WATER TREATMENT PLANT DESIGN

American Water Works Association
American Society of Civil Engineers

Edward E. Baruth, Technical Editor

Fourth Edition

McGRAW-HILL

New York Chicago San Francisco Lisbon London Madrid
Mexico City Milan New Delhi San Juan Seoul
Singapore Sydney Toronto

The McGraw·Hill Companies

Library of Congress Cataloging-in-Publication Data

Water treatment plant design / American Water Works Association, American Society of Civil Engineers ; Edward E. Baruth, technical editor.—4th ed.
  p. cm
  Includes bibliographical references and index.
  ISBN 0-07-141872-5 (alk. paper)
  1. Water treatment plants—Design and construction. I. Baruth, Edward E.
  II. American Society of Civil Engineers. III. American Water Works Association.
  TD434.W38   2004
  628.1′62—dc22   2004061109

Copyright © 2005, 1998 by American Water Works Association and American Society of Civil Engineers. All rights reserved. Printed in the United States of America. Except as permitted under the United States Copyright Act of 1976, no part of this publication may be reproduced or distributed in any form or by any means, or stored in a data base or retrieval system, without the prior written permission of the publisher.

Copyright © 1990 by McGraw-Hill, Inc.

The first edition was published in 1969 by American Water Works Association.

2 3 4 5 6 7 8 9 0   DOC/DOC   0 1 0 9 8 7

ISBN 0-07-141872-5

The sponsoring editor for this book was Larry S. Hager, the editing supervisor was Stephen M. Smith, and the production supervisor was Sherri Souffrance. It was set in Times Roman by Matrix Publishing Services. The art director for the cover was Handel Low.

At AWWA: Mary Kay Kozyra, editor; B. J. Crooks, administrative assistant.

Printed and bound by RR Donnelley.

McGraw-Hill books are available at special quantity discounts to use as premiums and sales promotions, or for use in corporate training programs. For more information, please write to the Director of Special Sales, McGraw-Hill Professional, Two Penn Plaza, New York, NY 10121-2298. Or contact your local bookstore.

Information contained in this work has been obtained by The McGraw-Hill Companies, Inc. ("McGraw-Hill") from sources believed to be reliable. However, neither McGraw-Hill nor its authors guarantee the accuracy or completeness of any information published herein and neither McGraw-Hill nor its authors shall be responsible for any errors, omissions, or damages arising out of use of this information. This work is published with the understanding that McGraw-Hill and its authors are supplying information but are not attempting to render engineering or other professional services. If such services are required, the assistance of an appropriate professional should be sought.

CONTENTS

Contributors ix
Preface xi

Chapter 1. The Challenge of Water Treatment Plant Design
Thomas J. Lane   1.1

Today's Challenges / *1.2*
Development of Water Treatment Design Projects / *1.4*
The Purpose of This Text / *1.5*
Bibliography / *1.6*

Chapter 2. Master Planning and Treatment Process Selection
George P. Fulton   2.1

The Master Plan / *2.1*
Water Quality Goals / *2.4*
Treatment Options / *2.9*

Chapter 3. Design and Construction   *R. Gary Fuller*   3.1

Preliminary Design / *3.1*
Final Design / *3.5*
Construction Assistance / *3.10*
Bibliography / *3.12*

Chapter 4. Intake Facilities   *Steven N. Foellmi*   4.1

Intake Features / *4.1*
Types of Intake Systems / *4.3*
Intake Design / *4.15*
Intake Design Considerations / *4.21*
Racks and Screens / *4.31*
Bibliography / *4.34*

Chapter 5. Aeration and Air Stripping   *John E. Dyksen*   5.1

Uses of Aeration/Air Stripping / *5.1*
Types of Aeration/Air Stripping Equipment / *5.2*

Principles of Air Stripping and Aeration / 5.8
Design of Aeration/Air Stripping Processes / 5.10
Pilot Testing / 5.22
Bibliography / 5.24

Chapter 6. Mixing, Coagulation, and Flocculation Paul J. Delphos and George M. Wesner 6.1

Definitions / 6.1
The Coagulation Process / 6.2
Coagulation and Membrane Technology / 6.5
Design of Chemical Mixing / 6.6
Flocculation Process Design / 6.14
Process Monitoring and Control / 6.23
Bibliography / 6.25

Chapter 7. Clarification John F. Willis 7.1

Conventional Clarification Design / 7.1
High-Rate Clarification / 7.22
Solids Contact/Slurry Recirculation Units / 7.28
Sludge Blanket Clarification / 7.32
Dissolved Air Flotation / 7.36
Contact Clarification / 7.40
Bibliography / 7.43

Chapter 8. High-Rate Granular Media Filtration Kevin Castro, Gary Logsdon, and Stephen R. Martin 8.1

Mechanism of Filtration / 8.2
Design Considerations / 8.2
Filter Design Criteria / 8.36
Filter Operation and Maintenance / 8.37
Bibliography / 8.41

Chapter 9. Slow Sand and Diatomaceous Earth Filtration Michael V. Broder and John C. Byron 9.1

Slow Sand Filtration / 9.2
Diatomaceous Earth Filtration / 9.16
Bibliography / 9.29

Chapter 10. Oxidation and Disinfection James C. Hesby 10.1

Regulatory Framework / 10.2
Chemical Oxidation Treatment / 10.10
Chlorination / 10.13
Chlorine Dioxide Systems / 10.34
Ozone Disinfection Systems / 10.39
Bibliography / 10.59

CONTENTS v

Chapter 11. Lime Softening *Michael B. Horsley, Doug B. Elder, P.E., and Leland L. Harms* 11.1

Water Hardness and Softening Treatment / *11.1*
Pretreatment / *11.5*
Lime Softening Processes / *11.6*
Softening Process Basins / *11.13*
Special Lime Softening Design Considerations / *11.16*
Future Trends in Softening / *11.19*
Bibliography / *11.19*

Chapter 12. Ion Exchange Applications in Water Treatment *Michael C. Gottlieb* 12.1

Introduction / *12.1*
The Ion Exchange Process / *12.6*
Cation Exchange Process / *12.9*
Anion Exchange Processes / *12.12*
Demineralization / *12.30*
Basics of Equipment Design / *12.32*
Regeneration Methods / *12.33*
Boundaries between Suppliers, Consultants, and Users / *12.53*
Future Trends in Ion Exchange / *12.54*
Bibliography / *12.54*

Chapter 13. Membrane Processes *Robert A. Bergman, P.E.* 13.1

Types of Membrane Processes / *13.1*
Membrane System Components and Design Considerations / *13.8*
Membrane Unit Design / *13.25*
Other Membrane Process Design Considerations / *13.41*
Bibliography / *13.48*

Chapter 14. Activated Carbon Processes *Richard D. Brady* 14.1

Characteristics of Activated Carbon / *14.1*
Design of Powdered Activated Carbon Facilities / *14.4*
Design of Granular Activated Carbon Facilities / *14.9*
Regeneration of Granular Activated Carbon / *14.33*
Bibliography / *14.37*

Chapter 15. Chemicals and Chemical Handling *Jerry L. Anderson, P.E.* 15.1

Receiving and Storing Process Chemicals / *15.1*
Chemical Feed and Metering Systems / *15.17*
Designing for Safety and Hazardous Conditions / *15.41*
Recent Trends in Chemical Handling and Use / *15.50*
Bibliography / *15.53*

Chapter 16. Hydraulics *Peter J. Barthuly* 16.1

Hydraulic Design / *16.1*
Head Loss Types and Calculations / *16.5*
Hydraulic Design Hints / *16.11*
Ancillary Hydraulic Design / *16.13*
Bibliography / *16.15*

Chapter 17. Process Residuals *Brian E. Peck and Jerry S. Russell* 17.1

Sludge Types, Quantities, and Characteristics / *17.4*
Process Residual Disposal Methods / *17.9*
Residuals Disposal Regulatory Requirements / *17.12*
Design Considerations and Criteria / *17.16*
Residuals Handling Considerations / *17.58*
Additional Information Sources / *17.67*
Acknowledgment / *17.68*
Bibliography / *17.68*

Chapter 18. Architectural Design *Philip C. Zimmerman* 18.1

The Role of the Architect in Water Treatment Plant Design / *18.1*
Facilities Design / *18.5*
Construction Alternatives and Building Material Selection / *18.14*
Design Standard–Promulgating Organizations and Organizations Offering Design
  Recommendations / *18.17*

Chapter 19. Structural Design *Ralph T. Eberts and Todd F. Schellhase* 19.1

The Design Process / *19.1*
Structural Design Considerations / *19.2*
Design of Basins, Vaults, Large Conduits, and Channels / *19.10*
Buildings and Superstructures / *19.12*
Bibliography / *19.13*

Chapter 20. Process Instrumentation and Controls *Dave Kubel* 20.1

Process and Operational Control Strategies / *20.2*
Control System Design / *20.15*
Instrumentation / *20.30*
Bibliography / *20.42*

Chapter 21. Electrical Systems *John H. Wiedeman* 21.1

Introduction / *21.1*
Codes and Standards / *21.1*
Electrical Distribution Systems / *21.3*
Transformers / *21.4*
Motors and Controllers / *21.5*

Lighting / 21.8
Lightning Protection / 21.9
Power Factor Correction / 21.10
Acceptance Testing / 21.10
Maintenance / 21.11

Chapter 22. Design Reliability Features   *John E. Spitko, Jr., P.E.*   22.1

Introduction / 22.1
Reliability and Redundancy Concepts / 22.1
Overall Water Supply System Reliability / 22.3
Design Concepts / 22.5
Design Principles and Practices / 22.8
Operation and Maintenance / 22.14
References / 22.15

Chapter 23. Site Selection and Plant Arrangement   *Darryl J. Corbin*   23.1

Water Treatment Plant Site Selection / 23.1
Arrangement of Water Treatment Plant Facilities / 23.22
Bibliography / 23.36

Chapter 24. Environmental Impact and Project Permitting
Philip C. Kennedy and Jane W. Wheeler   24.1

Environmental Issues Associated with Plant Construction / 24.1
Environmental Issues Associated with Plant Operations / 24.5
Project Permitting / 24.12
Integration of Environmental Issues and Regulatory Compliance into Plant Design, Construction, and Operation / 24.16
Bibliography / 24.18

Chapter 25. Operations and Maintenance Considerations during Plant Design   *Richard A. Rohan, P.E., and Stephen F. Baker*   25.1

Design Considerations for Plant Operations and Maintenance / 25.1
Preparing a Plant for Start-Up / 25.18

Chapter 26. Water Treatment Plant Construction Cost Estimating
Ronald F. Cilensek   26.1

Introduction / 26.1
Level of Estimates / 26.2
Estimating Methodologies / 26.4
Special Cost Considerations / 26.6
Finalizing the Cost Estimate / 26.8
Bibliography / 26.10

Chapter 27. Operator Training and Plant Start-Up *Thomas E. Arn and John Gasser*   27.1

Training and Start-Up Considerations / 27.1
Design-Phase Training / 27.2
Construction-Phase Training, Start-Up, and Post–Start-Up / 27.5

Chapter 28. Pilot Plant Design and Construction *Robert A. Stoops*   28.1

Pilot Plant Studies / 28.2
Pilot Plant Design / 28.3
Pilot Plant Construction / 28.21
Bibliography / 28.25

Chapter 29. Ultraviolet Disinfection *Paul D. Swaim, P.E., Christine A. Cotton, P.E., and Samuel S. Jeyanayagam, Ph.D., P.E., D.E.E.*   29.1

Principles of UV Disinfection / 29.2
Applications for UV Disinfection / 29.6
UV Disinfection Equipment / 29.8
Water Quality and Treatment Effects on UV Disinfection / 29.15
UV Disinfection Facility Planning, Design, and Project Implementation / 29.19
Construction, Testing, Start-Up, and O&M / 29.35
Potential Regulatory Requirements for UV Disinfection in North America / 29.38
Validation Testing / 29.43
Bibliography / 29.45

Chapter 30. Water Treatment Plant Security *John M. Winslow*   30.1

Water Quality / 30.2
Barriers / 30.2
Electronic Security Systems / 30.3
CCTV / 30.5
Procedures / 30.6
Bibliography / 30.7

Appendix A. Properties and Characteristics of Water Treatment Chemicals *Jerry L. Anderson, P.E.*   A.1

Appendix B. Abbreviations Commonly Used in the Water Industry   B.1

Index follows Appendix B

CONTRIBUTORS

Jerry L. Anderson, P.E.   *CH2M HILL, Louisville, Kentucky* (Chap. 15, App. A)

Thomas E. Arn   *Malcolm Pirnie, Inc., Phoenix, Arizona* (Chap. 27)

Stephen F. Baker   *MWH, Broomfield, Colorado* (Chap. 25)

Peter J. Barthuly   *Barthuly Hydraulics & Pumping Engineering, Inc., Burlington, Massachusetts* (Chap. 16)

Robert A. Bergman, P.E.   *CH2M HILL, Gainesville, Florida* (Chap. 13)

Richard D. Brady   *Richard Brady & Associates, San Diego, California* (Chap. 14)

Michael V. Broder   *Hazen and Sawyer, PC, New York, New York* (Chap. 9)

John C. Byron   *Stearns & Wheler, LLC, Cazenovia, New York* (Chap. 9)

Kevin Castro   *Stearns & Wheler, LLC, Cazenovia, New York* (Chap. 8)

Ronald F. Cilensek   *CH2M HILL, Alexandria, Virginia* (Chap. 26)

Darryl J. Corbin   *MWH, Dallas, Texas* (Chap. 23)

Christine A. Cotton, P.E.   *Malcolm Pirnie, Inc., Tucson, Arizona* (Chap. 29)

Paul J. Delphos   *HDR Engineering, Inc., Norfolk, Virginia* (Chap. 6)

John E. Dyksen   *Black & Veatch, Oradell, New Jersey* (Chap. 5)

Ralph T. Eberts   *Black & Veatch, Los Angeles, California* (Chap. 19)

Doug B. Elder, P.E.   *Black & Veatch, Olathe, Kansas* (Chap. 11)

Steven N. Foellmi   *Black & Veatch, Irvine, California* (Chap. 4)

R. Gary Fuller   *HDR Engineering, Inc., Denver, Colorado* (Chap. 3)

George P. Fulton   *Hazen and Sawyer, PC, New York, New York* (Chap. 2)

John Gasser   *Oakland, California* (Chap. 27)

Michael C. Gottlieb   *ResinTech, Inc., West Berlin, New Jersey* (Chap. 12)

Leland L. Harms   *Black & Veatch, Kansas City, Missouri* (Chap. 11)

James C. Hesby   *Black & Veatch, Concord, California* (Chap. 10)

Michael B. Horsley   *Black & Veatch, Kansas City, Missouri* (Chap. 11)

Samuel S. Jeyanayagam, Ph.D., P.E., D.E.E.   *Malcolm Pirnie, Inc., Columbus, Ohio* (Chap. 29)

Philip C. Kennedy   *CDM, Cambridge, Massachusetts* (Chap. 24)

Dave Kubel   *Westin Engineering, Rancho Cordova, California* (Chap. 20)

Thomas J. Lane   *Malcolm Pirnie, Inc., Jackson Heights, New York* (Chap. 1)

Gary Logsdon   *Retired, Lake Ann, Michigan* (Chap. 8)

Stephen R. Martin *CDM, Cambridge, Massachusetts* (Chap. 8)

Brian E. Peck *South Orange County Wastewater Authority, Dana Point, California* (Chap. 17)

Richard A. Rohan, P.E. *MWH, Houston, Texas* (Chap. 25)

Jerry S. Russell *Carollo Engineers, PC, Phoenix, Arizona* (Chap. 17)

Todd F. Schellhase *Black & Veatch, Kansas City, Missouri* (Chap. 19)

John E. Spitko, Jr., P.E. *Boucher & James, Inc., Doylestown, Pennsylvania* (Chap. 22)

Robert A. Stoops *Springfield Water & Sewer Commission, Springfield, Massachusetts* (Chap. 28)

Paul D. Swaim, P.E. *CH2M HILL, Englewood, Colorado* (Chap. 29)

George M. Wesner *Consulting Engineer, San Clemente, California* (Chap. 6)

Jane W. Wheeler *CDM, Cambridge, Massachusetts* (Chap. 24)

John H. Wiedeman *Wiedeman & Singleton, Inc., Atlanta, Georgia* (Chap. 21)

John F. Willis *CDM, Cambridge, Massachusetts* (Chap. 7)

John M. Winslow *AMRIC Associates, Ltd., East Syracuse, New York* (Chap. 30)

Phillip C. Zimmerman *Malcolm Pirnie, Inc., White Plains, New York* (Chap. 18)

PREFACE

Design of water treatment plants is evolving. New technologies and unit processes continue to emerge and have become much more common since publication of the third edition of *Water Treatment Plant Design* 7 years ago. Security issues, summarized in a new chapter in this fourth edition, have forever redirected many aspects of design. Additional reference materials for security design considerations are becoming available and should be consulted for further information.

Equipment design continues to broaden, yet sources of manufactured products have become more consolidated in the past 7 years. This new edition contains numerous revisions of illustrations and photography; however, not all new technologies or equipment offerings are represented. The American Water Works Association (AWWA) and the American Society of Civil Engineers (ASCE) are interested in obtaining input from readers on how to facilitate the future exchange of information on equipment. Alternatives such as Web-based links are being considered to better provide needed product information to the design community.

The first version of *Water Treatment Plant Design* was published in 1939 as a manual of engineering practice for the ASCE. In 1969, the manual assumed book form and was updated to include a discussion of developments in pretreatment and filtration processes. The 1969 edition was the result of a joint effort between committees of the ASCE, the AWWA, and the Conference of State Sanitary Engineers (CSSE).

The second edition was produced in 1990 through a joint effort of the AWWA and the ASCE. The material for each chapter was prepared by one or more authors and reviewed by a joint committee of AWWA and ASCE members.

The third edition, published in 1998, was a joint AWWA and ASCE effort and was essentially a complete rewrite of the previous edition. The information presented in the book was prepared as a guide and represented a consensus of opinion of recognized authorities in the field. A steering committee made up of members from both associations guided the revision process.

Updates to this fourth edition provide significant new information on many important topics. Authors from engineering firms and water utilities throughout North America have revised the chapters and written the two new chapters on UV technologies and security. Providing support to the chapter authors was a significant base of volunteer reviewers. Due to the ability of the authors to distribute drafts of their chapters electronically to large numbers of prospective reviewers, it became apparent that the attempt to accurately name all reviewers would result in inadvertent omissions. Therefore a general acknowledgment and thanks to all reviewers is hereby offered.

Guiding the development of the entire fourth edition was a dedicated joint AWWA/ASCE steering committee. The members of this committee were

C. Michael Elliott (Chair), Stearns & Wheler, LLC
Bruce E. Burris, HDR Engineering, Inc.
Anthony J. Geiss, Jr., Onondaga County Water Authority

Arthur D. Kney, Ph.D., Lafayette College
Hal Wiedeman, Wiedeman & Singleton, Inc.
Ian P. D. Wright, P.Eng., Associated Engineering Group, Ltd.

CHAPTER 1
THE CHALLENGE OF WATER TREATMENT PLANT DESIGN

Thomas J. Lane
*Malcom Pirnie, Inc.
Jackson Heights, New York*

When water treatment engineering first evolved in the early part of the twentieth century, its main goal was to ensure that infectious organisms in drinking water supplies were removed or inactivated. Chlorination and filtration practices were applied with tremendous success to the point that major death-causing waterborne disease outbreaks in the United States were virtually eliminated by the 1930s.

As a result, for engineers trained in the 1960s, 1970s, and 1980s, both education and industry belief was that all concerns of microbiological contamination in surface waters could be eliminated by providing filtration (with suitable pretreatment) to produce water of sufficient clarity (turbidity less than 1.0 or 0.5 ntu) and then chlorinating. Groundwater was thought to be already filtered, requiring only chlorination to maintain a distribution system residual. Any additional treatment was generally considered necessary only to address non–health-related parameters, such as excessive hardness or water discoloration caused by iron and manganese.

The principal challenge to water treatment engineers in the 1960s and 1970s was engineering cost-effectiveness: how to accomplish these simple treatment goals at the lowest total cost to the water utility. Thus, in these decades many new techniques and processes were developed to clarify surface water economically. These developments included improvements to sedimentation basin designs; high-rate clarification processes such as tube settlers, plate settlers, and dissolved air flotation; high-rate filtration processes; and proprietary preengineered or package equipment integrating flocculation, settling, and filtration processes.

In the 1970s and 1980s a new drinking water concern arose: the potential long-term health risks posed by trace amounts of organic compounds present in drinking water. A wave of regulations ensued with new maximum contaminant levels (MCLs) established for total trihalomethanes (TTHMs), pesticides, and volatile organic chemicals (VOCs). This trend continues today. In response to this concern and resulting treatment needs, water treatment engineers have successfully devised new methods of water treatment to remove organic compounds. These methods, such as air stripping, activated carbon adsorption, and enhanced coagulation, have been a primary focus of water treatment engineering over the last 20 years.

In the 1990s the old concern about microbiological contamination reemerged as a primary focus of water treatment engineers. The main driving forces behind this development were:

- The promulgation of the Surface Water Treatment Rule and Total Coliform Rule by the U.S. Environmental Protection Agency (USEPA) and the monitoring and enforcement actions that have occurred since 1989, when they went into effect
- Recent documented cases of contamination of drinking water supplies by waterborne diseases, mainly giardiasis and cryptosporidiosis, caused by cysts rather than bacteria

New approaches and processes began to be applied and continue to evolve to address these concerns. These approaches include renewed emphasis on source water protection, optimizing plant performance, and recycle stream management, plus consideration of new technologies, especially membrane treatment and ozonation.

Recently, new health effects research has led to reconsideration of the safe drinking water concentrations for several inorganic contaminents. In the United States the MCL for arsenic was reduced dramatically—from 50 to 10 ppb. This has created new treatment challenges for many utilities.

The effects of planned and unplanned recycling of municipal wastewater into raw water sources are gaining increased attention. A new class of constituents emanating from modern human life has been discovered in trace amounts in water supplies. These include pharmaceutical compounds and personal hygiene products. Additionally, new "emerging" pathogens are being discovered and their potential occurrence and treatability in water supplies evaluated. The potential impacts of these developments on the design requirements for water treatment facilities have yet to be determined.

In the last decade, new water treatment technologies have continued to be developed and older technologies improved. Notable developments have been

- The sudden emergence of ultraviolet (UV) radiation as a viable and economic process for achieving inactivation of cryptosporidium and other protozoan pathogens.
- The continued development and maturing of membrane technologies. Microfiltration and ultrafiltration are being used frequently for particulate removal in lieu of granular media filtration. Nanofiltration is being used increasingly for water softening and removal of organic carbon. Reverse osmosis has new emphasis for desalination applications and for removal of inorganic contaminents (e.g., arsenic).

Finally, since September 11, 2001, there has been increased recognition of the vulnerability of municipal water systems to acts of terrorism and vandalism. New approaches to siting and designing critical system components, including water treatment plants, are evolving to reduce this vulnerability.

TODAY'S CHALLENGES

Engineers who design water treatment systems today face many challenges. The most important of these are described as follows.

Integrated Treatment Systems

Traditional treatment engineering has focused on the treatment plant as the sole vehicle for controlling drinking water quality. The engineer's role was to characterize the quality of the source water entering the plant and devise treatment facilities to produce water meeting drinking water standards. The point of measurement for drinking water standards was the finished water exiting the plant.

Today's engineer must view the water treatment plant as only a major component in a multistep treatment process. This process includes consideration of the path that the water travels upstream of the plant in the watershed and the elements of the water transmission and distribution system downstream of the plant. Changing water quality must be managed in each of these steps, and new regulations require that drinking water standards be met at the customer's tap.

Regulatory Uncertainties

The definition of "safe" drinking water, which remained relatively fixed in the 1950s, 1960s, and 1970s, now seems to be constantly changing or under review as the water utility industry grapples to understand the potential health effects of trace amounts of an increasing variety of chemical compounds and infectious organisms. Today's treatment system engineer, in addition to addressing current drinking water standards, must anticipate potential future requirements. A water system designed today must be designed with sufficient flexibility to be modified to meet these potential requirements.

Regulatory uncertainties extend to other environmental concerns important to water treatment plant design, including waste management practices, chemical storage and feed operations, and workplace safety.

New Technologies

The state of the art of water treatment plant design is continually changing as new technologies emerge, offering new unit processes for water treatment or making currently used processes more efficient or economical. In addition, advances in computer technology and building materials are rapidly changing and improving the support systems associated with water treatment plants.

Multidiscipline Teams

A water treatment plant engineering design team traditionally consisted simply of a small group of civil engineers. This single-discipline team performed the majority of design work for virtually all plant components. Support disciplines of architects and structural, electrical, and mechanical engineers were used to execute the basic decisions made by the design team.

Today, the complexity of project and regulatory requirements dictates that a far more multidisciplined approach be used. Typically, a small group of civil engineers remains as the "project" engineers, but this group uses the expertise and resources of many different specialists to execute the design. In addition to traditional design support disciplines, these may include

- Process engineers
- Plant operations specialists
- Instrumentation and control engineers
- Health and safety specialists
- Environmental scientists
- Specialists in environmental permitting and public participation

Major design decisions today are no longer made unilaterally by the project team. Instead, a consensus is reached after participation by members of the design team and by individuals outside the team, including owners, operators, regulatory agencies, and the general public.

Project Delivery

The traditional procedure for construction of a new water treatment plant was for the engineering design and specifications to be prepared by an engineering firm or the owner's in-house staff. Bids were then taken, and the contract was awarded to the lowest responsible bidder. The design team usually monitored the construction to see that the design intent was carried out, and after construction was completed, the facilities were operated by the owner.

Today, a number of changes and variations to this traditional approach are being considered or implemented. Two of the principal alternatives are

- Design-build approaches, in which one entity is responsible for both design and construction
- Privatization approaches, in which the facility is owned by a private entity providing treatment service for the water utility

In addition, a global marketplace for water treatment engineering has developed. Ideas and practices are being exchanged among countries all over the world. In North America, there is increasing consideration of European treatment practices, technologies, and firms.

DEVELOPMENT OF WATER TREATMENT DESIGN PROJECTS

A water treatment design project passes through many steps between the time when the need for a project is identified and the time that the completed project is placed into service. The period before construction commences can generally be divided into the following phases:

1. *Master planning.* Treatment needs and feasible options for attaining those needs are established in a report. In subsequent phases, this report may be periodically updated to adjust to both system and regulatory changes.
2. *Process train selection.* Viable treatment options are subjected to bench, pilot, and full-scale treatment investigations. This testing program provides background data sufficient in detail to enable decisions on selecting the more advantageous options for potential implementation. These tests provide design criteria for major plant process units.
3. *Preliminary design.* In this "fine-tuning" procedure, feasible alternatives for principal features of design, such as location, treatment process arrangement, type of equipment, and type and size of building enclosures, are evaluated. In this phase, preliminary designs are prepared in sufficient detail to permit development of meaningful project cost estimates. These estimates help in evaluating and selecting options to be incorporated into the final design and allow the owner to prepare the required project financial planning.

4. *Final design.* Contract documents (drawings and specifications) are prepared that present the project design in sufficient detail to allow for gaining final regulatory approvals, obtaining competitive bids from construction contractors, and actual facility construction.

Many technical and nontechnical individuals must be involved, not only during the four phases of project development, but also between these phases to ensure that a project proceeds without undue delay. In addition to the engineer's design staff and the owner, these may include public health and regulatory officials, environmental scientists, and the public affected by both the proposed construction and the future water supply services to be provided.

The process train selection phase is only briefly covered in this book. Theory and procedures needed for this phase are the focus of *Water Quality and Treatment.* It is important that the interface between phase 1 and phase 2 and between phase 2 and phase 3 be carefully coordinated to allow uninterrupted continuity of design. In other words, viable options developed for consideration in phase 1, master planning, should provide a base for developing unit process test studies in phase 2. The process train selected in phase 2 provides the basis for phase 3, preliminary design, in which other factors influencing design are included in the evaluations before criteria for final design are developed and finalized.

Careful coordination of the various phases and entities involved provides the owner and the engineer with the opportunity to develop the most advantageous treatment solutions and designs, and helps avoid pitfalls in the schedule and decisions that might add to the cost of the project.

THE PURPOSE OF THIS TEXT

Water Treatment Plant Design is intended to serve as the primary reference for engineers who take on today's challenges of water treatment plant design. It covers the organization and execution of a water treatment plant project from planning and permitting through design, construction, and start-up.

The book is aimed at "project" engineers and managers: those professional engineers who lead the group of specialists who make up the design team. Generally, these individuals are graduates of civil or environmental engineering programs and are registered professional engineers.

For certain topics, especially the practical application of water treatment unit processes, this book aims to be an authoritative reference to design engineers. For other topics, only a general discussion of major concepts and issues is provided, and the reader is referred to more specialized references for detailed information.

Many books in circulation address the subject of water treatment engineering. As a joint publication of the American Society of Civil Engineers (ASCE) and the American Water Works Association (AWWA), *Water Treatment Plant Design* attempts to present an industry consensus on current design practices.

Organization of This Text

For the convenience of readers, chapters in this book have been organized as follows:

- Chapters 1, 2, and 3 examine the general preliminary and final design phases and engineering needs during project construction and initial operation.

- Chapters 4 through 14 address design practices for the major categories of unit processes applicable to water treatment plants.
- Chapters 15 through 21 cover the support systems associated with the design of water treatment facilities.
- Chapters 22 through 28 discuss general topics essential to developing a successful water treatment plant project.
- Chapters 29 and 30 are new in this edition and provide information on ultraviolet disinfection and security.

Because of the rapid changes that the water treatment engineering industry is undergoing, many chapters include a discussion of future trends. These sections attempt to inform the reader of new developments not yet in general practice, but that may soon become so.

BIBLIOGRAPHY

A list is provided at the end of each chapter of references specific to the chapter subject. In addition, a number of other texts are cited that the design engineer should be aware of for reference and additional details.

Relationship with *Water Quality and Treatment*

The AWWA publication *Water Quality and Treatment* is intended to be a companion reference to *Water Treatment Plant Design* and to serve as the primary reference for process engineers. *Water Quality and Treatment* provides details of water quality goals, with an emphasis on unit processes, standards, and the evaluation and selection of treatment process trains.

Other Design References

Other design references currently available include the following:

Kawamura, Susumu. *Integrated Design of Water Treatment Facilities.* New York: John Wiley & Sons, 1991.

Montgomery, James M. *Water Treatment Principles and Design.* New York: John Wiley & Sons, 1985.

Sanks, Robert L. *Water Treatment Plant Design for Practicing Engineers.* Ann Arbor, Mich.: Ann Arbor Science, 1978.

CHAPTER 2
MASTER PLANNING AND TREATMENT PROCESS SELECTION

George P. Fulton
Hazen and Sawyer, PC
New York, New York

The master plan is the first of many steps leading to a final water treatment plant design. In a strict sense, a public water system master plan is a very comprehensive document that includes many aspects of the ownership and operating features of that system. The principal intent of this chapter, however, is to address those issues of a master plan that affect water treatment plant design. Just as the master plan must recognize growth changes in a community as they may affect water supply demand and plant capacity, the plan must also observe trends in regulatory requirements that will influence treatment process selection. This chapter introduces the principal issues that provide the basis for design through its many phases with particular emphasis on treatment process selection that will not only permit compliance with current water quality goals, but also accommodate revisions and addition for future objectives.

THE MASTER PLAN

A master plan is the orderly planning of a water system's future improvement program. The initial step in preparing any water system design is to update the system's master plan. Many states and state utility commissions require all water systems to have an active master plan that anticipates system additions and improvements for many years into the future. The master plan for treatment should be periodically updated to reflect the improvements needed to compensate for changing system requirements imposed by facility wear, customer requirements, and changing water quality regulations.

Master Planning Issues

Master planning for water treatment facilities is often incorporated into the long-range capital improvement program for the water system. The master plan identifies present and future needs and the direction for developing the water system's facilities. Some specific items that should be covered are

- Identification of existing system components and service area
- Long-range projections of the area to be served by the water system
- Planning periods for the various water system facilities
- Present and future water demands
- Regulatory requirements for the ultimate approval and operation of the system
- Evaluation of alternative sources of supply

In addition, technical, environmental, institutional, financial, and operations and maintenance issues related to developing the recommended plan should be identified. This text is concerned with the design of water treatment plant facilities and with those aspects of master planning related to the physical features of design. Discussion of master planning, therefore, is essentially limited to those treatment considerations outlined by the chapter subjects. In addition to process and facility design issues, the scope would also include

- Site and facilities arrangement
- Environmental impact
- Construction costs
- Operations and maintenance (scope and costs)

Where design is to rehabilitate an existing plant, many of the data, such as operations and maintenance costs, may be provided from past experience. In this case, background site and environmental experiences may also exist to guide future direction.

More master planning difficulties are encountered in the design of a completely new facility. In this situation, background process and cost data are usually obtained from nearby operating facilities, from the experience of other water supplies with a similar water source, or from references such as those published by the U.S. Environmental Protection Agency (USEPA) and the AWWA on typical treatment costs. Other factors such as the influence of site location and environmental impacts would need to be developed from local knowledge.

The most important considerations in preparing a master plan are to

- Provide general guidelines for future water treatment action.
- Develop possible alternatives for further evaluation unless background data from existing experiences are so overwhelming that final decisions on treatment are obvious.
- Enable a liberal use of contingencies in developing cost estimates, with the magnitude of contingencies reflecting the confidence in the cost base. It is important that the owner not be led into quick acceptance of a treatment program that eventually turns out to be too costly. If the available information is questionable, it is best to delay program discussion until more meaningful background information is available.
- Include, as much as possible, potential features limiting site locations and environmental impacts in the early determination to resolve which alternatives obviously should be eliminated and what difficulties may be encountered with other options.

Principal Influences on the Master Plan

The principal influences and controls on developing water treatment process options in a master plan include

MASTER PLANNING AND TREATMENT PROCESS SELECTION

1. Drinking water regulations, including those that are current, pending, and anticipated
2. Treatment and residuals management options and limitations that produce quality consistent with regulations
3. Choices available where more than one treatment method or treatment train may be equal in cost and other features

Treatment method development should be evaluated in the above order. Rules and regulations establish what must be accomplished in treated water quality, and often more than one method of treatment may accomplish the required result.

The potential options and the limitations of each method are developed in the next step. Limitations may include factors such as cost, operating control, and size of facilities. Where several viable options are available, different treatment trains are evaluated, taking advantage of the multiple choices to determine the most advantageous option. Comparisons here may include factors such as best fit on the site, appearance, ease of operations and maintenance, and vulnerability to upset.

On June 12, 2002, the federal Public Health Security and Bioterrorism Preparedness and Response Act was signed. In accordance with the terms of this act, all water treatment plant designs now have to be evaluated in regard to vulnerability to terrorist attack and resulting potential damage to facilities and upset in treatment process. Incorporating the results of these evaluations, existing emergency response plans must be revised accordingly or new plans developed. Reducing facility vulnerability and improving emergency response capability must be considered in the future in both the development of master plans and the selection of a treatment process train.

When alternative source waters are available, the potential sources (surface water and groundwater) and specific intake or well locations should be evaluated to determine the water quality characteristics of each source. The vulnerability of source water to future contamination or water quality deterioration should receive particular consideration. The variability of the source water quality should be investigated because extreme water quality conditions often dictate treatment requirements. Once a new water supply is chosen, a protection program should be implemented (either watershed or groundwater protection) to maintain the integrity of the supply.

Planning Periods for the Master Plan

Major capital projects, such as large water treatment plants, generally require many years' lead time to develop meaningful design, construction, and financial arrangement plans. This, together with the expected long life of certain facilities, results in exceptionally long design periods for water systems. Master planning studies often develop the water supply and treatment needs for 30 to 50 years or more into the future.

Based on complexity, expendability, and cost, the various components of the water filtration plant are sized to meet the needs of varying periods of time. More difficult and expensive facilities, such as intake tunnels and major structures, are often designed for the life of the facility, which can be as great as 50 years or more. Other facilities, such as process treatment units, are often initially planned with a first-phase design of 10 to 25 years, with a plan to allow for future increments of expansion to accommodate the full life of the project.

Equipment such as pumps and chemical feed systems may have an expected life of 10 to 15 years. Therefore they are designed for shorter-term capacities with allowances for replacement to meet future needs of the facility.

WATER QUALITY GOALS

The Safe Drinking Water Act (SDWA) and its subsequent amendments provide the basic rules for water quality produced by a treatment system, but the design engineer cannot work solely from federal requirements because they are only minimum standards. Individual states have the option of making the standards more stringent or of expanding the basic regulations to include other quality standards. It is, therefore, important to work closely with state officials when one is considering process options and design details.

The USEPA is in a continuous process of modifying and expanding drinking water regulations under the SDWA and its amendments. This has become a complex process that has involved difficult scientific issues, as well as the political considerations that are inevitable to regulatory processes. A number of rule-making proceedings are involved in this process. Some of these have been completed, and others are in varying stages of development. These rules tend to fall into one of four categories:

- *Finalized rules that are in effect.* The rules have become established regulations. Where relevant, state public water supply enforcement agencies must incorporate these rules into their own regulations and determine how the rules will be administered.

- *Finalized rules that are not yet in effect.* These rules have completed the promulgation process, and provisions are known with certainty. An effective date for these rules, however, has not been reached, and they have yet to formally become a component of the established SDWA regulations. At this stage of rule development, state primacy agencies may be developing procedures for incorporating the rules into their respective state regulations and are assessing options for administering the rules once the rules become effective.

- *Proposed rules.* These rules have reached an intermediate stage that reveals specific USEPA intent. However, the provisions are still at a proposed level that allows for comment from interested parties. The USEPA is required to formally respond to all comments and may make modifications before promulgation of the final rule, depending on the availability of additional information and the impact of comments. It is significant to note that most rules have been modified during the time between the proposed and final stage as a consequence of this process.

- *Rules under development.* These rules have not been proposed, and USEPA intent is not fully developed. In some cases, draft rules may be developed; these have no formal status, and the USEPA has flexibility for changing the drafts without going through a formal response process to outside parties.

Rules not yet established in the Primary Drinking Water Standards and not yet enforceable may also be categorized as future regulations in progress. It is essential that the design engineer be aware, and stay abreast, of both the established rules and those in various stages of development.

Finalized Regulations

Finalized rules in the Primary Drinking Water Standards presently in effect include the Surface Water Treatment Rule (SWTR); the Total Coliform Rule; the Information Collection Rule (ICR); the Stage 1 Disinfectants/Disinfection By-products Rule (Stage 1 D/DBPR); the Interim Enhanced Surface Water Treatment Rule (IESWTR); the Filter Backwash Rule; the Phase I, II, and V Contaminant Rules; and the Lead and Copper Rule.

Rules that have been finalized provide two levels of standards to be regulated—primary and secondary standards.

Primary Drinking Water Standards are health-related criteria that require mandatory enforcement by the federal USEPA or by state primacy agencies. Existing primary standards, as established by rules currently in effect as of January 2003, are described in Table 2.1. It is noted that the table does not address all established standards, but rather those affecting design of the principal physical elements of treatment discussed later in the text.

Secondary Drinking Water Standards include criteria that are intended for control of aesthetic factors. Unlike primary standards, parameters developed as secondary standards are established as guidelines that are strongly recommended, but not required to be enforced. Some of these may be enforced at the discretion of individual state primacy agencies. Table 2.2 summarizes existing secondary regulations. While parameters governed by secondary standards are not generally health-related, they can have a significant effect on customer acceptance and can be a source of considerable customer complaint. Among these are water color and tastes and odors in the drinking water. While the secondary standards are presently observed in drinking water treatment more by customer needs than by regulatory control, future findings may result in a switch of some of these constituents to primary listing.

Significant Future Drinking Water Regulations in Progress

At press time, a number of new regulations were under development in various stages from proposed through final promulgation to anticipated effective dates (Table 2.3). The salient features of the more significant of these are as follows, with the anticipated effective dates shown in parentheses:

- Long-Term 1 Enhanced Surface Water Treatment Rule (LT1ESWTR-2004)—same requirements as Interim Enhanced Surface Water Treatment Rule extended to include systems serving under 10,000 people.
- Radon Rule (2005)—a limit of 300 pCi/L radon content for systems with a groundwater supply—higher limit may prevail with state multimedia mitigation programs.
- Sulfate Rule (undecided)—still under USEPA consideration (may not be included in list of contaminants).
- Stage 2 Disinfectants/Disinfection By-products Rule (Stage 2 D/DBPR-2005)—compliance based on compliance running average at individual locations rather than average of all locations.
- Long-Term 2 Enhanced Surface Water Treatment Rule (LT2ESWTR-2005) for all surface water systems or groundwater under the influence—at promulgations (2004) requires 2-year monitoring program of monthly samples for *Cryptosporidium* (systems over 10,000 people) or 1-year, biweekly *E. coli* samples (under 10,000). If results trigger remediation, measures to attain up to 2.5-log improvements may be selected from "tool boxes or bins" that may include additional treatment such as ozone or UV or even improved watershed protection programs, each measure having individual assigned log values.
- Ground Water Rule (GWR-2005)—initially will require treatment technology based on evaluation of sanitary survey conducted by primacy agencies. Depending on susceptibility of individual sources, up to 4-log treatment inactivation of viruses may be required.

TABLE 2.1 Presently Enforceable USEPA Primary Drinking Water Standards

Inorganics
A list of some 20 chemical elemental or compound materials. Note that the traditional treatment would have little or no effect in the reduction of many of the materials in this list.
Synthetic organic contaminants
A list of over 30 materials, for the most part manufactured for use as herbicides and pesticides. Many of these would require more than the usual filtration treatment to achieve the required reduction.
Volatile organics
A group of over 20 materials, essentially solvents used in manufacturing and other activities for cleaning.

Contaminant	Maximum contaminant level (except as noted), mg/L
Surface Water Treatment Rule (SWTR)	
Turbidity	
Not to exceed 5% of monthly samples	0.5 ntu
Microbiological	
Total coliform bacteria	More than 5% of samples positive
Disinfectant residual/heterotrophic plate count	More than 5% sample with less than detectable residual and HPC 500/mL
Radionuclides	
Radium 226 and 228	5 pCi/L
Gross alpha radioactivity	15 pCi/L
Beta and photon radioactivity	4 rem/yr*
Interim Enhanced Surface Water Treatment Rule (IESWTR)—system population 10,000 or more	
Turbidity	
Not to exceed 5% of monthly samples	0.3 ntu

Disinfectants-Disinfection By-products Rule			
Stage 1:			
Disinfection by-products			
Bromate	0.010		
Chlorite	1.0		
Haloacetic acid (five) (HAA5)	0.060†		
Total trihalomethanes (TTHM)	0.080†		
Total organic carbon removal			
Source water TOC, mg/L	Percent TOC removals for source water alkalinity, mg/L		
	0 to 60	> 60 to 120	<120
> 2.0 to 4.0	40.0	30.0	20.0
> 4.0 to 8.0	45.0	35.0	25.0
> 8.0	50.0	40.0	30.0
Alternatives to TOC reduction			
Typical example—reduction of TTHM and HAA5‡			
TTHM	0.040†		
HAA5	0.030†		

Note: Values in mg/L unless otherwise noted.

*The contribution of this category of radionuclides is to be computed. Tritium at 20,000 pCi/L and strontium 90 at 8 pCi/L, respectively, are computed to give no more than 4 rem/yr.

†Four quarterly samples; compliance based on running annual average of all sites.

‡One of the four alternatives permitted (see *Federal Register* for others).

TABLE 2.2 USEPA Secondary Drinking Water Standards

Contaminant	Maximum contaminant level
Aluminum	0.05–0.2 mg/L*†
Chloride	250 mg/L
Color	15 color units
Copper	1 mg/L
Corrosivity	Neither corrosive nor scale-forming
Fluoride	2.0 mg/L
Foaming agents	0.5 mg/L
Iron	0.3 mg/L
Manganese	0.05 mg/L
Odor	Three threshold odor numbers
pH	6.5–8.5
Silver	0.1 mg/L
Sulfate	250 mg/L
Total dissolved solids	500 mg/L
Zinc	5 mg/L

*Selected level for aluminum depends on the discretion of the state primacy agency.
†Under consideration as primary health issue.

TABLE 2.3 Future USEPA Drinking Water Regulations in Progress

Contaminant	Prospective maximum contaminant level, mg/L (except as noted)
Radon Rule	
Radon	300 pCi/L
Stage 2 Disinfectants/Disinfection By-products Rule (Stage 2 D/DBPR)	
Total trihalomethanes (TTHM)	0.080*
Haloacetic acids (HAA5)	0.060*
Long-term 1 Enhanced Surface Water Treatment Rule (LT1ESWTR)—systems under 10,000 population	
Cryptosporidium Giardia Viruses	Same as IESWTR
Long-term 2 of Enhanced Surface Water Treatment Rule (LT2ESWTR)	
Cryptosporidium	Treatment technique (removal or inactivation beyond conventional treatment)
Ground Water Rule	
Viruses	Treatment technology (4.0 log inactivation)

*Four quarterly samples at four system locations including two high-TTHM sites, one high-HAA5 site, and one other system site; compliance based on running annual average of all sites.

It is apparent that the EPA is now employing treatment technology and other mitigating measures in addition to maximum contaminant levels (MCLs) for several of these new rules.

Maximum contaminant level goals (MCLGs) have been established for some time for most of, if not all, the proposed new standards. In moving toward establishment means that the federal USEPA believes that the technology is available for treatment at a reasonable cost to the consumer.

The trends in new contaminant additions to the rules, tightening of existing standards, and expansion of regulatory application should continue as new health issue and treatment research indicates the need and potential resolutions. To make the problem of producing acceptable quality water even more complicated, the treatment plant designer now also has to contend with the possibility of unusual biological and toxic contaminants introduced into the water supply by terrorist activity. It is apparent that treatment facilities now must not only accommodate changes in regulatory requirements, but also continue to handle changes in influent raw water quality. As the rules become more stringent, it will become more difficult to handle rapid changes in quality due to natural causes such as intense rainfall and runoff. It will be even more difficult to handle "introduced" contaminants because of the unknown factor. Once it is introduced and contaminants are discovered, temporary shutdown of treatment facilities may be necessary. The tightening rules and the potential for unexpected contamination make water quality monitoring facilities even more essential features of water treatment planning design and operation.

It is emphasized that the primary purpose of this brief discussion of the new proposed rules is to illustrate the challenge presented to the design engineer in defining treatment goals. The established and proposed water quality rules and standards shown in the tables should serve only as a guide to the water treatment planner or designer. It is also essential that the details of water quality requirements be obtained from the appropriate state, provincial, or other agency.

Regulatory Trends—Treatment Obsolescence

In general terms, the goal of the SDWA is to produce the most contaminant-free drinking water that technology permits, limited only by the magnitude of treatment costs. As technology has improved, particular MCLs have become more stringent. Since 1970, for instance, turbidity limits have been reduced from 5.0 to 1.0 to 0.5 ntu and, under the Interim Enhanced Surface Water Treatment Rule, to 0.3 ntu. Some states may require even lower limits. The first-stage TTHM (total trihalomethane) and HAA5 (five haloacetile acids) limits recently became effective for average distribution system conditions. In the near future these limits will be required for worst-case locations in the distribution system. New regulatory requirements are continuously being added to the rules such as the reduction of organic content in treatment and the reduction of arsenic content.

As the regulatory requirements become more stringent, currently accepted treatment processes can and will become obsolete. If, for instance, turbidity levels in surface water treatment are reduced to 0.20 ntu, there are several types of filtration that may not be adequate operating in currently acceptable design modes. Such facilities may have to be replaced or, modified in operation, or supplementary filtration may have to be added. Some current filtration processes do not reduce organic content at all or reduce organics to a limited extent. These also may have to be replaced, modified, or supplemented in treatment. Potential treatment obsolescence must be recognized in treatment selection to avoid costly future changes.

TREATMENT OPTIONS

As illustrated in Table 2.4, a summary of more common drinking water treatment processes, there are many treatment options available to the designer to achieve the desired water quality results.

Principal Process Alternatives and Issues

Where pertinent, detailed unit process tests and comparisons should be conducted under the guidelines of the companion volume to this text, *Water Quality and Treatment,* to determine the more effective options. Bench and pilot-scale treatment operations would also be conducted in the water quality and treatment phase to determine the more critical process design and operating parameters.

Under the current regulations, some form of filtration must be provided for effective particulate solids removal. In general, filtration would be required for all source water supplies except perhaps for some groundwater sources. The particular type of filtration required would depend upon the particulate solids loading contributed by that in the source water in addition to that which may be generated in treatment such as in precipitation (i.e., cold lime softening, manganese dioxide) and coagulation. Filtration types and solids loading capabilities may be categorized in general terms as follows:

- Direct, slow sand, and diatomaceous earth filtration may be feasible for waters low in turbidity (5 ntu or less) and organic matter. Potential supply sources must also be stable, of high quality, and not subject to significant algal blooms or other major water quality changes.
- Conventional clarification and rapid sand filter plants would be needed for source waters higher in turbidity and organic matter content, and those where higher coagulant dosages may be required.
- The limitations of membrane filtration are more difficult to define. Closed-vessel applications would more likely be limited to source waters of higher quality. Immersed membranes can handle higher solids loading including that generated in precipitation and coagulation to improve solids removal.

In considering conventional clarification/filtration, the type of clarification selected would also be influenced by the type of source water solids. Where suspended and/or dissolved organic matter predominates, high-rate clarifiers including dissolved air flotation could be the more effective and more economic application. Where seasonal algal blooms occur, dissolved-air flotation may be the preferred alternative. Plain settling with greater detention would be desired where more profuse, denser solids predominate, such as in many river sources.

With few exceptions, most U.S. public surface water supplies are currently provided treatment the preponderance of which is some form of filtration. For this reason, a major water treatment plant design effort now and in the future will be devoted to retrofit rather than new design. New designs for the most part will be for new area development, major capacity expansion, or changing rules, such as for groundwater.

Another feature of water treatment plant process selection and design that must be observed is potential regulatory conflict. Scrutiny of Tables 2.1 and 2.3 will show that there are two groups of rules that currently apply or soon will apply to all public drinking water treatment. These are the control of microbiological contaminants and the formation of

TABLE 2.4 Most Common Drinking Water Treatment Processes

Water quality parameter	Process components

Turbidity-particulate reduction	Filtration
	• Rapid sand—conventional
	  Coagulation
	  Flocculation
	• Clarification
	  Plain settling
	  Plate settlers
	  Solids contact
	  Dissolved air flotation
	  Filtration
	• Rapid sand—direct mode
	  Coagulation/flocculation
	  Filtration
	  Slow sand filtration
	• Diatomaceous earth filtration
	• Membrane filtration
	  Ultrafiltration
	  Nanofiltration
	  Reverse osmosis

Bacteria, viruses, cyst removal	Partial reduction—filtration (above) Inactivation—disinfection • Chlorine • Chloramine • Chlorine dioxide • Ozone • UV
Color	Coagulation/rapid sand filtration Adsorption • Granular activated carbon (GAC) media • Powdered activated carbon (PAC) addition • Synthetic resins (ion exchange) Oxidation • Ozone • Chlorine • Potassium permanganate • Chlorine dioxide Nanofiltration
Taste and color control	Oxidation • Ozone • Chlorine • Chlorine dioxide • Potassium permanganate BAC adsorption
Volatile organic reduction	Air stripping • GAC adsorption • Combination of the above
Disinfection by-product control	Precursor reduction • Enhanced coagulation • GAC adsorption • Biologically activated carbon (BAC) media—preozonation • Nanofiltration

TABLE 2.4 Most Common Drinking Water Treatment Processes (*Continued*)

Water quality parameter	Process components
	By-product removal • GAC adsorption • Air stripping (partial)
Iron, manganese reduction/sequestering	Filtration of precipitators formed by preoxidation • Sand and/or anthracite media • Green sand media • Proprietary media Polyphosphate sequestering agent
Hardness reduction	Lime softening • Ion exchange • Nanofiltration
Inorganic, organic chemical reduction	Ion exchange Biologically activated carbon media Adsorption Reverse osmosis
Corrosion control	Posttreatment • pH adjustment • Inhibitors

disinfection by-products. New D/DBP Rules require defined reduction in water TOC content which may require enhanced or increased coagulation. It is possible that the increased solids generated could overload certain filters or filter types. While perfectly capable of producing low-turbidity water in conformance with the microbiological rules, solids overloading may cause turbidity upset, especially in single-pass direct filtration, slow sand and diatomaceous earth (DE) filtration, and ultrafiltration. Specific situations will vary. The D/DBP Rule does permit alternative options to TOC reduction (Table 2.1). One option is to maintain system running TTHM and HAA5 averages at half the Stage 1 D/DBPR requirements. In addition to TOC reduction, another feature that will influence the magnitude of TTHM and HAA5 formation is the time of travel or the residence time in the distribution system. There are many systems with low retention where the TTHM and HAA5 levels tend to be low, but where TOC levels require removal. This would be especially so in small to moderate-sized systems or in systems with an effective transmission network feeding the smaller distribution mains. In these situations, where no reduction (or a small reduction) in the TTHM and HAA5 levels may be required to meet the alternative standard, it might be more advantageous to concentrate treatment modifications on TTHM and HAA5 reduction than on TOC removal. This approach might result in reduced residual solids generation. In direct filtration and slow sand filtration plants, adding granulated activated carbon (GAC) media and/or adding oxidants to increase TOC removals in biologically activated media may be enough to meet the D/DBPR alternative to TOC removal.

If simple process supplementation does not achieve the new rule requirements, treatment addition should be considered. A possible addition to existing filtration processes to improve effluent quality would be nanofiltration. First stage filtration would improve the efficiency of nanofiltration, and the residual stream could be routed back to the initial filter (if appropriate) for recovery.

The purpose of the above discussion is simply to emphasize the need for the engineer to evaluate the effect of each treatment change or addition in process. It is also essential that options permitted in the regulations be explored to achieve both effective and economic water treatment plant designs.

Treatment Train Alternatives

In developing a water treatment train, the multiple treatment capabilities of the different methods and materials should all be considered to both simplify and reduce the cost of facility construction and operation. A treatment train should not be considered simply as a sequence of process steps. In essence, a treatment train encompasses a combination of processes that, when integrated, achieve the desired water quality changes and improvements.

Multiple capabilities of the different options in Table 2.4, Most Common Drinking Water Treatment Processes, are grouped in Table 2.5. Comparison demonstrates why rapid sand filtration with coagulation is the most common type of water treatment in use. This type of filter plant reduces the content of particulate matter, pathogens, disinfection by-product precursors, and color. If a filter is fitted with a GAC medium, the feature of taste and odor control is added, and color and precursor reduction are enhanced. With the addition of preozonation, the effective life of the GAC medium is increased because organics are removed more through biological action. In addition, preozonation, or application of another oxidant, will condition dissolved iron and manganese for removal in the filter media.

Options and supplementary features that may be considered in developing baseline rapid sand filtration trains are illustrated in Figure 2.1. Basic treatment facilities are shown in solid boxes with options shown dashed.

Figure 2.1 illustrates the flexibility and broad treatment capabilities of the several options that may be adapted for many treatment requirements. For instance, different coagulants may be considered that can be afforded high- to low-energy flocculation to produce a floc size and density best suited to the particular settling device or filter medium. Flocculation will also agglomerate some dissolved as well as suspended organic matter. Small, light, and floatable solids may best be removed by dissolved air flotation rather

TABLE 2.5 Common Multipurpose Treatment Measures

Treatment measure	Quality improvement
Filtration (all)	Particulate reduction
	Bacteria, virus, cyst reduction
Coagulation, rapid sand filtration (additional)	Precursors and by-product reduction
	Color removal
Oxidation	Pathogen inactivation
	Partial organics reductions
	Non- to biodegradable organics
	Color removal
	Taste and odor control
	Iron and manganese reduction*
GAC media	Rapid sand filter particulate removal
	Color removal
	Precursor and by-product reduction
	Additional reduction with preoxidation
	Taste and odor control

*With filtration.

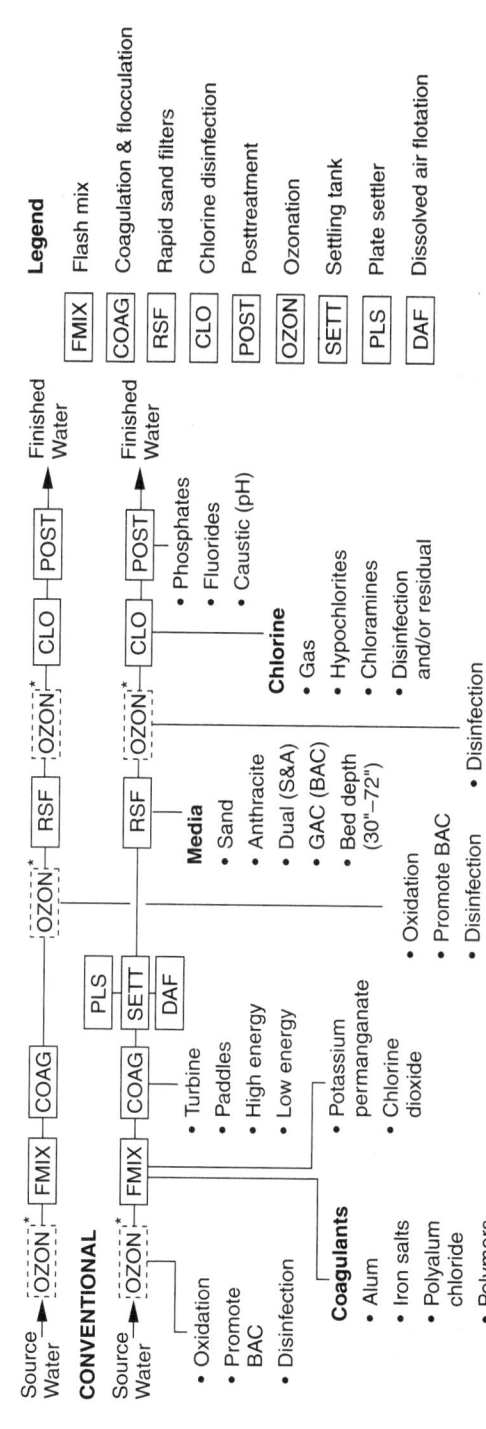

FIGURE 2.1 Baseline filtration options—rapid sand filters.

than by settling. Using GAC media preceded by ozonation provides the capability of additional organics reduction by biologically activating the media. Ozonation may also reduce or convert many organics to enhance disinfection by-product reduction and to reduce chlorine demand. The multiple advantages of the options shown in these figures will be discussed in greater detail later in this text.

Supplemental and additional treatment options that may be considered for normally single-pass filters are discussed below. Posttreatment and its options are the same as those shown in Figure 2.1.

Slow Sand Filter. Principal solids removal occurs in the top surface layer schmutzdecke, which is also biologically activated. Typical TOC reduction can range from 10% to greater than 20%. An intermediate GAC sandwich layer may be installed in the deep sand media to create an additional host layer. With preoxidation, using ozone, chlorine dioxide, or potassium permanganate, biological removal of organics will be enhanced in both the top and the sandwich layer.

Diatomaceous Earth Filters. This filter has superior cyst removal capabilities. However, there is no way to promote organics removal in this filter alone; a biologically activated GAC column would have to be added at considerable cost. Use of this filter might, therefore, be limited to waters of low organic content and low turbidity such as in many well supplies.

Membrane Filtration. Application of this type of filtration is developing. Little standardization has been established. Except for grading the pore size of membrane material from the finest for reverse osmosis to the coarsest for ultrafiltration and microfiltration, details are proprietary. Each manufacturer has individual arrangement details. For the most part, the most flexible use is with immersed-type ultrafiltration where coagulants may more readily be added to improve solids and organics removal. Preoxidation may also be employed to further enhance organics removal as well as promote the removal of iron and manganese. Note that the finer membrane types such as nanofiltration and reverse osmosis will directly remove these constituents, but at greater cost and with greater loss in the waste stream.

Additional-Stage Filtration. Mentioned before, existing filtration installations may not be able to meet the new requirements. In some instances it may be more practical than replacement to provide an additional stage of filtration to meet lower-turbidity standards. For instance, membrane filters could be added at the end of the train to further reduce turbidity and to remove other contaminants. With prior solids removal, membrane treatment becomes more efficient and waste of water may be eliminated by recycling to the original plant inlet. Closed-vessel membranes, however, may not be placed downstream of DE filters. The very fine siliceous DE particles in the effluent could damage and clog the membranes. For low-capacity DE filter units, it may be possible to add cartridge filters to further reduce turbidity. Membrane filtration can readily accommodate the effluent from all other granular media filters including the slow sand type.

While not exactly second-stage placement, immersed-type membrane units may be installed in existing rectangular settling basins of conventional filter plants while other pretreatment and posttreatment facilities may be kept in service.

Less Common Treatment

As mentioned, the above review of optional treatment trains included those more common or those more universal. Many systems must address the reduction or removal of less common or special contaminants such as those that may be found in the USEPA listing of inorganic, synthetic, and volatile organic contaminants. Where these exist, special treatment accommodation may be required. It is possible, however, that existing treatment for more common contaminants may be effective in reducing the amount of special contaminant or at the least provide necessary pretreatment. In very general terms, treatment methods that may be considered for contaminants in the three categories mentioned above are as follows:

- *Inorganics.* Oxidation or chemical reaction to produce *innocuous* compounds or precipitates and/or ion-exchange following filtration.
- *Synthetic organics.* Herbicides and pesticides, most of which may be removed in GAC columns. Some contaminants may be preconditioned by strong oxidants.
- *Volatile organics.* Removed by air stripping and/or in GAC columns.

There are a number of special or infrequently found contaminants that have not been discussed. Wherever special or less common contaminants may exist in source waters, it is always essential that detailed investigations, possibly including pilot studies, be conducted before treatment facilities are designed.

Treatment Comparisons and Evaluations

While the treatment rules become more demanding, the list of available "tools of the trade" is also expanding. It is up to the designer to take advantage of the many treatment resources discussed later in this text that are best adapted to the particular plant application.

The principal intent of these discussions is to emphasize that many treatment options and combinations of options are available to the designer and that all viable options must be investigated to achieve that treatment or combination of treatments best suited for the particular application. It is also essential that issues other than treatment capability be investigated for each option and each treatment train. These other issues may include the following (not necessarily in order of importance):

- Construction cost
- Annual operation costs
- Site area required
- Complexity of operation (required capability of operating staff and laboratory monitoring)
- Operation risk (most common causes, if any, of treatment failure)
- Flexibility of plant arrangement for future changes
- Waste disposal options

Consideration of viable options would also be critical to provide a flexible facility arrangement in which additions and modifications may be made for future treatment requirements. Drinking water treatment design is not static; it is a dynamic, ever-changing process.

CHAPTER 3
DESIGN AND CONSTRUCTION

R. Gary Fuller
HDR Engineering, Inc.
Denver, Colorado

This chapter provides general guidelines for managing the design and construction phases of a water treatment project. These projects involve addition of either entirely new facilities or the upgrade/expansion of existing facilities. The principal design and construction phases for both types of projects can typically be separated into preliminary design, final design, construction assistance, and facilities start-up. Traditionally, these phases of design and construction have been conducted sequentially, with a single entity conducting all phases. Growing interest in "design-build" projects has placed an emphasis on initiating construction prior to final design, accelerating the project schedule by combining the design and construction teams into one entity. Regardless of which approach is used, many of the design and construction issues remain the same.

PRELIMINARY DESIGN

Completion of the master planning and process selection phases establish the design concepts for the water treatment improvement project, forming the basis for preliminary design. The preliminary design phase must address a variety of design issues, concluding with concise recommendations for the principal components of the water treatment plant facilities. During preliminary design, the primary treatment processes and waste handling systems design parameters will be established to allow the design team to initiate development of the design concepts for all the other disciplines. Close coordination with the design team members, owner, and regulatory agencies is essential during this phase of the project.

Regulatory Agency Coordination

Regulatory agencies having approval authority over construction of a water treatment plant project may include local building departments, zoning commissions, and environmental commissions. In addition, there will be state and federal authorities having jurisdiction over water supply public health and environmental issues. Close coordination and communication with these agencies during the preliminary design phase is essential, as the plant design will eventually be reviewed and approved by these agencies.

Most regulatory agencies have design standards or guidelines that should be reviewed to identify any potential impacts to the project. As part of this review, agency permits required for the project should be identified. Permits may be required for a number of plant design issues that could affect project implementation, such as site approval, stormwater management, plant discharges to receiving streams, construction in waterways, environmental impacts, and water rights.

Although it may not be required, it is often beneficial to submit copies of preliminary engineering documents to regulatory agencies for review and comment, to avoid major changes during final design.

Plant Siting Analysis

Selection of potential sites for a new water treatment plant must take into consideration a number of factors. Some of the principal ones are

- Proximity of plant site to the source water and to customers being served
- Consideration of water transmission requirements to interconnect the plant to the source water and the water distribution system
- Proximity of plant site to ultimate treatment waste disposal location
- Environmental and land use concerns
- Subsurface and geotechnical considerations
- Land availability, cost, and zoning
- Storage requirements at plant site for raw water supply
- Compatibility with surrounding existing and planned developments
- Potential for flooding and stormwater handling requirements
- Availability of utilities (power, natural gas, sewer, telephone)
- Site topography and accessibility
- Vulnerability to security risks and natural disasters

Source water supply, treatment, and finished water transmission facilities must function as a complete system to provide a safe, reliable source of drinking water to the water distribution system. The water treatment improvements design must take all three types of facilities into consideration to ensure a proper transition between systems.

Upgrading or Expansion of Existing Facilities

Many water treatment plant design projects involve modifying existing facilities to provide increased capacity, to improve treatment performance, or to modernize and renovate equipment and structures. These projects require special considerations and investigations that are not required when one is designing a new plant on a new site.

Investigations made during preliminary design for these types of projects must answer questions such as the following:

- To what extent must the existing facilities be brought up to current design standards and building codes as part of the project?
- How can the original design, as-built facilities, and current condition be defined, and remaining useful life of existing facilities and structures be determined?
- How can the existing treatment facilities be kept in reliable service during the construction of plant modifications and be integrated with the new facilities?
- What safety or environmental hazards (such as lead paint, asbestos insulation, and underground petroleum spills) may be uncovered during the construction of the modifications?
- What provisions have been previously made in the plant to expand or upgrade the existing facilities?
- Are there any existing plant site or hydraulic constraints that must be taken into consideration with the facilities improvements?

Design decisions on these issues often require reasonable compromises that must be worked out among the design team, the owner, and regulatory agencies. The cost-effectiveness of modifying each existing plant component versus constructing a new component should be considered.

Custom-Built Equipment versus Preengineered versus Package

When the treatment plant under design has a capacity of less than about 10 million gallons per day (mgd), it is often possible to choose between custom-designed, preengineered, or "package" treatment units. In custom-designed units, major processes usually utilize reinforced concrete construction. The design engineer is responsible for all dimensions of the unit and the complete process electrical, mechanical, and structural design. Equipment suppliers provide and are responsible for design and performance of specific equipment such as weirs, gates, sludge collectors, and other equipment located in the process unit.

Preengineered units are generally made using steel or fiber-reinforced plastic tanks. An equipment supplier takes responsibility for the full process, structural, electrical, and mechanical design of the unit. The design engineer generally is responsible only for the foundation design and items such as influent water quality, hydraulic, and mechanical connections, electrical and instrumentation connections, and, if required, buildings or enclosures to house the process unit. Preengineered units are strictly field-assembled.

Package units are partially or completely factory-assembled and installed on the site. The package units are often skid-mounted, with the process components prepiped and prewired, requiring minimal fieldwork to install. The package units are often restricted to 1 to 2 mgd in capacity, as the size of the unit is constrained by shipping limitations.

Custom-designed process units tend to have higher construction costs but lower maintenance needs and longer useful lives. The best decision for a specific project can be made only after a careful analysis of the costs and benefits of each feasible alternative. In general, as the plant capacity becomes smaller, preengineered and package units become more cost-effective and require less time to construct.

Design Development

As part of the preliminary design phase, several key design parameters must be developed:

- Plant flow schematics indicating all treatment units and equipment for both the primary treatment processes and waste handling systems
- Conceptual plant layouts showing the arrangement of treatment process units and other related facilities with respect to site topography
- Plant hydraulic profile establishing operating water elevations through the plant at normal and peak flow rates, for both the primary treatment process and waste handling system
- Design criteria establishing treatment unit sizes and specific design factors for minimum, average, and maximum flow conditions of the major treatment processes and waste handling systems
- Requirements for ancillary facilities including space needs for plant administration, laboratory, maintenance, chemical handling and storage, low and high service pumping, and water storage
- Plant control concepts including plant control logic, control system architecture, and communications with off-site facilities
- Treatment waste handling and ultimate disposal scheme, including off-site waste transport requirements
- Architectural and structural concepts for the plant buildings and structures, including weather protection provisions for the treatment facilities
- Provisions for future expansions of the plant facilities, allowing for addition of equipment and treatment process units, with plant infrastructure oversized where appropriate
- Security measures required to provide the necessary protection of the plant facilities and ensure plant reliability

Financial Feasibility Evaluation

Financing for a water treatment plant project should be finalized during, or immediately after, the preliminary design phase, when sufficient information is available to establish a valid project budget. Preliminary estimated costs for plant facilities should be developed, including the implementation schedule, approximate cash flow, and operations and maintenance costs, that establish projected annual costs for the new facility. Potential sources of revenue for financing the project must then be investigated to establish a viable program to fund the planned facility. AWWA Manual M29, *Water Utility Capital Financing,* provides a comprehensive overview of traditional and innovative financing sources that should be evaluated by water utility management in determining short- and long-term strategies for meeting capital requirements.

Ongoing Coordination and Cost Evaluations

During the preliminary design phase, design development should be coordinated with all support disciplines to obtain input on the overall design. This coordination is necessary to make sure that preliminary engineering does not overlook any design issues that could affect critical design decisions. Although the process and civil disciplines are normally the primary focus of the preliminary engineering phase, the architectural, structural, mechanical, electrical, and instrumentation support disciplines should be included in these discussions.

Regular meetings should be held with the owner during preliminary design to discuss design issues and update the client on the process of design. Design disciplines pertaining to design issues being discussed should be involved in client meetings whenever possible. Involving the regulatory agencies that will need to approve the project may also be beneficial to the design effort. Quality control reviews should be conducted of the preliminary design by a team of design professionals independent from the design team.

Value engineering is often conducted before the beginning of the final design to evaluate the cost-effectiveness and constructability of the preliminary design. At this stage of design, value engineering focuses primarily on major design elements, such as number and size of treatment units, plant layout, design parameters, and support facilities. The value engineering comments should be reviewed with the owner and design team to determine which ideas will be incorporated into the design.

During preliminary design, a preliminary cost estimate for the water treatment plant project should be prepared to ensure that project cost is within the owner's budget established during earlier feasibility studies. Cost estimates are important for (1) making decisions between viable plant alternatives, (2) finalizing the treatment process train, and (3) planning the project's budgeting and financing. The cost estimate should provide a breakdown of costs for each major component of the project. A construction contingency of 15% to 25% is usually included in the cost estimate at this stage of design to provide for unforeseen costs not identified in a preliminary estimate.

FINAL DESIGN

The final design phase is initiated after completion of preliminary engineering and documentation of design decisions, obtaining approval of review agencies, and securing project financing. The design criteria and schematics developed in preliminary design will be used to guide the development of detailed drawings and specifications, modifying the preliminary design only as needed to accommodate the other design disciplines.

Production of Design Documents

Final design involves preparing detailed drawings and specifications within the project framework established during preliminary design. Contract documents must also be prepared that establish the bidding and contractual requirements for the construction of the plant facilities.

Drawings and Specifications. The final design drawings and specifications preparation must take into consideration a number of constraints and requirements not identified in the preliminary engineering phase. Treatment process design should be completed in greater detail before the support disciplines undertake final design. Critical design elements that should be completed early in final design include

- Selecting process equipment to be installed, including equipment sizes, weights, utilities required to support the equipment, equipment control requirements, access requirements for maintenance, and provisions for variations in the requirements of equipment from different manufacturers
- Delineation of ancillary facilities required, including piping, valves, chemical systems, controls, pumps, and other equipment

- Finalization of major plant piping and structures, including pipe size and routing, treatment structure sizes and elevations, and support facilities size and functions
- Development of process and instrumentation diagrams (P&IDs) for all treatment systems, including process units, piping, valving, instruments, and their interrelationships

Project Coordination. It is important to maintain close communications with the client and regulatory agencies through all phases of design. Regular meetings should be held with clients to solicit their input and apprise them of project status. Members of the project operations staff should be involved in the meetings, when possible, to provide input on design decisions and to become familiar with the new plant facilities. During design, all disciplines should be kept informed of design decisions by any discipline that could affect the design. Regular design team meetings should be conducted to exchange information and coordinate conflicts between disciplines.

Schedules of project work tasks and project milestones should be prepared for all stages of design, with the schedule updated as changes are necessary. Reviews of the schedule should be conducted periodically to confirm that work tasks and milestones are on schedule. As part of schedule development, project budgets can be prepared based on the identified work tasks and schedule. Input from each design discipline should be obtained during budgeting to establish the needed work tasks, schedule, and budgets for each design phase. Previous project budgetary information can be a particularly valuable source of information for developing budgets for a new project. AWWA Manual M47, *Construction Contract Administration,* provides information on the requirements for maintaining communications between all parties involved in the construction project.

Design Standards. It is critical that the plant facilities' design conform to both legally required and generally accepted design standards, including

- Treatment design criteria published by state and federal agencies
- Design codes, such as building, electrical, fire, and plumbing codes
- Applicable laws and ordinances, such as Occupational Safety and Health Administration (OSHA) safety standards and Americans with Disabilities Act (ADA) requirements
- Design standards developed by local agencies and by the owner

If the proposed design requires any deviations from legally required criteria, the changes must be discussed with the approving agencies and written variances or exemptions obtained. Any deviance from generally accepted practice should receive careful consideration and should be reviewed with the owner.

Coordination of Design Disciplines

Once final design is under way, it is essential that all design support disciplines be closely coordinated to avoid design conflicts and to make sure design information from all disciplines is received in a timely manner.

Civil Process Design. In developing the plant layout for construction of new facilities or modification of existing facilities, a number of design considerations should be addressed:

- Matching the plant facilities to the site topography to take advantage of physical features and grade changes across the site

- Providing plant access to facilitate chemical delivery and waste hauling, to allow plant operations and maintenance staff routine traffic, and to permit visitors to the plant facilities
- Siting of new facilities to avoid construction conflicts, allowing continual operation of existing facilities and providing easy access for plant operation
- Interfacing requirements with existing facilities, including treatment units, piping, buildings, controls, power supply, and chemical feed systems
- Providing ample site space for future additions or expansion to the plant facilities, including provisions for future tie-ins and vehicle/operator access

Plant hydraulic design should be conducted early in the project, with the hydraulic profile developed to establish structure elevations and the interrelationship between facilities. Plant facilities should be arranged on the site to take advantage of available physical relief across the site with respect to plant hydraulics, locating treatment structures to minimize earthwork where practical. Siting of plant facilities should also take into consideration a number of other site factors, including

- Topography elevation changes across the site
- Plant site buffer and security requirements
- Physical features, barriers, or constraints within the site
- Floodplain limits and flood protection requirements
- Drainage and stormwater detention provisions
- Piping and conduits between structures
- Vehicle access and parking for plant personnel, public, and deliveries
- Natural gas, power, sewer, and telephone utility routing
- Plant aesthetics and public acceptability
- Archaeological, historical, or cultural resource protection

The final process design should also address a number of factors that affect other design disciplines, including

- Final dimensions of treatment structures and buildings
- Utility requirements for process equipment
- Process equipment locations, dimensions, and operations requirements

Site Surveying. Topographic and property surveys should be made for the water treatment plant site to provide the baseline conditions for developing the plant location design and for obtaining many of the required permits. A permanent site benchmark should be established at the plant site a location that will not be disturbed by future construction. A legal description of the property is also needed, indicating all easements and rights-of-way at the site.

A site plan is then prepared indicating property limits, site contours, physical features, buried utilities and pipelines, structures and buildings, roadways, easements and rights-of-way, and other features.

Geotechnical Investigation. The geologic conditions at the water treatment plant site should be established by a geotechnical investigation. Any existing subsurface information available should be collected first and reviewed to identify potential problems. The location of soil and rock strata should then be established by test borings and excavation of test pits. This information is vital to establish the bearing capacity of soils, to establish rock elevations, and to determine groundwater levels. Location of soil borings or test pits should be indicated on the site topographic plan, with the site survey coordinated to accurately locate the geotechnical investigations. Seismic conditions at the site may also have to be investigated to determine any special design considerations that need to be addressed. In addition, it may be necessary to conduct hazardous materials investigations to identify any potential site contamination or subsurface concerns.

Architectural Design. Development of the architectural design for water treatment plant facilities must be closely coordinated with the process design as the treatment building requirements are normally based on the process requirements. Architectural concepts should be compatible with plant surroundings, with particular emphasis on minimizing building maintenance. Architectural design is also based on input from the other design disciplines, including

- Treatment unit and support facilities enclosure requirements
- Equipment and piping/conduit space requirements
- Operator access requirements and maintenance provisions
- Public access requirements to plant facilities

Where practical, architectural concepts used for plant additions or modifications should be compatible with existing plant buildings. Provisions for laboratory, office, maintenance shop, locker rooms, restrooms, storage, lunch rooms, meeting rooms, and control rooms are usually included in the plant design.

Structural Design. Structural design is normally most affected by the civil, process, and architectural design decisions. The size and location of the treatment structures are particularly critical to structural design. If unusual or difficult geotechnical conditions are encountered, it may be necessary to change structure concepts to adapt to the subsoil constraints. As a result, early coordination with the geotechnical investigations is necessary to identify any unusual conditions. Water containment structures will require special provisions to minimize water leakage and allow for variation of water level within the structure.

Mechanical Design. Mechanical design establishes heating, ventilating, and air conditioning (HVAC) requirements for buildings and enclosed structures. Chemical handling, storage, and laboratory areas normally require special attention because of their unique HVAC requirements. Mechanical and architectural designs must be coordinated closely to ensure that sufficient building space is allocated for the plumbing and HVAC equipment and ductwork. Fire protection systems may also be required for some plant areas, particularly the chemical handling and storage areas, and will need to be coordinated with the local fire department.

Electrical Design. Electrical design should address a number of issues related to other design disciplines, including

- Power supply needs for all plant equipment and ancillary facilities
- Standby power requirements for treatment and support systems
- Energy conservation measures for plant operation
- Interior and exterior lighting requirements for operation and security

Adequate building space must be provided for electrical switchgear and motor control centers, with major electrical equipment often isolated to protect and adequately cool the equipment. The electrical equipment location must protect the system from corrosive and dust-prone areas of the plant.

Instrumentation Design. Instrumentation design is closely interrelated with process design to provide necessary control and monitoring of treatment processes. Automation of plant control should be considered, where practical, which requires particularly close coordination with process designers. A central control room is normally provided for most treatment plants, with plant operations monitored and controlled from that location. The capability to access plant control and monitoring information from critical operation locations within the plant should be provided, particularly for plants with multiple buildings.

Design Reviews and Final Documents

It is extremely important that the design team continually keep in mind various reviews that must be periodically made of final documents as they are developed and completed. The review comments should be documented, with the action taken to address the comments noted.

Owner Reviews. Throughout final design, the owner should be involved in key design decisions. Periodic meetings to discuss plans and specifications should be conducted to ensure that design meets the owner's needs. Plans and specifications should be submitted to the owner for review as project design proceeds, including intermediate and final documents. The owner's comments should be discussed with design disciplines and incorporated into the documents. Participation of the owner's operations and engineering staff is extremely important, because they will be the ultimate users of the facility.

Quality Control Reviews. The final design drawings and specifications should be reviewed by an independent team involving all design disciplines represented on the project. The reviews should consider both the individual disciplines as well as the interrelationships between the various disciplines. The quality control review comments should be documented, with each design discipline providing a response on the action taken to address each comment.

Value Engineering. If value engineering is performed at this stage of the project, it normally focuses on design details. It would not ordinarily consider major design concepts, because changing the design at this stage involves significant changes to plans and specifications. Issues usually addressed include materials of construction, equipment selection, system redundancy, and constructability.

Regulatory Agency Reviews. Final plans and specifications must be submitted to various regulatory agencies for review, with the review comments incorporated into documents where necessary. If possible, meetings should be conducted with regulatory agencies during final design to keep them informed of project status and to obtain their input on critical design issues.

Final Cost Estimates. A detailed cost estimate should be prepared as soon as practical during final design. The estimate should be completed for each design discipline, indicating unit quantities and costs for all construction items. Cost estimates should be submitted for owner review. In addition, they must accompany applications for state or federal funding assistance.

Bid Documents. Final design plans and specifications can be formatted to allow construction using one of several contracting approaches. Normally, conventional bid documents are used that allow the owner to accept competitive bids and award the entire construction job, or job components, to the lowest qualified, responsible bidder. Under this approach all elements shown in the bid documents are the contractor's responsibility to construct, with the owner and engineer inspecting the project to ensure compliance with bid documents. In some instances, major equipment to be installed on the project is prepurchased by the owner and supplied to the contractor for installation. Prepurchased equipment is normally used to accelerate the construction schedule or to allow use of proprietary equipment.

Another bid approach gaining popularity is use of a "design-build" format. Under this approach, only preliminary engineering is completed for bidding, with final design prepared by the lowest, qualified, responsible bidder. The owner then monitors compliance with bid documents during construction, usually with assistance from the preliminary designer. The primary benefit of the design-build approach is the single source of responsibility provided to the owner, allowing a single point of contact for design and construction issues. Combining design and construction responsibilities provides a greater potential to shorten the project implementation period by overlapping design and construction. One of the primary concerns with using design-build is the possible loss of checks and balances between the designer and contractor, which is provided by the traditional contractual separation between both parties.

CONSTRUCTION ASSISTANCE

After completion of final design and incorporation of all pertinent review comments, the water treatment plant design can proceed into the construction phase. During construction of water treatment plant facilities, a number of design issues must be addressed to ensure the successful implementation of the contract documents. These relate to bid administration, construction administration, design disciplines coordination, and owner and regulatory review of change orders.

Bid Administration

The time frame required for bidding varies depending on the size and complexity of the project but is normally about 30 to 60 days. Bid administration activities generally include printing the bid documents, conducting prebid meetings, answering bidder questions, preparing addenda to bid documents, bid evaluation and award recommendation, and contract preparation. Bid administration concludes with the award of the construction contract and the issuance of a notice to proceed to the selected contractor.

Construction Administration

The administration of project construction consists of a number of tasks that must be provided to oversee the work in progress. Some of the principal duties include

- Field observation of the work as it progresses to verify the contractor's compliance with contract documents
- Clarification of contractor questions pertaining to the drawings and specifications
- Review of shop drawings submitted by the contractor
- Review of payment requests
- Monitoring compliance with the construction schedule
- Testing materials and systems for acceptability
- Coordination of field orders and change orders

Timely and relevant communication between all parties involved in the construction of the plant improvements is critical to the success of the project. On larger or more complex projects, formal "partnering" is sometimes utilized to facilitate communication between all parties and to minimize conflicts.

Coordination of Design Disciplines

All design disciplines should be actively involved in construction administration. Members of the design team should make periodic visits to the construction site to monitor construction activities related to their discipline. As part of plant construction administration, it is necessary to coordinate modifications and changes to design with the various design disciplines.

Construction Approvals

No matter how carefully a construction job is planned and designed, there are almost always some changes that must be made during construction as a result of unexpected problems or a change of mind by the owner, designer, or regulatory agency.

Owner Approval of Change Orders. To facilitate timely processing of changes to contract documents, the owner should be apprised of potential or upcoming change orders as they become apparent. Owner approval is normally required for most change orders unless the designer has been designated as the client's contractual authority. Documentation should be provided for each change order to ensure that all parties have a clear understanding of the changes it will cause in the project cost and construction schedule.

Regulatory Approval of Change Orders. In many instances in which regulatory approval of the design has been obtained, change orders must be approved by the agencies to ensure that the project still complies with agency regulations. Regulatory approval of change orders may also be necessary when funding for construction is provided by a regulatory agency.

Facilities Start-up

The final phase of a water treatment plant construction project involves activities associated with start-up and initial operation of the new facilities. The contractor is normally required to organize and conduct the start-up of new plant facilities, with the owner and designer overseeing the activities.

Operator Training

With any new or modified plant facility, it is important that plant staff receive training to operate the new equipment and systems. Operations and maintenance (O&M) manuals and formal training should be provided by manufacturers for all new equipment and systems. Electronic O&M documents are gaining popularity, and provide a valuable tool for accessing data from a variety of sources. Before start-up, equipment or systems manufacturers or suppliers should inspect equipment to verify proper installation, supervise any adjustments or installation checks, provide a written statement that equipment is installed properly and ready to operate, and instruct the owner's personnel on proper equipment operations and maintenance.

Equipment Start-up

Starting up a new or modified treatment plant must be carefully coordinated to provide a manageable sequence of operations that demonstrates acceptable operation of each system and piece of equipment. Depending on the type of facility, individual or multiple systems of the plant may undergo individual start-up if they will not interfere with normal operations of existing facilities. For new facilities, it may be necessary to operate the entire treatment process at start-up due to the close interrelationship of all treatment systems.

Construction Finalization

At the completion of construction, the design disciplines should conduct a final inspection of the plant improvements to confirm compliance with the drawings and specification. As-built drawings of the facilities should be prepared, incorporating all design changes made during construction. After completion of all contract requirements by the contractor, final acceptance of the project by the designer and owner should be issued.

BIBLIOGRAPHY

American Water Works Association. *Construction Contract Administration,* M47. Denver, Colo.: AWWA, 1996.

———. *Water Utility Capital Financing,* M29. Denver, Colo.: AWWA, 1998.

Corbitt, Robert A. *Standard Handbook of Environmental Engineering.* New York: McGraw-Hill, 1990.

Kawamura, Susumu. *Integrated Design of Water Treatment Facilities.* New York: John Wiley & Sons, 1991.

Water Environment Federation and American Society of Civil Engineers. *Design of Municipal Wastewater Treatment Plants,* vol. 1. Alexandria, Va.: WEF and ASCE, 1991.

CHAPTER 4
INTAKE FACILITIES

Steven N. Foellmi
Black & Veatch
Irvine, California

Intakes are structures built in a body of water for the purpose of drawing water for human use. As discussed in this chapter, intake systems include the facilities required to divert and transport water from a supply source, such as a river, lake, or reservoir, to a shore well or pumping station. For small water supplies, the intake system may be relatively simple, consisting of little more than a submerged pipe protected by a rack or screen. In contrast, for major water supply systems, intake systems can be extensive, with diversion accomplished by intake tower structures or submerged inlet works. An intake system may also include transmission conduits, screens, pumping stations, and, in some instances, chemical storage and feed facilities. This chapter presents a brief review of intake features, describes types of intake systems, and provides a discussion of intake design considerations. The design of racks and screens is also discussed.

INTAKE FEATURES

The purpose of an intake system is to reliably deliver an adequate quantity of water of the best available quality. Reliable intake systems are costly and may represent as much as 20% of the total water treatment plant investment. Pipeline construction associated with intakes may involve extensive underwater work and the use of specialized marine equipment. As a result of these and other factors, the cost of such work would be 2.5 to 4 times more than that for a similar land project. An intake system must possess a high degree of reliability and be able to supply the quantity of water demanded by a water utility under the most adverse conditions. Intakes are exposed to numerous natural and artificial perils, and it is important that the designer anticipate and make provision for operation under adverse conditions. Conservative structural and hydraulic design and careful selection of intake location should be priority considerations. Changing and increasingly stringent drinking water regulations present new challenges that potentially impact the design of intake facilities. Application of chemical treatment at the intake facilities may be advantageous to achieve water quality or treatment goals. Problems associated with zebra mussels provide additional challenges to intake system design.

FIGURE 4.1 Component block diagram.

Intake Design Objectives

Specific components that make up an intake facility are influenced by many factors, including characteristics of the water source, required present and future capacity, water quality variations, climatic conditions, existing and potential pollution sources, protection of aquatic life, water level variations, navigation hazards, foundation conditions, sediment and bed loads, required reliability, and economic considerations. In general, an intake facility consists of some combination of the components shown in Figure 4.1.

Each intake system presents unique challenges. Once the specific components are defined, the facility design must achieve the following:

- Reliability during water level fluctuations or channel instability
- Flexibility for water withdrawal at various depths, where desirable and feasible
- Protection against hydraulic surges, ice, floods, floating debris, boats, and barges
- Location to provide the best available water quality
- Prevention of entry of objects that might damage pumps and treatment facilities
- Adequate space for routine equipment cleaning and maintenance
- Facilities for removing pumps and other equipment for inspection and maintenance
- Location and design to minimize damage to aquatic life
- Adequate space and facilities for receiving, storing, containing, and feeding treatment chemicals

Intake Capacity

Unlike water system components such as pumping stations, basins, and filters, intake facilities are difficult to expand to provide additional future capacity. For this reason, long-range water supply needs must be carefully considered during the intake design. In general, intake requirements to meet water needs 20 to 40 years in the future (minimum) should be investigated. Although intake facility construction is generally costly, the incremental cost of incorporating provisions for expansion of capacity to meet future requirements into the original intake facility design and construction is relatively small.

Intake Reliability and Redundancy

Reliability is essential, especially for waterworks that depend on a single intake facility. For larger systems, construction of duplicate intake structures is becoming increasingly common, including multiple inlet ports and screens, screen chambers, pump wetwells, and

TABLE 4.1 Intake Location Considerations

Criteria	Remarks
Water quality	Local surface drainage
	Wastewater discharge points
	Lake and stream currents
	Wind and wave impacts
	Water depth and variation
Water depth	Maximum available
	Adequate submergence over inlet ports
	Avoid ice problems
Silt, sand	Locate to minimize impact
Navigation	Outside shipping lanes, designed for accidental impact
Trash and debris	Provisions for unrestricted flowby
W.S. elevation	Maximum practical hydraulic gradient
Treatment facility	Minimize conduit length to treatment plant
Cost	Lowest cost consistent with long-term performance and O&M requirements

discharge conduits. Where impact from zebra mussels can be expected, redundancy takes on increased emphasis because of the need to provide periodic inspection and maintenance of intake components. For systems served by a single intake, failure of the intake system means failure of the supply, an emergency condition that, if not promptly corrected, can become a water supply disaster.

Intake Location

Selecting the appropriate location for an intake facility must include an evaluation of the major factors presented in Table 4.1. A thorough study should be made of water quality data to help determine the best location for siting the intake structure. Where insufficient data are available, a water sampling and testing program may be warranted.

TYPES OF INTAKE SYSTEMS

A variety of intake systems have been employed and can be generally divided into two categories: exposed intakes and submerged intakes (Table 4.2). Additionally, many variations of these principal intake types have been used successfully. Intake systems are classified as either river intakes or lake/reservoir intakes because the circumstances, location, and types of structures used are generally quite different. Intake towers are commonly used in lakes and reservoirs and are usually located in the deepest water that can be economically reached. Intake towers may also be incorporated into a dam that creates a reservoir and used as part of a river intake system. Water is conveyed from the intake tower via a gravity pipeline or tunnel to the shore well or pumping station. Intake towers offer permanence, reliability, and flexibility in depth of draft, but their cost is substantial and accessibility can be problematic.

TABLE 4.2 Types of Intake

Category	Design type	Remarks
Exposed	Tower (integral with dam)	Applicable to larger systems, more expensive
	Tower (lake interior)	Navigational impact
	Shore well	Design for floating debris or ice
	Floating or movable	Improved access for O&M
	Siphon well	Increased flexibility, provisions for expansion
Submerged	Plain-end pipe or elbow	Applicable to smaller systems, less expensive
	Screened inlet crib	No navigational impact
		No impact from floating debris or ice
		Less flexibility
		Difficult O&M
	Gravel-packed wells	No navigational impact
		No impact from floating debris or ice
		Applicable to river systems with certain geologic conditions
	Horizontal collection systems	No navigational impact
		No impact from floating debris or ice
		Applicable to river systems with certain geologic conditions

River Intakes

Water systems utilizing rivers as a supply source can often combine an inlet structure and source water pumping into a single facility. As an example of this configuration, the bank intake system located on the Missouri River serving Water District No. 1 of Johnson County, Kansas, is shown in Figure 4.2. This facility, part of a long-range water supply program, has a capacity of 125 mgd (473 ML per day). The intake has six vertical, wet-pit pumps, each located in an individual cell and each protected by a removable bar rack and traveling screen. Source water enters through six rectangular ports located about 1 ft (0.3 m) above the bed of the river. Sluice gates enable each cell to be isolated for inspection, cleaning, or maintenance. The Johnson County intake also includes presedimentation facilities at the water source. Heavy solids (sands and silts) are separated from the flow stream to reduce the load on pumps and pipelines, as well as to improve overall performance of the treatment plant.

River channel degradation has become a significant challenge on some rivers, which has led to insufficient water withdrawal at intakes due to low flows. Emergency low-flow submersible pumps have been installed at some locations, including the Missouri River intake described above, as a temporary measure. A jetty was installed at a Kansas River intake to raise the river to a level where it would enter the existing intake ports. However, jetty installation is not always an acceptable remedy due to navigation, environmental, and recreational concerns.

The lock and dam system on the Ohio River consistently provides flow for barge traffic, ensuring sufficient depth for water withdrawal is present at the Northern Kentucky Water District's Ohio River Pump Station No. 1 (ORPS-1) (Figure 4.3). ORPS-1 withdraws water from the Ohio River for supply to the Fort Thomas Treatment Plant. ORPS-1 is a tower-type intake facility, with a firm pumping capacity of 60 mgd (227 ML per day). There are three interior chambers, or cells, inside the structure, each equipped with two

FIGURE 4.2 Missouri River intake (shore structures).

12 mgd (45 ML per day) vertical turbine pumps and a traveling screen. Raw water enters through stainless steel bar racks on one of six submerged ports (two per cell) and then passes through a traveling screen before reaching the pumps. The facility has the capability of feeding sodium hypochlorite and potassium permanganate at the port openings for zebra mussel control. The pump station is approximately 80 ft (24 m) in length and

FIGURE 4.3 Ohio River intake (shore structures).

78 ft (24 m) in width and is approximately 115 ft (35 m) high from the concrete foundation to the top of the station roof. Architectural enhancements were included to address aesthetic concerns due to the intake's location in a highly visible area and close proximity to the Greater Cincinnati Water Works' existing 1890s vintage tower intake.

Alternative intake systems should be considered to identify the most cost-effective solution that satisfies performance criteria. Examples of alternatives that may prove useful in overcoming low-water and flood difficulties as well as bed load problems include an exposed or submerged river inlet tower and shore pumping station, siphon intake, floating intake, movable intake, and well intake.

Siphon well intakes are usually installed in rivers and consist of a shore structure that receives water from the river through a siphon pipe (Figure 4.4). The siphon pipe inlet may be a submerged crib equipped with a trash rack or simply a screen section attached to the open pipe. Siphon well intakes have a record of satisfactory service and are generally less costly than other types of shore intakes.

Figure 4.5 shows the Louisville Water Company crib-type intake located on the Ohio River in Kentucky. The design capacity is 120 mgd (454 ML per day). The intake includes multiple screens mounted in a concrete crib structure constructed on the river bottom. Parallel 60-in. (1.5-m) source water conduits enable redundancy and increased hydraulic capacity to the pump station. Provisions are included to backflush each source water conduit from the pump station. The conduits also include chlorine feed capabilities to control zebra mussels, Asiatic clams, and freshwater sponges.

FIGURE 4.4 Siphon well intakes.

The groundwater potential of a river valley is also worthy of consideration. Gravel-packed wells or horizontal groundwater collection systems (collector wells) may, on occasion, offer an economical alternative to difficult, costly intake construction and the associated operating challenges when geologic conditions are favorable. Well intakes are not subject to the river intake hazards of navigation; flood damage; clogging with fish, sand, gravel, silt, ice, or debris; and extreme low water. An additional benefit associated with well intakes is the improved water quality due to riverbank filtration resulting from the surface water being subjected to ground passage prior to its collection. The water obtained from well intake systems along rivers is typically a mixture of groundwater and induced flow from the river. Gravel-packed wells are vertical collection systems with typ-

FIGURE 4.5 Louisville Water Company Ohio River intake (crib type).

INTAKE FACILITIES 4.9

FIGURE 4.6 Horizontal groundwater collection system. (*Source: Hydro Group.*)

ical diameters of 5 ft (1.5 m) or less with pipe casing. Collector wells are typically 13 to 20 ft (4.0 to 6.1 m) in diameter with a reinforced concrete caisson that has two or more horizontal screens extending into the saturated zone.

Figure 4.6 shows a typical collector well. A collector well, similar to Figure 4.6, has been installed in Lincoln, Nebraska, along the Platte River. A comparison between vertical and horizontal wells determined that life-cycle costs would be similar for the development of a 35 mgd (132 ML per day) well field. Two 17.5 mgd (66.2 ML per day) collector wells were designed for this installation due to reduced maintenance requirements and equivalent reliability compared with vertical wells. Extensive hydrogeologic testing was performed to design the collector wells and to identify the most favorable sites. The testing program included installation of test holes, observation wells, and test production wells. Long-term pumping tests were conducted to determine the aquifer properties and the amount of recharge available from the Platte River.

A 30 mgd (114 ML per day) well field along the Missouri River taps a 110-ft-deep (33.5-m) sand and gravel alluvial aquifer. This well field consists of seven gravel-packed vertical wells and one collector well. The vertical wells include an elevated platform, located above the flood of record, which houses the motor, controls, and valves. The collector well house is located above the flood of record and includes two floors, a bridge crane, and three 6 mgd (23 ML per day) pumps. One pump is outfitted with an adjustable-frequency drive. Vertical well capacities are 2.9 mgd (10.9 ML per day), and the collector well capacity is 15 mgd (57 ML per day).

River Intake Locations. The preferred location for a shore intake system provides deep water, a stable channel, and water of consistently high quality. In general, the outside bank of an established river bend offers the best channel conditions. The inside bank is likely to be troublesome because of reduced flow velocity, shallow water, and sandbar formations. The intake location should be upstream of local sources of pollution. Considerable variation in water quality can result from the entrance of pollution from tributary streams upstream of the proposed intake location, and water quality near one bank may be inferior to the quality encountered midstream or at the opposite shore.

Flood Considerations. It is essential to protect intake structures against flood damage. The intake structure must also be designed to prevent flotation and to resist the thrust of ice jams. Flood stages at the intake site should be considered carefully, and a substantial margin of safety provided. It is essential for the flood design criteria to be clearly marked on the design drawings. Due to watershed and channel alterations, future flood stages may exceed those of the past, and so the designer should consider the possibility that the intake will be exposed to flood stages in excess of those on record.

Silt and Bed Load Considerations. Many streams carry heavy loads of suspended silt at times, and heavy material move along the beds of many streams. The intake must be designed so that it will not be clogged by silt and bed load deposits. Silt, sand, and gravel can also cause abrasion of pumps and other mechanical equipment, leading to severe problems at the treatment plant. To help prevent such deposits, jetties may be built to deflect the principal flow of the river toward and past the face of the intake.

Lake and Reservoir Intakes

Both tower intakes and submerged intakes are employed for drawing water supplies from lakes and reservoirs. A tower intake may be designed as an independent structure located some distance from shore in the deepest part of the lake or reservoir. Access to these towers is typically provided by bridge, causeway, or boat. Towers must be designed to cope with flotation, wind, wave, and ice forces.

The Metropolitan Water District of Southern California (MWD) owns and operates the Diamond Valley Lake, which contains, at its maximum water level, about 800,000 acre · ft (990,000 ML) of imported raw water. Diamond Valley Lake serves as an emergency water supply for southern California. The intake tower, shown in Figure 4.7, is the structure by which water enters and exits the reservoir. The tower is 260 ft (79 m) tall, 100 ft (30 m) long in the direction of flow, and 80 ft (24 m) wide. It includes nine portal tiers, each equipped with two 84-in. (2.1-m) pipes with butterfly shutoff valves. Multiple tiers provide MWD the flexibility to selectively withdraw the highest-quality water as reservoir conditions change. The tower is equipped with four movable wire cloth screens to prevent the entry of fish during a drafting condition. The screens can be positioned over any of the intake ports selected for operation. Normal hydraulic withdrawal capacity of the intake tower is 1,100 mgd (4,200 ML per day) with a maximum capacity during emergency drawdown of the reservoir of 4,500 mgd (17,000 ML per day). The raw water is chlorinated within the lake's intake facilities to control algae and mussel growth, and the water is dechlorinated at the tower prior to entry into the reservoir.

The Charlotte-Mecklenburg Utilities Department intake on Lake Norman in North Carolina is an in-lake tower facility (Figure 4.8). The intake has an initial capacity of 54 mgd (204 ML per day) and is expandable to an ultimate capacity of 108 mgd (409 ML per day), which should satisfy projected demand for 35 years. The intake uses passive stainless steel screens mounted on the exterior of the structure's inlet pipes to exclude entry of debris and fish. The screens are designed with air backwash for cleaning, and redundancy is provided with two 60-in. (1.5-m) source water conduits to shore facilities. This facility incorporates a special "gazebo" architectural concept, which was specifically selected to blend the intake structure with the park setting.

INTAKE FACILITIES

FIGURE 4.7 Diamond Valley Lake inlet/outlet tower.

Submerged intakes generally do not obstruct navigation and are usually less costly than exposed towers. If properly located and designed, submerged intakes are not significantly impacted by ice. They consist of a submerged inlet structure, an intake conduit, and a shore shaft or suction well. Shore intakes are occasionally built on lakes and reservoirs, but in general, their effectiveness is reduced by ice, sand, and floating debris.

The intake shown in Figure 4.9 has an initial capacity of 44 mgd (167 ML per day) and includes provisions to add a fourth screen to provide a future capacity of 65 mgd (246 ML per day). This is a submerged intake with three T-shaped, stainless steel passive screens connected to a common header pipe. Each screen is equipped with an internally mounted air backwash connection used to remove debris from the screen surface. These screens will be located in the Woodward Reservoir in Stanislaus County, California.

An example of a floating intake can be found on Canyon Lake in southern California. Elsinore Valley Municipal Water District draws source water from Canyon Lake for its

FIGURE 4.8 Lake Norman in-lake tower.

INTAKE FACILITIES

FIGURE 4.9 Submerged intake screens.

FIGURE 4.10 Floating intakes.

9 mgd (34 ML per day) water treatment plant. The intake for this facility, shown in Figure 4.10, consists of four 3 mgd (11 ML per day) horizontal centrifugal pumps mounted on a floating platform anchored to the shore. The depth of water withdrawal is variable based on changeable suction piping length for each pump. A flexible hose is used for transmission of source water from the intake to the shore piping. An in-lake aeration system can be used to minimize lake stratification and to provide the best possible water quality to the plant for treatment. A design variation used to accommodate water level variations is a movable carriage intake, which is shown in Figure 4.11.

Lake and Reservoir Intake Location. The location of a lake or reservoir intake should be selected to obtain an adequate supply of water of the highest possible quality, with consistent reliability, economical construction, and minimal effect on aquatic life. To avoid sediment, sand, and ice problems, a submerged intake's inlet works should ideally be located in deep water. Water with a depth of at least 50 ft (15.2 m) is desirable. To achieve this depth in lakes where shallow water extends for a long distance from shore, a lengthy intake conduit may be required. In some locations, placement in shallow water may be acceptable if acceptable quality water is available.

Water Quality Impacts on Intake Location. In the past, pollution has forced the abandonment of some lake intakes, which has resulted in the need for careful evaluation of pollution potential at proposed new intake sites. When one is considering the pollution potential of a site, prevailing winds and currents are often significant. Review of seasonal water quality data provides further guidance during the process of site selection. Additionally, special water quality surveys over a period of several years may be required to define the optimum location.

FIGURE 4.11 Intake pump on movable carriage.

INTAKE DESIGN

Designing an intake system involves many different considerations including design capacity, intake layout, intake conduits, and pumping requirements. Intake design also requires consideration of several potential problems, as discussed below.

Design Capacity

Selecting design criteria and flow rates should reflect the longer planning period appropriate for major intake facilities. Hydraulic criteria to be evaluated are summarized in Table 4.3.

If the need for future increased capacity is likely, the potential for constructing an enlarged intake facility initially should be carefully evaluated. This may mean installing an oversized screen structure or parallel intake conduits (with one reserved for future capacity) to facilitate future expansion. Hydraulic analysis should also include calculation of a hydraulic overload condition to provide a safety factor against lost capacity caused by conduit deterioration over time, conduit fouling, or for emergency conditions.

Submergence Design

As intake capacity is directly related to inlet submergence, the designer should be aware of the substantial variations in water levels encountered in many lakes, reservoirs, and

TABLE 4.3 Hydraulic Criteria

Flow criteria	Capacity	Remarks
Design flow	Q	Present design capacity
Minimum flow	$0.10Q$ to $0.20Q$	System-specific
Ultimate flow	$2.00Q$ or higher	System-specific
Hydraulic overload	$1.25Q$ to $1.50Q$	Evaluate for all design conditions

rivers. The intake system must have adequate capacity to meet maximum water demand during its projected service life, and intake capacity equal to this demand must be available during the period of minimum water level in the source of supply.

A conservative approach should be used to establish the lowest water level for intake system design. Lack of adequate submergence during periods of extreme low water results in greatly reduced capacity. Deeper than normal submergence should be considered for rivers subject to floating and slush ice, or where extremely low water may occur during the winter season as a consequence of ice jams. On some rivers, it may be impossible to obtain adequate submergence under all anticipated operational conditions. For such situations, a low, self-scouring channel dam may be required for a reliable intake system.

Layout

Cellular or parallel component design of intake systems should be evaluated for major facilities. Cellular design divides the intake into two or more independent and parallel flow streams. This arrangement enhances reliability, provides flexibility, and simplifies maintenance activities. Individual cells can be taken out of service as required for inspection, cleaning, maintenance, or repair.

Inlet Works and Ports

For intakes located on deep reservoirs and rivers, gated inlet ports may be provided at several depths to facilitate selective withdrawal of water. In rivers, submergence may govern, and ports must be placed at as low an elevation as practical. However, the lowest ports of an intake should be sufficiently above the bottom of the channel to avoid clogging by silt, sand, and gravel deposits. Port inlet velocities should be selected to minimize entrainment of frazil ice, debris, and fish. Factors that affect selection of intake port configuration and locations are summarized in Table 4.4.

Development of an intake port design should include an overall operational strategy identifying potential problems that may be encountered and the capability of minimizing problems by selective withdrawal of water from the water source. It is likely that annual facility operation will require use of several intake ports in an effort to draw the highest-quality water. A water quality monitoring system capable of defining water quality pa-

TABLE 4.4 Intake Port Selection

Concern (to be avoided)	Required port location		
	Shallow	Intermediate	Deep
Organisms (requiring sunlight)		✓	✓
Warm water		✓	✓
Storm turbidity	✓	✓	✓
Plankton		✓	✓
Carbon dioxide	✓		
Iron and manganese		✓	✓
Color	✓	✓	

TABLE 4.5 Location and Spacing of Ports

Criterion	Suggested location
Vertical spacing	10–15 ft (3–5 m) maximum
Depth above bottom	5–8 ft (1.5–2.4 m)
Depth below surface	Variable
Intermediate ports	Best water quality, avoidance of pollution
Ice avoidance	20–30 ft (6–9 m) below surface
Wave action	15–30 ft (5–9 m) below surface

rameters throughout the depth of the water source should also be considered as an element of the intake facility design.

Experience has shown that the final number and spacing of inlet ports is affected by the specific conditions to be encountered, as shown in Table 4.5.

There is typically a wide variation of water quality, with respect to depth, in stratified lakes. At Shasta Dam in northern California, turbid water behind the dam became stratified within several weeks after a major storm. The most turbid water settled near the bottom, and the water became progressively less turbid toward the surface. These, and other, studies confirm the value of intake tower designs that include multilevel inlet ports. Occasional adjustment of the depth of draft can often substantially improve chemical, physical, and biological water quality parameters. These improvements are reflected in enhanced treatment performance and reduced treatment costs.

Intake ports should be selected to achieve reliable delivery of water while minimizing the inclusion of unwanted material or contaminants. Table 4.6 provides a summary of design criteria for exposed intakes.

The intake tower shown in Figure 4.12 is located in Monroe Reservoir, the water supply source for Bloomington, Indiana. Two intake cells, each equipped with three intake ports, provide variable depth withdrawal capability. Bar and traveling screens protect four vertical, wet-pit pumps. The design capacity of the intake system is 48 mgd (182 ML per day).

Various inlet designs for submerged intakes avoid sediment, sand, and ice problems. Inlets are best located in deep water with inlet velocities less than 0.5 ft/s (15 cm/s), with a preferred range of 0.2 to 0.3 ft/s (6.1 to 9.1 cm/s). Inlet structures are typically constructed of wood or other nonferrous materials with low heat conductivity that are less susceptible to ice deposits. Submerged intakes may be constructed as "cribs" surrounding an upturned, bell-mouth inlet connected to an intake pipeline. The crib is often constructed as a polygon, built of heavy timbers bolted together, weighted and protected by

TABLE 4.6 Exposed Intake Design Criteria

Criterion	Remarks
Port velocity	0.20–0.33 ft/s (6–10 cm/s); 0.50 ft/s (15 cm/s) maximum
Ports	Multiple; three minimum
Water level variation	Design capacity at minimum level; operating deck above 500-year flood level

FIGURE 4.12 Multiple inlet gates.

FIGURE 4.13 Lake intake crib.

concrete and crushed rock, and bedded on a crushed stone mat. The intake conduit conveys water to the shore well, which may also be the source water pump station. The shore well is typically designed to dissipate surges and may contain either fixed or traveling screens. Submerged intake systems using the wood crib arrangement have proved generally reliable on the Great Lakes when properly located.

Figure 4.13 shows a section of the Milwaukee, Wisconsin, intake crib located in Lake Michigan, with a rated capacity of 315 mgd (1,192 ML per day). The crib is an octagonal, coated steel structure, 11 ft (3.3 m) high and 52 ft (15.8 m) wide between parallel sides. It was floated into position and sunk by filling the air tanks with water. The horizontal baffle ensures relatively uniform flow through all parts of the intake screen. At the design flow rate, the average velocity through the screen openings is 0.31 ft/s (9.4 cm/s). Average water depth at the intake location is approximately 50 ft (15.2 m). The 108-in. (2.74-m) intake conduit extends 7,600 ft (2,320 m) to the pumping station located on the shore of the lake.

Other configurations for submerged intakes include hydraulically balanced inlet cones; screened, baffled steel cribs; and inlet drums. Hydraulically balanced inlet cones have been used at several Great Lakes intakes. The structure shown in Figure 4.14 consists of three groups of three equally spaced inlet cones connected to a cross at the inlet end of the intake conduit. This configuration provides essentially identical entrance velocities through all cones. The lower photograph shows a group of three inlet cones before placement in Lake Michigan.

Intake Conduit

The intake conduit, which connects the submerged inlet works with the shore shaft, typically consists of either a pipeline or a tunnel. Tunnels have a high degree of reliability, but are usually more costly to construct. For large water systems, a tunnel may provide an economical choice. The selection of the design velocity for intake conduits requires a balance between hydraulic headloss at high flow rates and the potential for sediment deposition at low flow. Velocities in the conduit should be sufficient to minimize deposition. If low flow rates are anticipated, provision for high-velocity backwash should be considered. Biological growths on the interior surface of the conduit may reduce its ca-

FIGURE 4.14 Hydraulically balanced inlet cones.

pacity, and this should be factored into the hydraulic design. Design criteria applicable to the design of intake conduits are presented in Table 4.7.

Subaqueous pipe is generally used for intake conduits, with pipe laid within a trench in the lake or reservoir bottom. It is desirable to have approximately 3 to 4 ft (0.91 to 1.22 m) of cover over the top of the pipe, plus an additional protective top layer of crushed rock.

TABLE 4.7 Intake Conduit Design Criteria

Number of conduits	One minimum; two or more preferred
Velocity	1.5–2.0 ft/s (46–61 cm/s) at design flow; 3–4 ft/s (90–120 cm/s) maximum
Type of construction	Tunnel or pipeline
Slope or grade	Continuous to drain or to an air release valve

The allowable drawdown in the shore shaft is a critical factor in the hydraulic design of the intake conduit. Drawdown results from friction loss in the conduit and equals the difference between the lake or reservoir level and the level of water in the shore shaft. Drawdown should be limited to avoid excessive excavation for the shore shaft.

The conduit should be laid on a continuously rising or falling grade to avoid reduction in conduit capacity from air accumulation at high points. When an undulating conduit cannot be avoided because of the profile of the lake bottom, provisions for air release should be made at the high points.

Pumping Facilities

The shore well serves as a screen chamber and a source water pump suction well. The depth of the shore shaft must be adequate to allow for drawdown when the intake operates at maximum capacity and the source water elevation is at its minimum. In addition, the well must provide an ample submergence allowance for the source water pumps. The well must also be capable of resisting and dissipating surges that occur during a power failure. Some shore wells are constructed with a shaft large enough to readily discharge the surge back to the supply source.

Many shore wells are equipped with fixed or traveling screens to remove objects large enough to cause pump damage. In general, traveling screens are preferable for most systems with the exception of small systems. Cellular construction of the shore well is advisable, and inlet control gates should be provided so that either all or a portion of the shore well can be taken out of service for inspection, maintenance, and repair.

INTAKE DESIGN CONSIDERATIONS

Designing an intake system requires consideration of a variety of potential problems that can be encountered. Regions with long, severe winters require significant consideration of potential impacts from ice formation. Zebra mussel impacts may need to be addressed in some regions. Fish protection is an especially important design consideration in the Pacific northwest and California. Design challenges applicable to most regions include flotation, geotechnical, mechanical, and chemical treatment considerations.

Ice Design Considerations

Intake systems located in regions with long, severe winters may be subjected to ice in its various forms including surface, frazil, and anchor ice.

Surface Ice. In some locations, surface ice and ice floes create a structural hazard to exposed intakes. On lakes, an accumulation of wind-driven ice floes near a shore intake can produce a deep, nearly solid layer of ice capable of restricting or completely blocking intake ports. Under such conditions, reliable intake operation is virtually impossible, and water supplies obtained from lakes and reservoirs subject to severe ice problems are typically served by offshore intakes.

Ice jams can cause partial or complete blockage of river intakes. Jams below a river intake can also cause extremely high river stages, and an upstream jam can produce low water levels at the intake location, reducing its capacity.

Frazil Ice. Frazil ice is small, disk-shaped ice crystals. Frazil ice has been aptly called "the invisible strangler." When conditions favor its formation, the rate of buildup on underwater objects can be rapid; frazil ice accumulation can reduce an intake's capacity substantially and can, in some cases, clog it completely in only a few hours.

Ice starts forming when water temperature is reduced to 32° F (0° C) and water continues to lose heat to the atmosphere. For pure water, supercooling to temperatures well below the freezing point is necessary to start ice formation, but with natural water, the required supercooling is much less. Two types of ice formation are recognized: static ice and dynamic ice (Table 4.8).

Static ice forms in quiet water of lakes and river pools. Dynamic ice formation occurs in turbulent water such as areas of high flow in rivers and in lakes mixed by wind action. Frazil ice formed under dynamic conditions adversely affects hydraulic characteristics of intakes.

When natural water loses heat to the atmosphere and a condition of turbulence exists, uniform cooling of a large fraction of the water body occurs. If the initial water temperature is slightly above the freezing point and cooling is rapid, a small amount of supercooling occurs, and small, disk-shaped frazil ice crystals form and are distributed throughout the turbulent mass. These small crystals are the initial stage of ice production. Other ice forms can develop from this initial ice production in sizable quantities.

Where there is little or no mixing, supercooled water and existing surface ice crystals are not carried to a significant depth, and the result is the formation of a layer of surface ice rather than a mass of frazil. Surface ice formation reduces heat loss from the water and usually prevents formation of frazil ice.

Two kinds of frazil ice have been identified, as shown in Table 4.9: active and inactive, or passive. Freshly formed frazil crystals dispersed in supercooled water and growing in size are in an active state. When in this condition, they will readily adhere to underwater objects such as intake screens or rocks. Frazil ice production and adhesiveness are associated with the degree of supercooling, which is related to the rate of cooling of

TABLE 4.8 Ice Formation Conditions

Type	Remarks
Static ice	Quiet waters
	Small lakes and river pools
Dynamic ice	Turbulent water
	Rivers in area of great flow
	Reservoirs with significant wind action
	Frazil ice formation occurs first
	Massive ice formations may follow

TABLE 4.9 Frazil Ice Characteristics

Type	Remarks
Active	Initial phase of formation
	Rapid growth in size
	Readily adheres to intake facilities
	Short-lived phase
	System clogging within a few hours
Inactive (passive)	Static or declining size
	Lost adherence and characteristics
	Less troublesome

the water mass. Frazil ice particles remain in an active, adhesive state for only a short time after their formation. With the reduction of supercooling and the return of the water to 32° F (0° C), frazil ice crystals stop growing and change to an inactive, or passive, state. Passive frazil ice is ice that has lost its adhesive properties and is therefore less troublesome.

Some confusion exists concerning the relationship between frazil ice and anchor ice. It has been suggested that anchor ice occurs rarely and consists of sheetlike crystals that adhere to and grow on submerged objects. Accumulations of frazil ice may closely resemble anchor ice. Some investigators designate all ice attached to the bottom as anchor ice regardless of how it is formed. Anchor ice may form in place on the bottom and grow by the attachment of frazil crystals. On the other hand, according to Giffen (1973), ice crystallization and growth directly on the surface of a shallow intake structure in open water is commonly termed *anchor ice*. Anchor ice normally does not form at depths greater than 40 to 45 ft (12.2 to 13.7 m), although the depth associated with anchor ice formation ultimately depends on water turbidity.

Predicting Frazil Ice Formation. The climatological conditions that encourage frazil ice formation are a clear night sky, an air temperature of 9.4° F (-12.5° C) or less, a day water temperature of 32.4° F (0.222° C) or less, a cooling rate greater than 0.01° F (0.0056° C) per hour, and a wind speed greater than 10 mi/h (16.1 km/h) at the water surface. Frazil ice generally accumulates in the late evening or early morning hours and seldom lasts past noon. Conditions favorable to frazil ice formation vary considerably from site to site, making it difficult to use weather data alone as a forecasting tool.

Frazil ice formation can be illustrated graphically by plotting water temperature variation versus time, as shown in Figure 4.15. When the original constant cooling rate (A to B) of water undergoing supercooling deviates, frazil ice begins to form. The process illustrated in Figure 4.15 takes only a few minutes. The rate of ice production at point C is equal to the cooling rate divided by the product of the ice density.

Design Features. Research and experience on the Great Lakes and elsewhere indicate that location and design features of submerged intakes can significantly reduce intake ice problems, but probably not completely eliminate them. Submerging lake intakes in deep water and sizing inlet ports for a velocity of 0.3 ft/s (9.1 cm/s) or less minimize the amount of frazil ice transported downward to the structure. However, during winter storms, strong wind and wave action can carry ice crystals and supercooled water to considerable depths, making accumulation of ice on and around the intake likely. Some of the procedures to alleviate frazil ice problems at submerged intakes are summarized in Table 4.10.

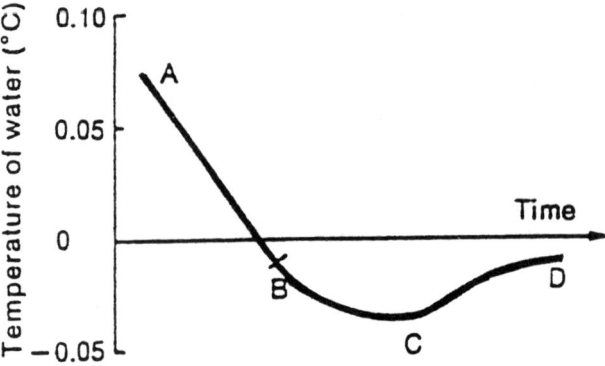

FIGURE 4.15 Frazil ice formation: water temperature sequence.

Creating a quiet body of water at the intake location promotes surface ice formation. At the Billings, Montana, waterworks, frazil ice was a severe winter problem in the turbulent Yellowstone River. In this case, the solution was to enlarge an off-river intake channel into an earthen forebay with a detention time of approximately 1 h. Surface ice formed on the nearly quiescent forebay. This insulating ice cover prevented the formation of additional frazil ice and provided the opportunity for river frazil carried into the forebay to combine with the surface ice and revert to a passive condition.

To prevent or at least minimize ice clogging, the structure can be built of low-heat-transfer materials with smooth surfaces not conducive to the accumulation of ice crystals. Metals such as steel are more susceptible to frazil ice formation because they have a high heat conductivity and act as a sink for the latent heat released when ice begins to form. This encourages ice buildup. In contrast, ice does not readily crystallize or grow rapidly on wood or plastic. Screens can be constructed using fiberglass-reinforced plastic with low thermal conductivity and a smooth surface. Any exposed metal surfaces can be coated with an inert material such as black epoxy paint to effect better thermal properties and to increase radiation heat gain. Screens have been developed that resist frazil ice formation.

Heated intake screens have been used successfully at power plant installations and may have application at waterworks intakes. Frazil ice does not adhere to objects with tem-

TABLE 4.10 Frazil Ice Mitigation

Criterion	Remarks
Location	Quiet waters away from turbulence Protected by ice cover Deep water
Inlet ports	Velocity Large bar rack opening [< 24 in. (0.6 m)] Low-heat-transfer materials Smooth surfaces Coating to improve thermal characteristics
Temperature	Heat to prevent ice adherence Raise local water temperature 0.18° F (0.1° C)

peratures slightly above the freezing point. For larger installations in cold climates, a prohibitive amount of energy would probably be required for heating intake screens.

Control Methods. An indication of icing problems at an intake is abnormal drawdown of the intake well. If the intake has a screen, excessive drawdown can rupture the screen.

Methods most commonly used to control frazil ice at an intake structure include injecting steam or compressed air at the intake opening, backflushing the intake with settled water, and manual or mechanical raking to remove the ice. If backflushing is to be routinely practiced, provisions must be included when the intake is designed. It is sometimes possible to clear partially clogged intake ports by a method termed *control drawdown,* which involves throttling the intake well pumps and maintaining reduced intake flow. Under some conditions, this flow may be sufficient to erode ice bridges at the ports and restore intake capacity.

If the plant has an alternate intake or sufficient storage capacity to operate without source water for a few hours, the simplest solution is to shut down the intake and wait for the ice to float off.

Zebra Mussel Considerations

The zebra mussel (*Dreissena polymorpha*) is a small bivalve mollusk native to Europe that has alternating light and dark stripes on its shell. Zebra mussels average about 1 in. in length and may typically live for 4 to 8 years. The mussels are most frequently found at water depths between 6 and 12 ft (1.8 and 3.7 m). However, they have been found at depths up to 360 ft (109.7 m) in the Great Lakes (Lake Ontario). The mussel grows filamentlike threads (byssus filaments) from the flat side of its shell that allow it to attach to hard surfaces, such as the bottoms of ships and the components of water facility intakes.

Zebra mussels have some natural predators, including some diving ducks and bivalve-eating fish such as the common carp and the freshwater drum. However, these predators appear to be of little help in controlling the zebra mussel population. The zebra mussel is well established in the Great Lakes region of the United States and Canada and proliferates under a relatively wide range of water conditions, as illustrated in Table 4.11. Projections indicate that they will eventually infest all freshwaters in two-thirds of the United States and the majority of southern Canada. The magnitude of the zebra mussel threat can be illustrated by the rate at which the mollusk can reproduce. A single female can produce between 30,000 and 40,000 eggs per year. Although biologists estimate that only

TABLE 4.11 Zebra Mussel Requirements for Reproduction

Water characteristics	Remarks
Water pH	Basic, 7.4 minimum
Calcium content of water	28 mg/L minimum
Water temperature Maximum Minimum	90° F (32.2° C) 32° F (0° C)
Velocity of flow	5.0–6.5 ft/s (1.5–2.0 m/s)
Location	Prefer a dark location

TABLE 4.12 Zebra Mussel Control

Treatment technique	Remarks
Thermal	95° F (35° C) for 2 h 100% effective Repeat 2 or 3 times per year
Chemical	Oxidizing chemicals such as chlorine, bromine, potassium permanganate, ozone, carbon dioxide Continue for 2 or 3 weeks or apply continuously Nonoxidizing chemicals are effective and being developed
Coating of components	Silicone-based coatings prevent attachment 80% successful Copper/epoxy paints
Construction materials	Copper Brass Galvanized steel
Mechanical	Shovel or scrape High-pressure hose Sandblasting Pipeline pigging
Other methods	Ultrasound High pressure Electrocution Oxygen depletion UV light

5% of the laid eggs may actually survive, population increases can be huge in areas where the environment is conducive to growth. The rapid reproduction rate can result in a zebra mussel problem changing from nonexistent to a water supply emergency in a few months.

A range of design techniques have been investigated as a means of controlling or preventing zebra mussel infestation of water intake facilities, as shown in Table 4.12. The most commonly used control method involves combinations of mechanical removal and chemical treatment with chlorine and potassium permanganate applications to control recurrence. Chemical treatment typically consists of a chemical feed point at the intake facility entrance to distribute treatment chemical across the full flow stream of the inlet. Mechanical removal may be necessary to clean bar screens or pipelines infested by adult mussels. If this is done, care should be exercised in planning the removal process. Disposal of large volumes of shells and rotting mussel can be problematic, and the rotting tissue has the potential of causing significant taste and odor problems in drinking water. Control technologies using microorganisms to prevent attachment by and to kill zebra mussels are being developed.

A second bivalve, the quagga mussel, has been discovered in the Great Lakes region. This mussel is named after an extinct African relative to the zebra and appears very similar to the zebra mussel to the untrained eye. The quagga tends to be slightly larger than the zebra mussel. The quagga does not have a flat side, which prevents it from anchoring to hard surfaces as tightly as zebra mussels. Both mussels have the ability to colonize hard, rocky surfaces and soft sediments. The quagga mussel appears to tolerate higher-salinity water than does the zebra and appears to survive at greater water depths. The quagga mussel is another problematic aquatic species that may need to be considered when one is designing an intake system.

Fish Protection Considerations

Design of intake systems may affect aquatic organisms present in the river, lake, or reservoir. An evaluation for endangered species should be conducted. If endangered fish species are present, the National Oceanic and Atmospheric Administration (NOAA) fisheries guidelines should be consulted. Fish protection is a significant design element in some areas such as California and the Pacific Northwest. Three basic types of devices used to prevent juvenile fish from entering flow diversions are (1) physical barrier screens, (2) behavioral guidance systems, and (3) capture and release systems. The most widely accepted and successful method of protecting juvenile salmonid passage is to use physical barriers. When one is designing a juvenile fish screen, the swimming ability of the species and the life stage targeted for protection should be considered. In the Pacific Northwest, the juvenile fish screen criteria developed by the fisheries agencies are designed to protect the weakest swimming species at its most vulnerable life stage. An effective approach for fish protection, which also minimizes the introduction of other contaminants, is to use low entrance velocities. The flow should approach the screen in a laminar mode. A screen can be placed on the end of the pump intake in a pressurized system. Cleaning of pump intake screens can be accomplished using fixed spray bar with rotating screen, fixed screen with rotating bar, or internal air-burst systems.

A positive method to reduce fish entry into submerged intake facilities is the velocity cap. Figure 4.16 illustrates a horizontal cap on the top of the intake structure, which forces all water to enter horizontally. Fish tend to swim against horizontal currents and avoid the intake.

The Clackamas River Waste District (CRW) in Oregon operates a filtration water treatment plant rated at 30 mgd (114 ML per day). Raw water is supplied to the plant via two concrete caisson intake structures in the Clackamas River connected to a raw water pump station inland. Original screening consisted of mesh traveling screens located at the pump station wet well. CRW modified the intake structures to meet NOAA fisheries guidelines set to protect salmonid fish listed under the Endangered Species Act. The modifications entailed the installation of 42-in.-diameter (1.07-m) stainless steel T screens on each of the intake caissons (Figure 4.17). The intake T screens are over 10 ft (3.0 m) long and sized to meet 0.4 ft/s (12.2 cm/s) approach velocity requirements for salmonid fish protection, for a flow of 18 mgd (68 ML per day) each. Screen openings are 0.069 in. (1.75

FIGURE 4.16 Velocity cap. (*Courtesy of USFilter, Envirex Products.*)

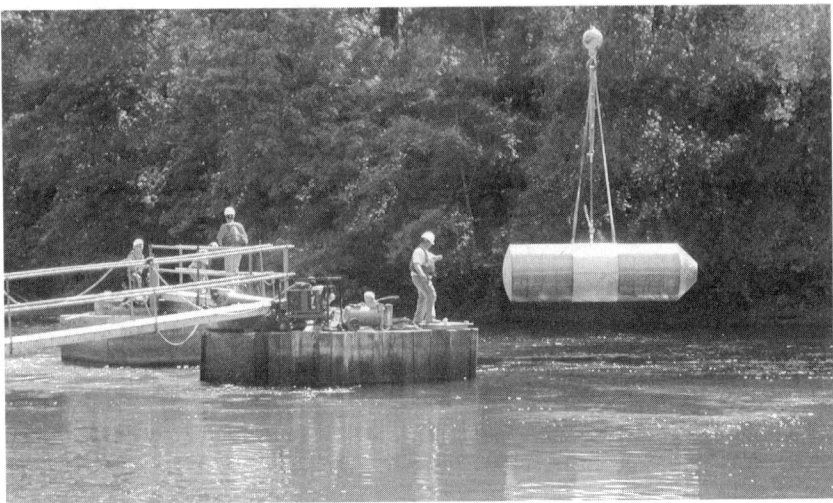

FIGURE 4.17 T screen used to modify intake to improve fish protection.

mm). A 150-psi (10.2-atm) air-burst cleaning system can be actuated by head loss across the screens, timed cycle, or manual operation.

The Geren Island Treatment Facility serves the city of Salem, Oregon. The intake, which is on the North Santiam River, is sized for a maximum flow of 147 mgd (556 ML per day). This intake facility, shown in Figure 4.18, conveys flows to the treatment facilities and includes a concrete intake structure, trash racks, twelve 36-in. (0.91-m) sluice gates, and two 84-in. by 60-in. (2.1-m by 1.5-m) control gates. A fish screening system prevents fish from entering the treatment system. The fish screening system consists of stainless steel fish screens with 0.069-in. (1.75-mm) slot openings, ultrasonic level sensors, and an air-burst cleaning system. Design was based on NOAA fisheries screening criteria, and meets the criteria with approach velocities of less than 0.4 ft/s (12.2 cm/s) and sweep velocities at least equal to the approach velocities.

Flotation Considerations

Designing intake facilities requires careful evaluation of uplift forces to ensure that the structure is stable over the full range of anticipated water surface elevations. Two general methods for resisting uplift forces are deadweight and foundation ties. The most common approach to stabilizing uplift forces is to design the structure with enough deadweight to resist all possible uplift force. Deadweight can be provided by tying the structure to a concrete mat or by filling cells of the structure with crushed rock or other heavy material. Foundation ties or rock anchors drilled into foundation rock are also occasionally used to tie down an intake structure. With either method, a safety factor against uplift of at least 1.2 should be provided.

Geotechnical Considerations

Geotechnical investigations must be completed as the first step in designing an intake. Investigations typically include borings unless adequate data are already available from a

FIGURE 4.18 Concrete intake structure including fish protection system.

previous installation in the same area. During the study phase, geotechnical data permit a more accurate comparison of the cost of viable alternatives to identify the most economical solution for the project. Geotechnical investigations should cover a sufficient area to account for adjustments in the intake location during final design. The cost of a few additional borings during the initial study is much less than the cost of mobilizing a barge for additional borings, should they be deemed necessary later in the design. Provisions should also be included in construction specifications for diver inspections to verify that the subgrades are clean of loose and soft material before the placing of tremie concrete.

Mechanical Considerations

The mechanical design for an intake structure includes many of the routine considerations found in other, similar structures. Special mechanical design considerations for intakes include adequate provisions for ventilation, insects, and flooding impacts.

Ventilation systems must be designed to accommodate heat generated by electric motors, motor control centers, and solar gain. Ventilation systems must also consider odorous gases that may be present, especially if screening material is temporarily stored in the building. With the trend toward more chemical treatment at the intake, ventilation must provide airflow required by applicable building codes for each stored chemical system. Ventilation of below-grade spaces that may occasionally be occupied for inspection or maintenance should also be provided.

Insects can be a serious problem at some intake sites, and insect screens on the air intake and exhaust louvers are one solution. However, large quantities of insects can block these screens, requiring increased maintenance by facility personnel. Another solution is to provide air filters with a large surface area on air handling units. Facility personnel may prefer the ease of changing a filter to the difficulty of cleaning louver screens.

Flooding conditions can affect the design of sanitary and storm drainage systems. If flooding is expected, backwater or knife gate valves should be provided.

Chemical Treatment Considerations

The intake facility provides the first opportunity in a water treatment system to impact, remove, or alter contaminants in the source water before subsequent treatment. A number of chemical treatment choices that may be provided at the intake should be considered during the intake facilities design (Table 4.13).

TABLE 4.13 Chemical Treatment Alternatives

Contaminant or criterion	Suggested treatment chemical
Debris and screenings	No treatment; discharge to water source or landfill
Taste and odor	Chlorine Chlorine dioxide Potassium permanganate Carbon
Coagulation	Alum or ferric chloride Cationic polymer
Zebra mussels	Chlorine and other oxidants

The choice of chemical treatment to be applied at the intake must be coordinated with overall optimization of the water treatment process, including the control of disinfection by-product formation. Historically, chlorine has been used as the standard treatment to oxidize taste and odor compounds at the intake structure. However, current practice is to avoid the early use of chlorine in the treatment process where the level of organics in the source water is highest because of the potential for forming chlorinated disinfection by-products such as trihalomethanes. Chlorine dioxide, potassium permanganate, and carbon are alternatives to chlorine that avoid or minimize the formation of chlorinated disinfection by-products.

RACKS AND SCREENS

Racks and screens remove suspended particulates from water, including leaves, debris, and other sizable clogging material. Racks and screens are essential to providing protection for downstream conduits, pumps, and treatment works. Properly designed intake racks and screens can also minimize the effect on fish.

Racks and screens can be divided into two broad categories: coarse screens (racks) and fine screens. A summary of characteristics associated with each type is presented in Table 4.14.

Design Considerations

Head loss through racks and screens is an important consideration to be evaluated as part of the design process. Racks and screens should be designed to minimize head loss by providing sufficient flow-through area to keep velocities low. The total area of clear openings in a screen typically ranges between 150% and 200% or more of the area or channel protected by the screen. The maximum head loss from clogging should be limited to between 2.5 and 5.0 ft (0.76 and 1.52 m), and the screen should be designed to withstand the differential hydraulic load. Head loss for mechanically cleaned screens of all types can be held nearly constant with proper operator attention. Curves and tables for head loss through screening devices are available from equipment manufacturers.

TABLE 4.14 Rack and Screen Characteristics

Type	Remarks
Coarse screens (racks)	
Trash racks	Clear opening 3–4 in. (8–10 cm)
	Inclined or vertical
	Manually or mechanically cleaned
Bar racks	Clear opening $3/4$–3 in. (2–8 cm)
	Inclined or vertical
	Manually or mechanically cleaned
Fine screens	
Traveling water screen	0.24–0.35 in. (6–9 mm) mesh cloth
Basket screen	$1/8$–$3/8$ in. (3–10 mm) most common
Disk or drum screen	Vertical
	Water spray cleaning

FIGURE 4.19 Course bar screen, mechanically cleaned.

Design of Racks

Coarse screens (also termed *trash racks* or *bar racks*) are commonly located at inlet ports to prevent entrance of large objects. Racks are generally constructed of ½- to ¾-in.-diameter (12.7- to 1.91-mm) metal bars, spaced to provide 1- to 3-in. (25- to 76-mm) openings.

Coarse screens are typically installed vertically or at an incline of about 30° from vertical. Both manual and automatic operation of the cleaning mechanism can be provided. Figure 4.19 illustrates one method for providing an economical grating system by mounting the rake on a traveling rail system so that a single rake can clean several coarse screens.

Design of Screens

Fine screens remove material that is too small to be deflected by coarse screens. Water velocity in net screen openings should be less than 2.0 ft/s (61 cm/s) at maximum design flow and minimum screen submergence. Hydraulically cleaned (backwashed) stationary screens have been used successfully at reservoir and river intakes. Numerous variations in design involving the use of backwashed, fixed screens have been utilized. It is generally recommended that the velocity through stationary screen openings not exceed about 0.6 ft/s (18 cm/s).

Traveling water screens are commonly installed after coarse bar screens. Figure 4.20 illustrates a typical traveling water screen. Screens can be made to any desired opening, but ⅛ to ⅜ in. (3.18 to 9.53 mm) is the most common size. Operation includes flushing from behind the screen with water jets to remove accumulated material from the screen and washing it into a disposal trough. Screenings may be returned to the water source or disposed of in a landfill, as required.

INTAKE FACILITIES 4.33

FIGURE 4.20 Traveling water screen.

BIBLIOGRAPHY

Ashton, G. D. "River Ice." *American Scientist* 67(1):38, 1979.

Barnes, H. T. *Ice Engineering*. Montreal, Canada: Renouf, 1928.

Baylis, J. R., and H. H. Gerstein. "Fighting Frazil Ice at a Waterworks." *Engineering News-Record* 140(16):562, 1948.

Burdick, C. B. "Water Works Intakes." *Journal AWWA* 38(3):315, 1946.

Clasen, J., and H. Bernhardt. "Experiences of Quality Control of Raw Water Storage Reservoirs." *Aqua—Journal of Water Supply Research Technology* 38:256, 1989.

Coughlan, Jr., F. P. "Locating the New Intake." *Journal AWWA* 50(5):668, 1958.

Cunningham, J. W. "Waterworks Intakes and the Screening of Water." *Journal AWWA* 23(2):258, 1931.

Donnan, J. A., and C. A. MacKay. "The Economic Implications of the Zebra Mussel in the Great Lakes." *Water Pollution Control—Canada* 129:12, 1991.

Driscoll, F. G. *Groundwater and Wells*, 2d ed. St. Paul, Minn.: Johnson Division, 1986.

Evans, G. P., and D. Johnson. *Detection of Pollution at Drinking Water Intakes, Environmental Protection: Standards, Compliance, and Costs*. Chichester, England: Ellis Horwood, Ltd., 1984, p. 239.

"Fine Screens." *Public Works* 126(4):c–12, 1995.

Foulds, D. M., and T. E. Wigle. "Frazil—The Invisible Strangler." *Journal AWWA* 69(4):196, 1977.

Giffen, A. V. *The Occurrence and Prevention of Frazil Ice Blockage at Water Supply Intakes*. Toronto, Ontario: Research Branch Publication No. W 43, Ministry of the Environment, 1973.

Gillespie, D. D. "Novel Construction Concept for Raw Water Intake." *Civil Engineering* 41:43, 1971.

Goda, Takesha. "Water Quality Management in the Yodo River Basin." *Water Science Technology* 23:65, 1991.

Hardenbergh, W. A. *Water Supply and Purification*. Scranton, Pa.: International Textbook, 1945, p. 325.

Heinzel, L. R. "Storm Effects on Turbidity in Trinity Project Waters." *Journal AWWA* 59(7):835, 1967.

Howson, L. R. "Saginaw-Midland Supply Project Begins Operations." *Water Works Engineering* 102(5):415, 1949.

Howson, L. R., and Gerald Remus. "Best Features of Four Existing Plants Highlighted in New Detroit Water Facility." *Water and Wastes Engineering* 7(12):50, 1970.

Huber, F. "Coquitlam Lake Water Tunnel Upgrading—Design and Construction. A Case History." *Canadian Geotechnical Journal* 26(1):90, 1989.

Ihling, H. M., and D. P. Proudfit. "New Facilities Add 100 mgd to Milwaukee's Water Supply." *Civil Engineering* 33(9):31, 1963.

Klauber, Avery. "Investigators Hunt for Great Lakes Enemy. New York Sea Grant." *Nor'easter* 2:14, 1990.

Kubus, J. J., and D. A. Egloff. "Monitoring Water Quality Parameters from Municipal Water Intakes." *Water Pollution Control Journal* 54:1592, 1982.

Lamarre, Leslie. "Invasion of the Striped Mollusks." *EPRI Journal* 16:12, 1991.

Lamarre, Leslie. "Zebra Mussels: The Assault Continues." *EPRI Journal* 18:22, 1993.

Lee, G. F., and C. C. Harlin, Jr. "Effects of Intake Location on Water Quality." *Industrial Water Engineering* 2(3):36, 1965.

Lischer, V. C., and H. O. Hartung. "Intakes on Variable Streams." *Journal AWWA* 44(10):873, 1952.

Logan, T. H. *Prevention of Frazil Ice Clogging of Water Intakes by Application of Heat*. REC-ERC-74-15. Denver, Colo: U.S. Department of Interior, Bureau of Reclamation, September 1974. Distributed by National Technical Information Service, U.S. Dept. of Commerce, Springfield, Va.

Matisoff, G., et al. "Controlling Zebra Mussels at Water Treatment Plant Intakes." *AWWA Surface and Ground Water Quality Symposium*. Cleveland, Ohio: AWWA, p. 79.

McDonald, Adrian, and Pamela Naden. "Colour in Upland Sources: Variations in Small Intake Catchments." *Water Services* 91:121, 1987.

Merriman, Thaddeus, and T. H. Wiggin. "Collection of Water Intakes." In *American Civil Engineers' Handbook,* 5th ed. New York: John Wiley & Sons, 1930.

Mills, E. L., G. Rosenberg, A. P. Spidle, M. Ludyanskiy, Y. Pligin, and B. May. "A Review of the Biology and Ecology of the Quagga Mussel (*Dreissena bugensis*), a Second Species of Freshwater Dreissenid Introduced to North America." *American Zoology* 36:271–286, 1996.

Mithcell, R., and J. D. Gu. "Use of Microorganisms and Their Metabolites for Zebra Mussel Control." *Zebra Mussel Research Technical Notes.* U.S. Army Corps of Engineers Technical Note ZMR-3-19, 1998.

Monk, R. D., Terry Hall, and Mohammed Hussain. "Real World Design: Appropriate Technology for Developing Nations." *Journal AWWA* 76:68, 1984.

Nordlund, Bryan. "Designing Fish Screens for Fish Protection at Water Diversions." National Marine Fisheries Service, 1996.

Payne, B. S., A. Miller, and G. Adams. "Effects of Elevated Carbon Dioxide Concentrations on Survivorship in Zebra Mussels (*Dreissena polymorpha*)." *Zebra Mussel Research Technical Notes.* U.S. Army Corps of Engineers Technical Note ZMR-2-21, 1998.

Peters, L. F. "Hydraulically Backwashed Well Screens Used as Intakes." *Water and Wastes Engineering* 6(3):52, 1969.

Richardson, W. H. "Intake Construction for Large Lakes and Rivers." *Journal AWWA* 61(8):365, 1969.

"Rise and Fall of a Pumping Station." *Water Works Engineering* 112(3):208, 1959.

Rutyna, Jason. "Zebra Mussel Control Methods at Water Intakes: Insights and Guidance Based on the Great Lakes Experience." *Michigan Section AWWA Annual Conference 2003, Bellaire, Michigan.*

San Giacomo, Richard, and M. E. Cavalcoli. "Opposing the Threat: Practical Application of Zebra Mussel Control." *Proceedings of the AWWA 1991 Annual Conference.* Denver, Colo.: AWWA, p. 137.

Schuler, V. J., and L. E. Larson. "Improved Fish Protection at Intake Systems." *Journal of Environmental Engineering Division (ASCE)* 101(EE6):897, 1975.

Shaefer, V. J. "The Formation of Frazil and Anchor Ice in Cold Water." *Trans-American Geophysical Union* 31(6):885, 1950.

Tobin, Patrick, and George Wesner. "Siting of Water Intakes Downstream from Municipal Wastewater Facilities." AWWA Research Foundation et al. *Water Reuse Symposium* 1:192, 1979.

Tone, Ralph, and E. T. Conrad. "Centrifugal Pumps Ride the Rails of Unique Water Intake." *Public Works* 100(12):64, 1969.

Tredgett, R. G., and D. R. Fisher. "New Waterworks Intake for Hamilton, Ontario." *Water and Wastes Engineering* 7(2):32, 1970.

U.S. Army Corps of Engineers. "Technical Instructions Water Supply." TI 814-01, 1998.

Weiss, C. M., and R. T. Oglesby. "Limnology and Quality of Raw Water in Impoundments." *Public Works* 91(8):97, 1960.

Weiss, J., E. J. Bouwer, W. P. Ball, C. R. O'Melia, M. W. Lechevallier, H. Arora, and T. F. Speth. "Riverbank Filtration—Fate of DBP Precursors and Selected Microorganisms." *Journal AWWA* 95(10):68–81, 2003.

Whitlock, E. W., and R. D. Mitchell. "Hydraulically Backwashed Stationary Screens for Surface Water." *Journal AWWA* 50(10):1337, 1958.

Wolman, Abel. "Ice Engineering." *Journal AWWA* 21(1):133, 1929.

CHAPTER 5
AERATION AND AIR STRIPPING

John E. Dyksen
Black & Veatch
Oradell, New Jersey

Aeration processes have been used to improve water quality since the earliest days of water treatment. In this process, air and water are brought into intimate contact with each other to transfer volatile substances to or from the water. Aeration processes generally are used in two types of water treatment applications. The first is the removal of a gas from water and is classified as desorption, or air stripping. The second is the transfer of a gas to water and is called gas absorption, or aeration. The U.S. Environmental Protection Agency (USEPA) has identified air stripping as one of the best-available technologies for the removal of volatile organic chemicals (VOCs) and radon from contaminated groundwater.

USES OF AERATION/AIR STRIPPING

Principal uses for aeration in water treatment include

- To reduce the concentration of taste- and odor-causing substances and, to a limited extent, for oxidation of organic matter.
- To remove substances that may in some way interfere with or add to the cost of subsequent water treatment. A prime example is removal of carbon dioxide from water for corrosion control and before lime softening.
- To add oxygen to water, primarily for oxidation of iron and manganese so that they may be removed by further treatment.
- To add ozone or chlorine gas to the water for oxidation and/or disinfection purposes.
- To remove gases such as radon, hydrogen sulfide, and methane.
- To remove ammonia from the water.
- To remove VOCs considered hazardous to public health.

TYPES OF AERATION/AIR STRIPPING EQUIPMENT

Structures or equipment for aeration or air stripping may be classified into four general categories: waterfall aerators, diffusion or bubble aerators, mechanical aerators, and pressure aerators.

Waterfall Aeration—Air Stripping

The waterfall type of aeration equipment utilizes air stripping and accomplishes gas transfer by causing water to break into drops or thin films, increasing the area of water exposed to air. The more common types are

- Spray aerators
- Multiple-tray aerators
- Cascade aerators
- Cone aerators
- Packed columns

Spray Aerators. Spray aerators direct water upward, vertically or at an inclined angle in a manner that causes water to be broken into small drops. Installations commonly consist of fixed nozzles or a pipe grid located over an open-top tank.

Spray aerators are usually efficient with respect to gas transfer such as carbon dioxide removal or oxygen addition. However, they require a large installation area, are difficult to house, and pose operating problems during freezing weather. To be most effective, spray aerators need a supplemental air exchange to minimize the buildup of contaminants in the air phase.

Multiple-Tray Aerators. Multiple-tray aerators consist of a series of trays equipped with slatted, perforated, or wire-mesh bottoms. Water is distributed at the top, cascades from each tray, and is collected in a basin at the base. It is important to have even distribution of water from the trays to obtain optimum unit efficiency. Coarse media such as coke, stone, or ceramic balls ranging in size from 2 to 6 in. (5 to 15 cm) are used in many tray aerators to improve the efficiency of gas exchange and to take advantage of the catalytic effects of deposited manganese oxides.

A type of multiple-tray aerator, the crossflow tower, has been extensively used in water cooling applications. Water is allowed to fall over the tray area while air is either forced or induced to flow across the slats, perpendicular to the water path. Tray aerators are analogous to cooling towers, and the problems encountered in design and operation are similar. Tray aerators must be provided with adequate ventilation. If they are placed in a poorly ventilated building, performance will be impaired by contamination of inlet air with the compounds being removed.

Artificial ventilation is provided to some types of tray aerators by supplying air from a blower at the bottom of the enclosure. These aerators exhibit excellent oxygen absorption and carbon dioxide, ammonia, and hydrogen sulfide removal.

Cascade Aerators. With cascade aerators, increases in exposure time and area-volume ratio are obtained by allowing water to flow downward over a series of steps or baffles. The simplest cascade aerator is a concrete step structure that allows water to fall in thin

FIGURE 5.1 View of a typical cone aerator. (*Courtesy of Ondeo Degremont.*)

layers from one level to another. The exposure time of air to water can be increased by increasing the number of steps, and the area-volume ratio can be improved by adding baffles to produce turbulence. In cold climates, these aerators must be housed, and adequate provisions must be made for ventilation. As with tray aerators, operating problems include corrosion and slime and algae buildup.

Cone Aerators. Cone aerators are similar to cascade aerators. They have several stacked pans arranged so that water fills the top pan and cascades down to each succeeding pan. A common type of commercial cone aerator is shown in Figure 5.1.

Packed Columns. Packed columns (also called packed towers or air strippers) are used extensively in the chemical industry and have been used in the water industry primarily for removal of volatile compounds such as VOCs from contaminated water. The extremely large surface area provided by packing in a column, combined with forced air flowing counter to the flow of water, provides considerably more liquid-gas transfer compared with other air stripping methods. A packed column consists principally of a cylindrical tower, packing material contained in the tower, and a centrifugal blower.

The quantity of air provided in relation to the quantity of water flowing through the column is known as the air-to-water ratio. This ratio is important in designing for removal of VOCs, as is the height of the packing material. A packed column installation is shown in Figure 5.2.

FIGURE 5.2 View of a packed column installation. (*Source: U.S. Environmental Protection Agency.*)

Diffusion-Type Aeration

Diffusion- or bubble-type aerators accomplish gas transfer by discharging bubbles of air into water by means of air injection devices. Compared with packed columns, diffused aeration provides less interfacial area for mass transfer but greater liquid contact time. On the other hand, packed columns provide a greater effective area but lower liquid contact time.

Diffuser Aerators. The most common type of equipment for diffusion aeration consists of rectangular concrete tanks in which perforated pipes, porous diffuser tubes or plates, or other impingement devices are inserted. Compressed air is injected through the system to produce fine bubbles, which, on rising through the water, produce turbulence resulting in effective water-air mixing.

This type of aeration technique is often adapted to existing storage tanks and basins. If porous tubes or perforated pipes are used, they may be suspended at about one-half tank depth to reduce compression head. Porous plates are usually located on the bottom of the tank. Static tube aerators are also used in a variety of applications and provide adequate aeration when properly designed.

A relatively new type of diffused aeration equipment is the low-profile system, which is a multistage, diffused bubble air stripping device. The device differs from conventional systems in that it operates with a water depth of about 18 to 36 in. (46 to 92 cm) compared with 10 ft (3 m) or greater for conventional installations. Shallow water depth permits the use of regenerative or centrifugal blowers instead of compressors and reduces overall height of the device to less than 4 to 5 ft.

In this system, water flows by gravity through a series of completely mixed staged reactors. Individual modules are designed and fabricated to include three, four, six, or eight stages, depending on the level of treatment required. Each stage is separated by a baffle wall and includes separate diffuser heads. Air is blown into each stage, creating a turbulent mixture of diffused bubbles and water to provide efficient mass transfer. A schematic of a typical device of this type is shown in Figure 5.3.

Individual modules are designed to treat water at flow rates ranging from 1 to 1,000 gpm (0.06 to 63 L/s). Modules used in parallel can treat water at higher flow rates. This system also removes carbon dioxide, VOCs, gasoline components, hydrogen sulfide, methane, and radon from contaminated groundwater and is particularly applicable to small water systems with low flow rates. Two drawings of low-profile systems are shown in Figure 5.4.

FIGURE 5.3 Schematic of a flow-through low-profile aeration system. (*Courtesy of Lowry Aeration Systems.*)

FIGURE 5.4 Two types of low-profile systems: (*a*) three-stage; (*b*) four-stage. (*Courtesy of Lowry Aeration Systems.*)

Draft-Tube Aerators. A draft-tube aerator consists of a submersible pump that rests on the bottom of a basin and is equipped with an air intake pipe extending to above the water surface. The partial vacuum created by the pump pulls air through the tube and mixes it with water at the pump intake. Aerated water is then directed outward along the floor of the basin. This type of aeration is an inexpensive and relatively effective means of adding aeration to an existing basin.

In-Well Aeration. A variation of diffused aeration to remove VOCs from groundwater supplies is in-well aeration. This technique has been investigated by the North Penn Water Authority, a public water utility located in Lansdale, Pennsylvania. The Authority received a grant from the AWWA Research Foundation to conduct detailed tests on the feasibility of in-well aeration.

Aside from obvious advantages of this type of treatment system, several disadvantages exist. One disadvantage is the dissolution of large quantities of air into water, causing water to appear milky. For practical use of this type of aeration system, water needs some atmospheric contact time to allow the milky appearance to disappear before it is pumped into the distribution system. The study found this treatment method to have relatively low efficiency.

Mechanical Aeration

Mechanical aerators employ motor-driven impellers alone or in combination with air injection devices.

Surface Aerators. Mechanical surface aerators are used extensively in wastewater applications for supplying oxygen to water. To a lesser extent, they are used to control taste and odor problems in water treatment and are commonly installed at a reservoir rather than at the treatment plant. They generally consist of an electric motor suspended on a float, with a driveshaft operating a propeller located a short distance below the water surface. The water is drawn up by the blade and thrown into the air in tiny droplets so that the water can pick up oxygen.

A variation in design is the surface aerator equipped with a draft tube extending below the propeller. With this design, water is drawn up from near the bottom of deeper basins.

Submerged Aerators. Submerged aerators operate in the reverse of surface aerators. The submerged blade draws water downward and, in the process, draws in air, which is diffused into the water. This type of aerator results in relatively calm water at the surface compared with surface aerators. Submerged aerators are best used for increasing dissolved oxygen levels.

Pressure Aeration

There are two basic types of pressure aerators, and the object of both is to aerate water that is under pressure. In one type, water is sprayed into the top of a closed tank while the tank is continuously supplied with compressed air. Aerated water leaves at the bottom of the tank. In the second type of pressure aerator, compressed air is injected directly into a pressurized pipeline and adds fine air bubbles to the flowing water.

With both systems, the higher the pressure used, the more oxygen dissolves into the water. Pressure aerators are primarily used for oxidizing iron and manganese for subsequent removal by settling, filtration, or both.

PRINCIPLES OF AIR STRIPPING AND AERATION

Transfer by aeration of a volatile material to or from water depends on a number of factors, including

- Characteristics of the volatile material
- Water and surrounding air temperatures
- Gas-transfer resistance
- Partial pressure of the gases in the aerator atmosphere
- Turbulence in gaseous and liquid phases
- Area-to-volume ratio
- Time of exposure

Equilibrium Conditions

The term *equilibrium* applied to gases dissolved in water signifies a steady-state concentration of dissolved substances. Aeration promotes the establishment of equilibrium between dissolved, volatile constituents in the water and the constituents in the air to which the water is exposed. For example, when water is exposed to air, oxygen and nitrogen dissolve in the water until a state of equilibrium is reached. The function of aeration is to speed up this natural process. True equilibrium may not be attained by aeration unless the air-water exposure period is relatively long. From a practical standpoint, however, it is generally not necessary to achieve absolute equilibrium.

Saturation Value

The concentration of a gas dissolved in a liquid at equilibrium is known as its *saturation value*. This value is an important characteristic of a dissolved gas. Saturation value is principally dependent on water temperature, partial pressure of the gas in the atmosphere in contact with the water, and presence of dissolved solids. The higher the partial pressure, the greater the dissolved gas concentration. This relationship is known as *Henry's law*. At a fixed partial pressure, the higher the temperature, the lower the solubility or saturation value of a gas. Gas solubility is also reduced by dissolved solids.

Saturation value has considerable practical and theoretical significance. It is the difference between the saturation value of a gas and its actual concentration in the water that provides the driving force for the interchange of gas between air and water. Water deficient in oxygen will absorb it when brought into contact with air, and the air-water equilibrium will be reached from the direction of oxygen deficiency. Prolonged aeration produces oxygen saturation.

On the other hand, if water contains more oxygen or, as is more commonly encountered, more carbon dioxide than the saturation amount, air stripping brings about release of the gas. In this instance, equilibrium is approached from the direction of supersaturation. The final result of prolonged air stripping, however, is the same—saturation.

Rate of Achievement

Equilibrium conditions are important in the aeration process, but of even greater significance to the design engineer is the rate of achievement of equilibrium. Equilibrium and rate of approach to it are not independent of each other. Under similar conditions, the further the air-water system is from equilibrium, the more rapid the interchange of gas in the direction of attaining equilibrium.

Significance of Films

Films at the air-water interface appear to have an important bearing on the rate of gas transfer. Both liquid and gas films can retard the rate of exchange of volatile material, but the liquid film is a more important factor in the transfer of gases of low solubility, such as oxygen and carbon dioxide.

Film resistance is influenced by many factors, but the most important are turbulence and temperature. High temperature and turbulence promote gas transfer by reducing film thickness. Increased temperature also increases the rate of molecular diffusion.

Rate of Transfer

The rate of transfer of a volatile substance from water to air is generally proportional to the difference between the existing concentration and the equilibrium concentration of the substance in solution. The relationship is expressed as follows:

$$M = K_L a \, (C_i^* - D_i)$$

where M = mass of substance transferred per unit of time per unit of volume (lb/h)/ft^3 [(kg/h)/m^3]
K_L = overall liquid mass-transfer coefficient, ft/h (m/h)
a = effective area for mass transfer, ft^2/ft^3 (m^2/m^3)
C_i^* = liquid-phase concentration in equilibrium with gas-phase concentration, lb/ft^3 (kg/m^3)
D_i = bulkhead liquid-phase concentration, lb/ft^3 (kg/m^3)

The driving force for mass transfer is the difference between actual conditions in the air stripping unit and conditions associated with equilibrium between the gas and liquid phases. Equilibrium concentration of a solute in air is directly proportional to the concentration of the solute in water at a given temperature.

Henry's law states that the amount of gas that dissolves in a given quantity of liquid, at constant temperature and total pressure, is directly proportional to the partial pressure of the gas above the solution. Thus Henry's law constant can be considered a partition coefficient describing the relative tendency for a compound to separate, or partition, between the gas and the liquid of equilibrium; it indicates a contaminant's volatility and its affinity for the aeration process. Substances with high Henry's law constants are easily removed by air stripping, and those with low constants are difficult to remove. Table 5.1 lists the Henry's law constants for several compounds. Vinyl chloride has an extremely high constant relative to the other VOCs.

The mass-transfer coefficient K_L is a function of the compound being stripped from water, the geometry and physical characteristics of the air stripping system, and the temperature and flow rate (contact time) for the liquid. It also incorporates the diffusion resistance to mass transfer in both liquid and gas phases and is related to local gas- and liquid-phase mass-transfer coefficients k_g and k_L, respectively.

For most stripping applications in water treatment, the bulk of resistance to mass transfer resides in the liquid phase. As a result, air stripping process design should be based on maximizing the liquid mass-transfer coefficient.

TABLE 5.1 Henry's Law Constants for Selected Compounds

Compound	Formula	Henry's constant atm[a,d]
Vinyl chloride	CH_2CHCl	3.55×10^5
Oxygen	O_2	4.3×10^4
Toxaphene[b]	$C_{10}H_{10}C_{18}$[c]	3.5×10^3
Carbon dioxide	CO_2	1.51×10^3
Carbon tetrachloride[b]	CCl_4	1.29×10^3
Tetrachloroethylene[b]	C_2Cl_4	1.1×10^3
Trichloroethylene[b]	$CHClCCl_2$	5.5×10^2
Hydrogen sulfide	H_2S	5.15×10^2
Chloromethane[b]	CH_3Cl	4.8×10^2
1,1,1-Trichloroethane[b]	CCH_3Cl_3	4.0×10^2
Toluene[b]	$C_6H_5CH_3$	3.4×10^2 (at 25° C)
Benzene[b]	C_6H_6	2.4×10^2
Chloroform[b]	$CHCl_3$	1.7×10^2
Bromodichloromethane	$CHCl_2Br$	1.18×10^2[e]
1,2-Dichloroethane[b]	CH_2ClCH_2Cl	61
Dibromochloromethane	$CHClBr_2$	47[f]
Bromoform[b]	$CHBr_3$	35
Methyl tertiary butyl ether	$C_5H_{12}O$	22[g]
Ammonia	NH_3	0.76
Pentachlorophenal[b]	$C_6(OH)Cl_5$	0.12
Dieldrin[b]	$C_{12}H_{10}OCl_6$	0.0094

[a] Temperature 20° C except where noted otherwise.
[b] Computed from water solubility data and partial pressure of pure liquid at specified temperature.
[c] Synthetic; approximate chemical formula.
[d] Kavanaugh and Trussell, 1980, except where otherwise noted.
[e] Warner, Cohen, and Ireland, 1980.
[f] Symons et al., 1981.
[g] Zorgorski et al., 1996.

The effective area a represents the total surface area created in the air stripping unit by producing numerous fine water droplets or by forming minute gas bubbles. The effective area is a function of air stripping equipment. In terms of effective area, an optimum treatment system is one that includes a high surface area for mass transfer per unit volume.

The mass-transfer coefficient K_L and the effective area for mass transfer a are usually evaluated as one constant, $K_L a$.

DESIGN OF AERATION/AIR STRIPPING PROCESSES

Air stripping equipment design has been developed extensively in the chemical industry for handling concentrated organic solutions. Procedures found in the chemical engineering literature can be applied to water treatment for trace organics removal. Based on principles described in the last section, the design of an air stripping installation is primarily dependent on the following factors:

- Temperature of the water and the surrounding air
- Physical and chemical characteristics of the contaminant to be removed
- The ratio of air to water being provided in the process
- Contact time between the air and water
- The water surface area available for mass transfer

The first two factors are fixed by source water quality and location of the installation. The other factors can be varied with the type of aeration equipment used. For all aeration and air stripping equipment, the water exiting the treatment equipment must be disinfected before the water is pumped to the distribution system.

Design of Diffused Air Equipment

When diffused aeration is employed, air stripping is accomplished by injecting bubbles of air into the water. Ideally, diffused aeration is conducted counter to the flow of water. Untreated water should be entering at the top and treated water exiting at the bottom while fresh air enters at the bottom and exhausted air exits at the water surface. Gas transfer can be improved by increasing the basin depth, producing smaller bubbles, improving the contact basin geometry, and incorporating a turbine to produce smaller bubbles and increase bubble holdup.

Diffused air aerators usually provide a longer aeration time than waterfall aerators, generally an advantage, but other factors influencing performance are the turbulence provided, air-volume ratio, and gas-transfer resistance. Because of these factors, comparison between the two types of equipment cannot be made solely on the basis of aeration contact time.

Basin Design. Tanks used for the diffused air process are usually made of concrete and are commonly 9 to 15 ft (3 to 5 m) deep and 10 to 30 ft (3 to 9 m) wide. The ratio of width to depth should not exceed 2, to achieve optimum mixing. Tank length is governed by the desired detention time, which usually varies from 10 to 30 min. Air diffusers are generally mounted along one side of the tank to impart a spiral flow to the water. A spiral flow pattern produces higher water surface velocities, which in turn promotes better gas transfer. In addition, with a spiral flow, a substantial number of bubbles do not escape immediately, but are carried across the basin where they are held in a more or less fixed position by the descending water.

Diffusers. Common types of diffusers are perforated pipes, porous plates or tubes, and various patented impingement or sparger devices. Compressed air is generally furnished by a rotary compressor sized to produce the correct volume and pressure. Diffusers produce small bubbles that rise through the water and cause turbulence and the opportunity for the exchange of volatile materials.

Diffusers are generally located near middepth in the tank, usually about the optimum efficiency point. Deeper location of the diffusers requires greater pressure head, which increases compressor power costs. The amount of air required ranges from 0.01 to 0.15 ft^3/gal (0.0008 to 0.012 m^3/L) of water treated. Sufficient diffuser capacity must be provided to supply air at the required rate without excessive pressure loss. Some installations include lateral baffles to prevent short-circuiting.

Air pressure requirements depend on submerging diffusers and friction loss through piping. Power requirements vary from 0.5 to 2.0 kW/mgd (0.00013 to 0.00053 kW/m² per day), with the average about 1.0 kW/mgd (0.00026 kW/m² per day). When porous plates or tubes are used, air should be filtered to avoid clogging of the diffusers. "Oil-free" compressors should be used to avoid contamination of the air.

Diffuser-type aerators require less space than spray aerators and generally more than tray aerators. They have practically no head loss through diffusion units, and this is usually an important aspect in overall plant design. Aeration units have few cold weather operating problems, and in moderate climates there is no need to house them. In some instances, diffusion aeration basins are used to provide chemical mixing.

Design of Spray Aerators

Exposure time for each drop from a spray aerator depends on its initial velocity and trajectory. Drop size, and the resulting area-volume ratio, is a function of the dispersing action of the nozzle. The initial velocity V of a drop emerging from an orifice or nozzle appears in the formula

$$V = C_v \sqrt{2gh}$$

and the discharge by the equation

$$Q = C_d A \sqrt{2g}$$

where h = total head on nozzles, ft
  g = acceleration from gravity, ft/s²
  A = area of opening, ft²
  C_v = coefficient of velocity
  C_d = coefficient of discharge ($C_d = C_v C_c$, where C_c is coefficient of contraction)

Coefficients of velocity, contraction, and discharge vary with the shape and other characteristics of the orifice or nozzle.

The trajectory of the spray used in an aerator may be vertical or inclined. If the angle between the initial velocity vector and horizontal is zero, theoretical exposure time t of the water drops is given by the formula

$$t = 2C_v \sin \theta \sqrt{2\frac{h}{g}}$$

The sine of an angle of less than 90° is less than 1.0, so a vertical jet gives the longest exposure time for a given value of h. But an inclined jet has the advantage of a longer path and less interference between falling drops. Wind also influences the path of the rising and falling drops, so an allowance must be made for its action.

Nozzle design is important in achieving optimum dispersion of water. Among special designs used are rifled nozzles, centrifugal (West Palm Beach) nozzles, Sacramento floating cones, impinging devices, and rotating reaction nozzles.

The size, number, and spacing of spray nozzles depend on the head of water being used, space available for aeration facilities, and interference between adjacent sprays. Theoretically, numerous small nozzles capable of producing atomized water would be the most efficient design. However, from a practical standpoint, very small nozzles should be avoided because of clogging and high maintenance requirements. Nozzles used in most spray aerators are 1.0 to 1.5 in. (2.5 to 3.8 cm) in diameter and have discharge ratings of

75 to 150 gpm (4.73 to 9.46 L/s) at about 10 psi (69 kPa). Nozzle spacing in most installations is between every 2 and 12 ft (0.6 and 3.7 m). The area allocated to spray aeration varies from 50 to 150 ft²/mgd [106 to 318 m²/(m³/s)] capacity, although much larger areas have been used at some treatment facilities.

Because interior and exterior corrosion can be serious problems in aerator piping, corrosion-resistant materials should be used wherever possible.

Spray aerators providing a high area-to-volume ratio are spectacular to see. They are rarely housed, so ventilation presents no problem. Gas transfer between water drops and air proceeds rapidly, and spray-type aerators usually have a relatively high efficiency. In general, spray aerators remove more than 70% of dissolved carbon dioxide, and removals as high as 90% have been documented. Disadvantages of spray aerators are principally the relatively large space requirements, freezing problems in colder climates, short exposure time between water and air, and high head requirements.

Design of Multiple-Tray Aerators

Multiple-tray aerators are generally constructed with three to nine trays and a spacing of 12 to 30 in. (30 to 76 cm) between trays. Space required for an aeration unit ranges from about 25 to 75 ft²/mgd (2 to 6 m²/ML per day) capacity, with 50 ft²/mgd (4 m²/ML per day) being about average. Water application rates range from roughly 20 to 30 gpm/ft² [17 to 20 (L/s)/m²]. These aerators have excellent oxygen adsorption and carbon dioxide removal capacities.

Ventilation Requirements. Tray aerators are, in many respects, analogous to cooling towers, and the design is similar. Ventilation and water distribution must be carefully considered in connection with location and design.

Multiple-tray aerators are usually housed, particularly in colder climates. A good example of an enclosed but well-ventilated installation is the Allen substation aerator at Memphis, Tennessee. Aluminum scroll panels are used to promote good cross-ventilation, and the roof is open except directly over the distributing trays. Carbon dioxide concentration in the source water exceeds 90 mg/L, and this aerator has consistently produced a 90% or greater reduction.

If a tray aerator must be enclosed and there is not sufficient natural ventilation, artificial ventilation must be provided. This is usually accomplished by supplying air with a blower at the bottom of the aerator so that it travels counter to water flow.

Important design considerations in designing tray aerators are the use of corrosion-resistant materials and methods of dealing with slime and algal growths. Aeration units are generally constructed using concrete, stainless steel, aluminum, and rot-resistant wood. Slime and algal growths may be controlled by treating the source water with chlorine or copper sulfate.

Carbon Dioxide Removal. Carbon dioxide removal by multiple-tray aerators can be approximated by the following empirical equation, developed by Scott (1955):

$$C_n = C_c 10^{-kn}$$

where C_n = concentration of carbon dioxide after passing through n trays, mg/L
C_c = concentration determined originally in distribution tray
n = number of trays including distribution tray
k = coefficient dependent on ventilation, temperature, turbulence, and other characteristics of installation; generally ranges from 0.12 to 0.16

Design of Packed Columns

The rate at which a volatile compound is removed by air stripping in packed tower aeration (PTA) depends on the following factors:

- Air-to-water ratio (A/W ratio)
- Height of packing in the column
- Available surface area for mass transfer
- Water loading rate
- Air and water temperatures
- Physical chemistry of the contaminants to be removed

The first four factors may be controlled in the design of an air stripping unit.

Airflow and Water Flow Requirements. Airflow required for a packed column depends on the Henry's law coefficient for the compounds to be removed from the water. Packing height is a function of the required VOC removal efficiency. In general, an increase in packing height results in higher VOC removal.

The air-to-water ratio used in a column is a function of water temperature and desired level of contaminant removal. This ratio determines the size of the blower, the primary component of operating costs for PTA systems. Air-to-water ratios typically range from 30 : 1 to 100 : 1. The water loading rate—the amount of water passing through the column—usually ranges from 25 to 30 gpm/ft^2 [17 to 20 (L/s)/m^2]. Compounds that are more difficult to remove, such as methyl tertiary butyl ether (MTBE), require higher air-to-water ratios (150 : 1 or greater) and are generally designed at lower liquid loading rates (15 to 17 gpm/ft^2) to minimize air pressure drop and reduce blower energy requirements. Column diameter is selected to accommodate the desired water loading on the column.

Packed column aeration removal effectiveness usually increases with an increase in water temperature, but it has been found that heating the influent water to increase removal effectiveness is not generally cost-effective.

Column Design. The relationship between packing height and column performance is derived from the basic mass-transfer relationship. In the following formula, packing height Z, in feet, is related to the *height of a transfer unit* (HTU) in feet, and the *number of transfer units* (NTU):

$$Z = (\text{HTU})(\text{NTU})$$

The HTU is a function of the liquid loading rate and $K_L a$. This relationship is expressed in the following equation:

$$\text{HTU} = \frac{L}{K_L a C_0}$$

where L = liquid flow, (lb mol/h)/ft^2 [(g mol/h)/m^2] and C_0 = molar density of water, lb mol/ft^3 (g mol/m^3).

The number of transfer units is a function of column performance and the substance to be removed. This relationship is expressed as

$$\text{NTU} = \frac{R_1}{R_1 - 1} \ln \frac{(X_i/X_o)(R_1 - 1) + 1}{R_1}$$

FIGURE 5.5 Effect of type of VOC on packed column design. (*Source: AWWARF/KIWA Report, 1983.*)

where X_i/X_o = ratio of influent to effluent liquid-phase concentration
R_1 = dimensionless stripping factor
$= \dfrac{(H)(G)}{(p_t)(L)}$ where G is gas flow, (lb mol/h)/ft² [(g mol/h)/m²]

The ability of a particular VOC to be stripped may be determined from Henry's law constant. The higher the constant, the more easily the VOC is removed by air stripping. The effect that the type of compound to be removed has on packing depth and air-to-water ratio is shown in Figure 5.5. An air-to-water ratio of about 20 : 1 is required to achieve 95% removal of trichloroethylene (TCE) with 15 ft (4.6 m) of packing medium that is 1 in. (2.5 cm) in diameter. For 95% removal of a less volatile compound such as 1,2-dichloroethane, an air-to-water ratio of about 120 : 1 is required for a column with the same size and depth of packing.

Desired removal efficiency also affects the design of a packed column. Figure 5.6 illustrates the relationship between air-to-water ratio and packing depth to achieve various efficiencies for TCE removal. About 6 ft (2 m) of 1-in. (2.5-cm) packing medium is required to achieve 80% removal of TCE with an air-to-water ratio of 20 : 1. To achieve 99% removal with the same packing and air-to-water ratio, about 20 ft (6.1 m) of packing would be required.

Water temperature must also be considered in designing a packed column. Most groundwater supplies have a water temperature of about 55° F (13° C). However, water temperature may be as low as 45° F (6° C) in northern regions and as high as 75° F (24° C) in some warm regions. The relationship between water temperature and removal efficiency is illustrated in Figure 5.7.

Design Considerations.   A diagram of a typical packed column installation is shown in Figure 5.8. In an installation to treat groundwater, the water is generally pumped directly

FIGURE 5.6 Effect of removal efficiency on packed column design. (*Source: AWWARF/KIWA Report, 1983.*)

FIGURE 5.7 Effect of water temperature on packed column design. (*Source: AWWARF/KIWA Report, 1983.*)

FIGURE 5.8 Packed tower aeration system. (*Source: Cook College Office of Continuing Professional Education/SOTC.*)

from the wells to the top of the column. Treated water is collected at the bottom of the column in a clearwell, from which it is usually pumped directly to the distribution system. The principal facility elements of a packed column include the following:

- Column and internal parts
- Packing material
- Blower
- Clearwell
- Booster pumps

The column is basically a tank usually constructed from fiberglass-reinforced plastic, aluminum, stainless steel, or concrete. Where stainless steel is used, chlorination of the incoming water should be avoided because of the aggressive nature of chlorinated, aerated water on stainless steel.

A demister is usually installed at the top of the tank to prevent objectionable clouds of moisture from coming off the column. Near the top of the column, piping is installed to distribute influent water evenly over the top of the packing material.

As water flows downward through the packing, it tends to migrate to the column wall, and redistributors are installed at intervals to support the packing material and redirect the water back toward the center of the column. Four commonly used distributor styles are the orifice plate and the trough (see Figure 5.9), orifice headers, and spray nozzles.

Packing. Packing materials are designed to simultaneously provide a low pressure drop for air passing through the column and maximum air-water contact area. Packing pieces for "dumped packing" are randomly dumped into the column. They are available in various shapes of ceramic, stainless steel, and plastic materials. Plastic is most commonly used in water treatment applications because it is durable, lightweight, and cost-effective. Some common shapes are illustrated in Figure 5.10.

Fixed packing comes in prefabricated sheets mounted at intervals inside the column. Although the initial cost is higher for this type of packing, manufacturers claim higher transfer efficiency.

Liquid Distribution. After the packing material, the distribution of the water over the packing material is the next most important design consideration for a packed column. Proper liquid distribution typically determines the successful operation of the packed column. Inadequate liquid distribution can be caused by improper design of the distribution tray or plate, or insufficient sealing between the distribution plate and the walls of the column. The result can be short-circuiting of the incoming water down the sides of the column and consequently reduced contact between the air and water. The distribution plate must be designed to provide even distribution of the water under all flow conditions. Some packed columns have been designed using spray nozzles for liquid distribution to overcome this potential problem. Also, wall wipers or redistribution rings have been used to reduce the effects of short-circuiting.

Air Blower. Airflow is provided at the base of the column by a centrifugal blower driven by an electric motor. Small towers are designed with blowers requiring a motor as small as 5 hp (3,700 W), but much larger blowers are required for larger units and when a high air-to-water ratio is required. Care must be taken in providing screens and locating the air inlet to prevent insects and airborne contaminants from being blown into the column. In some instances, clogging of the air inlet screens has caused a reduction in the air-to-water ratio and a resultant reduction in the treatment efficiency. Screens must be checked regularly or designed with pressure drop indicators to sense a drop in airflow through the intake.

Site Considerations. In general, water temperature in the column stays close to the temperature of the influent water. It has been demonstrated in several installations that there is no danger of freezing even under air temperatures well below freezing. However, in very cold conditions, this assumption should be confirmed. In cold climates, blowers and pumping equipment are usually housed for protection from ice and snow. Housing this equipment also provides increased security, reduced noise, and reduced maintenance.

Site considerations include zoning restrictions, height restrictions, and noise restrictions. There are instances where residents have opposed installation of a column in a residential neighborhood because of potential noise, visual impact of the relatively tall tower, and the perceived inhalation danger of the off-gas. Where there is public opposition to installation of a column, other means of treatment may have to be considered, or the source water may have to be piped to a more remote location for treatment.

Orifice-type distributor

Trough-type distributor

FIGURE 5.9 Types of distributors. (*Source: USEPA workshop on emerging technologies for drinking water treatment, 1988.*)

FIGURE 5.10 Some common shapes of plastic packing. (*Source: U.S. Environmental Protection Agency.*)

Exhaust Emission Considerations. The emission of contaminated exhaust air from packed column air stripping systems creates potential air quality problems. The transfer of VOCs from water to air might be a concern depending on site or local regulations, type of VOC, proximity to human habitation, exposure of treatment plant workers, local air quality, local meteorological conditions, daily quantity of water to be processed, and the contamination level. Also, sulfur compounds in the raw water can result in odor problems in the vicinity of a packed column.

The emission rate must be evaluated in the context of applicable air quality regulations and other site-specific factors. Air emission regulations are expressed in terms of permissible emission rates (lb/day or lb/h) or projected ground level concentrations (mg/m^3). If the treatment plant's emission rate is unacceptable, the column or plant process may be changed to bring the installation into compliance. Treatment options currently available to remove organics from off-gas include

- Packed column design modification
- Thermal incineration of the gas
- Catalytic incineration of the gas
- Ozone destruction
- Vapor-phase carbon adsorption

The least costly method of achieving compliance with air regulations is usually to modify the packed column design to dilute the emissions. Possible modifications include increasing the tower height, airflow rate, and exhaust gas velocity. If these steps are insufficient to achieve compliance, a vapor-phase treatment component may be required.

Thermal incineration of packed column off-gas has the disadvantage of high energy requirements. Catalytic incineration has lower temperature requirements but is currently

AERATION AND AIR STRIPPING 5.21

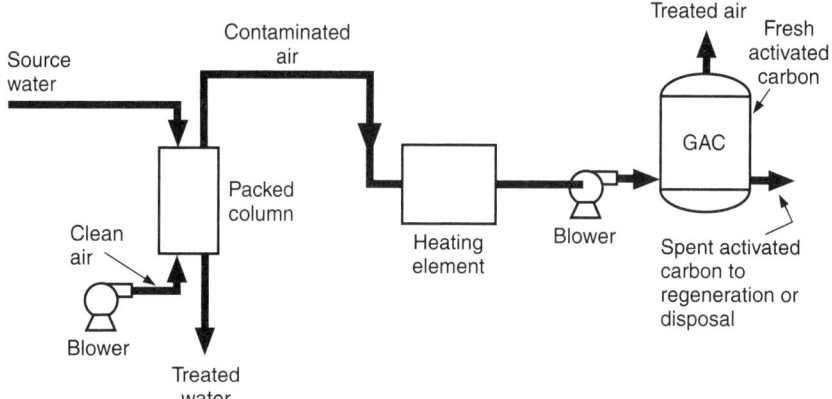

FIGURE 5.11 Schematic of vapor-phase GAC system. (*Source: AWWA WQD Committee Report, 1991.*)

not effective for removing low levels of chlorinated organics. Similarly, ozone destruction may not be cost-effective at this time.

Granular activated carbon (GAC) adsorption is generally the most cost-effective method to remove low-level organics from packed column exhaust air. Vapor-phase adsorption is attractive because the vapor-phase mass-transfer zone (MTZ) is much shorter than the liquid-phase MTZ, and the cross-sectional area requirement of the fixed bed is much smaller. Activated carbon usage is also less than that for liquid phase. A schematic of a vapor-phase GAC system is illustrated in Figure 5.11.

In operating a vapor-phase GAC system, the relative humidity of the off-gas should be reduced to prevent condensation of water vapor in the activated carbon pores by heating the air before it enters the GAC contactor. The competition of water vapor adsorption and gas-phase VOC adsorption onto GAC is minimized at an off-gas relative humidity of 40% to 50%.

Predicting contaminant breakthrough is a major concern with vapor-phase GAC systems because of the reliability of methods of estimating the vapor-phase GAC bed life. Possible approaches include monitoring GAC effluent air quality either continuously or intermittently, using a mass balance around the contactor, or combining these two approaches.

For example, a GAC bed from a pilot plant in Wausau, Wisconsin, treating off-gas containing TCE and PCE was regenerated with steam 3 times. The TCE capacity decreased from 80% to 60% of the initial capacity over the three cycles. This lessening in TCE capacity with successive adsorption/regeneration cycles was due to PCE buildup on the GAC. The PCE was not removed effectively under the existing regeneration conditions (100° C, 1 atm). Use of the equilibrium model indicated that regeneration with saturated steam 50° C above the boiling point of PCE (121° C) could improve PCE removal.

When one is considering installing a stripping column, as a first step, state and local officials should be contacted for information on air quality requirements.

Fouling of Packing. Packed column design must consider the possibility of scaling and fouling of the packing. Some installations experience few problems, but others have serious problems. Three main causes of fouling are carbonate scaling, iron oxidation, and microbial action. Fouling gradually causes a decrease in airflow through the column and if not corrected, can seriously reduce the column's performance.

Carbonate scaling of packing becomes a problem when influent water has a relatively high calcium carbonate hardness. Deposition occurs primarily because of pressure changes and a rise in the water's pH as CO_2 is released from the water. Carbonate scale deposits usually are a brittle, cementlike scale.

Ferrous iron in groundwater, another source of scaling, oxidizes easily in the presence of oxygen to form insoluble ferric compounds. The result is primarily iron hydroxide, which accumulates on the packing as a rust- or black-colored gel.

Microbial fouling is, in most cases, primarily from the presence of iron bacteria. Iron bacteria generally thrive in a dark environment, under aerobic conditions, and at temperatures between 40° and 70° F (4° and 21° C). These organisms derive energy from the oxidation of iron from the ferrous to ferric form. Some of the more common species of iron bacteria found in groundwater are *Gallionella, Crenothrix, Leptothrix,* and *Sphaerotilus.* Colonies of the bacteria can grow on the packing media, forming a slimy material that, if not controlled, can completely fill all void spaces in the packing.

One method of controlling fouling involves pretreating water with chlorine or permanganate and then filtering to remove oxidized solids. Another pretreatment method is to add chelating agents to inhibit formation of oxidation products. In any event, if the influent water has a high potential for fouling, the plant must provide facilities and a regular schedule of periodic cleaning of the buildup. Cleaning consists of circulating strong chlorine or acid solutions through the media.

PILOT TESTING

Pilot plant air stripping studies are usually conducted to determine the effectiveness of an aeration system in removing the contaminants of concern. Testing is performed using a laboratory or bench-scale prototype unit. Municipalities, private consultants, and the USEPA use pilot aeration tests to determine the usefulness of aeration in removing various VOCs.

Pilot tests are sometimes performed on samples "spiked" with the contaminant to be studied. However, it is generally best to use water directly from the contaminated source, because subtle differences in both the physical and chemical composition of the water could have some effect on the ability of an aeration unit to remove a contaminant.

Diffused Air Pilot Studies

The main components of a diffused air pilot plant are the tank, diffuser, and blower. The system is usually operated in a countercurrent fashion. Influent water is piped through a rotameter to measure flow and dispersed through a liquid distributor at the top of the tank. Water then flows down through the tank to be discharged at the bottom. Air is supplied by a compressor, piped through a rotameter to measure flow, and dispersed by a diffuser located at the bottom of the tank.

Packed Column Pilot Studies

A typical packed column pilot plant consists of the column with influent piping, valves, packing media, blower, and support structure, as shown in Figure 5.12. Influent is pumped through the metering valve and meter to the top of the column, where it is distributed by an orifice-type plate. Water trickles down through the packing and is discharged at the bottom of the column.

Support plates are placed at intervals in the column to provide intermediate support for media and to redistribute waterflow and airflow. Air from a compressor or blower is controlled by a metering valve and is introduced at the bottom of the column. Taps for sampling water are located on the influent line, at three locations along the side of the column, and at the bottom of the column.

Pilot plant size should be small enough to allow it to be easily transported to test sites, yet large enough to allow "scale-up" procedures to be meaningful. The diameter of a pilot column is commonly 12 in. (30.4 cm).

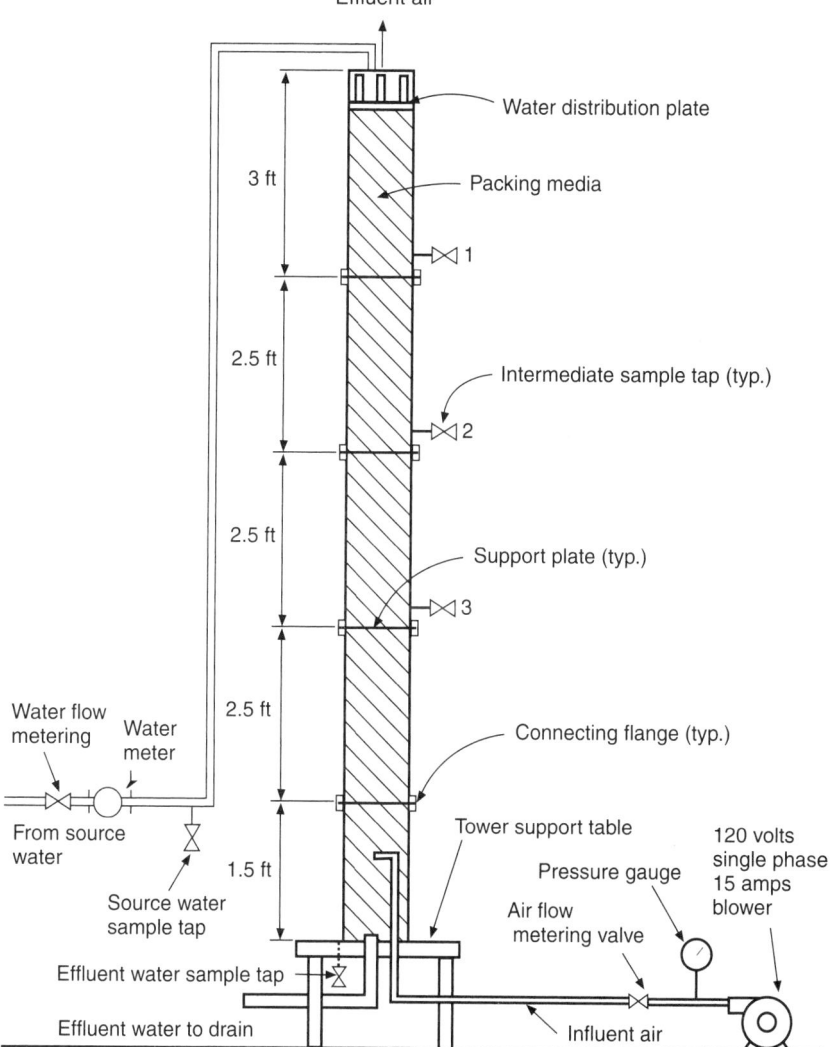

FIGURE 5.12 Schematic of a pilot air stripping column. (*Source: USEPA technologies for upgrading existing or designing new drinking water treatment facilities, 1990.*)

It is critical for measurements of both waterflows and airflows to be accurate in order to know the air-to-water ratio.

When one is using random packing in a column, the number of packing pieces per unit volume is ordinarily less in the vicinity of column walls. This packing distribution tends to produce a flow scheme in which water moves toward the walls and airflow is directed more toward the center of the column—a phenomenon referred to as *channeling*. Channeling is less pronounced when the diameter of the packing pieces d_p is smaller than one-eighth the column diameter d_c. It is recommended that, if possible, the ratio $d_p/d_c = 1 : 15$.

Testing Program and Evaluation of Results

A testing program for operating a pilot plant and evaluating results is usually necessary before one begins to design full-scale treatment facilities. Decisions must be made on factors such as the number of operations to be made, loading rates, air-to-water ratio that must be provided, and, in the case of packed columns, the type of packing to be used.

A critical part of operating a pilot plant is the sampling procedures. When samples are collected for VOC analysis, they must be collected without allowing any entrapped air to remain in the sample bottle. If even a very small bubble remains in the bottle, some of the volatile chemicals may leave the water, causing incorrect results when the sample is analyzed.

Other data that must be collected periodically during pilot plant operation are water and air temperatures, as well as chlorine, iron, and total organic carbon levels.

Once samples have been analyzed, test results can be evaluated. From these data, optimum full-scale operational and design conditions can be established. In a diffused air system, detention time, water depth, and air-to-water ratio are evaluated. In addition to these conditions, with a packed column system, KLa is determined and the best packing is chosen.

BIBLIOGRAPHY

Adams, J. Q., and R. M. Clark. "Evaluating the Costs of Packed Tower Aeration and GAC for Controlling Selected Organics." *Journal AWWA* 83(1):49, 1991.

Air Stripping for Volatile Organic Contaminant Removal. Denver, Colo.: American Water Works Association, 1989.

American Water Works Association. *Water Quality and Treatment,* 4th ed. New York: McGraw-Hill, 1990.

Atoulikian, Richard G. "Proper Design of Radon Removal Facilities for Cost-Effective Treatment." *Journal NEWWA* 111:4, December 1997.

Balla, John M. "MTBE Removal via Packed Column Air Stripping." *2001 AWWA Annual Conference Proceedings,* Washington, June 2001.

Balla, John M. "Is Your Packed Column Air Stripper Ready for the 21st Century?" *1999 AWWA Engineering and Construction Conference Proceedings,* Orlando, Fla., March 1999.

Boyden, B. H., et al. "Using Inclined Cascade Aeration to Strip Chlorinated VOCs from Drinking Water." *Journal AWWA* 84(5):62, 1992.

Controlling Radionuclides and Other Contaminants in Drinking Water Supplies: A Workbook for Small Systems. Denver, Colo.: American Water Works Association, 1991.

Dixon, K. L., et al. "Evaluating Aeration Technology for Radon Removal." *Journal AWWA* 83(4):141, 1991.

Duranceau, Steven J. "Removal of Hydrogen Sulfide from Ground Water Supplies and Residuals Control." *2002 AWWA Inorganic Contaminants Workshop Proceedings,* San Diego, Calif., February 3–5, 2002.

Dvorak, B. I., et al. "Uncertainty in Air Stripping Tower Design: Implications of Air-to-Water Ratio." IWA 1st World Congress, Paris, France, July 2000.

Dxombak, D. A., S. B. Roy, and H. J. Fang. "Air-Stripper Design and Costing Computer Program." *Journal AWWA* 85(10):63, 1993.

Existing VOC Treatment Installations: Design, Operation, and Cost. Denver, Colo.: AWWA Organic Contaminants Committee Report, 1991.

Jang, W., N. Nirmalakhandan, and R. E. Speece. *Cascade Air-Stripping System for Removal of Semi-Volatile Organic Contaminants: Feasibility Study.* Denver, Colo.: American Water Works Association Research Foundation and American Water Works Association, 1989.

Kavanaugh, M. C., and R. R. Trussell. "Design of Aeration Towers to Strip Volatile Contaminants from Drinking Water." *Journal AWWA* 73(12):684, 1980.

Lamarche, P., and R. L. Droste. "Air-Stripping Mass Transfer Correlations for Volatile Organics." *Journal AWWA* 81(1):78, 1989.

La Motta, Enrique J., et al. "Corrosion Control of Drinking Water Using Tray Aerators." *ASCE Journal of Environmental Engineering* 122:7, July 1996.

Langelier, W. F. "The Theory and Practice of Aeration." *Journal AWWA* 24(1):62, 1932.

Little, J. C., and R. E. Selleck. "Evaluating the Performance of Two Plastic Packings in a Crossflow Aeration Tower." *Journal AWWA* 83(6):88, 1991.

Schock, M. R., J. Holldber, T. R. Lovejoy, and J. Lowry. "California's First Aeration Plants for Corrosion Control." *Journal AWWA* 94:3, March 2002.

Scott, G. R. "Committee Report: Aeration of Water." *Journal AWWA* 44(4):873, 1955.

Spencer, Catherine M. "Aeration and Limestone Contact for Radon Removal and Corrosion Control." *Journal NEWWA* 112:1, 1998.

Staudinger, J., W. R. Knocke, and C. W. Randall. "Evaluation of the Onda Mass Transfer Correlation for the Design of Packed Column Air Stripping." *Journal AWWA* 82(1):73, 1990.

Symons, J. M., et al. "Treatment Techniques for Controlling Trihalomethanes in Drinking Water." U.S. EPA, Drinking Water Research Division, 1981.

VOCs and Unregulated Contaminants. Denver, Colo: American Water Works Association, 1990.

Warner, H. P., J. M. Cohen, and J. C. Ireland. "Determination of Henry's Law Constants of Selected Priority Pollutants." Wastewater Research Division, Municipal Environmental Research Laboratory, Cincinnati, Ohio, 1980.

Wen, Jason J., and Jeffrey B. Long. "Issues and Treatment Options to Comply with New Radon Rule." *2000 AWWA WQTC Proceedings,* Salt Lake City, Utah, November 2000.

Zogorski, J. S., et al. "Fuel Oxygenates and Water Quality: Current Understanding of Sources, Occurrence in Natural Waters, Environmental Behavior, Fate, and Significance." Prepared for the Interagency Oxygenated Fuels Assessment, Office of Science and Technology Policy, Executive Office of the President, 1996.

CHAPTER 6
MIXING, COAGULATION, AND FLOCCULATION

Paul J. Delphos
HDR Engineering, Inc.
Norfolk, Virginia

George M. Wesner
Consulting Engineer
San Clemente, California

Coagulation and flocculation may be broadly described as chemical and physical processes that mix coagulating chemicals and flocculation aids with water. The overall purpose is to form particles large enough to be removed by the subsequent settling or filtration processes. Particles in source water that can be removed by coagulation, flocculation, sedimentation, and filtration include colloids, suspended material, bacteria, and other organisms. The size of these particles may vary by several orders of magnitude. Some dissolved material can also be removed through the formation of particles in the coagulation and flocculation processes. The importance of dissolved material removal has become much more critical in recent years with increased regulatory emphasis on disinfection by-products and total organic carbon removal.

There are several excellent discussions on the theory of coagulation and flocculation in *Water Quality and Treatment* and other AWWA publications listed at the end of this chapter.

DEFINITIONS

Terms used in this chapter are defined as follows:

- *Coagulation* is the process in which chemicals are added to water, causing a reduction of the forces tending to keep particles apart. Particles in source water are in a stable condition. The purpose of coagulation is to destabilize particles and enable them to become attached to other particles so that they may be removed in subsequent processes. Particulates in source waters that contribute to color and turbidity are mainly clays, silts, viruses, bacteria, fulvic and humic acids, minerals (including asbestos, silicates, silica, and radioactive particles), and organic particulates. At pH levels above 4.0, particles or molecules are generally negatively charged. The coagulation process physically occurs in a rapid mixing process.

- *Mixing* is commonly referred to as *flash mixing, rapid mixing,* or *initial mixing.* The purpose of rapid mixing is to provide a uniform dispersion of coagulant chemical throughout the influent water.
- *Enhanced coagulation* is a phrase used by the U.S. Environmental Protection Agency (USEPA) in the Disinfectants and Disinfection By-products Rule. The rule requires that the coagulation process of some water supplies be operated to remove a specified percentage of organic material from the source water, as measured by total organic carbon (TOC). Enhanced coagulation (removal of TOC) can be achieved in most cases by either increasing coagulant chemical dosage or adjusting the pH during the coagulation reaction.
- *Coagulant chemicals* are inorganic or organic chemicals that, when added to water at an optimum dosage, cause particle destabilization. Most coagulants are cationic when dissolved in water and include chemicals such as alum, ferric salts, lime, and cationic organic polymers.
- *Flocculation* is the agglomeration of small particles and colloids to form settleable or filterable particles (flocs). Flocculation begins immediately after destabilization in the zone of decaying mixing energy following rapid mixing, or as a result of the turbulence of transporting flow. In some instances, this incidental flocculation may be an adequate flocculation process. A separate flocculation process is most often included in the treatment train to enhance contact of destabilized particles and to build floc particles of optimum size, density, and strength.
- *Flocculation aids* are chemicals used to assist in forming larger, denser particles that can be more easily removed by sedimentation or filtration. Cationic, anionic, or nonionic polymers are most often used in dosages of less than 1.0 mg/L.
- *Direct filtration* is a treatment train that includes coagulation, flocculation, and filtration, but excludes a separate sedimentation process. With direct filtration, all suspended solids are removed by filtration. In the process sometimes called in-line filtration, flocculation occurs in the conduit between the rapid mixing stage and the filter, in the volume above the filter media, and within the filter media.
- *Solids contact clarifiers* are proprietary devices that combine rapid mixing, flocculation, and sedimentation in one unit. These units provide separate coagulation and flocculation zones and are designed to cause contact between newly formed floc and settled solids.
- *Low-pressure membranes* are hollow-fiber membrane systems that provide micro- or ultrafiltration. These systems have pore sizes that are 10 to 100 times smaller than those of primary protozoa of concern (i.e., *Cryptosporidium* and *Giardia lamblia*). The membrane is a thin layer of polymer capable of separating materials based on size and chemical properties. These membrane systems typically operate in the range of -12 psi vacuum to 40 psi pressure.

THE COAGULATION PROCESS

Coagulation reactions occur rapidly, probably taking less than one second. Principal mechanisms that contribute to the removal of particulates when coagulating chemicals such as alum or ferric chloride are mixed with water include chemical precipitation, reduction of electrostatic forces that tend to keep particles apart, physical collisions between particles, and particle bridging.

Several factors affect the type and amount of coagulating chemicals required, including the nature of suspended solids and the chemical characteristics of the influent water.

Coagulant Chemicals

The most commonly used coagulants are

- Alum (aluminum sulfate), $Al_2(SO_4)_3 \cdot 14H_2O$. The most common coagulant in the United States, it is often used in conjunction with cationic polymers.
- Polyaluminum chloride, $Al(OH)_x(Cl)_y$. This is efficient in some waters, requiring less pH adjustment and producing less sludge.
- Ferric chloride, $FeCl_3$. This may be more effective than alum in some applications.
- Ferric sulfate, $Fe_2(SO_4)_3$. It is effective in some waters and more economical in some locations.
- Cationic polymers can be used alone as the primary coagulant or in conjunction with aluminum or iron coagulants.

Although alum is by far the most widely used coagulant chemical, ferric chloride or ferric sulfate forms a better-settling floc in some waters and may be more consistently effective in removing natural organic matter as compared to aluminum-based coagulants. Additionally, polyaluminum chloride often produces a better-settling floc in colder waters and often results in lower dosages, thereby producing less sludge than alum and ferric coagulants.

Flocculation Aids

Floc formed in many waters with alum is light and fragile and somewhat difficult to settle. Polymers and other additives can often help form a floc that is more efficiently removed by settling and filtration. Typical additives used for flocculation aids are

- High-molecular-weight anionic or nonionic polymers
- Activated silica
- Bentonite

These chemicals are normally added after the application of coagulants, from 5 to 600 s after mixing. If the water to be treated with a flocculent aid is already in the flocculation stage, the chemical should be added so that it can be spread across the flocculation basin.

Chemical Selection

The selection of coagulant chemicals and flocculation aids for use in a particular plant is generally based on economic considerations along with reliability, safety, and chemical storage considerations. The best method of determining treatability, the most effective coagulants, and the required dosages is to conduct bench-scale and, in some cases, pilot tests. Jar tests can be used to determine treatability and estimate chemical dosages. If possible, testing should cover all critical seasonal conditions. Pilot plant design and construction are discussed in Chapter 28.

When one is designing for coagulant application, as much flexibility as possible should be allowed, to accommodate changing conditions. Several points of addition for coagulant chemicals, particularly polymers, should be provided in the rapid mixing and flocculation processes. The order of chemical addition is also important in almost all waters.

Sludge quantity and disposal are important considerations in selecting the coagulant to be used. Metal-ion coagulants produce considerably larger volumes of sludge than polymers. The ability to predict the exact reaction and quantity of sludge that will be produced solely by the reaction formulas is limited. For this reason, predictions of treatability, chemical dosages, and sludge quantities must generally be determined by laboratory and pilot plant tests.

The coagulation process may, in some cases, be improved by preozonation. Ozone may significantly reduce coagulant requirements to the point where low residual solids (or filtration efficiency) make direct filtration feasible. However, due to the increasing power costs, in an effort to reduce ozone requirements, many recently constructed ozone facilities incorporate ozonation after clarification or filtration.

Oxidation with air and chemical oxidants such as chlorine and potassium permanganate may also aid coagulation by oxidizing iron and manganese, which can aid floc formation. Carbon addition, typically in the form of powdered activated carbon (PAC), may also improve coagulation, as it would remove a fair amount of organic matter prior to the coagulation process, thereby, reducing coagulant demand and the associated levels of sludge production as well as improving overall turbidity and organics removal. Similarly, new, specialty adsorbants/resins are actively being considered in the drinking water treatment community. One such adsorbant, a magnetic ion exchange (MIEX) resin by ORICA Watercare (Melbourne, Australia, and Englewood, Colorado), is specifically designed to remove low-molecular-weight organics, which are a primary contributor to many DBP precursors. As the DBP rules become more and more stringent in future years, there is little question more specialty-type coagulant aids will continue to be developed to further improve the removal of these precursors.

Adjustment of pH

Control of pH and alkalinity is an essential aspect of coagulation. The optimum pH for coagulation varies but is generally within the following ranges for turbidity removal:

- Alum: pH 5.5 to 7.5; typical pH 7.0
- Ferric salts: pH 5.0 to 8.5; typical pH 7.5

It can be necessary to adjust the pH of some source waters to achieve optimum coagulation. The pH is often lowered by adding carbon dioxide or an acid. Alum and ferric chloride consume alkalinity and can lower pH; however, reducing pH by adding more chemical than is required for coagulation should be avoided as it increases overall chemical costs and sludge production/costs. In some source waters with low pH or low alkalinity, it may be necessary to add caustic soda or lime to raise pH and to offset the acidity of metal-ion coagulants, even in an enhanced coagulation mode of operation. A thorough discussion of the effects of pH on coagulation appears in *Water Quality and Treatment.*

For waters that require enhanced coagulation to remove organic matter, the pH of coagulation should be lowered as compared to coagulation for turbidity removal only. Typically, the optimum pH for organics removal with alum is between 6.0 and 6.5, and between 5.5 and 6.0 for ferric coagulants. Often, polyaluminum chloride can provide organics removal without as significant a decrease in pH.

There are a number of secondary impacts of utilizing the higher coagulant dosages and lower pH values for enhanced coagulation. A few of these impacts include the following:

- *Increased solids.* The higher coagulant dosages directly result in increased sludge volumes. Chapter 17 discusses the relationship between coagulant dosages and sludge production.
- *Poorer dewatering characteristics.* The increased metal (Al^{3+} or $Fe^{2+ \text{ or } 3+}$) concentrations typically result in poorer dewatering characteristics. As a result, a change to enhanced coagulation may result in lower ultimate, dewatered solids concentrations.
- *Increased concrete/metal corrosion.* The lower pH of the coagulated water for TOC removal will be significantly more aggressive on concrete and metals as compared to the more neutral pH of water that has been coagulated for turbidity removal.

If pH is lowered to improve coagulation, it is typically necessary to raise the pH in the final effluent from the plant to provide a less corrosive finished water. The pH may be adjusted at one or more points in the treatment, including rapid mixing, prefiltration, and postfiltration. If the pH is lowered to improve coagulation and organics removal, it is often recommended to readjust the pH after the filtration process as compared to prefiltration. This is due to the fact that some organic matter may be adsorbed onto the floc that may carry over from the clarification process, and any prefiltration pH adjustment may then result in the "release" of this organic matter, which could pass through the filters and contribute to subsequent DBP formation.

For plants where only a small increase in pH is required, liquid caustic soda is most commonly used because of its ease of handling. When a large increase in pH is required, lime is normally the most economical choice. Lime, however, may add turbidity to a finished water; therefore, if lime is used for postfiltration pH adjustment, it is generally best to use a lime saturator to minimize the potential of turbidity addition. Also, in some waters, the utilization of soda ash for precoagulation alkalinity adjustment often helps the overall coagulation process.

COAGULATION AND MEMBRANE TECHNOLOGY

Low-pressure membrane technology (micro- and ultrafiltration) is becoming significantly more prevalent in the drinking water industry (see Chapter 13). Low-pressure membranes are purely size-exclusionary devices. As a result, anything smaller than membrane pore sizes (approximately 0.01 to 0.1 μm) will pass through the membrane. Therefore, membrane feedwaters with dissolved materials, such as organics and metals, require some form of additional treatment.

Often, in these cases, the most economical pretreatment process is simple coagulation. Potential coagulants for membrane pretreatment include those also used for conventional water treatment. Additionally, organic adsorption media such as PAC and MIEX, or oxidants such as potassium permanganate, chlorine, or chlorine dioxide can be applied upstream of a low-pressure membrane (assuming appropriate membrane compatibility) for enhanced dissolved material removal.

Similar to a direct filtration mode of operation for conventional technology, the goal of coagulation for membrane pretreatment is to produce a pinpoint floc that is capable of adsorbing dissolved matter, but minimizes solids loading onto the membrane filtration process. As noted briefly above, it is important to quantify membrane compatibility and performance with the coagulant of choice. Each commercially available membrane utilizes different membrane materials. As a result, the compatibility and performance of a coagulant for membrane filtration pretreatment will likely vary between membrane system and raw water supplies. As such, there are no specific guidelines for membrane system precoagulation except the general guidelines that are associated with conventional treatment.

DESIGN OF CHEMICAL MIXING

Chemical mixing can be accomplished by several different types of equipment designed to mix the applied chemicals with the source water as quickly as possible.

Mixing Intensity

The intensity of agitation required for optimum rapid mixing and flocculation is measured by the G value. The G value concept, developed by Camp and Stein in 1943, is widely used in designing rapid mixing and flocculation processes and is defined by the equation

$$G = \left(\frac{P}{\mu V}\right)^{1/2}$$

where G = root-mean-square velocity gradient, or rate of change of velocity, (ft/s)/ft
P = power input, ft · lb/s
μ = dynamic viscosity, lb · s/ft^2
V = volume, ft^3

Equations are also available to calculate G for various types of mixing arrangements, and manufacturers of mixing and flocculation equipment provide information on G values for their equipment. Another parameter used in designing mixing systems is Gt, which is the dimensionless product of G and detention time t in seconds.

Temperature Effects on Mixing

Rapid mix and flocculation systems design is temperature-dependent because water viscosity varies with temperature, as shown in Table 6.1. Guidelines for adjusting detention times in both rapid mix and flocculation basins are shown in Table 6.2.

Types of Rapid Mixing Systems

Coagulant chemicals can be mixed by several methods, including

- Mechanical devices in a dedicated basin
- In-line blenders

TABLE 6.1 Water Viscosity and Water Temperature

Temperature, °C	Temperature, °F	μ, cP	μ, lb · s/ft^2
0	32	1.792	3.75×10^{-5}
5	41	1.520	3.17×10^{-5}
10	50	1.310	2.74×10^{-5}
15	59	1.145	2.39×10^{-5}
20	68	1.009	2.10×10^{-5}
25	77	0.895	1.87×10^{-5}
30	86	0.800	1.67×10^{-5}

TABLE 6.2 Guidelines for Mixer Detention Times

Temperature, °C	Detention time factor
0	1.35
5	1.25
10	1.15
15	1.07
20	1.00
25	0.95
30	0.90

- Hydraulic methods
- Air mixing
- Induction mixing

Mechanical Mixers. Propeller- or paddle-type mechanical mixers in a dedicated basin are the most commonly used rapid mix system in water treatment plants. A typical arrangement for this type of rapid mixer is illustrated in Figure 6.1. Another mixer arrangement without stators and with a turbine-type blade is illustrated in Figure 6.2.

FIGURE 6.1 Propeller-type mechanical flash mixer.

FIGURE 6.2 Turbine-type mechnical flash mixer. (*Courtesy of Dorr-Oliver Eimco.*)

Rapid mixers attempt to provide complete mixing by near-instantaneous blending throughout the entire basin. As a result, an incoming volume of water immediately loses its identity as it is mixed with water that entered the basin previously. A mixer operating in this manner is also called a *backmix reactor* because basin contents are always blended backward with incoming flow.

Mechanical mixers are generally propeller- or paddle-type devices. More than one set of propeller or paddle blades may be provided on a shaft. Stators (baffles near the blades of the mixer or on the wall of the basin) may be provided to maximize energy transfer to the fluid and to minimize residual velocities at the outlet. Mechanical mixers are often constructed with a vertical shaft driven by a speed reducer and electric motor. Propeller-type mixers can be arranged so that flow is directed in any direction. With propeller-type

TABLE 6.3 Rapid Mixing Design Criteria from a 170-mgd Plant in San Diego, California

Number of stages: three

Type
    First stage: pump diffusion
    Second stage: mechanical
    Third stage: mechanical

Detention time
    First stage: 1 s
    Second stage: 30 s
    Third stage: 30 s

Number of basins: two
    Volume (each): 56,800 gal (214,900 L)
    Depth (each): 15 ft (4.6 m)
    Width (each): 22 ft (6.7 m)
    Length (each): 23 ft (7 m)

Mixing intensity G
    First stage (in-line): 1,000 s^{-1}
    Second stage: (basin): 150 to 300 s^{-1}
    Third stage: (basin): 150 to 300 s^{-1}

---

blades, coagulant chemical is generally directed to the eye of the propeller on the suction (upstream) side.

Mechanical mixers are not normally provided with variable-speed drives. If adjustments to energy input are necessary, they may be achieved by changing propellers or paddle blades or by mechanically adjusting the shaft speed.

Many treatment plant designs incorporate two or more rapid mix basins in series. The order in which coagulant chemicals are added is important in most waters, and more than one rapid mixer can provide the needed reaction time for each chemical. A plant in San Diego, California, uses three rapid mixers, as shown in the design criteria summarized in Table 6.3. Two other large water treatment plants in southern California use two vertical turbine mechanical mixers in series designed for G values of 440 to 670 s^{-1} with corresponding detention times of 30 to 10 s.

Typical design values for most mechanical rapid mix systems provide detention times of 10 to 60 s and G values of 600 to 1,000 s^{-1}.

In-Line Blenders. Although the traditional complete-mix type of basin with a mechanical mixer has performed satisfactorily in many plants for years, recent experience has shown that in-line blenders often provide more efficient rapid mixing. Advantages of this type of mixer are that it can approach nearly instantaneous dispersion of chemicals. In-line blenders operate at short detention times (less than 1 s) and at high G values.

An important consideration is that a short detention time and high G value may be a disadvantage in waters requiring more reaction time and the use of more than one chemical for floc formation. One answer to this problem may be an in-line mixer used in conjunction with a mechanical rapid mix basin (for example, installing an in-line blender as a first stage, followed by rapid mix basins to provide more detention time). The design summarized in Table 6.3 uses this concept.

In-Line Jet Mixers. Kawamura (1976) notes some problems with backmix-type rapid mixers and provides information on a design for several large water treatment plants. The system shown in Figure 6.3 was designed for an 82 mgd (310 ML per day) plant. The jet

FIGURE 6.3 Flash mixing facility. (*After Kawamura, 1976.*)

FIGURE 6.4 Section through pump mixer.

velocity at the nozzle is 24 ft/s (732 m/s) with a 950-gpm (60-L/s) capacity injection pump with a 10-hp motor. The G value is approximately $1,000$ s^{-1}.

Advantages of this system are that either source water without added chemicals or partially destabilized source water can be used in the chemical injection system. A valve installed in the pump discharge line can control pumping rate and vary energy input for various plant flows and types of coagulating chemicals.

Chao and Stone (1979) presented a recommended design for a variation of an in-line jet mixing system illustrated in Figure 6.4. Instead of a single orifice directed either upstream or downstream with the flow, this design uses multiple jets that inject perpendicular to the flow in the pipe. With two rows of eight jets per row, the duration of mixing is about 0.5 s and the G value about $1,000$ s^{-1}.

Mechanical In-Line Blenders. Mechanical in-line blenders provide rapid mixing of chemicals with water flowing in a pressure pipe. These devices consist of a propeller in the pipe and an electric drive system, as illustrated in Figure 6.5. These are normally proprietary items of equipment and can be specified to provide any required G value.

Static In-Line Blenders. Static in-line mixers, sometimes called motionless mixers, use energy of the flowing liquid to produce mixing. The design of this type of mixer attempts to create flow paths that result in consistent and predictable mixing performance. The units are available from more than one manufacturer and incorporate various arrangements of intersecting bars, corrugated sheets, and plates. A typical static mixer is shown in Figure 6.6.

Air Mixing. Air mixing can be a simple and reliable mixing method and has advantages where aeration of the source water is required anyway. In some instances, air mixing is incorporated into existing structures where it may not be convenient to install other mechanical equipment. It is especially applicable to deep conduits or vertical sections of pip-

FIGURE 6.5 Typical in-line blender (mixer).

ing. Energy applied in adding air may be computed as the volume of water displaced per unit time multiplied by the depth below the free surface, as shown in the following equation:

$$\text{hp} = \frac{Qh}{528}$$

where Q = free air discharge, ft^3/min
h = depth of air inlet nozzle below water surface, ft
hp = horsepower input

FIGURE 6.6 Typical static mixer. (*Courtesy of Komax Systems, Inc., Wilmington, Calif.*)

Energy applied by air injection may be varied directly by adjusting airflow. Air mixing is not widely used, and before it is incorporated into a design, an inspection should be made to determine whether scum and floatable material would be a problem. Limited quantities of floating scum (or sludge) may be accepted onto filters, but certain coagulants and algae may increase scums.

Hydraulic Mixing. Hydraulic mixing can be achieved by using V-notch weirs, Parshall flumes, orifices, throttled valves, swirl chambers, and simple turbulence caused by velocity in a pipe, fitting, or conduit. Hydraulic mixing is a nonbackmix method that can sometimes be highly efficient. The principal problem is that energy input varies with the flow. However, if a plant has relatively constant flows, energy variations may not be a concern. Seasonal flow variations can sometimes be overcome by varying the number of plant mixing modules in operation to maintain more or less constant flows on those modules in operation.

Total head loss across a throttled valve used for mixing coagulant chemicals should not exceed 4 ft (3.2 m). If head loss exceeds this amount, coagulants should be added to the flow downstream of the valve in the zone of decaying energy because excessive confined energy may shear polymers.

The energy provided by a weir with an effective fall of 1 ft (30 cm) provides a G valve of $1,000 \text{ s}^{-1}$ at 20° C. Such a weir mixer with a downstream baffle develops G values as a function of flow, as shown in Table 6.4. If the volume V where turbulence dissipates is assumed to be constant, G may vary significantly; but if turbulence volume is assumed to be proportional to the flow Q, there is a lesser variation in G.

Weir mixers require that coagulant chemicals be fed equally across the length of the weir at multiple points spaced at not more than the head distance of the weir. Because of maintenance problems with multiple-orifice chemical feed manifolds and other practical considerations, weir mixers tend to be used on plants of less than approximately 40 mgd (151 ML per day) capacity.

Induction Mixing. Induction mixing essentially serves as an all-in-one chemical feed, mixing, and control system. Induction mixing systems, which are primarily proprietary systems, have primarily been utilized for gaseous introduction of disinfectants in wastewater treatment facilities. However, the application of this process can be considered for metal coagulants in drinking water.

Induction mixers basically employ a vacuum via a propeller system to pull coagulant chemicals into the mixing system and then inject into the water stream. The chemical re-

TABLE 6.4 Hydraulic Weir Mixing

Flow Q (percent of maximum)	G relative (V constant)	G relative (VQ)
1.0	1.0	1.0
0.9	0.92	0.97
0.8	0.83	0.93
0.7	0.74	0.88
0.6	0.65	0.84
0.5	0.47	0.73
0.35	0.42	0.71

actions that are required for charge neutralization occur within seconds, and the induction system disperses the coagulant into the raw water stream extremely rapidly.

Induction mixing systems can be in-line in a piping system or submerged in a channel. The selection of the type of system is primarily governed by induction system design and the design water flow. By combining the chemical feed, mixing, and control system into one unit, induction mixing systems can also save power costs. Due to the uniqueness of these systems and the proprietary nature of the equipment, the evaluation and design of these systems should be closely coordinated with the equipment supplier.

FLOCCULATION PROCESS DESIGN

Building optimum size floc requires gentle mixing in the energy gradient range of 20 to 70 s^{-1} for a total period of approximately 10 to 30 min. Direct filtration requires a small, dense floc that can be formed at the higher end of the energy range. For settling in conventional basins and in units with settling tubes and settling plates, lower energy levels are applied to produce a large, dense floc that will resist breakup during contact with weirs and plates. Often, polymers are used to help form denser floc.

Floc begins to form within 2 s of coagulant addition and mixing. If high turbulence or shear is subsequently applied to the water, the formed flocs may be fragmented, and broken floc may not readily settle or re-form.

Optimum floc that is efficiently settled or filtered is usually formed under conditions of gradually reducing energy. In large plants, it may be difficult to distribute water to flocculation basins or filters without quiescent stages and high-energy stages. Conduits handling mixed water should minimize head losses, but may, on the other hand, include water jets or air mixing to maintain G at values of 100 to 150 s^{-1} before the water is transferred to the flocculation stage.

The gentle mixing process of flocculation is designed to maximize contact of destabilized particles and build settleable or filterable floc particles. It is desirable to maintain shear forces as constant as possible within the process. As a result, flocculator mechanisms tend to be slow and to cover the maximum possible cross-sectional area of floc basins.

It is desirable to compartmentalize the flocculation process by dividing the basin into two or more defined stages or compartments, as illustrated in Figure 6.7. Compartments prevent short-circuiting and permit defined zones of reduced energy input or tapered energy. To prevent short-circuiting, baffles are typically placed between each stage of flocculation. For mechanical (nonhydraulic) flocculation basins, baffles are designed to provide an orifice ratio of approximately 3% to 6% or a velocity of 0.9 ft/s (27 cm/s) under maximum flow conditions.

Incidental Flocculation

As coagulated water is transferred to flocculators in small plants, distances are short enough that incidental flocculation is negligible. But in large plants transfer may involve distances of more than 100 ft (30 m) through low-velocity conduits, weirs, or other means of distributing water equally to each flocculation basin or compartment. This travel in large plants involves turbulence, and flocculation and incidental flocculation take place. If velocities or levels between the conduits and the flocculation basin are not limited, floc may be fragmented and plant efficiency impaired. Higher coagulant feed rates may be required to overcome fragile floc problems.

FIGURE 6.7 Sections through horizontal shaft paddle flocculator.

Flow splitting to distribute the flow to the flocculation basins is normally by weirs or orifices. Weirs at the postmixing stage of the process should be low-velocity, submerged weirs. Typical velocities in conduits from the mixer to flocculation basins are 1.5 to 3.0 ft/s (46 to 91 cm/s). Distribution channels between the mixer and flocculation basins are often tapered, in either width or depth, to maintain constant velocity. Typical design criteria for a large plant are summarized in Table 6.5 for the facilities illustrated in Figure 6.8.

Flocculation Time

Most modern plants provide approximately 20 min of flocculation time (at 20° C) under maximum plant flows. Some references recommend flocculation times of 30 min or longer. Older references (including the first edition of this book) do not define *nominal flows*,

TABLE 6.5 Typical Mixing and Flocculation Design Criteria for System Shown in Figure 6.8

Plant capacity	150 ft^3/s
Design flow	97 mgd
Plant inlet pipe diameter	84 in.
Initial mixer	
Pump blender	2 units
Pump rating, each	10 hp
Energy input, G at 20° C	1,000 s^{-1}
Mixing zone	538 ft^2
Detention time	3.6 s
Distribution channel	
Depth	10 ft
Width	6.5 to 1.5 ft
Maximum velocity	1.2 ft/s
Flocculation	
Number of basins	2
Compartments, each basin	4
Depth, average	16 ft
Compartment, width × length	15 × 80 ft
Volume, per basin	76,800 ft^3
Detention time, total	18 min
Horizontal shaft paddles (each basin)	4
Maximum G, per compartment	50 s^{-1}
Maximum power, per compartment	2.0 hp
Chemical dosages	
Chlorine, refilter	5 mg/L
Alum, maximum, rapid mixer	20 mg/L
Cationic polymer, rapid mixer	2 mg/L
Nonionic polymer, second-stage flocculator	0.5 mg/L
Potassium permanganate, rapid mixer	2 mg/L

and it appears that earlier texts based detention times on mean, or nominal, flows rather than on maximum plant capacity.

In addition, temperature adjustments outlined in Table 6.2 should be considered. The design water temperature is the temperature most likely to be encountered under maximum flows.

For direct filtration plants, high-energy flocculation is typically in the range of 15- to 20-min detention. When clarification is required, lower energy input and detention times of 18 to 25 min are a guide. If compartmentalization is not provided, increased detention times should be provided in addition to adjustments for water temperature. In all cases, pilot plant or full-scale tests and economic evaluation offer the most reliable indicators.

For low-pressure membranes, the flocculation duration is typically in the range of 5 to 15 min, depending on the type of membrane and required organics removal.

Energy Requirements

Energy input (G values) for flocculation in plants using metal-ion and organic coagulant chemicals ranges from 20 to 75 s^{-1}. Typical G values and detention times for floccula-

FIGURE 6.8 Partial plan for mixing and flocculation facilities [368 ML/day (97 mgd)].

tion at 20° C are summarized in Table 6.6. Floc size and density should be matched to subsequent settling and filtration stages.

Types of Flocculators

Flocculation can be achieved by hydraulic methods or mechanical devices. Hydraulic methods are used most often in small plants. Mechanical flocculators cover a broad range of configurations.

TABLE 6.6 Flocculation Design Criteria

Process	G, s^{-1}	Detention time, s	Gt
Distribution channels mixer to flocculator	100 to 150	Varies	—
High-energy flocculation for direct filtration	20 to 75	900 to 1,500	40,000 to 75,000
Conventional flocculation (presettling)	10 to 60	1,000 to 1,500	30,000 to 60,000

Mechanical Flocculators. Mechanical flocculators are preferred by most design engineers in the United States because of their greater flexibility in varying G values and because they have low head loss. Typical arrangements for horizontal shaft, reel-type flocculators, and vertical paddle units are shown in Figures 6.7, 6.8, and 6.9. Another type of horizontal shaft flocculator, illustrated in Figure 6.10, oscillates with a back-and-forth motion. One advantage of this type of unit is that it prevents water from rotating continually

FIGURE 6.9 Vertical paddle flocculator compartment.

FIGURE 6.10 Oscillating flocculator. (*Courtesy of Dorr-Oliver Eimco.*)

in the same direction around the shaft. Leakage from packing glands has occurred in some horizontal shaft units.

The type of mechanical flocculator influences the shape of flocculation compartments. Vertical flocculators are often associated with square compartments with maximum dimensions of approximately 20 ft (6 m) square and depths of 10 to 16 ft (3 to 5 m). Horizontal shaft, reel, or paddle flocculator compartments are often 20 to 100 ft (6 to 30 m) long and 10 to 16 ft (3 to 5 m) wide.

Between each zone or stage of mechanical flocculation, baffles should be designed to prevent short-circuiting. For these baffles, typical orifice areas should provide a velocity of approximately 1.0 to 1.5 ft/s (30 to 46 cm/s). Baffles are normally constructed of wood but may also be concrete, brick, or woven stainless steel strips.

Vertical flocculators are often higher-speed devices than horizontal shaft flocculators, and the proportion of volume of the compartment that receives energy from the vertical flocculators may be less. As a result, a wider range of energy is applied to the flow in vertical flocculators, and for a portion of the time, some of the flow may be subjected to a higher G. Vertical flocculators are more applicable to high-energy flocculation situations such as direct filtration.

Where uniform floc is required, low-tip-speed flocculators may be more suitable. Equipment manufacturers should be consulted to ensure that appropriate paddle designs are specified for large plants using vertical flocculators.

Vertical flocculators are often specified because they have no submerged bearings, are usually higher-speed, and involve lower investments. High-speed flocculators, however, may not provide floc suitable for high-rate horizontal flow basins. Improved clarification may require increased coagulant doses or flocculent aids.

As a guide, for high-energy flocculators (G = 50 to 75 s^{-1}), maximum tip speed of mixer blades should not exceed 10 ft/s (30 m/s). For low-energy flocculators and paddle-type flocculators (G = 20 to 45 s^{-1}), blade tip speeds in the range of 1.0 to 2.5 ft/s (30 to 76 m/s) are appropriate. Some method of varying speed is normally provided. Variable-speed drives or provisions to change pulleys or gears for different shaft speeds are valuable features that should be provided on mechanical flocculators. As a rule, only the upper 25% of the speed range requires adjustment, and such adjustment will provide a variation of 65% to 100% of maximum G. The G output of a flocculator does not normally have to be varied frequently, but it may require adjustment after installation or on a seasonal basis.

Hydraulic Flocculation. Hydraulic flocculation methods are simple and effective, especially if flows are relatively constant. The assumed flocculation volume is the total volume of each compartment, even though in some cases there may be reduced turbulence in portions of the compartments. The disadvantage of hydraulic flocculators is that G values are a function of flow that cannot be easily adjusted.

Energy may be applied to water by means of maze-type baffles or cross-flow baffles, as illustrated in Figure 6.11. For maze-type baffles, optimum plug flow conditions prevail, and excellent results can be obtained. At velocities in the range of 0.7 to 1.4 ft/s (21 to 43 cm/s), adequate flocculation may be achieved from turbulence caused by the 180° turn at each end of the baffle. For lower channel velocities, it may be necessary to provide an orifice at the end of each channel to induce higher-energy input.

For cross-flow baffles, energy may be transmitted to the water in each compartment from the head loss across orifices in the entrance baffle. The G value in each compartment can be calculated from estimated head loss across baffles into each compartment with the following equation:

$$G = 62.5 \frac{hv}{t\mu}$$

where hv = head loss entering compartment, ft
t = detention time in compartment, s
μ = viscosity, lb · s/ft^2

The head loss through orifices in baffles may be computed from the square-edged submerged orifice formula where the discharge coefficient may be assumed to be 0.8. Many

FIGURE 6.11 Plan and section of maze and baffle flocculators.

texts give coefficients of 0.61 for square-edged orifices. In water treatment applications, head losses and velocities are lower than the velocities assumed in textbooks, less than 1.5 ft/s (46 cm/s), and coefficients are typically 0.8. If, in practice, head losses across baffles are greater than assumed, higher head losses and higher G values will result.

For cross-flow baffle flocculators, slots typically 4 to 6 in. (10 to 15 cm) high and 16 to 24 in. (41 to 61 cm) long should be provided in the bottom of baffles for cleaning purposes. Slots should be staggered to prevent short-circuiting. Normally the floors of hydraulic flocculation basins slope toward the discharge end to facilitate cleaning and to provide tapered energy.

Generally, flocculation basins are designed to be approximately the same depth as the adjacent clarifier.

End Baffles. Most flocculator systems require an end baffle between the flocculation zone and the clarifier, or some provision to prevent residual energy from the flocculation process from being transferred to the clarification stage. These baffles also minimize short-circuiting and reduce the effects of water temperature changes.

End baffles may be designed on the same basis as compartmentalization baffles described above. End baffles should not provide a barrier to removing sludge when the floc basin is cleaned, and limited openings in the bottom of the baffle are appropriate. Similarly, a small submerged section at the top of the baffle will allow scum to pass downstream.

Proprietary Designs. Several manufacturers provide proprietary designs that incorporate rapid mixing, flocculation, and settling in one unit. A typical configuration for this type of unit is illustrated in Figure 6.12. This equipment generally provides for upflow through the clarification stage, although in large-diameter units the flow may be more accurately described as radial.

FIGURE 6.12 Typical proprietary design of solids contact reactor unit.

The process may be designed to include intentional backmixing in the flocculation stage, and the clarification stage may be designed for "sludge blanket" conditions, where flow from the flocculation zone moves upward through a layer, or blanket, of sludge in the settling zone. Designs may include a conical section in the mixing and flocculation stage and increased area in the upflow zone in the clarifier stage.

A single-unit solids contact type of flocculator-clarifier may have advantages for some applications. These units perform best in waters that can develop a dense, fast-settling floc, but the process may be difficult to control in some waters. In general, they work very well on waters that are stable in water quality and with treatment systems that require steady water demands. "Flashy" river waters or water systems that require wide treatment rates can be problematic. Due to these issues and the potential variations between raw waters and system demands, at a minimum, bench-scale tests should be performed at all facilities considering these types of unit, and full-scale pilot tests are recommended for large plants.

Sludge Recirculation and Solids Contact. The intentional introduction of preformed floc or sludge into the mixing and flocculation stage is a feature of most proprietary water treatment equipment. High concentrations of suspended solids in the flocculation process (and in the sludge blanket of the clarifier) can provide improved efficiency in reducing particulates, colloids, organics, and certain ionized chemicals.

Recirculating sludge and reintroducing filter wash water into the mixing and flocculation stages of nonproprietary designs may improve efficiency and reduce chemical requirements. Sludge and wash water quality is an important consideration when one is determining the reuse of process residuals. Sludge recirculation may be difficult to optimize at the pilot plant stage of investigation. Providing flexibility for sludge recirculation and solids contact should be considered at the design stage of new plants.

Contact Flocculation. The ability of a coarse media bed to act as a flocculation system is well demonstrated. The time required to build optimum size and density of floc may be reduced where there is close contact with preformed flocs. The time factor is important for plants treating very cold water and for portable plants and pressure plants. Contact flocculation may be applicable for difficult high-suspended-solids water or water with low total dissolved solids that may not respond readily to metal-ion coagulants.

A typical application of the process is to pass coagulated water through a coarse media or gravel bed, either by gravity or under pressure. The gross detention time may be 3 to 5 min at 5° C. Flow may be either upflow or downflow, as illustrated in Figures 6.13 and 6.14. The system should provide for removing excess accumulated floc, usually by means of air scour, similar to filter air scour.

Pilot plant investigations should precede design, or the criteria for contact flocculation should be based on proven applications on similar water sources. Several manufacturers offer equipment or processes that use the principles of contact flocculation.

PROCESS MONITORING AND CONTROL

Design dosages for chemicals should be based on experience with similar types of waters or, preferably, with jar tests and pilot tests. Design dosages of coagulants can be determined from jar tests. The effectiveness of coagulation and flocculation can also be monitored by several parameters, including turbidity, zeta potential, streaming current, particle counts, and bench and pilot tests (discussed in other chapters of this volume).

FIGURE 6.13 Upflow gravel bed flocculator. (*Source: Schulz et al., 1994.*)

FIGURE 6.14 Downflow gravel bed flocculator. (*Source: Schulz et al., 1994.*)

BIBLIOGRAPHY

American Water Works Association. *Water Quality and Treatment,* 4th ed. New York: McGraw-Hill, 1990.

Amirtharajah, A., M. M. Clark, and R. R. Trussel (eds.). *Mixing in Coagulation and Flocculation,* AWWA Research Foundation Report. Denver, Colo.: American Water Works Association, 1991.

AWWA Coagulation Committee. "Committee Report: Coagulation as an Integrated Water Treatment Process." *Journal AWWA* 10(89):72, 1989.

Chao, J. L., and B. G. Stone. "Initial Mixing by Jet Injection Blending." *Journal AWWA* 10(79):570, 1979.

Coagulation and Filtration: Back to the Basics. AWWA Annual Conference Seminar Proceedings. Denver, Colo.: American Water Works Association, 1981.

Crozes, Gil, et al. "Enhanced Coagulation: Its Effect on NOM Removal and Chemical Costs." *Journal AWWA* 1(95):78, 1995.

Culp, Wesner, and R. L. Culp. *Handbook of Public Water Systems.* New York: Van Nostrand Reinhold, 1986.

James M. Montgomery, Consulting Engineers, Inc. *Water Treatment Principles and Design.* New York: John Wiley & Sons, 1985.

Kawamura, Susumu. "Considerations on Improving Flocculation." *Journal AWWA* 6(76):328, 1976.

Kawamura, Susumu. *Integrated Design of Water Treatment Facilities.* New York: John Wiley & Sons, 1991.

Operational Control of Coagulation and Filtration Processes, M37. Denver, Colo.: American Water Works Association, 2000.

Schulz, C. R., et al. "Evaluating Buoyant Coarse Media Flocculation." *Journal AWWA* 8(94):51, 1994.

Selection and Design of Mixing Processes for Coagulation, AWWA Research Foundation Report. Denver, Colo.: American Water Works Association, 1994.

Technologies for Upgrading Existing or Designing New Drinking Water Treatment Facilities. EPA/625/4-89-023. Cincinnati, Ohio: U.S. Environmental Protection Agency, March 1990.

Upgrading Existing Water Treatment Plants. AWWA Annual Conference Seminar Proceedings. Denver, Colo.: American Water Works Association, 1974.

Vrale, Lasse, and R. M. Jorden. "Rapid Mixing in Water Treatment." *Journal AWWA* 1(71):52, 1971.

CHAPTER 7
CLARIFICATION

John F. Willis
CDM
Cambridge, Massachusetts

Clarification has more than one application in water treatment. Its usual purpose in a conventional treatment process is to reduce the solids load after coagulation and flocculation. A second application, a process called *plain sedimentation,* is removal of heavy settleable solids from turbid water sources to lessen the solids load on treatment plant processes. Material presented in this chapter deals primarily with settling flocculated solids.

One way of designing the clarification process is to maximize solids removal by clarification, which generally requires lower clarifier loadings and larger, more costly units. Alternatively, the clarifier may be designed to remove only sufficient solids to provide reasonable filter run times and to ensure filtered water quality. This latter approach optimizes the entire plant and generally leads to smaller, less expensive facilities. Typical loading rates suggested in this chapter or by regulatory guidelines are generally conservatively selected to provide a high-clarity settled water rather than optimization of the clarifier-filter combination.

Clarifiers fall into two basic categories: those used only to remove settleable solids, either by plain sedimentation or after flocculation, and those that combine flocculation and clarification processes into a single unit. The first category includes conventional sedimentation basins (Figure 7.1) and high-rate modifications such as tube or plate settlers and dissolved air flotation (DAF). The second category includes solids contact units such as sludge blanket clarifiers and slurry recirculation clarifiers. Also included in this category is contact clarification in which flocculation and clarification take place in a coarse granular media bed.

CONVENTIONAL CLARIFICATION DESIGN

Most sedimentation basins used in water treatment are the horizontal-flow type in rectangular, square, or circular design. Both long, rectangular basins and circular basins are commonly used; the choice is based on local conditions, economics, and personal preference. Camp (1946) states that long, rectangular basins exhibit more stable flow characteristics and therefore better sedimentation performance than very large square basins or circular tanks. Basins were originally designed to store sludge for several months and were peri-

FIGURE 7.1 Typical conventional sedimentation tanks.

odically taken out of service for manual cleaning by flushing. Most basins are now designed to be cleaned with mechanical equipment on a continuous or frequent schedule.

Sedimentation Theory

A complete discussion of sedimentation theory and its application can be found in *Water Quality and Treatment* (AWWA, 1999). The designer is urged to become familiar with this theory before selecting and designing the clarification process.

When flocculated particles enter a basin and begin to settle, particles' settling velocities change as particles agglomerate to form larger floc. Because the settling properties of flocculent suspensions cannot be formulated, a basin's performance cannot be accurately predicted. However, for new plants, settling rates can be estimated from batch settling data developed with laboratory jar tests. For expanding existing plants, settling rates can be derived from evaluating the performance of existing basins during various influent water quality conditions. These evaluations often allow for increasing rates for existing basins and establishing higher rates, as compared to published guidelines, for new basins.

In an ideal continuous flow basin, sedimentation would take place as it does in the laboratory jar. However, in a real basin, wind, temperature density currents, and other factors cause short-circuiting, disruption of flow patterns, breakup of floc, and scouring of the settled sludge. The designer must learn as much as possible about the settling properties of the flocculated solids and then design basins to match these characteristics. When the designer does not have access to source water data, it is best to select design criteria known to have worked in similar applications, either from personal experience or from regulatory guidelines.

Design Approach

The primary approach in designing conventional sedimentation basins is to select a design overflow rate for maximum expected plant flow. This rate may be chosen based on all units being in service or on one unit being out of service, to allow for redundancy. After selecting an overflow rate, the designer should determine the number of units needed and select the type of sludge collection and removal equipment. The sludge removal equipment may limit basin dimensions, which could establish the size and number of units. With the number of basins selected, the designer should proceed to design inlet and outlet conditions and finalize dimensions to suit all design parameters and site conditions. The following are suggested guidelines for the various design parameters.

Overflow Rates. Hydraulic overflow rate is the primary design parameter for sizing sedimentation basins. This rate is defined as the rate of inflow Q divided by the tank surface area A. Units are typically rated in gallons per day per square foot, gallons per minute per square foot, or cubic meters per hour per square meter. Acceptable overflow rates vary with the nature of the settling solids, water temperature, and hydraulic characteristics of the settling basin.

Typical design overflow rates for sedimentation of solids produced through alum coagulation/flocculation are shown in Table 7.1. For the ranges shown, higher rates are typical for warmer waters with heavier suspended solids. Rates higher than these may be applicable for warm waters greater than 20° C. Lower rates should be used for colder waters with lower turbidity or that are high in organic color or algae. After evaluation of both cold and warm water loading rates, the design rate is based on whichever is more critical. Plant flow variations between cold and warm water periods often allow selection of higher rates for summer operation than the typical suggested loading rates.

Overflow rates can also be selected based on pilot studies. Piloting of conventional settling basins is not especially reliable, but it is often done using tube settlers (see Chapter 28). Data from such studies, along with jar testing, are often useful in design. Pilot testing of other types of settling, especially proprietary processes, is useful and is recommended.

TABLE 7.1 Typical Sedimentation Surface Loading Rates for Long, Rectangular Tanks and Circular Tanks Using Alum Coagulation

Application	(L/day)/m^2	gpd/ft^2
Turbidity removal	32,592 to 48,888	800 to 1200
Color and taste removal	24,444 to 40,740	600 to 1000
High algae content	20,370 to 32,592	500 to 800

Detention Time. Detention time (i.e., flow rate divided by tank volume) is usually not an important design parameter. Many regulatory agencies (e.g., Great Lakes, 2003), however, still have a requirement for detention periods of 4 h. It is likely that this detention requirement is a carryover from the days of manually cleaned basins designed to provide a sludge storage zone. These basins were often 15 to 16 ft (4.6 to 4.9 m) deep or greater and operated so that more than one-half the volume could be filled with sludge before being cleaned. Real detention time could vary from 4 h when clean to less than 2 h just before cleaning.

Modern designs with mechanical sludge removal equipment need not provide a sludge storage zone, and deep basins with long detention times are no longer required. Conventional basins with detention times of 1.5 to 2.0 h provide excellent treatment.

Basin Depth and Velocities. In theory, basin depth should not be an important parameter either, because settling is based on overflow rates. However, in practice, basin depth is important because it affects flow-through velocity. Flow-through velocities must be low enough to minimize scouring of the settled floc blanket. Velocities of 2 to 4 ft/min (0.6 to 1.2 m/min) usually are acceptable for basin depths of 7 to 14 ft (2.1 to 4.3 m), the shallower depths often used with multiple-tray basins. Single-pass basins are generally deeper, to offset the effects of short-circuiting from density and wind currents.

Basin depth may also play a role in allowing greater opportunity for flocculent particle contact. Additional flocculation that takes place as particles settle allows for growth of heavier floc and the formation of a sludge blanket that may be less susceptible to resuspension. The formation of this blanket helps increase the solids content of the residuals withdrawn by removal equipment. The blanket can, however, also contribute to the creation of a density current along the bottom of the tank, causing floc carryover to the effluent.

Number of Tanks. One important choice to be made is the number of basins. The minimum, and by far the least costly, plant would have only a single settling basin. However, that would make for poor operation, because tanks must periodically be taken out of service for maintenance. Two tanks would partially offset this problem, but unless plant flow can be reduced, the load on one tank could be excessive when the other is out of service. A minimum practical number of tanks would be three, allowing for a 50% increase on two tanks when one is out of service.

If the design overflow rate is conservative, the three-tank approach is acceptable. In general, however, a minimum of four tanks is preferred. The number of tanks may also depend on the maximum size tank that can accommodate the selected sludge removal equipment or on other factors, such as site constraints.

Factors to consider in selecting the number of tanks are their relationships to the flocculation basins and the filters. Are units to be lined up as consecutive processes, or is each process to be a separate unit? When processes are to be consecutive, a decision on the minimum or maximum number of filters or floc basins may determine the number of settling basins. The designer should refer to the chapters on flocculation and filtration design and should approach design of these units as a common process.

Rectangular Basins

Long, narrow basins have been used for sedimentation for many years and will be in operation for years to come. Such basins are not as affected by wind and density currents as are square or circular basins.

Basin Dimensions. Rectangular basins are generally designed to be long and narrow, with width-to-length ratios of 3:1 to 5:1. This shape is least susceptible to short-circuiting—the hydraulic condition in a basin when the actual flow time of water through the basin is less than the computed time. Short-circuiting is primarily caused by uneven flow distribution and density or wind currents that create zones of near-stagnant water in corners and other areas.

Basin widths are most often selected to match the requirements of the chosen mechanical sludge collection equipment. Chain-and-flight collectors, for example, are limited to about a 20-ft (6-m) width for a single pass, but it is possible to cover a wider basin in multiple passes. Traveling bridge collectors can be up to 100 ft (30 m) wide, limited only by the economics of bridge design and alignment.

Basin depths may be selected to provide a required detention time (although detention time is not a good design parameter) or may be selected to limit flow-through velocities and the potential for resuspension of settled floc. Basins with mechanical sludge removal are usually between 10 and 14 ft (3.0 and 4.3 m) deep.

Because settling is primarily based on area, multiple-tray basins have been developed as shown in Figure 7.2. A depth of about 7 ft (2 m) is typically provided between trays to allow access for cleaning and maintaining equipment.

Inlet Zone. A basin's effectiveness at any overflow rate can be greatly changed by short-circuiting. Short-circuiting reduces the actual area traversed by the flow, increasing the apparent overflow rate and reducing solids removal efficiency.

Many publications provide testimony to the importance of proper design of the basin inlet (Yee and Babb, 1985; Monk and Willis, 1987; Hudson, 1981; Kawamura, 1991). Good design of the inlet zone establishes uniform distribution into the basin and minimizes short-circuiting potential.

For long, narrow basins being fed directly from a flocculation basin, slots or a few individual inlets may suffice. To obtain uniform flow distribution through wider basins, perforated baffle walls should be provided; a typical arrangement is shown in Figure 7.3. For best results, flow from the flocculation basin should be in line with the basin axis.

Following hydraulic principles to ensure equal flow distribution, head loss through the perforations should be 4 to 5 times the velocity head of the approaching flow. The velocity gradient G should be equal to or less than that in the last flocculation compartment to minimize floc breakup (Hudson, 1981). The number of ports should be the maximum practical that will provide the required head loss. Port velocities typically must be about 0.7 to 1.0 ft/s (21 to 30 cm/s) for sufficient head loss. Ports should be arranged to cover as much of the basin's cross section as possible without creating high velocities in the sludge collection zone that might cause scouring action. Thus the lowest port should be about 2 ft (0.6 m) above the basin floor. Port spacing is typically 10 to 24 in. (25 to 61 cm) with a port diameter of 4 to 8 in. (10 to 20 cm).

Introducing flow across the entire inlet end of the basin reduces short-circuiting caused by density currents—created when water entering the top of the basin is colder and heavier than the water below. The cold influent settles quickly to the bottom, causing flow to move along the bottom of the tank and back across the top. When influent is introduced uniformly across the tank from top to bottom and side-to-side, water temperature remains more uniform, and density currents are less likely to form.

Outlet Design. Outlet design is also critical in reducing short-circuiting and scouring of settled solids. Outlet designs have undergone a number of transformations. Basins were originally designed with end weirs. This type of outlet causes an increase in horizontal and vertical velocity as flow is forced up the end wall to the weir, and the increased ve-

FIGURE 7.2 Two-tray sedimentation basin. (*Courtesy of CDM.*)

FIGURE 7.3 Typical inlet baffle showing orifices.

locities cause considerable floc carryover by scouring settled floc and removing floc that has not had time to settle.

In an effort to reduce velocities and carryover, long-finger weirs extending well down the basin were developed, with up to one-third of the basin covered in some installations. By increasing the surface area over which flow is collected, vertical velocity is reduced. Regulatory agencies have promoted this type of design by requiring weir loading rates of 20,000 gpd/ft [248,000 (L/day)/m] or less (Great Lakes, 2003).

However, one evaluation of the performance of long-finger weirs compared with end weirs showed that no benefit was achieved by the long weirs (Kawamura and Lang, 1986). Bottom density currents were still found to rise along the end wall, causing floc carryover to the end of the weirs. Finger weirs are, however, somewhat effective in breaking up wind-induced surface currents and thus may improve performance. A single end weir could be used, but the problem of high velocities at the end wall still exists.

One approach to address the vertical velocity problem is to use a perforated end wall, similar to the inlet distribution wall, to maintain parallel hydraulic flow along the length of the basin (Monk and Willis, 1987). Velocity approaching the wall does not increase except near the ports because the entire basin cross section is used. Velocity remains low along the floor, reducing the potential for scouring. Flow uniformity approaching the end wall helps to ensure that flow covers the entire basin surface to achieve the design overflow rate. The approach to designing the perforated wall is the same as for the inlet. A small head loss must be taken to obtain uniform distribution. Velocities through the ports may be higher than those for the inlet because the smaller floc carried over is less likely to be sheared. In some cases it is preferable to increase the head loss through the effluent ports to obtain hydraulic flow splitting to the basins rather than taking the head loss at the inlet. This avoids high inlet velocities that may cause breaking up of a fragile floc.

The use of effluent baffle walls may not be effective in controlling density currents in the settled solids. Such density currents may be as high as 2.5 to 6 ft/min (0.8 to 1.8 m/min), much faster than the tank average flow-through velocity that could carry solids through a baffle wall (Kawamura, 1991). Control of density currents in the sludge blanket may be best accomplished by frequent solids removal at several locations (Taebi-Harandy and Schroeder, 1995). Traditional mechanical sludge removal equipment, however, does not typically provide for draw-off at several locations. However, frequent or continuous solids removal should minimize density currents in the settled solids.

Manual Solids Removal. Although no longer used in many modern water treatment plants, basins can be designed for manual cleaning. In such cases basins must be designed to store sludge for a reasonable period of time. An extra depth of 4 to 5 ft (1.2 to 1.5 m) should be provided, basin floors must slope to a drain, and adequate pressurized water must be available for flushing. Manually cleaned basins are suitable for use in developing countries with low labor rates.

Basins can be designed with hoppers in the first half of the tank, where most of the sludge is likely to settle, and equipped with mud valves that can be frequently opened to waste the bulk of the sludge. This design reduces the frequency of removing the basin from service for complete cleaning. However, withdrawal of sludge from these hoppers should be frequent and controlled. Otherwise, if the solids become too thick, flow may "pipe" through to the drain, leaving the bulk of solids on the sides of the hoppers. Frequent removal of solids through these hoppers may also disrupt density currents and improve performance.

Mechanical Solids Removal. Most modern sedimentation basins are designed to be mechanically cleaned using a variety of mechanisms, most of which are proprietary. These include systems that drag or plow sludge along the basin floor to hoppers and systems that rely on hydraulic or siphon action to withdraw solids. Because each system has different design requirements for basin dimensions and solids draw-off, it is important that the designer research the available equipment or systems around which to design the tanks. The following text discusses various types of equipment commonly used. It is not a complete discussion, and new devices are being developed all the time. Good references for lists of currently available equipment and the names of manufacturers is the *Public Works Manual,* an annual publication of *Public Works Magazine* and the *AWWA Sourcebook,* annually published by AWWA.

Traditional Equipment. Traditional desludging equipment was mostly chain-and-flight drags made up of two strands of iron chain with wooden flights attached at 10-ft (3-m) intervals and operated at about 2 ft/min (0.6 m/min) to convey dense sludge to a hopper. Flights were usually made of redwood in lengths up to 20 ft (6 m). Cast iron or steel wearing shoes were attached to the wooden flights to prevent the wood from wear-

CLARIFICATION

FIGURE 7.4 Typical rectangular basin with chain-and-flight collectors with sludge hoppers.

ing, and these rode on steel T-rails cast into the concrete floor. A steel rail was also attached to the wall for the flight to ride on as it looped back. Figures 7.4 and 7.5 show two typical types of rectangular basin equipped with chain-and-flight collectors. Drags for continuous desludging have been popular with design engineers because they fit the long rectangular basin geometry.

New designs have been developed with ultrahigh-molecular-weight (UHMW) plastic or similar materials to replace the iron chain and sprockets. In addition, fiberglass flights have replaced the wooden boards, and plastic wearing strips are attached to the concrete floor and walls to replace iron rails. These new systems are corrosion-free and require less maintenance. However, when first installed, plastic chain tends to stretch. This requires adjusting the chain tension one or two times during the first year of operation. After that, the rate of creep reduces.

Circular Collector Equipment. Circular sludge collector units have been used in long, rectangular tanks to avoid using chain drag equipment. The circular units were generally installed at the influent end of the basin where most well-developed floc will settle, and a transverse barrier wall was added to stop the density current and drifting of sludge toward the unscraped effluent end. In these cases, the circular mechanism "pushed" sludge to a circumferential hopper at the center pier from which it was automatically discharged as sludge underflow. The remainder of the basin was periodically cleaned manually.

With a poorly settling floc or in basins with strong sludge density currents, placing the collector only in the first one-third to one-half of the basin may not be very effective, resulting in the need to frequently manually clean the basins. It is preferable to provide multiple collector mechanisms to improve sludge removal and eliminate the need for manual cleaning.

FIGURE 7.5 Typical rectangular basin with chain-and-flight collectors with chain-and-flight cross collectors.

In small rectangular basins, where corners not swept by the circular collector are relatively small, corner sweeps are generally not specified to minimize collector maintenance. When corner sweeps are not used, it is common practice to add steep corner fillets in basin construction to promote sludge movement to the area swept by the collector mechanism. However, in larger basins, where corner areas are large, or when sludge is of organic composition and must be continuously removed, corner sweeps are specified.

Corner sweeps can be subject to mechanical problems, and the larger the basin, the greater the likelihood of problems. Unfortunately, many corner sweeps fail or require an inordinate amount of maintenance because of poor mechanical details. The mechanical aspects of corner sweeps should be specified in great detail to obtain a high-quality, low-maintenance system.

In general, circular collector mechanisms are highly reliable and have low maintenance requirements because of their simplicity in providing positive sludge removal. Disadvantages include protecting their large submerged metal surfaces from corrosion and maintaining corner sweeps if they are required.

Carriage-Type Collectors. Oscillating-bridge collectors have become more popular during the past several decades. These top-of-wall running units span the width of one or more long, rectangular basins. One type cleans thick sludge by means of a single transverse vertical blade about 24 to 30 in. (0.61 to 0.76 m) deep that conveys dense sludge into a cross hopper. Another type uses a transverse suction header that discharges into a longitudinal trough along one side of the basin.

The transverse blade has an adjustable angle of attack and slides on flat bottom rails or sometimes on a heavy neoprene squeegee riding on the floor. The blade automatically adjusts to varying floor slopes and can be raised above the water level for maintenance. It pushes dense sludge at about 6 ft/min (1.8 m/min) toward the cross hopper on the cleaning run, and then it is hoisted about 3 ft (0.9 m) and travels at double speed back to the influent end to repeat.

Carriage units generally run on double-flanged iron wheels along heavy steel rails mounted on the long walls of the clarifier. Rubber tires running on top of concrete walls have been used occasionally in western Europe, but their use is discouraged in colder climates because of problems with snow and ice buildup. The units are traction-driven with automatic compensation to prevent "crabbing" or, preferably, driven on cog rails located adjacent to steel rails on either side of the basin.

Power to drive and hoist motors is typically supplied through a flexible power cable reeled in and paid out by a cable reel. Motor-driven cable reels work best because simple spring motor reels often overstress the cable when the carriage is at the far end. Another problem is that they commonly do not have enough reserve force to reel in the cable at the near end, and they may run over and sever the loose, kinked cable. Several cleverly designed power reels have been developed, either using a backstay cable to power the reel as it retrieves the cable or synchronizing the cable reel to carriage travel and employing a spring motor slave reel core to compensate for minor variations in cable length.

Feed rails are sometimes used for power feed, but they are vulnerable to vandalism and sometimes burn out or carbonize because of the slow speed of the brushes. Another form of power supply is an overhead power cable festooned and sliding back and forth on a taut steel carrier cable running a few feet above the surface for the length of the basin. This type of power system is unsightly and generally suitable only for short basins.

Cross Collectors/Cross Hoppers. A cross hopper is a trench, typically 3 or 4 ft wide by 2 or 4 ft deep (1 or 1.2 m wide by 0.6 or 1.2 m deep), running the width of one or more longitudinal sections of the sedimentation basin. Dense sludge falls into this cross trench and is scraped at about 2 ft/min (0.6 m/min) by chain-driven flights 8 in. (20 cm) deep, spaced 5 ft (1.5 m) on centers. These scraper flights deposit dense sludge into a deeper accumulating hopper at the end of the cross trench. The underflow is withdrawn hydraulically or by pumping from the hopper. Figure 7.6 shows one type of cross-collector arrangement.

A helicoid screw is sometimes used in the cross trench in place of chain-driven flights. The screw turns slowly, paced to have a theoretical capacity of 4 times the volume of sludge to be actually moved, to minimize bearing wear. The bottom of the cross trench is filleted to accommodate the outside diameter of the screw. In screw cross collector applications, instead of propelling dense sludge all the way to one end, the flights of the helicoid screw may be opposed so that dense sludge is carried only one-half the trench length to the center point where the accumulating hopper is placed.

Some designers like to use the traditional steep-sided (60°) hopper to remove sludge underflow, as illustrated in Figure 7.7. For basins greater than 10 ft (3 m) wide, more than one hopper must be used to keep the hopper depth within reason. Multiple hoppers can be more expensive than cross hoppers and not as satisfactory in operation; if they are joined with a manifold to a common pipe, only one hopper will be served. Sludge must

HELICAL-SCREW-TYPE CROSS COLLECTOR

CHAIN-AND-FLIGHT-TYPE CROSS COLLECTOR

FIGURE 7.6 Typical cross-collector arrangements. (*Courtesy of USFilter, Envirex Products.*)

be removed independently from each hopper. This is often done by using telescoping sludge valves, air lift pumps, or individually valved outlets. Frequent sludge removal is required to prevent consolidation and bridging in the corners. If the sludge become too dense, it will not flow to the outlet and will not be removed.

Indexing Grid System. The indexing grid system consists of a series of concave-faced, triangular blades rigidly connected to glide bars on the basin floor (Figure 7.8). The system operates beneath the sludge layer, sliding back and forth, indexing the sludge toward a sludge hopper at the end of the basin. The glide bars ride on top of ultrahigh-molecular-weight polyethylene wear strips anchored to the floor. The floor of the basin has no slope.

The stainless steel grid system is structurally reinforced by welded cross members to provide rigidity to the unit. Blade profiles are 2 in. (5 cm) tall, typically spaced 26 in. (66 cm) on center. The collector operates by gently pushing the grid system, including the sludge in front of each blade, in a forward direction at a speed of between 2 and 4 ft/min (0.6 and 1.2 m/min). The low profile height and slow operating speed help prevent resuspension of the sludge particles.

The system is driven by a low-pressure hydraulic unit using food-grade hydraulic fluid. The unit's speed can be adjusted to regulate sludge removal rates. The hydraulic unit is designed to oscillate the profiles. The stroke length of the actuator is longer than the spac-

FIGURE 7.7 Typical sludge hopper arrangements for rectangular basins.

ing of the profiles to create an overlapping effect. When the hydraulic cylinder reaches the end of its stroke, the grid system reverses at a speed 2 to 3 times the forward speed. This allows the triangular profiles to wedge their way under the sludge. This oscillating movement of the collector with the concave profile is said by the manufacturer to provide thickening of the sludge.

The low clearance of 10 in. (25 cm) on the bottom of the basin allows the system to work well under tube-and-plate settler systems.

FIGURE 7.8 Indexing grid sludge removal system. (*Courtesy of Parkson Corporation.*)

Track-Mounted Hydraulic Systems. As shown in Figure 7.9, one type of hydraulic removal system consists of a stainless steel header pipe with orifices sized and spaced for proper sludge removal. The collector pipe is attached to a pneumatically controlled drive assembly that travels on a stainless steel guide rail running the length of the tank. Collector pipes are generally a maximum width of 20 to 25 ft (6.1 to 7.6 m), and multiple units must be used to cover the width of wider tanks.

The collector pipe is attached to a sludge discharge pipe in the tank wall. The sludge discharge pipe contains a pneumatically actuated sludge valve located below water level. When the sludge valve is open, the water level in the basin creates a driving force to start flow into and through the collector system. The drive assembly is pneumatically powered to travel the length of the tank in both directions. As it moves along, sludge on the basin floor is picked up hydraulically at a travel speed of 1.5 ft/min (0.46 m/min). The number of times the collector traverses the tank must be determined from the volume of sludge produced and the flow capacity of the collector system. This is typically about 90 gpm (5.7 L/s), based on a differential head of 5 ft (1.5 m).

The system is fully automated through an electronic control system using a programmable logic controller to control how often the collector operates and the length of travel. For example, because most of the sludge typically accumulates in the first third of a rectangular basin, it is necessary to collect in that area on a frequent basis, with the collec-

FIGURE 7.9 Track-mounted hydraulic sludge removal system. (*Courtesy of Eimco Water Technologies.*)

FIGURE 7.9 (*Continued*)

tor traveling the full length of the basin less frequently. This method of operation avoids collecting a large volume of very low-solids water.

Instrument-quality compressed air at 100 psi (689 kPa) is supplied from a compressor system to a local electric/pneumatic interface panel mounted at the basin. Air is provided to the drive assembly and pneumatic sludge valves by means of umbilical hoses from the control panel to the drive assembly and valves. At one end of the basin is an extractor assembly consisting of vertical guide rails and a removable winch assembly to lift the collector header out of the tank for maintenance.

Because these collectors do not require expensive sludge hoppers, they are a low-cost option for retrofitting manually cleaned basins and since they do not require access from the surface, they can also operate effectively beneath plate or tube settler systems.

Options to the pneumatically driven collectors include continuous stainless steel tapes or chains, powered by a motor mounted at the top of one end of the basin, that pull the collector pipe back and forth along the bottom-mounted rail.

Floating Bridge Hydraulic Systems. Floating suction header solids removal systems were developed to provide a less expensive means of retrofitting existing manually cleaned basins than chain-and-flight units. The units may also be a lower-cost approach for new construction. Figure 7.10 shows a typical siphon desludging unit.

The suction unit is mounted on massive floats built of closed-cell Styrofoam encased in fiberglass-reinforced plastic (FRP). These floats are tied together as a rigid structure so that it is freestanding when the basin is drained. A header system supported by the floats draws dense sludge from the bottom of the basin, and by means of low-head siphon discharges it into a longitudinal trough attached to the side of the sedimentation basin. The cross header is broken into several subheaders, each carrying flow overhead to a control siphon freely discharging into the sludge trough. The effect of each lateral section can be observed and modulated. The siphon system is started with a portable low-differential suction source. Experience has shown that, for water treatment sludges, once started, the siphon can be maintained for months.

The floating bridge system is towed back and forth, either by a single, center-mounted, stainless steel flexible tow cable or by two cables acting on either end of the rigid floating structure. In either case, cables are powered by a geared motor drive and idler sheave arrangement mounted at either end of the basin on top of the wall in an accessible location. Because the floating system eliminates friction and most wind problems, remarkably little power is required to tow the bridges, even in basins 200 ft (61 m) long or longer. Dense sludge sucked up by the system is siphoned into the longitudinal trough. Valves control the amount of siphonage, with one valve for each siphon section forming the transverse header sludge pickup system.

Sludge suction pipes are of light-gage, epoxy-coated aluminum, stainless steel, or PVC and are equipped with fluidizing vanes. In operation, the suction system travels up and down the length of the basin at about 6 ft/min (1.8 m/min) as often as is required to vacuum the dense sludge layer (compression zone). When there is not enough sludge to require continuous suction up and down the basin, siphon discharge is temporarily arrested by programming the longitudinal trough discharge gate to close. The water level in the trough then rises to equal the basin level. When the next programmed desludging cycle begins, the trough valve opens and the siphon continues from where it stopped, again discharging dense sludge into the trough.

Both carriage and floating bridge collection mechanisms are constructed with little submerged metal to minimize corrosion problems. However, these units can be used only in temperate climates where ice accumulation is not a problem. Some installations have also experienced drive synchronization problems and incomplete sludge removal.

Underflow Control. Underflow drawoff must be carefully controlled. If underflow is removed at too low a rate, dense sludge accumulates in the basin, creating a sludge blan-

FIGURE 7.10 Floating bridge–type collector. (*Courtesy of Leopold.*)

ket that is too deep. If underflow is removed too quickly for too long an interval, the drawoff "postholes." That is, less viscous liquid breaks through, and dense sludge accumulates in the basin, overloading the desludging equipment.

Drawing off underflow at regular intervals is best, either manually, with a programmed blowdown valve, or using a transfer pump. Such a program must be completely adjustable and programmed by the design engineer during early operation to fit plant operation. Manual drawoff control by guess and "gut feeling" often leads to operating problems.

New facilities should be designed with sludge viewing pits that permit the operator to observe the consistency of the sludge during blowdown. Direct observation provides a means of optimizing withdrawal rates and reduces excessive loading of sludge handling facilities.

Circular Basins

Circular sedimentation basins became more prevalent in water clarification when periodic manual cleaning of long, rectangular basins became unpopular. The top-drive circular mechanisms used for sludge cleaning have no bearings under water, resulting in longevity with little maintenance. In reasonable sizes—not exceeding 125 ft (38 m) in diameter—the circular center-feed clarifiers perform as well as long, rectangular basins provided there is a reasonably well-balanced radial flow from the center well with substantial water depth maintained at the center.

Some circular basins are designed for rim feed with clarified water collected in the center. However, most circular basins used today are the center-feed type. Included in this category are square tanks with center feed that are used for their feature of lower cost by means of common wall construction. A typical circular clarifier is shown in Figure 7.11. A circular clarifier with a center flocculation zone is shown in Figure 7.12.

Basin Dimensions. Circular basins, like rectangular basins, are designed based on surface overflow rates, and rates used are typically the same as those for rectangular units. Circular basins may be of any diameter but are usually sized based on the commercially

FIGURE 7.11 Typical circular clarifier. (*Courtesy of Eimco Water Technologies.*)

FIGURE 7.12 Typical flocculator clarifier. (*Courtesy of Eimco Water Technologies.*)

available standard sludge removal systems. Circular tanks have been built as large as 300 ft (91 m) in diameter but more typically are less than 100 ft (30 m) in diameter.

Although settling theory is based on overflow rate, side water depth is an important consideration. Adequate depth mitigates hydraulic instability caused by wind currents, thermal currents, hydraulic scour, and random sludge blanket disturbances. Typical depths range from 10 to 15 ft (3 to 3.6 m). Because sludge is usually scraped to center hoppers, basin bottoms are sloped to the center. Large-diameter basins have two slopes, one steeper near the center to allow adequate depth to move the solids to central hoppers for removal.

Inlet Design. Flocculated water is usually introduced to the center of circular or square basins through a center riser into a circular feed well. Some clarifier designs allow the introduction of flocculated water into the side of the feed well. The intent of the feed well is to produce a smooth, radial flow outward toward the periphery of the basin.

The center feed amounts to a point source, because the feed well seldom represents more than 3% or 4% of basin area. For this reason, a great deal of flow mass is crowded into a small space and does not flow in an exactly radial pattern, leading to hydraulic imbalance and short-circuiting. This problem is accentuated in large-diameter shallow basins at high surface overflow rates in the 800 to 1,500 gpd/ft^2 [33,000 to 61,000 (L/day)/m^2] range.

One questionable feed well design involves using a small-diameter circular skirt of about 1% of the basin area extending only 3 to 4 ft (0.9 to 1.2 m) below the surface. The feed into this well is either from four ports discharging horizontally from a pier riser or from a horizontal pipeline discharging horizontally into the well from just below the surface. In the four-port design, variation of flow rate is accommodated with a design exit flow of about 2 ft/s (61 cm/s). This, however, does not ensure equal egress from each port.

A more controllable feed design includes using a distribution well inside a large feed well that is about 3% to 4% of the basin area. This distribution well has multiple ports hooded with adjustable biased gates. The gates balance tangential feed discharges by imposing about a 4-in. (10-cm) head loss through the ports. This type of discharge causes the homogenized mass within the large feed well to rotate around the vertical axis at about 2 ft/s (61 cm/s). The well-distributed, fine-scale turbulence within and below the feed well encourages floc aggregation, and the overall slow rotation ensures that flow from the bottom of the skirt into the hindered sludge mass moving radially across the floor has equal displacement vectors.

Density and displacement currents for circular basins are much the same as for long, rectangular basins. The vector system is influenced by well-flocculated influent mass sinking to the bottom adjacent to the feed well area, typically in the center one-third of the basin (about 10% of the total basin area). The vector system shows displacement radially along the bottom in the blanket zone and upwelling next to the peripheral wall. Clarified water generally flows across the surface toward the effluent.

Outlet Design. Clarified water collection must be uniform around the perimeter of the basin. This is accomplished by a circular trough around the perimeter with V-notch weirs or with submerged orifices. Some designs use a double-sided weir trough mounted inboard along at least 15% of the tank radius. This has the advantage of reducing wall flow disturbances and drawing overflow from a more widely distributed region to offset the effects of bottom density currents running up the peripheral wall. Inboard weir troughs also partially break up wind current stirring. Troughs should have small-diameter holes in the bottom to reduce buoyant uplift forces when they are empty.

Some designers prefer orifice troughs to overflow weir troughs because less floc breakup occurs. Others point out that the velocity gradient in a weir trough is no greater than in an orifice trough. Weir troughs are far easier to adjust for equal linear overflow, but if not properly adjusted, they have greater variation in flow than improperly adjusted orifices. Submerged orifice troughs reduce passage of floating trash to the filters and permit variation in basin water depths during operation. This capability is useful for balancing differences in plant inflow and discharge rates, such as when multiple filter washing occurs.

Regulatory agencies sometimes stipulate that weir rates should not exceed around 20,000 gpd per linear foot (248,000 L/m) of weir. Flood (1961) found that weir overflow rates several times this value could be used if the weirs were well distributed over a substantial portion of the surface. Placing a double-sided weir trough 1 ft (0.3 m) away from the peripheral wall satisfies the regulatory requirements, but still draws overflow from a narrow band of surface immediately in the path of the upwelling peripheral flow.

Sludge Removal. Sludge is removed from circular basins using circular collection equipment powered by a center turntable drive and plows that move sludge into a center sludge hopper.

Turntable Drives. The tried and true, relatively trouble-free drive for both bridge-supported and pier-supported circular collectors is the sealed-turntable drive with the gear and pinion running in oil. Properly lubricated and with automatic condensate overflow, these drives operate for years without major repair. Typical turntable drives rotate on renewable bearing strips, and the gear is split so that the ball bearings and strips can be replaced without dismantling the remainder of the equipment.

These drives are protected by an indicator and overload circuit breaker device actuated by the thrust of the primary worm gear driving the pinion and turntable gear. The indicator senses the torque load exerted on the collection arms by the sludge and turns off power if the load exceeds a preset limit.

Sludge Hopper and Bottom Slopes. Because a circular sludge hopper surrounding a central pier holds the greatest volume, it is preferable to the older style offset hopper design. A pair of heavy stirrups reach down from the arms of the circular scraper to move dense sludge around the hopper to the outlet, to prevent buildup of anchor sludge and grit. The sludge drawoff pipe should never be less than 6 in. (15 cm) in diameter, and it should be designed so that a rotor rodder, or "go-devil," can be placed into the line from outside the basin in case of clogging. In lime softening plants, this line should be given a short purge of clear water after each blowdown cycle to flush out residual slurry.

The slope of the basin bottom is important, especially when there is heavy or sticky sludge. Plow blades keep the bottom free of adhesions, literally plowing extremely dense sludge and grit to the center hopper. Otherwise, the thixotropic sludge flows along the bottom to replace the blown-down underflow. As the dense sludge approaches the basin center, the plow blade spacing reduces, and the shorter radius results in reducing tangential blade velocities.

In large basins, a second set of arms is typically employed to cover the center half (25% of the basin area) because the blade movement at this point is extremely slow. Deep blades formed into spiral sections bridge the main and auxiliary arms to push the crowded sludge into the hopper.

In basins larger than about 80 ft (24 m) in diameter, it is advisable to use a double bottom slope. The double bottom slope is essential for basins more than 125 ft (38 m) in diameter. The two-slope design gives steeper slopes (greater hydraulic gradient) at the sludge hopper without excessive basin depth. The greater center depth dissipates scouring currents and is needed because of the concentration of influent energy in this relatively small region.

Square Basins

Basins larger than 30 ft (9 m) square are typically equipped with circular sludge collection systems with corner sweeps, always in pairs, to clean sludge from corners. Corner sweeps eliminate the need for larger corner fillets, which are generally unacceptable except in very small basins. Corner sweeps should be avoided in basins larger than 100 ft (30 m) square because of structural problems and wear associated with large cantilevered corner sweep units.

Hydraulic problems usually occur with larger square basins. Radial density and displacement currents impinge on the peripheral walls at various angles, drift toward the corners, meet, and may cause a rising "corner floc" phenomenon that often contributes turbidity to effluent.

Where inboard weir troughs are used in square tanks, they are designed to cut across the corner to avoid rising corner floc. Where radial troughs are used (as is typical with upflow basins), they are always arranged to straddle the corner for the same reason. It is best not to run a single peripheral weir around the walls of a square basin because of the corner floc effect. Conventional square basins are not generally recommended.

HIGH-RATE CLARIFICATION

High-rate clarification refers to all processes that can be loaded at higher rate than is typically used in designing conventional clarifiers. The principal types of units currently being used are

- Tube settlers
- Plate settlers
- Solids contact units

- Sludge blanket clarification
- Dissolved air flotation
- Contact clarification

Tube Settlers

Tube settlers take advantage of the theory that surface overflow loading, which can also be defined as particle settling velocity, is the important design parameter. Theoretically, a shallow basin (i.e., short settling distance) should be effective. By using several shallow parallel tubes, surface area can be greatly increased and low flow-through velocity maintained in each tube to reduce scouring.

The first tube settlers were introduced in the 1960s by Microfloc. Typical tubes are 2 in. (5 cm) square, reducing the settling distance from several feet in a conventional basin to 2 in. (5 cm) or less inside the tube. The large effective settling surface results in a low overflow rate compared with the area of the tubes, which allows for a smaller basin, shorter detention time, and increased flow rates.

When tubes are placed at a 60° angle, they provide efficient settling and allow for settled solids removal from the tubes by gravity. As flocculated water rises through the tube, solids settle to the inclined surface, where they gradually gain mass and weight and eventually slide down the incline. In this way, a countercurrent flow pattern is developed. As the solids fall from the bottom of the tubes, they settle to the floor of the basin, where they can be removed by conventional sludge collection equipment. Localized velocities caused by thermal currents are damped by the tubes. Likewise, surface wind currents have little effect because settling occurs within the tubes.

Overall depth of a tube clarifier is usually the same as that of a conventional basin. This is necessary to provide room: below the tubes for sludge collection equipment, for uniform flow approaching tube inlets, for the tubes themselves, and for uniform flow distribution through the tubes up to the collection launders. Figure 7.13 shows a typical tube settler installation.

Design Criteria. Tube settler clarifiers are designed on the basis of the total projected surface area of tubes. A loading rate of about 0.5 gpm/ft^2 (1.2 m/h) is typical for aluminum- and iron-based coagulants, but the loading rate may range from 0.4 to 0.8 gpm/ft^2 (1.0 to 2.0 m/h). A 0.5 gpm/ft^2 (1.2 m/h) rate is equivalent to a loading rate of about 2 gpm/ft^2 (4.9 m/h) over the top area of tubes. Rates ranging from 1.0 to 3.0 gpm/ft^2 (2.4 to 7.3 m/h) over the top area may be used, depending on the settling characteristics of the flocculated solid (Neptune Microfloc, Inc., 1980).

Inlet Conditions. For tube settlers to operate with uniform loading, the hydraulics of the influent and effluent is very important. Influent turbulence adversely affects settling efficiency in two ways. First, high velocities do not allow even flow distribution into the tubes, important for ensuring equal loading on the tubes. Second, sludge falling from the tubes must be able to settle to the bottom of the basins. High velocities below the tubes break up and shear the falling floc, causing it to be resuspended and carried into the tubes. This overloads the tubes and affects operating efficiency.

To avoid inlet turbulence effects, tube settlers must be placed to create a stilling zone between the inlet and the settler modules. This stilling zone is usually at least 25% of the total basin area.

A minimum depth of 10 ft (3 m) is generally provided below the tubes to create low velocities approaching the tubes, to allow sludge to settle without breaking. This depth also allows for access to sludge collection equipment.

Rapid clarification results from reduced settling distance and laminar flow conditions.

FLOW DIAGRAM

FIGURE 7.13 Typical tube settler. (*Courtesy of USFilter, Microfloc Products.*)

Effluent Design. Flow leaving the tubes must be collected uniformly across the basin to equalize flow through the tubes. Flow is usually collected through submerged orifices in pipe laterals or launders, or in some cases by overflow weirs into the launders. A clear space of about 2 to 3 ft (0.6 to 1 m) above the tubes must be provided for transition distribution of flow to the collection laterals or launders. Launders are spaced at not greater than 5-ft (1.5-m) centers.

Solids Removal. Settled solids collect on the floor below the tubes and the plain settling area ahead of the tubes. These solids can be removed with the same types of equipment used in conventional basins. Although those types that travel along the basin floor are most applicable, tube modules can be designed and placed to allow for carriage-type collectors and circular-type units. Sludge equipment selection should be made before the basin is designed to accept the tubes.

Ballasted Flocculation

A proprietary process that takes advantage of the high rate settling characteristics of tube settlers is the USFilter Kruger "ACTIFLO" ballasted flocculation system, shown in Figure

FIGURE 7.14 ACTIFLO process for water treatment. (*Courtesy of Kruger Inc.*)

7.14. In this process, a microsand is added after chemical coagulation, but before flocculation, to act as a nucleus for floc formation. The sand adds weight to the floc, thus increasing its settling velocity and allowing for tube clarifier loading rates of up to 40 gpm/ft^2 (100 m/h), although more typically in the 15 to 20 gpm/ft^2 (23 to 50 m/h) range. The effect is to provide a substantially reduced settling area compared to tube settlers without the sand. The microsand also improves the speed at which floc formation takes place, which further reduces space requirements. Figure 7.14 shows the sequence of coagulation, microsand addition, maturation (flocculation), and clarification. Sludge is typically collected with a circular scraper and is pumped to a hydroclone for reclamation of the microsand. The solids side stream is very dilute and may require additional treatment before sending to thickeners or to a sewer system. If the stream is disposed to a lagoon, the dilute nature of the solids would not be a problem. The side stream also may be as high as 4% to 5% of the plant flow, so additional treatment is also recommended to reclaim the water.

Plate Settlers

Plate settlers were developed to improve the efficiency of conventional rectangular settling basins by taking advantage of the theory that settling depends on the settling area rather than detention time. Plate settlers date back to an English patent in 1886 (Purac) and were used primarily in the mining and mineral industries to separate heavy particles from slurries. In the late 1950s they were developed in Europe for treating drinking water.

Theory. Based on sedimentation theory, shallow basins provide the same settling as a deep basin. However, horizontal shallow basins are subject to the scouring action of the flow-through velocity, which lowers removal efficiency. Also, if horizontal shallow basins are stacked to reduce plant area, there is no easy method of sludge removal. For these reasons, plate settlers are designed to be vertically inclined, similar to tube settlers, to allow settled solids to slide down the inclined surface and drop into the basin below. Distance between plates is designed to provide an upflow velocity lower than the settling velocity of the particles, allowing particles to settle to the plate surface. The effective settling area is the horizontal projected area of the plate, calculated by multiplying the plate area by the cosine of the angle of the plate to the horizontal. Total settling area is the sum of the effective areas of each plate.

Design Criteria. The primary design criterion for plate settlers is the surface loading rate for each plate. Typical loading rates range from 0.3 to 0.7 gpm/ft^2 (0.7 to 1.7 m/h), depending on the settling characteristics of the solids, water temperature, and desired effluent quality. These loading rates allow for overall basin loadings from 2 to 6 gpm/ft^2 (5 to 15 m/h), several times those for conventional basins. This criterion allows for much smaller basins in new construction or for up-rating existing basins.

Basin Dimensions. Plate settlers are typically manufactured in modules. Dimensions of modules and plates vary by manufacturer and are proprietary. These proprietary dimensions require different basin geometries, and the system designer must work with manufacturers to establish appropriate dimensions for new construction or to decide how plates may be installed in existing basins. Some manufacturers provide standard-width plates of varying heights to fit custom designs.

The basic dimensions provided by each manufacturer are primarily established by how each manufacturer approaches the problem of flow control. Basin dimensions may also be controlled by how the sludge is to be removed. Sludge removal devices are typically the same as those used for conventional basins except those that travel on or above the surface of the basin. Two typical plate settler installations are shown in Figures 7.15 and 7.16.

FIGURE 7.15 Plate settlers. (*Courtesy of Parkson Corporation.*)

FIGURE 7.16 Plate settlers. (*Courtesy of Parkson Corporation.*)

Inlet Design. Inlet flow distribution is critical for efficient performance of plate settlers. In fact, the approach to distribution is a major difference between manufacturers. Some manufacturers feed the plates from side channels to prevent interference between the incoming flow and the falling settled solids. In one case the flow to each plate is controlled by head loss through an influent orifice. In another, flow is distributed from the side channel through slots and controlled by head loss taken through control orifices as flow leaves the plates.

A third approach provides a large space below the plates, similar to tube settler design, to minimize velocities and to design the plate outlet to induce sufficient head loss to ensure good flow distribution. The designer should evaluate each approach as it relates to the water to be treated and decide if one is more advantageous than another.

Outlet Design. Proprietary designs also provide for collection of clarified water in different ways. One design uses the upper half of the side inlet channel to collect clarified water laterally at the top of each plate section. Another design places a collection launder along the top of the plates with a control orifice located above each plate to induce enough head loss to ensure good flow distribution.

Other designs have uniformly spaced launders with weirs or submerged orifices that collect the flow along the entire surface. Launders should be spaced on the order of 6 ft (1.8 m). Submerged orifices should be designed to create sufficient head loss to ensure good flow distribution. An orifice velocity of 1.5 to 2.5 ft/s (46 to 76 cm/s) will generally be adequate.

Solids Removal. Settled solids that slide down the plates collect uniformly at the bottom of the basin, and they must be periodically removed. Typical removal equipment would be chain-and-flight collectors or bottom track units. Circular equipment with drive units above the plates has also been used.

SOLIDS CONTACT/SLURRY RECIRCULATION UNITS

Solids contact units, slurry recirculation, or sludge blanket types are designed to provide more efficient flocculation and greater opportunity for particle contact within the blanket, which also acts partially as a filter. The hydraulic design also provides for more uniform flow and is less subject to short-circuiting. For these reasons, solids contact units can handle 3 to 4 times the hydraulic loading of conventional basins. However, density currents created by differences in water temperature between the incoming flow or direct sunlight on the basin surface can be disruptive, by causing the blanket to "boil" with resulting rising of the floc particles into the effluent. In the most severe situations, the entire sludge blanket can be lost.

Solids contact units combine flocculation and sedimentation functions into a single basin. Some units operate with chemical feed directly to the inlet pipe, but a separate rapid mixer may provide better coagulation for turbidity or color removal applications. A large volume of previously settled solids is recirculated to the mixing zone to act as nuclei to form additional floc and to make more complete use of coagulation chemicals. Recirculation rates vary with the application, and may be up to 12 times influent flow for softening and up to 8 times influent flow for turbidity removal.

Slurry recirculation units were developed primarily for softening applications, where they have been extensively employed for many years. They have also been used in applications for turbidity or color removal.

Operation and Design Criteria

The basic theory of solids contact units is that contact of newly formed coagulation particles with previously formed floc enhances floc formation, creates greater opportunity for particles to make contact, allows for larger floc development, and allows higher loading rates. Providing this recirculation within a single basin, compared with pumping settled solids to a flocculator in a conventional plant, reduces equipment requirements and lowers facilities costs. In addition, recirculating settling floc within the basin is less destructive to the floc than recirculating after it has settled and had time to thicken and further agglomerate. A typical slurry recirculation unit is shown in Figure 7.17.

FIGURE 7.17 Typical solids contact slurry recirculation unit. (*Courtesy of Eimco Water Technologies.*)

Surface Loading Rate. The surface loading rate is the primary basis for sizing the clarification zone. Loading rates may range from 0.5 to 1.5 gpm/ft² (1.2 to 3.7 m/h) for color and turbidity removal, with even higher rates for turbid warm waters and lower rates for colored colder waters. Rates for softening are often higher. Surface settling rates used for solids contact units typically are higher, by as much as 2 to 4 times, than for conventional sedimentation in similar applications.

A number of factors affect settling efficiency, and loading rates should be adjusted accordingly. Specific gravity of particles in the settling zone often ranges from as low as 1.02 for alum floc to as high as 2.65 for sand particles. The designer should consider the nature of the flocculated material in selecting the loading rate.

Temperature is also an important consideration. Higher-temperature water allows for more efficient chemical coagulation and flocculation and has a lower viscosity, exerting less drag on settling particles. For these reasons, higher surface loading rates may be used for warm waters.

Flocculation/Contact Zone. A detention time of 30 min or more is provided in the reaction wells for units designed to remove turbidity to ensure complete chemical reactions and adequate time for flocculation. Shorter or longer times may be appropriate depending on water quality and temperature. Selection should be based on an understanding of flocculation principles. The designer should refer to Chapter 6 on flocculation for additional information.

Basin Dimensions. Slurry recirculation units are center-fed units with uniform collection of clarified water across the surface. To provide good flow distribution, units are generally circular with a diameter up to about 150 ft (46 m). Square tanks can take advantage of common wall construction, but they generally have problems associated with sludge removal from corners.

Side water depth, especially center depth, is an important design parameter to ensure adequate space for the mixing zone, uniform distribution of slurry to the basin clarification zone, and transition flow vertically to effluent collection troughs. Basin depth is usually established by proprietary equipment manufacturers' requirements.

Influent Design. Influent design is based on requirements for mixing chemically treated source water with previously formed flocculated and settled solids. To achieve this mix, flow is introduced into a recirculation drum or zone at the center of the unit.

Flow is then introduced to a flocculation reaction zone in a center well separated from the clarification zone by a hood. The design of this area varies depending on the manufacturer. The hood is designed to control flow uniformly to the clarification zone. The designer should work with the manufacturer to select equipment most suitable to the application and to obtain necessary dimensions for basin structural design.

Effluent Design. Clarified water is collected in radial launders to maintain a uniform vertical upflow within the settling area to help reduce short-circuiting. Launders are designed with low loading rates, typically about 20 gpm per linear foot [248 (L/min)/m], which sets the spacing. Spacing is usually about 15 to 20 ft (4.6 to 6.1 m).

Launders are designed with submerged orifices to overcome the problems of trying to level many weirs to the same elevation. Small changes in orifice elevation do not greatly affect flow distribution, but small changes in weir elevations have a major impact on flow.

Solids Removal. Most slurry recirculation units use a rake rotating around a center column to plow settled solids toward the center, where they are deposited in an annular hopper for discharge to waste. Some solids are picked up by the recirculating flow to be mixed with influent. The rake consists of a steel truss with squeegees and is supported from the rotating recirculation drum. Rake arms must be designed to withstand the torque caused by the weight of the solids being moved and by friction forces. The rake is driven by a center drive assembly similar to equipment for conventional circular clarifiers.

Equipment

Basin equipment consists of a conical reaction well, recirculation drum, influent pipe, and influent baffle. The equipment is generally mounted on a bridge extending across the basin and designed to act as a walkway for access to the center drive unit. Dual drives are provided, one to drive the sludge collection rakes, the other to drive the turbine mixer used to recirculate slurry and provide flocculation.

In large-diameter tanks, equipment is supported from a center column, and a bridge is provided only to access drive units. Some large-diameter tanks use peripheral drive units with rakes supported from the walls and a center column.

CLARIFICATION 7.31

FIGURE 7.18 Accelator solids contact unit. (*Courtesy of Ondeo Degremont.*)

The Accelator Design

The Accelator (Figure 7.18) is a special design of solids contact basin developed by Infilco (Vincent, 1991) and is now a trademark of Ondeo-Degremont Inc. This unit was initially used for softening applications but has widespread use for turbidity and color removal.

Influent, to which coagulating chemicals have been added, is introduced to a primary mixing zone where it is mixed with previously formed slurry. A large turbine mixer provides recirculation and mixing inside a central hooded area that separates the mixing zone from the clarification zone. Solids settling in the clarification zone drop below the edge of the hood and are picked up in the recirculating flow within the hood. The turbine impeller circulates slurry to a secondary zone located above the primary zone where continuous flocculation is taking place. Flocculated water overflows the top of the secondary zone baffle and flows downward and out along the slopes of the mixing zone hood.

In the clarification zone, solids settle while clarified water rises vertically to collection launders. Part of the clarification zone area contains hoppers with valved outlets. Solids settling over these hoppers can be retained and then periodically discharged, usually by a timed blow-off valve.

The volume of solids removed can be controlled by the number of hoppers in service. Solids not removed in the hoppers are drawn back under the hood to the mixing zone. The original Accelator did not have mechanical sludge removal equipment. This limited the diameter because of the depth required to maintain proper dimensions for the mixing zone and sludge collection hoppers. Depths greater than 20 ft (6 m) tend to be uneconomical compared with other types of slurry recirculation units.

Larger-diameter Accelerators operate with the same basic mixing zones but use mechanical scrapers to move sludge to a central hopper (Webb, 1993). As sludge moves under the hood of the mixing zone, some of it is resuspended, and the rest is periodically discharged to waste. The scraper may be structurally a part of the hood, supported by a separate scraper drive, or supported at the center by columns extending to the floor. Some Accelator designs use a peripheral drive with the scraper mechanism supported from a traveling bridge that extends from the outer wall to a center column and that rotates around the column.

The center mixing turbine operates at low tip speeds (variable-speed drives are typically used) to prevent floc particle shearing. The turbine circulates flow at pumping rates of up to 10 times the basin design flow rate.

SLUDGE BLANKET CLARIFICATION

Sludge blanket clarification is a variation of solids contact clarification in which coagulated water flows up through a blanket of previously formed solids. As the small coagulated particles enter the blanket, they contact particles within the blanket and flocculation takes place. Flocculated particles grow in size and become a part of the blanket. The sludge blanket grows in thickness and is suspended by the flow velocity passing through the blanket.

A blanket depth of several feet is required for efficient clarification. When the blanket depth has reached the design depth, the top of the blanket is above the level of sludge removal hoppers. Because there is no upward flow in the hopper area, the portion of the blanket next to the hopper settles into the hopper. Cohesion in the sludge blanket helps pull solids into the hopper as the blanket settles, creating a flow of solids along the top surface of the blanket. Sludge is periodically removed from the hopper by gravity.

Design Criteria and Application

Sludge blanket clarification may be used for applications where flocculent suspensions are formed, such as to remove turbidity, color, organic matter, tastes and odors, and iron and manganese. The process may be applied to highly turbid waters if the turbidity is colloidal in nature. However, heavy suspended solids should be removed by presedimentation because they may not be supported by the upward flow in the blanket and may settle out to create a maintenance problem. Waters with high levels of algae may be difficult to treat because algae may float to the surface, carrying flocculated solids with them. Otherwise, sludge blankets can be an efficient removal process operating at much higher loading rates than conventional clarifiers.

For a sludge blanket to perform efficiently, it must be designed to provide uniform upward flow, ideally with equal velocities across the entire cross section. This ideal is approached by introducing flow across the bottom of the basin from uniformly spaced distribution laterals, each with uniformly spaced orifices. Clarified water is then collected in equally spaced launders of either the overflow weir or the submerged orifice type.

One of the first applications of sludge blanket clarification was the pyramid type, shown in Figure 7.19, developed by Candy in the 1930s. In this unit, coagulated water is introduced to the bottom of the pyramid and flows upward through the sludge blanket with a reducing velocity resulting from the expanding area of the pyramid. Clarified water is collected in uniformly spaced launders, and sludge overflows to a hopper.

Because the hydraulic design reduces short-circuiting, and because the sludge blanket is heavier than the settling floc particles in a conventional basin, the surface loading rate can be increased. Recent application of plates and tubes in sludge blanket clarifiers has allowed even higher loading rates.

Pulsed Blanket Clarifier

The pulsed blanket clarifier (Hartman and Jacarrino, 1987), as shown in Figure 7.20, consists of the following subsystems:

- Vacuum chamber
- Inlet distribution system
- Effluent collection system

FIGURE 7.19 Pyramid-type sludge blanket clarifier. (*Source: Monk and Willis, 1987.*)

FIGURE 7.20 Cross section of the Superpulsator sludge blanket clarifier. (*Courtesy of Ondeo Degremont.*)

- Sludge extraction
- Drain, washdown, and sample system
- Controls

The treatment cycle begins when previously mixed influent water and coagulant chemicals enter the vacuum chamber. With the vacuum pump running and the vacuum release valve closed, pressure is reduced inside the chamber, causing the water level to rise to a predetermined level. This level is controlled by a timer that actuates the vacuum release valve. When the vacuum is released, the water level drops rapidly, causing a surge of water to flow into the distribution system. As the water level in the vacuum chamber drops to the hydraulic grade line of the influent water, the head is dissipated and the surge slows. A timer then closes the vent valve, and the vacuum pump, which runs continuously, begins evacuating the air to repeat the cycle. This rising and falling of the water level in the vacuum chamber create the rhythmic, controlled pulsing for which the pulsed blanket clarifier is known. Pulsations create and maintain the sludge blanket in homogeneous suspension critical for efficient solids removal. Pulsations also aid in flocculation created by the turbulence from the high-velocity flow leaving distribution pipe orifices.

As water enters the sludge blanket and passes upward through it under the forces of succeeding pulsations, the sludge blanket performs the double tasks of flocculation and filtration: It serves as a medium that agglomerates the newly formed floc, and it helps suspended matter and colloidal particles adhere to the floc.

Water exits the sludge blanket into the clarification zone. Because the flow rate is lower in the clarification zone than in the blanket, particles that escape the blanket settle back toward the blanket. In the Superpulsator, a trademark of Ondeo-Degremont Inc., clarification is enhanced by inclined plates that increase the settling area and efficiency of particle removal. At the top of the clarification zone, clarified water is collected in submerged orifice laterals or launders.

As solids are removed in the blanket and clarification zone, blanket volume increases and some solids must be removed to maintain stable conditions. A zone of sludge concentrators collects and concentrates excess solids. As these excess solids increase the volume of the blanket, they spill over into the concentrators. When the concentrator is full, thickened sludge is withdrawn through sludge removal headers usually on a timed basis by automatically controlled sludge valves.

Design Criteria. Pulsed blanket clarifiers are sized based on surface loading rate in the clarification zone. The area over the sludge concentrators is not included in this calculation. Typical design loading rates recommended by the manufacturer are as follows, but lower rates may be required by some regulatory agencies and for some water quality conditions (Webb, 1993):

Type of unit	*Loading rate*
Pulsed blanket	1.0 to 1.25 gpm/ft^2 (2.4 to 3.1 m/h)
Pulsed blanket with plate settlers	2.0 to 4.0 gpm/ft^2 (4.9 to 9.8 m/h)
Pulsed blanket with plate and tube settlers	3.0 to 5.0 gpm/ft^2 (2.3 to 12.2 m/h)

Basin Dimensions. Basin dimensions are set to provide uniform flow distribution. A typical Superpulsator consists of two 16-ft-wide (4.9-m) blanket sections with a central 10-ft-wide (3-m) source water influent chamber, sludge concentrator, and effluent section. Basin water depth is 16 ft (4.9 m). Basin length varies, limited by hydraulic considerations.

Inlet Design

Flow to the basin is controlled through the inlet distribution system. Flow from the vacuum chamber typically enters a conduit serving two halves of the basin. From this conduit, laterals extend to either side to provide uniform flow across the bottom of the basin. Laterals are spaced at about 3.5 ft (1.1 m) on center. Orifices in these laterals further serve to equally distribute flow across the basin. Orifices are designed to provide a certain entrance velocity based on the vacuum chamber hydraulic head, to ensure even flow distribution and create an energy level to enhance flocculation in the mixing zone as flow enters the basin. Baffles above the inlet laterals further ensure uniform flow to the blanket.

Outlet Design. Clarified water is collected at the clarifier surface in uniformly spaced laterals with submerged orifices. The lateral spacing is about 6.5 ft (2 m). Orifices are designed to induce enough head loss to aid in maintaining uniform upflow velocities in the blanket. Laterals discharge to an effluent channel located above the sludge concentrators. Submerged orifices are used because flow through them is not as affected by the pulsing action of the water surface as it would be over a weir.

Sludge Removal. As the sludge blanket builds in volume, it rises above the level of an overflow weir to sludge concentrators, which are hoppers located between the two halves of the basin. Sludge flows into concentrators, where it is allowed to partially thicken. It is periodically drawn off through timer-controlled valves, usually by gravity, to a sump from which it may be pumped or flow by gravity to sludge handling facilities. No mechanical equipment is used, which is one of the attractive features of this type of clarifier.

DISSOLVED AIR FLOTATION

In flotation, the effects of gravity settling are offset by the buoyant forces of small air bubbles. These air bubbles are introduced to the flocculated water, where they attach to floc particles and then float to the surface. Flotation is typically sized at loading rates up to 10 times that for conventional treatment. Higher rates may be possible on high-quality warm water.

Dissolved air flotation (DAF) is an effective alternative to sedimentation or other clarification processes. Modern DAF technology was first patented in 1924 by Peterson and Sveen for fiber separation in the pulp and paper industry (Kollajtis, 1991). The process was first used for drinking water treatment in Sweden in 1960 and has been widely used in Scandinavia and the United Kingdom for more than 30 years.

Previous uses of the process in the United States have been to thicken waste-activated sludge in biological wastewater treatment, for fiber separation in the pulp and paper industry, and for mineral separation in the mining industry. Only recently has this process gained interest for drinking water treatment in North America. It is especially applicable when treating for algae, color, and low-turbidity water. The first use in the United States was at New Castle, New York, in a 7.5 mgd (28 ML per day) plant that began operation in 1993. A typical DAF unit is shown in Figure 7.21.

Theory and Operation

Effective gravity settling of particles requires that they be destabilized, coagulated, and flocculated by using metal salts, polymers, or both. The same is true for DAF. In gravity settling the flocculation process must be designed to create large, heavy floc that settles

FIGURE 7.21 Typical dissolved air flotation tank. (*Courtesy of Parkson Corporation.*)

to the bottom of the basin. In DAF, flocculation is designed to create a large number of smaller floc particles that can be floated to the surface.

For efficient flotation, flocculated particles must be in contact with a large number of air bubbles. Three mechanisms are at work in this air/floc attachment process:

- Adhesion of air bubbles on the floc surface
- Entrapment of bubbles under the floc
- Absorption of bubbles into the floc structure

The size of air bubbles is important. If bubbles are too large, the resulting rapid rise rate will exceed the laminar flow requirements, causing poor performance. If bubbles are too small, a low rise rate will result and tank size may need to be increased.

In a typical DAF tank, flocculated water is introduced uniformly across the end of the tank, near the bottom, into the recycle dispersion zone. Recycle is continuously introduced through a distribution system of proprietary nozzles, valves, or orifices. When the recycle flow pressure is suddenly decreased from its operating pressure of 60 to 90 psi (414 to 620 kPa) to atmospheric pressure, saturated air within the recycle stream is released in the form of microbubbles with a size range of 10 to 100 μm, and averaging around 40 to 50 μm. These microbubbles attach to flocculated material by the mechanisms described previously, causing flocculated material to float to the surface.

At the surface, the bubble-floc forms a stable and continuously thickening layer of float, or sludge. If left at the surface, the float can thicken to as much as 3% to 6% dry solids. This can be an advantage if solids are to be mechanically dewatered, because solids may be suitable for dewatering without further thickening, or the thickening process can be reduced. Sludge thickness depends on the time it is allowed to remain on the surface and the type of removal system employed.

Recycle System

About 5% to 10% of clarified water is recycled into the air saturation system by recycle pumps. Recycle water is pumped through an eductor that introduces high-pressure air in the line from the air cushion of the saturation tanks. The air-water mixture is then discharged into the bottom of the saturation tank. The eductor increases the efficiency of air-to-water transfer by increasing contact time between the two media.

Air is supplied from an air compressor controlled by the saturation tank water level. Operating efficiency of this type system is in the 80% to 85% range and may be improved by using packed bed saturation tanks to increase the air-water interface. Packed bed saturators are more expensive, but their greater efficiency may make them more economical in large plants. Saturated water flows from the saturation tank through a system of headers to the proprietary recycle dispersion system, where the pressure is dropped to atmospheric and microbubbles are created.

Design Criteria

The size of DAF tanks is based on the surface loading rate. Standard practice has been to design in the range of 4 to 5 gpm/ft^2 (10 to 12 m/h), although higher rates may be possible. Edzwald et al. have reported that rates as high as 12 to 16 gpm/ft^2 (28 to 40 m/h) may be possible depending on water temperature (Edzwald et al. 1999). The use of pilot plant studies is recommended for rates higher than standard (Grubb, Arnold, and Harvey, 1994). Ondeo Degremont markets the AquaDAF system as being capable of handling loading rates of up to 20 gpm/ft^2 (50 m/h) or more. The efficiency of AquaDAF is said to be based on a proprietary effluent collection system.

Design should allow for a recycle rate of 5% to 10% of plant flow at a dissolved air pressure of 60 to 90 psi (414 to 620 kPa). Recycle rate and air pressure should be adjustable to allow for process optimization.

The recycle dispersion system must be selected to evenly distribute recycle flow across the width of the unit into the flocculated water flow.

One unique use of DAF is "stacked DAF," where the flotation and filtration take place in the same basin. In this case the water depth above the filter media is such as to allow flotation to take place with solids being removed at the surface. Flotation effluent goes directly to the filter. DAF loading rates in this case are limited to the rates the filter can handle.

Basin Dimensions

Basin dimensions may be flexible; the length-to-width ratio is not important. However, a maximum length of about 40 ft (12 m) is recommended, because in this distance, all the air bubbles would have typically risen to the surface. Tank depths are usually around 10 ft (3 m). If tanks are too shallow, forward velocities could carry air bubbles into the effluent. Tanks could be deeper, but there is no added benefit for increasing depth.

Influent Design

As noted above, flocculated water must be introduced uniformly along the width of the tank near the bottom. This can be done by means of a continuous slot or through uniformly spaced pipes. Recycle flow must also be uniformly distributed. When pipes are used for flocculated water, recycle flow is sometimes introduced into pipes just before the outlet.

Effluent Design

Clarified water is collected from the tank bottom by being forced under an end baffle wall or by pipe laterals or cast-in-place concrete flumes on the basin floor. In either case, flow must then be discharged over a control weir to maintain a stable water surface in the clarifier. The weir may be adjustable to fine-tune the water surface level. This is especially important when scrapers are used to remove the float.

Floated Solids Removal

Floating solids (called float) are removed from the surface by mechanical skimming or by hydraulic flooding. Mechanical skimming equipment must be maintained, but it allows the float to be removed at a dry solids content of 3% to 6%. At this concentration it may be possible to eliminate or at least reduce thickening equipment required for mechanical dewatering.

Flooding eliminates mechanical equipment in the flotation tank but results in a dilute solids stream of around 0.5% or less. This float removal method could create a need for larger thickening equipment.

Mechanical Skimming. Three common types of mechanical skimming equipment are the chain-and-flight skimmer, oscillating skimmer, and rotating skimmer.

Chain-and-Flight Skimmer. A typical mechanical skimming device is the chain-and-flight type, similar to those used to scrape settled solids in gravity sedimentation basins. In this case, however, flights are equipped with rubber squeegees or nylon brushes that drag the float along the water surface and up onto a dewatering beach. As solids are dragged up the beach, water is squeezed out and drains away, further thickening the float. The float then drops off the beach into a hopper from which it flows by gravity or is pumped to waste or thickening.

Solids content of the float is controlled by how often it is removed and by how much excess water is dragged along with the solids. Adjusting the flight depth into the water is critical to the amount of excess water and to treatment efficiency. If the moving flights are set too deep, they can create flow patterns in the basin that will cause poor particle removal and floc carry-through to the effluent. If the flight is set too shallow, it can disrupt the adhesive forces in the float blanket. This disruption causes solids to settle back into the tank, resulting in floc carry-through.

The skimmer is usually operated on a timer, with the frequency of operation based on solids production and the desired level of thickening. All mechanical components are above water, making maintenance easier than that for gravity sedimentation basins.

Oscillating Skimmer. This skimming device consists of a carriage-and-blade assembly that is pushed and pulled across the surface by a gear motor. Blade spacing is typically 5 to 7 ft (1.5 to 2.1 m), and the mechanism covers two-thirds of the flotation area. The lead carriage, closest to the sludge beach, is equipped with a profile-duplicating arm that allows the skimmer blade to exactly duplicate the shape of the beach. The beach can be either flat or curved.

All moving components are located above water, which allows easy inspection and maintenance. The gear motor oscillates a drive rod back and forth in 15-in. (38-cm) increments while engaging the skimmer carriages. When the carriage reaches the effluent end and the profile blade deposits the float into the hopper, the blades retract above the water surface. The carriage is then mechanically reversed to return to the influent end in an indexing manner.

The skimmer unit is constructed of stainless steel and plastic. High-hardness plastics provide low operating friction, allowing the mechanism to operate efficiently. The profile duplicating blade is typically a nylon brush that aids in sludge dewatering. The brush allows the sludge to thicken to a slightly higher solids percentage while the blade pauses inherently during operation on the sludge beach. Maintenance is minimal, consisting of checking the plastic wear blocks and regular servicing of the gear motor.

Rotating Skimmer. Another type of mechanical skimming device consists of a rotating shaft with curved or straight blades attached. The shaft is mounted at the effluent end of the basin above a sludge beach. As the blades rotate, they pull a portion of the float blanket onto the beach and into a hopper. The general flow of water toward the effluent causes the blanket to flow up to the beach, where the rotating blades continually remove a portion.

Hydraulic Removal. Float can be hydraulically removed from the surface of a DAF unit by flooding. This is accomplished by partially closing the basin effluent gate or in some cases raising a mechanical effluent weir, which causes increased head loss and a small rise in the water surface level. This increase in level brings the surface up above a weir that discharges to a sludge drain. Part of the flow through the basin is now diverted over this weir, dragging the float blanket along with it.

Adhesion of the sludge particles helps keep the blanket intact as it moves across the surface to the weir. Water is trickled down the basin walls to break adhesion to the wall so that the blanket can move freely. Float removal is relatively frequent to prevent the blanket from becoming too thick because a heavy blanket does not flow well. A solids content of about 0.5% is best.

When most of the blanket is removed, usually in about 10 min or less, the effluent gate reopens and operations return to normal. This mode of float removal eliminates all mechanical equipment in the basins. The disadvantage is that the removed solids are diluted and may require more extensive thickening for subsequent mechanical dewatering.

CONTACT CLARIFICATION

Contact clarification resembles filtration more than clarification. Coagulated water flocculates within the contact medium, may stick to the medium grains, and builds in size, eventually clogging the medium, at which point it is backwashed. Because this process is more filtration than gravity settling, it is capable of operating at rates of up to 20 times that for conventional settling on low-solids source waters requiring low coagulant dosages. High-solids waters and high-color waters that require high coagulant dosages quickly clog the medium, resulting in excessive backwashing. The process is generally marketed as a two-stage process, along with filters in package plants, but may be used separately to upgrade existing plants.

Roughing filters have been used for as long as filtration has been around. The contact clarification process, however, gained popularity with the development in the mid-1980s of the Adsorption Clarifier by Microfloc, now owned by USFilter. Similar processes are also marketed by Infilco Degremont, Inc., and Roberts Filter Co. The Adsorption Clarifier is shown in Figure 7.22.

FIGURE 7.22 Adsorption Clarifier. (*Courtesy of USFilter, Microfloc Products.*)

Application

The fact that clarification results from filtration limits the solids levels that can be removed by contact clarification without causing excessive cleaning. This process is most effective for source water having low turbidity, color, iron, and manganese. On such waters, it can be an economical alternative to more conventional types of clarification. It is especially applicable to upgrading overloaded direct filtration systems.

Design Criteria

Contact clarifiers are rated on the basis of surface loading rate and typically operate in the range of 8 to 10 gpm/ft^2 (20 to 24 m/h). To maintain effective flocculation, units must also be selected to operate in the range of 50% to 100% of design capacity.

Operation

Source water is chemically treated by the addition of metal salts or polymers in a rapid mixer, usually of the static mixer type. Chemically treated water is then introduced to the clarifier in an upflow or downflow mode, depending on the proprietary process selected. Clarified water is then filtered in a typical granular medium filter.

Inlet Conditions

Proprietary units are designed as upflow units. The Adsorption Clarifier uses a buoyant medium held in place by a screen. Chemically treated source water is introduced under the medium through a pipe lateral. Other types of units use a medium heavier than water that must be supported by a media-retaining support system typical of gravity filters. In all cases, source water flows up through the medium as it would for any upflow filter.

Outlet Conditions

Clarified water is collected above the medium in an effluent trough similar to backwash troughs in a filter and then flows to the filter through an outlet valve.

Sludge Removal

Because solids are removed by filtration, they must be removed by backwashing or flushing. The flushing cycle is initiated after a preset head loss has been reached or a preset time has passed.

The cycle is started by closing the effluent and influent valves. The medium is first air-scoured. Air is necessary in the adsorption clarifier to allow buoyant media to expand during the flushing cycle. Air reduces the apparent density of the water, causing media to lose buoyancy and opening pore spaces so that accumulated material can be flushed out. Air is also necessary with other types of media to assist water in scouring accumulated solids.

Source water is used to flush the bed, taking advantage of hydraulic head already available. Waste washwater is collected in the effluent trough but is directed to a waste drain instead of the filter. When the cycle is completed, the air is turned off, the waste drain valve is closed, the effluent valve is opened, and clarified flow once more is directed to the filter. No mechanical equipment is required for sludge removal, an advantage of these units.

BIBLIOGRAPHY

American Water Works Association. *Water Quality and Treatment*, 5th ed. New York: McGraw-Hill, 1999.

Camp, T. R. "Sedimentation and the Design of Settling Tanks." *Transactions of ASCE* 3(2285):895, 1946.

Camp, T. R. "Studies of Sedimentation Basin Design." *Sewage and Industrial Wastes* 25(1):1, 1953.

Dissolved Air Flotation: Field Investigations, American Water Works Association Research Foundation Report. Denver, Colo.: American Water Works Association, 1994.

Dissolved Air Flotation: Laboratory and Pilot Plant Investigations, American Water Works Association Research Foundation Report. Denver, Colo.: American Water Works Association, 1992.

Edzwald, James K., John E. Tobiasson, Tony Amato, and Lawrence J. Maggi. "Integrating High-Rate DAF Technology into Plant Design." *Journal AWWA* 91(12), 1999.

Flood, F. L. *Sedimentation Tanks*. Seminar Papers on Wastewater Treatment and Disposal. Boston: Boston Society of Civil Engineers, 1961.

Great Lakes–Upper Mississippi River Board of State and Provincial Public Health and Environmental Managers. *Recommended Standards for Water Works*. 2003.

Grubb, T. R., S. R. Arnold, and P. J. Harvey. "Recent Applications of DAF Pilot Studies, Retrofits and Conversions." Presented at AWWA Annual Conference, June 1994.

Hartman, R. H., and R. F. Jacarrino. "Superpulsator Clarifier for Potable Water Treatment." Presented at AWWA Annual Conference, 1987.

Hazen, A. "On Sedimentation." *Transactions of ASCE* 53:63, 1904.

Hudson, H. E. *Water Clarification Processes: Practical Design and Evaluation.* New York: Van Nostrand Reinhold, 1981.

Ives, K. J. "The Inside Story of Water-Treatment Processes." *Journal of Environmental Engineering* 121(12):846, 1995.

Kawamura, S. *Integrated Design of Water Treatment Facilities.* New York: John Wiley & Sons, 1991.

Kawamura, S., and J. Lang. "Re-evaluation of Launders in Rectangular Sedimentation Basins." *Journal WPCF* 58(12):1124, 1986.

Kollajtis, J. A. "Dissolved Air Flotation Applied in Drinking Water Clarification." Presented at AWWA Annual Conference, 1991.

Monk, R. D. G., and J. F. Willis. "Designing Water Treatment Facilities." *Journal AWWA* 79(2):45, 1987.

Neptune Microfloc, Inc. *60* Tube Application Guidelines for Water Treatment.* Technical Memorandum No. 1, Revision 2. Corvallis, Ore.: Neptune Microfloc, Inc., 1985.

Neptune Microfloc, Inc. *FTC and TC.* Technical Memorandum No. 14, Revision 0. Corvallis, Ore.: Neptune Microfloc, Inc., 1980.

Operational Control of Coagulation and Filtration Processes, M37. Denver, Colo.: American Water Works Association, 1992.

Taebi-Harandy, A., and E. D. Schroeder. "Analysis of Structural Features on Performance of Secondary Clarifiers." *Journal of Environmental Engineering* 121(12):911, 1995.

Vincent, J. "Effective Clarification and Softening with the Accelator Unit versus Conventional Treatment Methods." Presentation to the New York State Sixth Annual Water Treatment Technical Conference, Saratoga Springs, N.Y., April 9 and 10, 1991.

Webb, J. E. "Super Flocculation and High-Rate Clarification for Water Treatment." Presented at North Carolina Operators Association, September 29, 1993.

Yee, L. Y., and A. F. Babb. "Inlet Design for Rectangular Settling Tanks by Physical Modeling." *Journal WPCF* 57(12):1168, 1985.

CHAPTER 8
HIGH-RATE GRANULAR MEDIA FILTRATION

Kevin Castro
Stearns & Wheler, LLC
Cazenovia, New York

Gary Logsdon
Retired
Lake Ann, Michigan

Stephen R. Martin
CDM
Cambridge, Massachusetts

Filtration, as it applies to water treatment, is the passage of water through a porous medium to remove suspended solids. According to Baker (1948), the earliest written records of water treatment, dating from about 4000 B.C., mention filtration of water through charcoal or sand and gravel. Although a number of modifications have been made in the manner of application, filtration remains one of the fundamental technologies associated with water treatment.

Filtration is needed for most surface waters, to provide a second barrier against the transmission of waterborne diseases. Although disinfection is today the primary defense, filtration can assist significantly by reducing the load on the disinfection process, increasing disinfection efficiency, and aiding in the removal of precursors to disinfection by-product (DBP) formation. The Surface Water Treatment Rule (SWTR) and Enhanced Surface Water Treatment Rules (ESWTR) recognize three categories of granular filtration techniques:

- Rapid sand
- Slow sand
- Diatomaceous earth

This chapter covers the design of the first category of filters. However, in this instance the term *rapid sand* includes not only sand, but also other types of filter media such as crushed anthracite coal and granular activated carbon (GAC). Chapter 9 covers the other two categories of granular filtration techniques. Chapter 14 further discusses activated carbon processes, including GAC filters/adsorbers.

MECHANISM OF FILTRATION

Removing suspended solids by high-rate granular media filtration is a complex process involving a number of phenomena. Attempts to develop theories that quantitatively predict solids removal performance with sufficient precision and versatility to be of use in practical filter design have met with relatively little success. Consequently, filter media selection is often an empirical process. Pilot investigations are common tools for assessing the performance of a particular filter design (see Chapter 28).

In current high-rate granular media filtration techniques, solids removal occurs primarily as a two-step process (Cleasby, 1972). During the initial transport step, particles are moved to the surfaces of media grains or previously captured floc. Transport is believed to be caused largely by hydrodynamic forces, with contact occurring as stream lines converge in pore restrictions. The second step is particles' attachment to either grain or floc surfaces. Electrokinetic and molecular forces are probably responsible for the adherence of particles on surfaces within the bed (O'Melia and Crapps, 1964; Craft, 1966; O'Melia and Stumm, 1967). Physical straining through the surface layer of solids and biological growth (*schmutzdecke*) is the principal filtration mechanism of a slow sand filter, but it is generally a minor means of solids removal in high-rate granular media filters.

DESIGN CONSIDERATIONS

A number of interrelated components are involved in the overall design of a high-rate granular media filtration system:

- Regulatory requirements
- Pretreatment systems
- Filter media
- Filtration rates
- Depth of the filter box
- Mode of operational control
- Filter washing system
- Filter arrangements
- Underdrain system
- Filter performance monitoring
- Auxiliaries

These components are discussed in detail in the following sections.

Regulatory Requirements

The primary regulation related to the design of filters is control of effluent turbidity. The ESWTRs establish a maximum contaminant level (MCL) on combined filter effluent (CFE) at 0.3 ntu in 95% of samples collected in a month, for direct and conventional filtration systems (employing a clarification process). The CFE turbidity limit for slow sand and diatomaceous earth filtration is 1.0 ntu. Additionally, individual filter turbidity must be maintained below 1.0 ntu at any time and below 0.5 ntu 4 h into a filter run to avoid specific reporting and performance review action levels, based on continuous monitoring. Reporting and filter evaluation action is required if two successive 15-min individual filter turbidity measurements exceed 0.5 ntu after 4 h of filter operation and 1.0 ntu or 2 ntu at any point during the filter run. Generally, achieving the required turbidity limits will provide 2.5-log *Giardia* removal credit, 2.0-log virus removal credit, and 2.0-log *Cryptosporidium* removal credit for conventional filtration systems; and 2.0-log *Giardia* removal credit, 1.0-log virus removal credit, and 2-log *Cryptosporidium* removal credit for direct filtration systems. These credits vary from state to state, so consultation with your local regulators is encouraged.

Pilot investigations and filter design must consider the filter's ability to achieve these standards throughout the filter run. Filter-to-waste capabilities are not specifically required by drinking water regulations; however, their use is recommended to achieve compliance with individual filter turbidity limits. A short filter-to-waste period can "set" or ripen the filter to reduce spikes of turbidity at the beginning of a filter run. For DBP standards, the use of granular activated carbon filters may be considered to enhance removal of organic DBP precursors. While the regulatory limit for combined filter effluent turbidity is 0.3 ntu, general industry practice is to maintain filtered water turbidity at less than 0.1 ntu.

Pretreatment

Effective operation of a high-rate granular media filtration system requires pretreating the source water. The nature, as well as the quantity, of suspended material in the pretreated water is critical to filter performance.

Unflocculated water can be difficult to filter regardless of the type of medium in use (Cleasby, 1972; Hsiung, Conley, and Hansen, 1976). However, the work of Robeck, Dostal, and Woodward (1964) with dual-media filters showed that if the applied water is properly coagulated, filtration at rates of 4 or 6 gpm/ft^2 (10 or 15 m/h) produces essentially the same filtered water quality as filtration at a rate of 2 gpm/ft^2 (5 m/h). Subsequent investigations have shown similar results for mixed-media filters (Laughlin and Duvall, 1968; Westerhoff, 1971; Conley, 1972). Recent research has pushed the filtration rates for deep-bed filters to as high as 10 to 12 gpm/ft^2 (24 to 30 m/h); however, polymer filter aids and deeper coarse bed filters were required to achieve these rates and the raw water quality was excellent. Extended pilot studies and regulatory consultation are recommended when filtration rates above 4 gpm/ft^2 (10 m/h) are proposed.

Chemicals used in conjunction with high-rate granular media filtration are limited primarily to metal salts or cationic polymers as primary coagulants. Primary coagulants are ideally fed into rapid mixing basins preceding flocculation. Whether clarification is also required depends on the quantity of suspended solids, metals, and algae in the source water. Primary coagulants are intended to produce agglomerations of natural and chemical solids. Nonionic or anionic polymers are often added with the coagulant as a coagulant aid to assist in strengthening and growth of these agglomerations during flocculation. These same polymers can also be added as a filter aid to the filter influent water or to the washwater to increase the strength of adhesion between media grains and floc in coarse-to-fine filters. Proper pretreatment and mixing is essential to filter performance, especially at higher filtration rates (AWWA Research Foundation, 1992). Pretreatment may also include aeration or introducing an oxidant if an objective of water treatment is to remove iron or manganese.

A filter aid polymer can improve floc capture, provide better filtered water quality, and increase filter runs with higher head loss before turbidity breakthrough. Filter aid polymers are not generally used with fine-to-coarse filters because they promote rapid surface clogging. Filter aids are often fed at low dosages in dilute liquid form to allow dispersion without mechanical agitation just before filtration.

Filter aid polymer doses to gravity filters are usually low (0.02 to 0.05 mg/L). Doses required for pressure filters may be higher because of the higher operating head losses normally employed. Because water viscosity increases with decreasing temperature, breakthrough as a result of floc shearing is more likely at lower water temperatures. Consequently, increased polymer doses and a longer contact time before filtration may be required in cold weather.

Assuming that adequate coagulation is feasible, the designer must decide whether clarification is desirable. In the past, settling has been provided before high-rate granular media filtration when turbidities exceeded roughly 10 ntu (Culp and Culp, 1974). The increased storage capacities of dual- and mixed-media filters have made filtration of water with higher turbidities practicable. The primary advantage of providing direct filtration is the elimination of capital and operating costs associated with clarification. (For the purpose of this chapter, *direct filtration* is defined as high-rate granular media filtration directly following flocculation, without a clarification process.) The higher solids load on the filter will, however, shorten run times and increase the portion of product water required for filter washing. Although the point at which advantages outweigh disadvantages varies with local conditions, a number of investigators have suggested conditions that would justify consideration of direct filtration (Conley, 1965; Cleasby, 1972; Hutchison, 1976; Culp, 1977). Development of packaged or preengineered filtration systems in the 1970s has led to a hybrid process hereafter referred to as *two-stage filtration*. Two-stage filtration combines traditional high-rate granular media filtration (gravity or pressure) preceded by a high-rate clarifier or roughing filter, generally in an upflow configuration. Direct filtration and two-stage filtration are generally employed for higher-quality waters with lower and more consistent turbidity and lower organic content. It is imperative that pilot studies be conducted to determine the feasibility of direct filtration or two-stage filtration for each application.

Filter Media

Although the selection of filter media type and characteristics is the heart of any filtration system, selection is usually based on arbitrary decisions, tradition, or a standard approach. Pilot plant studies using alternative filter media and filtration rates can determine the most effective and efficient media for a particular water.

In drinking water applications in North America, the most commonly used filter media are natural silica sand, garnet sand or ilmenite, crushed anthracite coal, and GAC. Selecting appropriate filter media involves a number of design decisions concerning source water quality, pretreatment, and desired filtered water quality. Filter media cleaning requirements and underdrain system options depend on the filter configuration and filter media selected.

Media variables the designer can control include bed composition, bed depth, grain size distribution, and, to a lesser extent, specific gravity. In addition to media design characteristics, media quality can be controlled to some extent through specifications covering, where applicable, hardness or abrasion resistance, grain shapes, acid solubility, impurities, moisture, adsorptive capacity, manner of shipment, and other such factors. Suggested criteria and a discussion of the applicability of these parameters can be found in the AWWA *Standard for Filtering Material (B100)* and *Standard Granular Activated Carbon (B604)*.

In the United States, granular media have been traditionally described in terms of *effective size* (ES) and *uniformity coefficient* (UC). The ES is that dimension exceeded by all but the finest 10% (by weight) of the representative sample. It is also referred to as the "10% finer" size. The UC is the ratio of the "60% finer" size to the ES. Common practice in Europe is to express media sizes as the upper and lower limits of a range. These limits may be expressed either as linear dimensions or as passing and retaining sieve sizes (that is, 1.0 to 2.0 mm or $-10 + 18$ mesh).

The filtration process also affects the selection of the filter bed because of the special requirements of each type of process. The direct and in-line filtration processes must have filter beds with a large floc holding capacity. A reverse-graded filter bed, such as a dual-media or coarse deep bed, satisfies this requirement. In two-stage filtration, the filter bed of the first stage acts as a roughing filter and carries out the flocculation process. Data obtained from pilot filter tests and actual installations using the two-stage filtration process indicate that the first-stage filter bed may be designed in the same fashion as an ordinary filter (Kawamura, 2000).

Rapid sand filtration, with filtration rates ranging from 2 to 3 gpm/ft^2 (5 to 7.5 m/h), usually uses medium-sized sand (0.5-mm ES). High-rate filters of 5 to 10 gpm/ft^2 (12.5 to 25 m/h) always consist of a reverse-graded filter bed or a deep, large-sized monomedia.

Filter beds may be classified as graded fine-to-coarse, ungraded, graded coarse-to-fine, or uniformly graded, depending on the distribution of grain sizes within the bed during filtration. Transition from the ungraded media of a slow sand filter to the fine-to-coarse high-rate granular media filter resulted from dissatisfaction with the low loading rates and laborious cleaning procedure characteristic of slow sand filters. Filters with uniformly graded or coarse-to-fine beds are now operated at higher filtration rates and for longer run times than are feasible with conventional rapid sand filters.

Ungraded Media. The slow sand filter is a primary example of an ungraded bed. Because slow sand filters are not backwashed, no hydraulic grading of the media occurs. Distribution of the various grain sizes in the bed is essentially random. Typical slow sand filter beds contain 2 to 4 ft (0.6 to 1.2 m) of sand with an ES of 0.15 to 0.35 mm and a UC not exceeding 3.0. Refer to Chapter 9 for further information on slow sand filtration design.

Fine-to-Coarse Media. Fluidization and expansion of rapid sand filter beds during backwashing result in accumulating fine sand grains at the top of the bed and coarse grains at the bottom. Consequently, filtration occurs predominantly in the top few inches, and head loss increases relatively rapidly during operation. This sand medium typically has an ES of 0.35 to 0.60 mm (generally 0.5 mm) and a UC of 1.3 to 1.8. Grains passing a no. 50 sieve (0.3 mm) or captured on a no. 16 sieve (1.18 mm) are normally limited by specifications to very small portions of the medium. Bed depths are typically 24, 30, or 36 in. (0.6, 0.75, or 0.9 m), respectively.

Single-medium anthracite beds have been used in the same basic configuration as rapid rate beds. Because anthracite is more angular than sand, the porosity of an anthracite bed is higher than that of a sand bed containing media with the same ES. The porosity of a sand bed is generally 40% to 45% whereas a typical anthracite bed has a porosity of 50% to 55%. Consequently, anthracite does not perform in exactly the same manner as sand of equivalent size. Because of the lower specific gravity, anthracite beds are also easier to fluidize and expand than sand beds.

Coarse-to-Fine Media. In a coarse-to-fine bed, both small and large grains contribute to the filtering process. The presence of fine media in a filter is desirable because of the relatively large surface area per unit volume that fine media provide for particle adhesion. Fine media are instrumental in achieving the best-quality filtered water. Coarse media, when placed before fine media in the filtering sequence, decrease the rate of head loss buildup and increase available storage capacity in the bed.

Dual-media beds normally contain silica sand and crushed anthracite coal and are a very common filter media design. Triple-media beds contain an additional layer of garnet or ilmenite sand. Specific gravities of materials used in filtration are roughly as follows:

- Silica sand, 2.55 to 2.65
- Anthracite coal, 1.5 to 1.75
- Garnet, 4.0 to 4.3
- Ilmenite, 4.5

A typical dual-media bed contains 6 to 12 in. (0.15 to 0.3 m) of silica sand (ES 0.45 to 0.55 mm) overlaid by 18 to 30 in. (0.46 to 0.76 m) of anthracite (ES 0.8 to 1.2 mm). A typical mixed-media filter bed contains 3 to 4 in. (5 to 10 cm) of garnet (ES 0.15 to 0.35 mm), 6 to 9 in. (0.15 to 0.3 m) of silica sand (ES 0.35 to 0.5 mm), and 18 to 24 in. (0.5 to 0.6 m) of anthracite (ES 0.8 to 1.2 mm).

The degree to which media layers are intermixed in the bed depends on the sizes and shapes of the media used, the nature of the backwashing procedure, and the specific gravities of the different media. Disagreement exists over whether distinct layers or intermixed layers are most desirable. If layers mixed completely, the purpose of using more than one medium would be defeated. If no mixing occurs, individual fine-to-coarse layers would result, and the possibility of rapid clogging at interfaces would be raised.

Proponents contend that in a properly designed mixed-media filter, a gradual decline in pore sizes from top to bottom of the bed is established after backwashing. The original argument can be traced to Conley and Pitman (1960), Conley (1961), and Camp (1961, 1964) in the early 1960s. Brosman and Malina (1972) concluded that a slightly mixed bed was superior to a distinctly layered bed in terms of head loss development, filter run time, and filtered water turbidity. Cleasby and Sejkora (1975), however, disagree that superior performance can be attributed to interfacial intermixing in and of itself; rather, it is a result of differences in the media sizes required to construct mixed and separated beds. They found that to provide a relatively sharp interface in a dual-media bed, fairly coarse sand was required. The resulting bed would not provide the same filtered water quality as a bed using finer sand that mixed more readily with the coal.

The anthracite coal and silica sand used in dual-media filters inevitably result in some intermixing of layers. In a triple-media bed, intermixing of silica sand and garnet sand normally occurs more readily than mixing of silica sand and coal. Cleasby and Woods (1975) suggest that, as a rule of thumb, the ratio of the average particle size of coarse silica grains to the size of coarse garnet grains should not exceed 1.5, to ensure that some garnet remains at the bottom of the bed. They also suggest that a ratio of coarse coal grain size to a fine silica sand grain size of about 3 results in a reasonable degree of mixing in dual- or mixed-media beds. Brosman and Malina (1972) found that anthracite sand filter media with a size ratio at the interface of less than 3:1 exhibits little mixing and that the zone of mixing increases linearly as the size ratio increases above 3:1.

In a number of U.S. installations, taste and odor removal and filtration have been combined in a single unit using GAC (Hager, 1969; Hansen, 1972; Blanck and Sulick, 1975; McCreary and Snoeyink, 1977). GAC is sometimes added to existing rapid sand units from which some sand has been removed. GAC depths of 12 to 48 in. (0.3 to 1.2 m) over silica sand layers of 6 to 18 in. (0.14 to 0.5 m) have been reported. Typically, GAC with an ES of 0.5 to 0.65 mm has been used. This technique is usually applicable only where taste and odor, and not turbidity, are of primary concern. If turbidity levels are high, GAC pores become rapidly plugged, and carbon life is quickly reduced. If turbidity, taste, and odor are all significant problems, GAC beds should be preceded by conventional granular media filtration. If carbon adsorption is desired to remove organics, the depth of GAC that can be provided in a converted gravity filter is likely to be too shallow to provide adequate contact time; however, this may offer some additional carbon removal after conventional sedimentation.

Uniform Media. Uniformly graded deep-bed filters use relatively coarse media, ranging from 0.5 mm to as much as 6.0 mm. The UC is typically 1.2 to 1.3, but values as high as 1.5 may be found. Greater media depth is substituted for the lack of fine media in the bed. Such a substitution requires more vigilant operation of pretreatment systems to avoid breakthrough. Depths of 4 to 6 ft (1.2 to 1.8 m) are common, and in some cases media depths reach 8 ft (2.4 m). Filters of this type are not expanded during backwash, and stratification of grain sizes does not occur. These filters are generally designed to use air or air/water backwash.

There are many possible combinations of filter media size d and depth L. J. M. Montgomery (1985) presents a methodology for determining the optimum relationship between these two variables. The relationship between L and ES d_e (10% finer) of many high-rate filters is shown in Figure 8.1. In this figure, the average ES for a dual- or mixed-media

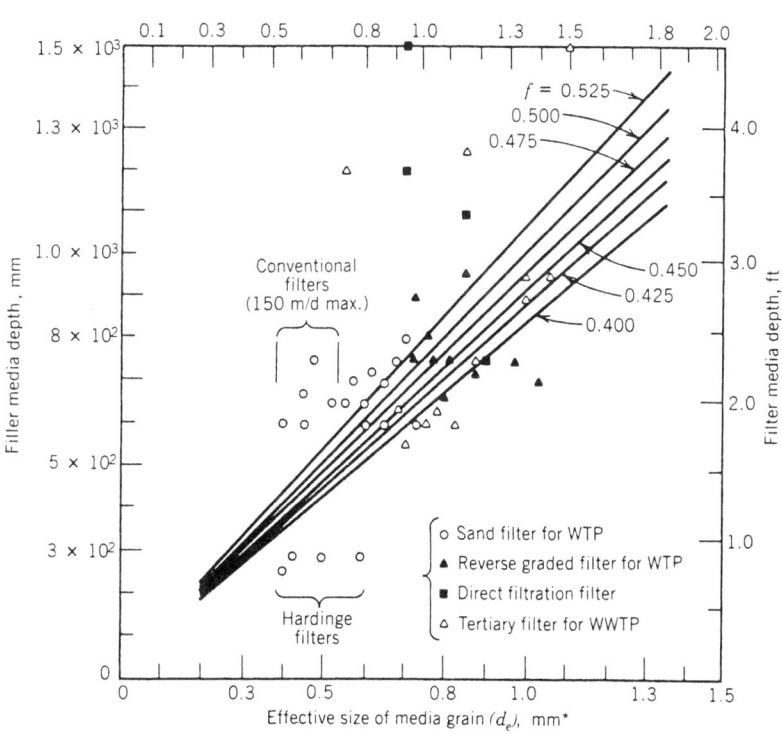

*Weighted average to be used for dual or multimedia bed

FIGURE 8.1 Relationship between depth and size of media. (*Courtesy of MWH Americas Inc.*)

filter was computed as a weighted average. Data in the figure indicate that as the media used become coarser, the required depth is increased, and as the media become finer, the depth required is reduced. Kawamura (1999) uses a length-to-diameter L/d ratio to select the proper depth and size of filter beds (with both L and d measured in millimeters). A value of the L/d ratio should be >1,000 in rapid sand filters, >1,250 in trimedia filters, and >1,300 in most coarse deep beds where d is 1.2 to 1.4 mm and >1,500 is most coarse deep beds where $d > 1.5$ mm.

Filtration Rates

Slow sand filters, designed for filtration rates of 3 to 6 mgd/acre at a rate of 0.05 to 0.10 gpm/ft^2 (0.1 to 0.2 m/h), were initially replaced by rapid sand filters that operated at rates of 1 to 2 gpm/ft^2 (2.4 to 5.0 m/h). The 2 gpm/ft^2 (4.9 m/h) rate became widely accepted as an upper limit in U.S. water supply practice for many years. For the past 30 years, it has been demonstrated that dual-media and mixed-media, as well as single-medium (sand or anthracite), filters can be successfully operated at much higher rates.

A number of investigators found dual- and mixed-media filters to operate successfully at rates from 3 to 8 gpm/ft^2 (8 to 20 m/h) in a variety of locations (Conley, 1961, 1965; Robeck, Dostal, and Woodward, 1964; Dostal and Robeck, 1966; Laughlin and Duvall, 1968; Tuepker and Buescher, 1968; Rimer, 1968; Westerhoff, 1971; Kirchman and Jones, 1972). The quantity of evidence of the practicality of high-rate filtration was such that in 1972 the AWWA Committee on Filtration Problems concluded that it had been amply demonstrated that filters could be designed and operated to produce water of acceptable quality at flows substantially higher than the rate of 2 gpm/ft^2 (5 m/h), once considered the maximum. Over the last 30 years, a number of pilot-scale and full-scale deep-bed uniformly graded anthracite filters have been operated reliably at rates of 10 to 15 gpm/ft^2 (24 to 37 m/h).

Average filtration rates of roughly 2 to 7 gpm/ft^2 (5 to 17 m/h) are reported for the upflow, biflow, and deep-bed filters discussed previously (Hamann and McKinney, 1968; Jung and Savage, 1974).

Logsdon et al. (1993) demonstrated that deep-bed monomedia direct filters could be operated at filtration rates up to 9 gpm/ft^2 on higher-turbidity waters (up to 60 ntu) with proper chemical pretreatment and polymer filter aid selection. Filtration rates are impacted by water temperature. Generally, when water temperatures drop below 45° F (8° C), water quality and filter run length deteriorate in high-rate filters (Kawamura, 1999). Many regulatory agencies will not approve rates in excess of 4 to 5 gpm/ft^2 (10 m/h) without successful pilot-scale testing. The designer should make every effort to obtain approvals for operation at higher rates. The quality of the raw water and extent of pretreatment will play a large role in the acceptable filtration rate.

Filter Operational Control

Filtration process control is critical to successful operation. Decisions regarding control methods must be made early during the design because they affect the physical layout of the filtering facilities and its performance. Filter control may be predicated either on head loss through the filter bed or on the rate of filtration. In either case, smooth transition during changes in filtration rate is highly desirable. The adverse effects of sudden flow surges on filtered water quality have been well documented (Cleasby, Williamson, and Baumann, 1963; Tuepker and Buescher, 1968).

Two basic modes of gravity filter control are commonly found: constant-rate and declining-rate. With the constant-rate mode, there are three ways to operate a filter: (1) a rate-of-flow controller in the filtered water piping; (2) influent flow splitting with the water level over the filter maintained at a constant level; and (3) influent flow splitting with the water level varying during the filter run.

Constant Rate with Rate-of-Flow Controller. With this type of control, water levels in all filters and the filter influent channel are maintained at a constant level. Flow is proportioned equally among the operating filters by means of a flow-measuring device (e.g., venturi or magnetic meter) and modulating valves incorporated in the effluent piping of each filter. Each filter controller receives a signal from the venturi meter and modulates the valves to ensure that each filter is filtering an equal portion of the influent flow.

As the water level in the influent channel rises or falls because of filter media clogging, filters being taken out of service for washing or maintenance, or variations in the plant flow, a level element in the filter influent channel (or settling basin, if employed) signals this movement to the controller, which, in turn, modulates the flow through the other filters. At the start of each filter run, the valve is almost closed to dissipate the surplus head caused by a high water level over the filters. As the filter media clogs, the water level rises and is sensed by the level element, and the controller compensates accordingly by opening the valves to recover an equal amount of head. When one of the filters is taken out of service for washing, the remaining filters in service must pick up additional flow. Operating variation in the water level is usually 6 in. (15 cm). Figure 8.2 shows a typical arrangement of this method of filter control.

Constant Rate with Constant Water Level and Influent Flow Splitting. This method incorporates individual inlet weirs in the influent channel entrance to each filter. Channel and weir lengths should be generously sized to ensure equal flow splitting. A level element in each filter sends a signal to a filter controller to maintain a constant level of water, accomplished by modulating a valve located in the effluent piping of each filter. At the start of the filter run when the head loss through the media is minimal, the valve is almost closed to dissipate surplus head. As head loss in the filter media increases, the water level rises, increasing the driving head. The level element signals this rise to the con-

FIGURE 8.2 Constant-rate filter with rate-of-flow controller. (*Source: Monk, 1987.*)

FIGURE 8.3 Constant-rate filter with influent splitting and constant water level. (*Source: Monk, 1987.*)

troller, which, in turn, commands the valve to open further, reducing head loss across the valve and maintaining a constant water level and flow. Figure 8.3 shows a typical arrangement of this method of filter control.

Constant Rate with Varying Water Levels and Influent Flow Splitting. This method is similar to influent flow splitting with constant level except that there are no level elements, controllers, or modulating valves. The water level at the start of the filter run is just above the top of the bed. Filters of this type normally discharge over an effluent weir, eliminating the possibility of bed dewatering. The relatively high discharge elevation requires an unusually deep filter box to provide filtering head. From 5 to 6 ft (1.5 to 1.8 m) of additional depth is typical. The filtration rate is determined by plant influent flow. The level in each filter rises as necessary to accept an equal portion of influent and indicates head loss. When the level rises to a fixed upper limit, filter washing is initiated. Figure 8.4 shows a typical arrangement of this method of filter control.

Declining-Rate Control

Declining-rate filters are equipped with effluent weirs rather than rate controllers. Flow is distributed on the basis of the relative conditions of the beds. Assuming that influent piping losses are roughly the same for all filters, a uniform operating water level in all filters is achieved. The filtration rate then becomes the highest in the cleanest bed and lowest in the dirtiest bed. In each bed, the filtration rate decreases as solids accumulate. An orifice plate or other simple flow-limiting device is used on each filter effluent line to limit maximum flow rate. To determine which bed is in greatest need of washing, some type of effluent rate indication must be provided. Advantages claimed for declining-rate filters include higher water production for a given run length and improved filtered water quality. A principal disadvantage is that after a filter has been backwashed, it immediately operates at a high rate, which can cause high turbidity to pass through for a period of time. Compliance with individual filter turbidity standards may be troublesome unless filter to waste is employed. Figure 8.5 shows a typical arrangement of this method of filter control.

FIGURE 8.4 Constant-rate filter with influent splitting and varying water level. (*Source: Monk, 1987.*)

Filter Media Washing

As the amount of solids retained in the filter media of a rapid sand filter increases, bed porosity decreases. At the same time, head loss through the bed and shear on captured floc increase. Before the head loss builds to an unacceptable level or turbidity breakthrough occurs, washing is required to clean the bed.

Failure to clean filter media adequately can lead to a multitude of problems. Initially, mudballs form and accumulate in the bed, causing clogging. Then the clogged areas contract as head loss increases. This shrinkage opens cracks in the filter media surface and

FIGURE 8.5 Declining-rate filter. (*Source: Monk, 1987.*)

sometimes at the filter walls. Cracks can cause short-circuiting of the bed during filtration, with subsequent decline in filtered water quality. Clogged areas also contribute to channeling of washwater, which can lead to bed upset. The mechanisms by which washing problems lead to filter failures are discussed in greater detail in Cleasby (1972).

The selection of a washing technique is closely tied to filter media and underdrain selection. In current practice, washing normally includes upflow water flushing. The rate and duration of water flushing are variable, however, and may be supplemented with air scour or surface water wash. Operational sequencing of combined washing systems and the source of the washwater introduce additional variations.

Washwater Source. Washwater source options include the following:

- Flow bled from high-service discharge and used directly for washing or to fill an aboveground washwater tank that is subsequently used for gravity washing
- Gravity flow from a separate elevated finished-water storage tank
- Direct pumping from a sump or belowground clearwell

Bleeding flow from a high-service discharge main results in energy loss because of pressure reduction required before washing to avoid media loss. For direct washing, a pressure-reducing valve or orifice is placed in the washwater supply line. For bleeding flow to fill a washwater tank, an altitude valve or other level control device is used to control the water level in the tank. In either case, the washwater supply line is often sized to restrict the maximum amount of water that can be delivered. Both options avoid provision of separate washwater pumps. Direct washing also avoids construction of a washwater tank but presents greater difficulty in controlling washwater flow. Because of the large pressure drop often involved in supplying washwater by high-service bleeding, the potential for cavitation in or following head-dissipating devices in the supply line is significant.

If elevated finished-water storage is not available to provide head for filter washing, washwater may be pumped to a separate washwater storage tank or directly to the filters. Use of a washwater tank permits pumping at a lower rate. Tank storage volume must be sufficient to permit filter washing at the maximum wash rate while the pump operates at the minimum run times.

A number of proprietary filters are available that obtain washwater by means other than those previously listed. One design uses vertical steel tanks divided into upper and lower compartments. Sufficient filtering head is provided so that following downflow filtration in the lower compartment, filtered water flows through a pipe into the upper tank. When terminal head loss in the filter bed is reached, washwater flows from the upper tank back through the filter.

Some filter control systems permit gravity flow washing of a filter using effluent from the filters remaining in service. Such filters are called *self-backwashing* filters. They do not use pumps or piping for backwashing; instead, all the filters discharge into a common channel. A filtered water weir controls the water level in the channel so that the water level is always higher than that of the filter washwater troughs or side weir. This difference in level must be sufficient to provide the head needed to deliver adequate water for backwashing. Also, there must be a sufficient number of filters in operation to meet the demand for backwash water. Figure 8.6 shows the configuration of a typical self-backwashing filter. To provide the required backwash driving head, the filter box must be substantially deeper than those required for more conventional types of filter backwash systems. But, because no equipment is involved, the capital cost is the lowest. One disadvantage of this design is that there is no way to control the backwash rate.

HIGH-RATE GRANULAR MEDIA FILTRATION

FIGURE 8.6 Self-backwashing filter. (*Source: Monk, 1987.*)

Washing Methods. The AWWA Subcommittee on Backwashing of Granular Filters has acknowledged four basic backwash methods:

- Upflow water wash without auxiliary scour
- Upflow water wash with air scour
- Upflow water wash with surface wash
- Continuous backwash

The application normally dictates the method selected. Filter bed expansion during upflow water washing results in media stratification. Air washing results in filter media mixing. If stratification is desired, air scour must be avoided or must precede fluidization and expansion with water. There are a number of conflicting opinions on the best method of backwash.

Upflow Water Wash without Auxiliary Scour. Upflow water wash alone may be sufficient in some filters receiving low solids loadings. In the absence of auxiliary scour, washing in an expanded bed occurs as a result of the drag forces on the suspended grains. Grain collisions do not contribute significantly to washing (Camp, Graber, and Conklin, 1971; Cleasby, Stangl, and Rice, 1975; Cleasby et al., 1977).

Maximum shear on the grains theoretically occurs (for typical filter sand) at a bed expansion of 80% to 100% (Cleasby et al., 1977). The increase in shear with increasing bed porosity is relatively slight beyond the point at which expansion begins. Optimal expansion may be less than 20% (Johnson and Cleasby, 1966). Normally, when water wash is applied exclusively, an expansion of 20% to 50% is used. Water wash at a sufficient rate to substantially expand (10% or greater) a granular bed is generally referred to as *high-rate water wash*. Water wash incapable of fully fluidizing a bed (i.e., less than 10% expansion) is generally referred to as a *low-rate water wash*.

Experience in the United States with high-rate water wash used alone is extensive. It is generally successful for applications that filter iron precipitates from groundwater or remove color from otherwise high-quality surface water. The relatively weak cleaning action of water wash without auxiliary scour of some type, however, generally renders it unsuitable for filters removing large quantities of suspended solids or for applications where polymers are used. Optimum expansion rate is influenced by the media size and porosity. Kawamura (1999) expresses the optimal expansion rate as

$$\text{Optimal expansion rate} = \frac{(0.1)^{0.22} - f}{1 - (0.1)^{0.22}}$$

$$= \frac{0.6 - f}{0.4}$$

where f = porosity ratio.

High-rate water wash tends to stratify granular media. In multimedia beds, this action is essential and beneficial, but it is not required for uniformly graded single-medium beds. In single-medium beds, high-rate water wash results in movement of the fine grains to the top of the bed, which has a negative effect on head loss and filter run length.

Upflow Water Wash with Air Scour. There are numerous approaches to using auxiliary air scour in backwashing filters. Air scour has been used alone and with low-rate water backwash in an unexpanded bed or slightly expanded bed. Each procedure takes place before either low- or high-rate water wash.

Air scour provides effective cleaning action, especially if used simultaneously with water wash. Cleaning is attributable to high interstitial velocities and abrasion between grains. On the other hand, air wash has substantial potential for media loss and gravel disruption if not properly controlled. Use of air scour can significantly reduce the quantity of water required for backwashing filters.

If more than one filtering medium is used and stratification of the bed is desired, high-rate water wash must follow air scour. In a single-medium bed, if a low-rate wash can adequately remove scoured solids, high-rate wash can be avoided.

If air scour occurs simultaneously with water wash, airflow must usually be stopped before washwater overflow into the washwater collection troughs to prevent media loss. For this reason, the permissible duration of air washing is short unless the concurrent water wash rate is low or the filter box is very deep.

Experience indicates that air scour essentially eliminates mudball formation. Difficulties have arisen, however, from failure to remove scoured solids from filter surfaces. Contributing factors probably include low water-washing rates, long horizontal-travel distances to backwash troughs, and a necessary lag between termination of air scour and initiation of higher-rate water wash.

Air scour complicates and increases the cost of wash systems as air blowers, air piping, and air and water backwash controls are required. Air backwash can also cause upsets to underdrain systems because of the potential to compress air in the underdrains. Pressure relief systems should be considered with air scour.

Upflow Water Wash with Surface Wash. Surface wash systems have been widely used for many years. Fixed systems distribute auxiliary high-pressure washwater from equally spaced nozzles in a pipe grid. Rotary systems have pipe arms that swivel on central bearings. Nozzles are placed on opposite sides of the pipes on either side of the bearing, and the force of the water jets provides the thrust required to rotate the pipe arms.

Rotary systems are more often utilized, and they generally provide better cleaning action, lower water requirements, and less obstruction for filter access. Possible problems with rotating surface wash units include failure to rotate, nozzle clogging, failure to clean in corners, abrasion of concrete walls near the point of closest passage of the arm, and locally high velocities caused when passing under washwater collection troughs. Either type of system may fail to provide auxiliary scour where it is most needed. This can be especially true in multimedia beds if substantial removals are occurring at media interfaces.

Surface wash systems are typically suspended about 2 in. (5 cm) above the surface of the unexpanded filter bed and employed prior to introducing washwater during low-rate backwash. Use of surface wash during high-rate backwash may cause media loss over the waste troughs. Systems have also been placed in the unexpanded bed of dual- and mixed-media filters where there is an accumulation of filtered materials deep within the filter bed. Dual-arm rotary systems that have one arm above and one arm below the unexpanded surface are also available. Nozzle plugging with media has been a problem with the submerged units.

Rotary surface wash systems may have either straight or curved pipe arms and generally have nozzle diameters of $\frac{1}{8}$ to $\frac{1}{4}$ in. (3 to 6 mm). Single-arm rotating units typically operate at 70 to 100 psi (480 to 690 kPa) and at flow rates of 0.5 to 0.7 gpm/ft^2 (1.3 to 1.8 m/h). Standard units are available up to approximately 14 ft (4 m) in diameter. Some models induct air into the washwater jets. A typical surface agitator and arrangements are shown in Figure 8.7.

FIGURE 8.7 Typical surface agitator and arrangements. (*Source. F. B. Leopold Co., Inc.*)

Advantages of auxiliary surface wash include proven effectiveness in alleviating dirty filter problems, improved cleaning (when compared with water wash alone) without a great change in system complexity, and possibly lessened danger of gravel upset if the quantity of washwater introduced through the underdrain is reduced.

Because surface wash systems constitute a possible connection between filtered and unfiltered water, backflow prevention devices must be provided in supply lines.

Continuous Backwash. An alternative to the automatic control of standard filters is the use of continuous backwashing filter beds, which eliminate the need to remove the beds from service for washing. Beds are divided into a series of narrow, contiguous cells, each containing its own underdrain system that allows it to be washed independently from remaining cells. Washing is accomplished by means of a traveling hood suspended above the bed. As the hood travels across the bed, each cell is isolated, and a small backwash pump draws clean water from the filter effluent and reverses the flow through that particular cell. Water is removed by a second washwater pump located in the traveling hood and discharged to waste.

The wash cycle time is controlled by preset adjustable timers to permit optimization of the automatic operation feature. Media depth varies with each application but is typically 30 to 36 in. (0.8 to 0.9 m).

In addition to automatic washing features, these filters have the capability of producing relatively constant washwater flow. In a properly sized system, this constant flow can eliminate the need to provide washwater equalization facilities and can permit direct recycle to the plant headworks. Consult with local regulators as to the feasibility of recycling washwater.

Wash Rates. In the United States, wash rates are expressed as volumetric flow per unit surface area (gpm/ft^2). In Europe, wash rates are expressed as the equivalent water rise velocity (ft/s, ft/min, in./min, mm/s, or m/h). Wash rates are generally variable and depend on washwater temperature, filter media characteristics, and washing method. Water viscosity decreases with increasing temperature. Consequently, as washwater temperature rises, drag forces on media grains are reduced, and higher wash rates are required to achieve bed expansion. Each degree Celsius increase in water temperature requires roughly a 2% increase in wash rate to prevent a reduction in bed expansion. Filter wash systems should be designed for the warmest washwater temperature that will be encountered.

Filter media characteristics also affect washing rate. Rate requirements increase with increasing grain size and density. Also, angular grains are more easily expanded than round grains. In filters using more than one type of filter medium, sizes of each type of medium must be selected carefully to ensure proper positioning after water wash. Recommended size ratios for dual- and mixed-media beds were discussed previously. Figure 8.8 displays the effect of media size on the water wash rate required to achieve 10% bed expansion for three common filter media. Figure 8.9 shows the effect of water temperature on the viscosity of water and on the wash rate for silica sand and anthracite coal. Specific backwash rate curves can be provided by media suppliers. A minimum rate of 15 gpm/ft^2 (37 m/h) is recommended, with typical rates ranging from 15 to 23 gpm/ft^2 (37 to 56 m/h). Rates as low as 10 gpm/ft^2 can be utilized for full-depth GAC or anthracite filters.

Characteristic washing rates and durations vary for each washing method discussed previously. The suitability of a washing method is related to influent water quality, filter media characteristics and bed configuration, and underdrain design. Consequently, not all washing methods are applicable in all cases, and different methods may or may not yield similar results in a particular case.

Upflow Water Wash without Auxiliary Scour. When water wash is used alone, a high-rate wash is employed. Generally a wash rate of 15 to 23 gpm/ft^2 (37 to 56 m/h) is ap-

FIGURE 8.8 Minimum fluidization velocity V_{mf} needed to achieve 10% bed expansion at 25° C. (*Source: Cleasby and Baumann, 1974.*)

FIGURE 8.9 Effect of water temperature on V_{mf} of sand and coal, and on absolute viscosity of water. (*Source: Cleasby and Baumann, 1974.*)

plied. After the water level in the filter has been lowered to the top of the washwater collection trough, the wash starts. It usually lasts from 3 to 15 min. A low-rate water wash is used at the end of the wash cycle in multimedia filter beds to restratify the filter media.

Upflow Water Wash with Air Scour. Three variations of air and water wash were discussed previously. The first—air scour alone followed by low-rate water wash—is commonly applied in Great Britain to single-medium sand filters with 0.6- to 1.2-mm ES media. After the water level in the filter is lowered to below the washwater overflow, air is injected at 1 to 2 ft^3/min/ft^2 (0.3 to 0.6 m^3/min/m^2) for 3 to 5 min. Water wash of 5 to 7.5 gpm/ft^2 (12 to 18 m/h) follows. Bed expansion and stratification are not achieved, although relatively cool water temperatures may result in fluidizing upper sand layers. Problems with gravel disruption have not been experienced if air and water are applied separately (Cleasby et al., 1977).

Air scour alone followed by high-rate water wash can be applied to dual-media or multimedia filters, because bed stratification occurs during water wash. This method has been used in the United States with air scour at 2 to 5 ft^3/min/ft^2 (0.6 to 1.5 m^3/min/m^2) followed by high-rate water wash at 15 to 23 gpm/ft^2 (37 to 56 m/h).

Concurrent air scour and water wash are generally limited to the deep, coarse-grained filters common in Europe. For 1- to 2-mm ES media, air scour rates of 2 to 4 ft^3/min/ft^2 (0.6 to 1.2 m^3/min/m^2) are used with a water flow of 6.3 gpm/ft^2 (15.4 m/h). For 2- to 6-mm ES media, 6 to 8 ft^3/min/ft^2 (1.8 to 2.4 m^3/min/m^2) and 6.3 to 7.5 gpm/ft^2 (15.4 to 18.3 m/h) are used. Concurrent air and water wash typically lasts 5 to 10 min and is followed by water wash alone for another 5 to 10 min. The rate of final water wash is generally 1 to 2 times that used with air scour. In some installations, concurrent air and water wash is used to improve transport of solids to the washwater collection troughs rather than to increase scour in the bed (Harris, 1970).

Upflow Water Wash with Surface Wash. Combined surface and water wash usually involves three phases. After the water surface level is lowered in the bed, surface wash is activated and operated alone for 1 to 3 min. Low-rate water wash is then applied simultaneously for an additional period of roughly 5 to 10 min. Termination of surface wash precedes a final phase (1 to 5 min) during which a higher washwater rate is used to expand the bed 20% to 50%. This usually requires a washwater rate of 15 to 23 gpm/ft^2 (37 to 56 m/h). Washwater flow during surface agitation is usually limited to that required to expand the bed only slightly. If anthracite makes up the top filtering layer, bed expansion above the surface-wash system may be desirable to reduce the likelihood of media loss. Rotary surface-wash systems typically add 0.5 to 0.7 gpm/ft^2 (1.2 to 1.8 m/h) to the washwater flow. Fixed-nozzle systems typically deliver 2 to 4 gpm/ft^2 (5 to 10 m/h).

Filter Arrangements

Filters can be configured in a number of ways in the overall plant layout. It is important to develop a layout that is the least costly and is operationally optimized (Begin and Monk, 1975). Amirtharajah (1982) shows how a minimum-cost filter design is obtained by differentiating the cost function in terms of length-to-width ratio as a variable.

Configuration of Filters. Filters are normally placed next to one another along one or both sides of a pipe gallery. This approach provides the most compact arrangement and also simplifies filter operation and maintenance. If possible, areas for future expansion should be provided at one end of the row (or rows) of filters, and piping in the gallery should be installed with blind flanges at the ends to make future filter additions easier.

HIGH-RATE GRANULAR MEDIA FILTRATION 8.19

Los Angeles (CA) Aqueduct Water Filtration Plant (300 mgd)

Hemlock Water Filtration Plant; Rochester, NY (48 mgd)

Attleboro (MA) Water Treatment Plant (12 mgd)

FIGURE 8.10 Typical filter configurations. (*Courtesy of CDM.*)

In larger plants, placing filters in rows on opposite sides of a pipe gallery is common practice. In smaller plants, a single row of filters results in simpler construction. Typical filter configurations are shown in Figure 8.10.

Location of the filtered water clearwell under the pipe gallery, and common walls between filtered and unfiltered water, should be avoided to prevent the possibility that leakage through the walls may contaminate filtered water. A false floor should be provided in the washwater gullet to prevent a common wall from existing between unfiltered water in the gullet and filtered water in the underdrain system. Drainage should be provided for spaces beneath false floors of this type.

Many designers favor construction of the conduits connecting pretreatment basins and filters in such a manner that floc destruction because of turbulence and high velocities is minimized. Drops, bends, and long runs should be minimized or avoided. The need to avoid turbulence, however, is not universally accepted (Cleasby, 1972). Low velocities may be dictated by head loss between basins and filters.

In warm climates, filters may be placed outdoors with precautions for controlling algae formation in the filter box. One method is to provide for shock chlorination of filter influent or washwater. In colder climates, filters are normally housed to prevent ice formation. Filter control consoles are usually provided adjacent to each filter or each set of filters to allow for local control of the backwash operation. Consoles should be provided with climate-controlled housing.

Number of Filters. From a cost point of view, one filter is the most ideal. Practically, however, four filters are the minimum number that should be used to allow for filter washing and the occasional need for a filter to be out of service for maintenance, without resulting in unreasonable rate increases in the other filters. If self-backwashing filters are used, allowing for the lowest rate of operation and for one filter out for maintenance, the remaining filters must be able to produce sufficient water to wash one filter effectively. Kawamura (1999) recommends the following formula as a guide to determining the required number of filters:

$$N = 1.2Q^{0.5}$$

where N is the number of filters and Q is the plant design flow rate in million gallons per day.

Size of Filters. The size of individual gravity filters is determined by plant capacity, filtration rate, and the number of filters desired. Hydraulic considerations and the effect of removing a filter from service limit maximum filter size. Additional considerations include the maximum area to which washwater or air scour can be evenly distributed, the maximum span length of washwater collection troughs, and available sizes of surface wash equipment, if used. Single gravity filters of up to 4,500 ft^2 (420 m^2) have been reported (Clark, Viesman, and Hammer, 1977), but units less than one-half this size are more typical, even in large plants. Large filters may be divided into two sections by using a central gullet, permitting one-half of the filter to be washed at a time, although influent and effluent piping is usually shared. In general, the practical maximum size of a filter is 1,000 ft^2 (90 m^2) provided the plant is not exceptionally large (Kawamura, 1999).

Pressure filters are usually limited by shipping constraints. The largest standard units typically available are 12-ft-diameter (3.7-m) tanks. This limits vertical filters to about 113 ft^2 (10 m^2) of filter medium. Horizontal filters are normally not longer than 40 ft (12 m). Larger units of both types can be specially fabricated on site.

Capital cost of filters can generally be minimized by designing for the minimum number of filters consistent with size limitations. Plant expansions are usually accomplished by adding filters of the same size as existing units, because installing larger filters may require extensive changes in the filter wash system.

Depth of Filter Box. A number of factors should be considered in designing a filter box, including

- Available head
- Depth of water over the filter media surface
- Fixed head losses
- Head losses through the filter media
- Shape of head loss curves
- Rate of filtration
- Elevation of the filtered water effluent control weir (if used)

Available head may be restricted by site conditions and plant layout, by the designed maximum length of filter run, by optimizing head loss, or by terminating turbidity breakthrough. Ideally, allowable head loss is determined from pilot plant studies. Fixed head losses through piping, venturi meters, and throttling valves can be calculated from manufacturers' literature and by hydraulic analyses. Head losses through filter media are determined by pilot plant studies or are calculated from Darcy's formula.

Once these factors are known, water depth and weir elevation can be established to achieve a cost-effective design. Objectives in designing a filter box should be to minimize the cost of construction and to avoid the possibility that the filter will develop negative pressures. Monk (1984) presents the methodology required to determine optimum filter box depth.

Underdrain Systems

An underdrain system has two purposes: to collect water that passes through the filter media and to distribute washwater (and air, if used) uniformly across the filter bed. Support gravel is required when openings in the underdrain system are larger than the filter medium directly above it. Although the support gravel or other support method does not contribute to particulate matter removal, it aids in distributing washwater. For this reason, it should be considered part of the underdrain system. Uneven distribution of washwater can displace support gravel, eventually requiring removal of the filter media to be regraded or replaced.

Four basic types of underdrain systems are common: pipe laterals, blocks, false bottom, and porous bottom.

Traditionally, the greatest difficulty in underdrain design has been in providing a barrier to the finest medium that does not clog during filtration or filter washing. Early attempts to use fine screens or strainers were largely unsuccessful, leading to the use of gravel layers below filter sand. The position of gravel layers may, however, be disrupted during filter washing. Jet action, which is discussed in greater detail elsewhere (Cleasby, 1972), causes sand and gravel mixtures to be more easily disrupted than gravel alone. If auxiliary air scour is used, even greater gravel disturbance may occur. Fine gravel, usually placed at the sand-gravel interface, is most easily dislocated. A possible solution to this problem is the use of gravel in a coarse-to-fine-to-coarse, or "hourglass," gradation, which has been shown to be highly stable at high washwater rates (Cleasby, 1972). Fine media penetrate the upper coarse gravel layer without apparent ill effect.

Mixed-media beds with very fine garnet at the bottom of the bed are generally constructed with a layer of coarse garnet on top of the silica support gravel. Coarse garnet prevents leakage of the fine garnet and also helps stabilize the underlying silica gravel.

European-type deep-bed filters use relatively coarse and uniformly graded media. As a result, bed stratification is not required, and air scour presents less of a hazard to proper bed operation. Also, the use of strainers is more likely to be feasible because of the larger permissible openings. Consequently, false-bottom underdrains with nozzles designed for both air and water distribution and without support gravel are commonly used in deep-bed filters.

Pipe Laterals. Pipe lateral underdrains were once popular because of their relatively low cost and adaptability for use in pressure filters. Problems with relatively high head loss and poor washwater distribution resulted in a general decline in their use. They are still encountered, however, when older filters are upgraded.

Pipe underdrain systems generally consist of a centrally located manifold pipe to which smaller, equally spaced laterals are attached. Lateral pipes usually have one or two rows of $\frac{1}{4}$- to $\frac{3}{4}$-in.-diameter (6- to 19-mm) perforations on their bottom sides. The lateral pipes may be fitted with nozzles as illustrated in Figure 8.11. Guidelines for lateral design include the following ratios:

FIGURE 8.11 Pipe lateral underdrain with nozzles. (*Courtesy of Paterson Candy Ltd.*)

- Total area of orifices (surface area of bed): 0.0015 to 0.005:1
- Cross-sectional area of lateral (total area of orifices served): 2 to 4:1
- Cross-sectional area of manifold (total area of laterals served): 1.5 to 3:1

Orifices are normally spaced at 3 to 12 in. (8 to 30 cm) and laterals at roughly the same spacings as the orifices. Approximately 18 in. (45 cm) of support gravel is required to cover a lateral network. Three to five graded layers are usually involved, with sizes varying from 1½ to ⅛ in. (38 to 3 mm). The bottom layer should extend 4 in. (10 cm) above the highest washwater outlet.

Blocks. A commonly used block underdrain consists of vitrified clay blocks with ¼-in.-diameter (6-mm) dispersion orifices located across the top of each block. Support gravel is required with this type of underdrain system. The size and arrangement of these blocks and typical support gravel layers are shown in Figure 8.12. In mixed-media applications, the third gravel layer is replaced by garnet of similar size. This type of underdrain system is suitable only for water washing. However, auxiliary air scour may be provided by adding an air piping grid at the sand-gravel interface, as shown in Figure 8.13.

Another type of block underdrain is designed for concurrent air/water wash. Blocks are constructed of polyethylene and consist of a primary feeder lateral (lower) and a secondary compensating lateral (upper), shown in Figure 8.14. Small control orifices open from the feeder lateral directly into the compensating lateral. Washwater and air are admitted to and flow through the feeder lateral and rise to discharge from the control orifices into the compensating lateral. The triangular shape of the primary lateral distributes incoming washwater and air uniformly along its length. Support gravel is typically used with this type of underdrain, graded in an hourglass configuration. As a replacement for

FIGURE 8.12 Typical gravity filter. (*Source: F. B. Leopold Co., Inc.*)

8.24 CHAPTER EIGHT

FIGURE 8.13 Vitrified clay tile underdrain with auxiliary air scouring system. (*Source: Roberts Water Technologies, Inc.*)

FIGURE 8.14 Plastic block underdrain designed for use with air/water wash. (*Source: F. B. Leopold Co., Inc.*)

support gravel, an integral media support (IMS) cap made of plastic beads sintered together may be installed on top of the plastic block underdrain, as shown in Figure 8.15. Water quality should be reviewed closely prior to use of an IMS cap.

False Bottoms. One of the most widely used false-bottom underdrains is constructed of precast or cast-in-place reinforced concrete supported on concrete sills. This underdrain system contains uniformly spaced inverted pyramidal depressions. Unglazed porcelain spheres are placed in the depressions to distribute flow. Each depression is filled and leveled with 1- to 1½-in. (25- to 38-mm) gravel before placement of overlying gravel support layers. A typical arrangement including the gravel layers is shown in Figure 8.16. The last silica gravel layer should be replaced by coarse garnet in a mixed-media filter bed.

Other false-bottom underdrains have impervious bottoms penetrated by nozzles. Nozzle-type underdrains are used primarily in filters employing air/water wash systems. Fine openings in the nozzles eliminate the need for support gravel, which, accordingly, reduces

FIGURE 8.15 Integral media support (IMS) cap designed for use without support gravel. (*Source: F. B. Leopold Co., Inc.*)

filter box depth. Other varieties of false-bottom underdrains are constructed of concrete, polyethylene, or tile blocks; monolithic concrete; or steel plates. Nozzles are equipped with plunge pipes for air wash and are usually constructed of stainless steel, plastic, or brass. Plunge pipes usually adjust to allow for leveling after installation. Nozzle orifices are sometimes smaller on the filter side to prevent clogging during filtration, even though such a configuration can contribute to clogging during filter washing. Some failures resulting from plugging and breakage have been experienced with this type of underdrain. A typical false-bottom underdrain using monolithic concrete is shown in Figure 8.17.

Porous-Bottom Underdrains. Porous-bottom underdrains constructed of porous aluminum oxide plates have been used in both block and false-bottom configurations. They

HIGH-RATE GRANULAR MEDIA FILTRATION 8.27

FIGURE 8.16 Concrete wheeler bottom with gravel support. (*Source: Roberts Water Technologies, Inc.*)

FIGURE 8.17 Underdrain layout. (*Courtesy of Ondeo Degremont.*)

are constructed of plates mounted on concrete or steel piers or on clay tile saddles to form blocks. Very small pore sizes make porous-bottom underdrains susceptible to plugging and therefore unsuitable for use in softening or iron and manganese removal plants or other plants where plugging by chemical deposition may occur. They may also clog with rust or debris during backwashing.

Additional problems that may occur include breakage because of the brittle nature of the porous material and failure of caulked joints between plates. Porous bottoms have been used successfully in a few locations, but they are less widely accepted than the block or false-bottom underdrain systems discussed previously.

Filter Performance Monitoring

Filter design should include instrumentation to monitor filtered water turbidity (individual and combined), filtration rate, head loss, backwash rate, and filter run length. If auxiliary air scour backwash is used, airflow monitoring should also be included. Particle counting may also be afforded for an additional operational tool. Pilot filters may help determine primary coagulant, coagulant aid, and filter aid dosage.

Turbidity. Turbidity is the most common measure of filter performance. Turbidity is an indication of the amount of suspended material in a water sample. The nephelometric technique, the current standard in the United States, compares the intensity of light scattered by a water sample with a standard reference suspension under the same conditions. Turbidimeters can be either discrete sample or on-line type. On-line turbidimeters include a light source to illuminate the water sample, one or more photoelectric detectors, and an analog indicator of turbidity based on the intensity of light scattered 90° to the path of the incident light. Turbidity is reported as *nephelometric turbidity units* (*ntu*).

Each filter's effluent turbidity should be monitored and recorded continuously (maximum 15-min intervals) using an on-line turbidimeter to detect variances from normal operation immediately. Observations of the effects of fluctuations in source water quality, rate changes, equipment malfunctions, chemical feed variations, filter washing, and other such occurrences contribute to the operator's understanding of the plant's performance and increase the ability to deal with such situations.

Turbidity measurement is sometimes used to automatically initiate a filter wash cycle or to actuate an alarm whenever the filter effluent reaches a preset maximum turbidity level. Most water treatment plants establish a filtered water turbidity goal (say, 0.1 ntu) well below current regulatory standards.

Turbidity of the waste washwater and filter-to-waste water can also be monitored to assess the performance of filter washing and ripening, respectively. A different type of turbidimeter, capable of measuring higher turbidity levels, can monitor waste washwater.

Sample piping should be designed so that it does not collect air bubbles that can distort readings (Letterman, 1994). Air bubble traps, available from most turbidimeter suppliers, can be used for this purpose. On-line turbidimeters should be located where they are readily accessible because they require periodic cleaning and calibration to ensure accurate readings.

Additional information on turbidity measurement and methods can be found in the USEPA *Guidance Manual for Compliance with the Interim Enhanced Surface Water Treatment Rule Turbidity Provisions* (1999).

Particle Counting. Particle counting is rapidly gaining acceptance to monitor filter performance (Lewis, Hargesheimer, and Ventsch, 1992). Particle counters are instruments that can quantify and size particles in water by light-scattering techniques and can be either discrete sample or on-line type. Particles ranging in size from approximately 1 to 500 μm can be quantified by particle counters. Particle counters can provide a direct measurement of the number of particles in a particular size range and serve to verify the log removal of particles achieved.

Particle counters are particularly useful in determining the log removal of particles in the *Giardia* and *Cryptosporidium* size ranges. *Giardia* cysts generally fall in the 5- to 15-μm range, and *Cryptosporidium* oocysts fall in the 4- to 7-μm range; however, the correlation between particle counts and *Giardia* and *Cryptosporidium* concentration has not been fully established. Filtration theory indicates that 1- to 3-μm particles are the most difficult to remove, but Moran et al. (1993) found earliest breakthrough of 3- to 7-μm particles in test filters. The 3- to 7-μm size range includes cysts and oocysts currently of primary concern.

Current standards require a 2- to 2.5-log removal of *Giardia* cysts and a 2-log removal of *Cryptosporidium* oocysts by a filtration process, depending on the type of filtration. Future regulations may increase the *Cryptosporidium* oocyst log removal requirement. For very clean source waters, particle counters may not demonstrate proper log removals because of the low required filtered water particle counts. Even in these situations, fluctuations in filtered water particle counts can still give an excellent indication of filter performance.

Engineers and plant operators should be aware that at present no standard method of analysis has been established, so particle counts can vary from instrument to instrument. Sensors associated with particle counters may have different operating principles such as particle size detection limit, particle concentration range, sample flow rate, or pressure requirements. Bends and flow-altering devices should be avoided upstream of particle counters because they can alter performance. Chemical and physical cleaning of particle counters is required and should be considered in their installation. The operator should be aware that stable particle counts are difficult to obtain when samples are switched periodically between filter effluent lines.

Filtration Rate. Flow-measuring devices are recommended for monitoring flow through individual filters. Filtration rates can be monitored and controlled by comparing the metered flow rate through a filter to its surface area. Meters can be combined with modulating valves to automatically control filtration rates. The most common type of measuring device is a venturi tube because it can be easily checked in the field with manometers. It is often impossible to provide the ideal length of pipe preceding the flow-measuring device, particularly when it is mounted in the filter's effluent piping. However, this is usually not a concern if the total plant flow is being split equally between all operating filters. Local and remote indicating and recording devices are usually included. Orifice plates can also be placed in filter effluent lines to limit maximum filtration rates. If flow splitting by weirs or other such device is employed, the filtration rate may be determined by dividing the total plant flow by the number of operating filters.

Head Loss. Head loss in a filter bed is a valuable indicator of filter bed condition and may be used to automatically activate filter washing. Head loss through the filter media is normally monitored by differential pressure-cell devices that measure the water pressure above and below the filter media.

Aside from head loss developed by particle retention within the filter media, operating head loss depends on the filtration rate, the clean bed head loss through the filter media, and head losses through the filter underdrain system and the effluent rate controller. *Terminal head loss* is the difference between the static head "available" between the water elevations in the filters and the filtered water effluent control weir, less the operating head losses through the clean media, underdrain system, effluent piping connections and bends, and the effluent rate controller. Modern plants typically have a terminal head loss of 8 to 10 ft (2.4 to 3 m), while many older plants have significantly less. Clean bed head losses range from 1 to 2 ft (0.3 to 0.6 m) depending on media specifications and filtration rate.

Filters should be washed when terminal head loss is reached; otherwise, turbidity breakthrough may occur. Also, a vacuum can result if head loss at any level in the filter bed exceeds static head. This situation is referred to as *negative head* and can cause air binding of the filter media. When pressure in the filter bed drops below atmospheric levels, dissolved gases are released from the water being filtered. Gas bubbles trapped in the bed further increase head loss and aggravate the problem. They may also result in media displacement during filter washing.

This problem is particularly acute when filtering is done with insufficient water depth over the media or when surface waters are saturated with atmospheric gases because of rising temperatures in the spring. Remedies for air binding in gravity filters include increasing washing frequency, maintaining adequate static head above the media surface, and keeping the clearwell water level above the top of the filter media to keep it submerged. Pressure filters normally discharge well above atmospheric pressure and are not subject to air binding.

The head loss sensor connection to the filter box should be located approximately 4 in. (10 cm) above the top of the washwater collection trough to prevent washwater from entering the sensor. A sediment trap with drain installed on the sensor line will capture any sediment that may enter the line. The end of the sensor should be turned up, keeping a full column of water in the line at all times to minimize air entrainment. A fine-mesh stainless steel screen installed on the end of the sensor will prevent clogging with filter media. However, this screen requires periodic cleaning to prevent a buildup of material that may cause a false head loss reading.

Another valuable monitoring method is to measure head loss at points within the filter bed by installing several pressure taps at various depths of the filter bed. These pressure taps can be connected to transparent tubes, creating a piezometer board. The pressure taps can be monitored and recorded continuously for better observation and control

(Monk and Gagnon, 1985). Although this unit is a good performance monitoring tool, it is not necessary to install one on each filter.

Filter Wash Rate. Because filter wash flow requirements may vary with the seasons owing to differences in water temperature and pretreated water quality, operator knowledge of the filter washwater rate in use at a particular time is essential. Flow tubes are usually employed as the monitoring device and are usually matched with a downstream flow control device to control the washing rate. Recording meters are generally not necessary, but a totalizing device is desirable to determine the overall volume of water used in washing. An alarm can be provided that is actuated if the wash rate exceeds a predetermined maximum.

Length of Filter Run. Filter performance is often judged by the length of the filter run, but too long a filter run may not be good for filter operation. Long filter runs make washing a filter much more difficult because of particulate matter compaction in the filter media. In addition, a long filter run indicates that the filter is not working at its most cost-effective capacity.

Rapid sand filters are generally operated with run lengths between 12 and 72 h, typically with 24-h runs. Pressure filters may have somewhat longer filter runs than gravity filters if they can be operated at higher head losses without turbidity breakthrough.

Pilot Filters. Pilot filters are bench-scale models of full-scale plant filters that can be used to determine optimum coagulant dosage. Coagulated water is diverted to pilot filters from the full-scale pretreatment units. Monitoring the pilot filter effluent turbidity provides an indication of the adequacy of coagulant feed. Pilot filters greatly reduce the lag time in coagulant feed system adjustment and improve plant performance. Parallel pilot filters are usually provided to ensure continuous control. Because of the nature of connecting piping to pilot filters, higher filter aid polymer dosages are usually required than in the full-scale plant. The effect of increasing the polymer dose is to shorten filter run times. Consequently, pilot filters are generally not used to predict run lengths or polymer dosage. In most cases, however, these variations do not affect determination of optimum coagulant dosage.

Filter Operation and Control

Several types of filtration operation and control are used. Many plants have filter control consoles located immediately adjacent to the filters they serve to observe filters for malfunction during backwashing. Remote operation from the plant's central control console is also practiced in some plants. This allows a single operator to wash filters and still observe other plant processes.

In the past, all major valves were controlled by individual manual controls, and all filter operations were operator-directed. However, advances in sensing and control equipment have made the use of remote automatic or semiautomatic control commonplace. In semiautomatic operation, filter washing is initiated by the operator but consists of a predetermined sequence that requires no additional attention.

Fully automatic filters are washed without operator input on the basis of loss of head in the filter bed, filtered water turbidity, or a fixed maximum run time. Automatic systems permit operation of all filters from a central location, reducing personnel requirements. However, remote automatic operation may not permit the operator to observe the wash cycle directly. Automatic systems should allow easy modification of the washwater and auxiliary scour rates and sequence by the operator. Process instrumentation for monitoring and alarming to indicate when problems arise is essential for all modes of operation.

Auxiliaries

Auxiliary equipment that the engineer must consider in filter design includes piping and conduits supplying and removing water from the filter, valves and actuators used to control filter operations, and washwater troughs.

Conduits. Filters require hydraulic connections for influent water, filtered water, washwater supply, auxiliary scour, washwater drain, and (if used) a connection to allow filtered water to be wasted. Influent water is usually delivered to a gravity filter through the washwater gullet. Influent to a pressure filter is generally distributed by a tapped pipe serving as a manifold or by a baffle plate. Influent conduits should be designed to deliver water to the filters with as little disturbance as possible. Free fall or turbulence, which can disturb the filter media, is undesirable. Delivering influent beneath the water surface in the filter or baffling the incoming stream prevents media disturbance. Necessary measures depend on the control strategy used.

Typical piping serving a gravity filter is shown in Figure 8.18. Influent conduits should be sized to limit velocities to 2 ft/s (0.6 m/s). This may result in the use of an influent flume rather than a pipe in large plants. Hydraulic considerations generally result in velocities of 3 to 6 ft/s (0.9 to 1.8 m/s) in washwater and filtered water piping. At higher velocities, head losses often become excessive, and undesirable effects such as water hammer are more likely to occur. Cement-lined ductile iron or steel pipe or stainless steel pipe is commonly used for filter piping. Flanged, grooved-end, or mechanical joints and connections should be used for ease of maintenance.

Design considerations include avoiding high points and including air release valves in washwater supply piping, ensuring adequate backpressure within the throat of the filtered water venturi meter, and providing accessible piping to permit proper maintenance.

Valves and Actuators. A typical filter is equipped with at least five valves: influent, filtered water, washwater supply, washwater drain, and surface wash or air wash supply. A filter-to-waste valve may also be included; however, waste lines constitute a potential cross-connection and must be equipped with air gap protection against backflow from the drain to the filter.

Rubber-seated butterfly valves are most common in filter pipe galleries because of their relatively short laying length. In larger plants, the influent and washwater drain may be gates rather than valves. These valves and gates can be operated either manually or automatically from a local filter operating console or from a remote central control console. The filter rate control valve and its actuator must provide for stable and accurate flow control. It typically should operate between 0° and 45° open. It should also automatically close during power failure to prevent dewatering the filter media. Valve placement should permit easy access for maintenance, with valves and gates equipped with easily distinguishable position indicators.

Valve actuating systems may be hydraulic, pneumatic, or electric. Hydraulic systems were developed first but are generally no longer installed in new plants because of problems with leakage and with plugging of orifices in the lines by deposition from the fluid. Pneumatic or electrical systems are used in most new construction. Pneumatic systems are generally less expensive, but they require oil- and moisture-free air. Electrical systems offer greater reliability, but initial cost is usually higher. In the event that maintenance is required, electric controls may require greater technical skill to service than pneumatic controls. Flow control valves must be carefully designed to provide required accuracy and to avoid cavitation.

Washwater Troughs. In the United States, washwater troughs are suspended at even spacings above gravity filter beds to provide uniform removal of washwater during back-

FIGURE 8.18 Typical gravity filter piping. (*Courtesy of CDM.*)

washing. These same troughs also normally distribute influent flow uniformly across the filter media's top surface. This limits horizontal travel required and equalizes static head on the underdrain system. In contrast, European designs often feature narrow beds with overflow walls on one or both sides, but not suspended over the media. In these designs, tilting side weirs, horizontal water jets, and a procedure allowing influent water to enter the filter on the side opposite the overflow wall are sometimes used to aid the movement of scoured solids to waste. This method is termed *cross-wash*. Spacing troughs in U.S. practice are usually at 5- to 7-ft (1.5- to 2-m) centers to limit horizontal travel distances to 2.5 to 3.5 ft (0.8 to 1 m).

If troughs are placed too close to the surface of the unexpanded bed, media may be lost during backwash. The design elevation of the weir edge of the trough may be determined by adding the depth required for maximum bed expansion (usually 50%) and the overall depth of the trough, plus a small margin of safety of 6 to 12 in. (0.15 to 0.3 m). If air scour is practiced, additional care must be taken. In a conventional trough, simultaneous use of washwater and air must be stopped when the level of water rises to the bottom of the trough, or media will be lost. In a limited number of plants, special baffled plates on either side of the trough minimize the loss of media during concurrent water/air wash.

French (1981) has given the following criterion for the location of the top of the troughs based on trough spacing:

$$H = 0.34S$$

where H = height of the top edge of the trough above the fluidized bed and S = center-to-center spacing of the troughs. Troughs are usually made of fiberglass-reinforced plastic (FRP), stainless steel, or reinforced concrete. Troughs made of FRP and stainless steel usually have semicircular bottoms, and concrete troughs have V-shaped bottoms. Trough bottoms should not be flat because froth and sludge tend to accumulate and fall back onto the surface of filter media. Typical trough cross sections, including two with special baffled plates for concurrent air/water wash, are shown in Figure 8.19.

a. Fiberglass-reinforced plastic trough.   b. Reinforced concrete trough.

c. Engineered Stainless Steel Design (ESSD™) wash trough with Type II MULTIWASH^R separator baffles. *(Source: General Filter Company.)*

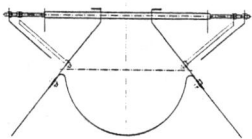

d. ScourGuard^R baffled wash trough. *(Courtesy of Eimco Water Technologies.)*

FIGURE 8.19 Typical washwater trough cross sections.

The required cross-sectional area of the trough for a given washwater flow can be estimated from Figure 8.20. A more rigorous analysis can be obtained by referring to derivations provided by Fair, Geyer, and Okun (1968) and Brater and King (1976).

After troughs are installed in a filter, weir edges must be leveled to uniformly match a still water surface at the desired overflow elevation. It is critical that troughs be properly supported both vertically and horizontally so that their weirs remain absolutely level

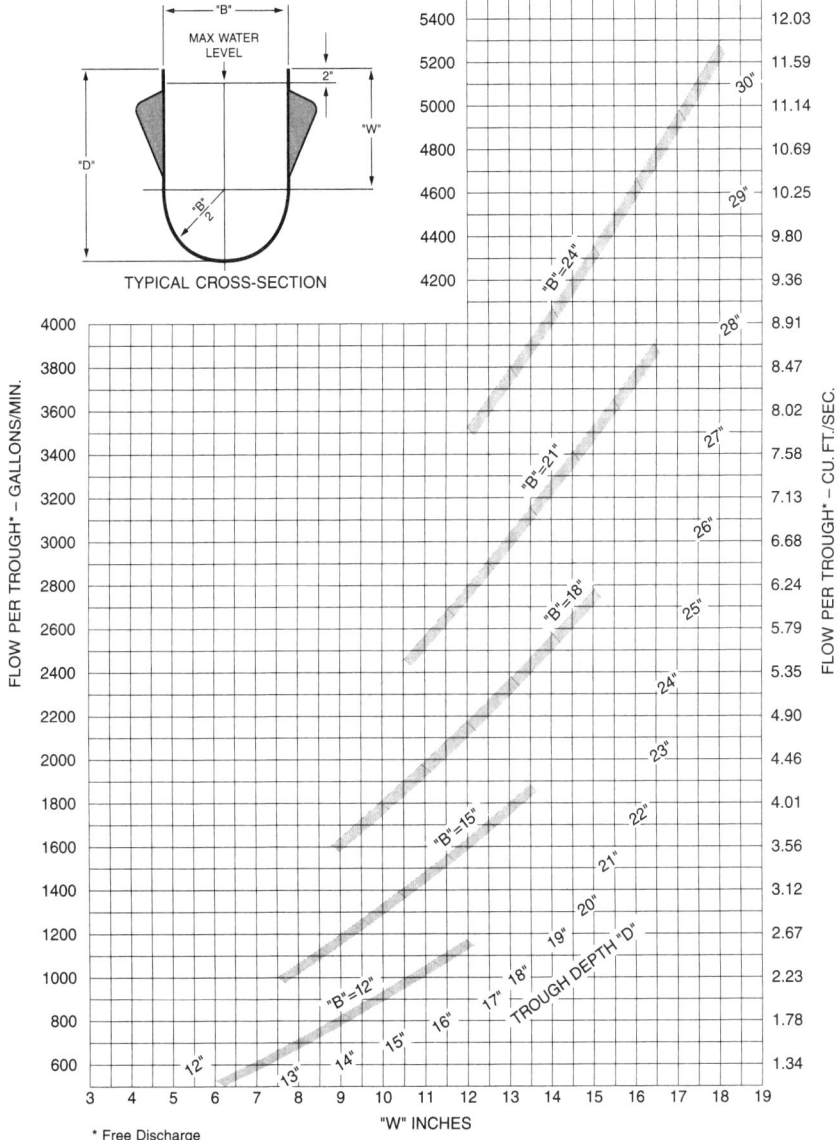

FIGURE 8.20 Washwater trough-sizing diagram. (*Source: F. B. Leopold Co., Inc.*)

during backwashing. Center supports are typically used whenever trough length exceeds 14 ft (4.3 m).

FILTER DESIGN CRITERIA

The first step for the design engineer when designing filtration facilities should be to review all current federal, state, and local laws and regulations that may be applicable. The second step should be to review all applicable standards prepared by various organizations and associations.

Standards Set by Regulatory Agencies

The first design criterion to be considered for a filtration process is its ability to meet applicable water quality standards. The U.S. Environmental Protection Agency (USEPA) has established the Surface Water Treatment Rule (SWTR) and the Enhanced Surface Water Treatment Rules (ESWTR) as the controlling standards for filtration of surface water and groundwater under the direct influence of surface water. The SWTR and ESWTR require a public water supplier using a surface water or groundwater source deemed "under the influence of surface water" to achieve 3-log (99.9%) removal and/or inactivation of *Giardia*, 4-log (99.99%) removal and/or inactivation of viruses, and 2-log (99%) removal of *Cryptosporidium*.

Credit for log removal is given to filtration processes based on their type, with the remaining required log removal to be achieved by disinfection. Conventional filtration is usually given a 2.5-log credit for *Giardia* removal; direct, slow sand, and diatomaceous earth filtrations are usually given a 2-log credit; and all filtration processes are usually given a 2-log credit for *Cryptosporidium* removal, provided turbidity limits are met.

Maximum turbidity levels are established for filtration performance standards. For conventional and direct filtration, filtered water turbidity must be less than 0.3 ntu in 95% of samples collected in a month. For slow sand and diatomaceous earth filtration, less than 1.0 ntu must be achieved in 95% of the samples. A level of 1 ntu must not be exceeded at any time in conventional and direct filtration, and 5 ntu for slow sand and diatomaceous earth filtration. These water quality limits must be considered in designing filtration facilities, as well as design of the entire treatment train.

In addition to drinking water quality standards, most states have adopted minimum standards for designing filtration facilities, either with specific criteria or reference to other design standards. The most widely known and adopted standard is *Recommended Standards for Water Works,* commonly known as the "Ten States Standards" (Great Lakes, 2003). This standard contains specific criteria on filter size, arrangement, allowable filtration rates, structural details, hydraulics, materials of construction, filter washing, and control systems. In some instances, states require pilot testing of treatment processes to establish their validity and design criteria.

Standards that limit the recycle of waste washwater to the plant headworks and discharge to surface water sources have also been established by the USEPA. Recycle should generally be provided with some minimum level of pretreatment and be returned ahead of the introduction of coagulants. Waste discharges to a waterway usually require the implementation of a federal National Pollutant Discharge Elimination System (NPDES) permit. If a sanitary sewer is available, local sewer use ordinances (based on wastewater treatment plant discharge limits set by an NPDES permit) usually require pretreatment of the waste washwater. State or local regulations may also require recycling of waste washwater within the water treatment plant. State regulatory officials should be consulted during the planning phase to determine recycle and/or discharge requirements for waste washwater. The Filter Backwash Rule places specific requirements on the recycle of backwash water within the plant.

Standards Recommended by Others

The American Water Works Association (AWWA) has published various manuals and handbooks that provide useful and up-to-date information for planning and designing filtration facilities. The AWWA Research Foundation has published many research reports and design manuals related to filtration, generally specific to a particular type of filtration or new advances in filtration technology.

The AWWA standards for filtering materials are a particularly useful tool because they establish minimum quality standards for the purchase and installation of filter media of various types.

NSF International has established a certification program for all direct and indirect drinking water additives. Standard 60 relates to drinking water treatment chemicals, such as coagulants and coagulant aids. Standard 61 relates to drinking water system components to ensure that these additives are compatible with drinking water systems, including filter media and coatings and linings for valves and piping systems.

Standards such as building, fire, and plumbing codes and other local ordinances must be considered in the design of filters and their housing. However, most codes and ordinances do not specifically mention water treatment plants, so local building code and fire department officials should be consulted during the planning phase to obtain clarification of their specific requirements. Occupational Safety and Health Administration (OSHA) standards for worker safety also must be considered.

FILTER OPERATION AND MAINTENANCE

When designing filtration equipment, the engineer must keep in mind how the facilities will be operated and maintained, including common filter behavior and filter operating problems.

Common Filter Behavior

In between filter backwashes, filters "ripen," improving their ability to remove flocculated particulate matter from the water, until turbidity breakthrough, terminal head loss, or the next scheduled filter wash occurs. Filters remove floc and larger particles by straining at the top surface of the filter media and smaller particles by the mechanisms of transport (i.e., interception, sedimentation, and diffusion) and attachment to the deeper filter media. In general, increasing the filtration rate tends to decrease the time to reach terminal head loss or turbidity breakthrough. Increasing the ES of the filter media tends to decrease the time to reach breakthrough but increases the time to reach terminal head loss. J. M. Montgomery (1985) includes a thorough discussion of filtration theory and mathematical models for particle removal and hydraulics.

Filter Ripening. Clean filter media require a period of conditioning before particle attachment mechanisms take effect. The mechanisms of particle transport apply during and

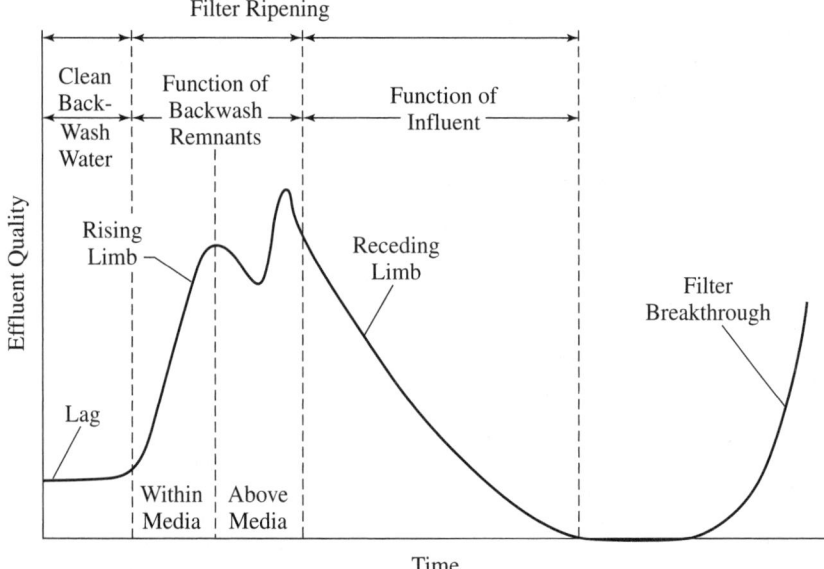

FIGURE 8.21 Characteristics of initial effluent quality. (*Source: Amirtharajah and Wetstein, 1980.*)

after filter ripening, but the mechanisms of particle attachment require attraction to the filter media of opposite charges or coagulation with charge-neutralized particles previously attached to the filter media. Consequently, inadequate particle and turbidity removal may occur for a period of time after a filter is put into service following backwash. This time period typically lasts from 5 to 30 min. Amirtharajah (1980) thoroughly explores and describes the filter ripening process (Figure 8.21). Moran et al. (1993) found that for smaller particles, ripening continues well into the filter run, whereas for larger particles, ripening occurs early in the filter run, but removal efficiency decreases afterward.

Methods successfully used to minimize the problem of filter ripening include a short period of wasting the first portion of filtered water (i.e., filter-to-waste), a slow increasing of the filtered water flow rate after a backwash, resting a filter, and adding a polymer filter aid to the washwater supply.

Filter Efficiency. The computation of filter efficiency relates the effective filtration rate to the operating filtration rate as in

$$\frac{R_e}{R_o} = \frac{\text{UFRV} - \text{UBWV}}{\text{UFRV}}$$

where R_e = effective filtration rate
 R_o = operating filtration rate
 UFRV = unit filter run volume
 UBWV = unit backwash volume

As an example, a filter operating at 4.0 gpm/ft^2 (10 m/h) for 48 h and requiring 300 gal/ft^2 (12,200 L/m^2) for an adequate backwash would have an effective filtration rate of 3.9 gpm/ft^2 (9.5 m/h) and an efficiency of 97%.

Net Filter Production. A filter operating at 4.0 gpm/ft^2 (10 m/h) for 48 h and requiring 300 gal/ft^2 (105 L/m^2) for an adequate backwash would have a net filter production of 4 gpm/ft^2 × 60 min × 48 h − 300 gal/ft^2 = 11,220 gal/ft^2 (3,946 L/m^2). Properly designed and operated filters should exhibit net filter production volumes ranging from 7,500 to 12,500 gal/ft^2 (306,000 to 509,000 L/m^2). A net filter production volume of 5,000 gal/ft^2 (204,000 L/m^2) or less could indicate inadequate pretreatment, filter-clogging algae in the influent water, excessive fines or mudballs in the filter media, mineral precipitates in the underdrains, air binding, or hydraulic restrictions causing inadequate head between filters and clearwell.

Common Filter Problems

Common filter problems include inadequate pretreatment or filter washing, gravel bed upset, air binding, restart after shutdown, and filter media replacement.

Inadequate Pretreatment. The pretreatment process (i.e., coagulation, flocculation, and clarification) in a conventional plant generally should produce pretreated waters with turbidities no greater than about 4 ntu with levels less than 2 ntu preferred. Pretreated waters with turbidities much greater than 4 ntu are indicative of floc carryover that tends to cause short filter run lengths. Conversely, pretreated waters with turbidities of 1 ntu or less in a conventional plant may result in inefficient filter operation or inadequate particulate removal.

In a direct filtration plant, pretreated water turbidity typically exceeds source water turbidity. The flocculation process should be operated to minimize floc size and allow penetration of the floc deep within the filter media, effectively using its entire depth.

Jar tests or on-line pilot filters, zeta potential, or streaming current instrumentation can greatly assist the operator in optimizing the coagulation process and ensuring adequate pretreatment for effective filtration.

Inadequate Filter Washing. Inadequate filter washing can result in poor filtered water quality and mudball formation. Cracks can occur in filter media when compressible solids remaining from previous filter runs pull filter media together and away from the filter box wall. Pretreated water can then travel through the cracks and bypass much of the filter media. Mudballs result from residuals remaining from previous filter runs sticking to filter media and forming agglomerations that grow too large to reach the washwater collection troughs during washing. As they grow heavier, mudballs can sink to create impassable regions within the filter media, typically at the anthracite-sand or sand-gravel interface. The impassable regions result in higher effective filtration rates, poorer filtered water quality, and shorter filter runs.

Air scour and surface wash systems can prevent the formation of cracks and mudballs, but previously formed mudballs may have to be removed manually or by soaking filter media with acidified water. Some plants include waste washwater turbidimeters as an operational tool to monitor the waste washwater quality and to minimize the volume of washwater usage.

Support Gravel Upset. Filter media support gravel upset may occur from operational errors such as washing a dry or drained filter. The initial rate for filling the filter should not exceed 5 gpm/ft^2 (12.2 m/h). Otherwise, rising water can compress air within the filter media pores to short-circuit the washwater, channel through the filter media, and disturb the support gravel.

Opening the washwater rate control valve too fast can cause gravel upset. Also, trapped air in the washwater header piping or in the underdrain system may be released in an uncontrolled manner, creating a visible boiling action at the top of the fluidized media. If the support gravel is disturbed, it requires manual regradation after the overlying filter media are removed.

Improper specifications for the gradation and thickness of gravel layers can also create support gravel upset. Refer to *AWWA Standard B100—Filtering Material* and consult with filter media and filter underdrain suppliers for guidance on support gravel for various filter media and underdrain combinations. Design errors relating to the washwater supply or rate controller, improper programming of an automatic filter wash sequence, and improper installation of the gravel layers can also lead to support gravel upset.

Air Binding. Filter influent waters, particularly from surface water sources, typically contain significant concentrations of dissolved gases. Depending on water temperature, the dissolved gas concentration may reach saturation point. Surface water supplies typically reach saturation point during algal blooms, during seasonal changes when temperatures increase, or where there is significant cascading and aeration of source water.

When head loss exceeds the available head at some elevation within the filter media, pressure falls below atmospheric and air escapes from solution. Air binding occurs when the accumulation of air bubbles blocks the water's path. An excessive effective filtration rate and significantly increased head loss result. Air binding most often occurs a few inches into a single-medium filter or just below the anthracite-sand or GAC-sand interface in a dual-media filter. Air escapes from the media upon closure of the filtered water rate control valve, and the resulting agitation can cause loss of media. By design, air binding does not occur in pressure, declining-rate, or self-backwashing filters.

Recommended Standards for Water Works (Great Lakes, 2003) recommends a minimum water depth of 3 ft (1 m) above filter media in gravity plants. Modern plants that operate at relatively high filtration rates typically require a water depth of 5 ft (1.5 m) or more to prevent air binding within the media. The designer should consider the rate of head loss development along with the desired filter run length to develop design criteria for the filter media, filter box depth, and freeboard.

Restart after Shutdown. Water treatment plants sometimes operate with individual filters removed from service during low-demand periods. Additionally, some plants shut down at night or cycle the filters on and off based on distribution storage tank levels. Some plants place individual filters or entire plants on standby for periods of time and even drain the filters when they have sufficient redundancy or more than one plant. Preferably, all filters should be operated continuously at a reduced filtration rate. Continuous operation prevents formation of undesirable biological activity caused by water standing in the filter box and avoids loss of ripened filter media in a drained filter. In any case, inactive filters require special measures for restart.

Filters that are shut down with standing water require a brief washing or filter-to-waste period to ensure adequate filter performance before filtered water flows to the clearwell. The restart wash or filter-to-waste period should be programmed into the centralized control system for the plant.

Drained filters require manual restart with close supervision, including gradual refilling with washwater to prevent gravel upset or media separation, a complete backwash, and a longer than normal period of filter-to-waste to ensure adequate performance before opening the filtered water control valve. Filters should always be filled from the bottom to prevent air entrainment.

Filter Media Replacement. Rapid sand filters with dual and mixed media can lose 5% to 7% of the media per year from air binding, excessive washing, air scour, or surface wash. The loss can be particularly serious when there is cold water, mismatched dual or mixed media, low washwater collection troughs, and leakage through the support gravel or underdrains. Filter performance and the effectiveness of surface wash decline when media loss exceeds about 20% of the original depth (Kawamura, 1991).

Lost media should be replaced with appropriately specified material following the guidelines presented earlier to ensure that the entire bed of media approximates original specifications. Replacement media for dual- and mixed-media filters require special attention. If a layer of fines has accumulated on the surface of the filter, it may be wise to scrape off a thin layer of top media before adding new media. After replacing the new media on top of the filter, the operator should initiate two complete filter wash cycles to fluidize the media, wash off fines, and ensure restratification.

Filter Evaluation. Filter evaluation procedures can be found in Nix and Taylor (2003).

BIBLIOGRAPHY

Amirtharajah, A., and D. P. Wetstein. "Initial Degradation of Effluent Quality during Filtration." *Journal AWWA* 72(9):518, 1980.

Amirtharajah, A. "Design of Granular-Media Filter Units." In *Water Treatment Plant Design*. Ann Arbor, Mich.: Ann Arbor Science, 1982.

AWWA/ANSI. *Standard for Filtering Material—B100*. Denver Colo.: American Water Works Association, 1996.

AWWA/ANSI. *Standard for Granular Activated Carbon B604*. Denver Colo.: American Water Works Association, 1994.

AWWA Research Foundation, *Design and Operation Guidelines for Optimization of the High-Rate Filtration Process: Plant Demonstration Studies*. Denver, Colo.: AWWARF and AWWA, 1992.

Baker, M. N. *The Quest for Pure Water*. Lancaster, Pa.: Lancaster Press, 1948.

Begin, E. E., and R. D. G. Monk. "Improvements to New Bedford's Water Supply." *Journal of the New England Water Works Association* 89(2):145, 1975.

Blanck, C. A., and D. J. Sulick. "Activated Carbon Fights Bad Taste." *Water and Wastes Engineering* 12(9):71, 1975.

Brater, E., and H. King. *Handbook of Hydraulics,* 6th ed. New York: McGraw-Hill, 1976.

Brosman, D. R., and J. F. Malina, Jr. *Intermixing of Dual Media Filters and Effects on Performance*. Austin, Tex.: University of Texas Center for Research in Water Resources, 1972.

Camp, T. R. "Discussion: Experience with Anthracite-Sand Filters." *Journal AWWA* 53(12):1478, 1961.

Camp, T. R. "Theory of Water Filtration." *Journal Sanitary Engineering Division, American Society of Civil Engineers* 90(8):1, 1964.

Camp, T. R., S. D. Graber, and G. F. Conklin. "Backwashing of Granular Water Filters." *Journal Sanitary Engineering Division, American Society of Civil Engineers* 97(12):903, 1971.

Clark, J., W. Viesman, Jr., and M. Hammer. *Water Supply and Pollution Control,* 3d ed. New York: Donnelly, 1977.

Cleasby, J. L. "Filtration." In W. J. Weber, Jr. (ed.), *Physicochemical Processes for Water Quality Control*. New York: Wiley-Interscience, 1972.

Cleasby, J. L., J. Arboleda, D. E. Burns, P. W. Prendiville, and E. S. Savage. "Backwashing of Granular Filters." *Journal AWWA* 69(2):115, 1977.

Cleasby, J. L., and E. Baumann. *Wastewater Filtration: Design Considerations*. EPA 625/4-74-007. Washington: U.S. Environmental Protection Agency, Office of Technology Transfer, 1974.

Cleasby, J. L., and G. D. Sejkora. "Effect of Media Intermixing on Dual Media Filtration." *Journal Environmental Engineering Division, American Society of Civil Engineers* 101(8):503, 1975.

Cleasby, J. L., E. W. Stangl, and G. A. Rice. "Developments in Backwashing of Granular Filters." *Journal Environmental Engineering Division, American Society of Civil Engineers* 101(10):713, 1975.

Cleasby, J. L., M. W. Williamson, and E. R. Baumann. "Effect of Filtration Rate Changes on Quality." *Journal AWWA* 55(7):869, 1963.

Cleasby, J. L., and C. F. Woods. "Intermixing of Dual Media and Multimedia Granular Filters." *Journal AWWA* 67(4):197, 1975.

"Committee Report. State of the Art of Water Filtration." *Journal AWWA* 64(10):662, 1972.

Conley, W. R. "Experience with Anthracite-Sand Filters." *Journal AWWA* 53(12):1473, 1961.

Conley, W. R. "Integration of the Clarification Process." *Journal AWWA* 57(10):1333, 1965.

Conley, W. R. "High Rate Filtration." *Journal AWWA* 64(3):205, 1972.

Conley, W. R., and R. W. Pitman. "Test Program for Filter Evaluation at Hanford." *Journal AWWA* 52(2):205, 1960.

Craft, T. F. "Review of Rapid Sand Filtration Theory." *Journal AWWA* 58(4):428, 1966.

Craft, T. F. "Comparison of Sand and Anthracite for Rapid Filtration." *Journal AWWA* 63(1):10, 1971.

Culp, R. L. "Direct Filtration." *Journal AWWA* 69(7):375, 1977.

Culp, G. L., and R. L. Culp. *New Concepts in Water Purification.* New York: Van Nostrand Reinhold, 1974.

Dostal, K. A., and G. G. Robeck. "Studies of Modifications in Treatment of Lake Erie Water." *Journal AWWA* 58(11):1489, 1966.

Fair, G., J. Geyer, and D. Okun. *Water and Wastewater Engineering,* vol. 2. New York: John Wiley & Sons, 1968.

French, J. A. "Flow Approaching Filter Washwater Troughs." *Journal Environmental Engineering Division, American Society of Civil Engineers* 107(2):359, 1981.

Great Lakes–Upper Mississippi River Board of State and Provincial Public Health and Environmental Managers. *Recommended Standards for Water Works.* Albany, N.Y.: Health Education Services, 2003.

Hager, D. G. "Adsorption and Filtration with Granular Activated Carbon." *Water and Wastes Engineering* 6(8):39, 1969.

Hamann, C. L., and R. E. McKinney. "Upflow Filtration Process." *Journal AWWA* 60(9):1023, 1968.

Haney, B. J., and S. E. Steimle. "Upflow Filter for Potable Water Production." *Journal Environmental Engineering Division, American Society of Civil Engineers* 101(8):489, 1975.

Hansen, R. E. "Granular Carbon Filters for Taste and Odor Control." *Journal AWWA* 64(3):176, 1972.

Harris, W. L. "High Rate Filter Efficiency." *Journal AWWA* 62(8):515, 1970.

Hsiung, A. K., W. R. Conley, and S. P. Hansen. "The Effect of Media Selection on Filtration Performance." Paper presented at Spring Meeting of Hawaii Section, AWWA, April 1976.

Hutchison, W. R. "High Rate Direct Filtration." *Journal AWWA* 68(6):292, 1976.

Johnson, R. L., and J. L. Cleasby. "Effect of Backwash on Filter Effluent Quality." *Journal Sanitary Engineering Division, American Society of Civil Engineers* 92(2):215, 1966.

Jung, H., and E. S. Savage. "Deep Bed Filtration." *Journal AWWA* 66(2):73, 1974.

Kawamura, S. "Design and Operation of High-Rate Filtration. Part 1." *Journal AWWA* 67(10):535, 1975.

Kawamura, S. "Design and Operation of High-Rate Filters." *Journal AWWA* 91(12):77, 1999.

Kawamura, S. *Integrated Design of Water Treatment Facilities,* 2d ed., New York: John Wiley & Sons, 2000.

Kirchman, W. B., and W. H. Jones. "High Rate Filtration." *Journal AWWA* 64(3):157, 1972.

Laughlin, J. E., and T. E. Duvall. "Simultaneous Plant-Scale Tests of Mixed Media and Rapid Sand Filters." *Journal AWWA* 60(9):1015, 1968.

Letterman, R. D. "What Turbidity Measurement Can Tell Us." *Opflow AWWA* 20(8), 1994.

Lewis, C. M., E. E. Hargesheimer, and C. M. Ventsch. "Selecting Particle Counters for Process Monitoring." *Journal AWWA* 84(12):46, 1992.

Logsdon, G. S., D. G. Neden, A. Mark, D. Ferguson, and S. D. LaBonde. "Testing of Direct Filtration for Treatment of High Turbidity Water," *Journal AWWA* 85(12):39, 1993.

McCreary, J. J., and V. L. Snoeyink. "Granular Activated Carbon in Water Treatment." *Journal AWWA* 69(8):437, 1977.

Monk, R. D. G. "Improved Methods of Designing Filter Boxes." *Journal AWWA* 76(8):54, 1984.

Monk, R. D. G. "Design Options for Water Filtration." *Journal AWWA* 79(9):93, 1987.

Monk, R. D. G., and A. P. Gagnon. "A New Method of Filter Monitoring." *Public Works* 116(10):68, 1985.

J. M. Montgomery Consulting Engineers. *Water Treatment Principles and Design.* New York: John Wiley & Sons, 1985.

Moran, D. C., M. C. Moran, R. S. Cushing, and D. F. Lawler. "Particle Behavior in Deep-Bed Filtration. Part 1: Ripening and Breakthrough." *Journal AWWA* 85(12):69, 1993.

Nix, D. K., and J. S. Taylor. *Filter Evaluation Procedures for Granular Media.* Denver, Colo.: AWWA, 2003.

Oeben, R. W., H. P. Haines, and K. J. Ives. "Comparison of Normal and Reverse Graded Filtration." *Journal AWWA* 60(4):429, 1968.

O'Melia, C. R., and D. K. Crapps. "Some Chemical Aspects of Rapid Sand Filtration." *Journal AWWA* 56(10):1326, 1964.

O'Melia, C. R., and W. Stumm. "Theory of Water Filtration." *Journal AWWA* 59(11):1393, 1967.

Rice, A. H., and W. R. Conley. U.S. Patent 3,343,680. 1967.

Rimer, A. E. "Filtration through a Trimedia Filter." *Journal Sanitary Engineering Division, American Society of Civil Engineers* 94(6):521, 1968.

Robeck, G. G., K. A. Dostal, and R. L. Woodward. "Studies of Modifications in Water Filtration." *Journal AWWA* 56(2):198, 1964.

Tuepker, J. L., and C. A. Buescher, Jr. "Operation and Maintenance of Rapid Sand and Mixed Media Filters in a Lime Softening Plant." *Journal AWWA* 60(12):1377, 1968.

Westerhoff, G. P. "Experience with Higher Filtration Rates." *Journal AWWA* 63(6):376, 1971.

CHAPTER 9
SLOW SAND AND DIATOMACEOUS EARTH FILTRATION

Michael V. Broder
Hazen and Sawyer, PC
New York, New York

John C. Byron
Stearns & Wheler, LLC
Cazenovia, New York

Although rapid sand filters predominate, the two types of filters covered in this chapter may be effective in many applications where source water quality permits. In general, the combined costs of constructing and operating slow sand and diatomaceous earth (DE) filters may be considerably less than the cost of rapid sand filtration plants for the same capacity.

The principal mechanisms for separation of solids in all porous media filters are attachment and straining or entrapment. Because of the relationship of the somewhat large pores in rapid sand filter media compared with particulates, the primary mechanism for separation is attachment. In the case of slow sand and DE filters, however, the pore-particulate size relationship results in more substantial separation by entrapment.

During the initial operation period of slow sand filters, the separation of organic matter and other solids generates a layer of biological matter on the surface of the filter media. Once established, this layer is the predominant filtering mechanism. Solids are removed from water by a DE filter as the source water flows through a precoat layer of powderlike DE.

For all practical purposes, most solids are separated at the surface of the media in both actions. Because of the small pore size of the media, particulate separation is ideal for removing the cysts of *Giardia* and *Cryptosporidium*. In most situations, neither type of filter requires previous conditioning of the raw water.

In general, application of slow sand and DE filtration should be limited to source waters with turbidity levels less than 5 ntu. Where particulates are dominant (rather than organic matter) both types of filters may be used with water of up to 10-ntu turbidity. If either type of filter is used on water with higher turbidity, filtrate quality will generally be acceptable, but the more rapid buildup of solids on the filter results in rapid loss of head and shortened length of filter runs.

With establishment of the Stage 1 Disinfectants/Disinfection By-product Rule, filtration of surface water must achieve specified reductions in total organic content (TOC) and/or reductions in disinfection by-product levels [total trihalamethanes (TTHM) and sum of five haloacetic acids (HAA5)]. The Interim Enhanced Surface Water Treatment Rule calls for lower drinking water turbidity. This means that use of slow sand and DE filters may be limited further without the use of supplemental measures or additional treatment.

SLOW SAND FILTRATION

Slow sand filtration was the first type of porous media filtration used in water treatment. The first recorded installations occurred in Scotland and England in the early 1800s. By the mid-1800s, legislation was passed in London, England, requiring filtration of water to be consumed. The first recorded installation of slow sand filtration facilities in the United States was in Poughkeepsie, New York, in 1872. Subsequent development of rapid rate filtration then slowed the pace of construction of slow sand plants in the United States in the early 1900s.

Around 1980, interest in using slow sand filtration was rekindled as the U.S. Environmental Protection Agency (USEPA) conducted research to develop treatment options that are simple to operate for use by small communities and that produce high-quality effluent. Research reconfirmed that, at recommended filter rates and with appropriate media and source water quality, slow sand filtration can produce a low-turbidity effluent and can effectively remove microbiological contaminants.

When the USEPA passed the Surface Water Treatment Rule (SWTR) in 1989, further pressure was placed on communities that were not filtering surface water supplies to add filtration. Slow sand filters were rated along with rapid sand and DE filters as baseline treatment in the regulations. As a result, slow sand filtration has once again become a treatment method routinely considered in evaluating filtration options in many U.S. communities. Abroad, it is used to provide safe drinking water to many poor or rural communities. A major portion of the city of London water supply is treated by slow sand filtration.

Renewed interest in the slow sand process has generated new research into improving treatment performance. The focus has been on expanding use of slow sand filtration in treating poorer-quality source water, especially with higher turbidity and organic content.

Slow Sand Treatment Mechanism

As its name implies, slow sand filtration is accomplished by passing water at a relatively low rate through a sand medium. The filtration rate is on the order of one-hundredth of the rate used in a typical rapid sand filter.

Because of the relatively low filter rate, head loss across the bed occurs gradually over a much longer time. Average filter run length is normally between 45 and 60 days. In some newer installations, filter run lengths in excess of 6 months and even greater than 1 year have been reported.

Slow sand filtration accomplishes its treatment primarily through biological activity, with the bulk of this activity taking place on the surface of the sand bed. A layer develops on the sand surface that is called *schmutzdecke*, an accumulation of organic and inorganic debris and particulate matter in which biological activity is stimulated. It has been

TABLE 9.1 Source Water Quality Limitations of Slow Sand Filters

Parameter	Recommended limit
Turbidity	5 to 10 ntu*
Algae	200,000/L†
True color	15 to 25 platinum color units
Trihalomethanes	50 µg/L
Nonpurgeable dissolved organic carbon	2.5 mg/L
UV absorbance‡	0.080 cm^{-1}
Dissolved oxygen	>6 mg/L
Phosphorus (PO$_4$)	30 µg/L
Ammonia	3 mg/L
Iron	<1 mg/L
Manganese	<1 mg/L

*Presence of clay particles must be determined.
†Dependent on identification of algal species and assumes covered filter.
‡Absorbance at wavelength of 254 nm.

found that some biological activity also extends deeper into the bed, where particulate removal is accomplished by bioadsorption and attachment to the sand grains.

Source Water Quality Considerations

In considering whether slow sand filtration is an appropriate treatment method, source water quality must be carefully evaluated. If source water quality data are not available, pilot testing of the source water is essential to determine the applicability of the slow sand treatment option. Table 9.1 lists source water quality parameters with recommended limits (Spencer and Collins, 1991).

Turbidity. Both the level and type of turbidity in source water must be considered. In general, most existing slow sand plants successfully treat source water turbidity of less than 10 ntu (Slezak and Sims, 1984), which is recommended for an upper limit in designing new facilities. Also of some importance is the stability of the water. Slow sand facilities operate more efficiently if source turbidity is relatively constant and generally ≤ 5.0 ntu.

Of equal importance is the nature of particulates. Source waters that normally contain clay particulates or that pick up clay after storm events will cause problems for slow sand filters. This difficulty for slow sand filters occurs because clay penetrates deep into the bed or may even carry through the filter, causing an immediate problem of elevated filtered water turbidity and a long-term problem of filter clogging and reduced length of filter runs.

Algae. In a few instances, it has been found that the presence of certain types of algae actually enhances the filtration process by providing greater surface area for biological activity. In general, however, the presence of algae in the source water reduces filter run

TABLE 9.2 Algal Species Classification for Slow Sand Filtration

Filter clogging	Filamentous	Floating
Tabellaria	*Hydrodictyon*	*Protococcus*
Asterionella	*Oscillaria*	*Scenedesmus*
Stephanodiscus	*Cladophora*	*Synura*
Synedra	*Aphanizomenon*	*Anabaena*
	Melosira	*Euglena*

lengths. Table 9.2 presents a list of commonly found algal species, divided into categories related to their effect on filter performance (Spencer and Collins, 1991). Filter-clogging species are detrimental to filter performance, while filamentous species may actually enhance filter performance by providing greater surface area. Floating species would not result in direct clogging of the filter, but may shorten run lengths based on poorer-quality raw water.

Algae may be present in source water delivered to the filter and may also occur in an uncovered filter bed open to sunlight. In general, it is prudent to reduce algal content in source water to as low a level as possible to limit its effect on filter performance. Observation of algal growths, as well as identification, will aid with assessing the need for pretreatment, such as copper sulfate, and in determining when filter run lengths may be shortened. Some researchers have suggested the measurement of chlorophyll at concentrations of 5 mg/m^3 as a limit in source water (Cleasby et al., 1984).

Color. Color in treated water is currently categorized by USEPA as a secondary contaminant in drinking water supplies, with the focus being aesthetic concerns. As identified by Christman and Oglesby (1971), the yellow to brown color of many source waters can be the result of microbial breakdown of lignins from woody plants. True color removals of 25% or less were reported by Cleasby et al. (1984). Other research has indicated a removal range between 15% and 20% for total organic carbon (Fox et al., 1994; Collins et al., 1989).

When one is evaluating the applicability of slow sand filtration for a specific source water, a review of historical trihalomethane (THM) data can reveal whether the expected low removal efficiency of aquatic organic substances by the process is a concern. Where historical color and THM data are unavailable, a sampling program can be initiated to aid in evaluating whether slow sand filtration is an appropriate treatment method.

Iron and Manganese. Slow sand filters remove iron and manganese through precipitation on the sand surface in a scaling-like action, but an upper limit of 1 mg/L of iron is suggested to avoid forming an iron precipitate that could clog filters. A similar limit for manganese would also appear to be acceptable. Collins et al. (1989) showed that iron precipitate on a slow sand filter enhanced the removal of organic precursors.

Dissolved Oxygen. The presence of dissolved oxygen in source water is critical for stimulating a healthy schmutzdecke for proper slow sand filter operation. Some slow sand plants use aeration of the water as a pretreatment. Reduction of dissolved oxygen levels commonly occurs following algal blooms, so that the importance of dissolved oxygen in the source water is another reason to control algal growth in the source. Potential problems resulting from dissolved oxygen deficiencies include tastes and odors, redissolving of precipitated metals, aesthetics, and increased chlorine demand (Ellis, 1985).

Nutrients. The proper operation of the schmutzdecke is somewhat dependent on the presence of sufficient concentrations of carbon, nitrogen, phosphorus, and sulfur. Carbon and sulfur (in sulfate form) are prevalent in most source waters. However, protected reservoir systems may have limited concentrations of nitrogen and phosphorus present.

It has been reported that, for every 1 mg of carbon removed by the schmutzdecke, 0.04 mg of nitrogen and 6 μg of phosphorus are required (Skeat, 1961). Slow sand filters have also shown the ability to remove up to 3 mg/L of ammonia from source water under the right conditions. Ammonia can be used as a source of nitrogen for the filter.

Effluent Water Quality

Slow sand filtration has been shown to be effective in achieving removal of *Giardia* and viruses. Effluent turbidities in the range of 0.1 to 0.2 ntu are typical for high-quality source waters, while turbidities of up to 1.0 ntu may be considered an upper limit. Removal of organic substances is generally in the range of 15% to 25%. Recent research has focused on improving removal because of disinfection by-product formation considerations.

Typical treatment performance of conventional slow sand filtration plants is listed in Table 9.3. Current regulations require effluent turbidities of less than 1.0 ntu. Pilot testing of the source water is recommended for determining the operational parameters and possible need for supplemental treatment to meet the established turbidity requirements. Limited data are available on removal capabilities with respect to *Cryptosporidium*, but research is continuing.

Design of Slow Sand Filters

The slow sand filter is relatively simple in arrangement, having only three basic elements in addition to a control system. Typical of any filter design, the complete train includes clearwell storage, disinfection, and posttreatment. Figure 9.1 presents a cross-sectional view of a typical filter bed.

Filter Box Design. The filter box contains all the filtering components of the system. These include source water storage (above the sand bed), filter sand, the underdrain

TABLE 9.3 Typical Treatment Performance of Conventional Slow Sand Filters

Water quality parameter	Treatment performance or reduction capacity
Turbidity	<1.0 ntu
Coliforms	1 to 3 log units
Enteric viruses	2 to 4 log units
Giardia cysts	2 to 4+ log units
Total organic carbon	<15% to 25%
Biodegradable dissolved organic carbon	<50%
Trihalomethane precursors	<25%

FIGURE 9.1 Typical covered slow sand filter installation.

system, and, in some cases, facilities for collecting wastewater generated during the cleaning process. The box floor and sides are generally constructed of concrete. Roof designs for covered filters vary and may include wood truss, steel, precast concrete, or cast-in-place concrete.

If the filter unit is to be covered, the height of the box must be adequate to provide for the depth of sand and support media, underdrain system, source water storage above the media, and headroom for cleaning and resanding operations. The filter box area is determined by the unit rate of flow and required supply flow.

Slow sand filters may also be uncovered. There are currently many operating facilities that are uncovered in the U.S. Pacific Northwest, Europe, and South America. Besides lower initial cost, an advantage of uncovered filters is the far greater ease of using mechanical equipment for cleaning and maintenance. If filters are to be covered, major considerations include providing headroom for equipment during cleaning and repair, lighting, and ventilation.

A 2- to 3-ft (0.6- to 0.9-m) freeboard depth should be added over the normal water surface to provide for fluctuations in water depth within the filters without reaching ceiling height. This also provides room to install permanent lighting to improve the efficiency of cleaning and resanding. Generous headroom also allows the use of larger mechanical equipment within the filters, which can significantly reduce the time required for cleaning or resanding operations, particularly in large filters.

Filter Sand. Guidelines for filter sand characteristics and proper media depth vary between those of the International Research Center (IRC) and the *Recommended Standards for Water Works* (commonly known as the 10 State Standards). The IRC manual recommends sand with an effective size of 0.15 to 0.30 mm with a uniformity coefficient between 3 and 5. The 10 State Standards recommend an effective size of 0.30 to 0.45 mm and a uniformity coefficient of ≤ 2.5.

A finer effective size may improve particulate removals but generally results in shorter filter run lengths. Media that are too large allow deeper bed penetration and may even result in filter breakthrough or clogging. A deeper penetration of particles in the filter bed also means that more sand must be removed during a scraping cycle. It has been suggested that a better approach is to increase the depth of the sand rather than to reduce media size if a more conservative design is desired.

Pilot testing of the process using different media sizes provides data on removals and filter run lengths and can serve as a basis for media selection.

Sand depth should generally be between 18 and 35 in. (460 and 890 mm), but some plant operators have reported satisfactory treatment with sand depth as low as 12 in. (300 mm). Most slow sand plants in the United States are designed with a minimum sand depth of 30 in. (760 mm).

If filters are cleaned by manual scraping, about $\frac{1}{2}$ in. (1 cm) of sand is removed during the scraping. Final sand depth should be determined based on cleaning method, anticipated filter run lengths, number of scrapings desired before resanding, sand availability and expense, and impact of downtimes on plant capacity. The minimum depth before resanding should be 18 in. (460 mm).

Underdrain and Support System. One common type of underdrain consists of a manifold and perforated laterals installed below the sand bed. Most new designs use a plastic piping system for filter underdrains. Piping material must be certified for contact with potable water. Typical lateral sizes range from 4 to 8 in. (100 to 200 mm) with the underdrain system header in the range of 8 to 16 in. (200 to 400 mm). Figure 9.2 is a view of an installation in progress of a perforated PVC underdrain system.

The underdrain system must be designed to cause minimal head loss within the system. Head losses through the individual perforations of the laterals must be a fraction of head loss through the lateral itself to provide a balanced flow across the system. The design engineer should refer to hydraulic textbooks for guidance with respect to piping manifold system designs.

Other underdrain systems use prefabricated plastic or clay filter blocks or a false floor of concrete blocks or brick with gravel media above. Because of the large area of a slow sand filter, the prefabricated type is normally expensive to install and is used infrequently. The hydraulics of a false floor system must be similar to those of the piped system.

Gravel support media usually consist of multiple layers of graded gravel. The gravel layers are coarsest on the bottom and become finer with each layer. Gravel supports the

FIGURE 9.2 Underdrain installation.

sand, and the fine layer prevents sand from migrating down to clog underdrain openings. The 10-State Standards recommend gravel support layers similar to those required for a rapid sand installation with a media depth between 18 and 24 in. (460 and 610 mm) and gravel sizes in a range from $\frac{3}{32}$ to $2\frac{1}{2}$ in. (2 to 64 mm) in a five-layer system.

Source Water Storage. Source water storage is the depth and volume of water overlying the sand surface within the basin. This volume varies in different designs (between 3 and 24 h of plant capacity). As storage capacity increases, there are some benefits with respect to equalizing source water quality, sedimentation of larger particles, and even biological action within the water column itself.

The primary purpose of the water level above the sand, however, is to provide the driving head across the filter bed. A typical terminal head loss for a slow sand filter is in the range of 4 to 5 ft (1.2 to 1.5 m). Therefore typical depths of water above the sand should range between 6 and 7 ft (1.8 and 2.1 m) to provide for the additional driving force required for head loss through the clean sand bed and through the piping systems. If a filter is to be covered, the height of the filter box above the sand is governed primarily by space requirements for cleaning and resanding, so provision of a 6- to 7-ft (1.8- to 2.1-m) depth of water can easily be accomplished.

Access for Cleaning and Maintenance. Early sand filters were constructed either with no covers (roofs) or with earthen embankments over cast-in-place concrete roofing systems. Filters constructed with a roof typically used access hatches or ports spaced at intervals above the bed surface. Cleaning the beds was cumbersome and normally accomplished either by manually hoisting the scraped schmutzdecke up through the access way or by hydraulic transfer of a slurry of removed material through a piping system.

New covered filter installations typically use some sort of structure with an access ramp into the basins or "ship's doors" to allow direct access into the filter box at the elevation of the sand surface. Installation of access ramps is generally more costly (when installed with a covered filter) but may result in reduced maintenance when compared with a ship's door. An access ramp and filter are depicted in Figure 9.3.

FIGURE 9.3 Access ramp and filter.

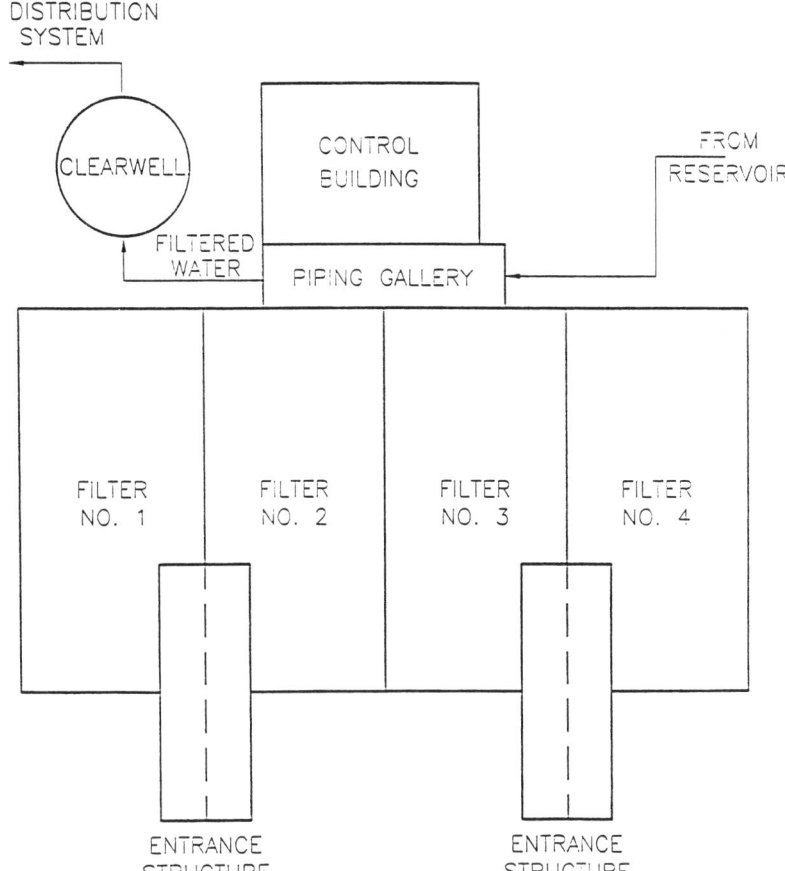

FIGURE 9.4 Typical slow sand plant layout with exterior entrance structures.

A typical ramp is constructed of concrete (or earth for uncovered installations). Ramp widths vary, but consideration should be given to the width of equipment expected to be used in the box. The slope should be 1 ft (0.3 m) vertical to 5 ft (1.5 m) horizontal or shallower. A steeper slope could result in equipment slipping on the ramp surface when sand or other debris is tracked on it.

A cross section of a typical plant using a filter entrance structure is shown in Figure 9.4. Entrance doors at the top of the ramp must be corrosion-resistant because of the prevalent high humidity from the water within the filters.

Ship's doors are constructed of steel or aluminum. Steel is stronger but generally requires more maintenance, such as painting, and corrodes faster than aluminum. Design of aluminum doors must consider anticipated loads carefully. Hinge lubrication is critical with either material to ensure smooth operation.

Lighting. Older, covered slow sand installations relied on portable or minimal lighting when working within the filters or used natural light through skylights or from open hatches on the roof. If funds are available, a permanent lighting system increases visibility and eliminates hazards associated with portable lighting.

Because of high-humidity conditions, the lighting design must incorporate corrosion-resistant materials such as aluminum and stainless steel. Lighting units must include integral gaskets and methods of sealing wire penetrations to minimize moisture within the units. Conduit systems must use corrosion-resistant materials.

Lighting units must be mounted in a manner that minimizes damage from high water levels or from operating cleaning equipment. Lighting levels on the order of 10 fc (100 lm/m^2) provide sufficient light to allow efficient and safe operations.

Ventilation. Little consideration was given to ventilation in older, covered slow sand filters. Usually ventilation consisted of open hatches with portable ventilators installed during maintenance operations. Although portable units increase ventilation, the actual rates are generally below those required by code when engine-driven equipment is being operated inside the filter unit.

For newer installations, ventilation rates for motorized equipment must be designed to protect operators and should be similar to those provided in parking garages. Rates must be set according to expected emissions of motorized equipment operated within the basin. The number of air changes required is generally governed by state or federal regulations.

Filter Roofing System. Selection of a filter roofing system must consider capital expense, long-term reliability, headroom, availability of material, and site conditions. Cast-in-place systems are generally the most expensive but are the most durable. Precast concrete slabs are less expensive than cast-in-place but have a slightly shorter life span. Systems constructed of wood or steel experience significant deterioration or corrosion problems and require replacement within a relatively short time. Obviously, the replacement expense versus initial capital expense must be evaluated in design.

Methods of covering the roof system include shingles, metal roofing, single-ply membrane, built-up roofing, and earthen fill. Again, each system has specific capital costs, and not all are usable with each roof support system. An earth cover system can help blend a facility into its surrounding but requires a strong support structure to carry the load.

Filter-to-Waste System. The design of a slow sand filter should provide the ability to discharge filtered water to waste. The filter-to-waste system allows the operator to check effluent quality after a filter cleaning and before bringing the filter back on line. It also provides a method to "cleanse" filter media after resanding or other major reconstruction work.

The filter-to-waste system must be designed to avoid a potential cross-connection. The ideal system includes an air gap on the filter effluent to the drain pipe. However, many regulators allow a hard piped connection to a drain system, with valving to direct water to finished water storage or to waste. In the latter case, it is normally required to install the filter-to-waste line at an elevation below the filtered water line and to provide an air gap at the ultimate discharge point to minimize potential contamination.

Filter Draining. At some point in the filter cycle, head losses increase to such a level that filter production is unacceptably low, and the filter must be taken off-line for cleaning. Because the rate is so low through the filter and the hydraulics between adjoining filters may not allow significant drop in filter level to occur, some form of separate drain should be provided.

This drain should generally be installed to discharge filter supernatant either back to the source or to a lagoon. The drain can also be connected to the filter-to-waste system through an air gap, or it may be installed as a separate system. Cross-connection to the

FIGURE 9.5 Piping gallery.

filtered water system must be considered when one is designing this drain. After the supernatant is discharged, full drainage of the filter box can be accomplished through the filter and into the filter-to-waste system. Figure 9.5 shows a typical piping gallery for a slow sand facility.

Flow Control. Flow through a slow sand filter can be controlled on either the influent or effluent side. One method of influent control provides a constant level over the filter, and declining-rate filtration results. This means that as the head loss increases across the bed, the filter rate is reduced. In the second method, the filtered water level is adjusted to increase as filter head loss increases to provide the same flow, or constant-rate filtration.

Effluent is controlled either by a control valve or by fluctuating finished water storage levels in response to head loss changes across the filter bed. The intent is to maintain a constant filter rate by adjustments in available head as head loss changes.

With either system, finished water storage must ensure that water levels are maintained at a minimum of 1 ft (0.3 m) above the top of the filter sand to avoid problems associated with negative pressures within the sand bed. Control systems should seek to maintain constant flow through the plant and minimize the surge of filter starts and stops that could affect effluent quality.

Storage of finished water after slow sand filtration serves two functions. First, it can provide a method of maintaining submergence of the filter media under all conditions. This limits problems that could develop from air binding within the sand bed. Second, storage provides contact time for disinfection after filtration. The volume and form of storage vary according to system requirements.

Filter Rates. Typical filter rates are usually in the range of 0.04 to 0.10 gpm/ft^2 (0.09 to 0.24 m/h). Filter rates can be established through pilot testing. In some cases, it has been demonstrated that higher rates are possible while maintaining acceptable effluent quality and filter run lengths. However, based on concerns about microbiological contaminants, it is prudent to design filter rates conservatively to allow flexibility for future increased treatment requirements and fluctuating demands for water quantity and quality.

One design approach is to allow a moderate to low rate, even when one filter is off-line and the treatment system is under maximum water demand conditions. This allows some flexibility to meet future increased demands without capital expenditures and provides a buffer under poor source water quality conditions. This approach also increases initial capital expenditures because of larger filter area requirements.

Some cost savings may be gained by allowing on-line filters to operate at high rates when one unit is off-line, but research has shown that even though the schmutzdecke has been established, an increase in filter rate can have an adverse impact on treated water quality. The higher rate also results in reduced length of filter runs.

A compromise approach is to provide moderate flow rates when all units are in service. Higher rates result with one unit off-line, but only for short periods. This provides flexibility when dealing with water quality and demand issues and reduces capital expenditures.

Automation. Slow sand filtration plants can require much simpler automation than other filtration processes. But to some degree, the level of automation is driven by regulations and the client. For automation, the design must consider

- State and federal regulations
- Initial cost and operating costs
- Level of complexity desired by client and operator
- Work shifts to be used
- Source water quality fluctuations
- Record-keeping needs and requirements
- Critical plant parameters and equipment

The recommended minimum location points for recording and monitoring include

- Source water turbidity
- Source water flow
- Filter level
- Filtered water flow and head loss (individual)
- Filtered water turbidity (individual or combined)
- Chlorine residual (before and after storage)
- Finished water flow
- Finished water storage level

Other monitoring points may be specific to each site's requirements and should be discussed with the client.

Pilot Testing

The most advantageous way of determining the applicability of a treatment process to a specific source water and the performance that may be expected is through pilot studies. In fact, reliable prediction of treatment results is difficult without pilot testing. The results of studies provide background data to establish design parameters and estimate filter run lengths. Data can also be used as an introduction of the process for orienting operations staff.

Planning is critical to the successful implementation of the pilot program. The protocol must establish

- Pilot plant design and construction
- Conditions for runs
- Duration of tests
- Run parameters
- Data to be collected
- Laboratory analyses required

Some suggested water quality parameters to be monitored on the source and filtered water include

- Turbidity
- Color
- Temperature
- Algae
- Bacteria
- Dissolved oxygen
- Iron and manganese
- Particle counts
- Filter head loss
- Flow rate
- Filter run length

It is important to collect all samples in accordance with established procedures, to minimize potential for contamination. Whenever possible, filter sand used for pilot studies should be the same sand that will be used in the full-scale plant. It has also been suggested that sand be installed in a "dirty" condition (similar to that delivered in construction) to determine the washing-out period during plant start-up. When tests are completed, pilot filters can be moved into the full-scale plant for future research.

Pilot plant construction varies. Previous designs used many types of materials, including fiberglass or concrete pipe and PVC pipe. The intent is to provide a watertight container with room for installing sand and support media, underdrain piping, sample ports, flow measurement equipment, and source water storage. Typical pilot plant setups are 8 to 12 ft (2.4 to 3.7 m) high. The diameter of the pilot should be as large as possible to approximate future flow rates and facilitate operation and cleaning.

New Developments

Improvements and innovations to expand slow sand filtration treatment capabilities are discussed in the following paragraphs.

Roughing Filters. Roughing filters serve as a pretreatment method to reduce sediment loading to the sand filters. Many varieties are currently in use; in general they consist of gravel-type media with different gradations. The filters commonly are designed in stages and for either vertical upflow or downflow or horizontal flow.

Recent designs used gravel filter material that decreases in size with flow direction. The gravel size range is between 0.2 and 2 in. (5 and 50 mm), and flow velocities are in the range of 0.02 to 0.08 ft/min (0.3 to 1.5 m/h). Pardon (1991) showed that greater than 90% removal can be obtained for particles 10 μm and greater, and 72% removal of particles between 2 and 5 μm, through use of vertical roughing filters. Roughing filters are cleaned by flushing at high rates.

Other pretreatment methods for extremely turbid waters use single-stage gravel filters located next to or within the source (mostly rivers or canals). These filters operate similarly to roughing filters. Cleaning is normally accomplished manually because most of the removal is at the top surface of the filter.

Preozonation. Concern with disinfection by-products in finished water has increased the need to improve organic precursor removals through the treatment process. Research has been conducted on ozone as a preoxidant ahead of slow sand filtration as a means of improving organics removal.

In general, ozone use is greater in Europe than in the United States, and many plants in Europe have ozone preoxidation and GAC adsorption before slow sand filtration. Malley et al. (1991) reviewed past research and performed pilot studies to evaluate the effect on treatment performance of preozonation before slow sand filtration. Ozonation converts nonbiodegradable organic matter to biodegradable forms, to enhance biological activation of the filter media.

In this ozone treatment scheme, reductions occurred in ultraviolet (UV) absorbance and trihalomethane formation potential (THMFP). It was also observed that ozone enhanced conditions in the filter water columns for removal of other objectionable matter. However, the breakdown of organic matter also reduced filter run times. The studies showed that ozonation by-products were removed through the biological slow sand process.

Using ozone ahead of slow sand filtration may allow many communities to meet new disinfection by-product regulations. However, treatment improvements may result in increased operating and maintenance costs relating to shorter filter run times and operating costs of the ozone system.

Granular Activated Carbon. The addition of granular activated carbon (GAC) to slow sand filter media was initially tested in England at the Thames Water Utilities. Thames Water currently operates seven slow sand treatment facilities with a combined capacity of about 700 mgd (2,500 ML per day). Because of pesticide levels in the source waters and strict regulations for pesticide removals, the utility determined that adding a GAC treatment step could allow water quality goals to be met.

To avoid the relatively high cost of constructing GAC adsorbers, the utility explored installing GAC within the filter bed. The "sandwich" bed used a 3- to 8-in. (75- to 200-mm) layer of GAC installed 4 to 6 in. (100 to 150 mm) below the sand surface. The performance of the system was compared with that of conventional slow sand filtration with respect to head loss, color removal, TOC removal, and THMFP. The results are presented in Table 9.4. Chlorine demand was also reduced with the use of GAC, and pesticide levels were reduced to below standards.

Based on test results, Thames Water installed the GAC sandwich in several of its treatment facilities.

Biologically activated carbon (BAC) with use of both preoxidation with ozone or other oxidants and biological activation of the sandwich layer will enhance reduction of both

TABLE 9.4 Comparisons of the Operation of Slow Sand and GAC Sandwich Filters

Parameter	Conventional slow sand	GAC sandwich filter
Cleaning frequency	30 days	No change
Color removal	20%	50%
Total organic carbon reduction	20%	35% to 40%
Total trihalomethane formation potential reduction (24-h contact time)	130 µg/L	60 g/L

TOC and disinfection by-products even further. When used primarily for the purpose of organic reduction, the life of the GAC sandwich layer may be extended to 5 years or more.

Filter Fabric. In research conducted in the United States by Collins et al. (1989) and in England by Graham et al. (1991), a synthetic, nonwoven fabric installed on the sand surface was shown to increase filter run lengths while maintaining effective treatment. The benefit is that removal of suspended particles occurs on the fabric with the intent of simplifying the cleaning process.

When one is specifying a fabric, parameters to be considered include porosity, specific surface area, and fabric thickness. Graham et al. (1991) showed that a fabric thickness between ¾ and 1¼ in. (20 and 30 mm) was necessary to allow adequate fabric removal and cleaning and that fabrics with a specific surface area between 4,000 and 4,500 ft^2/ft^3 (13,000 and 15,000 m^2/m^3) provided optimal results. They also demonstrated that use of fabric meeting this criterion increased filter run times by 400% over those of a conventional filter.

Issues to resolve before filter fabric is used involve the mechanism for removal and cleaning of the fabric, especially for large installations. Further pilot studies may be warranted to determine the applicability of this modification with source waters of varying quality.

Filter Harrowing. The process of harrrowing filters was initially developed as a filter cleaning method at the West Hartford, Connecticut, slow sand filtration facility. The plant currently has 22 slow sand beds in operation with a capacity of 50 mgd (189 ML per day). Beds vary in size between ½ and ¾ acre (0.2 and 0.3 ha). To reduce the time and expense for cleaning, the utility developed the process of *harrowing* filters in the 1950s. In the process, a tractor with a mounted spring-tooth harrow operates within the bed while water about 6 in. (15 cm) deep is flowing across the sand surface. The harrow breaks up the top of the filter surface, and water carries away the debris. The process removes the accumulated source water particulates while maintaining an active biological material in the top several inches of the sand.

The process creates wastewater as the bed is being cleaned. The volume generated depends on source water quality, filter area, and cross-flow velocity and depth and is generally in the range of 40 to 60 gal/ft^2 (163 to 244 L/m^2) of filter area.

The experience at West Hartford has shown that using the harrowing process reduces cleaning time to one-quarter of the time required for conventional cleaning. The harrowing process requires the following facilities that may not be provided for a conventionally cleaned slow sand filter:

FIGURE 9.6 Harrow tractor.

- Access ramp for harrowing equipment to enter the filter
- Harrowed water influent distribution system
- Harrowed wastewater collection system
- Holding lagoon for the harrowed wastewater

The filter box must be designed with enough headroom to permit tractor operation, and support columns should be minimized to improve maneuverability. Figure 9.6 shows a harrow tractor operating in a filter basin. Influent and waste headers can consist of either perforated piping or channels with adjustable weirs.

Wastewater lagoons should be designed to carry one to two harrowings with the ability to allow time for settling and to discharge the decanted water in stages. Because wastewater has generally not received previous chemical pretreatment, discharge requirements are usually less rigorous.

The process is now currently used at several operating plants in the United States to improve plant performance and reduce cleaning times. Designers of larger-capacity plants should consider using this cleaning process.

DIATOMACEOUS EARTH FILTRATION

Diatomaceous earth (DE) filtration has been used effectively for drinking water treatment since 1942 when it was adopted as a standard method for the U.S. Army. The DE filter was selected because of its portability and effectiveness in removing *Entamoeba histolytica* cysts (Black and Spaulding, 1944). These cysts are pervasive in some parts of the world and are difficult to control with disinfectants alone. The capability of DE filtration to effectively remove particulates applies equally well to the latter concerns of *Cryptosporidium* and *Giardia* cysts where cyst removals of approximately 6 logs have been achieved (Ongerth and Hutton, 1997). DE filters are commonly called precoat filters because of the precoat of the filter leaves that initiates every operating cycle. Although DE is the most common precoat material used, other precoat material such as ground perlite performs as well in many applications.

For many years, the type of equipment available limited the use of DE filters for municipal drinking water treatment. The use of stainless steel and plastics in the fabrication of equipment has significantly changed the performance capability of the filters by improving their ease of operations and maintenance.

DE is mined from the fossilized remains of microscopic plants called diatoms, deposited in what were the beds of ancient oceans. A powdered medium is manufactured from the diatomite deposits that is almost pure silica. One of the more common diatomite media used for drinking water treatment has a mean particle size of 22.3 μm with 80% of the particles ranging in size from 5 to 64 μm. This medium, when deposited on the filter septum, has an average pore size of about 7.0 μm. Specifications for DE material is covered in AWWA Standard B101 (2001).

DE Filter Operation

As illustrated in Figure 9.7, DE filter operations occur in three steps:

1. A precoat of about ⅛ in. (3 mm) is deposited on the filter.
2. After the precoat has been deposited, filtering begins, and at the same time a small amount of DE material (called *body feed*) is added to the source water to maintain the porosity of the media.
3. Particulates in the source water are trapped in the precoat layer until maximum head loss is reached, at which time the filter run is terminated and media material is cleaned from the septa.

Porosity Control. The principal requirement for maintaining effective DE filter runs is to maintain the porosity of the filter cake. Source water solids generally vary in size and

| Precoat Formation | Filter Cake Formation | Filter Cake Removal |
| a. | b. | c. |

FIGURE 9.7 Precoat filtration cycles.

FIGURE 9.8 Effect of varying body feed addition.

are a mixture of relatively inert matter and solids that are predominantly organic. If source water is filtered through the precoat alone, the buildup of solids and compression of the accumulated cake quickly reduce filter cake porosity, and head loss increases at an exponential rate. This may be avoided by adding body feed to the source water in sufficient amounts to produce a constant flow versus head loss relationship (Figure 9.8).

Although the rate of flow does not affect effluent quality or turbidity breakthrough, the flow rate for precoat filters should generally be limited to about 2 gpm/ft^2 (7.8 m/h). The shape of the precoat filter head loss curve that reflects both feed and flow conditions is, therefore, an important feature to control effective filter run performance.

Supplementary Treatment. Supplementary measures may be added to the basic DE filter process to enhance the filtration process and to expand the process to remove some nonparticulate constituents.

Natural color in source water supplies can be caused by either organic or mineral matter. Color can result from the decay of plant matter or from the solubilization of iron in the soil, and in many instances the mineral and organic matter may be bound together. Therefore color can be present either in particulate form or in solution. Particulate color consists mostly of negatively charged colloids, and even though the precoat medium has low pore size, charged colloids pass through unless the charge is neutralized. The use of a strong oxidant such as ozone has been demonstrated to be effective in conditioning color for removal.

When color is particulate rather than dissolved, DE filters reduce source water color of about 25 color units (CU) and less to below 5.0 CU. With source color between 25 and 60 CU, filter effluent is generally no higher than 10 CU. Supplemental treatment such as preozonation or alum-coated media may be required to improve removal of particulate color and to reduce dissolved color.

Dissolved iron may be precipitated by aeration or by adding a strong oxidant so that the iron may be removed as a particulate in DE filtration. The use of magnesite (magnesium oxide) has been found to facilitate removal of some forms of iron. Magnesite mixed along with body feed is held for about 10 min to form a negatively charged suspension of magnesium oxide (MgO) that gradually undergoes hydration and solution.

Manganese may be removed in DE filtration with potassium permanganate ($KMnO_4$) added to the body feed, followed by flow detention, usually from 10 to 20 min. Detention time is important and should be determined in bench and pilot tests. The rate of $KMnO_4$ addition and body feed rate depend on the amount of manganese in the source water and other water quality characteristics.

Where iron and manganese are both present in source water, supplementary conditioning must usually be accomplished in separate steps, with iron treatment preceding the $KMnO_4$ addition. When there is a large amount of iron to be removed, it may be necessary to have two filters in series, with iron conditioning preceding the first filter and the addition of $KMnO_4$ and detention between the first and second filters.

Practically the only carryover of solids in DE filter effluents would be very fine DE particles used in precoat and body feed. Although the presence of these particles is innocuous to health, effluent water turbidity must meet the established requirements. Use of a finer DE for precoat can sometimes achieve lower turbidity levels. A slight reduction in the applied flow rate can also help to improve effluent turbidity. Where only slight additional turbidity reduction is needed, the use of a simple cartridge filter following the DE filter accomplishes the desired DE removal. Use of cartridge filters to polish the effluent will be more effective for the lower-capacity DE installations when solids carryover is minimal.

Note that without the addition of separate, additional treatment processes, DE filtration will not reduce the organic content of source water.

Filter Design Considerations

The principal elements of a flat-leaf DE filter are shown in Figure 9.9. Several options are available in designing each of these elements, as well as in the integrated assembly design.

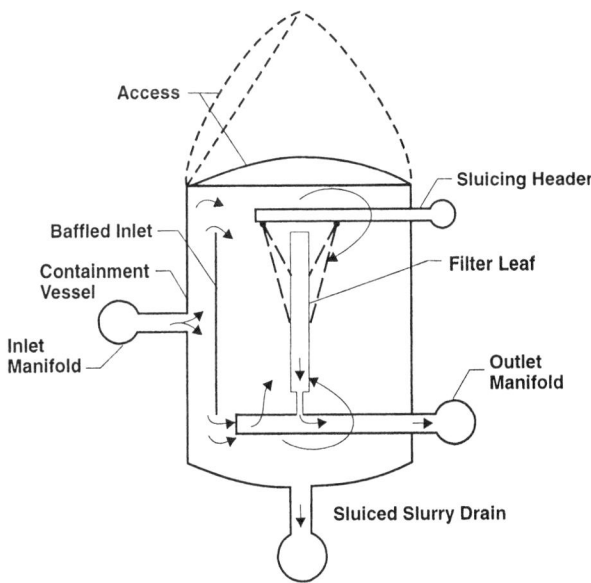

FIGURE 9.9 Principal elements of a flat-leaf filter.

Types of DE Filters. Two basic groups of DE filters are available. If source water is to be forced through the filter under pressure, the containment vessel must be closed, as illustrated in Figure 9.9. Filters operated under a vacuum, on the other hand, may be open vessels.

Although there is theoretically no limitation to what pressure may be applied to a pressure-type filter, practical considerations of pumping costs have limited head loss to a maximum of 35 psi (241 kPa). Most systems used for drinking water filtration are designed for a maximum head loss of 25 psi (172 kPa).

Filter Construction. Pressure filters are always constructed as cylindrical pressure vessels mounted either vertically or horizontally. Most units fabricated today are made of stainless steel for the shell and most internal parts. The type of stainless steel used depends on the corrosivity of the water being treated.

Vacuum filters are built as rectangular tanks. Because of the low differential heads they are subject to, vacuum filter containments and internal parts, except certain structural supports, are most often fabricated of plastics for their chemical resistance and reduced maintenance. For larger units, the containment vessel may be constructed of concrete.

Filter Elements. Inlet water flow is introduced to a DE filter through the containment wall, fitted with an internal baffling device to prevent disturbance of the filter cake. Filter cake may be cleaned from the filter by scraping, vibration, hydraulic bumping (surging), or manually hosing down the septa from the top of an open vessel.

Many arrangements of filter elements are available, constructed in both a tubular and flat form, with the flat (or "leaf") design being by far the most common. Some typical filter element designs are illustrated in Figure 9.10. Filter elements may be mounted either horizontally or vertically, and they may be either fixed in position or able to rotate.

Vertical mounting of the leaves is used almost exclusively for water treatment applications. Most pressure and vacuum filters constructed today have fixed leaves mounted by means of spigot-type "push-on" outlets installed in sockets on a manifold and sealed by O rings or flat gaskets. The outlet manifold is usually located below the leaves to provide them with support while allowing gravity to assist in seating the push-on connections.

Variations for both leaf connections and manifold location are selected depending on operating conditions and requirements for inspection and maintenance. Fixed-leaf pressure filters may also be divided into retracting shell and retracting bundle types for internal access. Both options may be used for any size filter, but the retracting bundle type

Rectangular Leaf   Circular Leaf   Tubular

FIGURE 9.10 Filter elements.

FIGURE 9.11 Common vertical leaf filters.

is generally preferred for larger units. As illustrated in Figure 9.11, the retracting shell or tank design has the shell mounted on wheels on rails. An electric motor or hydraulic piston opens and closes the unit.

In the retracting bundle design, the shell head is suspended from an overhead monorail. The bundle, attached by the manifold and frame to the head, is retracted by means of the monorail, which is usually motor-driven. Internal rails attached to the shell support the end of the leaf bundles when the shell is opened.

Thorough cleaning of the septa at the end of a filter run is important in maintaining peak efficiency. Most units used for water treatment sluice the cake with water sprays, which creates a slurry that can be easily handled and treated and does not require opening the filter vessel. Fixed-leaf filters are usually cleaned with high-pressure spray jets mounted on oscillating spray heads, with single or multiple jets directed between the filter leaves.

Rotating filter leaves usually have a stationary spray header, and coverage is obtained as the leaves rotate past the sprays. Open filters may be cleaned manually using high-pressure sprays and may require covers over the units to contain the spray. Additional devices that may assist in the complete removal of the cake slurry from the filter containment include spray jets in the invert of the vessel or an air scour to suspend the material before the vessel is drained.

Filter Leaf Design

A section of a typical flat leaf showing the principal construction elements is shown in Figure 9.12. The flat filter leaf with a broad surface and limited thickness should be designed with the following goals:

FIGURE 9.12 Typical construction of flat-leaf filter element.

- Leaf and outer frame must be stiff enough to resist warping under the force exerted at maximum differential pressure.
- The unit must have a backing screen to prevent the cloth septum from flexing under gradually increasing pressure.
- The path provided for filtrate flow must not restrict flow and must create minimal head loss through the leaf.

It is essential that the filter cake remain undisturbed. An adequate filter leaf design prevents the possibility of cake movement due to warping of the frame or flexing of the septum.

Central Drainage Chamber. There are three basic types of filter leaf drainage chambers: heavy wire mesh with wire spacing up to 1 in. (2.5 cm); expanded metal sheets that provide a deeper chamber with increased rigidity; and the Trislot, a proprietary design having thin metal bars with welded transverse round or wedge-shaped wires.

Backing Screen. The backing screen is an intermediate screen used when the irregular surface of the central drainage chamber may permit flexing of the cloth septa under conditions of varying pressure.

Filter Septum. Filter septum materials are cloth weaves made with either stainless steel wires or plastic monofilaments. The principal purpose of the septum is to retain the precoat, which must bridge the openings in the weave. Because openings in the weave are larger than the major portion of particulates in precoat material, the precoat is retained by bridging. The cloth septum must be uniformly woven to produce an even precoat that reduces the extent of the recirculation required to deposit the material. The weave must also be designed so that it sluices cleanly, drops the cake freely, and resists plugging and damage.

One of the more common wire cloths for water filtration is a standard 24 × 110 Dutch weave. Another type of weave is the multibraid, composed of bundles of wire in both directions. This weave is less vulnerable to the entrapment of particles and blinding than the standard weave. Woven wire cloth may also be "calendered," which involves passing the cloth through compression rollers to flatten the rounded wire at the surface of the weave. Calendering improves precoat retention characteristics and generally strengthens the cloth against rough treatment.

Plastic cloth is used predominantly for vacuum-type DE filters and is available in a variety of weaves using either polyester or polypropylene monofilament. Plastic cloth may be supplied as a bag to envelop a filter leaf or as a cloth caulked into a leaf frame.

Binding Frame Closures. The binding frame surrounds the filter leaf to prevent leakage around the septum. The outside frame is also the principal structural element to provide rigidity and prevent warping. Depending on the shape, the outside binder may also collect flow from the central chamber and supplement flow routing to the outlet nozzle.

Vacuum Filter Leaves. Vacuum filter leaves used for drinking water treatment are often made of plastic. As shown in Figure 9.13, the central drainage chamber and outlet spigot are molded in a single piece, usually of high-impact styrene. Ridges or other raised patterns provide the required flow path. The raised pattern is spaced so that intermediate screens are not required. The septum, in the form of an envelope with zipperlike closures, is sealed at the bottom outlet by a gasket that also provides tight closure for the manifold connection. Figure 9.14 shows two ways that filter leaves may be mounted in a vacuum filter with supplementary support to improve rigidity.

Bottom Outlet   Detail of Leaf Outlet

FIGURE 9.13 Vacuum filter leaf.

Bottom Outlet   **Vertical Edge Outlet**

FIGURE 9.14 Vacuum leaf mountings.

Outlet Connections. Fixed-leaf filter leaves usually have spigot-type outlet connections made of castings machined to fit into the sockets of the outlet manifold. Figure 9.15 shows two types of outlet construction. The central drain hub shown in the figure is used for rotating leaves and is of two-piece construction, clamped to the center of the circular leaf by bolts. The leaf outlet connection must be of sufficient size to allow full flow of the filtrate collected by the leaf at a minimum head loss. As leaf size and loading increase, the distribution of flow within the leaf and the transition to the outlet connection increase in importance.

Comparison of DE Filter Systems

If DE filtration has been selected for use in a water treatment system, a selection must be made between vacuum and pressure filter options. The length of filter run expected from either type of filter should be determined in pilot testing preceding design evaluation. The two types of filter systems may be evaluated in regard to costs, including the following items:

- Capital cost of installation of the complete filter system and ancillary equipment

FIGURE 9.15 Outlet connections.

FIGURE 9.16 Pressure system schematic.

- Capital costs of structures to house the equipment
- Average length of filter runs
- Cost of DE material required for operation
- Pumping power costs

Pressure Filter Systems. The principal elements included in a pressure filter system are shown in Figure 9.16. In the filtering mode, source water is pumped into the filter shell, and a baffle plate distributes flow and prevents it from eroding the cake on the septa. A flow-measuring device installed in the filter outlet line activates a valve that maintains a constant flow rate.

In the precoat cycle, the source water inlet and filtrate outlet valves remain closed. If the system includes a precoat tank, as shown in the schematic, first the system is filled with previously filtered water to a depth sufficient to reach the top of the shell. Then the pump recirculates the water as precoat slurry is added. In an alternative arrangement, a recirculation tank is not used. With level controls in the filter shell, the unit is first partially filled with water, then filled completely along with the addition of precoat slurry.

Precoat accumulates on the septa as the slurry is recirculated. Because the flow tends to seek the path of least resistance, recirculation tends to concentrate on those leaf areas not previously covered, until a reasonably uniform precoat thickness is eventually formed. After full cover is achieved, the remaining suspended precoat material is intercepted by the filter, and the turbidity of recirculated flow quickly clears up. At this time, in a simultaneous operation, the recirculation valve closes, and the filter inlet and outlet valves open to initiate the filter cycle.

When the maximum available head differential is reached, the filter run is terminated and the filter is cleaned. In the cleaning cycle the inlet valve is closed, the waste valve is opened, and after the unit is fully drained, sluice water is turned on to flush away the filter cake.

9.26 CHAPTER NINE

The principal advantages of the pressure filter system are as follows:

- The ability to operate at application rates of up to 3 gpm/ft^2 (7.8 m/h) requires fewer filter units to process a given amount of water.
- The enclosed filter units generally make a neater installation with minimum housekeeping maintenance.
- A wide range of sizes of pressure units are available, from small up to about 3,000 ft^2 (280 m^2).

The principal disadvantages of the pressure-type precoated filter system are the relatively high initial cost of the filter units and somewhat higher power costs for operation because of the higher operating pressure.

Vacuum Filter Systems. As illustrated in Figure 9.17, a vacuum precoat filter system operates essentially the same as a pressure filter except that the filter is open to the atmosphere and the pump pulls the flow of water through the filter cake. Vacuum-type DE filters operate at flow rates of from 0.5 to 1.0 gpm/ft^2 (1.2 to 2.4 m/h) of leaf area, with 0.75 gpm/ft^2 (1.8 m/h) about average. Total head available for operating the filter system is limited by the suction capability of the pump, generally about 18 in. (460 mm) of mercury.

Because the vacuum filter is an open unit and operations can be easily observed, precoat and cleaning cycles are simplified. Sluicing may be done by fixed or rotating sprays, or the leaves can be manually cleaned with a high-pressure hose.

One limitation of vacuum filters sometimes occurs during operation with cold water. In winter some water is saturated with air, and if the velocity through the cake generates a low enough hydraulic pressure, the air is released from solution to form bubbles that "pop" particles from the precoat, producing turbidity. However, many vacuum filters operated with cold water do not experience this problem.

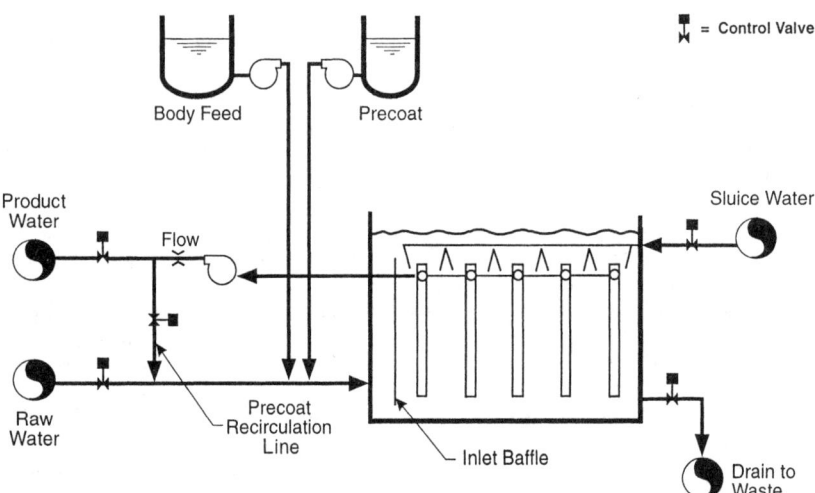

FIGURE 9.17 Vacuum filter system schematic.

Principal advantages of the vacuum filter system are as follows:

- For small plants, the initial cost is about one-third the cost of a stainless steel pressure filter of the same capacity.
- It is easy to observe the precoat and cake accumulation because the unit is open.
- Precoat, filter, and sluicing cycles may be operated manually because the unit is open.
- Operation and maintenance of the units generally requires fewer skills than does pressure filter operation and maintenance.

The principal disadvantages are as follows:

- The relatively low filtration rate requires more or larger units than for pressure units.
- Filter units must be protected from the weather because wind, rain, or ice disturbs the filter cake, and sunlight promotes algae growth.

Ancillary Facilities

DE is a fine powder that has abrasive properties when handled both in dry form and in slurries. Special precautions must be taken to protect plant staff against dust, and equipment and pipelines must be protected against erosion. Because of its abrasive properties, DE breaks down in size under turbulent operating conditions.

DE Delivery. DE can be delivered in 50-lb (22.7-kg) plastic-lined paper bags, in 900-lb (408-kg) woven plastic bags, or in bulk truck or rail car loads. The 50-lb (22.7-kg) bags are usually stacked about 12 to a pallet and are handled with a forklift truck. Bags are then individually lifted by plant staff and discharged through a bag breaker equipped with a dust collector that empties into a dry feeder hopper or into a slurry tank.

The 900-lb (408-kg) bags are delivered, two to a pallet, in an "over-and-under" stack. Special forklift trucks and unloading frames are used for handling and unloading the bags. Bags are transferred from the pallet with a sling to a monorail and then moved into place over a discharge hopper. A spout, tucked into a flap in the bottom of the bag, is unrolled and attached to a discharge spout, and the bag is then unloaded with the assistance of a vibrator mounted on the holding frame. Discharge may be to a pneumatic conveyor for dry bulk storage or to a wetting chamber for pumping and storage as slurry. Because bags are completely enclosed, dust collectors are not required for unloading facilities.

Bulk delivery by truck or railcar may be made directly to storage silos with a pneumatic system similar to equipment used for handling bulk carbon or lime.

Slurry Storage. For small operations, DE slurry is usually stored in fiberglass-reinforced plastic tanks and, for large volumes, in reinforced concrete basins. All storage units must be equipped with continuously operating mixers to keep DE in suspension. Depending on slurry concentration, it may be difficult to resuspend DE once it has been allowed to settle. To limit attrition of DE in the turbulence of mixing, slurries should be used within 3 days.

Slurry Conveyance. Steel pipelines convey slurry either by gravity or by pumping. The flow must be continuous, and at the end of a conveying cycle, the pipeline must be completely flushed. The line should not be flushed to the point of delivery unless flushing water volume is measured and considered in determining the concentration of DE in the slurry at the point of use. Another method used is to deliver flushing water to a holding tank when DE concentration is determined before supplementary DE is added for use in mixing a new batch of slurry.

For DE concentrations of 4% or less, case-hardened centrifugal pumps may be used for slurry transfer. Abrasion-resistant chemical feed metering pumps are also acceptable for these concentrations. Thicker slurries must be conveyed with either peristaltic-type pumps or rubber-lined centrifugal pumps. To minimize breakdown of DE material, pipeline velocities should be limited to no more than 8 ft/s (2.4 m/s), but must be maintained at over 3 ft/s (0.9 m/s) to prevent settling.

Waste Disposal. Because it is predominantly inert, waste DE slurry may be easily treated by thickening with polymers and dewatering on belt filters. DE slurry may also be settled and concentrated in lagoons.

At large installations, DE may be recovered from wastes for reuse as body feed, but it should never be used for precoat. Virgin DE should be used for precoat to provide a final barrier for any pathogens that may pass through the upper levels of the cake. DE recovery saves costs of DE purchases and reduces the costs of waste disposal.

In the recovery process, separation of the lighter source water solids from the denser DE is accomplished by pumping wastes through a series of vertical hydrocyclones. The heavier DE material gravitates to the bottom discharge to exit into a collector tank, while most of the source water solids and the finer fraction of the DE remain in the center core of the unit. The finer material is then removed hydraulically from the top of each unit. Because of the erosive nature of DE slurry, it is necessary to use hydrocyclones constructed of ceramic material and pumps designed for abrasion resistance.

A typical recovery system with connections to the filtration process is shown in Figure 9.18. Recovery rates generally range between 85% and 90%. Because of the possible carryover of some bacteria, viruses, and cysts that have been separated in precoat filtration, it is advisable to disinfect the recovered material before use as body feed. One simple method of disinfection is the application of a heavy dose of potassium permanganate to the material before it is reused. Where *Cryptosporidium* cysts may exist, ozonation or UV may be the only suitable methods of disinfection.

System Controls. Most pressure filter systems are provided with automatic operating systems similar to those for rapid sand filters. Vacuum filter systems may also be auto-

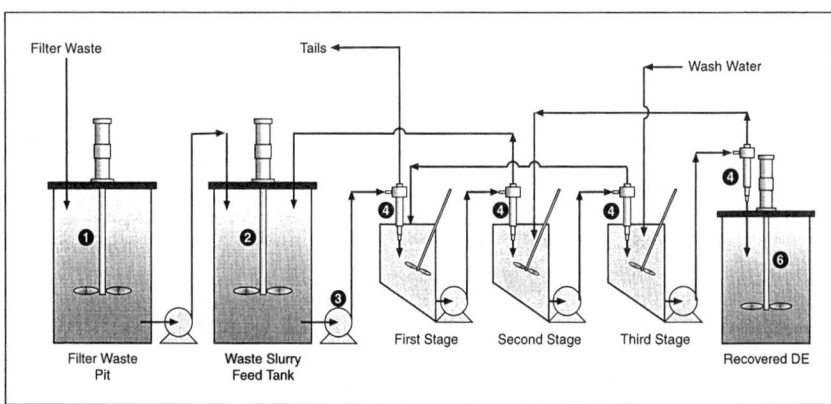

FIGURE 9.18 Schematic of a diatomaceous earth recovery system.

mated or operated manually. Where automation is used, the precoat cycle, body feed, and washing cycles may be initiated and performed automatically. The washing cycle may be automatically initiated when maximum head loss is reached, or the staff operator may be alerted by an alarm so that the cycle may be manually initiated.

A flow-measuring device on the outlet of pressure filters controls system flow with a throttling valve on the product water line. Most installations use a supply pump that operates at a fixed head, and the flow is controlled by the throttling valve. In some installations, it may be advantageous to use a variable-speed supply pump instead of the control valve to maintain required flow. For vacuum filters, a level control may be added to maintain a relatively fixed water level in the filters while the flow device controls outlet flow.

BIBLIOGRAPHY

Slow Sand Filtration

Allen, R. E. "A Method of Cleaning Slow Sand Filters Developed at the West Hartford, Connecticut, Water Treatment Plant." Slow Sand Filtration Workshop, University of New Hampshire, October 1991.

Biologically Enhanced Slow Sand Filtration for Removal of Natural Organic Matter. American Water Works Association Research Foundation. Denver, Colo.: American Water Works Association, 1993.

Calgon Carbon Corporation. *GAC Sandwich Filter.* Pittsburgh, Pa.: Water Services, February 1994.

Christman, R. F., and R. T. Oglesby. "Microbiological Degradation and the Formation of Humus." In K. V. Sarkanen and C. H. Ludxig (eds.), *Lignin: Occurrence, Formation, Structure, and Reactions.* New York: Wiley-Interscience, 1971.

Cleasby, J. L., D. J. Hilmoe, and C. J. Dimitracopoulos. "Slow Sand and Direct In-line Filtration of a Surface Water." *Journal AWWA* 76(12):44, 1984.

Collins, M. R., T. T. Eighmy, J. M. Fenstermacher, and S. K. Spanos. *Modifications to the Slow Sand Process for Improved Removals of Trihalomethane Precursors.* American Water Works Association Research Foundation Report. Denver, Colo.: American Water Works Association, 1989.

Ellis, R. V. "Slow Sand Filtration." *CRC Critical Reviews in Environmental Control* 15(4):315, 1985.

Fox, Kim, M. R. Collins, and N. J. D. Graham. *Slow Sand Filtration: An International Compilation of Scientific and Operational Developments.* Denver, Colo.: American Water Works Association, 1994.

Graham, N. J. D., T. S. A. Mbwette, and L. Di Bernardo. "Fabric Protected Slow Sand Filtration: A Review." Slow Sand Filtration Workshop, University of New Hampshire, October 1991.

Huisman, L., and W. E. Wood. *Slow Sand Filtration.* Geneva, Switzerland: World Health Organization, 1974.

Leland, D. E. "Slow Sand Filtration—Piloting Requirements." Slow Sand Filtration Workshop, University of New Hampshire, October 1991.

Malley, J. P., M. R. Collins, and T. T. Eighmy. "The Effects of Preozonation on Slow Sand Filtration." Slow Sand Filtration Workshop, University of New Hampshire, October 1991.

Manual of Design for Slow Sand Filtration. American Water Works Association Research Foundation. Denver, Colo.: American Water Works Association, 1991.

Mbwette, T. S. A. "The Performance of Fabric Protected Slow Sand Filters." Doctoral dissertation, University of London, United Kingdom.

McNair, D. R., R. C. Sims, D. L. Sorenson, and M. Hulbert. "Schmutzdecke Characterization of Clinopilolite-Amended Slow Sand Filtration." *Journal AWWA* 79(12):74, 1987.

Pardon, Maurice. "Removal Efficiency of Particulate Matter through Vertical Flow Roughing Filters." Slow Sand Filtration Workshop, University of New Hampshire, October 1991.

Recommended Standards for Water Works. Albany, N.Y.: Health Education Services, 1997.

Skeat, W. O. *Manual of British Water Engineering Practice,* 3d ed. "The Biology of Water Supply," Chapter 20. Cambridge, England: Institution of Water Engineers, W. Heffer and Sons Ltd., 1961.

Slezak, L. A., and R. C. Sims. "The Application and Effectiveness of Slow Sand Filtration in the United States." *Journal AWWA* 76(12):38, 1984.

Spencer, C. H., and M. R. Collins. "Water Quality Limitations to the Use of Slow Sand Filtration." Slow Sand Filtration Workshop, University of New Hampshire, October 1991.

Wegelin, M. "Roughing Filters for Surface Water Pretreatment." Slow Sand Filtration Workshop, University of New Hampshire, October 1991.

Whipple, G. C. *The Microscopy of Drinking Water,* 3d ed. "Purification of Water Containing Algae," Chapter 17. New York: John Wiley & Sons, 1914.

Diatomaceous Earth Filtration

Black, H. H., and C. H. Spaulding. "Diatomite Filtration Developed for Field Troops." *Journal AWWA,* November 1944.

Fulton, G. P. *Diatomaceous Earth Filtration for Safe Drinking Water.* Reston, Va.: American Society of Civil Engineers, 2000.

Ongerth, J. E., and P. E. Hutton. "DE Filtration to Remove *Cryptosporidium.*" *Journal AWWA,* December 1997.

Precoat Filtration, M30. Denver, Colo.: American Water Works Association, 1995.

Schuler, P. F., and M. M. Ghosh. "Diatomaceous Earth Filtration of Cysts and Their Particulates Using Chemical Additives." *Journal AWWA,* December 1990.

Standard for Precoat Filter Media, B101. Denver Colo.: American Water Works Association, 2001.

CHAPTER 10
OXIDATION AND DISINFECTION

James C. Hesby
Black & Veatch
Concord, California

The principal health risk from drinking water in most locations is waterborne diseases from microbial contamination. According to the World Health Organization, an estimated 1.7 million deaths a year can be attributed to unsafe water supplies. Within the United States, waterborne diseases are a lower risk, but serious outbreaks still occur with some regularity. The Centers for Disease Control and Prevention (CDC) reported that for the 24-month period 1999–2000, there were 39 outbreaks of waterborne disease associated with drinking water in 25 states. Of the outbreaks for which the cause was identified, 90% were associated with pathogens, and the remainder were associated with chemical poisoning. Most outbreaks in community systems are a result of improper treatment or breaks in the distribution system that allowed for system contamination. Consequently, recent regulations have emphasized improved microbial controls through multibarrier techniques including improved watershed protection to minimize contamination at the source, filtration for contaminant reduction, disinfection for inactivation of the remaining pathogens, and residual disinfectant for the distribution system. Also, recognizing that encysted waterborne protozoans are more resistant to traditional disinfection practices, the regulations are forcing higher levels of disinfection and use of alternative technologies. Concurrently, other regulations seek to reduce the levels of disinfectant chemicals and their reaction by-products, which are considered long-term health risks. This, then, is the designer's challenge, to provide comprehensive microbial reduction to minimize the risk of waterborne disease while also minimizing the formation of disinfection by-products.

This chapter provides a discussion of chemical oxidants that can be used for disinfection, as well as other treatment objectives. Disinfection by ultraviolet light (UV) is covered in Chapter 29.

The concept of waterborne disease was first realized in the 1850s during a cholera epidemic in London. But it wasn't until almost 20 years later that Louis Pasteur and Robert Koch developed the germ theory of disease, and it was another 30 years before the regular use of disinfectants to kill the germs. Continuous chlorination was used for the first time in Lincoln, England, in 1905 to arrest a typhoid outbreak. The first regular use of disinfection in the United States was at the Bubbley Creek Filtration Plant in Chicago in 1908, about the same time that Dr. Harriette Chick first advanced her theory of disinfec-

TABLE 10.1 Physical Constants for Common Disinfecting Agents

Name	Symbol	Molecular weight	Solubility in water at 1 atm and 25° C, g/L	Boiling point, °C	Melting point, °C	Heat of vaporization, cal/g
Chlorine	Cl_2	70.91	7.29	−34.5	−101	68.7
Ozone	O_3	48.00	0.006*	−112	−192	54.0
Chlorine dioxide	ClO_2	67.45	8.0†	10.9	−59	96.6

*190 O_3 by weight.
†Assumes equilibrium with 10% ClO_2 gas phase.

tion. The first use of ozone for disinfection was at Nice, France, in 1910. Since that time, disinfection has become an accepted water supply practice throughout the world. Chlorination has been the dominant method employed, but ozonation has been widely used also. There has also been increasing use of chlorine dioxide as a disinfectant in the United States and Europe.

Some of the physical constants for the three most common disinfecting agents, chlorine, ozone, and chlorine dioxide, are listed in Table 10.1. Note that all three normally exist as gases, although chlorine dioxide liquefies at a temperature near 10° C. Chlorine is available as a compressed liquid, but ozone and chlorine dioxide must be manufactured on-site (ozone because it decomposes, chlorine dioxide because it is dangerous to store in a concentrated compressed form). Chlorine can also be provided as one of the hypochlorites available as a bulk liquid or generated on-site.

Oxidants are used in water treatment to accomplish a wide variety of treatment objectives besides disinfection, including mitigation of objectionable tastes and odors, removal of color, removal of iron and manganese, and oxidation of organic chemicals. Oxidation of contaminants in water by means of aeration is covered in detail in Chapter 5. Information on the chemicals used in water treatment is provided in Appendix A. Information on chemical handling and chemical feed equipment is covered in Chapter 15.

For additional information on the theory and chemical reactions involved in oxidation and disinfection, refer to the companion AWWA text *Water Quality and Treatment*.

REGULATORY FRAMEWORK

A number of regulations have been promulgated by the U.S. Environmental Protection Agency (USEPA) under the Safe Drinking Water Act (SDWA) that affect how or when a water treatment system performs oxidation or disinfection. In addition to the requirements under the Primary and Secondary Regulations, several specific rules have a great impact on oxidation and disinfection process and are described in the following sections.

Secondary Regulations

The National Secondary Drinking Water Standards apply to drinking water contaminants that may adversely affect the aesthetic qualities of water, such as odor and appearance. These qualities have no known adverse health effects, and thus secondary regulations are not mandatory. However, the quantities listed in the Secondary Standards do seriously affect acceptance of water by the public, and for this reason, compliance with the limits established by USEPA is strongly recommended. In addition, some states have enacted more stringent regulations that require compliance with some of the maximum contaminant levels (MCLs) in Secondary Standards.

The contaminants most commonly treated for removal by oxidation under the Secondary Standards are taste and odor, iron, and manganese. These oxidation processes are discussed in greater detail later in this chapter.

The Surface Water Treatment Rule

The Surface Water Treatment Rule (SWTR) was promulgated by USEPA on June 29, 1989. This regulation applies to every public water system in the United States that uses surface water as a source. It also applies to groundwater systems that the state determines might become contaminated by surface water; these systems are labeled *groundwater under the direct influence of surface water,* or *GWUI.*

The purpose of the regulation is to protect the public from waterborne diseases that are most commonly transmitted by contamination of surface water. Because it is difficult to monitor for particular pathogens, such as *Giardia lamblia* and viruses, the SWTR emphasizes treatment techniques as the condition for compliance instead of having MCLs for pathogens. Because of the wide variety of water qualities, local conditions, and methods of treatment, the rule does not prescribe a particular method of treatment but instead offers several alternatives. Any of these methods may be used by a water system to meet the overall goal, which is removal or inactivation of essentially all disease-causing organisms.

To ensure that water quality goals are met, the SWTR contains many operation and monitoring requirements. Studies indicate that viruses and the encysted protozoans (*Giardia* and *Cryptosporidium,* which is addressed in the Enhanced Surface Water Treatment Rule) are among the most resistant waterborne pathogens; therefore, water systems that attain adequate removal or inactivation of these organisms will, to the best of current knowledge, provide adequate protection from other waterborne disease organisms.

Disinfectant Contact Requirements. Most water systems using a surface water source must use sedimentation and filtration to ensure adequate removal of pathogens. Under the SWTR requirements, only water systems with extremely low-turbidity source water may be allowed to operate without filtration, and then it is under very stringent operating and monitoring conditions. All surface water and GWUI systems, whether they provide filtration or not, must practice disinfection under highly specific conditions. Disinfection requirements must be met prior to the water reaching the first customer.

The effectiveness of a chemical disinfectant in killing or inactivating pathogens depends on

- The type of disinfectant used
- The disinfectant residual concentration (denoted by C)
- The time the water is in contact with the disinfectant (denoted by T)
- Water temperature
- The pH of the water, which has an effect on inactivation if chlorine is used

The residual concentration C of a disinfectant in milligrams per liter (mg/L), multiplied by the contact time T in minutes, is called the *CT value.* The *CT* values required by the SWTR to guarantee the necessary reduction in pathogens by various disinfectants may be obtained from tables in publications referenced at the end of this chapter. Each water

TABLE 10.2 Surface Water Treatment Rule Disinfection Requirements

Process	Log removals	
	Giardia cysts	Viruses
Minimum log removal inactivation	3	4
Conventional treatment credit	2.5	2
Remaining for disinfection	0.5	2
Direct filtration credit	2	1
Remaining for disinfection	1	3

system's treatment must be sufficient to ensure that the total process of removal plus inactivation achieves at least 99.9% (3-log) inactivation or removal of *Giardia* cysts and 99.99% (4-log) inactivation or removal of viruses. Source waters that are particularly vulnerable to microbial contamination may require greater log reductions, at the discretion of the primacy agency.

Credit for physical removal of pathogenic organisms is given to properly operated filtration processes, as indicated in Table 10.2. The remaining log inactivation is required to be achieved by the disinfection process. Application of the CT concept is discussed further under the design consideration section for each oxidant.

Approval of Lower CT Values. The CT values presented in the tables provided by USEPA are generally considered to be conservative. Each primacy agency may allow lower CT values for individual systems based on on-site studies showing that adequate inactivation is achieved under all flow and raw water conditions. Protocols and requirements are extensive but may be justified for systems that have unusual circumstances warranting the studies.

Single Point of Disinfection. Systems with only one point of disinfectant application may calculate the CT that is being achieved by the entire system by measuring the disinfectant residual at the *exit of the contact volume*. The multiplication of this residual concentration C and the contact time T through all basins and piping from the application point to the measurement point will provide a conservative CT value. This is the simplest calculation, but as indicated in Figure 10.1, this simple CT calculation does not take credit for the higher disinfectant residual that exists in the contact volume prior to the exit.

An alternate method to calculate the level of disinfection is to use segregated flow analysis (SFA). SFA estimates disinfection in a contact volume by calculating disinfection in a number of theoretical "packets" of water that enter the contact volume. Some packets have very short contact times T but are exposed to high disinfectant concentrations C, while other packets have very long contact times T but are exposed to low disinfectant concentrations C. The CT values for each packet are estimated by multiplying the disinfectant residual with time curve C and the "F curve" T from the tracer test of the contact volume. The CT value for each packet is used to calculate the log inactivation, and then all the log inactivations are summed to calculate the overall level of disinfection in the contact volume. SFA is most applicable to disinfectants with rapid decay rates, such as ozone, because of the conservative nature of the T_{10} concept. (T_{10} is discussed below.) However, SFA must be approved for individual systems by each primacy agency. A similar comprehensive approach is the Integrated Disinfection Design Framework (IDDF). Information regarding the IDDF can be found in AWWARF's publication *Implementation of the Integrated Disinfection Design Framework* (2001).

Multiple Disinfectants or Application Points. Systems that apply disinfectant at more than one point will have to profile the system by computing the CT for each section be-

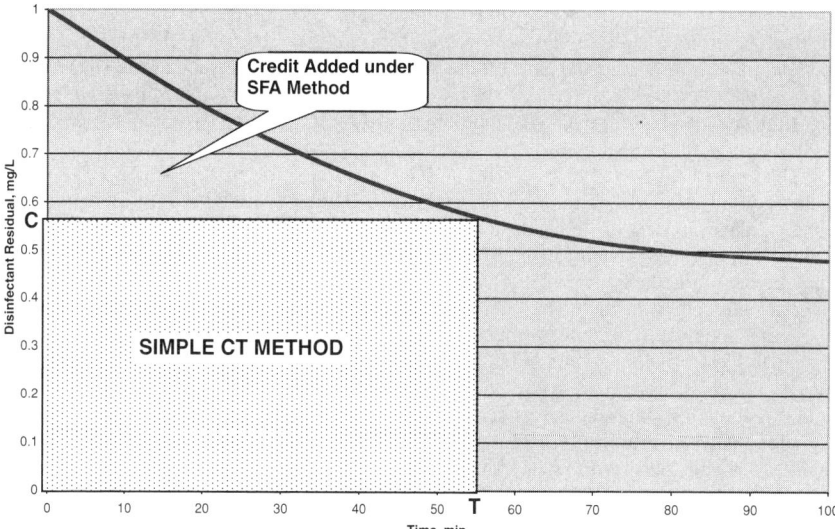

FIGURE 10.1 Disinfection CT.

tween application points. For some systems it may also be advantageous to divide the treatment train into additional sections between the disinfection application points to achieve the greatest CT credit.

Performance Ratio. The concept of performance ratio is a convenient way to monitor and report CT compliance because the required CT often varies with pH and temperature. The *performance ratio* (also called *inactivation ratio*) is the actual CT divided by the required CT and must always be at least 1.0. Many water treatment plants target a performance ratio in excess of 1.0 (for example, 1.10) to operate at an increased level of safety and ensure compliance.

Variations in Peak Hourly Flow. The inactivation credit in each disinfection section of a system is to be determined under the conditions of peak hourly flow. However, in some systems with large reservoirs, peak hourly flow may not occur at the same time in all sections.

To simplify determination of peak hourly flow, USEPA suggests that CT values for all sections be calculated during the hour of peak flow through the last section. This is best determined by a flowmeter immediately downstream of the last section.

Determination of Contact Time. The contact time T used in calculating CT values is the time it takes water to move from the disinfectant application point to the point at which the residual is measured. This time T varies with the configuration and physical characteristics of each individual basin or conveyance, as short-circuiting will occur more in some basins or conveyances than in others.

Contact Time in Pipes. The time during which water is in contact with a disinfectant while flowing through pipes is straightforward. It assumes that water moves in a relatively uniform manner between two points and can be calculated on the basis of uniform plug flow as

$$T = \frac{\text{internal volume of pipe}}{\text{peak hourly rate through pipe}}$$

Contact Time through Reservoirs. Under most conditions, water does not move through reservoirs, tanks, and basins in a uniform manner. Therefore the time T used to compute CT in reservoirs depends on the design of the reservoir, such as the shape, inlet and outlet design and locations, and the baffling. In general, reservoirs with a large length-to-width ratio and with good inlet and outlet baffling minimize short-circuiting and provide the most uniform flow.

The contact time used to calculate the CT is the detention time at which 90% of the water passing through the reservoir is retained within the reservoir—in other words, the time it takes for 10% of the water to pass through the reservoir. This detention time, or contact time, is designated as T_{10}. The value of T_{10} for a reservoir at various flow rates may be determined experimentally by tracer studies or theoretically by approximation.

The most accurate method of determining contact time through reservoirs is by experiments using tracer chemicals such as fluoride or lithium. The studies are performed by feeding controlled amounts of the tracer chemical at the reservoir inlet and making repeated analyses of samples collected at the outlet. Unfortunately, both the contact time and the detention time under various flow rates are not linear functions, so it is recommended that tracer studies be performed using at least four flow rates that span the normal flow range. This information can then be used to construct a curve of detention time versus flow rate that can be used to determine T_{10} at any flow with fair accuracy.

Under certain conditions, the state primacy agency may allow the contact time for a reservoir to be determined by an approximation. The method involves multiplying the theoretical contact time (plug flow) of a reservoir by a rule-of-thumb factor that takes into consideration the reservoir design. Examples of reservoirs with poor, average, and superior baffling conditions are shown in Figures 10.2, 10.3, and 10.4. The shaded areas on the figures indicate areas with little or no flow (dead space) in both a horizontal and vertical perspective, which causes much of the flow to short-circuit directly from the inlet to the outlet. Table 10.3 summarizes the baffling conditions and the proportion of T_{10} to the theoretical contact time for each classification.

FIGURE 10.2 Examples of poor baffling conditions in basins. (*Source: Guidance Manual.*)

FIGURE 10.3 Examples of average baffling conditions in basins. (*Source: Guidance Manual.*)

The detention time and contact time in reservoirs will also vary with the water level, the depth of sludge, and whether the water level is rising or falling, so these factors must be included in the CT calculations.

Contact Time through Filters. Filters usually have good flow distribution, so they can be considered as having conditions similar to superior baffling. The T_{10} for filters can therefore be calculated by subtracting the volume of filter media, gravel, and underdrains

FIGURE 10.4 Examples of superior baffling conditions in basins. (*Source: Guidance Manual.*)

TABLE 10.3 Baffling Classifications

Baffling condition	T_{10}/T	Baffling description
Unbaffled	0.1	No baffling, agitated basin, very low length-to-width ratio, high inlet and outlet velocities
Poor	0.3	Single or multiple unbaffled inlets and outlets, no baffles
Average	0.5	Baffled inlet or outlet with some intrabasin baffles
Superior	0.7	Perforated inlet baffle, serpentine or perforated intrabasin baffles, outlet weir or perforated launders
Perfect (plug flow)	1.0	Very high length-to-width ratio, perforated inlet, outlet, and intrabasin baffles

from the volume of water in the filter box; dividing by the flow rate; and multiplying by the superior factor of 0.7. Experience has shown that the T_{10} for typical filters operating at 4 gpm/ft^2 (10 m/h) is about 15 min.

Disinfectant Monitoring Requirements. The SWTR requires that the disinfectant residual of water entering the distribution system be continuously monitored by water systems serving a population of more than 3,300. The residual cannot be less than 0.2 mg/L for more than 4 h during periods when water is being served to the public. Anytime the residual falls below this level, the water system must notify the state. Systems serving a population of less than 3,300 may take grab samples on an ongoing basis in place of continuous monitoring. Samples must be collected at least as frequently as given in Table 10.4.

The disinfectant residual must be measured at the same points in the distribution system and at the same time as total coliforms are sampled. Disinfectant residuals must not be undetectable in more than 5% of the samples each month for any 2 consecutive months that water is served to the public.

Water systems must submit special reports to the state detailing the monitoring required by the SWTR.

SWTR Changes. The 1989 requirements of the Surface Water Treatment Rule have now largely been superceded by the Interim Enhanced Surface Water Treatment Rule (IESWTR) and the Long-Term 1 Enhanced Surface Water Treatment Rule (LT1ESWTR).

TABLE 10.4 Required Disinfectant Residual Sampling Frequency for Small Systems

Population served	Samples per day*
< 500	1
501 to 1,000	2
1,001 to 2,500	3
2,501 to 3,300	4

*Samples must be taken at intervals prescribed by the state.

The requirements of these two rules are similar, with the IESWTR applying to public water systems (PWS) serving more than 10,000 people and the LT1ESWTR specifying requirements for systems serving fewer than 10,000 people. Both rules are applicable only for systems using surface water or groundwater under the direct influence of surface water, and they establish a limit on finished water turbidity of 0.3 ntu as well as turbidity criteria for individual filters. Also for systems that filter, at least a 2-log removal of *Cryptosporidium* was mandated; but if the system complies with the turbidity limit, the 2-log removal requirement will be met. The rules also include triggers for disinfection benchmarking, disinfection profiling, and filter profiling.

The pending Stage 2 Long-Term Enhanced Surface Water Treatment Rule will increase the minimum removal/inactivation requirement for *Cryptosporidium* to 3-log. Raw water monitoring will determine if additional removal/inactivation beyond 3-log will be necessary. If so, the results of the monitoring dictate the additional treatment and/or inactivation required. USEPA will provide guidance regarding credit received for various treatment processes including inactivation from ozone, chlorine dioxide, and UV. Conventional, and probably direct filtration, plants will receive a 3-log removal credit if turbidity limits are met.

Regulation of Disinfection By-products

All oxidation-reduction reactions will form by-products from the reaction, and the by-products formed from many common oxidants used in water treatment are now subject to regulation. In 1979 trihalomethanes, a group of by-products of chlorination, were regulated at 0.10 mg/L. Further by-product regulation is proceeding under Stages 1 and 2 of the Disinfectants/Disinfection By-product Rules (Stage 1 DBPR and Stage 2 DBPR).

The Stage 1 DBPR was promulgated at the end of 1998, and it set limits for the concentrations of both disinfectants and disinfection by-products, as shown in Table 10.5. If

TABLE 10.5 Regulated Levels for Disinfectants and Disinfection By-products

Parameter	MCLG/MRDLG,* mg/L	MCL/MRDL,* mg/L
Trihalomethanes (total THMs)		0.080
Chloroform	0	
Bromodichloromethane	0	
Bromoform	0	
Dibromochloromethane	0.06	
Haloacetic acids (HAA5)		0.060
Dichloroacetic acid	0	
Trichloroacetic acid	0.3	
Bromate	0	0.010
Chlorite	0.8	1.0
Chlorine	4.0	4
Chloramines	4.0	4
Chlorine dioxide	0.8	0.8

*MCLG = maximum contaminant level goal, MRDLG = maximum residual disinfectant level goal, MCL = maximum contaminant level, MRDL = maximum residual disinfectant level.

sufficient TOC removal was not accomplished through the treatment process, the rule required water systems using surface water or groundwater under the direct influence of surface water to practice enhanced coagulation, i.e., additional removal of organic precursor materials, unless the system could meet at least one of six alternative compliance criteria. The MCLs for the chlorinated DBPs, total trihalomethanes (TTHMs), and sum of five haloacetic acids (HAA5) are computed as a running annual average (RAA) of all monitoring stations under the Stage 1 DBPR.

The Stage 2 DBPR will maintain the MCLs established under the Stage 1 Rule, but the chlorinated DBPs, TTHMs, and HAA5 will now be regulated on a locational running annual average (LRAA) basis. Each location, selected under a new monitoring plan intended to use monitoring sites where high levels of DBPs occur, must meet MCLs of 80 and 60 μg/L for THMs and HAA5, respectively, 6 years after the rule is promulgated. Additionally, after the first 3 years, the Stage 1 DBPR monitoring locations must meet limits of 120 and 100 μg/L for THMs and HAA5, respectively, as LRAA in addition to meeting Stage 1 limits of 80/60 as RAA.

Groundwater Disinfection Rule

Federal regulations require all surface water and GWUI systems to practice disinfection and maintain a chlorine residual in the water entering the distribution system; but there is no federal requirement for groundwater systems to practice disinfection. Many states have gone beyond the federal requirements and currently require all or certain classes of public water systems to practice chlorination.

The proposed Ground Water Rule (GWR) was published in the *Federal Register* on May 10, 2000, and is expected to be promulgated in final form during the fall of 2003. This rule is intended to protect the public from consuming contaminated drinking water from groundwater sources by identifying which source waters are subject to fecal contamination and then specifying corrective action. The rule will apply to approximately 157,000 public water systems using groundwater. Disinfection is not mandatory but may be required for vulnerable systems.

The GWR relies on five key components to accomplish its objectives:

- Sanitary surveys that will identify deficiencies in water systems
- Hydrogeologic sensitivity studies for systems that do not disinfect
- Source water microbial monitoring for some systems
- Corrective action if deficiencies are identified
- Compliance monitoring for groundwater systems that disinfect to ensure they achieve a minimum of 4-log inactivation and/or removal of viruses

CHEMICAL OXIDATION TREATMENT

Oxidation-reduction (redox) reactions form the basis for many water treatment processes addressing a wide range of water quality objectives. These may include removal of iron, manganese, sulfur, color, tastes, odor, and synthetic organics (herbicides and pesticides). A redox reaction consists of two half-reactions: the oxidation reaction, in which a substance loses, or donates, electrons; and the reduction reaction, in which a substance gains, or accepts, electrons. An oxidation reaction and a reduction reaction must always be coupled because free electrons cannot exist in solution and electrons must be conserved.

Oxidizing agents, or oxidants, used in water treatment include chlorine, chlorine dioxide, permanganate, oxygen, and ozone. The appropriate oxidant for achieving a specific water quality objective depends on a number of factors, including raw water quality, specific contaminants, and local chemical and power costs. For critical applications, the designer should insist on bench- or pilot-scale evaluations of treatment alternatives to select the best approach and determine appropriate design criteria.

Iron and Manganese Removal

Iron and manganese have secondary MCLs (SMCLs) of 0.3 and 0.05 mg/L, respectively. These SMCLs have been considered as safe limits to avoid the staining of plumbing fixtures and laundry, but experience shows that lower levels are desired to avoid difficulties. Targets of < 0.1 mg/L iron and < 0.02 mg/L manganese should be normal water quality goals.

Dissolved iron and manganese are normally in the reduced state (Fe II and Mn II) and can be removed by oxidizing to Fe III and Mn IV, where they will precipitate as $Fe(OH)_3$ and $Mn(OH)_2$. Precipitates are subsequently removed in sedimentation and/or filtration steps. Several oxidants are available for this process, namely, chlorine, chlorine dioxide, ozone, and potassium permanganate. They are also removed through conventional lime softening treatment. The stoichiometric requirements for each oxidant are given in Tables 10.6 and 10.7. Most oxidants also react with organics in the water, so some testing is normally required to determine the appropriate dose.

Small well water systems with excessive levels of iron and manganese often apply an oxidant, provide a period of detention for the reaction to take place, and then remove the precipitated iron and manganese with a pressure filter. Figure 10.5 illustrates such a system.

Chlorine is an effective oxidant for iron, and oxidation to $Fe(OH)_3$ proceeds rapidly. It is not as effective for manganese removal, however, at normal pH conditions. Chlorine

TABLE 10.6 Oxidation of Manganese

Reaction	Oxidant, mg/mg Mn^{2+}	Alkalinity used, mg/mg Mn^{2+}	Sludge,* lb/lb (kg/kg) Mn^{2+}
A. Oxygen $2MnSO_4 + 2Ca(HCO_3)_2 + O_2$ $= 2MnO_2 + 2CaSO_4 + 2H_2O + 4CO_2$	0.29	1.80	1.58
B. Chlorine $Mn(HCO_3)_2 + Ca(HCO_3)_2 + Cl_2$ $= MnO_2 + CaCl_2 - 2H_2O + 4CO_2$	1.29	3.64	1.58
C. Chlorine dioxide $Mn(HCO_3)_2 + 2NaHCO_3 + 2ClO_2$ $= MnO_2 + 2NaClO_2 + 2H_2O$ $+ 2CO_2$	2.46	3.60	1.58
D. Potassium permanganate $3Mn(HCO_3)_2 + 2KMnO_4$ $= 5MnO_2 + 2KHCO_3 + 2H_2O$ $+ 4CO_2$	1.92	1.21	2.64

*Sludge weight based on MnO_2 as the precipitate. It is highly probable that portions of the sludge will consist of MnOOH and $MnCO_3$.

TABLE 10.7 Oxidation of Iron

Reaction	Oxidant mg/mg Fe^{2+}	Alkalinity used, mg/mg Fe^{2+}	Sludge,* lb/lb (kg/kg) Fe^{2+}
A. Oxygen $4Fe(HCO_3)_2 + O_2 + 2H_2O$ $= 4Fe(OH)_3 + 8CO_2$	0.14	1.80	1.9
B. Chlorine $2Fe(HCO_3)_2 + Ca(HCO_3)_2 + Cl_2$ $= 2Fe(OH)_3 + CaCl_2 + 6CO_2$	0.64	2.70	1.9
C. Chlorine dioxide $Fe(HCO_3)_2 + 2NaHCO_3 + ClO_2$ $= Fe(OH)_3 + NaClO_2 + 3CO_2$	1.21	2.70	1.9
D. Potassium permanganate $3Fe(HCO_3)_2 + KMnO_4 + 2H_2O$ $= 3Fe(OH)_3 + MnO_2 + KHCO_3$ $+ 5CO_2$	0.94	1.50	2.43

*Sludge weight based on $Fe(OH)_3$ as the precipitate. It is highly probable that portions of the sludge will consist of $FeCO_3$.

is also used to maintain $MnO_2(s)$ coatings on filter media, which allows for the continuous removal of manganese through the filter bed.

Potassium permanganate, $KMnO_4$, has often been used at treatment plants for oxidation of these two dissolved species. Potassium permanganate is best used at the front of the treatment works to allow contact prior to the introduction of other chemicals. The re-

FIGURE 10.5 Chlorination, detention, and filtration (iron and manganese removal).

actions have been found to be both pH- and temperature-dependent. Permanganate is also used for the regeneration of manganese greensand filter media.

Ozone has been effectively used for iron and manganese oxidation, often in conjunction with protection against tastes and odors. Iron is almost instantaneously oxidized by ozone while manganese is somewhat slower, but even this reaction is normally complete within about 30 s.

Oxidation using chlorine dioxide is generally complete within a few seconds. Chlorine dioxide is also normally applied at the head of the treatment process. It has the advantage over chlorine of not forming chlorinated DBPs, but a regulated DBP, chlorite, is one of its by-products. Depending on the concentrations of iron and manganese in the raw water, the chlorite formed may limit the use of chlorine dioxide for this application unless the chlorite is subsequently removed by another process.

Lime softening is also an effective way of removing iron and manganese. Again, the reduced forms are oxidized to insoluble precipitates, and they are removed along with the lime sludge. The chemical reaction that describes this oxidation process is as follows:

$$4Fe(HCO_3)_2 + 4Ca(OH)_2 + 2H_2O + O_2 \rightarrow 4Ca(HCO_3)_2 + 4Fe(OH)_3$$

Taste and Odor Control

Most objectionable tastes and odors that occur in raw water, particularly those of organic nature, can be mitigated by judicious application of a preoxidant. Surface waters in particular are prone to taste and odor problems from the presence of algae, other odor-causing organisms, and decaying vegetation.

The most well-known and common odor-causing compounds associated with algae are methylisoborneol (MIB) and geosmin. Both are produced by actinomycetes and various blue-green algae and are particularly resistant to oxidation. Because of the competing reactions with an oxidant and the potential for creating new taste and odor problems, or of increasing toxin levels through oxidation of blue-green algae, bench- or pilot-scale studies should be conducted to establish design criteria and select the appropriate oxidant for specific taste and odor problems. Particularly tough applications may require both oxidation and an adsorption step to lower tastes and odors to acceptable levels.

CHLORINATION

Although a good deal of work on modeling disinfection has recently been done, the principal disinfection theory in use today is still the Chick-Watson theory, which was developed by Dr. Harriette Chick and refined by H. E. Watson in 1908. In simplified form, this theory states that the rate of destruction of pathogens by a disinfectant is proportional to the number of pathogens and the concentration of the disinfectant. At present the maintenance of free chlorine residuals is the principal means by which water utilities ensure that the water they provide is properly disinfected. *Free residual chlorine* is a term used to refer to a chlorine residual that is not combined with ammonia or organic nitrogen.

Chlorine is most often available commercially in pressurized vessels containing both liquid and gas fractions, as sodium hypochlorite (household bleach), and as calcium hypochlorite. Gaseous chlorine is most often employed by larger utilities because of its significantly lower cost; however, transportation and storage of gaseous chlorine do impose a certain risk of serious accidents, and the Uniform Fire Code now treats gaseous chlorine as an acutely hazardous material (AHM). Consequently, many utilities have switched to hypochlorite, to minimize safety problems in densely populated areas.

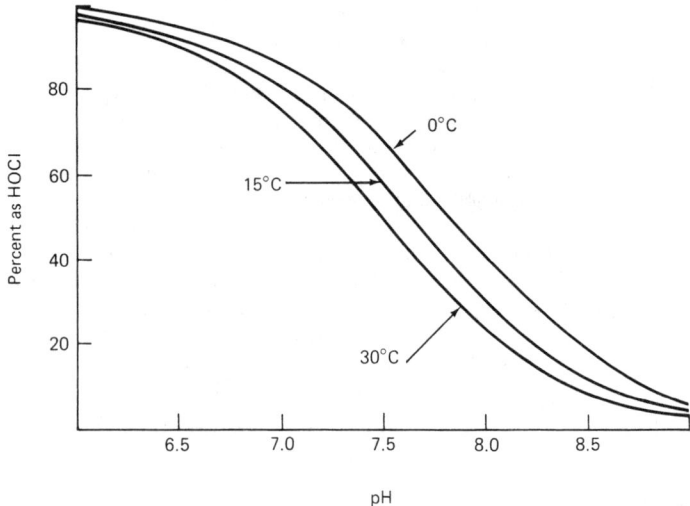

FIGURE 10.6 Influence of pH and temperature on fraction of aqueous chlorine as HOCl.

Chlorine Chemistry

When chlorine gas is dissolved in water, it quickly forms hypochlorous acid (HOCl), which then disassociates to form a mixture of the acid and hypochlorite ion (OCl^-). Although both forms are disinfectants, the acid is more effective, so the effectiveness of free chlorine as a disinfectant depends primarily on the amount of hypochlorous acid available to react with the pathogens. This fraction is most important because the HOCl is nearly 1,000 times more effective than the ionized form of OCl^-.

The relative proportion of these two forms depends on the pH of the solution. The fraction of chlorine present as HOCl at various pH levels and at various temperatures is shown in Figure 10.6. Generally speaking, the HOCl species dominates at pH levels greater than 8. Between pH 7 and 8 the speciation of aqueous chlorine is highly pH-dependent. Note that although pH is the dominant factor, changes in temperature also result in a modest change in the proportion.

Chloramination

If free chlorine is added to completely pure water, the free residual will be the same as the amount of chlorine added. However, if there are any contaminants in the water, the chlorine will react with them. The contaminants could be microorganisms, organic or inorganic compounds that can be oxidized (such as iron and manganese), or ammonia. The amount of chlorine used in the reaction with these contaminants is called the *chlorine demand.*

When chlorine reacts with ammonia, chloramine compounds are formed. The ammonia may be naturally occurring in the water or may be added in the treatment process to purposely form chloramines. Chloramines are useful disinfectants in some situations and are referred to as *combined residual chlorine.*

Understanding of the reactions between chlorine and ammonia is important to an adequate understanding of chlorination chemistry. These reactions are complex, and the temperature, pH mixing regimen, and Cl_2/NH_3 weight ratio all influence both the rate and the products of the reaction. When a small amount of chlorine is added ($Cl_2/NH_3 < 4$), monochloramine is the dominant species formed. As more chlorine is added, dichloramine and trichloramine are formed, along with other products such as NO_3 and N_2 gas. The following are some of the typical reactions that may occur:

Monochloramine: $NH_4^+ + HOCl = NH_2Cl + H_2O + H^+$
Dichloramine: $NH_2Cl + HOCl = NHCl_2 + H_2O$
Trichloramine: $NHCl_2 + HOCl = NCl_3 + H_2O$
Nitrogen: $2NH_4^+ + 3HOCl = N_2 + 5H^+ + 3Cl + 3H_2O$

The relationships between these reactions are shown in Figure 10.7. Monochloramine is the major compound formed until the mg Cl_2/mg NH_3 ratio exceeds 4. This is the compound desired when chloramines are used for disinfection in potable water. The recommended Cl_2/NH_3 weight ratio for the formation of monochloramine is 3 to 4 because this will minimize the concentration of unreacted ammonia remaining in the water.

Weight ratios of Cl_2/NH_3 greater than 4 but less than 8 should be avoided because dichloramine is formed in this region. Dichloramine is a disinfectant, but it also produces undesirable tastes and odors. Formation of nitrogen gas is not shown in Figure 10.7, but it also starts to occur in this region.

The low point in the total chlorine residual curve shown at a Cl_2/NH_3 weight ratio of 8 in Figure 10.7 is called the *breakpoint*. Increasing the Cl_2/NH_3 ratio above this value will produce a free-chlorine residual. Nitrogen gas is also produced in the breakpoint reaction. A dose of 9 to 10 mg Cl_2/mg NH_3 should be used for designing breakpoint facilities.

Increasing the free-chlorine residual to values well in excess of the breakpoint is counter productive because nitrogen trichloride will start to form. This is a sparingly soluble, foul-smelling gas that will cause consumer complaints. Excessive chlorine doses should be avoided.

Saunier (1976) has extensively modeled the chloramine reactions in both potable water and wastewater. Additional insights to the reactions between these compounds are available in his paper.

Combined residual chlorination was first used at Ottawa, Ontario, Canada, in 1916. The process enjoyed modest success as a way to eliminate tastes and odors produced by the use of free chlorine until 1939, but was widely discontinued during World War II because ammonia was not available for civilian use. Combined residual chlorination did not return to widespread use until the concentrations of chlorinated disinfection by-products (TTHMs and HAA5) in the water delivered to consumers were regulated by the USEPA. Use of a monochloramine residual in the distribution system became common at that time because combined chlorine did not produce these by-products. However, the use of this compound is not a panacea. Its germicidal action is substantially less than that of free chlorine, and the ammonia released during decay of this compound is a food source for nitrifying bacteria.

The USEPA permits the use of chloramine as a primary disinfectant if the *CT* requirements published by this agency are met. However, chloramine is much more commonly used as a secondary disinfectant in the distribution system after primary disinfection is achieved by the use of free chlorine, ozone, chlorine dioxide, or UV in the treatment plant. The concentration of chloramine required will depend upon the size of the distribution system and the decay rate of the residual. The maximum chloramine concentration allowed by USEPA is 4 mg/L as Cl_2. Most facilities provide an initial residual from 2.5 to 3.5 mg/L to minimize taste and odor complaints from consumers. The minimum chlo-

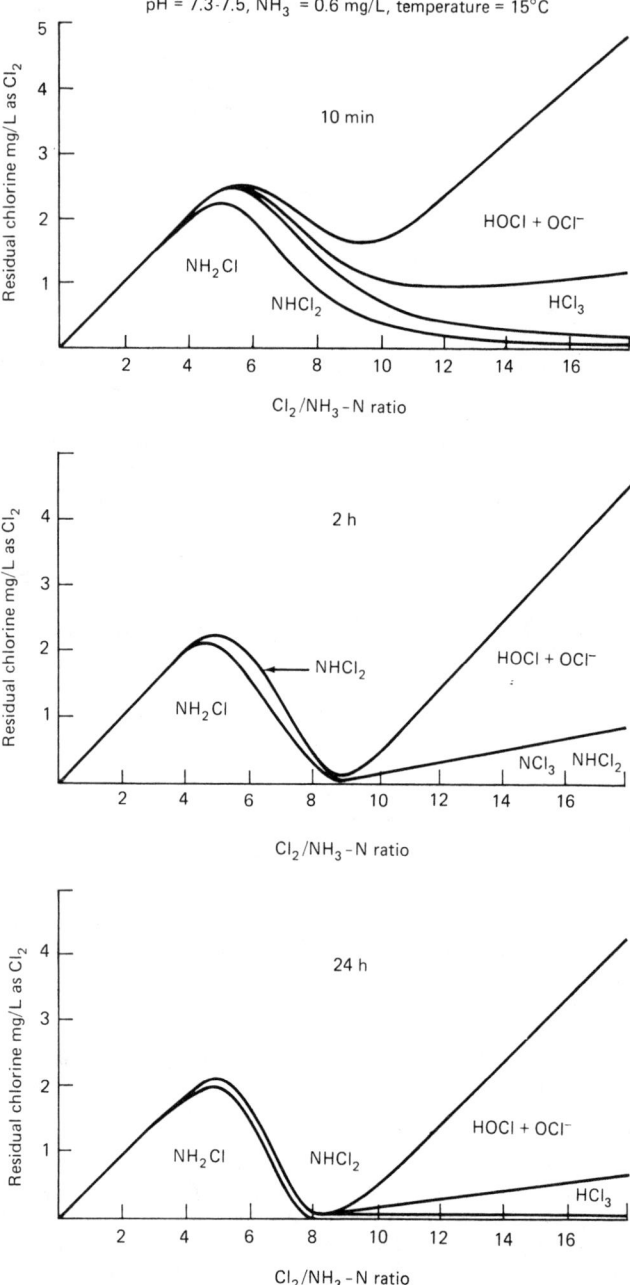

FIGURE 10.7 Free residual chlorine curves. (*Source: A. Palin, "A Study of the Chloro Derivatives of Ammonia," Water Engineering, vol. 54, December 1950, p. 151.*)

ramine residual permitted at the far end of a distribution system depends upon individual state regulations. However, a residual of less than 0.5 to 1.0 mg/L will make the system vulnerable to the onset of biological nitrification. Once nitrification starts to occur, the only way to correct this problem is to change to the use of a free-chlorine residual until the nitrifying microorganisms have been inactivated. Some utilities switch from a combined to a free-chlorine residual at regularly scheduled intervals to prevent the onset of nitrification. The placement of rechlorination stations at strategic locations in the distribution system can also be used to reform monochloramine from the ammonia released during the decay process.

Adding refrigeration-grade anhydrous ammonia, ammonium sulfate, or ammonium hydroxide to water containing free chlorine will form chloramine. Both the chlorine and the ammonia must be very rapidly mixed into the complete volume of water to prevent the formation of trace amounts of di- or trichloramine. The chemical feed system must be designed to maintain a constant ratio between the free-chlorine residual and the quantity of ammonia-nitrogen being added to the system.

A comprehensive water quality monitoring program should be conducted throughout distribution systems using a chloramine residual. Parameters measured should include combined chlorine residual, free ammonia, nitrite and nitrate ion, as well as the bacteriological and physical measurements routinely performed on these samples. The most sensitive indicator of the onset of biological nitrification will be the presence of nitrite ion in the sample.

Monochloramine provides a long-lasting residual in distribution systems, but it is inherently unstable even in the absence of reactive substances. The net reaction, simplified from the approximately 14 individual reactions that govern it, can be written as

$$3NH_2Cl = N_2 + NH_3 + 3Cl^- + 3H^+$$

A kinetic model describing chloramine formation and auto-decomposition has been developed by Jafvert and Valentine (1993) and Vikesland et al. (1998). Use of this model is relatively complicated. However, in the absence of chlorine demand reactions, monochloramine decay can be estimated by a simple second-order equation. The integrated form of this relationship is

$$\frac{1}{[NH_2Cl]} - \frac{1}{[NH_2Cl]_0} = {}^kVCSC^t$$

where ^kVCSC is the Valentine chloramine stability coefficient. This coefficient increases with decreasing pH and the initial chloramine concentration that determines the total free ammonia present in the system. It also increases with increasing total inorganic carbon and temperature of the water. Utilities may use this coefficient to calculate the effects of water quality and chloramination practices (say, mg Cl_2/mg HN_3) on monochloramine stability. Observed decay rates that exceed those predicted by the stability coefficient can be used to locate problems in the distribution system, such as the presence of oxidizable iron or organic slime on pipe walls. Observed decay rates that are less than the stability coefficient may point to the existence of relatively stable organic chloramines in the water.

The chemistry of chloramination in the presence of bromide is not completely understood. However, it is well documented that several compounds may be produced by the reactions between chlorine, ammonia, and bromide. Both bromamines (NH_2Br, $NHBr_2$, and NBr_3) and bromochloramine ($NHBrCl$) have been found (Wajon and Morris, 1980; Symons et al., 1998). The Metropolitan Water District of Southern California reported an increase in TTHMs from 10 to 20 μg/L while using chloramines during a drought when the bromide level in the raw water increased to 0.5 mg/L.

In addition to affecting TTHM concentrations and speciation, water containing bromide will interfere with chlorine residual measurements. A common method of determining chlorine residuals utilizes DPD (N,N-diethyl-p-phenylenediamine). Valentine (1986) studied the use of DPD in the presence of bromochloramine. He concluded that bromochloramine reacts as half bromamine and half chloramines. It is therefore measured as both free chlorine and total chlorine.

Chlorine Gas Systems

Each chlorine system consists of the following system components (see Figure 10.8): storage media, vacuum regulator, gas feeder, and an injector or gas induction unit. For some systems evaporators may be used in the system. Chlorine in the system upstream of the vacuum regulator (storage systems and evaporators) is pressurized, and chlorine downstream of the vacuum regulator is maintained at a vacuum.

Storage, Evaporators, and Vacuum Regulators

Chlorine is delivered to the site in the storage media, which can be 150-lb (68-kg) cylinders, 1-ton (1,016-kg) containers, tank trucks, or railroad tank cars. In all these vessels, liquid chlorine occupies approximately 85% of the volume when the product is delivered. This is to provide room for the expansion of liquid chlorine if the container should be heated.

Chlorine cylinders and containers should never be directly heated. As a safety precaution, their outlet valves are equipped with a small fusible plug that melts at approximately 158° F (70° C) and releases some chlorine to cool the cylinder before a more serious accident can occur.

Small to medium-size water systems generally withdraw gas from the top of the container. The maximum withdrawal rate with this method is about 40 lb (18 kg) per day for a 150-lb (68-kg) cylinder and 400 lb (180 kg) per day for a 1-ton (1,016-kg) container. Higher feed rates can be obtained by connecting two or more cylinders or containers to feed simultaneously. The temperature of the chlorine feed room should be maintained at about 65° F (18° C).

Provision must also be made for a weighing device. Weighing devices are usually a small platform scale on load cells for cylinders and trunnions on load cells for ton containers. Frequently, two weigh scales are provided to facilitate the determination of chlorine consumption when empty containers need replacing.

If the containers being used are 1 ton (1,016 kg) or larger and the withdrawal rates exceed those available with the direct evaporation method described earlier, chlorine evaporators may be used. Evaporators are available in several capacities, but they are all on the same 10,000 lb (4,535 kg) per day chassis. Smaller evaporators simply have smaller heaters. When an evaporator is used, liquid chlorine is withdrawn from the bottom of the container and transported to the evaporator, where it is converted to a gas. The most common type of evaporator uses an electric resistance heater in a hot water bath surrounding a vessel in which the liquid chlorine is converted to gas.

The heat of vaporization of chlorine is very low, approximately 69 cal/g, compared with 540 cal/g for water. Commercially available chlorine evaporators are designed with extra capacity to ensure that the existing gas is superheated and does not recondense on the downstream side. When an evaporator is being used beyond its capacity, misting occurs. Misting is a severe problem in a chlorine system as the pipe materials generally used downstream of the vacuum regulator (and some vacuum regulators) cannot withstand contact with liquid chlorine and will rapidly degrade and fail under those conditions. A chlorine gas filter should be installed on the exit gas line from the evaporators to remove im-

FIGURE 10.8 Chlorine system. CL, chlorine; CLV, chlorine under vacuum; CLS, chlorine solution; SA, sample line; UW, utility water. (*Courtesy of MWH Americas Inc.*)

purities in the chlorine that would be detrimental to the chlorine gas feeder. Evaporators should be equipped with an automatic shutoff valve interlocked to the evaporator operation to prevent liquid chlorine from passing to the chlorine feeder.

When possible, all portions of the chlorine feed system that contain liquid chlorine should be designed and operated with all the liquid in the system as a continuous medium. To shut down the evaporator, it is necessary only to close the effluent valve on the evaporator. No other valves between the evaporator effluent valve and the liquid chlorine container should be shut. Long liquid chlorine lines should be avoided, but if they are unavoidable, chlorine expansion chambers should be provided. It should be emphasized that liquid chlorine has a high temperature expansion coefficient. Unless expansion is permitted, the temperature increase in trapped liquid will result in pressure high enough to rupture the pipes. As such, any sections of liquid chlorine piping that can be isolated should be fitted with an expansion chamber. Economical design should limit the number of expansion chambers, and as such isolation valves should be used only as strictly necessary.

In chlorine system design, the storage media, evaporators, and vacuum regulators are generally kept in a dedicated room and the balance of the equipment in a separate room. Because of the extreme health hazard associated with exposure to concentrated chlorine, individual exposure to chlorine should be limited as much as possible. Chlorine storage rooms should have controlled entry points, and access to these rooms should be limited to qualified individuals. To prevent chlorine release to the surrounding area and to limit exposure to individuals, some building codes require either that the chlorine storage area have a chlorine gas scrubber or that the cylinders or ton containers be kept in a secondary vessel or "coffin."

For very large chlorine systems, tank trucks are often used. In this case, automatic shutoff actuators should be installed to isolate the truck by closing the truck valves in the event of a chlorine leak. This requirement is dictated by recent changes to the law enforced by the Department of Transportation. This requirement is only enforceable if the truck cab is connected to the trailer; however, it is still recommended that these automatic shutoff systems be installed. These automatic shutoff systems are also recommended if the storage medium is railroad tank cars. Equivalent shutoff systems are available for cylinders and ton containers. The systems for tank trucks and railroad tank cars must be pneumatically driven. The systems for cylinders and ton containers are generally electrically driven. While chlorine leaks or releases from 150-lb cylinders and 1-ton containers can be contained or scrubbed relatively easily, releases from tank trucks and railroad tank cars are difficult to contain. This should be considered when addressing security needs.

Vacuum regulators come in various sizes and configurations. In general, they are usually wall-mounted, but for small systems where gas withdrawal is used, vacuum regulators can be mounted directly on the cylinder or ton container. These regulators are suitable only if a single cylinder or ton container is used at a time. Because of small differences in the physical construction when multiple vacuum regulators are used in parallel, they will open to different degrees. As a result, it becomes possible to pull chlorine at unequal rates from the multiple containers, and liquid chlorine can be pulled through the vacuum regulators. As stated above, the piping materials generally used downstream of the vacuum regulators cannot withstand contact with liquid chlorine and will rapidly degrade and fail under those conditions.

Gas Feeders and Injectors/Gas Induction Units

Chlorine gas feeders are commonly called *chlorinators*. Often this includes the injector (shown in Figure 10.9), but sometimes it does not. For clarity, rather than the term *chlorinator,* the term *chlorine gas feeder* is preferred.

OXIDATION AND DISINFECTION   10.21

FIGURE 10.9 Flow diagram for conventional chlorinator. (*Courtesy of Wallace and Tiernan Division of USFilter.*)

A conventional chlorine gas feeder consists of an inlet pressure-reducing valve, a rotameter, a metering control orifice, and a vacuum differential regulating valve. A simple schematic is shown in Figure 10.9. The chlorine gas comes to the chlorine gas feeder and is converted to a constant pressure (usually a mild vacuum) by the influent pressure-reducing valve. The chlorine then passes through the rotameter, where the flow rate is measured under conditions of constant pressure (and consequently constant density), and then through a metering or control orifice.

A vacuum differential regulator is mounted across the control orifice so that a constant pressure differential (vacuum differential) is maintained to stabilize the flow for a particular setting on the control orifice. The flow through the control orifice can be adjusted by changing the opening on the orifice. The control orifice has a typical range of 20 to 1, and the vacuum differential regulator has a range of about 10 to 1. Thus the overall range of these devices combined is about 200 to 1. On the other hand, a typical rotameter has a range of about 20 to 1.

Chlorine feeders are available in standard overall sizes (500, 2,000, and 10,000 lb/day) and then are fitted with rotameters, which either match the capacity of the overall feeder or are somewhat smaller. For instance, if the required flow rate necessary for an application is 5,000-lb/day, a 10,000-lb/day model would be selected and a 5,000-lb/day rotameter would be installed within the unit. Thus the chlorine should be selected based on future design capacities, and the rotameter installed at any particular time should be appropriate for current demands. In addition, note that while the rotameter has a 20-to-1 turndown capability, this is only in manual mode. In automatic mode, a chlorine feeder only has a 10-to-1 turndown capability.

The driving force for the system comes from the vacuum, which is created by the chlorine injector or gas induction unit. As the chlorine is dropped to a vacuum at the vacuum regulator, the chlorine must be pulled through the vacuum regulator, piping, and chlorine feeder. As a result, if the injector or gas induction unit is not operating correctly, the chlorine feeder will not be able to operate. In extreme cases, there may be insufficient vacuum to open the vacuum regulator, in which case no chlorine would be fed at the feed point. As such, proper design and selection of injectors and gas induction units is critical to the application of a chlorine system.

Injectors rather than gas induction units are used predominantly to generate the vacuum required for chlorine systems. Injectors serve two purposes. First, they generate the vacuum necessary for the system. Second, they put the gaseous chlorine into a liquid solution so that it can be easily dispersed into the process flow. Injectors (eductors, ejectors) operate on the Bernoulli principle wherein a volume of water is passed by a small orifice generating a vacuum. The predominate variables that affect the proper operation of an injector are the quantity of flow (gpm) and pressure (psig) of the source water, which acts as the motive force for the injector and the backpressure of the generated chlorine solution. If any of these parameters is insufficient for the selected injector, the injector may generate only a light vacuum or may not function at all.

Gas induction units operate similarly to injectors but rather are immersed in the process water and use a motorized impeller to pull water past the orifice to generate a vacuum. Gas induction units have the additional benefit of providing very efficient mixing of the chlorine solution with the process water.

It should be mentioned that both injectors and gas induction units must be matched to the capacity of the chlorine feeder. For instance, a chlorine feeder with a 10,000-lb/day chassis and 5,000-lb/day rotameter would require a 5,000-lb/day injector or gas induction unit. A 10,000-lb/day injector or gas induction unit could be used but would result in a more expensive unit and a larger water or power demand.

A final issue in regard to injectors is location. Injectors can be positioned either local to the chlorine gas feeders or remote of the chlorine feeders (local to the feed point). Both have benefits and disadvantages. The advantages of injectors local to the gas feeders is that if there are multiple feed points, only one water source is required and all the chlorine equipment is in a single location (resulting in simplified operations and management (O&M)). The disadvantages of injectors local to the gas feeders are that a pressurized chlorine solution must be distributed throughout the plant and that backpressure on the injectors may be excessive, resulting in large water demands. The advantages of injectors remote of the chlorine feeders include the elimination of distribution of a pressurized chlorine solution and less backpressure on the injectors. The disadvantages of injectors remote of the chlorine feeders include distributed equipment (resulting in increased operations and management) and the requirement of an extensive plant water system. The choice between remote and local installation should be made based on project requirements and client preferences.

Chlorine Piping and Valving

There are six distinct areas of concern regarding piping and valving in a chlorine system: ton container to the piping system (flexible connections), pressurized liquid chlorine piping, pressurized gas chlorine piping, vacuum piping between the vacuum regulator and chlorine feeder, vacuum piping between the chlorine feeder and the injector or gas induction unit, and the chlorine solution piping between the injector and the feed point.

Sizing of the pressurized chlorine piping is trivial and will result in very small-diameter piping for pressure drop requirements. One-inch piping minimum should be used for these installations. It is very unlikely that larger piping will be required.

Sizing of the vacuum lines is more complicated. As the chlorine is under vacuum and the pressure is very small, any pressure drop of any magnitude results in a significant change in the density of the compressible chlorine gas. As a result, the use of the Darcy equation is unsuitable for the calculation of pressure drop in vacuum lines, as the chlorine gas is *not* incompressible (an assumption built into the Darcy equation). Rather the use of either the isothermal or adiabatic equations is recommended for this application as both are suitable for compressible flow. The isothermal equation is the simpler of the two and can be found in Crane's Technical Paper No. 410, 1988, p. 1-8 (equation 1-6). However, manufacturers of chlorine feed equipment typically supply charts for selection of pipe sizes for different lengths of system piping. These tables should be used preferentially.

Another concern for chlorine system design is appropriate material selection. In general, on the pressure side of the vacuum regulator, metallic piping materials should be selected; and on the vacuum side, thermoplastic piping materials should be selected.

Dry *gaseous and liquid chlorine* will not attack carbon steel at normal temperatures; as a consequence, liquid chlorine is packaged in steel containers. On the other hand, like liquid oxygen, liquid chlorine will sustain combustion of steel once any portion of the steel-chlorine contact surface has been heated to the kindling point of 438° F (225° C). Because of this potential danger, heat should never be applied to a chlorine container or piping containing chlorine. If a steel pipe containing liquid chlorine or even chlorine gas at reduced pressure is accidentally cut with a welder's torch, the pipe will ignite and continue to burn as long as there is a chlorine supply available. Small amounts of moisture will also cause chlorine to attack steel. Because a trace amount of moisture is unavoidable, some of the corrosion product ($FeCl_3$) is always found in chlorine containers and in chlorine lines. Some may wish to use enhanced metallurgies for pressurized chlorine service. Stainless steels are acceptable for this service but supply very little advantage. Titanium piping will spontaneously ignite in the presence of dry chlorine (although not with wet chlorine) and so should be avoided at all costs.

As stated earlier, there are three distinct sections of pressurized chlorine piping: the flexible connection, the pressurized liquid chlorine piping, and the pressurized gaseous piping. Often as a chlorine container is emptied, the vacuum generated at the injector is pulled all the way back to the chlorine container. It must be noted that as the chlorine drops in pressure, the temperature of the chlorine will drop below $-20°$ F, which is the lower limit at which carbon steel can be used because of brittle fracture phenomena. For the gaseous chlorine piping, the mass of the piping is so much greater than the mass of the gas that the piping never gets sufficiently cold for brittle fracture to be a concern. On the liquid side, the liquid will get much colder than $-20°$ F, and the relative mass of the liquid and piping is such that the piping will achieve these low temperatures as well. There are two solutions to this problem: Either use an enhanced metallurgy, or design the liquid piping such that it is either free draining to the evaporator or chlorine container. If the second methodology is selected, a 2% minimum slope should be maintained in the piping.

For the metallic piping in the pressurized section, Schedule 80 seamless steel piping should be used (as appropriate). Reducing fittings should be used rather than bushings, and ammonia-type unions with lead gaskets should be used rather than ground joint unions. All parts should meet Chlorine Institute standards. Isolation valves should be ball type or rising stem type, having cast iron or steel bodies, Hastalloy trim, Teflon seats, and stainless steel balls (as required). Ball valves are preferred because it is easier to operate and because the lever indicates the position of the valve at a glance. Piping systems should be assembled by welding. Only diaphragm-type pressure gauges should be installed in chlorine systems, and these should have silver diaphragms and Hastalloy-C housings.

Once a vacuum is created downstream of the vacuum regulator or once chlorine solution is created downstream of the injector, carbon steel pipe is no longer appropriate for chlorine service. If a leak occurs in piping in vacuum service, air and water will get into the piping. The water will mix with the chlorine in the line to form hypochlorous acid, which will rapidly destroy the steel. As stated earlier, rather than metallic piping, thermoplastic piping materials should be used. The thermoplastic material used most often is Schedule 80 polyvinyl chloride (PVC). There are two types of PVC pipe available: type I and type II. Type I is PVC without any components added for property control. Type II PVC has other components present. These components are readily attacked by chlorine, and this attack can result in premature pipe failure. The same argument applies to Chlorinated PVC (CPVC) pipe. As such only PVC type I should be used for chlorine service.

Anhydrous Ammonia Systems

Ammonia is available on the commercial market in three useful forms: anhydrous ammonia, aqueous ammonia, and ammonia sulfate. Aqueous ammonia and anhydrous ammonia are the forms most commonly used.

Anhydrous ammonia is available as a compressed liquid in containers nearly identical to those used for chlorine. Most of the equipment used in connection with chlorine can also be used for feeding ammonia with minor modification. Chlorine gas feeders can be used as ammonia gas feeders through the use of a simple modification kit. Ammonia has a heat of vaporization of 328 cal/g and may be fed from the top of the cylinder in the same manner as chlorine, although not at the same rate. The capacity of a 1-ton (1,016-kg) ammonia cylinder by the evaporation method is about 84 lb (38 kg) per day. Like chlorine, ammonia liquids should be transported in black iron pipe with welded steel fittings. Note that the turndown for ammonia gas feeders is much less than for chlorine gas feeders.

Although anhydrous ammonia can be fed by an injector system, it is highly soluble in water, and simple direct feed ammonia gas feeder designs are common (pressure feed). Although an ammonia injector-type system eliminates the need for transporting ammonia under pressure to distant locations in the plant, ammonia is alkaline, unlike chlorine which is acidic, and the ammonia will soften the water at the point of injection, producing a precipitate of calcium carbonate. This can cause severe scale problems in the injection system. Users of anhydrous ammonia report that carbonate precipitates also build up at diffuser ports when anhydrous ammonia is fed directly.

These problems are particularly troublesome in hard water, so ammonia injector systems may require that soft water or a special self-cleaning diffuser with rubber sleeve be used.

Aqueous Ammonia Systems

Aqueous ammonia is delivered in a solution that is 19% ammonia by weight. Aqueous ammonia is often fed through PVC lines. Over time, resin formulations of PVC change. The PVC pipe currently available is not compatible with aqueous ammonia. As such, carbon or stainless steel pipe should be used in aqueous ammonia feed systems.

In concentrated form, ammonia reacts chemically with copper, so under no circumstances should any brass, bronze, or other copper alloy be used in any ammonia feed system (aqueous ammonia and anhydrous ammonia feed systems). The vapors above the solution in an aqueous ammonia tank are extremely potent. Therefore provision should be made for disposing of the displaced vapors in the tank in a safe manner, such as through transfer back to the delivery vehicle when the storage tank is being refilled. Pressure relief valves on these tanks should pass through a water-type scrubber before going to the atmosphere.

Delivery of aqueous ammonia to the feed point is typically performed with diaphragm metering pumps. The ideal control system would permit a feed rate proportional to the product of the flow and the chlorine residual such that the appropriate ammonia-to-chlorine ratio for chloramine formation can be maintained.

Aqueous ammonia is often fed through stainless steel end-type diffusers with corporation stops for ease of cleaning. If carrier water is used, the water should be softened and the diffuser should have $3/16$-in.-or-larger (4.5-mm) orifices designed with a significant backpressure such that any scale formation is knocked off the diffuser.

When one is ordering aqueous ammonia, debris-free chemical should be specified, because aqueous ammonia is sometimes delivered with considerable debris present. Alternatively, facilities may be installed to strain undesirable debris from the product before it is used.

Ammonium Sulfate Systems

Solid ammonium sulfate is usually fed into a simple mixing tank using a gravity or volumetric feeder. Once mixed, the solution can be transported using the same methods described earlier for aqueous ammonia. If the local water is hard, scaling problems may occur, and softening of this carrier water should be considered.

Sodium Hypochlorite Systems

Sodium hypochlorite is used by many small and medium-size water systems to feed chlorine because of the greater ease of handling and much greater safety as compared to chlorine gas. In recent years, some larger water systems that are located in metropolitan areas have also changed from chlorine gas to the use of sodium hypochlorite because of an increased emphasis on safety.

Purchasing Hypochlorite

Sodium hypochlorite (liquid bleach) is formed by combining chlorine solution and sodium hydroxide. In other instances, it is manufactured from chlorine and sodium hydroxide that have been separately shipped to the manufacturing site. The reaction that proceeds in this instance is as follows:

$$Cl_2 + H_2O \rightarrow HOCl + HCl \text{ (gas chlorine to chlorine solution)}$$

$$HOCl + HCl + 2NaOH \rightarrow NaOCl + NaCl + H_2O$$

A slight excess of sodium hydroxide is often added to increase the stability of the sodium hypochlorite product.

When the sodium hypochlorite is added to water, it hydrolyzes to form hypochlorous acid (HOCl), the same active ingredient that occurs when chlorine gas is used. The hypochlorite reaction slightly increases the hydroxyl ions (pH increase) by the formation of sodium hydroxide, whereas the reaction of chlorine gas with water increases the hydrogen ion concentration (pH decrease), forming hydrochloric acid.

In most waters, these differences are not significant, but when high chlorine doses are used in poorly buffered waters, these effects should be considered. They can be evaluated by calculation or by simple laboratory tests.

In the commercial trade, the concentration of sodium hypochlorite solutions is usually expressed as a percentage. The *trade percent* is actually a measure of weight per unit volume, with 1% corresponding to a weight of 10 g of available chlorine per liter. Common household bleach, at a trade concentration of 5.25%, has approximately 5.25 g/100 mL, or 52.5 g of available chlorine per liter. Sodium hypochlorite available for municipal use usually has a trade concentration of 12% to 17%. These are approximate concentrations, and they should always be confirmed for a particular shipment by laboratory procedures.

When one is purchasing bulk sodium hypochlorite, purchasing specifications should be used delineating the acceptable ranges for available chlorine (12% to 17%) and pH (11 to 11.2), as well as maximum contaminant limits for iron (2 mg/L) and copper (1 mg/L). Specifications should also require that shipments be free of sediment and other deleterious particulate material. Shipments should be analyzed for the concentration of chlorine, the pH, and the concentration of metal contaminants.

Hypochlorite Degradation

Sodium hypochlorite solutions are subject to degradation during storage. The rate of degradation is accelerated by high hypochlorite concentrations, high temperature, the presence of light (UV radiation), low pH, and the presence of heavy metal cations such as iron, copper, nickel, and cobalt.

During storage, sodium hypochlorite continually degrades to salt and oxygen through the following reaction pathway:

$$2NaOCl \rightarrow 2NaCl + O_2$$

Because of the generation of diatomic oxygen, sodium hypochlorite continually effervesces. This causes several distinct problems in sodium hypochlorite systems. The first problem is that metering pumps will air-bind (strategies to limit or eliminate this problem are discussed later). The second problem is that if sodium hypochlorite is trapped in a section of piping (between two valves, for instance), the chemical will degrade and attempt to effervesce. Because the air bubbles do not have anywhere to go, the pressure in the trapped section increases. The pressure will continue to increase until the pipe ruptures. Cases of pipe rupture (and valve failure) have been reported in the literature.

Several strategies should be followed in the design of a sodium hypochlorite system to limit the danger inherent in the system. First, valves should be kept to a minimum in the system so that there are limited sections of pipe that can be isolated. Second, any section of pipe that can be isolated must be provided with a relief valve. Third, operations and maintenance procedures that limit the possibility of sodium hypochlorite being trapped within piping should be put in place. These procedures can include lockout/tagout of valves and flushing of systems prior to shutdown and isolation.

Hypochlorite Storage and Pipe and Valve Materials

The storage of sodium hypochlorite must be carefully managed to limit degradation and the formation of chlorate. The quicker that the chemical is used, the less time it will have to degrade, so storage time should be limited. Because degradation is a strong function of concentration, the chemical will degrade more in absolute terms on day 1 than on day 10 or 20.

In that the rate of degradation is a function of temperature and concentration, storage time should be limited. Often storage is limited to less than 28 days and temperatures to 70° F (21° C) or less. The chemical may be delivered warm, that is, at temperatures up to 85° F (30° C), which must be considered in the system design. Alternative approaches to sodium hypochlorite management include installation of the storage tanks in a climate-controlled room to maintain the chemical at the desired temperature, provision for additional storage capacity to allow for dilution of the chemical upon delivery, or installation of immersion coolers in the storage tanks connected to a chilled water source to cool the chemical. The first and second approaches are far more common than the third.

Materials that are used for bulk storage of sodium hypochlorite include FRP (fiberglass-reinforced plastic), polyethylene (polyethylene resins vary, so specific selections must be specified with a tank supplier), or lined carbon steel. FRP tanks are the most common selection. They are superior in construction as compared to polyethylene and less expensive than lined carbon steel. Sodium hypochlorite tanks should be vented, and provision should be made for sampling the contents.

Schedule 80 PVC and CPVC are the most commonly used materials used for sodium hypochlorite piping. Historically, many PVC and CPVC piping systems have failed due to leaks at the solvent welded connections. Special attention must be paid to the glues used to put the PVC and CPVC pipe together. Glues generally available are not suitable for sodium hypochlorite service. The glues consist of solvent, native material (PVC or CPVC), and additives. Sodium hypochlorite attacks the additives and results in leaks at the solvent welded connections. As a result, only glues that consist solely of solvent and the native material can be used.

Special attention must be paid to the valves used in sodium hypochlorite systems. Common choices include ball and diaphragm valves. Ball valves are unsuitable for sodium hypochlorite service for two reasons: Chemical can be trapped in the interstitial volumes of the valves, resulting in exploding valves, and the presence of sodium hydroxide results in valve sticking. Special attention must also be paid to the specifics of the diaphragm valves selected. Sodium hypochlorite is especially aggressive to elastomers, and as such, common elastomers such as EPDM or Viton cannot be used in direct contact with sodium hypochlorite. Rather a two-piece diaphragm consisting of a Teflon diaphragm contacting the chemical with an elastomeric backing (usually EPDM) should be used. The Teflon and elastomeric backing must be independent of the diaphragm and cannot be directly bonded to each other.

Hypochlorite Feed

Most small to medium-size water systems feed hypochlorite with positive displacement diaphragm metering pumps. Because of the effervescing experienced in sodium hypochlorite systems due to the degradation characteristics, metering pumps are subject to air binding. When metering pumps air-bind, the pumps stall and are incapable of pumping chemical, which results in interruption of chemical feed (and hence lack of chlorination and potential plant shutdown). As such, special provisions should be followed to limit pump air binding:

- The metering pumps should be located as close to the storage tanks as possible. The suction lines should be oversized.
- The pumps should be located well below the level of the liquid in the storage tank.

FIGURE 10.10 Metering pump suction piping detail. *Notes:* (1) Sodium hypochlorite metering pump suction piping shall be routed to slope continuously downward from the day/bulk storage tank. (2) Valve shall remain open except when day/bulk storage tank level is below the pump centerline. (3) Locate tee for vent as close as possible to pump suction connection. The vent piping shall be connected to the run of the tee in the vertical position. Vent piping shall not be connected to the branch of the tee.

- The piping should be sloped up toward the tanks from the pumps such that the bubbles can flow back to the tank.
- Suction piping at the pump should be configured in a specific manner to allow for disengagement of the bubbles prior to entering the pump head (see Figure 10.10).

In addition, some pump manufacturers supply a special *hypo pump,* which bypasses a small amount of hypochlorite from inside the pump head to the pump discharge to allow any accumulated air to exit the pump head. This results in a small decrease in pump accuracy.

On-site Generation of Chlorine

Although on-site generation of disinfectants has been possible for many years, it has not generally been found to be practical. However, new concerns over accidental release of chlorine gas and the relatively high cost of scrubbers that may have to be installed by many plants to prevent release of gas into adjoining neighborhoods have raised new interest in on-site generation. Several systems are now available that use new technology to generate disinfectants and use only salt and electric power. They can even be designed to operate unattended at a remote location and are intended to meet all USEPA standards for disinfection and maintaining a chlorine residual in the finished water.

There are two types of systems available for on-site generation. The first and most common generates a very dilute (0.8%) sodium hypochlorite solution directly from saltwater, and the second generates pure chlorine gas directly and sodium hydroxide as a by-product. In the second method the chlorine can be fed in two distinct ways. Either the generated chlorine is fed directly via a conventional chlorine feed system wherein the generation effectively replaces local chlorine storage; or the generated chlorine is remixed with the generated sodium hydroxide to generate a concentrated (12.5%) sodium hypochlorite solution, which is stored on-site and fed with conventional metering equipment. The

FIGURE 10.11 Schematic of ClorTec™ on-site hypochlorite-generating process. (*Courtesy of Severn Trent Services.*)

second system is considerably more complex than the first and has seen infrequent use in the United States, so it is not discussed further. The first type of on-site generation system has seen widespread use.

The first type consists of the following pieces of equipment, as shown in Figure 10.11: water softener, salt storage and brine generation tank, brine metering, brine mixing, sodium hypochlorite storage tanks, and sodium hypochlorite metering pumps.

The heart of these systems is the electrolytic cells. The electrolytic cells generate dilute sodium hypochlorite solutions by electrifying brine solutions through the following reaction pathway:

$$NaCl + H_2O + electricity \rightarrow NaOCl + H_2$$

The cells come in distinct sizes, so multiple cells are usually arranged in racks to attain a set capacity. The cells are subject to scaling, corrosion, and plugging from contaminants in the incoming water, so all the water used in the systems must be softened. Also, diatomic hydrogen is generated as a waste product and must be disposed of. Hydrogen is extremely explosive, so vapor spaces must be purged with air to keep the concentration of hydrogen below the lower explosive limit (LEL).

Power to the cells is supplied by a transformer/rectifier, which controls how much power is sent to the cells. This allows control of the amount of sodium hypochlorite generated. Also, the quantity and quality of the power used are powerful indicators of the efficiency and condition of the electrolytic cells. When the power draw becomes excessive, the cells must be removed from service and cleaned in a mild acidic solution to remove scale and regenerate the cell.

Key to the design of these systems is how much chemical storage to maintain on-site and what equipment to back up. In general, the chemical storage is as salt with only small amounts of stored hypochlorite. Because the generated sodium hypochlorite is very dilute, the volume of storage required for sodium hypochlorite is very large as compared to the salt storage. Usually 30 days of salt storage is supplied, and 3 days of sodium hypochlorite storage is supplied. This keeps the system reasonably economical.

The electrolytic cells are the most expensive part of the system. Sometimes installed spares are supplied, but more frequently they are not. Rather, the sodium hypochlorite storage tanks are designed to handle concentrated sodium hypochlorite (12.5%), so that in the event of cell failure, sodium hypochlorite can be delivered to the site. The disadvantage of this is that multiple metering pumps must be supplied because of the large size variation.

Choosing between Chlorine, Delivered Sodium Hypochlorite, and On-site Generation

When a facility is researching what chemical to use for disinfection, the most common choices will be chlorine, delivered sodium hypochlorite, and on-site generation of sodium hypochlorite.

If a net present-value analysis is performed, chlorine will almost always be the best choice. Chlorine will have a relatively low installed cost and the lowest chemical cost. Delivered sodium hypochlorite will typically have the lowest installed cost, but the highest chemical cost. On-site generation will have the highest installed cost, but a relatively low chemical and power cost. Of course, in all these instances local variations can affect the most economic choice, but this is usually how they fall out.

If a facility is considering delivered sodium hypochlorite and on-site generation, it is likely that the facility is concerned with safety or otherwise concerned with having chlorine on-site. As such, often new and existing chlorine systems are being eschewed in favor of delivered sodium hypochlorite and on-site generation systems.

The specific project requirements, including sparing philosophy, local delivered chemical cost, and power cost, will greatly influence the choice between the two options.

Chlorination Control

Proper design of the control system for chlorination facilities is as important as any other aspect of their design. Methods currently used are manual setting, based on flow and periodic or continuous residual measurements; continuous feedforward control, based on continuous flow measurements; feedback control, based on continuous residual measurements; and compound, closed-loop control, based on continuous measurements of both flow and chlorine residual. Each of these methods is illustrated in Figure 10.12. Details of control signal manipulation are not shown.

Control Methods. The manual control method is limited by the diligence of the operator. If either the flow or the chlorine residual changes, the operator must make adjustments; consequently, this method gives a wide range of performance. The feedforward (flow-pacing) method is a significant improvement because the flow has been eliminated as a variable, provided the operator is diligent in maintaining the control system. For water supplies where chlorine demand is very stable, feedforward control is preferred.

Feedback control is theoretically superior to feedforward control because the chlorine residual is being directly used to control chlorine addition. Unfortunately, this is not al-

FIGURE 10.12 Alternative chlorination control methods.

ways the case; because chlorine residual analyses drift off course unless they are regularly maintained, feedforward control is better than feedback control alone.

Whenever chlorine demand is variable, compound-loop control is preferred. This system is sometimes accomplished by controlling the differential vacuum regulator on the chlorinator with the flow signal and setting the chlorine gas-metering orifice with the signal from the residual analyzer. Alternatively, it can be provided by electronically adding both control signals and using the results to control the differential vacuum. The first approach allows the chlorinator to operate over a dynamic range of 200 to 1, whereas the differential vacuum will only allow a dynamic range (maximum feed/minimum feed) of 20 to 1. However, in most installations, a dynamic range of 20 to 1 is satisfactory.

With either feedback or compound-loop control, lag time is one of the principal design parameters. *Lag time* is the time between the moment when the chlorine is added to the effluent and the time when the residual analyzer signal comes to the chlorinator. Lag time includes the transit time from the point where the chlorine is initially mixed until it reaches the sampling point, the transmission time between the sample point and the chlorine residual analyzer (in the sample line), and the analysis time.

The analysis time is usually a minor factor. If the lag time is too much longer than the response time of the analyzer, the level of the chlorine dose will sawtooth. White (1999) suggests that the lag time be maintained at an average level of 2 min, with a maximum of 5 min. Low-flow conditions should be considered.

It should be understood from the beginning that the purpose of the chlorine residual analyzer discussed here is to control the chlorine dose. If continuous monitoring of the chlorine residual after the chlorine contact period is desired, another chlorine residual analyzer is required.

The following are some of the most common design errors: poor chlorine sample conditions, analyzer located too far from sampling point, and effluent chlorine dose paced to influent flow.

Sample Point Location. The principal consideration in locating the sampling point for the control analyzer is that there must be good mixing. If the sample is taken before adequate mixing has occurred, the result will be erratic readings unsuitable for control. For the majority of initial mixing designs, sampling should be provided immediately downstream of the initial mixing device. Ordinarily, chlorine residuals are stable enough for control measurements after just a few seconds of contact.

If no initial mixing device is present, the sampling point should be far enough downstream to ensure that good mixing has occurred. For turbulent flow, 10 pipe diameters is usually sufficient; however, low-flow conditions should be considered, and if adequate mixing cannot occur in a reasonable time, an initial mixing device will be necessary for control purposes.

Analyzer Location. Chlorine residual analyzers should always be located as near as possible to the sampling point, even if special housings are required. Sample lines should be designed for velocities of about 10 ft/s (3 m/s), and the transit time between the sampling point and the residual analyzers should be minimized.

Chlorine Dose Pacing. The chlorine dose should always be paced to the flow most representative of the point of addition. A common error in design is an arrangement in which the effluent chlorine dose is paced using influent flow measurements. Too many events occur between a plant's influent and its effluent, and such a design often results in an erratic chlorine dose and an unmanageable operating system.

Chlorine Residual Analyzers

Two methods for continuous chlorine residual analysis are currently available: the automatic amperometric titrator and the ion-selective probe. In an automatic amperometric titrator, the cell has an indicating electrode made of copper concentrically mounted around a platinum reference electrode. Water flows into the space between the two, and a potential is imposed between the electrodes, resulting in a current flow that is proportional to the amount of chlorine in the sample. Ordinarily, a pH 4 buffer is used, and the free chlorine is measured. The use of a buffer with excess potassium iodide will cause the unit to titrate the total chlorine residual, and an excess of combined chlorine will interfere with attempts to measure the residual. One type of amperometric titrator is shown in Figure 10.13.

FIGURE 10.13 Flow diagram of amperometric chlorine residual analyzer. (*Courtesy of Wallace and Tiernan Division of USFilter.*)

Chlorination System Design for SWTR Compliance

The following simplified steps can be taken to determine probable CT required and the probable credit:

1. Determine the total removal/inactivation required.
   - Determine whether the usual 3-log *Giardia* and 4-log virus removal/inactivation applies.
   - Determine whether higher removal/inactivation is required because of a vulnerable source.
2. Determine the credits that will be allowed for physical removal.
   - For conventional treatment—2.5-log *Giardia* and 2-log virus.
   - For direct filtration—2-log *Giardia* and 1-log virus.
3. Determine credits required for disinfection.
   - For conventional treatment—minimum 0.5-log *Giardia* and 2-log virus.
   - For direct filtration—minimum 1-log *Giardia* and 3-log virus.
4. Determine CT required for design conditions, using tables provided in publications detailing the SWTR requirements. Note that the worst case would be a combination of high pH, low temperature, and high flow; however, these may not occur concurrently, and several combinations of representative conditions should be considered.
5. Using Figures 10.2, 10.3, and 10.4, determine the hydraulic efficiency of the basins that the water will pass through from the point of disinfectant application until it enters the distribution system (or at the first customer, if appropriate).
6. Using Table 10.2, compute the total effective detention time through the plant.
7. Multiply the detention time by the chlorine residual at the first customer, and compare it with the required CT.

If primary disinfection can be applied as the raw water enters the plant, there is usually not much problem in getting adequate detention time through the process basins in a full conventional plant. However, many plants now practice zero or minimal prechlorination to control disinfectant by-product (DBP) formation, so the CT credit must be achieved in the clearwell or a separate contact basin with optimized hydraulic efficiency. The designer must be particularly careful when relying on a clearwell, or any treated water storage basin, for CT credit. Most existing clearwells are poorly baffled or have none at all. In addition, their basic function as useful storage volume, which may be withdrawn, may conflict with the need to maintain more than a minimum volume to satisfy contact time requirements.

CHLORINE DIOXIDE SYSTEMS

In certain circumstances, chlorine dioxide is an excellent choice among disinfectants. In particular, it is effective in destroying phenols, yet it does not form trihalomethanes in significant amounts. Chlorine dioxide's disinfectant properties are not adversely affected by a higher pH, as those of a free chlorine residual are. Consequently, chlorine dioxide is a much-quicker-acting disinfectant than chlorine is at higher pH levels. In western Europe, use of chlorine dioxide is increasing, particularly in Holland, Germany, France, and

Switzerland, regions where source water is of lower quality. In these locations, chlorine dioxide is used for disinfection, often as an adjunct to ozonation.

Chlorine Dioxide Disinfection

Chlorine dioxide does not dissociate or disproportionate as chlorine does at normal drinking water pH levels. Like chlorine, chlorine dioxide exerts a demand when it is first added to a water supply, which must be overcome if a persistent residual is to be maintained. Like chlorine, chlorine dioxide is photosensitive (light-sensitive), and because it is a gas at temperatures above 11° C, its residuals are easily removed by aeration.

Although chlorine dioxide has recently been used at more treatment plants in the United States because it does not form trihalomethanes or haloacetic acids, there is still concern that there are other organic by-products of chlorine dioxide that are not yet well understood, and it may have other undesirable reaction products. Information presently available indicates that the reaction products include aldehydes, carboxylic acids, and ketones.

It is generally considered that chloroorganic by-products are not produced by reaction between chlorine dioxide and organic compounds, but may be present in practical applications as a result of free chlorine present in the chlorine dioxide solution. The principal inorganic by-products of chlorine dioxide reactions within water treatment are chlorite ion (ClO_2^-), chloride ion (Cl^-), and chlorate ion (ClO_3^-), in the order listed. Both chlorate and chlorite ions, particularly the chlorite ion, have been implicated in the formation of methemoglobin. Consequently, most European countries limit the level of chlorine dioxide that can be used. The recent Stage 1 Disinfectant Disinfection By-products Rule (DDBPR) (1998) sets limits of 0.80 mg/L for chlorine dioxide and 1.0 mg/L for the chlorite ion. No maximum contaminant level has yet been proposed for the chlorate ion.

Chlorine Dioxide Generation

Chlorine dioxide cannot be stored once it is generated because it is not safe. Numerous stimulants may cause the pure gas to explode, including an increase in temperature, exposure to light, changes in pressure, and exposure to organic contaminants. As a result, chlorine dioxide is usually generated on-site.

All chlorine dioxide for drinking water treatment is generated from sodium chlorite. Most generation techniques use the two-chemical oxidative process, in which chlorine, either as a gas or in solution, is mixed with a sodium chlorite, $NaClO_2$. Chlorine dioxide can also be generated by direct electrolysis of sodium chlorite.

The goal in generating chlorine dioxide from chlorine and sodium chlorite is to maximize the chlorine dioxide *yield,* defined as the molar ratio of chlorine dioxide produced to the theoretical maximum. The term *conversion* is also used in referring to chlorine dioxide generation reactions; this is the molar ratio of the amount of chlorine dioxide formed to the amount of sodium chlorite fed to the system. For other reactions that produce chlorine dioxide, such as the hydrochloric acid–sodium chlorite reaction, yield and conversion will have different values.

Studies of the mechanism and kinetics of the chlorine–sodium chlorite reaction have shown that conditions favoring the formation of chlorine dioxide are those in which the reactants are present in high concentrations and the chlorine is present as either hypochlorous acid or molecular chlorine (Cl_2). Three methods for the generation of chlorine dioxide from chlorine and sodium chlorite are commercially available: the aqueous chlorine–sodium chlorite system, the gas chlorine–aqueous sodium chlorite system, and the solid sodium chlorite–gas chlorine system.

Aqueous Two-Chemical System. The earliest systems produced chlorine dioxide by simply pumping a sodium chlorite solution into a chlorine solution, followed by a short reaction time. Acceptable yields were achieved by feeding 200% to 300% more chlorine than the stoichiometric requirements. The chlorine dioxide solution from a generator of this type contains high levels of chlorine in addition to chlorine dioxide.

With the discovery that potentially toxic chlorinated organics are generated by the reaction of chlorine and naturally occurring humic substances in water supplies, plus the growing interest in chlorine dioxide as a replacement for some chlorination practices, generation methods were sought that would produce a chlorine-free chlorine dioxide. One of the most common methods for chlorine dioxide generation currently in use that strives to meet this requirement is the pH-adjusted method.

The *pH-adjusted* system uses hydrochloric acid fed into the chlorine solution before reaction with the sodium chlorite. The acid feed serves to shift the chlorine solution equilibrium and the chlorine hydrolysis equilibrium, favoring hypochlorous acid and molecular chlorine. The acid feed must be carefully controlled so that the pH of the chlorine dioxide solution can be maintained between 2 and 3. Yield is reduced at both higher and lower pH values. Yields of more than 90% have been reported from the pH-adjusted system, with approximately 7% excess (unreacted) chlorine remaining in the solution.

Another modification that produces high yields of chlorine dioxide, with minimal amounts of chlorine remaining in the chlorine dioxide solution, requires that the chlorine solution used for generation have a chlorine concentration greater than 4 g/L. The exact relationship of excess chlorine required for 95% yield and initial chlorine solution concentration is shown in Figure 10.14.

Because this concentration of chlorine in solution is near the upper operating limit of commercial chlorine ejectors, and because these ejectors operate at constant water flow rates, the yield of this method of generation depends on the production rate, with lower production rates resulting in lower yields. This type of generator is normally operated on an intermittent basis to maintain high yield when less than maximum production capacity is required. Chlorine dioxide solutions in the 6- to 10-g/L concentration range are prepared and immediately diluted to about 1 g/L for storage and subsequent use as needed.

A schematic of the aqueous chlorine–sodium chlorite system is shown in Figure 10.15.

Vapor-Phase Two-Chemical System. The vapor-phase two-chemical system is a more recent development to improve efficiency and reacts chlorine gas with a concentrated sodium chlorite solution under vacuum. The chlorine dioxide produced is removed from the reaction chamber by a gas ejector, which is similar to the common chlorine gas vacuum feed system.

This generation technique produces chlorine dioxide solutions with yields in excess of 95%. The chlorine dioxide solution concentration is 200 to 1,000 mg/L and contains less than 5% excess chlorine, which is defined as the amount of unreacted chlorine remaining in the chlorine dioxide generator effluent. The system is operated on a continuous basis and achieves a high yield over the entire production range (Figure 10.16).

Electrochemical Systems. Chlorine dioxide (ClO_2) can be generated by direct electrolysis of a sodium chlorite ($NaClO_2$) solution. A major advantage of this type system for drinking water treatment is that only one chemical (sodium chlorite) and a power source are required, thus eliminating the need for on-site storage of chlorine gas. In principle, the process is quite simple, as illustrated by the following equation:

$$ClO_2^- \ (aq) \rightarrow ClO_2 \ (aq) + e^-$$

The $NaClO_2$ solution is fed into the reaction cell, and ClO_2 gas is produced at the anode by direct electrolysis; two by-products, sodium hydroxide (NaOH) and hydrogen

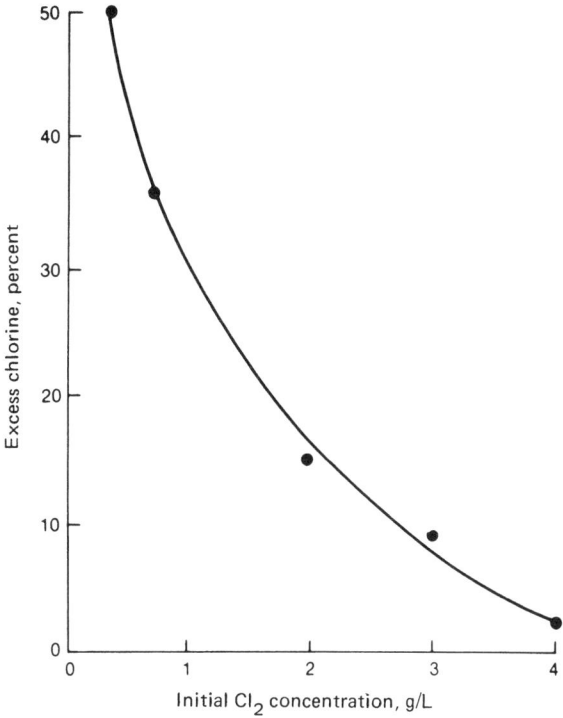

FIGURE 10.14 Excess chlorine necessary for greater than 95% conversion of chlorite to chlorine dioxide. (*Source: M. Aieta and J. Berg, "A Review of Chlorine Dioxide to Drinking Water Treatment," Journal AWWA, vol. 78, no. 6, June 1986, p. 62.*)

FIGURE 10.15 Aqueous chlorine–sodium chlorite system schematic with optional acid food. (*Source: M. Aieta and J. Berg, "A Review of Chlorine Dioxide in Drinking Water Treatment," Journal AWWA, vol. 78, no. 6, June 1986, p. 62.*)

FIGURE 10.16 Gas chlorine–sodium chlorite system schematic. (*Source: M. Aieta and J. Berg, "A Review of Chlorine Dioxide in Drinking Water Treatment,"* Journal AWWA, *vol. 78, no. 6, June 1986, p. 62.*)

gas (H_2), are produced at the cathode. The overall reaction is shown by the following formula:

$$2NaClO_2 + 2H_2O \rightarrow 2ClO_2 + 2NaOH + H_2$$

The caustic solution is dilute and can be used as a feedstock chemical for pH adjustment in the process water, and the H_2, while extremely explosive, poses no particular problem because it is diluted to low concentrations (one-fourth the explosion limit) with air and vented to the atmosphere (Gates, 1998). The ClO_2 is removed from the anode portion of the reactor to avoid further oxidation to chlorate ion (ClO_3^-) and, possibly, perchlorate ion (ClO_4^-) (Gordon, 2001). One commercial electrochemical system separates the ClO_2 from the anolyte stream by stripping it from solution with air while another system separates it by a patented membrane process called *perstraction* (Figure 10.17). Gordon (2001) described perstraction as a *membrane distillation* process whereby "a gas-permeable hydrophobic membrane is used to separate a donor solution (reaction chamber) from an acceptor solution (water flow)." Additional details regarding the membrane and the perstraction process are provided by Cowley et al. (1996).

Commercial electrochemical generators used by the paper pulp industry produce several hundred pounds of ClO_2 per day, but those used to treat drinking water are typically much smaller, most commonly producing less than 50 to 150 lb (23 to 68 kg) per day.

Sodium Chlorite

Sodium chlorite is available as a solid, orange-colored powder that is about 80% $NaClO_2$ by weight and in the form of liquid solutions that are 25% to 32% $NaClO_2$ by weight. If granular sodium chlorite is used, it should be stored in a separate building equipped with sloped floors, drains, and facilities for hosing down spills. The building should be con-

FIGURE 10.17 Electrochemical ClO_2.

structed of materials with maximum fire resistance, and it should be designed with the material's explosive potential in mind. Furthermore, $NaClO_2$ must be kept from contacting organic materials (leather boots, cloth, mops, etc.) because rapid oxidation reactions can ensue and cause fires.

OZONE DISINFECTION SYSTEMS

Although the first use of ozone in water treatment coincided with the first use of chlorine for disinfection, ozonation has been a common practice in only a small number of countries, notably France, Germany, and Switzerland. In many other countries, including the United States, England, and Japan, interest in ozonation in water treatment increased substantially in the 1990s because of suspected carcinogenic properties of the trihalomethanes that are formed when organic compounds, naturally present in water, react with chlorine. By the end of the millennium there were known to be more than 250 ozone systems in drinking water treatment service in the United States at plants with capacity ranging from less than 1 mgd to well over 100 mgd.

Ozone Chemistry

Ozone is a highly reactive gas formed by electrical discharges in the presence of oxygen. Its most distinguishing characteristic is a very pungent odor. In fact, the word *ozone* is derived from a Greek word that means "to smell." The use of this gas in water treatment requires an understanding of its physical and chemical behavior. The physical chemistry of ozone is important because a number of complex factors affect its solubility, reactivity, autodecomposition, and stability.

Ozone is an allotrope of oxygen. Substantial amounts of energy are required to split the stable oxygen — oxygen covalent bond to form ozone, and the ozone molecule readily reverts to elemental oxygen during the oxidation reduction reaction. Ozone is more

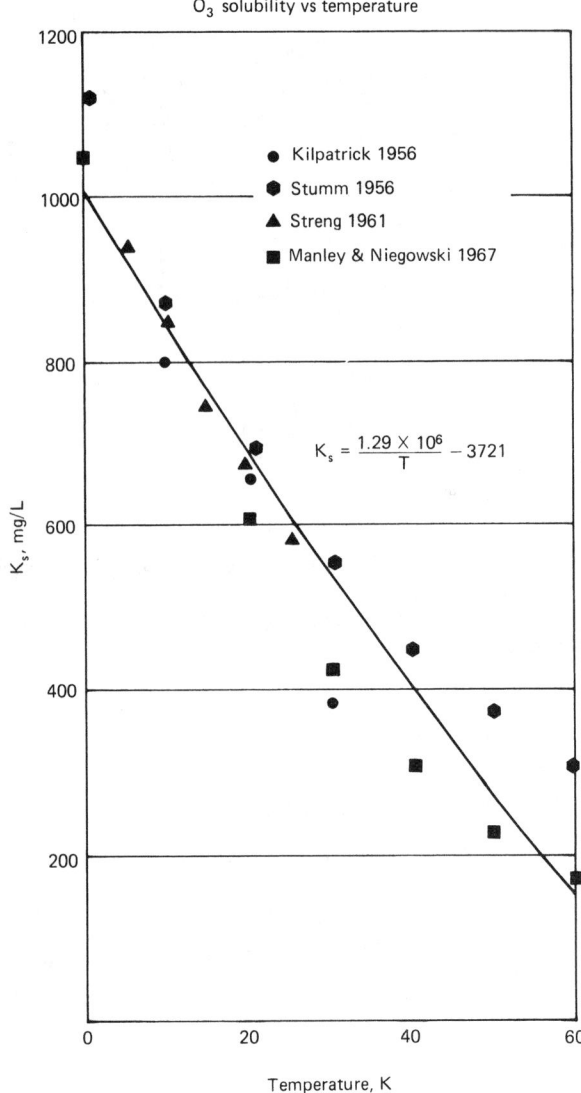

FIGURE 10.18 Effect of temperature on solubility of ozone.

soluble in water than oxygen. The effect of temperature on ozone solubility is shown in Figure 10.18.

As illustrated in Figure 10.19, once ozone enters solution, it follows two basic modes of reaction: direct oxidation, which is rather slow and extremely selective, and autodecomposition to the hydroxyl radical. Autodecomposition to the hydroxyl radical is catalyzed by the presence of hydroxyl radicals, organic radicals, hydrogen peroxide, UV light, or high concentrations of hydroxide ion. The hydroxyl radical is extremely fast and

OXIDATION AND DISINFECTION

FIGURE 10.19 Reaction pathways of ozone in water. (*Source: J. Hoigne and H. Bader, "Role of Hydroxyl Radical Reactions in Ozonation Processes in Aqueous Solutions,"* Water Resources Bulletin, *vol. 10, 1976, p. 377.*)

nonselective in its oxidation of organic compounds, but at the same time, it is scavenged by carbonate and bicarbonate ions to form carbonate and bicarbonate radicals.

These carbonate and bicarbonate radicals are of no consequence in organic reactions. Furthermore, the hydroxyl radicals and organic radicals produced by autodecomposition become chain carriers and reenter the autodecomposition reaction to accelerate it. Thus low-pH conditions favor the slow, direct oxidation reactions involving O_3, and high-pH conditions or high concentrations of organic matter favor the autodecomposition route. High concentrations of bicarbonate or carbonate buffer, especially carbonate buffer, reduce the rate of autodecomposition by scavenging hydroxyl radicals. This means that ozone residuals last longer at low pH and in highly buffered waters.

The rate of autodecomposition is highly dependent on many factors, including pH, UV light, temperature, ozone concentration, and the presence of radical scavengers. Decomposition can be expressed, assuming pseudo-first-order kinetics and in terms of ozone residuals, as

$$C_t/C_0 = e^{-kt}$$

where C_t = ozone concentration at time t, mg/L
C_0 = ozone concentration at time 0, mg/L
k = decay rate constant, L/min
t = time, min

For design purposes, the decay rate constant must be determined for the water to be treated under expected operating conditions, including temperature, pH, and water quality.

Ozone Systems

Ozone systems are typically comprised of four basic subsystems: ozone generation, feed gas preparation, contacting, and off-gas disposal.

FIGURE 10.20 Basic ozonator configuration. [*Source: H. Rosen, "Ozone Generation and Its Relationship to the Economical Application of Ozone in Wastewater Treatment," in F. L. Evans III (ed.), Ozone in Water and Wastewater Treatment, Ann Arbor Science Publishers, Ann Arbor, Mich., 1972.*]

Ozone Generation. The basic configuration of an ozone generator (ozonator) is shown in Figure 10.20. An electromotive force (voltage) is impressed across two electrodes with a dielectric and discharge gap in between. Oxygen or air is passed through the corona discharge between the two electrodes, and some of the oxygen or oxygen in the air is converted to the ozone allotrope. Design principles suggest that the voltage necessary to produce ozone is a function of the product of the gap pressure and the gap width.

$$V = K_1 p g$$

where V = necessary voltage
p = gap pressure
g = gap width
K_1 = constant

In a similar sense, the yield of the ozonator is directly proportional to the frequency, the dielectric constant, and the square of the voltage applied and is inversely proportional to the thickness of the dielectric.

$$\text{Yield} = \frac{K_2(feV^2)}{d}$$

where f = frequency
V = voltage
e = dielectric constant
d = thickness of dielectric
K_2 = constant

This equation suggests some problems that are inherent in the design of ozonators. One is that the dielectric should be made as thin as possible. Very thin dielectrics are, however, more susceptible to failure. In the same sense, the yield is related to the square of the voltage, indicating that high voltages are desirable. On the other hand, dielectric failure also occurs when high voltages are used.

Equipment designs are currently available as low-frequency (i.e., line frequency of 60 Hz in the United States), medium-frequency (400 to 1,000 Hz), and high-frequency (up to 2,000 Hz). The medium-frequency design has become the most common, with each manufacturer optimizing operating conditions to balance the interacting factors. Recent design improvements involving dielectric materials, gap width, and better methods of removing heat from the ozone cell have drastically improved the performance and capabilities of ozone generation.

Three basic types of ozone-generating systems are now in use: the Otto plate, the conventional horizontal tube, and the Lowther plate. Each of these designs is sketched in Figure 10.21. The Otto plate was designed in 1905. Although inefficient, this design is still

FIGURE 10.21 Alternative ozonator configurations. (a) Otto plate–type generator unit; (b) tube-type generator unit; (c) Lowther plate generator unit. [Source: H. Rosen, "Ozone Generation and Its Relationship to the Economical Application of Ozone in Wastewater Treatment," in F. L. Evans III (ed.), Ozone in Water and Wastewater Treatment, Ann Arbor Science Publishers, Ann Arbor, Mich., 1972.]

being used in parts of western Europe. Its principal disadvantages are that it is inefficient and that only low pressures can be used within the unit. High pressures are desirable so that the ozone can be bubbled through deep ozone contact chambers.

The tube-type generator is composed of a number of stainless steel tubes fitted into a large vessel and surrounded by cooling water. A concentric glass tube with a conducting coating on the inside is placed inside each stainless steel tube. A potential is applied between the inside coating of the glass tube and the outside steel tube, and air or oxygen is then passed through the gap in between. Variations of this design are by far the most common ozone generators in use today.

Discounting improvements in the tube-type generator, the Lowther plate unit is the most recent development. Whereas the other two units are usually water-cooled, this unit is air-cooled. It is made up of a gastight arrangement of an aluminum heat dissipater, a steel electrode coated with a ceramic dielectric, a silicone-rubber spacer, and a second ceramic-coated steel electrode with inlet and outlet. The silicone-rubber spacer sets the width of the discharge gap. Several of these units can be pressed together and manifolded to increase generator production. This design has been most effective in smaller capacities used for aquariums, cooling water, spas, and similar applications.

Feed Gas. The function of a feed gas system is to supply oxygen to the ozone generators of appropriate quality and in quantities sufficient for the process. Feed gas may be either air or high-purity oxygen from bulk liquid (LOX) or on-site generation. Due to the high efficiency achieved with high-purity oxygen and the simplicity of operations, many recent installations have been based on LOX feed systems. Design of a feed gas preparation system should address the following parameters of the gas:

- Oxygen concentration
- Moisture content
- Particulates and other contaminants
- Temperature
- Pressure
- Mass flow rate

Oxygen Concentration. Because ozone generation is the conversion of molecular oxygen to ozone, increasing the oxygen concentration of the feed gas will increase the production of ozone. Air-based systems provide a feed gas with approximately 22% oxygen. Increased oxygen concentration can be achieved through supplementing, or replacing, the airstream with a high-purity oxygen source using one of the following:

- Purchased liquid oxygen (LOX) delivered to the site and stored in refrigerated thermos tanks at 95% to 99%+ oxygen
- On-site generation of gaseous oxygen using cryogenic air separation technology at 90% to 99%+ oxygen
- On-site generation of gaseous oxygen using pressure or vacuum swing (PSA/VSA) technology at 90% to 95% oxygen

Moisture Content. The most important gas quality parameter is moisture content. Moisture content of gases is usually expressed as dew point, which is the temperature at which moisture begins to condense out of the gas. To understand gas preparation systems, the designer should remember that the moisture-holding capacity of a gas decreases with decreasing temperature and with increasing pressures.

In ozone applications, the feed gas dew point must be extremely low. Excess moisture will adversely affect ozone production and may react with nitrous oxides within the ozone generator to form nitric acid, causing subsequent damage to the generator. Current practice is to design for −60° C as the minimum acceptable dew point. Most systems operate at below −80° C dew point.

Particulates and Other Contaminants. Particulates can potentially cause problems at a number of locations within an ozone system. Compressors with close running clearances must be protected from particles that would score the impeller, lobe, or rotor surfaces. Desiccant dryers and molecular sieve material must be protected from dust and hydrocarbons that would block the active pores, thereby reducing the material's effectiveness. Because the ozone generator acts as an electrostatic precipitator, particulates and hydrocarbons will attach to the dielectric surface. In some cases, if particulates accumulate in the unit, hot spots will develop and eventually lead to dielectric failure. Fine bubble ozone diffusers must be protected from particulates that would eventually clog the diffuser.

Particulates may be inorganic or organic and may be present in the ambient air or generated from within the ozone system itself. Typical inorganic materials include sand, dust, lime dust, coal dust, construction debris, and moisture droplets. Common organic materials include pollen, cottonwood seeds, and the like. From within an air feed ozone system, the most common particulate is desiccant dust.

Filters are the best method of controlling particulates and in air feed systems should be placed ahead of the compressor, ahead of the desiccant dryers, and after the desiccant dryers. In LOX-based systems particulate filters should be placed downstream of the vaporizers. The goal in any system should be a final feed gas entering the ozone generator free of all particulates larger than 0.3 μm.

In addition, hydrocarbons may be present in the atmosphere in large metropolitan areas and in the oil- and gas-producing areas of the country. Although not commonly used, lubricated-type compressors will also introduce unwanted hydrocarbons into the system. Hydrocarbons in the feed gas can be controlled with coalescing and carbon-absorber filters. For installations with lubricated screw-type compressors, the coalescing filter will typically remove oil droplets larger than 0.05 μm. However, a carbon absorber is necessary for capturing and removing hydrocarbon vapors. LOX purchase specifications should limit total hydrocarbons to less than 20 ppm.

Temperature. At elevated temperatures, the rate of ozone decomposition increases. Consequently, feed gas temperatures should be relatively cool to avoid rapid decomposition of ozone as it is produced within the generator. Current practice limits entering gas temperature to below 90° F (33° C).

Pressure. The system pressure at the point of delivery from the feed gas preparation system to the generator is dictated by two factors. First is the pressure required to overcome all downstream pressure losses through the ozone contact basin, including those from control valves, the generator, line losses, diffusers, and the static head of the water above the diffuser.

The second is the design pressure at which the manufacturer has optimized generator performance. If the manufacturer's optimum design pressure exceeds the downstream losses, a pressure-maintaining valve will be placed downstream of the generator. Low-frequency generators typically operate at low pressures, 8 to 12 lb/ft^2 (55 to 83 kPa); medium-frequency generators operate at 18 to 25 lb/ft^2 (124 to 172 kPa).

Mass Flow Rate. Feed gas preparation systems must be designed to provide sufficient mass flow to achieve the design ozone production at the desired ozone concentration. The mass flow required can be calculated as follows:

$$\text{Required feed gas mass flow (lb/h)} = \frac{\text{design ozone production (lb/d)}}{24 \times \text{ozone concentration}}$$

The mass flow rate can be converted to a volumetric flow rate for compressor selection. However, although compressors typically used in ozone systems are constant-volume machines, the mass flow rate will vary with inlet air temperatures because of changes in inlet gas density. Because the lowest mass flow rate output from a constant-volume compressor occurs when the inlet temperature is highest, summer will usually be the critical condition for compressor selection.

Air Feed Systems. Feed gas preparation systems can be classified according to the general operating pressure within the feed gas subsystem as follows:

- Low pressure: < 15 psig (< 103 kPa)
- Medium pressure: 15 to 65 psig (103 to 448 kPa)
- High pressure: > 65 psig (> 448 kPa)

Low-pressure systems are designed for very low pressure losses and usually operate as constant-volume systems (i.e., constant gas flow rate). This allows the use of rotary lobe-type blowers, which are reliable, low-maintenance units. Because of the high moisture-holding capacity of the low-pressure air, drying always includes both refrigerative and desiccant dryers, as presented in Figure 10.22. Desiccant dryers are limited to thermally regenerated types.

Medium-pressure systems are configured much as the low-pressure systems are, as illustrated in Figure 10.23, except the rotary lobe blower is not an acceptable compressor type, and a pressure reduction step is incorporated after the dryers. Many systems are being designed today for a broad range of system pressures to allow for additional pressure losses from dirty filters, aged desiccant, and fouled diffusers. Although "normal" system operating pressure is expected to be 22 to 27 psig (152 to 186 kPa), the systems are being designed to maintain operations at pressures up to 35 psig (241 kPa). Like low-pressure systems, both refrigerative and desiccant drying types will normally be provided in a medium-pressure system. Again, only thermally regenerated desiccant dryers can be used.

High-pressure systems can usually be differentiated from the other types by the use of multistage, positive displacement compressors and pressure swing-type desiccant dryers, as presented in Figure 10.24. Pressure swing desiccant dryers will operate at pressures down to 65 psig (448 kPa), but with higher pressures, purge requirements for regeneration are reduced. Typical operating pressures are 85 to 100 psig (586 to 690 kPa). High-pressure systems have been used successfully in many small systems using less than 750 lb/day (340 kg/day) ozone production capacity. Only a few larger installations use high-pressure drying.

High-Purity Oxygen. High-purity oxygen (LOX), purchased and delivered to the plant site, has gained in popularity because of its simple operation. This system has only two major components, the storage tank and the vaporizers.

Liquid Storage Tanks. Liquid-oxygen storage tanks are double-walled thermos-type tanks designed to minimize evaporation losses, and they can be either vertical or horizontal. Although custom-made tanks can be fabricated, it is less expensive to purchase standard-size tanks. If LOX is supplied from an outside commercial source, the storage tanks are normally rated at 200 to 300 psig (1,379 to 2,068 kPa). In special cases, pressurized storage tanks can be built to operate at 600 psig (4,137 kPa). Although the classification of the tanks by operating pressure varies from manufacturer to manufacturer, the following general guidelines can be used:

- Operating pressure < 50 psig (< 345 kPa): low pressure

FIGURE 10.22 Low-pressure air preparation ozonation system.

FIGURE 10.23 Medium-pressure air preparation ozonation system.

FIGURE 10.24 High-pressure air preparation ozonation system.

10.48

- Operating pressure = 200 to 300 psig (1,379 to 2,068 kPa): medium or high pressure
- Operating pressure = 600 psig (4,137 kPa): high pressure

LOX storage tanks absorb ambient heat, which causes some of the stored liquid product to evaporate. The evaporated oxygen gas must be vented or routed to an ozone generator to prevent overpressurizing the tanks. Depending on tank size, about 0.2% to 0.5% per day of stored liquid oxygen will vaporize.

On-site storage capacity varies depending on the relative availability of LOX. Most installations should be provided with at least twice the expected delivery size, which is typically 5,000 lb (2,268 kg), so that it will not be necessary to deplete the supply on hand before taking a full delivery.

Vaporizers. The stored LOX must be vaporized to be used. Vaporization is accomplished by using water, electricity, steam, liquid petroleum gas (LPG), natural gas, or an ambient-type vaporizer. The water-type vaporizer uses a water-to-LOX heat exchanger. Filtered and disinfected water should be used to prevent plugging of the vaporizer. To reduce water use, cooling water returned from the ozone generator can be used to supply the water-type vaporizer. To meet instantaneous oxygen demands with stored LOX, an uninterrupted water supply is required.

The electric vaporizer uses an immersion-type electric heater submerged in a concrete basin filled with water. The heater raises the water temperature, and the heat is transferred to a submerged stainless steel or copper coil where LOX is vaporized to oxygen gas (GOX). This method requires about 80 kWh to maintain the safer temperature and about 120 kWh per ton of oxygen vaporized.

The steam-type vaporizer uses a steam-to-oxygen heat exchanger and requires approximately 350 lb (159 kg) of steam per ton of oxygen produced.

The ambient-type vaporizer is a radiator-type vaporizer using ambient heat to vaporize LOX. This type of vaporizer consumes the least amount of energy but requires more space. At 100,000 ft^3/min (2,830 m^3/min) vaporization capacity, the space required for the ambient-type vaporizer is about 36 by 14 ft (11 by 4 m). However, a fan can be provided to increase ambient airflow and reduce the size of the vaporizer. In addition, a heater is sometimes provided to prevent the vaporizer from freezing.

The LPG/natural gas vaporizer is a direct-fire vaporizer using LPG (such as propane) or natural gas to power a water heater and a water recirculation system. The heated water vaporizes LOX to GOX. This type of vaporizer can provide an uninterrupted oxygen supply from the storage tanks during plant power outage. If LPG is used as the primary fuel, a storage vessel is required. For safety reasons, this type of vaporizer should be located remote from the LOX storage tanks. The different types of vaporizers are compared in Table 10.8.

On-site Generation of Oxygen. Oxygen can be generated on-site as it is needed, using either the cryogenic separation or the adsorptive separation process.

Cryogenic Separation Process. Cryogenic air separation is the oldest of the air separation technologies. Since it was introduced in 1902, the technology has been improved many times and is the most widely used in both industrial and municipal applications to separate oxygen from air. Since its inception, it has been used in more than 2,000 installations worldwide. Municipal applications in the United States alone include more than 70 installations, mostly in the treatment of wastewater. It is a sophisticated process, offered only as a preengineered package. However, alternative components (e.g., air compression, cooling, prepurification, and system operating packages) are still available.

The cryogenic air separation process uses the principles of gas liquefaction followed by fractional distillation to separate air into oxygen and nitrogen. Components of air will

TABLE 10.8 Comparison of Vaporizers

Type of vaporizer	Advantages	Disadvantages
Water-type vaporizers	Low space requirement Simple to operate Reliable	Require continuous supply of plant water Require pumps
Electric vaporizers	Low space requirement Simple to operate	High energy consumption Heating element prone to corrosion Heater must be continuously energized High cost
Steam-type vaporizers	Low space requirement Low cost	Require a continuous supply of steam
Ambient-type vaporizers	Low cost Low energy consumption	Large space requirement
Liquid petroleum gas (LPG) natural gas–type vaporizers	Does not depend on plant power supply Low space requirement	Require LPG or natural gas Noisy operation

liquefy at a temperature below $-300°$ F ($-184°$ C) and at a pressure of 14.7 psia (101 kPa). Liquefaction is achieved by expanding the compressed air rapidly from a high-pressure stage to a low-pressure stage. This sudden expansion causes the air temperature to drop drastically and allows the air to liquefy.

Nitrogen and oxygen gases can be separated from liquefied air by using their different boiling points at various pressures. The boiling points for oxygen and nitrogen are

	Boiling points at different pressures		
	14.7 psia (101 kPa)	22 psia (152 kPa)	90 psia (620 kPa)
Nitrogen	$-320°$ F ($-196°$ C)	$-314°$ F ($-192°$ C)	$-286°$ F ($-177°$ C)
Oxygen	$-297°$ F ($-183°$ C)	$-290°$ F ($-179°$ C)	$-259°$ F ($-162°$ C)

As indicated in the table, nitrogen always has a lower boiling point than oxygen under the same pressure. If the temperature is maintained at the boiling point of nitrogen, the nitrogen will vaporize, while oxygen remains liquid. In addition, the boiling point of nitrogen at 90 psia (620 kPa) is higher than that of oxygen at 22 psia (152 kPa), so nitrogen gas at the high-pressure side of the condenser will condense and release heat to vaporize liquid oxygen at the low-pressure side. By using two separate distillation columns, one operated at the high pressure and the other at the low pressure, the nitrogen and oxygen can be separated.

In addition to the production of gaseous oxygen (GOX) and liquid oxygen (LOX), by-products such as gaseous nitrogen (GAN) and liquid nitrogen (LIN) can be produced. Cryogenic air separation is the only method that can produce GOX, LOX, LIN, and GAN at the same time.

A simplified schematic diagram of the cryogenic air separation system is presented in Figure 10.25. The major components of the system provide air compression, air purification, gas expansion, distillation, and product storage.

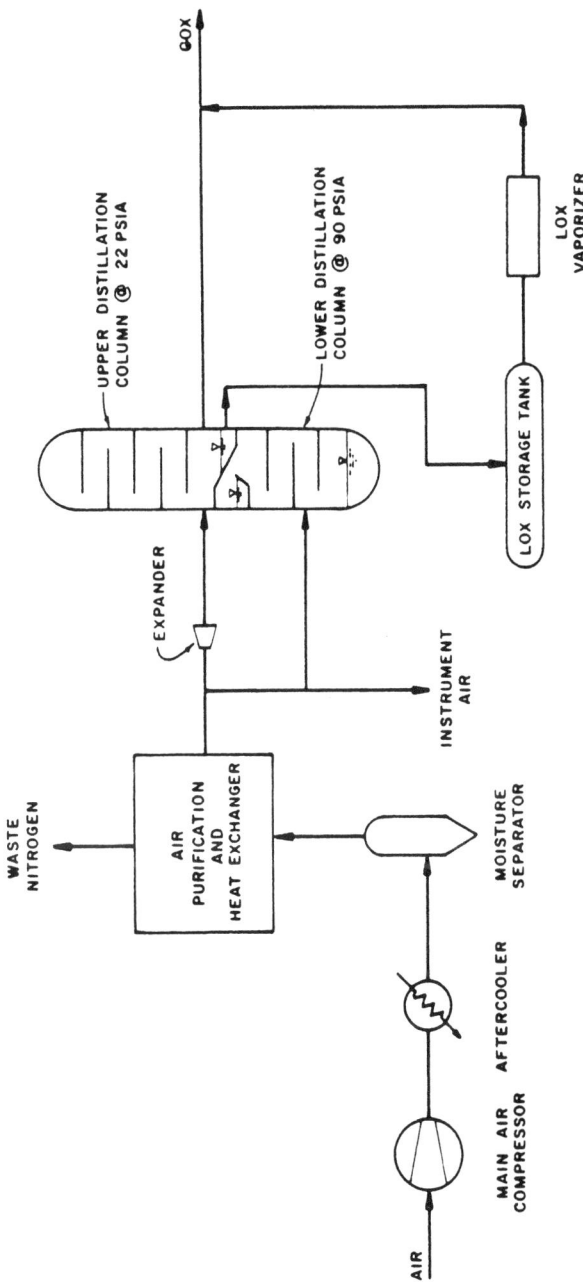

FIGURE 10.25 Typical schematic of cryogenic system.

A cryogenic air separation system consumes about 260 to 340 kWh (936 to 1,224 MJ) of energy per ton (1,016 kg) of oxygen produced, depending on the plant capacity, with about 90% of it used by the main compressor. Because of its high power requirements, high-efficiency compressors and drives are required to reduce the total operating cost.

Turndown of a cryogenic air separation system is limited primarily by the main air compressor and the distillation column. The maximum turndown for a single main air compressor is about 70% to 75% of the rated capacity. Below that limit, the compressor will become unstable and must be vented. The turndown of a cryogenic system is also limited by the distillation column to about 50%. Below 50% of system capacity, GOX must be vented or LOX must be produced in order to maintain system stability.

The turndown range can be increased by using multiple trains. For a two-train system with each train rated at 50% of capacity, the turndown can extend to 30% of total rated capacity. However, because of high capital costs and operating complexity, the selection of the number of trains must be based on process requirements rather than turndown range.

The response rate to demand changes of a cryogenic system is relatively slow, normally about 2% to 3% per minute. Therefore, from 100% to 70%, it will take about 10 to 15 min to reduce the production rate. If one train is shut down but maintained in a cold condition (sometimes referred to as "bottled up") for less than 12 h, the "cold start-up" time to reach full production is about 1 to 2 h. If the system is maintained in cold condition for more than 12 h, the cold start-up time will increase to 2 to 3 h. If the cryogenic system is restarted from ambient temperature, LOX is required, and the "warm start-up" time is about 2 to 3 days.

Providing redundancy of equipment to ensure uninterrupted operation is a common practice for municipal agencies. Redundancy normally includes multiple trains of equal-capacity equipment, with one or two more trains on standby. However, manufacturers indicate that with a sound operation and maintenance program, a cryogenic plant can be on-line over 98% of the time without a second train. They also indicate that a standby cold box is not needed, because there are no rotating parts in this unit. However, this would require adequate LOX backup storage in the event of a cold box problem. Redundancy of rotating equipment and product storage tanks is desirable.

Noise levels from an air separation plant can be high unless special consideration is given during the design to reduce the noise. Noise levels from the main air compressors in general exceed 90 dBa (decibels a-weighted) at a distance of 3 ft (0.9 m). Therefore it is common to house this equipment in a sound-insulated enclosure. A separate silencer in addition to the inlet filter/silencer assembly may be needed to reduce the noise at the inlet of the compressors. Silencers should be provided at the outlet of the compressor surge vent line and the product and waste vent lines. Switching valves would generate a noise level over 100 dBa at a distance of 3 ft (0.9 m) and should be provided with an enclosure and silencer to reduce the noise.

Depending on the equipment size, sheet metal enclosures with an acoustic treatment may be required for the booster compressors and gas product compressors. Most liquid product pumps are small and should not cause concern about the noise level. The cooling water pumps are usually centrifugal and should have noise levels below 85 dBa.

Depending on the size, the main air compressor may require a power supply of 480, 2,400, or 4,160 V. The remaining equipment normally requires either a 120- or 480-V supply. The main air compressor is started unloaded and may take 5 to 15 min to reach its operating status, depending on compressor size. If the standby unit is started before the duty unit is shut down to maintain uninterrupted air supply, the power requirement should include the standby unit.

The cryogenic plant requires large quantities of cooling water equal to 8 to 10 gpm (30 to 38 L/min) per ton (907 kg) of oxygen production.

Adsorptive Separation. Separation of air into high-purity oxygen and nitrogen gas streams can be accomplished by preferential adsorption of nitrogen onto a solid adsorbent as air is passed through a column (or bed). Nitrogen is retained in the adsorbent bed while oxygen (being less preferentially adsorbed) passes through the column as the product gas at an oxygen purity between 90% and 95%.

When the column becomes saturated with nitrogen, oxygen production from that column is discontinued and the adsorbent is regenerated. Regeneration occurs by elevating the temperature or dropping the pressure in the adsorption column, which reduces the capacity of the adsorbent. The nitrogen is disturbed (released) from the adsorbent, and the highly concentrated nitrogen stream is purged from the system. At this point, the column is returned to the adsorption mode, and production of enriched oxygen is resumed.

Adsorptive separation processes are classified by the method of regeneration. The adsorbent can be regenerated by raising the temperature or by decreasing the pressure. Raising the temperature of the bed in the regeneration process is known as thermal swing adsorption (TSA). When the pressure is lowered during regeneration, the process is called either pressure swing adsorption (PSA) or vacuum swing adsorption (VSA). The use of PSA or VSA technology for air separation is more popular than the TSA process because of lower capital and operating costs.

The difference between PSA and VSA systems is the operating pressure. A PSA system utilizes a compression step of between 30 and 60 psig (209 and 419 kPa) before the adsorbent beds, and the VSA system operates at between 0.5 and 15 psig (3 and 105 kPa) for adsorption. Regeneration takes place at near-atmospheric pressure in a PSA system; the VSA columns are regenerated under vacuum. Thus a pressure change occurs in both PSA and VSA systems to desorb nitrogen. One characteristic common to both PSA and VSA is the inability to produce liquid oxygen. This is in contrast to cryogenic air separation, which is based on the liquefaction process.

For typical municipal water treatment operations, separation capacity will be used during most of the year. Both VSA and PSA oxygen generation have the capability to operate at low utilization rates. Turndown can be nearly 100%, but with a substantial power penalty.

Most of the energy losses in an adsorptive system occur during the regeneration of the adsorbent bed. It is at this time that the energy expended to raise the pressure in a PSA system is lost (from depressurization of the column) or a pressure drop is induced in a VSA system (with a vacuum pump). Because the adsorption process utilizes multiple beds whose operation is set by the timing sequence for adsorption, desorption, and purging, provisions are made to alter the timing sequence during turndown operation. By extending the duration of each step, the regeneration step occurs less often and less energy is lost. This improves the energy efficiency of the process at turndown. These projections indicate that the VSA system would provide the most energy-efficient operation at capacity production. The lower energy requirements of the single-train VSA versus the single-train PSA continue until approximately 50% of production capacity, at which time the energy requirements are about equal. Accordingly, the magnitude of the energy savings offered by the VSA system decreases as the production rate is lowered.

The response time to changes in production rate is almost instantaneous for both types of adsorptive separation processes. If less oxygen is required, the flow rate through the system is reduced by adjusting a valve on the outlet of the column. The compressor on a PSA system would then go into unload mode to maintain the same pressure in the columns, which lowers the flow rate. For a system utilizing a preset volume of gas through the column to initiate regeneration, the microprocessor controlling the timing cycle for the beds would automatically adjust for the lower production rate. This allows quick response to changes in oxygen production.

The main sources of noise in a PSA separation process are the main air compressors. Noise emissions from these compressors can range from 80 to 90 dBa. Noise attenuation should be provided for these units. Typical installations are in an enclosed area, with inlet, surge, and purge silencers.

The sources of noise in a VSA system include the main air blower, the product compressor, the switching valves, and the vacuum pump used for regeneration. The vacuum pump is the major source of noise for this system, with noise levels above 90 dBa. Exhaust silencers are usually necessary.

Cooling water requirements range from 2 to 5 gpm (8 to 19 L/min) per ton (907 kg) of system capacity for a VSA process. Several manufacturers have suggested that high-quality water (total dissolved solids less than 200 ppm) be used as the seal water for the vacuum pumps. The manufacturer of the most popular vacuum pump used in VSA systems has indicated that city tap water is generally used for seal water, with no other special requirements. A PSA process requires between 4 and 5 gpm (15 and 19 L/min) per ton (907 kg) of system capacity, with the major portion being used by the compressor's intercooler and aftercooler.

Space requirements for VSA and PSA systems are roughly the same as for a cryogenic facility.

Ozone Contactors

Ozone contact basins provide for transfer of ozone gas into the liquid, promote ozone contact throughout the liquid, and serve to retain the ozonated liquid for a period of time as required to accomplish the desired reactions. The specific process objective and corresponding reactions should dictate contact basin design. Reactions that are rapid relative to the ozone mass-transfer rate from gas to liquid phase are best served by contactors that promote the maximum transfer of ozone in the shortest time. For these applications, such as oxidation of iron, manganese, or simple organics, contact time is often less important, and contactors that rely on single points of application may be suitable. For reactions that are slow relative to the ozone mass-transfer rate, such as disinfection or oxidation of complex organics (including the very persistent herbicides and pesticides), contact time is critical and favors contactors with extended detention time and multiple application points, such as the conventional multistage fine bubble diffuser design.

Factors Affecting Transfer Efficiency. The mass transfer of ozone into water has been described by the two-film theory of gas transfer. However, the calculations are complex, and designers have usually avoided them in favor of conservative estimates for transfer efficiency based on past experience with similar designs. With the continued development and use of ozone technology, it becomes important for the designer to understand the basic factors that affect transfer efficiency, including contactor characteristics, feed gas characteristics, and source water characteristics.

Contactors have been developed in many configurations, and mass transfer will vary with any characteristic that affects the driving force between the gas and liquid. For the conventional fine bubble diffuser design, the essential factor is depth of water over the diffusers, with efficiency increasing with increasing depth. Additional factors of less importance include hydraulic detention time, liquid flow direction relative to gas flow direction, and number of stages.

Feed gas characteristics that influence mass transfer include ozone dose, feed gas concentration, and bubble size. Mass-transfer efficiency will decrease with increasing ozone dose or bubble size but increase with increased ozone concentration. Recent developments in ozone generation technology with resulting ozone concentrations well over 10% have improved transfer efficiency. However, the corresponding decrease in gas flow rates may significantly change the hydrodynamics within the contact basin, making uniform contacting (distribution of the ozone throughout the liquid) much less certain.

Certain characteristics of the water itself can also influence the ozone transfer rate, including temperature, pH, and water quality. The solubility of ozone increases with increasing temperatures, and thus the transfer rate will increase at higher temperatures. As pH increases, the transfer rate will increase. The presence of ozone-reactive materials, including organics, iron, or manganese, will increase mass-transfer efficiency.

Common Types of Contactors. Types of contactors that are commonly used include

- Conventional fine bubble
- Turbine
- Packed column
- Injectors
- Deep U-tube

The selection of contactor type depends on many factors, including site considerations and economics, but most often the contactor should be selected based on the specific treatment objective. Appropriate choices are indicated in Table 10.9.

Fine Bubble Contactors. The fine bubble diffuser contactor consists of a series of over/underbaffled cells. Ozone is applied to some or all cells through a grid of ceramic diffusers at the bottom of the basin. The fine bubble diffuser contact basin is the most widely used contactor. Its prevalence is justified by many factors, including

- No moving parts
- Adaptable to both rapid and slow reactions
- Adaptable to high transfer efficiency
- Adaptable to high hydraulic efficiency (T_{10}/HRT, where HRT is the theoretical hydraulic retention time)

In an effort to improve transfer efficiency, the depth of bubble contactors has increased over the years to the current practice of 20 to 22 ft (6 to 7 m) over the diffusers. Further increases in depth yield only marginal improvement in transfer efficiency, especially at low ozone doses.

Distribution of gas throughout the liquid must be carefully considered, especially in the design of disinfection applications where successful operation cannot be directly mea-

TABLE 10.9 Contactor Selection

Treatment objective	Suitable contactor choices
Primary disinfection	Multistage, fine bubble diffuser type
Iron and manganese oxidation	Injectors, multistage fine bubble
Color removal	Multistage fine bubble, injectors, deep U-tube, turbine
Taste and odor control	Multistage fine bubble, deep U-tube
Algae removal	Turbine
Oxidation	Deep U-tube, injectors, multistage fine bubble

sured. Several guidelines have been developed to assist the designer in ensuring adequate contact between the ozone gas and the liquid:

- Maintain a gas-to-liquid ratio between 0.05 and 0.20. This guide comes from the chemical engineering field for effective transfer and contacting in a two-phase (liquid and gas) system. However, this may be difficult to maintain under a wide range of flow and dose conditions.
- Maintain a minimum gas floor loading rate of 0.12 ft^3/min/ft^2 (0.037 m^3/min/m^2). This guide comes from studies on activated sludge systems and represents the minimum gas flow rate for proper mixing of a liquid mass.
- Maintain uniform floor coverage with diffusers spaced at no more than 3 ft (0.9 m). This guideline was developed in recognition of opportunistic flow of the liquid through areas low in or devoid of gas bubbles.

If followed, these guidelines will yield a design that effectively promotes uniform contacting and distribution of ozone throughout the liquid. In disinfection applications, this means the designer can be assured that all the liquid has been equally exposed to the disinfectant. These guidelines work well with conventional air feed systems operating at typical water treatment doses.

However, recent developments in ozone generation, resulting in ozone concentrations in excess of 10% when using high-purity oxygen as the feed gas, have made it almost impossible to meet these guidelines when operating at typical disinfection doses of 1 to 3 mg/L. Solutions to these applications have included supplemental mixing with air or water injectors. One manufacturer recommends using more, small [4-in.-diameter (10-cm)] diffusers with smaller bubbles (1 to 2 mm) to improve the floor coverage. Ultimately, the higher ozone concentration available from the latest generation of generators is forcing the industry to consider modifications, and alternatives, to the conventional multistage fine bubble diffuser contactor.

The search to optimize the hydraulic efficiency T_{10}/T in fine bubble contactors has been investigated by using tracer dye studies of existing basins, on computer modeling based on finite element analysis, and using computational fluid dynamic techniques. While complex in development, this work provides practical results for the designer.

By using D, L, and W to represent the depth, length in the direction of flow, and the width perpendicular to the direction of the flow, respectively, it has been shown that hydraulic efficiency is very closely related to D/L. While previous sources have recommended basin configurations of 1.5 to 1.0 to 1.0 (D, W, L), the T_{10}/T can be improved by 50% or better if D/L is increased to 4.0 or higher. Hydraulic efficiency appears to be independent of width. Consequently, to optimize T_{10} for contactors with a depth of 20 ft (6.1 m), the cell length should be less than 5 ft (1.5 m).

Appendix O of the SWTR Guidance Manual provides extensive discussion in evaluating CT for existing basins. For most applications, the designer may use the T_{10} method for estimating basin performance. However, the determination of CT for ozone contact basins is complicated by the time required to establish a residual concentration and the relatively rapid decay of ozone residual in water. Consequently, a cell-by-cell evaluation is necessary, and the CT for an ozone contact basin is the sum of the CT values for all cells.

The designer may conservatively estimate the T_{10}/T for the entire contactor. The D/L value may be estimated by

$$\frac{D}{L} = \frac{D \times \text{number of cells}}{\text{total contactor length (excluding baffle thickness and counting chimneys as cells)}}$$

TABLE 10.10 First Cell Log Flat Inactivation Credits

Outlet ozone residual	Credit
0.1 mg/L	1-log virus
0.3 mg/L	0.5-log *Giardia*, 1-log virus

The T_{10} for each cell may be estimated by allocating the total T_{10} based on a linear extrapolation by the following:

$$T_{10} \text{ cell} = (V \text{ cell})(T_{10} \text{ total})(V \text{ total})$$

Each cell may be evaluated as follows:

First Cell. In the first cell, the initial ozone demand is being satisfied, and the residual is just being established. This cell may be considered for flat inactivation credits in accordance with Table 10.10. Or C may be taken as the average for the contactor. For the designer taking T_{10} as one-half the allocated T_{10} and C as one-half the target outlet, C will give a conservative estimate of CT equivalent to assuming the residual is initially established mid-depth of the contactor.

Subsequent Cells. For subsequent cells, T_{10} is taken as the allocated T_{10}. The value of C may be taken as the average for the cell (that is, C value in $+$ C value out, divided by 2), or as the C at the outlet.

For the designer, basin operation will usually be established to achieve a target residual in the outlet from the first cell of 0.1 mg/L or more, although long-established European practice has been to achieve a residual of 0.4 mg/L. In subsequent cells, additional ozone is applied to maintain the residual for the desired contact time, the last cell(s) provide for ozone decay, but may also be used for CT credit.

Diffuser contactors should be designed with 2 to 3 ft of headroom to provide for unimpeded gas flow to the off-gas exit. In addition, many operations will exhibit foaming within the basin, which should not be allowed to enter the off-gas piping.

Turbine Contactors. Turbine contactors have been used widely for water treatment around the world, including several installations in the United States. The aspirating turbine draws ozone gas into the contactor and mixes it with the water. Major advantages of this design include effective mixing, reduced opportunity for clogging from particulates or oxidation products, and high transfer efficiency without deep structures. Major disadvantages, which limit expanded use, include the additional energy input and limited turndown capability from fixed gas flow rates. Because the turbine contactor functions as a completely stirred reactor, in disinfection applications it should be followed by detention chambers or placed in multiple stages.

Packed Column Contactors. Packed column contactors have seen limited use in water treatment, but interest may increase with increasing ozone concentrations, which are less compatible with conventional designs. In this design, the reactor is filled with ceramic rings; the liquid flows down through the reactor while ozone gas is applied at the bottom. This design provides effective contacting for high liquid flow/low gas flow conditions as experienced with low doses of high ozone concentration. Hydraulic efficiency approaches plug flow characteristics. The primary disadvantages of packed columns are the high cost of the packing material and potential for scaling on the packing.

Direct Injection Contactors. Direct injection of ozone into the liquid stream has received renewed interest in the United States for several reasons, including the increasing number of smaller applications. In addition, direct injection offers potential advantages when operating with high ozone concentrations. Direct injection works on the venturi principle and involves pressurized water flow past a small orifice, which creates a partial vacuum, drawing in the gas. In most applications, this is conducted in a sidestream, which is subsequently blended with the remaining liquid flow in a static mixer. In small applications, this may all occur in a pipeline; for larger applications, injection would normally be followed by a reaction chamber. The primary disadvantages of this system are the energy input required to move the liquid in the sidestream and poor turndown characteristics.

Off-Gas Disposal

One of the principal design problems in ozone contact systems is the disposal of off-gases from ozone contactors. Assuming that ozone contactors use from 90% to 100% of the ozone that is applied, the air exiting from the contactor may have ozone concentrations as high as 0.5% by volume. This compares with a threshold odor level of 0.05 ppm for ozone and an 8-h OSHA standard of 0.1 ppm.

To date, regulations have not been established on the levels of ozone that may be discharged to the atmosphere, but there is no question that large volumes of air containing 0.5% ozone cannot be casually discharged. Five principal methods of off-gas disposal may be considered:

- Reinjection
- Heating to cause autodecomposition
- Chemical reduction with a reducing agent
- Catalytic reduction with a metal oxide
- Dilution

Reinjection generally involves the construction of two ozone contact basins. The fresh ozone is introduced into the downstream contact basin, and the off-gases are then re-pumped and reinjected into the upstream contact basin. Given the efficiencies of ozone consumption in each contact stage and the loss of ozone during the repumping process, the ozone residual in the air exiting from the reinjection stage can be as low as 0.001% or 10 ppm. Thus injection alone does not completely solve the problem. Rather, reinjection must be used in tandem with some of the other techniques described.

Chemical reduction is another method for removing ozone residuals from off-gases. The chemical reduction could be accomplished by passing the off-gases from the ozone contact chamber in countercurrent flow with an ozone-specific reducing agent in a scrubber much like those used for removing fumes from industrial off-gases. The key to this method is the selection of an inexpensive reducing agent that is not also oxidized by the oxygen present in the air. No uniformly satisfactory reducing agent has been developed to date.

Ozone rapidly dissipates when it is heated. Consequently, in some designs the ozone contactor off-gases are heated to a temperature at which decomposition of the ozone is nearly instantaneous. Temperatures as high as 250° C have sometimes been indicated. The obvious disadvantage of this method is the amount of heat required. In some European designs, the hot air exiting from the ozone decomposer is recycled to a preheater to warm the air that is about to enter the decomposer. This reduces energy requirements but increases capital costs.

Most recent designs have used thermal/catalytic destruct units for off-gas treatment. These consist of a vessel containing the catalyst preceded by a heater. Catalytic reduction involves passing the ozone off-gases across a surface that catalyzes the decomposition of ozone to elemental oxygen. Most catalytic compounds shown to be effective for ozone reduction are proprietary and are based on iron or manganese oxides.

The catalytic reaction is very rapid, and empty-bed contact times are on the order of 1 min. The preheater primarily functions to eliminate moisture from the incoming gas, and is typically designed for a 20° F ($-8°$ C) temperature rise to reduce the relative humidity from 100% to about 50%. The thermal/catalytic destruct systems are capable of reliably reducing off-gas ozone concentrations to well below 0.1 mg/L.

Construction Materials for Ozone Equipment

In designing systems for ozonation, the highly aggressive character of ozone should be kept in mind. All rubber, most plastics, neoprene, ethylene-propylene-diene rubbers, and aluminum are unacceptable materials for use with ozone. The only acceptable materials are 316 stainless steel, 305 stainless steel, glass, Hypalon, Teflon, and concrete. There is some dispute about the usefulness of type 1 PVC. Manufacturers have often recommended type 1 PVC, but its quality does not seem to be uniform from place to place, and incidents of PVC failure occur regularly.

BIBLIOGRAPHY

Aieta, E. M., and J. D. Berg. "A Review of Chlorine Dioxide in Drinking Water." *Journal AWWA* 77(6):64, 1986.

Automated Measurement of Aqueous Ozone Concentration. Denver, Colo.: American Water Works Association Research Foundation, 1993.

Case Studies of Modified Disinfection Practices for Trihalomethane Control. Denver Colo.: American Water Works Association Research Foundation, 1990.

Chick, H. "An Investigation of the Laws of Disinfection." *Journal of Hygiene* 8:92, 1908.

Chlorination By-Products: Production and Control. Denver, Colo.: American Water Works Association Research Foundation, 1986.

Chlorine Dioxide: Drinking Water Issues. Denver, Colo.: American Water Works Association Research Foundation, 1993.

Controlling Disinfection By-Products. Denver, Colo.: American Water Works Association, 1993.

Development and Validation of Rational Design Methods of Disinfection. Denver, Colo.: American Water Works Association Research Foundation, 1995.

Disinfectant Residual Measurement Methods. Denver, Colo.: American Water Works Association Research Foundation, 1992.

Disinfection Dilemma: Microbiological Control versus By-Products. Denver, Colo.: American Water Works Association, 1993.

Duguang, Z., C. Ramirex Cortina, and M. Roustan. "Modeling of the Flow Patterns in Ozonation Chambers Contactor by Residence Time Distribution Studies and by Computational Fluid Dynamics Approach." In *Proc. 12th World Congress of the International Ozone Association*, vol 2. Tours: Instaprint S.A., 1995, pp. 215–226.

Experimental Methodologies for the Determination of Disinfection Effectiveness. Denver, Colo.: American Water Works Association Research Foundation, 1993.

Federal Register. Part III, Environmental Protection Agency, *List of Regulated Substances and Thresholds for Accidental Release Prevention and Risk Management Programs for Chemical Accident Release Prevention, Final Rule and Notice.* 40 CFR Parts 9 and 68, vol. 59 (no. 20), 4493–4499, January 31, 1994.

Federal Register. National Primary Drinking Water Regulations: Monitoring Requirements for Public Drinking Water Supplies; Final Rule. 40 CFR Part 141, vol. 61, no. 94, pp. 24354–24388, May 14, 1996.

Federal Register. *National Primary Drinking Water Regulations: Disinfectants and Disinfection By-products; Final Rule.* 40 CFR Parts 9, 141 and 142; vol. 63 (no. 241): 69389–69476, December 16, 1998.

Finch, G. R., and M. Belosevic. "Controlling *Giardia* spp. and *Cryptosporidium* spp. in Drinking Water by Microbial Reduction Processes." *Journal Environmental Engineering Science* 1:17–31, 2002.

Gates, D. *The Chlorine Dioxide Handbook.* Eds. B. Cobban, P. Murray, and A. Livingston. Denver, Colo.: AWWA, 1998.

Gordon, G. "Is All Chlorine Dioxide Created Equal?" *Journal AWWA,* 93(4):163–174, 2001.

Gordon, G., and A. Rosenblatt. "Gaseous Chlorine-Free Chlorine Dioxide for Drinking Water." In *Chemical Oxidation: Technology of the Nineties. Fifth International Symposium.* International Chemical Oxidation Association, Vanderbilt University, Nashville, Tenn., February 15–17, 1995.

Griese, Mark H., J. J. Kaczur, and G. Gordon. "Combining Methods for the Reduction of Oxychlorine Residuals in Drinking Water." *Journal AWWA* 84(11):69–77.

Guidance Manual of Compliance with the Filtration and Disinfection Requirements for Public Water Systems Using Surface Water Sources. U.S. Environmental Protection Agency. Denver, Colo.: American Water Works Association, 1990.

Henry D. J., and E. M. Freeman. "Finite Element Analysis and t10 Optimization of Ozone Contactors." In *Proc 12th World Congress of the International Ozone Association,* vol 2. Tours: Instaprint S.A., 1995, pp. 201–214.

Hill, A. G., and H. T. Spencer. "Mass Transfer in a Gas Sparged Ozone Reactor." In *Proceedings First International Symposium Ozone for Water and Wastewater Treatment.* Edited by R. G. Rice and M. E. Browning. Norwalk, Conn.: International Ozone Association, 1975, pp. 367–381.

Hoehn, R. C., B. L. Long, and D. Gates. "Status of Chlorine Dioxide Disinfection Technologies." *Proceedings AWWA Annual Conference* (on compact disk), Denver, Colo., June 11–15, 2000, Paper 6 in Session THU9 "Getting Alternative Disinfection Strategies Approved for General Use."

Hoigne, J. "Mechanisms, Rates, and Selectivity of Oxidation of Organic Compounds Initiated by Ozonation of Water." In *Handbook of Ozone Technology and Applications,* vol. 1: *International Ozone Technology and Applications.* Edited by R. G. Rice and A. Netzer. Norwalk, Conn.: International Ozone Association, 1982, pp. 341–379.

Identification and Occurrence of Ozonation By-Products in Drinking Water. Denver, Colo.: American Water Works Association Research Foundation, 1993.

Impact of Ozone on the Removal of Particles, TOC, and THM Precursors. Denver, Colo.: American Water Works Association Research Foundation, 1989.

Implementation of the Integrated Disinfection Design Framework, report 90848. Denver, Colo.: American Water Works Association Research Foundation, 2001.

In-Line Ozone and Hydrogen Peroxide Treatment for Removal of Organic Chemicals. Denver, Colo.: American Water Works Association Research Foundation, 1992.

Kilpatrick, M., et al. "The Decomposition of Ozone in Aqueous Solutions." *Water Research* 10:377, 1976.

Lee, S. H., et al. "Surveillance for Waterborne-Disease Outbreaks—United States, 1999–2000." Atlanta, Ga.: Centers for Disease Control, 2002.

Manley, T., and S. Niegowski. "Ozone." In *Encyclopedia of Chemical Technology,* 2d ed., vol. 14. New York: John Wiley & Sons, 1967, p. 410.

Masschelein, Willie J. *Chlorine Dioxide: Chemistry and Environmental Impact of Oxychlorine Compounds.* Ann Arbor, Mich.: Ann Arbor Science, 1979.

Morris, C. "The Acid Ionization Constant of HOCl from 5 to 35° C." *Journal of Physical Chemistry* 70:3798, 1966.

Morris, C. "Aspects of the Quantitative Assessment of Germicidal Efficiency." In *Disinfection in Water and Wastewater.* Ann Arbor, Mich.: Ann Arbor Science, 1975.

Ozone and Biological Treatment for DBP Control and Biological Stability. Denver, Colo.: American Water Works Association Research Foundation, 1994.

Ozone Disinfection of Giardia *and* Cryptosporidium. Denver, Colo.: American Water Works Association Research Foundation, 1994.

Ozone and Ozone-Peroxide Disinfection of Giardia *and Viruses.* Denver, Colo.: American Water Works Association Research Foundation, 1992.

Palin, A. "A Study of the Chloro Derivatives of Ammonia." *Water and Water Engineering* 54:151, 1950.

Pilot-Scale Evaluation of Ozone and Peroxone. Denver, Colo.: American Water Works Association Research Foundation, 1991.

Race, J. *Chlorination of Water.* New York: Wiley, 1918.

Rakness, K. L., R. C. Renner, B. A. Hegg, and A. G. Hill. "Practical Design Model for Calculating Bubble Diffuser Contactor Ozone Transfer Efficiency." *Ozone Science and Engineering* 10:173, 1988.

Roustan, M., and J. Mallevialle. "Theoretical Aspects of Ozone Transfer into Water." In *Ozonation Manual for Water and Wastewater Treatment,* edited by W. J. Maschelein. New York: John Wiley & Sons, 1982, pp. 47–52.

Saunier, B. "Kinetics of Breakpoint Chlorination and of Disinfection." Ph.D. thesis in civil engineering, University of California—Berkeley, 1976.

Sources, Occurrence, and Control of Chlorine Dioxide By-Product Residuals in Drinking Water. Denver, Colo.: American Water Works Association Research Foundation, 1994.

Streng, A. "Solubility of Ozone in Aqueous Solution." *Journal of Chemical and Engineering Data* 6:431, 1961.

Taste and Odor Problems Associated with Chlorine Dioxide. Denver, Colo.: American Water Works Association Research Foundation, 1991.

Trussell, R. R., and M. D. Umphres. "The Formation of Trihalomethanes." *Journal AWWA* 70:604, 1978.

Von Huben, H. *Surface Water Treatment: The New Rules.* Denver, Colo.: American Water Works Association, 1991.

Water Chlorination Principles and Practices, Manual M20. Denver, Colo.: American Water Works Association, 1973.

Watson, H. "A Note of the Variation of the Rate of Disinfection with Change in the Concentration of the Disinfectant." *Journal of Hygiene* 8:536, 1908.

White, G. C. *Handbook of Chlorination and Alternative Disinfectants,* 4th ed. New York: John Wiley & Sons, 1999.

CHAPTER 11
LIME SOFTENING

Michael B. Horsley
Black & Veatch
Kansas City, Missouri

Doug B. Elder, P.E.
Black & Veatch
Olathe, Kansas

Leland L. Harms
Black & Veatch
Kansas City, Missouri

Cold lime softening uses chemical precipitation with lime and other chemicals to reduce a water's hardness and, in some cases, to enhance clarification before filtration. Hot-process softening is predominantly used in industrial applications and is not discussed here. Ion exchange softening is covered in Chapter 12, and membrane processes that may be used to soften water are covered in Chapter 13.

Designing plants with lime softening processes is somewhat different from the design of other types of water treatment plants. Factors responsible for these differences include types of chemicals used, the relatively large quantity of some of these chemicals, the special chemical handling considerations, and the nature of the corresponding chemical reactions. These factors influence process and equipment design and selection, plant layout, and other design considerations.

WATER HARDNESS AND SOFTENING TREATMENT

Hardness in water is the sum of the concentrations of multivalent ions, principally calcium and magnesium. Other ions that produce hardness include iron, manganese, strontium, barium, zinc, and aluminum; but these ions are generally not present in significant quantities. Hardness is generally expressed in terms of equivalent milligrams per liter (mg/L) of calcium carbonate. The sum can also be expressed in milliequivalents per liter. Another expression, used commonly in the past, is grains per gallon, where 17.1 mg/L (as calcium carbonate) is equal to 1 gr/gal.

Total hardness is usually defined as simply the sum of magnesium and calcium hardness in milligrams per liter as $CaCO_3$. Total hardness can also be differentiated into carbonate and noncarbonate hardness. Carbonate hardness is the portion of total hardness present in the form of bicarbonate salts [$Ca(HCO_3)_2$ and $Mg(HCO_3)_2$] and carbonate compounds ($CaCO_3$ and $MgCO_3$). Noncarbonate hardness is the portion of calcium and magnesium present as noncarbonate salts, such as calcium sulfate ($CaSO_4$), calcium chloride ($CaCl_2$), magnesium sulfate ($MgSO_4$), and magnesium chloride (MgCl). The sum of carbonate and noncarbonate hardness equals total hardness.

Acceptable Level of Hardness

Sawyer (1994) classified the degree of hardness as follows:

Hardness	mg/L as $CaCO_3$
Soft	0–75
Moderate	75–150
Hard	150–300
Very hard	300+

The degree of hardness acceptable for finished water varies with the consumer or industry served. In 1968, the American Water Works Association (AWWA) established a water quality goal for total hardness of 80 to 100 mg/L expressed as calcium carbonate, but current AWWA water quality goals do not include hardness.

Magnesium hardness of not more than 40 mg/L is also often proposed as a goal to minimize scaling at elevated temperatures in water heaters. The magnesium concentration present before precipitation of magnesium salt is a function of pH and temperature of the finished water. In recent years, many utilities using a softening process have allowed total hardness in finished water to approach 120 to 150 mg/L to reduce chemical costs and residuals production and, in some cases, to produce a less corrosive water by increasing alkalinity.

Desired hardness is often stated as a policy by a utility that softens water. If a policy exists, the designer must provide a treatment plant design meeting that policy. If no policy exists but softening is desired, it is up to the designer to arrive at a reasonable hardness for design purposes. At present, total hardness of 120 mg/L or less and magnesium hardness of 40 mg/L or less appear to be acceptable design criteria for softening facilities for most applications.

In practice, total hardness produced and the pH of the treated water may be adjusted during operation by varying chemical feed to meet other treatment goals, including modification of hardness goals, stability and corrosivity requirements, and turbidity removal under varying source water quality conditions. Treated water pH of different softening facilities typically varies between 8.5 and 10.0, depending on all these factors.

Benefits of Softening

Potential benefits of softening water at a central treatment plant include the following:

- Reducing dissolved minerals and scale-forming tendencies
- Reducing consumption of household cleaning agents

- Removing radium 226 and 228
- Removing heavy metals, including arsenic
- Removing certain organic compounds and reducing total organic carbon (TOC)
- Removing silica and fluoride
- Removing iron and manganese
- Reducing turbidity of surface waters in conjunction with the hardness precipitation process

The degree of removal of constituents usually depends on the treatment process used. Benefits to the consumer depend on source water quality and user requirements.

Softening Plants in the United States

Currently more than 1,000 domestic-use water plants in the United States use lime softening processes. Although water plants using lime softening are found in most sections of the United States, the majority of large plants are located in the midwestern states and in Florida. These plants range in size from less than 1 mgd to more than 100 mgd.

Chemistry of Lime Softening

Water softening involves a number of complex and dynamic chemical interactions. The chemical reactions involved and calculation of chemical feed requirements are discussed in detail in AWWA's *Water Quality and Treatment*. The discussion that follows in this text simplifies the chemistry involved, highlighting only the predominant reactions.

Lime, the primary chemical used for water softening, reacts with carbon dioxide and carbonate hardness to precipitate calcium carbonate and magnesium hydroxide. Quicklime, CaO, is first slaked to produce calcium hydroxide:

$$CaO + H_2O = Ca(OH)_2 \tag{11.1}$$

Chemical Reactions. Reactions between calcium hydroxide and carbon dioxide and bicarbonate alkalinity are shown in Equations (11.2) and (11.3). The reactions convert the bicarbonate alkalinity present to carbonate alkalinity, which precipitates as insoluble calcium carbonate.

$$CO_2 + Ca(OH)_2 = \underline{CaCO_3} + H_2O \tag{11.2}$$

$$Ca(HCO_3)_2 + Ca(OH)_2 = \underline{2CaCO_3} + 2H_2O \tag{11.3}$$

The optimum pH to produce minimum soluble calcium carbonate is about 10.3, depending on water temperature, total dissolved solids, and other factors affecting the solubility of calcium carbonate. In precipitating the calcium ion, 2 mol of calcium carbonate is formed for every 1 mol of calcium ion removed from the water, as shown in Equation (11.3).

Magnesium hardness, present as magnesium bicarbonate, is removed in a stepwise fashion, as shown in Equations (11.4) and (11.5).

$$Mg(HCO_3)_2 + Ca(OH)_2 = \underline{CaCO_3} + MgCO_3 + 2H_2O \tag{11.4}$$

$$MgCO_3 + Ca(OH)_2 = \underline{CaCO_3} + \underline{Mg(OH)_2} \tag{11.5}$$

Magnesium hydroxide does not precipitate quantitatively, as suggested by Equation (11.5), because the solubility of magnesium hydroxide depends on pH. Generally a pH of 11.0 to 11.3 is necessary to reduce the magnesium ion concentration to low values. Excess lime must be added to raise the pH for precipitation of magnesium hydroxide. The resulting excess hydroxide alkalinity must be converted later to carbonate alkalinity to produce a water of minimum calcium hardness. This process, generally termed *recarbonation,* requires carbon dioxide:

$$Ca(OH)_2 + CO_2 = \underline{CaCO_3} + H_2O \tag{11.6}$$

Once calcium carbonate is formed, its properties are such that resolubilization takes place only at a very low rate. To remove noncarbonate hardness—calcium or magnesium hardness present in excess of the alkalinity—requires soda ash. Equations (11.7) and (11.8) illustrate noncarbonate hardness removal.

$$MgSO_4 + Ca(OH)_2 = \underline{Mg(OH)_2} + CaSO_4 \tag{11.7}$$

$$CaSO_4 + Na_2CO_3 = \underline{CaCO_3} + Na_2SO_4 \tag{11.8}$$

No softening occurs in Equation (11.7), as magnesium hardness is only exchanged for calcium hardness. Soda ash is used in Equation (11.8) to remove the calcium noncarbonate hardness either originally present or formed as a result of the reactions in Equation (11.7).

Chemical Requirements. These equations allow reasonably good approximations of the amounts of lime and soda ash required to soften a water. The lime required to remove carbonate hardness can be calculated as shown in Equation (11.9):

$$CaO \text{ (lb/mil gal)} = 10.6[CO_2 \text{ (mg/L)}] + 4.7[\text{alkalinity (mg/L)} \\ + \text{ magnesium hardness (mg/L)} + x] \tag{11.9}$$

where CaO is 100% pure, CO_2 is expressed as CO_2, alkalinity and magnesium hardness are expressed as $CaCO_3$, and x is the required excess hydroxide alkalinity in milligrams per liter as $CaCO_3$. The magnesium hardness shown is the amount present in the water to be treated. Desired excess alkalinity can be determined from the magnesium hydroxide solubility relationship; it is typically in the range of 30 to 70 mg/L and can be estimated at 65 mg/L, expressed as $CaCO_3$, for minimum magnesium hardness.

Equation (11.10) shows the calculation for the quantity of soda ash required to remove noncarbonate hardness

$$Na_2CO_3 \text{ (lb/mil gal)} = 8.8[\text{noncarbonate hardness (mg/L)} - x] \tag{11.10}$$

where Na_2CO_3 is 100% pure, noncarbonate hardness is expressed as $CaCO_3$, and x is the noncarbonate hardness left in the water.

Because CaO is usually 88% to 95% pure, results from Equation (11.9) must be divided by actual chemical purity. Soda ash is usually essentially pure, so no adjustment to the calculation is required. If $Ca(OH)_2$ is used instead of CaO, the required amount of CaO should be calculated by dividing by 56/74, the ratio of the molecular weights.

Use of Caustic Soda. Caustic soda, NaOH, can be used in place of lime or soda ash. Fewer residuals are produced, and caustic soda is easier to handle, store, and feed. Caustic soda is generally purchased as a 25% or 50% aqueous solution. Softening reactions with caustic soda are shown in Equations (11.11) through (11.15).

LIME SOFTENING 11.5

$$CO_2 + 2NaOH = Na_2CO_3 + H_2O \qquad (11.11)$$

$$Ca(HCO_3)_2 + 2NaOH = \underline{CaCO_3} + Na_2CO_3 + 2H_2O \qquad (11.12)$$

$$Mg(HCO_3)_2 + 2NaOH = \underline{Mg(OH)_2} + 2Na_2CO_3 + 2H_2O \qquad (11.13)$$

$$MgSO_4 + 2NaOH = \underline{Mg(OH)_2} + Na_2SO_4 \qquad (11.14)$$

$$CaSO_4 + Na_2CO_3 = \underline{CaCO_3} + Na_2SO_4 \qquad (11.15)$$

The sodium carbonate formed in Equations (11.11) through (11.13) is available to precipitate calcium noncarbonate hardness, as shown in Equation (11.15). A combination of lime and caustic soda can be used, the ratio depending on the calcium noncarbonate removal required. This combination provides some savings in chemical cost compared with the use of caustic soda alone because caustic soda is more expensive than lime. Using caustic soda may be a good option for low-alkalinity water, because alkalinity reduction with caustic soda is one-half that of lime softening. A disadvantage of using caustic soda is the increase in finished water sodium concentration.

PRETREATMENT

The principal types of pretreatment used before lime softening are aeration and presedimentation.

Aeration

Aeration may be used to remove carbon dioxide from the source water before softening. This is usually only applicable to groundwaters where carbon dioxide concentrations are relatively high. Lime removal of carbon dioxide in source water adds to operations costs because of chemical expenses and increased calcium carbonate residuals, in accordance with Equation (11.2).

Induced-draft or open-tray aeration is often used and may reduce the carbon dioxide level to 10 mg/L or less. Aeration also oxidizes iron that may be present. For some groundwaters containing substantial iron, clogging of aeration trays is a problem. The aerator should be designed to minimize clogging and provide ready access for periodic cleaning. The lime dosage required to react with carbon dioxide may be estimated by using part of Equation (11.9), as follows:

$$CaO \text{ (lb/mil gal)} = 10.6 \times CO_2 \text{ (mg/L)} \qquad (11.16)$$

where CO_2 is expressed as CO_2 (mg/L).

Reduced lime consumption and residuals production associated with aeration must be weighed against the capital cost, as well as the operating and maintenance costs, of aeration equipment. Aeration is primarily used where carbon dioxide levels are high enough to justify the cost. Residuals produced by the reaction of lime and carbon dioxide are in accordance with Equation (11.2) and may be estimated as

$$\text{Dry weight } CaCO_3 \text{ residuals (lb/mil gal)} = 19.0 \times CO_2 \text{ (mg/L)} \qquad (11.17)$$

where CO_2 is expressed as CO_2 (mg/L).

Aeration is often used to oxidize iron in iron and manganese removal plants, but the elevated pH of the softening process together with chemical oxidation, if needed, can remove iron and manganese without the need for aeration.

Presedimentation

Presedimentation is used primarily by those plants treating high-turbidity surface waters, such as sources on the Missouri and Mississippi rivers. Some of these plants use metal salt or polymer coagulants to enhance suspended-solids removal.

Presedimentation provides a more uniform water quality at the treatment plant, removes a major portion of the suspended solids with little chemical cost, and offers potential cost savings in residuals treatment and disposal. Because of the cost of dewatering and land disposal of these solids and the minor impact on highly turbid rivers, state and federal agencies have often allowed return of presedimentation residuals directly to the river.

Presedimentation also provides an opportunity for pretreatment and removal of tastes and odors and other organic compounds with powdered activated carbon and oxidizing agents before the elevated pH associated with the softening process. In some cases, this provides more efficient and effective treatment.

LIME SOFTENING PROCESSES

The common lime softening processes are (1) the straight lime or lime–soda ash process, (2) the excess lime or excess lime–soda ash process, and (3) the split-treatment excess lime or excess lime–soda ash process. The following paragraphs present formulas that can be used to estimate chemical feed requirements for these processes. A design factor of at least 10% is typically added to the chemical feed dosages indicated by the formulas for design of the feed systems. A more detailed presentation can be found in AWWA's *Water Quality and Treatment*.

Straight Lime or Lime–Soda Ash Process

For the straight lime softening process, only carbonate hardness in the form of calcium is to be removed, and only lime is added to the softening stage. The lime dosage required to react with calcium hardness may be estimated by using part of Equation (11.9) as follows:

$$\text{CaO (lb/mil gal)} = 10.6[\text{CO}_2 \text{ (mg/L)}] + 4.7[\text{alkalinity (mg/L)}] \quad (11.18)$$

where alkalinity is equal to the portion of carbonate hardness associated with calcium to be removed.

For the straight lime–soda ash process, both carbonate and noncarbonate hardness are removed, and soda ash is added to the softening stage. The dosage of lime required to react with calcium hardness may be estimated by using Equation (11.18), where alkalinity is the portion of carbonate hardness associated with calcium to be removed. The dosage of soda ash required may be estimated using Equation (11.10).

The straight lime or lime–soda ash process removes only calcium hardness down to a minimum of about 35 mg/L.

LIME SOFTENING 11.7

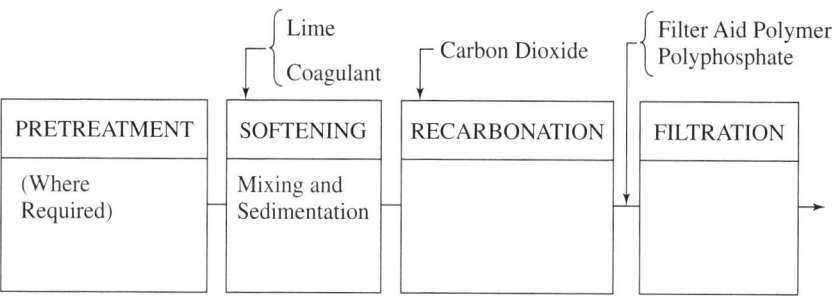

FIGURE 11.1 Single-stage lime process.

Recarbonation. Recarbonation with carbon dioxide is required to convert carbonate alkalinity to bicarbonate alkalinity and stabilize the water before filtration. An often used guideline is 40 mg/L of bicarbonate alkalinity in the finished water, although this could vary depending on stability requirements for the particular water. The reaction of carbonate alkalinity with carbon dioxide to produce bicarbonate alkalinity is

$$CO_2 + CaCO_3 + H_2O = Ca(HCO_3)_2 \qquad (11.19)$$

The dosage of carbon dioxide required to react with the carbonate alkalinity to produce bicarbonate alkalinity may be estimated as follows:

$$CO_2 \text{ (lb/mil gal)} = 3.7 \times \text{carbonate alkalinity (mg/L)} \qquad (11.20)$$

where carbonate alkalinity is that amount to be converted to bicarbonate alkalinity.

The straight lime or lime–soda ash process is typically a single-stage softening process where a single basin is used for softening followed by a recarbonation basin prior to filtration. A flow diagram for a single-stage lime or lime–soda ash process is shown in Figure 11.1.

Straight Lime or Lime–Soda Ash Process Examples. The following examples demonstrate the use of the above equations to estimate chemical feed requirements.
Given:

Surface water source (carbon dioxide is negligible for this source)
Total hardness = 270 mg/L as $CaCO_3$
Alkalinity = 180 mg/L as $CaCO_3$
Calcium hardness = 240 mg/L as $CaCO_3$
Magnesium hardness = 30 mg/L as $CaCO_3$
Noncarbonate hardness (NCH) = 270 − 180 = 90 mg/L as $CaCO_3$
Available lime is 90% CaO
Available soda ash is very close to 100% Na_2CO_3

EXAMPLE 11.1 Reduce calcium carbonate hardness as much as possible, using lime only. (*Note:* In this example, the magnesium hardness is less than the selected target maximum of 40. Therefore, lime addition for magnesium hardness removal is not required.)

Using Equation (11.18) for approximate lime feed requirements, we find

CaO (lb/mil gal) = 10.6[CO_2 (mg/L)] + 4.7 [alkalinity (mg/L)]

CaO (lb/mil gal) = 10.6(0) + 4.7(180) = 846 lb/mil gal, as 100% pure CaO

Correcting for purity, we get

$$CaO = \frac{846}{0.9} = 940 \text{ lb commercial lime/mil gal}$$

Adding selected design factor of 10% for minimum feeder size gives

1.1(940 lb commercial lime/mil gal) = 1,030 lb/mil gal commercial lime

Approximate resulting hardness = 35(approx. minimum solubility of $CaCO_3$)

+ 30(Mg hardness) + 60(remaining calcium NCH) = 125 mg/L as $CaCO_3$†

Approximate remaining alkalinity = 35 mg/L as $CaCO_3$

(approximate minimum solubility of $CaCO_3$)†

Using Equation (11.19) for approximate CO_2 feed requirements gives

CO_2 (lb/mil gal) = 3.7(40 mg/L finished water bicarbonate alkalinity)

= 148 lb/mil gal

Adding selected design factor of 10% for minimum feeder size gives

1.1(148 lb/mil gal) = 160 lb/mil gal

EXAMPLE 11.2 Reduce the calcium hardness in the finished water, using both lime and soda ash to reach a target finished water hardness of 100 mg/L.

As shown in Example 11.1, addition of lime at 1,030 lb/mil gal will yield a finished water with total hardness = 125 mg/L. Removal of a portion of the remaining calcium NCH will be required to reduce hardness to 100 mg/L.

Noncarbonate hardness in water = 270 − 180 = 90 mg/L

Noncarbonate hardness to be left in water = 90 − (125 − 100) = 65 mg/L

Using Equation (11.10) to estimate the soda ash feed requirements gives

Soda ash (lb/mil gal) = 8.8(NCH − x) = 8.8(90 − 65) = 220 lb/mil gal

Adding selected design factor of 10% for minimum feeder size gives

1.1(220 lb/mil gal) = 242 lb/mil gal

Approx. resulting hardness = 35(approx. minimum solubility of $CaCO_3$)

+ 35(remaining calcium NCH) + 30(remaining Mg NCH) = 100 mg/L as $CaCO_3$†

---

†*Note:* Results are approximations based on theoretical equilibrium conditions. These conditions are seldom duplicated in practice, and some variations should be expected at the particular plant. Alkalinity in the finished water may be increased to 40 mg/L or more by reducing the amount of carbonate hardness removed.

Approx. remaining alkalinity = 35 mg/L as $CaCO_3$
(approx. minimum solubility of $CaCO_3$)†

Approximate CO_2 feed requirements remain the same as for Example 11.1.

Excess Lime or Excess Lime–Soda Ash Process

In lime softening, the term *excess lime* refers to the removal of magnesium, in addition to calcium, by providing excess lime in the primary softening process. The excess lime process can generally remove calcium hardness to a level of about 35 mg/L and magnesium hardness to about 10 mg/L. Chemical dosage requirements can be estimated by using Equation (11.9) for the excess lime process and Equations (11.9) and (11.10) for the excess lime–soda ash process.

Single-Stage Excess Lime or Excess Lime–Soda Ash Process. The single-stage excess lime or excess lime–soda ash process is similar to Figure 11.1. Excess lime leaving the softening basin is reacted with additional carbon dioxide in the recarbonation basin to lower the pH of the water and to stabilize the water before filtration.

Although single-stage excess lime softening plants have been operated successfully, several problems are commonly encountered. It is more difficult to achieve process stability with excess lime softening in the typical recarbonation basin, resulting in potential deposition of calcium carbonate on the filter media. In addition, residuals buildup in the recarbonation basin creates periodic basin cleaning requirements. Particulate loading to the filters is high during periods of process instability in the softening basin.

Two-Stage Excess Lime or Excess Lime–Soda Ash Process. A two-stage excess lime softening process is shown in Figure 11.2. The two-stage process has several advantages over the single-stage excess lime process. By limiting recarbonation at the second stage to enough carbon dioxide to react with the excess lime, additional softening by removal of calcium carbonate may be performed in the second stage.

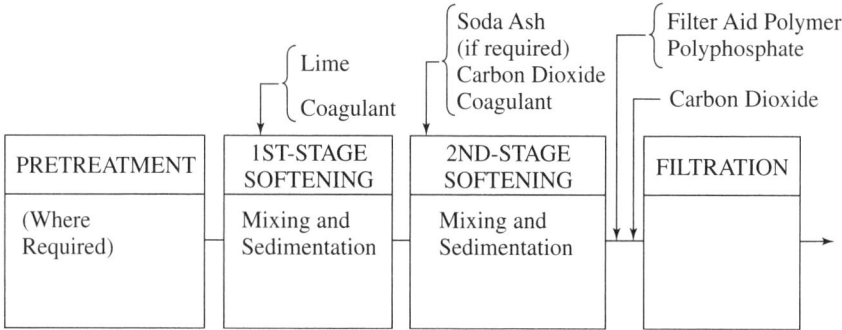

FIGURE 11.2 Excess lime softening process.

---

†*Note:* Results are approximations based on theoretical equilibrium conditions. These conditions are seldom duplicated in practice, and some variations should be expected at the particular plant. Alkalinity in the finished water may be increased to 40 mg/L or more by reducing the amount of carbonate hardness removed.

The second stage provides additional process stability before filtration and is equipped with residuals collection equipment to eliminate residuals buildup. A coagulant added at the second stage further clarifies the water before filtration. The second stage also provides an opportunity for further treatment by adsorption or oxidation after primary softening and before filtration. After the second stage, carbon dioxide is added to convert carbonate alkalinity to bicarbonate alkalinity for stability, similar to single-stage treatment.

Recarbonation. Recarbonation is required for the excess lime process to convert excess hydroxide alkalinity to carbonate alkalinity and also to convert carbonate alkalinity to bicarbonate alkalinity for stabilizing the treated water.

The reaction of carbon dioxide with hydroxide alkalinity is shown in Equation (11.6). The dosage of carbon dioxide can be estimated as follows:

$$CO_2 \text{ (lb/mil gal)} = 3.7 \times [\text{hydroxide alkalinity (mg/L)}] \quad (11.21)$$

where hydroxide alkalinity is the excess hydroxide alkalinity required for magnesium removal.

In addition, carbon dioxide is required to convert carbonate alkalinity to bicarbonate alkalinity for stabilizing treated water in accordance with Equation (11.20).

Excess Lime or Excess Lime–Soda Ash Process Examples. The following examples demonstrate the use of the above equations to estimate chemical feed requirements. Given:

Groundwater where carbon dioxide is present
CO_2 = 20 mg/L as CO_2 (mg/L)
Total hardness = 305 mg/L as $CaCO_3$
 Alkalinity = 224 mg/L as $CaCO_3$
 Calcium hardness = 235 mg/L as $CaCO_3$
 Magnesium hardness = 70 as $CaCO_3$
 Noncarbonate hardness (NCH) = 305 − 224 = 81 as $CaCO_3$
Available lime is 90% CaO
Available soda ash is very close to 100% Na_2CO_3

EXAMPLE 11.3 Reduce hardness as much as possible, using lime only. (*Note:* In this example, the magnesium hardness is greater than the selected target maximum of 40. Therefore, lime addition for magnesium hardness removal is required.)

Using Equation (11.9) for approximate lime feed requiremments, we have

$$CaO \text{ (lb/mil gal)} = 10.6 CO_2 \text{ (mg/L)} + 4.7 [\text{alkalinity (mg/L)}$$
$$+ \text{ magnesium hardness (mg/L)} + x]$$

Assume 65 mg/L of excess hydroxide alkalinity (expressed as $CaCO_3$) is required to reduce Mg hardness to ~10 mg/L.

$$CaO \text{ (lb/mil gal)} = 10.6(20) + 4.7(224 + 70 + 65) = 1,899 \text{ lb/mil gal,}$$

as 100% pure CaO

LIME SOFTENING   11.11

Correcting for purity gives

$$CaO = \frac{1,899}{0.9} = 2,110 \text{ lb commercial lime/mil gal}$$

Adding selected design factor of 10% for minimum feeder size gives

1.1(2,110 lb commercial lime/mil gal) = 2,320 lb/mil gal commercial lime

Approx. resulting hardness = 35(approx. minimum solubility Ca) + 10(approx. minimum solubility of Mg as $CaCO_3$) + 81(NCH) = 126 mg/L as $CaCO_3$†

Approx. remaining alkalinity = 45 mg/L as $CaCO_3$

(approx. minimum solubility of Ca plus Mg as $CaCo_3$)†

Using Equations (11.19) and (11.21) for approximate CO_2 feed requirements gives

CO_2 (lb/mil gal) = 3.7(40 mg/L bicarbonate alkalinity)

+ 3.7(65 mg/L excess OH alkalinity)

= 389 lb/mil gal

Adding selected design factor of 10% for minimum feeder size gives

1.1(389 lb/mil gal) = 428 lb/mil gal CO_2

EXAMPLE 11.4 Reduce the noncarbonate hardness for the water shown in Example 11.3 enough to reach a target finished water hardness of 100 mg/L.
The required lime dosage is calculated as in Example 11.3.

Noncarbonate hardness to be removed = 126 − 100 = 26 mg/L

Noncarbonate hardness to be left in water = 81 − 26 = 55 mg/L

Using Equation (11.10) to estimate soda ash feed requirements gives

Soda ash (lb/mil gal) = 8.8(NCH − x) = 8.8(81 − 55) = 229 lb/mil gal

Adding selected design factor of 10% for minimum feeder size yields

1.1(229 lb/mil gal) = 252 lb/mil gal

Approx. resulting hardness = 35(approx. minimum solubility Ca)

+ 10(approx. minimum solubility of Mg as $CaCO_3$) + 55(remaining NCH)

= 100 mg/L as $CaCO_3$†

---

†*Note:* Results are approximations based on theoretical equilibrium conditions. These conditions are seldom duplicated in practice and some variations should be expected at the particular plant.

Approx. remaining alkalinity = 45 mg/L as $CaCO_3$ (approx. minimum solubility of Ca plus Mg as $CaCo_3$)†

Approximate CO_2 feed requirements remain the same as for Example 11.3.

Split-Treatment Excess Lime or Lime–Soda Ash Process

Split-treatment softening is a variation of the two-stage excess lime or lime–soda ash process that can be advantageous under some conditions. A schematic of the process is shown in Figure 11.3.

In split-treatment softening, a portion of the flow to the first-stage softening process (generally 10% to 30%) is bypassed to the second stage. Carbon dioxide and bicarbonate alkalinity in the bypassed flow react with excess lime leaving the primary softening process, both to soften the bypassed flow with excess lime and to stabilize water entering the secondary basin.

The net result of split treatment softening is a savings in lime and carbon dioxide feed requirements and a reduction in residuals production. In addition, the first-stage softening unit may be smaller because it receives reduced flow. Chemical reactions involved and calculation of chemical feed requirements and flow split are covered in detail in *Water Quality and Treatment*.

Because a portion of the flow is not treated in the first stage of softening, it is important to determine whether adequate treatment of bypassed flow is provided in the second stage. Split-treatment softening is most often used for groundwaters but has also been used for some surface waters with pretreatment or coagulation of the bypassed flow.

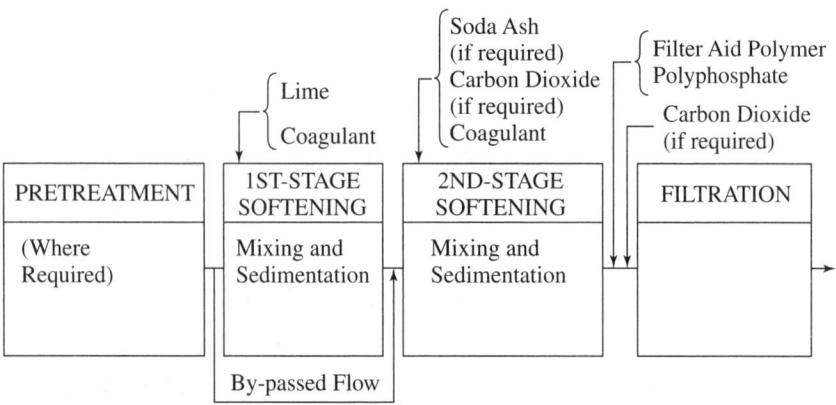

FIGURE 11.3 Split-treatment excess lime process.

---

†*Note:* Results are approximations based on theoretical equilibrium conditions. These conditions are seldom duplicated in practice and some variations should be expected at the particular plant.

SOFTENING PROCESS BASINS

Conventional Softening Basins

Conventional softening basins are similar to conventional basins used for coagulation and clarification. They consist of rapid mixing, flocculation, and sedimentation.

Lime, soda ash, coagulant, and a polymer coagulant aid (if required) are mixed with the water prior to flocculation. Often the mixing is done hydraulically in a flume. Mechanical mixers may be problematic due to encrustation with calcium carbonate.

Flocculation typically consists of horizontal paddle wheel or vertical turbine flocculators with total detention times of 30 to 45 min and up to three-stage tapered flocculation. Flocculators are usually equipped with variable-speed drives. The flocculation basin should be designed to facilitate periodic cleaning because of residuals buildup.

Sedimentation loading rates are in the range of 0.4 to 1.0 gpm/ft^2 (0.98 to 2.4 m/h) with detention times of 2 to 4 h. The higher loading rates are generally used for waters not requiring coagulation for turbidity removal. Continuous residuals collection equipment should be provided, and there must also be a means of draining the basin for periodic cleaning.

Recycling previously formed calcium carbonate residuals from the sedimentation basin to the mixing zone prior to flocculation is beneficial in the softening process. Recycling accelerates the precipitation reactions, and the process more closely approaches true solubility when the mix is seeded with these previously formed crystals. Recycling residuals also allow precipitation to occur on the recycled residuals that serve as nuclei, reducing precipitation on the mechanical equipment.

In addition, recycling calcium carbonate residuals promotes growth of larger calcium carbonate crystals that settle and dewater more rapidly. A study has indicated that 50% to 100% residuals recycle, based on solids produced, controls particle size optimally for this type of process (Burris et al., 1976). This same study found that particle growth approached an equilibrium value after about 4 cycles.

Figure 11.4 illustrates the effect of residuals recycle on particle growth. Curve A is a sample of residuals taken before recycling was practiced. Curve B represents a sample of sludge taken after several cycles at 25% recycling based on solids produced. Curve C represents 300% sludge recycling. Curve D was a sample of sludge taken from a solids contact clarifier with an extremely high recycle rate operating at a high solids concentration. Residuals recycle for conventional basins may be accomplished by a separate basin sump and solids handling pumps that recycle sedimentation residuals to the mixing zone.

Conventional softening basins are mostly found at older facilities. They provide a high degree of process stability, but the relative size and number of the basins, when compared with high-rate solids contact processes, substantially increase the initial cost. In addition, the high-rate solids contact processes are generally more effective in driving the softening reaction to completion and reducing chemical costs.

Solids Contact Softening Basins

Solids contact softening basins combine mixing, residuals recirculation, and sedimentation functions in one basin. Rapid mixing may be provided ahead of the solids contact unit but generally is not needed. Lime and other chemicals are applied directly to the mixing zone of the solids contact unit. This type of unit provides a high degree of continuous sludge recirculation and contact, and it produces more stable water and larger cal-

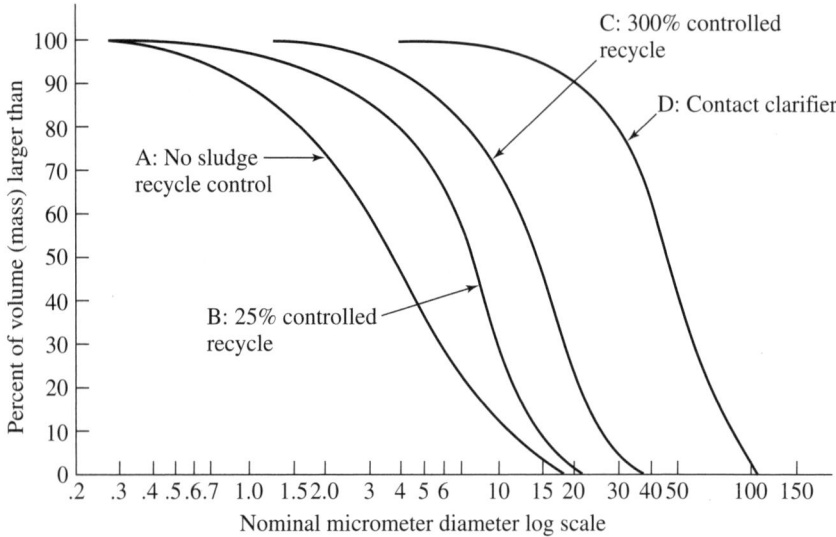

FIGURE 11.4 Particle size distribution comparison.

cium carbonate crystals. The depth and density of the residuals in the basin can be controlled by wasting residuals and by the degree of recirculation. The mixing and recirculation zone is separated from the sedimentation zone by a conical baffle wall within the basin. An example of a solids contact unit is shown in Figure 11.5.

Effective solids contact units pull settled residuals from near the floor at the center of the basin with a large-diameter turbine or impeller and recirculate the residuals at a high rate with the water entering the basin. The maximum recirculation rate is typically 10 to 1 based on incoming flow. The mixer is provided with a variable-speed drive for adjusting the recirculation rate to meet treatment conditions.

Where multiple solids contact units are installed, small recirculation pumps, operated to recirculate solids from one unit to another, are often used during start-up of a unit that has been out of service for cleaning. This is especially helpful during periods of low water temperature.

Side water depth of solids contact basins usually varies from 14 to 18 ft (4.3 to 5.5 m) and depends on basin size and the equipment manufacturer's requirements. Contact time in the mixing zone is typically 15 to 30 min, measured by the volume of water within and directly under the baffle wall.

Surface loading rate in the sedimentation zone is generally measured 2 ft (0.6 m) below the water surface, based on the surface area between the baffle wall and the basin wall. Surface loading rates are usually in the range of 1.0 to 1.75 gpm/ft^2 (2.4 to 4.3 m/h). Where coagulation for turbidity removal is required, surface loading rates are generally in the lower end of this range or less. Where possible, design requirements should be based on successful plants using the same or similar source water.

For the straight lime and lime–soda ash process, residuals are composed primarily of heavy calcium carbonate crystals without the more gelatinous and lighter magnesium hydroxide component. The heavy calcium carbonate crystals create some difficulty in recirculating residuals. In any case, the manufacturer should be advised of the nature of the process so that the basin design will function properly. Equipment specifications should establish a minimum percentage of solids to be maintained in the mixing zone. A mini-

FIGURE 11.5 Solids contact clarifier.

mum percentage of solids of 1% by weight is recommended by the *Recommended Standards for Water Works* (commonly called the 10 State Standards).

Solids contact units offer a smaller footprint because of the combined, higher-rate process. They also provide a high degree of residuals recirculation and contact to complete the softening reaction and are typically used instead of conventional basins for softening. However, solids contact units can require greater operator attention to monitor and control the solids inventory in the basin, and these units are more prone to upset from varying inlet water temperatures, flows, and turbidity conditions.

Helical flow solids contact units are an alternative to conventional solids contact units. Helical flow solids contact units are typically constructed of steel and have an inverted conical shape. These units are designed to induce a circular flow pattern within the clarifier, with the incoming water flowing upward through a blanket of previously formed precipitates. Lime (and soda ash if necessary) is added to the incoming water at the base of the clarifier. The conical shape of the clarifier and the helical flow pattern provide the required influent/solids mixing and contact without the need for mechanical mixing equipment within the clarifier.

Recarbonation Basins. For single-stage softening processes, a recarbonation basin with a contact time of at least 20 min is typically provided to stabilize water before filtration. The contact basin should be designed for plug flow, and mixing should be provided ahead of the basin to introduce carbon dioxide.

SPECIAL LIME SOFTENING DESIGN CONSIDERATIONS

Many special design and operations and maintenance considerations are associated with lime softening processes, including problems with encrustation and removing, handling, and disposal of the large quantities of residuals resulting from treatment processes.

Encrustation Problems

A significant problem in operations and maintenance of lime softening processes is encrustation of calcium carbonate on the process and hydraulic components. Encrustation can occur anywhere but is generally greater and harder near application points of lime and soda ash and where water velocity is increased, such as on mixing equipment. Encrustation can also occur in flumes and pipelines after lime or carbon dioxide addition or where split-treatment flow is introduced before a secondary process.

The burden of designing for encrustation on equipment is usually placed on the basin equipment manufacturer. An allowance should also be made in designing pipelines and flumes, with additional hydraulic capacity and provisions for ease of cleaning. Open flumes are generally preferred over pipelines for this reason. Calcium carbonate buildup may be removed with high-pressure jet equipment or by mechanical means. Basins and pipelines are usually cleaned once per year during seasonal low-demand periods. If the plant has no low-demand periods, additional treatment capacity in the form of more treatment units will have to be provided.

Recirculating residuals to the point of addition of lime or, for second-stage processes, to the point of addition of carbon dioxide or split-treatment flow provides nuclei for calcium carbonate deposition and may reduce encrustation problems.

Additional information on sludge thickening, dewatering, and handling is provided in Chapter 17.

Residuals Removal and Handling

Residuals from softening reactions are dense when compared with coagulation residuals, with solids concentrations varying from 3% to 15% or greater by weight. Straight softening processes removing only calcium carbonate produce the highest-density residuals. Residuals from excess lime processes are generally less dense because of the influence of magnesium hydroxide, which is lighter and more gelatinous. For excess lime softening, an average residuals blowdown solids concentration of 5% by weight is often used as a rough guideline. For split-treatment softening with coagulation of the bypass flow, design consideration must be given as to whether to combine residuals from softening and coagulation units.

Residual scrapers for softening basins require higher-torque units than for conventional treatment basins because of the high density of residuals. The design of the load on scraper mechanisms depends on the type of softening and the stage of the process. Equipment manufacturers and experience with similar treatment facilities should be consulted for guidance.

Residuals blowdown from softening basins is usually controlled by a repeat cycle timer that allows periodic, timed blowdowns at full pipe flow. Backflushing piping with clear water after each blowdown must be provided to reduce the possibility of lines clogging from settled residuals. Provisions for sampling the residuals blowdown should be included. Some facilities have residuals blowdown viewing pits where the residuals can be observed and sampled.

Residuals pumps for recirculation or disposal are typically solids handling centrifugal units, although positive displacement pumping units may be required for extremely dense residuals. Recirculation pumps should be provided with variable-speed drives.

Residuals Production

The quantity of residuals produced by the softening process can be estimated by performing a mass balance on the residuals-producing calcium and magnesium, which represent the hardness removed in the process. One or both of the substances are present as hardness in the water entering the plant. In addition, calcium is added to the plant flow by the addition of lime.

Some hardness also leaves the plant in the finished water. The rest of the hardness leaving the plant is in the residuals produced. The residuals are in the form of calcium carbonate for the calcium hardness removed and in the form of magnesium hydroxide for the magnesium hardness removed.

The calcium balance is as follows:

Dry weight $CaCO_3$ residuals (lb/mil gal) = 20.9

$$\times [(Ca\ in) + (Ca\ added\ by\ lime)$$
$$- (Ca\ out),\ mg/L] \qquad (11.22)$$

where Ca is calcium in milligrams per liter as calcium (calcium carbonate equivalent divided by 2.5).

For quicklime (CaO), Ca added by lime is $0.71 \times CaO$ (mg/L) $\times$ percent purity/100. For hydrated lime [$Ca(OH)_2$], Ca added by lime is $0.54 \times Ca(OH)_2$ in mg/L $\times$ percent purity/100.

The magnesium balance is

Dry weight $Mg(OH)_2$ residuals (lb/mil gal) = 20.0

$$\times [(Mg\ in) - (Mg\ out),\ mg/L] \quad (11.23)$$

where Mg is magnesium in milligrams per liter as magnesium (calcium carbonate equivalent divided by 4.1). Impurities in lime are also a source of residuals produced in the softening process. The amount can be estimated as follows:

Dry weight lime impurities (lb/mil gal) = lime dose (lb/mg)

$$\times (100 - \text{percent purity})/100 \qquad (11.24)$$

Total dry weight solids of softening residuals for the excess lime treatment process are usually about 2.5 times the hardness removed, expressed in milligrams per liter or pounds per milligram. The straight lime process produces total dry weight solids for softening residuals approaching 2.0 times the hardness removed.

Residuals are also produced by source water turbidity removed in the process, by the precipitation of iron and manganese that may be removed in the process, and by coagulants and powdered activated carbon that may be used in the process. Guidelines for determining the quantities of residuals produced by these items may be found in Chapter 17.

Filtration

The design of filters for softening plants is similar to that for conventional clarification plants. Many modern softening plants use dual-media filters with bed depths of 36 (0.91 m) or more, although other types of media have also been used successfully. Design loading rates typically range from 3 to 5 gpm/ft^2 (7.3 to 12.2 m/h). Available head should be 8 ft (2.4 m) or more at design rate, measured from the water surface in the filter to the filter effluent line as it leaves the filter.

Particulate loading on softening plant filters consists, to a large extent, of fine calcium carbonate that makes softening plant filters behave somewhat differently than coagulation plant filters. Softening plant filters are generally characterized by long filter runs, on the order of 36 to 48 h or more, and are often washed based on time rather than head loss or turbidity breakthrough. However, effluent turbidity must be carefully monitored and controlled because it is not unusual for the fine calcium carbonate to pass through the filter. Filter aid polymer should be provided to assist in controlling turbidity breakthrough.

Water is typically under a calcium carbonate plating condition before filtration to maintain stability and to reduce corrosion in the distribution system. Therefore a polyphosphate is usually fed before the filters at a dosage of 0.25 to 1.0 mg/L to control encrustation of the filter media. Annual inspection of the media for cementation and inspection of the underdrain every 4 to 5 years through careful excavation of the media may also be performed to monitor encrustation.

Surface wash or air-water backwash is essential to assist filter backwashing. The supplemental agitation breaks up cementation of filter media and helps remove solids in the filter. The design of backwash systems is similar to that for conventional plants.

Pellet Reactors

Pellet reactors have been used for softening in the Netherlands for many years and have been installed at a number of locations in North America. A pellet reactor consists of an inverted conical tank where calcium carbonate crystallizes on a suspended bed of fine sand.

Advantages of the pellet reactor are its small size and low installation cost. Residuals consist of small pellets that dewater readily, minimizing residuals volume. However, pellet reactors should not be considered for systems high in magnesium content because of magnesium hydroxide fouling of the reactor.

Pellet reactor systems should be designed cautiously. Where the application is not proved, pilot testing is advisable. Design should be carefully coordinated with the equipment manufacturer.

Chemical Feed and Layout Considerations

The design of lime softening plants is influenced, to a large degree, by the need to handle and feed large quantities of lime. Lime solution readily encrusts solution pumps, pipelines, and troughs and presents major maintenance considerations. Pumping lime solutions should be avoided if possible; gravity flow in open troughs or hoses readily accessible for cleaning are preferable. Soda ash feed systems also have encrustation problems.

The lime and soda ash feed systems should be located as close to the point of feed as possible. On the other hand, feed systems for these chemicals, particularly lime, require frequent attention and should be located as close as practical to the operator's station. Because of these factors, it is generally preferable to bring the water to be treated as close as practicable to central chemical feed facilities readily accessible to the operator.

Provisions should be made to feed alum or ferric coagulants and coagulant polymer at the mixing facilities ahead of softening basins. It is often necessary to use coagulants to enhance clarification within softening units. Provisions should also be made to feed polyphosphate and filter aid polymer ahead of the filters.

FUTURE TRENDS IN SOFTENING

With the technological advances in membrane process manufacturing, it is anticipated that lime softening may gradually be replaced by membrane processes for some applications. This option appears to be particularly applicable for smaller installations where the potential for reduced operator attention and remote monitoring of membrane processes has a significant cost impact. In addition, membrane processes produce residuals containing only the constituents removed from the source water, which may make membranes more attractive to regulatory agencies where residuals disposal to source waters is a consideration. Membrane processes may also remove regulated contaminants in conjunction with softening.

Lime softening uses large amounts of chemicals (primarily lime) and produces large amounts of residuals in comparison to coagulation treatment only. Because of the cost of lime, other chemicals, and residuals disposal problems, some utilities have considered reducing the degree of hardness removal. This idea has raised the question as to what degree of hardness removal for a particular utility is beneficial and cost-effective to the consumers. Greater attention to this area is expected in the future.

BIBLIOGRAPHY

Aieta, E. M., E. J. Singley, A. R. Trussell, K. W. Thorbjarnarson, and M. J. McGuire. "Radionuclides in Drinking Water: An Overview." *Journal AWWA* 79(4):144, 1987.

American Water Works Association Trace Inorganic Substances Research Committee. "A Review of Solid-Solution Interactions and Implications for the Control of Trace Inorganic Materials in Water Treatment." *Journal AWWA* 80(10):56, 1988.

Berg, G., R. B. Dean, and D. R. Dahling. "Removal of Poliovirus 1 from Secondary Effluents by Lime Flocculation and Rapid Sand Filtration." *Journal AWWA* 60(2):193, 1968.

Burris, M. A., K. W. Cousens, and D. M. Mair. "Softening and Coagulation Sludge—Disposal Studies for a Surface Water Supply." *Journal AWWA* 68(5):247, 1976.

Chaudhuri, Malay, and R. S. Engelbrecht. "Removal of Viruses from Water by Chemical Coagulation and Flocculation." *Journal AWWA* 62(9):563, 1970.

Corrosion Control by Deposition of $CaCO_3$ Films: A Handbook of Practical Application and Instruction. Denver, Colo.: American Water Works Association, 1978.

Internal Corrosion of Water Distribution Systems, 2d ed. Cooperative research report. Denver, Colo.: American Water Works Association Research Foundation and DVGW–Technologiezentrum Wasser, 1996.

Introduction to Water Treatment, vol. 2: *Principles and Practices of Water Supply Operations.* Denver, Colo.: American Water Works Association, 1996.

Liao, N. Y., and S. J. Randtke. "Removing Fulvic Acid by Lime Softening." *Journal AWWA* 77(8):78, 1985.

Liao, N. Y., and S. J. Randtke. "Predicting the Removal of Soluble Organic Contaminants by Lime Softening." *Water Research* 20(1):27, 1986.

Lime Handling, Application and Storage. Denver, Colo.: American Water Works Association, 1995.

Manual of Treatment Techniques for Meeting the Interim Primary Drinking Water Regulations. U.S. Environmental Protection Agency, 600/8-77-005, Washington, D.C., 1978.

"Quality Goals for Potable Water: A Statement of Policy Adopted by the AWWA Board of Directors on January 28, 1968, Based on the Final Report of Task Group 2650P—Water Quality Goals, E. L. Bean, Chair," *Journal AWWA* 60(12):1317, 1968.

Randtke, S. J., et al. "Removing Soluble Organic Contaminants by Lime-Soda Softening." *Journal AWWA* 74(4):192, 1982.

Recommended Standards for Water Works. Great Lakes–Upper Mississippi River Board of State Public Health and Environmental Managers, 2003.

The Removal and Disinfection Efficiency of Lime Softening Processes for Giardia *and Viruses.* Denver, Colo.: American Water Works Association, 1994.

Riehl, M. L., H. H. Weiser, and R. T. Rheins. "Effect of Lime Treated Water upon Survival of Bacteria." *Journal AWWA* 44(5):466, 1952.

Sawyer, C. N., P. L. McCarty, and G. F. Parkin. *Chemistry for Environmental Engineers.* New York: McGraw-Hill, 1994.

Sproul, O. J. *Critical Review of Virus Removal by Coagulation Processes and pH Modifications.* EPA-600/2-80-004. Washington, D.C.: EPA, 1980.

Waste Sludge and Filter Washwater Disposal from Water Softening Plants. Cincinnati, Ohio: Ohio Department of Health, 1978.

Water Quality and Treatment: A Handbook of Community Water Supplies, 5th ed. Denver, Colo.: American Water Works Association, 1999.

Wentworth, D. F., R. T. Thorup, and O. J. Sproul. "Poliovirus Inactivation in Water-Softening Precipitative Processes." *Journal AWWA* 60(8):939, 1968.

CHAPTER 12
ION EXCHANGE APPLICATIONS IN WATER TREATMENT

Michael C. Gottlieb
ResinTech, Inc.
West Berlin, New Jersey

INTRODUCTION

Ion exchange processes are widely used in water and wastewater treatment to remove objectionable ionic contaminants. The nature of ion exchange, as used by industry, has changed dramatically over the last 80 years. The first ion exchange processes used inorganic zeolites mined from natural deposits as a cation exchanger for water softening. Modern synthetic polymer-based exchange media are today used as cation and anion exchangers. They have largely displaced the natural zeolites, primarily due to higher capacities, except in specialized applications. By the mid-1990s, in response to more stringent discharge limits and pending implementation of tighter potable water guidelines, there was increased development activity of ion exchanges with high selectivities for specific substances. Some of these are organic-based and others are inorganic-based.

Ion exchange processes primarily treat waters with low mineral content, to reduce undesirable ionic contaminants. By far, the largest single application of ion exchange is in softening. Softening is the process of removing calcium and magnesium ions from water. Ion exchange is widely used in the power and utility industries to produce softened or demineralized water and in-home water softeners. The use of ion exchange in these applications has been widely practiced for several decades and is well defined.

Since September 11, 2001, a greater awareness of possible terrorist threats to potable water systems has created new challenges for water treatment operators. So-called "dirty bombs," made from conventional explosives and packed with radioactive substances, are of concern because they can spread radioactive substances to public water supplies. Fortunately, ion exchange technology already exists to remove these substances and some of the poisons that may have to be dealt with. A separate section on terror is included, describing the most likely substances to be involved. The individual treatment approaches for their removal will be listed separately according to the substances themselves.

Ion exchange is also used in many specific applications, for example, to reduce or remove potentially harmful ionic contaminants from potable water supplies in chemical processes and for product purification and recovery for specialty separations, such as chromatographic separations by size, valance charge, and as catalysts. A full discussion of ion exchange process theory can be found in *Water Quality and Treatment.*

Ion exchange, as its name implies, is the exchange of ions from one phase to another. In water treatment, the exchange of ions occurs between the solid phase of the ion exchanger and influent water. For example, in a water softener, a cation exchange resin operating in the sodium form exchanges sodium ions for an equivalent amount of calcium and magnesium ions in the influent raw water. The result is that most of or all the calcium and magnesium ions are removed from the raw water by the resin, and an equivalent number of sodium ions are added to the water from the resin.

Ions, such as calcium, magnesium, barium, copper, lead, zinc, strontium, radium, ammonium, fluorides, nitrates, humates, arsenates, chromates, uranium, anionic metallic complexes, phosphates, perchlorates, hydrogen sulfide, bicarbonates, sulfates, and many others, are routinely removed from water by ion exchange resins. Proper certifications are required for resins used in potable waters. The original regulations that applied to ion exchange resins are stated in paragraph 21CFR173.125 of the Food Additives Regulations of the Food and Drug Administration (FDA). The common cation exchange resins used in water softening and their strongly basic anion exchange resins counterparts, as well as other commercially available resins at the time the regulations were issued, are listed by composition in the CFR. The listed types of resins are therefore acceptable for use in potable water applications. However, the CFR requirement specifies that the end user test the resins for compliance with maximum allowable extractables in order to be in compliance.

Several new classes of resins have been developed since the FDA-sponsored CFR was issued. However standards and protocols for resins that are used in potable water applications are available through the American National Standards Institute (ANSI). These include toxicology review of the ingredients used to manufacture the resin, operating capacity and effluent quality testing, and extraction tests under simulated start-up conditions to make sure the effluent will be safe and the resin will perform as stated. Third-party certification against the ANSI standards for resins used for potable water treatment is often mandated. Organizations such as NSF International and Water Quality Association Gold Seal are certified to issue certifications against the ANSI adopted standards. Certification seals are issued with the name of the certifying organization and the number of the ANSI adopted standard. Several more organizations are expected to be certified to perform this service in the near future.

It is impossible to cover the subject adequately in a single chapter. In writing this chapter, the authors hope to provide the reader with a general overview of the ion exchange process, general information about how ion exchange systems are designed and built, and impart enough information that the reader will be able to decide whether ion exchange is an appropriate technology to solve a specific problem and to estimate the size and scope of the equipment required. At the end of this chapter is a comprehensive Bibliography organized by topic for those wanting more details.

Chemical Structure of Ion Exchange Polymers

Almost all modern day ion exchange resins consist of an organic polymer, chemically bonded to an acidic or basic functional group. Most ion exchange polymers are polymerized polystyrene cross-linked with divinylbenzene (DVB). Other polymers are used, for example, acrylic or methacrylic acids, phenol formaldehyde, epoxy polyamine, and pyridine-based polymers. The acrylics and epoxies are the most widely used in this group, which, in total, comprise only a small percentage of the ion exchangers produced commercially.

In general, the polymerization reaction is carried out by suspending the reacting monomers in a nonmiscible liquid containing a dispersant. The mixture is stirred, and the monomers polymerize into cross-linked polymers in the form of spherical droplets, which later become solid spherical hydrophobic copolymer beads. The beads are screened to the proper size range and then converted to hydrophilic ion exchange resins by adding functional groups through additional chemical reaction steps. Most ion exchange resins are hydrated to the extent that chemically bound water comprises one-half the weight of the final product after all surface moisture has been drained away.

Cation resins are created by attaching negatively charged functional groups to the copolymer structure. Strongly ionized (strongly acidic) cation exchange resins have sulfate functional groups. These sulfate groups have a permanent negative charge which is strongly attracted to the positively charged cations in water. A second group of cationic exchangers is made by attaching carboxylic acid functional groups. These are not fully ionized and are limited to specific applications. Another broad group of cation exchange resins have functional groups that act as chelants; some examples are those based on iminodiacetate, aminophosphonic, and thiol groups.

Anion resins are functionalized by attaching chemical groups with positive charges to the copolymer. While there are several chemicals that can be used to functionalize anion resin, only a few are actually used in commercial products, and these are all based on amines.

Type I (strong base) anion resins are functionalized with trimethylamine. Type II (strong base) anion resins are functionalized with dimethylethanolamine. The resulting structures of the amines in type I and type II anion resins are quaternary. These are fully ionized. These resins are referred to as being strongly basic due to their ability to react with neutral salts such as NaCl to create NaOH (salt splitting). Less strongly ionized resins have tertiary, secondary, and primary amine functional groups. These are referred to as weakly basic. The weakly basic anion resins are generally limited to neutralization reactions as they are unable to react effectively with salts. Most weakly basic resins exhibit some strong base characteristics. Those with appreciable strong base capacity are sometimes called intermediate basic resins.

There is also a group of anionic chelating resins of the weakly basic type. Some examples are those based on picolylamine, thiouronium, and methylglucamine. These perform effectively at low pH and exhibit high selectivity for certain metals or metal containing anionic complexes.

Size of Resin Beads

Ion exchange resins are homogeneous solids; most of the exchange groups are below the surface of the resin bead. The exchange of ions from the liquid to the solid phase occurs at the surface of the beads. The ions then diffuse through the solvated polymer from one ion exchange site to another toward the center of the bead. To create a balance between pressure loss and kinetics, the size of resins chosen for most water treatment applications is a range between 0.3 and 1.2 mm in diameter. This size range has been proved to provide the best combination of flow equalization, pressure drop, and kinetics. Resins smaller than 0.3 mm are in used in specialty applications. They have enhanced kinetic properties that make them superior to larger resins for certain types of slow chemical reactions. However, their small size makes them more expensive to produce, requires special containment equipment, and creates higher pressure drops.

Gel versus Macroporous Resins. The physical structure of the commonly available ion exchange resins is classified as macroporous or gelular. In the gelular types, the pores are in the copolymer structure and are due to molecular spacing. The macroporous resins have actual physical pores in the resin bead. The macroporous types have high surface area and a higher surface-to-volume ratio than the gel types. Macroporous resins are currently made in two different ways. The original (first-generation) method agglomerated microspheres into large spherical beads. A second process (second generation) is based on extraction technology in which an inert substance is added to the reacting monomers mixture. After polymerization the inert substance is removed from the resulting copolymer beads, leaving behind discrete pores.

Historically, macroporous resins were originally made to provide physical stability to very highly cross-linked cation exchange resins. By their nature, macroporous resins are more resistant to stresses brought on by changes in operating conditions. Highly cross-linked macroporous resins are basically better able to withstand severe and rapid changes in operating environments such as sudden and very high flow rates, sudden temperature and pressure changes, *very* high temperature and oxidative environments. Increasing cross-linkage in gel resins to gain thermal and oxidative resistance also decreases kinetics and increases brittleness, which leads to reduced physical stability. These effects can be reduced by introducing macroporosity. The volume of the discrete pores provides stress relief but also reduces the amount of copolymer within a bead and therefore reduces the volumetric capacity of the resin. This reduced volumetric capacity makes macroporous resins less desirable in most bulk water treatment applications.

However, macroporosity has been proved superior for the high-capacity weakly ionized resins, due to the macroporous resins' ability to accommodate the very significant size change that these resins undergo between the exhausted and regenerated ionic forms.

Inorganic Ion Exchangers and Zeolites. Inorganic zeolites are no longer commonly used in softening or in bulk deionization. However, they exhibit high affinities for certain substances which makes them ideal to remove specific contaminants. Usually these products are not stable at all pH values, so pH control is required. The major categories of commercially available inorganic exchanges and zeolites are green sand, activated alumina, and aluminosilicates. They are pH-sensitive.

Green sand and some aluminosilicates are manufactured from mined deposits. Aluminosilicates can also be synthesized. Green sand is primarily used for removal of iron and manganese. Some of the aluminosilicates such as clinoptilolite and chabasite are used for the removal of ammonia from wastewater and in the nuclear power industry for removal of specific radio nuclides such as cesium and rubidium. Activated alumina is a highly processed form of aluminum oxide. It has a high selectivity for fluoride, arsenic, and lead.

Adsorbents. In certain applications, the nonfunctional polymers, especially those with macroporous structures, are used as adsorbents. They generally have limited or no ion exchange functionality. Instead, they have certain characteristics such as static charge, copolymer structure, or pore size that make them useful for some types of chemical separations. They are not widely used in water treatment.

Nature of the Ion Exchange Process. When salts dissolve in water, they separate into charged ions. Cations carry positive charges, and anions carry negative charges. With this in mind, there are two classes of ion exchange resins: those that exchange cations and those that exchange anions. Each of these classes of exchangers is further divided into strongly ionized and weakly ionized, according to the nature of the functional groups. Almost all ion exchange materials can be classified in this manner.

The general ion exchange reactions shown below are for cation softening and anion softening. The following equations show the exchange of sodium ions from the sodium form of a strong acid cation exchange resin for calcium and magnesium (cation softening) and also the exchange of chlorides for other anions by the chloride form of a strong base anion exchange resin (anion softening). Cation softening is so much more common than anion softening that the term *softening* by itself refers only to the cation softening process.

Cation softening:

$$4R - Na + Ca^{2+} + Mg^{2+} \rightarrow R_2Ca + R_2Mg + 4Na^+$$

Anion softening:

$$3R - Cl + SO_4^{2-} + HCO_{3-} \rightarrow R_2SO_4 + RHCO_3 + 3Cl^-$$

As a result of these two ion exchange processes, all the salts in the raw water would be changed by the ion exchange resins to an equivalent amount of sodium chloride.

The Nature of Resin Degradation

Ion exchange resins degrade both physically and chemically. In practical terms, the *degradation* of a resin can be defined as any change in the physical or chemical properties that adversely affects the operating performance or the ability of the resin to attain desired results. A change in the resin properties is not automatic grounds for replacement. Degradation must be evaluated in light of the way the resin is used. However, it is safe to say that most resin replacements occur as a result of degradation. A physically degraded ion exchange resin that continues to perform well chemically without high pressure loss or sporadic quality due to flow channeling may be acceptable for certain low-flow-rate applications. However, the same resin would not be acceptable in high-flow-rate applications such as condensate polishing. A strong base anion resin that has chemically degraded may give good results in separate bed applications on certain kinds of waters, but the same resin would retain only a small fraction of its original operating capacity in a mixed-bed demineralizer operating on the same water.

Physical. Ion exchange resins degrade physically by breaking. Usually the resin develops cracks, which then fracture. The result of this fragmentation is a reduced average particle size. The void volume (space between beads) also decreases as the irregularly shaped fragments fill spaces between the beads, which increases pressure loss and can lead to channeling (irregular flow distribution) and clogged flow distributors. New resins typically have more than 90% whole, perfect uncracked beads. Over time, physical and osmotic stresses will gradually crack and break beads.

Once placed in service, resins are often exposed to oxidants such as dissolved oxygen and chlorine. The chemical bond between the DVB cross-linker and the copolymer is the primary place where oxidative attack occurs. Cleavage of the divinylbenzene copolymer bond reduces the level of cross-linking. As the cross-linkage is reduced, the resin becomes more hydrated and the beads swell.

Oxidative. Oxidation occurs first on the outside of the beads and works its way inward. This results in an uneven distribution of cross-linking and swelling, which creates an osmotic stress. This is a major cause of bead breakage in cation resins used in domestic water softeners. Both macroporous and gel types are subject to oxidative degradation.

Chlorine in potable water is present as the OCl^- anion. The negatively charged anions are excluded from the interior of the cation resin due to charge repulsion by the ion exchange groups. The lower the DVB level, the higher the water retention and the more space between the ion exchange groups, which in turn reduces change repulsion and lowers the resistance to penetration by the hypochlorite ion into the resin. Therefore, lowering the DVB cross-linker level increases susceptibility to attack by chlorine in two ways. The life expectancy under oxidizing conditions of a water softening resin is proportional to the square of the cross-linking level. Reducing the DVB level by 50%, for instance, will decrease resin life due to oxidation by chlorine by a factor of 4. Most domestic softener manufacturers use cation resins with 8.0% DVB.

Anion resins are similarly affected by chlorine except that the negatively charged hypochlorite ion is not excluded from the polymer. Therefore, anion resins are more susceptible to chlorine attack. Since it is an anion, the hypochlorite ion is exchanged onto the anion exchange resin. The amine group of the anion resin is the primary point of attack by the hypochlorite ion. Although de-cross-linking does occur, the primary indicator of degradation is loss of functionality.

Chlorine attack on the amine groups of an anion exchange resin can cause either partial oxidation of the amine groups, which lowers the basicity (more weakly basic), or cleavage of the entire amine group from the resin, which destroys the ion exchange group. In either case the result is lower water retention values. This tends to counter the effect of DVB cleavage. Because of the double route that oxidation takes in the anion resins, the water retention may not immediately be affected by chlorine attack. Therefore, the best way to measure anion resin degradation from oxidation is to track the changes in strongly basic, weakly basic, and total capacities.

Chemical Degradation. The ion exchange groups can become impaired by an ion exchange reaction that is not easily reversible, or by surface clogging, which occurs when a resin becomes fouled with oil, rust, or biological slime. When any of this happens, the resin is referred to as being *fouled*. Fouling due to coating by water-soluble polymers from clarification upsets also occurs.

The most common form of cation resin fouling is due to chemical precipitants such as iron or barium. These can either coat the resins or form clumps of precipitated salts that cannot be removed by the normal regeneration process.

Anion resins are commonly fouled by naturally occurring organic substances which, though exchangeable as ions, have such slow diffusion rates that they are not fully removed by the normal regeneration process unless special resins or special regeneration procedures are used. This often makes their exchange onto the resins an irreversible process. When this continues over a long time the resin will become impaired and performance will drop off. This condition is commonly referred to as *organic fouling.*

Proper regeneration techniques and chemical procedures reduce the tendency to foul. Some types of resins are more resistant to fouling than others, but usually there is a trade-off in other properties so that selection of the type of anion resin for a given application depends on several factors, such as water temperature ranges, operating efficiency capacity requirements, regeneration temperatures, and required quality levels.

THE ION EXCHANGE PROCESS

It is important to keep in mind that the ion exchange process only works with ions. Substances that do not ionize in water are not removed by ion exchange. Each type of ion exchange resin exhibits an order of preference for various ions. This can be stated quantitatively through selectivity coefficients. These are similar to the equilibrium constants in ionic equilibria. In a similar manner, the equilibrium concentrations of ions in the resin phase and in the water phase can be calculated from the selectivity coefficient.

Each ion pair has a unique selectivity value for each ion exchange resin. The higher the selectivity coefficient, the higher the relative affinity of the ion for the resin. The higher the affinity, the easier it is to load the ion, and conversely the more difficult it is to remove during regeneration. The operating performance, capacity, and leakage data for ion exchange resins for the common ions found in water are usually provided by the resin supplier.

In general, the ion exchange resin is used in and regenerated to an ionic form, which will exchange ions that are acceptable in the treated water. As the untreated water passes through the resin, the undesirable ion or ions exchange for the unobjectionable ion on the resin. For example, a cation exchange resin is regenerated with sodium chloride and operated in the sodium cycle. The resin exchanges sodium ions for all the positively charged ions (cations), including hardness-causing calcium and magnesium ions, and the water is thereby softened. Likewise, if an anion resin is regenerated with sodium chloride, it operates in the chloride cycle. It will exchange chlorides for the anions, i.e., bicarbonates, sulfates, nitrates, etc., and the effluent water will have a chloride concentration equivalent to the total concentration of the anions in the raw water.

Although salt exchanges are common, so are other ion exchanges that involve acids and bases. In the process of demineralization, for example, the cation resin is operated in the hydrogen form and exchanges hydrogen ions. This converts all incoming salts to their equivalent acids. After the cation exchange, the water passes through a hydroxide form anion exchanger where anions are exchanged for hydroxides, which converts the acids to water molecules. In this process the net result is that an equivalent amount of water molecules is added to the water in exchange for the salts that are removed.

The weakly acidic and weakly basic ion exchange resins are generally incapable of converting salts to acids and bases and are generally limited in use to neutralization reactions. The hydrogen form weakly acidic cation resins are usually used to neutralize alkalinity, and weakly basic resins are likewise used to neutralize acidity.

Ion exchange is a dynamic equilibrium-driven process, and regeneration is never 100% complete. There is always some, albeit trace, level of the undesirable ionic constituent left on the resin and in the product water. The most complete ion exchange reactions are those in which the resulting products disappear, such as by H^+ and OH^- neutralization which drives the reaction to completion by the formation of water molecules.

When the equilibrium is favorable, the exchange of ions occurs in a narrow band within the resin bed, and a large portion of the resin bed can be used before significant leakage occurs due to kinetic slippage. When the equilibrium is not favorable, the exchange zone ion is bigger and more diffuse. It can be as long as or longer than (in terms of the amount needed to reach the desired level of purification) the entire resin bed. The leakage of the undesirable ion is almost immediate and gradually increases throughout the service run. When the exchange zone is larger than the entire bed, the desired quality cannot be achieved, even in the initial portion of the service cycle.

Resin Capacity and Regeneration

Ion exchange resins have a finite capacity. When this capacity is used up, the resins are exhausted and leakage of the unwanted ions increases. The exhausted resin can be regenerated with a salt, acid, or base solution containing the ion whose "form" the resin will be operated in. This is passed through the resin bed in sufficient quantity and at a sufficiently high concentration to reverse the exchange, desorb, and replace the previously exchanged ions from the resin with the ions from the regenerant solution. The high concentration of the regenerant minimizes waste volumes and changes the equilibrium relationships to make the regeneration more efficient.

The most common regenerant used in potable water applications of ion exchange is sodium chloride. It is used in softening, dealkalization, barium, radium, uranium, selenium, arsenic, and nitrate removal. Potassium chloride, though more expensive, is also used with very similar results in cases where low sodium levels are desired.

In demineralization, the resins are regenerated with acids and bases. Cation exchange resins are regenerated with acid, the most common acids being sulfuric and hydrochloric. Strongly basic anion resins are usually regenerated with caustic soda (sodium hydroxide), but potassium hydroxide, though more expensive, can also be used. Weakly acidic cation resins can be regenerated with weak acids such as carbonic, citric, acetic, or spent acids from acceptable sources such as the regeneration of strong acid resins. Weakly basic resins can be regenerated with weak bases such as sodium carbonate or ammonia or spent caustic such as leftover from regeneration of strong base resins. In most cases the weak acid and weak base resins are regenerated with the same strong acids and bases used to regenerate strong cation and strong anion resins, especially when they are used together. However, it is not unusual to use the waste regenerants from the strongly ionized resins to regenerate the weakly ionized resins. In this manner, regeneration efficiencies can approach 100% whereas 10% to 50% is the norm for strongly ionized resin systems alone.

In general, for the ion exchange process to be effective, the volume of treated water must be greater than the volume of waste generated by the regeneration process, i.e., the backwash, regeneration, and rinse cycles. As the total ionic strength of the solution increases, the exhaustion (service) cycle throughput is reduced proportionately. When the ionic concentration is greater than 500 mg/L, ion exchange may become impractical or less attractive than other processes. However, even at 1,000 mg/L, ion exchanger is still an effective technology for softening and removal of selective ions. At 10,000 mg/L, it is usually not practical for most applications except trace ion removal. The removal of trace ions usually involves resins that are specially formulated and that are highly selective for the ions to be removed.

Ion exchange resins are generally limited to processing waters that are relatively free of oxidants, physical contaminants, or oily substances that could coat the resin beads. The organic polymers used to make ion exchange resins have upper temperature limits in the neighborhood of 300° F (149° C). That establishes the maximum temperature level for any organic-based adsorbent, ion exchanger or not. The functional groups in strongly basic anion resins are thermally less stable than the polymer and have lower temperature limits whereas the functional groups in cation exchanges are more stable. The ionic form of the resin affects its chemical stability. The functional groups of anion resins are less stable in the alkaline form. The functional groups of the cation resins are less stable in the acidic form. Sodium form cation resins are routinely used at temperatures approaching 300° F (149° C). Salt form anion resins are sometimes operated at temperatures up to 175° F (79° C) and occasionally as high as 200° F (93° C). In the alkaline form, the type I strong base resins are best operated below 140° F (60° C); the type II resins should not be operated over 100° F (37° C). The weakly basic resins are more stable and can be operated at 70° F (21° C). Strong base resins based on acrylic polymers should not be operated at temperatures above 95° F (35° C).

Types of Ion Exchange Processes

In municipal and domestic applications, ion exchange resins are normally used in single beds for the removal of specific substances, such as hardness, nitrates, naturally occurring organics (color), and alkalinity. In most cases the resins are regenerated with salts; therefore the total dissolved solids and pH of the treated water are not significantly altered. Acid and caustic regenerations are used in deionization processes, which are primarily used in industrial applications.

CATION EXCHANGE PROCESS

When only specific ions need to be removed, it is convenient to use the appropriate type of ion exchanger operated in an acceptable ionic form. Any environmentally and economically acceptable salt can be used to regenerate the exhausted resins. Softening is the most widely practiced of all ion exchange processes. Calcium and magnesium ions are referred to as *hardness* because they react with soap to form curds. This makes it "hard" to wash in waters containing these substances. Ion exchange softening, of course, involves all cations; for example, copper, iron, lead, and zinc are all exchanged along with the calcium and magnesium. A strongly acidic cation exchange resin is regenerated with sodium chloride. This places the resin in the sodium form. The resin can then exchange its sodium ions for calcium, magnesium, and other cations in the raw water. The result is a water containing essentially all sodium cations.

Ions with higher valences such as Ca^{2+}, Mg^{2+}, $Fe^{2+ \text{ or } 3+}$, Cu^{2+}, and Pb^{2+} are more highly preferred by the resins at the TDS levels typical of potable water. During the service cycle, the multivalent ions are removed by the resin in exchange for an equivalent amount of sodium ion; the sodium ion is replenished by the sodium chloride used to regenerate the resin.

Salt Selection

A good regenerant should be inexpensive and contain an exchangeable ion that can effectively displace the unwanted ions from the resin. It should be soluble so as to be delivered at reasonably high concentrations in order to keep waste volumes low, and its co-ion should remain soluble when paired with the displaced ions. Most chloride salts are soluble and chlorides are nontoxic, so it is not surprising that sodium chloride is the most common salt used to regenerate softeners. Potassium chloride, though more expensive, can be used with similar results.

Operating Capacity

The operating capacity of the resin is defined by the number of "exchangeable" ions it removes from the raw water each cycle. In the case of a water softener, calcium, magnesium, and other divalent ions constitute the exchangeable ions. There are several ways to express this. The most common expression of capacity in the United States and the rest of North America is as kilograins (as $CaCO_3$) per cubic foot of resin. The higher the concentration of exchangeable ions, the lower the volume-based throughput capacity.

The operating capacity of the resin varies according to the regenerant level. The regenerant level is normally stated in terms of pounds of regenerant per cubic foot of resin. Higher dosages give higher operating capacities and lower hardness leakage (higher quality) but at reduced salt efficiency. Generally, the selection of regeneration levels is based on using the minimal amount of salt that will operate the softener at the required degree of hardness leakage.

Operating efficiency depends on several factors including the choice of resin, regeneration level, method of regeneration, flow rate during the service cycle, and quality level used to determine start and end of the service cycle.

There are a variety of process schemes to maximize regenerant effectiveness. These usually require more complex and costly equipment, trading reduced chemical operating

Ammonia

Ammonia is a colorless gas that dissolves readily in water. It is commonly present in most water supplies in trace quantities as the ammonium ion due to degradation of nitrogenous organic matter. It may also be present due to the discharge of industrial wastes. Ammonia reacts with water to form ammonium hydroxide, which dissociates into ammonium (NH_4^+) cations according to the following reaction:

$$NH_3 + H_2O \rightleftarrows NH_4^+ + OH^-$$

$$NH_4 + (OH) \rightleftarrows NH_4^+ + OH^-$$

When ammonia is present in concentrations of a few parts per million, the degree of ionization of ammonia is a function of the pH. At a pH below 8, ammonia is present primarily as the cationic ammonium ion. At a pH above 9.0, ammonia is essentially unionized and is not efficiently removed by salt cycle ion exchange.

Strong acid cation exchange resins, such as the kind typically used for softeners, will exchange sodium for ammonium ions. The process is in fact a standard softener. The cation resin has a higher affinity for divalent hardness ions; therefore hardness ions displace the ammonia, which in turn displaces more sodium. As the service cycle continues, the hardness ions form a band in the upper portion of the resin, and ammonia forms its own band just beneath the hardness. Since ammonia removal is required, ammonia is included as an exchangeable ion along with the hardness ions in sizing the system. The service cycle is ended when ammonia levels in the softened water rise. If the service cycle is allowed to continue past the ammonia break, the hardness will continue to load on the resin and displace ammonia. Ammonia levels could reach concentrations in the effluent equal to the total hardness plus ammonia concentrations in the raw water. This could be much higher than influent ammonia levels. When this occurs, it is known as *dumping*.

Ammonia can also be removed by inorganic zeolites such as chabasite, mordenite, and clinoptilolite. Some zeolites are more highly selective for ammonia than hardness and are able to remove it throughout the acceptable pH range for potable water (6 to 9) without the danger of dumping. Not having to remove hardness gives them very high throughput capacities compared to strong acid cation resins for ammonia removal from waters containing appreciable levels of hardness.

Barium

Barium is an alkaline earth metal which is occasionally found in groundwater primarily in Arizona, Texas, Michigan, Vermont, and Florida. Barium is also present in trace amounts in some surface waters with the highest levels occurring in the lower Mississippi basin. Barium is very toxic when ingested as a soluble salt—a dose of even 550 mg is considered lethal.

Barium Removal by Softeners. Barium is readily exchanged on to the strong acid cation exchange resins in softeners. Barium has a higher selectivity for cation resins than either calcium or magnesium (hardness), and is removed throughout the entire softening cycle. It does not begin to leak until after hardness. The increase in barium caused by hardness breakthrough, though small compared to the hardness leakage, may be sufficient to warrant termination of the service cycle. Therefore, barium removal by ion exchange is achieved by ordinary softening, and the process is designed as an ordinary softener.

When sulfate is present, barium solubility is limited to the parts per billion range. If sulfates are present in the raw water, it is probable that barium is present only as a suspended solid and therefore cannot be removed by ion exchange at levels above a few parts per billion.

The barium form of the resin is more difficult to regenerate than the hardness form. Therefore barium will tend to accumulate on the resin, especially at lower salt dose levels. During the initial (first time use) exhaustion cycle, barium continues to load on the resin after hardness breakthrough by displacing previously exchanged hardness (calcium and magnesium) ions. Later, when the resin bed is regenerated, the barium is less efficiently displaced from the resin than the hardness ions. The ratio of barium to hardness left on the resin after regeneration will be substantially higher than that in the influent water. The barium will be pushed toward the bottom of the resin bed, where it could cause leakage in the next service cycle. The regenerant dose level should be high enough to prevent barium buildup in the resin bed.

If a significant amount of barium is exchanged onto the resin during the service cycle, there is the potential for barium fouling from precipitation of barium sulfate. Appreciable amounts of sulfate either in the dilution water or in the regenerant salt itself will cause precipitation of the barium in the resin, which will foul the resin. Barium-fouled resins can be cleaned, but the cleaning process is slow and involves corrosive chemicals such as hydrochloric acid. It is usually necessary to use external vessels to treat the fouled resins to avoid corrosion of the softener. In any case, the cleanup of barium-fouled resins is difficult and not usually practical. One method of cleaning barium-fouled resins is to soak the resin bed in 10% hydrochloric acid. This converts sulfates to bisulfates, which are more soluble. The process usually requires several hours and vigorous agitation before giving measurable improvement. It is often less expensive to discard and replace the resins when performance drops below acceptable levels due to barium fouling.

Combined Softening and Decationization. Waters containing hardness and appreciable levels of alkalinity can be partially demineralized and fully softened by having two columns of strong acid cation resins operated in parallel. One column is in the hydrogen form, and one is in the sodium form, the two effluents are blended at a ratio determined by the untreated water composition. The overall effect is partial demineralization, alkalinity reduction, and softening. Through proper control of the blend ratios, the pH is maintained at acceptable levels.

Barium Removal by Hydrogen Form Weak Acid Dealkalizers. Barium can also be removed by using a weak acid cation resin. The resin can be used in either the hydrogen or the sodium form. The use of weak acid resins in the hydrogen form for softening is limited to removing hardness associated with alkalinity. This process is usually employed for alkalinity reduction and limited to waters that have high hardness-to-alkalinity ratios. The effluent will have a reduced and somewhat variable pH. Initially the resin will remove all cations in exchange for hydrogen and will produce a very low pH. Very quickly the resin will cease exchanging for monovalent ions such as sodium. Then more gradually, some alkalinity will start to slip through, eventually followed by hardness, and then barium. The pH will rise gradually. The reaction of the hydrogen form weak acid resin is usually limited to divalent cations associated with the alkalinity content of the water. Alkalinity in the water will be converted to carbon dioxide, which is a gas, by the hydrogen ions. The carbon dioxide will have to be removed. The most common methods of CO_2 removal are air stripping and deaerating heaters.

Hydrogen form weak acid resins combine readily with the hydrogen ion and are very efficiently converted back to the hydrogen form. Only about 20% excess regenerant above the theoretical dose is needed. Hydrochloric acid as a regenerant has the advantage that it can be used at high concentrations, which means lower waste volume. Hydrochloric acid regenerant concentrations of 6% to 10% are typical. Concentrated hydrochloric acid is hazardous and corrosive and gives off toxic fumes. Proper precautions must be taken when using this chemical. Sulfuric acid cannot be used except at low concentrations, < 0.7%, because it would form insoluble calcium (and barium) sulfate, which would foul the resin bed.

A weak acid resin can be used in the sodium form to remove barium and hardness. This involves a two-stage regeneration: first acid, to remove hardness, then an alkali, to neutralize the bed using sodium carbonate or sodium hydroxide. Sodium bicarbonate can be used but is less effective.

The operation of a weak acid resin whether in the hydrogen or sodium cycle is more complex and usually more expensive than the use of salt regenerated strong acid cation resins because of higher chemical costs, the need for acid-resistant construction, materials, wastewater neutralization, and the need to deal with or strip CO_2 from the product water.

Radium

Radium 226 and radium 228 are natural groundwater contaminants that usually occur at trace levels. A strong acid cation exchange resin operated in the sodium cycle is a very effective method of radium removal. Like barium, radium has a higher selectivity for cation exchange resins than hardness and will be removed during the normal water softening cycle. In a manner similar to barium, on the first cycle of exhaustion, radium will continue to load on the resin bed until well after hardness breakthrough by displacing all the other ions previously loaded, including calcium and magnesium. This effect is valid only for the first cycle and can be misleading.

Point-of-use cartridges, which use ordinary softener resins on a one-time basis for radium removal, will provide decent radium removal for over 5 to 15 times (depending on the particular resin) as long as they will produce softened water. In regenerable systems, however, radium like barium is much harder to regenerate off the resin, and more of it will remain in the resin after regeneration. The radium will be pushed toward the exit (bottom) of the resin bed during the regeneration cycle. As the softener becomes exhausted, hardness leakage reaches the radium-rich end of the resin bed. The hardness, which is less preferred by the resin, will displace only a small, but nevertheless, significant amount of radium from the resin, causing radium levels to increase to unacceptable levels. Therefore, in systems that are regenerated it is necessary to limit the service cycle to the softener capacity for hardness. Since the amount of radium is insignificant compared to hardness, the softener design calculations are made for an ordinary softener.

ANION EXCHANGE PROCESSES

Dealkalization, sulfate, nitrate, selenium, arsenic, and uranium can all be removed by anion exchange. These substances are all present as anions. The most common process is to use salt (NaCl) regenerated strongly basic anion exchange resins operating in the chloride cycle. The design of the equipment is essentially the same for any of these substances. Only the throughput capacities and the resin volumes are different because of the differences in relative affinities and concentrations. Those substances with high affinities can continue to load to higher concentrations on the resins by displacing others previously exchanged but with lower relative affinities.

The gel-type strongly basic resins whose functional groups are based on either trimethylamine (these are called type I) or dimethylethanolamine (these are called type II), have the approximately same order of affinities for the common ions found in potable water except for the hydroxide ion, which is not a factor here, and perchlorate. When operated in the salt cycle, the regeneration efficiency is about the same for both types.

The general order of affinities for both types of strongly basic resins for the common ions found in water at less than 500 ppm total dissolved electrolytes is

$$PO_4 > SeO_4 > SO_4 > NO_3 = AsO_4 > SeO_3 > CO_3 \rightarrow AsO_3 > HCO_3$$

Ion exchangers using strong base resins that are regenerated with NaCl are commonly referred to as *anion softeners* because of the similarity of the equipment to the traditional cation-based water softeners. When the pH of the resin is at or above 7, anion resins tend to give off fishy odors. This is due to the (slow) decomposition of their functional groups, which are amine-based. The alkanol amine group of the type II resin gives off less offensive odors than the type I resins and is sometimes preferred over the type I resins in potable water applications for this reason.

When a strong base resin is operated in the chloride cycle, it will exchange chlorides for all the incoming anions. When all the exchangeable ions, in this case chlorides, have been used up, the resin bed will contain only the ions that it removed from the raw water. These ions will have distributed themselves in order of their relative affinities.

The bicarbonate ion is the least strongly held and is positioned nearest the exit end of the vessel. If the vessel is operated as a dealkalizer, the service cycle would be ended at this point and the vessel regenerated with a salt (NaCl) solution. However, if the vessel is used for nitrate removal and there is no desire to remove bicarbonates, the service cycle would continue. The nitrates and other ions would continue to be exchanged onto the resin bed, and the bicarbonates would be driven off. The service cycle would end when the nitrates on the resin were pushed down to the exit end of the vessel. Regardless of the application, the vessel, resins, and regeneration equipment are identical. The operating capacity of course depends on the composition of the untreated water, the concentration of the impurity being removed, and the specific goal of the treatment. The specific applications are discussed separately in the following sections.

Dealkalization

Dealkalization is the removal of bicarbonates, carbonates, and hydroxides. In potable waters the hydroxides could be controlled simply by injection of acid. This leaves bicarbonates and carbonates which are not amenable to acid injection because of the increased corrosivity at lower pH. Strongly basic anion exchange resins operated in the chloride cycle can remove bicarbonates and carbonates in exchange for chlorides. This also eliminates "temporary hardness," which is an effective way of minimizing scale buildup in water heaters and boilers. Figure 12.1 shows a softener followed by a salt cycle dealkalizer (12.11) (strong base anion resin dealkalizer).

Carbonates are usually a very small percent of the overall alkalinity. The bicarbonate ion, which predominates, has the lowest affinity of the common anions for the exchange resins. It is the first ion to appear in the effluent, leaving the exhausted bed filled with sulfates at the top, just above nitrates, which are above chlorides, and the bicarbonates at the bottom.

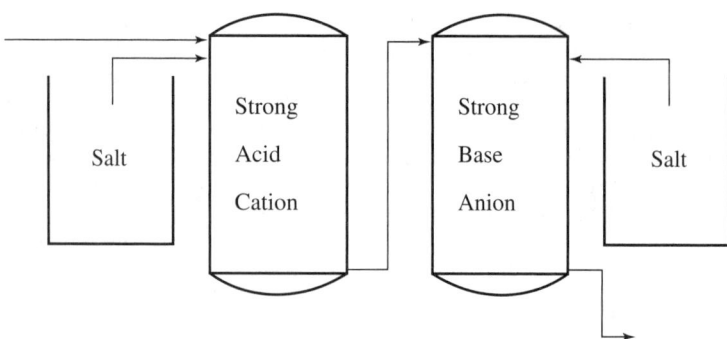

FIGURE 12.1   Strong base dealkalizer.

Although the sulfate has a higher affinity for the resin during the exhaustion cycle, the sulfate/chloride affinity relationship is affected by the concentration and is reversed at the regenerant concentrations. Approximately 5 lb of salt per cubic foot of resin, applied at 4% to 6% concentration, is sufficient to fully regenerate the resin bed to the chloride form. Higher dose levels offer very little gain.

Since the bicarbonate ion is the least strongly held of the common ions by the resin, the operating capacity depends on the total concentration of ions. The operating capacity for most gel type II anion exchange resins can be estimated at 12,500 gr/ft^3 resin. Throughput capacities are calculated based on all the anions in the inlet water including chlorides as being exchanged. First, calculate the sum in milligrams per liter of all the anions as their calcium carbonate equivalents, and divide that by 17.1 to convert the concentration to grains per gallon. Bicarbonates will be reduced by about 80% to 90% throughout most of the service cycle, and the treated water pH will be about 5. When alkalinity leakage reaches 30%, the service cycle is terminated. The operating capacity can be increased to about 15,000 gr/ft^3 and the effluent pH increased to above 7, by adding about 0.25 lb of sodium hydroxide to each 5 lb of salt used for regeneration.

pH Effects of Dealkalization.   The pH of ordinary tap water is determined by the equilibrium between carbon dioxide, bicarbonate, and carbonate between pH values of 5.0 and 10.6. In the absence of hydroxides, the removal of the bicarbonate and carbonates automatically reduces the pH of tap water to 5.0, which is the pH of carbon dioxide in water. This is what happens when salt is used alone as the regenerant. Carbon dioxide is only slightly ionized, so it is not removed by the resin. This scenario is changed by adding a small amount of caustic to the regenerant salt, usually 0.25 lb of NaOH to each 5.0 lb of salt. This places a very small amount of hydroxide on the resin. Its main effect is that it converts the nonregenerated bicarbonate left on the resin to carbonates. The carbonates on the resin will combine with carbon dioxide to form bicarbonates, which are held by the resin. By this mechanism, CO_2 is removed and the pH is shifted back to 7.0. The effluent is free of bicarbonates and carbon dioxide.

Dealkalization with Weak Acid Resins Regenerated with Acid.   Weakly acidic resins operated in the hydrogen form are, by their nature, unable to react with cations except in the presence of a neutralizing agent such as alkalinity. They exchange hydrogen for cations but are only effective in removing hardness and other divalent ions that are associated with alkalinity. Therefore, weakly acid resins are able to work effectively as dealkalizers, but only on certain waters. Because they are weakly acidic, they have extremely high affinities for the hydrogen ion and are fully converted to the hydrogen form with little more than the stoichiometric amount of acid during regeneration. The degree of exchange during the service cycle, or operating capacity, is limited by the amount of alkalinity, the ratio of divalent cations to alkalinity, and total ionic concentration. The key factors in determining the potential operating capacity are the hardness-to-alkalinity ratio and the leakage endpoint. The base operating capacity can vary from about 10 to 60 kgr/ft^3 of alkalinity as the hardness-to-alkalinity ratio varies from 0 to 1.2 or higher. The regeneration dose is usually calculated as 120% of the operating capacity. Flow rate and temperature during the service cycle also affect the operating capacity and can reduce the operating capacity by 35% as the temperature drops from 70° to 35° F (21° to 2° C) or as the flow rate doubles from 2 to 4 gpm/ft^3.

Hardness is removed and bicarbonates are converted to carbonic acid, which can be removed by a degasifier. In effect the effluent is partially demineralized. The degree of softening depends on the hardness-to-alkalinity ratio, and any hardness in excess of alkalinity remains. The regeneration equipment and all process and tanks, piping, and valves need to be acid-resistant. Sulfuric acid is commonly used as the regenerant because it is less expensive, but calcium sulfate precipitation is a very real concern. When sulfuric acid is used, the regenerant concentration should not be allowed to go above 0.75%. It is usually held between 0.5% and 0.75%. Hydrochloric acid can be used at higher concentrations, with 6% to 10% the common range. This gives much lower waste volumes. Hydrochloric acid is not as widely used because it is more expensive and more corrosive and gives off hazardous fumes. Equipment costs are usually more expensive also, because of the increased material costs associated with the increased corrosivity of HCl.

pH Effects. Weakly acidic resins have a small amount of strong acid capacity and will initially convert all salts to acids in the beginning of the service cycle. Initial pH values can be as low as 2 and rise gradually, but remain under 7 for over three-fourths of the service cycle. When this process is used to treat boiler feedwater, the carbon dioxide generated by the exchange reaction has to be removed by aeration or vacuum degasification; otherwise, it will be converted back to bicarbonate when caustic is added to raise the pH.

Dealkalization by Dual Strong Acid Columns Operated in the Sodium and Hydrogen Cycles. In this scheme two vessels, each loaded with strong acid resins, operate in parallel; one is regenerated with salt and the other with an acid. The two effluents are combined in a ratio based on the composition of the untreated inlet water so that acid generated in the hydrogen cycle vessel is just sufficient to neutralize the alkalinity in the effluent from the salt regenerated vessel. This process produces partially demineralized, fully softened water with zero alkalinity.

The sodium cycle vessel is sized and designed as a traditional salt regenerated softener. The hydrogen cycle vessel is typically designed to operate at the highest acid efficiency, usually about 3 lb/ft^3 of sulfuric acid. The throughput capacity and leakage and blending ratios will depend on the water analysis. When it is practical, the two vessels are sized to exhaust simultaneously.

Arsenic

Arsenic occurs widely in the earth's crust, usually in the form of insoluble complexes with iron and sulfides. Another source of arsenic in potable water supplies comes from its extensive use in the past as a pesticidal agent. It also exists in soluble form, primarily as arsenites (AsO_3) and arsenates (AsO_4). Ingestion of as little as 100 mg of arsenic per day can cause severe poisoning in humans.

There have been recent changes regarding arsenic levels. The proposed maximum contamination level (MCL) for arsenic has been lowered to 10 ppb, from the current 50-ppb acceptable level. The effective date for compliance with the rule is January 2006. From the time the MCL review discussions began in earnest in the late 1990s, many efforts to develop the product technology to meet the new standard economically have been undertaken. Reports of new media for removal of arsenic have appeared and continue to appear frequently in technical journals. Until recently, activated alumina (AA) and granular ferric hydroxide (GFH) had been identified as the two most likely adsorbents to be employed for arsenic removal. However, other media have been developed with promise of improved performance. Iron oxide–based media and most recently titanium oxide–based media are showing what appears to be remarkable performance in producing very low levels of arsenic at low operating costs. These media appear to be designed for one-time use, are able to treat in excess of 10,000 and even 20,000 bed volumes before exhaustion, and can operate over a wide pH range. In some cases the new materials may be able to remove both the pentavalent (arsenate) and more difficult to remove trivalent (arsenite) forms of arsenic. Recent reports suggest that some of the inorganic-based media are able to pass the leach tests required for nonhazardous disposal once they become exhausted.

Development work is ongoing and at a rapid pace. It is prudent to conduct literature searches at the start of any new project as technology discussed here will likely have been improved on or even made obsolete. Review what's available in technical approaches when the time comes to select the technical approach for a new or upgraded water system. For existing facilities, chemical addition, clarification, and filtration may already be in place, requiring only feed rate and composition changes to meet the new requirements. Conventional treatment will most likely remain the treatment method of choice for surface water supplies.

The soluble forms of arsenic can be removed by strongly basic ion exchange resins, activated alumina, or other adsorbent media. In either case, the process works best when arsenic is present in the arsenate form. Proper pretreatment and testing should be employed to ensure that all the arsenic is oxidized to the arsenate form.

In oxygen-deprived waters, such as in well water, the arsenite species is usually 20% to 50% of the total arsenic. Arsenite is a weak acid, and at pH levels below 9 it is unionized and therefore not readily removed by ion exchange. However, arsenate is a relatively strong acid and fully ionized at a pH levels above 6. Oxidative pretreatment with chlorine is an effective way of converting arsenite to arsenate. Activated alumina exhibits as much as an 80-times increase in throughput capacity for arsenate as for arsenite at the same concentrations on otherwise identical waters.

Strongly basic anion exchange resins have the advantage over activated alumina of being regenerable in a single step and with common salt in much the same manner as a softener. Activated alumina needs to be regenerated in two steps, first with an alkali such as sodium hydroxide and then the bed is neutralized with an acid rinse. However, anion resins have a disturbing tendency to dump arsenic in preference for sulfate and have recently fallen out of favor for small unmonitored systems.

The presence of iron or high levels of sulfate will interfere with and impair the removal efficiency of arsenic in any form, as the sulfate ion is removed along with the arsenic and therefore increases the exchange load. Iron can be used to remove arsenic by coprecipitation and may be used as a first-stage process with second-stage polishing via ion exchange. Any remaining iron from the first stage can physically foul the polishing media. Throughput capacities and resin volume calculations should include sulfates as exchangeable ions.

Arsenic is not detectable by normal human senses; toxic levels could exist without warning. It is important that the treated effluent be properly monitored to ensure that arsenic removal is occurring and to prevent unsafe levels of arsenic in the treated water from being delivered to the distribution system.

Oxidation of arsenite to arsenate with chlorine occurs at a very rapid rate. Test results from an EPA-sponsored study showed that over 95% of the arsenite was converted to arsenate in less than 5-s contact time with 1 ppm residual chlorine. These tests showed that the reaction was insensitive to pH in the range of 6.5 to 9.5. However, the reaction rate decreased substantially outside this range.

The presence of other oxidizable substances must be accounted for, as they will consume chlorine and may act as reverse catalysts for the oxidation of arsenite. For example, the presence of 5 ppm of TOC in one case slowed the reaction rate so that almost 50% of the arsenite remained unoxidized after 30 s. In this case more than 30 min was required to reach 80% oxidation. The presence of chloramine has also been shown to reduce the rate of oxidation substantially. Since both TOC and chloramines can be present (ammonia is converted to chloramine by chlorine), each contaminated water being considered for this type of treatment should be thoroughly evaluated at the pilot plant stage to ensure that this critical step is carried out effectively and that all arsenites are converted to arsenates.

The relative affinity and operating capacity of strongly basic anion exchange resins is markedly higher for arsenate than for arsenite. Sulfates and chlorides are preferred over arsenites, but not over arsenates regardless of the type of media. This means that they are removed before the arsenites, and therefore both of these ions would be included as exchangeable ions, which increases the loading. The arsenate ion, however, has a somewhat higher affinity for the strongly basic anion resins higher than chloride but lower than sulfate. This means that arsenate *dumping* can occur if the resin column is inadvertently overrun. Surface fouling due to naturally occurring organics reduces the affinity of the resins for all forms of arsenic, resulting in reduced run lengths when fouling occurs. Improved ion exchange processes using multiple beds in round-robin fashion show promise in avoiding arsenic dumping and increasing regenerant efficiency (also reducing brine discharge). They work on the principle of overrunning the first stage past the sulfate break and using the second stage to catch the arsenic. The sulfate-loaded resin is regenerated and placed into the second-stage position while the arsenic loaded resin takes on the primary position. Since the arsenic is only present at trace levels, the regeneration of both beds to remove the arsenic is not required on a regular basis. This approach, while interesting, is not yet offered commercially.

Activated Alumina. Activated alumina has been used successfully to remove both forms of arsenic. However, control of pH within a fairly narrow range and careful monitoring of water chemistry are required for good success. Preoxidation of arsenic to the arsenate form and pH control to the 5.5 to 6.0 range are highly recommended.

Regeneration of activated alumina is a two-step process. Following the backwash cycle, it is first regenerated with sodium hydroxide and rinsed; next it is washed with a dilute acid such as sulfuric, and then the bed is rinsed down to the proper pH before being returned to service.

Strong base ion exchange resins remove arsenic rapidly and to low levels, and can operate over a wide pH range. However, only the arsenate can be removed without interference from other ions commonly found in potable water, so oxidative pretreatment to convert arsenite to arsenate whenever arsenite is present is a must. Additionally, strong base resins have a higher affinity for sulfates than arsenates, and sulfates are typically present at concentrations orders of magnitude greater than that of arsenic. This requires close attention to effluent monitoring to avoid the potential chromatographic dumping of arsenic at concentrations much higher than that in the influent. If an ion exchange resin is run past the arsenic breakthrough all the way to the sulfate breakthrough, no arsenic will be left on the resin.

Sulfate

Sulfates at high concentrations in potable water supplies cause diarrhea in humans and livestock. They can be removed by strong base chloride cycle anion exchange. Type II resins are typically used because of their lower odors. Removal rates are typically in the range of over 95% to 100%. A salt dosage of 5 lb/ft^3 is sufficient, and operating capacities of 20 to 25 kgr of sulfate, expressed as calcium carbonate per cubic foot of resin, are typical for gel-type strong base resins. Sulfate displaces the other common ions including chlorides and nitrates during the exhaustion cycle, so the operating throughput is calculated based only on the sulfate concentration in the raw water.

Nitrate

Nitrates are considered to be toxic to human infants and ruminant animals such as cows. They occur naturally as a result of the decomposition of nitrogen-containing waste matter, also as a result of farm runoff from fertilizer. The maximum level of nitrates in potable water is usually limited to 10 ppm as nitrogen. At levels above 50 ppm nitrates as nitrogen (N) can cause methemoglobinemia in infants, also known as blue baby syndrome.

Type I or type II strong base gel anion resins are often used for nitrate removal. They are selective for nitrates over all the common ions in tap water with the exception of sulfates. During the service cycle, sulfates will load first and can displace nitrates at concentrations equal to the sulfates plus nitrates if the vessel is overrun. This is discussed below under "Nitrate Dumping" (see Figure 12.2).

Sulfate, being divalent, undergoes selectivity reversal at the higher ionic concentrations during regeneration. Therefore sulfates are more fully removed from the resin than nitrates during regeneration.

FIGURE 12.2 Nitrate removal capacity.

During the service cycle, all the anions are converted to chlorides at the entrance to the resin bed. At the exit end of the bed, the chlorides displace some of the remaining nitrates left over from the previous cycle, and a small but significant amount of nitrate leakage occurs as the service cycle begins. How much nitrate leakage occurs depends on the total ionic concentration and degree of regeneration. To ensure nitrate levels below 10 ppm as nitrogen, about 10 lb/ft^3 of NaCl or higher is often considered as a minimum dose for cocurrently regenerated columns.

Waste Discharge. Regenerant discharge is often a problem in rural areas because of the potential of nitrates reentering the groundwater.

Nitrate Dumping. The standard type I and type II resins are more selective for sulfates at the low ionic concentrations of the service cycle. This means that sulfates will dominate the resin bed, occupying the first layer of resin nearest the inlet distributor. Nitrates will occupy the resin layer directly after sulfates and will be displaced down the column by sulfates throughout the service cycle. Nitrates will likewise push the chlorides, which in turn will do the same to bicarbonates. This system is similar to a dealkalizer except that it is run past the alkalinity breakthrough, until the nitrate breakthrough. Initially the pH will drop to 5.0 as the resin bed operates as a dealkalizer. After the bicarbonates have been displaced from the bed, the pH will rise again, and finally the nitrates will begin to increase in the effluent. The service cycle is usually terminated when the effluent nitrates reach 20% to 30% of the influent or 10 ppm as nitrogen. If the resin bed is allowed to continue to process water, the sulfates will continue to load on the resin by displacing nitrates, thus increasing the level of nitrates in the effluent to a level that could become as high as the sum of sulfate plus nitrates in the raw water. In many waters, sulfate levels are much higher than nitrate levels, so dumping could result in effluent NO_3 levels reaching several times as high as the influent levels.

The nitrate dumping phenomenon is shown for a nonnitrate selective resin, in this case a type II resin, in Figure 12.3. Figure 12.4 shows the breakthrough pattern of a nitrate se-

FIGURE 12.3 NO_3 breakthrough pattern for a type II, nonselective resin.

FIGURE 12.4 NO$_3$ breakthrough pattern for a nitrate-selective resin.

lective resin. Notice that with the nitrate selective resin the sulfate breaks through before the nitrate. Also, the sulfate level will rise above the influent level if the resin is run beyond its usable capacity. When a nitrate selective resin is run past its usable capacity, the nitrate level will gradually rise to the influent level, but not above.

Nitrate Removal by Selective Resins. Nitrate removal by so-called nitrate selective anion exchangers is becoming widely practiced around the world and in the United States. These ion exchangers are similar in composition and structure to type I strong base resins. However, their functional groups are based on larger amine compounds than the trimethylamine used in ordinary type I resins. Ion exchange resins that use triethylamine, tripropylamine, and tributylamine have been offered commercially as *nitrate selective*. The larger amine groups are more widely spaced compared with the trimethylamine groups used in standard resins. This makes it more difficult to exchange with multivalent ions such as sulfate. Actually, the term *nitrate selective* is a misnomer. These resins are effective because they are less selective for multivalent ions such as sulfates. Because of this, sulfates are not able to cause the massive dumping of nitrates when these resins are overexhausted as is the case with ordinary resins. These resins therefore offer greater safety in high-sulfate waters. At levels of about 25% or less sulfates, the standard resins will have a higher operating capacity than the so-called selective types. Due to the larger size of the amine groups, the selective resins have lower total capacities so the overall operating capacity is smaller than for the standard-type resins. When sulfate levels are greater than 25% the selective resins will give higher throughputs. The difference in operating capacity with changing sulfate levels, comparing selective and nonselective nitrate resins, is shown in Figure 12.5.

Sodium chloride is very effective in regenerating both standard and selective types of resins. Average nitrate leakages are about the same for either type of resin at similar regenerant levels. The biggest difference between the selective and nonselective resins is that sulfates are dumped by nitrates from the selective resins instead of the other way round in standard resins.

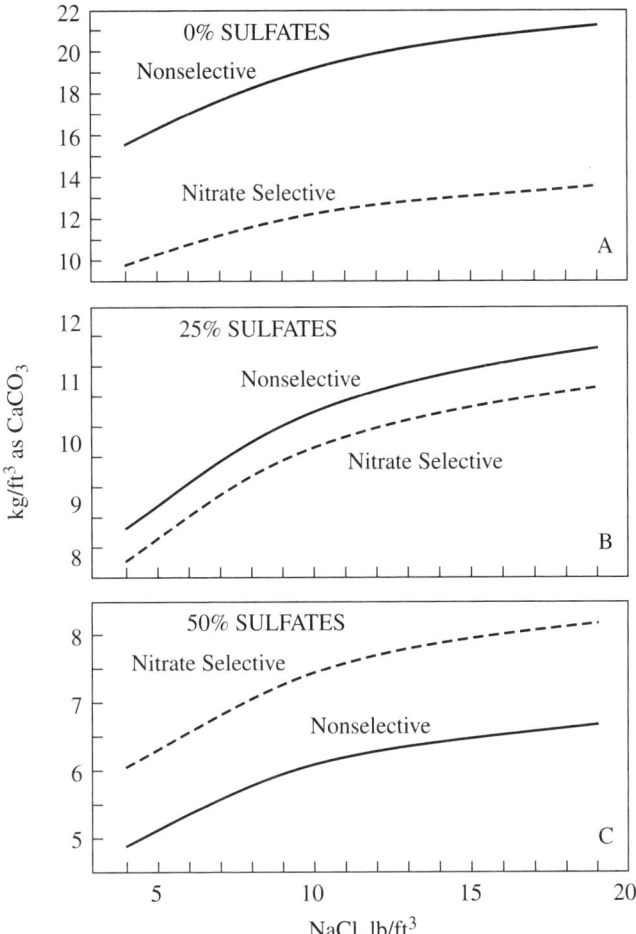

FIGURE 12.5 Removal capacities.

Lead

Common ion exchange softening also removes lead and is the preferred method when softening is already in practice. Although cation exchange resins remove lead preferentially, lead salts are not very soluble in most water supplies. Of course, this also means that lead (as a cation) is not present in most water supplies. When lead is present, it is generally present as a suspended solid. When lead is removed by cation exchange resin, it is usually by a combination of filtration and ion exchange.

Fortunately, ionic lead levels in softener resins remain low enough to not interfere with the softening process, and lead removal becomes an added benefit.

If a 100% guarantee for lead removal has to be made, it is very important to perform bench-scale tests. The treatment system must include filtration, and the filtration must be an absolute cutoff size significantly smaller than 1 μm.

It is important to remember that an ion exchange column can only remove ions and that with wastewater contaminants, such as lead, the real story is that these types of contaminants are frequently not in an ionic form. Therefore, ion exchange alone may not be suitable for their removal.

Fluoride

Bone char and activated alumina are both effective for fluoride removal and are the preferred exchange media. They are highly selective for fluorides over the other ions in water. Strongly basic anion exchange resins are not used, because the fluoride ion is weakly held and therefore not well removed. Activated alumina is a semicrystalline inorganic adsorbent. Bone char is made from charred and steam-activated animal bones. It consists basically of calcium hydroxy appetite, which bonds strongly to fluoride. Bone char is not as readily available or as widely used as activated alumina, but both have been successfully used to reduce fluoride levels.

Defluoridation with activated alumina is similar to other ion exchange processes. The alumina is generally in a column configured for downflow exhaustion and regeneration. Because of limited kinetics, it's best to use bed heights of at least 3 ft. The pH of the water being treated should be adjusted to 5.5 to 6.0. Unlike for organic-based ion exchangers, the breakthrough curve is very gradual. Fluoride leaks from the column throughout the service cycle. The endpoint of the service cycle is the maximum fluoride concentration to which the cycle can run that is acceptable to the user. Because of the ever-changing breakthrough pattern, it is very difficult to blend as a technique to reduce treatment costs. A more practical approach is to store the entire exhaustion cycle volume and to run the system until the average over the entire cycle meets or is just below the maximum acceptable level.

Perchlorate

Perchlorate consists of an atom of chlorine surrounded by four atoms of oxygen. It occurs as ammonium, potassium, magnesium, or sodium salts. These salts bind weakly to soil particles and are not significantly broken down in the environment. Perchlorate salts are extremely soluble and highly mobile and migrate faster and farther than other water contaminants. These properties make perchlorate particularly persistent and problematic. The primary source of perchlorate in the western United States is due to ammonium perchlorate, an oxidizing agent used in solid propellants. Its source is primarily from the previously legal discharges of wastewater from military installations and defense contractors. Perchlorates are also used in a variety of manufacturing processes, such as car airbags, leather tanning, fireworks, and electronic tubes. Prior to the 1960s, perchlorate was also used as a medical treatment for patients with hyperthyroidism. Today, over 90% of the perchlorate produced goes into solid rocket fuel.

Perchlorate impairs the thyroid function, because it is taken up in preference over iodide. As far back as 1966, perchlorate was identified as a suspected carcinogen. Although the initial focus was on California, because of concentration of defense-related activities, it became clear that other locations with defense activities were beginning to show up as sources of significantly higher levels of perchlorate. In 1992 and again in 1995, the USEPA-recommended concentrations for perchlorate in potable water were 1 to 5 ppb (micrograms per liter) for children and 4 to 18 ppb for adults. The USEPA draft recommendation in 2002 was lowered to 0.3 ppb for children and 1 ppb for adults. Later in

2002, the Office of Environmental Health Hazard Assessment (OEHHA) published 6 ppb for adults and 2 ppb for children.

The removal of perchlorate from potable water supplies is somewhat similar to the removal of nitrate. In both cases, ion exchange technology is the best currently known method. Perchlorate is even more strongly held by ion exchange resins than nitrates are. Sulfate ions are the only major competitor besides nitrates for exchange sites on strong base anion resins. When nitrate removal is also required, nitrate capacity becomes the limiting operation and posttreatment for perchlorate removal may be required. However, perchlorate is so strongly held by ion exchange resins that it is difficult to regenerate using normal regeneration methods and traditional regeneration chemicals. Sulfates are less strongly held by the resin than perchlorates, but they are almost always present at concentrations several orders of magnitudes higher and, therefore, significantly reduce the throughput capacity per cycle.

To control the amount of regenerant waste, a regenerable ion exchange resin must have high capacity for the contaminant of interest and be efficiently regenerated. The very high affinity of perchlorate for strong base anion exchange resins goes against this. Type II strongly basic anion exchange resins and acrylic strong base anion exchange resins can be regenerated with lower amounts of brine than can type I strong base anion resins. However, the problem remains of disposal of the brine and/or the destruction of the perchlorate. As of March 2003, much work is being done to address this problem of brine reuse and perchlorate destruction from the waste brines.

Recent developments in ion exchange technology have resulted in resins that are extremely deselective of divalent ions. Such resins were first commonly referred to as *nitrate selective*, because they allowed resins to remove nitrates without interference from sulfates. Although perchlorates are much more strongly held than nitrates, the presence of sulfates in the water can reduce the operating capacity of type I, type II, and acrylic strong base anion resins. Further developments in the field of ion exchange technology have now resulted in resins with even further reduced selectivity for divalent ions such as sulfate, reducing sulfate interference in perchlorate removal by orders of magnitudes below the original nitrate selective resins. The result is that today there exist several resin types that are able to treat over 100,000 bed volumes before perchlorate breakthrough, based on typical inlet perchlorate concentrations well below 1 ppm.

Typical selectivity coefficients for perchlorate compared to chloride have been reported in various literature. There are several ways of stating relative affinities between ions and ion exchange resins. Distribution coefficients, separation factors, and selectivity coefficients are commonly used. Each defines different mathematical relationships, but they have in common that they represent ratio relationships for a specified pair of ions. Values above 1 mean the first ion is preferred, and less than 1 that it isn't. These relationships depend strongly on the solution conditions in which they were measured; therefore they should be used only for relative comparisons and not as absolute preference of a resin for one ion compared to another. Table 12.1 shows the selectivity coefficients for perchlorate versus chloride.

With this development comes the potential use of resins on a nonregenerable basis. Because of their high affinity for perchlorate compared to sulfate, these resins can be used without being regenerated on a competitive basis when compared against regenerable, less expensive resins, including the cost of the regenerant chemicals, the equipment for the regeneration, the disposal cost, and the treatment cost of the regenerant brines. Because of their high affinity, these resins pass the leach test for hazardous material so that they can be disposed of, in most cases, as ordinary landfill. However, specific tests for the resins intended for use must be conducted to ensure compliance with local laws at the time the plant is being designed to build, as those laws may change from the date of this publication.

TABLE 12.1 Perchlorate Selectivity Coefficients of Various Anion Resins

Type of resin	Selectivity coefficient ClO_4 versus Cl
Acrylic strong base, gel	5
Type I styrenic, gel	100–150
Type II, styrenic, gel	50–100
Tributylamine	2,000–4,000
Triethylamine macroporous styrenic	500–1,000

Recently, efficient regeneration schemes for these superhigh-perchlorate-capacity resins are being developed that reduce chemical waste. One of the more promising schemes is being patented (U.S. patent pending, 09/491,242). This process involves the use of ferric chloride complexed with hydrochloric acid to form tetrachloroferrate ($FeCl_4^-$), an ion with exceptionally high affinity for anion resin (much higher than perchlorate). By washing with water, the complex is broken, leaving the resin in the chloride form, free of perchlorate and ready for the next service cycle. As little as two bed volumes of the tetrachloroferrate solution are needed for complete regeneration of even the bifunctional resins. When coupled to a perchlorate destruct system, the regenerant solution can be reused without any perchlorate or regenerant sourced chloride being returned to the environment.

There are several possible destruct mechanisms for the perchlorate in waste brine and reuse of the waste brine. However, most involve significant heat and/or pressure and are therefore somewhat expensive and complicated. A simple process using a ferrous salt has been developed to convert perchlorate back to chloride ions and oxygen, while at the same time converting ferrous ions to ferric ions. This process is a low-temperature, inexpensive adjunct to the technique of using tetrachloroferrate ion as a regenerant. This process is also "patent applied for." The time interval between regenerations is much longer than that for typical ion exchange systems. This should allow development of energy-efficient approaches and smaller multicycle regeneration systems. These developments could tip the balance in favor of using resins on a regenerable basis.

Flow sizing criteria for these resins tend to be the same as for normal regenerated resins, 2 to 8 gpm/ft^3 of ion exchange resin (1- to 4-min EBCT). It has been reported that some of the larger, very highly selective functional groups used in the bifunctional resins are more flow-rate-sensitive than others. The second functional group is added to compensate for this. All the selective resins can be operated at flow rates approaching 15 gpm/ft^3 (30-s EBCT). Some capacity is lost at higher flow rates. Flow rate sensitivity will probably vary greatly among the different resin types. Also, extremely high flow rates in traditional forms of ion exchange applications such as water softening have led to bed compaction, plugging, and other problems; thus long-term design at such high flow rates carries a significant risk of poor reliability and frequent maintenance. As with all ion exchange systems, limits of oxidants, suspended solids, and organic contaminants are necessary for long-term operation without fouling. Table 12.2 describes some of the resins and their suggested roles on perchlorate removal applications.

TABLE 12.2 Perchlorate Removal Applications for Various Resin Types

Name of functional group	Chemical symbol	Polymer background	Comment
Trimethylamine	$N-(CH_3)_3$	Acrylic	Sufficiently low affinity for perchlorate that it can be regenerated with common salt. Best used with countercurrent technology. Typical capacity—highly influenced by sulfate content in the water.
Dimethylaminoethanol	$(CH_3)_2-N-C_2H_5OH$	Styrenic type II	Sufficiently low affinity for perchlorate that it can be regenerated with common salt. Best used with countercurrent technology. Typical capacity—highly influenced by sulfate content in the water. Is more stable than the acrylic resin and less likely to give off a fishy odor.
Triethylamine	$N-(C_2H_5)_3$	Styrenic	Nitrate selective commonly available, very high capacity, very difficult to regenerate, but may be usable with reclaimed regenerant schemes using countercurrent technologies.
Tripropylamine	$N-(C_3H_6)$	Styrenic	Higher ClO_4 selectivity than triethylamine type, one-time use applications. Offer excellent throughput capacity at relatively low cost on a single-use basis, and can be used with countercurrent technology and regenerant reclaim schemes in regenerable applications.
Tributylamine	$N-(C_4H_7)$	Styrenic	Very highly selective. Offers excellent throughput capacity at relatively low cost on a single-use basis, and can be used with countercurrent technology and regenerant reclaim schemes in regenerable applications.
Bifunctional	Various	Styrenic	Combinations of highly and very highly selective groups. Much slower kinetically than other choices, but higher throughput capacity, intended for single-use applications, and may be usable with regenerant reclaim schemes and countercurrent technology. Much higher cost than other types of resins.

Selenium

The MCL for selenium is 0.01 mg/L. When selenium is found in potable water, it is usually as a divalent anion. The selenate ion (SeO_4^{2-}, Se^{6+}) is more highly preferred than the selenite form (SeO_3^{2-}, Se^{4+}) by strongly basic anion exchange resins. The operating capacity of a strongly basic anion exchange resin will be greatest when all the selenium is present as selenate, SeO_4. The selenate ion is more preferred than any of the ions commonly found in potable water including sulfate, whereas selenite ion is poorly exchanged and is less preferred than sulfates.

When a resin loaded with selenite is exhausted, the sulfate will continue to be loaded on the resin and displace previously loaded selenite ions. As this happens, the concentration of selenite in the effluent can approach the sum of the sulfate plus selenite ions in the raw water. The sulfate level in the raw water is usually many times higher than the allowable limit of selenium. Therefore, the danger exists of selenium dumping and appearing in the effluent at levels much higher than in the influent, if the resin bed is overrun. Dumping does not occur if the selenium is all converted to the selenate ion prior to the ion exchange vessel because selenate is preferred over sulfate.

Oxidation of selenites to selenates by chlorine occurs quite readily in the pH range of 6.5 to 7.5. A retention time of 5 min will ensure that more than 70% of the Se^{4+} is converted to Se^{6+} when the free chlorine level is maintained at 5 mg/L or more. Free chlorine is a much more effective agent than either potassium permanganate or hydrogen peroxide. Over 99% of the selenite can be converted to the selenate in 15 min with a 5-ppm free chlorine residual. But a 2-ppm chlorine residual takes 4 times as long. Both pH and the chlorine residual level must be controlled to maintain stable and effective operation. For example, by letting the pH rise to 8.3 and the chlorine residual drop to only 1 ppm, only 80% of the selenite may be converted to selenate in 30 min. Each installation should be evaluated on its own to determine the necessary parameters for proper chlorination.

Even though the selenate ion has a higher affinity for strongly basic resins than sulfates, their relative affinities are sufficiently close that when the sulfate breaks through, it causes an increase in the selenate level in the effluent. Selenates will begin to rise gradually once sulfate breaks through, usually within 10% of the throughput at which sulfate breakthrough occurs. The effluent concentration of selenium will soon rise above maximum allowable levels. This could happen without notice unless the effluent is carefully monitored. For this reason it is considered standard practice to end the service cycle and regenerate the resin at or before the sulfate breakthrough.

Operating Capacities and Parameters. Since selenium is only present in trace quantities, its concentration alone will have little effect on the resin's throughput capacity. It is prudent that any anion exchange system designed for selenium removal be designed to run as a sulfate removal system to a sulfate leakage endpoint. The system should be sized on the basis of the water analysis with the highest sulfate concentration. The ion exchange vessel and regeneration equipment for removing sulfates and selenium are the same as used for dealkalization, except for the monitoring equipment.

The relative affinities of divalent ions such as sulfates and selenates against monovalent chlorides drops at the higher ionic concentration during regeneration. To carry out a proper regeneration, the anion resin should be regenerated with sodium chloride at concentrations of at least 5% at a flow rate that will allow at least 30-min contact time. This can be achieved with a salt dose of 5 to 10 lb/ft^3 injected at a concentration of at least 5%, to a maximum concentration of 15%. The selenium and sulfate are pushed from the resin bed. Any residual selenium will be found at the bottom of the bed after the regeneration cycle. It will resist leakage until another divalent ion such as sulfate begins to leak through at the end of the next service cycle.

Operating Cycle. During the service cycle, all the anions are loaded on the resin and exchanged for an equal amount of chlorides. During the early portion of the service cycle, the effluent will contain only chloride ions. As the service cycle continues, selenates continue to load onto the inlet portion of the resin and displace the previously loaded sulfates down the column. The sulfates do likewise to the nitrates and chlorides, which in turn displace the bicarbonates, which have the lowest relative affinity. The bicarbonates appear first in the effluent, followed by chlorides and nitrates and then the sulfates. At these low concentrations, the divalent ions are much more strongly held. The presence of high levels of chlorides or bicarbonates in the inlet water will have only a negligible impact on the capacity of the resin for selenium. Selenium leakage will remain low throughout the service cycle until sulfate breaks through. The sulfates will displace a portion of the selenium from the bottom of the bed, left over from the previous regeneration, and selenium levels will rise soon after the sulfate levels increase.

Radioactive Substances (Terrorist-Related)

Ion exchange solutions to terrorist acts will likely be looked at primarily for removal of radioactive substances. Although some poisons, such as cyanide, can be removed by ion exchange, they are usually better dealt with by already in-place treatment methods such as pH adjustment and chlorination. Radioactive substances could come in small quantities from using "dirty bombs" (conventional explosives with radioactive substances attached) as the delivery/dispersant system. The short-term radioactive dose and exposure may be very small, even within acceptable limits at the municipal level. Longer-term solutions will probably involve ion exchange. It's believed that radioactive hydrogen, cesium, and iodine might be involved, because they are commercially available for industrial/medical uses. Other possible candidates are those present in waste streams and spent fuel rods of nuclear generating systems, such as tritium; cobalt; cesium; strontium; plutonium; radioactive isotopes of iron, zinc sodium, manganese, and zinc plus others; and strontium, rubidium, and iodine.

Ion exchange can easily remove these substances. The same strongly basic resins that are used for nitrate and perchlorate removal are also highly selective and useful for iodine and uranium removal. The biggest problem is the disposal of the resins or the treatment of the waste regenerant that could be 100,000 times more concentrated than the untreated influent. Specific regeneration schemes, based on existing technology and resin selections will vary depending on the ionic composition of the water. Likewise, so will the process scheme and resin selection. Single use/replacement/disposal versus regeneration with or without postprocessing and reclaim are just some of the scenarios to be considered. Most of, if not all, these can be accomplished with existing technology, but lead times for delivery are typically measured in months and, in the aftermath of a terror incident, could become years. Preplanning using current treatment processes and water supplies should be looked on as insurance. Information on removal of several radioactive substances is listed below.

Radioactive Cobalt. Cobalt is not soluble in potable water. If cobalt is present, it will be in solid form and best removed by filtration.

Radioactive Cesium. This can be removed with the same resins used for ammonia. In the case of waters containing ammonia or where chloramination is practiced, the water should be treated prior to adding ammonia because ammonia is a major competitor (as also potassium) for the cesium selective resins (see ammonia).

Radioactive Iodine. Radioactive iodine has a very high affinity for strong base resins and can be removed easily and efficiently. In nonradioactive scenarios, the strongly basic resins would be regenerated. In a radioactive scheme, a nonregenerable approach would be taken using the same kind of selective resin used for nitrate, perchlorate, and uranium. This could be the better economic choice depending on the water analysis, disposal, and resin purchase price costs. Fortunately, in almost all cases, these resins can be interchanged in the same equipment that, except for disposal-related processes, is a standard design.

Radioactive Uranium. The concentration in water would likely be so small and the throughput capacity of the ion exchange resin so high that a nonregenerable system is the most likely treatment process, using either ordinary strongly basic nitrate selective resins or recently developed specialty resins with even higher preference for monovalent ions like nitrate, perchlorate, and uranium carbonate. Throughput capacities would be similar or identical to those of "ordinary radioactive" uranium removal for ordinary resins and potentially up to several thousand times higher with the latest developments. However, the radiation level in the resin could become a limiting factor so that ordinary resin, in some cases, may well have the same capacity as the superselective ones when operated under limited-dose scenarios. The biggest difference lies in the added cost of disposal of the spent resin or regenerants as hazardous/radioactive wastes. For radioactive resins, disposal costs are approximately 3 times their purchase price (as of March 2003) for nitrate selective anion exchangers. Uranium from nuclear wastes, such as reactors, is several orders of magnitude more radioactive than the uranium loaded into the reactor. The normal isotope mix of enriched uranium is sufficiently low that the workers who loaded most of the original uranium into nuclear reactors wore no protective garments except gloves. That was to protect the fuel rods from human oil deposits that could later char and become a corrosion site. The same fuel rods, when spent, could easily kill the worker who tries unloading it the same way due to the different isotopes formed during the nuclear service cycle. Uranium and plutonium are not normally present in wastewater from nuclear plants, but would be a problem if spent fuel was stolen and used to make a dirty bomb.

Uranium

At pH values above 6, uranium exists in potable water primarily as an anionic carbonate complex that has a tremendous affinity for strongly basic anion exchange resins. Strongly basic anion exchange resins can be used to remove uranium. The process has been tested and found to be very effective at pH of 6 to 8.2. Higher pH values could result in uranium precipitation, which makes the problem one of physical removal. Lower pH values change the nature of uranium to a nonionic and/or cationic species. Tests have shown effective removal (over 95%) of uranium at pH as low as 5.6. But after the pH was reduced to 4.3, the removal rate dropped to 50% and the run lengths (throughput capacities) were reduced by over 90%! It has been shown that sudden changes in pH of the influent water to values below 5.6 results in dumping of previously removed uranium. Therefore, it is important to keep the inlet water pH above 6 at all times. In situations where the pH cannot be maintained above 5.6, other treatment methods should be considered.

Throughput Capacity. The uranium carbonate complex has a relative affinity for strongly basic anion exchange resin that is over 100 times greater than any common ions, including the divalent ions such as carbonates and sulfates. At the pH levels associated with potable water applications (6.0 to 9.0), the carbonate ion is negligible as it exists primarily as the bicarbonate species, which is monovalent. Therefore, the sulfate ion is the only potential competitor. The throughput capacity of a strong base anion resin for ura-

TABLE 12.3 Difference in Uranium Removal with Different Sodium Chloride Concentrations

NaCl concentrations, %	Uranium removed, %
4	47
5.5	54
11	75
16	86
20	91

Regenerant level approximately 22 lb/ft^3 of type I gel anion resin.

nium removal can conservatively be estimated as being 100 times the throughput to sulfate breakthrough. To maintain this capacity over many cycles, it is necessary to use sufficient salt and at sufficiently high concentrations to regenerate the resin back to the chloride form.

Regeneration. To regenerate the uranyl carbonate ion from the resin, it is important that the concentration of the regenerant at the resin bed be sufficiently high to reverse or reduce its relative affinity (compared to chloride) to workable levels, and to use enough regenerant and contact time to complete the process. Sodium chloride is the most common regenerant. Table 12.3 shows the difference in effectiveness of various concentrations of salt.

At concentration above 20% a regenerant level of 15 lb/ft^3 is sufficient to ensure better than 90% uranium removal. Leakage will remain low through the service cycle even without complete regeneration because of the very high selectivity at the low ionic concentrations during the service cycle. Leakages are normally well below 1% for regeneration levels of 15 lb/ft^3 of sodium chloride applied at concentrations of 10% or higher.

Other Regenerants. The chloride ion at neutral or low pH is the most commercially effective ion for the regeneration of uranium from anion resins. Neutral salts are usually preferred because of environmental concerns and materials of construction considerations. Regeneration with pure hydrochloric acid, though not recommended because of the nature of hydrochloric acid and the added expense for corrosion-resistant equipment, shows an even better efficiency than sodium chloride because of its low pH. Higher pH, such as through the addition or use of alkalis such as sodium hydroxide or sodium bicarbonate or sodium carbonate, would result in severely decreased uranium regeneration. At high pH values, uranium tends to form $Na_2U_2O_7$, a nonionized precipitate that stays in the resin and interferes in subsequent service cycles.

TOC, Color, and Trihalomethane (THM) Precursors

These are usually caused by naturally occurring organics, which are the result of the natural decay of vegetable matter, primarily leaves, grasses, roots, and fallen branches. The resulting soluble or colloidal dispersion of semisoluble matter is a mixture of fulvic and humic acids, tannins, and lignins. These are primarily aromatic hydrocarbons with carboxylic groups, which therefore behave as anions. Carboxylic acids are moderately to weakly ionized. Although they are exchanged as anions, they are kinetically slower and bulkier than common inorganic ions. Therefore the ion exchange processing equipment and regenerant delivery systems have to be designed for lower flow rates and longer regenerant contact times than systems designed for exchanging inorganic ions.

The molecular weight distribution of naturally occurring organics varies widely, usually between 200 and 80,000 or more, depending on the source, age, and environmental conditions. Typically over 80% of the naturally occurring substances that cause TOC have molecular weights under 10,000. They are thus larger than inorganic ions but still small enough to enter the gel phase of the ion exchange resins. The TOC values are typically expressed as ppm as carbon. This can be converted to approximate ion exchange concentrations as calcium carbonate equivalents by multiplying the TOC values, ppm as carbon, by 1.5 to get approximate ppm as calcium carbonate.

The lower cross-linked, more highly porous "standard" gel resins designed for inorganic exchange are somewhat better then the standard 8% DVB type but may still lack enough gel-phase porosity to be effective. Special grades of resins, with extraordinary high gel-phase porosities, are far more successful in this kind of service.

Macroporous versus Gel Resins. Macroporous resins were the first types widely used for organics because initially it was believed that surface adsorption played a significant role in the removal process. It has been shown that the best performing macroporous resins are those that have the highest level of gel-phase porosity. It has also been shown that gel-phase porosity is the most significant predictor of performance. Since over 95% of the naturally occurring TOC matter is removed only by ion exchange and not adsorption, macroporous structures cannot offer advantages unless combined with low cross-linked gel phase. Activated carbon is effective in removing much of the 5% of naturally occurring TOC substances that are nonionic. For best removal, activated carbon should be used as a posttreatment, after the water passes through the resin.

The loading capacity and degree of breakthrough during the service cycle are affected by competing ion ratios especially sulfates to TOC. Once sulfates begin to leak about 50% of the influent, TOC will appear in the effluent. Therefore the service cycle should be terminated when sulfate leakage increases if complete TOC removal is required. If the service cycle is run past the sulfate break, the organics with the highest affinities will continue to load by displacing those with lower affinities and also sulfates. They will accumulate on the resin and make it more difficult to regenerate the resin, which could lead to premature fouling.

The naturally occurring TOC substances have affinities for the resins that are quite similar to sulfates. Theoretically, they can be regenerated from a strong base resin in the same type of process, as used for sulfates, i.e., sodium chloride. However, the regeneration process is limited by kinetics. Salt regeneration can be made more effective by using a 10% concentration of warm brine and adding 1% to 2% sodium hydroxide to it. A dose level of 10 lb/ft^3 applied over a 2-h contact time at 70° F is the minimum recommended. Shorter contact times will produce proportionately lower organic removal; longer contact times will add only about 3% per hour to the 2-h organic removal amount. The TOC operating capacity is calculated based on sulfates and TOC as the exchanging ions.

DEMINERALIZATION

Demineralization, which is also called deionization, is the name given to the removal of ionic substances. It is the result of exchanging all the positively charged ions for hydro-

gen ions and all the negatively charged ions for hydroxide ions. The hydrogen and hydroxides then react to form water molecules:
Cation exchange for hydrogen ions:

$$M^+ + R - H \rightarrow R - M + H^+$$

Anion exchange for hydroxide ions:

$$A^- + R - OH \rightarrow R - A + OH^-$$

Hydroxides and hydrogen ions react to form water:

$$H^+ + OH^- \rightarrow H_2O$$

Separate Beds

In acid exchanges, cation resins are used in the hydrogen form to exchange hydrogen ions for all other ions. Strongly acidic cation resins are normally used, although weakly acid cation resins are sometimes used in conjunction with the strong acid resins to increase regeneration efficiency in larger systems where the increased capital cost can be justified. The cation bed is normally used as the first step in the deionization process. Otherwise, the divalent cations would precipitate as hydroxides in the anion vessel. Some systems used in pharmaceutical applications do practice what is called *reverse deionization*, usually including a softener that precedes a hydroxide form anion exchanger followed by a hydrogen form cation exchanger.

Generally speaking, demineralization always requires cation and anion exchange. Several varieties of each kind of resin can be selected depending on the specific water analysis, quality requirements, and operating conditions. The cation and anion resins can be in separate vessels (separate beds or two beds) or in a single vessel (mixed beds). In the latter, the reaction products disappear by forming water molecules, which drives the exchange reactions and produces a greater level of purity.

In deionization systems, anion resin is used in the hydroxide form. Most demineralizers employ strongly basic anion resins, which remove all ions, including silica and carbon dioxide. There are two kinds, type I and type II. The type II resins are used more frequently for drinking water applications because they are functionalized with an amine that does not impart as much of a fishy odor to the water. Type II anion resins tend to have higher operating capacity and are more efficiently regenerated, but are less thermally stable and do not have as high a life expectancy as type I resins. Type I gelular anion resins are most widely used in large demineralizers. The more highly porous, 4% DVB gel type I anion resins are more resistant to organic fouling than are the 8% DVB type I resins and are more efficiently regenerated. Their operating capacities and fouling resistance are not as good as those of type II. Type II's better fouling resistance is due at least in part to its failure to load organic ions in the first place.

In some systems weakly basic anion resins are used. This type of resin is only able to neutralize acidity from mineral acid salts such as sulfates and chlorides that are created in the cation exchange vessel. Weakly basic exchangers in general do not remove weakly ionized substances such as carbon dioxide and silica, which is okay for some applications. They are very efficiently regenerated and are often used in conjunction with strong base resins in large-scale plants where the extra capital cost is more than offset by reduced chemical costs.

Various combinations of cation and anion resins are used in the process of demineralization. They can be as simple as separate vessels of cation and anion resins in a two-bed system, to more complex systems consisting of several different vessels with various combinations of weak acid, strong acid, weakly basic, and strongly basic resins.

When the bicarbonate alkalinity is more than 50 ppm, it is sometimes, depending on the flow rate, more practical to employ a forced-draft decarbonator to remove carbon dioxide in the effluent of the cation exchange vessel by blowing air through the water. This reduces the load on the anion exchange vessel, which follows the decarbonator.

Mixed beds of cation and anion resins give the effect of multiple stages of two-bed systems. Most mixed resin systems consist of strongly acidic and strongly basic cation and anion resins. Mixed beds are most widely used for polishing water that has previously been deionized either by a membrane process or by a two-bed demineralizer. The main advantage of mixed bed demineralization is the improved water quality.

BASICS OF EQUIPMENT DESIGN

This section covers the basics of how ion exchange systems are sized, the requirements for ancillary systems, and the selection of materials of construction. These vary depending on the type of equipment and the equipment suppliers' preferences. Therefore, the subject is presented as a brief introduction rather than as a design manual.

Flow Rates

Most ion exchange systems are limited by flow rate for physical reasons such as pressure loss or the physical characteristics of the resin rather than by any particular chemical requirements. With respect to linear flow rates, if the linear flow rate exceeds about 20 gpm/ft^2, the pressure exerted on the resin bed, together with other stresses such as osmotic and thermal, is sufficiently large to begin to break some of the resin beads. The strength of the individual beads is actually much higher, but when pressure loss across the bed routinely reaches about 50 psi, excessive bead breakage can be expected. In general, the maximum allowable pressure loss across the cation beds should always be less than 20 psi and across the anion beds less than 20 psi. It is common practice to allow for a 5- to 10-psi increase in pressure drop across the resin bed over time. Therefore, the design value for the maximum flow rate across a resin bed is usually in the neighborhood of 15 gpm/ft^2.

On the low end, the minimum flow rate is that which the liquid distributors can accommodate. Typically this is the flow rate during the regeneration cycle. The maximum practical turndown is about 5:1. A liquid distributor designed for a maximum flow rate of 5 gpm may not give good distribution at flow rates below 1 gpm. Some distributors can do somewhat better than this with 10:1 being about the maximum, while other distributors, such as those used in some of the less expensive commercial type tanks, do not do this well. Rinse water flow rates of less than $\frac{1}{4}$ gpm/ft^3 should generally be avoided due to the potential difficulty in rinsing the regenerant out of the resin bed and the consequent prolonged leakage of regenerant into the product water during service.

Regeneration Frequency and Media Depth

Resin volume requirements are generally selected based on the user preference for cycle times between regenerations within the limits defined by flow rate and bed depth requirements. Regeneration frequencies are usually kept at less than 2 or 3 times per 24 h.

The service cycle times can normally vary from a minimum of approximately 8 h to several days, depending on the process, the volume of resin used, and the mix and total concentration of ions in the feed water. Systems that cycle more frequently are generally less reliable. On the other hand, systems that regenerate infrequently can encounter difficulties due to ions migrating deeply onto the resin beads, foulants that harden into the resin over time, and bacterial growth that occurs in stagnant resin beds.

For the purpose of roughing out a system, a reasonable starting point is to use a flow rate of 10 gpm/ft^3 and a resin bed depth of 3 ft. The final resin volume should be fairly close to the first estimate. Most ion exchange processes have very fast kinetics, and the space velocity (or gallons per minute per cubic foot) is not normally an important factor in determining the capacity or efficiency of the ion exchange bed. However, flow rates in excess of 5 gpm/ft^3 may lead to linear flow rates greater than 15 gpm/ft^3.

Resin bed depths of less than 24 in. are generally not recommended, primarily to avoid inefficient operation. Even though the theoretical height of the ion exchange zone is usually less than 6 in., the "actual" height of the exchange zone is distorted by imperfections in the distribution and collection systems within the vessel. A considerable fraction of the total resin bed may be lost to these imperfections. Also, between 2 and 4 in. of the total bed is usually lost simply because it lies beneath the bottom distributions. This is a significant fraction of the total resin volume when the bed depth is less than 24 in. Bed depths greater than 6 ft are generally avoided because of concerns of exceeding the previously mentioned pressure drop limitations of the resin.

REGENERATION METHODS

There are several different methods for regenerating resins. The method chosen will largely determine how efficiently the ion exchange resins will operate and how complex the regeneration will be. The most commonly employed regeneration method is called coflow or cocurrent regeneration. The resin is regenerated in the same direction as the service flow. This method is used in almost all salt cycle exchangers, i.e., softening, dealkalization, etc. The usual cycle consists of a backwash to purge the resin bed of suspended solids and resin fines and fragments, followed by chemical injection of a solution of the regenerant salt, acid, or base through the resin bed, followed by a rinse cycle to flush the regenerant from the resin bed. Since the regenerant flow is in the same direction as the service flow, the ions at the top of the bed have to be pushed downward all the way through the resin bed before they can be purged. This makes the coflow method somewhat inefficient. It also leaves a portion of the exchanged ions remaining in the resin at the bottom of the bed, where they can cause leakage in subsequent service cycles. The leakage is highest at the beginning, and as the service cycle progresses, the leakage becomes less. The effect is more noticeable at low regenerant dose levels. This leakage phenomenon is very slight in softeners but very noticeable in nitrate removal salt exchange units. The nitrate leakage at the beginning of the cycle is significantly higher than the leakage in mid-cycle. The overall leakage can vary by over 2 to 1 from beginning to end of the service cycle.

Various means of improving the efficiency of cocurrent regeneration can be employed. One of the least complicated is called *thoroughfare regeneration*. This is commonly used in systems that have two or more exchangers in series that are regenerated with the same regenerant chemical. The dilute regenerant first is passed through the last polishing column and then is passed through the preceding columns before it flows to waste. The regenerant chemical is more efficiently used and gives lower and less variable leakage in the final effluent. Another method of improving efficiency in coflow exchangers is to re-

FIGURE 12.6 Regeneration method versus leakage.

claim the last portion of the regenerant and reuse it as the first portion of regenerant during the next cycle. This is surprisingly difficult to do effectively, especially at low dose levels. If too much is reclaimed, then the ratio of the unwanted ion will be high in the reclaimed portion of the regenerant and will compete with the regenerant ions. When this happens, it can virtually eliminate the effectiveness of the reclaim. When done correctly, reclaiming regenerant can save up to one-third of the total regenerant dose.

Various methods of countercurrent regeneration have been devised to overcome the limitations of coflow regeneration. Countercurrent regeneration is primarily used in demineralization, and except for being occasionally used in nitrate removal, it is almost never used in salt cycle exchange. In all countercurrent regenerations, the dilute regenerant solution is introduced in a flow direction that is opposite to the service flow such that the least used portion of the resin (polishing zone) bed is contacted by the fresh regenerant first. This results in very low leakages even at the lowest regenerant dose levels, which provides higher regeneration efficiency. Countercurrent systems save on chemical operating costs, and waste discharge is reduced. These advantages are offset by higher capital costs and a more complex operating system. Maintenance costs are usually higher than for cocurrent regenerated systems. A comparison of leakages at equal regenerant levels is shown in Figure 12.6.

Cocurrent versus Countercurrent

The design of a demineralizer, at least in the United States, is often based on customer preference. It may also be based on the skill of the company building the equipment and its experience with various types of demineralizer designs. The method of regeneration chosen will most likely be made based on effluent water quality requirements and on the inlet and outlet water quality. The cleaner the raw water and the higher the effluent quality required, the more likely it is that countercurrent regeneration will be employed. The dirtier the raw water and the less stringent the effluent quality requirement, the more likely a coflow unit will be employed.

Countercurrently regenerated demineralizers do not do well when there are significant amounts of foulants in the feedwater. Suspended solids, iron, and organic content are the primary culprits. Suspended solids accumulate in the resin beads. Countercurrently regenerated units do not employ a full bed backwash as a normal portion of the regeneration procedure. There is no means of purging the suspended solids from the bed on a routine basis. In fact, if a means was provided of doing so, this would interfere with the countercurrent principle and would make a countercurrently regenerated unit operate less effectively with lower quality, somewhere in between a coflow unit and a countercurrent unit.

Iron and other types of foulants accumulate on resin beds and require periodic cleaning procedures. This is more of a disadvantage for countercurrently regenerated units as it is generally more difficult to clean the resin in a vessel that has limited freeboard and most countercurrently regenerated exchangers have limited freeboard. Any movement of the resin bed, which is unavailable during a cleaning procedure, will impair quality over several subsequent cycles.

Organics are less well purged from countercurrently regenerated units primarily because most of these units employ smaller chemical doses. Regeneration is the primary means of purging the organics from the resin. When the regeneration is conducted with a smaller chemical dose, it is less able to remove the organics. Therefore, at lower dose levels the resin tends to become organically fouled more rapidly. The effects of organic fouling are felt more strongly by countercurrently regenerated units because they tend to make better water quality to start with.

The very best water quality in terms of resistivity that can be produced by a countercurrently regenerated two-bed demineralizer is about 10 MΩ or 0.1 μmho. However, most countercurrent systems do not do this well. The typical quality produced by a properly designed and carefully operated countercurrent demineralizer is usually in the neighborhood of 1 to 2 MΩ or 0.5 to 1 μmho. Many of the countercurrently regenerated systems that are not perfectly designed operate in the quality range of 1 to 5 μmho or 200,000 kΩ to 1 MΩ. Countercurrently regenerated systems that are poorly designed or fouled may produce worse quality than a coflow exchanger that is properly operated.

Coflow exchangers do not typically produce as high a water quality as a countercurrently regenerated unit. The typical reduction in conductivity from inlet to effluent in a two-bed system is approximately 90%. The conductivity is caused by sodium hydroxide due to sodium leakage from the cation vessel and hydroxide not neutralized after being generated in the anion vessel. On a chemical equivalent basis, the sodium ion concentration is equal to the hydroxide ion concentration. Since sodium hydroxide has about twice the equivalent conductivity of neutral salts such as sodium or calcium chloride, the reduction in concentration across the two-bed unit is typically better than 95%, that is, 5% leakage or less.

Countercurrently regenerated units also produce lower silica leakage than do coflow units. The best silica leakage that can be produced is in the range of 5 to 10 ppb. It is more likely to see silica in the range of 20 to 50 ppb. Silica leakage depends on the effectiveness of the regeneration following the previous service cycle. Silica polymerizes onto the anion resin, and the regeneration involves depolymerization. This process is dependent on contact time and temperature. There is a potential problem in countercurrently regenerated units with respect to silica in that the caustic dose is frequently chosen to optimize chemical efficiency. In some cases there may not be sufficient excess basicity to depolymerize and dissolve the silica off of the anion resin. Therefore, when countercurrently regenerated units have a high percentage of silica in the influent, it is desirable to avoid anion resin loading with more than 5 kgr/ft^3 of silica. In cases where the silica load is 50% or greater of the total anion load, extraordinary regeneration measures may be required.

Degasification

Another factor in the selection of a demineralizer design is whether to use a degasifier. CO_2 is present as a gas. It is also created by the reaction of the bicarbonates with the hydrogen ions generated in the cation exchange vessel. CO_2 can be removed by a forced-draft degasifier or by vacuum deaeration in between the cation and anion exchangers. This reduces the ionic load on the anion vessel. A degasifier complicates the design for several reasons:

- It may make the anion exchanger much smaller in size than the cation exchanger, and this creates potential problems in flow rates and regeneration times. However, the smaller size means lower chemical operating costs.

- In most cases two or more trains of exchangers share a common degasifier. Simultaneous regeneration becomes more complicated as the degasifier has to stay in service while a particular train is regenerating.

- The employment of rinse recycle becomes far more complicated in cases with a degasifier. In fact, it is generally not practical to rinse-recycle the cation exchanger although in many cases, even with a degasifier, the anion regeneration can finish with rinse recycle.

- Although the reason to use a degasifier is to reduce operating costs, a certain amount of caustic may be required to neutralize the waste acid from the cation regeneration. This should be considered.

Cocurrent Exchangers

Cocurrent ion exchange is the oldest of all the designs and is the simplest. Although they are inherently less efficient and do not provide as good a water quality as other designs, they are a forgiving design and can be used in dirty water applications with high turbidity.

The cocurrent exchanger consists of a tank that contains a bed of ion exchange resin. At the top there is an upper distributor (a means of distributing the water over the surface of the resin bed), and at the bottom there is an underdrain collector (a means of removing the water from the resin where it exits the vessel). The raw water enters at the upper distributor and flows down through the resin bed (Figure 12.7). During regeneration the regenerant chemical also flows downward through the resin bed. In smaller units the regenerant chemical and feedwater are introduced through the same upper distributor. In larger units there is generally a separate regenerant distributor located just above the resin bed. The advantage of a regenerant distributor is that it saves water during regeneration.

Coflow exchangers have freeboard or empty space between the upper distributor and the resin bed, which allows for expansion of the resin bed during the backwash portion of the regeneration cycle. During backwash, the inlet distributor becomes the outlet collector.

The regeneration of a coflow ion exchanger consists of four steps:

1. *Backwash.* During backwash a flow of water is introduced through the underdrain and flows up through the resin bed at a rate sufficient to expand the resin bed by about 50%. The purpose is to relieve hydraulic compaction, move the finer resin material such as resin fragments to the top of the bed, and remove any suspended solids from the bed that have accumulated during the service cycle.

2. *Chemical (regenerant) injection.* A dilute solution of regenerant chemical flows down through the resin bed, stripping the ions off the resin that were collected during the service cycle and restoring the resin into what is called the regenerated form.

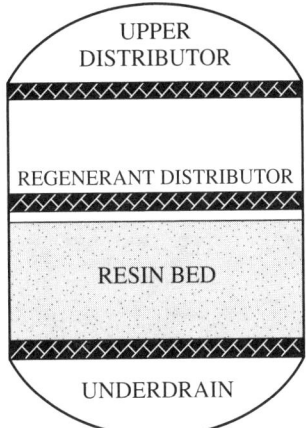

FIGURE 12.7 Coflow exchanger.

3. *Displacement rinse or slow rinse.* The displacement rinse is conducted at or close to the same flow rate as the injection step. Its purpose is to slowly push the regenerant chemical through the resin bed, allowing for the necessary contact time to complete the regeneration process. The slow rinse is an efficient way of removing the bulk of the regenerant solution from the exchange tank.
4. *Fast rinse.* This is the final rinse step and is usually conducted at or close to the service flow rate. Its purpose is to flush the last traces of regenerant chemical out of the exchanger and prepare the exchanger for return to service.

Split-Flow Exchangers

Split-flow (sometimes called cocounter) exchangers look similar to coflow exchangers except that the regenerant collector is located in the resin bed, usually about one-third of the way down (Figure 12.8). The regenerant collector is always screened and must be braced similar to a mixed-bed interface collector due to the hydraulic forces on the collector during service and regeneration cycles. Most of the regenerant comes in through the bottom and flows upward through the bottom portion of resin; but some of the regenerant is introduced through the top and flows downward through the top portion of the resin. Both portions of regenerant chemical exit through the regenerant collector.

The regeneration of a split-flow exchanger generally consists of the following steps:

1. *Subsurface backwash.* During this step, water is brought in through the regenerant collector and flows upward through the top portion of the resin bed. Since ion exchange media are good filter media, most of the dirt that might be trapped in the resin bed will be in the top portion. However, as in any countercurrently regenerated design, it is not a good idea to backwash the entire resin bed. Doing so would counteract the countercurrent advantages. The subsurface backwash relieves compaction of the upper resin bed and reduces the stresses against the regenerant collector. Subsurface backwash is conducted at the same flow rate for a coflow exchanger of the same diameter. However, the backwash time is generally limited to just a few minutes, as the main purpose is fluffing the bed and relieving hydraulic compaction.

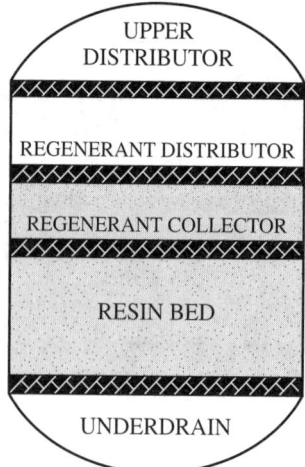

FIGURE 12.8 Split-flow exchanger.

2. *Chemical (regenerant) injection.* In this step the diluted regenerant is split into two portions. One is fed into the top of the bed and flows downward, and the other is fed into the underdrain and flows upward.

When sulfuric acid is used as the regenerant, there are advantages to using a two-step acid injection as a means of minimizing calcium sulfate precipitation. The regenerant flow rates and times are the same as in other designs using sulfuric acid; however, the upward flow should not exceed 1 gpm/ft^2 as any flow in excess of this might fluidize the bottom portion of the resin bed.

3. *Displacement rinse or slow rinse.* This step is usually carried out by continuing the dilution water from the chemical injection step, but is sometimes performed at a higher rate to minimize the regeneration time. The necessary volume of displacement water depends on whether there will be a recycle rinse. If, in the final rinse step, the bed will be rinsed to waste, then approximately two bed volumes of displacement water are employed (15 gal/ft^3 of resin). If the final rinse will be performed in a recycle manner, then it is prudent to plan to use at least three bed volumes and maybe as many as six bed volumes.

4. *Final rinse.* This can be conducted in a straight through to waste manner or by recycling through the accompanying exchanger (cation to anion or anion to cation). Rinse recycle is a water-saving feature and is commonly employed in countercurrently regenerated two-bed demineralizers. The rinse volume required depends on the required quality to begin the service cycle. The cation requires less water than the anion resins. Cation rinse volumes are in the neighborhood of 6 to 10 bed volumes. The anion rinse volumes are in the neighborhood of 6 to 15 bed volumes. The larger volumes are required for the lower conductivity endpoints or for systems where the resin has deteriorated and the anion resin might be organically fouled. The flow rate most commonly used is the service cycle flow rate. However, it can be any convenient rate not less than approximately 2 gpm/ft^2 or 0.5 gpm/ft^2 and generally is not higher than the service flow rate.

Countercurrent Regeneration (CCR) Exchanges

CCR units with downflow service and upflow regeneration are the most common form of countercurrently regenerated ion exchange systems currently manufactured (Figure 12.9). Water block CCR regeneration is very similar to the split-flow regeneration except that instead of splitting the regenerant flow top and bottom, the regenerant all goes through the bottom. The service flow is downward, and the regenerant flow is upward. Because the upflow regenerant tends to expand and fluidize the resin bed, defeating the purpose of countercurrent regeneration, there must be some means employed to keep the bed packed during regeneration. The simplest method is to bring water in from the top of the bed. Air is used instead of water as the blocking flow in some systems.

Water block and air block exchangers both require a regenerant collector located just below the surface of the resin bed. It has been found necessary to bury the regenerant collector in the resin bed to a depth of approximately 6 in. to prevent fluidization of the resin during the upward regeneration. This 6-in. (roughly) layer of resin is considered inactive and is not included in the ion exchange capacity or throughput calculations.

The required flow rate of the blocking water must be sufficient to prevent excessive movement of the resin (to prevent fluidization) and must be compatible with the chemical solution employed during regeneration. This means that blocking water used for anion regeneration must be hardness-free.

Another method of preventing bed fluidization is called the *air block system*. The water above the resin bed is drained to the regenerant collector, and pressurized air is brought in through the top of the tank. An air dome prevents water buildup above the regenerant collector and prevents the resin bed from expanding and fluidizing. The air block method of countercurrent regeneration is tricky and more complicated than the water block method. The regeneration procedure is complicated by the necessity to drain down and then refill the bed before and after chemical injection and displacement. Another complication of air block units is that a portion of the regenerant tends to flow up above the regenerant collector. Some fluidization of the resin inevitably occurs. The air dome cannot keep the water level or chemical level from rising above the bed. To the extent that the regenerant solution rises above the collector, the regenerant does not displace out of the bed.

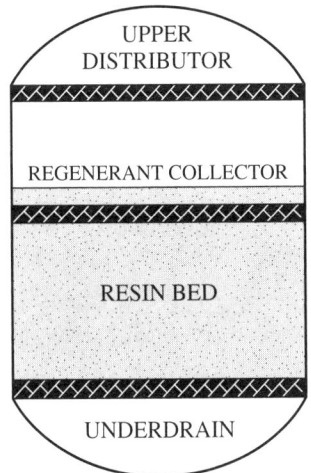

FIGURE 12.9 CCR exchanger.

1. *Subsurface backwash.* Subsurface backwash is performed at the normal backwash rate for a reduced period of time compared to other designs, by bringing water in through the regenerant collector and out the top of the tank.
2. *Drain to bed level.* Normally air is brought in through the vent and is used to pressure-drain the water to the level of the regenerant collector.
3. *Chemical (regenerant) injection.* Chemical injection is employed at the same flow rate, concentration, and time as in other designs. The limitation of upward regenerant flow rate is about the same as for other upwardly regenerated units (approximately 2 to 3 gpm/ft^2 maximum).
4. *Displacement or slow rinse.* Displacement is normally employed at the dilution water rate, although it is sometimes advantageous to displace at a higher rate. The requirement is at least three bed volumes.
5. *Refill.* Refill is generally performed from the top through the rinse water valve.
6. *Optional rinse to waste.* Rinse to waste depends on the water quality requirements. The rinse to waste is approximately three to five bed volumes.
7. *Rinse recycle.* Rinse recycle is 6 to 10 bed volumes for the cation and 10 to 15 bed volumes for the anion, normally performed at the service rate. The volume depends on the endpoint desired.

Packed Bed Design

Another factor in the choice of demineralizer is the size required. The packed bed design offers the very smallest possible size. The coflow design offers the widest potential variety of sizes and does not require that the size be fitted exactly to the resin volume. One of the potential problems with all countercurrently regenerated units is that the resin volume is fixed by the vessel size and must be very close to an exact volume (Figure 12.10). Resins expand and contract during service and exhaustion. Resins are shipped by weight

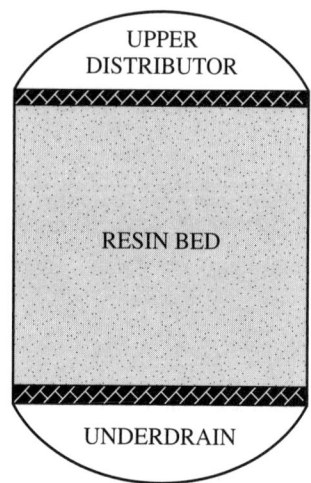

FIGURE 12.10 Packed bed exchanger.

rather than volume even though it is common practice to label and purchase resins by volume. It is therefore necessary to fit the resin volume to the tank. If an inexact amount of resin is blindly put into a tank that is going to employ countercurrent regeneration, the chance of failure is pretty high. A packed bed demineralizer consists of a tank containing an upper distributor and an underdrain. The resin bed fills the entire tank except for the amount of freeboard required to allow for the resin swelling and contraction of the resin bed between the regenerated and exhausted forms. There is no separate regenerant collector. The regenerant flow is performed countercurrently to the service flow such that upflow service units have downflow regeneration. The direction of the service flow rate may be upward or downward, but most of the packed bed designs built to date have upflow service and downflow regeneration.

External Regeneration

External regeneration is performed in a vessel other than the service vessel. This type of regeneration is widely practiced by the service exchange companies and is the preferred method in condensate polishing demineralizers in the electric power industry. The resin becomes fully mixed as it is transferred from the service vessel, so there is no advantage to CCR. Externally regenerated units are always regenerated in a downflow fashion. Resins used in mixed beds must be separated prior to regeneration so that the acid regenerant can contact the cation portion of the resin and the sodium hydroxide regenerant can contact the anion portion of the resin. If the wrong chemical contacts the wrong resin, it can cause fractured resin beads due to osmotic shock and precipitation of various salts, and contamination of the resin bed.

Multiple-Step Regeneration

These types of regenerations are used in specialty applications, for example, where the resin is neutralized after the acid or base regeneration. The salt forms of weakly ionized resins have better kinetic properties and are often used for special purposes. The sodium form weak acid cation resin is very selective for divalent and trivalent metals such as lead, copper, and chromium. Weakly basic resins are used in the salt form for specialized exchanges such as chromate removal. It is extremely difficult to convert weakly ionized resins directly to the salt form, while an acid or base step followed by neutralization is relatively easy.

Regeneration

When the resin's capacity is exhausted and the leakage of undesirable ions rises in the effluent to an unacceptable point, the resin is regenerated. The regeneration process reverses the exchange reaction under controlled conditions, thus leaving the resin in the desired ionic form.

Salt Regeneration. High ionic concentrations are more effective regenerants and are more efficient. On the other hand, sudden changes in ionic concentration cause ion exchange resins to rapidly change size. This sudden volume change osmotically shocks the resin and can cause bead breakage. So it is necessary to control the rate of change and total concentrations during regeneration. For salt regenerated exchangers, the best range of salt concentration is between 8% and 15%, with 10% being the most commonly employed concentration.

Acid and Caustic Cycles. For acid base exchanges, the concentrations used are generally somewhat lower than for salt regenerations. For acids the typical concentration range is 2% to 10% for strong acids, but concentrations of 0.5% to 1.0% are used with sulfuric acid regeneration of weak acid resins in high-calcium waters. Hydrochloric acid is usually injected at concentrations of 5% to 10%. For caustics the typical concentration range is 4% to 6%.

Sulfuric acid concentration must be controlled when calcium is present. Higher concentrations favor more complete regeneration, but calcium sulfate will precipitate if the concentration is too high. Multiple steps with varying concentrations are sometimes used with sulfuric acid to reduce waste volumes and maintain regenerant efficiency.

While contact times as short as 10 to 15 min have been employed successfully, in most cases a contact time of 30 to 40 min is used for cation resins and weakly basic anion resins. Strongly basic resins may require 60-min injection time and heated regenerant where silica removal is critical. Even longer contact times are necessary for organic traps, and systems experiencing organic fouling.

In many units following the introduction of the regenerant chemical, there is a slow rinse or displacement rinse that is used to help push the regenerant chemical through the resin and out at approximately the same flow rate as that used during the chemical introduction. This slow rinse displaces the regenerant, ensures adequate regenerant contact time, and decreases the overall rinse requirement. The final rinse is at the service cycle rate and is used to purge the last traces of chemical from the ion exchanger and prepare the system for the next service cycle.

Rinse volume requirements vary considerably depending on the type of resin used, the adequacy of the internal design, the age of the resin, and the presence of foulants. Most systems require 7 to 10 bed volumes of final or fast rinse. Rinse requirements can vary from a minimum of about 2 bed volumes to over 20 bed volumes depending on the type of system and the endpoint requirements.

Rinse recycle instead of the final fast rinse is sometimes employed as a method of reducing wastewater volume in demineralizers. When the effluent purity during the slow rinse reaches about the same level as in the raw water, the rinse is recycled to the front end of the system rather than discharged to waste. Since the effluent quality is changing rapidly, there is very little load on the resins. Rinse recycle can usually be employed when the regenerant concentration has dropped to less than about 0.1% in the slow rinse effluent. Rinse recycle cannot be employed with salt regenerated units as there is no mechanism for the removal of the salt left in the rinse water; however, final rinse waters can be saved for use in the next backwash or regenerant dilution. This is usually not practical except on large systems.

Vessel Design

The ion exchange vessel must contain the ion exchange resin beads while allowing the liquid to flow through them in such a fashion that the resin bed remains packed. The design must ensure that the various chemical solutions used during regeneration flow through the media properly. Most ion exchange systems employ pressurized tanks, although there are a few gravity flow ion exchange systems in use. Gravity flow type of systems are generally applicable only to very large systems and systems that have sufficient surface area to allow the necessary flow rates without large pressure drops.

There are two basic choices of tank materials for pressure vessels: fiberglass and steel. Fiberglass tanks are generally less expensive, but due to the way the fiberglass tank is made, there are fewer options for accessories and connections that can be made part of the system design. Fiberglass tanks are lightweight and have an advantage over steel tanks in that they are inherently noncorrosive. However, they are not available in very large sizes and are generally limited to tank diameters of 8 ft and less. They are not generally used for pressures above 150 psig or temperatures above 100° F (under pressure). Steel tanks can be made in almost any configuration. However, steel tanks are heavier than fiberglass tanks. They are almost always lined and, therefore, more expensive. Where high temperatures are employed, stainless steel is sometimes preferred over carbon steel. The cost of building a stainless steel vessel makes it prohibitively expensive for all but a few specialty applications.

Distributor Design

As a bare minimum, the tank should be equipped with an inlet distributor, to get the water into the tank, and an underdrain collector, to get the water out of the tank. It is advisable to have if there are separate connections for the introduction of the resin into the tank and the removal of resin from the tank. For tanks that are larger than 3 ft in diameter, it is necessary to have an access manway, so that a worker can climb inside the tank to install or repair the internal distributors and collectors. Manways are best located at the top of the tank. Larger manways are preferable to small manways. If the manway is located on the side of the tank, it can create distribution problems. View ports are a feature used in more sophisticated ion exchange systems. They allow the operator to visually inspect resin to make sure that resin has not been lost due to excessive backwash or problems with the underdrain collector, and to watch the resin during the regeneration and backwash cycles to ensure that the resin bed remains packed and does not fluidize when it is not supposed to.

To ensure the proper flow of liquid through the resin bed, the water must enter the tank in such a fashion as to provide minimum turbulence and laminar flow into the resin bed. Any protuberance will create a disturbance and an area of maldistribution. The resin bed should present a flat surface that the water can flow straight down through and into the underdrain collector. The path length should be equal for all portions of the bed. Most pressure vessels have dished heads. The dish creates a disturbance that distorts the flow and reduces the efficiency of the ion exchange resin. Large tanks benefit from the addition of either flat, false bottoms or subfills such as concrete or sometimes gravel that fill the bottom head and provide a flat surface for the resin bed to lay upon. By far the most important factor in the efficiency and quality produced by an ion exchange system is the design of the internal distributors and collectors. The inlet distribution system must allow the water and chemical to enter the vessel in such a way as to spread out equally throughout the entire tank diameter with equal velocity. The velocity must be sufficiently low as to not disturb the resin bed below. Distributor design is much more crucial in tanks with limited freeboard. Underdrain collectors must be located as close to the bottom of the vessel as possible with the least amount of resin underneath the collector as possible, and should have collection spacing of about 6 in. Hub and lateral collectors have a problem in the larger tank sizes for two reasons: (1) The radials are far apart at the edges of the tank. (2) The relative surface area close to the edge of the tank is large; however, the number of collection points in a radial is small. Therefore, it is very difficult in large tanks to get even distribution with the hub and lateral design. Another problem is that the hub itself causes an area of poor distribution right in the center of the tank.

If the underdrain is not located within 1 to 2 in. of the bottom of the tank, the rinse requirement will increase significantly due to the difficulty of removing the regenerant chemical that lays out and hides below the collector. The underdrain must also have sufficient open area to collect the water without becoming clogged with resin. The slotted pipe type of collectors sometimes used in the less expensive and smaller systems have the smallest amount of open area and are highly prone to plugging. One common system, which works well, consists of a lateral pipe with drilled orifices pointed down toward the bottom of the vessel with a layer of coarse plastic mesh protector wrapped over the lateral and a layer of plastic or stainless screen of the desired size covering the mesh protector. Another common system is similar to the first except that the drilled orifices are covered with a continuous wedge wire sleeve. These two lateral systems are among the most widely used and are considered by some to be superior to strainer designs. Strainers are reasonably good collectors, but they have too much open area to be inherently good distributors. They do not create enough pressure drop to create equal distribution at the low flows used during regeneration. A second problem is that to obtain reasonable open area in the strainer itself, it is generally higher than $1\frac{1}{2}$ to 2 in. high. The extra height creates an area of poor distribution toward the bottom of the strainer. Various manufacturers have tried to overcome this problem by the addition of check valves within the strainers, putting shrouds over the strainers, or using a large number of very small strainers, with some success.

Large systems with large amounts of freeboard generally benefit from a separate chemical distributor. There are two advantages. The first one is that the chemical distributor can be designed for good distribution at the regenerant flow rate. The second advantage is that by locating the chemical distributor close to the resin bed, the volume of liquid in the freeboard area does not have to be displaced during the regeneration and rinse steps. This reduces the rinse requirement. Some systems, such as mixed beds and certain types of countercurrently regenerated exchangers, have liquid collector distributors that are buried within the resin bed. These distributor/collectors generally are designed in a similar fashion to the underdrain. There is significant hydraulic force against these distributors during the service cycle and potentially during the regeneration cycle, and if they are not very firmly braced, they are prone to breakage.

Piping Design

Smaller systems, and in particular the single home type softeners, generally have multiport valving systems. Multiport valves are low in cost and are fairly reliable up to about 2 in. in size. They are widely available. Their main disadvantage is that they have a fixed cycle. Although several options are available in some designs, they are generally considered not suitable for most countercurrently regenerated systems. Larger systems almost always employ a valve nest and a piping tree to deliver and collect the various flows.

Many piping system materials are available; however, PVC is most commonly used for smaller systems, and either plastic-lined steel or stainless steel is most commonly used for larger systems. The most important feature of the valves used is that they be capable of tight shutoff over many cycles. The diaphragm-type valve has proved itself to be a very good choice for this purpose, albeit it has higher pressure drop and costs more than other types of valves. Two other valve-type choices are commonly used. Ball valves and, in larger systems, butterfly valves are frequently used.

Chemical Storage and Dilution

Some small systems simply employ open-top tanks to mix and store the chemical solution at the required concentration prior to regeneration. Larger systems generally use some type of bulk chemical storage and dilution stations. In these systems, the concentrated chemical is pumped usually to a mixing T and blended with dilution water prior to entering the exchange tank. The smaller systems with dilution tanks generally require that an operator manually fill the tank with dilution water and concentrated regenerant. This can expose the operator to dangerous chemicals. The larger, in-line dilution stations do not routinely expose the worker to the chemical. However, the chemicals are pumped under pressure, and there is the potential danger to workers should the chemical line rupture. Salt exchanges are not considered hazardous, but acid and base exchanges are.

The use of bulk chemical storage generally requires a containment structure in the area surrounding the dilution station, where the pressurized chemicals are used, or some type of enclosure to prevent a worker from being accidentally sprayed, should a leak or rupture in the pipe line occur.

For small systems the eductor type of pump is ideal because it can be used for pumping and diluting the regenerant in one step. Larger systems generally employ positive displacement pumps such as diaphragm pumps and gear pumps, although sometimes centrifugal pumps are used. All the various types of chemical pumps have advantages and drawbacks. Metering pumps have problems with pulsation and are not particularly reliable; gear pumps are far better in this respect, but are not free of maintenance problems. During regeneration the resin bed changes size, and the pressure drop across the resin bed will vary. It is harder to keep a constant flow rate with a centrifugal pump than with the other types of pumps used. The chemical dilution station is one of the highest maintenance areas of an ion exchange system.

Waste Collection and Disposal

In the past it was not uncommon for the wastewater leaving an ion exchange system to go to the sewer with no treatment or monitoring. On the whole, this practice killed innumerable fish and helped to significantly pollute waterways. Today we are still paying for these tragedies. The unwanted ions that are regenerated off the resin are present in the wastewater in concentrated form. In addition, the wastewater contains either concentrated salt solution or the concentrated acid and base solutions from the regenerant chemicals used. The average ionic strength of the wastewater often exceeds 10,000 ppm, and it can be as high as 60,000 ppm or more. The concentration factor is usually in the range of 10 to over 100. Most ion exchange systems today employ a wastewater tank where the regenerant waste flow is placed prior to treatment and disposal. As a minimum, the wastewater is neutralized and monitored to ensure that it is neutral prior to being discharged. In some areas of the country, and for some types of wastewater, it is necessary to impound the spent regenerant portion of the waste and to send it off-site to a place where it can be further treated. Depending on the nature of the undesirable constituent and the location within the country, this cost of waste collection and disposal can exceed the operating cost of the ion exchange equipment itself and is an increasingly important concern in the design of ion exchange systems.

Disposing of Regenerant Wastes. State and municipal codes should be examined to see what limitations are for discharging regenerant wastes and for specific objectionable substances such as barium, radium, and suspended solids. Community or municipal size systems should include procedures for environmentally safe waste disposal. Regenerant treatment can include brine reclamation in softening installations, precipitation and removal of a particular contaminant, or collecting the contaminated spent brine and hauling it to a central treatment facility that is equipped to safely dispose of the waste. In cases where municipal, district, or sanitation district codes exist, they should be followed, providing they do not exceed regulations with higher authority.

The expected level of a specific contaminant in the waste can easily be estimated at the design stage. It takes about 80 gal/ft^3 to regenerate an ion exchange resin, including the backwash, regeneration, and rinse cycles. The concentration ratio is found by dividing the throughput capacity in gallons per cubic foot of resin by the regeneration volume. Multiply the concentration ratio by the concentration of the contaminant in the raw water level to determine the average concentration to be discharged.

For example, since each cubic foot of softener resin requires about 80 gal for regeneration, let's assume for a specific installation that the throughput is 2,400 gal/ft^3 of resin during the service cycle. The concentration factor is 2,400 divided by 80, or 30. This means that the hardness level in the regenerant waste will be 30 times as high as in the influent water. In most cases as of 1995, most regenerant wastes are discharged into the local sanitary sewer. Many softeners are already in use on individual homes and are discharging such wastes even in areas that contain objectionable substances such as barium or radium. In the future, it is likely that these discharges will be restricted or prohibited.

Monitoring and Validating System Performance

The ion exchange process is quite predictable provided that the feedwater conditions do not change and that the condition of the resin bed remains reasonably constant, free of fouling and degradation. For this reason many ion exchange systems are designed with nothing more than throughput monitoring and regeneration initiation based on throughput. In the case of home water softeners, time demand regeneration based on an average expected usage is often used. For treatment systems such as home softeners, the only consequence of overrunning the ion exchange unit is hard water getting into the downstream piping and perhaps affecting the ability of soap to suds in a shower. When a contaminant is potentially harmful, failure to monitor could be unsafe. Centralized treatment systems should have periodic monitoring of objectionable constituents to verify that the effluent water quality from the ion exchange system remains within the desired limits. In certain cases there is the possibility of an objectionable contaminant being dumped by the resin if the system were to be overrun past exhaustion. In these cases, the need for periodic monitoring becomes much more significant as there is the potential for the contaminant level to reach significantly higher levels in the effluent than its level in the raw water. Treated water reservoirs can store enough water to significantly reduce the impact of dumping. However, continuous overruns can result in complete discharge and therefore total negation of the treatment process, i.e., zero average removal.

Certain ion exchange processes depend on the substance being exchanged to be in a certain ionic form before it can be successfully removed by ion exchange. Such is the case of ions like barium, silicate (silica), arsenate (arsenic), selenate (selenium), and others. In such cases, it is necessary to validate the effluent quality and monitor the feedwater to be certain that the impurities are in the proper ionic state for removal. If pretreatment equipment is provided to alter the ionic form, it too must be monitored and validated. Beyond these requirements, other parameters that should be monitored depend on the availability of workforce, size of the system, and intent of the user to operate the system in the most economical possible fashion. For instance, chemical concentration and dosage used during regeneration are very necessary for determining the economics of the operation and may be of little value for any other purpose.

Resin Replacement

The usual rate of resin deterioration in properly pretreated clean water systems is quite slow and is somewhat predictable. Nevertheless, it is helpful to periodically monitor the condition of the resin bed itself to verify the condition of the resin. Such monitoring enables the user to determine when resin replacement is necessary and to perform maintenance and cleaning procedures before the bed deteriorates to the point where poor performance occurs. It is beneficial to periodically perform complete water analyses of the raw and treated waters and to verify that the overall water quality remains within the desired limits. This is not usually practical for point-of-use systems or home systems but is routinely done at central treatment facilities. Since ion exchange resin beds in some situations are fertile breeding grounds for bacteria, it is prudent to include occasional bacterial analysis as part of the validation of system performance.

Automation

Many systems are best operated manually. Systems with only one or two vessels that regenerate no more than once a day most likely will have more maintenance time required to keep up with the automation than to regenerate manually. However, there are advantages to having automatic regenerations. For point-of-use systems, this frees the homeowner from a tedious chore. The automatic home system regeneration controller is usually set to regenerate on weekends or evenings when minimal water use is expected. In large treatment facilities where multiple vessels or trains are used, automation can significantly reduce the amount of workforce required. Automation has the advantage over manual operation in that regenerations are performed in a predictable fashion time after time in the same way. This reduces the possibility of human error; however, it does increase the possibility of machine error. In sophisticated automated systems with program logic control (PLC) and computer data logging, the cost of automating a system can exceed the equipment cost itself. This is an important point to consider in the planning stage for a new ion exchange system. The more complicated control systems require more highly skilled computer technicians to properly maintain the automatic control hardware and software. Because the state of the art in computer automation is changing so rapidly, it is likely that any automation system purchased today will be outdated within a few years.

Point-of-Use (Single-Home) versus Centralized Systems

Point-of-use systems have advantages for very small communities because they have lower capital costs and because they can be applied to a smaller portion of the total water flow and be used where consumers require or desire the removal of a specific contaminant. As an example, a water supply that has a relatively high nitrate level, but one which is within the USP guidelines for potable water, might be used by an adult couple without treatment and without danger of illness. Another couple with young children might choose to treat the water to remove the nitrates to protect the infant children against the possible illness associated with high nitrate levels. Another example is a water supply with very hard water. A homeowner might purchase a softener to avoid the nuisance of having hard water. Yet another homeowner on a salt-restricted diet might prefer not to have a softener or to limit its use to nonpotable outlets, to avoid the increase in sodium caused by the hardness exchange. Most homeowners do not soften the water used for lawn sprinkling, which would be wasteful but unavoidable in a central system. Individual home systems in some cases may not have the same restrictions on waste discharge that centralized treatment facilities have. However, they have a significant drawback, in that there is seldom any periodic monitoring of the effluent quality, and the homeowner may not be aware when the softener or purifier has malfunctioned, or that the treated water is not as purified as desired. Centralized treatment systems, however, have economy of scale on their side. The cost per unit of water produced is usually significantly less than that of the smaller systems. However, they have a disadvantage in that they must usually treat the entire water supply—even portions that are not necessary, such as for watering lawns and flushing toilets. Centralized treatment allows for more effective monitoring of the effluent quality, and in cases where the contaminant must be removed from the wastewater prior to discharge, the centralized treatment facility provides an easier pathway to collect and treat the regenerant waste.

Exchange Tank or Lease Services

Exchange tank services are becoming more widely available and are one of the larger future trends in the way that ion exchange is used. Exchange tank services are available for all flows, from point-of-use systems up to large centralized treatment facilities. There are advantages in the capital cost saved by not installing the treatment system, and there are advantages in the freedom of not having to regenerate and dispose of the wastewater produced by the regeneration process. The exchange tank service companies are experienced in performing regenerations and are, in some cases, able to charge a lower unit water cost than can be provided by a municipality or a homeowner. Most of these companies will perform the monitoring and validation requirements of the ion exchange system. These services are far more readily available in urban communities than in rural communities as they are transportation-sensitive (location-sensitive).

Operation and Maintenance of Equipment

Operation. The main duty of an operator is to operate the equipment and to document the procedures followed and the effluent parameters of the ion exchange system. In manual systems the operator must also operate valves and pumps during the regeneration procedure and in some cases even make up the chemical solutions used during regeneration. As a bare minimum, the operator should log the flow rate, throughput, effluent quality, and time and date of each regeneration. It is also helpful to record the flow rates, concentration doses, and inventory levels of chemicals used during regeneration. In highly automated systems these functions would be performed by PLC. The objective of taking these data is to be able to determine trends in the ion exchange system performance and to spot malfunctions or events that might cause malfunctions and to prevent them from happening.

Most ion exchange systems require between 1 and 2 h to regenerate. Where manual operation is employed, the operator must be present during the entire regeneration, although she or he will have time between steps to perform other duties, such as filling out the logbook. Even with automated systems, if the process is critical and the effluent is critical, an operator should be present during the regeneration to watch over the automatic system and to verify that it performs the way that it is supposed to. In larger systems automation has other advantages. It frees the operators from the necessity of climbing to reach valves that may be in rather inaccessible locations. It generally ensures that regenerations will be carried out in a predictable fashion and performed the same way each time in every unit within the system. A single operator may be able to control several automated processes simultaneously from a central control room, thereby reducing the overall workforce requirements. Most large industrial ion exchange plants are automated and have operators whose duties include watching over the automated regeneration. The strategy of having operators who are dedicated to the ion exchange plant is not always economically feasible. The second-best choice would be to have a chemist or a lab technician on a part-time basis perform the necessary operation of the ion exchange system, as these types of personnel are generally more familiar with the chemistry involved in the process.

Maintenance. General maintenance of an ion exchange system includes cleaning the equipment periodically, particularly in the area of the dilution stations, as the regenerant chemicals are corrosive. If the equipment area is not kept clean, rapid deterioration of the hardware will occur. Other routine maintenance requirements such as replacing valve diaphragms and repairing pump seals are generally the same as for other systems that employ this type of hardware. With respect to the ion exchange system, the maintenance requirements tend to be less predictable and are performed at irregular, long intervals.

Although ion exchange plants are not designed with the intent that the resin will foul, it is not uncommon for one or more contaminants in the feedwater to cause resin fouling over time. One of the more significant maintenance requirements is cleaning the ion exchange resin. Resin cleaning may be nothing more than an occasional (yearly) extended backwash under manual supervision to remove excessive suspended solids and broken resin beads that have accumulated, to something as complicated as repeated chemical cleanings with various acids, salts, and caustics. The specific cleaning process depends on the nature of the contaminant or foulant. When the foulants are not able to be sufficiently removed, it necessitates resin replacement. Many of the chemicals used to clean resin are not compatible with the materials of construction of the equipment. Therefore, it is not uncommon to have to clean the resin externally. Resin cleaning procedures generally involve a minimum of 8-h outage time and must be followed by extraordinary regeneration before the resin bed can be put back into service. Some of the resin cleaning chemicals are hazardous and must be properly stored, handled, and disposed of. For this reason it is usually impractical for small systems to be cleaned, and it is most common to see the resin replaced when it becomes fouled, rather than cleaned. Even in large treatment systems, the economics involved may favor resin replacement over resin cleaning. Today some of the exchange tank or water treatment service companies will perform resin cleaning services for a fee. This avoids tying up operating vessels and allows time for more vigorous treatments, which may eliminate the need to change out the resin beds.

Resin Removal and Replacement. Resin removal and replacement is a periodic maintenance requirement that occurs in all ion exchange systems sooner or later. If the system operates in an unfouled manner, the interval between resin replacements may be in excess of 10 years for water softeners and cation exchangers and 3 to 6 years in anion exchangers. However, the average is 3 to 5 years, and in cases where the water fouls the resin, it can be even more often. For small household systems or with small exchange tanks that can be picked up, it is relatively simple to unscrew the top, pick the tank up, turn it upside down, and shake the resin out, using a hose to sluice out any resin that clings to the inside wall of the vessel. For larger systems, the resin can be sluiced or siphoned out of the vessel. In urban areas there are vacuum truck operators that, for a fee, will remove the resin by vacuuming it out of the tank. It is important to clean out the tank as well as possible before refilling with fresh resin, particularly if the resin is being removed because of fouling. After the resin has been removed, while the tank is still empty, the maintenance personnel should always take advantage of the opportunity to inspect the tank and internals for signs of wear or damage. In small systems this can be accomplished by peering into the tank with a flashlight. However, there is no substitute for closer examination by a maintenance worker actually climbing inside the tank. In the inspection the worker should look for signs of bent or broken parts, excessive corrosion, gaskets that may have slipped, etc. The internals are not indestructible, and the worker who enters the vessel must be careful not to step on and thus damage the internals by his or her entrance. If the tank is lined, the lining should be examined for signs of blistering, tears, corrosion showing through, or other indications that lining failure could occur in the foreseeable future. If there is damage to the internals or vessel lining, this damage should be identified and either repaired prior to putting the vessel back into service or scheduled for repair.

Once the unit has been emptied of resin and examined, it is ready to have the replacement resin installed. For small systems, it is practical to pour the resin through a funnel into the open top of the tank. For larger systems, it is feasible but labor-intensive and time-consuming to pour the resin in the top of the tank or to remove the resin from its drum with buckets and pour the buckets into the tank. Various methods have been devised for putting resin into tanks. One of the simplest is to use a water-powered eductor to suck the resin straight out of the barrel and into the exchange tank. This method requires a source of pressurized water to operate the eductor. Another convenient method is to fill the resin drums with water and to use an air-operated diaphragm pump to pump the resin as a slurry. It is relatively fast and can be performed by a single worker without much labor. For very large systems, it is sometimes practical despite the extra cost to have the resin delivered in a tanker truck and pumped directly from that truck into the service vessel.

Disposal. Ion exchange resins themselves are nonhazardous and are generally suitable for disposal as ordinary trash in a landfill. However, you must take into consideration the ions held by the resin in determining its classification as a waste material. Resins that have been used to remove toxic substances become toxic. If the regeneration is not able to completely remove the contaminant from the resin, then the resin may have to be disposed of as though it were the toxic substance contained therein. Local requirements vary, and some disposal sites may not take ion exchange resins or may take ion exchange resins only in limited quantities or only after leach tests confirm acceptability. Ion exchange resins that are regenerated with acids or bases take on the characteristics of those chemicals and therefore may be considered hazardous as acids or bases. To avoid this, prior to disposal, these resins should be thoroughly exhausted into the salt form such that their pH is neutral. It is usually better to remove the excess water from the resin prior to disposal. This reduces the weight and handling difficulty.

Reliability and Availability

Ion exchange systems are not known for having high availability. This is due to a number of causes. First, in single-tank systems, the equipment is periodically out of service for regeneration. It is not uncommon to purchase three 50% capacity exchangers; thus while one unit is in regeneration or off-line, for maintenance, the other two handle the load. It is prudent in larger systems to have at least one spare exchange tank that is ready to be placed in service, should an unforeseen problem occur with a unit in service.

By far the biggest factor in the unscheduled outage of ion exchange systems is the chemical dilution systems for the regeneration flows. All regenerant chemicals are corrosive; some are very corrosive, and if they are not carefully handled or if spills are not promptly cleaned up, the area of the chemical dilution stations deteriorates. Although most equipment vendors will provide spare chemical pumps, they usually do not provide the spare dilution stations. Since the dilution station is one of the most likely parts of the system to fail, this failure will cause an unscheduled outage of the equipment. For centralized ion exchange systems, the addition of a very large treated water storage tank holding 8 to 24 h of treated water volume at the normal flow rate helps to improve the availability of treated water and overcome reliability problems. Overall, a single ion exchange tank will probably not be available more than 90% of the time and perhaps as little as 50% of the time. Even ion exchange systems with multiple exchange tanks and spare tanks seldom exceed 98% availability.

Resin Deterioration

Ion exchange resins are in general very stable polymers, and their chemical deterioration occurs at a very slow pace. Anion resins are generally less stable than cation resins. The acid form cation resin and hydroxide form anion resin forms are less chemically stable than their salt forms. Ion exchange resin beads are also susceptible to physical damage. In systems that have high pressure drop, the hydraulic force of the water, especially in conjunction with other stresses, can crack the resin beads. In general, cation resins are stronger than anion resins. As a general guideline, cation resin beds should not be subjected to pressure drops in excess of 25 psi, and anion resin beds should not be subjected to pressure drops in excess of 20 psi. Pressure drops in excess of 50 psi across the resin bed may lead to rapid physical deterioration of the resin beads. In poorly designed ion exchange systems, where the water flow is so turbulent that it causes resin movement, physical damage will also occur. The movement of the beads causes surface erosion as the beads collide against one another and chips are broken from the surface of the resin. Another type of physical stress is due to osmotic changes. This together with physical and thermal stresses causes much of the damage that occurs during regeneration because the resin bead changes size rapidly when subjected to differing ionic concentrations and temperatures and pressures. The very worst effect occurs in acid base cycling because heat, osmotic, and physical stresses all act at the same time. For example, when an acid form anion resin is be exposed to sodium hydroxide, this type of cycling causes internal cracks within the resin that ultimately cause the resin beads to break apart. Weakly acidic and weakly basic exchange resins are more prone than their strongly ionized counterparts due to the larger volume change that occurs with these resins when they are regenerated from the exhausted to the regenerated forms. The consequence of physical deterioration of a resin bed is increased pressure drop and erratic quality and run lengths due to maldistribution and channeling. There is also a problem with the small fragments from the broken resin beads clogging the underdrain or escaping the underdrain collector and contaminating the treated water.

By far the largest cause of deterioration of ion exchange resin is oxidation. The vast majority of drinking water supplies are chlorinated. Chlorine is a strong oxidant that attacks the polymer structure of ion exchange resins. This causes the resin bead to swell, to increase in moisture, and to become softer as the oxidant attacks the polymer structure of the resin. It can also attack the functional groups and cause loss of capacity. Capacity loss is limited to the anion exchangers. Eventually the oxidant can cause the resin bead to begin to dissolve, which contributes organic contaminants to the treated water. The effects of oxidation are variable and depend strongly on the degree of cross-linkage used in the polymerization of the resin bead. Polymer structures with high levels of cross-linkage are more resistant to oxidation. Oxidation occurs more rapidly at elevated temperatures and in the presence of transition metal catalysts such as iron or copper. Although the most common oxidant is chlorine, hydrogen peroxide, ozone, and even oxygen in the feedwater, especially at elevated temperatures, can contribute to the oxidation deterioration of ion exchange resins.

Although oxidation is the prime cause of loss of resin functionality, there are several other causes. Because ion exchange resins have different selectivities for varying ions, and some ions exchange much more slowly than others, some ions with extremely high selectivity or slow diffusion rates can build up and foul the resin. Such is the case with ions such as barium on cation resins, and naturally occurring organic acids, and organic sulfonates, on anion resins. These ions go on to the resin and do not readily come off. They permanently occupy a portion of the resin's exchange capacity; this leads to loss of available capacity.

All strong base anion resins are thermally unstable and undergo a gradual loss of functionality under all conditions. This rate of loss is directly affected by the operating temperature and the internal pH. The weak base resins are the most stable. Of the strong base resin, the type I styrenic resins are the most resistant, followed by the type II styrenic resins, and then the strong base acrylic resins which are the least thermally stable. Anion resins used in the salt form have lower internal pH values and are significantly more stable than those used in the hydroxide or basic form.

Physical fouling can take one of two directions. Either the foulant can plug up the flow channels between resin beads and prevent equal distribution of service water or regenerant flows, or the foulant can actually coat the resin bead and prevent or retard the exchange of ions into and out of the resin. In the former case, the effect of fouling may or may not be noticed as increased pressure drop because the effect of channeling is uncertain. In cases where the physical foulant coats the surface of the resin beads, the resin exhibits a marked kinetic impairment. At very low flows, the water quality may be somewhat acceptable, but as flow is increased, the water quality deteriorates. Certain types of chemical fouling also cause this type of flow response. The extent to which a physical foulant is present in the resin bed can be determined by looking at the resin under a microscope of 20 to 30 power, or in many cases simply by taking a handful of the resin, feeling it, and looking at it closely. Chemical foulants are more difficult to identify and are frequently not visually apparent. Most chemical foulants are ions with very high selectivities for the resin. They are exchanged onto the resin but are not easily removed. Such is the case with respect to cation resin fouling by barium, lead, or aluminum. Similarly, anions such as phosphates can significantly foul anion resins and are quite difficult to remove during the normal regeneration with NaOH.

Another type of chemical fouling is precipitation. Precipitation can occur on the surface of the resin bead, in which case it resembles other forms of physical fouling, but can also occur inside the resin beads. Internally precipitated salts are the most difficult to remove from ion exchange resins and usually require successive long soaks in fairly high concentrations of acids or bases.

Silica fouling of anion resins occurs in deionization systems where the ratio of silica is high and the caustic regeneration dosage and temperature and contact time are not sufficient to fully remove the silica from resin during the regeneration cycle. Silica is only slightly soluble and polymerizes on the resin during the service cycle. If the regeneration process fails to remove as much silica as was exchanged, a coating of silica builds up in a shell around the resin bead. Silica depolymerizes at high pH, and this is accelerated with high temperatures. So long contact times and elevated temperatures during regeneration are helpful. Elevated regenerant dosage levels are required to ensure that enough excess caustic, above and beyond the amount consumed by the regeneration of the resin, is present to redissolve and remove the silica from the resin during regeneration.

Resin Leachables

Chemical discharge by ion exchange resins occurs when they begin to oxidize. In the case of anion resins, this discharge can also be caused by thermal degradation of the ion exchange functional groups. In the case of oxidation, the leachables are oligomers of polystyrene and may not have any discrete chemical structure, except that they will generally contain at least one aromatic group and may also contain portions of the functional groups that happen to be attached to the part of the polymer that breaks away from the bead it-

ION EXCHANGE APPLICATIONS IN WATER TREATMENT 12.53

self. In the case of cation resins, the two most likely leachables are benzene sulfonic acid and sulfa benzylic acid. In the case of anion resins, the most likely leachables are the functional groups and are generally classified as amines. The dominant leachable from type I strong base resins is trimethylamine, and the dominant leachable from type II anion resins is acetaldehyde. In cases of severe oxidation, the anion resin can also have leachables that are made up of portions of the polymer.

Leachable levels are always highest when the resin is new or when it has been allowed to stand for a period of time; this decreases with age. The concentration drops when the flow increases. This is so because the resin is deteriorating at a relatively constant rate and the leachables diffuse into and are diluted by the surrounding water. This is one reason why an ion exchange bed that has sat idle for more than full cycle time should be rerinsed before its return to service. Under normal steady-state operation, and in the absence of fouling, the leachable concentrations will decline to the low parts per billion range. To avoid the unnecessary deterioration of an ion exchange resin bed, the bed should be protected from strong oxidants such as chlorine, chloramine, ozone, hydrogen peroxide, potassium permanganate, or even oxygen if the feedwater temperature is above 140° F.

Guidelines for Resin Replacement

Generally speaking, if no specific guidelines exist, ion exchange resins should be replaced when their useful capacity has deteriorated by 25% from the performance when the resin was new or if the total volume of resin has decreased by 25% or more and the resin is not in "like new" condition. This is the point in many cases where economics favor the replacement of the resin and the cost of the resin is offset by savings in regenerant and waste treatment costs. Somewhere around 25% to 50% loss of total capacity usually results in a deterioration of water quality in addition to the reduction in throughput. Other criteria for resin replacement include failure to make sufficient quantities of good-quality water, failure to rinse up to service quality in a reasonable time.

It is generally better not to top off ion exchange beds simply because a small percentage of the total resin volume has been lost, particularly if the capacity of the resin is significantly degraded. It is generally more efficient to operate with reduced throughput until such time as the entire bed can be replaced. This is not always practical as the loss of throughput capacity may necessitate increased numbers of regenerations, which may overtax the wastewater system or the operator's ability to maintain continuous service.

BOUNDARIES BETWEEN SUPPLIERS, CONSULTANTS, AND USERS

The process of creating an ion exchange system usually involves a resin supplier, an equipment supplier, a consultant or design engineering firm, a contractor (both mechanical and electrical), and an end user or operator of the finished plant. Except for relatively small systems, it is unlikely that the role of these various suppliers and users will be handled by a single company. These roles will almost always be handled by various companies who must all work together and cooperate to have a successful ion exchange plant. The end user must begin the process of creating the plant by collecting information about needed water quality and quantity. The end user must determine that the plant will be built, must economically justify it, and must secure capital for its construction. The end user will then generally contract with a design engineering consultant to create a specification for the finished ion exchange plant. The consultant will then work with equipment suppliers and resin suppliers to determine the size and shape of the plant necessary to meet a certain objective. The finished design is generally put out to bid, and the consultant and end user will select an equipment supplier.

The equipment supplier builds the equipment, the installation contractor installs the equipment, the end user and the equipment supplier commission the plant, and the end user then takes over the operation of the finished exchange plant. The traditional roles played by the various companies involved can lead to certain types of difficulties if the various parties do not communicate well. The bidding process, particularly if the selection is made solely on the basis of the lowest price, may lead to shoddy workmanship and subquality components. Some types of partnering arrangements have been proved to provide a better finished product at an overall price that is little or no more expensive than the total cost provided by the traditional bid method of supply. Regardless of how the various roles are played, it is extremely important that the end user take an active interest in the design and construction of the ion exchange plant. The user who believes that by hiring a consultant his or her work is finished, is very likely going to be unpleasantly surprised to receive the finished product, as it will not likely be what the user had hoped to get.

FUTURE TRENDS IN ION EXCHANGE

It has become common to see ion exchange used in conjunction with membrane technology, especially in applications with high salinity. In the future the role of ion exchange will shift from bulk removal of ions toward selective ion removal and polishing. The use of cation resins for water softening will continue as the largest single role of resins, even as pretreatment for membrane processes. In some applications, the role of ion exchange for bulk ion removal will diminish as membrane processes become more efficient. Ion exchange polishing of membrane-treated waters will continue to grow. Ion exchange will become more widely used for selective ion removal and product recovery, and purification. Currently the areas of rapid growth in ion exchange are selective removal of contaminants such as nitrates, heavy metals, radioactive ions, and organic acids from potable water supplies and for specific metal removal from wastewater streams.

There will be increasing reliance on the companies that provide exchange tank services. Exchange tank services are ideally suited for point-of-use applications, especially in polishing membrane effluents. They allow the regeneration of the ion exchange material in bulk at centralized facilities. This includes the treatment of the regenerant wastewater. Another recent trend is for owners to contract out the operation and maintenance of the plants. This eliminates workforce, training, and management costs.

BIBLIOGRAPHY

Arsenic (See Also Selenium)

1. *Arsenic (III) Oxidation and Removal from Drinking Water*, Phyllis Frank and Dennis Clifford, EPA, Project Summary, April 1986.
2. *Arsenic, Arsenite (AsO_2^{-1}), Arsenate (AsO_2^{-3}), Organic Arsenic Complexes, Recognized Treatment Techniques for Meeting the Primary Drinking Water Regulations for the Reduction of Arsenic (AsO_2^{-1}, AsO_4^{-3} and Organic Arsenic Complexes) Using Point-of-Entry (POU/POE) Devices and Systems*, Technical Application Bulletin, Water Quality Association, February 1992.

3. *Arsenic: New EPA Regs to Trigger Extensive Remediation Efforts*, Dr. Robert V. Russo and Bruce Saaski, Water Conditioning & Purification Magazine, January 1994.

4. *The Equilibrium Arsenic Capacity of Activated Alumina*, Eric Rostenblum and Dennis Clifford, EQA, Project Summary, February 1984.

5. *Pilot Study of Fluoride and Arsenic Removal from Potable Water*, F. Rubel and F. S. Williams, USEPA 600.2-80-100, August 1980.

6. *ResinTech Product Bulletin SIR-900, Lead, Fluoride and Arsenic Selective Media*, 1 Resintech Plaza, 160 Cooper Rd., West Berlin, N.J.

7. *Arsenic Removal from Water Utilizing Natural Zeolites*, Dagmar Bonnin, 1997 Annual Conference Proceedings; American Water Works, pp. 421–441, June 15–19, 1997

8. *Barium Arsenic Removal during Precipitative Softening*, Laurie S. McNeill and Marc Edwards, Journal of Environmental Engineering, vol.123, no. 5, pp. 453–460, May 1997.

9. *Toxics Leaching from Residuals of Arsenic Treatment under Landfill Conditions*, Amlan Ghosh, Dhananjay Mishra, and Wendell Ela, 2002 Annual Conference Proceedings; American Water Works, June 16–20, 2002.

10. *Arsenic Removal by Enhanced Coagulation and Membrane Processes*, Janet G. Hering and Menachem Elimelech, Publisher AWWA—AWWARF, 1996.

11. *Arsenic Occurrence and Speciation in United States Drinking Waters: Implications for Water Utilities*, Hsiao-wen Chen and Marc Edwards, Proceedings Water Quality Technology Conference, November 17–21, 1996.

12. *Arsenic Removal from Water Utilizing Natural Zeolites*, Dagmar Bonnin, 1997 Annual Conference Proceedings; American Water Works, pp. 421–441, June 15–19, 1997.

13. *Arsenic Behavior in Packed Bed Adsorption Reactors: Media Performance vs. Water Quality Composition*, Federico G. A. Vagliasindi and Mark M. Benjamin, 1997 Annual Conference Proceedings; American Water Works, pp. 443–456, June 15–19, 1997.

14. *Arsenic Removal Using Potassium Permanganate Generated Greensand*, Gregory F. Lauf, Proceedings 1994 Water Quality Technology Conference, Part I, November 6–10, 1994.

15. *Toxics Leaching from Residuals of Arsenic Treatment under Landfill Conditions*, Amlan Ghosh, Dhananjay Mishra, and Wendell Ela. 2002 Annual Conference Proceedings; American Water Works Association, June 16–20, 2002.

16. *A Novel Ion Exchange Process for As Removal*, Kim et al., Journal of American Water Works Association (AWWA), pp. 77–85, March 2003.

Barium and Radium

17. *Ion Exchange and Inorganic Adsorption*, Dennis Clifford, Water Quality and Treatment, F. Pontius, ed., pp. 561–640, 1990.

18. *Exhausting and Regenerating Resin for Uranium Removal*, Zhihe Zhang and Dennis A. Clifford, Journal American Water Works Association, April 1994, vol. 86, no. 4, pp. 228–241.

19. *Modifying Ion Exchange for Combined Removal of Uranium and Radium*, Dennis Clifford, and Zhihe Zhang, Journal American Water Works Association, April 1994, vol. 86, no. 4, pp. 214–227.

20. *Ind. & Engineering Chemistry—Fundamentals*, A. K. Sengupta and Dennis Clifford, 25, pp. 249–58, 1986.

21. *Removing Barium and Radium through Calcium Cation Exchange*, Anthony G. Myers, Vernon L. Snoeyink, and David W. Snyder, Journal AWWA, Research and Technology, May 1985.

22. *Removal of Barium and Radium from Groundwater*, Vernon L. Snoeyink, Carl C. Chambers, Candace K. Schmidt, Rick F. Manner, Anthony G. Myers, Julie L. Pfeffer, Sharon K. Richter, and David W. Snyder, EPA/600/M-86/021, Environmental Research Brief, February 1987.

23. *Weak Acid Ion Exchange for Removal of Barium, Radium and Hardness*, D. W. Snyder, V. L. Snoeyink, and J. L. Pfeffer, Journal American Water Works Association, 78: 98, September 1986.

24. *Strong-Acid Ion Exchange for Removal of Barium*, V. L. Snoeyink, C. C. Chambers, and J. L. Pfeffer, Journal AWWA 79(8):66–72, 1986.

25. *Radium Removal from Groundwater by Ion Exchange Resin*, Lucius Cole, P.E., and Jack Cirrincione, Technical Paper, Water Quality Association, March 1987.

26. *Determination of Radium Removal Efficiencies in Iowa Water Supply Treatment Processes*, The U.S. Environmental Protection Agency Office of Radiation Programs, Technical Note, ORP/TAD-76-1, June 1976.

27. *Strong-Acid Ion Exchange for Removing Barium, Radium, and Hardness*, Vernon L. Snoeyink, Carl Cairns-Chambers, and Julie L. Pfeffer, Research and Technology, Journal AWWA, August 1987.

28. *Evaluating Ion Exchange for Removing Radium from Groundwater*, Suresh Subramonian, Dennis Clifford, and Winston Vijjeswarapu, Research and Technology, Journal AWWA, May 1990.

29. *Weak-Acid Ion Exchange for Removing Barium, Radium, and Hardness*, David W. Snyder, Vernon L. Snoeyink, and Julie L. Pfeffer, Research and Technology, Journal AWWA, September 1986.

Fluoride (See Also Arsenic)

30. *Exposure to Excessive Fluoride during Hemodialysis*, William J. Johnson and Donald R. Taves, Technical Note, Sidney International, vol. 5, pp. 451–454, copyright 1974.

31. *Defluoridation of Drinking Water in Small Communities*, Final Report, National Sanitation Foundation Assessment Services, Gordon Bellen, Marc Anderson, and Randy Gottler.

32. *Design Manual Removal of Fluoride from Drinking Water Supplies by Activated Alumina*, Frederick Rubel, Jr., P.E., EPA-600/2-84-134, August 1984.

33. *Fluoride Radium Removal from Groundwater by Ion Exchange Resin*, Lucius Cole, P.E., and Jack Cirrincione, Technical Paper, Water Quality Association, March 1987.

General

34. *Fundamentals of Ion Exchange*, Michael C. Gottlieb, Water Quality Association, 1989.

35. *Water Quality and Treatment: A Handbook of Community Water Supplies*, 5th ed., AWWA, 1999.

36. *Ion Exchange Resins*, Robert Kunin, Robert E. Krieger Publishing Company, 1985.

37. *Demineralization by Ion Exchange in Water Treatment and Chemical Processing of Other Liquids*, Samuel B. Applebaum, Academic Press, 1968.

38. *Ion Exchange for Pollution Control*, vol. 1, Calvin Calmon and Harris Gold, CRC Press, 1979.

39. *Management of Point-of-Use Drinking Water Treatment Systems*, Final Report, National Sanitation Foundation Assessment Services, Gordon Bellen, Marc Anderson, and Randy Gottler.

40. *Estimation of Ion Exchange Process Limits by Selectivity Calculations*, Robert E. Anderson, Journal of Chromatography, 201, 1980.

41. *Performance Projects for Anion Resins in Two Bed Demineralizers*, Michael Gottlieb, International Water Conference, October 1989.

42. *The Practical Aspects of Ion Exchange in the Service D.I. Industry*, Michael Gottlieb, Water Technology, February 1989.

43. *Practical Principles of Ion Exchange Water Treatment*, D. L. Owens, Tall Oaks Publishing, Littleton, Colo., 1985.

Mercury

44. *Mercury (II) Removal from Aqueous Solution Using Ion Exchange and Carbon Adsorption*, Keith W. Fife, Joel D. Williams, and H. G. Folster, Wastewater Treatment, Mercury, Ion Exchange, Adsorption, Ann Arbor Science Publishers, Inc., P.O. Box 1425, Ann Arbor, MI 48106, pp. 643–654.
45. *Removal of Mercury from Drinking Water Using Activated Carbon*, Leon Thiem, Diane Badorek, and John T. O'Connor, Journal American Water Works Association, August 1976, vol. 68, no. 8, pp. 447–451.

Nitrate

46. *Nitrate Removal from Drinking Water in Glendale, Arizona*, Dennis Clifford, C. C. Lin, L. L. Horng, and J. V. Boegel, PB 87-129 284/AS, NTIS, Springfield, Va., 1987.
47. *AWWA Journal*, S. Subramonian, Dennis Clifford, and W. Vijjeswarapu, 1990, pp. 61–70.
48. *Removal of Nitrate from Contaminated Water Supplies for Public Use*, G. A. Guter, EPA-6005-81-029, U.S. EPA, Cincinnati, Ohio, 1981.

Organics (Naturally Occurring)

49. *Removal of Dissolved Organic Carbon by Coagulation with Iron Sulfate*, Robert L. Sinsabaugh, III, Robert C. Hoehn, William R. Knocke, and Arthur E. Linkins, III, Research and Technology, Journal AWWA, May 1986.
50. *Organic Traps*, Application Bulletin, ResinTech, Inc., 615 Deer Road, Cherry Hill, N.J.
51. *The Role of Organics in Water Treatment Parts 1, 2 and 3*, Amber-hi-lites, Rohm and Haas Company, Fluid Process Chemical Department, no. 169 (Part 3), Winter 1982.
52. *Reducing Cancer Risks by Improving Organic Carbon Removal*, Bryan D. Black, Gregory W. Harrington, and Philip C. Singer, Disinfection By-Products, Journal AWWA, June 1996.
53. *The Use of Anion Exchange Resins for the Removal of Natural Organic Matter from Municipal Water*, J. M. Symons, P. L-K. Fu, and P. H-S Kim, International Water Conference, Proceedings Book, 92-120, Pittsburgh, Pa., October 1992.
54. *The Reversible Removal of Naturally Occurring Organics Using Ion Exchange Resins: Part 1*, Michael Gottlieb, Industrial Water Treatment, vol. 27, no. 3, May 1995.
55. *The Reversible Removal of Naturally Occurring Organics Using Resins Regenerated with Sodium Chloride, Part 2*, Michael Gottlieb, Ultrapure Water, vol. 14, no. 8, November 1996. Also published in Ion Exchange Developments and Applications, Proceedings of IEX '96, Edited by J. A. Greig, copyright 1996, by The Royal Society of Chemistry.
56. *Operating Experiences with a New Organic Trap Resin*, International Water Conference Prodeedings, IWC95-1, Pittsburgh, Pa., October 1995.
57. *Removal of Aquatic Organics by Anion Exchange Resins*, P. L. K. Fu and J. M. Symons, Proc. AWWA Annual Conf., Orlando, Fla., June 1988.

Perchlorate

58. *Perchlorate in the Environment*, E. T. Urbansky, Kluwer Academic/Plenum Publishers, Proceedings of the American Chemical Society Division of Environmental Chemistry Composium on Perchlorate in the Environment, August 22–24, 1999.
59. *ATR-FTIR Detection of >/= 25 µg/L Aqueous Cyanide, Perchlorate, and PFOS*.

60. *Evaluating Ion Exchange Technology for Perchlorate Removal*, I. Najm, R. R. Trussell, L. Boulon, B. Gallagher, R. Bowcock, and D. Clifford, Proceedings of the 1999 AWWA Annual Conference, Water Research, American Water Works Assoc., Denver, Colo.

61. *Efficient Treatment of Perchlorate (ClO_4)—Contaminated Groundwater by Bifunctional Anion Exchange Resins*, B. Gu, G. M. Brown, S. D. Alexandratos, R. Ober, J. A. Dale, and S. Plant, Chapter 16, "In Perchlorate in the Environment," Kluwer Academic/Plenum, New York, 2000.

62. *Pilot Scale Tests to Remove Perchlorate from the RLWT Effluent by Ion Exchange*, P. Worland, FWO-WFM, January 2001.

63. *Abiotic and Biotic Perchlorate Removal in an Activated Carbon Filter*, Jess C. Brown, Vernon L. Snoeyink, and Mary Jo Kirisits, Journal AWWA, February 2002.

64. *Elimination and Minimization of Ion-Exchange Brine Disposal during Perchlorate and Nitrate Treatment*, Thomas Gillogly, Stewart Lehman, Lee Aldridge, Dennis Clifford, and Deborah Roberts, 2002 Annual Conference Proceedings; American Water Works, New Orleans, La., June 2002.

65. *Assessment of the Electrochemical Reduction of the Perchlorate Ion*, Thomas L. Theis, Amy K. Zander, Xiang Li, Marc A. Anderson, and Jeosadaque Sene, American Water Works Association Research Foundation, Denver, Colo., January 2002.

66. *Modeling Ion-Exchange Treatment of Perchlorate in Groundwater*, Anthony R. Tripp and Dennis A. Clifford, AWWA 2002 Inorganic Contaminants Workshop, San Diego, Calif., February 3–5, 2002.

67. *Biological Treatment of Perchlorate-Contaminated Brines*, Tina M. Gingras and Jacimaria R. Batista, AWWA 2002 Inorganic Contaminants Workshop, San Diego, Calif., February 3–5, 2002.

68. *A New Design for a Low Waste, Low Cost Ion Exchange System for Removal of Nitrate, Arsenic, Perchlorate and Chromium: Brine Recycling Methods*, Gerald A. Guter, Peter L. Jensen, and Dan Ziol, AWWA 2002 Inorganic Contaminants Workshop, San Diego, Calif., February 3–5, 2002.

69. *Electrodialysis Reversal (EDR) and Ion Exchange as Polishing Treatment for Perchlorate Treatment*, Vincent Roquebert, Stephen Booth, Robert S. Cushing, Gil Crozes, and Ed Hansen, The International Journal on the Science and Technology of Desalting and Water Purification, December 2000.

70. *Dual Removal of Perchlorate and Nitrate from Groundwater through Biological Reduction*, N. L. Burns, D. D. Schulte, and M. F. Dahab, AWWA Proceedings Water Quality Technology Conference, Nashville, Tenn., November 11–15, 2001.

71. *Advanced Ion Chromatographic Methods for the Determination of Bromate, Bromide, Chlorite, and Perchlorate in Drinking Waters*, Shree Karmarkar, Jessi Ammerman, and Tim Bahowick, AWWA Proceedings Water Quality Technology Conference, Nashville, Tenn., November 11–15, 2001.

72. *A Zero-Valent Iron (FE[0]) Packed-Bed Treatment Process*, Paul Westerhoff and Paul Johnson, American Water Works Association Research Foundation, Denver, Colo., August 2001.

73. *Occurrence of Perchlorate in Drinking Water Sources*, Richard W. Gullick, Twila S. Barhorst, and Mark W. LeChevallier, Journal AWWA, January 2001.

74. *The Treatability of Perchlorate in Groundwater Using Ion Exchange Technology*, Anthony R. Tripp and Dennis A. Clifford, Proceedings Water Quality Technology Conference, Salt Lake City, Utah, November 5–9, 2000.

75. *Application of Ion-Exchange Technology for Perchlorate Removal from Drinking Water*, Issam Naim, Rhodes Trussell, Lina Boulos, Brian Gallagher, Robert Bowcock, Carol Williams, and Dennis Clifford, AWWA 1999 Annual Conference Proceedings; American Water Works Association, Chicago, June 20–24, 1999.

76. *Investigation of Perchlorate Removal in Drinking Water Systems*, Kvehee Kim and Mirat D. Gurol, AWWA 1999 Annual Conference Proceedings; American Water Works Association, Chicago, June 20–24, 1999.

77. *Improved Analysis of Perchlorate in Water*, Philip Harrington and Yvonne Shen, Proceedings 1998 Water Quality Technology Conference, San Diego, California, November 1–4, 1998.

Selenium (See Also Arsenic)

78. *Selenium Oxidation and Removal by Ion Exchange*, J. V. Boegel and Dennis Clifford, PB 86-171 428/AS, NTIS, Springfield, Va., 1986.

Softening

79. *Information Facts on Home Water Softening, Total Dissolved Solids, Chlorides, & Water Usage*, Joseph F. Harrison, P.E., CWS-V, Water Quality Association, March 1991.

Terrorism (General)

80. *Minimizing the Vulnerability of Water Supplied to Natural and Terrorist Threats*, Robert M. Clark and Rolf A. Deininger, Proceedings of 2001 Information Management & Technology Conference Proceedings, Atlanta, Ga., April 8–11, 2001.
81. *Water Treatment Plant Chemical Security & Protection* (Video, VHS, AWWA, Denver, Colo.), 2002.
82. *A Risk Based Approach to Terrorism Preparedness*, Proceedings of 2002 Annual Conference; American Water Works, New Orleans, La., June 16–20, 2002.
83. *Strength in Numbers: The Benefits of Large San Francisco Bay Area Water Utilities Joining Forces to Address Post 09/11 Security Needs*, Steve Dennis, 2002 Annual Conference Proceedings; American Water Works Association, New Orleans, La., June 16–20, 2002.
84. *Vulnerability Assessment of Water Utilities and Protective Measures*, Sam Irrinki, 2002 Annual Conference Proceedings; American Water Works Association, New Orleans, La., June 16–20, 2002.
85. *Water Terrorism: How Vulnerable Is Your Utility?*, Kelvin L. Baker and Albert L. Perez, Annual Conference Proceedings; American Water Works Association, New Orleans, La., June 16–20, 2002.
86. *Deter and Detect with Technology's Help*, Alan Manning, Annual Conference Proceedings; American Water Works Association, New Orleans, La., June 16–20, 2002.

Uranium

87. *Minimizing the Vulnerability of Water Supplied to Natural and Terrorist Threats*, Robert M. Clark and Rolf A. Deininger, Proceedings of 2001 Information Management & Technology Conference, Atlanta, Ga., April 8–11, 2001.
88. *Legislation/Regulation—Defining a Guideline for Uranium*, Frederick W. Pontius, Journal American Water Works Association, August 2000, vol. 92, no. 8, pp. 18, 20, 22, 24.
89. *Methods for Removing Uranium from Drinking Water*, Thomas J. Sorg, Journal American Water Works Association, July 1988, vol. 80, no. 7, pp. 105–111.
90. *Methods for Removing Uranium from Drinking Water*, Thomas J. Sorg, Journal AWWA, July 1988.
91. *Operation of Small Scale Uranium Removal Systems*, Robert T. Jelinek and Randy J. Correll, Proceedings AWWA Seminar on Radionuclides in Drinking Water, Kansas City, Mo., June 14, 1987, pp. 99–117.

92. *Modifying Ion Exchange for Combined Removal of Uranium and Radium*, Dennis Clifford and Zhihe Zhang, Removing Radionuclides, Journal AWWA, April 1994.

93. *Uranium Removal from Groundwater*, in preparation, Dennis Clifford and Z. Zhang.

94. *Uranium, Recognized Treatment Techniques for Meeting the Primary Drinking Water Regulations for the Reduction of Uranium from Drinking Water Supplies Using Point-of-Use/Point-of-Entry Devices and Systems, Technical Application Bulletin,* Water Quality Association, February 1992.

95. *Exhausting and Regenerating Resin for Uranium Removal*, Zhihe Zhang and Dennis Clifford, Journal AWWA, April 1994.

96. *Removal of Uranium from Drinking Water by Ion Exchange and Chemical Clarification*, Steven W. Hanson, Donald B. Wilson, Naren N. Guanji, and Steven W. Hathaway, Project Summary, EPA/600/S2-87/076, December 1987.

97. *Operating a Small Full-Scale Ion Exchange System for Uranium Removal*, Robert T. Jelinek and Thomas J. Sorg, Journal AWWA, July 1988.

Zeolites

98. *Design and Testing of Natural/Blended Zeolite Ion Exchange Columns at West Valley*, P. Burn, D. K. Ploetz, A. K. Saha, D. C. Grant, and M. C. Skriba, Adsorption and Ion Exchange, AIChE Symposium Series, no. 259, vol. 83.

99. *SORBPLUS Adsorbent Product Data Sheet*, Alcoa Industrial Chemicals, December 1992.

100. *Ion Exchange Separations with Molecular Sieve Zeolites*, John D. Sherman, Linde Molecular Sieves, Ion Exchange Bulletin, presented at the 83d National Meeting, American Institute of Chemical Engineers, Houston, Tex., March 20–24, 1977.

Miscellaneous

101. *Economic Considerations in the Replacement of Ion Exchange Resins*, M. C. Gottlieb and F. J. DeSilva, I.S.P.E. EXPO 92, May 7, 1992, Philadelphia, Pa.

102. *Economic Considerations in the Selection of Ion Exchange Resins*, M. C. Gottlieb, 24th Liberty Bell Corrosion Course, September 1986.

103. *Ion Exchange Resin Selection and Performance Projections with a Personal Computer*, M. C. Gottlieb, 24th Liberty Bell Corrosion Course, September 1987.

104. *The Expected Life of Anion Exchangers*, L. Wirth, Jr., Nalco Reprint no. 40.

105. *Silica Removal with Highly Basic Anion Exchange Resins*, R. Osmun and L. Wirth, Jr., Industrial and Engineering Chemistry, May 1951.

106. *Performance Projections for Anion Resins in Two Bed Demineralizers*, M. C. Gottlieb, International Water Conference, October 1989.

107. *Symmetry Modeling of Ion Exchange Processes with Resin and Zeolites*, F. X. McGarvey and M. C. Gottlieb, Second Chemical Congress of the North American Continent, August 1980.

108. *The Relationship between Resin Operating Capacity and Static Analysis Results*, F. J. DeSilva and M. C. Gottlieb, Ultrapure Water Conference East, April 1991.

109. *A Software Model for Optimizing Resin Selection*, F. J. DeSilva and M. C. Gottlieb, 28th Liberty Bell Corrosion Course, September 1990.

110. *Ion Exchange Resins*, R. Kunin, Robert E. Krieger Publishing Co., 1958, p. 16.

111. *The Significance and Limitations of Laboratory Resin Analyses in Evaluating Ion Exchange Operating Performance*, G. J. Crits and D. H. Valentine, International Water Conference, October 1988.

112. *Thermal Degradation of Strongly Basic Anion Exchange Resins in Caustic Regenerants*, F. X. McGarvey, E. W. Hauser, B. Backs, and J. Stellitano, International Water Conference, October 1987.
113. *Technical Notes Amberlite IRA-93,* Rohm and Haas Bulletin IE-106-74/76, 1977.
114. *Ten Years Experience with Reverse Flow Dual Stratified Bed Systems in China*, Zhang Shuchun, P. A. Yarnell, and J. C. Konen, International Water Conference, October 1990.
115. *Fate of Plutonium Released from the Nagasaki A-Bomb, Japan,* A. Kudo, Y. Mahara, T. Kauri, and D. C. Santry, Water Science and Technology, 1991, vol. 23, no. 1, pp. 291–300.

CHAPTER 13
MEMBRANE PROCESSES

Robert A. Bergman, P.E.
CH2M HILL
Gainesville, Florida

A wide variety of membrane processes can be categorized according to driving force, membrane type and configuration, and removal capabilities and mechanisms. Membrane processes in the drinking water industry are used for desalting, softening, dissolved organics and color removal, particle and microbial removal, and other purposes. Although membrane technologies became commercially available more than 30 years ago, they are experiencing rapid development and improvements.

Membranes are increasingly cost-effective with better performance characteristics, and their applications continue to grow. For example, membrane filtration systems are now used to remove *Giardia, Cryptosporidium,* and other particles from surface water supplies and to treat backwash return waters from conventional water treatment plants.

This chapter is based on current state-of-the-art membrane system design criteria. Some of the design data will undoubtedly change as new developments occur. The reader should keep abreast of new developments in membranes and system components by attending training seminars, reviewing publications and ongoing research reports, and contacting manufacturers, consultants, and other professionals practicing in the field.

TYPES OF MEMBRANE PROCESSES

Membrane processes can be classified by the driving force used to promote the water treatment

- Pressure
- Electrical voltage
- Temperature
- Concentration gradient
- Combinations of more than one driving force

13.2 CHAPTER THIRTEEN

Pressure-driven and electrically driven membrane processes are the only commercially available and commonly used membrane processes for water treatment.

Pressure-Driven Membranes

The pressure-driven membrane processes are

- Reverse osmosis (RO)
- Nanofiltration (NF)
- Ultrafiltration (UF)
- Microfiltration (MF)

Membranes are permeable to water but not to substances that are rejected and removed. All membrane processes separate feedwater into two streams (see Figure 13.1). The permeate (for RO, NF, or UF) or filtrate (for MF) stream passes through the membrane barrier. The concentrate (or retentate) stream contains the substances removed from the feedwater after being rejected by the membrane barrier. The driving force for these membrane processes may come from (1) a pressurized feedwater source with the membranes installed in pressure vessels, called modules, or (2) a partial vacuum in the filtrate/permeate flow stream caused by use of a filtrate/permeate pump or gravity siphon. The vacuum-driven processes typically apply to MF and UF only and have membranes submerged or immersed in nonpressurized feedwater tanks.

Pressure-driven membrane processes can be designed for cross-flow or dead-end operating mode. In the cross-flow mode, the feed stream flows across the membrane surface, and permeate (or filtrate) passes through the membrane tangential to the membrane surface. Cross-flow operation results in a continuously flowing waste stream. Sometimes a cross-flow system is designed with a concentrate recycle with a reject stream (feed-and-bleed mode). Many MF and UF systems treating relatively low-turbidity waters are de-

FIGURE 13.1 Pressure-driven processes using feed or permeate pumps.

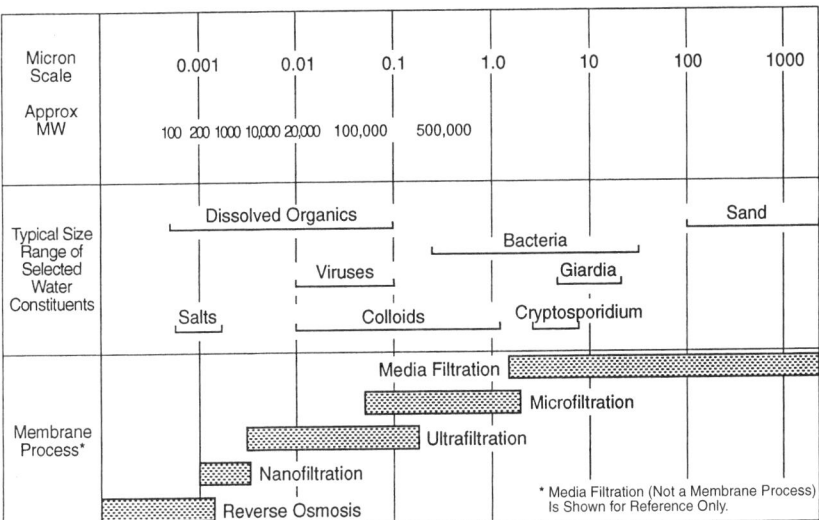

FIGURE 13.2 Pressure-driven membrane process application guide.

signed to operate in a dead-end flow pattern where the waste retentate stream is produced by an intermittent backwash.

Figure 13.2 shows the relative removal capabilities for pressure-driven processes and compares these processes with media filtration. MF and UF separate substances from feedwater through a sieving action. Separation depends on membrane pore size and interaction with previously rejected material on the membrane surface. NF and RO separate solutes by diffusion through a thin, dense, permselective (or semipermeable) membrane barrier layer, as well as by sieving action. The required membrane feed pressure generally increases as removal capability increases. Table 13.1 presents typical feed pressures for pressure-driven membrane processes.

Electrical Voltage–Driven Membranes

The electrical voltage–driven membrane processes are

- Electrodialysis (ED)
- Electrodialysis reversal (EDR)

Alternating anion and cation transfer ion exchange membranes in flat-sheet form are placed between positive and negative electrodes (see Figure 13.3). Applying a voltage across the electrodes causes a direct current (dc) to flow, resulting in positively charged cations moving toward the negative electrode (cathode) and negatively charged anions moving toward the positive electrode (anode). This causes alternating compartments to become demineralized and the other compartments to become concentrated with ions.

A typical electrodialysis system has many anion and cation transfer membranes and spacers (membranes with spacers are called *cell pairs*) stacked vertically between electrodes. Feedwater enters the stack under a pressure of about 50 psi (340 kPa), ions are re-

TABLE 13.1 Typical Feed Pressures for Pressure-Driven Membrane Processes

	Typical feed pressure range	
Membrane process	psi	kPa
Reverse osmosis		
Brackish water application		
Low pressure	125 to 300	860 to 2,070
Standard pressure	350 to 600	2,410 to 4,140
Seawater application	800 to 1,200	5,520 to 8,270
Nanofiltration	50 to 150	340 to 1,030
Ultrafiltration and microfiltration		
Pressure type	3 to 50	21 to 340
Vacuum type	−1 to −12	−7 to −83

moved, and the dilute (demineralized) stream leaves as product water. A typical design includes recycling the concentrate stream and discharging concentrate (blowdown) (feed-and-bleed mode) to waste. Electrodialysis does not remove electrically neutral substances such as silica or particulate matter because product water does not pass through a membrane barrier as it does in pressure-driven membrane processes.

Electrodialysis reversal is a variation of the electrodialysis process. The electrical polarity of the electrodes is reversed on a set frequency, typically about every 15 min, reversing the direction of ion movement to "electrically flush" the membranes for membrane scale and fouling control (Meller, 1984).

Selecting a Membrane Process

Membrane processes that satisfy treatment objectives at the lowest possible life-cycle cost should be selected. The first step is to identify overall project goals and define the cor-

FIGURE 13.3 Voltage-driven electrodialysis process.

rect treatment objectives. Source water quality should be compared with product water quality goals to determine the degree of removal required for various constituents. Removal may be by membrane process or associated pretreatment and posttreatment processes.

Historical, current, and expected future water source quality data and product criteria should be considered.

The following factors should be considered in membrane selection:

- Source water characteristics and availability
- Pretreatment and posttreatment requirements
- Product water quality and quantity requirements and blending options (for example, split-flow treatment)
- Waste residuals disposal
- Need for bench or pilot testing
- Capital and operation and maintenance (O&M) costs

Table 13.2 provides a starting point for a preliminary determination of which membrane processes, if any, apply to the treatment objectives. Membrane characterizations conducted during bench and pilot tests may be necessary to determine which process will provide the desired degree of removal (Bergman and Lozier, 1993).

Membrane Composition

Most membranes used in processes for municipal water treatment are prepared from synthetic organic polymers. Pressure-driven RO and NF processes use either cellulosic or noncellulosic membranes. Cellulosic types include cellulose acetate, cellulose acetate blends, and cellulose triacetate. Noncellulosic types include polyamides, polyurea, sulfonated polysulfone, sulfonated polyfuran, polypiperazides, polyvinyl alcohol derivatives, and other composites. MF and UF membranes use polyvinylidene fluoride (PVDF), polypropylene, polyethersulfone, polysulfone, cellulose acetate derivative, polyacrylnitrile, and other materials. Inorganic UF membranes, such as ceramic membranes with an alumina barrier layer, are available but are used to a lesser extent than are organic membranes.

ED and EDR membranes are essentially ion exchange resins in flat-sheet form. These are synthetic polymers consisting of either cross-linked sulfonated copolymers of vinyl compounds (cation transfer type) or cross-linked copolymers of vinyl monomers with quaternary ammonium anion exchange groups (anion transfer type).

Membrane Configurations

Most pressure-driven membranes for municipal water treatment are arranged in spiral-wound and hollow-fiber configurations and, to a lesser extent, tubular and plate-and-frame configuration.

Flat-sheet membranes for pressure-driven processes are most commonly assembled into a spiral-wound element (module) in which multiple membrane "leaves," each composed of two membrane sheets separated by a permeate carrier, are connected to a central permeate collector tube. A feed concentrate spacer is placed between each leaf, and the leaves and spacers are rolled around the central permeate collection tube (Figure 13.4a).

TABLE 13.2 Typical Applications of Membrane Processes

Process	Application

Reverse osmosis and electrodialysis	Total dissolved solids reduction
	   Seawater desalting (RO favored)
	   Brackish water desalting (RO is typically more cost-effective than ED or EDR for greater than 3,000 mg/L TDS)
	   Brackish water desalting of high-silica waters (ED and EDR are favored)
	Inorganic ion removal
	   Fluoride, calcium and magnesium (hardness)
	   Nutrients (nitrate, nitrite, ammonium, phosphate)
	   Radionuclides (RO only)
	   Others listed under drinking water regulations
	Dissolved organics removal (RO only)
	   THMs, other DBP precursors
	   Pesticides, synthetic organic chemicals (SOCs)
	   Color

Nanofiltration	Hardness removal
	Dissolved organics removal
	   THMs, other DBP precursors
	   Pesticides (SOCs)
	   Color

Ultrafiltration and microfiltration	Particulate removal
	   Suspended solids
	   Colloids
	   Turbidity
	   Bacteria
	   Viruses (UF only; MF can remove some viruses if they are attached to larger particles)
	   Protozoan cysts
	Organics removal (dissolved organics removal by UF only; dependent on molecular weight size of organics and UF membrane pore size; however, MF and UF can remove dissolved organics if coagulants or powdered activated carbon is used upstream of the membrane barrier)
	Inorganic chemical removal (after chemical precipitation or pH adjustment)
	   Phosphorus
	   Hardness
	   Metals (e.g. iron, manganese, arsenic)

One or more spiral-wound elements are placed inside each pressure vessel in a series arrangement (Figure 13.4b).

RO hollow-fiber membrane modules (also called *permeators*) are commonly constructed by forming the hollow, fine fibers into a U shape, "potting" the open ends in an epoxy tube sheet, and placing the potted assembly inside a pressure vessel, as shown in Figure 13.5. The flow direction through RO hollow fibers is typically outside-in, with feedwater on the outside of the fibers and permeate within the fibers' central bore. For UF membrane modules, larger-diameter hollow fibers are commonly potted at both ends and assembled into modules having a straight-through flow path for the feed-concentrate

FIGURE 13.4 Typical spiral-wound RO membrane module for pressure-driven processes. Vessels with side ports near the end caps for feed and concentrate connections are also available. [*Adapted from* The U.S.A.I.D. Desalination Manual *(Buros et al., 1980) and is used courtesy of the U.S. Agency for International Development.*]

FIGURE 13.5 Typical hollow fine-fiber RO membrane module. [*Adapted from* The U.S.A.I.D. Desalination Manual *(Buros et al., 1980) and is used courtesy of the U.S. Agency for International Development.*]

FIGURE 13.6 Representation of hollow-fiber UF module.

stream within the fibers and with the permeate passing on the outside of the fibers (inside-outside flow pattern), as shown in Figure 13.6.

Hollow-fiber MF often is configured in a straight-through flow pattern, but typically uses an outside-in flow with the filtrate within the fibers. Vacuum-type hollow-fiber MF and UF place the fibers directly in an open feed tank (not in a pressure vessel) and use negative pressure on the filtrate side of the membrane (central bore of the fibers) as the driving force.

Electrically driven ED and EDR processes typically use flat-sheet membranes assembled into stacks of up to 500 membrane pairs with a single pair of electrodes, as shown in Figure 13.7. ED/EDR stacks can have multiple electrical and hydraulic stages.

MEMBRANE SYSTEM COMPONENTS AND DESIGN CONSIDERATIONS

Some of the more important considerations in membrane system design are the system components, feedwater characteristics, and pretreatment that may be required.

Definitions

Applied pressure: feed pressure minus product pressure.

Concentrate pressure P_c: hydraulic pressure of the concentrate or retentate flow stream leaving the membrane modules or circulating in an ED/EDR membrane stack.

Concentration polarization: The phenomenon whereby retained solutes and solids accumulate at a membrane surface in concentrations greater than in the bulk flow stream.

Delta-P or differential pressure (ΔP) for pressure-driven membrane systems: feed pressure minus concentrate pressure.

Delta-P or differential pressure (ΔP) for electrically driven membrane systems: pressure difference between the demineralized stream and the concentrate stream.

Demineralized stream pressure: dilute (product) stream pressure in an ED/EDR system.

MEMBRANE PROCESSES 13.9

FIGURE 13.7 Typical ED membrane stack. (*Adapted from* Electrodialysis and Electrodialysis Reversal, *AWWA M38, 1995.*)

Feed pressure P_f: hydraulic pressure of the feedwater flow stream entering the membrane modules.

Flux (water flux): rate of product water flow through a pressure-driven membrane, commonly expressed in units of gallons per day per square foot of active membrane area (gpd/ft^2) or meters per second (m/s).

Interstage pressure P_i: in a multistage membrane system, concentrate pressure from one stage and feed pressure to the following stage.

Net driving pressure (NDP): the net driving force for a pressure-driven membrane system.

$$\text{NDP} = P_{tm} - \pi_{tm}$$

where NDP = net driving pressure, psi
P_{tm} = transmembrane (hydraulic) pressure differential, psi
π_{tm} = transmembrane osmotic pressure differential, psi

Osmotic pressure (π): a natural pressure phenomenon exhibiting a force from a low-concentration stream (e.g., product) to a high-concentration stream (e.g., feed and concentrate). Osmotic pressure is related to the solution's ionic strength and must be offset by feed pressure. Osmotic pressure for saline water is about 10 psi (70 kPa) per 1,000 mg/L of total dissolved solids (TDS) (see Figure 13.8). Osmotic pressure can generally be ignored in UF and MF water treatment applications.

Product pressure P_p: hydraulic pressure of the permeate or filtrate flow stream at the outlet of the membrane modules.

Recovery Y: permeate or filtrate flow rate divided by the feedwater flow rate, usually referred to as permeate or product water recovery for RO and NF and feedwater recovery for UF and MF systems.

$$Y = \frac{Q_p}{Q_f} \times 100$$

where Y = recovery, %
Q_p = product flow rate (volume)
Q_f = feed flow rate (volume)

Solute passage: solute passage SP is the fraction of solute present in the feed that remains in the permeate, typically expressed as a percentage. When the solute considered is TDS, it is usually called salt passage.

$$\text{SP} = \frac{C_p}{C_f} \times 100$$

where SP = solute (salt) passage, %
C_p = solute (salt) concentration in permeate, mg/L
C_f = solute (salt) concentration in feed, mg/L

FIGURE 13.8 Simplified concept of osmosis and reverse osmosis.

Solute rejection (removal): solute rejection or removal SR (or for TDS, salt rejection or removal) is the fraction of solute in membrane feedwater that remains in the concentrate (retentate) stream and does not enter the product stream, expressed as a percent. For ED/EDR systems, *removal* is the term used instead of *rejection*.

$$SR = \frac{C_f - C_p}{C_f} \times 100 = 1 - \frac{C_p}{C_f} \times 100 = 100 - SP$$

where SR = solute (salt) rejection or removal, %
SP = solute (salt) passage, %
C_p = concentration of solute in permeate or filtrate stream, mg/L
C_f = concentration of solute in feed stream, mg/L

Transmembrane osmotic pressure π_{tm}: feed stream (average feed/concentrate) osmotic pressure of a pressure-driven membrane minus permeate osmotic pressure, mathematically expressed as

$$\pi_{tm} = \frac{\pi_f + \pi_c}{2} - \pi_p$$

where π_{tm} = transmembrane osmotic pressure, psi
π_f = feed osmotic pressure, psi
π_c = concentrate osmotic pressure, psi
π_p = permeate osmotic pressure, psi

Transmembrane pressure P_{tm}: feed stream (average feed/concentrate) pressure (crossflow operating mode) or feed pressure (dead-end operating mode) minus the permeate (product) pressure, mathematically expressed as

$$P_{tm} = \frac{P_f + P_c}{2} - P_p = P_f - \frac{\Delta P}{2} - P_p$$

where P_{tm} = transmembrane pressure, psi
P_f = feed pressure, psi
P_c = concentrate pressure, psi
P_p = permeate pressure, psi
ΔP = feed/concentrate pressure differential $P_f - P_c$, psi

Membrane System Components

A simplified membrane system flow schematic is shown in Figure 13.9. Membrane systems typically include pretreatment, the membrane unit(s), product (permeate or filtrate) posttreatment, and possibly concentrate (or retentate) waste stream posttreatment. Membrane systems are commonly designed with multiple parallel process "trains" to give greater flexibility in production output rates and to allow cost-effective incremental expansions. The flow schematic also shows that in some cases it is possible to bypass a portion of source or pretreated water around the membranes and blend it with the product to produce finished water, before disinfection (i.e., split-flow treatment). In some applications, two membrane units may be placed in series such as in the dual-membrane system, with MF or UF as pretreatment for NF or RO (Figure 13.10), or two-pass RO desalting (Figure 13.11). In the two-pass seawater desalting process, the second-stage concentrate is typically less saline than the raw seawater and commonly is recycled to lower the TDS of the membrane feed.

FIGURE 13.9 Typical membrane system flow schematic.

Membrane System Feedwater Characteristics

Membrane system design must consider source water quality and temperature. Design feedwater composition should encompass current source water quality, anticipated changes in quality, and the effects of membrane pretreatment processes. Historical records are useful for determining a range of values and averages for specific water quality parameters and in estimating long-term trends. Seasonal variations for surface waters should also be considered.

Table 13.3 presents recommended items that should be analyzed for the feedwater to a proposed pressure-driven membrane system plant.

The feedwater flow rate required for a system depends on membrane system recovery. Maximum possible recovery of a membrane system is usually controlled by feedwater quality. For RO, NF, and ED/EDR, maximum allowable recovery often depends on the concentration of sparingly soluble salts and, except for ED/EDR, silica in the source water supply.

FIGURE 13.10 Example dual-membrane system for treating a saline surface water.

FIGURE 13.11 Example two-pass RO seawater desalination system.

Sometimes small flat-sheet membranes with varying pore sizes are used with bench test equipment to characterize feedwater in terms of solute fractions in various molecular weight size ranges. This is sometimes called *molecular weight fingerprinting* and can be used for membrane selection.

Feedwater Pretreatment

Most membrane treatment systems require some pretreatment for source water. The type of pretreatment system required depends on feedwater quality, membrane type, and design criteria for the membrane unit. Pretreatment may be used to

- Condition the feedwater to allow membrane treatment to be effective, for example, using coagulants to create particles large enough to be removed by microfiltration
- Modify the feedwater to prevent membrane plugging, fouling, and scaling; maximize the time between cleanings; and prolong membrane life

Design of specific components needed to pretreat membrane feedwater is discussed in other chapters of this book. In general, surface waters require more extensive pretreatment than groundwaters because of the presence of significant levels of suspended solids and biological matter.

For example, Figure 13.12 shows typical groundwater and surface water pretreatment for RO and NF processes. In some situations, pretreatment can be relatively simple, such as adding chemicals, such as acid or scale inhibitor, and cartridge filtration. Other waters may require pretreatment as extensive as a full conventional water treatment plant with coagulation, flocculation, sedimentation, and filtration.

Suspended Solids Control. Essentially all membrane systems require control of suspended solids in the feedwater to protect the membranes from excessive solids loading and/or to improve overall system performance.

Pretreatment for UF and MF systems typically includes strainers or bag filters rated in the 100- to 500-μm size range. Membrane manufacturers usually specify the maximum feedwater suspended solids concentration or turbidity allowed for their particular products. If the raw water solids (and organic) content is high, pretreatment coagulation and clarification are sometimes used. The cost of this more extensive pretreatment is at least partially offset by cost savings in the MF/UF system cost, because it can be designed at increased flux rates as a result of the improved feedwater quality.

TABLE 13.3 Recommended Feedwater Analyses for Pressure-Driven Membrane Systems

RO, NF, and ED systems	MF and UF systems
Temperature (field)	Temperature (field)
pH (field)	pH (field)
Alkalinity	Alkalinity
Hardness	Hardness
Conductivity (at 25° C)	Iron (total and dissolved)
Evaporative TDS (at 180° C)	Manganese (total and dissolved)
TDS by summation of major ions*	Turbidity
Calcium	Total suspended solids
Magnesium	Chlorine residual and other strong oxidants
Sodium	Total organic carbon
Potassium	Bacterial analysis
Barium	Silica†
Strontium	
Iron (total and dissolved)	
Manganese (total and dissolved)	
Bicarbonate*	
Carbonate*	
Hydroxide*	
Carbon dioxide*	
Sulfate	
Chloride	
Fluoride	
Nitrate and nitrite	
Phosphate (total)	
Silica (total)	
Silica (reactive)	
Ammonium	
Bromide	
Boron	
Turbidity	
Total suspended solids	
Silt density index (15 min) (field)	
Hydrogen sulfide	
Chlorine residual and other strong oxidants (field)	
Total organic carbon	
Radionuclides	
Bacterial analyses (total plate count)	

*Calculated value.
†May react with certain MF/UF cleaning solutions.

a. Groundwater source

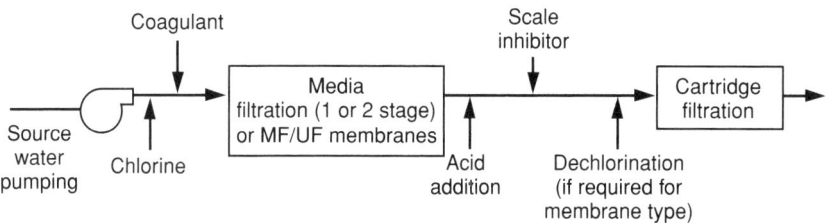

b. Low-turbidity surface water source (includes seawater)

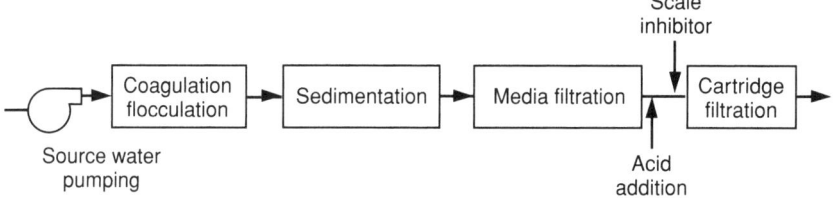

c. High-turbidity surface water source

d. Excessive calcium water (surface or ground) source

FIGURE 13.12 Typical RO or NF pretreatment processes.

Feedwater must be relatively free of suspended matter for RO, NF, and, to a lesser degree, EDR systems. Small suspended particles (colloids) must be removed to the extent determined by the type of membrane. Improved suspended solids removal can significantly lower the frequency rate of required membrane cleaning.

Nearly all RO and NF membrane systems use pretreatment cartridge filters nominally rated at 1 to 25 μm, depending on the type of system. Groundwater RO and NF systems commonly use 5-μm-rated cartridge filters. ED/EDR systems typically use 20- to 25-μm cartridges. Where solids loading would result in frequent filter cartridge replacement, or

TABLE 13.4 Feedwater Turbidity and SDI Limits Recommended by Manufacturers for RO, NF, and ED/EDR Systems

	RO and NF		ED/EDR
	Spiral wound	Hollow fiber	
Maximum turbidity, ntu	1*	—	2 to 3
Maximum SDI (15 min)	3 to 5	3 to 4	—
Maximum SDI (5 min)	—	—	15

*Recommended turbidity less than 0.2 ntu.
Sources: Spiral-wound RO data based on product literature from Dow-FilmTec, Koch Fluid Systems, Hydranautics, and TriSep. Hollow-fiber RO data based on product literature from the DuPont Company. EDR data based on Ionics, Inc., Bulletin No. 121-E, *EDR—Electrodialysis Reversal,* March 1984.

where cartridge filters are desired for backup protection only (the preferred method), additional pretreatment, such as granular media filters with or without chemical addition, is used. For waters with high suspended solids loading (usually surface water sources), a coagulation-flocculation-sedimentation pretreatment process may be used. MF or UF systems can be used as pretreatment for NF or RO (dual-membrane systems).

The two most common indicators of feedwater suspended solids content used today for RO, NF, and ED/EDR membrane systems are turbidity and the silt density index (SDI), although the use of particle counters is increasing. SDI is determined from the rate of plugging of a 0.45-μm filter under a feed pressure of 30 psig (207 kPa) as described in ASTM D4189. The ED/EDR, RO, and NF manufacturers usually specify maximum allowable turbidity or SDI limits. Typical turbidity and SDI limits for the various membrane desalting processes, depending on the particular membrane, are listed in Table 13.4.

Scaling Control. Scaling control is applicable to RO, NF, and ED/EDR. Design for all three processes must consider calcium carbonate and sulfate scaling control. RO and NF system design must also consider the need for silica control. Because electrodialysis does not remove silica from the feedwater and does not concentrate it in the concentrate flow channels, silica does not limit ED design recovery.

Depending on hydraulic recovery, the concentration of salt ions and silica in feedwater can be increased during the treatment process by as much as 10-fold. If concentrations exceed the solubility product of the compound at ambient conditions of temperature and ionic strength, scale can form within the modules, decreasing productivity and deteriorating permeate quality. More important, it can also cause failure of the membrane module. The sparingly soluble salts of concern for drinking water systems are calcium carbonate ($CaCO_3$); the sulfate salts of calcium, barium, and strontium ($CaSO_4$, $BaSO_4$, and $SrSO_4$, respectively); and silica (SiO_2). Other salts, such as calcium fluoride (CaF_2) or calcium phosphate, may also limit recovery in some waters.

Calcium Carbonate Control. The pH of calcium carbonate ($CaCO_3$) solubility can be estimated by

$$pH_s = pCa + pAlk + K$$

where pH_s = solubility pH
 pCa = negative logarithm of calcium concentration
 $pAlk$ = negative logarithm of alkalinity (bicarbonate, HCO_3 concentration)
 K = constant related to ionic strength (and TDS) and temperature

MEMBRANE PROCESSES 13.17

Example:

Temp = 20°C
pH = 7.6
Ca^{++} = 800 mg/L as $CaCO_3$
HCO_3 = 774.8 mg/L as $CaCO_3$
TDS = 7853.6 mg/L (use 5000 mg/L)
pCa = 2.10
$pHCO_3$ = 1.81
K_{LSI} = 2.37
pH_s = 2.10 + 1.81 + 2.37 = 6.28
pH_c = 8.19
LSI = 8.19 - 6.28 = +1.91

FIGURE 13.13 Langelier saturation index. (*Adapted from* DuPont Permasep Products Engineering Manual, *1992*.)

The tendency to develop $CaCO_3$ scale during the treatment of fresh and brackish waters can be determined by calculating the Langelier saturation index (LSI) of the concentrate stream.

$$LSI = pH_c \text{ (of concentrate stream)} - pH_s$$

Figure 13.13 presents a nomograph that can be used to determine LSI. For seawater desalting, the Stiff and Davis saturation index (SDSI) is often used. Figure 13.14 shows graphs with which the SDSI K value can be determined.

Control for $CaCO_3$ scale can be achieved by using the following methods:

- Acidifying to reduce pH and alkalinity
- Reducing calcium concentration by ion exchange or lime softening
- Adding a scale inhibitor chemical (antiscalant) to increase the apparent solubility of $CaCO_3$ in the concentrate stream
- Lowering the design recovery

Feedwater acidification is the most widely used method for controlling $CaCO_3$ scaling. Sulfuric acid is generally used because of its low cost and less hazardous (nonfuming) characteristics. When scaling from sparingly soluble sulfate salts is a concern and adding sulfate ions is undesirable, hydrochloric acid should be considered.

The need for acidification is lessened with the introduction of sodium hexametaphosphate (SHMP) or polyacrylic acid or other types of antiscalants. Polyacrylic acid antiscalants at dosages of less than 5 mg/L typically can provide scale control up to an LSI (or SDSI) value of about +2.0. Also, polyacrylic acid antiscalants are often used at plants

FIGURE 13.14 Stiff and Davis K versus ionic strength and temperature. (*Adapted from* DuPont Permasep Products Engineering Manual, *1992*.)

employing noncellulosic membranes, and pH control is not required to minimize membrane hydrolysis—a concern with the use of cellulose products. When feedwater LSI is initially high, softening may be required.

Sulfate Scale Control. Scaling potential of sulfate salts can be estimated by calculating the ion product for each salt in the concentrate stream and comparing it with the solubility product (K_{sp}) of the salt at the temperature of interest. Figures 13.15 and 13.16 present calcium sulfate ($CaSO_4$), strontium sulfate ($SrSO_4$), and barium sulfate ($BaSO_4$) solubility graphs.

The most widely used method of control for sulfate scale is the addition of SHMP or polyacrylic acid or other appropriate types of antiscalants. Polyacrylic acid antiscalants are commonly used for RO and NF applications because they permit a greater degree of supersaturation of critical ions in the concentrate stream than does SHMP.

Silica Control. The silica (SiO_2) scaling potential of the concentrate stream usually can be estimated by the following formula (DuPont Company, 1992):

$$SiO_{2(max)}(mg/L) = SiO_{2(temp)}(mg/L) \times pH \text{ correction factor}$$

where $SiO_{2(max)}$ = maximum concentrate stream silica, mg/L
$SiO_{2(temp)}$ = silica solubility concentration at pH 7.5 as a function of water temperature, mg/L
pH correction factor = silica solubility factor at various pH values

Concentrate silica as a function of temperature is shown in Figure 13.17, and concentrate stream pH correction factors are shown in Figure 13.18.

In some cases, RO and NF systems have successfully operated with concentrate stream silica levels exceeding theoretical solubility limits, especially with some applications us-

FIGURE 13.15 K_{sp} for $CaSO_4$ versus ionic strength. (*Adapted from* DuPont Permasep Products Engineering Manual, *1992*.)

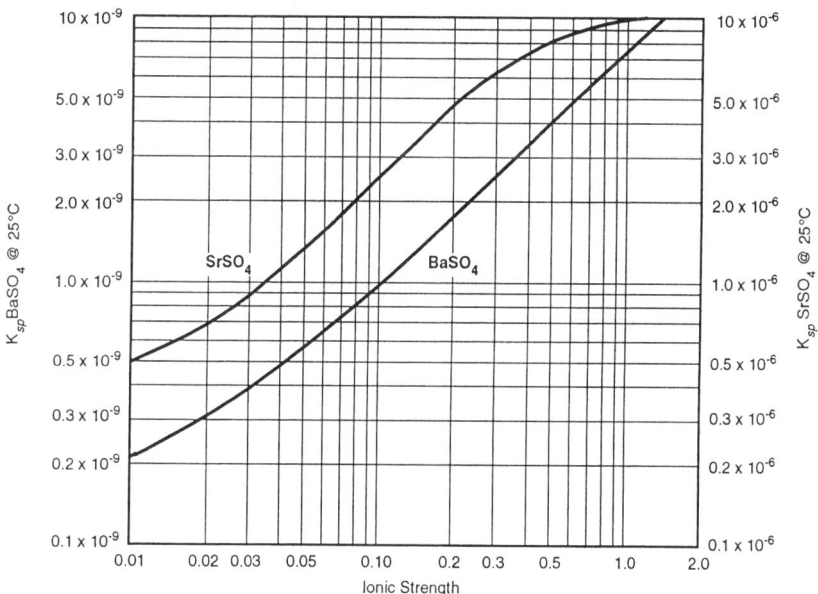

FIGURE 13.16 K_{sp} for $BaSO_4$ and $SrSO_4$ versus ionic strength. (*Adapted from* DuPont Permasep Products Engineering Manual, *1992*.)

FIGURE 13.17 Effect of temperature on SiO_2 solubility. (*Adapted from* DuPont Permasep Products Engineering Manual, *1992.*)

ing new silica-specific antiscalants. Silica scaling has occurred in other systems at concentrations much less than theory would predict because of complexation and precipitation of silica with trivalent ions such as oxidized iron or aluminum. Where feedwater silica concentrations approach solubility limits, it is essential that the presence of these metal species be minimized.

The following pretreatment methods can be used to control silica scale:

- Reducing hydraulic recovery to reduce concentrate stream silica concentrations
- High lime softening to reduce feedwater silica concentrations
- Increasing the temperature of the feedwater
- Increasing feedwater pH to 8.5 or higher (taking into account the impact of increased pH on $CaCO_3$ scaling potential)
- Adding specific silica antiscalant chemical

Concentration of Sparingly Soluble Salts and Silica. Concentrations of sparingly soluble ions and silica (for RO and NF) in the concentrate stream can be estimated by dividing feedwater concentrations of each by 1 minus the desired hydraulic recovery:

$$C_c = C_f \times \frac{1}{1 - Y}$$

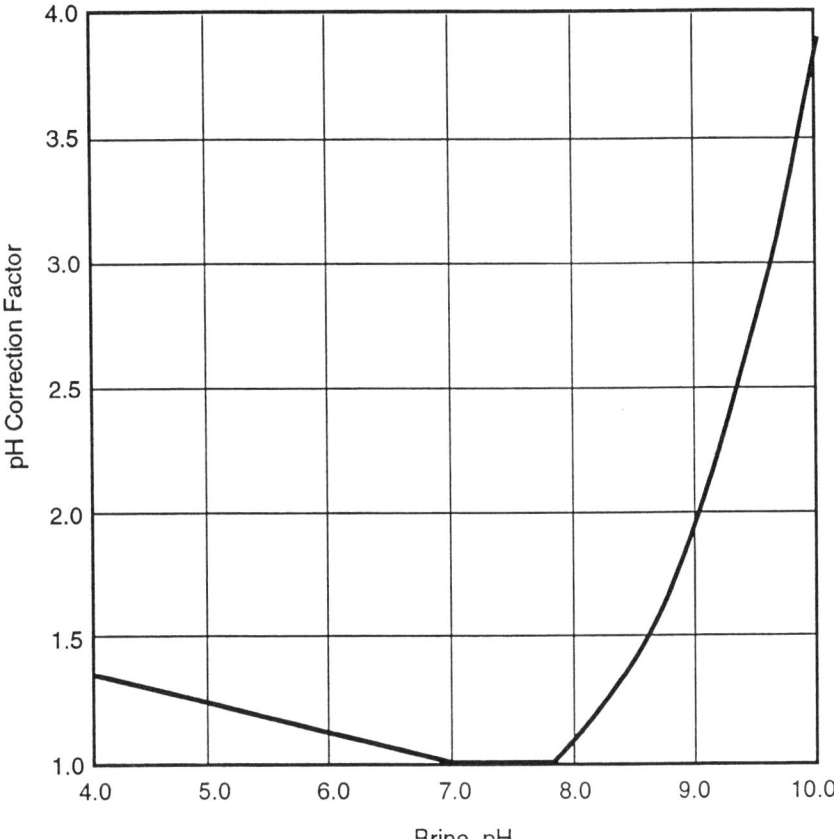

FIGURE 13.18 SiO$_2$ pH correction factor. (*Adapted from* DuPont Permasep Products Engineering Manual, *1992.*)

where C_c = concentration of solute in concentrate stream, mg/L
C_f = concentration of solute in feedwater stream, mg/L
Y = hydraulic recovery (permeate flow rate divided by feed flow rate), expressed as decimal

This expression overestimates bulk stream concentrate ion concentration because it assumes that membrane rejection of each ion is 100%, which is never the case. A more accurate method uses a formula taking into account the membrane's percent rejection value for each ion, if known. This formula is

$$C_c = C_f \times \frac{(1 - Y) \times \text{SP}_i}{1 - Y}$$

where SP$_i$ is the solute passage for the specific ion, expressed as a decimal.

With available computer programs for the projection of RO, NF, and EDR system performance, the need to manually calculate scaling potential has declined. It is more con-

TABLE 13.5 Typical RO and NF Concentrate Stream Solubility Design Criteria Used by Some Manufacturers

Compound	Concentrate stream criteria[a,b]		
	Typical (without scale inhibitor)	Typical (with scale inhibitor)[c]	Maximum (with scale inhibitor)[c]
Calcium carbonate ($CaCO_3$)	$LSI^d < -0.2$	$LSI < +1.8^h$	$LSI < +2.3^h$
Calcium sulfate ($CaSO_4$)	$IP^e < 0.8 \times K_{sp}^{\ f}$	$IP < 1.5 \times K_{sp}$	$IP < 2 \times K_{sp}$
Strontium sulfate ($SrSO_4$)	$IP < 0.8 \times K_{sp}$	$IP < 8 \times K_{sp}$	$IP < 8 \times K_{sp}$
Barium sulfate ($BaSO_4$)	$IP < 0.8 \times K_{sp}$	$IP < 40 \times K_{sp}$	$IP < 60 \times K_{sp}$
Silica (SiO_2)[g]	$IP < 0.8 \times K_{sp}$	$IP < 1.5^g$	<200 mg/L[g]

[a]The membrane and antiscalant manufacturers should be contacted to confirm design criteria for the specific products for the specific application.
[b]In some cases, maximum calcium fluoride and calcium phosphate levels are specified.
[c] Assumes addition of an appropriate scale inhibitor and that, when shut down, the system is flushed within the specified time period, commonly less than 4 h.
[d]LSI is the Langelier saturation index. Positive LSI values indicate potential for calcium carbonate precipitation. Negative LSI values indicate corrosive tendencies.
[e]IP is the ion product of the compound based on the concentrations in the concentrate stream.
[f]K_{sp} is the solubility product. If the ion product (IP) is greater than the K_{sp}, precipitation can occur. If the IP is less than K_{sp}, precipitation usually does not occur.
[g]Presence of polyvalent metal oxides also affects criteria. Silica precipitation usually occurs from polymerization, which is a relatively slow process, but is significantly affected by other ions in the water.
[h]For brackish water greater than about 10,000 mg/L TDS and seawater applications, the Stiff and Davis saturation index (SDSI) is typically used instead of LSI. SDSI values typically are not greater than +0.5 for RO design.
Sources: DuPont Company: Wilmington, Del. Hydranautics: Oceanside, Calif. Dow–FilmTec: Midland, Mich. Koch Fluid Systems: San Diego, Calif. TriSep: Goleta, Calif.

venient to send the feedwater quality analysis to the membrane manufacturer, system supplier, or membrane treatment consultants and request scaling predictions. The desired recovery and the type of membrane to be used are also variables in the analyses. Table 13.5 presents typical concentrate stream solubility criteria used in RO and NF design.

Microbial Control. Microbial growth can foul membrane systems and cause performance declines. In addition, certain types of cellulose acetate RO and NF membranes can be degraded by some biological enzymes. Microbial control methods for membrane systems are dependent on the source water quality, type of membrane system, and membrane tolerance to potential disinfection chemicals, especially oxidants such as chlorine. Only chemical biocides and biostats that are acceptable to the membrane manufacturers and approved for potable water applications (by independent testing agencies such as NSF International) should be used for drinking water treatment plants.

Many groundwater systems do not have microbial fouling problems because of the relatively low bacterial concentrations and microbial activity. Surface water and other source waters with high microbial populations may need continuous or intermittent on-line microbial control methods to minimize membrane fouling.

Essentially all membrane systems use periodic (off-line) chemical cleaning to remove foulants, including microbial matter. MF and UF systems also use backwashing, sometimes with chemical addition (such as hypochlorite, if the membrane is chlorine-tolerant), to remove foulants, including microbes. Some MF and UF membranes (such as PVDF membranes) can be exposed to high concentrations of free chlorine.

Chlorination is sometimes used for disinfection, although disinfection by-product formation potential may preclude the use of free chlorine. Chlorine dosage rate must be closely controlled. Most types of cellulose acetate RO and NF membranes can tolerate up to 1.0 mg/L free chlorine. However, most composite RO and NF membranes, such as polyamide (PA) membranes, cannot tolerate chlorine or other strong oxidants. If free chlorine is used prior to membranes intolerant to free chlorine, dechlorination processes, such as sodium bisulfite or sulfur dioxide addition, can be used upstream of the membranes. Granulated activated carbon (GAC) is also sometimes used for dechlorination, but usually only for small systems. Microbial regrowth downstream of chemical dechlorination systems or microbial matter released from GAC may also cause problems in the membrane system. The chlorination and dechlorination system, if used, must be designed to be very reliable to prevent unacceptable oxidant concentrations from reaching the membranes.

Chlorination-dechlorination is also sometimes used for ED systems because most ED membranes cannot tolerate chlorine. Some EDR membranes may have relatively long life with a continuous free chlorine exposure of up to 0.3 mg/L.

Limited concentrations of chloramines are also sometimes used for bacterial control in membrane systems where the membranes can tolerate the combined chlorine concentration. The membrane manufacturer should be contacted to obtain maximum acceptable oxidant exposure criteria, if any. The chloramination system (feeding ammonia and chlorine) must be designed to very reliably maintain the target combined chlorine residual and preclude any free chlorine from entering the membrane system.

Biological control through periodic shock treatments with sodium bisulfite or other nonchlorine compounds is also used for membrane systems. Ultraviolet light disinfection can be used as pretreatment for membrane systems, but the lack of a residual may result in regrowth problems. MF or UF membranes in dual-membrane systems also reduce biological matter prior to the NF or RO system.

Bacterial control should also be considered for membrane units not in service. When off-line for extended downtime, membranes are commonly placed in preservative and storage solutions of various formulations to retard microbial growth using the cleaning system.

Hydrogen Sulfide Control. Where hydrogen sulfide is present, such as in some well waters, no chlorine or exposure to air can be allowed; colloidal sulfur will form and foul an RO or NF membrane system. Because chemical cleaning is typically ineffective in removing sulfur from the membrane surface, membranes that become fouled are usually replaced. When low levels of hydrogen sulfide exist in feedwater, the system can be designed to be airtight and hydrogen sulfide can be removed in posttreatment degasifiers. Hydrogen sulfide greater than 0.1 to 0.3 mg/L can cause problems with ED/EDR systems and must be removed in pretreatment. RO and NF membranes can tolerate relatively high concentrations of hydrogen sulfide without damage.

Iron and Manganese Control. Iron and manganese may cause problems with RO, NF, and ED/EDR membrane systems. However, as long as these metals are kept in their reduced state and dissolved in water, they cause little problem with RO and NF systems. Because iron and manganese form precipitates in the presence of oxygen or at high pH, acid addition is often used for metal oxide control (in addition to its other uses, previously discussed).

Sequestrants such as sodium hexametaphosphate are also commonly fed to inhibit iron and manganese deposition. Some types of polyacrylic acid antiscalants react with iron and manganese, causing fouling problems in downstream membranes, and should not be used. Proprietary antifoulants (scale inhibitor/dispersants) are also commonly used to control fouling and scale in the presence of iron and manganese.

Iron and manganese can foul ED/EDR membranes even when they are in the reduced state. In addition, manganese can plate out on the electrodes, decreasing their efficiency. For ED/EDR systems, iron removal is usually recommended if feedwater iron concentration exceeds 0.3 mg/L or if the manganese level exceeds 0.1 mg/L.

If iron or manganese precipitates, or if either metal is present from corrosion products, ED/EDR, RO, and NF systems can be fouled. Appropriate materials should be used for construction of feedwater supply or pretreatment systems.

Three pretreatment processes can be used for iron or manganese removal:

- Oxidation using air, chlorine, or another oxidant, followed by granular media filtration, MF, or UF
- Oxidation using potassium permanganate, followed by a manganese greensand filter
- Cation ion exchange softeners

Residual oxidant levels must be monitored and controlled before entering the membrane system. If lime softening is used for scale control, it has the additional benefit of lowering iron and manganese levels in feedwater.

Organics Control. Organic matter may be classified as either suspended or dissolved. Suspended (and colloidal) organic solids and microbial matter (previously discussed) adversely affect membrane systems.

Naturally occurring dissolved organics cause fouling in ED/EDR systems and generally, to a lesser degree, in RO or NF systems. In fact, some NF and RO systems are used to remove natural and synthetic dissolved organics and color. Membrane manufacturers are aware of many organic compounds that cause problems in their systems and can be contacted for specific information. There is no definitive correlation between the quantity of organics present, such as measured by total organic carbon (TOC) analyses, and performance decline of RO or NF membranes. However, if TOC exceeds 10 mg/L in feedwater, the potential effects of fouling usually warrant further investigations. Pilot testing is often used to determine the effects of dissolved organics on membrane systems.

The organic content of the feedwater to MF and UF systems may control the acceptable design flux and the backwash and cleaning frequencies. Sometimes, MF/UF membrane systems are designed in a direct filtration arrangement, where metal coagulant addition and flocculation (often less than 5 min) are provided upstream of the MF/UF unit to remove organic matter (and solids) and improve performance. In some systems, powdered activated carbon (PAC) is fed prior to the membrane filter to aid in organic removal and lessen fouling.

Oil, greases, hydrocarbons, and various organic solvents and other chemicals can damage or foul membranes and should not be allowed to enter the feedwater supply.

Conventional coagulation, flocculation, and sedimentation processes and lime softening can reduce the organic content of feedwater before it enters membrane units. Activated carbon filters are sometimes used to remove organics or for dechlorination of feedwater entering RO, NF, and ED systems; but problems caused by carbon fines or biological microorganisms released from carbon filters have occurred in some systems. Surfactants and organic polymers, especially the cationic type, can foul membranes and should not be used without membrane manufacturer approval.

Control of pH. All membrane types have specified pH ranges in which they should operate. Many types of ED and polyamide RO and NF membranes can tolerate continuous exposure to feedwater in the pH range of 4 to 10 and intermittent exposure to an even wider pH range. The various types of cellulose acetate RO and NF membranes normally require pH between approximately 4 and 6.5 to minimize degradation caused by hydrolysis. Many types of polysulfone, polypropylene, and PVDF MF and UF membranes tolerate a pH range of at least 2 to 13. Inorganic MF and UF membranes can generally accept feedwater with nearly any pH.

Temperature Control. Water temperature significantly affects membrane systems in several ways:

- It alters membrane material characteristics and membrane life.
- Water viscosity and density affect membrane hydraulic performance and required membrane area.
- It changes solubility of sparingly soluble salts and silica, which limits the design recovery of membrane desalting processes.

Membrane systems must operate within the manufacturer's temperature guidelines to maximize membrane life. The various types of cellulose acetate membranes deteriorate from hydrolysis at increasing rates as temperatures rise. Flux decline caused by compaction of RO and NF membranes also is greater at higher temperatures.

The maximum temperature limit for the various brackish water cellulose acetate RO and NF membranes ranges from 35° to 40° C, depending on the membrane. The various types of polyamide RO membranes typically are rated at maximum temperatures ranging from 40° to 50° C, but their useful life can be significantly reduced at these elevated temperatures.

ED/EDR membranes commonly can operate at temperatures up to 45° C. Typically, PVDF, polysulfone, polypropylene, and several other types of MF and UF membranes tolerate a temperature up to at least 40° C. Inorganic (ceramic) MF and UF membranes can tolerate temperatures well over 100° C.

Viscosity and density of water increase at colder temperatures, and for pressure-driven systems greater membrane area is required to produce a specified product flow at a given feed pressure; or for a given system and feed pressure, the production rate declines at lower temperatures.

Temperature affects membrane system performance not only directly by influencing product water flow rates or salt removal, but also indirectly by affecting solubility of compounds, which can precipitate and foul the system. Some compounds, such as calcium carbonate, have a greater tendency to scale at higher temperatures. However, most compounds normally encountered in natural waters have improved solubilities as the temperature rises.

MEMBRANE UNIT DESIGN

Once treatment objectives have been identified, design criteria can be developed for the membrane process. The design considerations are grouped for

- RO and NF units
- UF and MF units
- ED and EDR units

Reverse Osmosis and Nanofiltration Unit Design

Because of the high solute rejection of RO and NF membranes, which concentrate inorganic ions among other constituents, there are many special considerations in the treatment unit design.

Design Equations. The basic behavior of permselective (semipermeable) membranes can be described by the following two diffusion model equations. Permeate flow through the membrane may be expressed as

$$F_w = A \times (P_{tm} - \pi_{tm})$$

where F_w = water flux, g/(cm^2 · s)
A = water permeability coefficient, g/(cm^2 · s · atm)
P_{tm} = hydraulic pressure differential applied across membrane, atm
π_{tm} = osmotic pressure differential across membrane, atm

The solute (or salt) flux through the membrane may be expressed as

$$F_s = B \times (C_1 - C_2)$$

where F_s = solute (or salt) flux, g/(cm^2 · s)
B = solute (or salt) permeability constant, cm/s
$C_1 - C_2$ = concentration gradient across membrane, g/cm^3

Water and solute permeability coefficients are characteristic of the particular membrane type.

Water flux depends on applied pressure, but solute flux does not. As pressure of membrane feedwater increases, water flow through the membrane (water flux) increases, while solute flow remains essentially unchanged. Permeate quantity increases with increased applied pressure, as does the quality (decreased solute concentration).

Water flux decreases as the salinity of the feed increases because of increased osmotic pressure differential resulting from increased salinity. As increasing amounts of water pass through the membrane system, the salinity of the remaining feedwater (concentrate) increases. Concentrate osmotic pressure increases, resulting in a lower water flux with increasing overall percent water recovery.

Finally, because salinity of the feed concentrate stream increases with increasing permeate production from a given volume of feed, and the membrane rejects a fixed percentage of solute, product water quality decreases (higher concentration) with increasing recovery. Table 13.6 presents typical feed pressures and recoveries for RO and NF systems.

TABLE 13.6 Typical RO and NF System Feed Pressure and Recovery

Application (with example TDS)	Typical operating pressure		Typical product water recovery,* %
	psi	kPa	
Reverse osmosis			
Seawater (35,000 mg/L TDS)	800 to 1,200	5,520 to 8,270	30 to 45
Brackish water (5,000 mg/L TDS)	350 to 600	2,410 to 4,140	65 to 80
Brackish water (1,000 mg/L TDS)	125 to 300	860 to 2,070	70 to 85
Nanofiltration			
Freshwater (500 mg/L TDS)	50 to 150	340 to 1,030	80 to 90

*Maximum allowable recovery is site-specific and depends on the feedwater's scaling potential, membrane rejection and product water quality requirements, concentrate stream osmotic pressure, and the use of scale-inhibiting chemicals.

Source: Adapted from Bergman, Robert A. "Anatomy of Pressure-Driven Membrane Desalination Systems." *1993 Annual AWWA Conference Proceedings—Engineering and Operations.* Denver, Colo.: American Water Works Association, 1993.

MEMBRANE PROCESSES   13.27

Recovery Considerations. As recovery increases, the following factors must be considered in designing an RO or NF system.
  Scaling. The concentration factor and potential for scaling increase as recovery increases. The source feedwater composition must be evaluated to estimate maximum operating recovery and the necessary pretreatment requirements (for example, pH adjustment or scale inhibitor addition). The concentrations of solutes are greater near the membrane surface than in the bulk stream due to *concentration polarization*.
  Hydraulics. Optimal performance requires adhering to minimum concentrate and maximum feed flow conditions for membranes. Feed flow to the first element in a pressure vessel and concentrate flow from the last element in a pressure vessel must satisfy the manufacturer's stated requirements.
  System design must provide adequate membrane concentrate flow. Concentrate staging of membranes and pressure vessels is typically used for recoveries greater than 50% to 60% (see **Membrane Module Arrays and Staging**). Some small systems are designed with concentrate recycle to produce flows above the minimum specified by the membrane manufacturer.
  Source Water Use. The required volume of source feedwater necessary to produce the same volume of permeate decreases as the recovery rate increases. Maximizing recovery rates minimizes both the source water requirement and the volume of concentrate generated.
  Permeate Water Quality. Feed-concentrate average salinity increases as recovery increases. Because the flow of solutes through the membrane is a direct function of their concentration in the feed concentrate stream, permeate quality decreases as recovery increases.

Solute Rejection and Solute Passage. The removal, rejection, or passage of solutes in a membrane system requires consideration of several variables.
  Manufacturer's Specifications. RO and NF membranes are rated for nominal and minimum rejections based on a specific test condition. Each RO membrane manufacturer typically provides both design and minimum specifications relating to percent rejection for sodium chloride (NaCl). With NF modules, specifications are also given for selected divalent salts, for example, magnesium sulfate ($MgSO_4$), and possibly organics in terms of general molecular weight cutoff in daltons.
  For example, low-pressure spiral-wound RO membrane elements are commonly rated to have 96% to 99% salt rejection at 150 to 225 psig (1,035 to 1,550 kPa) feed pressure; 25° C feed temperature; 8% to 15% recovery; 1,500 to 2,000 mg/L NaCl feed; and a pH of 5.7 to 7.0.
  Seawater spiral-wound RO elements are typically rated at 99% to 99.7% salt rejection at 800 psig (5,520-kPa) feed pressure; 25° C feed temperature; 7% to 10% recovery; 32,000 to 35,000 mg/L NaCl feed; and pH of 5.7 to 7.0.
  NF membranes used for membrane softening, THMFP, and color removal typically have 95% to 98% $MgSO_4$ rejection with a 1,000 to 2,000 mg/L $MgSO_4$ feed solution (or greater than 70% salt rejection with 1,500 mg/L NaCl feed) at 70 to 100 psig (483 to 689 kPa) feed pressure; 25° C feed temperature; 10% to 15% recovery; pH of 6.5 to 7.0; and a molecular weight cutoff in the 150- to 400-dalton range.
  Inorganic versus Organic Solutes. Both RO and NF membranes reject ionic and many nonvolatile organic solutes to a high degree. Composite membranes typically reject organic compounds better than do cellulose acetate or polyamide hollow fine-fiber membranes.
  In general, volatile organic compounds are poorly rejected by all membrane types (less than 50%), although certain composite formulations have considerably higher rates. NF membranes reject multivalent ionic and many nonvolatile organic solutes to a high de-

gree (90% or greater); monovalent solute rejection is much lower (75% or less) and is strongly dependent on the types of co-ions present in the feed and the feed pH.

Types of Membranes. Membrane composition may affect solute rejection over time. Composite membranes are generally stable and maintain their rejection properties over long periods of operation. Cellulosic membranes continuously undergo hydrolysis, which gradually diminishes their rejection properties. The extent of rejection loss depends on the rate of hydrolysis, but often the salt passage is assumed to double within 3 years of operation.

Increased temperature also increases the rate of cellulose acetate membrane hydrolysis. Cellulose acetate membrane module manufacturers typically recommend that feed pH be controlled between 5.0 and 6.2. The useful life of a membrane is also highly dependent on external (nonmembrane) factors associated with the application such as pretreatment operation, type of cleaning chemicals, and frequency of cleaning.

Operating pH. Feedwater pH may affect the rejection properties of composite membranes and the degree of rejection for certain ionic constituents. Composite membranes generally have an optimum pH at which rejection is maximized, although it is not always a published parameter. Rejection of certain ions such as fluoride, bicarbonate, and boron can vary with pH. As pH increases above 5.5, fluoride and bicarbonate ion rejection increases. Boron rejection increases significantly as pH is increased from neutral to high.

Flux. For a system containing more than one stage, the flux value can be expressed as a system average or as individual values for each stage. The individual module flux rate within a system generally decreases from beginning to end. Lead modules operate at a higher flux rate than trailing modules because feed TDS (and osmotic pressure) is lowest and feed pressure highest at the lead end of the system.

System capacity is determined by the amount of membrane area provided (number of elements in the system) and the flux rate. The number of membrane elements required can be calculated, given an average membrane flux and active membrane area per element. Most large-capacity NF and RO systems use 8-in.-diameter $\times$ 40-in.-long elements containing 350 to 440 ft^2 (33 to 41 m^2) of membrane area per element and are rated at 8,000 to 15,000 gpd (30 to 57 m^3/day) each.

As flux rate increases, the loading rate of potential foulants at the membrane surface increases. Theoretically, there is an optimum flux rate for every water supply and membrane combination. In general, groundwaters allow higher design flux rates, and surface water applications are restricted to lower design flux rates.

The water production capacity of a membrane module specified by the manufacturer, typically expressed in gallons per day per square foot (gpd/ft^2), is based on a membrane flux achieved during laboratory testing at standard test conditions. Typically, manufacturer product specifications base performance on fluxes from 20 to 30 gpd/ft^2 (0.034 to 0.051 m/h). Rarely, if ever, can this rate be achieved during system operation because of fouling constraints and the impact of recovery. Typical average system flux rates for groundwater are 13 to 18 gpd/ft^2 (0.022 to 0.031 m/h) and for surface waters between 8 and 12 gpd/ft^2 (0.014 and 0.020 m/h).

Low-pressure RO and NF systems are sometimes designed with flux balancing to better balance the flux rates of the various stages. This is an attempt to minimize fouling rates and extend the intervals between membrane cleanings, or to improve the overall hydraulics of the system. Where flux balancing is desired, either permeate backpressuring the first stage of the system or boosting the pressure of the interstage feeds can be used. These techniques lower transmembrane pressure of the first stage to limit its production, shifting a greater proportion of the production to the later stages.

New "low-fouling" RO membranes are now available that have favorable properties: low membrane surface roughness, little or no surface charge, and less hydrophobicity.

The degree to which fouling would be lessened, if any, for a given membrane element type and application, should be determined with pilot testing.

Temperature. Water temperature is a factor that must be considered in all membrane systems. The important effects of temperature on membrane system design are these:

- Membrane operations using high-temperature waters require lower operating pressures to achieve a given flux, compared with operation on low-temperature waters.
- Membrane permeate quality degrades as water temperature increases and membrane flux is held constant (water flow through the membrane is constant and solute flow increases). Minimum temperature should be used to determine maximum anticipated feed pressure (and thus pump pressure and motor horsepower) at the design flux rate because the net driving pressure (NDP) is less at a higher temperature.
- If the design flux rate is held constant, maximum temperature dictates the worst permeate quality condition because solute passage is greater at higher temperatures.

Assuming all other factors (including membrane area and feed pressure) remain constant, the permeate productivity of a pressure-driven membrane system is about 30% to 40% less at 15° C than it is at 30° C. For RO and NF systems, an approximation of permeate flow at any temperature relative to flow at 25° C, assuming all other factors are constant, is as follows:

$$Q_p = Q_{p(25° C)} \times 1.03^{(T - 25)}$$

where Q_p = permeate flow at temperature T
$Q_{p(25° C)}$ = permeate flow at 25° C
T = water temperature, °C

Typical membrane system design necessitates evaluation of performance at minimum and maximum temperatures assuming new (initial start-up) and "used" (often 1, 3, or even 5 years of operation) membranes. A maximum design flux or feed pressure is assumed, and the minimum temperature is used to determine the required membrane area for "used" membranes. Given the membrane area, the performance at maximum temperature is then evaluated to verify that product water quality goals are also met.

All polymeric membranes have maximum operating temperatures. The use of high-temperature waters may exclude the use of cellulose acetate membrane, and if temperatures are too high, membrane feedwater cooling will be required for all types of membranes.

Feed Pressure Requirement. The required feed pressure depends on the following:

- Membrane type
- Flux
- Recovery
- Osmotic pressure
- Temperature
- Permeate pressure
- Changes over time

All the above points have been discussed previously except changes over time. To ensure that a system continues to produce the desired quantity and quality of permeate over the expected life of the membrane, the feed pump must provide sufficient membrane feed pressure throughout the design life of the system. The three main causes of time-related performance changes are membrane fouling and scaling, membrane compaction, and membrane degradation.

To illustrate the effect of these factors on water flow through the membrane, system productivity is typically plotted as a function of operating time (Figure 13.19), referred to as a *flux decline curve*. Flux decline is defined as the loss in system productivity, expressed as a percentage of initial productivity, that occurs with operating time, assuming constant feedwater quality, pressure, temperature, and recovery. Flux decline causes an increase in required feed pressure. The principal causes of flux decline are membrane compaction and the effects of fouling and scaling that cannot be reversed by cleanings (referred to as *irreversible* fouling or scaling). Additional design pumping pressure, typically at least a 15% to 20% increase in available net driving pressure, must be included in the design to offset flux decline.

Membrane compaction occurs when the membrane compresses under the applied pressure of operation. The degree of compaction depends on the specific membrane type. This compression increases membrane resistance to water flow and increases NDP. A majority of compaction (and flux loss) occurs during the first 100 h of operation. Membrane projections allow for decreases in productivity over a period of time, usually 3 to 5 years.

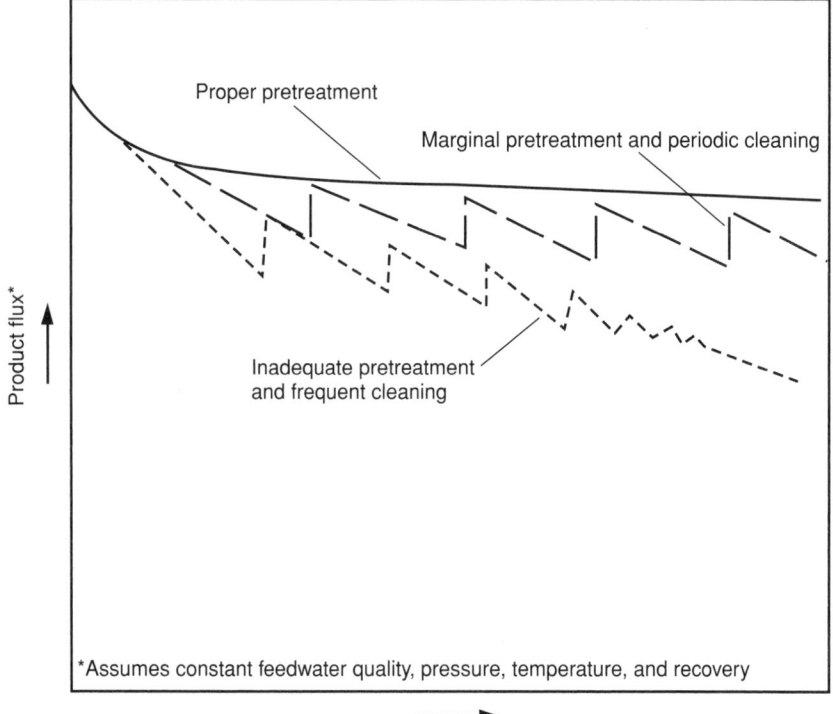

FIGURE 13.19 Membrane flux decline curves under different pretreatment conditions. (*Adapted from* DuPont Permasep Products Engineering Manual, *1992.*)

Fouling or scaling occurs when inorganic scales, suspended solids, organics, or biofilms collect on the membrane surface or the membrane module. Proper pretreatment system design should minimize membrane fouling. A membrane cleaning system should be included in the design to periodically clean and restore membrane performance.

Product (or permeate) pressure must be offset by feedwater pressure to produce the proper net driving pressure. Feed pump requirements are minimized by designing the facility for minimum permeate backpressure. For some low-pressure RO and NF systems, a permeate backpressure valve is installed for the first stage to help balance flux rates throughout the system in an attempt to minimize fouling. Some membrane manufacturers specify maximum allowable permeate pressures for their products to prevent module damage.

Membrane Module Arrays and Staging. The arrangement of membrane modules used to achieve the desired process flow or optimum hydraulic configuration is called the *membrane module array*. Arrays are based on the number and location of pressure vessels (modules or permeators). For example, a 2:1 array would have two parallel pressure vessels feeding one. Depending on the quality of the feedwater or permeate quality requirements, RO and NF system arrays are either concentrate-staged or permeate-staged. Sometimes permeate staging is referred to as multiple "passes."

Concentrate-Staged Design. Standard 40-in.-long (102-cm) spiral-wound RO or NF membrane elements have a maximum individual operating recovery of 8% to 15%. To achieve higher recoveries, up to eight elements are loaded in series into pressure vessels, with the concentrate flow from the first element becoming the feed flow to the second element, and so on.

In a typical large, spiral-wound system design, six or seven elements are loaded into each pressure vessel (referred to as a 6M or 7M vessel). Each 6M vessel in the design provides for an operating recovery of approximately 50%. In a concentrate-staged design (Figure 13.20) containing up to three stages, pressure vessels are arranged so that the concentrate flow from the first stage serves as the feed to the second stage, and the concentrate flow from the second stage serves as the feed to the third stage. In this flow configuration, concentrate staging maximizes system recovery.

As an example, to achieve 75% recovery, two parallel 6M vessels in the first stage can be arranged to feed a single vessel in the second stage (called a 2:1 array). If feed flow is 100 gpm (6.3 L/s) to the first stage using 6M vessels operating at 50% recovery, first-

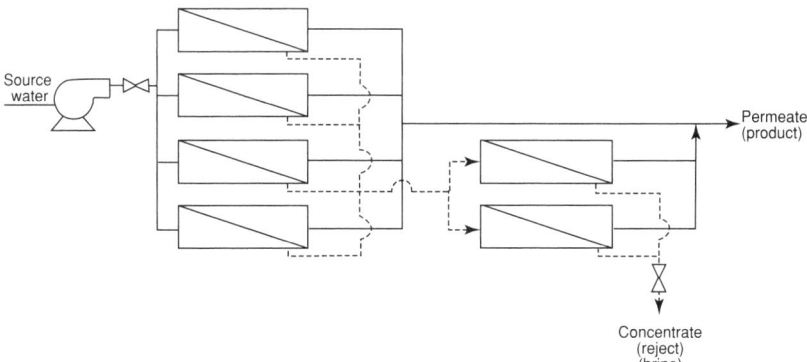

FIGURE 13.20 Concentrate staging (two-stage design shown).

stage permeate flow will be 50 gpm (3.2 L/s) and the concentrate flow will be 25 gpm (1.6 L/s) from each vessel, or a total of 50 gpm (3.2 L/s). This concentrate flow would then be fed into a single 6M vessel in the second stage, where 50% recovery for this stage would yield a concentrate flow and a permeate flow of 25 gpm (1.6 L/s) each. The total system produces 75 gpm (4.7 L/s) of permeate and 25 gpm (1.6 L/s) of concentrate.

Commonly, two-stage and three-stage design using 6M vessels can yield recoveries of 75% and 85%, respectively. Designs using 7M vessels can achieve a recovery of 65% in one stage and 85% in two stages. Typically, design involves an iterative process considering various combinations of pressure vessel staging. There are usually several concentrate-staged arrays that meet the treatment objectives and hydraulic criteria, especially if flux balancing (discussed under *Flux*) is employed. In many cases, the array selected is not a "perfect" mathematical taper (i.e., each stage does not have exactly 2 times as many pressure vessels and membrane elements as the following stage).

For hollow-fiber RO membrane modules (permeators), each pressure vessel contains a "bundle" of membrane fibers. Modules in each stage are placed in parallel and, for some systems, with designed pressure loss in each module's concentrate outlet piping or tubing (before concentrate header piping). This pressure drop helps balance flows between each module. For example, a pressure drop of 35 psi (240 kPa) or more has been used for many single-stage systems and the final stage of multistage systems. Commonly, maximum recoveries up to 50%, 75%, and 90% are used for one-, two-, and three-stage hollow-fiber membrane systems, respectively.

Permeate-Staged Design. In applications in which the TDS of the feedwater is too high to produce a permeate of sufficient quality with a single pass through the membrane, system design incorporates permeate staging (Figure 13.21). This design is common to RO systems treating seawater with very high salinities (45,000 mg/L or greater); when treating typical seawater salinity (35,000 mg/L), where permeate quality requirements are more stringent than typical drinking water standards; or when an extra safety factor for product water quality is desired. In this type of design, permeate from the first RO train becomes feedwater to the second RO train. The first-pass system uses high-pressure seawater RO membranes, and the second-pass system uses low-pressure brackish water membranes.

Normally, not all first-stage permeate requires treatment in a second pass, and some can bypass and be blended in finished water. This offers the advantage of reducing the size of the second pass. Concentrate from the second-pass system is generally recycled back to serve as feedwater to the first-pass system.

FIGURE 13.21 Permeate staging (multiple-pass design).

MEMBRANE PROCESSES   13.33

Blending Membrane Permeate and Bypass Water. In cases in which feedwater quality is such that membrane treatment provides a permeate with significantly better than needed finished water quality, it may be possible to blend the permeate with source or pretreated water. The advantages of blending are the

- Reduction of the required membrane capacity for a given finished water flow rate
- Reduced concentrate flow needing disposal for a given finished water flow rate
- Corresponding reduction in plant capital and O&M costs, including concentrate disposal costs
- Reduction in the amount of permeate posttreatment required for corrosion control

The opportunity for blending depends on the relative concentrations of critical constituents in both the membrane bypass and permeate relative to treated water goals and the composition of the membrane feed (bypass water).

The potential for blending can be determined using mass balance techniques for each product water quality constituent to determine the allowable blend ratios. Generally, the designer can easily determine the most critical constituents that need to be evaluated for a given application. The constituent requiring the permeate flow to be the highest percentage of the finished water flow is the limiting constituent and controls the overall allowable blending rate. The potential for blending can be calculated as follows:

$$Q_p/Q_{fin} = \frac{C_{source} - C_{goal}}{C_{source} - C_p}$$

where Q_p = membrane permeate flow rate
Q_{fin} = finished water flow rate
C_{source} = concentration of solute in source (feedwater)
C_{goal} = concentration of finished water (after blending) solute quality goal
C_p = concentration of solute in permeate

The resulting bypass flow rate is

$$Q_{byp} = Q_{fin} - Q_p$$

where Q_{byp} is the membrane bypass flow rate that is blended with the membrane permeate to form the finished water (before subsequent posttreatment processes).

The required source water flow rate can be calculated as follows:

$$Q_{source} = Q_{fin} \times \left(\frac{BR}{Y} + 1 - BR\right)$$

where Q_{source} = source water flow rate
BR = blend ratio = Q_p/Q_{fin}
Y = membrane system recovery = Q_p/Q_f
Q_{fin} = finished water flow rate

Using Membrane Computer Performance Projections for System Design. An RO or NF system is seldom designed by hand calculations. Most major membrane and system manufacturers, as well as some consultants, have developed computerized modeling programs that generate theoretical feed pressures and permeate quality at both initial and future operating conditions. However, designers must use caution because many projections may be accurate for initial operating conditions but can be inaccurate in predicting future performance, particularly if the effects of site-specific feedwater membrane foulants on membrane performance have not been determined. The following variables are required to predict system performance:

- Design feedwater composition (including temperature)
- Process train capacity and desired recovery
- System array, number of elements per vessel, and membrane module type
- Fouling characteristics of feedwater
- Membrane age (number of operating years)

Other Design Considerations. Depending on the type of membrane process and the specific application, the following additional items need to be considered.

Energy Recovery. RO and NF systems commonly include feedwater pumps with adjustable-frequency (variable-frequency) drives to minimize energy usage. For many applications, it is cost-effective to include energy recovery devices in the design, especially when the waste concentrate pressure is high and the system recovery is relatively low. Such devices are often included when the potential for energy savings using a particular energy recovery device exceeds associated capital and maintenance costs for the device. The selection of a specific type of energy recovery depends on the flow rates and variability, recovery available concentrate pressure, costs, and other factors. Some commonly used energy recovery devices are

- Impulse turbine
- Integrated turbopump
- Turbocharger
- Pressure exchanger

Automatic Flushing Systems. The membrane system design should include a means of automatically flushing membrane modules after planned or unplanned shutdown of the membrane train. The flushing system is used to remove from the modules concentrate that would otherwise remain if flushing were not performed. Flushing is particularly important when concentrate contains sparingly soluble constituents in excess of their theoretical solubility (supersaturated) that may precipitate and scale the modules. A scale inhibitor may be used to prevent this under normal operating conditions. However, in many cases the inhibitor will not permanently prevent scaling if concentrate remains in the modules for long periods. Flushing displaces the sparingly soluble constituents to eliminate this scaling potential.

Permeate is often used for flushing because it is generally the best-quality water (lowest in foulants and scalants) available on-site, although in many cases membrane feedwater is acceptable or possibly even more desirable. For example, many systems treating groundwaters with hydrogen sulfide have been designed with feedwater flushing because of the potential for membrane fouling when flushed with permeate containing colloidal sulfur formed while in storage. If permeate is used, design should include provisions for storing a sufficient amount of permeate to flush the system and to displace the concentrated water. If chlorine-intolerant composite membranes are used, only unchlorinated permeate should be stored and used for flushing. If cellulosic membrane modules are used, low-chlorine permeate is often preferred because of its ability to inhibit bacterial action during the downtime period. If multiple membrane trains are provided, a portion of total permeate flow can be diverted to flush an individual train and limit the amount of storage needed.

Permeate Drawback Tank. Hollow-fiber seawater RO systems commonly have the permeate drawback tank piped such that, after feed pump shutdown, an immediate backflow of permeate water is available to flow back through the membranes by means of the natural osmotic pressure differential across the membranes. The water prevents potential damage from this "osmotic drawback."

Ultrafiltration and Microfiltration

UF and MF membrane water treatment system design should involve communication with manufacturers and others familiar with the products. Many systems have proprietary design features that the designer should know. MF and UF systems are relatively new to the municipal water treatment marketplace, and new products continue to become available. Facility design requirements can vary greatly between the commercially available MF and UF systems. The following design parameters, however, are applicable to most, if not all, membrane filtration systems.

Design Equations. Based on Darcy's law, the flux through a membrane can be determined as follows:

$$J = \frac{P_{tm}}{\mu(R_m + R_t)}$$

where J = filtrate or permeate flux (flow rate/membrane area)
 P_{tm} = pressure across the membrane (transmembrane pressure TMP)
 μ = absolute viscosity
 R_m = clean membrane resistance (reciprocal of length)
 R_t = total resistance from foulants (reciprocal of length)

The flux is directly proportional to the transmembrane pressure (TMP). For a given membrane area, the filtrate or permeate flow rate is also directly proportional to the TMP. The flux is inversely proportional to the water viscosity. Therefore, as the water temperature decreases, MF and UF systems must increase the TMP to maintain production (see *Temperature* below).

The clean membrane resistance varies from one membrane type to another. The resistance from foulants can be from materials deposited on the membrane surface (sometimes called "cake" resistance), concentration polarization, blocking of the pores, and adsorbing to the membrane. The "cake" can improve the removal rates of the system until it is removed by backwashing and re-formed in the next cycle.

Configuration. There are two basic types of UF and MF processes used in water treatment: pressure-type systems where the membranes are housed in pressure vessels and vacuum-type systems where the membranes are submerged or immersed in nonpressurized tanks. There are significant differences in the facilities to be designed for these two configurations; therefore, it is common to select the configuration (and even the specific membrane product) before detailed design is started on the facility. The specific membrane product is often determined from a competitive proposal process considering initial and life-cycle costs based on design criteria determined during pilot testing.

Recovery. The recovery of a UF or MF system, typically called feedwater recovery, is the system's final product volume over a given time period divided by the feedwater flow volume. The quantity of source water or permeate/filtrate used for backwashing and flushing is considered in calculating feedwater recovery. Typical recoveries for UF and MF systems range from 85% to greater than 95%. Overall recovery of a system can sometimes be improved by recycling backwash water after solids removal. In some cases, recoveries exceeding 98% can be attained with backwash recycling. The maximum recovery can be identified by pilot testing.

Solute and Particle Rejection. In MF and UF systems, solute or particle rejection is primarily a function of membrane type and pore size rating. However, previously rejected substances remaining on the membrane surface also can act as a "dynamic membrane" and affect solute removal, especially for UF systems. Typical MF membrane pore size ratings range from 0.1 to 0.5 μm. Commonly used UF membranes range from about 2,000- to 150,000-dalton molecular-weight-cutoff (MWCO).

Flux. The permeate or filtrate flux through UF and MF membranes depends significantly on transmembrane pressure and water temperature. Design flux rate used for a system is also strongly defined by membrane feedwater quality because the rate of plugging and frequency of backwashing and cleaning are affected by the operating flux rate. Design flux rate for a specific application is often determined by bench or pilot testing. Typical flux rates for UF and MF membranes vary widely, depending on membrane product and specific application, and range from about 20 to 100 gpd/ft^2 (0.034 to 0.170 m/h).

Temperature. Temperature affects the required driving pressure for a UF or MF system because of changes in feedwater viscosity. Figure 13.22 shows the viscosity of water as a function of temperature. At 20° C, the absolute viscosity is approximately 1.00 cP. Because the filtrate/permeate flux through clean membrane is inversely proportional to viscosity, about 50% more TMP is needed to maintain a constant flux at 5° C than at 20° C. Assuming a constant TMP and membrane area, the product flow at 5° C would be approximately 35% less than at 20° C.

For UF and MF systems, an approximation of permeate flow at any temperature relative to flow at 20° C is as follows:

$$Q_{pT} = \frac{Q_{p(20° C)}}{e^{-0.0239(T-20)}}$$

where Q_{pT} = permeate flow at temperature T
$Q_{p(20° C)}$ = permeate flow at 20° C
T = water temperature, °C
e = 2.71828

In the design process, it is critical to define the desired production rate at a specific water temperature. In many cases, less production is needed in the colder winter months because of decreased water demands from the utility customers. The MF or UF system could then be designed to produce the rated capacity at a warmer temperature, thus reducing the required membrane area and saving capital costs.

FIGURE 13.22 Absolute viscosity of water versus temperature.

Pumping Requirement. UF and MF systems typically require several pumps (depending on the specific system and application): feed or permeate pump, recirculation pump, backwash pump, cleaning pump, chemical transfer and feed pumps, and vacuum pump (for vacuum-type systems). In a few systems, the feed or permeate pumps have been eliminated by the use of gravity head or siphons.

The feed pump for a system is sized to meet the maximum required pressure and flow rate considering the membrane type, temperature, flux, piping and other pressure losses (separate from the transmembrane pressure), and degree of fouling over time. Additionally, some systems are designed for increased feed flow rates during part of the operating cycles. For example, one major MF manufacturer with a proprietary gas backwash system uses an increased-capacity feed pump to provide the desired feedwater flow rate for flushing out suspended solids removed from membranes by the gas backwash. Typical feed pressures for pressure-type MF and UF systems used in municipal water treatment range from 15 to 60 psi (100 to 414 kPa), and from -1 to -12 psi (-6.9 to -83 kPa) for vacuum-type systems.

If the design includes retentate stream recirculation (common in feed-and-bleed cross-flow operating mode), recirculation pumps are sized to overcome pressure losses through the system while providing the desired recirculation flow rates, typically 3 to 6 times greater than the source water flow rate.

Backwash pumps, if required, are sized for the specific needs of the membrane system and commonly provide targeted flow to the filtrate or permeate side of the membranes at low pressure.

Membrane Module Arrays. MF and UF systems are typically designed using multiple parallel units. For large-capacity systems, multiple groups (each containing several membrane units) form modular arrays or trains. Each array has common manifold piping to control service flows, backwashing, and cleaning-in-place (CIP).

Cross-Flow Velocity. Many hollow-fiber membrane filtration systems operating with "inside-out" flow patterns require specific minimum flow rates to provide sufficient cross-flow velocity to control fouling and to flush retentate from the fibers. Typical cross-flow velocities range up to 3 ft/s (0.9 m/s). Commonly, recirculation pumps recycle the fluid on the feed side of the membranes to provide adequate velocities (see Figure 13.23).

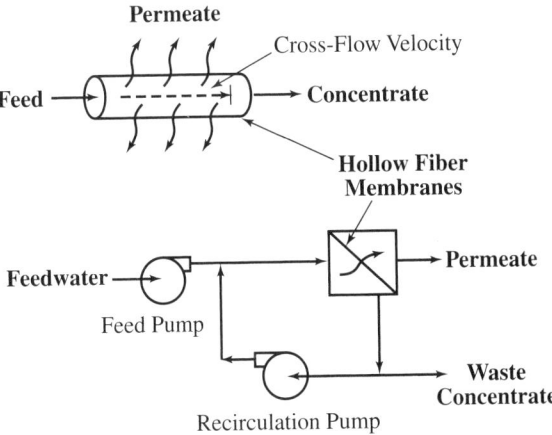

FIGURE 13.23 Use of recirculation pumping to provide needed cross-flow velocities.

FIGURE 13.24 MF/UF TMP control with periodic backwashing and cleaning.

Backwashing. MF and UF systems require periodic backwashing and chemical cleaning (cleaning-in-place, CIP) to maintain permeability, minimize the needed driving pressure, and prolong the useful life of the membranes (see Figure 13.24). Depending on the specific membrane system, backwashing uses unchlorinated water, chlorinated water, or gas (e.g., air). Gas-backwashed systems first remove foulants from the membrane by transferring gas through the membrane from the filtrate to the feed side and then flushing the foulants to waste using feed flow. Backwash is accomplished using relatively frequent (5-min to several-hour) and short duration (3- to 180-s) backwashes, depending on the specific application and membrane system. For example, it would not be unusual to have a system that backwashes for 15 s every 15 min or 90 s every 20 min. For municipal water systems, backwashing is typically an automatically controlled process initiated on a set frequency and set for a specific duration. Backwash can also be automatically initiated based on TMP rise and/or other monitored parameters.

Chemically Enhanced Backwash (CEB). Some MF and UF systems use *chemically enhanced backwashes* (CEBs) or "mini-cleans" to maintain permeability, control TMP, and extend the cleaning frequencies. CEBs are usually less frequent than typical backwashes and may include a short time for the membranes to soak in the chemical solution. Common CEB chemicals are sodium hypochlorite, citric acid, and caustic soda, and their concentrations and use are dependent on the specific membrane product and the targeted type of foulant.

Cleaning System. MF and UF systems are typically cleaned without removing the membranes from the system—called *cleaning-in-place*. MF and UF systems commonly include chemical CIP systems used for periodic major cleanings when routine backwashes and/or CEBs are not adequate. Most municipal systems have manually initiated, automatically controlled chemical cleaning systems. Chemicals used, their concentrations, and CIP protocols are specific for each membrane product and the type of foulants expected to be present. Common cleaning chemicals are sodium hypochlorite, citric acid, caustic soda, and detergents. CIPs are often performed in multiple steps and typically include soaking time and can take several hours to perform. Typically, a CIP system consists of one or more chemical solution tank(s), a transfer and/or recirculation CIP pump, and other facilities and equipment.

Membrane Integrity Tests. Most systems include instrumentation that continuously monitors filtrate quality and membrane performance. This instrumentation may include turbidity monitors, particle monitors and counters, and other devices. For municipal systems, state regulations should be reviewed to determine specific requirements for system monitoring and membrane integrity testing.

Most MF and UF membrane systems are equipped for periodic off-line integrity testing using air, based on the bubble point theory. In the pressure decay test, air is applied to the membranes at a pressure less than needed to be forced through the pores of intact membranes, and decline in air pressure is measured. If the rate of pressure decay is too large, the membrane unit is considered to have failed the test. In another test, the diffusive airflow test, the volume of water displaced by air passing through the membranes is measured and evaluated for acceptance. These tests are conducted with the membrane unit out of service, although the testing can be automated and scheduled to take place on a timed cycle. Membrane units not passing the tests can then be diagnosed for the source of the leakage, such as broken fibers or valves not sealing properly, and remedial measures can be taken. To meet the requirements of the Long-Term 2 Enhanced Surface Water Treatment Rule (LT2ESWTR), sufficient air pressure must be used in the test to achieve a resolution of 3 μm (USEPA Membrane Filtration Guidance Manual, Proposal Draft, June 2003).

Electrodialysis and Electrodialysis Reversal

Electric current through an ED/EDR system can be calculated by using Faraday's law (*Electrodialysis and Electrodialysis Reversal*, AWWA M38, 1995):

$$I = \frac{F \times F_d \times \Delta N}{e \times N}$$

where I = direct current, A
F = Faraday's constant (96,500 A · s/equivalent/cell pair)
F_d = flow rate of demineralized stream through membrane stack, L/s
ΔN = change in normality of demineralized stream between inlet and outlet of the membrane stack
e = current efficiency
N = number of cell pairs in membrane stack

Electrical resistance can be calculated by Ohm's law:

$$R = \frac{E}{I}$$

where R = resistance, Ω
E = voltage, V
I = current, A

ED/EDR current efficiency is defined as follows:

$$e = \frac{F \times F_d \times \Delta N \times 100}{I \times N}$$

Recovery. Recovery of an ED/EDR system is often limited by the concentration of sparingly soluble salts, similar to RO and NF units. However, silica scaling is not a concern because silica is not concentrated (or removed) by ED/EDR. Pretreatment processes, such as those described previously for RO, can be used to control the concentration of sparingly soluble salts (for example, $CaCO_3$, $CaSO_4$, $SrSO_4$, $BaSO_4$) to allow higher recoveries. One major EDR manufacturer allows LSI values up to +2.2 and $CaSO_4$ concentrations up to 175% of saturation without chemical addition (Meller, 1984). Higher values are allowed by the manufacturer with appropriate antiscalant addition.

For EDR systems, recovery is lowered somewhat by discharge of concentrate off-specification water immediately after polarity reversal when the concentrate compartments become the dilute (demineralized) water compartments.

System recovery is also affected by electrical water transfer. Typically 0.5% of demineralized stream flow passes through to the concentrate stream per 1,000 mg/L TDS removed electrically (not hydraulically) along with ions (Meller, 1984).

Solute Removal. The following factors affect solute removal of an ED/EDR system.

Desalination Ratio. Salt removal in an ED/EDR system is directly proportional to current flow across the stack. Typically, TDS is concentrated about 3.0 to 3.6 times (i.e., the desalination ratio) in a single pass through the system (Wolfe, 1993). The desalination ratio varies with different membrane types and varying ions. For instance, the desalination ratio may be as high as 10 to 1 for nitrate and only 1.5 to 1 for sulfate using monovalent selective membranes (Wolfe, 1993).

Limiting Current Density. There is a limit to the rate at which ions can be transferred through an ED/EDR membrane. Polarization occurs when too few ions are present to allow proper current flow and the resulting high electrical resistance causes water molecules to split. Typically, 70% of the limiting current density is used as the maximum design value.

Current Leakage. There is also a limit to the voltage that can be applied to an ED/EDR membrane stack. At that limit, excessive electric current leaks through an adjacent membrane to the concentrate stream, causing heat and resulting in damage. Maximum design voltage is typically 80% of the voltage where current leakage occurs.

Back Diffusion. If the concentrate is too concentrated, ions can diffuse (against the force of the dc potential) back to the demineralized stream. Typical design practice lowers membrane stack efficiency when the concentration of the concentrate exceeds 150% of the concentration in the demineralized stream.

Types of Membranes. A number of membrane products are available. Some membranes are designed to maximize specific ions, such as monovalent-specific membranes that do not concentrate calcium sulfate as much as conventional membranes. This may allow higher recovery operation, if calcium sulfate solubility is the limiting criterion.

Temperature. Temperature is one of the most important factors affecting ion removal. Ion removal increases about 2% per degree Celsius temperature rise.

Product Water Quality. Because electrical resistance increases as the demineralized stream becomes more and more dilute, ED/EDR systems often are designed to limit demineralized flow streams to a minimum of about 200 mg/L (although lower concentrations are possible).

Power Consumption. Power is required for pumping power and current flow across membrane stacks. The pumping power for a single-pass system can be estimated, assuming a flow rate of about 1.3 times the feed flow rate (for dilute and concentrate), and the pressure loss through the stack at about 50 psi (340 kPa).

An estimate of the stack power requirement can be made from the following formula (Wolfe, 1993):

$$\frac{\text{Desalting kWh}}{1{,}000 \text{ gal}} = \frac{F \times \Delta C \times 3.788}{1{,}000 \times \text{eq wt} \times e}$$

where F = Faraday constant, 26.8 A · h/equivalent
ΔC = ion concentration removal, mg/L
eq wt = equivalent weight for ion removed (58.5 g/eq for NaCl)
e = average current and rectifier efficiency (typically about 0.83)

Temperature and pH. Maximum water temperature limitation for ED/EDR stacks is generally about 45° C. Commonly used membranes can tolerate a pH range from 1 to 10. Temperature is one of the most important factors affecting ion removal of an ED/EDR system. Ion removal increases about 2% per degree Celsius temperature rise. A two-stage ED/EDR system can have TDS removal approximately 25% greater at 30° C than at 15° C.

Staging. ED/EDR systems have electrical stages and hydraulic stages, sometimes in a single stack. Multiple hydraulic stages within one stack are created by using special interstage membranes designed to accommodate elevated pressure differentials. An electrical stage is created by using a pair of electrodes (anode and cathode).

Typically, 40% to 67% salt removal is possible in one hydraulic stage (M38, American Water Works Association, 1995). Greater salt removals are accomplished by using additional hydraulic stages in series. In an ED/EDR system, salt removal is directly proportional to electric current and inversely to flow rate because of less available detention time. Because current flow can be regulated independently for each electrical stage, multiple hydraulic and electrical stages optimize hydraulic and electrical parameters.

Differential Pressure. Demineralized stream pressure is usually controlled to be about 0.5 to 1.0 psi (3 to 7 kPa) greater than concentrate stream pressure to ensure that any cross-leakage moves toward the concentrate stream and does not lower product water quality (M38, American Water Works Association, 1995). This differential pressure is maintained slightly positive.

Operating Mode. An ED/EDR membrane system can be designed for batch or continuous flow operating mode. Municipal water treatment systems commonly use the continuous mode when the feed stream is demineralized and passes from the system as demineralized water. Some of the concentrate is recycled through the stack, and some leaves as blowdown (feed-and-bleed design). Chemicals are sometimes added to the recycle stream for scale control.

Electrode Compartments. Electrodes are usually made of platinum-coated titanium. Chlorine gas, oxygen gas, and hydrogen ions are produced at the anode, creating an acidic condition. Hydrogen gas and hydroxide ions are generated at the cathode, raising the pH and increasing the conditions for scaling. Water from electrode compartments is usually mixed together and transferred to a degasifier before final disposal or possible recycle.

OTHER MEMBRANE PROCESS DESIGN CONSIDERATIONS

Other considerations necessary in membrane process design include the need for posttreatment, membrane cleaning, disposal of waste residuals, instrumentation and control, efficient O&M, and building design.

Posttreatment

Membrane product flow streams usually require some form of posttreatment before distribution. Posttreatment provides disinfection, corrosion control, and removal of dissolved gases and volatile compounds.

Posttreatment Disinfection. Most pathogens are removed by RO, NF, UF, and MF membranes, and microbiological quality of the permeate (or filtrate) is usually excellent. However, it is possible for microbes to pass through during the process when a membrane barrier is compromised (such as a broken fiber) or when other imperfections are present in the membrane system (such as an O-ring leak or contamination before or during operation of the permeate side of the membrane module or permeate piping). In ED/EDR systems, the product water does not pass through a membrane barrier. Consequently, product water posttreatment usually includes disinfection, typically with chlorine.

Posttreatment disinfection design is similar to that of conventional water treatment plants and is described in Chapter 10.

Posttreatment for Corrosion Control. Corrosion control is not usually required for UF and MF processes. However, RO, ED/EDR, and to a lesser extent NF product streams are commonly corrosive because of the lack of calcium and alkalinity and the acidic pH. Posttreatment operation for corrosion control can include the following activities:

- Releasing carbon dioxide in a degasifier
- Adding caustic (with or without previous CO_2 addition), sodium bicarbonate, or sodium carbonate to increase bicarbonate alkalinity and pH
- Adding a corrosion inhibitor chemical
- Adding calcium chloride to increase calcium levels
- Adding lime to increase both calcium ion and alkalinity and pH (but this may cause unacceptable turbidity)
- A combination of these processes

Selection of the most appropriate posttreatment method for corrosion control should be site-specific and depend heavily on membrane product water quality.

Removal of Gases and Volatile Organic Compounds. Water treatment membrane processes do not remove dissolved gases and, in general, provide poor removal of most volatile organic compounds (VOCs). Therefore these components must be removed during permeate posttreatment.

Hydrogen sulfide (if present in the source water) and carbon dioxide are usually the predominant gases present in membrane product water. Sulfide is often present in groundwater source waters, and carbon dioxide can be present in groundwaters or generated by feedwater acidification. Unless the permeate has sufficient alkalinity (which is generally not the case), carbon dioxide removal is not desirable because the gas can be used as a source of bicarbonate alkalinity for corrosion control. VOCs are often a problem when treating groundwaters contaminated by industrial processes.

The most common method of treatment is the use of a degasifier, commonly a packed tower with a blower. Tower design should be a function of the critical gas or VOC present in the product and the degree of removal required. When carbon dioxide and hydrogen sulfide are both present, sulfide removal usually controls the degasifier design.

If the pH of the permeate is greater than 6.5, hydrogen sulfide removal may require that the permeate be acidified before degasification. Acidification ensures that a large proportion of the sulfide is present in the gaseous form, in contrast to sulfide ions. Degasifier off-gases containing sulfides may also require scrubbing to minimize odor and corrosion problems and for safety reasons. Scrubbing is typically achieved by conveying the gas through chlorine and/or caustic solutions to convert the sulfide to an ionized form. Degasification for VOC removal may also require off-gas treatment with granular activated carbon (GAC), depending on state or local air quality regulations.

Membrane Cleaning System

With few exceptions, all membrane systems are subject to fouling by one or more source water components and therefore require periodic cleaning. Membrane cleaning is usually performed without removing membranes from the pressure vessels or the system (i.e., cleaned in place). A cleaning system is designed to prepare and recirculate chemical solutions through some of or all membrane modules at low pressure.

The cleaning system can also be used to feed special membrane posttreatment chemicals (not to be confused with membrane system posttreatment) that sometimes are used to improve membrane performance. The cleaning system also serves to prepare and transfer membrane storage solutions, or preservatives, used to prevent microbial growth and in some cases to prevent freezing when the membrane system is shut down for extended periods, typically more than a week.

The cleaning system for an RO or NF system should be designed to accommodate all cleaning and membrane storage solutions expected to be used at the plant. Cleaning systems typically consist of the following basic components:

- Tanks with mixers for holding cleaning solutions.
- A pump providing sufficient head to circulate the solution from the tank through the membrane system and back to the tank.
- A cartridge filter to intercept any suspended solids present in the cleaning solution and prevent them from clogging the module flow passages.
- Piping and valves to transfer the solution to and from the train feed, permeate, and concentrate headers. The design should consider the use of removable piping spools or hoses to provide positive air break in the cleaning system connection points. An air break prevents chemical solutions from entering the product header piping and being transported to downstream product facilities when the membrane system is being put back into service or when one train is being cleaned while other trains remain in operation.
- Instrumentation and control for proper system operation, including tank level, temperature switches, and a pH monitor.
- A cleaning solution heating device (not always required), such as an immersion heater, to increase solution temperature and improve the efficiency of cleaning, particularly when cleaning with detergents.
- A cleaning solution cooling device (usually not required), such as a heat exchanger external to the cleaning solution tank or cooling coils inside the tank, if it is determined that the maximum solution temperature needs to be controlled. The need for cooling increases when the cleaning solution makeup water is at relatively high temperature or the heat energy expected to be transferred to the cleaning solution from the cleaning pump while recirculating is too great.

Typical design allows for separate cleaning of each stage of a multistage train. With large-capacity trains, individual sections of a stage can be designed to allow separate cleaning, minimizing cleaning tank volume and pumping requirements.

Residuals Disposal

The designer must consider the least expensive, acceptable method of disposing of wastes generated in both pretreatment processes and residuals resulting from membrane treatment processes.

Concentrate from membrane processes is typically considered a waste stream that must be disposed of in accordance with applicable local discharge regulations. Because these regulations vary considerably from one geographic location to another, it is not practical to provide more than general information on the proper methods of concentrate disposal for a certain treatment. Disposal alternatives are easier for MF and UF systems, which do not concentrate salts, than for RO, NF, and ED/EDR systems. Potential concentrate disposal options for RO, NF, and ED/EDR include (Mickley et al., 1993)

- Discharge to surface water
- Land application (e.g., irrigation, possibly after blending with other low-TDS waters)
- Injection wells
- Evaporation ponds
- Evaporators
- Wastewater collection system
- Wastewater treatment plant effluent

Spent (used) cleaning and membrane storage solutions are typically disposed to a sanitary sewer system after pH neutralization. Often spent solutions are transferred to storage tanks with provisions for adding acid and caustic to neutralize low- and high-pH cleaning solutions. Neutralized spent solution is then transferred to the disposal location.

Instrumentation and Control

Instrumentation and control (I&C) facilities vary greatly depending on membrane process type, application, and degree of automatic monitoring and control desired. Some membrane systems are fully automatic for normal operation and are designed to be monitored remotely by means of modems connected to computer-based controls.

A minimum of I&C equipment is necessary for all systems to protect equipment from damage. For example, low-suction pressure switches or other devices automatically shut off the system on loss of adequate flow. It is also common to provide on-line instrumentation to measure feed and product water qualities.

If source water temperature and turbidity are variable, the membrane system should be provided with instrumentation to continuously measure water temperature and feedwater turbidity. Product water turbidimeters or particle counters usually monitor UF and MF systems. For RO, NF, and ED/EDR systems, conductivity monitors installed on the feed and product flow streams monitor salt removal performance. Where acid is fed for feedwater pH adjustment, an on-line pH meter with high/low alarms is commonly used. ED/EDR systems typically provide for pH measurement of the feed, product, and concentrate flow streams.

Pressure-driven membrane systems typically have either feed and product or product and concentrate flowmeters. Pressure gauges indicate all critical pressures and pressure differentials, such as membrane feed, interstage, concentrate, permeate, and, unless determined solely by calculations, differential pressure across each stage.

ED/EDR systems typically have flowmeters for the dilute in, concentrate makeup, and electrode flow. Pressure gauges are usually provided to measure inlet and outlet pressures of the dilute and concentrate flow streams, as well as the pressure of the electrode flow stream of each membrane stack. Additionally, ED/EDR systems continuously monitor current flow and voltage; for EDR systems, measurements are needed for each electrical polarity.

Elapsed time meters provided on the power supply to the membrane system track accumulated run time. Sampling stations should be provided at all important process locations in the piping system for periodic sampling and analyses.

Most drinking water membrane systems designed today have controls based on a programmable logic controller (PLC). These systems can be designed to control flows, pressures, and other parameters. Many RO and NF systems have controls to provide automatic flushing upon shutdown, and most MF and UF systems have automatically controlled backwash/backflush facilities. Many large RO and NF plants are designed with computer systems that compute normalized performance data for operation monitoring and determine when to perform membrane cleanings.

Operations and Maintenance (O&M) Design Issues

Membrane systems require proper O&M to perform according to design and to extend membrane life. Depending on the type of membrane system and application, O&M requirements vary greatly. As a minimum, operations staff should monitor and record all critical parameters, operation incidents (such as feedwater quality upsets and shutdowns), and maintenance performed. The membrane plant designer should provide all needed instrumentation and representative sampling points for operations staff.

System design should consider adequate spacing and access for equipment maintenance, as well as devices for lifting and moving equipment. Proper storage facilities should be provided for membranes, chemicals, other equipment replacements, and spare parts.

Membrane Process Building Design

Floor plans of membrane process buildings vary greatly depending on the type of membrane process, the site-specific conditions and design constraints, and the number and types of nonprocess areas included. Process areas include the following pieces of equipment, depending on the membrane process and application:

- Pretreatment filters, such as cartridge filters, basket strainers, and bag filters
- Chemical feed systems for pretreatment and posttreatment chemicals
- Membrane feed pumps
- Membrane treatment units
- Membrane cleaning and flushing systems
- Posttreatment equipment
- High service pumps
- Holding tanks (where applicable) such as feed and spent backwash water tanks for an MF or UF system
- Electric switch gear and motor control center
- Emergency generator

Typical nonprocess areas commonly incorporated into building designs are as follows:

- Control room
- Laboratory
- Chemical storage area
- Maintenance shop
- Spare parts storage room

- Restrooms and locker rooms
- Offices
- Mechanical room for heating, ventilation, and air conditioning equipment
- Lunch/break room
- Training room

Particular attention should be given to the size and location of doorways. In some cases removable wall panels should be provided to allow large equipment to be removed and installed.

Materials of Construction

It is particularly critical to use proper construction materials in membrane systems because corrosion products can foul membranes, resulting in increased cleaning frequency and shortened membrane life. In addition, process flow streams and chemicals used are commonly highly corrosive, and for pressure-driven desalting systems, high pressures are often required.

Where pressures and temperatures allow, nonmetallic (PVC, CPVC, FRP, etc.) piping and valves are commonly used. Nonmetallic piping is common throughout MF, UF, ED/EDR, and in some cases NF systems. For applications requiring higher pressures and temperatures, such as most RO systems, appropriately selected stainless steels are typical. For example, a 5,000 mg/L TDS brackish groundwater RO system commonly has the following components (other available material options can also be considered):

- Piping from supply well to pretreatment cartridge filter—PVC or FRP
- Cartridge filter housing—316L stainless steel or FRP
- Piping from cartridge filter to membrane feed pump suction connection—PVC or FRP
- Membrane feed pump discharge piping—316L stainless steel
- RO pressure vessels—FRP
- RO membrane unit interstage and concentrate lines (up to and including the concentrate control valve)—316L stainless steel
- Concentrate line downstream of the concentrate control valve—PVC or FRP
- Permeate piping—PVC or FRP

For low-salinity (freshwater) applications, 304L stainless steel is sometimes used for low- and high-pressure piping. For seawater RO system high-pressure lines, higher-alloy stainless steels, with a molybdenum content of 6% or more, are commonly used. Alternatively, polyethylene-lined carbon steel piping has been used for seawater RO high-pressure piping, although piping failures have been reported when the lining was improperly manufactured or damaged after manufacturing.

Because many membrane systems require concentrated chemicals, storage and handling systems must be designed with appropriate materials and with operator safety in mind.

Process Reliability and Redundancy Issues

In designing membrane systems, the same reliability and redundancy issues considered for conventional treatment facilities should be evaluated. Membrane treatment systems have been used for water treatment for more than 30 years, and when properly manufactured, installed, operated, and maintained, the systems are very reliable.

Membrane treatment facilities commonly have multiple parallel process trains. Where required, additional membrane units can be installed to provide needed redundant capacity when one or more units are out of service for maintenance or membrane cleaning. In many cases, particularly where the membrane system provides expansion to an existing water supply system, only critical system components are provided with redundancy, rather than the entire membrane unit. This not only minimizes the financial investment, but also allows the membrane system to operate as continuously as possible. If there is no complete standby membrane unit, the need to prepare and feed membrane storage solutions is minimized among other advantages.

If a product water quality problem is detected in a membrane system, the specific membrane modules within the system having the problem can be identified for remedial action. With many membrane systems, the deficient membrane modules can be isolated and the rest of the system operated until there is a convenient time for maintenance.

Bench and Pilot Testing

Bench testing is often performed in conjunction with a feasibility study to determine which, if any, membrane process is applicable to treatment requirements. If a particular process has been selected, bench testing results can develop basic characterization data on the process. Bench testing is typically 1 day to 2 weeks long and uses a simple membrane characterization apparatus consisting of a feed pump; a pressure assembly containing a flat-sheet, hollow-fiber, or tubular membrane; and valves to set flow rates, pressures, and recovery. Typical operating data derived during bench testing include initial membrane feed pressure to produce a given flux and initial permeate quality (representative of low-recovery operation) at a given membrane flux.

Compared with bench testing, membrane pilot testing is lengthy, often taking 1 to 4 months of operation. Test duration is generally governed by the amount of time the proposed membrane and pretreatment systems must be operated to correctly quantify longer-term trends in membrane performance, such as flux decline rate and rate of change of solute transport.

Bench and pilot testing both have limitations that the designer must consider. For example, source water may not be representative of water quality over the life of the membrane system. The designer should consider extra allowances, or safety factors, for critical design parameters.

Designer and Vendor Interface Issues

Many membrane products are available for water treatment applications, and new products continue to enter the marketplace. It is important that the membrane facility designer contact membrane manufacturers and system suppliers to identify the best candidate membrane products for the application and to identify all critical design issues applicable to each product. This not only minimizes potential problems with the constructed facility and allows the best candidate membranes to be considered, but also provides for the designer the benefit (directly or indirectly) of the latest computer design software to minimize design calculation time.

BIBLIOGRAPHY

"Alternative Water Supplies—Desalting and Reuse." *Proceedings of AWWA 1992 Annual Conference.* Vancouver, B.C. Denver Colo.: American Water Works Association, 1992.

American Water Works Association. *Water Quality and Treatment,* 4th ed. New York: McGraw-Hill, 1990.

American Water Works Association. *Proceedings 1991 Membrane Technology Conference.* Orlando, Fla. Denver, Colo.: American Water Works Association, 1991.

American Water Works Association. *Proceedings 1993 Membrane Technology Conference.* Baltimore, Md. Denver, Colo.: American Water Works Association, 1993.

American Water Works Association. *Electrodialysis and Electrodialysis Reversal,* M38. Denver, Colo.: American Water Works Association, 1995.

American Water Works Association. *Reverse Osmosis and Nanofiltration,* M46. Denver, Colo.: American Water Works Association, 1999.

American Water Works Association. *Microfiltration and Ultrafiltration Membranes in Drinking Water, M53.* Denver, Colo.: American Water Works Association, 2004.

American Water Works Association Research Foundation, Lyonnaise des Eaux, and Water Research Commission of South Africa. *Water Treatment Membrane Processes.* New York: McGraw-Hill, 1996.

Applegate, L. E. "Membrane Separation Processes." *Chemical Engineering* (6):64, 1984.

Applegate, L. E. "Post-treatment of Reverse Osmosis Product Waters." *Journal AWWA* 78(5):59, 1986.

AWWA Water Desalting and Reuse Committee Report. "Membrane Desalting Technologies." *Journal AWWA* 81(11), 1989.

Baker, R. W., et al. *Membrane Separation Systems. A Research Needs Assessment.* Final Report by the Department of Energy Membrane Separation Systems Research Needs Assessment Group. Springfield, Va.: National Technical Information Service NTIS-PR-360, U.S. Department of Commerce, 1990.

Bergman, R. A. "Anatomy of Pressure-Driven Membrane Desalination Systems." *Proceedings AWWA 1993 Annual Conference—Engineering and Operations.* San Antonio, Tex. Denver Colo.: American Water Works Association, 1993.

Bergman, R. A., and Lozier, J. "Membrane Process Selection and the Use of Bench and Pilot Tests." *Proceedings of the Membrane Technology Conference.* Baltimore, Md. Denver, Colo.: American Water Works Association, 1993.

Buros, O. K., et al. *The U.S. Agency for International Development Desalination Manual.* Denver, Colo.: CH2M Hill International, 1980.

Byrne, Wes. *Reverse Osmosis: A Practical Guide for Industrial Users.* Littleton, Colo.: Tall Oaks Publishing, Inc., 1995.

DuPont Company. *PEM Permasep Products Engineering Manual.* Wilmington, Del.: DuPont Company, 1992.

Duranceau, Steven J. (ed.). *Membrane Practices for Water Treatment.* Denver, Colo.: American Water Works Association, 2001.

"Engineering Considerations for Design of Membrane Filtration Facilities for Drinking Water Supplies." *Proceedings of AWWA 1992 Annual Conference,* Vancouver, B.C. Denver, Colo.: American Water Works Association, 1992.

Jacangelo, J. G., N. L. Patania, and J. M. Laine. *Low Pressure Membrane Filtration for Particle Removal.* Denver, Colo.: AWWA Research Foundation, American Water Works Association, 1992.

James M. Montgomery, Consulting Engineers, Inc. *Water Treatment Principles and Design.* New York: Wiley, 1985.

Katz, W. "Electrodialysis for Low TDS Waters." *Industrial Water Engineering.* June/July 1971.

Laine, J. M., and J. G. Jacangelo. "System Components and Process Design Considerations for Ultrafiltration of Untreated Drinking Water Supplies." *Proceedings AWWA 1992 Annual Conference—*

Engineering and Operations, Vancouver, B.C. Denver, Colo.: American Water Works Association, 1992.

Meller, F. H. (ed.). *Electrodialysis (ED) and Electrodialysis Reversal (EDR) Technology.* Watertown, Mass.: Ionics, Inc., 1984.

Mickley, Mike, R. Hamilton, L. Gallegos, and J. Truesdall. *Membrane Concentrate Disposal.* Denver, Colo.: AWWA Research Foundation and American Water Works Association, 1993.

Sourirajan, S. *Reverse Osmosis.* New York: Academic Press, 1970.

Sourirajan, S., and T. Matsuura. *Reverse Osmosis/Ultrafiltration Process Principles.* Ottawa, Ontario: Division of Chemistry, NRC Canada, 1985.

Taylor, J. S., S. J. Duranceau, W. M. Barrett, and J. F. Goigel. *Assessment of Potable Water Membrane Applications and Research Needs.* Denver, Colo.: AWWA Research Foundation and American Water Works Association, 1989.

U.S. Congress, Office of Technology Assessment. *Using Desalination Technologies for Water Treatment.* OTA-BP-O-46. Washington: Government Printing Office, 1988.

U.S. Environmental Protection Agency. *ICR Manual for Bench- and Pilot-Scale Treatment Systems.* Cincinnati, Ohio: Technical Support Division, Office of Ground Water and Drinking Water, U.S. Environmental Protection Agency, 1996.

U.S. Environmental Protection Agency. *Membrane Filtration Guidance Manual, Proposal Draft.* Office of Water, U.S. Environmental Protection Agency, June 2003.

Wale, R. T., and P. E. Johnson. "Microfiltration Principles and Applications." *Proceedings AWWA 1993 Membrane Technology Conference,* Baltimore, Md. Denver, Colo.: American Water Works Association, 1993.

"Water Desalting: A Viable Technology for Water Supply." *Proceedings of AWWA 1993 Annual Conference,* San Antonio, Tex. Denver, Colo.: American Water Works Association, 1993.

"What Membranes Can Do for You?" Seminar S5. AWWA 1993 Annual Conference. San Antonio, Tex., 1993.

Wolfe, T. D. "Electrodialysis Design Approaches." *Proceedings AWWA 1993 Membrane Technology Conference,"* Baltimore, Md. Denver, Colo.: American Water Works Association, 1993.

CHAPTER 14
ACTIVATED CARBON PROCESSES

Richard D. Brady
Richard Brady & Associates
San Diego, California

Activated carbon is an adsorbent material that provides a surface on which ions or molecules in the liquid or gaseous phase can concentrate.

The use of activated carbon in water treatment in the United States has been limited primarily to removing taste- and odor-causing compounds, pesticides, and other organic contaminants.

CHARACTERISTICS OF ACTIVATED CARBON

Activated carbon has a random structure that is highly porous, with a broad range of pore sizes ranging from visible cracks and crevices down to molecular dimensions. Intermolecular attractions in the smallest pores create adsorption forces. These forces cause large and small molecules of dissolved contaminants to be condensed and precipitated from solution into the molecular-scale pores. Activated carbon is an effective adsorbent because it provides a large surface area on which the contaminant chemicals can adhere.

Activated carbon is available in two different forms: powdered (PAC) and granular (GAC). Adsorptive properties of GAC and PAC are fundamentally the same because they depend on pore size, the internal surface area of the pores, and surface properties independent of overall particle size. Each brand of commercially available PAC or GAC has properties making it most suitable for particular applications. Besides adsorptive capacity and selectivity in removal, these properties include the ability to withstand thermal reactivation and resistance to attrition losses during transport and handling. Powdered activated carbon is added to water, mixed for a short time, and removed. Adsorption of molecules occurs while the PAC is in contact with the water. In both Europe and the United States, PAC is typically fed as a powder using dry feed equipment or is batched as a slurry and fed with metering pumps or rotodip feeders. PAC is usually added early in the treatment process and then either settles out with the floc in the pretreatment basins or is removed from the filter beds during backwashing.

Granular activated carbon is generally used in adsorption beds or tanks through which the water passes for treatment. As GAC is used for treatment, surfaces within the pores gradually become covered with chemical molecules until the carbon is no longer able to adsorb new molecules. At that point, the old, or spent, carbon must be replaced with new, virgin, or fresh reactivated carbon. The adsorptive capacity of the spent GAC can be restored by thermal reactivation.

One method of using GAC is to install it as a partial or complete replacement for the conventional granular media in conventional gravity filters. In this case, GAC acts as a filtering medium and provides limited adsorbance as water passes through the filters. The other common type of installation is to locate the GAC contractor as the last step in the treatment process, commonly with a preozonation step.

The use of PAC adsorption is most appropriate in situations where taste and odor or organic contaminant, loading episodes are moderate, seasonal, and/or infrequent. The use of GAC adsorption beds should be considered whenever a system experiences moderate to severe taste and odor problems or organic contaminant loading. PAC may also be considered for small to midsized plants to address these conditions.

Physical Properties of Activated Carbon

Surface area within porous carbon structure provides the capacity to adsorb dissolved organic materials such as natural organic matter (NOM), disinfection by-products (DBPs), or taste and odor compounds. Carbons used for adsorption in drinking water applications have a minimum surface area of 73 acre/lb (650 m^2/g), with typical surface areas on the order of 112 acre/lb (1,000 m^2/g). A series of tests developed for testing the suitability of various types and brands of activated carbon are detailed in American Water Works Association (AWWA) Standard B604, *Granular Activated Carbon* and B600, *Powdered Activated Carbon.*

Iodine Number and Molasses Number. The quantity of small and large pore volumes in a sample of activated carbon is described by the iodine number and molasses number. Adsorption tests are also used to approximate the distribution of pores available for adsorption. Using standard reference adsorbates (such as iodine and a molasses solution) allows the activity characteristics of different carbons to be compared. Iodine's small molecular size can characterize the small pore volume of a carbon and its ability to adsorb contaminants small in molecular size. The iodine number is the mass of iodine adsorbed (in milligrams) from a 0.02 N bulk solution by 1 g of carbon. The molasses number is a measurement comparing the color (optical density) of the filtrate from a standard activated carbon with the color of the filtrate from the carbon being investigated.

When activated carbon is in use, the iodine and molasses numbers decrease with time as adsorption occurs and available adsorption sites are filled. Some water treatment systems use iodine or molasses numbers to determine when to replace carbon. To use the iodine number as a surrogate measure (replacing actual bench-scale testing of the GAC with the water needing treatment), a correlation between the iodine number and the degree of carbon exhaustion should be developed for each specific adsorption application.

The AWWA standard indicates a performance requirement that the adsorptive capacity of the granular activated carbon, as measured by the iodine number, shall not be less than 500 mg/g carbon. The standard procedure for determining the iodine number of activated carbon is ASTM D4607.

Carbon Weight. The weight of carbon in air is the apparent density. For water treatment applications, carbon density is described by bulk density or as backwashed and drained. For a filter of known dimensions, density determines the weight (or mass) of GAC required to fill that filter. Typically, carbon density should be determined on an as-received basis, and calculations can then be made to correct for moisture content. The apparent-density test apparatus and the procedure for determining the apparent density are described in the AWWA standard.

Apparent density is important because, for new GAC systems, the initial quantity of activated carbon is typically specified either by volume or by weight. Volume (cubic feet or cubic meters) for first-time installation is specified as backwashed, drained, and in place. After the initial installation, additional GAC needed to replace lost carbon may be specified by volume or by weight. The standard specifies that the apparent density of GAC shall be at least 0.25 g/cm^3.

Moisture Content. The AWWA standard indicates that moisture content of GAC shall not exceed 8% by weight as packaged or at the time of shipment in the case of bulk shipments.

Abrasion Resistance. Abrasion resistance is a property describing carbon durability. Activated carbon is exposed to abrasion during shipping, installation, backwashing, and regeneration. If the carbon is not durable enough, abrasion can generate undesirable fines or crushed carbon. Increased fines can result in increased head losses across the filter bed, increased loss of carbon, or degradation of water quality. Abrasion resistance is expressed in terms of the abrasion number; the greater the number, the more resistant the carbon is to abrasion.

Durability. Although the industry has not yet agreed on a standard test for predicting carbon durability, the AWWA standard suggests two tests. The stirring abrasion test is recommended for lignite-based GAC, and the Ro-Tap abrasion test is recommended for bituminous-based GAC. For either test, the retention of average-size GAC shall be at least 70%.

Ash Content. Ash content reflects the purity of the carbon. In the United States, most activated carbons are manufactured from coal. Higher-quality coals, such as metallurgical-grade bituminous coals, produce carbons with ash contents of approximately 5% to 8%. Subbituminous coals produce carbons with ash contents of approximately 10% to 15%. Lower-grade coals, such as lignite, produce activated carbons with the highest ash content of approximately 20%. Ash found in these coal-based carbons can contain calcium, magnesium, iron, and silica. These constituents can form precipitates in areas with hard water supplies. According to AWWA Standard B604, water-soluble ash in GAC should not exceed 4%.

GAC Particle Size. The particle size of GAC used in a filter affects pressure drop, filtration abilities, requirements for backwash rate, and the rate at which adsorption equilibrium is reached. Smaller particle sizes increase the pressure drop across the carbon bed and necessitate lower backwash rates. However, small GAC particles reach equilibrium more rapidly than large particles because of the smaller (shorter) distance organics must diffuse to reach the center of the particle and the larger surface area–to–volume ratio of smaller GAC particles. Mesh size describes the range of particle sizes to be used in a filter.

The effective size of GAC is defined in the AWWA standard as the size opening through which only 10% of a sample of representative filter material will pass. For example, if the size distribution of the media grains has 10% finer than 0.600, the effective size of the GAC is 0.600 mm.

The uniformity coefficient is a ratio of the size opening that will pass 60% of a representative sample of the filter material to the size opening that will pass just 10% of the same sample. The maximum uniformity coefficient recommended by AWWA is 2.1 after backwashing and draining a GAC filter. If two types of GAC media have the same effective size, the one with a larger uniformity coefficient (that is, 1.7 to 2.4) will have a greater number of fines in the upper layers of a stratified bed and larger particles in the lower layers. Typical uniformity coefficients for GAC produced in the United States are less than or equal to 1.9.

Manufacture of Activated Carbon

Activated carbon is made from organic materials having a high carbon content. A wood product such as sawdust, coconut shells, or wood may be used, or a coal product such as bituminous coal, lignite, or peat is converted to activated carbon by heating the material to between 300° and 1,000° C. The resulting carbon material provides large surface areas and, accordingly, a large number of adsorption sites.

DESIGN OF POWDERED ACTIVATED CARBON FACILITIES

PAC is used by surface water treatment plants either full-time or as needed for taste and odor control or removal of organic chemicals.

Characteristics of PAC

The apparent density of PAC ranges from 23 to 46 lb/ft^3 (0.36 to 0.74 g/cm^3) depending on the type of material and the manufacturing process. Iodine and molasses numbers are often used to characterize PAC. Information needed to evaluate different PACs is shown in Table 14.1, which summarizes the manufacturers' specifications for some commercially available PACs. In general, the following criteria should be used to evaluate different PACs:

TABLE 14.1 Manufacturers' Specifications of Some Commercially Available PACs

Parameter	PAC 1[a]	PAC 2[b]	PAC 3[c]	PAC 4[d]	PAC 5[e]	PAC 6[f]
Iodine number, mg/g	800	1,199	600	900	1,000	550
Molasses decolorizing index	9	—	—	14	18	—
Moisture as packed, %	5	3	5	10	10	4
Apparent density, g/cm^3	0.64	0.54	0.74	0.38	0.38	0.50
Ash content, %	—	6	—	3 to 5	3 to 5	—
Passing 100 mesh, %	99	—	99	95 to 100	95 to 100	99
Passing 200 mesh, %	97	—	97	85 to 95	85 to 95	95
Passing 325 mesh, %	90	98	90	65 to 85	65 to 85	90

[a] Aqua-Nuchar, Westvaco, Covington, W.Va.
[b] WPH, Calgon Corp., Pittsburgh, Pa.
[c] Aqua, Westvaco, Covington, W.Va.
[d] Nuchar S-A, Westvaco, Covington, W.Va.
[e] Nuchar SA-20, Westvaco, Covington, W.Va.
[f] Hydrodarco B, American Noril, Jacksonville, Fla.
Source: Adapted from I. N. Najm et al., 1991.

ACTIVATED CARBON PROCESSES 14.5

- The AWWA standard for PAC specifies a minimum iodine number of 500. Higher iodine numbers indicate greater adsorbancy.
- Higher molasses numbers indicate greater adsorbancy.
- Moisture content should be minimized and should not exceed 8%.
- AWWA standard specifies a minimum apparent density of 0.25 g/cm^3.
- Ash content, which is an impurity that reduces adsorptive capacity, should be minimized.
- Smaller PAC particles adsorb organic compounds more rapidly than larger particles, so PAC is manufactured as a very fine powder. Typically, 65% to 95% of commercially available PAC passes through a 325-mesh (44-μm) sieve.

Selecting a PAC Application Point

PAC is primarily used to control tastes and odors. Important factors that should be considered in selecting a PAC application point include the following:

- The contact time between PAC and organics in the source water is important and depends on the ability of the carbon to remain in suspension. Providing a minimum of 15 min of contact time is generally sufficient for most taste and odor compounds. Considerably longer contact times may be needed for 2-methylisoborneol (MIB) geosmin removal.
- The surface of PAC particles should not be coated with coagulants or other water treatment chemicals before the PAC has had adequate contact time with the source water.
- PAC should not be added concurrently with chlorine or potassium permanganate because PAC will adsorb these chemicals. The best location for PAC addition is usually at the head of the plant, either in the source water pipeline or in a basin dedicated to rapid or flash mixing chemicals. Applying PAC at the earliest point in the treatment process allows the longest contact time possible before the application of other chemicals. If PAC must be added later in the treatment process, dosages may be higher to account for shorter contact times and interference by other chemicals, such as coagulants and chlorine.

If PAC is added to the sedimentation basin effluent or filter influent, particular care must be taken in filter operation. Because of its small carbon particle size, PAC can pass through a filter and cause black water complaints from consumers. An approach used by some systems is to add carbon at two or more application points, with part of the carbon added to the source water and smaller doses added before filtration to remove remaining taste- and odor-causing compounds. Table 14.2 summarizes some of the important advantages and disadvantages of PAC addition at each of these points.

PAC is increasingly being used in conjunction with microfiltration or ultrafiltration membrane systems. Not all membrane filtration products are compatible with PAC so the types of PAC that can be used may be limited. Therefore, it is vital that the use of PAC be discussed with the membrane filtration supplier when considering this application.

PAC dosage depends on the type and concentration of organic compounds present. Common dosages range from 2 to 20 mg/L for nominal taste and odor control, but doses can exceed 100 mg/L to handle severe taste and odor episodes or spills of organic chemicals.

TABLE 14.2 Advantages and Disadvantages of Different PAC Application Points

Point of addition	Advantages	Disadvantages
Intake	Long contact time, good mixing	Some substances may adsorb that otherwise would be removed by coagulation, thus increasing the activated carbon usage rate
Slurry contactor preceding rapid mix	Excellent mixing for the design contact time; no interference by coagulants; additional contact time possible during flocculation and sedimentation	A new basin and mixer may have to be installed; some competition may occur from molecules that otherwise would be removed by coagulants
Rapid mix	Good mixing during rapid mix and flocculation; reasonable contact time	Possible reduction in rate of adsorption because of interference by coagulants; contact time may be too short for equilibrium to be reached for some contaminants; some competition may occur from molecules that otherwise would be removed by coagulation
Filter inlet	Efficient use of PAC	Possible loss of PAC through the filters and into the distribution system

Source: Adapted from I. N. Najm et al., 1991.

Basis for Applying Powdered Carbon

Although some treatment plants feed PAC continuously at a low dosage, a far greater number of plants keep a supply available for use during emergencies, such as organic pollution, that may last only a day or so. Other plants may apply PAC for only a few weeks out of the year during periods of poor source water quality, such as during algae blooms. A major dilemma facing the designer is how to provide a PAC feed and storage system to meet the full range of feed rates. PAC system design must also deal with how best to provide for a chemical that is potentially dusty, difficult to handle, and difficult to store for long periods, but is needed on a moment's notice.

The designer must first consider the specific needs of the owner, from which a suitable PAC feed strategy can be developed. The following questions should be asked before the design is begun:

- How often, over the course of a typical year, is carbon needed?
- What is a reasonable range of PAC feed rates to meet historical source water quality characteristics?
- What is the worst-case scenario for PAC feed?
- How quickly can a shipment of PAC be delivered to the plant site?

A PAC system must be ready when needed, quick to respond, and flexible over a wide range of needs.

Powdered Carbon Application Equipment

Powdered activated carbon can be fed as a powder using dry feed equipment or as a slurry using metering pumps or rotodip feeders. A dry feed system should be considered if PAC feed is infrequent and the maximum feed rate is less than a few hundred pounds per hour. The feed system usually includes a bag-loading hopper, an extension hopper, a dust collector, a dissolving tank or vortex mixer tank, and an eductor.

A much better way to feed PAC is to mix it with water to form a slurry. A slurry system (Figure 14.1) should be considered if PAC is frequently used and the maximum feed rate exceeds several hundred pounds per hour. The metering pump system usually includes a PAC slurry storage tank and can either be a hydraulic diaphragm pump or rotary volumetric feeder. However, if PAC is not required regularly, the slurry tank sits either empty or full. If the tank is full, the carbon and water mixture must be kept continuously stirred to keep the carbon in suspension. If the tank is kept empty with the intent of mixing slurry when it is needed, the taste and odor episode may be over before the slurry system is operable.

PAC is typically available in 50-lb (23-kg) bags or in bulk form from trucks or railroad cars. PAC is usually purchased in bags if dry feed is to be used and in bulk if it is to be fed as a slurry. If delivered in bulk form, carbon is removed from the tank car or truck using an eductor, or it is blown into the slurry tank directly from the delivery truck. The slurry can be formed by the eductor and then transferred to a storage tank, or slurry can be batched directly inside the tank. In either method, slurry should be batched at about 1 lb of PAC per 1 gal of water (0.1 kg/L). A batch meter should be provided on the water supply line to accurately measure the amount of water added to a specific load of carbon. The system illustrated in Figure 14.1 incorporates the following design features:

- A 30,000-gal (113,500-L) square concrete activated carbon slurry tank
- A 30-hp vertical turbine mixer, with type 316 stainless steel shaft and blades
- A bag loader and dust collector
- An air bubbler and stilling well for slurry tank level measurement, with a water flush connection for periodic purging of the stilling well chamber
- A pneumatic fill connection directly to the bulk delivery truck

FIGURE 14.1 Activated carbon feeding system.

- A batch meter and flush connection, with gauges for monitoring supply pressure
- A single rotary volumetric feeder (because carbon is not fed constantly at this facility, only a single feeder has been provided)
- An eductor, equipped with a rotameter for measuring water supply
- A solenoid valve on the water supply line that keeps the water flowing for a period of time after the feeder is turned off to flush the discharge line
- Sample connections
- Cleanouts

For the PAC system illustrated in Figure 14.1, the maximum plant flow rate is 90 mgd (340 ML per day). At maximum flow and an average dosage of 1.5 mg/L, the tank provides 26 days of storage. At a maximum feed rate of 10 mg/L, only 4 days of storage is available.

Plants using PAC on a regular basis should consider at least two slurry tanks so that a shipment of carbon can be placed into one tank before the other is empty. However, most applications can be satisfied by a single storage tank. A carbon slurry storage tank can be circular or square and made of steel or concrete. Tank storage capacity should be approximately 20% greater than the maximum carbon load delivered by railroad car or truck. A protective lining can be provided to inhibit corrosion, but unlined concrete tanks are usually sufficient. Mechanical mixers should be provided in the storage tank to keep the PAC slurry in suspension. All metal surfaces should be type 316 stainless steel.

Slurry can be pumped from the main storage tank to a day tank that holds the volume of slurry to be fed over a period of several hours, or slurry can be fed from the main storage tank directly to the chemical feeder. The day tank, if provided, should be plastic or steel with a corrosion-resistant lining. The tank must be equipped with a mixer to keep the slurry in suspension. Slurry should then flow by gravity to the volumetric feeder.

An eductor usually moves the slurry from the feeder to the application point. Piping should slope downgrade to the application point, with provisions for flushing any carbon that may settle out and clog the pipe. Avoid any vertical drops of pipe down off a header unless the drops are used continuously. If they are not used continuously, PAC will settle in the bottom of them and plug the pipe. Because of the abrasive nature of carbon slurry, piping should be corrosion- and erosion-resistant, such as rubber, plastic, or stainless steel. Pump impellers and mixing blades in the slurry tank and day tank should also be of stainless steel or fiberglass. To avoid feed pipe clogging, a minimum flow velocity of 5 ft/s (1.5 m/s) should be maintained in the pipelines. Care must be taken to provide long-radius elbows.

Wet activated carbon removes oxygen from the air, and as a result, slurry tanks or other enclosed spaces containing carbon may have seriously reduced oxygen levels. Personnel who must enter these spaces should use an oxygen meter to check the atmosphere and also have attached safety belts and another worker present to pull them from danger if necessary.

Common PAC System Operating Problems

The most common operating problem with PAC is chemical handling. Most dry feed systems currently in use can be labor-intensive. Dust is a major problem if a dry feed system is used, but the amount of dust allowed to become airborne can be minimized by good design practice.

A problem that is not uncommon is PAC passing through the filters and entering the distribution system, provoking complaints from consumers. Black water is usually caused by inadequate coagulation or sedimentation or high doses of PAC added just before the filters. The problem is usually eliminated if carbon is fed into the source water intake or into the rapid mix basin.

Careful attention must be paid to the interaction of PAC with other water treatment chemicals. Activated carbon chemically reduces substances such as free and combined chlorine, chlorine dioxide, ozone, and potassium permanganate, and the demand for these chemicals will be substantially increased. Mixing PAC with chlorine also reduces the adsorption capacity of the activated carbon for selected compounds. Competition between the two chemicals must be avoided. Adding PAC to a source water supersaturated with $CaCO_3$ or other precipitates, or after lime softening, may lead to particle coating of the PAC and a corresponding decrease in adsorption efficiency.

Carbon dust is potentially explosive. Explosion-proof motors should be used if contact with carbon dust is a possibility. Because PAC adsorbs organic compounds, including gases that could reduce its effectiveness, PAC storage must be carefully located. Carbon should be stored in a separate, climate-controlled storage area. The stock of bagged PAC should always be stored in a manner that allows stock to be rotated. Storing carbon outside on pallets under canvas or plastic sheets for long periods is not recommended.

PAC slurry solidifies if not mixed frequently or, preferably, continuously. Feed lines become clogged, as do the bags of filter dust collectors. Hydraulic eductor lines carrying slurry also become clogged if they are shut down without being thoroughly flushed. It is recommended that mixers be set on timers to allow scheduled, unattended operation.

If a slurry feed system is used, carbon should be fed below the water surface in the slurry tank so that only a small amount of dust is produced during the loading period. A dust collector should be provided on top of the slurry tank to prevent any dust from escaping into the plant.

Some bag storage should be maintained to supplement a slurry-based system for use if sudden, extremely heavy use exceeds the slurry tank capacity. A bag loader can be provided on top of the slurry tank to allow the tank to be recharged with PAC without calling for a full tanker truck or to meet lower feed rates. Although this is not recommended except for worst-case conditions, carbon bags can be dumped directly into the intake well or elsewhere at the head of the plant.

The PAC feed system should be located as close to the application point as possible. It is suggested that the system be periodically operated, even without a need for carbon feed. Working parts of the system should be checked to be sure the system is always ready for use. Periodic operation also ensures that feed lines are cleaned, carbon is mixed, and the system is flushed. The discharge piping for any feed device should include a flushing line, automatically set to flush the pipelines for a period of at least 60 s after carbon feed is stopped. Because moist PAC is highly corrosive, all metal parts that come in contact with carbon should be type 316 stainless steel.

DESIGN OF GRANULAR ACTIVATED CARBON FACILITIES

As illustrated in Figure 14.2, three basic options are available for locating a GAC treatment step for both new and existing water treatment plants:

- Prefiltration adsorption, ahead of the conventional filtration process (prefilter adsorber)
- Postfiltration adsorption, after the conventional filtration process (postfilter adsorber)
- Filtration/adsorption, combining water filtration and GAC filtration into a single process (filter-adsorber)

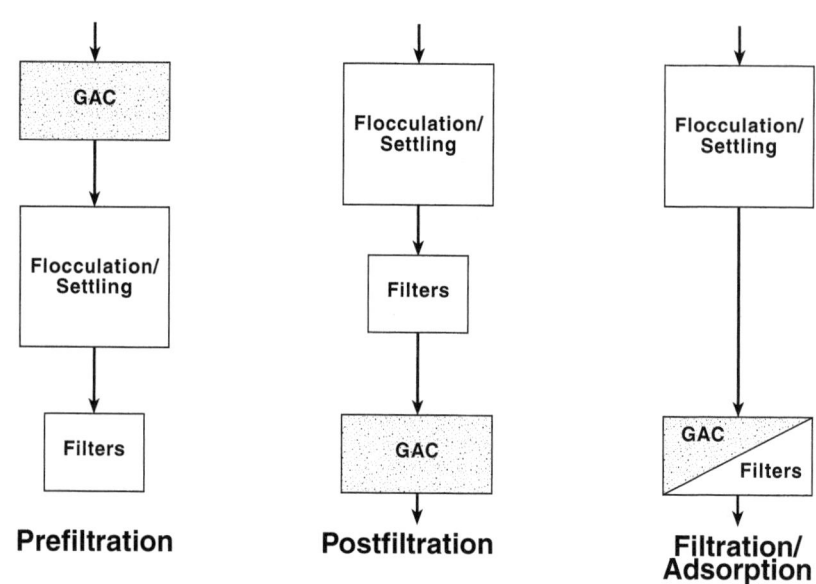

FIGURE 14.2 Basic options for locating GAC treatment.

Under certain plant configurations and site-specific conditions, prefiltration adsorption may be feasible. However, this option has limited applications and benefits and therefore is not discussed further in this chapter.

The more conventional option for most applications is postfiltration adsorption in which the water receives complete treatment before GAC application. In this option, the highest-quality water is applied to the GAC system, the only objective being to remove dissolved organic compounds. As a result, the carbon usage rate is minimized, as is the frequency of regeneration, providing for an efficient GAC design.

General Considerations

General design considerations in assessing the feasibility of GAC for a specific application include

- Design flow rate and system hydraulics
- Types and amounts of organics to be removed
- Type of GAC
- Breakthrough and empty bed contact time (EBCT)
- Adsorber volume and bed depth
- Surface loading rate
- Backwashing requirements
- Carbon usage rate
- Type of contactor units
- Gravity or pressure units

ACTIVATED CARBON PROCESSES   14.11

Design Flow Rate and System Hydraulics. GAC contractor design is similar to the design for conventional gravity dual-media filtration. Hydraulically, the design must accommodate the maximum flow of the plant. GAC volume is determined based on design flow for the GAC facilities and the EBCT established for the flow. Because flow rates and organic concentrations often vary seasonally, careful analysis is needed to prevent undersizing or oversizing the contactors.

Types and Amounts of Organic Compounds to Be Removed. Both the type and the amount of organic compounds to be removed strongly affect the type and level of pretreatment prior to GAC adsorption, as well as the rate of adsorption of the contaminant molecules.

Type of GAC. GAC particles are similar in size to filter media, typically ranging between 1.2 and 1.6 mm in diameter. General properties of typical types of GAC produced in the United States are presented in Table 14.3. Some commonly manufactured mesh sizes available for GAC and their associated size ranges appear in Table 14.4.

Breakthrough and EBCT. Primary factors the designer must consider in determining the volume required for a GAC adsorber or contactor are breakthrough and EBCT.

Breakthrough. Breakthrough can be defined as the point at which the concentration of a contaminant in the effluent adsorption unit exceeds the treatment requirement. Total organic carbon (TOC) is often used as the performance standard defining GAC exhaus-

TABLE 14.3 Properties of Granular Activated Carbon Produced in the United States*

	The Carborundum Company	Calgon Corporation		NICIT	Witco Chemical
Product name	GAC 40	Fitrasorb	Hydrodarco HD-1030	Hydrodarco 83 Plus	Witcarb 950
Base material	Western bituminous coal	Bituminous coal	Lignite	Bituminous coal	Petroleum coke
U.S. standard sieve size	12 × 40	12 × 40	10 × 30	8 × 30	12 × 40
Effective size, mm	0.6	0.8 to 0.9	0.8 to 0.9	0.8 to 0.9	0.8 to 0.9
Uniformity coefficient	≤1.9	≤1.9	≤1.7	≤1.9	≤1.7
Apparent (or vibrating feed) density, g/cm^3	0.47	0.4 to 0.5	0.40 to 0.50	0.47	0.46 to 0.53
Washed density, lb/ft^3	25	25	23.5	26	29 to 33
Iodine number, mg/g	1,050	1,050	600	900	1,050
Surface area, m^2/g	1,000 to 1,100	1,050	650	1,000	1,000
Available in:	60-lb bags, bulk	60-lb bags, bulk	40-lb bags, bulk	50-lb bags, bulk	50-lb bags, 200-lb drums, hopper trucks, tote bins

*AWWA Standard for Granular Activated Carbon [AWWA B604-74, Journal AWWA 66(11):672, 1974] provides minimum specifications on the properties of granular activated carbon used as an adsorption medium for treating drinking water.

TABLE 14.4 Commonly Manufactured GAC Mesh Sizes

Mesh size (U.S. standard sieve size)	Particle diameter range (mm)
8 × 16	1.18 to 2.36
8 × 20	0.850 to 2.36
8 × 30	0.600 to 2.36
10 × 30	0.600 to 2.00
12 × 40	0.425 to 1.70
14 × 40	0.425 to 1.40
20 × 40	0.425 to 0.85
20 × 50	0.300 to 0.85
Sand	0.38 to 0.65
Anthracite	0.45 to 1.6

tion. As a rule, if the GAC filter effluent concentration is greater than the performance standard for 3 or more consecutive days, it is time to regenerate or replace the GAC.

Breakthrough depends on the quality of both the influent stream and the design of the carbon bed. Figure 14.3 illustrates a typical breakthrough curve for a GAC adsorber from the time of start-up to exhaustion. As the adsorptive capacity of a contactor becomes exhausted at any location in the bed, the zone of adsorption migrates farther down the bed. Eventually, when there is very little adsorptive capacity left, the contaminant begins to break through in the effluent. As more and more leakage occurs, the bed becomes ex-

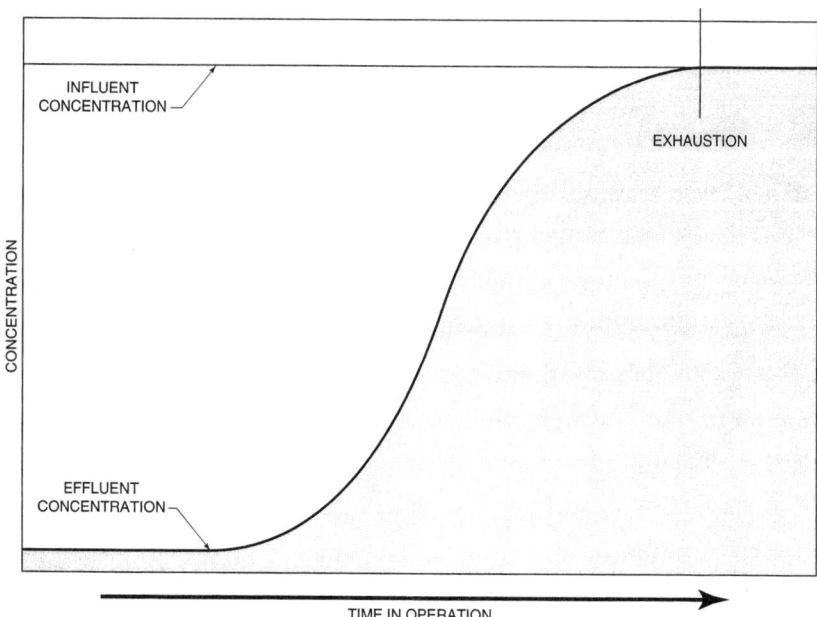

FIGURE 14.3 Typical breakthrough pattern for GAC filter.

hausted. Breakthrough curves are important to the designer because they define the relationship between the physical and chemical parameters of the carbon system (e.g., flow rate, bed size, carbon exhaustion rate), the determination of the number of beds or columns, their arrangement (either series or parallel), and treatment plant effluent requirements.

EBCT. EBCT is calculated as the volume of the empty bed (occupied by the GAC) divided by the volumetric flow rate of water through the carbon. Alternatively, *EBCT* can be defined as the depth of GAC divided by the linear velocity of water flowing through the carbon bed. Note that EBCT is a false-residence time.

EBCT is used instead of detention time because of the ease of calculation. An actual detention time would have to account for the porosity of the bed, a variable that changes with carbon size and type. EBCT can be varied by changing the bed depth at constant flow or by changing flow with constant bed depth. Together, the design EBCT and the design flow rate define the amount of carbon to be contained in the adsorption units.

Longer EBCTs can delay breakthrough (to a point) and improve carbon usage rate; shorter EBCTs can expedite breakthrough. Thus the time of GAC operation between replacement or regeneration depends on the EBCT. For most water treatment applications, EBCTs range between 5 and 25 min. In addition, a factor having a greater influence on operating costs than EBCT is volume throughput, which is the number of bed volumes of water processed before the breakthrough concentration is reached.

Adsorber Volume and Bed Depth. After the EBCT has been determined, carbon volume and bed depth can be selected. The adsorber design volume depends on bed volume and the amount of freeboard (excess vessel capacity beyond design operating levels). Freeboard can range up to about 50% for fixed and expanded bed systems. If bed expansion is unnecessary, a freeboard of 20% to 30% may be adequate to allow for proper bed expansion during backwashing.

No freeboard is needed for upflow pulsed beds. An economic evaluation is usually made of capital and operating costs to compare carbon columns of various depths.

Surface Loading Rate. The surface loading rate for GAC filters is related to the design flow of a particular treatment plant. Surface loading rates are defined in the same manner as for conventional granular media filters. The surface loading rate is the rate of a volume of water passing through a given area of GAC filter bed, usually expressed as cubic meters per square vector (m^3/m^2) or gallons per minute per square foot (gpm/ft^2). Surface loading rates for GAC filters range from 2 to 10 gpm/ft^2 (5 to 24 m/h), although rates of 2 to 6 gpm/ft^2 (5 to 15 m/h) are more commonly used as design criteria.

Surface loading should be kept high for those compounds where the mass transfer rate is controlled by the rate of transfer of the chemical from the bulk liquid to the interior pores of the GAC. Typically this is the case for highly adsorbable compounds (e.g., many SOCs). When mass transfer is controlled by the rate of adsorption (and transport) within the GAC particle, surface loading is not important. This is the case with most less-adsorbable compounds. Figure 14.4 illustrates the relationship between hydraulic loading and pressure drop for several brands of GAC.

Backwashing Requirements. A GAC filter bed is backwashed using the same general procedures typically used for backwashing conventional granular gravity filters. If GAC is installed as a sand filter replacement, a redesign of the backwash supply system, including the rate of flow control and washwater troughs height, is often necessary. This redesign is necessary because of the difference in particle density between GAC and sand—about 1.4 and 2.65 g/cm^3, respectively. If GAC is used as a simple replacement for anthracite coal as a filter medium and the granular sizes are similar, the backwash system may be adequate because particle densities are nearly identical.

FIGURE 14.4 Carbon bed pressure drop versus hydraulic loading.

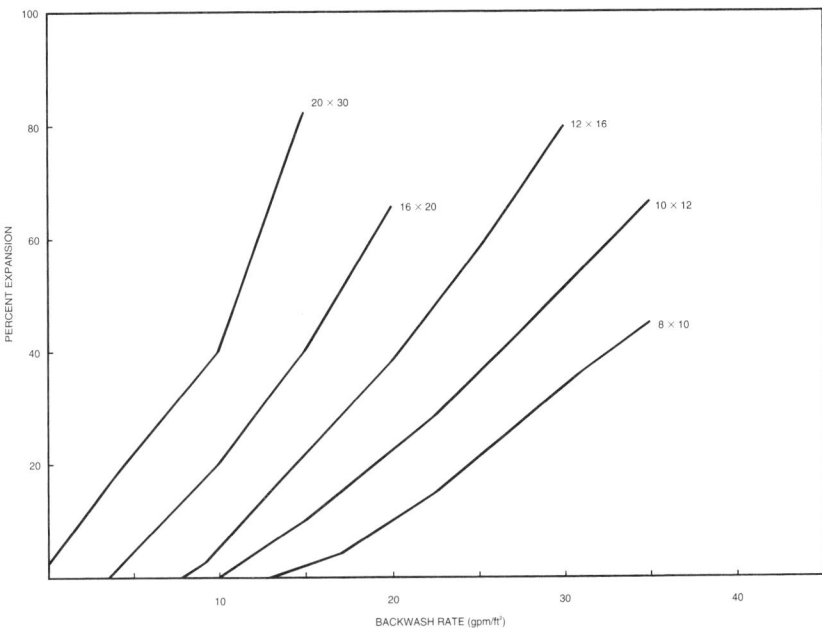

FIGURE 14.5 Backwash characteristics of various mesh sizes at 25° C.

GAC particle size distribution and wetted density vary among different carbon brands and even among different deliveries of the same carbon. Appropriate backwash rates can be obtained from the manufacturer for each type of carbon (Figure 14.5). Backwash rates must be adjusted to account for specific media characteristics and for changes in backwash water temperature (Figure 14.6). Installing a surface wash or air scour system to assist with filter cleaning may be necessary to control mudball formation.

A good conservative design should allow for 75% to 100% expansion of light GAC media, but 50% is generally considered to be adequate. The design should provide for sufficient freeboard to reduce media losses during the backwashing cycle.

Carbon Usage Rate. The carbon usage rate (CUR) determines the rate at which carbon will be exhausted and how often the carbon must be replaced. The CUR essentially determines the size of the entire regeneration system. The CUR for GAC systems removing organic compounds may be determined by using physical models or adsorption isotherm models. A pilot-scale test is often used to evaluate the complexity of multiple chemical interactions. A quicker and more economical method for evaluating GAC column performance is the *rapid small-scale column test* (RSSCT). Small systems can be designed to best simulate the performance of a full-scale system using dimensional analysis. Dimensionless parameters for the full-scale system and the small system are designed to be equal.

Carbon treatment effectiveness improves as contact time increases. The percentage of total carbon in a bed that is exhausted at breakthrough is greater in a deep bed than in a shallow bed. At a point beyond the optimum bed depth, the additional adsorber volume provided acts primarily as a storage capacity for spent carbon. The actual selection of bed

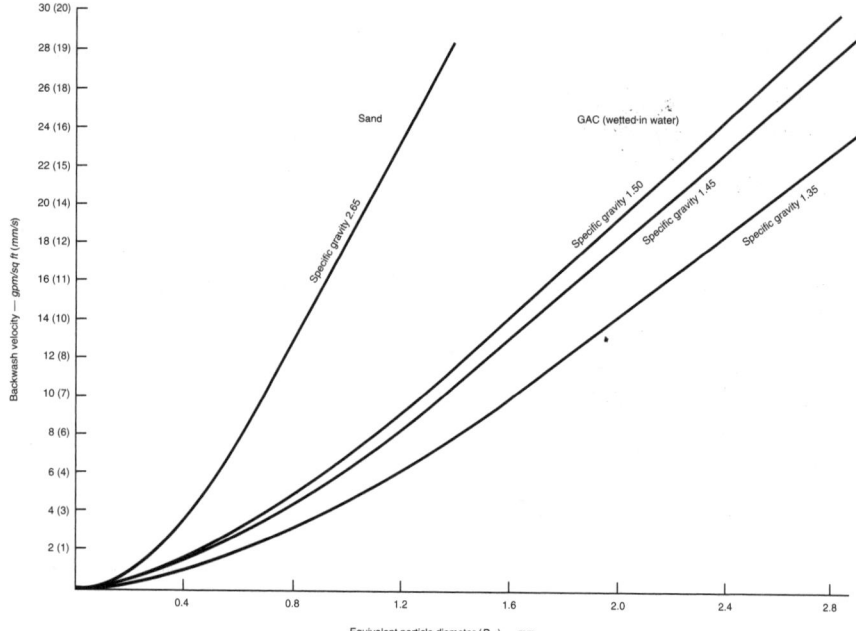

FIGURE 14.6 Determination of appropriate medium size and backwash rate for a water temperature of 68° F (20° C) (backwash velocity = $1.3 \times V_{mf}$ for D90).

depth and corresponding adsorber volume also depends on reactivation frequency and contactor construction cost. The relationship between increased adsorber volume and reduced reactivation frequency can be compared against a reduced adsorber volume and increased reactivation frequency to determine the optimum characteristics for design. Together, CUR and EBCT have the greatest effect on capital and operating costs for GAC processes.

GAC adsorption units can have one of three basic configurations, with single or multiple adsorbers operated in series or in parallel:

- Downflow fixed bed
- Upflow fixed (packed) or expanded (moving) bed
- Pulsed bed

The upflow configuration is not suitable for application in a potable water treatment process unless media filters are located downstream of the adsorption units (Kawamura, 1991).

Type of Contactor Units

Gravity or Pressure Units. Larger installations are likely to be gravity fed. Pressure adsorption units are cost-effective for smaller installations. Pressure flow can be used for either downflow or upflow beds. Pressure flow achieves higher hydraulic loadings than would be economically feasible with gravity flow. This higher loading reduces the required adsorber cross-sectional area. Pressure flow also permits operation at higher suspended solids concentrations with no backwashing or with less frequent backwashing than is possible with gravity flow. Pressure adsorbers can be less expensive to design and construct because they can be prefabricated and shipped to the site.

For many systems, the decision between pressure or gravity adsorbers is based solely on initial installed cost. Most existing and many new water treatment plants require that the total plant flow be pumped to postfiltration GAC facilities because of insufficient hydraulic head downstream of the existing or new conventional filters. Pressurized adsorbers (commonly called contactors) may be more cost-effective in these installations.

Design of GAC Filtration/Adsorption Beds

For filtration/adsorption, either as a filter cap or full-depth GAC, major design considerations include:

- Filter run lengths
- Means for cleaning the GAC bed as a part of routine operations
- Life expectancy of the GAC
- Chlorine, polymer, and ozone feed points
- Size and depth of GAC as filter media
- Matched GAC and sand barrier layers during backwash (GAC cap)
- Head loss restrictions
- Backwash restrictions
- Flow rate restrictions—inlet and outlet piping and conduits

Retrofitting Existing High-Rate Granular Media Filters to Filter-Adsorbers

One option readily available for most water treatment facilities is to replace conventional filter media with GAC. Existing filter boxes can be converted to GAC filter-adsorbers simply by removing a portion of or all the granular media from the filter box and replacing them with GAC. Alternatively, an entirely new filter box, underdrain, and backwashing system for the GAC can be designed and constructed. In these applications, GAC is used for turbidity and solids removal, biological stabilization, and dissolved organics removal by adsorption.

Filter-adsorbers must balance the constraints of both adsorption and filtration. They typically require backwashing at about the same frequency as conventional filters. A filter-adsorber design can be easily installed as a retrofit for an existing conventional granular filter, but the designer must recognize and account for the fact that filter-adsorbers

- Have shorter filter run times and must be backwashed more frequently than postfilter adsorbers, which in some applications do not require backwashing at all
- May incur greater carbon losses than postfilter adsorbers because of more frequent backwashing
- May cost more to operate than postfilter adsorbers because carbon usage is less efficient

Washwater Trough Considerations. In retrofitting existing filters, modifications may be necessary to position the washwater collection troughs to allow for proper bed expan-

L — Filter media depth at rest

P — Height of backwash troughs

S — Spacing between backwash troughs

H_o — Height of top of backwash trough over filter media

Where $0.75 \times L + P < H_o < L + P$

And $1.5 H_o < S < 2 H_o$

FIGURE 14.7 Design guidelines for filter washwater troughs.

sion to take place. Although some loss of GAC is expected over time as a result of the backwash process, the trough must be located at the correct elevation above the top of the filter bed to ensure that particle losses are not excessive. The height and spacing guidelines of washwater troughs are illustrated in Figure 14.7. Given the high cost of GAC as compared with anthracite, excessive and unnecessary GAC loss during backwash is costly.

GAC Support System. For support, GAC can be placed directly on sand or graded gravel. In either case, particular attention must be paid to the type of underdrain system used. Underdrain products are available that allow GAC to be placed directly on top of the underdrain. The most common types are the stainless steel or plastic wedge wire screen and block underdrains. In the block underdrain design, a permeable membrane is placed over the block underdrain. The membrane prevents GAC loss through the underdrains and provides a surface for the uniform application of filter washwater. A cross-sectional view of

FIGURE 14.8 Wedge wire underdrain for GAC filters.

a wedge wire underdrain is shown in Figure 14.8. The block underdrain with a porous membrane is shown in Figure 14.9. If air scour is proposed, ensure the underdrain is compatible with use with air scour.

Converting Existing Dual-Media Filter to GAC Postfilter Adsorber. To convert a dual-media filter to a GAC postfilter adsorber, the process is a little more complex. Many larger gravity dual-media filters—usually greater than 1,000 ft^2 (about 100 m^2)—have center filter gullets that split the filter into two physically separate but hydraulically connected cells. The separation is designed to separate the total filter area into two manageable areas that can be backwashed separately, one cell at a time.

This design reduces the size of the backwash facilities (pumps and storage tanks can be one-half the size) and ensures that the rate of backwash flow is not disproportionate to the total instantaneous plant flow. The center filter gullet supplies source or settled water to the filter in the upper gullet, collects the filtered water, and supplies filter backwash water (reverse flow) to the underdrains in the lower gullet.

Many current dual-media filters at one time were sand filters. Converting sand media to dual media, to attain higher filtration rates without the need for physical plant expansion, has been found to be beneficial for many treatment plants.

However, in many of these applications, the filter box is barely deep enough to allow the media change to sand and anthracite coal. Excessive media losses, from losing an-

FIGURE 14.9 Block underdrain with porous membrane.

thracite over the backwash troughs of shallower filter boxes, are common. These filters typically have very shallow filter box depths relative to the depth that would be provided for either a new dual-media filter design or a GAC filter or postfilter adsorber design. Providing empty bed contact times for GAC filtration or adsorption presents physical demands that cannot be easily met with this type of design.

A simple method of converting dual-media gravity filters to GAC contactors is illustrated in Figure 14.10. The depth of GAC needed to provide the empty bed contact time can be accommodated with minor physical modifications. The filter can be retrofitted to provide downflow in one cell and upflow in the adjacent cell. As a result, the depth of GAC needed for EBCT can be provided with a shallow box. In many applications, converting existing dual-media filters to GAC, just as these same filters were once converted from sand to dual media, can be accomplished at a fraction of the cost of constructing entirely new contactors. When one is contemplating a conversion to GAC postfilter adsorbers, the value and cost benefit of retrofitting the existing filters should not be overlooked.

The conversion from sand or dual media to GAC can be accomplished as follows:

- Remove all filter media from the filter (sand or sand and anthracite coal and the support gravel) either physically or by eduction.
- Inspect the underdrain system.
- Increase the height of the backwash troughs if required for the filter media.
- Alternatively, replace existing underdrains with a "gravel-less"-type underdrain system, such as a type that uses a permeable membrane layer, to achieve the depth needed for the new or deeper media bed.
- Check existing backwash supply facilities. Redesign may be necessary because GAC may require more or less backwash flow.
- Inspect the existing surface wash or air scour system. A system should be considered if none is currently in place.
- Train plant staff in new backwash procedures.

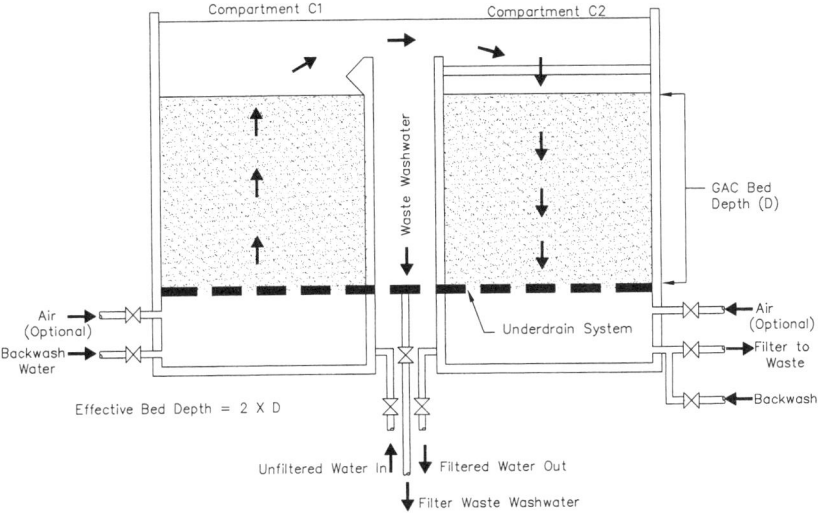

FIGURE 14.10 Conversion of split-cell filters to GAC contactors.

Capping Existing Slow Sand Filters with GAC

Slow sand filter operations can be improved by adding a GAC filter cap. Slow sand filtration is a process of filtering water through a sand layer at a very low flow rate. Backwashing is not required. Instead, the filters are typically scraped after 1 to 6 months of operating time. Scraped sand is eventually replaced with new sand to restore the sand filter to the desired depth. The depth is usually 24 to 40 in. (625 to 1,000 mm), supported by 12 to 18 in. (300 to 450 mm) of graded gravel. Underdrains typically consist of perforated pipes, porous concrete slabs, or channels formed by double layers of loose brick. Filter areas may be as large as 50,000 ft^2 (4,645 m^2). Water height above the sand surface is usually maintained between 3 and 5 ft (1 and 1.8 m).

Slow sand filter performance can also be improved by adding a GAC sandwich layer, placed 4 to 6 in. (100 to 150 mm) below the top of the sand layer, as shown in Figure 14.11. The GAC layer is typically 3 to 8 in. (75 to 200 mm) thick, depending on water quality conditions. Operation is similar to that of the normal slow sand filter. At the time of sand scraping, sand is removed down to the GAC level. Sand is then added again if the GAC is determined to be active (not exhausted or spent). When the GAC layer is spent, it can be removed and reactivated, then returned to the sandwich filter for reuse.

Design of Postfiltration Adsorbers

Perhaps the most important consideration facing the designer is to choose the proper EBCT and process flow configuration (adsorbers in series or parallel) to maximize GAC effectiveness. The type of regeneration system used for reactivating spent carbon is also an important process design element.

Separate GAC adsorbers after conventional filtration offer increased flexibility for handling GAC and for designing specific adsorption conditions. Considerably longer contact times can be obtained than usual for the filter-adsorber.

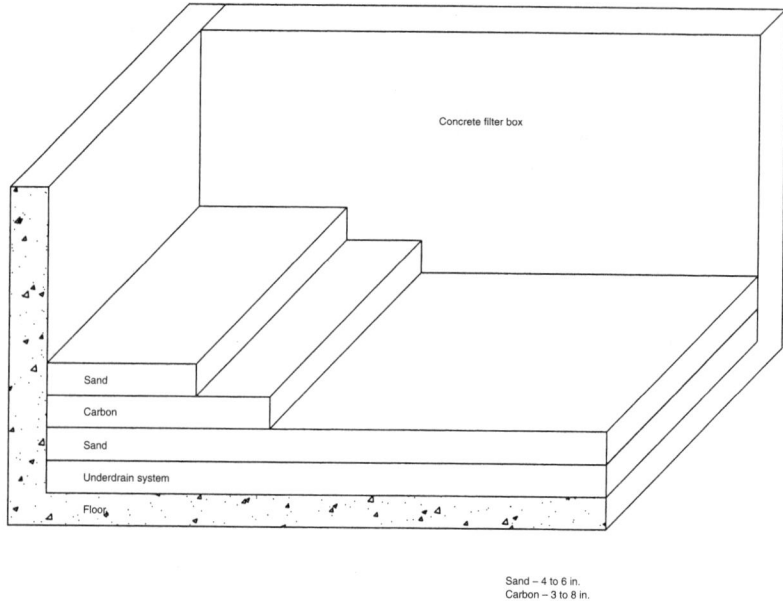

Sand – 4 to 6 in.
Carbon – 3 to 8 in.
Sand – 8 to 10 in.
Underdrain system – see Figures 14.8 & 14.9

FIGURE 14.11 Media arrangement in a GAC sandwich slow sand filter.

Postfilter adsorbers typically use GAC with small effective sizes and large uniformity coefficients to promote rapid adsorption of organic compounds, better stratification, and reduction of overall carbon loss. These adsorbers also can provide an improved barrier against microbial penetration, are more compatible with other advanced treatment processes such as ozonation (or ozone and hydrogen peroxide), and use more of the adsorptive capacity of the carbon.

Adsorber Configurations. Various combinations of the three GAC contactor configurations and flow patterns can be made. Some of the most common contactor configurations are shown in Figure 14.12. Table 14.5 provides a summary of each configuration described below.

Fixed Beds in Series. In this configuration, flow is downward through the carbon bed for each unit connected in series, with the EBCT divided between the number of adsorbers. The first in the series receives the highest contaminant loading and is thus the first to be exhausted. The last in the series receives the lightest contaminant loading and serves as a polishing step. When carbon is removed for reactivation, the first adsorber in the series is removed from service, with the next adsorber in line becoming the lead unit. This system can be constructed with an extra adsorber on standby to become the first adsorber when the lead adsorber is taken out of service. The configuration can be arranged and controlled so that the freshest GAC is always the last step in the series.

Capital cost considerations usually limit the number of adsorbers in series to four or fewer, because the increased cost of piping and valving counters the cost benefit of reduced carbon usage. Countercurrent flow provides highly efficient usage of the carbon by maximizing carbon exhaustion in the lead adsorber before it is removed for reactivation.

ACTIVATED CARBON PROCESSES   14.23

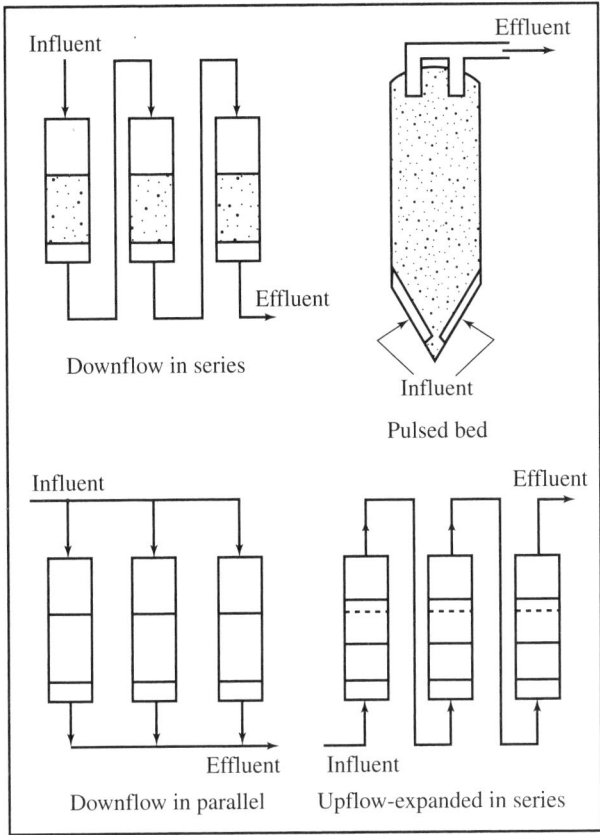

FIGURE 14.12 Commonly used GAC contactor configurations.

TABLE 14.5 Guide to Selection of GAC System Configuration

Adsorption system configuration	Applicable if:
Single adsorber	Low carbon exhaustion rate (usage rate) occurs.
Fixed beds in series	High carbon exhaustion rate occurs. High effluent quality must be ensured at all times.
Fixed beds in parallel	High system pressure drop is minimized. Large total flow rate is required.
Moving beds, upflow	High carbon exhaustion rate occurs. High effluent quality is required. Either some carbon fines can be tolerated in the effluent, or carbon adsorption is followed by filtration.
Expanded beds, upflow	High suspended solids are present in the influent, and some suspended solids can be tolerated in the effluent.

Premature breakthrough of organics may also be reduced. Carbon usage rates under series operation can approach one-half the rate under parallel operation, reducing operating costs accordingly. The lead absorber can be backwashed to remove suspended solids that accumulate in the carbon bed.

Fixed Beds in Parallel. The most common GAC application in drinking water treatment is the downflow fixed bed in parallel operation. Downflow operation is appropriate when the carbon bed is to be used as a suspended solids filter as well as an adsorber. In this design, carbon contained in the adsorber remains stationary. Flow is divided equally to each contactor, and each contactor is sized for the design EBCT. Suspended solids are periodically removed by backwashing in a manner nearly identical to that used for sand and dual-media (sand and anthracite) filters. When the carbon adsorber is preceded by conventional filtration, downflow operation can sometimes be used with reduced or no backwashing at all. In a parallel configuration, each carbon bed receives essentially the same quantity and quality of flow. Start-up of individual units is staggered so that exhaustion of the carbon occurs sequentially. This allows removal of all carbon from each adsorber, one at a time, for reactivation. For systems operating at full design capacity, a spare adsorber can be provided to bring online when an adsorber is taken out of service.

In this arrangement, the level of carbon exhaustion when an adsorber is discharged is not as high as that of fixed beds in series because no adsorber is operated completely to breakthrough. However, carbon use in a given adsorber can be increased by blending effluent from all adsorbers. One or more adsorbers can be run slightly beyond the breakthrough point while other adsorbers produce water with concentrations below the breakthrough point.

Because effluent from each of the units is blended, each unit can be operated until it is producing a water with an effluent concentration in excess of the treated water goal. Only the composite flow must meet the effluent quality goal. For example, if 10 adsorbers are used in parallel, each adsorber can process 10,000 bed volumes of throughput if the effluent TOC criterion is 50% of the influent, compared with 5,000 bed volumes if only a single contactor were used or if all contactors were operated in parallel but were replaced at the same time. This method of operation may be most appropriate for large plants.

Upflow Expanded Beds. Upflow expanded (moving) beds are best suited for waters with high suspended solids concentrations when suspended solids are to be removed by subsequent processes (filtration). For high suspended solids concentrations, upflow beds may be preferred, because solids accumulation and corresponding head losses would be excessive in downflow adsorbers. For low suspended solids concentrations, upflow adsorbers can be considered, because the carbon bed is not needed as a solids filter.

In an upflow bed, the upward movement of water causes the carbon bed to expand slightly (approximately 10%). A higher CUR is expected for expanded beds because mixing of the carbon creates a longer mass transfer zone. Mixing may allow the release of carbon fines into the effluent flow. Expanded beds should not be used where downstream contamination by suspended solids or carbon fines passing through the bed would be a problem.

Pulsed Beds. A pulsed bed operates in an upflow mode, with water and carbon flow moving countercurrent. Pulsed bed adsorbers permit intermittent or continuous removal of spent carbon from the bottom of the bed while fresh carbon is added at the top without system shutdown. The chief advantage of this system is better carbon use because only thoroughly exhausted carbon is removed. In contrast with fixed beds, a pulsed bed cannot be completely exhausted. This prevents any contaminant breakthrough into the effluent that may cause effluent water quality standards to be exceeded.

The performance of pulsed beds is affected by suspended solids or biodegradable compounds that cause extensive biofilm growth on the activated carbon, so that the bed must be backwashed. Backwashing leads to mixing of the fresh activated carbon at the top of the bed with spent activated carbon deeper in the bed and destroys some of the beneficial countercurrent effect. Additionally, some activated carbon fines may be produced during upflow that may require removal by a subsequent process.

Another characteristic of the pulsed bed system is that a steady-state constant effluent concentration (assuming a constant influent concentration) is achieved. In fixed beds, effluent concentration gradually increases with time.

Adsorber Design. Various adsorption vessels can be considered by the designer. The design depends on the size of the plant, type of adsorber selected, mode of operation, and number of adsorber units. Adsorption units are typically cylindrical steel vessels (pressure or gravity flow) or cast-in-place concrete rectangular structures (gravity flow only). The type of vessel used depends on plant size and specific site constraints. Basic characteristics of typical adsorbers are presented in Table 14.6. The four fundamental adsorber types are

- Open-top steel, gravity flow
- Enclosed steel, gravity flow
- Enclosed steel, pressure flow
- Open-top concrete, gravity flow

Treatment plants of 10 mgd (38 ML per day) or less usually select pressure steel vessels, although many designers prefer open-top beds to allow the operator to see the backwash. Larger plants usually have open-top concrete tanks. Other features that differentiate the individual adsorbers are the details of internal hardware, such as liquid distributors and collectors, carbon bed support methods, underdrains, and backwashing apparatus.

Adsorber sizing is based on flow rate, hydraulic loading, and EBCT. These variables determine the adsorber volume, depth, cross-sectional area, and number of individual adsorber vessels. Once these quantities are known, adsorber design can proceed.

TABLE 14.6 Basic Characteristics of Adsorbers

Adsorber type	Material	Volume* range, ft^3	Diameter†	Remarks
Steel, gravity flow	Lined carbon steel	6,000 to 20,000	20 to 30 ft	Field-fabricated by welding or bolting preformed steel plate; mounted on concrete slab foundation
Steel, pressure flow	Lined carbon steel; stainless steel	2,000 to 50,000	Up to 12 ft	Shop-fabricated; over-the-road transportation constraints limit size
Concrete, gravity flow	Standard reinforced	1,000 to 200,000	Usually rectangular	Field-constructed; designs vary; 2:1 length-to-width ratio is common

*To convert ft^3 to m^3, multiply by 0.0283.
†To convert ft to m, multiply by 0.3048.

TABLE 14.7 Standard Diameters for Circular Steel Vessels (ASME)

Shop-fabricated	Field-fabricated
7'11"	15'6"
9'6"	21'6"
10'0"	29'9"
11'0"	38'8"
12'0"	55'0"

Circular Adsorbers. Circular adsorbers include both pressure and gravity flow steel vessels. Adsorbers in parallel or in series affect equipment layout because of the differing needs for piping and valving.

The selection of the number of adsorbers for any given design depends on vessel size and permissible hydraulic loading. Commercially available steel adsorbers have standard diameters. Shop-fabricated steel vessels are limited to a diameter of 12 ft (3.7 m) and a length of 60 ft (18.3 m) because of trucking transportation constraints. Larger vessels can be installed by shipping partial pieces (circular rings and end pieces) and then fabricating the complete vessel in the field. However, this approach is not likely to be cost-competitive with cast-in-place concrete alternatives for any but the largest installations. American Society of Mechanical Engineers (ASME) standard diameters for unfired pressure vessels for shop-fabricated and field-fabricated steel vessels are given in Table 14.7.

Rectangular Adsorbers. Rectangular adsorbers are usually constructed of conventionally reinforced concrete. Multiple vessels are generally built into a single large concrete structure with individual treatment units sharing common walls, pipe galleries, and operating areas. The major design features subject to variation are the length-to-width ratio of the individual unit, the design of the influent and effluent channels and conduits, bed depth, and the type of underdrain system.

A typical length-to-width ratio is 2 : 1, similar to the ratio used for conventional gravity filtration. Influent and effluent conduits are usually oriented to the inside of the common pipe gallery to facilitate placement of isolation valves, meters, and instrumentation and control equipment. Bed depth is determined from other factors, but it is usually not less than 600 mm and can be as deep as 6,000 mm or deeper. A cross section of a typical open-top concrete, rectangular, gravity flow contactor is illustrated in Figure 14.13.

Number of Adsorbers. With bed volume and surface area established, the number and size of contactors can be determined. Design should allow for the full plant flow to be treated with one contactor out of service for maintenance or carbon replacement. As with the design of conventional filters, fewer large contactors are generally more cost-effective than constructing a greater number of smaller contactors. Whether or not operation of the units is staggered, the method of effluent blending may also affect the number of contactors.

The number of contactors is influenced by the following:

- Initial capital costs
- Long-term operating costs
- Ease of operation—generally the fewer the units, the easier to operate
- Backwash system size required

FIGURE 14.13 Cross section through a typical, open-top concrete, rectangular, gravity flow GAC contactor. (*Adapted from Westerhoff and Miller, 1986.*)

- Rate at which backwash water must be applied because this is a function of contactor surface area
- Ability to provide uniform distribution of backwash water over the effective maximum area of a single unit
- The ability to accommodate a reduction in capacity with one unit out of service for backwash, regeneration, or maintenance

- Number and size of valves required
- Structural considerations

For purposes of determining the optimum number of contactors, estimates of capital costs should be developed for various contactor configurations. A plot of the relative capital cost as a function of the number of contactors aids in selecting the best configuration.

Perhaps the most important factor in the design of a GAC adsorption facility is the type of underdrain system to be used. Underdrains must perform several important functions:

- Collect water at a uniform rate to ensure GAC retention in the contactor
- Prevent binding or plugging of the collection system
- Distribute water evenly over the entire contactor area
- Provide uniform fluidization of GAC at a low backwash rate during transfer of spent carbon out of the contactors so that carbon moves to the points of discharge

Common Operating Problems

Operating problems encountered when activated carbon is used as a filter medium or adsorber are similar to those found with most rapid sand or dual-media filters. The most common difficulties faced in operating a GAC facility that the designer should be aware of when preparing a design include

- Effectively monitoring the carbon bed depth
- Properly backwashing the carbon, as carbon bed density and depth can change during backwashing
- Removing carbon from the contactors and storage tanks
- Transporting the carbon
- Regenerating the carbon
- Maintaining a good-quality contactor influent
- Controlling contactor flow rate
- Controlling biological activity in the carbon adsorber

Monitoring the Carbon Bed Depth. A means must be provided for continuously monitoring and measuring the levels in all storage vessels, including contactor and storage tanks (virgin, spent, and regenerated carbon). The designer must recognize that slurry concentrations and flow rates change during the filling process. When carbon is placed in an adsorber, it typically has a density of approximately 90% of the apparent density for random-packed graded media. When a carbon column is first backwashed, density decreases from approximately 90% to approximately 83%, resulting in a permanent expansion of 8%.

An 8% expansion in a 24-in. (60-cm) filter is only 1.9 in. (4.8 cm) and does not present any major operating problems. However, an 8% expansion in a 15-ft-deep (4.6-m) carbon bed entails a permanent expansion of 14 in. (35 cm). This can create a major problem when one is trying to closely control the surface elevation of a deep bed.

The problem of controlling surface elevation of deep bed, graded media is complicated further by the fact that the bed also compacts during operation. The same 15-ft (4.6-m) column could see a compaction of 5 to 15 in. (13 to 38 cm). If surface wash sweeps are used, they must be placed to accommodate this expansion and contraction of the bed. As noted, conventional surface wash sweeps are not recommended.

Backwashing the Carbon Properly. The basic objectives of the backwashing process are to expand the bed for cleaning and to restratify the bed after washing to maximize the carbon usage rate. These simple objectives are not easily attained.

GAC density when wetted in water is appreciably lower than sand. Thus, the carbon bed can easily rise as a plug. The force required to lift any bed is the difference between the force of gravity and the buoyant force. For a carbon bed, an upward pressure of 3 psi (21 kPa) is sufficient to lift the bed. This problem is compounded if the entire carbon bed has been drained, because air in the bed prevents the upper bed from expanding when it is not submerged in water. Although only a small pressure is acting on the bed, the cross-sectional area is so large that a tremendous force can develop. This situation can result in damage to the carbon bed and tank internal equipment.

To prevent the carbon bed from rising as a plug, the backwash rate should be gradually increased to ensure that the bed expansion occurs from the top downward. Slowly ramping or stepping up the backwash rate from approximately 3 to 4 gpm/ft^2 (8 to 10 m/h) to 14 to 15 gpm/ft^2 (34 to 37 m/h) should be sufficient to ensure proper bed expansion. Conversely, ramping the backwash rate down following backwash ensures proper bed restratification. Because of the potential for damage of the system components that can result from a rising carbon plug, and the difficulty of maintaining a constant bed depth due to expansion and compaction of the carbon, conventional surface sweeps are not recommended for deep bed carbon systems.

Removing Carbon from the Adsorbers. Because granular carbon flows as a slurry when diluted with transport water, it is most easily removed from a contactor in the form of a slurry. Dilution or transport water can be added in a number of ways, the most common of which is through the underdrains or tangential to the vessel wall at the bottom of the tank. GAC slurry can be withdrawn through ports at the base of the wall or in the vessel sidewall. Multiple ports are very efficient, but single ports are also effective.

Controlling Biological Activity. Biologically active carbon must be controlled to avoid undesirable effects. The larger surface area provided by the GAC media provides an excellent attachment location for microorganisms. If anaerobic conditions develop, odor problems within the system will result. Chlorine application to GAC filter-adsorbers or postfilter adsorbers is not recommended because chlorine does not prevent growth and can increase the concentration of adsorbed chloroorganics. Chlorine can make activated carbon more brittle because it destroys some of the activated carbon by chemical reduction. Brittle carbon is then susceptible to increased breakup during filter backwash.

Backwashing for filter-adsorbers and periodic media replacement (for both filter-adsorbers and postfilter adsorbers) are necessary to control biological growth. Backwashing should not be excessive because of its possible detrimental effect on adsorption efficiency. Mixing may also take place during backwashing. As a result, GAC with adsorbed molecules near the top of the bed moves deeper into the bed, leading to possible early breakthrough. The large uniformity coefficient of most commercial activated carbons promotes restratification after backwash. If the underdrain system does not properly distribute the washwater or if backwashing procedures do not aid restratification, substantial mixing of the activated carbon can occur with each backwash.

Biological activity can be beneficial for removing some compounds. Biological treatment reduces assimilable organic carbon (AOC) and other biodegradable compounds, resulting in a biologically stable water. If provided, an upstream oxidation step (ozone or ozone with hydrogren peroxide) renders more of the AOC biodegradable, resulting in a greater biological activity on the GAC, rather than in the distribution system. Bacterial growth in a distribution system is undesirable because it can accelerate pipeline corrosion, produce tastes and odors, and increase the amount of disinfection needed to maintain a residual throughout the distribution system. Maximizing the production of AOC within the treatment plant itself, and then providing for a biological treatment step with GAC, can be an effective method for reducing the concentration of AOC delivered to the distribution system.

Any disinfectant residual applied to a biologically active GAC adsorber is quickly consumed, wasting chemicals and inhibiting biological activity on the GAC. Adding chlorine ahead of GAC produces chlorinated disinfection by-products that are adsorbed. Thermal regeneration oxidizes these by-products to dioxins and furans. For this reason, primary disinfection with chlorine should be provided after contact with GAC. The design engineer should also be aware that microbial growth on filter-adsorbers can cause increased rates of head loss buildup and shorter filter runs if growth is excessive and the backwash system is not able to control the growth. These possible conditions should be considered in the design.

Biological activity is lower for cold water than for warm water. As a result, biological oxidation cannot be counted on throughout the year if temperatures are low enough to affect biological activity.

The designer must also be aware that it is possible for activated carbon particles to migrate through and penetrate the filter-adsorber or postfilter adsorber underdrain system, providing a habitat for microbial growth that cannot be controlled by disinfection. Zooplankton and other undesirable organisms can also grow in GAC filters that are biologically active. Increased backwashing may be necessary (once every 5 days) to keep filter beds clean and free of these organisms before they have an opportunity to flourish.

As noted previously, if the system becomes anaerobic, odor problems may develop. This can occur if large concentrations of ammonia are allowed to enter the adsorber, insufficient dissolved oxygen is present in the water, or the bed is removed from operation for an extended period. Proper operation should ensure that sufficient oxygen is present at all times.

Maintenance Requirements

In activated carbon adsorption, maintenance requirements fall into preventive and repair categories for the main components of the activated carbon system.

Adsorbers. Steel adsorbers are lined with rubber, painted with epoxy, or constructed of type 316 stainless steel. The life of a rubber lining is a function of the frequency and extent of backwashing and the frequency and changeout of the activated carbon. Minimal maintenance should be involved, and it is expected that the lining will need to be replaced every 10 to 15 years. Underdrains should be inspected after each carbon replacement.

Transfer Equipment. Pipelines should be designed for suitable velocities to subsequently minimize the effects of erosion and corrosion. Lines should be thoroughly flushed to keep any residual carbon from plugging the lines and accelerating corrosion. The life of an eductor is generally a function of the amount of carbon transferred. A properly designed and operated eductor can handle 2 to 4 Mlb (0.9 to 1.8 Mkg) of carbon before being replaced. Slurry pumps should be avoided because of the initial costs and the amount of subsequent maintenance required.

Carbon Losses. In a facility using GAC and on-site reactivation, there are three areas where carbon losses can occur:

- Within the adsorbers
- In the transport system
- In the reactivation furnaces and ancillary equipment

ACTIVATED CARBON PROCESSES   14.31

For a new shipment of GAC, fines are typically specified in the range of 4.0% maximum, although requiring less than 2.0% is common. The majority of initial carbon fines are backwashed out of the system within 1 to 2 days after the GAC is placed into service. During regular backwashing operations, abrasion of particles occurs. The mean particle diameter of typical coal-based GAC products is usually reduced by 0.1 to 0.2 mm during the on-line time of 2 to 4 years.

A loss of GAC bed depth of 0.2 to 1.0 in. (0.5 to 2.5 cm) per year is common. For a 24-in.-deep (0.6-m) bed, this translates to 1% to 4% loss per year. Losses beyond these values indicate that there may be a problem with the underdrain system or that the applied backwash rate is excessive, resulting in the migration of carbon over the tops of the washwater troughs.

Medium for Biological Treatment. Both filter-adsorbers and postfilter adsorbers become biologically active when disinfection occurs after the GAC process. When ozone or an advanced oxidation process (AOP), such as ozonation and hydrogen peroxide, precedes the GAC treatment step, the GAC also becomes biologically active. Ozonation makes some nondegradable compounds biodegradable by breaking the compounds into smaller, more biodegradable products and thereby provides an additional means of organics removal.

Practical Design Suggestions

Carbon adsorption system design is not complex. In fact, most of the design elements and features are similar to those found in the design of a conventional filter system. However, there are several design considerations worth noting that should be taken into account when one is preparing a carbon adsorption design.

Design of Adsorber Vessels

- For gravity downflow, open-top concrete adsorbers, consider dual cells to reduce the backwash rate. In an attempt to limit the total number of contactors, the subsequent backwash flow rate required for the larger filter area may be too great or require too much of the instantaneous plant flow for backwashing. Providing dual cells allows each filter cell to be backwashed separately, reducing the instantaneous backwash flow rate to one-half. Dual cells reduce the size of the washwater system but increase the number of valves. In large systems, the number of valves to open, close, operate, and maintain is an important consideration.
- If a stainless steel wedge wire underdrain system is used, underdrain laterals should be at least 1 ft off of the vessel floor to facilitate carbon removal. Sloping the vessel bottom from the back to the front assists in flushing.
- Underdrain systems with nozzles are not recommended, because they are subject to plugging. Systems with false bottoms or plenums in which carbon fines can accumulate also are not recommended. Where nozzles and false bottoms are used in combination, plugging of the nozzles may create uplift pressures during backwashing, leading to structural failure of the false bottoms. Allowing carbon fines to collect in the underdrain system is also undesirable because these fines can provide a habitat for the growth of microbes and other undesirable organisms.
- Provide a method for periodic evaluation of the carbon bed depths by placing a permanent reference mark such as a stainless steel plate, laminated plastic staff gauge, or painted gauge marks directly on the vessel wall.

- For conventionally reinforced concrete structures in contact with carbon, a minimum concrete cover of 15 to 40 mm (depending on reinforcement) should be provided over reinforcing steel. Carbon is extremely corrosive and attacks reinforcing steel through cracks in concrete.
- Provide a washdown hose site on the vessel deck to facilitate cleaning.
- Provide air release valves on the backwash water supply piping to ensure that air has been completely purged.
- Provide adsorber-to-waste connections (similar to conventional filter-to-waste) to permit additional removal of GAC fines remaining after backwashing.

Design of Carbon Storage and Transport Facilities. Carbon replacement is a major expense associated with carbon adsorption systems. Because slurry storage and transport can cause major attrition losses, facilities must be designed with care. Carbon deliveries in excess of 9,000 kg (10 tons) usually justify an eductor/carbon slurry transport system. Design guidelines include the following:

- To minimize carbon-to-carbon abrasion, avoid air-assisted transfer of carbon.
- Transfer all carbon in a water slurry form, designed for a slurry of approximately 1 to 3 lb/gal (0.12 to 0.34 kg/L).
- Minimize the length of carbon slurry transfer pipes through optimum arrangement of structures and equipment.
- Pipelines delivering carbon slurry to the tanks should discharge below the water surface to minimize attrition and overflows from the dewatering screw. The tank overflow pipeline should be screened or diverted to a collection point where carbon can be recovered.
- Provide a metering device to measure carbon slurry flow rate in the piping system.
- Limit velocities in carbon slurry transfer lines to 3 to 5 ft/s (0.9 to 1.5 m/s).
- Avoid the use of throttling valves on carbon slurry lines and on potable water dilution lines that are ahead of the point of carbon introduction.
- Use a minimum diameter of 50-mm type 316 stainless steel pipe for carbon slurry transfer lines and long sweep bends to minimize pipe erosion and improve hydraulics. Mitered bends should not be used for piping carbon slurry.
- Minimize the number of movements (valve changes, flow direction changes, pump starts) to effect a complete carbon transfer.
- Limit the speed of recessed impeller, rubber-lined, centrifugal carbon transfer pumps to 900 rpm or less.
- Design carbon storage bins to receive the entire wet contents of carbon transferred from a single carbon vessel, plus at least 20% for expansion.
- Install pressure gauges with diaphragm seals throughout the eduction system to monitor system performance.
- Use double-seated valves for in-line shutoff and isolation throughout, because the system may see carbon flow in both directions. Double-seated knife gate valves, plug valves, or ball valves fabricated from materials suitable for carbon contact can be used.
- Design the carbon slurry system to move slurry through the system continuously until all carbon has been flushed from piping. The process must not be allowed to stop until all carbon has been moved through the piping, or else carbon will settle and plug the lines. Provide cleanouts and flushing connections throughout the piping system.

- Provide a vacuum cleaning system for picking up loose carbon.
- Provide a method for automatically and accurately measuring carbon levels in storage bins and vessels. A backup manual system should be provided to track bed and bin levels. Depths can be monitored by placing a permanent reference mark such as a stainless steel plate, laminated plastic staff gauge, or painted gauge marks directly on the inside wall of each bed and bin.

Materials of Construction. Materials of construction for carbon adsorption plants vary. Epoxy-coated steel, type 316 stainless steel, concrete, and fiberglass are all used for vessels; stainless steel, lined carbon steel, fiberglass, and plastic have been used for piping. Problems have been reported with plastic due to abrasion and scouring from the transport of carbon slurries.

Dry carbon and carbon in water slurries are not corrosive. Damp or wet carbon, however, is extremely corrosive. Precautions must be taken to ensure that only proper materials are used in contact with damp carbon. All metal in contact with damp carbon must be corrosion-resistant (type 316 stainless steel) or noncorrosive (fiberglass). Tanks, vessels, and fittings may be fabricated of epoxy or rubber-coated steel, stainless steel (type 316), glass-reinforced polyester (fiberglass), or concrete. At existing plants, embedded metals can be field-lined with rubber, glass-reinforced polyester, vinyl ester, or epoxy. Valves and instruments such as flowmeters and eductors must be protected from corrosion.

REGENERATION OF GRANULAR ACTIVATED CARBON

Facilities for on-site regeneration of GAC are expensive, making it generally impractical for smaller water systems which are generally designed to waste spent carbon and replace it with new. Larger systems must carefully look at the economics of installing regeneration equipment.

Regeneration Facilities

GAC is expensive when compared with sand and anthracite filter media. As a result, it is often cost-effective to regenerate and reuse GAC. Two basic approaches to regenerating GAC are off-site regeneration and on-site regeneration.

The rate at which contaminants break through the carbon bed determines the size of the regeneration system. The primary design criterion for a reactivation system is the rate of carbon regeneration (mass per unit time, e.g., kg/h). A complete regeneration or reactivation process typically consists of a furnace system, including a feed system, a drying or dewatering scheme, and a reactivation process. A simple regeneration system schematic is presented as Figure 14.14.

Typical carbons used in water treatment require regeneration from every 6 months to 5 years, depending on the application. On-site regeneration is generally not cost-effective unless the carbon exhaustion rate is over 910 kg per day. Current U.S. applications fall in the range of off-site regeneration (225 to 700 kg per day). For facilities using less than 225 kg per day, off-site disposal should be considered.

For off-site disposal applications, virgin carbon is first purchased from a carbon supplier. Once carbon becomes exhausted, it is transported in slurry form by gravity to a drainage collection tank where the supernatant is routed to the plant headworks for treat-

FIGURE 14.14 Activated carbon regeneration system schematic. (*Adapted from Westerhoff and Miller, 1986.*)

ment. Recovered carbon is then manually packaged or conveyed to hopper trailers and shipped to a landfill or an incineration facility, depending on the chemicals that have been adsorbed.

The off-site disposal concept is a simple solution for applications with small carbon exhaustion rates. Because toxic or hazardous materials are likely to be absorbed on the spent carbon, landfill disposal may not be a feasible long-term option if the disposed carbon is considered a hazardous waste. Fortunately, this is not often a concern for GAC used in drinking water treatment. However, incineration may be necessary to satisfy environmental concerns regarding proper ultimate disposal.

Off-site regeneration is similar to off-site disposal, possessing many of the economic benefits of carbon reuse. However, the number of handling steps and resulting carbon attrition involved in off-site regeneration must be considered. In most cases, GAC manufacturers are capable of retrieving, transporting, and regenerating the spent GAC at their own facility on a contract basis, but most regeneration facilities combine every type of carbon they receive.

Customized regeneration is expensive. Before one enters into an agreement with a carbon supplier for off-site regeneration, assurance should be given that only virgin GAC will be returned to the plant. Because it may not be possible to track and monitor, receiving regenerated activated carbon obtained from municipal or industrial waste treatment processes should be avoided.

On-Site Regeneration Alternatives

The three most common methods of GAC regeneration are steam, thermal, and chemical. Of the three methods, thermal regeneration is the most widely used. Chemical and steam regeneration techniques are used primarily in industrial systems designed to recover the adsorbate. Only thermal regeneration is discussed in this chapter.

Thermal Regeneration

Thermal regeneration is typically carried out in five steps:

1. Dewatering to 40% to 50% moisture content by draining. It is more efficient to dewater moisture than to evaporate extra water in the drying stage.
2. Drying at temperatures of up to 200° C.
3. Baking or pyrolysis of adsorbates at 500° to 700° C.
4. Activating carbon at temperatures above 700° C.
5. Quenching regenerated carbon.

Reactivation Rate. The primary consideration in designing a carbon reactivation system is the reactivation rate. This rate depends on carbon loading and carbon usage or exhaustion rate in the adsorbers. Carbon loading is the amount of organic matter removed from the water treated per unit quantity of carbon and is usually expressed in terms of milligrams per gram (mg/g) or pounds per pound (lb/lb) of organic materials to carbon. Carbon usage is often expressed as carbon dosage or kilograms of carbon per megaliter of water treated and is inversely related to carbon loading.

Carbon usage (exhaustion) rate is the mass of carbon exhausted per unit time. It is usually expressed as kilograms or pounds per day and is inversely related to reactivation frequency. Reactivation frequency measures the time between the reactivation of a given bed of carbon. The CUR is the minimum rate at which carbon would be reactivated in a continuous system with no intermediate storage. The designer should provide sufficient carbon storage between the adsorbers and reactivator so that the furnace is sized above the minimum reactivation value. Operation at a higher rate takes into account times when furnaces are out of service for maintenance or repair, while still satisfying the carbon usage rate of the adsorbers.

Types of Furnaces. Four types of regeneration furnaces may be used to remove adsorbed organics from activated carbon:

1. Electric infrared oven
2. Fluidized bed furnace
3. Multiple-hearth furnace
4. Rotary kiln

In selecting the type of furnace to be used, the following criteria should be considered:

- Capital cost
- Operating experience and service record
- Ability to change feed rates
- Individual control of the drying, pyrolysis, and activation zones
- Uniformity and quality of the regenerated carbon
- Design and operating parameters that may lead to potential problems such as explosions, fires, and dust
- System efficiency with respect to plant downtime during carbon regeneration
- Downtime
- Energy requirements

- Operating and maintenance costs
- Experience of system suppliers and availability of service

Electric Infrared Oven Furnaces. An infrared furnace consists of an insulated enclosure through which carbon is transported on a continuous metal conveyor belt. Heat for reactivation is supplied to the furnace by a number of electric infrared heating elements mounted in the top of the tunnel. Spent carbon is fed to the reactivator through an airlock which controls the carbon feed rate and minimizes the air intake. Carbon is spread evenly onto the belt conveyor and transported slowly through discrete chambers or zones of the furnace.

Chambers provide increasingly higher temperatures. Atmosphere and temperature can be controlled with this type of furnace. Carbon discharges into a conventional water-filled quench tank for cooling. Modules are made of a steel shell, lined with thermal shock-resistant ceramic fiber insulation, and include support rollers for the conveyor belt.

Fluidized Bed Furnaces. A fluidized bed furnace suspends carbon particles by an upward-flowing gas stream. Gas velocity is controlled so that the weight of particles in the bed is just balanced by the upward force of the gas. The use of a fluidized bed in this application offers the advantages of uniform temperatures within the bed and high heat and mass transfer rates. Steam must be injected to control the temperature in the furnace.

Dewatered carbon is first dried to a moisture content of about 1%. It then flows by gravity to a lower chamber which is maintained at about 982° C. As the carbon temperature increases to the temperature of that zone, adsorbed organic materials are pyrolyzed and gasified. After leaving the furnace, carbon is cooled in a water-filled quench tank.

Air is injected into the space between the drying and heating zones that serves as a second combustion zone. This zone burns organic compounds and the hydrogen and carbon monoxide released during reactivation in the lower chamber. Part of the heat produced dries wet carbon in the upper chamber, with the remainder recycled to the lower chamber reactivation zone.

Multiple-Hearth Furnaces. A multiple-hearth furnace consists of a refractory-lined steel shell containing five to eight circular hearths. Burners are generally located on the bottom hearth to provide heat necessary for reactivation. Burners can also be located on higher hearths to provide improved temperature profiles for increased performance and flexibility of operation.

Spent carbon enters the furnace through a dewatering screw conveyor onto the top hearth. A rotating center shaft with supporting arms and rabble blades moves carbon in a spiral path so that it drops from hearth to hearth and finally out of the bottom of the furnace into a quench tank for cooling. Residence time in the furnace is controlled by the rotation speed of the center shaft. The hollow rabble arms are cooled by ambient air blown through them. The atmosphere within the furnace is tightly controlled to prevent excessive carbon oxidation. Steam is usually added in the lower hearths for temperature control.

Rotary Kiln Furnaces. Multiple-hearth furnaces are the traditional method of granular activated carbon reactivation, although fluidized bed and infrared furnace installations are becoming more common, on the contrary, rotary kilns are more energy-intensive and are becoming less competitive with other technologies.

Furnace Size. Furnace size is based on the rate at which carbon is charged per unit of furnace dimension and varies with the type of reactivation furnace chosen. For multiple-hearth furnaces, values are usually given in pounds of carbon per square foot of hearth per day (or kg/m^2 per day). For multiple-hearth furnace design, typical values are 40 to 115 lb/ft^2 per day (145 to 560 kg/m^2 per day). Values for fluidized bed furnace loadings range from 1,460 to 1,700 lb/ft² per day (7,125 to 8,300 kg/m² per day). These numbers are based on the cross-sectional area perpendicular to the flow of combustion air. In an infrared furnace, carbon flow depends on the width of the conveyor belt and residence time.

Regeneration By-products

Activated carbon must be regenerated in a way that does not pollute the environment. For this reason, scrubbers and afterburners are used to minimize particulate and gaseous emissions, respectively. The afterburner oxidizes organic compounds, and the scrubber removes particulate matter and any soluble chemical species from the gas stream. Off-gas control is important in controlling dioxins and furans produced during regeneration. For these reasons, local air quality regulations should be reviewed early in the design process. A supplementary dust collector is used only where it is necessary to collect particulate matter not normally removed by scrubbers.

Transporting Carbon from the Quench Tank

Eductors typically transport carbon away from the quench tank. The quench tank is normally a small tank that receives hot regenerated carbon from the furnace, and the water level must be controlled to prevent hot carbon from being exposed to air. When a method other than an eductor is used to transport carbon, the volume of makeup water must be greatly increased. The increased volume of makeup water causes turbulence that keeps carbon in suspension and increases carbon losses. With an eductor, most of the transport and dilution water is supplied by the water treatment system.

BIBLIOGRAPHY

Adsorption of Pesticides by Powdered Activated Carbon. Denver, Colo.: AWWA Research Foundation, American Water Works Association, 1994.

Clark, R. M., and B. W. Lykins, Jr. *Granular Activated Carbon Design, Operation and Cost.* Boca Raton, Fla.: Lewis, undated.

Control of Organic Compounds with Powdered Activated Carbon. Denver, Colo.: AWWA Research Foundation, American Water Works Association, 1991.

Crittenden, J. C., D. W. Hand, H. Ariham, and B. W. Lykins, Jr. "Design Considerations for GAC Treatment of Organic Chemicals." *Journal AWWA* 79(1):74, 1987.

Culp, R. L., and R. M. Clark. "Granular Activated Carbon Installations." *Journal AWWA* 75(8):398, 1983.

"Design and Use of Granular Activated Carbon: Practical Aspects." *Proceedings Technology Transfer Conference.* May 9 and 10, 1989. Denver, Colo.: American Water Works Association, 1989.

The Effect of Metals on Thermal Regeneration of Granular Activated Carbon. Denver, Colo.: AWWA Research Foundation, American Water Works Association, 1994.

GAC Filter Adsorbers. Denver, Colo.: AWWA Research Foundation, American Water Works Association, 1987.

Graese, S. L., V. L. Snoeyink, and R. G. Lee. "Granular Activated Carbon Filter-Adsorber Systems." *Journal AWWA* 79(12):64, 1987.

The Hazardous Potential of Activated Carbons Used in Water Treatment. Denver, Colo.: AWWA Research Foundation, American Water Works Association, 1994.

J. M. Montgomery Consulting Engineers. *Water Treatment Principles and Design*. New York: John Wiley & Sons, 1985.

Kawamura, Susumu. *Integrated Design of Water Treatment Facilities*. New York: John Wiley & Sons, 1991.

Knappe, D. R. U., V. L. Snoeyink, G. Dagois, and J. R. DeWolfe. "Effect of Calcium on Thermal Regeneration of GAC." *Journal AWWA* 84(8):73, 1992.

Najm, I. N. "Effect of Initial Concentration of a SOC in Natural Water on Its Adsorption by Activated Carbon." *Journal AWWA* 83(8):57, 1991.

Najm, I. N., V. L. Snoeyink, B. W. Lykins, Jr., and J. Q. Adams. "Using Powdered Activated Carbon: A Critical Review." *Journal AWWA* 83(1):65, 1991.

Optimization and Economic Evaluation of Granular Activated Carbon for Organic Removal. Denver, Colo.: AWWA Research Foundation, American Water Works Association, 1989.

Organics Removal by Granular Activated Carbon (GAC). Denver, Colo.: American Water Works Association, 1989.

Oxenford, J. L., and B. W. Lykins, Jr. "Conference Summary: Practical Aspects of the Design and Use of GAC." *Journal AWWA* 83(1):58, 1991.

Standardized Protocol for the Evaluation of GAC. Denver, Colo.: AWWA Research Foundation, American Water Works Association, 1992.

Technologies for Upgrading Existing or Designing New Drinking Water Treatment Facilities. Cincinnati, Ohio: U.S. Environmental Protection Agency, 1990.

Weisner, M. R., J. J. Rook, and F. Fiessinger. "Optimizing the Placement of GAC Filtration Units." *Journal AWWA* 79(12):39, 1987.

Westerhoff, G. P., and Richard Miller. "Design of the GAC Treatment Facility at Cincinnati." *Journal AWWA* 78(4):146, 1986.

CHAPTER 15
CHEMICALS AND CHEMICAL HANDLING

Jerry L. Anderson, P.E.
CH2M HILL
Louisville, Kentucky

A principal issue in modern water treatment plant design is the decision of what chemicals are to be used for treatment processes, how they are to be shipped and stored, and what type of chemical feed equipment should be used. This chapter summarizes alternatives and issues associated with gaseous, dry, and liquid chemicals and the associated equipment and piping necessary for their use. Other issues for the designer to consider involve the many regulations that must be met and safety considerations in planning the chemical feed systems. Finally, the importance of securing and hardening of chemical handling facilities is a part of modern design.

Additional discussions of the theory of chemical treatment may be found in detail in *Water Quality and Treatment*. Details of specific chemicals commonly used in water treatment processes are provided in Appendix A.

RECEIVING AND STORING PROCESS CHEMICALS

Many alternatives are available for receiving and storing process chemicals, depending on plant size and location, treatment processes to be used, and many other factors.

Sizing Storage and Feed Systems

Sizing of storage and feed systems begins with an investigation of dosage requirements for each chemical used. Chemical feed rates (lb/day, ft³/h, gal/day, kg/day, etc.) can be computed from dosages (mg/L) and plant flow rates [million gallons per day (mgd), ML per day]. For chemicals in dry form:

$$\text{Feed rate (lb/day)} = 8.34 \text{ lb/gal} \times \text{dosage (mg/L)} \times \text{flow (mgd)}$$

For chemicals in solution form:

$$\text{Feed rate (gal/h)} = \frac{8.34 \text{ lb/gal} \times \text{dosage (mg/L)} \times \text{flow (mgd)} \times 100\%}{24 \text{ h} \times \text{concentration (\%)} \times \text{density (lb/gal)}}$$

Chemical storage or inventory requirements can be computed from feed rates and the number of days of storage required.

Capacities and Feed Rates. Many different conditions must be evaluated to determine the range of feed rates used to select feed equipment capacities for each chemical. For an existing treatment plant, or where there is another plant using the same water source, historical records should be investigated to obtain peak- and minimum-hour dosages for the plant's chemicals. For a new plant without records, bench- and pilot-scale testing is required. Table 15.1 provides guidelines for chemical feed and storage capacities, but these guidelines should be integrated with consideration to regional and state standards, operating records of similar plants, and experience.

Minimum feed rate capacities and the feed equipment's capability to accurately feed at low rates are as important as maximum capacities. Once dosage and feed rates are calculated, these usually must be adjusted upward to account for chemical purity or percent availability for reaction.

Sizing the Inventory. Inventory size is based on factors that vary from plant to plant, even within the same system. First, determine the maximum monthly feed rate of the chemical, accounting for dosage and plant flow rate. Second, the appropriate number of days of chemical storage should be established for each chemical, typically between 7 and 30 days. Major considerations include the general availability of the chemical, location of chemical suppliers, reliability of suppliers during periods of shortage, normal delivery time, and delays in shipment that may be caused by weather conditions. Another consideration is the possibility of delays in receiving shipments because of transportation strikes or natural or human-caused emergencies. Some state regulatory agencies specify a minimum number of days for certain chemicals.

Once the number of days of storage for each chemical has been identified, it is possible to determine whether bulk deliveries are practical. In smaller plants using bulk delivery chemicals, the quantity stored may be dictated more by the size of the bulk shipping containers in use in the area than by the desired number of days' supply. Many regulators require minimum storage requirements equal to 1.5 truckloads of a bulk delivery.

Excess inventory is undesirable if the chemical has a limited shelf life. Chemicals such as soda ash and quicklime that either possess hygroscopic properties or are otherwise affected by moisture can become difficult to handle after several months in bulk storage unless precautions are taken to control the environment. Chemicals furnished in return-

TABLE 15.1 Suggested Chemical Feed and Storage Design Parameters

Design parameter	Dosage condition	Plant flow rate condition
Maximum feeder capacity	Peak hour	Peak hour
Minimum feeder capacity	Minimum hour	Minimum hour
On-site storage	Maximum month	Maximum month
Day tank volume	Maximum day	Maximum day

able containers, such as chlorine cylinders or ton containers, may be subject to demurrage charges unless these are waived by agreement in the purchase contract.

Delivery, Handling, and Storage of Chemicals

Modes of transporting chemicals include railroad tank cars, trucks, and, in rare instances, barges. The type of unloading facilities depends on whether chemicals arrive in bags, drums, or bulk; the form in which the chemical will be fed to the process; the type of carrier (rail or truck); and the location and type of storage silo, tank, or other storage facility.

Receiving Shipments. Whether the chemical is liquid or solid, truck delivery is usually preferred because of its simplicity, maneuverability, and generally prompt and predictable delivery. In contrast to rail shipments, truck delivery makes more load size options available while still preserving the economies of bulk purchasing. Modern air-slide and pneumatic unloading equipment on these vehicles permits quick and easy delivery of bulk loads below grade and to overhead silos.

Unloading platforms or docks must be provided in all but the smallest plants to accommodate truck deliveries of containerized chemicals, including bagged material, drums, small gas cylinders, and the like. Horizontal transport of nonbulk chemicals from the unloading dock is normally accomplished by hand or power trucks, conveyors, or a monorail system. Storage in areas above or below the unloading area requires installation of an inclined conveyor, dock leveler, hoist, or elevator designed for the maximum anticipated loads.

Receiving and Storing Pressurized, Liquefied Gases. Gases commonly used in water treatment are chlorine and ammonia for disinfection processes and carbon dioxide for stabilization and pH control after softening. Oxygen is sometimes used for the gas feed to ozonation generators. All these gases are shipped as pressurized liquids. Carbon dioxide and oxygen are refrigerated in their liquid state.

Chlorine. Chlorine used in the water industry is seldom stored in on-site receivers. Instead, the shipping container is used for storage. If chlorine is received at a treatment plant in a tank truck or barge, a sufficient pressure differential must be maintained between the shipping container and the storage container during transfer.

Gaseous chlorine is commercially available in containers of the following sizes:

- 150-lb cylinders
- 1-ton cylinders
- 15- to 17-ton tank trucks
- 16- to 90-ton railroad tank cars

Cylinders of 100-lb (45-kg) weight are also available, but are rarely used. The dimensions of the 150-lb (68-kg) and 1-ton (907-kg) cylinders are shown in Figure 15.1. Because chlorine tank trucks do not have standard sizes, dimensions should be obtained from local chlorine suppliers.

In all the containers listed above, liquid chlorine occupies a maximum of approximately 85% of the volume when the product is delivered. This 15% allowance provides room for liquid chlorine to expand if the cylinder becomes warm. No chlorine container should ever be directly heated. If the liquid were to become warm enough to expand and fill the entire container, tremendous hydrostatic pressure would result, and the container would rupture. As a safety precaution, cylinder outlet valves are equipped with a small

FIGURE 15.1 Dimensions of standard chlorine cylinders. (*a*) A 150-lb cylinder. (*Courtesy of PPG Industries, Inc., Pittsburgh, Pa.*) (*b*) A 1-ton cylinder. (*Courtesy of the Chlorine Institute, Inc.*)

fusible plug that melts at approximately 158° F (70° C) and releases some liquid chlorine to cool the cylinder before a more serious accident can occur.

Chlorine is most often fed by withdrawing gas from the top of the container, and the reduced pressure above the liquid then causes some of the liquid to evaporate, providing additional gas. The maximum withdrawal rate with this method is about 40 lb per day (18 kg per day) for a 150-lb cylinder and 400 lb per day (181 kg per day) for a 1-ton cylinder. Maximum continuous withdrawal rates for containers of other sizes can be approximated by comparing the surface area available to absorb the heat required to replace the heat of evaporation lost. If chlorine is being used in this manner, the containers should be maintained in an environment that can be heated to 65° F (18° C). The design-maximum 24-h withdrawal rate of a system designed to withdraw gas from the top of the container should not exceed the continuous withdrawal rate of the containers on-line.

Bulk Shipment of Chlorine. Chlorine tank barges have either four or six tanks, each with a capacity of 85 to 185 tons (77 to 168 t), and must comply with Coast Guard regulations for piping, loading, and unloading arrangements. The valves are similar to those on tank cars, but the arrangement is not standard, and on some barges the valves are larger.

Tank trucks consist of cargo tanks for chlorine permanently attached to a motor vehicle. U.S. Department of Transportation (DOT) regulations are explicit as to tank and accessories design and maintenance, marking, and emergency handling. Only a few chlorine tank trucks are currently being operated in the United States. Railcar deliveries are practical only for the largest plants and are subject to the economic feasibility of constructing a siding from the nearest rail line.

Ton containers of chlorine are authorized for rail shipment, but only on a special tank car frame known as a multiunit tank car designed to hold 15 containers. The initial rail shipment of these containers is unloaded from the car for use at the consumer's plant. In subsequent shipments, the full containers are exchanged for empties, which are returned for refilling. No freight is charged on the return carload of empty containers because tank cars are entitled to free return movement.

It is impractical to ship less than 15 containers because the transportation fee is figured for 15 full containers at prevailing carload rates. Multiunit tank cars must be consigned for delivery and unloading on a private track. If a private track is not available, containers may be removed from the car frame on carrier tracks with previous written permission. Regulations provide that one or more ton containers may be transported on trucks or semitrailers under special conditions.

If logistics show that purchase of chlorine by tank truck or railroad tank car is feasible, stationary storage facilities should be considered. The user can purchase chlorine for a better price if the tank car or trailer is on-site only for the time it takes to unload it. On the other hand, the use of tank truck trailers and railroad cars for on-site storage is also common.

Stationary chlorine facilities should be designed in complete accordance with the recommendations of The Chlorine Institute as described in *Facilities and Operating Procedures for Chlorine Storage.* Provision must also be made for a weighing device, either a lever scale system or, more commonly, load cells.

An air padding system is recommended for unloading the tank car and removing gas from the tank before inspection. The air should be dried with a heat-reactivated, desiccant-type air dryer. Facilities must be provided to vent chlorine gas or chlorine-and-air mixtures from the storage tank to the consuming process or other disposal system. When chlorine is transferred from the storage tank to the consuming process, air padding may be necessary. The procedure is essentially the same as that required for emptying tank cars.

Tank cars should be emptied through a suitable-metal flexible connection that accommodates the rise of the car as its springs decompress. Tank cars are almost invariably emptied by discharging liquid. Liquefied gases may be unloaded by their own vapor pressure. Cold weather usually decreases the unloading rate. Sometimes it is desirable to place an air pad over the chlorine vapor in the car to facilitate unloading. The air pad may be provided by the chlorine supplier or at the point of use. Weight-measuring devices are preferred to determine tank contents; gauge glasses should not be used. Adequate lighting, including auxiliary power sources, should be provided for night operations.

The location, design, maintenance, and operation of chlorine bulk storage tanks may be subject to local or state regulations and to insurance requirements. The number and capacity of storage tanks should be consistent with the size of shipments received and the rate of consumption. Receiving and unloading areas and safety precautions applicable to the handling of single-unit cars, cargo trucks, and barges are subject to DOT, Coast Guard, and other regulations.

Handling Chlorine Ton Containers and Cylinders. Cylinders should be stored upright and secured in a manner that permits ready access and removal. Ton containers should be stored horizontally, slightly elevated from ground or floor level, and blocked to prevent rolling. A convenient storage rack for ton containers can be constructed with rails or I-beams to support both ends of the containers.

Ton containers should not be stacked or racked unless special design provisions are made for easy access and removal. Storage areas should be clean, cool, well ventilated, and protected from corrosive vapors and continual dampness. Cylinders and ton containers should preferably be stored indoors, in a fire-resistant building, and away from heat sources, other compressed gases, and flammable substances. If containers are stored outdoors, the area should be shielded from direct sunlight and accumulations of rain and snow. If natural ventilation is inadequate, storage and use areas should be equipped with suitable mechanical ventilators.

All storage areas should be designed so that personnel can quickly escape in emergencies. If possible, two exits should be provided from each separate room or building in which chlorine and other gases are stored, handled, or used.

Various mechanical devices, such as skids, troughs, or upending cradles, should be provided to facilitate the safe unloading of chlorine cylinders. Specially designed cradles or carrying platforms are recommended if it is necessary to lift the cylinders by crane or derrick. Chains, lifting magnets, and rope slings that encircle cylinders are unsafe. Hand trucks are preferred for lateral movement.

Special provision should be made for lifting ton chlorine containers from a multiunit tank car or truck by means of hooks designed to fit on a chain sling or lifting clamp, in combination with a hoist or crane with at least a 2-ton (1,800-kg) capacity. Containers should be moved to point of storage or use by truck or by a crane and monorail system.

Ammonia. Anhydrous ammonia is most commonly shipped in 150-lb (68-kg) cylinders and 1-ton (907-kg) containers similar to those used for chlorine. The guidelines described for handling and storing chlorine containers also apply to ammonia. Ammonia and chlorine gases should not be stored or fed in the same room.

Carbon Dioxide and Oxygen. Carbon dioxide and oxygen gas can be manufactured on-site for use in water treatment processes. More commonly, it is delivered by tanker truck and pumped into refrigerated, pressurized, outdoor storage tanks on the plant site. The liquids are vaporized and sometimes heated to the desired temperature for process use. The storage tanks are equipped with fill lines having quick-connect couplings located for easy access by delivery trucks. Tanks and all piping segments capable of isolation must be provided with pressure relief safety valves.

Evaporators. Chlorine, ammonia, oxygen, and carbon dioxide are usually stored as liquid and must be evaporated to a gas for metering and flow control to the point of application. Depending on ambient temperatures and rates of withdrawal, some of these gases must be evaporated by warming in vaporizers.

If chlorine containers are 1 ton or larger and the withdrawal rates exceed the quantity available by direct evaporation, evaporators may be required. Evaporators are available in capacities of 4,000, 6,000, and 8,000 lb per day (1,800, 2,700, and 3,600 kg per day). When an evaporator is used, liquid is withdrawn from the bottom of the container and transported to the evaporator, where it is converted to gas. The most common type of evaporator uses an electric resistance heater in a hot-water bath surrounding a vessel in which the liquid is converted to gas.

Sometimes the bath is heated with steam or with a separated, recirculating hot-water system. Figure 15.2 shows a cross section of a typical evaporator. The heat of evaporation of chlorine is low, approximately 69 cal/g, compared with 540 cal/g for water. However, evaporators should be designed with extra capacity to ensure that the gas is superheated and does not recondense on the downstream side.

When an evaporator is being used beyond its capacity, misting occurs. A chlorine gas filter should be installed on the exit gas line from the evaporators to remove impurities

FIGURE 15.2 Cross section of typical evaporator. (*Courtesy of Wallace and Tiernan Division of USFilter.*)

in the chlorine that would be detrimental to the chlorinator. Evaporators should be equipped with an automatic shutoff valve to prevent any liquid from passing to the chlorinators.

When possible, all portions of the chlorine feed system that contain liquid chlorine should be designed and operated with all liquid in the system as a continuous medium. To shut down the evaporator, it is necessary only to close the effluent valve on the evaporator. No other valves between the evaporator effluent valve and the liquid chlorine container should be shut. Very long liquid chlorine lines should be avoided; if they are unavoidable, expansion chambers should be provided. It should be emphasized that liquid chlorine has a high temperature expansion coefficient. Unless expansion is permitted, the temperature increase in trapped liquid will increase the pressure enough to rupture the pipes.

Ammonia feed systems are similar to those used for chlorination. Unlike chlorine and ammonia, carbon dioxide is usually vaporized and maintained with positive pressure during the metering and flow rate control process.

Delivery, Handling, and Storage of Dry Chemicals. Dry chemicals may be purchased and stored in bags or barrels if relatively small quantities are to be used. For larger use, economy of bulk delivery and storage must be considered.

Bulk Delivery by Truck or Railcar. When large quantities of the product are consumed, bulk delivery provides not only economic benefits, but also protection against supply shortages and transportation difficulties. Obviously, the type of storage facilities provided varies with the chemical.

Flow is the key word in chemical handling of bulk solids. In general, the flow of material increases with particle size, uniformity, hardness, smoothness, absence of fines, and lower moisture content. These physical parameters can be controlled to some degree by purchase specification.

Solids unloading is usually accomplished with pneumatic equipment (blower or vacuum), air slides, or mechanical screw conveyors or bucket elevators. The latter equipment is satisfactory for unloading lumpy or coarse material where excessive dust is not a problem.

Pneumatic truck unloading of dry bulk chemicals is done through the user's pipe conveying system, consisting of a truck inlet panel, interconnecting piping to silos or storage bins, safety release valve, and dust collector mounted on top of the storage bin. Piping diameter is usually standardized at 4 in. (10 cm), with bends having a minimum radius of 4 ft (1.2 m).

Maximum length of piping between the inlet panel and storage depends on the nature of the material being transported. Pebble lime, for instance, may be blown as much as 100 ft (30 m) vertically, providing the total length of run does not exceed 150 ft (46 m). Lightweight powdery material can easily be transported up to 300 ft (90 m) over a combined vertical and horizontal distance. Pneumatic trucks equipped with air compressors are generally available in capacities of 700 to 1,300 ft^3 (20 to 40 m^3).

Covered railroad hopper cars used for delivery of dry bulk chemicals may have capacities up to about 3,700 ft^3 (105 m^3). Some of these cars are constructed with two to four compartments, each provided with its own bottom discharge gate. Material is withdrawn through these gates positioned over undertrack hoppers. Air vibrators are used where necessary to facilitate the movement of fine, powdery material through the hopper.

A more widely used type of railcar permits a method of pneumatic unloading similar to that used on pneumatic trucks. Others employ air slide conveyors to move the material out to the truck hopper. Canvas connectors, or stockings, generally connect to the truck hopper that feeds the conveying system. The user must provide air compressors to pressurize the cars or the vibrators and air slides, because compressors are not part of the car equipment, as is the case with pneumatic trucks. Hopper cars may also be unloaded using a vacuum system consisting of suction pump, filter receiver, and discharge air lock or rotary gate at the top of the silo or storage bin.

Storage of Bulk Dry Chemicals. Once the desired material is delivered, proper bin or silo capability must be provided. Dry chemical flow in the bins must be maintained by installing vibrating or pulsating devices, live bin bottoms, and internal devices to control packing and arching. The height of bins or silos is usually approximately 2.5 to 4 times the diameter, with the discharge area as large as possible.

Hopper bottoms should slope at least 60° from the horizontal; for storage of hydrated lime, an even greater slope is desirable. Offset hopper sections are commonly used, with the outlet on one side of the vertical bin axis. The distortion produced by the varying slope angles tends to prevent arching in the conveying section. The use of vibrators to maintain flow requires caution and consideration of the type of material being handled; the worst possible situation occurs when fine materials (such as hydrated lime) are overvibrated and packed. Such material can only be vibrated intermittently, for instance, by a 2- to 4-s pulse repeated several times a minute. By contrast, lumpy material such as pebble lime can be vibrated continuously during discharge. The interrupter unit used to control the vibrator must be interlocked electrically with the process feeder to allow vibration only during discharge.

Air jets and pulsating air pads are also commonly used to fluidize light materials such as hydrated lime. Numerous other devices are available, the most popular being the "live" bin bottom. These units operate continuously during discharge and use gyrating forces or upward-thrusting baffles within the hopper to eliminate bridging and rat-holing. Less sophisticated devices include double-ended cones supported centrally within the hopper, rotating chains or paddles, and horizontal rods run from wall to wall.

Quicklime and hydrated limes are abrasive but not corrosive, and steel or concrete bins and silos can be used for storage. It is imperative that the storage units be airtight as well as watertight to reduce the effect of air slaking; this includes relief valves, access hatches, dust collector mechanisms, and so on, all of which are normally exposed to the weather. Bins and silos can be designed with rectangular, square, hexagonal, or circular cross sections; the first three make optimum use of plant space, but the circular silo is less susceptible to sidewall hangups, which tend to occur in corners of bins of other shapes. Regardless of the cross-sectional configuration of the vertical storage unit, the bottom is always designed with a hopper or conical base right up to the discharge gate. The design volume of any silo or bin should be based on the average bulk density of the chemical, with an allowance for 50% to 100% extra capacity beyond that required to accommodate a normal-size delivery.

Because there is a tendency for water of crystallization from alum to partially slake lime, it is necessary to avoid mixtures of alum and quicklime. In a closed container, this may lead to a violent explosion. For the same reason, equal care should be taken to avoid mixtures of ferric sulfate and lime.

Powdered activated carbon, a finely ground, low-density material, is capable of producing copious quantities of black dust with the least disturbance. As a result, large users of this material who can accept air slide truck or rail bulk shipments prefer to have the carbon unloaded directly into slurry tanks with a dust collector or scrubber on the vent. In this way, further handling of the powder into and out of storage is avoided, and dust problems are minimized. If dry storage is necessary or preferred, a vacuum transfer system minimizes dust. Suppliers should be consulted about trucks or railcars that can accommodate vacuum hookups.

Storage of Materials in Bags and Drums. Areas used for dry chemical storage of bagged material, drums, and other chemical containers must be fireproof, dry, and well ventilated. Compressed gases in cylinders should be stored separately in an area provided with separate mechanical ventilation or exhaust fans.

Containment of Dry Chemical Dust. Any unloading or transfer of dry chemicals creates dust, especially when airflow equipment is used to unload the material. Operation of this type of equipment requires discharging dust-free conveying air to the atmosphere, a job for which a bag-type filter is best suited. Because these filters collect chemical dust that can be returned to the process, it is common practice to mount the collector on top of the silo or dry chemical feeder, so vibration of the bags drops the chemical back into the original container.

Dust collector operation is remotely controlled from the unloading point or truck inlet panel. It is possible to have one dust collector serve more than one storage receiver, provided the same chemical is being handled, but under no circumstances should a mixture of different chemical dusts be allowed to accumulate within one collector. A dust collector's size depends on the volume of air to be handled. Air blowers mounted on pneumatic trucks commonly have the capability of producing up to 750 ft^3/min (21 m^3/min) of air.

Bag filters work for most dust collection applications, but vortex scrubbers should be considered for powdered activated carbon slurry tanks. The unloading rate for carbon into a slurry tank potentially could overwhelm a bag filter and require excessive maintenance. Properly specified vortex scrubbers can provide adequate removal of carbon dust with relatively minimal maintenance. When mounted on top of the carbon slurry tank, vortex scrubbers operate with a water spray entering near the top and traveling countercurrently to the carbon-laden air entering from the bottom. Water and trapped carbon discharge directly into the slurry tank.

Delivery, Handling, and Storage of Liquid Chemicals. Some of the chemicals most commonly used in liquid form in water treatment are listed in Table 15.2. When one is computing storage and feed rate requirements, the effective density must be used. The effective density is determined by the bulk density, active component of the chemical molecule, and solution strength, as shown in the table.

Liquid chemicals, such as caustic soda, liquid alum, acids, or corrosion inhibitors, are generally delivered by tank truck. To facilitate safe unloading, it is the user's responsibility to provide appropriate fill-pipe connections, clearly labeled and equipped with protective caps. A concrete drip sump protected with a chemically resistant coating should be provided beneath all fill-pipe connections.

Local and state regulations may require secondary containment for the entire unloading area. Storage tank overflow pipes should terminate into the secondary containment. If the stored chemical has fuming or corrosive properties, the overflow must be sealed with a vapor check valve.

TABLE 15.2 Characteristics of Common Liquid Chemicals

Chemical name	Formula	Specific gravity	Density of solution (lb/gal)	Total molecular weight	Active molecular weight	Active (%)	Typical solution strength (%)	Effective density (lb/gal)
Sodium hypochlorite	NaOCl	1.21	10.1	74.5	51.5	69.1	12.5	0.87
Fluorosilicic acid	H$_2$SiF$_6$	1.21	10.1	144.1	144.1	79.2	23.0	1.84
Alum	Al$_2$(SO$_4$)$_3$	1.34	11.2	342	342.0	100	50.0	5.6
Ferric chloride	FeCl$_3$	1.44	11.9	162.5	162.5	100	40.0	4.76
Phosphoric acid	H$_3$PO$_4$	1.57	13.1	98	95.0	96.9	75.0	9.52
Caustic soda	NaOH	1.54	12.8	40	40.0	100	50.0	6.4
Ammonia, anhydrous	NH$_4$OH	0.9	7.48	35	35.0	100	29.0	2.17

CHEMICALS AND CHEMICAL HANDLING   15.11

Vent pipes should terminate outdoors with screened elbows facing downward. Vent and fill lines must be run without traps. Overflow lines may interconnect sealed tanks containing the same material, but the lines should not be allowed to terminate in the open air unless they are inside the secondary containment area. Such a line would otherwise become the primary vent and permit a spill to occur in the event of an overfill. As with any chemical, but particularly with liquids, the user must be certain that the quantity ordered can be accommodated in storage at the time of delivery.

Accidental overflows are best prevented by providing dependable level indicators and alarms on all tanks. To further guard against overfilling a storage tank, a high-level audible alarm should be provided as part of the in-plant chemical tank-level indicating system. This alarm should be mounted outside at the unloading station to alert the vehicle operator. The alarm can be common to all tanks served by the particular unloading station.

Several level-sensing systems are available with local or remote indicating and alarm capability. Pneumatic level sensors using bubble pipes are effective and accurate, but before they are used, first it must be determined that air bubbles will not deteriorate the chemical whose level is being measured.

Types of Storage Tanks. Storage tanks for liquid chemicals are commonly fabricated from steel, stainless steel, fiberglass-reinforced plastic (FRP) [also referred to as reinforced thermoset plastic (RTP)], and various forms of polyethylene, such as high-density, cross-linked polyethylene (HDXLPE).

For very corrosive chemicals, such as ferric salts, sodium hypochlorite, and sodium hydroxide, the nonmetallic materials are normally most appropriate. Polyethylene tanks should be designed in accordance with ASTM D-1998. Whenever possible, nozzles should be constructed as integrally molded flanged outlets (IMFOs) to avoid stress points and potential leak sites. Polyethylene is more appropriate for smaller day tanks.

The use of FRP for chemical service began in the mid-1950s. ANSI Standard RTP-1, *Reinforced Thermoset Plastic Corrosion Resistant Equipment,* should be the basis for design of FRP tanks having internal pressures less than 15 psi (103 kPa). FRP has greater structural strength and flexibility for installation of nozzles and other appurtenances in the sidewalls. A variety of polyesters, vinyl esters, and other materials are available for the resin or double laminate construction, depending on chemical resistance requirements.

The choice of tank material or lining, in addition to its chemical resistance, depends on the concentration of chemical, pressure, and temperature conditions, and tank dimensions. The tank manufacturer must be consulted about the types of materials to use for the proposed chemical. Improper material selections could lead to tank failure.

Open-top tanks can be used for stable, nonvolatile liquids, but their use is restricted. Buried tanks must be strapped to anchor blocks with sufficient mass to prevent flotation when the tank is empty and must meet U.S. Environmental Protection Agency (USEPA) leak prevention criteria. These criteria mandate the use of double-walled tanks with leak detection alarms. Buried tanks must also be equipped with manholes, vent lines, and other connections.

The majority of storage tanks are located in the lower or basement areas of buildings, where air is naturally cooler, or outdoors if winters are not too severe. As a result, heating and insulation may be required for those tanks that contain chemicals subject to crystallization. Installing insulation and thermostatically controlled hot-water or steam coils, or electric immersion heaters within or around the tank, ensures positive protection against slushing or freezing. If the problem is marginal or occurs only during severe seasonal weather, applying electrical heating tape on the tank exterior usually offers sufficient protection.

Spill Containment. Chemical storage and feed facilities require adequate provision for spill containment with consideration for worker safety; water quality; customer protection; local, state, and federal regulations; and general environmental protection. Depending on the capacity of chemical storage facilities, the strictest regulations require secondary containment for chemical unloading areas, tertiary containment for chemical storage tank and feed equipment areas, and double-walled piping with leak sensors in the annular space for outside chemical piping runs.

Liquid chemical storage areas should have dike walls, or other impermeable structures, providing adequate secondary containment capacity based on the volume of the single largest storage vessel, a safety factor (say, 10%), and freeboard. The model facility provides common secondary containment for bulk and day tanks, pumping equipment, safety valves, and immediate piping associated with each chemical. Consequently, the structure can contain leaks from the most vulnerable areas, including storage tank fittings and pump connections, without the risk of mixing incompatible chemicals. For corrosive chemicals, secondary containment structures require a chemical-resistant coating or liner to protect and prolong the structure's useful life.

Secondary containment areas generally should not include any floor drains or other floor or wall penetrations that could compromise the integrity of the containment structure. Design should provide access to critical isolation valves without requiring the operator to enter the containment structure, or motorized isolation valves with remote control or automatic interlocks. As an alternative, the facility can include a floor drain connected to a separate secondary containment vessel at a lower elevation.

The design also should provide a leak sensor in a sump or low point within the containment structure, local and remote alarms, isolation valve interlocks, and other instrumentation and controls necessary to enhance safety and ensure rapid response to emergency conditions.

Secondary containment provisions for liquid chemicals should extend to unloading areas with tank filling connections, as well as inside-outside chemical piping runs. Inside piping should run in troughs or channels that allow access for repairs. As a minimum, exposed inside piping should run along walls where a leak would not endanger personnel or other passersby. Outside piping or tubing should run inside conduit or concrete encasement, allowing access for repairs. Some plastic pipe manufacturers provide integrated double-walled piping for secondary containment purposes.

Alum Storage Facilities. Alum is probably the most commonly used liquid chemical in water treatment plants. In liquid form [$Al_2(SO_4)_3 \cdot 14H_2O$], it contains 48.5% alum by weight, the balance being water. As a liquid it has a light green to light yellow appearance. Specific gravity of the bulk liquid is 1.33.

A tank truckload contains between 3,300 and 5,500 gal (12,500 and 20,800 L), and a railcar contains 8,000 to 18,000 gal (30,300 to 68,100 L). At 48.5% strength, its freezing point is 2° F ($-17°$ C), but it can begin localized crystallizing at warmer temperatures. The freezing point is highly dependent on the percent strength. Both higher and lower concentrations have higher freezing temperatures, so alum is rarely handled at other concentrations. For example, at the slight increase to 50% strength, the freezing point rises dramatically to 30° F ($-1°$ C).

In freezing climates, storage tanks should be installed indoors or in a heated enclosure to maintain the liquid temperature between 45° and 60° F (7° and 16° C). Storage and feed lines exposed to low temperatures should also be insulated and equipped for flushing with water if crystallization could occur. Storage tanks should be sized for at least 7,500 gal (28,400 L) if alum is delivered by truck.

Storage tanks are most commonly constructed of FRP and polyethylene. Alternatively, rubber-lined steel and type 316 stainless steel can be used. Piping can be schedule-80 polyvinylchloride (PVC) with socket end joints if liquid temperatures do not exceed 120° F (49° C). The temperature of liquid alum when delivered may occasionally exceed 120° F (49° C), so the unloading line for the tank should be chlorinated polyvinylchlo-

CHEMICALS AND CHEMICAL HANDLING   15.13

ride (CPVC), 3-in. (80-mm) diameter, as a precaution. Plug, needle-and-ball valves, 150-lb integral flanges with PVC or type 316 stainless steel construction, and tetrafluoroethylene (TFE) or similar material sleeves or retaining rings are recommended. A schematic layout of an alum feed system is shown in Figure 15.3.

Caustic Soda Storage. Liquid caustic soda is delivered by railcar and tanker truck. Tanker trucks can haul up to 3,000 gal (11,400 L) per load. Storage tanks should be constructed of stainless steel, FRP, polyethylene, or steel lined with rubber or polypropylene. As with alum, schedule-80 PVC with socket end joints is the recommended piping material if liquid temperatures do not exceed 120° F (49° C). Caustic soda generates heat when mixed with water, resulting in temperatures approaching this level, so use of CPVC piping should be considered for this application. Valves can be steel or ductile iron with Teflon components, rubber-lined, or PVC.

The freezing point of caustic soda solution is highly dependent on the solution strength (Figure 15.4). Freezing points for solutions greater than 50% quickly elevate, making it impractical to handle and store. Because 50% caustic soda begins to crystallize at approximately 54° F (12° C), storage tanks must be indoors or insulated and heated to avoid crystallization. A typical chemical feed schematic for caustic soda appears in Figure 15.5.

Sodium Hypochlorite Storage. Commercial sodium hypochlorite (NaOCl), or liquid bleach, is marketed in carboys and rubber-lined drums holding up to 50-gal (190-L) volume, and in trucks. Storage tanks should be constructed of FRP, polyethylene, or steel lined with rubber or polypropylene. As with alum, schedule-80 PVC with socket end joints is the recommended piping material if liquid temperatures do not exceed 120° F (49° C).

Valves may be a plug type made of steel (lined with PVC or polypropylene), or PVC diaphragm valves. Although PVC ball valves would not react with the chemical, hypochlorite releases small amounts of gas as it decomposes, and PVC ball valve failures due to gas buildup have been reported. A similar type of failure has occurred with stainless steel (SST) ball valves in hydrogen peroxide service. Any type of ball valve, especially 1-in. and larger sizes, may allow hypochlorite to weep through the seal and crystallize. To avoid these problems, a diaphragm valve constructed of composite material using a Teflon diaphragm can be used.

When hypochlorite is added to water, it hydrolyzes to form hypochlorous acid (HOCl), the same active ingredient that occurs when chlorine gas is used. The hypochlorite reaction slightly increases the hydroxyl ions (pH increase) by the formation of sodium hydroxide, whereas the reaction of chlorine gas with water increases the hydrogen ion concentration (pH decrease), forming hydrochloric acid. In most waters, these differences are not significant, but when high chlorine doses are used in poorly buffered waters, these effects should be considered. They can be evaluated by calculation or by simple laboratory tests.

In the commercial trade, the concentration of sodium hypochlorite solutions is usually expressed as a percentage. The *trade percent* is actually a measure of weight per unit volume, with 1% corresponding to a weight of 10 g of available chlorine per liter. Common household bleach, at a trade concentration of 5.25%, has approximately 5.25 g/100 mL or 52.5 g/L of available chlorine. Hypochlorite available for municipal use usually has a trade concentration of 12.5% to 17%. These are approximate concentrations and should always be confirmed by laboratory procedures.

Because increasing the concentration of any salt lowers the freezing point of a solution, the freezing points of various solutions of sodium hypochlorite are a function of their concentrations, with the more dilute concentrations approaching the freezing point of pure water. Figure 15.6 shows the freezing temperature of hypochlorite solutions as a function of concentration in the concentration ranges normally experienced.

The chlorine concentration in hypochlorite solutions is adversely affected by high temperature, light, low pH, and the presence of certain heavy metal cations. Iron, copper,

FIGURE 15.3 Coagulant alum/ferric chloride system. (*Courtesy of MWH Americas Inc.*)

CHEMICALS AND CHEMICAL HANDLING

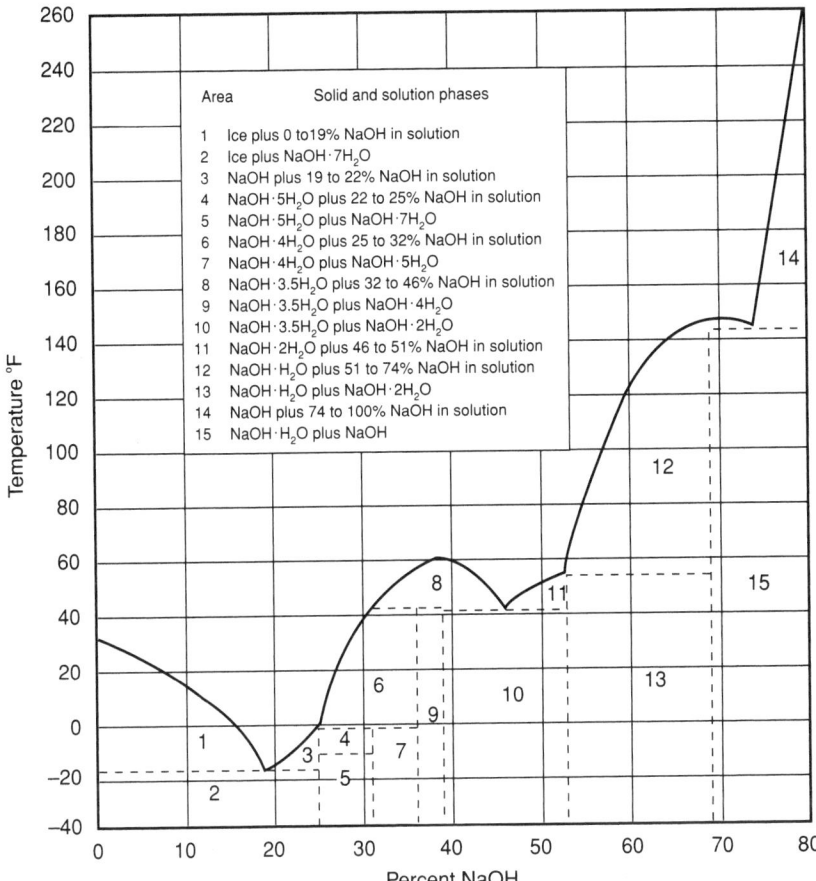

Area	Solid and solution phases
1	Ice plus 0 to 19% NaOH in solution
2	Ice plus $NaOH \cdot 7H_2O$
3	NaOH plus 19 to 22% NaOH in solution
4	$NaOH \cdot 5H_2O$ plus 22 to 25% NaOH in solution
5	$NaOH \cdot 5H_2O$ plus $NaOH \cdot 7H_2O$
6	$NaOH \cdot 4H_2O$ plus 25 to 32% NaOH in solution
7	$NaOH \cdot 4H_2O$ plus $NaOH \cdot 5H_2O$
8	$NaOH \cdot 3.5H_2O$ plus 32 to 46% NaOH in solution
9	$NaOH \cdot 3.5H_2O$ plus $NaOH \cdot 4H_2O$
10	$NaOH \cdot 3.5H_2O$ plus $NaOH \cdot 2H_2O$
11	$NaOH \cdot 2H_2O$ plus 46 to 51% NaOH in solution
12	$NaOH \cdot H_2O$ plus 51 to 74% NaOH in solution
13	$NaOH \cdot H_2O$ plus $NaOH \cdot 2H_2O$
14	NaOH plus 74 to 100% NaOH in solution
15	$NaOH \cdot H_2O$ plus NaOH

FIGURE 15.4 Freezing points of caustic soda solutions.

nickel, and cobalt are the most common problem-causing cations. The concentration of hypochlorite itself also has a major impact on hypochlorite degradation. Table 15.3 shows the half-life of a hypochlorite solution as a function of chlorine concentration and temperature.

When one is purchasing bulk sodium hypochlorite, purchasing specifications should be used, delineating the acceptable ranges for available chlorine (12.5% to 17%) and pH (11 to 11.2), as well as maximum contaminant limits for iron (2 mg/L) and copper (1 mg/L). Specifications should also require that shipments be free of sediment and other deleterious particulate material. On arrival, shipments should be analyzed for the concentration of chlorine, pH, and the concentration of metal contaminants.

Sodium hypochlorite storage must be carefully managed to limit degradation and the formation of chlorate. The rate of degradation is a function of storage time, temperature, and chlorine concentration, so storage time should preferably be limited to less than 28 days. The chemical may be delivered at temperatures up to 85° F (30° C), which must be considered in the system design.

FIGURE 15.5 Caustic soda system. (*Courtesy of MWH Americas Inc.*)

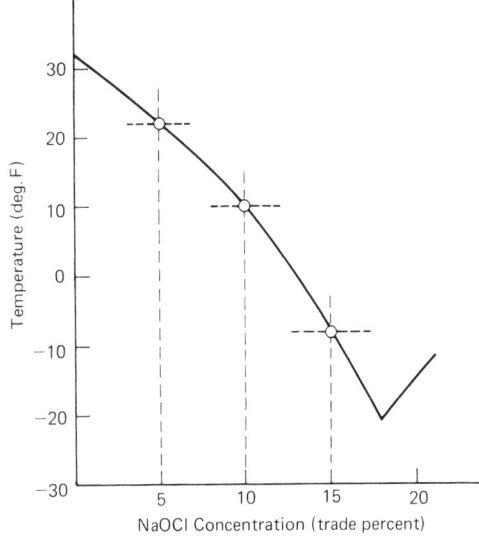

FIGURE 15.6 Freezing temperatures of hypochlorite solutions.

TABLE 15.3 Influence of Temperature and Strength on Decay of Hypochlorite Solutions

% Cl$_2$	Half-life (days)			
	100° C	60° C	25° C	15° C
10.0	0.079	3.5	220	800
5.0	0.25	13	790	5000
2.5	0.63	28	1800	—
0.5	2.5	100	6000	—

Storage temperatures should not exceed about 85° F (30° C) because above that level, the rate of decomposition increases rapidly. Although storage in a cool, darkened area greatly limits the deterioration rate, most manufacturers recommend a maximum shelf life of 60 to 90 days.

All hypochlorite solutions are corrosive to some degree and can affect the skin, eyes, and other body tissues. Accordingly, rubber gloves, aprons, goggles, and similar suitable protective apparel should be provided for preparing and handling hypochlorite solutions. Areas of skin contact should be promptly flushed with large quantities of water. Every precaution should be observed to protect containers against physical damage, to prevent container breakage, and to minimize accidental splashing.

CHEMICAL FEED AND METERING SYSTEMS

The types of chemical feed equipment for the water treatment plant must be part of the decision of whether gaseous, dry, or liquid chemicals are to be used.

Gas Feed Equipment

Chlorine and ammonia gases are used in disinfection treatment processes and require careful selection of equipment because of the aggressiveness and dangers associated with the gases.

Chlorinators. A conventional chlorinator consists of

- An inlet pressure-reducing valve
- A rotameter
- A metering control orifice
- A vacuum differential regulating valve

A simple schematic is shown in Figure 15.7. The driving force for the system comes from the vacuum created by the chlorine injector. The chlorine gas flows to the chlorinator and is converted to a constant pressure (usually a mild vacuum for safety reasons) by the influent pressure-reducing valve. Present design practice locates the influent pressure-reducing valve as close as possible to the storage containers to minimize the amount of pressurized chlorine gas piping in the plant.

FIGURE 15.7 Flow diagram for conventional chlorinator. (*Courtesy of Wallace and Tiernan Division of USFilter.*)

The chlorine then passes through the rotameter, where the flow rate is measured under conditions of constant pressure (and consequently constant density), and then through a metering or control orifice. A vacuum differential regulator is mounted across the control orifice so that a constant pressure differential (vacuum differential) is maintained to stabilize the flow for a particular setting on the control orifice. Flow through the control orifice can be adjusted by changing the opening on the orifice. The control orifice has a typical range of 20 to 1, and the vacuum differential regulator has a range of about 10 to 1. Thus the overall range of these devices combined is about 200 to 1. On the other hand, a typical rotameter has a range of about 20 to 1. Thus the chlorinator should be selected based on design capacities, and the rotameter installed should be appropriate for current demands.

Chlorine Gas Piping. Between the chlorinator and the injector, the chlorine gas flows in a vacuum. Although the head loss of the gas flow is usually small, it is critical that the vacuum created by the injector be transmitted to the chlorinator without significant dissipation. As a consequence, the diameter of the chlorine vacuum lines should always be designed rather than arbitrarily selected. Lines should be sized to limit the total pressure drop over the pipe length to between 1.5 and 1.75 in. of mercury (1.7 to 1.9 ft of water) under maximum injector vacuum levels.

Dry Chemical Feed Equipment

Feeders for powders and granules may be classified as gravimetric (accuracy range 0.5% to 1% of set rate) and volumetric (accuracy range 1% to 5% of set rate depending on material fed). Gravimetric feeders are preferred when accurate feeding of chemicals with varying bulk densities is important.

Loss-in-Weight Gravimetric Feeders. Gravimetric feeders use a feeder hopper suspended from scale levers, a material feed control mechanism, and a scale beam with motorized counterpoise. The rate of weight loss of the hopper equals the weight loss equivalency of a traveling counterpoise when the feeder is in balance. If it does not, the scale beam deflects, and the feed mechanism increases or decreases the feed accordingly. Although these feeders are highly accurate, their capacity is usually less than 1,000 lb/h (454 kg/h). The total amount of material fed may be recorded or read directly off the weight beam at any time.

Belt-Type Gravimetric Feeders. These feeders are available in numerous forms and are usually designed to handle specific types of material. Weight belts can be of the pivoted type for heavy feed rates of 250 tons/h (227 t/h) and up, or the rigid belt type passing over a live or weigh-deck scale section, with a feed hopper at one end of the belt and a control gate to regulate the flow and depth of material placed on the belt. A scale counterpoise is adjusted to establish the desired belt loading, and the control gate is automatically repositioned in proportion to the error signal. Various gate control systems are available depending on the material to be handled, response time desired, feed range, and capacity. Normally belt speed is varied to produce the desired flow of material. Total quantity of material fed can be read directly on a totalizer or similar device.

Volumetric Feeders. Although more than a dozen types of volumetric feeders are available, all operate on the principle of feed rate control by volume instead of weight. Advantages of volumetric feeders include low initial cost, good overall performance at low feed rates, and acceptable accuracy for materials with stable density and uniformity. Disadvantages include unresponsiveness to density changes, fixed orifices or openings subject to clogging, and calibration by manual sampling (which must be done regularly).

The many types of volumetric feeders available permit a good choice based on capacity requirements and the nature of the material to be fed. The roll feeder forms a smooth ribbon of material of adjustable thickness and width; this feeder is unique in its ability to handle very low feed rates of fine-ground materials such as hydrated lime. It cannot be used for coarse granular material.

The screw feeder is a popular unit employing rotating or reciprocating feed screws that can handle most dry chemicals. Most of these require hopper agitation or vibration to maintain screw loading. The range of feed is good (at least 20 to 1), with capacities up to 600 ft^3/h (17 m^3/h) using 6-in. (15-cm) helical screws. Minimum feeds on certain models are very low when using fine powdery material through a small-diameter screw.

On belt feeders, the material is deposited on a moving belt from an overhead hopper and passes beneath an adjustable vertical gate. The speed of the belt and the position of the gate establish the volume of material passing through the feeder. These are high-capacity feeders that can handle anything from powder to 1.5-in. lump materials at rates from 600 to 3,600 ft^3/h (17 to 100 m^3/h) depending on belt width. A schematic of a typical volumetric feeder system for potassium permanganate is shown in Figure 15.8.

Rotary paddle feeders consist of a paddle or series of compartments revolving within an enclosure that receives material from the hopper and releases it through a discharge chute as rotation proceeds. Feed rate is normally controlled by using a sliding gate or varying the speed of the paddle shaft. These feeders have a unique application in that they can deliver material into vacuum or pressure systems because they form an air lock. They are also commonly used to feed chemicals that tend to flow or gush out of control through a fixed orifice.

Vibrating feeders employ a vibrating mechanism attached to a slightly inclined feed trough. Flow is controlled by regulating the depth of material and the intensity of vibration. These feeders are used only on dry, nonhydroscopic, free-flowing materials. They are generally used in smaller installations, and their accuracy is acceptable as long as the material is of consistent quality and there are no large voltage fluctuations that would affect the amplitude of the vibrator.

An oscillating hopper feeder consists of a main hopper fitted with an oscillating apex section that discharges to a stationary tray or plate. Oscillation of the hopper pushes previously deposited material off the tray in one or more directions. Capacity is controlled by adjusting the depth of chemical deposited and regulating the length of stroke. Because these feeders can handle a variety of chemicals from powder to pebble lime, they are popular in smaller plants.

Virtually all dry feeders, gravimetric or volumetric, can be equipped to operate automatically in proportion to a flow or other process signal. The means for accomplishing this varies according to the nature of the feeder's control mechanism. In its simplest form, time-duration control using a resettling time can provide proportional feed using a manually adjusted feeder, provided a flow-proportional pulse or contact signal is available from the flowmetering system.

Lime Slakers. Slaking means combining water with quicklime (CaO), in various proportions, to produce milk-of-lime, a lime slurry, or viscous lime paste. Slakers operate at elevated temperatures, with or without auxiliary heaters, because of the exothermic reaction between CaO and water.

Aside from capacity, operational flexibility, and the desired concentration of the slaked lime product, the most appropriate method of and equipment for slaking quicklime depend on the characteristics and quality of the quicklime supply. Lime slakers come in two basic varieties: slurry (detention) and paste slakers. Elements common to both varieties

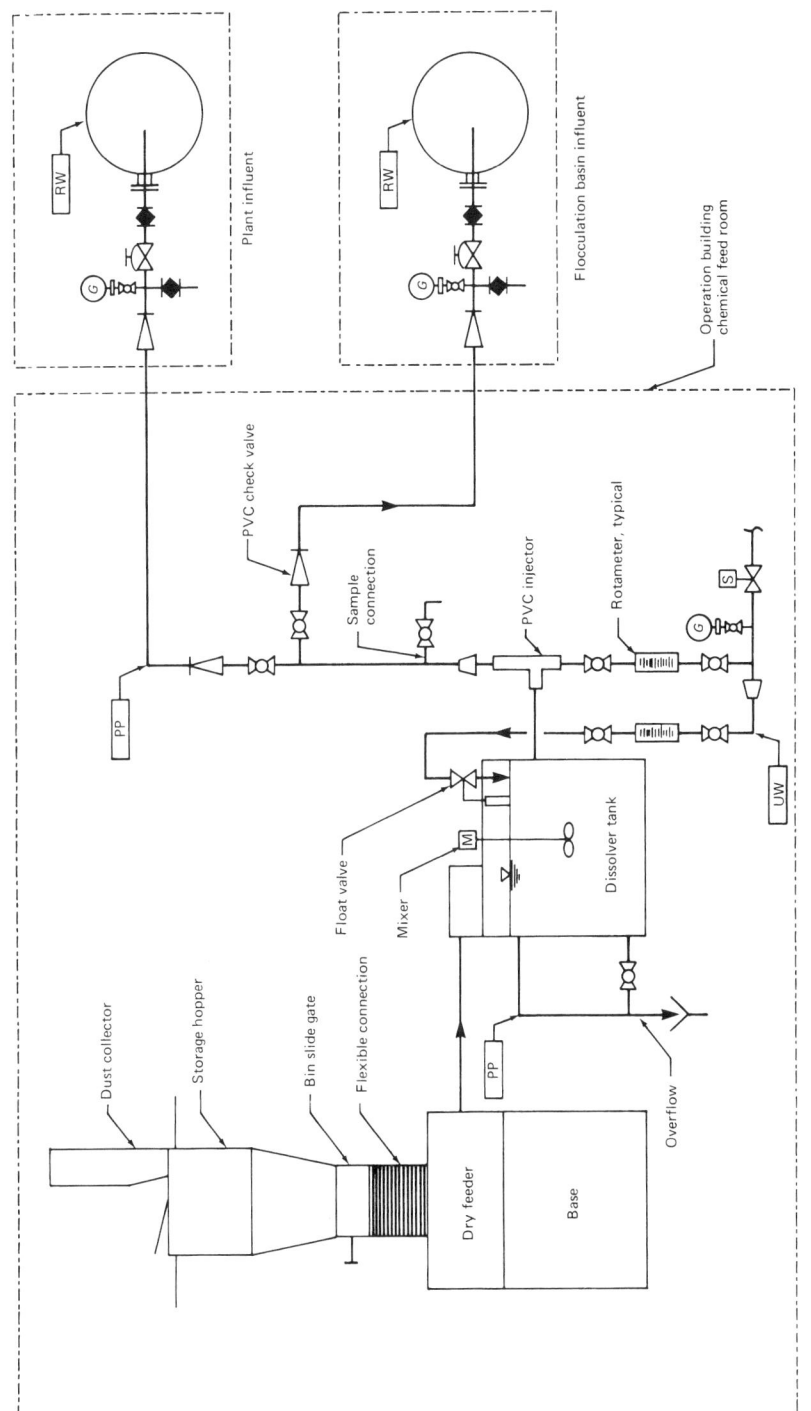

FIGURE 15.8 Potassium permanganate system. (*Courtesy of MWH Americas Inc.*)

include a quicklime (dry) feeder, a water flow control valve, temperature controls, a grit removal device, a dilution chamber, and a final reaction vessel. All slakers require an integral water vapor and dust collector to maintain a slight vacuum within the slaker and discharge clean air.

The difference between the two varieties relates to the temperature and consistency of the lime-water mixture as it passes through the slaker. A slurry (detention) slaker typically mixes lime and water at a weight ratio between 1:3 and 1:4. Paste slakers, and some slurry slakers with auxiliary heaters, mix the lime and water at a ratio of about 1:2. The slurry slaker uses a mechanical (typically an impeller) mixer and maintains the slurry level in the slaker by regulating the water flow rate. A paste slaker uses a pug mill type of agitator and regulates the water flow rate based on the torque imposed on the agitator. In this manner, the paste slaker adds only enough water to achieve and maintain the desired consistency.

Batch Mixing. In most water treatment facilities today, batch mixing typically applies only to potassium permanganate, polymer, and powdered activated carbon. Batch mixing of potassium permanganate in *saturation tanks* provides a constant-strength solution for application to multiple points at variable rates via positive displacement pumping equipment. For continuous operation, the design should include at least two tanks to maintain consistent permanganate dosages while preparing the solution in one of the tanks.

Batch mixing can work in a similar fashion to prepare solutions from dry or emulsion polymers. Again, the design should include at least two tanks to maintain consistent polymer dosages while preparing and aging the solution in one of the tanks. Special automatic batch mixing equipment for dry polymers is available.

Activated Carbon. Powdered activated carbon can be stored in bins and extension hoppers to supply dry feeders. This type of installation is appropriate in smaller plants and in plants where carbon usage is intermittent and the carbon is received and stored in paper bags. Activated carbon is combustible and will burn when ignited. The ignition point of the activated carbons varies from 600° to 800° F (315° to 426° C). After ignition, activated carbon does not burn with a flame, but glows or smolders until all carbon material is oxidized. At least one U.S. utility has experienced spontaneous combustion in its bulk storage bins. The combustion was smothered by injecting carbon dioxide in the bottom of the bin and placing dry ice on the top of the carbon.

Storing carbon in paper bags presents a hazard in that the paper burns more rapidly. Bags of powdered carbon should be stacked in rows with aisles between so that each bag is accessible for removal in case of fire.

In the event of an activated carbon fire, the safest procedure, if possible, is to place the smoldering material in a metal container and haul it outside the building. A smoldering carbon fire may be extinguished by means of a very fine spray or mist of water from a hose or by a foam-type chemical extinguisher. Do *not* attempt to extinguish the carbon by a direct stream of water, because this causes the light, smoldering particles to fly into the air and spread the fire.

Installing an overhead sprinkler system in the storage and feeding rooms is a practical precautionary measure. Activated carbon should not be stored where it can come into contact with gasoline, mineral oils, or vegetable oils. These materials, when mixed with carbon, slowly oxidize until the ignition temperature is reached. Never mix or store carbon with such materials as chlorine, lime hypochlorites, sodium chlorite, or potassium permanganate. Such mixtures are known to be spontaneously combustible.

Activated carbon is an electrical conductor and should not be allowed to accumulate as dust near or on open electric circuits. Some activated carbons are subject to deterioration in storage, so carbon storage areas should be relatively free of such air contaminants as sulfur dioxide, chlorine, hydrogen sulfide, and organic vapors. Normal safety equipment, such as protective clothing, respirators, neck cloths, gloves, and goggles, should be provided for workers handling powdered activated carbon.

Sodium Chlorite. Sodium chlorite is a dry, flaked salt, which, because of its powerful oxidizing nature, is shipped in steel drums bearing an Interstate Commerce Commision (ICC) yellow label classification. It is stable when sealed or in solution, but it is highly combustible in the presence of organic material. For this reason, the solution should not be allowed to dry out on floors, but should be hosed down with minimum splashing. Technical-grade sodium chlorite is an orange flaked salt.

Sodium chlorite should be stored in an enclosed space specially prepared for the purpose, and removed from the storage room only as needed for immediate use. Empty containers should be returned to the storage room immediately after each use unless they are shipping containers, in which case they should be thoroughly flushed with water (to the sewer) as soon as they are empty and should be immediately disposed of well away from any building. Shipping containers should never be used for any other purpose after they are empty.

Calcium Hypochlorite. Although calcium hypochlorite is a stable, nonflammable material that cannot be ignited, contact with heat, acids, or combustible, organic, or oxidizable materials may cause fire. It is readily soluble in water, varying from about 21.5 g/100 ML at 32° F (0° C) to 23.4 g/100 ML at 104° F (40° C). Tablet forms dissolve more slowly than granular materials and provide a fairly steady source of available chlorine over an 18- to 24-h period.

Granular forms usually are shipped in 35- or 100-lb (16- or 45-kg) drums, cartons containing 3.75-lb (1.7-kg) resealable cans, or cases containing nine 5-lb (2.3-kg) resealable cans. Tablet forms are shipped in drums and in cases containing resealable plastic containers.

Because of its strong oxidizing powers and reactivity with organic materials, calcium hypochlorite should be segregated from other chemicals or materials with which it can react, or stored in a separate location. To minimize the loss in available chlorine content that occurs with elevated temperature, cool storage areas should be provided. Containers should be kept dry and located in a darkened area unless the containers themselves keep out excessive light. Their size should be consistent with use requirements. Stored containers should be arranged such that they can be easily moved from the storage area in the event of leaks.

Sodium Carbonate (Soda Ash). Soda ash used to soften water is a grayish-white powder containing at least 98% sodium carbonate. It may be shipped in bulk, in bags, or in barrels. Soda ash is noncorrosive and may be stored in ordinary steel or concrete bins or silos and fed using a conventional chemical dry feeder. Its solution may be transmitted through conventional pipelines or troughs. Hazards associated with soda ash are primarily those of a chemical dust. Protective clothing and devices such as gloves, respirators, and goggles should be provided.

Sodium Chloride. Sodium chloride has a tendency to absorb moisture and to cake under certain conditions. It should be protected from moisture and is best stored in concrete bins. Sodium chloride is highly soluble in water, may be readily made up to a desired concentration, and may be fed by a standard liquid chemical metering device. The solution may be transmitted through rubber or bronze lines. Exposure of the skin to large amounts of the dry salt would have a tendency to cause the skin to dehydrate. Protective clothing and devices such as gloves and face shields should be provided. Large users of brine solution prefer to receive delivered salt directly into a saturator tank and avoid the problems of dry storage.

Liquid Chemical Feed Equipment

Liquid chemical feed equipment includes various types of positive displacement pumps, centrifugal pumps, dipper-wheel feeders (discussed in the section on slurries), and eductors. The designer must select appropriate pumping equipment for the intended application and pump components compatible with the intended chemicals.

Positive displacement pumps handle a nearly constant rate of flow regardless of the back pressure they pump against, making them ideal for metering precise chemical flow rates. Several types of positive displacement pumps are manufactured; those most commonly used in water treatment plants are piston and diaphragm and progressive cavity.

Piston and Diaphragm Pumps. Piston and diaphragm pumps are accurate, their capacity range dependent on stroking speed and length of stroke. Diaphragms pump the liquid so that the mechanical components of the pump do not come into direct contact with corrosive chemicals. Diaphragms are set in oscillating motion by a piston either directly connected to the diaphragm or indirectly connected through a hydraulic fluid.

Flat-faced diaphragm and tubular diaphragm pumps come in different configurations, with some advantages and disadvantages to each, but all operate based on the same principle, with an elastomeric diaphragm, compression chamber, and inlet-outlet check valves. Diaphragm pumps can accommodate a wide range of capacities and operate against relatively high back pressures. These pumps are driven by motors or are solenoid-driven.

Air-operated diaphragm pumps are also available, the discharge rate controlled by regulated air admission and exhaust to the power side of the diaphragm. Pump stroking speed should be limited to less than 120 strokes per minute. Flow rate adjustments are made by varying stroke length, speed, or both.

Although manufacturers sometimes claim the capability of high turndown ratios by using a combination of stroke length and speed, pumps should be sized so that the turndown ratio does not need to exceed 10 to 1, to help ensure accuracy at low feed rates. For adjustable-speed drives, the turndown ratio should not be expected to exceed 5 to 1, to maintain stability at low speeds. In addition, a tachometer for feedback to the control system should be considered to help provide stable speeds.

Progressive Cavity Pumps. Progressive cavity pumps, with their unique rotor and stator elements, are capable of pumping thick pastes, gritty slurries, or viscous shear-sensitive fluids. These pumps can wear more rapidly than other pumps handling slurry under similar conditions unless care is exercised in selecting proper construction materials. Progressive cavity pumps are specialized service pumps with a relatively high initial cost; with preventive maintenance, they can be expected to give trouble-free service.

Other Pumps. Properly selected peristaltic pumps have some applications for liquid chemical feed applications, particularly at lower capacities. Operating similarly to peristaltic pumps, hose pumps can meet high-capacity liquid chemical feed requirements. Gear pumps operate similarly to other positive displacement pumps but use gear teeth as the motive elements. Gear pumps have particular applications for highly viscous liquid chemicals such as polymers.

Valves and Appurtenances. With the possible exception of peristaltic pumps, all liquid chemical feed pumping equipment requires certain valves and other accessories to provide a safe and reliable system. All positive displacement pumps should have a suitable pressure relief valve installed in the discharge piping to prevent overpressurizing of piping due to inadvertently closed valves or other obstructions. In some cases these valves are integral with the pump. Calibration chambers for calibrating the discharge from metering pumps should be installed on the suction piping with valves and fittings to isolate it from the chemical supply tank, so that measurements can be made with the pump operating against normal discharge conditions.

Back pressure valves are used to prevent gravity flow or siphoning to a low-pressure application point and to ensure consistent operation. A typical set point for back pressure valves in this application would be at least 30 psig. Even when chemical feed pumps normally discharge to a high-pressure application point, an accidental loss of pressure in the system could result in an overdose by gravity flow or siphoning action, which back pressure valves would mitigate.

Figure 15.9 shows a typical liquid chemical feed system, including isolation valves, strainers, and a calibration chamber on the suction side and a series of safety valves (pressure relief, check, back pressure, antisiphon) along with isolation valves on the discharge side of the chemical feed pumps. Vents for calibration chambers should extend above storage tanks or whatever the maximum level of the liquid is in the piping system.

Many positive displacement pumps have integral check valves, and some are equipped with combination pressure relief, priming, back pressure, and antisiphon valves. Even when a pump has its own combination of safety valves, the design should provide at least a redundant antisiphon valve to help protect the water supply from an overdose of chemical, should a system upset occur.

Low pressures, which could encourage gases to come out of solution and form bubbles, should be avoided. This requires larger suction lines that are as short as possible. Suction piping should have a positive static pressure when possible. A standpipe (or calibration column) near the pump suction should be considered to divert bubbles that may be present in the suction line.

Positive displacement pumps typically require some type of pulsation dampener on the discharge (and sometimes suction) piping to reduce intermittent high pressures and partly equalize the flow of liquid chemical to the application point. Pulsation dampeners should be placed upstream of back pressure valves to work properly. Even with pulsation dampening, the piping pressure rating should be at least 150% of average working pressure.

Valves for isolation should be used often on chemical feed piping. Even long lines with no attachments or equipment should be equipped with isolation valves for maintenance and cleaning. Unions should be installed around pumps and valves to facilitate their removal.

Eductors. Eductors, also referred to as ejectors or jet pumps, are useful for introducing gases, solutions, or slurries into a diluted water solution followed by application to the process water. A stream of water passing through a venturi in the eductor creates a subatmospheric pressure, which draws the chemical into the eductor. Because eductors are incapable of flow rate control, the chemical must be metered in some fashion before entering the eductor. The chemical introduced at the eductor can be monitored by a flowmeter and its flow rate controlled by a valve in response to a reading from the meter. This practice should be used only when adequate suction pressure is available from the chemical supply source, to allow the control valve to accurately control the chemical entering the eductor.

Centrifugal Pumps. Centrifugal pumps are used primarily in chemical feed systems for chemical transfer from bulk tanks to day tanks. Centrifugal pumps for chemicals operate similarly to centrifugal pumps for water. Speed variation affects both the discharge flow

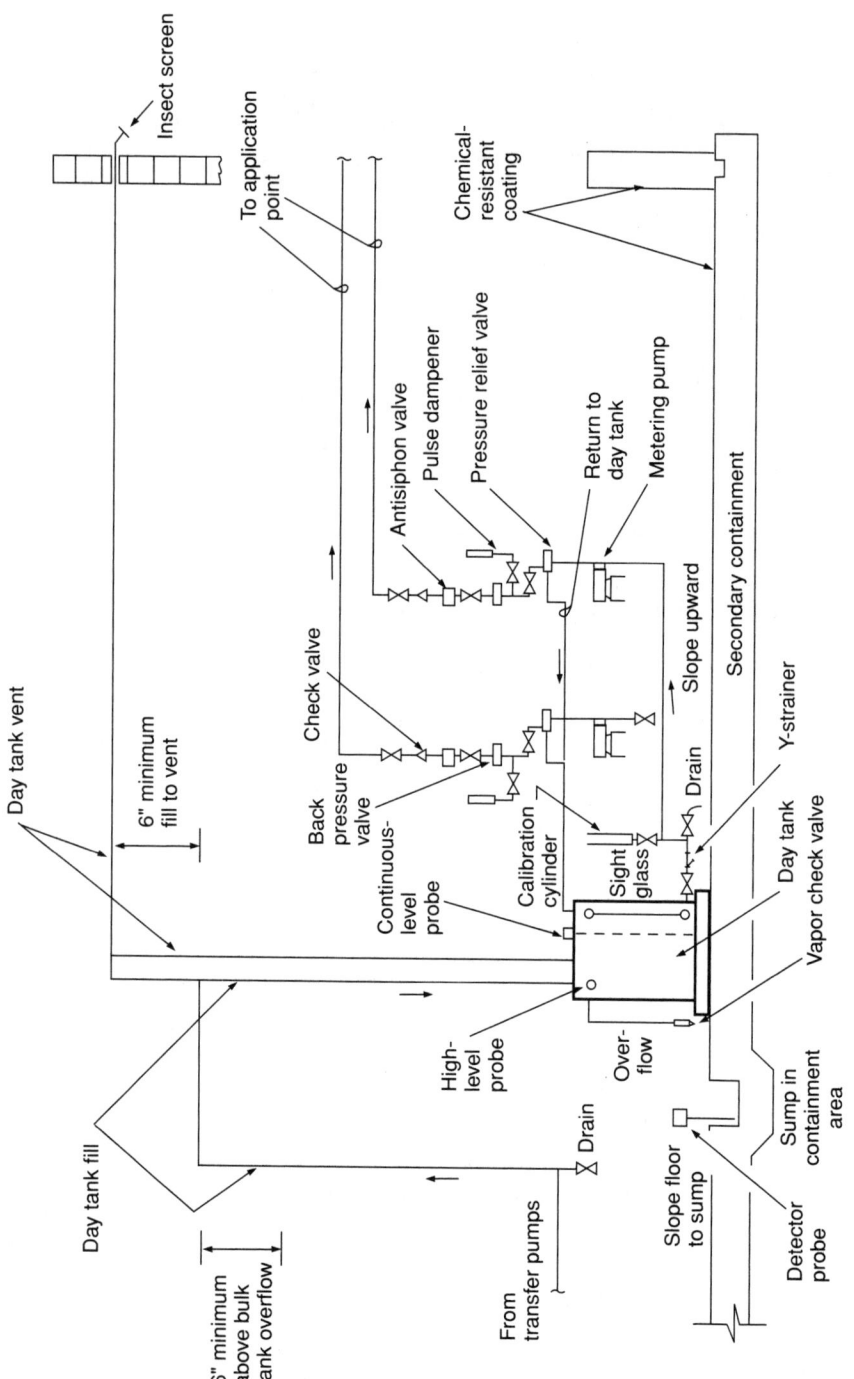

FIGURE 15.9 Typical arrangement for valves and apparatus.

rate and the head. For this reason, centrifugal pumps typically are not used to deliver chemical directly to the injection point, except in a recirculating system with rate control valves at the application points. Centrifugal pumps come with mechanical seals that contain the liquid chemical within the pump body, or with sealless magnetic drives that eliminate the need for mechanical seals. Centrifugal pumps also have important applications for supplying high-pressure water to eductors.

Feed Systems for Slurries. As used in water treatment, a slurry can be described as a suspension of a relatively insoluble chemical in water. Powdered activated carbon and lime (calcium hydroxide) are the most common slurries handled in water treatment. Most questions about these slurries deal mainly with handling and mixing. Once a slurry (carbon or lime) has been prepared, a pumping and piping system conveys the material to the point or points of application. Different techniques are required to handle carbon and lime slurries.

Carbon Slurries. Numerous tests have demonstrated that there is no loss of adsorptive activity when carbon is held in water as a slurry over long periods (up to 1 year). Mixing equipment used to wet and slurry activated carbon should operate with the agitator revolving at a minimum of 60 to 70 rpm. It is preferable that the agitator motor be dual-speed, providing about 80 rpm for initial wetting and 40 rpm to maintain suspension. Agitators normally consist of two sets of stainless steel paddles, one set near the bottom of the tank and the other set placed approximately 18 in. (0.46 m) from the top. With such an arrangement, activated carbon can be slurried almost as rapidly as it is discharged from the delivery vehicle. With the proper equipment, tests have shown that an air slide car containing 21 tons (19 t) of carbon can be unloaded in less than 1 h. Water treatment plants purchasing less than carload quantities of powdered activated carbon may also find a slurry system desirable.

Tank sizing is based on a slurry concentration of 1 lb/gal (2 g/L). It is common practice to size the tank to have a total volume equal in gallons to the maximum load (in pounds) of carbon to be received plus 20% for freeboard.

Carbon slurry tanks are usually square concrete structures with a bitumastic or epoxy lining. If steel tanks are used, the surface must be cleaned to bare metal before lining. The mixer agitator shaft, impeller, assembly bolts, pump suction piping, and other such parts must be made of stainless steel or rubber-covered. Sufficient horsepower should be available to handle carbon slurry concentrations up to 1.5 lb/gal (180 g/L). A schematic drawing of a carbon slurry feed system is shown in Figure 15.10.

Mechanical failure of the mixing equipment renders the slurry system inoperative. For this reason, those installations that are critically dependent on uninterrupted carbon feed could be equipped with an air agitation system for backup. Although it is not as efficient as mechanical mixing, adequate suspension can usually be maintained until mixer repairs are made. Obviously, large plants that use carbon routinely should consider the feasibility of installing duplicate slurry tanks, not only for mechanical standby but also because it is almost impossible to maintain a constant-strength slurry during the recharge and wetting period after a new delivery of carbon.

Tank level readings are best accomplished using an air bubbler system with a stainless steel bubble pipe. Slurry is controlled by measuring makeup water to the tank through a water meter.

Carbon, unlike lime, does not react chemically when diluted in water, and therefore it does not cause scaling and salt precipitation problems. On the other hand, construction materials for piping, pumps, and valves must be resistant to the corrosive nature of powdered carbon slurries. Suitable materials for this service include type 316 stainless steel, rubber, silicon bronze, monel, Hastelloy C, Saran, and fiberglass-reinforced plastic.

FIGURE 15.10 Activated carbon system. (*Courtesy of MWH Americas Inc.*)

On smaller installations, where feed control is achieved by regulating the flow of dry carbon through a dry feeder equipped with a slurry pot, the feed line carrying the carbon suspension from the machine should be of ample size to handle the volume of water required for dilution. If possible, the line should be installed with a continuous downgrade to the point of application. Provision should be made for cleaning out any carbon that may settle in the line and cause clogging.

Whenever feeding is stopped for any reason, the line should be flushed with clear water. In many smaller plants, a rubber hose is used for the entire feed line, so that stoppages may be easily eliminated by manipulation. If an ample volume of flushing water is not available through the slurry pot, dilution water should be added downstream.

Larger installations using agitated slurry storage of constant strength employ metering pumps for feed control. These pumps should discharge to a dilution box located at an appropriate elevation to permit gravity feed. Dilution water is added through a rotameter to maintain a velocity of at least 1.5 ft/s (46 cm/s) to prevent settling. The point of application should be above the water surface, and unavoidable high spots in the line must be vented.

Carbon slurry concentrations generated by dry feeder–slurry pot systems naturally vary, because dry carbon feed is variable. On the other hand, agitated slurry storage systems are commonly batch-mixed and maintained at 1 lb of carbon per gallon of water for convenience in computing feed rates. For this reason, more dilution water should be added after the metering pump.

Lime Slurries. The concentration of lime slurry suspension is commonly expressed as pounds of CaO or $CaOH_2$ per gallon of water. It is also expressed as percent lime solids, although some confusion can arise unless the dry-weight basis (oxide or hydroxide) is specified. Lime suspensions are commonly referred to in three categories of concentration:

- *Paste* (as prepared by paste-type slakers). A wet, lime, puttylike mass containing about 30% to 35% quicklime with sufficient body to exhibit some angle of repose.

- *Slurry* (as prepared by slurry-type slakers). A creamy suspension containing about 20% to 25% quicklime that may be poured or pumped.

- *Milk-of-lime* (ready for application). Waterlike consistency containing up to 1.5 lb/gal (180 g/L) hydrated lime. Milk-of-lime, the final diluted product from paste and slurry slakers and dry feeder slurry pots, is readily pumpable.

A typical waterworks milk-of-lime ready for application contains between 1 and 1.5 lb/gal (2 and 3.2 g/L) of hydrated lime (10.7% to 15.2%). At 59° F (15° C) these suspensions exhibit specific gravity readings of 1.07 and 1.11, respectively. Lime concentration can readily be checked by using a hygrometer and specific gravity tables.

The formation and handling of lime slurries involve both chemical and mechanical considerations. Lime slurries result from adding controlled amounts of water to either slaked quicklime or hydrated lime. A slurry of reasonable purity is not corrosive and is relatively easy to keep in suspension, provided that the slurry has been stabilized. Stabilization occurs when all chemical reactions between the water used and the calcium hydroxide have been completed. This essentially is a softening reaction.

If slurries are transported before stabilization is complete, dissolved solids contained in the dilution water will precipitate on piping, valves, and pumps as a scale. A detention time of 15 min in an agitated slurry storage tank is usually sufficient for maximum development of salt crystals, which tend to grow on themselves. By maintaining suspension, these crystallized scale-forming compounds can be pumped and transported to the process with a minimum of problems. Conversely, attempts to pump directly from hydrated lime-wetting tanks or from slaker sumps can produce intolerable maintenance problems.

Agitated detention or stabilization of lime slurry does not resolve all problems. Quicklime slaked on-site contains fine settleable impurities that can be highly abrasive. This material is usually too fine to remove by slaker grit removal systems, so purchase specifications should emphasize or reward purity of product. Water quality for slaking and slurry dilution is also important, particularly with respect to the amount and type of dissolved solids present.

Water of medium or high hardness or containing sulfates is particularly objectionable in that it is likely to interfere with the hydration process itself. In such cases it may be feasible to consider installing water-softening equipment of ample capacity to satisfy water supply needs for slaking and slurry dilution. In less severe cases, the addition of 5 to 10 mg/L of hexametaphosphate to the supply line sequesters the mineral hardness present when a 2- or 3-min detention time is provided before reaching the slaker.

Once the quicklime has been slaked or the hydrate wetted to form slurry, it is essential that it be discharged as directly as possible into an agitated storage tank, where mixing and stabilization occur. Mixers must be designed to maintain the impurities and precipitated salts, as well as the lime, in suspension. As a general rule, about 0.5 to 1 hp (370 to 746 W) per 1,000 gal (3,785 L) of tank capacity is required for cylindrical tanks of less than 3,000-gal (11,400-L) capacity and slurry concentration not exceeding 1 lb/gal (120 g/L). Smaller tanks are usually equipped with propeller agitators; tanks larger than 2,000-gal (7,570-L) capacity resort to turbine-type mixers. Speeds vary from approximately 350 rpm for propeller mixers to 100 rpm or less on larger turbine units.

Because most slurry storage tanks are agitated continuously, care must be taken to position mixing impellers in the lower portion of the tank when slaker operations are automatically controlled by level controls within the tank. Under no circumstances should the impeller be located so that exposure above the liquid surface is possible. Tanks should be equipped with peripheral baffles set 90° or 120° apart to control vortex formation. Baffles should terminate above the bottom and be attached to the sides with intermittent spaces to prevent lime buildup in corners. Level sensors for slaker control, probes or sonic devices, and alarms are usually required. Float-actuated controls are usually impractical, even with stilling wells, because of heavy solids buildup that results.

One of the principal problems encountered with small systems using gravity feed lines for unstabilized slurry is the periodic line clogging because of low velocity. As a result, a compromise solution to the problem is to design the line for easy cleanout, removal, or flushing. Other measures commonly used to convey lime slurry include

- Polyphosphates in dilution water
- Flexible hose to allow breakup of scale
- Open troughs or channels
- Duplicate lines and passing chlorine solution through these lines on an alternating schedule

On larger installations it is preferable to employ pumps that circulate the slurry to the point or points of application and back to storage. A properly designed system has the pump and piping sized to produce a minimum velocity of 3 to 4 ft/s (0.9 to 1.2 m/s) in the return line. At each application point, several methods can be used to control the flow of slurry to process:

- Control valve, manually operated or automatically modulated by an external controller analyzer
- Discharge into a local day tank equipped with agitator and metering pump; tank refills through an automatic valve actuated by level sensors

The recirculating line must be run as close as possible to the points of application to minimize any additional length of piping where optimum velocities cannot be maintained. Takeoffs that are periodically closed, as in refilling day tanks, must be tapped and valved off the top of the pipe to prevent settling in the static line behind the closed valve. Throttling valves in continuous service may be tapped from the side or bottom of the line. Flow stability through these diversion valves requires some back pressure in the recirculating line; this is normally accomplished by installing a throttling valve in the line as it returns to storage.

Although ball valves and plug-type valves perform reasonably well in open-close service, they are not recommended for throttling service. Similarly, diaphragm valves tend to collect compacted sludge behind their weirs. Pinch valves operate well in throttling service, provided they are properly sized and not subject to a vacuum. These valves are available with manual, electric, hydraulic, or pneumatic operators. Flow modulation is possible with each type of power operator.

Mild steel piping is satisfactory for most slurry lines where rigid piping is preferred. Tees and crosses should be used as elbows to facilitate cleaning. A freshwater flushing system should be installed to flush out piping, pumps, and valves when the system is shut down for any reason. Reduction of line diameter must not be so abrupt that it causes a violent hydraulic disturbance that can result in dewatering and compaction of the lime.

Pipeline designers usually use a coefficient of $C = 100$ for slurry lines carrying up to 3 lb/gal (594 g/L) of hydrated lime. Piping should not be excessively oversized initially to accommodate estimated future system capacity. The penalty for oversizing is usually increased maintenance problems during the early years. Dissolved solids tend to be precipitated out of the process water where lime slurry or any other alkaline substance is applied. For this reason, an air gap is preferable between the end of the feed line and the surface of the water being treated. If a submerged application cannot be avoided, some means must be provided for periodic cleaning of the end of the slurry pipe to break up the precipitated mass, which may eventually plug the feed line.

Slurry Pumps. Slurry pumps generally fall into two categories: centrifugal pumps and positive displacement, or controlled-volume, pumps. Centrifugal pumps are generally employed for low-head transfer or recirculating service. With proper selection of casing and impeller material and an appropriate shaft seal, satisfactory service can usually be attained at a reasonable cost. Replaceable liners and semiopen impellers are preferred. It is important that the pump design allow easy dismantling for cleanout and repair. Lime slurry requires the lowest speed of rotation (1,725 rpm or less) consistent with hydraulic requirements to control impeller plating. Using water-flushed seals on centrifugal lime slurry pumps is not recommended, because it usually results in localized scaling.

Controlled-volume pumps are typically used where metering or positive control of slurry flow is required, such as at the point of application to the process. Several types are available for slurry service. Piston-type pumps where the slurry is in direct contact with the cylinder walls are not recommended for slurry service because of uncontrolled wear and abrasion. Similarly, peristaltic or squeeze-type pumps are subject to wear and excessive tubing replacement and are not normally used for slurry pumping.

Dipper-Wheel Feeders. The rotating dipper feed has been a longtime favorite for feeding slurries where gravity feed is possible between the feeder and the point of application. The feeder consists of a tank in which the slurry level is maintained by a float valve (or overflow weir if gravity return to slurry storage is practical), a dipper wheel with variable-speed drive, and a totalizer to register wheel revolutions. The dipper wheel is usually divided into eight segments, or dippers, each containing about 500 mL of slurry liquid. As the wheel rotates, an agitator bar maintains the slurry in suspension. The inlet float valve to the tank must be routinely cleaned, particularly if it is connected to a pressurized slurry recirculating system. An overflow connection must be provided when a float valve is used; the selection of an appropriate discharge point for this overflow requires planning for each project. Discharging the overflow to the process is not recommended. The overflow line is not required where gravity return of excess slurry to storage is possible using an overflow weir instead of the float valve.

Flow Control Valves. Sensitive flow control valves are capable of being positioned or modulated by an external control signal. The valve functions to control the flow of chemical to the process in proportion to the signal output analyzer. For this type of feed system to operate satisfactorily, chemical must be supplied to the valve under reasonably constant head, free of suspended material that could clog the valve. Because this constitutes what is referred to as a closed-loop system, it is imperative that the response and sample detention time of the overall control loop be properly designed to prevent cyclic overfeeding and underfeeding.

Splitter Boxes. Custom-designed splitter boxes can provide a constant ratio of slurry discharge rates by gravity flow to multiple application points. For example, one splitter box with three weirs discharging to separate troughs can supply lime slurry to three different basins. The weir design, which can provide for adjustable weirs, dictates the flow split ratios.

Feed and Metering Systems for Polymers. Many treatment plants feed commercial-strength liquid polymer direct from shipping containers or storage tanks, or manually prepared dry polymer solutions from batch-mixing tanks. The relatively high chemical cost requires maximum activation and minimum waste of polymers in solution, and preengineered polymer feed equipment can often achieve both goals.

Several preengineered equipment packages automatically mix and dilute commercial-strength liquid polymers and deliver a diluted (and in some cases, aged) polymer solution to the point of application. These packages typically include a positive displacement pump, check, pressure relief and back pressure valves, a dilution water flow control valve, an integral mixing chamber, and feed rate control instrumentation. Many of these equipment packages require the operator only to connect hoses to the inlet and outlet ports and the power plug to an electrical outlet.

Semiautomatic and fully automatic batch-mixing and solution feed equipment is also available for dry polymers. Skid-mounted equipment packages typically include a dry polymer storage hopper, dispenser and conveyor (pneumatic or hydraulic), dust collector, mix tanks and agitator, an aging tank, a water flow control valve, positive displacement pumping equipment to deliver the polymer solution to the point of application, and all necessary instrumentation for the batching process and solution feed rate control.

Equipment manufacturers also provide hybrid equipment packages that can accommodate liquid or dry polymer stock. These have manual, semiautomatic, or fully automatic control features and in some cases include redundant mix tanks or storage tanks for aging and activation of the polymer solution. These equipment packages typically include solution transfer and feed pumps in addition to the stock liquid polymer pump or the dry polymer eductor.

Ancillary Feed Equipment. This section describes several equipment items necessary for a complete chemical feed system.

Transfer Pumps. Transfer pumps recirculate during dilution of chemical solutions in the storage tanks, move the chemicals from storage to day tanks, or, for some polymer batching systems, transfer solutions from a mixing tank to an aging tank. These pumps are generally the low-head centrifugal type, with a capacity several times greater than the maximum application rate of the process. Progressive cavity or controlled-volume units are sometimes used for transfer pumping where liquid polymers or other shear-sensitive liquids are handled.

Materials for fabrication of pump liners, body, impeller, and shaft depend on the corrosive and abrasive nature of the material pumped. In some instances, temperature can be a factor if nonmetallic parts are involved. When one is ordering a pump for chemical service, it is best to consult with the pump supplier and check with the chemical manufacturer. Total reliance on published tables of corrosion- or abrasion-resistant materials for various chemicals can be misleading.

In itself, speed is not a major design factor where true solutions are being handled. However, slurries, viscous liquids, and polymer solutions require special handling and use of low-rpm pumps. In no case should a centrifugal-type slurry pump operate in excess of 1,750 rpm. Liquid, undiluted polymers may be transferred using a 1,750-rpm pump; however, once dilution and aging have taken place, the product is subject to molecular shearing if it is transferred in a centrifugal pump operating at any speed. Screw-type progressive cavity pumps are recommended for this purpose.

Generally, pumps designed to use flushing water on shaft seals should not be used for chemical transfer. Chemical dilution, scaling in lime slurries, and the expense of furnishing a non–cross-connected seal water system are some of the problems encountered. Pumping of carbon slurry is one of the few exceptions.

Modern technology has made available a variety of dry-mechanical-shaft seals to handle most types of liquids and slurries. In addition, indirect magnetic drives are available for smaller pumps.

Day Tanks. Day tanks generally minimize the amount of hazardous chemicals that may be stored within a chemical feed area or minimize the amount of chemicals that will be lost if there is a rupture in the suction lines between the storage tank and the chemical feeders. The use of day tanks is primarily a safety issue and applies to hazardous chemicals such as acids or caustic soda.

In addition, day tanks locate stored chemicals closer to pulsating diaphragm metering pumps to minimize losses in suction piping, provide a near-constant suction head, and allow for proper operation of the metering pump. Locating day tanks close to pumps is useful when bulk storage tanks are relatively remote because of chemical delivery and unloading constraints. Day tanks are typically sized to provide 24 h of chemical storage based on the maximum chemical volumes metered to the application points. Smaller day tanks can limit the volume of hazardous chemicals and subsequently prevent the area from receiving a more stringent hazardous classification under building and fire codes.

The filling mechanism between the bulk storage tank and the day tank should be carefully designed to prevent overfilling the day tank. Controls include automatic shutdown of transfer pumps or valves controlling the flow between bulk storage tanks and the day tank, and alarms on high levels in the day tanks. Visual and electronic level measurements are recommended, and secondary containment of day tanks should also be provided. Some systems require an operator to hold a safety switch to continue the transfer operation.

Day tanks can be mounted on scales to monitor feed rates and calibrate chemical metering equipment. Materials used for day tanks are generally the same materials used for bulk storage.

Piping and Conduits. Design of piping systems to convey chemicals must include provisions for redundancy, isolation, and maintenance. These provisions are essential when one considers problems that can occur from deposits and precipitates when eductors and positive displacement pumps are used to convey slurries. Eductors and piping that handle slurries should be sized to maintain minimum velocities to minimize solids deposition, especially for powdered activated carbon slurries. Flushing provisions are desirable for systems with slurries or polymer solutions to evacuate chemicals in the piping and prevent deposition or solidification of chemical solids when the system is off for any length of time.

Suction lines between supply tanks and pumps must be kept as short as possible, and the entire system should be located close to the application point. Where possible, it is important to avoid a layout that permits siphoning by simultaneous existence of positive suction pressure and low or negative discharge pressure on the pump. Back pressure valves installed on the discharge line prevent this condition and can help maintain the accuracy of discharge on diaphragm metering pumps. Back pressure valves should not be used for slurry applications.

Piping system design for use with oscillating or reciprocating pumps must be carefully prepared to avoid cavitation in the pump chamber and vapor lock in the suction piping from low suction pressures below the net positive suction head (NPSH) required. These effects occur because of the inertia effect on the liquid column from the movement of the piston or diaphragm as it changes direction on either end of the stroke.

The NPSH required must be established for the chemical being conveyed and then compared with the available NPSH based on piping head losses calculated for both viscosity head loss and acceleration head loss. The higher of these two head losses is used to calculate available NPSH. Suction chambers located close to metering pumps can minimize head losses in suction piping due to pulsating flows. Discharge pressures based on piping head losses must also be determined using the higher of these two types of head loss.

Table 15.4 is a condensed listing of common water treatment chemicals and the corrosion resistance of various types of piping materials at temperatures up to 104° F (40° C). It is important to remember that temperature and operating pressures are equally important parameters in piping materials selection, particularly when using plastic materials. With plastic materials, temperatures in excess of 104° F (40° C) generally reduce the maximum safe working pressure rating of the pipe.

The designer should consult manufacturers' data for special plastics acceptable for temperatures up to 176° F (80° C). All chemical piping should be schedule-80 thickness, particularly if threaded joints are used. Double-walled pipe with an annular space between the carrier pipe and the containment pipe should be considered for aggressive chemicals such as acids or caustic soda. This design minimizes the release of aggressive and hazardous chemicals to the environment in the event of a leak. In many cases, a leak detection mechanism can be included to provide warning that a leak of the chemical into the annular space has occurred.

Unions or flanges and suitable isolation valves must be provided at each pump or feed-controlling device to permit removal for routine maintenance. Plastic pipe fittings sized 3 in. (80 mm) and smaller are typically the solvent-welded type. Plastic pipe fittings 4 in. (100 mm) and larger are similar, except that pipe joints may use solvent-welded flanges instead of sleeves. Threaded pipe fittings are not generally recommended.

Valves used for open-close service should be the straight-through pattern, maintaining full-line size. Plug, ball, and diaphragm valves fabricated of the appropriate material are normally used for solutions free of suspended matter. Pinch valves are capable of handling slurries in both shutoff and modulating-control service. The relatively complicated internal parts of globe and gate valves would be exposed to chemical flow, so their use is not recommended.

It is good practice to support rigid plastic piping on hangers at intervals of 4 ft (1.2 m) or less. Piping must not be clamped by the hanger, because movement resulting from expansion and contraction should not be restricted. Flexible piping and rubber tubing are best supported by channel troughs or sections of steel piping supported on hangers or brackets. In these cases, gaps should be left in the support piping at bends and changes of direction to facilitate tubing installation and removal.

Where rigid metallic piping is used for the solution line, support hangers or brackets should be provided at intervals that prevent sagging for the size of pipe used. In any case,

TABLE 15.4 Piping Material Recommendations

	Piping material								
Chemical	Iron or steel	Type 316 stainless	Type 304 stainless	Copper	PVC—type 1	Fiberglass-reinforced polyester (FRP)	Polypropylene	Rubber tubing	Glass
Activated carbon (slurry)	NR	X				X		X	
Alum	S	S	NR		S	X	X		X
Ammonia, aqua	S	X						X	
Calcium hydroxide (slurry)	S	X	X		X		X	X	
Calcium hypochlorite					X		X		X
Carbon dioxide (dry)	S	X	X	X	X	X	X	X	
Chlorinated copperas					X			X	
Chlorine (dry gas)	S			X	NR		NR		
Chlorine solution	NR	NR	NR		S	X		X	
Chlorine dioxide (3% soln.)					X				X
Coagulant aids	Consult manufacturer—generally not corrosive								
Copper sulfate		X	X		S	X	X	X	
Dolomitic lime (slurry)	X	X	X		X		X	X	
Ferric chloride	NR	NR	NR	NR	S	X	X	X	X
Fluosilicic acid	NR	NR	NR	NR	X		X		NR
Hydrochloric acid	NR	NR	NR		X	X	X	X	
Potassium permanganate (2% soln.)	X	X	X		X	X	X		

Key: S = Industrial standard or excellent for handling
X = Suitable for handling
NR = Not recommended

TABLE 15.4 Piping Material Recommendations (*Continued*)

Chemical	Iron or steel	Type 316 stainless	Type 304 stainless	Copper	PVC—type 1	Fiberglass-reinforced polyester (FRP)	Polypropylene	Rubber tubing	Glass
Sodium carbonate (soln.)	S				X	X	X	X	
Sodium chloride		X			X	X	X		X
Sodium chlorite					X	X	X		X
Sodium fluoride (1% to 5% soln.)		X			X	X	X	X	
Sodium hexametaphosphate (soln.)		X			X	X		X	
Sodium hydroxide (to 50% soln.)	X	X	X		X	X		X	
Sodium hypochlorite (to 16% soln.)					S		X	X	X
Sodium silicate	S	X	X		X	X	X	X	
Sodium silicofluoride			X		X			X	
Sulfur dioxide (dry gas)	X	X	X						X
Sulfur dioxide (soln.)		X							
Sulfuric acid (conc.)	S								
Sulfuric acid	NR				S	X	X	X	X

Key: S = Industrial standard or excellent for handling
X = Suitable for handling
NR = Not recommended

supports should be provided at least every 10 ft (3 m). Burying chemical piping, particularly under concrete slabs, is not desirable due to the possibility of undetected leaks. Piping trenches can be used to convey piping below grade while still allowing access to the pipe for inspection and repairs.

Rotameters. Rotameters indicate flow in conjunction with flow control valves and can become part of a feed control system. Certain types of rotameters can be equipped with rotor position transmitters to permit remote recording of flow rates. Rotameters more generally provide a visual indication of flow, of either water or chemical solutions, to assist in manually setting flow rates for dilution or carrying water.

Visual rotameters, where the float position can be seen relative to a calibrated scale, are practical for clear liquids free of iron or other impurities that obscure scale calibrations. Where liquids do not allow visual indication, magnetically actuated indicators may be used. Branching chemical feed streams through more than two or three rotameters to different application points usually requires continuous manual flow rate adjustment. A more reliable method of feeding multiple application points is to use separate feed pumps dedicated to each application point.

Miscellaneous Feed Systems. Numerous other methods and devices, some of which date back to the 19th century, can be used to feed liquid chemicals or to feed dry chemicals that are dissolved or slurried. Many of these devices are still in operation at smaller plants throughout the country, in testimony to their simplicity and adequacy for the job at hand. Some of the more common chemical feed devices are described in the following paragraphs.

Dissolving Tanks. Dissolving tanks have a perforated wooden or metal basket filled with a weighed quantity of the chemical to be dissolved and hung in the upper part of a water-filled tank. Mechanical mixers can be installed to assist in dissolving the chemical. When all material is dissolved, the batch is ready for use, and it is applied by pumps or other flow control devices. This method is popular for handling glassy polyphosphates in smaller installations and for other lumpy materials that dry feeders do not handle well.

Pot Feeders. Pot feeders are normally used with coarse chemicals, such as lump alum. The pot is charged manually through a pressure-tight cover. An orifice plate or gate valve in the pressure line to be treated generates a differential pressure that is tapped off the line and permitted to flow through the pot via a flow control valve. In theory, the chemical dissolves and maintains a constant saturated solution in the pot. As the chemical is consumed through the pot, it is displaced and replenished in proportion to the flow. However, the system is subject to clogging with chemical impurities.

Gravity Orifice Feeders. Gravity orifice feeders generally consist of a constant-head supply tank and a fixed or adjustable orifice to deliver the desired rate of flow. These devices can be started and stopped remotely, but they do not lend themselves to slurry feeding or to proportional feed control.

Displacement or Decanting Feeders. A decanting feeder consists of a tank equipped with an overhead mechanism that lowers a displacement cylinder or decanting pipe arm into a prepared batch of chemical solution. The rate of feed flowing over the overflow weir or through the decanting pipe is related to the speed at which the cylinder or pipe is lowered. Mixers can be used if slurries are to be fed.

Utility Water. High-pressure, or utility, water has several uses in chemical feeding systems, including diluting chemical solution, flushing chemical feed systems, providing carrying water for neat chemicals, and washing down chemical feed areas. This water is generally obtained from plant-pumped finished water systems or from the distribution system served by the water treatment plant. Use of this water within the chemical feed area requires the precautions described below to maintain the drinking water status of the plant utility water system. It may be useful to install a flowmeter on this water supply to track in-plant water usage.

Dilution of Chemicals. Many chemicals, such as caustic soda and polymers, are delivered in concentrations inappropriate for direct use in unit processes. Utility water reduces the concentration of the chemical to an acceptable level before delivery of the chemical to the application point. Polymers are unique in that they must be properly activated in the dilution water before reaching the application point. Many methods of activating polymer are available, but they all generally involve adding dilution water to neat polymer.

Because caustic soda has a high freezing point, it is diluted in many areas of the country to prevent freezing under normal seasonal temperatures. Care must be taken with dilution water used for caustic soda (or other liquids with a high pH) because of its tendency to precipitate calcium and the exothermic nature of the dilution reaction. It is recommended that dilution water used with high-pH solutions be softened before use.

The amount of dilution water used must be carefully calculated to know the concentration of the diluted chemical and to calculate the feed rate required for the desired dosage. In many cases a flowmeter for batch mixing is used on the dilution water to provide the exact quantity of water necessary for the desired dilution ratio.

Flushing Feeders and Chemical Feed Lines. For most chemicals, it is not desirable to allow a concentrated solution to remain for long periods in chemical metering equipment and chemical lines. For polymers, slurries, some coagulants, and chemicals that cause precipitates or crystals to form in the pumps and piping, a flushing water system is recommended. This system flushes chemical metering equipment and downstream chemical pipelines whenever chemical metering is disrupted for any length of time.

Control systems can be designed to automatically flush the feed equipment and piping whenever a feed system is turned off. Flushing time should be sufficient to completely evacuate the discharge pipeline of all chemicals to the application point.

Conveying Water. It is sometimes desirable to meter a chemical dose into a stream of water to convey the chemical to the application point. The amount of water used to carry the chemical is not necessarily important, but a flushing velocity of 2 to 5 ft/s (0.6 to 1.5 m/s) may be desirable. Chemical is metered in its neat form with the dose dependent on the plant flow rate and the amount of chemical fed, not the amount of carrying water used.

A typical example of this condition is the use of an eductor or injector (jet pump) to provide the motive force for moving the chemical from its metering device to the application point. Care must be taken if carrying water is to be used with alum because the hydrolyzing chemical reaction begins to occur as soon as the alum is added to water. In most cases, the carrying water is metered to determine the volume of water so that it can be kept to a minimum while still maintaining adequate velocities in the solution pipeline.

Carrying water is almost always used with gaseous chlorine. The carrying water, in passing through an injector, creates a vacuum that draws gaseous chlorine from the storage tanks through a vacuum regulator. Chlorine is not controlled through carrying water but by other means, such as controlling gas flow in the chlorinator. Carrying water provides the means of creating a vacuum and then mixing gaseous chlorine with water to create a chlorine solution. The quantity of carrying water for chlorine is dictated by the amount necessary to create a vacuum in the injector against the back pressure created in the downstream piping from the carrying water flow.

Conditioning and Heating Requirements. Whenever water is mixed with chemicals, a reaction occurs that may or may not create a reason for concern. In the case of ammonia or caustic soda, water should be softened before adding the chemical to prevent precipitation of calcium or magnesium carbonates and ultimately clogging of the line downstream of the addition point. Softened water should also be used for any chemical that significantly elevates solution pH or alters the chemicals' solubility. Commercial ion exchange equipment can generally soften utility water adequately for mixing and conveying most chemicals used in water treatment.

Some chemicals go into solution more readily or in higher concentrations in warm water than cold water. These chemicals include dry polymers, ferric sulfate, sodium silicofluoride, soda ash, and some quicklime (slaking rather than dissolving). Blenders may combine hot water with utility water to achieve a preset temperature before mixing with the chemical. Care must be taken to adequately size hot water heaters in this application to ensure that an adequate quantity of hot water is available for all chemicals that need water with an elevated temperature. Hot water is also desirable for use in washing down polymer feed areas. Spilled polymers are more easily flushed with hot, rather than cold, washwater.

Cross-connection Control. Whenever utility water is used in direct connection with chemical feed systems, cross-connections are a concern. If negative pressures are created in the plant water system, chemicals can be drawn back into the potable water system and contaminate the drinking water. Most health agencies require reduced-pressure backflow prevention devices to be installed in all utility water systems that are connected to non-potable water systems, such as chemical feed systems. A careful analysis will identify all potential cross-connections and ensure installation of acceptable backflow prevention devices between the potential cross-connection and the potable water system.

Diffusion of Chemicals at Application Points

It is critical to the efficient use of chemicals in treatment processes to adequately disperse the chemical throughout the flow. When a chemical is applied in turbine rapid mixers or static mixers, the chemical should be injected into the center of the flow stream immediately upstream of the mixing devices. The mixing devices then disperse the chemical evenly throughout the flow. However, in many cases a mixing device is not used, and the chemical must be dispersed evenly throughout the flow via other means such as diffusers and the hydraulics of the conduit or basin where the chemical is added.

With the exception of carbon dioxide and ozone, most chemicals are brought to the application point in the form of liquid solutions or slurries. Although these gases can be injection-fed, carbon dioxide and ozone are usually direct fed as fine bubbles to the process water through relatively deep contact tanks. Typically, a counterflow (water flowing downward while the bubbles rise) is used with fine bubble diffusers to provide efficient transfer of the gas to the water.

For applying solution chemicals to both pipes and channels, the chemical is usually dispersed in the flow using diffuser pipes with multiple orifices. Figure 15.11 shows several of the more common diffusers used in dispersing chemicals. If chemicals are in slurry form, dispersion is usually accomplished using an open-ended pipe for pipe flow or a trough with notches or holes at the bottom for channel flows.

When disinfection is entirely by free chlorine, it is important to distribute the chlorine dose uniformly across the flow cross section. Traditionally, good designs have accomplished this with open-channel or pipeline diffusers. It can be shown, however, that truly complete mixing across the flow cross section only takes place some distance downstream from diffusers of this type. When free chlorine residuals are used, this may not be a critical problem.

When water contains significant amounts of ammonia (10% of the chlorine dose or more), initial mixing takes on new importance. Under these conditions there is evidence that chlorine is more effective during the first few seconds after its addition. With ammonia present, failure to mix the chlorine solution rapidly with all the water can result in

Type A: In-line Diffuser

Type B: In a Tank Inlet or in a Channel

Type C: In-line (Removable)

Type D: In a Tank Inlet or in a Structure

Type E: Gravity Feed from a Trough Above Water Surface

Type F: Fine-bubble Gas Diffusers

FIGURE 15.11 Types of chemical diffusers.

a break-point reaction with part of the water and little or no reaction with the remainder. The break-point reaction is not reversible. Thus part of the water receives excess chlorine, and the remainder receives little or none. Consequently, good initial mixing becomes a much more important process step, and initial mixing devices should be designed to spread chlorine throughout the flow cross section in the shortest possible time with the least possible backmixing. This can be accomplished by in-line mixers, gas aspirator devices, a hydraulic jump, or a pump diffuser system. Flash-mixing chambers such as those used for coagulants in a conventional water treatment plant are not satisfactory because they result in backmixing.

Careful attention must be given to the dispersion of scale-producing chemical solutions that can clog diffuser pipe orifices. These chemicals include metal salt coagulants, such as alum or ferric chloride, and alkali chemicals, such as caustic soda, soda ash, lime-saturated solutions, and ammonia solutions. For these types of chemicals, provisions must allow cleaning of the diffusers.

Redundant diffusers may also be included to allow one diffuser to be taken out of service for cleaning while the other diffuser remains in operation. A relatively unique self-cleaning design for ammonia solution diffusers uses short lengths of Teflon tubing at each orifice. These lengths of tubing move in the flow stream and can break off scale products that form at the end of the tubing where the solution contacts the main process water.

DESIGNING FOR SAFETY AND HAZARDOUS CONDITIONS

A general overview of safety requirements and suggested protective measures for handling various chemicals is presented in Table 15.5. Suggested protective measures shown in the table apply to the various chemicals used whether they are in dry, liquid, or gaseous state. Refer to the chemical supplier's most current Material Safety Data Sheet (MSDS) for up-to-date chemical safety requirements and protective measures. See below for a complete description of an MSDS. Another useful general reference for chemical safety, exposure limits, and incompatibilities is the *NIOSH Pocket Guide to Chemical Hazards* (National Institute of Occupational Safety and Health, 2003).

With all safety procedures, a written program should be developed and reviewed regularly with all personnel. This establishes what should be done in emergency situations before they occur. Every new employee who may be expected to use the equipment should be instructed in safety emergency procedures. Documented periodic review and training sessions are recommended. These may be mandatory under worker safety, transportation, or environmental regulations.

Federal, State, and Local Regulations and Codes

At present, regulatory requirements related to worker safety, public safety, and environmental protection are in a state of flux, as are many basic standards for drinking water safety. The use of many water treatment chemicals is subject to overlapping regulation by local, state, and federal statutes and codes. Many water treatment chemicals are classified as hazardous materials and regulated accordingly.

Early in the design process it is highly advisable to consult with appropriate agencies to determine which have input or enforcement authority over the project. These typically include local and state public health agencies, local and state environmental agencies, state or regional federal Occupational Safety and Health Administration (OSHA) offices, and local fire marshals. Federal OSHA's jurisdiction is the safety and health of private as opposed to public works, sector workers and employees, and related workplace conditions; however, some states have adopted and enforce federal OSHA regulations for municipal and other government employees. Local fire departments are concerned with acute effects of a chemical release both on-site and off-site. Failure to address concerns of these authorities early in the treatment plant design could lead to costly changes later in the design process.

Some states have their own OSHAs authorized to enforce state regulations instead of federal OSHA regulations. These state regulations may be more restrictive than the federal OSHA requirements cited in this chapter. Sources of assistance in identifying pertinent agencies include staff environmental and safety personnel available to the design

TABLE 15.5 Protective Measures for Water Treatment Chemicals

Chemical (D = dry; L = liquid; G = gas)	Positive ventilation	Protective clothing	Neck cloths	Gloves	Rubber boots	Rubber gloves	Goggles	Face shields	Rubber aprons	Respirator	Gas mask	Avoid skin contact	Safety shower and eye baths	Remarks
Activated alumina (D)	■													Store away from gasoline, mineral or vegetable oils, calcium hypochlorite (HTH), lime, sodium chlorite, or potassium permanganate
Activated carbon														
Powder (D)	■	■	■											
Granulate (D)	■	■	■											
Alum sulfate (D)				■			■			■		■		
Alum sulfate (L)				■			■			■		■		Similar to other acid
Ammonium hydroxide (L)	■					■	■	■	■		■	■	■	Moist NH$_3$ reacts with many metals and alloys—liquid contact produces burns
Ammonium sulfate (D)			■									■	■	See alum sulfate above
Anhydrous ammonia (G)	■	■	■											Fire sprinklers and water hoses effective in removing gas
Bauxite (D)	■													
Bentonite (D)	■													
Calcium carbonate (D)	■													
Calcium hypochlorite (D)	■					■		■		■		■		
Carbon dioxide (G)	■													
Chlorine (G)	■													Avoid contact with hydrogen or organic compounds or other flammable materials
Chlorine dioxide (G)		■			■	■		■		■		■	■	Solution is corrosive
Copper sulfate (D)		■				■		■		■		■	■	Very corrosive
Ferric chloride (D)		■				■		■		■		■		Very corrosive
Ferric sulfate (D)		■	■			■		■		■		■		
Ferrous sulfate (D)		■	■			■		■		■		■		
Ferrous sulfate (L)		■	■			■		■		■		■	■	

TABLE 15.5 Protective Measures for Water Treatment Chemicals (*Continued*)

Chemical (D = dry; L = liquid; G = gas)	Positive ventilation	Protective clothing	Neck cloths	Gloves	Rubber boots	Rubber gloves	Goggles	Face shields	Rubber aprons	Respirator	Gas mask	Avoid skin contact	Safety shower and eye baths	Remarks
Fluorosilicic acid (L)	■	■	■			■		■	■			■	■	
Fluorspar (D)	■	■	■					■				■	■	Have lime slurry on hand
Hydrated lime (D)	■	■	■	■				■		■		■	■	Etches glass when moist
Hydrochloric acid (L)	■	■	■			■		■	■	■		■	■	Can burn eyes or skin
Iron-exchange resins (D)														Hydrogen cation resins acid
Ozone (G)	■											■	■	
Potassium permanganate (D)	■	■	■	■		■	■	■				■	■	Large quantities present fire hazard
Quicklime (D)	■	■		■			■							Can burn eyes or skin
Sodium aluminate (D)	■	■		■			■							
Sodium aluminate (L)	■	■		■			■							
Sodium bisulfate (D)	■	■		■			■					■		
Sodium carbonate (D)	■	■	■					■	■	■		■	■	Can dehydrate skin
Sodium chloride (D)	■		■				■		■					
Sodium chlorite (D)	■	■	■						■	■		■	■	Rinse any spills immediately with water
Sodium fluoride (D)	■						■					■		
Sodium polyphosphate, glassy (D)	■													
Sodium hydroxide (D)	■						■	■		■		■	■	
Sodium hydroxide (L)	■				■	■						■		
Sodium hypochlorite (L)	■	■		■			■					■	■	
Sodium silicate (D)	■	■		■			■	■				■		
Sodium fluorosilicate (D)	■	■	■									■		
Sodium sulfite (D)	■													
Sodium dioxide (G)											■	■		
Sulfuric acid (L)	■	■				■		■	■			■	■	

TABLE 15.6 EPCRA Threshold Planning Quantities

Chemical	Threshold planning quantity (lb)
Chlorine	100
Chlorine dioxide	Not listed
Anhydrous ammonia	500
Aqua ammonia	Not listed
Hydrogen peroxide (conc. 52% or greater)	1,000
Sulfuric acid	1,000
Ozone	100

team; state-specific compliance guides developed by state agencies, chambers of commerce, or private publishers; and outside consultants.

Disclosure requirements may be necessary under the Emergency Planning and Community Right-to-Know Act (EPCRA), also known as Title III of the Superfund Amendments and Reauthorization Act (SARA) of 1986 (42 USC 11,001). Sections of EPCRA require facilities with chemicals above the thresholds given in Table 15.6 to report to their State Emergency Response Commission (SERC) and coordinate with their appropriate Local Emergency Planning Committee (LEPC). This can include providing an annual copy of their chemical inventory to the SERC, the LEPC, or the local fire department. Construction of a new facility exceeding these amounts will require notification.

Transportation, loading, and unloading of hazardous materials are also regulated by the U.S. Department of Transportation (49 CFR 171 through 180).

An additional subcategory of hazardous materials mandated by the Clean Air Act Amendments of 1990 [Section 112(r), 42 USC 741(r)] receives additional regulatory requirements because of their particularly acute ability to injure workers, the public, or the environment. Table 15.7 lists typical water treatment chemicals so regulated. All these laws stem from the 1984 toxic cloud in Bhopal, India, which killed or injured thousands of individuals, and from subsequent petroleum refinery and chemical transportation accidents.

TABLE 15.7 Selected Accidental Release Program Regulatory Thresholds

Chemical	USEPA threshold quantity (RMP)(lb)	OSHA threshold quantity (PSM)(lb)
Chlorine	2,500	1,500
Chlorine dioxide	1,000	1,000
Anhydrous ammonia	10,000	10,000
Aqua ammonia	20,000 (conc. 20% or greater)	15,000 (conc. 44% or greater)
Hydrogen peroxide	Not listed	75,000 (conc. 52% or greater)
Sulfuric acid	Not listed	Not listed
Ozone	Not listed	100

Storage or use of chemicals above these threshold quantities may be subject to OSHA's current Process Safety Management (PSM) regulation (29 CFR 1910.119), to the USEPA Risk Management Plan (RMP) regulation (40 CFR 68), or to both.

OSHA requires evaluation and minimization of risks to workers (within the facility itself, or "inside the fence") through safety programs; USEPA scrutinizes public health and environmental effects (beyond the facility boundaries, or "outside the fence") using air dispersion modeling studies. Both regulations share an identically defined requirement for a detailed failure and release evaluation of the entire system containing the listed chemical. This evaluation is called a Process Hazard Analysis (PHA).

Because of the potential costs and effects of these regulations, in 1995 the American Water Works Association Research Foundation funded development of a water industry–specific guidance document entitled "Compliance Guidance and Model Risk Management Program for Water Treatment Plants" (Project Number #272), published in 1998. General technical guidance publications are also available from the American Institute of Chemical Engineers.

If a facility is required to comply with PSM or RMP, the design team should schedule project time and funds to conduct a PHA on the design. A PHA can be accomplished using one of several alternative techniques and normally requires a group approach with both engineering design and facility operations expertise. The results and recommendations of the PHA can directly and indirectly affect plant design and operations, equipment options, treatment chemical management, and even community relations.

Before the process equipment is started up for the first time, a formal Pre–start-up Safety Review is also required to ensure that all required design elements, training, and management systems are in place. Subsequent changes to process equipment must follow a documented Management of Change procedure.

AWWA Chemical Standards

AWWA publishes a comprehensive list of standards covering many chemicals used for water treatment processes. The standards, ANSI/AWWA B200 through B700 series, cover purity issues, physical characteristics, chemical characteristics, sampling, testing protocols, packaging information, current transportation information, and support on preparing purchase specifications for water treatment processes.

NSF Listings and Certifications

Drinking water standards commonly cite the publications of NSF International (NSF—formerly the National Sanitation Foundation). NSF develops voluntary standards for drinking water system components and drinking water chemicals. These can include chemicals added to the treatment process, piping, lubricants in water treatment equipment, and so on. Purchased process chemicals may have to be certified as meeting NSF maximum contaminant levels. NSF standards may also be adopted by reference in state drinking water regulations. If new chemicals are added to the facility water treatment process, finding certified suppliers may become an issue.

Material Safety Data Sheets

All chemicals sold should be accompanied by a Material Safety Data Sheet (MSDS). Use of MSDSs is mandated by OSHA as part of the Hazard Communication Standard (29

TABLE 15.8 Required Contents of Material Safety Data Sheets (ANSI Z400.1)

1. Chemical product/company product identification
2. Composition/information on ingredients
3. Hazard identification
4. First aid measures
5. Firefighting measures
6. Accidental release measures
7. Handling and storage
8. Exposure controls/personal protection
9. Physical and chemical properties
10. Stability and reactivity
11. Toxicological information
12. Ecological information
13. Disposal considerations
14. Transport information
15. Regulatory information
16. Other information

CFR 1910.1200). ANSI Standard Z400.1 details the recommended contents, as shown below. The MSDS contains important basic information for the design team (Table 15.8). Current copies should be obtained from potential suppliers.

Confined Spaces

A *confined space* is an area large enough for an employee to enter that has limited or restricted means of exit and is not designed for occupancy (29 CFR 1910.146). Examples include tanks, silos, storage bins, vaults, and pits. They are regulated by OSHA because

1. They have the potential to create an oxygen-deficient or hazardous atmosphere.
2. A worker could become engulfed or suffocate in the storage material.
3. The space has converging walls that could trap a worker.

Because of the extensive management program mandated for confined spaces, the design team should work with the operations staff to identify and minimize confined spaces in the plant. This will significantly improve the facility's long-term ease of operability.

Safety Equipment

Safety equipment may be used by many people and must be easily cleaned and disinfected. A regular inspection and maintenance program will ensure that all protective equipment is kept clean and in good repair. Protective clothing must include foot protection to prevent injury from falling objects. This is especially important in receiving and transferring inventory. In general, regulatory agencies prefer engineering controls on potential health and safety dangers and use of personal equipment when such controls are not feasible. Engineering controls include enclosure or confinement of the operation, general and local ventilation, and substitution of less toxic materials.

Protective Equipment and Clothing. Protective clothing used in handling chemicals should be impervious and should cover exposed areas of the body, including arms and legs. Where gloves are used, they must be appropriate to the exposure. They may be standard work gloves of heavy canvas or leather construction or rubber or synthetic gloves impervious to various liquid chemicals. Gloves should generally protect the forearms as well as the hands.

Hard hats are normally required where workers may be subject to injury from falling or flying objects and must meet ANSI Standard Z89.1. Safety shoes should conform to the ANSI *National Standard for Safety Toe Footwear,* Z41.1.

Eye protection is generally provided by protective goggles fitted to each individual worker. Where skin exposure is a concern, the face should be protected by an 8- to 10-in.-high (20- to 25-cm) face shield that normally covers the full face. Personnel performing emergency response duties may require higher levels of chemical protection as required by OSHA (29 CFR 1910.120).

Respirators and Masks. Respirators or gas masks should be provided, as indicated in Table 15.5 or by the product's MSDS. Respirators may be of the particulate filter type, commonly referred to as filter respirators. These should be properly fitted. They are generally used for short, intermittent, or occasional dust exposures. Respirators must be approved for protection against the specific type of dust that may be encountered. Respirators should be selected in accordance with *American National Standard Practices for Respiratory Protection,* Z88.2. Normally, respirators are used as a protection from particulate matter, dust, or mist. Breathing air can be supplied to the airline-type respirator from cylinders or air compressors. Oxygen must never be used with airline-type respirators.

For gas or gas and particulate matter not immediately dangerous to life, the airline-type respirator is normally satisfactory. This can be a hose mask without a blower, or it can be a chemical-cartridge respirator provided with a special filter for the specific contaminant present in the atmosphere.

For gas or gas and particulate matter that are immediately dangerous to life, a self-contained respirator or gas mask should be used. A hose mask with blower or a gas mask with a special filter can also be used. In unknown conditions, self-contained breathing apparatus and the pressure-demand-mode positive pressure should be employed.

Personnel using respirators should be certain that they experience minimum face mask leakage and that the respirator is fitted properly. A semiquantitative fit test should be set up annually for each user of a nonpowered particulate respirator. Previous medical clearance or periodic medical review may also be necessary.

Wearing contact lenses with a respirator in a contaminated atmosphere is not advisable. Individuals wearing corrective glasses may find that they cannot achieve a proper seal with a gas mask or full face mask. In such cases, corrective lenses or lens inserts should be provided in full face pieces and should be clearly identified. Facial hair can also prevent a proper seal and may require a "clean-shaven" policy.

Respirator use is governed by OSHA regulation 29 CFR 1910.134, Respirator Protection. Requirements such as a monitoring program, a buddy system, and confined-space restrictions may apply.

Safety Showers and Eye Washes. Where indicated for specific chemicals, suitable facilities for quick drenching or flushing eyes through the use of an eye or total body wash must be provided. Both facilities must use potable water from a reliable source and be located within the work area when a person may be exposed to injurious or corrosive chemicals. The water supply pipe to these devices can be equipped with a flow switch that activates an alarm when used. Safety showers installed outdoors must be freeze-proof.

Safety Screens. Safety screens are a simple engineering control that can prevent worker injuries resulting from accidental contact with process equipment and chemicals. Appropriate locations should be identified during design through a combined effort of the design team and the facility operations staff. General guidelines can be found in OSHA 29 CFR 1910.23.

Protection from Toxic Gases

As previously discussed, typical gaseous treatment chemicals such as chlorine, anhydrous ammonia, and ozone are classified as hazardous or toxic under worker safety, public health, environmental, and transportation regulations. They are also governed by regional model building and fire codes.

Effective water treatment chemicals can be powerful and dangerous materials and must have the full respect and attention of the designer. Treatment chemicals that can become airborne, such as chlorine gas, are of particular concern. Chlorine is classified by the U.S. Department of Transportation as a toxic and corrosive gas. It is also a strong oxidizer and can support combustion, including steel combustion. When chlorine gas is inhaled or contacts mucous membranes, such as the eye or lung, it reacts to form acids. At sufficient concentrations, these acids can maim or kill the victim by causing severe damage resembling a burn to body tissue. Direct contact with liquid chlorine can also cause freeze burns.

Stored under pressure in the liquid phase, chlorine liquid evaporates to fill 460 times its original volume. Consequently, even a small indoor release of chlorine liquid can flood an enclosed room with vapor and overcome the occupants. Chlorine gas is heavier than air and tends to remain concentrated at ground level. Consequently, a large outdoor release of liquid or gaseous chlorine can be life-threatening miles downwind from the leak. It is extremely important for the safety of plant staff and neighbors that designers follow recognized codes and practices for chlorine systems. Where design varies from standard practice, it must be rigorously analyzed for system reliability and integrity.

Building and Fire Code Requirements

Three regional organizations publish model building and fire codes, as well as the National Fire Protection Association (NFPA), publisher of model fire codes and standards. Each organization takes a different approach to hazardous materials control. A code does not have the force of law until it is adopted as written or with local modifications by an enforcement body, such as a city, county, or state fire marshal.

It is consequently difficult to generalize about code requirements. Codes may be specific and used to enforce requirements, such as containment and treatment systems for toxic gases. At the beginning of the design process, it is necessary to determine which agencies enforce building and fire code requirements and what versions of the respective codes have been adopted.

Ventilation Requirements

Positive ventilation indicated for specific chemicals is essential to remove gases or dust that may develop from materials handling. Proper ventilation design can also avoid some areas becoming designated as *confined spaces*. The contaminated area must be isolated from the rest of the plant, and the exhaust systems must be run for a sufficient time to remove all contaminated air before that area is reopened.

The rate of exhaust must be sufficient to promptly clear the laden air from the contaminated area. Doors to the area should be flanged and sealed tightly when closed. The construction, installation, inspection, and maintenance of the exhaust system should conform to requirements in American Society of Heating, Refrigeration, and Air Conditioning (ASHRAE) and *Recommended Standards for Water Works* (2003), together with state and local codes governing design and operation of exhaust systems.

The static pressure in exhaust ducts leading from the ventilation equipment should be periodically checked to ensure continuing satisfactory operation. If there is an appreciable change in pressure drop, the system should be cleaned.

It is normally good procedure in handling dust-laden air to have the ventilation equipment discharged through dust-collecting equipment. This equipment must be set up so that accumulated dust can be removed without contaminating other working areas.

Heating, ventilation, and air conditioning (HVAC) of rooms containing equipment regulated under PSM or RMP, such as chlorination equipment, must be carefully designed. In the event of a leak it may be desirable to automatically shut down the room ventilation to contain the leak. It may also be desirable to automatically shut down air conditioning systems in adjacent buildings to prevent fugitive vapors from being drawn inside. Either way, the HVAC system for chlorine storage and feed rooms should be separated from the HVAC systems serving other areas of the same building.

Containment and Treatment Systems

Containment and treatment systems are design approaches intended to prevent the release of process chemicals that could cause employee, public, or environmental injury. These can be *active*, meaning that they have a human, mechanical, or other energy source, or *passive*, meaning that they do not have an energy source. Passive systems can include dikes, catch basins, and drains for liquids and enclosures for both liquids and gases. Containment for liquids is discussed in the section on secondary containment for liquids. Passive systems are often preferred by regulatory agencies because there is less opportunity for failure during emergency operation. Active systems include chemical scrubbers that neutralize a released chemical with another chemical and water sprays that can partially lower an airborne chemical.

No national consensus approach exists at present specifically addressing containment and treatment systems for water treatment plants. One regional model fire code promotes containment and neutralization of specified chemicals, although opinions differ on the value of such systems. General references on accidental chemical release prevention are available. When considering neutralization systems, the designer must balance risks introduced by storage of the neutralizing agent.

The chlorine scrubbing system, or scrubber, represents existing technology for controlling accidental or routine releases of chlorine gas. A scrubber is basically a process equipment unit that collects chlorine vapor and mixes it with alkaline solution (caustic) to remove the chlorine. The main reaction product is hypochlorite. The most common chemical used for scrubbing is sodium hydroxide (caustic soda), although others such as sodium carbonate and calcium hydroxide may be used. Scrubbers are often associated with neutralizing accidental releases, but they are also used for neutralizing routine process releases.

Proper sizing and design of the scrubber are important. Significant heat is generated by the caustic-chlorine reaction, so materials must be chosen for heat and chemical resistance as well as for strength.

Scrubbing systems are a tested and proven technology for chlorine storage vessels up to 1 ton in size, but they have not been reliably tested beyond this range. Nevertheless, scrubbers and enclosures have been built for 90-ton chlorine railcars, although at this time performance of these larger systems has not been verified by a full-container release.

Scrubbers require storage of significant quantities of caustic, and neutralization capacity is also critical. If the neutralizing caustic becomes saturated, chlorine gas, along with steam and oxygen, may evolve, causing foaming, which could become a significant emergency situation. Caustic storage imposes its own chemical management requirements, such as secondary containment, as well as additional handling risks for facility personnel. Caustic can also react with some metals to form hydrogen, creating a potential fire or explosion hazard.

Scrubber designs include venturi systems, horizontal sprayers, and packed media and usually recirculate caustic solution. Efficiency can be improved by increasing the caustic/gas interface ratio. Proper attention must be paid to container enclosure design, such as the storage room. The structure must provide for any expected pressurization, as well as wind, snow, and seismic loadings. Ventilation must also be sized for optimum air transfer between the enclosure and the scrubber unit, including any recirculation requirements.

Enclosing a leaking chlorine container generally limits options for direct access by response personnel because current chemical protective clothing is not designed for the concentration of chlorine that can be reached in containment rooms. For this reason, accidental release scrubbers generally should be sized to neutralize the complete contents of the largest container. Local fire codes may specify performance parameters, such as the permissible concentration level of chlorine in scrubber exhaust.

Reliability is essential for accidental release scrubber systems. Appropriate chlorine sensors, alarms, and control systems are needed. The designer should also consider using redundant systems and conservative structural safety factors. For example, scrubbers may be designed for either dry or wet chlorine gas. For emergency purposes it is more conservative to assume the gas is wet and select materials of construction for the more corrosive operating environment.

Fire and Explosion Prevention

Although generally rare, fire and explosion are potential risks at treatment facilities. For example, chlorine supports combustion in steel at 483° F (251° C). Fuels such as gasoline and propane may be present at support operations. Chemicals used to adjust pH levels can react with one another. It is important that plant design be carefully evaluated to avoid the potential for mixing incompatible materials during transportation, storage, and use.

The National Fire Protection Association develops model standards to address fire and explosion issues. Among these is the ANSI-NFPA publication *NFPA 101—Life Safety Code*, which discusses fire design considerations for a variety of building types.

Once facility personnel are familiar with emergency plans, it is advisable to conduct an orientation session for the local fire suppression agency. Joint facility and fire personnel drills may also be useful.

RECENT TRENDS IN CHEMICAL HANDLING AND USE

New trends in chemical use at water treatment plants include increased use of sodium hypochlorite, new safety requirements for gaseous chlorine installations, new chemicals for water treatment, and increased use of bulk chemicals. Securing and removing hardening of chemical handling facilities are also inherent in modern design.

Sodium Hypochlorite Systems

Chlorine has been successfully and safely used at municipal water and wastewater treatment facilities for many years. In recent years, however, utilities have been considering the use of hypochlorite rather than chlorine gas because of an increased emphasis on safety where large amounts of gas are stored in or transported through urban areas.

If cost were the only criterion for selecting the chemical form, gas would be chosen rather than sodium hypochlorite. Hypochlorite costs 2 to 4 times the cost of the equivalent quantity of gaseous chlorine. Local code and regulatory requirements for employee training and safety equipment for gaseous chlorine in some communities, however, cause hypochlorite and gaseous chlorine to be more comparable in total chlorination costs. For this reason, many large and small utilities have converted from gas to hypochlorite, and more are expected to do the same.

Chlorine Gas Leak Containment and Treatment Systems

Another reason for the recent emphasis on safety issues associated with chlorine handling has been brought about by code officials and public leaders concerned about protecting those who operate the facilities, those who reside and work in the surrounding area, and those who respond in case of an emergency. Heightened awareness of safety issues and the desire to provide as safe a system as possible have made it necessary to consider emergency chlorine scrubbers and other available safety measures in any new or rehabilitated facility that handles chlorine or other hazardous or toxic compressed gases.

Providing the impetus for this trend are the Uniform Fire Code, Superfund Amendment and Reauthorization Act (SARA Title III), Clean Air Act Amendments (CAAA), and regional and local building codes. OSHA regulations for process safety management of highly hazardous chemicals apply in those states where OSHA has jurisdiction over employees operating and maintaining municipal treatment plants and in all states over employees of privately owned plants.

New Water Treatment Chemicals

Water treatment chemicals that are relatively new include liquid oxygen, polyaluminum chloride, phosphoric acid, and several forms of phosphates, hydrogen peroxide, and polymers.

Liquid oxygen is used for generating ozone and results in more efficient production than an air source. It is conveniently delivered and stored on site. Polyaluminum chloride is a primary coagulant, and its chief advantage is a residual, which generates a smaller volume of sludge that is easier to dewater than some other coagulants.

The use of phosphoric acid and other phosphates is steadily increasing to help control internal corrosion in the distribution system piping. Designers should be aware that some phosphate products lower in phosphates accomplish the same purposes, and chemical selection should factor into environmental considerations. Hydrogen peroxide is sometimes used together with ozone for oxidation purposes. Although polymers have been used in water treatment for years, the increased interest in enhanced coagulation and maximizing particle removal through filters will probably increase the use of polymers as coagulant, flocculent, and filter aids.

Dechlorinating Agents. All water utilities regularly dispose of chlorinated water associated with routine operation and maintenance activities, such as water main flushing and

TABLE 15.9 Dechlorination Agents

Name	Formula	Grams required to neutralize 1 g of chlorine	Remarks
Ascorbic acid (vitamin C)	$C_6H_8O_6$	3	Will depress pH in large doses
Calcium thiosulfate	CaS_2O_3	1–1.3	Weak oxygen scavenger. Reaction is pH-dependent, and reaction time is slower than that of other agents, especially with chloramines. Has a relatively high toxic threshold
Hydrogen peroxide	H_2O_2	0.48	Not recommended for combined chlorine due to slowness of reaction
Sodium ascorbate	$NaC_6H_7O_6$	3.3	Minimal effect on pH
Sodium bisulfite	$NaHSO_3$	1.46	Strong oxygen scavenger
Sodium metabisulfite	$Na_2S_2O_5$	1.34	Strong oxygen scavenger
Sodium sulfite	Na_2SO_3	1.7	Moderate oxygen scavenger; the only agent that is available in tablet form
Sodium thiosulfate	$Na_2S_2O_3 \cdot 5H_2O$	1.6–2.2	Weak oxygen scavenger. Reaction is pH-dependent and will depress pH. Reaction time is slower than that of other agents, especially with cloramines. Has a relatively high toxic threshold
Sulfur dioxide	SO_2	1.1	This toxic gas is hazardous and requires careful handling. Strong oxygen scavenger

repairs and construction of new water mains. Sanitary sewers would normally be the preferred point of disposal. Large discharges to sanitary sewers can potentially overload sewers, small treatment plants, and pump stations and can even cause dry weather combined sewer overflows (CSOs) and sanitary sewer overflows (SSOs). Where sanitary sewers are not available or undersized, chlorinated water must be discharged to storm drainageways and could potentially harm aquatic life. As a result, water utilities are finding it necessary to dechlorinate not only highly chlorinated water used for disinfection after new main construction and repairs, but also potable water disposed during routine water main flushing.

A summary of the more common dechlorination agents showing chemical formulas, reaction ratios for neutralization of chlorine, and special remarks is presented in Table 15.9. A description of each agent is presented in Appendix A.

Bulk Storage and Handling

In the past, bulk storage and handling facilities were often limited to plants consuming large quantities of chemicals. Now smaller plants use bulk facilities to avoid potentially hazardous unloading and handling procedures and to simplify labor-intensive demands of handling containers and bags of chemicals. Before deciding on bulk service, plants using small quantities of chemicals should determine whether a long shelf life (which may result with large, bulk storage tanks) would affect the quality of stored chemical. If excessive shelf life is a concern, it may be practical to use smaller storage tanks and order partial truckloads of chemicals, although the delivery cost would be higher.

Acknowledgments

The following manufacturers contributed information on chemical properties and characteristics: Air Products and Chemicals, Inc.; Albright & Wilson Industrial Chemicals Group; American Cyanamid Co.; American Norit Company, Inc.; Calgon Carbon Corp.; Delta Chemicals, Inc.; Dow Chemicals U.S.A.; General Chemical Corp.; HVC, Inc.; International Dioxide, Inc.; Interox America; Mayo Chemical Company, Inc.; Monsanto Company; Occidental Chemical; Olin Corporation; PVS Chemicals, Inc., Sternson Water Treatment Chemicals; and US Filter.

BIBLIOGRAPHY

American Institute of Chemical Engineers. *Guidelines for Hazard Evaluation Procedures,* 2d ed. New York: Center for Chemical Process Safety, 1992.

ANSI/NSF. *Standard 60. Drinking Water Chemicals—Health Effects.* Ann Arbor, Mich.: NSF International, 2002.

AWWA. *Water Quality and Treatment,* 5th ed. New York: McGraw-Hill, 1999.

AWWA Standards B200 through B703, which cover the topics of softening, disinfection chemicals, coagulation, scale and corrosion control, taste and odor control, and prophylaxis. Denver, Colo.: American Water Works Association.

The Chlorine Institute. *The Chlorine Manual,* 6th ed. Washington, D.C.: The Chlorine Institute, 1997.

The Chlorine Institute. *Chlorine Scrubbing Systems,* 2d ed. Pamphlet 89. Washington, D.C.: The Chlorine Institute, 1998.

The Chlorine Institute. *Non-refrigerated Liquid Chlorine Storage,* 6th ed. Pamphlet 5. Washington, D.C.: The Chlorine Institute, 1998.

Kawamura, Susumu. *Integrated Design of Water Treatment Facilities,* 2d ed. New York: John Wiley & Sons, 2000.

NAS Chemicals Codex, 2d ed. Washington D.C.: National Academy of Science, 1985.

National Fire Protection Association. *NFPA 101—Life Safety Code.* Quincy, Mass.: National Fire Protection Association, 1985.

National Institute of Occupational Safety and Health. *NIOSH Pocket Guide to Chemical Hazards.* New York: Nostrand Reinhold, 2003.

National Lime Association. *Lime Handling, Application, and Storage.* Bulletin 213. Arlington, Va.: National Lime Association, 1990.

Recommended Standards for Water Works. Report of the Great Lakes–Upper Mississippi River Board of State and Provincial Public Health and Environmental Managers. Albany, N.Y.: Health Education Services, 2003.

Uniform Fire Code. Article 80. Whittier, Calif.: International Conference of Building Officials, 1994.

CHAPTER 16
HYDRAULICS

Peter J. Barthuly
Barthuly Hydraulics & Pumping Engineering, Inc.
Burlington, Massachusetts

The calculation of a water treatment plant hydraulic grade line involves much more than the simple summation of head losses through the processes. Coincident with determining the hydraulic profile, obviously needed to prevent spillage over the channel and tank walls, are other important hydraulic considerations. One important consideration is the necessity for equal distribution of flow among the various unit processes. There is also the need to prevent negative pressures where air is admitted into the plant's pipelines, causing unexpected head losses.

HYDRAULIC DESIGN

Good hydraulic design is a foundation of a well-designed water treatment plant. However, although hydraulics is a common element in the education of every environmental engineer, the nuances of plant hydraulics are not. This chapter provides advice on the types of head losses that need to be calculated, reference sources for critical head loss data, and information on equal flow distribution.

Preliminary Hydraulic Design

The first step in the hydraulic design of a new treatment plant is to prepare the preliminary hydraulic grade line. The preliminary hydraulic grade line is not by definition the definitive solution, but rather serves to expose areas of concern in developing the plant design.

Establishing Design Criteria. The criteria used in preparing the hydraulic design depend on such factors as the type of treatment system to be used, design flow of the plant, plant layout, and any special criteria and limitations of the site or equipment that may affect the hydraulics of the total system.

Type of Treatment Plant. The type of plant that is basic to the public water supply field is the conventional rapid sand filter treatment plant. There are, however, many varieties of water treatment plants, including packaged treatment with pressure filters, ozonation treatment, air stripping, reverse-osmosis membrane process, and water softening ion exchange. The conventional rapid sand filter plant is the focus of discussion for this chapter.

A conventional plant consists of rapid mixing, flocculation, clarification, and granular media filters. Pretreatment basins are sometimes installed before the mixing process. Filter effluent is usually stored in a clearwell that provides both equalization of the incoming and outgoing plant flow and detention time for the disinfectant, if clearwell is baffled properly, before the finished water is pumped to the distribution system. The conventional plant is sometimes modified where the level of treatment allows the exclusion of the clarification step. Such a plant is referred to as a direct filtration plant.

Coordinating with the Project Design Team. It is essential to a well-designed plant that the members of the team designing the various plant components meet regularly to coordinate the interaction of design functions. The initial meeting is important, to set a firm foundation from which the design can proceed; for example, at the initial meeting the hydraulic designer receives required information. It is presumed for the purposes of this chapter that hydraulic design is a separate function; however, the project team may have members that provide dual design functions for a project.

Essential information to be determined by the hydraulic designer before the beginning of hydraulic computations for a water treatment plant are maximum and average daily design flows, water level at plant inlet and outlet, number of units out of service during maximum flow, and the required filtering head and required freeboard at tanks and channels. Design average flow and maximum hydraulic flow must be established by the project team before hydraulic calculations proceed.

Maximum hydraulic flow can be slightly higher than maximum day demand of the distribution system to accommodate filter backwash water requirements when adequate system storage is provided. Design flow can be the peak hour demand of the distribution system at the extreme, where no storage is provided. Any value between the maximum day and peak hour may be the design maximum flow.

Design Flow and Available Plant Hydraulic Head. Maximum plant flow must be known to compute the maximum *hydraulic grade line* (HGL). This, in turn, is necessary to design wall elevations to contain water levels within tanks and channels. Average flow is needed to determine the average flow HGL, which sets average depths and volumes in the plant's unit processes. Inlet and outlet water levels are needed to determine available plant hydraulic head.

The outlet level, usually the maximum clearwell level, sets the start of the HGL and, hence, the elevation of the treatment plant units. The outlet level must be adjusted during and even after the calculation of the initial HGL to reconcile the physical profile of the plant with the surrounding plant site.

The hydraulic designer must also be provided with the anticipated head needed for the filter to trap material. This head is referred to as the *available filter head* (AFH) and is the difference in head between clean filter media and dirty filter media. To fail to account for the head loss of a dirty filter in addition to that of the clean filter causes excessive backwashing and leads to failure of the plant to provide the design volume of treated water. The required AFH is usually based on the results of filter runs determined during pilot studies. Pilot studies are recommended unless good data are available from an existing plant treating the same or similar source water.

Filter box elevation is typically set so that the top of the filter media is slightly above the clean water *hydraulic grade line elevation* (HGLE) at the maximum flow condition (with the required number of units out of service). The clean water HGLE may be up to 1 ft (0.3 m) below the media surface. This minimizes the required depth of the filter box and prevents excessive negative pressures within filter media as the maximum amount of solids is trapped. Setting the filter box too high could cause excessive negative pressures as the media become dirty. This in turn causes air to come out of solution and leads to hydraulic binding of the media with air bubbles (Steel, 1960).

The freeboard required to safely contain flow within plant facilities during the maximum flow condition must be discussed with the project design team before the initial HGL is calculated. Alternatively, the freeboard can be adjusted to the appropriate value as a result of the initial HGL calculation. For basins and channels with calm water surfaces, a freeboard of about 12 in. (25 cm) is considered adequate. Basins with turbulent water surfaces, such as rapid mix basins, require a higher freeboard because of splashing. Ozonation basins require a freeboard of 2 to 3 ft (0.6 to 1 m) or more to allow drawoff of the ozone gas.

Plant Layout. Before the preliminary or initial plant HGL can proceed, a scaled layout of plant facilities must be provided by the project team. The hydraulic design engineer can then use this layout to determine the length of connecting pipes and channels between the various plant facilities. The path of original connecting pipes and channels can be provided by the project team or by the hydraulic designer. The final setting of the pipe and channel routes is often a result of the give-and-take of the project team and the hydraulic designer as the plant's design develops.

Special Criteria and Limitations. During operation at maximum design flow, a number of plant process units could be out of service. To provide for such an eventuality, the hydraulic designer must obtain from the project team the number of units that can reasonably be expected to be out of service during maximum flow.

Although it is crucial that the plant be designed to be hydraulically adequate to meet the design year conditions, it is also important to plan for stages of expansion beyond the design year. For some treatment plant layouts, future staging will govern the size of pipes and channels.

Minimum velocity is not usually a concern in itself when one is dealing with relatively clean water through a treatment plant. A minimum velocity must be maintained in source water lines to prevent solids from settling in the piping. Some regulatory agencies, however, may establish a maximum velocity limit for all piping. The limit does not usually come into play as a critical hydraulic factor because it is the available hydraulic head that sets the allowable velocities. That is, because high velocities go hand in hand with high head losses, hydraulic design naturally progresses toward reasonable velocities.

The exceptions are the permissible velocities for flocculated water. Floc particles must not be severely agitated, or they may shear, causing flocculation problems and inhibiting clarification. To prevent shearing of floc, flow velocity must be kept low from the outlet of the flocculation basins, where floc has formed, to the inlet of the clarification basins, where floc is removed.

Actual maximum velocity that floc particles can tolerate depends on the type of floc generated and its strength. Maintaining velocities at approximately 0.5 ft/s (15 cm/s) or less prevents shearing of the floc particles under most conditions. Maximum velocity should correlate with maximum flow conditions through channels, pipes, or gates. For direct filtration plants where clarification is not used, flocculated water is directly applied to filters, and floc shear may not be a concern.

Determining the Preliminary Hydraulic Profile. Preparing a preliminary HGL is usually required only for the worst-case condition—the maximum flow condition (hydraulic capacity) with the prescribed number of units out of service. All remaining HGLs can be prepared after resolving all issues and problems and after calculation of the final HGL to the maximum flow condition.

Before one can proceed with the preliminary HGL, the location of the process units and the routes of the connecting pipes and channels must be provided, as well as the downstream control water surface elevation. Downstream control water elevation is needed because all treatment plant HGL evaluations must proceed from a hydraulic control point and in an upstream direction.

Ongoing discussions with the project team can often resolve problem issues as they arise. Other issues will be resolved after the initial HGL is completed. The problems that arise are often interconnected, with the solution to one problem affecting others. For this reason, all areas of concern must be reviewed and eventually resolved by the project design team. For example, simply raising the walls of process units to contain the flow may affect the type of pump selected, or raising walls to accommodate the HGL may be too costly or may not be a viable alternative because of incompatibility with the ground profile.

Areas of concern must be fully documented for eventual resolution. Documentation may take the form of a list of notes or a comprehensive design memorandum. The objective is to compile a list of concerns as the hydraulic analysis proceeds, not letting minor problems delay the completion of the HGL. Minor problems can be resolved after the HGL is completed.

Of course, any fatal flaws in the design must be brought to the attention of the project design team as soon as possible for resolution. The most common problem issues are related to plant head limitations that require a tight HGL and flow distribution where the flow stream is not equally split to process units. Flow distribution is an especially insidious problem because geometry alone, even with symmetry of the supply channels and pipes, does not ensure equal flow splitting to the process units. Instead, flow distribution requires deliberate design and often additional head loss.

Presentation and Adjustment of the Preliminary HGL. Once the preliminary HGL is calculated and concerns are highlighted, the resulting problems must be addressed and resolved by cooperation between the hydraulic design engineer and project design team. The hydraulic design engineer must present the results of the HGL to the project team. The presentation and related meetings should be an open discussion in which problem solving is required from all parties.

Effects on project scheduling and on each of the engineering functions must be considered. The most difficult issues are in areas where multiple engineering functions have valid considerations that conflict. Creative thinking and team cooperation must be used to resolve these issues. All issues that are resolved during the discussions with the project design team must then be incorporated into the hydraulic design and the HGL adjusted accordingly.

Intermediate and Final HGL Design

After modifications resulting from determining the preliminary HGL are incorporated into the plant design, the next round of HGL calculations is performed. The result is the intermediate HGL. This level of HGL may be sufficient for the final HGL if no additional issues or problems are uncovered.

For large, complex treatment plants, another round or two of hydraulic analyses may be required to resolve all hydraulic issues. As for the preliminary HGL, presentation of results to the project design team and ensuing discussions are required to resolve problems. All modifications decided on are incorporated into the design, and the final HGL is calculated.

Once the final HGL is completed for maximum flow condition with units out of service, other HGLs can be calculated. Other HGLs typically required are the maximum flow condition with all treatment units operating and the average HGL with all units operating. The minimum HGL is not normally required but can be done at this time if desired.

HYDRAULICS

FIGURE 16.1 Typical hydraulic grade line maximum day flow (50 MGD) conventional rapid sand filter water treatment plant.

After completion of the required HGLs, the hydraulic analyses should be fully documented in a hydraulic design report and presented to the project design team. All major changes and issues should be summarized in the report. The report should contain all piping and channel sizes and elevations, as well as HGL elevations throughout the treatment plant. All HGLs are presented on the hydraulic profile sheet of the design drawings. Figure 16.1 presents an example of a typical hydraulic grade line profile.

HEAD LOSS TYPES AND CALCULATIONS

Head losses through piping, valves, filters, channels, gates, and weirs all have an important bearing on determining the plant hydraulic grade line.

Pipelines

The hydraulics of pipelines and the corresponding head loss determinations are the most understood of all head loss calculations by engineers. Many references (Daugherty, Franzini, and Finnemore, 1985; WEF, 1991) adequately present the topic of pipeline losses.

In general, the Hazen-Williams equation is traditionally used for pressurized pipes and the Mannings equation for free surface pipe and channel flow. The Darcy-Weisbach equation is the most theoretically correct of the pipeline head loss equations, but because it requires multiple steps, it is not commonly employed in the design of water treatment plants or in the environmental field as a whole. For this reason, the Darcy-Weisbach equation is not presented here but can be found in other references (Daugherty, Franzini, and Finnemore, 1985; WEF, 1991). The respective head loss equations are as follows. The Hazen-Williams equation:

$$\text{HL} = \left(\frac{V}{K \times C \times R^{0.63}}\right)^{1.85} \times L$$

where HL = head loss, ft (m)
  V = velocity, ft/s (m/s)
  C = head loss coefficient
  R = hydraulic radius, A/P, ft (m)
  A = pipe area, ft^2 (m^2)
  P = pipe perimeter, ft (m)
  L = pipe length, ft (m)
  K = 1.318, English units (0.85, metric units)

Mannings equation:

$$\text{HL} = \left(\frac{V \times n}{K \times R^{2/3}}\right)^2 \times L$$

where HL = head loss, ft (m)
  V = velocity, ft/s (m/s)
  n = head loss coefficient
  R = hydraulic radius, A/P, ft (m)
  A = wetted area, ft^2 (m^2)
  P = wetted perimeter, ft (m)
  L = conduit length, ft (m)
  K = 1.49, English units (1.0, metric units)

Typical head loss coefficients for the most commonly used materials employed in treatment plant design are $C = 120$ for cement-lined ductile iron pipe and $n = 0.013$ for concrete and channels. For other channel and pipe lining materials, the designer should consult the manufacturer's literature.

Valves and Fittings

Valve and fitting head loss calculations are familiar to most engineers in the environmental engineering field. Although some engineers use the equivalent pipe length method, the method that is more representative of the theory behind fitting and valve head loss computations is based on the use of K values according to the following equation:

$$\text{HL} = \frac{K \times V^2}{2g}$$

where HL = head loss, ft (m)
  K = loss coefficient

V = velocity, ft/s (m/s)
g = gravitational acceleration constant, 32.2 ft/s² (9.81 m/s²)

The K values to determine fitting and valve head loss more closely represent the cause of the head loss—turbulence from flow contraction or expansion. The K values can be thought of as related to coefficients of contraction C based on the following equation:

$$K = \frac{1}{C^2}$$

The K value method also has the advantage of wider availability of values. The main confusion in using the K value or equivalent pipe length method to determine head loss is the wide variability of values listed in the various references. It should, however, be kept in mind that if pipeline velocities are relatively low, any errors resulting from overstated or understated K values rarely cause failure in the hydraulic design. Although it is not critical to use the "correct" value, whatever that may be, it is critical that a reasonable head loss estimate be made for each valve and fitting, especially where many valves and/or fittings occur, such as in a filter gallery.

Many references (Daugherty, Franzini, and Finnemore, 1985; Heald, 1988; King, 1963; Miller, 1990) list K values for various valves and fittings. The most commonly used K values are as follows:

Fitting type	K
90° elbows	0.2 to 0.3 (where K = 0.2 for 12-in.-diameter pipes or greater)
45° elbows	0.1 to 0.2 (where K = 0.1 for 12-in.-diameter pipes or greater)
Pipe entrance with tank	0.5 (flush entrance) to 1.0 (reentrant)
Pipe exit	1.0
Gate valve (full open)	0.1
90° miter bend	1.2

The greatest difficulty usually comes in selecting the proper K values for complex fittings such as tees with combining or dividing flow. Miller (1990) is an excellent source of K values for tees.

The most accurate information on head loss through valves can be obtained from valve manufacturers. The manufacturers commonly represent the head loss coefficient as a Cv value where this value is defined as the flow in gallons per minute associated with a 1-psi pressure drop through the valve. By using the fitting head loss equation, the Cv value can be transformed into a K value to determine head loss on the same basis used for fittings.

Of special concern are valves in rate controlling systems. The K value used must correspond to the value when the valve is about 60° to 70° open to allow for good hydraulic control characteristics. For valves used for isolation purposes only, the full open K value is appropriate.

Filter Media

The major concentration of head loss in a conventional water treatment plant using rapid sand filters occurs at the filtering media and the filter box effluent piping. This location can typically account for over 10 ft (3 m) of head loss through dirty filter media and effluent piping, including the rate controller. The clean media head loss is affected by the types of media installed in the filter box. Pilot studies commonly done in conjunction with the design of the treatment plant provide not only important data on the expected head loss through the media selected, but also information on required filter run time and head loss to be expected for the dirty filter. The difference between dirty filter head loss and clean filter loss is the AFH. Typical values of AFH can run from an extreme low value of about 5 ft (1.5 m) to a high value of about 7 to 10 ft (2.1 to 3 m).

The types of media used in filters are garnet sand, silica sand, granulate anthracite coal, and activated carbon. Head loss through sand depends on characteristics of the sand, such as porosity and gradation of media particles. A rough initial estimate for head loss through sand is 1 ft (0.3 m) per 1 ft (0.3 m) of sand layer thickness at 6 gpm/ft^2 [0.004 (m^3/s)/m^2].

By using established equations (Fair, Geyer, and Okun, 1968), head loss can be calculated, given the physical properties of the media. To calculate filter media head loss, either physical properties of the sand media or head loss characteristics of the sand media must be obtained from suppliers. Head losses of proprietary media such as activated carbon also are best obtained from the supplier.

Sand and other types of filtering media are supported by gravel or other layers of coarse materials. As with media, the physical characteristics of the material or head loss characteristics of the coarse support material must be obtained from the material's supplier for head loss calculation. Head loss through the coarse support material is much less than head loss through the media. For example, head loss through coarse gravel material is often a nominal value of about 0.1 ft (0.03 m) at nominal flow rates.

Filter bottoms are designed primarily to distribute the much higher backwash flows. Therefore the head losses they cause during filtering operation may be so low as to be negligible. However, it is best to check head loss characteristics with the underdrain system manufacturer.

Filter Piping

Much of the head loss at the filters is caused by the many fittings and valves in the piping. Head loss through the effluent piping immediately adjacent to the filter box may be as high as 2 to 4 ft (0.6 to 1.2 m) at maximum flow conditions. Piping size is driven by the limited space available and the allowable head loss. Figure 16.2 shows a typical piping arrangement at the outlet of a rapid sand filter box.

The principal feature of effluent piping is the rate controller that maintains a constant level in the filter as treatment plant flows fluctuate and as filters become dirty with trapped material. The typical rate controller is composed of a flowmeter followed by a downstream butterfly valve. The rate controller must be sized to allow travel of the butterfly valve disk over a reasonable range of operation to provide good hydraulic control characteristics. If possible, the butterfly disk should be sized to limit operation between 20° and 70°. For operation outside these limits, advice of a control valve and instrumentation specialist should be obtained. The rate controller is often sized smaller than the adjacent pipe size.

The flowmeter component of the rate controller must be sized to allow flow measurement over the range of flows to be conveyed. The flowmeter used in some rate controllers is a shortened variation of the venturi tube called a flow tube. For adequate flow measurement at low flow, a flow tube must be sized to permit at least 0.5 in. (1.3 cm) of head differential to be developed at extreme low flow.

The flow tube size must also accommodate measurement at maximum flow without excessive head loss. Maximum head differential is limited to about 5 ft (1.5 m). Actual head loss through the flowmeter depends on the size and type of flowmeter and is usu-

FIGURE 16.2 Typical filter effluent piping.

ally a small percentage of the head differential. Information on appropriate limits and head loss should be obtained from the flowmeter manufacturer.

Channels and Troughs

The hydraulics of channels and troughs is governed by the hydraulics of open channel flow. References are available (Chow, 1959; Henderson, 1966) that discuss in detail the methods used to calculate backwater profiles or head loss through open channels or free-surface flowing pipes. Commercial computer software programs are available to calculate head loss through open channels.

There often are channels in a treatment plant that collect flow as the flow proceeds downstream, such as the effluent collector channel of a number of parallel sedimentation basins or the filter trough when backwashing is occurring. This type of open channel flow is referred to as *spatially varied flow* and must be analyzed using spatially varied flow methodology (Chow, 1959) or computer software programs using such techniques.

Gates and Ports

Many gates and ports are used in a water treatment plant design. At the inlet of unit processes such as sedimentation basins, gate openings through the inlet wall not only allow flow into the basin but also distribute flow across the basin width through multiple gates or ports. The required head loss through ports can be high.

Determining head loss for gates and ports depends on the degree of contraction and expansion of the flow through the opening. The more contraction and expansion that occur, the higher the head loss will be. The degree of contraction and expansion depends on the submergence of the opening, the thickness of the wall opening, and the number of sides causing the contraction. A submerged gate causes contractions at four sides; an unsubmerged gate causes contraction at three sides.

Hydraulic handbooks (Davis and Sorensen, 1969) provide contraction coefficients based on these parameters. Once the appropriate contraction coefficient is obtained, head losses through submerged gates and ports can be calculated by using equations previously presented for fittings and valves. For unsubmerged gates and ports, the equations still apply, with velocity based on the depth through the opening relative to its downstream side.

Weirs

Weirs are employed at the outlet of process basins to maintain a relatively constant level within the basins. Many hydraulic textbooks and handbooks (Davis and Sorensen, 1969; Henderson, 1966; King and Brater, 1963; WEF, 1991) present the theory behind the operation of weirs as well as the coefficients that correspond to the types of weirs available. The most commonly used weirs within a treatment plant are sharp, broad-crested, and V-notch weirs. The basic form of the equation for sharp and broad-crested weirs is

$$Hw = \left(\frac{Q}{C \times L}\right)^{2/3}$$

where Hw = weir head, ft (m)
Q = flow, ft³/s (m³/s)
C = weir coefficient (in either English or metric)
L = weir length, ft (m)

For sharp-crested weirs at the effluent end of relatively deep basins, the weir coefficient is approximately 3.27 (1.8, metric). For broad-crested weirs, however, the weir coefficient C depends on the height of weir head and the thickness of the weir wall (King and Brater, 1963).

For short-length weirs where flows at the ends of the weir contract and reduce the effective hydraulic length of the weir, the weir length L must be adjusted accordingly. Effective weir length is determined by subtracting a weir length equal to 0.1 times the weir head for each end where contracted flow occurs.

The 90° V-notch weir is the most commonly used type of weir. It is described by the following equation:

$$Hw = \left(\frac{Q}{C \times N}\right)^{1/2.48}$$

where Hw = weir head, ft (m)
Q = flow, ft³/s (m³/s)
N = number of notches
C = 2.48, English units (1.34, metric units)

Venting of the weir nappe (the area just below the weir) must be done to prevent fluctuation of the weir head. The V-notch weir is naturally vented; but for sharp- and broad-crested weirs, care must be given to the venting of the weir nappe. For weirs with contracted ends, venting is ensured. For weirs where the flow is not contracted, small pipe vents should be installed.

The above weir equations presume that the weir is free-flowing (i.e., not submerged by the downstream water level). For submerged weirs, the effect on weir head can be determined from equations and empirical data presented in hydraulic handbooks (King and Brater, 1963). For weirs submerged above their critical depth, a good approximation of the change in head across the weir can be obtained by multiplying the velocity head relative to the downstream side by a K value of 1.5. This K value presumes that head loss across a submerged weir can be treated as an abrupt contraction and expansion.

HYDRAULIC DESIGN HINTS

This section presents a number of design hints that the hydraulic designer should be aware of in designing water filtration plants.

Use of Computer Programs

Proprietary hydraulic computer programs can be used to calculate the HGL through water treatment plants, but most of the programs are available only to the personnel of the engineering consulting firm that developed the program. Undoubtedly, many more hydraulics programs, some more reliable than others, will be developed in the future.

Unfortunately, computer programs may be incapable of capturing the nuances of a design that lead to its success or failure. Therefore, no matter how good the program is, the most important component in the calculation of any hydraulic profile is the user. The use of any computer program to calculate HGL must be accompanied by an engineer familiar with treatment plant hydraulics. In short, use computer programs with caution, and always check their output for reasonableness.

Available Filter Head

As discussed in preceding sections, AFH is the difference between the dirty filter and clean filter head loss at maximum design flow. The reason that AFH is mentioned again here is to reiterate its importance. An uncommon but serious design error is to account only for the clean water head loss through the filter. Without accounting for the head loss caused by the filter as it becomes dirty, the entire design and operation of the water treatment plant become severely compromised.

Flow Distribution

To minimize chemical usage and to maximize water treatment, the distribution of flow to and across each unit process must be addressed. Symmetry alone does not ensure proper flow distribution because locally generated velocity currents often adversely affect the presumed flow distribution attributed to symmetry. Also, flow distribution can be adversely affected by uneven setting of weir elevations. Flow distribution must be maintained either by rate controllers, as is done at filters, or by inducing head loss at points of the desired distribution.

Distribution of flow by rate controllers to unit processes other than filters is not often desirable because of equipment cost and additional maintenance required. Where feasible, flow distribution should be obtained by inducing head loss. The principle behind flow splitting is to provide a high enough head loss at points where equal flow distribution is required so that the head differential acting against equal flow distribution is made relatively small in comparison. An inlet header channel or pipe that feeds flow to a number of parallel process units is usually a location where equal flow distribution is required. The following equation (Camp and Graber, 1968; Fair, Geyer, and Okun, 1968) describes flow distribution along an inlet header to parallel units fed by gates or ports:

$$m = \left(1 - \frac{H}{HL}\right)^{1/2}$$

where m = ratio between lowest and highest flows
 H = head differential along channel between first and last port, ft (m)
 HL = head loss across port, ft (m)

The head differential H is based on the maximum velocity head and the friction loss in the channel or pipe. For a conduit where flow is taken off uniformly along its length, friction loss is one-third of the head loss that would occur for a conduit conveying the entire flow along its length, with no flow taken off. The following equation allows calculation of the differential head H along a conduit with uniform discharge of flow along its length:

$$H = \frac{V_o^2}{2g} \times \left(1 - \frac{hl}{3}\right)$$

where H = head differential along inlet header channel, ft (m)
 V_o = velocity head at upstream end of channel, ft/s (m/s)
 g = acceleration constant, 32.2 ft/s² (9.81 m/s²)
 hl = frictional head loss along inlet header channel with no flow taken off, ft (m)

For conduits that approach uniform discharge of flow along their length to a reasonable degree, the one-third friction loss presumption still yields a good approximation of head loss. Frictional head loss along the inlet header channel is usually computed by using the Mannings equation. For conduits where only a few takeoff ports are located, the presumption of uniform discharge of flow along its length is not valid. For such cases, manual analysis using detailed open-channel flow techniques is required. Flow distribution is discussed in further detail in several references (Camp and Graber, 1968; Fair, Geyer, and Okun, 1968; James M. Montgomery, 1985).

Another item related to distribution of flow is the placement of perforated baffle walls within sedimentation basins. Baffle walls are installed to prevent local velocity currents within the basin that could short-circuit flow and upset formation of the sludge blanket. Both occurrences would adversely affect settled water quality. The location of perforated baffle walls is empirical and is therefore based on past successful designs or established design standards (James M. Montgomery, 1985).

Where head loss cannot be tolerated at the inlet to sedimentation basins to distribute flow, multiported baffle walls can be installed at the downstream end of the basin to affect the required head loss. This arrangement avoids disruption of the floc.

Filter Effluent Piping

To minimize the head loss through filter piping, the designer should be particularly careful in laying out multiple 90° bends, butterfly valve alignment, and conditions that may allow air entrainment.

Compound 90° Bends. The layout of effluent piping must avoid using combined 90° bends in series and at right angles unless they are spaced sufficiently apart. Reference material (Padmanabhan and Nystrom, 1985) has presented research data indicating that closely spaced 90° combined bends cause swirling flow, leading to head losses much higher than losses normally expected with two 90° bends. The reference concludes that typical spacing required for combined 90° bends should be about 6 pipe diameters.

Butterfly Valve Alignment. When a butterfly valve is placed near a bend or tee where the flow turns, the disk of the valve should be aligned in the same plane as the turning flow to offer minimum resistance to the turning flow. This minimizes additional head loss that would result from the disk encroaching into the turning flow stream.

Avoid Air Entrainment. Air entrainment in filter effluent piping can be prevented by ensuring that piping, especially the packing boxes for valve stems, is not subjected to negative pressure under any flow condition. To prevent this from happening often requires that a minimum HGL elevation be maintained downstream from effluent piping. To buoy up the HGL elevation, a weir is sometimes installed at an effluent box immediately at filter effluent discharge piping or at an inlet box at the downstream clearwell.

Unless a pressure disk diaphragm is installed at the low-pressure tap of a flow tube, the low-pressure tap should be installed below the minimum HGL elevation to always ensure positive pressure. Proper calculation of the HGL elevation at the low-pressure tap requires that the velocity head at the throat of the meter be subtracted from the associated energy grade line elevation.

Rapid Mixing Basins

Conventional rapid mixers usually add energy to, or at least do not deplete energy from, the flow stream, so no head loss is typically presumed for action of the mixers themselves. Other types of mixers, such as in-line units and jet mixers, are used in many plants, and these units may cause some head loss. The manufacturers of these units should be contacted for information on the head loss that they will cause.

If conventional mixing basins are used, head loss associated with the basins is caused by their respective inlet and outlet openings. Because the configuration and opening sizes affect the mixing process, the layout and design of mixing basins should be reviewed by the mixing equipment manufacturer.

ANCILLARY HYDRAULIC DESIGN

In addition to establishing the hydraulic grade line through the treatment plant, a number of individual pump and piping systems must be considered and designed by the hydraulic designer.

Filter Backwash System

The filter backwash system must be hydraulically designed to ensure proper backwashing. Required backwashing flow rates must be obtained from the project team before the hydraulic analysis can proceed. Some systems pump backwash water, and other systems are designed for complete gravity operation using a storage tank that furnishes water at the required level. Both types require a slow opening and closing valve to prevent surging in the filter that could cause media washout or failure of the filter bottom.

Whether the filter backwash is a gravity or pumped system, backwash piping proceeds from the supply to the filter backwash system. The control valve should be designed to operate within a range of 20° to 70° open. Filter bottoms and supply pipe should be designed to allow enough distribution of flow over the filter bottom to achieve adequate washing of the media while at the same time preventing media mounding and washout. Filter bottoms must be selected with good flow distribution characteristics, and the supply pipe and inlet gullet preceding the filter bottom must be configured to enhance the distribution of flow through the filter bottoms.

Information on head loss through filter bottoms must be obtained from the manufacturer. Head loss through the media depends on the degree of expansion required. For fully expanded beds, head loss through the media bed is equal to the weight of media (Fair, Geyer, and Okun, 1968; Steel, 1960) and can be calculated according to the following equation:

$$HL = (s - 1)(1 - f)l$$

where HL = head loss through media, ft (m)
s = specific gravity of media
f = porosity of media
l = media depth, ft (m)

Head loss through the supporting gravel depends on the gravel's physical characteristics such as effective size and porosity. Given its characteristics, head loss can be calculated with equations describing flow through media (Fair, Geyer, and Okun, 1968). One reference (Steel, 1960) states that head loss can be approximated at 0.1 ft per 1 ft (0.1 m/m) of gravel depth at a backwash vertical rise rate of 12 in./min (30 cm/min). Head loss is directly proportional to gravel depth and backwash rise rate. This head loss presumes a typical support bed of coarse gravel. Where support beds include other materials such as garnet, the material's supplier must be consulted to obtain its physical characteristics or, if known, its head loss characteristics.

Filter Waste Discharge

Dirty backwash water must be directed to waste. The waste stream is conducted from filter troughs to a waste gate that is opened to direct flow to a water reuse basin or to a waste lagoon or sewer. If directed to a reuse basin, the decanted reuse water is pumped to the head of the treatment plant, and the residuals are conveyed to a waste lagoon or sewer. Residuals may be conveyed by gravity or may require pumping. If filter waste discharge is conveyed directly to a waste lagoon or sewer, the system may be gravity-driven or may require pumping. For any of these cases, the filter waste discharge system must be hydraulically designed. Backwash waste is discharged at a high flow rate but has a relatively low volume because of short duration. The designer should evaluate the benefits of flow detention on the sizing (and cost) of the plant's hydraulic units.

Water that initially passes through a filter after backwashing is often initially diverted to the filter inlet channel for reuse or to waste. If directed to the filter inlet channel, in-line pumping is required. If directed to waste, the system may be gravity-driven or require pumping. In either case, the filter-to-waste system must be hydraulically designed. A significant hydraulic concern is the potential of "shocking" the filter in the process of switching from filter-to-waste and back to filter effluent. Special modulating valves should be considered to perform these functions gradually without disrupting the filter.

Pumps

A variety of pumps and pumping systems are needed for the operation of water treatment plants. Most treatment plants require pumping the source water to the treatment plant, as well as pumping treated water to the transmission and distribution system. The types of pumps normally employed for these services are horizontal split-case centrifugal and vertical turbine pumps.

Pumping systems within the treatment plant include filter backwash, filter waste discharge, and filter-to-waste systems. Horizontal split-case centrifugal and vertical turbine pumps are usually found in backwash systems. The smaller ancillary filter pumping systems are often serviced by end suction pumps. All pumping systems require hydraulic analyses for proper design and operation. Some of the references that amply cover pump design are Karassik et al. (1976) and Sanks et al. (1989).

BIBLIOGRAPHY

Camp, T. R., and S. D. Graber. "Dispersion Conduits." *Journal Sanitary Engineering Division, ASCE* 94(2):31, 1968.

Chow, V. T. *Open-Channel Hydraulics.* New York: McGraw-Hill, 1959.

Daugherty, R. L., J. B. Franzini, and E. J. Finnemore. *Fluid Mechanics with Engineering Applications,* 8th ed. New York: McGraw-Hill, 1985.

Davis, C. V., and K. E. Sorensen. *Handbook of Applied Hydraulics,* 3d ed. New York: McGraw-Hill, 1969.

Fair, G. M., C. G. Geyer, and D. A. Okun. *Water and Wastewater Engineering,* vol. 2: *Water Purification and Wastewater Treatment and Disposal.* New York: John Wiley & Sons, 1968.

Heald, C. C. *Cameron Hydraulic Data,* 17th ed. Woodcliff Lake, N.J.: Ingersoll-Rand, 1988.

Henderson, F. M. *Open Channel Flow.* New York: Macmillan, 1966.

James M. Montgomery Consulting Engineers, Inc. *Water Treatment Principles and Design.* New York: John Wiley & Sons, 1985.

Karassik, I. J., W. C. Krutzsch, W. H. Fraser, and J. P. Messina. *Pump Handbook.* New York: McGraw-Hill, 1976.

King, H. W., and E. F. Brater. *Handbook of Hydraulics,* 5th ed. New York: McGraw-Hill, 1963.

Miller, D. S. *Internal Flow Systems,* 2d ed. Houston, Tex.: Gulf, 1990.

Padmanabhan, M., and J. B. Nystrom. *International Conference on the Hydraulics of Pumping Stations.* September 17–19, Manchester, England. Cranfield, Bedford, England: British Hydromechanics Research Association, 1985.

Sanks, R. L., et al. *Pumping Station Design.* Boston, Mass.: Butterworth, 1989.

Steel, E. W. *Water Supply and Sewerage.* New York: McGraw-Hill, 1960.

WEF Manual of Practice No. 8. *Design of Municipal Wastewater Treatment Plants,* vol. 1. Joint Task Force of the Water Environment Federation and American Society of Civil Engineers. Brattleboro, Vt.: Book Press, 1991.

CHAPTER 17
PROCESS RESIDUALS

Brian E. Peck
South Orange County Wastewater Authority
Dana Point, California

Jerry S. Russell
Carollo Engineers, PC
Phoenix, Arizona

Regulatory changes have caused dramatic changes in the handling of residuals from water treatment plants since the early 1970s. The Clean Water Act placed limitations on the ability of drinking water treatment plants to discharge residuals to sanitary sewers or natural waterways. This placed a greater emphasis on the on-site handling of process wastes. Changing regulations with regard to constituents such as arsenic have also impacted the range of opportunities for the ultimate disposal of residuals. Over the past decade, growing concerns about the level of protozoa in drinking water have spurred further changes to the handling of backwash waste and the recycling of waste streams within treatment plants.

The impact of regulations has complicated the design process for residuals handling systems. The design process now involves a balancing of the options for disposal treatment. The general steps in designing residuals handling, recovery, and disposal are as follows:

- Determine the types, characteristics, quality, and quantity of waste flows.
- Evaluate treatment and disposal options that are available.
- Review the regulations and restrictions affecting each disposal method.
- Review treatment changes that could reduce the quantities of wastes, including economies and advantages of recycling some of the wastes.
- Review the economics, advantages, and disadvantages of all alternatives.
- Select the best treatment or disposal alternatives.

The flow of these design steps is illustrated in Figure 17.1.

The types of wastes produced from different types of treatment processes are illustrated in Figure 17.2. The primary focus of this chapter is residuals from conventional treatment plants, direct filtration plants, and lime softening plants as these facilities are the largest source of water treatment in the United States. However, the growth in the use

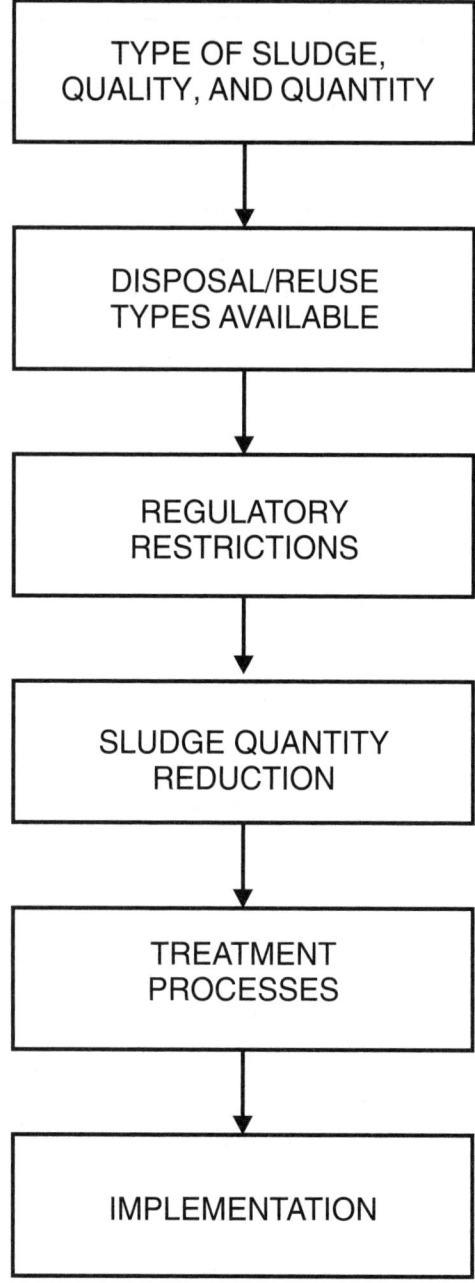

FIGURE 17.1 Steps in development of a sludge handling system.

PROCESS RESIDUALS 17.3

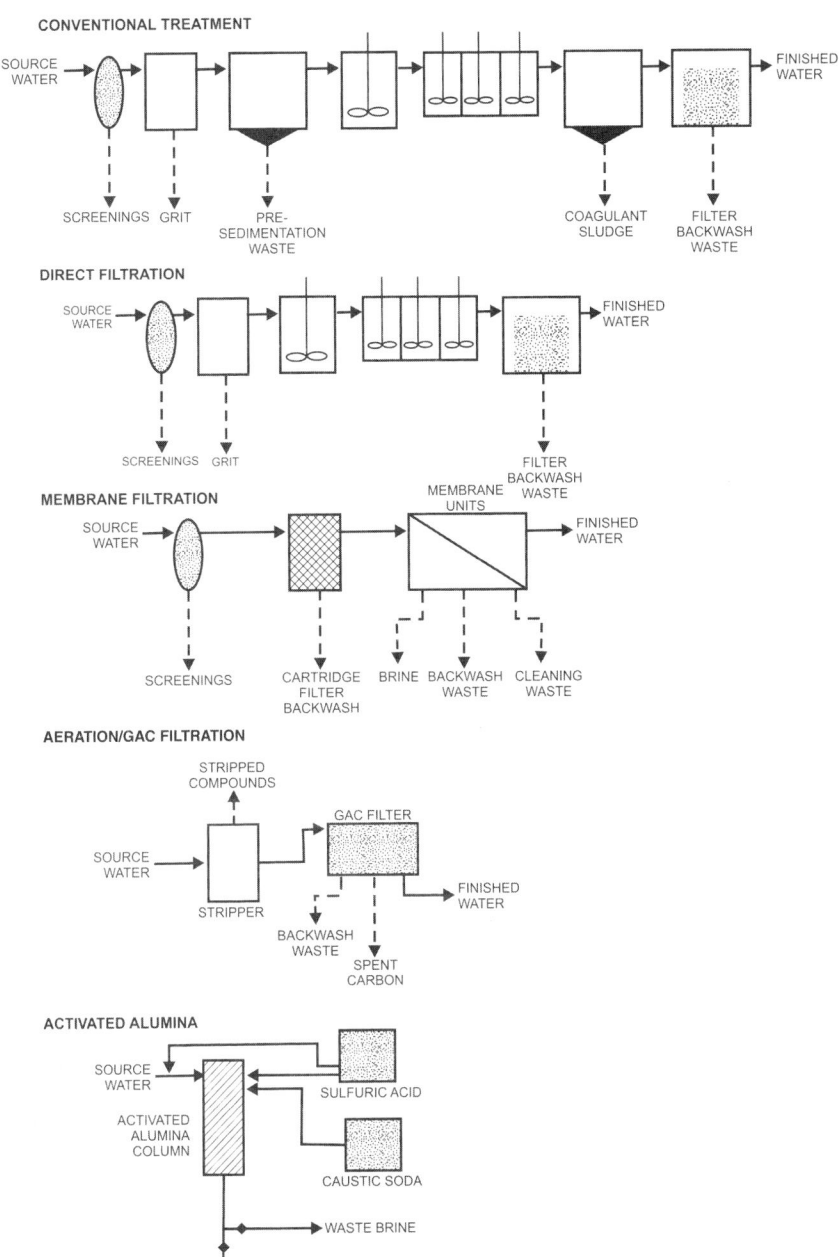

FIGURE 17.2 Water treatment plant wastes.

of new processes such as granular activated carbon, membrane filtration, and ion-exchange processes has created new types of wastes that require unique handling and disposal techniques.

The array of residual disposal and treatment options is so wide that it is not possible to provide detailed information on each element. Direction is provided to sources of more detailed analysis, where available. This chapter also provides ranges of data including loading rates and process discharge qualities. The reader is cautioned that the design of the residual handling processes is very sensitive to source water quality and the operation of the main treatment facilities. Values provided in this chapter are guides. The designer is encouraged to review the operation of like facilities and, where appropriate, to perform bench and scale testing to confirm design parameters.

SLUDGE TYPES, QUANTITIES, AND CHARACTERISTICS

Wastes from water treatment plants have traditionally been referred to as sludge or solids; however, there are also nonsolid wastes that must be treated or controlled, and the broader term *residuals* is becoming more common. The term *waste* itself may be "politically incorrect" because it fails to recognize potential beneficial reuse of some of the by-products of treatment processes.

Sludge generated from water treatment processes includes suspended solids removed from source water and chemical precipitates created by the treatment process. The various types of sludge resulting from water treatment processes generally fall into the following categories:

- Solids in filter backwash water
- Aluminum or iron coagulant sludges
- Iron and manganese precipitates
- Softening plant sludges

In deciding what treatment or disposal methods are appropriate for each type of sludge, consideration must be given to the

- Type of solids
- Quantities of sludge being generated
- Sludge characteristics that may influence sludge dewaterability

Filter Backwash Water Sludge

Filter backwash water can be troublesome to handle. It has low solids content even after thickening, it usually dewaters poorly, and most of the solids are difficult to separate by gravity without using coagulant aids. Filter backwash water deserves careful consideration not only with regard to solids disposal, but also because of potential recovery of a significant volume of water.

Filter backwash water typically represents 2% to 5% of total water processed. The quantity of solids depends on filter efficiency and the amount of solids applied to the filter, but the concentration generally varies from 50 to 400 mg/L. The amount of solids applied to the filter depends on the pretreatment provided.

Where sedimentation precedes filtration, typical suspended solids of water entering the filters range from 4 to 10 mg/L, and the backwash contains a solids loading of about 35 to 85 lb/mil gal (4 to 10 kg/ML). Attainable sedimentation basin effluent turbidity can be less than 1 nephelometric turbidity unit (ntu); however, typical sedimentation effluent turbidities are 2 to 6 ntu. In the direct filtration process, flocculated water is applied directly to the filter. As a result, solids loading is a function of coagulant dosage and source water turbidity.

Filter backwash solids are typically difficult to separate from liquid. Washwater recovery ponds sized to hold backwash water for 24 h or more may recover up to 80% of solids with the use of polymers or other coagulant aids. Reclaimed water is then usually reprocessed through the treatment plant. Flow through washwater clarifiers equipped with V-notch weirs and undercurrent baffle can recover 90% or more of solids.

Coagulant Sludges

Aluminum and iron coagulants result in inorganic sludges containing compounds such as aluminum hydroxide and ferric hydroxide along with clay, silts, and organic and inorganic matter precipitated by the coagulant. The nature of sludge produced is highly variable, depending on source water quality. Seasonal variations in source water also affect such characteristics of the sludge as its thickening density and dewaterability.

The characteristics of coagulant sludges also vary with the proportion of material removed from the water. High-turbidity waters usually result in sludges that are more concentrated and less difficult to dewater; low-turbidity waters present a more difficult sludge processing problem. In general, settled iron sludges have a higher solids concentration than alum sludges, and the addition of polymer or lime increases the solids concentration of both. Coagulant sludges are essentially biologically inert, having low biodegradable organic content and a near-neutral pH. Iron and alum coagulant sludges may be characterized as follows:

Solids content	Sludge character
0% to 5%	Liquid
8% to 12%	Spongy, semisolid
18% to 25%	Soft clay
40% to 50%	Stiff clay

Alum Sludge. Alum sludge is gelatinous with poor compactability. It generally concentrates to 0.5% to 2.0% (5,000 to 20,000 mg/L) in sedimentation basins. When filter alum [$Al_2(SO_4) \cdot 14H_2O$] is added to water, it forms aluminum hydroxide [$Al(OH)_3$]. For every 1 mg/L of alum (as 17.1% Al_2O_3) added, 0.44 mg/L solids is formed. The quantity of aluminum escaping the filters and appearing in the finished water depends on the pH of the water.

The optimum coagulant dosage to minimize sludge quantities should be developed from a series of jar or pilot filter tests. Repeat tests and sludge quantity analyses should be run at least seasonally to account for variations in water quality.

Suspended matter in the source water is usually reported in nephelometric turbidity units, but there is no absolute correlation between turbidity units and total (dry-weight) suspended solids (TSS). However, based on observed values where both parameters have been measured, the ratio of TSS to ntu normally varies from 1.0 to 2.0 and can often be as high as 10.0. Two examples of estimating solids residue from alum coagulation are shown below.

EXAMPLE 1 Alum sludge formed in treatment of a low-turbidity source water:

Source water turbidity	10 ntu
Alum dose	30 mg/L
Aluminum hydroxide sludge	30 mg/L × 8.34 × 0.44 = 110 lb/mil gal (13 kg/ML)
Raw water solids	10 ntu × 1.5 (assumed) × 8.34 = 125 lb/mil gal (15 kg/ML)
Total solids	110 + 125 = 235 lb/mil gal (28 kg/ML)

EXAMPLE 2 Alum sludge formed in treatment of a high-turbidity source water:

Source water turbidity	150 ntu
Alum dose	60 mg/L
Aluminum hydroxide sludge	60 mg/L × 8.34 × 0.44 = 220 lb/mil gal (26 kg/ML)
Source water solids	150 ntu × 1.5 (assumed) × 8.34 = 1,876 lb/mil gal (225 kg/ML)
Total solids	220 + 1,876 = 2,096 lb/mil gal (251 kg/ML)

Iron Salts Coagulation Sludge. Iron coagulants used in water treatment include ferric sulfate [$Fe_2(SO_4)_3$], ferrous sulfate ($FeSO_4 \cdot 7H_2O$), ferrous chloride ($FeCl_2$), and ferric chloride ($FeCl_3$). The precipitate formed is ferric hydroxide, $Fe(OH)_3$. Like alum sludges, ferric hydroxide is hydrophilic and thickens poorly. The amount of sludge formed can be estimated as 2.9 mg/L of solids formed for every 1 mg/L of iron added. The solids may also be estimated as shown in Example 3.

EXAMPLE 3 Iron sludge formed in the treatment of a low-turbidity source water:

Source water turbidity	10 ntu
Ferric sulfate dose (as iron)	15 mg/L
Ferric hydroxide sludge	15 × 2.9 × 8.34 = 363 lb/mil gal (43 kg/ML)
Source water solids	10 ntu × 1.5 (assumed) × 8.34 = 125 lb/mil gal (15 kg/ML)
Total	363 + 125 = 488 lb/mil gal (59 kg/ML)

Iron and Manganese Precipitates

Water treatment to remove iron and manganese consists of first oxidizing soluble iron and manganese by means of aeration or by adding a chemical such as permanganate, chlorine, or ozone. Precipitates formed are principally ferric hydroxide, ferric carbonate, or manganese dioxide. Then these precipitates are removed in sedimentation or filtration processes. The sludge produced is inert and typically red or black.

For each milligram per liter of iron or manganese in solution, 1.5 to 2 mg/L of sludge production may be anticipated. However, because concentrations of iron and manganese found in most natural waters are typically low, overall iron and manganese sludge volume is generally much lower than the volume of coagulation or softening sludge.

Iron and manganese sludge removed from filters by backwashing generally settles sufficiently in 2 h to allow decanting and recycling of backwash water to the head of the water plant. Concentration of the remaining sludge varies considerably, with typical values of 10% to 30% of the total backwash water volume.

Softening Plant Sludge

Water softening with lime or soda ash produces sludge containing precipitates such as calcium carbonate, calcium sulfate, magnesium hydroxide, silica, iron oxides, aluminum oxides, and unreacted lime. Coagulated organic and inorganic substances typically constitute a small fraction of the sludge mass. When highly turbid waters are to be softened, turbidity is normally removed by coagulation before softening.

Softening sludge is relatively inert and stable, and biologically inert as a result of the high pH caused by unspent lime and high alkalinity. Softening sludge is normally easier to concentrate than coagulant sludges. The solids content of lime sludges typically ranges between 2% and 15%.

The quantity of lime softening sludge produced at water treatment plants varies greatly depending on the hardness of the source water, the source water chemistry, and the desired finished water quality. The volume of softening sludge produced is typically much greater than that produced by coagulation processes. Solids in lime softening sludge normally concentrate in the sedimentation basin to 10%.

Lime softening sludges are primarily calcium carbonate with varying amounts of other constituents. The dewaterability of the sludge varies with the concentration of magnesium hydroxide that has been captured, ranging from a few percent to as much as 30%. Sludges low in magnesium hydroxide may be dewatered to cakes having 60% solids, whereas cake solids may be as low as 20% to 25% with higher magnesium hydroxide concentrations. Magnesium hydroxide solids are gelatinous and similar in nature to aluminum and iron coagulant solids. The calcium carbonate solid is more discrete and crystalline and more readily dewatered. The character of lime sludges at varying moisture contents is generalized as follows:

Solids content	*Sludge character*
0% to 10%	Liquid
25% to 35%	Viscous liquid
40% to 50%	Semisolid, toothpaste consistency
60% to 70%	Crumbly cake

Lime sludge cakes in the 50% to 65% moisture content range are generally sticky and difficult to discharge cleanly from dump trucks.

Brine Wastes

Methods of disposal of wastes from membrane and ion exchange processes are growing in importance as these processes are increasingly used for drinking water treatment. At the same time, government regulations on acceptable methods of disposal are becoming stricter.

Membrane Process Waste. Membrane processes used in water treatment include microfiltration (MF), ultrafiltration (UF), nanofiltration (NF), electrodialysis reversal (EDR), and reverse osmosis (RO). These processes involve separation of particles or ions from source water by passing the flow through a semipermeable membrane. Waste from a membrane treatment unit consists of constituents that do not pass through the membrane and are termed *reject, concentrate,* or *brine.* The volume of the reject can be calculated by using the following formula:

$$Q_c = Q_f(1 - R)$$

where Q_c = reject flow
Q_f = feedwater flow
R = recovery rate

Recovery rate depends on various factors, including the source water quantity, fouling potential, feed rate, operating pressure, and type of membrane. Typical recovery rates are as follows (Mickey, 1993):

Membrane process	Typical system recovery rates (%)
Microfiltration	90 to 95
Ultrafiltration	90 to 95
Nanofiltration	75 to 90
Electrodialysis reversal	70 to 90
Brackish water reverse osmosis	60 to 85
Seawater reverse osmosis	20 to 50

The quantity of the reject water is highly dependent on the type of membrane process and source water quality. Reject water from large-pore processes has water quality comparable to backwash waste from a conventional water treatment process. RO and EDR are typically applied to remove total dissolved solids (TDS), and the concentration of specific ions or salts is relatively high in the reject water. The relative concentration of ions in a reject water may be determined by using the following equation:

$$I_c = I_f \times \text{CF}$$

where I_c = ion concentration in reject water
I_f = ion concentration in feedwater
CF = concentration factor = $1/[(1 - R)r]$
r = fractional rejection rate for an ion

If an ion is completely rejected by a membrane, the fractional rejection rate r is 1.0 and the following equation applies:

$$\text{CF} = \frac{1}{1 - R}$$

For example, a membrane process with an 80% recovery rate would concentrate an ion in the reject water 5 times the concentration in the source water.

One of the advantages of membrane processes is the decreased use of chemicals such as coagulants. Chemicals added before the membrane process generally include acid, antiscalant, and, in some cases, low concentrations of disinfectant. These chemicals generally pass through larger-pore-size membranes.

TABLE 17.1 Typical Chemical Constituents of Ion Exchange Wastes

Constituent	Concentration range (mg/L)
TDS	15,000 to 35,000
Ca^{2+}	3,000 to 6,000
Mg^{2+}	1,000 to 2,000
Hardness (as $CaCO_3$)	11,600 to 23,000
Na^+	2,000 to 5,000
Cr	9,000 to 22,000

Source: Mickey, 1993.

The water treatment industry has evolved through the increased utilization of low-pressure membrane processes (microfiltration and ultrafiltration). Two types of residual streams are typically generated by these processes:

- Backwash
- Cleaning-in-place (CIP)

Surveys have found that the backwash waste generally represents 95% to 99% of the residual waste generated from the low-pressure membrane waste. CIP waste involves unique handling due to the use of chemical cleaning constituents such as sodium hypochlorite, citric acid, and caustic soda. Surveys have further shown that many of these chemical wastes are disposed through the sanitary sewer system. State and local regulations must be reviewed in new facilities to determine if this discharge is appropriate or if an alternate disposal method is needed (AWWA Residuals Management Research Committee Subcommittee on Membrane Residuals Management, 2003).

Ion Exchange Process Waste. Brines are also produced as a waste from ion exchange processes. Ion exchange has primarily been used as a softening process, but other applications include removal of specific ions such as nitrate and barium. Waste flow from the ion exchange process is produced during backwashing, regeneration, and rinsing. The quantity of brine waste is generally between 1.5% and 10% of the water treated by the process unit (ASCE/AWWA/EPA, 1997).

Ion exchange waste brine typically has high TDS and low suspended solids. The concentration depends on source water hardness, target ion concentration, regeneration rate, rinsing procedure, and resin capacity. Typical ranges of values are shown in Table 17.1.

Cationic ion exchange resins are usually regenerated with an acid solution. The low-pH backwash water may require neutralization before disposal depending on the disposal method. Similarly, anionic ion exchange wastes may require neutralization because they are regenerated with a basic solution of salt.

PROCESS RESIDUAL DISPOSAL METHODS

Disposal of treatment plant wastes includes removal of the wastes from the treatment plant or reuse of some of the residuals.

Sludge Disposal Alternatives

Five methods of disposal are typically considered for process wastes:

- Discharge to a natural waterway
- Discharge to a sanitary sewer system
- Discharge to permanent lagoons
- Burial in a landfill
- Reuse of all or a portion of the wastes

Each of these disposal methods involves different regulatory requirements and may require varying levels of pretreatment before disposal.

Discharge to a Natural Waterway. Although discharge to natural waterways has traditionally been the predominant form of both handling and discharging process wastes, requirements of the National Pollution Discharge Elimination System (NPDES) have essentially terminated all direct discharges. The primary concern has been the introduction of pollutants into the aquatic environment. The impact of aluminum toxicity from alum coagulant wastes on aquatic biota has been a chief concern. There is also concern about the impact on wildlife and on the environment of wastes containing high levels of solids, total dissolved solids, pH, various trace metals, nitrates, and chlorine, as well as the potential for creating excessive flow rates. Each of these items may be included in the NPDES permit if a plant is granted permission to discharge wastes to a waterway.

Discharge to a Sanitary Sewer. The practice of disposing of water treatment plant solids to sanitary sewers has become increasingly common. The economies of scale provided by treating water and wastewater treatment solids together are attractive, and dilution of the inorganic sludges with organic sludges makes the resulting sludge more acceptable for land disposal.

Discharge of sludge to sanitary sewers must be coordinated with the sewer authority operation and maintenance department and wastewater treatment plant authorities. The impact of both the chemical nature and the volume of the sludge on the wastewater facility needs particular consideration. Additional solids loading needs to be assessed, because it affects the waste treatment plant solids handling capacity. The additional liquid and solids load also increases operational and maintenance costs to some extent. Discharges to a sewer should be monitored and controlled to minimize the possibility of large quantities of relatively inert sludge filling the digesters and upsetting the wastewater treatment process.

Managers of some wastewater utilities are often concerned that water treatment plant solids will adversely affect their treatment processes. However, experience has shown that controlled addition of water treatment plant sludge does not cause disruption, and there may even be benefits to the waste treatment operation as a result of enhanced sedimentation.

Culp and Wilson (1979) studied the effect of adding alum sludge to an activated sludge wastewater treatment facility. They reported no identifiable benefit or detriment to the treatment process or to the anaerobic digester. The increase in wastewater sludge quantities was reported to be in proportion to added water treatment solids.

Discharge of lime sludge to sanitary sewers should be considered carefully because sludge may produce encrustations on weirs, channels, and piping. Because the volume of softening sludge disposed of is also typically large, it may be more than the waste treatment plant facilities can handle.

Discharge to a Lagoon. If adequate land is available, dilute sludge may be diverted directly to lagoons where coagulant solids concentrate to 6% to 10% over time. If a water treatment plant operating at an average rate of 1 mgd (3.8 ML per day) produces 2,000 lb/mil gal (240 kg/ML) of solids, 25 to 50 acre · ft (3,080 to 6,160 m^3) of storage capacity is required for every decade of operation.

Softening sludge, on the other hand, can be expected to concentrate to 20% to 30% and may attain a 50% concentration over a period of years. For a plant with a 1 mgd (3.8 ML per day) treatment rate that produces 2,000 lb/mil gal (240 kg/ML) solids, about 10 acre · ft (1,230 m^3) of storage capacity is required every decade. This option may be attractive for small treatment plants, but it is impractical for most larger treatment plants.

Lagoon storage of water treatment plant solids may be an attractive alternative for the short term. However, the lagoon will eventually be filled. Reuse of the land may require varying degrees of reclamation of the lagoon.

Burial in a Landfill. Water treatment plant wastes disposed of in a sanitary landfill must first have solids concentrated to a semisolid or cake form. When properly dewatered, sludge may be disposed of using sludge-only trenches or area fill techniques. Alternatively, sludge may be codisposed with refuse. Because of new environmental laws, the number of acceptable landfills has been greatly reduced and the cost of opening and closing landfills is much higher than it was just a few years ago.

Beneficial Reuse. The beneficial reuse of water treatment residuals has not been developed to the same degree as the reuse of biosolids in municipal wastewater treatment. However, a number of utilities have developed programs for the preparation and marketing of water treatment plants for reuse. Some of the examples of beneficial reuse include the following:

- *Land application.* Land application is the most prevalent form of beneficial reuse of water treatment residuals. Applications have included cropland, sod farming, forest lands, citrus groves, and arid climatized plant nurseries. The typical value of the residuals is as an additional source of moisture. Certain types of residual can also provide mineral requirements such as iron. Recent studies have shown that land application of alum or iron water treatment plant residuals can be a successful method of protecting surface water quality by reducing the amount of dissolved phosphorus from agricultural runoff water. The water treatment residuals act as a sorbent.
- *Composting.* Water treatment residuals have been mixed with biosolids and other organic wastes in compost piles. The water treatment residuals offer moisture, pH modification, and bulking.
- *Mixing with biosolids.* It may be possible to combine dewatered residuals with dewatered biosolids cake to produce a product with potential agricultural benefits. Based on information obtained through researching previous case studies and consulting professionals in the biosolid residual disposal field, a minimum ratio of 3 parts biosolids to 1 part residual may be required to produce a product that could potentially be marketed for land application. Additional evaluation and testing, with consideration for the specific projected biosolids and residuals cake properties, would be required to determine the appropriate ratios (Marie et al., 2003).
- *Cement manufacturing.* Both lime and coagulant residuals can serve to a limited degree as substitute materials in the cement manufacturing process.
- *Brick making.* Coagulant solids provide silts and clays that are used in the manufacture of bricks. Experimentation with this process has been occurring for over 30 years. This process has not proved economically effective; however, research has shown that

the water treatment residuals can be used in the process without affecting the quality of the brick.

- *Coagulant recovery.* Recovery of the coagulant in the treatment process has been proposed as one method of reducing wastes at treatment facilities. This approach is addressed in further detail elsewhere in this chapter.
- *Nutrient control.* Nutrient control is a relatively new form of land application wherein water treatment residuals are mixed with polluted soils to bind soil phosphorus and nitrogen. This can reduce nutrient concentrations in the runoff from affected areas.
- *Road subgrade.* Lime softening wastes have been effectively used for road subgrade after mixing with rock.
- *Landfill cover.* Residuals can be used in combination with other fill materials to serve as cover material for landfill cells.

Each of these forms of reuse has minimum requirements for the quality of the processed sludge including but not limited to solids concentration, pH, and total organic carbon. A series of texts have been prepared to provide guidance for the development of beneficial reuse programs including *Commercial Application and Marketing of Water Treatment Plant Residuals* (Cornwell et al., 2000) and *Land Application of Water Treatment Plant Sludge* (Elliott et al., 2000).

Disposal of Brine

Brine disposal options for an RO or EDR process plant include discharge to natural waterways, evaporation, and discharge to an injection well.

Discharge to freshwater generally requires a permit from the state and is limited by the TDS concentration of the brine in comparison with the freshwater flow rate. Coastal water treatment systems have an advantage in that there is less objection to discharging the brine to the ocean.

In areas where surface discharge is not possible, evaporation ponds have been used. However, groundwater protection laws in most areas now require lining the pond system to prevent contamination of groundwater. Industrial processors have used several mechanical forms of evaporation, including brine concentrators and crystallizers, but these technologies are generally considered cost-prohibitive for water treatment facility use.

Pumping water treatment brine wastes into injection wells is a possible option if wastes do not contain any hazardous constituents. Injection wells are now closely controlled by state and federal regulations, and the permit to allow injection includes showing that contaminants will not migrate to or adversely affect an underground source of drinking water.

RESIDUALS DISPOSAL REGULATORY REQUIREMENTS

This section provides a general discussion of federal regulations affecting disposal of water treatment plant wastes. However, each disposal situation may face different requirements because of the wide variation in state and local regulations.

Most federal regulations do not address water treatment discharges specifically. Broad categories of wastes that are regulated include solid, hazardous, and radioactive wastes. Requirements for each should be reviewed to determine applicability for water treatment plant residuals. Most water treatment plant wastes do not fall into either the hazardous or the radioactive category.

Regulation of Discharges to Natural Waterways

The 1972 Clean Water Act established the NPDES, which can affect water treatment process waste discharges into waterways. However, the application of this program has differed throughout the United States. Some states ban water treatment plant discharges completely; other states have set discharge limits based on total suspended solids and pH. There are some regions where direct discharges may be allowed for filter backwash waste and overflow from residuals handling processes such as lagoons and thickeners.

The Clean Water Act originally set procedures and prepared guidelines for discharges from water treatment plants, but the guidelines were never implemented. Guidelines identified best practical technology (BPT) control for different types of water treatment, and the requirements were established in terms of total suspended solids and pH. The approximate discharge limit suggested for total suspended solids was between 30 and 60 mg/L; this limit was based on the assumption that the discharge was 2% of the stream flow. Because these guidelines were not fully adopted, it has fallen to U.S. Environmental Protection Agency (USEPA) regional offices and individual states to set specific limits for discharges.

Regulation of Discharge into Sewer Systems

Discharging of water treatment residuals to a sanitary sewer system is generally covered by the Clean Water Act, which requires that each utility with a wastewater treatment plant capacity over 5 mgd (19 ML per day) develop a pretreatment program. The program is to establish minimum pretreatment standards for all industrial customers. State and local agencies may develop more restrictive standards that may limit flows and contaminant levels in water treatment discharges into sewerage systems.

Waste treatment plants that accept water treatment plant wastes must be sure there are no contaminants in the waste that will pass through the treatment systems and cause the wastewater plant to exceed one of the monitoring parameters of its NPDES discharge permit.

Regulation of Permanent Lagoons and Landfills

Subtitle D of the Resource Conservation and Recovery Act (RCRA) applies to the management of nonhazardous solid waste. The USEPA sets the regulatory direction and provides technical assistance. Planning, regulation, and implementation of solid waste programs are generally carried out by state and local agencies.

Regulations include location restrictions, facility design criteria, operation criteria, groundwater monitoring requirements, corrective action requirements, financial assurance requirements, and closure and postclosure care requirements. Lagoons and water plant residual landfills usually must meet these requirements under state direction.

Regulation of Land Application

Regulations for land application of water treatment residuals are usually found at the state and local levels. Some states specifically regulate the land application of alum coagulation residuals (ASCE/AWWA/EPA, 1997).

Federal regulations for land application are typically not issues because of the following:

- Standards for the Use or Disposal of Sewage Residuals, 40 CFR 503. Water treatment residuals are specifically excluded from these criteria.
- Criteria for Classification of Solid Waste Disposal Facilities and Practices, 40 CFR 257. Pollutant concentrations listed in these rules are typically higher than contaminant levels found in water treatment solids.

Regulation of Deep Well Injection

Underground injection control (UIC) rules were developed by the USEPA after the 1980 amendments to the Safe Drinking Water Act. The regulations are compiled in 40 CFR 122, 144, and 146. Some states have accepted primary enforcement responsibility (primacy) for the program and have developed their own injection well requirements. Other states have relinquished regulatory control to the USEPA.

Well injection is defined by federal regulations to include bored, driven, drilled, or dug wells. Injection wells are further separated into classes I through V. Water treatment waste disposal injection wells are generally described as class I wells. In this category, nonhazardous industrial or municipal wastes may be injected beneath the lowest formation containing a source of drinking water, and must be at least 0.25 mi (402 m) from the nearest drinking water well. There are also a variety of additional siting, permitting, and monitoring requirements placed on a permit for an injection well.

Impact of Evolving Water Quality Regulations on Solids Handling

Arsenic. The discussion of characteristics indicated the presence of various trace metals. The concentration of these metals is typically not high enough to have a significant impact on residuals handling (with the possible exception of land application). However, the establishment of a lower standard for arsenic could change both main stream treatment processes and residuals handling.

The current maximum contaminant level (MCL) for arsenic is 10 μg/L. The EPA has been reviewing a proposed lowering of this level to 5 μg/L. Processes that may be typically used to reduce arsenic to a lower level include the following: lime softening, enhanced alum and ferric chloride coagulation, activated alumina, iron oxide–coated sand filtration, and nanofiltration/reverse osmosis. The higher arsenic concentration in the plant residuals have an impact on disposal options as regulatory limits are considered:

- *Land application.* As set forth by Part 503 regulations, 41 mg/kg for mixtures with biosolids.
- *Discharge to sanitary sewer.* There are specific technically based local limits.
- *Discharge to natural waterway.* There are permit-specific limitations; one example for a specific site was a standard of 20.5 μg/L for human health protection with respect to fish consumption.
- *Landfill disposal.* Landfill disposal is allowable if the toxic characteristic leaching potential (TCLP) of the applied solids is not exceeded. The current TCLP associated with the 50 μg/L MCL is 5 mg/L. Solids must be shipped to a licensed hazardous waste facility if the TCLP is exceeded.

TABLE 17.2 Arsenic Treatment Residual Disposal Alternatives

Treatment	Discharge to receiving stream	Discharge to sanitary sewer	Land application	Landfill*
Alum/ferric coagulation	No	Not likely	Possible	Yes
Lime softening	No	Not likely	Possible	Yes
Ion exchange	No	Not likely	Possible	Yes
Activated alumina	No	Not likely	Possible	Yes
Iron oxide–coated sand filtration	No	Not likely	Possible	Yes
Nanofiltration/reverse osmosis	Possible	Possible	No	No

*Haul to hazardous waste fill as alternative.

The AWWARF report "Arsenic Treatability Options and Evaluation of Residuals Management Issues" performed an analysis showing the disposal options for residual from an arsenic removal system, as presented in Table 17.2.

Nanofiltration and reverse osmosis generate a reject stream with a low concentration of arsenic that creates possibilities for discharge in a stream or sanitary sewer. However, it may not be feasible to concentrate this flow sufficiently to allow disposal for land application or landfilling (Amy et al., 2000).

Recycle Streams. It is generally desirable to recover as much water as possible through an on-site waste handling process. The recovery process achieves both water conservation and minimization of the cost of alternative disposal. The main concern regarding recycling water is its impact on treatment plant finished water quality. The parameters of concern are

- Increased concentration of *Giardia* and *Cryptosporidium* cysts
- Increased concentration of assimilable and total organic carbon
- Increase in total trihalomethanes and trihalomethane precursors
- Increase in turbidity and particle counts
- Increase in aluminum concentration

Recent concern has focused on the potentials for cyst concentration and filter breakthrough because of recycling recovered water. The Environmental Protection Agency issued the Filter Backwash Recycle Rule in June 2001, in response to this concern. This rule impacts utilities that recycle one or more of the following side streams: spent filter backwash water, thickener supernatant, and wastes from the dewatering process. In general, the rule requires existing utilities to submit a date to their respective state agencies regarding recycle streams at existing facilities. The rule impacts new facility designs in that recycled flows must be returned to the beginning of the conventional treatment or direct filtration process. Each state may also develop additional recycle treatment requirements based on the data received from existing facilities. The rule does not specifically require that the recycle flow rate be limited to less than or equal to 10% of the plant flow rate. However, several states do set this limit (Pontius, 2003).

DESIGN CONSIDERATIONS AND CRITERIA

Methods for processing and disposing of residuals from a water treatment plant should be investigated early in the planning phase. Sludge disposal may represent a substantial portion of the investment and operating costs of providing treated water and may influence the source water selection and the method of treatment. When one is evaluating design alternatives for processing and disposing of waste solids, the following items should be considered:

- Sludge handling and transport
- Sludge dewatering techniques
- Sludge disposal
- Recovery and reuse of coagulants or lime

Minimizing Sludge Quantities

One of the first considerations for a designer beginning to design a new water treatment plant is how sludge handling and disposal costs can be kept to a minimum. There are two general methods of reduction: reducing sludge production through process modifications and recovering spent coagulants and lime for reuse.

One method of reducing the quantity of sedimentation wastes is by using polymers to enhance the performance of coagulants such as alum or ferric chloride. For some source water, the ratio of alum dose to polymer dose required for effective coagulation is approximately 50:1; this results in a sludge production ratio of aluminum hydroxide to polymer of 7:1.19, realizing a significant reduction of sludge volume.

In lime softening plants where a significant fraction of hardness is attributed to magnesium, split-flow lime softening can reduce total sludge production compared with excess lime softening. A high pH is necessary for magnesium hardness removal. In addition, most of the calcium hardness must be removed before magnesium hardness will precipitate.

In the split-flow process, the lime dose required to treat the entire flow is added in the first softening basin. Magnesium and calcium hardness precipitate in the primary softening basin as a result of the elevated pH. Unreacted lime enters the recarbonation basin or second softening basin, where it removes additional calcium hardness from the bypassed source water. Total sludge production is usually lower because less lime is used than in excess lime softening.

If more than one water source is available, possible differences in residuals created from the treatment of different qualities of source water should be considered. For example, selecting softer water may result in smaller sludge quantities.

Where softening of a water supply is required, alternative processes can be selected. For instance, the softening plant can be operated to selectively remove only the calcium fraction total hardness. However, the resulting higher magnesium concentrations may reduce the life of water heaters installed on the water system. Ion exchange softening may also be considered if there is a feasible method of disposing of the waste brine.

Recovery of Coagulants

Recovery of coagulants for reuse has been examined as a means of helping resolve the waste disposal problem of water treatment plant solids. Recalcination of spent lime is a proven technology in many locations, but coagulants recovery and reuse have been more elusive. Recent technology in alum recovery shows promise, and research efforts are being made to recover and reuse spent iron coagulants.

Alum Recovery. The traditional scheme for alum recovery consists of thickening, reducing pH, and separating residual precipitates from the dissolved aluminum decant. The recovered alum solution is decanted from the separation stage and reused. Aluminum recoveries of 60% to 80% have been reported at pH levels of 3.0. However, some sludge may require pH values as low as 1.0. When 1.9 lb (0.9 kg) of sulfuric acid is reacted with 1 lb (0.5 kg) of aluminum hydroxide, 1.9 lb (0.9 kg) of alum is formed. This acidic alum recovery process was eagerly pursued in several Japanese water treatment plants, but concern about the potential buildup of heavy metals in the recovered alum stopped further use of the process in 1972. Metals such as chromium and iron can be converted to a soluble form during acidification. Other metals and impurities may also be present in the sulfuric acid, and these impurities may become concentrated in the recovered alum. Further detractions for the process included the expense and critical operating control required.

Another process that may be considered for alum recovery is the liquid ion exchange process (Westerhoff and Cornwell, 1978). Thickened alum sludge is acidified to a low pH (about 2.0), separated from remaining precipitates in a sedimentation tank, and subjected to liquid ion exchange. The aluminum is extracted into a liquid carrier that is immiscible in water and separated. Aluminum in the separated liquid carrier is stripped with sulfuric acid, and the carrier liquid is recycled. The extractant is selective for aluminum, creating no buildup of impurities or heavy metals. A 95% recovery of alum is achievable under laboratory conditions.

If sodium aluminate proves to be a good coagulant for a specific water, the alkaline method of alum recovery may be applicable. Aluminum can be redissolved from alum sludge by raising the pH to 12 to 12.5 with sodium hydroxide. This converts the aluminum hydroxide to sodium aluminate. Aluminum recoveries of 90% to 95% are reported.

Iron Coagulant Recovery. Recovering iron coagulants involves acidification of the ferric hydroxide and a recovery technique similar to that described for the acidic alum recovery process. The pH must be reduced to 1.5 to 2.0 to attain 60% to 70% recovery of iron. Because of the expense and the poor dewatering characteristics of the sludge, there has been little interest in this process.

Lime Recovery. A coagulant recovery technique for lime softening plants has also been developed (Black and Thompson, 1975). It is based on a combination of water softening and conventional coagulation procedures that can be applied to all types of water. Magnesium carbonate is used as the coagulant, with lime added to precipitate magnesium hydroxide as the active coagulant. The resulting sludge is composed of $CaCO_3$, $Mg(OH)_2$, and the turbidity removed from the source water. Sludge is carbonated by injecting CO_2 gas, which selectively dissolves the $Mg(OH)_2$. Carbonated sludge is filtered, with the magnesium recovered as soluble magnesium bicarbonate in the filtrate. The magnesium bicarbonate coagulant is then recycled to the point where chemical is added to the source water. At that point, it is precipitated as $Mg(OH)_2$, and a new cycle is initiated. Filter cake produced in the separation step contains $CaCO_3$ and the turbidity removed from the source water.

The process may be expanded to recover lime. The filter cake ($CaCO_3$ plus turbidity particles) is slurried and processed in a flotation unit to separate turbidity particles from the $CaCO_3$. The purified $CaCO_3$ can then be dewatered and recalcined to quicklime.

Residuals Handling and Transportation

Methods for transporting water treatment plant residuals are similar to other solids and sludge materials handling methods. Residuals in liquid form may be pumped through pipelines or transported by tanker trucks. Alternatively, residuals in cake form may be pumped or trucked to a loading facility or final disposal location.

Hydraulics. The design of solids handling facilities requires hydraulic calculations for the sizing of pumps and piping. The Hazen-Williams formula is an empirical equation that is valid only for water (around room temperature and in the turbulent flow regime) and is not appropriate for process streams with solids concentrations higher than 2% to 3%. At solids concentrations above 2% to 3%, the flow stream becomes nonnewtonian. Under these conditions the system head loss is no longer proportional to flow, and fluid viscosity is not a constant. Various texts, including *Pumping Station Design* by Sanks et al. (1998) and *Wastewater Engineering: Treatment, Disposal, and Reuse* by Metcalf & Eddy, Inc. (1991), have provided graphs, based on collected data, comparing the relationship of head loss to flow at various solids concentrations. The graphs provide multipliers for calculated clear water system head loss based on a given sludge solids concentration. However, these graphs were developed for sewage sludges as opposed to the chemically coagulated residuals generated in water treatment plants. These graphs can still provide a useful tool for estimating system head loss for water treatment residuals with the application of the appropriate safety factors. In general, these graphs show that as solids concentrations increase, head loss can increase exponentially by factors of 5 to 500 over that for clear water. For a given solids concentration, as the solids flow stream velocity within a pipe increases, head loss spikes initially (thixotropic behavior) and then decreases after flow has been initiated (Sanks et al., 1998; Metcalf & Eddy, Inc., 1991). Some designers have used the Hazen-Williams equation to predict system head loss for water residuals by measuring head loss in a similar system and then calculating the C value. This C value (that is, 80) is then applied to the new system design. If pumping over long distances is required, pilot testing with the specific process residuals streams may be required to properly size the pumps and piping system.

Pumping. Nonclog centrifugal pumps are often used to pump residuals flows with concentrations less than 4%. Recessed impeller and combined screw-centrifugal pumps may also be used for pumping residuals flow streams from gravity thickeners and clarifiers (usually less than 5% solids) commonly found at water treatment plants. However, these pumps have not been used as frequently in potable water applications as in sewage treatment applications.

Positive displacement pumps are generally the most common pumps used at water treatment plants for pumping residuals flow streams. Progressive cavity pumps and diaphragm pumps have been used effectively for pumping residuals flow streams from clarifiers and gravity thickeners. Progressive cavity pumps, along with high-pressure models of piston pumps, have also been used for moving dewatered residuals (cake) as an alternative to screw or belt conveyor systems. Positive displacement pumps can be more expensive to maintain than centrifugal pumps when pumping residuals with high grit contents. Grit tends to wear out the contacting, rotating parts on positive displacement pumps. However, positive displacement pumps provide linear flow to pump speed relationships, simplifying system design, and can generally provide much higher system pressures than centrifugal pumps.

Piping. Ductile iron piping is commonly used at water treatment plants in diameters between 4 and 8 in. for conveying settled or thickened residuals. Large-diameter steel pipe is often used for carrying less concentrated flow streams such as recycled or recovered water returned to the head of the plant. Lining of ductile iron and steel pipe is commonly done with cement mortar. Glass-lined piping is uncommon due to the expense and the satisfactory performance of cement mortar lining. Plastic pipe should not be used with any flow stream that has the potential for carrying significant quantities of grit due to the potential excessive wear that may occur. Plastic piping has been used effectively for gravity flows and low-solids applications such as drying bed decant. Care should also be used in selecting the class of piping materials based on the higher pressures that may be seen in overcoming the thixotropic nature of residuals. For example, clearing a pipe full of high concentrations of solids at rest generally requires a high initial pressure head, which decreases as the velocity of the flow stream is increased.

Pipe layout and design should minimize the numbers of pipe fittings (including elbows and tees) where possible, to reduce the system head loss. The design should also include pipe cleanouts to allow manual flushing with water, compressed air, or pipe pigs, as appropriate, to remove obstructions and settled solids. Compressed air has been found to be useful for dislodging settled solids in gravity flow pipelines.

Trucking. Because trucking residual cake to remote sites (i.e., landfills) compresses solids due to the vibration during travel and can release free water, the associated truck container should be watertight. Lime sludge is commonly sticky and can adhere to the truck container, making it difficult to dump. Special surfaces such as high-density polyethylene liner attached to the inside container body can be used to reduce adherence of sticky residuals. Drier cakes also tend to dump more easily from hauling containers than do wet cakes. Properly conditioned residuals cakes tend to release little free water during shipping and tend to be less sticky.

Sludge Treatment Techniques

Economics, regulatory requirements, and various other factors may result in the need for utilities to provide treatment of residuals before transport and disposal. The purpose of these processes is usually to reduce the quantity of liquids being transported with the solids. Figure 17.3 shows treatment categories that generally handle coagulant sludges, softening sludges, and filter backwash waste. The three primary categories of residuals treatment processes are thickening, dewatering, and drying. These categories are typically delineated according to the ranges of solids concentration of the cakes produced as the end product of each process. The minimum acceptable cake dryness for residuals will de-

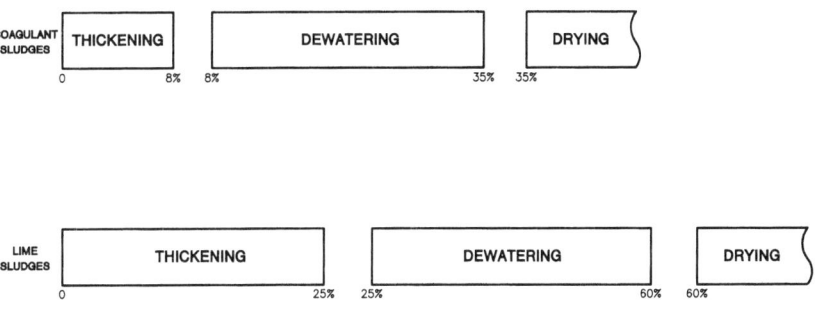

FIGURE 17.3 Percent dry solids for treatment of coagulant and lime sludges.

pend on the ultimate disposal site requirements. Municipal landfills generally require that cake pass the paint filter test. At cake solids less than 15% to 20%, the cake may not pass the paint filter test.

Following are the normal ranges of concentration of alum or metal hydroxide sludge by various processes:

Process	Solids concentration
Thickening	$\leq 8\%$
Dewatering	8% to 35%
Drying	$> 35\%$

The processes generally achieve a higher percent solids concentration for lime sludge than for a metal hydroxide sludge. For example, a centrifuge may dewater a metal hydroxide sludge to 15% while lime sludge is dewatered to 50%. Following are the normal ranges of concentration of lime softening water treatment plant sludge by various processes:

Lime sludge process	Solids concentration
Thickening	$\leq 25\%$
Dewatering	25% to 60%
Drying	$\geq 60\%$

Handling Backwash Wastewater. The handling of process wastes at conventional water treatment plants typically focuses on processes to handle both settled solids from sedimentation basins and backwash waste from filters. Treatment facilities that utilize on-site solids treatment processes typically use a more routine thickening-dewatering configuration for solid streams discharged from the sedimentation process. The handling of the filter backwash waste requires a modified or alternative approach due to the following characteristics:

- High flow rate
- Frequency of flow associated with backwash procedure
- Lower solids loading (as compared to sedimentation waste)

Direct mixing of the filter backwash waste with sedimentation basin solids can result in a significant upsizing of the thickening or dewatering process used at a facility. This was typically avoided in the past by one of the following three approaches to handle filter backwash waste.

1. *Direct discharge to a stream or adjacent body of water.* This was often acceptable due to the low solids content of the discharge. The opportunity for this type of discharge has become more limited due to the NPDES permitting process.

2. *Discharge to the sanitary sewer system.* Sanitary sewer discharge of backwash wastes was somewhat less common due to the high hydraulic loading involved. The sizing of the sanitary sewer system was often restrictive. Various local permitting requirements also impact this type of discharge.

3. *Equalization and recycling within the main treatment process.* The equalization and recycle flows have been historically popular due to the benefit of water conservation. This approach is being reconsidered due to the impact of new regulations on recycling within treatment facilities.

Treatment Processes for Filter Backwash Wastes. The treatment of filter backwash wastes begins with the concept of concentration. Various methods are used to separate as much water as possible from the waste stream so that the size of further treatment can be reduced. This is very important for processes with high costs for structure (thickening), equipment (mechanical dewatering), or land (drying beds, filters). Concentration is often achieved with either sedimentation or thickening. This sizing of these processes can often be reduced through equalization. Figure 17.4 shows how equalization may be used in conjunction with different treatment processes.

EQUALIZATION. Equalization basins are typically circular or rectangular concrete structures with an open top. However, many different designs have been used ranging from steel tanks to lined ponds. The sizing of the equalization basin begins with determining the downstream limitation or requirement including one of the following:

- *Recycle limitation.* Many states have set the recycle limit to be less than or equal to 10% of the plant flow rate. This criterion would set the maximum discharge rate from the equalization basin.
- *Loading on downstream process.* This approach would define the downstream hydraulic design loading rate for a subsequent process such as thickening. The preliminary design for a facility typically considers different combinations of sizing the equalization basin and the downstream process to determine the most cost-effective design of the combined process. The sizing of the sanitary sewer system was often restrictive. Various local permitting requirements also impact this type of discharge.

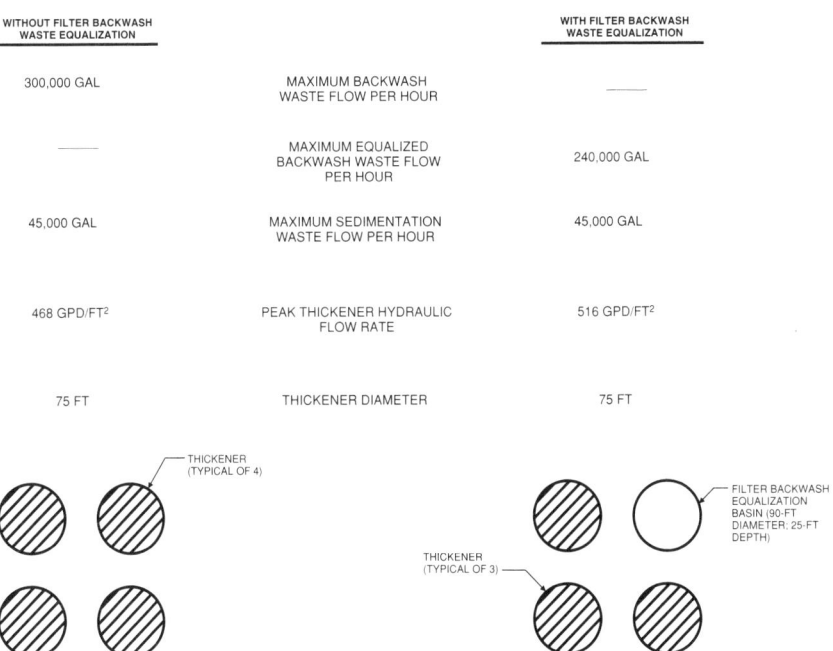

FIGURE 17.4 Impact of equalization on unit process sizing.

The design of an equalization basin then requires setting the following parameters:

- Basin discharge rate (gallons per minute)
- Total volume per backwash (gallons)
- Length of backwash (minutes)
- Backwash frequency (number per day)
- Spacing of backwash (e.g., how many backwashes can be done in a row, amount of time in between)

The setting of backwash length, volume (based on a multiple-step rate), and frequency for a new treatment plant is usually based on evaluation of those same parameters on a like facility using similar raw water quality.

One recent paper described using the parameters above for the development of a 24-h storage curve to aid in the sizing of the equalization basin. An example of a storage curve is presented in Figure 17.5. This curve tracks the fill and draw rate of an equalization basin as a sizing tool (Cornwell et al., 2001).

One key factor in the design of an equalization basin is the configuration of the facility within the overall configuration of the plant hydraulic profile. It is desirable to have the backwash waste flow from the filter by gravity. This often results in a deep excavation to construct the basin.

Equalization basins are typically designed without any internal mechanism for collecting solids. The intent of the basin is generally to minimize the holding time of the backwash waste. Solids will tend to settle during periods with low backwash frequency. Filter media

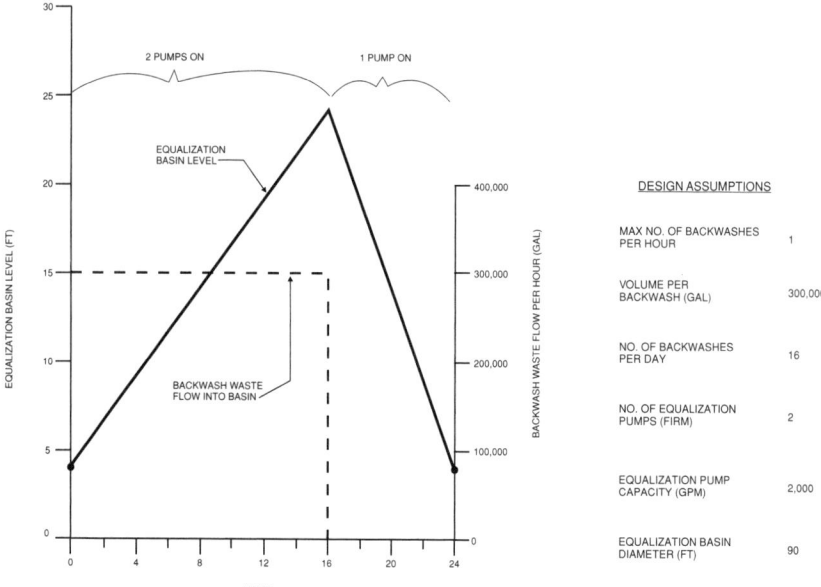

FIGURE 17.5 Equalization storage curve.

FIGURE 17.6 Typical equalization basin layout.

carried over the filter launders will also tend to collect in the basin. There are several mechanisms that can be used to help control solids buildup within an equalization basin:

- Minimize the surface area compared to the height of the basin to avoid dead spots.
- Slope the floor of the basin to a sump that can be pumped out (see Figure 17.6).
- Utilize pumps capable of removing the solids.
- Utilize multiple basins or a basin with compartments that facilitates cleaning.

SEDIMENTATION AND THICKENING. Planning for recycle treatment has been driven in large part by the focus on the potentials for cyst concentration and filter breakthrough because of recycling recovered water. Studies have indicated that this potential is significantly reduced by equalization and limitation of the rate of return of recycle flows back to the main treatment stream. Further treatment of the recycle stream is often performed through sedimentation and thickening. Sedimentation and thickening are similar processes largely differentiated by the overflow rates. Sedimentation is covered elsewhere in this text. The application of this process to filter backwash recycle is varied with overflow rates ranging from 100 to 1500 gpd/ft^2 (4,100 to 61,000 L/m^2 per day). Bench-scale testing for the city of Cleveland indicated that an alum polymer fed at rate of 2 to 4 mg/L at an overflow rate of 1000 gpd/ft^2 (41,000 L/m^2 per day) reduced total suspended solids from 100 to 5 mg/L.

The design of thickening as a concentration method is addressed in the next section. Thickening is also a very effective method of recycle treatment. A recent study in Salt Lake City evaluated the effectiveness of backwash waste treatment in the removal of *Giardia*- and *Cryptosporidium*-size particles. Tests for the removal of 2- to 5-μm particles through settling at overflow rates between 160 and 550 gpd/ft^2 (6,500 to 22,400 L/m^2 per day) indicated particle removal rates greater than 90% (Ashcroft et al., 1997). Various studies have also shown the effectiveness of coagulant aid, such as the addition of an anionic polymer before thickening, in significantly improved particle removal.

The design of these processes is addressed elsewhere in this text and in this chapter. Note that the regulations have driven the modification of recycle streams at many treatment facilities to utilize treatment such as sedimentation and thickening. Space is often at a premium for these retrofit facilities. Therefore, processes with a small footprint, such as plate settlers, are often considered for this application.

Additional Treatment Processes for Sidestreams. As discussed throughout this chapter, a wide variety of potential residual sidestreams can be generated by water treatment processes. As technologies such as membranes continue to emerge and become more widely accepted, additional sidestream flows including concentrate and chemical cleaning flows will become more prevalent. It is frequently desirable to recycle these sidestream flows back to the head of the treatment process to conserve water, improve facility operation, or provide a method of handling when alternative disposal options are not available.

In most conventional water treatment plants, a majority (in terms of both volume and solids loading) of the residual sidestream flows can be attributed to spent filter backwash water and return of recovered water. The promulgation of the Filter Backwash Recycling Rule (FBRR) by EPA on August 7, 2001, provided regulations that govern the recycling of filter backwash water within the potable treatment process. Generally stated, the FBRR ensures that the 2-log *Cryptosporidium* removal requirement established in the Interim Enhanced Surface Water Treatment Rule (IESWTR) and proposed in the Long Term 1 Enhanced Surface Water Treatment Rule (LT 1 ESWTR) is not negatively impacted by recycle practices. The rule essentially requires that recycle sidestreams be returned through the processes of a system's existing conventional or direct filtration treatment process that is recognized to be capable of achieving the required 2-log *Cryptosporidium* removal.

Sidestream flows can be treated prior to recycle by a variety of processes. The most appropriate process is often based on the types of contaminant to be removed and the degree of removal required. Although conventional treatment processes such as sedimentation are currently widely and successfully utilized for residual sidestream treatment, alternative treatment technologies, including membranes, ultraviolet (UV) disinfection, ozone, and ultrasound, are emerging for residual sidestream treatment. The following sections provide a brief overview of these alternative treatment technologies and their potential application in treating spent filter backwash and associated residual sidestream wastes generated by conventional treatment processes.

ULTRAVIOLET LIGHT. Ultraviolet sterilization is a physical process by which electromagnetic energy is transferred from a source (specifically a lamp) to an organism's genetic cellular material, thereby destroying the organism. UV light has been successfully utilized for disinfection for many years. However, a variety of factors including cost and limited institutional knowledge resulted in the use of UV light on a very limited basis. Due to recent advances in equipment testing, design, and manufacture, and progressively more stringent water quality regulations, large-scale application of UV light is becoming significantly more feasible.

UV light has been proved to be effective against a variety of microorganisms including bacteria and viruses. Recent studies have also shown it to be effective against *Giardia* and *Cryptosporidium*.

With the emergence of UV light as an effective and economical disinfection technology, the merits of UV light as a residual sidestream treatment process are being more closely considered. The effectiveness of UV light is heavily influenced by a variety of factors including color, solids (turbidity), and iron salts in the water. High concentrations of all these constituents can effectively reduce the disinfection potential of UV light, thereby impacting the effectiveness of the process. UV light systems typically require a relatively clean influent flow stream (< 0.5 ntu) to optimize the efficiency and effectiveness of the process. Sidestream residual flows (specifically spent filter backwash and recovered water) have a relatively high solids content and associated turbidity. Consequently, to maximize the effectiveness of the UV light unit in treating residual sidestream flows, it would be necessary to implement a physical separation process, such as sedimentation, upstream of the UV light reactor to reduce the turbidity of the influent water to acceptable levels. Overflow from the separation process could then be fed to the UV light disinfection process for subsequent treatment.

Depending on the specific application, a standard sedimentation process can remove between 75% and 90% of the solids in the sidestream. Numerous studies have indicated that solid-liquid separation processes effectively reduce pathogens including *Cryptosporidium* and *Giardia* in the post solid-liquid separation flow stream. The microbial fraction removed in the settled sludge is typically bound within the sludge matrix and has no method of reproduction (i.e., excestation potential). As such, the solid-liquid separation process has effectively met the requirements of the FBRR, and no additional treatment is required.

Based on the ability of conventional (sedimentation) treatment techniques to reduce pathogens, it may be difficult to justify the capital and O&M costs associated with installation of a dedicated UV light system downstream of the solid-liquid separation process. However, in specific applications where high concentrations of *Cryptosporidium* exist in the residual flow stream (thereby increasing the likelihood that *Cryptosporidium* may be carried over into the decant/return water flow), the application of UV light may become a reasonable supplementary treatment solution for the residual sidestream. This configuration would provide a multiple-barrier approach to pathogen reduction, thereby further reducing the potential risks associated with pathogen return in the recycled/recovered water flow stream. Studies have indicated that UV light disinfection of clarified spent filter backwash water (SFBW) is capable of achieving greater than 4-log inactivation of *Cryptosporidium* (Cornwell et al., 2001).

MICROFILTRATION (MF) AND ULTRAFILTRATION (UF) MEMBRANES. Similar to UV light disinfection, recent advances in equipment testing, design, automation, and manufacture, coupled with progressively stringent water quality regulations, have made large-scale application of membrane processes substantially more feasible as a water treatment technology. Membranes are capable of producing high-quality water, making them an attractive technology for water treatment applications.

Membrane systems may be classified as either pressure-driven, or electricity-driven, based on the type of driving force that promotes separation of particles from the water. Pressure-driven membranes (more common in water treatment applications) can be further classified according to the amount of pressure required to achieve solid-liquid separation. High-pressure systems, including nanofiltration (NF) and reverse osmosis (RO), use semipermeable membranes designed to promote the removal of dissolved substances from the water. High-pressure membranes require operating pressures in the range of 75 to 1,200 psig. Low-pressure systems, such as microfiltration (MF) and ultrafiltration (UF), use more porous membranes designed to remove suspended particles from the feedwater. Low-pressure membranes are typically operated at pressures less than 40 psig (although low-pressure vacuum systems can operate at pressures less than 12 psig). Low-pressure membranes can be further classified into two categories according to their flow configuration:

- *"Inside-out" membranes.* Feedwater is applied to the inside of a hollow-fiber membrane, and permeate is collected on the outside of the fiber.

- *"Outside-in" membranes.* Feedwater is applied to the outside of the hollow-fiber membrane, and permeate is collected on the inside of the fiber.

With continuing improvements in membrane technology and flexibility, the potential use of low-pressure membranes (MF/UF) as a sidestream treatment technology is becoming more commonly considered. However, several factors should be taken into account when one is evaluating the use of membranes for treatment of residual sidestreams. First, the characteristics of the sludge (specifically TSS/turbidity) may mandate the membrane configuration available for implementation. High-turbidity waters can result in rapid fouling of the membrane and subsequent decreased efficiency. Inside-out membranes will generally plug more rapidly under high solids loading conditions while outside-in membranes typically perform better in these applications. Consequently, an outside-in membrane configuration would likely be more conducive to use in residual sidestream treatment applications. It may also be possible to install a preliminary solid-liquid separation process (such as sedimentation), upstream of the membrane units to address high feedwater turbidities. However, it would be difficult to justify the capital and O&M costs associated with installation of a dedicated membrane system when a typical upstream solid-liquid separation process can produce acceptable decant/return water quality without subsequent treatment. Finally, although the cost of membrane treatment is continually becoming more economical, it is still significantly more costly than conventional sidestream treatment processes such as sedimentation and thickening (which meet the requirements of the FBRR). Consequently, economic considerations should be carefully weighed against specific water quality benefits and process preferences when one is evaluating the potential use of membrane technology for residual sidestream treatment.

OZONE (OZONATION). Ozone, an oxidizing biocide, has been used for drinking water treatment on a municipal scale for numerous years. Ozone, as a disinfectant, has been proved to be effective in destroying hazardous pathogens, including *Cryptosporidium,* in bench- (laboratory), pilot-, and full-scale experiments. Since ozone gas breaks down rapidly, treatment plants must generate the gas on-site. Ozone's successful track record makes it an appropriate technology to be evaluated for treatment of spent filter backwash and associated residual sidestream wastes (generated by conventional treatment processes).

Ozone has a high oxidation potential which increases its reactivity with other elements and compounds. This high reactivity results in high kill rates of fungus, bacteria, and viruses. Ozone causes destruction of the organism's cell wall. When the cell wall ruptures, it exposes the organism to the external environment, which destroys the cell. This method of disinfection is in contrast to chlorine, a common disinfectant used in municipal water treatment, which diffuses into the cell protoplasm, inactivating cell enzymes, and killing the organisms.

Ozone is very reactive and corrosive and requires the use of corrosion-resistant materials. The rate of ozone production is a function of the oxygen concentration and impurities in the gas. Two important components of an ozone generator include a clean oxygen supply and a high alternating current that produces voltages from 6 to 20 kV. Following generation, the ozone-enriched gas is contacted with the water to be disinfected (treated) in a bubble or countercurrent tower contactor.

Ozonation also has several disadvantages compared to other potential treatment technologies, which must be considered as part of an evaluation of ozone as a sidestream treatment process. When water containing bromide is treated with ozone, bromate is formed. Bromate is regulated by EPA as a disinfection by-product. Additionally, ozonation of a high-turbidity sidestream (such as SFBW) increases the ozone demand, which subsequently increases sidestream treatment (O&M) costs.

ULTRASOUND. High-intensity (power) ultrasound, for disinfection, is a new and upcoming technology in the water and wastewater industry. Although, the industry has seen limited use of this technology in the past few years, currently high-intensity ultrasound is being applied in the areas of disinfection, decontamination, and dewatering. The use of low-intensity ultrasound, for flow and level measurement, is already well established in the water and wastewater industry.

Ultrasound has been successful in the deactivation of *Cryptosporidium* oocysts found in drinking water. In Germany, ultrasound is being used to treat wastewater sludge to shorten the digestion period, increase gas yield, and reduce the amount of sludge for disposal. Other applications currently being researched include the use of ultrasound to clean ceramic membrane filters and as a pretreatment process, prior to UV light disinfection, to increase the efficiency of UV irradiation. Initial results suggest that ultrasound is an appropriate technology to be evaluated for treatment of spent filter backwash and associated residual sidestream wastes (generated by conventional treatment processes).

Ultrasound can be used to create high-frequency energy that causes liquids to vibrate. When liquids are exposed to these high-frequency vibrations, the physical and chemical changes result in cavitation. *Cavitation* can be defined as the rapid formation and collapse of microscopic gas bubbles in liquid as the molecules in the liquid absorb ultrasonic energy. This cavitation effect ruptures the cell walls of organisms, resulting in the death of the associated organism.

High-intensity ultrasound is generated when a high-voltage current actuates an ultrasonic transducer, which in turn generates vibration in a liquid. This treatment technique has several general disadvantages including the requirement for large amounts of power, and the inefficiencies related to converting electric energy to sound energy, and transferring ultrasound energy from the transducer to the liquid. Typically, piezoelectric and magnetostrictive transducers are used for ultrasound applications. These transducers vary in energy conversion (electrical to sound) efficiency and useful life. The potential applications, advantages, and disadvantages associated with the use of high-intensity ultrasound for sidestream treatment are not well known due to the limited use and testing of this technology in the water and wastewater industry. Consequently, care should be taken to fully and completely evaluate (research, pilot-test, etc.) high-intensity ultrasound prior to consideration as a sidestream treatment alternative.

FURTHER RESEARCH REQUIRED. Under most conditions, sidestream residual flows are returned to the head of the associated treatment process following treatment. In some cases, the recycled water may have poorer water quality than the raw water, depending upon the level of treatment received by the recycled water. As the number of available drinking water supplies diminishes and finished water regulations become more stringent, water conservation and reuse practices will continue to increase among water treatment facilities. These factors will likely cause many facilities to evaluate or reevaluate their existing sidestream treatment processes to consider alternative treatment techniques.

Several research studies have been conducted to evaluate the effects of recycle streams on the primary treatment process. In addition, the performance of several residual sidestream treatment processes (typically conventional processes) has also been assessed. Although most facilities are currently using conventional treatment processes, such as sedimentation, for residual sidestream treatment, the use of emerging treatment technologies including UV light, membranes, ozone, and ultrasound should be considered as potential alternatives under certain applications.

Significant research has been, and is continuing to be, conducted on the use of these emerging technologies for primary treatment applications. However, based on the fact that conventional sedimentation processes are capable of economically achieving the required sidestream treatment, little research has been conducted specifically on the use of these technologies for sidestream treatment. As these alternative treatment technologies become more widely available and potential supplementary benefits (beyond conventional treatment). For instance, membranes are capable of producing a very high-quality product from the sidestream flow. Consequently, it may be possible to return the decant/recovered water from the membrane sidestream treatment process directly to the finished water flow stream, in lieu of return to the head of the main treatment process. This configuration

17.28 CHAPTER SEVENTEEN

could save the costs of additional treatment (chemicals, power, etc.) associated with processing the flow through the entire treatment process. Further research and site/application-specific analyses may ultimately determine additional benefits associated with specific alternative sidestream treatment technologies not formerly recognized.

Thickening Processes. Thickening is typically the first step toward reducing the quantity of water treatment residuals. The products of the process are a low solids return flow and a thickened solids by-product flow. Return flow is generally returned to the water treatment processes, termed *used water* recovery. The thickened solids flow is either further thickened or transported off-site for disposal. Thickening has traditionally been used at water treatment plants in tandem with a dewatering or drying step. The function of thickening is to reduce land area requirements or mechanical needs of dewatering. There are three types of thickening:

- Gravity
- Flotation
- Mechanical

Gravity Thickening. Gravity thickening is the most predominantly used technique at water treatment plants. It reduces sludge bulk, provides a more consistent feed material, and reduces the size of subsequent dewatering units. Polymer is often added to increase particle size and reduce solids carryover in reclaimed water. A section of a typical gravity thickener is shown in Figure 17.7, and a picture of a gravity thickener is shown in Figure 17.8.

Sedimentation basin sludge is usually less than 1% solids when drawn off and can often be thickened to 2% solids. Aluminum and iron hydroxides may be conditioned with the aid of polymers to provide improved thickening. Typical design parameters reported

FIGURE 17.7 Gravity thickener section. (*Source: U.S. EPA*, Process Design Manual: Sludge Treatment and Disposal, *1979*.)

FIGURE 17.8 Gravity thickener. (*Courtesy of City of Phoenix Water Services Department.*)

for alum sludge thickening are 100 to 200 gpd/ft² (240 to 490 m/h) when conditioned with polymers. Alum sludges mixed with clay or lime have exhibited thickened concentrations ranging from 3% to 9% at higher overflow rates than sludges without clay or lime.

Lime sludge thickening provides more concentrated solids and a more consistent feed material for dewatering units. Solids loading levels from 60 to 200 lb/ft² of solids (290 to 980 kg/m²) of thickener surface area per day are common. High magnesium hydroxide concentrations reduce sludge dewaterability and thus reduce the density of thickened sludge.

Flotation Thickening. Flotation thickening has been used with a significant amount of success in the wastewater industry. Dissolved air flotation and recycle flotation are the two forms of flotation thickening.

In the dissolved air flotation thickening process, air is added at pressures in excess of atmospheric pressure either to the incoming residual stream or to a separate liquid stream. When pressure is reduced and turbulence is created, air in excess of that required for saturation at atmospheric pressure leaves the solution as very small bubbles of 50- to 100-μm diameter. Bubbles adhere to the suspended particles or become enmeshed in the solids matrix. Because the average density of the solids-air aggregate is less than that of water, the agglomerate floats to the surface. Water drains from the material that floats to the surface, called *float*; the float is continuously removed by skimmers.

Recycle flotation is most often used to concentrate metallic hydroxide residuals. In this type of system, a portion of the clarified liquor, or an alternate source containing only minimal suspended matter, is pressurized. Once saturated with air, it is combined and mixed with the unthickened residual stream before it is released to the flotation chamber. The advantage of this system is that it minimizes high-shear conditions, an extremely important factor when one is dealing with flocculent-type residuals (metallic hydroxides).

Coagulant residuals can be thickened in flotation thickening systems to a concentration of about 2,000 to 3,000 mg/L using a solids flux rate of 10 to 30 lb/ft^2 per day (50 to 150 kg/m^2 per day). This is higher than can be achieved by simply settling flocculated surface water in a clarifier, but less than can be achieved by gravity thickening. Flotation units are usually best operated with a continuous feed. Optimum operation requires adding polymer or other conditioning agent at a dosage of about 2 kg/dry ton (2 kg/907 kg dry).

Mechanical Thickening. In situations where it is not possible to obtain adequate thickening of sludge by using gravity or flotation thickening, mechanical thickening equipment may be used. The most commonly used type is gravity belt thickeners, but other mechanical thickening processes may also be considered.

GRAVITY BELT THICKENERS. A gravity belt thickener concentrates solids by letting gravity pull the free water through a moving porous belt. Free water passing through the belt is called *filtrate*. Figure 17.9 illustrates a typical gravity belt thickener. A polymer solution is normally injected and mixed into the solids in a chamber that feeds the gravity thickening zone. Fixed or adjustable plows guide the solids as water drains through the moving belt. Thickened solids are then discharged from the end of the thickener as the belt reaches the end roller. A scraper or an adjustable ramp is used to assist removal of the solids to a sludge hopper. A belt wash station then cleans the belt before it rotates back to the front roller.

Gravity belt thickeners have maximum hydraulic and solids loading rates that vary according to the manufacturer's design and the character of the sludge to be thickened. Actual loading rate is usually determined by pilot or bench testing. Concentration of 2.5% to 4.5% solids of metal hydroxide sludges can usually be achieved.

Advantages of a gravity belt thickener include simple design, low operating cost, limited operator attention, minimal chemical conditioning, and suitability for rapidly settling sludges such as lime. Disadvantages include two waste streams, filtrate and belt washwater, labor required for operation and maintenance, and usually chemical conditioning of the sludge.

FIGURE 17.9 Gravity belt thickener schematic. (*Source: WEF Manual of Practice No. 8, ASCE Manual and Report No. 76,* Design of Municipal Wastewater Treatment Plants. *Joint Task Force of the Water Environment Federation, Alexandria, Va., and American Society of Civil Engineers, Reston, Va., 1998. Reprinted with permission.*)

Gravity belt thickeners are usually equipped with 1-, 2-, or 3-m-wide belts, with 2-m most common. Depending on the manufacturer's design, equipment is furnished with a belt tracking system that may be hydraulic, pneumatic, or mechanical. Before the purchase of a gravity belt thickener, plant operations and maintenance staff should be consulted to see if they have a preferred design.

Provisions must be made to furnish equipment with a washwater supply of about 45 to 60 gpm (3 to 4 L/s) at the pressure specified by the manufacturer, and for receiving and disposing of the spent washwater. Other design considerations are the necessity of furnishing a polymer dosing system, splash control, and providing building humidity and ventilation control.

OTHER MECHANICAL THICKENING PROCESSES. Gravity belt thickening is the most common type of mechanical thickening. However, other mechanical devices have been used or studied for sludge thickening at water treatment plants, and these processes are discussed later in this chapter.

Dewatering Processes. Dewatering processes are typically separated into two groups: natural and mechanical. The difference between thickening, dewatering, and drying processes at times can blur. Natural dewatering processes, in particular, can receive concentrated waste flows directly from the main treatment stream and process these flows to a solids concentration that corresponds to the drying process.

Natural Dewatering Processes. Natural dewater refers to those methods of sludge dewatering that remove moisture by natural evaporation, gravity, or induced drainage. These processes are less complex, are easier to operate, and require less energy to operate than mechanical systems. However, they require a large land area, the operation depends on climatic conditions, and they are labor-intensive.

Natural dewatering typically uses the following mechanisms for the concentration of the solids stream:

- Sedimentation
- Drainage, or percolation
- Evaporation
- Decanting

These mechanisms are illustrated in Figure 17.10.

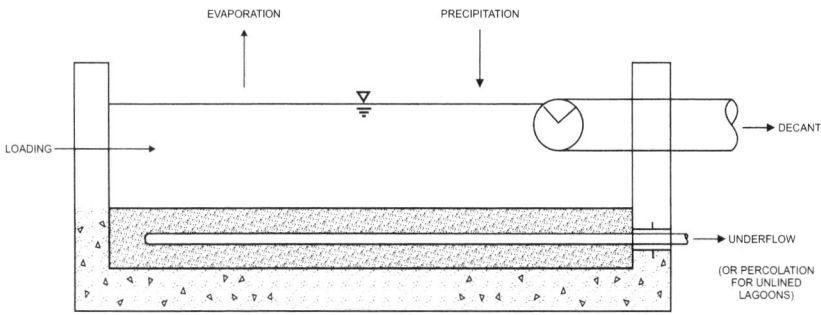

FIGURE 17.10 Natural dewatering sizing parameters.

Local groundwater protection regulations impact the design of natural dewatering processes. The regulations may require lining of a dewatering basin to protect an underlying aquifer. The state of Arizona, for instance, requires that a liner with a minimum equivalent permeability of 10^{-6} cm^{-1} be provided beneath waste lagoons. Lining increases the cost of the waste basin by both the liner installation cost and additional area and volume required due to the loss of the percolation function.

Natural dewatering processes are typically broken down into two categories: drying beds and lagoons. The two processes are typically differentiated by the depth of loading in the basin. Drying beds are typically loaded between 1 and 3 ft (0.3 and 0.9 m) in depth; lagoons are typically loaded between 5 and 20 ft (1.5 and 6.1 m) in depth.

DRYING BEDS. The lower loading depth can generally translate to both a more rapid cycle period and a greater surface area or land requirement depending on how the dewatering mechanisms are utilized. There are four basic categories of drying beds:

- *Sand drying beds.* These rectangular beds with sand or gravel bottoms contain a piped underdrain system to remove percolated water.
- *Solar drying beds.* These rectangular beds have the bottoms sealed with asphalt pavement or concrete to prevent percolation (to comply with local groundwater regulations).
- *Wedgewire beds.* These use a wedgewire septum with the goal of utilizing a filtration mechanism through sludge packed along the septum.
- *Vacuum-assisted drying beds.* These beds utilize a mechanical system to generate a vacuum over the bottom of the drying beds to enhance natural percolation.

Sand Drying Beds. The sizing of an individual sand drying bed is typically chosen with a goal of evenly distributing the solids loaded onto the bed. One reference recommends a length-to-width ratio of 1.7 to 2.5 with a length less than 75 ft (23 m). Each bed is divided by a wall constructed of concrete, unlined earthen berm, or shotcreted earthen berm. The depth of the bed is a function of the underdrain layer, sand layer, operating depth, and freeboard. A minimum freeboard of 12 in. (0.3 m) is recommended.

A typical underdrain system is shown in Figure 17.11. Underdrain pipes are commonly constructed with either vitrified clay pipe or plastic pipe. The plastic pipe is generally easier to work as the need arises to modify the configuration or perforation pattern. A minimum size of 4-in. diameter is recommended with pipes spaced 8 to 20 ft (2.4 to 6.1 m) apart. The underdrain pipes should be installed with a minimum 1% slope. The underdrain pipes are installed in gravel that is approximately 18 in. (0.46 m) deep. The gravel is graded from approximately 1.0 in. at the bottom to 0.1 in. at the top. The gravel is over-

FIGURE 17.11 Sand drying bed underdrain system.

lain with 12 to 18 in. (0.3 to 0.46 m) of sand. There are many different views on sand sizing. A typical sand might exhibit an effective size between 0.3 and 0.75 mm with a uniformity coefficient not greater than 3.5.

Some sand drying beds are installed with concrete runners to facilitate cleaning of the basins. These runners allow loaders or other mechanical equipment to clean the bed without digging too far into the sand and gravel with the resultant damage to the underdrain system. The layout of the runners requires knowledge of the equipment size and type that will be used to clean the beds. Many facilities operate with marked indicators on the dividing walls as guides to the equipment operator; other facilities operate with the knowledge that periodic replacement of sand and repair of underdrains will be necessary. Beds should have access ramps for cleaning equipment constructed of concrete or asphalt pavement.

The design of the inlet is an important feature for a drying bed, to facilitate even distribution of the solids application while reducing the incoming velocity. An example of an inlet distribution box is shown in Figure 17.12. An alternate approach involves the use

FIGURE 17.12 Drying bed inlet structure.

of pipe nozzles distributed around the perimeter of the drying bed. Splash blocks are located at the inlet to prevent undermining of the sand and gravel layers.

Various methods for dewatering can be used. The decant method should be designed with sufficient length to avoid disrupting the settling pattern of the bed contents due to high, localized velocities. Rotating weirs have been used effectively due to the shallow depth of the bed (Vandermeyden et al., 1998).

A variation on the standard sand drying bed is the freeze-assisted sand bed. Alum residuals have a jellylike consistency that makes them extremely difficult to dewater. By freezing and then thawing the sludge, the bound water is released from the cells, changing the consistency to a more easily dewatered granular type of material. Freezing alum residuals changes both the structure of the residuals slurry and the characteristics of the solids themselves. In effect, solids tend to be compressed into large discrete conglomerates surrounded by frozen water. When they thaw, drainage occurs instantaneously through the large pores and channels created by the frozen water. Cracks in the frozen mass also act as conduits to carry off the melt water.

Freezing can be done mechanically or naturally. Because of the high cost associated with mechanical systems, natural systems are most common. The optimum effects of both the freezing and thawing portions of the cycle can be obtained by exposing solids on uncovered beds. Water may drain during thawing at a faster rate and produce a greater volume when compared with applying the same unconditioned solids to a conventional sand bed.

The critical operational requirement is to ensure complete freezing of the solids layer before the next layer is applied. Hand probing with a small pick or axe usually helps make this determination.

Solar Drying Beds. Solar drying bed construction shares many common characteristics with drying beds. However, the sealing of the bed bottom and loss of the percolation mechanism impact the overall design. The bed bottom is designed with either concrete or asphalt pavement. The divider walls are typically concrete. These walls are usually placed monolithically or installed with waterstop to prevent leakage into either the subsurface or adjacent basins.

The loss of the percolation mechanism places a greater emphasis on evaporation and decanting. Some facilities utilize a mechanism to turn over the sludge. Facilities can use a tractor-mounted horizontal auger or other device to regularly mix and aerate the sludge. Mixing and aeration break up the surface crust that inhibits evaporation. Bed sizing must take into consideration the ability of a mixing vehicle or loader to navigate within the bed.

The decant mechanisms are similar to those for the sand drying bed. However, many facilities install an additional slide gate flush with the basin bottom to facilitate basin draining and cleaning.

A solar drying bed is pictured in Figure 17.13.

Wedgewire Beds. The wedgewire, or wedgewater, process is physically similar to the vacuum-assisted drying beds. The medium in this case consists of a septum with wedge-shaped slots about 0.01 in. (0.25 mm) wide. This septum supports the sludge cake and allows drainage through the slots. Through a controlled drainage process, a small hydrostatic suction is exerted on the bed, removing water from the sludge.

Vacuum-Assisted Drying Beds. In vacuum-assisted drying, a vacuum is applied to the underside of rigid, porous media plates on which chemically conditioned sludge has been placed. The vacuum draws free water through the plates, and essentially all sludge solids are retained on top, forming a cake of fairly uniform thickness. Solids can be concentrated to between 11% and 17%, depending on the type of solids and the kind and amount of conditioning agents used.

One problem encountered with this system involves improper conditioning of the sludge. The wrong type of polymer, ineffective mixing of polymer and solids slurry, and incorrect dosage result in poor performance of the bed. In addition, overdosing polymer

FIGURE 17.13 Solar drying bed. (*Courtesy of City of Phoenix Water Services Department.*)

may lead to progressive plate clogging and the need for special cleaning procedures to regain plate permeability.

Plate cleaning is critically important. If it is not performed regularly and properly, media plates will clog and the beds will not perform as expected. The special cleaning measures then required are costly and time-consuming.

LAGOONS. Lagoons are one of the oldest processes used to handle water treatment residuals. Lagoons can be used for storage, thickening, dewatering, or drying. In some instances, lagoons have been used for final disposal of residuals.

The traditional lagoon consists of either an earthen berm built on the ground surface or a large basin excavated in the ground. Lagoon depth typically varies from 4 to 20 ft (1.2 to 6.1 m). The surface area of lagoons ranges from 0.5 to 15 acres (AWWARF, 1987). Lagoon bottoms often have been installed unlined to allow dewatering by percolation. This type of approach can have operational problems due to the blinding of the soil pores with sludge with the subsequent reduction in the infiltration rate. Lagoons can also be designed with underdrain systems similar to the approach used for sand drying beds. State and local regulations have become more stringent with regard to preventing pollution of groundwater and may affect the design of water treatment residual lagoons. Liners using materials such as high-density polyethylene (HDPE), leachate collection systems, and monitoring wells are becoming common features of lagoon designs. A typical section for a lined lagoon is shown in Figure 17.14.

Inlet and decant facilities for lagoons are different than those used for drying beds due to the greater area and depth for a lagoon. The greater surface area of the lagoon typically requires multiple inlet points around the lagoon. Various designs have also used in-

FIGURE 17.14 Lagoon lining profile.

let pipes with feed valves at different operating levels. The need to decant from variable depths is also important for a lagoon. Figure 17.15 illustrates one approach using a decant pipe mounted on a swivel joint.

SIZING NATURAL DEWATERING PROCESSES. The design of a natural dewatering process requires the sizing of relatively large basins. Typical dimensions were discussed in the previous section. The sizing of the overall dewatering process requires the identification of the ultimate method of operation. Natural dewatering processes differ from thickening and mechanical dewatering in that the loading is projected over a long time (weeks or months). Natural dewatering systems go through two phases:

- *Loading phase.* This is the phase during which the basin is filled with the waste stream. The primary concentration mechanisms during this phase are settlement and decanting. Each day or week of the loading phase can be subdivided into three repeating periods:
  - *Fill.* The waste stream is being pumped or drained into the bed or lagoon.
  - *Settling.* The waste stream is shut off. A period of time is provided to allow the contents of the bed or lagoon to settle.
  - *Decant.* The operating level in the bed or lagoon is lowered through decanting. This is the straightforward approach. Some beds and lagoons are long enough that all three of these steps are performed continuously.

- *Holding phase.* This is the phase during which the contents of the basin are held to allow dewatering through the mechanisms of evaporation and, if applicable, percolation. Additional operations may be conducted within the bed or lagoon to turn or till the contents.

FIGURE 17.15 Lagoon decant structure.

Each bed or lagoon has a series of operating layers, as shown in Figure 17.16. The primary layers are the solids holding layer and the liquid operating layer. The depth at which sludge may be applied ranges from 8 to 30 in. (20 to 75 cm) for coagulant sludge and from 12 to 48 in. (30 to 122 cm) for lime sludge.

The following parameters must be established to determine the size and number of drying beds or lagoons:

V = maximum daily volume of waste flow to beds or lagoons, gal

L = loading of solids to beds or lagoons, lb

FIGURE 17.16 Drying bed operating layers.

The area of the drying beds used during the loading phase can be estimated from the following equation:

$$A = \frac{V}{N \times D_o \times 7.48}$$

where A = drying bed area, ft^2
  N = number of uses of beds each year
  D_o = operating depth of drying bed, ft
  V = annual volume of sludge for disposal, gal

The length of loading time for a bed or lagoon is a function of the solids loading rate. Typical loading criteria for lagoons and drying beds are not well documented, but coagulant sludges are typically applied to sand drying beds at 2 to 5 lb/ft^2 (1.0 to 2.4 g/m^2). The sludge loading rate (SLR) can be calculated based on the area being loaded by the application of the following formula:

$$\text{SLR} = \frac{L}{A \times N}$$

The term D_o (the operating depth) must be carefully considered in the design of the drying bed or lagoon. The timing of the fill, or loading, cycle must be compared to the design intent for the decant. The area calculation above is based on the assumption that a portion of the volume is removed through either percolation or decant. These values can also be used to establish the sludge concentration in the sludge holding layer:

$$\text{DS}_i = \frac{\text{SLR} \times 100}{62.4 \times D_s}$$

where DS_i = sludge holding layer concentration, %
  D_s = depth of sludge holding layer, ft
  SLR = sludge loading rate

These data are then used to evaluate the necessary holding time for the bed or lagoon.

The required time to achieve a desired solids concentration on the bed can be obtained from the following equation:

$$T = D_i - (D_i \times \text{VU}) - \frac{D_i \times \text{DS}_i/\text{DS}_f}{E}$$

where T = net evaporation time, months
  VU = % volume reduction through underflow/100
  DS_f = finished solids concentration, %
  E = seasonal net pan evaporation rate, in./month

The relationship of depth and concentration can also be described through the following equations (Vandermeyden et al., 1998):

$$D_d = \frac{D_i \times \text{SS}_i}{\text{SS}_d}$$

where D_d = depth in bed or lagoon after draining or decanting, ft
  D_i = initial depth of bed or lagoon, ft
  SS_i = initial solids concentration in bed or lagoon, %
  SS_d = solids concentration in bed or lagoon after draining or decanting, % (note that this assumes that all solids remain within the bed or lagoon)

This equation is further modified to note the change in depth and concentration in the bed due to evaporation after a long holding period:

$$D_f = \frac{D_d \times SS_d}{SS_f}$$

where D_f = final depth of bed or lagoon, ft, and SS_f = final solids concentration in bed or lagoon, %. The change in depth due to evaporation is then noted as

$$D_e = D_f - D_d$$

where D_e = change in depth, ft, due to evaporation.

The design of the drying beds is complicated in that the loading and evaporation rates vary according to time of the year. The development of the basis of design is an iterative process. An example is provided below.

EXAMPLE 4 Determine the number and sequencing of solar drying beds. Given:

- One drying bed is filled at a time.
- The loading is 500 lb/day year-round.
- Each bed is to be filled to a maximum loading of 4 lb/ft^2.
- The fill cycle is expected to last 1 month.
- The initial concentration of the discharge to the drying bed is 1%.
- The bed is filled each day to depth of 2 ft.
- The solids concentration after settling and decanting in the bed is 2.75%.
- One solar bed must be vacant at any given time for peak solids loading events and other contingencies.

Determine the number of solar drying beds and the operational sequence of drying beds.

Table 17.3 shows how the equations used above may be applied to a problem to determine the holding (evaporation) time needed to achieve the target final concentration of solids. After the holding times have been established, a sample loading chart can be assembled, as shown in Figure 17.17, to determine the number of drying beds needed. In this example, six drying beds are needed to maintain one vacant drying bed at any given time.

The sizing of a natural dewatering system requires a significant number of assumptions. It is common at many existing facilities to find drying bed systems that have been oversized. This is probably due to the inexact nature of the design methodology and the desire to maintain conservative assumptions in the calculations.

The development of different parameters used in the mathematical modeling and sizing of natural dewatering systems is discussed in detail in the AWWARF text *Nonmechanical Dewatering of Water Plant Residuals*.

Mechanical Dewatering. Mechanical equipment commonly used for dewatering water treatment plant sludge includes filter presses, belt presses, and centrifuges.

FILTER PRESSES. Pressure filters are typically divided into two categories: plate and frame presses and diaphragm presses. Plate and frame presses are also called fixed-volume presses and are the older category of units. Presses use a series of rectangular or circular frames with a filter cloth spanning the edges of the frame. A hydraulic drive locks the plates together at the beginning of a cycle, and solids are pumped into each chamber formed by the presses.

Pressures within each chamber range between 100 and 225 psi (690 and 1,550 kPa). Two pumps are sometimes used in this procedure: one with a relatively high pumping

TABLE 17.3 Drying Bed Loading Table

Maximum solids loading rate (lb/ft²): 2.5. Drying bed surface area (ft²), each: 5,000. Thickened sludge concentration (%): 3.

Month	Solids loading (lb/day)	No. of drying beds to be loaded	Thickened solids concentration (%)	Volume (ft³)	Initial depth D_i (ft)	Decant rate (%)	Net evaporation rate (in./month)	Holding time (months)	Total evaporation during holding time (in.)	Final solids concentration (%)
January	500	2	3	8,280	0.83	40	2	2	5	31.0
February	700	2	3	10,470	1.05	30	2	2	7.5	29.1
March	700	2	3	11,592	1.16	22.5	3	2	9.5	32.6
April	500	2	3	8,013	0.80	37.5	4.5	1	5	28.6
May	500	2	3	8,280	0.83	35	5	1	5.5	31.1
June	400	1	3	6,410	1.28	47.5	5.5	1	6.5	29.3
July	400	1	3	6,624	1.32	45	6.5	1	7	27.4
August	400	1	3	6,624	1.32	50	7	1	6.5	32.9
September	400	2	3	6,410	0.64	30	6.5	1	4.5	26.1
October	500	2	3	8,280	0.83	40	4.5	2	5	31.0
November	500	2	3	8,013	0.80	47.5	3	2	4	27.5
December	500	2	3	8,280	0.83	50	2	2	4	30.8

DRYING BED NO. 1		DRYING BED NO. 2		DRYING BED NO. 3	
Fill:	Jan	Vacant:	Jan	Vacant:	Jan, Feb
Hold:	Feb, Mar	Fill:	Feb	Fill:	Mar
Vacant:	Apr - Dec	Hold:	Mar, Apr, May	Hold:	Apr, May
		Vacant:	Jun	Vacant:	Jun, Jul, Aug
		Fill:	Jul	Fill:	Sep
		Hold:	Aug, Sep	Hold:	Oct, Nov, Dec
		Vacant:	Oct, Nov, Dec		

DRYING BED NO. 4		DRYING BED NO. 5		DRYING BED NO. 6	
Hold:	Jan, Feb, Mar	Hold:	Jan, Feb	Hold:	Jan, Feb, Mar
Fill:	Apr	Vacant:	Mar, Apr	Vacant:	Apr, May
Hold:	May, Jun	Fill:	May	Fill:	Jun
Vacant:	Jul	Hold:	Jun, Jul	Hold:	Jul, Aug
Fill:	Aug	Vacant:	Aug, Sep	Vacant:	Sep, Oct
Hold:	Sep, Oct, Nov	Fill:	Oct	Fill:	Nov
Fill:	Dec	Hold:	Nov, Dec	Hold:	Dec

FIGURE 17.17 Drying bed loading sequence.

rate to fill the chambers and one lower-rate pump to maintain pressure in the chambers after the fill is completed. At the end of the cycle, the plates are disengaged, allowing the dewatered solids to drop from the unit. Some units are termed *recessed chamber filter presses,* which generally denotes the method of mounting the cloth on the frame.

Diaphragm filters began to be used at water treatment plants during the 1980s; *variable-volume presses* is another term for these units. Operation is an enhanced version of the traditional plate and frame unit. After the fill and filter steps are completed, flexible diaphragms along the plate are expanded with compressed air or water to further compress solids in each chamber. This action is shown in Figure 17.18. Diaphragm filters generally provide higher filter cake solids and decreased cycle times compared with the traditional plate and frame press. Filtered water from the press, termed *filtrate,* flows to the end of the units for drainage away from the dewatering equipment. Chemicals used to condition sludge in filter presses include lime, ferric compounds, and polymers. Precoat materials have also been used to minimize the potential for blinding of the filter media. It is becoming more common for filter presses to use polymer as the sole conditioning chemical.

The primary operating issue with the filter press is the accumulation of solids on the filter media after each cycle. High-pressure rinse water is incorporated into the operation of most models as a sequence at the end of each cycle. Some manufacturers also recommended periodic washing of the filters with an acid-based cleaning fluid. A plate and frame press is shown in Figure 17.19.

FIGURE 17.18 Dewatering mechanism of plate and frame filter press. (*Courtesy of Eimco Water Technologies.*)

Although the filter press is applicable for dewatering lime softening sludges, it is primarily used for metal hydroxide coagulant sludges. Filter presses are often used for difficult sludges because the batch operation can maintain solids under pressure for extended periods until the desired consistency is attained. With alum or iron salt sludges, polymer conditioning, lime conditioning, or diatomaceous earth precoat is often required. Fly ash

FIGURE 17.19 Filter press. (*Courtesy of City of Tempe, Johnny G. Martinez Water Treatment Plant.*)

may also be appropriate. Cake solids contents of 20% to 50% result with pressing cycle times of 2.5 to 22 h, with approximately 8 h being average. Filter cloth life is usually about 12 to 18 months.

Filters can be provided in a variety of materials, weaves, and air permeabilities. The material should be selected based on durability, cake release, blinding, and chemical resistance. Selection is usually guided by either pilot testing or experience with similar solids streams.

Layout and profile of a filter press system are shown in Figure 17.20 with recommended clearances around the unit and head space above the filter for crane access. A traveling bridge crane is recommended for assistance in removing plates and drivers from the device.

Advantages of pressure filter dewatering include the following:

- Generally highest cake solids of any mechanical dewatering process
- High-quality filtrate
- Good mechanical reliability
- Adaptable to varying influent solids conditions

Disadvantages include relatively high construction and operating costs and the problem of some sludges blinding the media.

BELT PRESSES. Numerous types of different belt filter presses are manufactured at present. Although each of these units uses a unique design, the three schematic stages of operation in common are as follows:

- *Chemical conditioning zone.* Solids entering the press are flocculated with a chemical additive.

FIGURE 17.20 Typical filter press layout. (*Source: AWWARF Handbook of Water Treatment Plant Waste Management.*)

- *Gravity drainage zone.* Flocculated solids flow onto a porous traveling belt where free water drains from the solids by gravity.
- *Compression zone.* Solids are compacted between two belts.

The last stage is sometimes differentiated into a low-pressure zone and a high-pressure zone. Solids are initially wedged together as belts come together in the low-pressure zone, and a cake is formed at this point that withstands the shear forces of the high-pressure zone. In the next stage, forces are exerted by the upper and lower belts passing through a series of decreasing-diameter rollers to further dewater the sludge. Chemical conditioning for most belt filter press systems consists of polymer injection.

The cake has a tendency to collect on belts, creating a long-term effect of blocking the porous openings. Solids are removed in one of two ways. On some models, cake discharge blades scrape solids from the belt. On other designs, a spray wash system rinses both belts after the cake has dropped from the belts.

Belt presses used to dewater coagulant sludges generally produce cake solids of 15% to 20% when aided by optimum polymer dosage.

The belt filter press involves a large number of key components critical for proper operation. Research should be conducted to determine the right configuration for a given application. The two belts for the unit are typically constructed of woven synthetic fibers. Belts can be obtained in either seamed or seamless configuration. Seamed belts are more prone to wear at the joint; seamless belts avoid this wear but have a higher installation cost.

Rollers on belt filters are generally subject to failure caused by the high compression forces imposed on them. Rollers must be corrosion-resistant, and some manufacturers use perforated rollers to improve drainage. Roller bearings should be self-aligning and should be sealed appropriately against exposure to drainage and washdown. They should also have an L-10 life of 300,000 h (the number of hours of operation at which 10% of the bearings could be expected to fail). Accessibility of replacement parts should be reviewed for each manufacturer's design.

Filter belts are prone to tearing by abrasives in the incoming solids stream, but in most plants all abrasives have been removed by the treatment processes. However, there remains the possibility for sticks to blow into sedimentation or thickener basins. If this is a possibility, it may be desirable to add grinders to the suction side of the system feed pumps.

Layout and profile of a filter press system are shown in Figure 17.21. An elevated walkway may be desired to access the top of the unit.

Some of the principal advantages of using a filter press system are

- Low energy consumption compared with other mechanical dewatering techniques
- The most experience of available mechanical dewatering processes
- Price competitive from large number of manufacturers

Principal disadvantages are that they generally require a large amount of operator attention and that odor control is more difficult than for other mechanical dewatering techniques.

CENTRIFUGES. A centrifuge performs solids separation by applying a centrifugal force to the contents of a spinning bowl. The centrifugal force applied in municipal sludge dewatering is typically over 1,500 times that of gravity.

Several types of centrifuges have been used for solids separation in industry. The solid bowl centrifuge has developed into the principal unit used for large-scale municipal water treatment residuals dewatering. It has also been referred to as the *scroll* or *decanter centrifuge*. A schematic view of a solid bowl centrifuge is shown in Figure 17.22. The centrifuge uses two rotating elements: the bowl and the scroll. The bowl provides the solids separating force; the scroll moves the solids toward discharge from the unit. The bowl and the scroll operate at different speeds.

Solid bowl centrifuges operate continuously rather than in a batch mode. Liquid separated from the solids is termed *centrate*. Inside the spinning bowl the liquid level is a

FIGURE 17.21 Typical belt filter press installation. (*Source: AWWARF Handbook of Water Treatment Plant Waste Management.*)

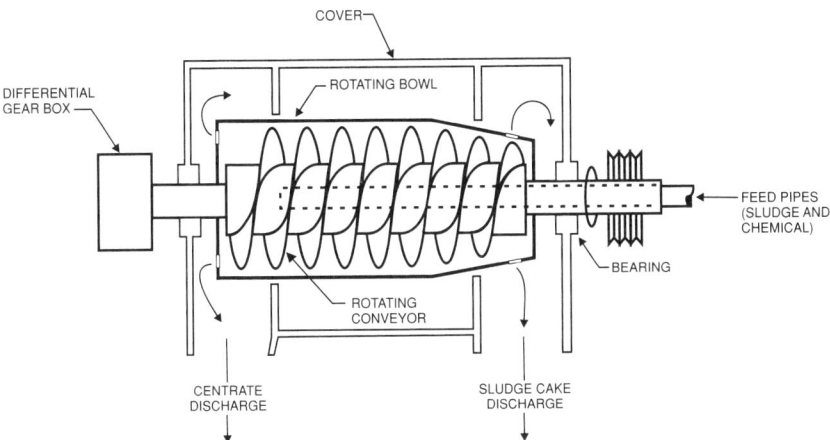

FIGURE 17.22 Centrifuge schematic. (*Source: AWWARF Handbook of Water Treatment Plant Waste Management.*)

constant distance from the rotating axis. Liquid level is controlled by weirs or dams at the end of the machine.

The centrifuge shown in the figure is called a *countercurrent* centrifuge. In this design the solids and the centrate are discharged from opposite ends of the machine. An alternative design approach, called a *cocurrent* centrifuge, diverts centrate into a series of conduits that travel parallel to the direction of the solids discharge. Manufacturers publish the relative advantages and disadvantages of each machine type. A centrifuge facility design must either predetermine a centrifuge type or be flexible enough to handle alternate solids and centrate discharge points. Centrifuges are rarely operated without a conditioning chemical.

Centrifuge loading is generally defined in terms of hydraulic loading (gallons per minute or liters per second) and solids loading (pounds or kilograms per day). Hydraulic capacity is generally defined by the centrifuge supplier. Increasing hydraulic loads affect the ability of the unit to separate solids. Using a hydraulic loading higher than the unit capacity generally leads to a decrease in centrate quality. Higher hydraulic loads may also require a higher chemical conditioning dosage.

Use of a bowl centrifuge to dewater alum sludge produces widely varying results. Some installations report that as little as 1 to 2 lb (0.5 to 1 kg) of polymer per ton of solids yields 98% recovery with a 30% cake. In other locations with low-turbidity water, as much as 4 lb (1.8 kg) of polymer per ton of solids may be required to produce a 15% cake. When the bowl centrifuge is used to dewater lime softening sludge, it yields a cake solids concentration ranging from 33% to 70%.

The bowl centrifuge may be used to selectively separate magnesium hydroxide from calcium carbonate or to produce a cake. Separating magnesium hydroxide precipitate from calcium carbonate precipitate is desirable when waste lime sludge is to be recalcined. However, separation is difficult to achieve with many waters. Magnesium hydroxide may also be recovered by carbonation of the calcium carbonate–magnesium hydroxide sludge. This process solubilizes magnesium, resulting in a high magnesium content in the centrate. Separation of the magnesium and calcium precipitates with a centrifuge results in 60% to 75% of the magnesium in the centrate. With the recarbonation/magnesium carbonate recovery process, magnesium recovery of 80% in soft water and nearly 100% in hard water is achievable.

Specifications for centrifuges should be tailored to protect the unit from corrosion, abrasion, and wear. Areas prone to abrasion include the bowl wall and the scroll blades. Stainless steel or ceramic liners should be considered for the inside of the bowl. Hardened materials or replaceable units may be considered for certain centrifuge components such as the scroll blades. Research on each type of centrifuge should include anticipated life of components, ease of parts replacement, and availability of replacement parts.

Layout and profile of a centrifuge system are seen in Figure 17.23, which shows recommended clearances around the units. A typical design issue is the removal of the centrifuge cover for access to the adjustable weirs or dams. It is standard to provide some lifting mechanism to remove the centrifuge cover, and the designer should consider installing a traveling bridge crane to move the cover out of the way while centrifuge adjustments are being made. Figure 17.24 shows a centrifuge installation.

Advantages of centrifuges include

- Continuous feed operation
- Minimal operator attention required compared with other mechanical dewatering techniques
- Excellent safety record of units
- Cake produced with relatively high percent solids
- Maintenance requirements relatively low compared with other mechanical dewatering techniques

Drying. Drying of water treatment plant residuals has historically revolved on the issue of economics—how to reduce disposal transportation costs through reducing solids volume. Drying to solids concentrations greater than 35% is becoming a regulatory issue in many areas. For instance, the state of California requires that solids concentration of a water treatment plant waste be at least 50% before disposal in a landfill. Similar to the dewatering process, the drying process may be carried out either through open-air means or through mechanical devices.

Open-Air Drying. Any of the solar drying or lagoon processes presented in the discussion of dewatering may be used for sludge drying. Drying depends only on evaporation, so it may require years to achieve the desired solids concentrations. One method of accelerating the drying is to use a tractor to periodically furrow and mix the solids to increase exposure to sun and air.

Mechanical Drying. Mechanical, or thermal, drying provides an alternative to open-air drying in locations where space is limited or environmental concerns are prohibitive. Thermal drying uses heat to evaporate water from sludge. There are no currently known applications of thermal drying at municipal water treatment plants within the United States. However, the technology has been used to a limited degree in municipal wastewater treatment in this country.

There are two basic categories of thermal drying processes:

- *Direct.* Direct thermal drying processes work on the principle of convection with direct contact between the heating medium and the sludge being processed.
- *Indirect.* Indirect drying processes work on the principle of conduction with the heating medium (water, steam, air, oil) physically separated from the sludge being processed. Figure 17.25 shows an example of an indirect drying process.

Many manufacturers can supply a drying unit with either a direct or an indirect heating mode.

FIGURE 17.23 Typical centrifuge layout. (*Source: AWWARF Handbook of Water Treatment Plant Waste Management.*)

FIGURE 17.24 Centrifuge system. (*Courtesy of City of Phoenix Water Services Department.*)

FIGURE 17.25 Heat drying process.

PROCESS RESIDUALS 17.51

Design Considerations. There are several issues that should be addressed in the design of a drying facility.

- *Feed solids moisture content.* The higher the feed solids concentration, the less energy will be necessary to achieve the drying goal.
- *Solids storage.* Thermal drying is typically operated as a batch process. The operating facility will need some storage unit or bin to contain the feed solids between cycles.
- *Containment.* Water treatment residuals typically have a low organic concentration. However, heat applied to the drying of the residuals is apt to break down any present organic material. This process can tend to generate odors. Therefore, some form of containment or housing with odor control may be necessary.
- *Equipment spacing and area ventilation.* Operating equipment and pipelines will be warm. Insulation and ventilation should account for this issue. Proper spacing around equipment will also reduce the possibility of operator injury.
- *Product moisture content.* Drying can typically achieve solids contents of 90% to 95%. However, the resultant product may tend to be dusty to the point that handling, transport, and disposal are difficult. A lower concentration will both yield an easier product to handle and lower energy consumption.

Auxiliary Solids Handling Systems. The preceding sections of this chapter have focused on the individual processes for waste treatment. Two additional elements are common components of any waste handling system: transport and chemical conditioning.

Handling and Transport. For purposes of considering handling alternatives, process residuals can be divided into two classes: low-concentration wastes, having less than 15% solids; and high-concentration wastes, with over 15% solids.

LOW-CONCENTRATION WASTES. Sludge in liquid form may be pumped through pipelines or transported in tank trucks. Dilute concentrations of coagulant sludges may flow by gravity or may have to be pumped via centrifugal pumps with nonclog impellers. Pumps and pipelines should be protected from both corrosion and abrasion. Abrasion of pump impellers is a particularly common problem in pumping sludge. Some treatment plants use pneumatic ejectors to transport sludge to avoid pump abrasion problems.

When sludges contain from 4% to 15% dry solids, positive displacement pumps are generally used. The most common type is the progressive cavity pump with a coil-shaped metallic rotor in a synthetic stader chamber. Important details in the installation are to provide pressure relief and flushing connections. If the pump is to be used to feed a mechanical dewatering device such as a belt filter press, the pump will need to be equipped with a variable-speed drive.

HIGH-CONCENTRATION WASTES. Cake solids are typically trucked away from the treatment plant site. Three traditional methods for truck loading are

- A truck drive-through directly under the dewatering equipment
- A series of conveyors from the dewatering equipment to the truck loading dock
- Use of special cake handling pumps

Designing a truck driveway requires previous knowledge of the truck type and dimensions. The unloading system shown in Figure 17.26 uses a protected chute to reduce spillage of material around the truck. The truck unloading pad should be sloped to facilitate cleanup of spilled material. A truck wash-down area may also be necessary, either at the plant site or at the dump location.

FIGURE 17.26 Truck unloading system. (*Courtesy of City of Phoenix Water Services Department.*)

Conveyor belts or screw conveyors are commonly used to transport thickened lime sludge (30% or greater) and thickened coagulant sludge (15% or greater). Conveyor manufacturers should be consulted about anticipated sludge quality before selection of equipment. The rise rate of the conveyor is dictated not only by building space, but also by the angle of repose of the particular sludge. Special belts may be specified to minimize backflow along the belt. Provisions for cleaning the belts include flexible scrapers with adjustable tension and rinse water sprays. Figure 17.27 shows a conveyor system used to lift dewatered solids.

Positive displacement pumps are available that can handle caked sludge. One manufacturer uses a piston-type pump fed by a screw conveyor.

Trucks used for transporting sludge should be watertight because free water usually results from the solids compression caused by vibration. Lime sludge is sticky and adheres to the truck body, so special surfaces in the container shell should be considered to reduce adherence.

Chemical Addition. Sludge thickening and dewatering often require a chemical for conditioning. Polymers are most commonly used for this purpose. Design of polymer feed systems is complicated by the wide variety of polymer types and brands available. Bench testing and pilot testing are almost always used to determine the best type of chemical and dosage. Dewatering equipment suppliers may also provide information about chemicals that work best in their equipment. If any sidestream from the dewatering process is recycled into the treatment process, only chemicals meeting the standards of the National Sanitation Foundation for addition to drinking water can be used.

FIGURE 17.27 Conveyor system. (*Courtesy of City of Phoenix Water Services Department.*)

Polymers can be used in dry or solution form. Solution-based polymers can be made up of varying degrees of water, oil, and polymer. Some equipment, such as the unit shown in Figure 17.28, is equipped to feed either liquid or dry chemicals.

Solids handling systems typically utilize dedicated polymer feed systems, as the type of polymers used can be different from those used as coagulant or filter aids. Separate systems may also be necessary for polymers used in the thickening and the dewatering processes due to the different attributes of those polymers. The method of activating the polymer is an important consideration in plant design. Figure 17.29 shows a schematic of two approaches to polymer feed systems. The traditional approach has been to provide a batch tank for aging the diluted polymer. Several equipment suppliers now have devices that activate polymers through the addition of energy, and they have been found to be successful with certain types of polymers.

Polymer system sizing requires parameters including pounds of neat polymer per hour, dilution to be used, quantity of dilution water required per hour, and quantity of diluted polymer used per hour. *Polymer usage* is typically defined as ratio of pounds of polymer to pounds of dry solids in the treatment process. This ratio can be defined, typically as a range, through the pilot-scale testing of the dewatering system.

Common features of a polymer feed system used for a dewatering system include the following:

- Polymer feeders are typically dedicated to the dewatering units (if mechanical dewatering is used) for simplicity of control.

- A standby polymer feeder is typically provided due to the critical nature of the polymer feed in the effective function of the dewatering system.

FIGURE 17.28 Polymer feed equipment. (*Courtesy of City of Phoenix Water Services Department.*)

Polymer feed systems are typically designed with schedule-80 PVC piping and ball valves. Liquid neat polymer is stored in either fiberglass-reinforced plastic or polyethylene tanks. The storage sizing for solids handling systems is more complicated than that for typical chemical feed systems. The shelf life of the polymer is balanced with the anticipated polymer delivery time. Polymer tanks for storage of emulsion or mannich polymers are sometimes equipped with recirculation pumps.

The indoor storage of polymer will have special requirements if the proposed brand is rated as flammable. This may trigger the need to provide a fire protection system for all or a portion of the storage building. Polymer systems are typically installed within concrete containment areas regardless of the hazard rating of the chemical. Polymer spills are extremely slick and pose a hazrard to the operation and maintenance staff.

Odor Control. Odor control at drinking water treatment plants has not been developed to the same degree as at wastewater treatment plants due to the lower level of odor-generating compounds. Odors in water treatment residuals typically can be divided into four categories:

- Odors associated with compounds in the raw water such as geosmin and MIB
- Odors resulting from disinfection (most significantly where the source of backwash water has been disinfected)
- Odors resulting from the biological degradation of treatment residuals
- Odors resulting from the growth, death, and biological degradation of aquatic biota in relatively stagnant open basins such as lagoons

* = TYPICAL FOR EMULSION OR MANNID POLYMER SYSTEM
** = PROVIDE HOT WATER CONNECTION IF AVAILABLE

FIGURE 17.29 Polymer feed system schematics.

Control of odors generated within the residuals handling system has typically been dealt with through operational control. Operators have used methods such as reduction of detention time within the sedimentation and thickening processes and direct application of chemicals such as lime to open-air dewatering/drying processes. The growing emphasis in on-site treatment and mechanical dewatering has led to more attention regarding odor control at water treatment solids handling facilities.

Characterizing Potential Odors. Minimal information regarding odor levels emanating from solids handling facilities has been published. One study by the city of Phoenix, Arizona, sampled air quality around the solids handling facilities of two existing plants. This brief (2-day) analysis recorded the following levels:

Constituent	Range (ppb)	Average (ppb)
Hydrogen sulfide	2–40	7.4
Reduced sulfur compounds	12–47	23
Amines	Below detection limits	Below detection limits

These values may vary significantly depending on the raw water quality, process configuration, and operational mode of a treatment facility. For instance, the use of powdered activated carbon (PAC) in the main treatment process may have a significant effect on odor potential. Extended storage of PAC-laden residuals may result in the desorption of compounds into the water with subsequent discharge of odors to the atmosphere. However, these values do provide an example of constituents that may be considered in an odor control design.

Odor Control Options. The following odor control options typically may be included as part of a solids handling facility design:

- For open-air dewatering and drying systems: chemical stabilization
- For thickening and solids storage basins: covering, capture, and treatment; and chemical stabilization
- For mechanical dewatering and treatment: enclosure, capture, and treatment

CHEMICAL STABILIZATION. This type of system functions on the principle of chemical oxidation of the odor-causing compounds. The goal of chemical stabilization is generally to raise the pH of a residual process stream or holding basin to 11. Most odors associated with residual process streams result from hydrogen sulfide and other reduced sulfur compounds, which have higher solubility at high pH. By maintaining a pH of 11, these compounds are kept in the liquid phase and therefore are not released to the atmosphere to cause odors. Adjustment of pH has traditionally been conducted through the manual application of bags of lime to either a drying bed or a lagoon. The in-process addition of either lime or caustic soda to achieve the pH increase is becoming more common.

Some facilities have also included the ability to add a chlorine compound to "shock" the inflow into a thickening basin or an open-air basin. This addition is typically done in environments that are conducive to aquatic growths such as algae. This approach has had mixed operational results, as the necessary level of chlorine may result in additional odor problems.

Potassium permanganate and copper sulfate addition to open-air basin has also been used as a method to control algal growths. The use of copper sulfate may be limited due to metal limits imposed by the final disposal of the dewatered and/or dried residuals.

ENCLOSURE, CAPTURE, AND TREATMENT. The enclosure or covering of a process is typically limited to areas with either long detention times or concentrated solids, including the following:

- Solids storage or equalization basins
- Mechanical dewatering facilities
- Residuals truck loading bays
- Filtrate or centrate collection tanks

Tanks, basins, and other nonoccupied spaces are typically ventilated at 4 to 6 air changes per hour to ensure a slight vacuum is induced on the headspace at all times. Occupied spaces such as rooms, dewatering facilities, truck bays, and the like are typically ventilated in accordance with applicable codes and guidelines, which frequently require 12 air changes per hour.

Collected air is typically collected through fiberglass ducting due to the potential for corrosion in most metallic ducting systems.

The treatment of collected odor-bearing airstreams is generally achieved through one of the following scrubber systems:

- *Water scrubbers.* This type of system considers the solubility of odor-causing compounds in water. Hydrogen sulfide and reduced sulfur compounds tend to have a low solubility in water. A water scrubber uses packing material to increase the area of interface for air-liquid contact. This process requires the determination of a water recirculation rate based on the estimated concentration of compounds.

- *Chemical scrubbers.* This type of system functions on the principle of chemical absorption. Caustic soda and sodium hypochlorite are typically used for the removal of hydrogen sulfide, reduced sulfur compounds, and amines. The chemical scrubber design focuses on chemical injections rates, recirculation rates, and the number of passes based on volume of the airstream and the concentration of odor-controlling compounds.
- *Granular activated carbon scrubbers.* Granular activated carbon scrubbers utilize surface adsorption for the removal of odor-generating compounds. Three types of GAC scrubbers are commonly available:
  - *Catalytic GAC:* generally found to be more effective for hydrogen sulfide removal than for reduced sulfur compound removal
  - *Unimpregnated GAC:* typically effective for reduced sulfur compound removal
  - *Impregnated GAC:* generally more effective for the removal of hydrogen sulfide with limited capacity for the removal of reduced sulfur compounds

One design approach that has been utilized on the wastewater treatment handling system approach to odor control is to use a chemical scrubber with the addition or potential future addition of a GAC scrubber to increase the removal rate of potential odor-generating compounds.

Recycling/Disposal of Liquid Wastes. As Figure 17.30 illustrates, each type of waste treatment produces a liquid sidestream such as decant or filtrate. These may fit into either of two definitions:

- *Recoverable water* is water that may be recirculated to the main water treatment process.
- *Nonrecoverable water* is water that may not be recirculated to the plant and must be disposed of through an alternative means, such as to a sanitary sewer or by evaporation.

It is generally desirable to recover as much water as possible, both for water conservation and to minimize the cost of alternative disposal. Some sidestreams are automatically

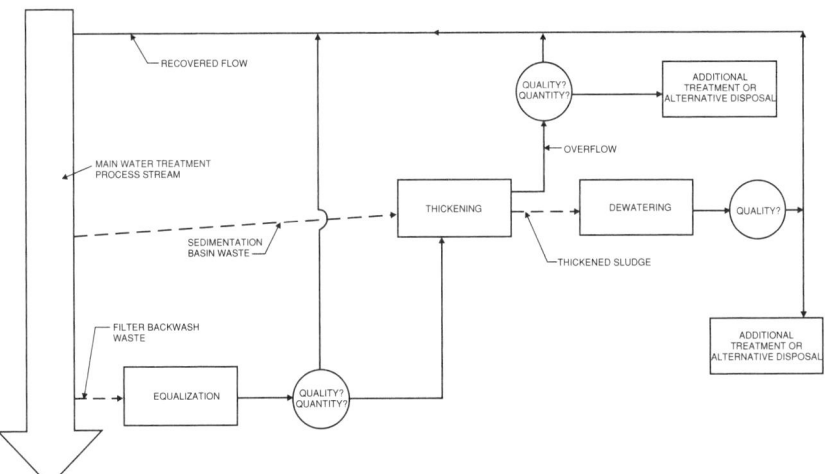

FIGURE 17.30 Concentration and sidestreams.

defined as nonrecoverable. This category includes filtrate from a mechanical dewatering process where a nonapproved polymer has been used, but most sidestreams have the potential for recovery.

The main concern regarding recycling recoverable water is its impact on treatment plant finished water quality. The parameters of concern are

- Increased concentration of *Giardia* and *Cryptosporidium* cysts
- Increased concentration of assimilable and total organic carbon
- Increase in total trihalomethanes and trihalomethane precursors
- Increase in turbidity and particle counts
- Increase in aluminum concentration

RESIDUALS HANDLING CONSIDERATIONS

Among the decisions that must be made by the water treatment plant designer regarding residuals disposal are the selection of the disposal method and equipment best suited for the plant, methods of transporting residuals, and the final disposition of liquid and solid wastes.

Residuals Handling Process Performance

The residuals handling process for a treatment plant is selected after a thorough comparison of available treatment and disposal alternatives.

Comparison of Thickening Processes. Factors that affect performance of the various types of thickening processes include the type of solids, their characteristics, influent concentration of the solids to the process, climatological conditions for open-air processes, variations in influent flow rates, and type and dosage of chemical conditioner.

This section provides typical ranges of values for percent solids that each thickening and dewatering process may generate. The comparison includes solids loading on the units and the solids concentration of the various types of residuals often experienced in these processes.

A comparison of thickening processes is shown in Table 17.4, and a summary of dewatering process performance provided by *Water Treatment Plant Waste Management* (AWWARF, 1987) is shown in Table 17.5.

Definition of Residuals Processing Requirements. To identify the ideal combination of unit processes necessary to form a preliminary residuals processing alternative, it is im-

TABLE 17.4 Comparison of Thickening Processes

Type of process	Residual	Percent solids
Gravity	Carbonate	1 to 30
Gravity	Hydroxide	1 to 3
Flotation	Hydroxide	0.2 to 0.3
Gravity belt	Hydroxide	2.5 to 4.5

TABLE 17.5 Range of Cake Solid Concentrations Obtainable

	Percent solids concentration	
Process	Lime sludge	Coagulant sludge
Gravity thickening	15 to 30	3 to 4
Scroll centrifuge	55 to 65	10 to 15
Belt filter press	—	10 to 15
Vacuum filter	45 to 65	—
Pressure filter	55 to 70	35 to 45
Sand drying beds	50	20 to 25
Storage lagoons	50 to 60	7 to 15

perative that specific requirements of the residual processing system be defined. To define specific requirements of a system, the following fundamental information must be obtained (or generated if it does not already exist) (ASCE/AWWA/EPA, 1997):

- Residuals disposal limitations
- Quantity and quality of residuals sources
- Resource recovery potential limitations
- Residuals mass balance

Disposal Limitations. In most cases the ultimate method of residuals disposal determines limitations, which in turn define process design requirements. The percent solids content of the residual is considered the primary criterion to define acceptable limits of a disposal option. Landfill and land application disposal options can, of course, be accomplished with a wide range of solids content, but both disposal options have different equipment requirements at the low and high solids limits of their ranges. The following tasks will help establish disposal limitations that a water utility must consider for further evaluation of residual process alternatives.

1. *Identify the disposal limitations of available residual disposal alternatives.* The range of normal residual solids content of the six most common methods used in the water industry are:
   - Land application: <1% to 15% solids
   - Landfilling (codisposal): >15% to 25% solids
   - Landfilling (monofill): >25% solids
   - Discharge to sewer: <1% to 8% solids
   - Direct stream discharge: <1% to 8% solids
   - Residuals reuse: <1% to >25% solids

2. *Identify external disposal costs, if they apply.* Disposal by landfilling in a codisposal municipal landfill can be associated with a per-unit-weight tipping fee levied by the landfill authority. Discharges to a sewer also have a user charge that can be assessed based on flow rate units or the combination flow rate and suspended solids concentrations. In some instances residuals reuse requires the water utility to pay a fee to the

user to cover additional costs of special handling, or the difference in production costs associated with accommodating the sludge over a more traditional material.

Quantity and Quality of Residuals. Many water utilities have not closely monitored the quantity and quality of the treatment plant residuals. This deficiency of data is typical and requires additional effort and time to do the following:

1. *Inventory residual sources.* Typical residual sources associated with conventional water treatment plants are
   - Headworks bar screens
   - Grit basins
   - Presedimentation basins
   - Sedimentation basins
   - Filter backwash
   - Filter to waste

2. *Quantify flow rates and quantities of residual produced.* Flow rate data collected for each residuals source should include the following:
   - Maximum, minimum, and average flow rates
   - Maximum, minimum, and average volume at each release
   - Frequency of events
   - Average event duration
   - Maximum number of events per day
   - Potential for concurrent events

3. *Determine the quality of each residual source.* Parameters that should be analyzed are
   - Aluminum and iron
   - pH
   - Total dissolved solids
   - Total organic carbon
   - Trihalomethane formation potential
   - Trace metals
   - Toxic characteristic leachate procedure (TCLP)

Resource Recovery Potential. Although standard water conservation strategies support the recovery of reusable residuals, consideration must also be given to potential risks of continuous recycling of infectious organisms. Typical recycling of untreated residual streams such as backwash and filter-to-waste residuals back to the plant influent can result in accumulation of organisms in treatment processes. In light of these concerns, the value of water continues to increase with time, and recovery potential for reusable residual streams should be evaluated using the following procedure:

1. *Assess the value of recovered water.* Essentially the value of recovered water depends on the value of source water and the relative abundance of supply. Establish whether the cost of recovery is justified.

2. *Identify residual sources that are easily recovered.* Backwash and filter-to-waste residual streams generally constitute the largest volume of residuals generated with the lowest percent solids concentration. Treating these residual streams usually is justified

PROCESS RESIDUALS

for the value of the water that is recovered. Other residual streams should be evaluated closely with regard to their potential impact on finished water quality.

3. *Identify residual sources unsuitable for recovery.* The most difficult residual streams for which to justify recovery are the sidestreams of mechanical dewatering processes. They usually contain many components that are undesirable to return to the plant.

Residuals Mass Balance. The most valuable exercise in evaluating residuals generated at a water plant is the development of a mass balance. A mass balance is a theoretical accounting of the water and solids contents that enter the plant, solids that are added to the water, the water and solids contents that leave the plant in the finished water, and the residual streams. This assessment must include the assessment of the effects of normal variations in source water quality and the associated chemical feed rates necessary to treat the water on the residual production rates. The objective of the mass balance is to specifically define the following parameters:

1. Establish maximum, minimum, and average solids production rates for the source water and for each residuals sidestream in terms of percent solids, flow rates, and total solids produced per million gallons of water treated.
2. Define the allowable return residuals stream limits in terms of hydraulic and percent solids limits (e.g., maximum backwash returned to the plant influent should not exceed 5% of operating production capacity).
3. Define the allowable solids content in the finished water.
4. Define the allowable flow rate and solids content for available residuals disposal options.
5. Based on the allowable return limits, effluent limits, and disposal limits for each residual stream, determine a required residuals processing efficiency for each residual stream. The objective of this exercise is to identify potential solids content objectives for each residual stream process.

Residual Handling Process Design Issues

Methods used in handling water treatment plant residuals depend on the character of wastes produced, the quantities of materials to be produced, and the type and size of treatment equipment to be used.

Estimating Solids Production. Solids mass balances are prepared to develop estimates of hydraulic and solids loading on residuals handling unit processes. Solids production from a water treatment plant may be estimated based on source water suspended solids removal and quantity of process chemical added. The following formulas may be used in developing solids loadings (AWWARF, 1987):

$$S = \begin{cases} Q(0.44\text{Al} + \text{SS} + P)[8.34 \text{ lb/(mil gal/mg/L)}] & \text{aluminum sulfate coagulant sludge} \\ Q(2.9\text{Fe} + \text{SS} + P)[8.34 \text{ lb/(mil gal/mg/L)}] & \text{ferric chloride coagulant sludge} \end{cases}$$

where S = solids produced, lb dry solids per day
 Q = water production, mil gal
 Al = alum dosage, mg/L (alum as 17.1% solution)
 SS = suspended solids concentration, mg/L
 Fe = ferric chloride dosage, mg/L (dosage as Fe)
 P = polymer dosage, mg/L

A similar equation has been developed for solids produced from a lime softening plant:

$$S = Q(2.0\text{Ca} + 2.6\text{Mg})[8.34 \text{ lb/(mil gal/mg/L)}]$$

where Ca = calcium hardness removed as $CaCO_3$ and Mg = magnesium hardness removed as $CaCO_3$. Note that factors (0.44Al, 2.9Fe, P) should be added when metal hydroxide coagulants or polymers are used in a lime softening facility.

A series of issues must be addressed in estimating solids production for the preliminary design of a facility:

1. Historical trends of source water suspended solids loading must be developed to determine the average day, maximum day, and peak hour loadings. These values are typically used to evaluate the following design issues:
   a. Average day: used in analyzing annual solids disposal fees, land availability for open air dewatering, and drying processes
   b. Maximum day: used to size thickening, dewatering, and drying processes
   c. Peak hour: used for the sizing of piping and pumping
2. If equipment such as centrifuges or belt filter presses is sized based on maximum day loadings, the residuals handling systems must be designed with the capacity to absorb the difference between peak hour and maximum day loading. This may be done by incorporating equalization basins into the design. An alternative approach is to consider the ability of a given process to handle short-term spikes in the solids loading rate.
3. Most water treatment facilities record solids loadings in terms of turbidity (ntu) rather than suspended solids. A sampling program should be established to define a statistical correlation between turbidity and suspended solids.
4. Estimating source water suspended solids loadings is inherently more difficult for new water treatment plants as opposed to the design of residuals handling facilities for existing plants. The following options are available for estimating loadings for a new plant:
   a. Use historical records for a water treatment plant with similar processes and a comparable water source.
   b. Develop a water quality database through a program of sampling, jar testing, filtration, and analysis. Care must be taken with this program in that a short-term program represents conditions for only one period in time and may not represent the long-term water quality for the site.

Determining Equipment Sizing. Water treatment plant residuals handling systems vary from simple to complex. The mass balance provides the starting point for the design process. The next step is to generate a schematic based on refined flow rates and individual equipment loadings.

Issues that must be determined at this preliminary stage in design are the number of anticipated shifts, coordination with main plant operations, and provision for redundant process trains. Figure 17.31 shows an example of how these factors can impact the design of a mechanical dewatering system. Projected equipment suppliers also play a key role in this design phase as specific equipment capabilities and requirements are identified.

The open-air dewatering and drying system is one of the most common forms of residuals handling processes. Although the process approach and system equipment are relatively simple, the actual design depends on local climatological conditions, regulatory requirements, and projected use.

FIGURE 17.31 Impact of staffing and design on contingencies.

Hydraulic Profile. The hydraulic profile of a solids handling system is typically more complex than the mainstream treatment process as the flow schematic is more complicated. Figure 17.32 shows the profile of a typical facility. One of the first design factors in the process is the attempt to provide gravity flow conditions for the relatively high waste flows from the sedimentation basins and filters. The flexibility of shifting streams to different processes may also impact the design of the hydraulic profile. Setting the grade of a mechanical dewatering facility is generally simplified as thickened sludge pumping is always necessary for the controlled feed of the dewatering equipment. However, a

FIGURE 17.32 Waste handling system hydraulic profile. (*Courtesy of City of Phoenix Water Services Department.*)

two-story building with dewatering equipment above a truck bay can be a very high-profile structure. The architectural cost of the building can be significantly impacted by the visibility to off-site neighbors.

Contingency Planning. Contingency planning design issues that should be incorporated into a water treatment plant residuals handling facility are listed in Table 17.6, along with some potential design resolutions.

Backup systems and storage spaces are key contingency planning items often incorporated into residuals handling facility designs. For example, for a facility that uses cen-

TABLE 17.6 Residual Handling Facility Contingency Planning Issues

Potential problems	Potential resolutions
Equipment breakdowns	Provide redundant units and piping.
	Allow for solids storage during repair.
	Base design on single shift; second or third shift can be used on emergency basis.
Solids disposal disruption	Plan for a secondary disposal approach.
	Allow for solids storage during disruption.
Solids production greater than expected	Provide for safety factors in process design.
	Allow for solids storage during peak periods.
	Base design on single shift; second or third shift can be used on emergency basis.

trifuges as the primary mode of dewatering, solar drying beds may also be included in the facility for the following purposes:

- To handle peak solids loadings the centrifuges may not be able to effectively treat
- To serve as a dewatering process in the event of a centrifuge failure
- To provide a dewatered solids storage area in case off-site solids hauling is disrupted

Figure 17.31 shows how contingency planning can be utilized in the design of a solids handling facility.

Facility Drainage. The need to drain portions of the treatment plant for annual maintenance should also be addressed in the layout of a solids handling facility. Equalization basins, redundant clarifiers or thickeners, drying beds, and lagoons can all be used as a destination for drainage water. However, the design must avoid directing floor or storm drains that could carry potentially contaminated water into the solids handling system process stream.

Residuals Disposal Unit Process Selection

The decision to install or modify solids handling processes at a water treatment plant can result from a variety of forces ranging from state or local regulation of waste stream discharge to minimizing operating costs. In any event, the decision is accompanied by a selection procedure to identify the residuals handling processes appropriate for the plant.

Selecting residuals handling processes is a more complicated task than selecting traditional liquid stream water treatment processes for the following reasons: There is much less existing operating experience with residuals handling processes to use as a basis of comparison, and residuals handling processes are more difficult to test with a procedure such as a jar test or a pilot filter because of the interdependence of solids handling with other treatment processes.

The following selection criteria are typically used in screening and selecting residuals handling processes:

- Discharge limitations and the effective operating range of the residuals handling process
- Similar operating experience with unit process
- Bench- and pilot-scale testing of unit processes
- Environmental impact of unit process
- Economics

Environmental Impacts. Environmental impacts have become a standard criterion in developing desired alternatives for any public facility. However, environmental criteria can be frustrating as a basis for comparison because of their subjectivity. The environmental impact of a water treatment plant residuals handling process varies greatly depending on whether the viewpoint is that of an operator, a neighbor, or a migrating egret. However, environmental criteria are typically part of the selection process. Some environmental criteria are described below:

- *Effectiveness in meeting discharge requirements.* The ultimate criterion for selecting a process or a combination of processes is whether the process will reliably and consistently meet the regulatory discharge requirements. This depends on many factors, including the sizing of the application and process loading conditions.

- *Effect on groundwater quality.* Any process with the possibility of contaminating groundwater is a possible culprit in this category. Unlined lagoons and drying beds are typically of greatest concern.
- *Creation of noise.* This criterion is a concern to plant employees and site neighbors. Noise is a typical concern for thickening and dewatering equipment using mechanical equipment. Open-air processes may also involve noise in the form of front-end loaders or other devices used to move and remove solids. Off-site transport of solids is another solids concern. Mitigation techniques include supplying hearing safety equipment for workers, enclosing the operation in acoustically sealed buildings, and specifying operating hours for front-end loaders and other vehicles.
- *Creation of odors.* Many operators emphatically state that there are no odors associated with water treatment residuals, and in most cases these individuals are correct. However, the possibility always exists for odors with open-air processes and sometimes even with mechanical systems. The traditional technique for odor mitigation is an operational change in the loading and removal of solids from process units. Chemicals such as caustic and chlorine may have to be added to stabilize and destroy odor-causing organisms in the sludge. Odor control units may be required for systems installed indoors.
- *Energy use.* Energy use is a function of both direct usage by equipment and the cost of removing solids from the site. Mechanical dewatering techniques involve the highest energy consumption, although there is a wide variation among different processes. Different applications of processes can also affect the percent solids concentration of the residual before disposal. A higher concentration translates to a lower volume for disposal. This lower volume then relates to less transportation energy.
- *Attraction of insects and other pests.* Insects are a potential concern with open-air processes. Submerged lagoons or gravity thickeners may serve as breeding grounds for mosquitoes and other flying insects. Flies and gnats may be a concern with drying residuals in the open air.
- *Impact on the neighborhood.* This broad category encompasses many other environmental concerns that result in the lowering of local property values.
- *Creation of air pollution.* Air pollution is a more typical concern in evaluating wastewater processes. Concerns about stripping volatile contaminants to the air with water treatment systems are minimal. Possible concerns may arise from the potential use of gas- or diesel-driven engines with mechanical equipment.
- *Space requirements.* Many plant sites do not have the benefit of a large property area or the possibility of expansion onto adjacent vacant land. Limited space typically drives process selection away from open-air processes to mechanical processes.
- *Employee and public safety.* Mechanical handling processes typically involve the highest safety concern of potential employee accidents. All processes involve some potential hazard, ranging from suffering heatstroke while working in a solar drying bed to drowning in a thickener.

One issue to note in comparing environmental criteria is the interrelationship of these criteria with both construction and operations and maintenance costs. For example, groundwater quality concerns associated with lagoons can be resolved by installing a relatively impermeable liner beneath the lagoons. Thus the liner mitigates the environmental concern at a significant increase in construction cost. This should be considered in the analysis of different residuals handling approaches.

Bench Tests. Bench and pilot tests serve two functions in selecting residuals handling processes: They provide additional data for the sizing of the full-scale equipment, and they provide an indication of the process performance (in other words, what percent solids can be achieved).

Bench and pilot testing should be considered after the number of alternative residual handling schemes have been winnowed to a few options requiring further study. Testing may not be needed if there is similar experience with process at a comparable facility, such as a plant of similar size with approximately the same source water quality and unit processes.

The following bench-scale tests are commonly used for thickening and dewatering:

- *Thickening:* settling tests, flotation test, capillary suction time test
- *Dewatering:* time-to-filter test, filter leaf test, capillary suction time test, settling tests

Bench testing does not provide any direct correlation to the performance of a given process. Bench tests are typically used in conjunction with full-scale operation to develop correlation, for instance, between the results of capillary suction time tests and the performance of centrifuges. The application of the bench test is then to analyze the effect of different sludge conditioners. Bench tests may have application to selecting a residuals handling process if the bench tests can be correlated to process performance based on previous testing. Equipment suppliers should be asked for the availability of past bench test comparisons.

Pilot Testing. Pilot testing involves the use of test equipment similar enough in size to the full-scale operating equipment that testing results can be directly compared between pilot test and full-scale operation. This testing may involve constructing a scaled-down version of a process basin such as a gravity thickener or a sand drying bed. Suppliers of mechanical thickening and dewatering equipment typically have test units available for rent or loan.

The use of bench and pilot tests has many pitfalls. Testing must be as representative as possible of real-world conditions. Testing procedures must simulate all processes involved with an alternative. This is a particular concern with thickening and dewatering, where incoming solids must represent the solids concentrated and transported from the previous process. Even when the bench testing and pilot testing successfully model process performance, the results may not reveal operational problems that are part of the process.

The duration of bench and pilot testing is of particular concern. Year-round operation may involve significant changes in source water quality that may greatly affect the operation of the residuals handling process. One week's worth of pilot testing may not reflect this impact. It is generally not feasible to perform year-round pilot testing. The use of short-term pilot testing to correlate results with bench tests may allow long-term testing on a lesser scale.

Analysis of pilot-testing results should consider the limitations of the test setup and protocol. If only limited testing was performed, a lower emphasis should be placed on the test results.

ADDITIONAL INFORMATION SOURCES

Regulatory restrictions have placed an increased focus on the design of water treatment plant residuals handling facilities in the 1980s and 1990s. This design field is complicated by regulations, site neighbor concerns, and a growing number of technical options for treatment and disposal. The water treatment plant designer also has a limited number of reference materials to utilize.

Two texts are particularly suggested for a more detailed coverage of this subject. The handbook *Water Treatment Plant Waste Management* (AWWARF, 1987) provides a comprehensive overview of residuals management issues. It also presents a series of cost curves that are useful in the preliminary screening of treatment and disposal alternatives.

The design procedures presented in this chapter originate from *Technology Transfer Handbook: Management of Water Treatment Plant Residuals* (ASCE/AWWA/EPA, 1997). This text provides a broad background into the regulatory requirements and solids charateristics associated with residuals management. This text provides additional data on cost estimating of residuals handling systems.

ACKNOWLEDGMENT

The authors would like to gratefully acknowledge the efforts of Jason Marie and Dave Sobeck in the preparation of texts and figures for this chapter.

BIBLIOGRAPHY

Amy, G., et al. *Arsenic Treatability and Options and Evaluation of Residuals Management Issues.* Denver, Colo. American Water Works Association Research Foundation, 2000

Arora, H., et al. "Spent Filter Backwash Water Contaminants and Treatment Strategies." *Journal AWWA* 93(5), 2001.

ASCE/AWWA/EPA. *Technology Transfer Handbook: Management of Water Treatment Plant Residuals.* New York: American Society of Civil Engineers, 1997.

Ashcroft, Craig, et al. "Modifications to Existing Water Recovery Facilities for Enhanced Removal of *Giardia* and *Cryptosporidium*." *Proceedings WEF and AWWA Joint Residual Conference.* Alexandria, Va.: Water Environment Federation, 1997.

AWWA Residuals Management Research Committee Subcommittee on Membrane Residuals Management. "Committee Report: Residuals Management for Low Pressure Membranes." *Journal AWWA,* 95(6), 2003.

AWWARF. *Water Treatment Plant Waste Management.* Denver, Colo.: American Water Works Association, 1987.

Benjes, H. H., Jr. "Treatment of Overflows from Sanitary Sewers." Presented at the 9th Texas WPCA Conference, Houston, Tex., July 1970.

Black, A. P., and C. G. Thompson. *Plant Scale Studies of the Magnesium Carbonate Water Treatment Process.* Publication EPA-660/2-75-006. Cincinnati, Ohio: U.S. Environmental Protection Agency, 1975.

Carr, S., et al. "Why Dry?" *Proceedings of the WEF/AWWA Joint Residuals and Biosolids Management Conference,* San Diego, Calif., 2001.

Cornwell, D. A. "An Overview of Liquid Ion Exchange with Emphasis on Alum Recovery." *Journal AWWA* 71(12):741, 1979.

Cornwell, C. A., and R. G. Lee. *Recycle Stream Effects on Water Treatment.* Denver, Colo.: American Water Works Association Research Foundation, 2003.

Cornwell, D. A., and R. M. Lemunyon. "Feasibility Studies on Liquid Ion Exchange for Alum Recovery from Water Treatment Plant Sludges." *Journal AWWA* 72(6):64, 1980.

Cornwell, D. A., and M. J. Macphee. "Effects of Spent Filter Backwash Recycle on *Cryptosporidium* Removal. *Journal AWWA* 93(4), 2001.

Cornwell, D. A., and J. A. Susan. "Characteristics of Acid Treated Alum Sludges." *Journal AWWA* 71(10):604, 1979.

Cornwell, D. A., et al. *Commercial Application and Marketing of Water Treatment Plant Residuals.* Denver, Colo.: American Water Works Association Research Foundation, 2000.

Cornwall, D., et al. "Considerations for Sizing Equalization Basins for SFBW Handling: Cleveland Case Study." *Proceedings of the WEF/AWWA Joint Residuals and Biosolids Management Conference,* San Diego, Calif., 2001.

Cornwell, D. A., et al. *Treatment Options for Removal of* Giardia, Cryptosporidium, *and Other Contaminants in Recycled Backwash Water.* Denver, Colo.: American Water Works Association Research Foundation, 2001.

Culp, R. L., and W. I. Wilson. "Is Alum Sludge Advantageous in Wastewater Treatment?" *Water and Waste Engineering* 16(79):16, 1979.

Edzwald, J. K. *Impacts of Filter Backwash Recycle on Clarification and Filtration.* Denver, Colo.: American Water Works Association Research Foundation, 2001.

Elliott, H. A., et al. *Land Application of Water Treatment Plant Sludge.* Denver, Colo.: American Water Works Association Research Foundation, 2000.

Gumerman, R. C., R. L. Culp, and S. P. Hansen. *Estimating Water Treatment Plant Costs.* Publication EPA 600/2-79-162. U.S. Environmental Protection Agency, 1979.

Hagstrom, L. G., and N. A. Mignone. "Centrifugal Sludge Dewatering Systems Can Handle Alum Sludge." *Water and Sewer Works* 125(5):54, 1978.

J. M. Montgomery, Consulting Engineers, Inc. *Water Treatment Principles and Design.* New York: John Wiley & Sons, 1985.

Knocke, R. W. "Thickening and Conditioning of Chemical Sludges." *Proceedings ASCE Environmental Engineering Conference,* San Francisco, Calif., July 1978.

Marie, J. R., et al. "Joint Solids Handling Facility Feasibility Study for Water Treatment Residuals and Biosolids in Phoenix, Arizona." Paper presented at the Water Environment Federation 76th Annual Technical Conference and Exposition, Los Angeles, Calif., October 2003.

Marinelli, F. *Impact of Backwash Recycling on Water Treatment Plant Performance.* Denver, Colo.: American Water Works Association Research Foundation, 2001.

Metcalf & Eddy, Inc. *Wastewater Engineering: Treatment, Disposal, and Reuse.* New York: McGraw-Hill, 1991.

Mickey, M. "Membrane Concentrate Disposal Issues." *Proceedings of the 4th Joint WEF and AWWA Conference on Biosolids and Residuals Management.* Alexandria, Va.: Water Environment Federation, 1993.

Pigeon, P. E., et al. "Recovery and Reuse of Iron Coagulants in Water Treatment." *Journal AWWA* 10(7):397, 1978.

Pizzi, N., et al. "Evaluation of Backwash Recycling Strategies for Cleveland, Ohio." *Proceedings of the WEF/AWWA Joint Residuals and Biosolids Management Conference,* Charlotte, N.C., 1999.

Pontius, F. "Update on USEPA's Drinking Water Regulations." *Journal AWWA,* 95(3), 2003.

Reddy, S., et al. "Odor Control at a Water Treatment Plant?" *Proceedings of the WEF/AWWA Joint Residuals and Biosolids Management Conference,* Charlotte, N.C., 1999.

Sanks, R. L., et al. *Pumping Station Design,* 2d ed. Boston: Butterworth Heineman, 1998.

U.S. Environmental Protection Agency. *Process Design Manual: Sludge Treatment and Disposal.* Cincinnati, Ohio: USEPA, 1979.

Tobiason, J. E., et al. "Full-Scale Assessment of Waste Filter Backwash Recycle." *Journal AWWA,* 95(7), 2003.

Vandermeyden, C. "Design and Operation of Nonmechanical Dewatering Systems." *Proceedings of the 4th Joint WEF and AWWA Conference on Biosolids and Residuals Management.* Alexandria, Va.: Water Environment Federation, 1993.

Vandermeyden, C., et al. *Nonmechanical Dewatering of Water Plant Residuals.* Denver, Colo.: American Water Works Association Research Foundation, 1998.

WEF/ASCE. *Design of Municipal Wastewater Treatment Plants,* 4th ed. Alexandria, Va.: Water Environment Federation, 1998.

Westerhoff, G. P., and D. A. Cornwell. "A New Approach to Alum Recovery." *Journal AWWA* 70(12):709, 1978.

Wilhelm, J. H., and C. E. Silverblatt. "Freeze Treatment of Alum Sludge." *Journal AWWA* 68(6):312, 1976.

CHAPTER 18
ARCHITECTURAL DESIGN

Philip C. Zimmerman
Malcolm Pirnie, Inc.
White Plains, New York

Successful treatment plant design must address an array of issues, many outside of the traditional area of water quality engineering. The facility must fulfill its primary objective, which is to provide a quality product; but it is also important how the facility is perceived by those who work in it, visit it, live near it, and see it every day. It is often the architectural and landscape design forms, the visual portions of the project, that mold public perception.

Careful consideration must be given during the planning and design process to address the needs of plant personnel for a well-thought-out, conveniently arranged, pleasant working environment. These features will engender positive employee response and provide an efficient and functional workforce.

Construction systems and materials must also be selected that will combine to create a facility that is cost-effective, as well as resource- and energy-efficient.

THE ROLE OF THE ARCHITECT IN WATER TREATMENT PLANT DESIGN

The design team for a water treatment plant design project generally includes an architect. In a team, the whole is more than the some of its parts. Associated with the benefits inherent in the early involvement of the architect is the additional benefit of team synergy. Team synergy occurs when professionals focusing on the same problem, but approaching it from different perspectives, offer solutions that spark additional ideas that could be realized only with the cooperation of all team members. It is this kind of complementary and cooperative effort that will produce the best work and that should be cultivated. Only when all design professionals collaborate does the owner truly receive the highest-quality service. It is through such collaborations that the best water treatment engineering is combined with the highest-quality building design concepts.

Aesthetics and Contexturalism

The architect is trained to view the buildings and environment of a project to establish an aesthetic approach for the design of the facility. The architect will often point out the need to "visually relate" or "respect the scale of" the proposed buildings with the communities that are adjacent to the plant. Even where the location of the facility is relatively isolated, the design should be visually pleasing, because, as major works of substantial expense, they are reflections of community pride and achievement. The architect will balance these considerations against available design and construction budgets and suggest construction materials appropriate to each building. The architect is also trained to propose the proper design of spaces for "people functions" and to relate the various plant functions to one another and to the site.

The architect should preferably begin meeting with the design team during the time of facility planning and will need to determine factors such as the number of personnel, scope of laboratory services, and extent of on-site maintenance that is being planned. The architect should particularly be involved in coordinating the placement of external building elements that affect the architectural design, such as louvers, light fixtures, and stack penetrations. If these decisions are to be made with the participation of the owner, it is generally best to meet as a team with the owner so that all perspectives are considered when important decisions are made.

Codes and Standards

It is the responsibility of the architect to investigate, understand, and implement applicable provisions of building construction codes, design standards, and regulations such as the American National Standards Institute, National Fire Protection Association, and Americans with Disabilities Act. Many trade organization standards define the level of acceptable workmanship and materials for construction systems. A list of the sources of many design standards has been included at the end of this chapter.

State and local building codes often reference some of these standards by name, and in such cases, the standard becomes part of the construction code. If a particular standard is not referenced, it may mean that the local code has different or more specific requirements, and those requirements must be followed. If a construction code is mute on a particular topic and there is no higher governing code, national standards can be used as the basis for prudent design.

A standardized approach to dealing with building construction and safety issues has long been sought by the design professional and the code enforcement communities alike. Many governing jurisdictions have begun to adopt IBC 2000, a standardized construction code published by the International Code Council, Incorporated. This construction code is the work of five subcommittees representing Building Officials and Code Administrators (BOCA), International Conference of Building Officials (ICBO), and Southern Building Code Conference International (SBCCI), first drafted in 1997. The intent of this effort was to create a comprehensive set of regulations for building systems consistent with, and inclusive of, the scope of the existing model codes published by each representative that adequately protects public safety, health, and welfare. Many states have begun to accept this standardized approach.

Standards specific to water treatment process design should remain the responsibility of the engineers, but these standards often impose construction and fire protection requirements of which the engineer will need to be made aware. Codes and national standards represent minimum standards, and the design team should always consider improvements that supersede code requirements if they will improve safety, lower operating and maintenance costs, or provide the owner with a more functional design.

It is important to determine at the beginning of the design process what the owner's code compliance responsibilities are. Some large municipal clients may be able to work with state and local planning, zoning, and building code officials in shaping code requirements to the project needs. The design team should also involve the local code enforcement officer early in the project. Valuable insight on local requirements can often be provided to the design team, which expedites the entire design process. Knowing exactly what is required may also save on costly design changes that would be caused if uncovered toward the completion of design or, worse, during construction.

The architect must also be familiar with the requirements of local architectural review boards, planning and zoning departments, and any possible effects of other governing authorities on the design of the project.

Environmental Design

One important duty of the architect is to attempt to minimize the impact that a new facility has on the natural environment. The architect typically has little control over site selection and process considerations, but can play a critical role in overcoming negative public perception of a new project. This can often be accomplished through inventive architectural design and site planning. The public must also be shown that it is necessary to see the total environmental benefit of the project, not just the facilities alone.

Such phrases as *green architecture* and *sustainable architecture* are now making their way into design vocabulary to indicate resource- and energy-efficient buildings. This philosophy seeks to take advantage of regional construction opportunities and methodologies, site orientation, natural daylighting, and resource-efficient recycled materials in order to save energy, conserve resources, and improve the quality of the indoor environment.

The U.S. Green Building Council has introduced a "green" building rating system known as LEED, for Leadership in Energy and Environmental Design, in an effort to provide a national standard for what constitutes a green building. Through its green building rating system it provides design guidelines and third-party review that allow a project to achieve "points," leading toward certification at various achievement levels. Benefits to the project include verifiable focus on the energy efficiency, resource efficiency, and indoor environmental quality of the project and the benefits that accrue through positive community exposure and publicity.

Sealing of Documents

Many states now require that certain building types involve the services of a licensed architect, and the completed contract documents for applicable portions of the work must bear the architect's seal and signature. To provide this, most jurisdictions require that the work be done under the architect's personal direction. Many states restrict the corporate practice of architecture regardless of an individual licensee's status. The issue of who may seal what types of documents should be established at the outset of a project.

Consideration of the Site

Architectural form is greatly affected by regional differences in area topography, climate, and individual site details. Plant site design should include adequate space for visitor, employee, and delivery parking that is accessed from a clearly defined traffic pattern that splits off traffic according to its type and destination. It is preferable that employee parking be isolated from the main entrance and be near an employee entrance that is close to lockers and other personnel facilities. This will reduce traffic near administrative and public areas.

If possible, administrative areas should be nearest the main entrance, not only for the convenience of visitors, but also for control of visitor access. The public areas must comply with the requirements of the Americans with Disabilities Act, and because compliance will require additional space, this should be considered early in the site planning process.

Landscaping offers a powerful mechanism with which to shape a site and goes far beyond obvious aesthetic benefits. Careful use of deciduous and evergreen plant material can bring clarity to the service road network by highlighting directional approaches while concealing less desirable or confusing views. Landscape material also gives scale to structures by integrating them with the ground plane and by softening visually hard edges and long expanses of blank walls. Plant material can make important contributions to energy conservation. Deciduous plant material allows winter sun to warm the building while providing shade from the heat of summer sun, thus reducing the energy consumption of structures.

In colder climates, evergreen windbreaks and earth berms can be used to buffer buildings from winter winds. A similar approach can be used in warm climates to channel breezes, which will provide natural cooling.

Architectural Programming

The client often does not fully know, or has not thought about, many of the factors that will affect internal architectural design and will need to be guided by the consultant. These are some of the questions that can be asked:

- What will be the staffing of the plant?
- What special administration or billing needs are anticipated?
- What accommodations should be provided for visitors?
- How much and what kind of laboratory analysis will be performed at the facility?
- What degree of employee training is anticipated?
- How many offices and conference rooms should be provided?
- How extensive should the maintenance facilities be?
- Should vehicle storage be provided?

The answers to these and other questions will uncover important planning considerations, which then can be discussed in greater detail.

It is also important to know if the client is composed of more than one bureau or responsible entity. Does your immediate client contact need to seek input from other departments in order to determine planning choices for the plant? Once the full scope of facility programming is understood and agreed on, it is important to establish how each programmatic element interrelates.

The information gathered from the client can next be presented as *adjacency and circulation diagrams* that show the relationship between the various work and storage areas in the building. On more complex projects, it may be necessary to present secondary relationships between spaces, which may affect their locations within buildings. These areas can then also be related to site orientation, topography, views, and other important external planning criteria because these factors will also influence the location and architecture of the buildings. The information that has been gathered will determine how compact or open the plan of buildings should be.

Planning for Persons with Physical Disabilities

The Americans with Disabilities Act (ADA) of 1990 represents landmark civil rights legislation that is the culmination of a series of related laws that began with the Civil Rights Act of 1964. The law essentially requires that individuals with life-limiting handicaps be afforded the same opportunities of access to places and jobs as able-bodied individuals enjoy. Although the law itself is complicated and is still being interpreted by the courts, the important point for designers to focus on is the need to provide adequate access for the public and employees to certain spaces within the project. The designers should remember to check with both the requirements of the state building code and the federal ADA accessibility guidelines to determine the requirements for the project, because the more restrictive of the two standards will apply.

In general, it is the owner who is responsible for complying with the law, so it is important that the engineer and architect review requirements with the owner. It is also important to have the interpretation arrived at by the design firm and owner in writing. In any case, it may be desirable for the owner to hire qualified physically handicapped individuals at some time in the future, and it is much less expensive to accomodate the necessary facilities initially than to add them later. Typical accommodations include elements such as ramps, elevators, accessible toilets, doors with adequate latch side clearances, hallways wide enough for wheelchairs to pass, and special design guidelines for stairs, elevators, and various equipment.

FACILITIES DESIGN

The architectural design of water treatment plant facilities generally involves consideration of administrative areas, facilities for the plant staff, laboratory spaces, and maintenance facilities.

Administrative Facilities

The design of administrative facilities must include consideration of not only offices, but also areas that are available to the public and areas for conferences and training.

Lobby and Reception Area. Most plants will need a lobby or reception area for use on occasions when the public visits the plant. Heightened public interest in both public health and environmental issues is creating increasing requests for visits by groups of children, civic groups, and individuals, especially in larger facilities. These activities require greater attention to public spaces and may increase the required lobby or reception area considerably beyond that which would otherwise be necessary.

Office Area. Office space for a treatment plant may vary from only a superintendent's or operator's desk to many offices for a large number of employees. Any planning for offices must include space for plant records. These records comprise contract documents, drawings, operations and maintenance manuals, and operating records, including recorded charts and logs. In larger treatment plants, it may be advantageous to provide a records storage room or technical library and training room dedicated to this function. It is not recommended that a lunchroom serve as a library and training facility.

Incorporating a technical library as part of a public room or conference room often will work well. If a room large enough to serve as a lecture room is provided, it may be equipped with a projection screen, visual aid panels, a lectern, and demonstration tables. Consideration should also be given to providing an assistive listening system if an audio system is installed in the space in response to the requirements of ADA accessibility guidelines. Figure 18.1 shows a plan for a water treatment plant administration building in which lobby space is used to provide a small public exhibit area where visitors can learn how the plant functions and view some typical equipment used in the processing of the water.

Personnel Facilities

The design of new water treatment plants gives increased attention to making the facilities comfortable for employees, and in particular, consideration must be given to providing facilities for women as well as men.

Planning for Change. Employment at water treatment facilities offers opportunities for job descriptions to be filled by either male or female applicants, and because of this, it is important to consider planning approaches that permit flexibility in accommodating the shifting proportions of male and female workers. Facilities that must be made available to workers of either sex include water closets, lavatories, lockers, shower facilities, drying areas, and laundry rooms. Unfortunately, current practice is to simply double the facilities that are planned to accommodate the shifting number of male and female employees. Even though most of one of the modules may not be used at present, the situation could change in the future. One alternative is to locate relatively fixed facilities such as toilet rooms and showers at opposite ends of the area that will be dedicated to lockers. The area between them could be allowed to shift in use through the relocation of partition walls, if required in the future. In smaller plants it may be possible to make only a reasonable assumption on future male/female employment.

Locker Rooms. Many water treatment plants are designed to assign two or sometimes three lockers to each employee. A smaller locker is used for street clothes, and a larger locker is provided for work clothes. Some employees also require an equipment locker for assigned tools, boots, and rainwear.

A food locker for each employee is provided in some treatment plants. These lockers are usually located in the lunchroom and are used to store personal food items. Handicap access to locker rooms is required by ADA. The type of locking mechanisms to use on lockers should be discussed with the client. Figure 18.2 shows a modular personnel area for a small treatment plant. The layout accommodates a staff of nine males and nine females and provides each employee with three lockers.

Lunchroom. Providing a separate lunchroom facility in a treatment plant contributes to the hygiene and health of plant employees by providing a central location for food storage and preparation, by fostering proper wash-up procedures, and by discouraging consumption of food in process areas. Providing a separate lunchroom space also serves to minimize disruption of process and administrative activities while employees are eating.

The lunchroom space should preferably be located near an accessible route, away from administrative activities and conference rooms, and near the toilet and locker facilities so employees will find it easy to wash before eating. Modular seating and tables should be provided so that they can be reconfigured for various activities, and durable kitchen equipment should be installed for employee use. The room should be open and light-filled and, preferably, with good views out to the building exterior or process areas of the plant.

FIGURE 18.1 Administration building.

ROOM LIST

1	GAS CHROMATOGRAPH	15	PANTRY
2	BOTTLED GAS STORAGE	16	LUNCH ROOM
3	CORRIDOR	17	OFFICE
4	LAB STORAGE	18	SEATING AREA
5	PILOT ROOM	19	ELECTRICAL / TELEPHONE ROOM
6	BACTERIOLOGICAL	20	COFFEE ROOM
7	ATOMIC ABSORPTION	21	MEN'S TOILET
8	CLOSET	22	MEN'S SHOWER / LOCKERS
9	LAB OFFICE	23	VESTIBULE
10	LABORATORY	24	WOMEN'S TOILET
11	LOBBY	25	WOMEN'S SHOWER / LOCKERS
12	EXHIBITION	26	JANITOR'S CLOSET
13	MECHANICAL ROOM	27	RECORDS / STORAGE
14	CONFERENCE ROOM	28	BUNK ROOM
		29	PERSONNEL COMMON SPACE

The architect should also acknowledge that a growing number of states and local governments are instituting regulations governing smoking in areas that are not private offices, and this needs to be addressed in the treatment plant design. Some plants provide special smoking areas with separate ventilation systems; otherwise employees may have to leave the building to smoke. Signs prohibiting or limiting smoking as required by law or directed by the client should be furnished as part of the building design.

FIGURE 18.2 Locker room, toilet, and shower room plan.

Laboratories

Since passage of the Safe Drinking Water Act, water treatment regulations, quality standards, and monitoring requirements have become increasingly strict, which, in turn, has caused the laboratory requirements for a plant to continuously escalate to perform more and increasingly sophisticated analyses.

Laboratory Planning Standards. In terms of total regulations, laboratories are the most regulated portion of the water treatment plant. They are usually regulated by the state's Department of Health and are regularly audited for strict compliance. They also must conform to Occupational Safety and Health Administration (OSHA) standards and may occasionally be audited by OSHA. Laboratories are also regulated by many states for compliance with clean air standards. Many cities and local jurisdictions have developed their own regulations governing laboratories.

The National Fire Prevention Association (NFPA) Standard 45, *Fire Protection for Laboratories Using Chemicals,* and other standards referenced as part of this document are used by most local fire marshals and inspectors as a basis for their own standards. There are also other NFPA standards that most states or local authorities refer to in determining compliance. Finally there are the Environmental Protection Agency Testing Methodology Standards, which describe required testing, equipment, and facilities that laboratory personnel must use in order to maintain their laboratory certification. The designer must be thoroughly familiar with all applicable regulations and standards before beginning design.

Many larger laboratories must be separated into organics, inorganics, and bacteriological testing areas. These distinct areas must be kept environmentally isolated from one another, or cross-contamination of test results can occur. Even the use of the same laboratory coat by a worker who performs tests in both the organics and inorganics testing areas of the laboratory can affect results, because the solvents used in one area are the very chemicals being tested for in another.

Although these areas must be separated, it is important for the safety of laboratory personnel that visual connection be provided between the laboratory rooms by the use of glazed safety windows in walls and large vision panels in doors, where they are allowed by the building code.

Safety must be a prime consideration in the laboratory layout. First aid kits, fire extinguishers, fire blankets, eye washes, and emergency showers are usually required by codes and governing authorities. Even if they are not mandated by code, they should be provided as a matter of good practice.

If possible, fume hoods should be located farthest from room entry doors, which will allow employees to exit away from an accident that occurs in a fume hood. At the same time, it also permits fresh air to be introduced at the door side of the room so that occupants will be moving toward the source of fresh air as they exit the room. Contaminants that have temporarily spilled into the room will eventually be exhausted through the fume hood. If the laboratory is using sophisticated equipment, such as gas chromatographs and atomic absorption units, it is wise to include an automatic fire suppression system to protect equipment from damage by fire.

Laboratory chemicals should be stored in approved cabinets and vented to fume hoods or in an approved manner. Solvents and flammables should be stored in specially designed rooms complying with NFPA 30, *Flammable and Combustible Liquids Code* (or other standard referenced by local governing authorities). In some situations, flammable liquids may be stored in individual cabinets in compliance with NFPA 30.

Treatment plant laboratories should be provided with their own independent air supply and exhaust systems. Many codes require this, but in the interest of safety, it should never be overlooked.

Estimating the Laboratory Requirements. The principal factors in estimating needs for the treatment plant laboratory are the type of testing to be done, the quantity of tests to be run each day, and the type of equipment that will be used within the foreseeable future. Establishing the requirements for these will help determine some of the following items that must be furnished:

- Number and type of fume hoods
- Size of chemical storage rooms
- Types and quantities of gas cylinder to be stored
- Size of laboratory personnel support areas
- Quantity and types of bench space and laboratory furniture
- Electrical, gas, vacuum, and compressed air systems that will be required
- HVAC requirements for the laboratories
- Need for a deionized water system

Understanding these parameters will also help clarify the choice of finish materials for walls, ceilings, and floors and establish the importance of various safety issues as they affect the choice of partitions, doors, and glazing materials. It is best, if possible, to interview the owner's laboratory staff and learn firsthand the kind of testing and equipment that must be provided. Establishing this at the early phases of a project will lead to far fewer changes and a much happier client when the job is completed.

A number of factors affect the laboratory staffing requirements, such as

- Type of laboratory equipment to be used
- Extent of automatic monitoring and recording instruments
- Amount of work that will be outsourced to other laboratories
- Level of training of laboratory personnel

The staffing needs should be projected as far into the future as possible. There should be enough built-in flexibility in the design of the laboratory to accommodate a reasonable amount of changes in operation caused by regulatory changes, increase in the size of the plant or water system, and more extensive testing due to technological changes.

Estimating the Necessary Floor Area   Once the annual number of person-hours has been estimated, the number of persons involved can be determined by considering the actual on-the-job hours per person. A guideline provided by OSHA suggests 2.5 ft (0.75 m) of fume hood per person working in the laboratory. Other sources suggest 4 ft (1.2 m) of workbench per person. However, planning rules of thumb need to be checked once additional staffing and equipment information is available.

A methodology has been developed to derive laboratory area requirements from known bench space requirements. A completely efficient room would provide bench space along all four walls without any interruptions for circulation or openings in walls. An extremely large room can generally be laid out to be the most efficient. Smaller rooms are less efficient because the same amount of space dedicated to circulation and openings makes up a larger percentage of the total area available.

Some rules of thumb use 200 to 300 ft^2 (19 to 29 m^2) per staff member to determine the area of the laboratory, with benchtop working surface assumed to be 30% to 40% of the room area. Sinks should be provided at locations that are convenient but out of main circulation paths, each serving about 20 ft of benchtop work area. Sinks should not be adjacent to instruments.

A modular layout of electrical outlets should be established along the tops of all work surfaces in the laboratory. Some types of benchtop equipment must be hardwired, and some equipment also requires power conditioning. If some equipment must be furnished with an uninterruptable power source, space for battery units may have to be considered as part of the laboratory design.

Most laboratory areas should be furnished with compressed air, burner gas, and vacuum outlets. The compressors and pumps necessary for these systems should be located outside of the laboratory to avoid vibration. Most laboratories also require quantities of deionized water. Systems for producing relatively small amounts of deionized water can be located within the laboratory, or central systems can be provided for laboratories using larger quantities.

Equipment and supplies in addition to those required by the state should be selected based on the frequency of tests, the sophistication of the unit processes, cost-effectiveness, and the desire to achieve optimum plant efficiency. The selection of major equipment items can be made from the equipment requirements suggested in *Standard Methods for the Examination of Water and Wastewater* and AWWA Manual M12, *Simplified Procedures for Water Examination.* Figure 18.3 shows a typical laboratory layout for a small plant.

Laboratory Operation Considerations.   If the laboratory under design is to perform analyses from other treatment plants, it will need to be provided with a way of receiving and properly storing samples that are brought in, as well as an orderly arrangement for tracking and preparing the samples for analysis. This is usually done by providing a separate sample receiving room with refrigerators, benchtops, and fume hoods, as well as a generously sized storeroom. Larger laboratories now use a computer program designated LIMS (laboratory information management system) to record, track, and report on the analyses of samples that are processed by the laboratory.

Laboratory lighting must receive particularly careful consideration. Glare can render computer screens unreadable, and extreme variations of light and shadow can be annoying and disorienting. Opening large areas of the laboratory to northerly light will provide

FIGURE 18.3 Process laboratory.

good, uniform lighting during daylight hours, but in cold climates it will cause unnecessary energy loss and possibly even discomfort to personnel. Methods of overhead skylighting with shade control or a south-facing clerestory with light shelves to distribute light along the plane of the ceiling should be considered as part of a well-designed laboratory.

Laboratory balances are highly sensitive to vibration, so balance tables are commonly placed against a bearing wall, next to a column, or at a location that is least likely to pick up vibrations from the heavy equipment operating in the plant. If there is no placement within the laboratory that is satisfactory, various types of vibration-isolated floor construction are available and should be considered.

Adequate space should be provided between back-to-back work areas. The minimum distance is usually considered to be 4 ft (1.2 m), and 5 ft (1.5 m) is generally better. The requirements of fume hood dimensions will usually determine clear ceiling heights in laboratory spaces and cause them to be higher than code-required minimums.

Maintenance Facilities

When a new water treatment plant is designed, careful consideration should be given to providing the necessary repair shops for the types of work that will be performed by the plant staff, storage rooms for tools and spare parts, and basic tools for the shops.

Repair Shops. The size, number, and function of repair shops are, to some degree, related to the size of the plant. It is also important to determine early in the design stage whether the plant will serve as a maintenance or materials receiving hub for other plants. Some clients choose to use outside facilities at other treatment plants or have access to a central maintenance complex off-site, and this will be equally important to establish at an early point in the project design.

Some large plants have the maintenance functions located in a separate building to isolate noise, to allow for free movement in handling large equipment, and for truck delivery. In addition, maintenance facilities often store costly equipment, and security is improved by the provision of a centralized location. Wherever the maintenance shop is located, there should be provision for truck delivery of large equipment. Doors should be made wide enough and high enough to allow equipment to pass in and out of the shop area.

Where plant size and anticipated maintenance workloads warrant, a separate electrical repair shop and a paint shop may be provided. Electrical repair shops require a high degree of cleanliness. Paint solvents are flammable, and most building codes will require strict standards of safety for these spaces.

The type and complexity of work that will occur in the maintenance shop must be determined early in the design. This will, in turn, dictate the type and size of repair machinery that should be provided. Clearances required around shop equipment must also be considered so that long, heavy, or bulky items can be handled and safely worked on. It is important to check all shop design layouts for safe clearance around each piece of shop equipment.

Adequate bench space is necessary in all shops. Benches should be sturdy and selected for heavy-duty service, allowing rough work to occur on top surfaces selected of stain- and scratch-resistant material. Drawers should be installed in workbenches to hold small, frequently used parts and hand tools. Tool storage shelving and storage systems should be selected to permit easy retrieval of tools near the point of work.

Attention should be given to bringing generous amounts of controlled natural lighting into the space, supplemented with task lighting at the equipment. The environment of the shops should be buoyant but not distractive. This can often be achieved by the use of high ceilings and clerestory or sawtooth monitors to bring light deeply into spaces. The use of skylights is less desirable because as the sun moves across the sky, it will be difficult to control glare and hot spots in the shop and could, during certain times of the day, make the use of some areas impossible or at least unsafe.

One item commonly overlooked in the design of repair shops is the need to use water and to dispose of wastewater. Testing of solution tanks for feed machines, the injectors of chlorinators, or electric mixers commonly results in spilled water. It is valuable for a shop to have a drainage area in a corner or an end of the room, set off by a 4-in.-high concrete curb, for use in wet testing equipment. A floor drain should also be installed at the low point of the shop floor so that the entire room can be hosed down if desired. If possible, it is also useful for a shop to have an area that can be used for solvent washing or steam cleaning equipment. Figure 18.4 shows a typical layout for a small maintenance shop.

Maintenance Shop Tools. A wide variety of tools are required for a water treatment plant maintenance shop. Heavy tools typically include drill presses of various types, bench

FIGURE 18.4 Shop and storage building.

and pedestal grinders, hydraulic presses, a portable hoist, a portable pump, pipe threaders, milling machines, an arbor press, welding rigs, lathes, band saws, and cutoff saws. Light tools should include those for plumbing, automotive, electrical, painting, carpentry, sheet metal, and masonry work. Also included should be grease guns, oilers, and trouble lamps.

For housekeeping, mops, pails, brooms, and vacuum cleaners are needed and will require dedicated storage areas. For outside maintenance, wheelbarrows, rakes, shovels, hoses, lawn sprinklers, and a power lawn mower should be provided and given adequate storage space. In cold climates, snow removal equipment is needed. Safety equipment should include first aid equipment, a resuscitator, and portable breathing equipment.

Storage, Spare Parts, and Security. Spare parts should generally be kept in a separate, secure storage area that can be locked for inventory control. The room should include a section of drawers, bins, cabinets, racks, a large amount of shelf space, and some peg racks on which roll material such as gasket stock, screen wire, and hardware cloth may be stored. For ease of inventory control, the spare parts storage area should preferably be located near the maintenance shop.

CONSTRUCTION ALTERNATIVES AND BUILDING MATERIAL SELECTION

There are a great number of alternative building materials available to water treatment plant designers today. The selection of which one will be best for a particular situation must be made on a consensus of the design team and client of what will be durable, economical, and aesthetically pleasing.

Structures over Process Units

The decision of whether to place superstructures over process units generally depends on the climate. In areas where there is minimal danger of freezing, structures such as flocculation and sedimentation basins can be left uncovered, which greatly reduces the capital costs. In addition, the visual impact of the plant is easiest to handle when the area and height of structures are kept to a minimum. In very hot climates it is often best to cover equipment such as filter operating consoles, both to protect the equipment and to make operations more comfortable for plant personnel.

Preengineered Structures

For the smallest and simplest facilities, preengineered, or even premanufactured, structures may offer a viable option for reducing capital costs because less detailing of construction systems will be required. Preengineered structures are primarily used for warehouses and as enclosures for manufacturing facilities, but they have been used for covering some water treatment facilities.

They are best used when the function and floor plan of the building are straightforward and there is no need to support heavy overhead pipelines, monorails, or cranes. Some manufacturers of preengineered structures can accommodate some limited support of overhead loads by altering the design of the structure. These buildings are generally designed to be economical, so they offer little opportunity to accommodate unforeseen future additional loadings.

When a preengineered building is selected to house chemical facilities, care must be exercised that there are no adverse effects of the chemicals on the wall panels and structural supports. A review of local building codes should be made to determine whether there are any restrictions on noncombustible, unprotected structures, based on building area, height, and occupancy type.

Planning for the use of preengineered buildings must start early, and it is best to consider only standard systems available from the manufacturers. The architect should keep in mind that although these buildings are preengineered, compliance with building code requirements still must be ensured by the project designer; so a careful and complete building code study should be made before full commitment is made to this direction as the most cost-effective building alternative.

Often the use of an engineered concrete masonry building will be no more expensive, when all costs are evaluated, and perhaps will be more durable and long-lived than a metal building. This, of course, must be determined individually for each project because there are many factors to be considered. Although preengineered structures can be erected quickly after they arrive on-site, the project engineer must remember that a fabrication lead time must be considered.

Building Panel Systems

The use of panelized building modules will speed the enclosure of structures and will usually permit erection to continue regardless of weather conditions. There are many panel system options to chose from. The architect may suggest the use of precast concrete, insulated steel or aluminum (either field- or factory-assembled), translucent fiberglass, or even preassembled panels of thin masonry or ceramic tile applied to strong-backs and metal decking. Advancements in structural silicones and acrylic additives make these preassembled panels relatively lightweight and long-lasting. A number of exterior insulation and stucco wall systems are now available that share the advantages of fast erection. When building codes allow their use, and the detailing, specification, and installation supervision of these systems are done carefully, they can be highly effective for construction of water treatment plant facilities.

Environmental Considerations in Design

In all designs of water treatment plant facilities, no matter what the size or budget of the project is, there is an opportunity to build in a sustainable manner. Creating healthy indoor environments requires careful selection of durable, easily maintained, and stable materials that do not continue to off-gas over their life span.

In a world of high energy costs, the architect should focus on how well the materials that are selected insulate from hot or cold, as well as how much energy was consumed in the manufacture, transportation, and erection of the materials—their so-called embodied energy. Many governing jurisdictions have adopted one or another of the model energy conservation codes, and all designers should address the requirements. Where does our structure get its energy? Is it renewable? Can we obtain some of the energy we will need to power, light, heat, and cool our structures from alternative energy sources? Some states offer incentive programs through one or another of their agencies or through the National Energy Conservation Act of 1992 and the Department of Energy, and the designer needs to find out if monetary or other incentives will benefit the project.

Materials that close the recycling loop or that contribute to environmental cleanup should be explored. Many construction materials are now being offered that use all or some recycled materials in their manufacture. Some use currently unrecycled wastes and turn them into usable products. The designer should appraise these. Some are undoubtedly using the current interest in the environment for gain while others truly are assisting environmental cleanups by recycling waste material.

Insulating glass with wavelength-selective indium oxide and silver-coated polyester interlayers that reject unwanted heat in summer and retain indoor heat in winter is available. These units can be purchased with inert gas-filled (argon or krypton) "dead-air" spaces, which further improves their resistance to heat flow, as will systems of insulated edge sealants available from some glass manufacturers. The designer should never fail to specify insulated doors and thermal-break windows, which offer significant improvements over their counterparts. Weather stripping and seals save tremendous energy, whether in a hot or cold climate.

Included in the list of available standards listed at the end of this chapter are some sources where designers can learn about sustainable design. Contacting even just a few will provide a wealth of environmental building knowledge.

Selection of Interior Finishes

In general, the selection of interior finishes in process spaces should be based on durability and low maintenance, whereas in administrative and office environments, more attention must be given to providing a pleasant working environment. Considerations also include any possible effects on indoor air quality by avoiding materials that tend to off-gas or absorb pollutants and give them off over time, or are difficult to fully clean such as carpeting. The designer should refrain from using any phenol formaldehyde adhesives and minimize the use of laminates within occupied spaces. Zero-VOC paints are now available. Materials that require frequent cleaning or strong cleaning solutions to maintain should be avoided.

Materials such as ceramic tile or vitreous glazed structural tile make excellent choices for many types of spaces found at the plant. They are available in many styles, and the installation system should be as is suggested in publications provided by the Tile Council of America. Other recommended selections include regular or thin-set terrazzo for floors and gypsum board containing recycled gypsum and recycled paper for walls in areas not subject to dampness. For areas subject to dampness, such as shower and locker rooms, cement board or concrete block is usually used as a base for ceramic tile, or glazed structural tile may be used as both wall and finish in one system.

Areas such as basements, boiler rooms, and pipe galleries are generally not given a special finish unless the walls are concrete block, in which case a high-durability paint can be applied to facilitate maintenance and improve light distribution.

Selection of Colors. The American National Standards Institute, in ANSI A13.1, *Safety Color Code,* and in ANSI Z535.1, *Scheme for the Identification of Piping Systems,* has tried to standardize the identification of piping using color coding. Many water systems have their own standards, which have served them well for years, and the designer needs to be aware of this. There are no restrictions on color schemes except that care should be used in the laboratory. The determination of small concentrations of minerals is often done colorimetrically, and light reflected from variously colored areas in the laboratory may interfere with these tests. Laboratories should be painted in subdued colors such as off-white, light gray, or light blue.

Selection of Surface Coatings. Steel surfaces at a water treatment plant that particularly require protective coating include clarifier mechanisms, structural members, storage tanks, and piping supports. Most governing jurisdictions have maximum allowable volatile organic compound (VOC) content for paint, and most manufacturers have responded with low-VOC paint systems. Any paint in contact with potable water must be approved in conformance with National Sanitation Foundation International Standard 60.

The durability of an applied coating is generally only as good as the preparation that allows it to bond to the steel or material being coated. For this the recommendations of both the manufacturers and the Steel Structures Painting Council should be adhered to. Some surfaces, such as galvanized steel and fiberglass, are generally difficult surfaces on which to obtain adequate adhesion, and on these surfaces it is best to specify a light 60- to 80-mesh sandpaper finish.

Painting can be avoided by the selection of materials that do not require protection, such as stainless steel or ceramic tile, but there are many locations throughout the plant where it is not economically feasible to use corrosion-resistant materials. In these cases, there is no way to avoid the use of paint. It is generally best for future maintenance if the designer selects as few painting systems as possible for the plant. Paint systems that can be applied to a broad variety of substrates and still offer excellent wearability and longevity are preferred. Materials such as cured epoxies and polyurethane enamels are state of the art in current plant painting technology and offer many variations to meet specific needs.

DESIGN STANDARD–PROMULGATING ORGANIZATIONS AND ORGANIZATIONS OFFERING DESIGN RECOMMENDATIONS

Adhesive and Sealant Council, Inc.
7979 Old Georgetown Rd., Suite 500
Bethesda, MD 20814
http://www.ascouncil.org

Aluminum Association
900 19th St., NW, Suite 300
Washington, DC 20006
http://www.aluminum.org

American Architectural Manufacturers Association
38800 County Club Drive
Farmington Hills, MI 48331
http://www.aci-int.org

American Forest and Paper Association
1111 19th St., NW, Suite 800
Washington, DC 20036
http://www.woodcom.com

American Hardboard Association
1210 W. Northwest Highway
Palatine, IL 60067
http://www.corporate.bricsnet.co.uk

American Industrial Hygiene Association
2700 Prosperity Ave., Suite 250
Fairfax, VA 22031
http://www.aiha.org

The American Institute of Architects
  Committee on the Environment
AIA Environmental Resource Guide
1735 New York Ave., NW
Washington, DC 20006
http://www.aia.org/cote

American Institute of Steel Construction
One East Wacker Dr., Suite 3100
Chicago, IL 60601-2001
http://www.aisc.org

American Institute of Timber Construction
7012 S. Revere Parkway, #140
Englewood, CO 80112
http://www.aitc-glulam.org

American Iron and Steel Institute
2000 Town Center, Suite 320
Southfield, MI 48075
http://www.autosteel.org

American Lumber Standards Committee
P.O. Box 210
Germantown, MD 20875
http://www.alsc.org

American National Standards Institute
1819 L Street, NW
6th floor
Washington, DC 20036
http://www.ansi.org

American Plywood Association
P.O. Box 11700
Tacoma, WA 98411
http://www.apawood.org

American Society for Testing and Materials
100 Barr Harbor Dr.
West Conshohocken, PA 19428-2959
http://www.astm.org

American Society of Heating, Refrigerating
 and Air-Conditioning Engineers
1791 Tullie Circle, NE
Atlanta, GA 30329
http://www.ashrae.org

American Wood-Preservers' Association
P.O. Box 286
Woodstock, MD 21163-0286
http://www.awpa.com

Architectural Woodwork Institute
1952 Isaac Newton Square West
Reston, VA 20190
http://www.awinet.org

Asphalt Roofing Manufacturers Association
1156 15th St., NW, Suite 900
Washington, DC 20005
http://www.asphaltroofing.org

Brick Industry Association
11490 Commerce Park Dr.
Reston, VA 20191-1525
http://www.bia.org

Builders' Hardware Manufacturers Association
355 Lexington Ave., 17th floor
New York, NY 10017
http://www.buildershardware.com

Carpet and Rug Institute
P.O. Box 2048
Dalton, GA 30722
http://www.carpet-rug.com

Center for Maximum Potential Building
 Systems, Incorporated
8604 FM 969
Austin, TX 78724
http://www.cmpbs.org

Center for Resourceful Building Technology
P.O. Box 100
Missoula, MT 59806
http://www.crbt.org

Ceramic Tile Institute of America, Inc.
12061 Jefferson Blvd.
Culver City, CA 90230-6219
http://www.arcat.com

Consumer Product Safety Commission
4330 East-West Highway
Bethesda, MD 20814-4408
http://www.cpsc.gov

Decorative Laminate Products Association
13924 Braddock Rd., Suite 100
Centreville, VA 22020
(703) 222-2300

Door and Hardware Institute
14150 Newbrook Dr., Suite 200
Chantilly, VA 20151-2223
http://www.dhi.org

EIFS Industry Members Association
3000 Corporate Center Dr., Suite 270
Morrow, GA 30260
http://www.eima.com

Factory Mutual Systems
P.O. Box 7500
Johnston, RI 02919
http://www.fmglobal.com

Glass Association of North America
2945 SW Wanamaker Dr., Suite A
Topeka, KS 66614-5321
http://www.glasswebsite.com

Gypsum Association
810 1st St., NE, Suite 510
Washington, DC 20002
http://www.gypsum.org

Hardwood Manufacturers Association
400 Penn Center Blvd., Suite 530
Pittsburgh, PA 15235
http://www.hardwood.org

Indiana Limestone Institute of America
400 Stone City Bank Building
Bedford, IN 47421
http://www.iliai.com

Marble Institute of America
28901 Clemens Rd., Suite 100
West Lake, OH 44145
http://www.marble-institute.com

National Association of Architectural Metal
 Manufacturers
8 S. Michigan Ave., Suite 1000
Chicago, IL 60603
http://www.arcat.com

National Building Granite Quarries
 Association, Inc.
1220 L. Street, NW, Suite 100-167
Washington, DC 20005
http://www.nbgqa.com

National Center for Appropriate Technology
P.O. Box 3838
Butte, MT 59702
http://www.ncat.org

National Concrete Masonry Association
13750 Sunrise Valley Dr.
Herndon, VA 20171-4662
http://www.ncma.org

National Fire Protection Association
One Batterymarch Park
Quincy, MA 02169-7471
http://www.nfpa.org

National Recycling Coalition, Inc.
1325 G. Street NW, Suite 1025
Washington, DC 20005
http://www.nrc-recycle.org

National Renewable Energy Laboratory
1617 Cole Blvd.
Golden, CO 80401-3393
http://www.nrel.gov

National Roofing Contractors Association
10255 W. Higgins Rd., Suite 600
Rosemont, IL 60018
http://www.nrca.net

National Sanitation Foundation International
P.O. Box 130140
789 N. Dixboro Road
Ann Arbor, MI 48113-0140
http://www.nsf.org

National Terrazzo and Mosaic Association
201 North Maple Ave., Suite 208
Purcellville, VA 20132
http://www.ntma.com

Occupational Safety and Health Administration
200 Constitution Ave., NW
Washington, DC 20210
http://www.osha.gov

Resilient Floor Covering Institute
401 E. Jefferson St., Suite 102
Rockville, MD 20850
http://www.rfci.com

Sheet Metal and Air Conditioning Contractors
  National Association
4201 Lafayette Center Dr.
Chantilly, VA 20151-1209
http://www.smacna.org

Single Ply Roofing Institute
77 Rumford Ave., Suite 3B
Waltham, MA 02453
http://www.spri.org

The Society for Protective Coatings
40 24th Street, Suite 600
Pittsburgh, PA 15222-4656
http://www.sspc.org

Southern Pine Inspection Bureau
4709 Scenic Highway
Pensacola, FL 32504-9094
http://www.spib.org

Steel Door Institute
30200 Detroit Rd.
Cleveland, OH 44145-1967
http://www.steeldoor.org

Tile Council of America, Inc.
100 Clemson Research Blvd.
Anderson, SC 29625
http://www.tileusa.com

Underwriters Laboratories
333 Pfingsten Rd.
Northbrook, IL 60062
http://www.ul.com

West Coast Lumber Inspection Bureau
P.O. Box 23145
Portland, OR 97281
http://www.wclib.org

Western Wood Products Association
522 SW 5th Ave.
Portland, OR 97204-2122
http://www.wwpa.org

U.S. Department of Commerce
1401 Constitution Ave., NW
Washington, DC 20230
http://www.commerce.gov

U.S. General Services Administration
1800 F. Street, NW
Washington, DC 20405
http://www.gsa.gov

U.S. Green Building Council
1015 18th Street, NW, Suite 805
Washington, DC 20036
http://www.usgbc.org

CHAPTER 19
STRUCTURAL DESIGN

Ralph T. Eberts
Black & Veatch
Los Angeles, California

Todd F. Schellhase
Black & Veatch
Kansas City, Missouri

The structural design of a modern water treatment plant is a dynamic, interactive process that requires specialized expertise. Because most structural engineers are typically more familiar with the design of conventional buildings, this chapter focuses on the design features that are unique to treatment plant structures.

THE DESIGN PROCESS

The engineer responsible for the design of structures at a water treatment plant must be intimately familiar with the general principles and codes governing structural design. The designer should work with the project team during preliminary design to establish the configurations of the facilities based on economy, structural efficiency, and performance criteria. The design should then be developed and detailed in a manner that accounts for the special considerations required for a water treatment plant.

Responsibility of the Structural Engineer

Historically, the civil, process, architectural, and structural designs of a water treatment plant were prepared by the same individual and shown on the same set of drawings. Although this comprehensive approach resulted in a well-coordinated design and an efficient use of construction documents, the additional complexities of a modern treatment plant make this approach impractical.

Engineering, like most technical professions, has advanced to the point where no one engineer can adequately address all the considerations associated with the design of a large water treatment plant. Today's design team, therefore, requires an engineer who specializes in structures to support the project team in developing structural drawings and specifications depicting the design of the structures.

The structural drawings should include all the information critical to the structural design, including dimensions, details, and materials that form the basis of the design. The final construction drawings should be stamped and signed by the professional engineer responsible for the structural design.

Interaction with the Design Team

The engineer who leads the design of a water treatment plant establishes the basic process and plant arrangements. The structural engineer should provide input early in the design process to optimize the efficiency of the structures. The civil and structural engineers and architect should work closely together in establishing the distance between structures, the materials of construction, and the foundation lines. Each party should appreciate the fundamental principles governing the other's design.

The mechanical engineer, who also supports the project team, is usually responsible for the design of plumbing, HVAC, and special mechanical systems. The detailed mechanical design usually begins after the basic civil and structural arrangements have been set. The structural and mechanical engineers coordinate the equipment support, the locations of pipe penetrations, and the reactions and support of the piping systems.

The electrical engineer usually designs the power, lighting, and instrumentation and control systems. This design generally follows the process design and often parallels the detailed structural design. The electrical and structural engineers coordinate the support of electrical equipment and the locations of electrical conduits.

Use of Codes and Standards

The specific codes that govern structural design should be established early in the design process. Most of the structures at a water treatment plant do not conform to traditional building codes because their function and behavior differ widely from those of conventional buildings. Consequently, the structural engineer must often base the design of treatment plant structures on standards or reference documents used for nonbuilding structures.

Although many water treatment facilities are exempt from local building codes, the structural design should conform to nationally recognized codes and standards and exceed the minimum requirements of local building codes. Local codes should also be used to establish basic design criteria such as wind and earthquake loadings.

STRUCTURAL DESIGN CONSIDERATIONS

Among the many basic considerations that the structural engineer must evaluate before any design can begin are the types of materials that will be used, how the facilities are laid out, and what local site conditions will influence the design.

Material Selection

The materials selected for use in the design of a treatment plant are influenced by many factors, including the size and shapes of the structures, the relative costs of materials, soil and other local conditions, architectural design, and environmental conditions.

Concrete. Concrete is extensively used for water-containing and water-conveying structures because it is generally durable and resistant to corrosion. The type of cement specified should take into account the quality of the source and treated water, the treatment chemicals, and the exposure of the site to atmospheric pollutants and seawater. The arrangement of reinforcing, jointing, and crack control should be thoroughly addressed by the structural engineer. Concrete alone should not be considered a watertight material, and the owner should be informed about the amount of cracking to be expected.

Steel. Steel is often used in the construction of plant superstructures, tanks, platforms, and walkways. The high strength and modulus of elasticity of carbon steel make it an efficient and economical construction material. Primary steel shapes and plates can be made corrosion-resistant by applying protective coating systems.

Masonry. Many water treatment plant buildings and partition walls are constructed with masonry such as concrete block, brick, and stone. Masonry materials are durable and have good corrosion resistance. Masonry structures should be designed in accordance with local building codes, using materials that are available in the region.

Aluminum. Aluminum is often used for walkways, grating, and handrails in areas exposed to a mildly corrosive environment. The structural components should be designed and specified to nationally recognized standards. Connections should be welded in the shop, and dissimilar materials should be separated. The engineer must check and limit deflections in aluminum structures during design.

Stainless Steel. Stainless steel is used in water treatment plant construction at locations where corrosion resistance is critical and where the design requires a high-strength material. Stainless steel is available only in a limited number of shapes and sizes and is significantly more expensive than steel and aluminum. When possible, stainless steel components should be welded in the shop rather than on the job.

Fiberglass. Fiberglass is often used for fabrication of equipment and parts that will be located in a highly corrosive environment. The designer must consider the deflection and fire resistance of fiberglass, and the material and performance requirements must be specified in detail in the contract documents.

Configuration of the Structures

The structural arrangements should be defined during preliminary design, based on process requirements and efficient use of structural elements.

Layout of Structures. The configuration of a structure often establishes its relative economy and long-term performance. The project team generally proposes a conceptual layout based on functional requirements. The structural engineer should then review the proposed arrangements and coordinate with the project engineer to economize the cost of the structures. Once set, the structural layout should not be changed during detailed design.

The structural engineer needs to first review the conceptual layout for horizontal and vertical symmetry. The configuration should be simple, and discontinuities that might cause a buildup of stress should be avoided. Units or components should be single, continuous structures, or control joints should be provided between components with different performance characteristics. If asymmetric structures are unavoidable and cannot be separated into balanced units, additional strengthening should be provided at the discontinuities.

The stiffness of each structural element needs to be considered individually. Where incompatibilities exist, structural elements should be isolated from one another. The cost-effectiveness of vertical offsets needs to be estimated and scrutinized.

The designer must use good judgment when varying the thickness of cast-in-place concrete members. The relative cost of additional concrete must be compared with the added costs of reinforcing bar splices and forming, if the thickness of concrete members is varied.

Wherever possible, foundation lines should remain relatively consistent. The construction costs of excavation and soil preparation for foundations must be evaluated, along with the ability of the soil to support the imposed loads.

Once the preliminary design is substantially complete, the structural arrangements should be reviewed by a senior structural engineer who was not involved in the preliminary design. This review will primarily focus on cost-effectiveness and overall structural integrity. Establishing the proper configuration of the structures during preliminary design will minimize initial construction cost and reduce long-term operating costs for the owner.

Efficiency of Structures. Many design firms establish general layout criteria for conceptual design. By developing economical member size-to-span ratios, each project can use cost-effective structural components. Because structures at a water treatment plant are modest in size, a highly refined design of individual members is seldom justified.

Interaction between Structures. When establishing the basic geometry of structures, the designer must consider the relative rigidity between structural components. Common walls separating structures of different types must be carefully considered. Double-walled construction may be appropriate where structural isolation is required and where all process systems can be readily isolated between structures. A constructability review should be performed on all structures located immediately adjacent to each other.

The interaction between individual building components should be considered during preliminary design. A diaphragm's ability to deform must be evaluated against its boundary conditions. The architectural, mechanical, and electrical systems must be compatible with the drift of a flexible superstructure, and the reactions from intermediate platforms and walkways need to be applied to the building's superstructure during final design. Breaks should be provided in catwalks and ladders spanning independently moving structures.

The different types of equipment in water treatment plants often create dynamic forces on a structure. Oscillating equipment should be located as low as possible in the structure, and its effects on individual structural components must be analyzed. Dynamic or heavy equipment should preferably be supported on separate foundations. Equipment base isolators may also be used.

In seismic areas, if equipment such as pump motors, conveyors, blowers, and hoists must be located on upper floors, the supporting structure must be rigid and the equipment rigidly attached or adequately guyed. The structure and equipment must respond as a unit with provisions to resist amplification of seismic loadings by the building structure.

Jointing of Structures. Where necessary, structures should be subdivided by joints. Control joints can be either expansion or contraction type and are generally used to isolate structures or structural elements from each other.

An expansion joint permits relative expansion and contraction between elements. The use of shear transfer devices or waterstops at these joints depends on the degree of isolation and watertightness required.

A contraction joint permits relative contraction, but not expansion, between concrete elements. It often serves as an intentionally weakened plane between concrete elements to encourage shrinkage cracking at the joint instead of at random locations. Contraction joints are typically detailed to transfer shear across the joint. Waterstops may or may not be required at contraction joints depending on the need to prevent the passage of water. Reinforcing is not usually continued across control joints, but the designer may choose to continue a small percentage of reinforcing for shear transfer.

Construction joints are normally used to facilitate construction and not to isolate elements from one another. They are normally used to bond two adjacent concrete elements together as though the joint did not exist. Such joints are not usually considered in the design calculations, although special detailing at construction joints should be indicated on the drawings. Construction joints are not contraction joints unless so detailed.

Concrete joints that affect the design and performance of a structure should be clearly indicated on the drawings and described in the specifications. Ambiguous joint requirements on the drawings and in the specification will create problems in the field that may eventually affect the performance of the structure.

Alternating or delaying pours between joints or installing delayed pour strips can accommodate some initial shrinkage. However, shrinkage within concrete will continue at a decreasing rate for several years after placement. Minimum shrinkage reinforcement should be in accordance with American Concrete Institute (ACI) Standard 350 and should be determined by the spacing of the control joints, not construction joints.

Waterstops are specified wherever it is necessary to inhibit the passage of moisture across a control or construction joint. Waterstops should be included in walls and slabs of concrete structures containing treated water and in walls where one side is wet and the other side is dry. To comply with ACI Standard 350.1, waterstops should be included in divider walls between reservoir or basin chambers.

Public Access Considerations. Some structures at a water treatment plant are exempt from building code requirements with regard to public access. This should not deter the project team from using good judgment when they are establishing the access in and out of a structure. Buildings that are regularly occupied should meet building code requirements for access. Provisions of the Americans with Disabilities Act (ADA) should also be satisfied to the degree desired by the owner and required by the governing authorities. Structures that are seldom occupied should meet OSHA standards and should include provisions for safe access and ready evacuation. Caution should be used when structures are designed with only one means of egress.

Environmental Considerations

Depending on where a water treatment plant is to be located, a number of environmental considerations might be important in the design.

Freezing Considerations. Structures built in locations subject to freezing should be designed to meet the local building code requirements pertaining to minimum depth of foundations. In addition, a geotechnical report should identify a minimum depth of foundation based on site-specific soil characteristics. Basins that are constructed with slabs on grade must carefully consider the potential of frost heave if the basin may at any time be completely emptied during the winter. Unless the soil is resistant to frost heave, the structure must be surcharged to resist upward forces. Pile foundations exposed to frost heave should be designed to resist uplift forces or should be isolated through the frost zone.

Structural Watertightness. A unique consideration for structures at a water treatment plant is that many of the structures must be watertight. This criterion influences much of the structural design. The design load factors, shrinkage reinforcing, and reinforcing cover should comply with ACI Standard 350.

Almost every concrete structure is subject to some shrinkage cracking, a phenomenon that the owner should be made aware of. The designer's challenge is to control and minimize shrinkage cracking consistent with the use, cost, and importance of the structure. Shrinkage cracking can usually be effectively controlled by proper location and installation of joints and shrinkage reinforcing. The percentages of shrinkage reinforcement should vary with the spacing of control joints as identified in ACI Standard 350. It should also be kept in mind that the effects of concrete quality and curing, as well as member restraints (intentional or otherwise), on shrinkage cracking can be significant.

The special load factors for liquid containment structures indicated in ACI Standard 350 serve to reduce service load tensile stresses in the reinforcement and to limit the width of shrinkage and flexural cracking. Minimizing crack width in watertight structures is important for both leak prevention and protection of the reinforcing steel. The load factors should be used for members in direct contact with liquid, members directly above large areas of free water, and members exposed to high-humidity or potentially corrosive environments. The structural design memorandum for the project should clearly indicate where the additional load factors will be used. The reinforcing selected for concrete structures should be checked for crack control in accordance with ACI Standard 318. Reinforcing for liquid containment structures should have a Z value less than that specified in ACI Standard 350.

Dampproofing should be applied to concrete walls exposed to water or backfill, if the opposite side is above grade or inside of a room. The specified dampproofing should be compatible with the liquid to which it will be exposed.

Corrosion Protection. The choice of protective coating systems for corrosion control of structural members depends on the degree of exposure to corrosive environments. Concrete members in water treatment plants rarely need protective coatings, but epoxy-coated reinforcing may be specified for concrete members exposed to high concentrations of chlorides. Steel components in a water treatment plant can usually be adequately protected by application of an epoxy coating. The specified epoxy system must be compatible with the structural steel primer.

Aluminum, stainless steel, and fiberglass are often used in highly corrosive environments. However, even these materials may require a protective coating system suitable for particularly corrosive conditions.

Foundation Considerations

There are often many deep foundations located adjacent to one another when a water treatment plant is constructed. The ability to construct these structures and the sequence of construction must be carefully considered by the structural designer. Shoring systems are usually required for buried structures and individual components constructed below grade. Buried pipes adjacent to or beneath a structure must be given special consideration when the designer is establishing the bearing capacity of the foundations.

The Geotechnical Report. A soils report stamped and signed by a professional geotechnical engineer should be prepared for the site of a new water treatment plant or for an expansion or modification to an existing plant. Plant site conditions are particularly significant in the design of concrete basins and vaults because these massive structures require stable bearing. The report should include the following information:

- Level of groundwater
- Active, passive, and at-rest soil pressures
- Allowable bearing pressures
- Minimum depth of foundations
- Overexcavation and backfill recommendations
- Corrosivity readings
- Estimates of settlement
- Recommendations for appropriate foundation systems
- Soil information necessary to construct the plant
- For areas of seismic activity, site-specific seismic information

In addition, the report should identify the potential presence of chemically reactive soils or shrinking and swelling clays and provide recommendations on how to mitigate their effects. After the plans are partially completed, the geotechnical engineer should review the final foundation design for compliance with the report's recommendations.

Use of Mat Foundations. Mat foundations are often used in the design of liquid-containing structures and to minimize differential settlement of large structures. Where a mat is constructed on fill, the fill materials should be of uniform thickness and compacted to a degree that will ensure comparatively high soil shear strength, little future consolidation, and no differential settlement.

A mat foundation may be subject to uplift forces from external water pressures acting on the mat. Structures that extend below the groundwater table should be checked for a minimum factor of safety against buoyancy of 1.1. The groundwater level assumed in the calculations should be based on the design flood level, which typically has a recurrence interval of 100 years or more. No live load or backfill friction should be assumed to be acting on the structure. The stiffness of the structure should be considered when relying on the full structure deadweight. Long-span slabs, in particular, may be too flexible to mobilize the deadweight of interior walls.

Buoyant forces causing flotation can be controlled in several ways:

- By providing a positive tie-down mechanism
- By tying the foundation to supporting piles designed to resist uplift
- By increasing the weight of the structure by using mass concrete, adding to the overburden on top of the tank, or keeping the tank at least partially filled
- By providing a positive drainage system to lower the groundwater table in the vicinity of the structure

Pressure-reducing valves (PRVs) are sometimes used to relieve groundwater pressure but should be used only to alleviate a temporary high-water condition, because continuous flows through the valve will eventually cause it to clog. In addition, the soil surrounding the valve must be carefully selected to establish a drainage path for the groundwater and to maintain the valve operation.

Use of Spread Footings. The geotechnical engineer should specify the allowable bearing pressure and the maximum settlement potential for spread footings. The fact that spread footings have a higher potential for differential settlement than mat foundations should be considered in the structural design. Because of the multiple foundation depths normally required in the design of a water treatment plant, the designer should verify the feasibility of locating spread foundations adjacent to deeper foundations and, where necessary, extend the footing down to native soils or design the higher footing for bearing on compacted fill.

Use of Deep Foundations. Some plant sites may require the use of pile or pier foundations. Water treatment plants are typically located in low-lying areas and on alluvial plains. It is usually necessary in these locations to support structures on deep foundations or to otherwise stabilize the underlying soil. Deep foundations will significantly add to the cost of a treatment plant and should be used with discretion. Basins or structures constructed on different foundation systems should be separated by a flexible joint.

Piles installed to resist seismic-induced bending stresses must be provided with ductility. The tops of piles should be embedded in the structure as deeply as feasible in order to transfer forces. Individual pile caps and piers should be interconnected by ties. The effects of driving new piles on the support of foundations at an existing treatment plant should be evaluated before selecting pile foundations.

Lateral Soil Pressures. The structural engineer should carefully consider the ability of a structure to deflect before selecting the applicable lateral soil pressure. Walls must be free to rotate or significantly deflect before active soil pressures can develop. A basin wall laterally restrained by a top slab or sidewalls will be subject to at-rest soil pressures. A structure must undergo substantial movement before passive pressures can develop.

Construction below the Water Table. The groundwater elevation should be carefully considered when the depth of structures is set. Dewatering a site for construction is usually relatively costly. In addition, the long-term risks of uneven settlement, flotation, and water leakage when a building extends below the water table should be a significant consideration in the structural design.

Seismic Design

The design of treatment plants in seismic areas requires careful attention to detail. Geotechnical investigation is a logical first step in determining foundation conditions and structural stability issues. The second step should be to properly identify all the pertinent seismic forces and to design the structures accordingly. Anchorage of the components and equipment also warrants careful consideration so that a structural system may survive a sizable seismic event with minimal damage.

Danger of Soil Liquefaction. Depending on the soil profile, liquefaction of soil can occur in a high-seismic region. Liquefaction is the loss of shear strength or shearing resistance in loose, saturated, cohesionless soils as an aftermath of rapid shaking and reconsolidation of ground caused by moderate to strong earthquakes. Liquefaction may cause lateral movement of soil masses above liquefied soil layers along a downward slope or unrestrained surface. Surface fissures and sand boils may occur when liquefied material attempts to migrate upward, pushing through shallow overlaying strata of unliquefiable material near the ground level.

In addition to causing lateral movement or spreading, liquefaction may result in extensive, unanticipated ground settlement. Consequently, it is essential to consult a geotechnical engineer experienced in facilities built on soils that have the potential to liquefy.

If a treatment plant must be built over a stratum susceptible to liquefaction or densification, the designer may alter the soil characteristics or design the structure to overcome earthquake effects. Depending on the local topography, an effective drainage system may be constructed to prevent liquefaction. Sensitive clays may be removed and replaced with stable fill material, or structures may be supported on piles extending through the sensitive clay layer. Studies have shown that the costs of soil stabilization for a complete facility are high; therefore proper siting can be critical to mitigating seismic damage.

Special Tank Design Considerations. Another potential problem associated with earthquakes is failure of tank foundations. One possible reason is the increased localized loading caused by the tank overturning moment. The earthquake motion may cause the soil structure to liquefy, lose shear strength, or simply consolidate (settle), depending on the soil conditions. This may allow the tank to tip or settle unevenly, causing the tank shell or roof to buckle and sometimes fail. Proper design includes provisions for friction or other restraint of surface basins and vaults.

Seismic-induced forces must also be properly identified and applied to the structure and its components. If a structure is to contain liquid, the sloshing effects of the liquid must be taken into account by using accepted and recognized analytical methods in designing the containment structure. The analysis may involve the effective mass method described in American Water Works Association (AWWA) Standard D100. The designer may also refer to ACI Standard 350.3 or to Haroun and Housner (1980) for a complete, detailed analysis of earthquake-induced loadings on water-holding basins.

Another consideration in the design of tanks and basins located in seismic areas is that adequate freeboard must be provided so that the contents will not spill out and flood the immediate area or damage the roof of the structure.

Equipment Design Considerations. All the essential components of a water treatment plant should be designed according to seismic risk analysis. The process units that are critical to the plant's serviceability after an earthquake are identified first. The unit processes that are not essential to short-term continuation of service are then identified, and the acceptable duration of their downtime is established. Nonessential facilities might include carbon regeneration, lime regeneration, sludge collection, sludge disposal, and similar functions that are not directly related to producing safe treated water for short periods.

Anchorage of all components and equipment to the structure must be designed for the appropriate seismic forces as required by the governing building code and standards. Piping connections should be flexible enough to allow for differential movements of adjacent components. Special care must be taken in locating equipment, storage tanks, and feed lines containing hazardous materials.

Most regulatory agencies require that chlorination facilities be located in a separate, well-ventilated room, away from the rest of the treatment facility. Chlorine feeders should also be separated from chlorine storage rooms. Similar precautions should be taken with other hazardous materials. All gas bottles and containers of hazardous materials must be restrained in appropriate racks. Chlorine cylinder scales must be equipped with hold-down devices so that the cylinders will move with the scales or will be restrained.

Vertical end suction pumps, often used for solids handling, may be tall, slim structures, so their bases and anchorages must be designed to resist overturning. Horizontal pumps and their motors should be mounted on a single structure so that they respond together. This concept also applies to a battery of pumps connected to a single header. Pumps and piping should always be separated by a flexible connection to allow them to respond independently to earthquake movement.

Heavy electrical equipment must be anchored to base pads or buildings. Precision equipment, such as residual analyzers, recorders, indicators, electronic instrumentation, switch gear, equipment instrumentation, and communications systems, should be mounted rigidly to avoid amplification of seismic acceleration.

Anchor bolt embedments should be used in preference to expansion anchors where continuous dynamic loads are encountered (e.g., rotating equipment). Expansion anchors are acceptable for static loads and infrequent dynamic loads such as earthquakes.

Seismic design factors for basins are usually determined as specified in Chapter 9 of ACI Standard 318 rather than Chapter 21, because the basin is not a building frame system. The load factor for hydrostatic forces is typically reduced for the seismic case as allowed by ACI Standard 318, because the amount of water is clearly defined.

Loading Criteria

The various types of structures used in a water treatment plant must each receive careful consideration of the loading under all possible conditions.

Types of Structural Loads. Unlike buildings, tanks and basins must continually resist both vertical and lateral loads while in service. Lateral loads can be separated into external loads typically resulting from the surrounding soil and internal loads resulting from the contents of the structure.

External soil loads consist of either at-rest or active soil pressure, compaction pressures, surcharge from trucks or adjacent structures, and groundwater. Internal service loads include the hydrostatic pressure exerted by the contained water. As a minimum, the structure should be analyzed for the hydrostatic loads resulting from an overflow condition. A load commonly overlooked is the tensile forces in slabs and walls resulting from the internal pressures on the structure.

Vertical loads can include soil overburden, equipment, construction loads, and roof or floor dead and live loads. The roof slab of a tank or basin may also be subject to vacuum pressures created by the treatment process.

Load Factors and Load Combinations. Ultimate load factors and load combinations for concrete structures are defined in ACI Standards 318 and 350. The increased load factors for flexure, tension, and shear applicable to tanks and basins are intended to reduce the stress in the reinforcing to a level comparable with the working stress design used in the past. In addition, load factors and load combinations from the local building code should be reviewed and used for design if more severe than the national standards.

DESIGN OF BASINS, VAULTS, LARGE CONDUITS, AND CHANNELS

The basins, vaults, large conduits, and channels used in water treatment plants are special structures that must be designed to withstand the worst-case loading that may be imposed on them.

Loading Considerations

Many treatment facility structures are rigid and box-shaped, often buried in the ground, and sometimes without a superstructure. These types of structures are unique in many respects, and their design is normally based on a few simplifying assumptions. The walls of the structures are designed as plates to loads derived from U.S. Bureau of Reclamation tables, from Portland Cement Association (PCA) tables, or from finite element computer programs. Walls with aspect ratios greater than 2:1 are generally designed as one-way slabs. The designer should be aware that the moment in the center of the plate may be underestimated because of moment redistribution at the corners of the walls, and that the walls intersecting the plate must be designed for the same moment as determined for the end conditions of the plate.

The footings of a structure may also be subject to an increase in positive moment at the center of the span because of rotation of the footing on elastic soil. This phenomenon applies to both strip footings and mats. A strip foundation is normally used for basins that are particularly wide. A mat foundation is commonly used for vaults, conduits, and channels or where needed to resist groundwater uplift or to distribute bearing pressures over a wider area.

The stiffness and strength of the support should be evaluated for adequacy. For example, a basin wall could appear to be braced by several floor slabs of connecting channels; however, the slabs may not be adequately restrained to function as a brace.

Support conditions for unique structures should be thoroughly examined for variations. Normally, the roof slab of a conduit or channel is supported by the two sidewalls, except where the channel turns and the support conditions change.

The structural design should separately consider the internal and external loads acting on treatment plant structures. Lateral soil pressures should never be used to counteract internal hydrostatic forces, because the backfill is likely to be excavated at some time when the plant must remain in operation. The design of environmental structures should consider at least these factors:

- Settlement
- Earth-retaining forces
- Flotation
- Appurtenant items such as baffles, sludge collectors, water troughs, and filter surface washers
- In high-seismic areas, the internal effects of sloshing

Design Recommendations

Specific design recommendations for tanks, vaults, and similar structures are as follows:

- Specify the appropriate type of backfill used behind tank walls during construction. A noncohesive soil is normally used, because it is easier to attain a high compaction density with minimum effort.
- Use flexible connections between storage facilities and all inlet and outlet pipelines. These connections or joints can be mechanical, restrained expansion, rubber, ball-and-socket, or gimbel restrained-bellows-type couplings.
- Design walls serving as shear walls or diaphragms with expansion and construction joints keyed to carry the shear forces. To maintain watertightness at joints where some movement is anticipated, use joints with flexible waterstops.
- Design the top slab of buried structures for pattern loading of the overburden soil to accommodate construction. If the overburden soils above a structure are subject to truck traffic and are less than 2 ft (0.6 m) thick, the top slab should be designed for the actual truck loading applied directly to the slab as required by the American Association

of State Highway and Transportation Officials (AASHTO) *Standard Specification for Highway Bridges,* in addition to the overburden soil.

BUILDINGS AND SUPERSTRUCTURES

The performance requirements for structures at a water treatment plant are unique in many respects. While the design of these structures must comply with the local building codes, it must also account for the particular service conditions the facilities are subject to. Crack control, corrosion resistance, and unusual structural geometry all play an important role in the design of water treatment plant structures.

Design Criteria

Control buildings, buildings housing equipment, and other superstructures for treatment facilities should generally be designed in accordance with the local building codes. The design approach used for these types of buildings is similar to that used for conventional buildings, with a few exceptions.

Phantom loads are often added to floor and roof members of treatment plant buildings to simulate small, miscellaneous mechanical and electrical loads imposed either as part of the initial facility or by future modifications. The building may also be subject to unique live loads not covered by the building codes. The designer should consult American Society of Civil Engineers (ASCE) Standard 7 for a comprehensive listing of live loads that should be applied to industrial structures such as those found at a water treatment plant.

In high-seismic areas, a dynamic analysis will be required for structures with plan or vertical irregularities. The design engineer should particularly keep in mind that the seismic design criteria prescribed in the building code are considered *minimum* requirements to ensure life safety. The important structures at a water treatment plant should be designed to remain functional during and after an earthquake.

Analysis and Design of Buildings

The structural framing system for a building should be simple and symmetric. Vertical and plan irregularities should be avoided when at all possible to minimize construction costs and to improve long-term performance of the structure.

The selection of the structural framing system should be based on functional requirements and construction costs. In general, the gravity framing system and the lateral load-resisting system are analyzed separately. For concrete framing, the gravity system is usually analyzed and designed as a string of continuous beams supported by hinges or columns, depending on the restraint condition. The effects of patterned live loads also should be taken into account during design. For steel framing, the gravity system is typically analyzed and designed as a series of simply supported members.

In buildings with rigid diaphragms, the lateral forces should be distributed to each frame based on their relative rigidity. For a building with flexible diaphragms, the lateral forces are distributed to each frame based on their tributary width. Whether a diaphragm is rigid or flexible is based on its relative rigidity as compared with the vertical shear-resisting element. For simple symmetric buildings or asymmetric buildings with flexible diaphragms, only a two-dimensional analysis is required for the lateral design.

Other Design Considerations

The fire resistance rating of a structure should be determined early in the design and taken into account in selecting the framing system. Many buildings in a treatment plant are not entirely enclosed or air-conditioned. This exposes structural components to corrosive environments that may affect the fireproofing or coating systems. The design details must address special concerns and issues in water treatment plants, such as the inability of protective coatings to protect the top flange of a beam after welding the metal deck to the beam.

The potential effects of temperature change and shrinkage should be considered in the design of large superstructures. Many industrial-type buildings at treatment plants are subject to large temperature fluctuations caused by the equipment they house.

The structural framing system and the building enclosure should be compatible. A building cladding system is often initially selected based on aesthetic considerations and durability. The preferred system may or may not be compatible with the framing system chosen solely for structural efficiency. The structural engineer and the architect should work together to identify the optimal combination of building systems and adjust their individual designs accordingly.

The building enclosure and other attachments to the structure must be detailed to accommodate the anticipated vertical and horizontal movements. If masonry walls are used to enclose a building with a moment framing system, expansion joints must be provided in the masonry walls near the corners to allow for building movement. The structural framing system should also be compatible with the building cladding and partition details, equipment limitations, and the HVAC and piping systems it supports.

BIBLIOGRAPHY

American Association of State Highway and Transportation Officials. *Standard Specification for Highway Bridges*. Washington, D.C.: American Association of State Highway and Transportation Officials, 2002.

American Concrete Institute. *Building Code Requirements for Structural Concrete and Commentary*, ACI 318-02. Farmington Hills, Mich.: American Concrete Institute, 2002.

American Concrete Institute. *Code Requirements for Environmental Engineering Concrete Structures and Commentary*, ACI 350-01. Farmington Hills, Mich.: American Concrete Institute, 2002.

American Concrete Institute. *Seismic Design of Liquid-Containing Concrete Structures and Commentary*, ACI 350.3-01. Farmington Hills, Mich.: American Concrete Institute, 2002.

American Concrete Institute. *Tightness Testing of Environmental Engineering Concrete Structures and Commentary*, ACI 350.1-01. Farmington Hills, Mich.: American Concrete Institute, 2002.

American Society of Civil Engineers. *Minimum Design Loads for Buildings and Other Structures*, ASCE 7-02. Reston, Va.: American Society of Civil Engineers, 2003.

American Water Works Association. *AWWA Standard for Welded Steel Tanks for Water Storage*, ANSI/AWWA D100-96. Denver, Colo.: American Water Works Association, 1997.

Haroun, M. A., and G. W. Housner. *Dynamic Analysis of Liquid Storage Tanks*, Report EERL 80-04. Pasadena, Calif.: California Institute of Technology, Earthquake Engineering Research Laboratory, 1980.

Moody, W. T. *Moments and Reactions for Rectangular Plates*. Engineering Monograph No. 27. Denver, Colo.: U.S. Bureau of Reclamation, 1963 (reprinted 1990).

Portland Cement Association. *Circular Concrete Tanks without Prestressing*. Skokie, Ill: Portland Cement Association, 1992.

Portland Cement Association. *Rectangular Concrete Tanks*. Skokie, Ill.: Portland Cement Association, 1998.

CHAPTER 20
PROCESS INSTRUMENTATION AND CONTROLS

Dave Kubel
Westin Engineering
Rancho Cordova, California

The degree of instrumentation and automation in today's water treatment plants varies significantly. Many plants are fully automated and are staffed only during a day shift. Others employ minimal technology and rely on manual operation. Increasingly, automation is part of the design philosophy for new plants and upgrades of older plants. Some of the business reasons driving this change include

- *A need for improved treatment quality.* Modern automation and control allows closer and more consistent water treatment than is practical using manual control. The control system can be designed to alarm the operator whenever abnormal conditions require operator intervention.
- *Tighter regulatory requirements.* Increased regulatory requirements lead to a heavier dependence upon automation and instrumentation. Most regulations require grab samples for compliance testing. As the sampling intervals and record keeping increase, significant labor savings can result from adding on-line instrumentation and data collection. For example, the recent Interim Enhanced Surface Water Treatment Rule requires turbidity reporting at 15-min intervals. This makes on-line monitoring economically attractive.
- *A need to reduce costs.* There are many areas for cost savings through automation. This can be done through reduced chemical usage, reduced equipment wear and tear, and optimized energy usage. A few examples are sequencing pumps to minimize energy use; backwash turbidimeters to reduce energy and washwater use by terminating backwash cycles based on turbidity levels; and optimizing chemical control loops to minimize chemical costs.
- *Improved labor efficiency.* Labor savings can be achieved by automating time-consuming tasks, such as filter backwashes. A modern control system can allow operators to monitor and control an entire plant from workstations located strategically around the plant or off-site by using laptop computers.
- *Better operational, maintenance, management, and engineering decisions.* With extensive operational information being collected and stored in a relational database, staff

can analyze and report information in new ways. This allows decisions to be made based upon the most accurate and up-to-date information possible. In addition, historical operational information can assist in evaluating long-term operational trends and answer questions that require detailed historical information. Current advances include using historical information to develop expert systems or artificial neural networks that use fuzzy logic to further assist in making operational decisions.

- *Earlier detection of problems.* Most control systems include an alarming capability that detects and notifies an operator of operation or process problems. In addition, on-line instruments can detect many types of water quality problems in real time as opposed to waiting for the results of grab sample analysis.

- *Quicker response to emergencies.* Centralized control can dramatically reduce the time required to make changes in water production during a crisis.

Because of the rapid advances in computer technology, the designer needs to be not only knowledgeable about process control strategies and instrumentation but also familiar with areas such as computer systems architecture, communications, and network design.

PROCESS AND OPERATIONAL CONTROL STRATEGIES

The design of a treatment plant needs to reflect how each process of the plant will operate and the appropriate level of automation. Water treatment plants often use a combination of manual, semiautomatic, and fully automatic controls. For example, flocculator drives can be manually activated when a sedimentation basin is in service. Filter backwash sequencing may be manually initiated, with operator intervention required to advance through subsequent steps. Automatic controls for water treatment can be simple or complex, depending on the application. The designer needs to balance the increase in complexity and long-term maintenance costs against the automation benefits. This section provides an overview of typical plant control hierarchies and common strategies for control of subprocesses within the water treatment plant.

Hierarchy of Control

General control philosophies should be developed and agreed to early in the design process. The control hierarchy should include both local manual controls that will operate independently of any computer system and one or more computer-based modes. An example approach (as shown in Figure 20.1) could be as follows:

Local Control. Local control refers to controls that are located close to plant equipment and processes and that operate independently of the plantwide computer control system. These are typically backup control modes that override computer control or for equipment that rarely requires a change in state.

- *Local equipment controls.* These are manual field controls mounted on motor starters or valve actuators to allow maintenance personnel to place the equipment out-of-service, make it unavailable, or locally control the equipment for testing.

- *Hardwired interlocks.* These interlocks are implemented to protect personnel and equipment or to avoid operational mishaps such as overflowing a reservoir. They are

FIGURE 20.1 Example of control hierarchy.

hardwired for reliability and also so that they are effective in both local and computer modes. They operate independently of any controller or computer system.

- *Local manual controls.* This type of control is typically located at an MCC or control panel and is selected by putting a Hand-Off-Auto switch in the Hand position. Local manual control allows the operator to run equipment by pushing local start and stop buttons; set a variable-frequency drive (VFD) speed by turning a potentiometer; position a valve via a digital panel-mounted interface; or set a chemical flow by adjusting a setpoint on a local controller. This mode can have complex relay logic that controls proper sequencing of equipment and interlocking. When in this mode, control from the computer control system is generally locked out.

Computer Control. Computer control includes all control modes implemented in the computer control system, and it is not necessarily specific to the location where the operator station is located.

- *Auto.* When facilities or unit processes are selected to operate in this mode, the controller controls the process and equipment. The controller is programmed to fully control all aspects of the process or facility. Equipment at a site can be placed in this control mode only by an operator at the site. It is typically selected at a motor control center (MCC), control panel, or the equipment by putting a Hand-Off-Auto switch in the Auto position. Most control strategies described later in this chapter are programmed into the controller to operate in auto mode.
- *Computer manual.* This control mode is initiated by the operator through the graphical screens at the control center or other workstation. This allows an operator to remotely override the controller Auto controls and force a piece of equipment to run or stop (open or close) or maintain a manually entered setpoint.
- *Optimization control.* Some systems might include more sophisticated systemwide optimization control. This type of control allows computer optimization of operational setpoints based upon real-time demand forecasts, energy and water quality optimization models, etc.

Types of Control Algorithms

This section discusses the types of automatic control that are commonly implemented in the field controller. The following control descriptions are conceptual and are included to give a general understanding of the types of algorithms used in water treatment plants.

Sequential Control. Sequential control is used to control two-state devices such as fixed-speed motors and open-close valves. It may coordinate the operation of a group of devices or a single device. A sequential control scheme commonly comprises boolean logic (for example, ladder logic) and timers. It may be used for sequencing the start-up or shutdown of a plant, controlling the pumps in a pump station, or backwashing a filter. When one is designing sequential controls, it is important to thoroughly consider all possible conditions that can occur, including power loss and recovery and equipment failures. Methods for managing failures can range from simply halting the sequence and issuing an alarm to very complex failover and recovery algorithms.

Continuous Control. Continuous control is used to maintain a process variable, such as flow or chlorine residual, at a setpoint value. Continuous control includes both feedback and feedforward (ratio) control. A feedback *flow* controller, for example, adjusts a control valve or feed pump continuously to reduce the magnitude of the error between the measured process flow and the desired flow setpoint. There are numerous algorithms for implementing continuous control, each with different applications.

  On-Off Control. On-off control is the simplest and most common automatic feedback control mode. This approach will maintain the process variable by turning on/off equipment as needed. As an example, a heater thermostat operates in on-off control. If the thermostat setpoint is 80°, the thermostat will turn the heater on when the temperature drops below 80°, and off when the temperature rises above 80°. On-off control typ-

FIGURE 20.2 Feedforward control example: polymer dosing.

ically includes a dead band which will prevent the temperature from sitting exactly at 80° and constantly turning the heater on and off. With a 2° dead band the heater will turn on at 80° and off at 82°. Wet-well level control commonly uses on-off control. A wet-well level is maintained while the influent flow varies by using high- and low-level float switches to turn fixed-speed discharge pumps on and off.

Feedforward Control. Feedforward control measures one or more process input parameters and calculates the necessary final control element setting based on these values. Ratio control is the most common water treatment application of feedforward. This concept is shown in Figure 20.2. Here it is used for flow pacing polymer into a main line water flow. The polymer flow is maintained at a fixed *ratio* or proportion of the main line flow. The ratio controller F(flow) continuously computes the needed metering pump settings (speed or stroke) based on the measured main line flow and the operator-entered ratio. Hence the polymer flow into the mixing tank is always maintained at the correct proportion of the main flow. This provides excellent results where water quality and polymer concentration are fairly constant.

Feedback Control (Including PID Control). Feedback control is the easiest way to automate process control. Feedback control measures the value of a controlled process variable and compares this with the value of the desired setpoint. The feedback controller then adjusts the manipulated variable (usually a valve position or pump speed) to reduce the error and maintain the process at setpoint. Level control is a common application of feedback control. An example is shown in Figure 20.3. The controller continuously compares the measured tank level with an operator-entered level setpoint. If the tank level is higher than setpoint, the controller automatically adjusts the valve to reduce the flow into the tank. Similarly, if the tank level is low, the valve position is adjusted to increase flow into the tank. A commonly used control algorithm for adjustment based upon the calculated error is a *proportional/integral/derivative* (PID) control algorithm. Feedback control is often referred to as *closed-loop control.*

Control Equation
Proportional Control: $M(t) = K_c[E(t)]$
PI Control: $M(t) = K_c\left[E(t) + \dfrac{1}{T_i}\int E(t)dt\right]$
PID Control: $M(t) = K_c\left[E(t) + \dfrac{1}{T_i}\int E(t)dt + T_d \dfrac{d}{dt}E(t)\right]$
Where: $E(t)$ = Error = SP - PV K_c = proportional gain T_i, T_d = integral and derivative time constants

FIGURE 20.3 Feedback control example: tank level control.

These are several control algorithms commonly used for feedback control:

PROPORTIONAL CONTROL. Proportional control simply adjusts the controlled device in a manner proportional to the measured process variable. For example, a proportional control algorithm for Figure 20.3 might be set up to have the pump fully closed when the level is above 10 ft and fully opened when the level is below 8 ft and to adjust the valve position linearly when the level is between 8 and 10 ft.

The difference between the minimum and maximum levels (8 and 10 ft in the example) is known as the *proportional band*. If the proportional band is set too small, the loop gain will be too high, resulting in level oscillations. Thus, proportional control is not recommended when tight control is desired.

PROPORTIONAL PLUS INTEGRAL CONTROL. An integral term can be added to the proportional control algorithm to improve control accuracy while avoiding loop instability.

The integral term (or *reset*) in this equation is affected by both the magnitude of the error and the time that the error is present. Over time, a small error will cause the integral term to generate a change in the controller output to reduce the error. The rate at which the controller integral term increases is the integral gain. The integral term ramps up and down as the process variable exceeds or falls below the setpoint.

In the above tank level control example, this means the valve position will adjust to whatever position is required to maintain the level at the desired setpoint. Integral control often allows system setpoint to be achieved with lower proportional gains than with proportional-only control. *Reset windup* can occur when the controlled device has been taken out of service, so the accumulated error times the duration eventually ramps the controller output to maximum. When the device is returned to service, the controller has to "unwind" the output back down, using the proportional term. Digital control systems usually monitor the status of the controlled equipment, so a simple solution is to program the controller to force the integral term to a minimum if the equipment is off.

PROPORTIONAL-SPEED FLOATING CONTROL. Another integrating process element is a motorized valve, often used for filter effluent flow control. The valve actuator responds to open and close commands and stops when neither command is present. The valve travel speed is fixed by motor speed and gearing. For large errors, proportional control response is hindered by the ramping *process lag* of the relatively slow actuator. For small errors, the valve moves too far before its travel can be stopped. Reset (integral) control is not required, because the valve holds its last position when the error is zero. To minimize valve hunting for small controller errors, the duration of the open and close commands can be made proportional to the error. For small errors, the valve may be jogged only 0.25 s every 10 s. For large errors, the valve may be moved 8 s every 10 s.

PROPORTIONAL PLUS INTEGRAL PLUS DERIVATIVE CONTROL. Proportional plus integral plus derivative (PID) control has a derivative control term added to adjust controller output in proportion to the rate of change of the controlled process variable. If the process variable changes rapidly, controller output is reduced by an amount proportional to the rate of the change. This action is useful in compensating for integrating-type process lags, such as the tank level in the above example. Derivative action is commonly used in temperature control applications where thermal inertia is a significant process lag.

More Advanced Process Control Algorithms. More advanced control approaches exist that can improve process control beyond conventional PID and feedforward control. Some of these include the following:

ADAPTIVE GAIN CONTROL. Adaptive gain control is used in conjunction with other feedback control algorithms. A typical control loop is tuned to provide the best control possible while maintaining stability at all times. However, the *best possible control* will vary with plant conditions. Thus, the controllers are detuned from optimum performance to ensure stability over varying plant conditions.

A digital controller can be programmed to monitor the effect produced by the last control change on the process and to adjust the controller gain accordingly. For example, when gravity filters are clean, their flow control valves are only partially open, and a small valve movement produces a large flow change. The controller adjusts the duration of motorized filter effluent flow control valve commands, depending on the flow change produced during the last control interval. Adaptive control is relatively common in water plant environments.

NONLINEAR CONTROL. An automatically controlled process may have a highly nonlinear response to changes in controller output. A good example is a pH control loop, where the titration curve for the process is steeper on both sides of the neutralization pH. Thus, the loop gain changes dramatically over the control range. To compensate for this and to provide the best control and stability, nonlinear process models in the controller loop are included. A controller with reduced gain near the neutralization point is more stable than a fixed-gain controller and can provide improved control.

MODEL-BASED CONTROL. One of the most difficult control problems is process dead time, which prevents the process variable from being measured in sufficient time to make feedback control stable. For example, additional chlorine must be fed to an open reservoir to compensate for solar decay, which varies with the solar intensity and transport time in the reservoir. A mathematical model of solar decay can be included in a programmable chlorination controller that also measures solar radiation intensity, reservoir flow rate, and level.

FUZZY LOGIC CONTROL. Fuzzy logic uses fuzzy set theory, which tests overlapping logical conditions for degrees of membership and makes appropriate control decisions. Fuzzy logic emulates the way humans think, in shades of gray rather than in black and white. Several manufacturers provide controllers with fuzzy logic support or tuning. Although fuzzy logic control has been utilized on several processes in water treatment plants, its use has not yet been widely adopted. However, considerable development work is being done.

Water Treatment Plant Control Strategies

This section discusses control strategies employed to operate water treatment plant processes. To limit the discussion, a *generic* treatment plant will be used, as shown in Figure 20.4. This is not intended to be comprehensive, but to illustrate a few approaches to many of the more common processes. Also see Table 20.1.

Raw Water Pump Controls. Flow is one of the most commonly controlled plant parameters. Changes to plant flow are usually kept to a minimum to avoid disruption to downstream processes. In most water treatment plants, a single flowmeter measures the plant influent flow rate. Multiple flowmeters can be used to provide redundancy and a wider range of measurable flows. Other meters are installed at places throughout the process to measure other important flows. Typically the operator sets the desired plant influent flow. This becomes a flow setpoint to a feedback flow controller. In cases where flow can be gravity-fed, the controller would typically modulate an influent valve to control the flow. In other cases, constant-speed and/or variable-speed pumps are commonly used. Constant-speed pumps may be sequenced to change the flow in steps or may be used in conjunction with variable-speed pumps or throttling valves to achieve a reasonably smooth flow range.

Pump control of the raw water includes pump sequencing and speed control. A PI or PID flow controller is used to adjust the speed of the on-line pumps and maintain plant flow at setpoint. The sequence control logic starts and stops pumps as necessary.

FIGURE 20.4 Example of water treatment plant—primary processes. (*Source:* Instrumentation and Computer Integration of Water Utility Operations, *AWWARF, 1993, revised by the author.*)

TABLE 20.1 Common Monitored and Controlled Items

	Raw water pumping	Mixing/ coagulation/ flocculation	Clarification/ sedimentation	Filtration	Disinfection	RWW pumping/ sludge handling	Treated water pumping/ storage	Miscellaneous
Monitor	• Water temperature • Level • Flow • Turbidity • pH • Conductivity • Chlorine demand	• Level • Flow • Turbidity • Chlorine residual • pH • Mixer status	• Valves • Sludge level • Effluent turbidity	• Level • Differential pressure • Valve positions • Flow • Turbidity • Particle counts	• Disinfection feed • Flow • Chlorine residual • Turbidity	• Sludge level • Tank level • Recycle flow • Turbidity	• Clearwell level • Chlorine residual • Turbidity • Flow • Pump status/speed	• Power usage/ quality • Chemical preparation and storage • Sump pump systems
Control	• Raw water pumps • Flow control	• Chemical feed • Flocculator control • Blowers • Flow valving	• Sed. basin sludge collection control • Sludge removal valves/ pumps	• Flow, level • Backwash valves • Backwash pumps/ storage • Filter-to-waste pumps • Surface wash pumps • Air scour blowers	• Disinfectant feed	• Sludge pumps • Scraper • Sludge level • Recycle pumps • Drying beds	• Effluent pumps • Discharge psi • Discharge flow	• Plant start-up • Plant shutdown • Chem. prep., storage, and pumping

Source: Instrumentation and Computer Integration of Water Utility Operations, AWWARF, 1993, revised by the author.

Mixing, Coagulation, and Flocculation

Coagulation. Coagulants are chemicals added to water to help suspended particles agglomerate during the flocculation process. Coagulant and coagulant aid chemicals are generally fed by positive displacement chemical metering pumps, using flow or flow/turbidity feedforward control, or feedback from a streaming current detector. The coagulant flow signal can also be used as feedback for the feed pump speed. Coagulant dosing is sometimes controlled using flow feedforward control combined with feedback control from the plant effluent turbidity. The chemical feed rate is often monitored with a flowmeter.

Control of pH. The pH adjustment is sometimes used to improve removal of inorganic contaminants, such as minerals and heavy metals, and for proper coagulation in low-alkalinity waters. Lime or caustic soda is added to raise the pH of the water or acid or carbon dioxide is added to lower the pH of the water. The pH control is highly nonlinear, and specialized adaptive gain control algorithms can be used for improved control. The pH control can include feedforward flow control adjusted with feedback control based upon downstream pH measurement. The use of both feedforward and feedback elements is known as *compound control.* This is shown in Figure 20.5.

Flash Mixing and Flocculation. Control of flocculation and mixing equipment is typically limited to turning on or off the equipment. This can be done manually through the control system or automatically, based upon sections of the plants that are in service. Optimum flocculation requires gentle mixing at the appropriate intensity of agitation, as defined by the G value (described in prior chapters). The speeds of mechanical mixers and flocculators are commonly controlled to allow the level of agitation to be adjusted. Control is typically manual through the control system, but can be based on water temperature or pollution level.

Clarification and Sedimentation. There are several approaches to clarification and sedimentation that can be employed, as described in Chapter 7. Often they require no control of hydraulics, with the level being set by an overflow weir. Sludge is collected by means of mechanical scrapers or rakes, which are manually turned on or off locally or

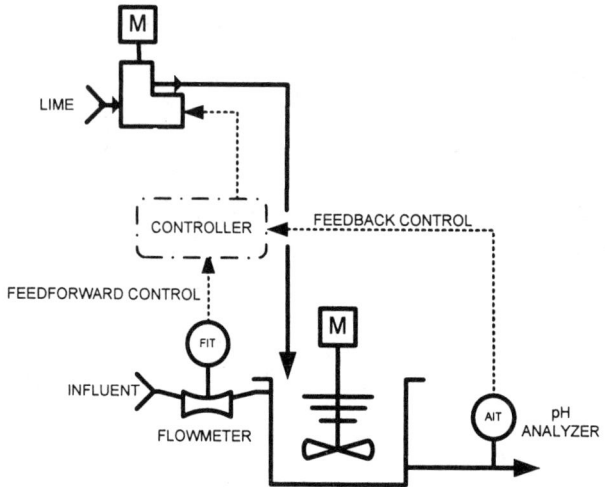

FIGURE 20.5 Compound pH control loop.

through the control system. Sludge is removed by opening a valve and/or turning on a pump for a set time period and set frequency. Monitoring of this process includes monitoring of valve positions, pump, and motor status.

Filtration

Filter Flow Controls. The number of filters required on line is determined by the plant flow and the optimum flow per filter. Constant-rate filters equipped with flow controllers adjust automatically as the plant inflow rate changes. A commonly used method is cascade level flow control (Figure 20.6). A level signal from the filter influent channel is transmitted to a PI level controller. The controller output adjusts the flow setpoints of the individual filter flow controllers. The individual filter flow controllers maintain flow through a PI controller with feedback from a filter effluent flowmeter. As plant flow rate increases, the filter influent channel level increases above the level controller setpoint, increasing the level controller output. This raises the filter flow controller setpoints, so the filter flow control valves open to dispose of more water. The filter influent channel level drops, eventually reducing the level error to zero, and the level controller integral term maintains the increased flow control setpoint. The level controller can be proportional-only control if the channel level can be allowed to rise a few inches at maximum plant flow.

If integral control is used, reset windup in the level controller must be prevented by suitable interlocks to force the flow to zero on plant shutdown and to the minimum flow setpoint on plant start-up. The highest value of the level controller flow rate must always be limited to the maximum allowable filter bed flow rate. Chapter 8 includes descriptions of additional filter types.

Filter Backwash Controls. The operation of filters eventually progresses to the point where backwashing is required to clean the filter media. Backwashing can be initiated by using any combination of the following:

- High head loss (constant-rate filters)
- Low flow rate (variable declining rate filters)

FIGURE 20.6 Example of cascade level/flow control.

- Elapsed run time
- High effluent turbidity
- High effluent particle count
- Predetermined schedule
- Operator judgment

High filter effluent turbidity can also be caused by other factors besides filter solids breakthrough, such as improper coagulant feed or improper filter aid feed, so turbidity is usually combined with other filter parameters to initiate backwash. Once it is determined that a filter is in need of backwashing, it is placed in a queue to be backwashed.

As discussed in Chapter 8, several backwashing techniques are available. The backwash control strategies involve significant sequential control strategies. A typical valving layout is shown in Figure 20.7. (*Note:* Water quality instruments are not shown.) An example backwash sequence proceeds as follows.

1. The filter's inlet valve is closed. This is coordinated with bringing on a standby filter, thus maintaining a relatively constant plant flow. The filter level is allowed to drop to a preset level just above the media, and then the effluent (modulating) valve is closed.
2. The drain valve is opened, the surface wash pump is started, and the surface wash valve is opened. After a setpoint time, the backwash pump is started and the backwash valve slowly ramps open to a low rate of flow and holds.
3. After a setpoint time, the surface wash valve is partially closed, reducing the surface wash flow. The backwash valve ramps to a high flow rate for a setpoint time, then ramps to a low rate.
4. The surface wash valve closes. The drain valve closes and refills the filter with clean backwash water.
5. The backwash valve closes.
6. After refill, the filter enters a filter-to-waste cycle. The inlet valve slowly opens, and the filter-to-waste valve opens. Water is run through the filter for a setpoint volume or

FIGURE 20.7 Example of filter backwash layout.

until desired water quality is obtained, as measured by on-line turbidimeters or particle counters. The filter is then either returned to service or placed out of service.

In some plants this procedure is manually controlled, usually from a control console on the filter deck where the operator can watch the process. Automation of the backwash sequence is complex, but can save tremendous operator time. Automated filter backwash scheduling can minimize plant flow disturbances as filters are placed in and taken out of service. In addition to proper programming of the backwash sequence, the strategy should properly handle unexpected failures in the sequence. Instrumentation such as valve limit switches and flowmeters is utilized with watchdog timers to verify proper completion of each step of the sequence and to detect failed sequences.

Given consistent washwater quality, and steady-state process conditions, a filter usually cleans in a fixed length of time. However, backwash turbidimeters can be added to the automatic controls to limit the amount of washwater pumped.

Disinfection

Chlorination Controls. The most commonly used disinfectant for water treatment is chlorine. Where the plant flow rate is constant and influent water quality is consistent, manual control of chlorination can be satisfactory. However, plants typically provide for some level of control of chlorine injection. Of prime importance with disinfection is maintaining a minimum *contact time* to ensure proper disinfection. Chlorine contact time requirements are discussed in Chapter 10.

Most chlorination controls employ ratio control (feedforward) for flow-pacing the chlorine. This allows for automatic adjustment due to changes in flow but not water quality. To adjust for water quality, manual or automatic (described below) adjustment to the ratio can be done. Flow between different parts of the plant can vary significantly due to filter backwashing, plant water usage, and the time it takes for changes in plant flow to propagate through the plant. Thus, chlorination should be controlled using a flowmeter that measures flow as close as possible to the point of chlorine addition. This will provide the most accurate flow-paced control.

A constant ratio of chlorine at the injection point does not ensure a consistent downstream chlorine residual or consistent contact time. Changes in chlorine demand are affected by changes in raw water quality as well as sunlight, which are generally slow-changing. To address this, automatic chlorine residual feedback control can be added to provide improved control. Chlorine injection is automatically adjusted based upon the measured downstream chlorine residual and an operator-entered setpoint. If the measured residual is higher than the setpoint, the chlorine being injected is reduced. Likewise, if the measured residual is lower than the setpoint, the amount injected is increased.

Feedback control for chlorination must consider the process lag times. *Process lag time* refers to the time from when a change in chlorine injection is made to when the measured chlorine residual reflects the change. Process lag time includes three elements. The first is hydraulic transport lag, between the point of chlorine addition and the analyzer sample piping inlet. This time is inversely proportional to the plant flow rate. The sample piping inlet must be far enough away from the point of chlorine addition to allow mixing of chlorine with water at all plant flows. The second element is the fixed transport lag in the sample piping between the process stream and the analyzer. This delay is dependent upon the analyzer's sample flow rate and the sample pipe diameter and length. If the analyzer is located away from the sample point, pumping the sample in a loop can be used to reduce the transport delay. In this situation, most of the sample is bypassed back to the process (Figure 20.8). The third element is the reaction time of the analyzer.

To account for lag time, the feedback control loop must be tuned to react slowly in comparison to the process lag time; otherwise, loop instability can result. However, by

FIGURE 20.8 Compound chlorination control loop with time lags.

slowing down the loop response, control performance is reduced. Thus, long process lag times can result in controller settings that give unacceptable performance. It is often impractical to use a single feedback control loop where the chlorine residual feedback is measured too far downstream of the injection point. Ideally, the total lag time at lowest plant flows should not exceed 3 to 5 min. This may require the use of two or more sampling points.

Although feedback control alone can be used to adjust for changes in process flow as well as water quality, it is typically coupled with feedforward flow pacing. The primary control is feedforward flow pacing, which can react very quickly to changes in flow since there are not the significant process delays associated with the feedback loop. The feedforward ratio is then slowly trimmed by the feedback loop. This is shown in Figure 20.8.

Another approach is to use feedforward control in conjunction with feedback from two chlorine residual analyzers. The first analyzer is located a couple of minutes downstream of the injection point, which is used for a conventional PI controller. The second analyzer is placed at the discharge of the contact basin. This is used to "trim" the setpoint for the second controller and is only updated at a period greater than the dead time.

Sludge Handling and Return Water. Sludge treatment is typically broken into three primary processes: thickening, dewatering, and drying. Each of these processes can be accomplished in a variety of manners, as described in detail in Chapter 17. Each approach has differing control requirements.

Mechanical dewatering equipment, such as filter presses, belt presses, and centrifuges, are provided with local panels that control the equipment and provide a relatively simple interface to the plant control system. Some of this type of equipment, such as filter presses, operates in a batch process mode as opposed to a continuous mode. This can require coordination with other plant processes.

The example in Figure 20.4 utilizes a sedimentation basin, a gravity thickener, and drying beds. The sedimentation basin and gravity thickener are controlled in the same manner as sedimentation basins previously described, with rakes or scrapers running continuously and thickened sludge removal controlled based upon time. Drying beds are sequenced to fill the beds in a given order, with each bed filling for a set time. Provisions for flushing the lines can be manual or operator-initiated.

Return water pumping is generally based upon level in the basin or tank that it pumps from. Pumps should be sequenced to even out the run times.

Treated Water Pump Controls. Treated water pumps can be either manually or automatically controlled, based on transmission or distribution system pressure or reservoir level. Multiple pumps can be automatically staged based on failure of the running pumps to maintain a certain pressure control range. Equally sized variable-speed pumps are usually all run at the same speed, because the pumps supply a common discharge header, and the discharge pressure reduces the flow produced by slower pumps. Pump speed is usually controlled using PI or PID control. Pumps supplying a pressurized pipeline can be stopped based on sustained periods of reduced speed or flow rate.

Miscellaneous Plant Controls

Start-up and Shutdown. In a small treatment plant, plant start-up and shutdown can be accomplished by manually turning equipment on and off, adjusting chemical feeders, and opening and closing valves. In a medium to large plant, automatic controls can make these operations relatively simple, with minimal staff. In addition, automatic shutdown of the plant can be done for emergency events such as overflows or power failures.

Chemical Delivery, Preparation, and Storage. Chemical delivery and chemical preparation are generally batch processes that include a number of specific steps to fill tanks with chemicals of the appropriate concentrations and age. When chemicals are provided of higher concentration than is needed, it is common to prepare a batch of diluted chemicals a few times a day, storing it in day tanks for continual injection into the plant.

Electric Power Monitoring. Electric power is a major cost associated with treating and pumping water. Billing is commonly based upon peak demands, energy, and power factor. Peak demands are tracked by the power company and are used to set *demand charges*. These can have large impacts over extended periods, such as the next year.

Energy monitoring can be included to track overall and process energy consumption synchronized with the power company demand intervals. Depending upon the system, operators can be alarmed through the control system when electricity usage is approaching the current demand peak usage and can take action, if possible.

Energy monitoring can also be used to trigger plant shutdowns based upon loss of power or to initiate backup power systems.

CONTROL SYSTEM DESIGN

Control system design should consider more than just the treatment plant. Operators control an entire water system, which may include multiple plants and distribution and transmission systems. The trend is toward greater integration of disparate plants into one common control system platform, leading to improved operational efficiency, potential optimization, and reduced maintenance costs.

Control System Architectural Issues

Many questions arise when one is considering the entire system that would not if the focus were limited to one plant (Figure 20.9). These systemwide questions can easily have the greatest impact on the final design architecture.

FIGURE 20.9 Example of water system control system architecture.

Assuming that there are existing plants and remote sites, some of the questions that should be considered when one is building a new plant include:

- Will the plant be capable of being operated remotely from another facility? Will other plants be controlled from the new plant?
- From where will the existing and new remote sites, such as treatment plants, pump stations, and ROFCs, be controlled? From just one plant or from all plants? How much integration is desired between existing control system software and new software? This will impact real-time and historical database integration, alarm management, configuration scope and requirements, performance requirements, etc. It may require the software at the new plant to be of the same make/model as that at existing facilities, allowing for the most integrated system, but limiting the competitiveness of the bid process. Where do the alarms for the remote sites go, and how is that managed? How will alarms for the new plant be handled when the plant is unattended?
- Should the communications to the existing remote sites all be routed through the existing plants, or should there be independent communications paths to minimize single points of failure?
- Will the existing control systems be due for replacement at the time the new plant is being brought on-line? Should the existing control system software be replaced or upgraded at the same time to maximize integration?
- What future water system needs are anticipated that will impact the new control system? This could include future CIP projects that should be accounted for in system sizing and performance, or planned optimization projects which might have instrumentation or data requirements.

There are a number of core issues associated with the new plant that will also impact the final control system architecture.

Unattended Plant Operation. Many plants are attended only by roving operators or are unattended nights and weekends. Control systems designed for fully or partially unattended operation should incorporate comprehensive process and equipment monitoring, automatic alarm paging for the operator, and the ability for an operator to monitor the plant remotely.

Redundancy and Availability. Sufficient redundancy and backup should be provided so that single failures that can seriously cripple the continuous operation and/or control of the plant are appropriately accounted for. Availability calculations can be made to validate the design. This should account for expected downtimes of devices and computers as well as the mean time to repair. System availability requirements should be specified and tested. Policies for on-site spare requirements should be driven from the availability calculations. Typical devices to consider for redundancy include historians, operator stations, local-area networks, remote terminal units (RTUs) and programmable logic controllers (PLCs), communication interfaces, and critical instrumentation. Reliability philosophies for each process unit should be coordinated with plant designers. Backup power for each control system device should be considered.

System Sizing and Performance Issues. When one is procuring a control system, it is preferable to clearly define the intended size and performance requirements of the system for the vendors. Each vendor knows its product and its limitations best, and each product has different limitations. By setting requirements for ultimate sizing and performance and allowing the vendor the flexibility to detail the design to meet the requirements, significant performance issues are more likely to be avoided. Sizing and performance items to specify include

- Process I/O quantities (AI, AO, DI, DO, serial)
- Historical data storage—quantity and frequency of points stored (by type), and duration maintained on-line
- Remote and in-plant data acquisition rates
- Response times—control output actuation, display generation and update, screen data refresh, trending rate, alarm response

Location of Operator Interfaces and Control Centers. A large plant should be designed so that the fewest staff possible can operate it easily and efficiently. Small plants often require more design features than larger ones, because there may be only one plant operator on duty. Some plants may be unattended on nights and weekends.

To allow operators to leave the control room, workstations can be strategically located throughout the plant, allowing complete monitoring and control. Optionally, some control systems have interfaces that allow an operator to carry a notebook computer or small pocket computer (PDA) for monitoring and control.

If there is no convenient location for a process area control room, design can specify process controllers with local operator interface (LOI) displays. These LOIs do not necessarily require full graphic displays, but they should have an alphanumeric display area large enough to display all local process variables, such as valve positions, flows, levels, and automatic commands, being issued. The LOIs should also have access to plantwide data from the local controller. Password protection should be included in the LOI for critical control actions.

Local Control Panels. Sooner or later any automatic control system will fail, possibly at an inconvenient time, such as the middle of an automatic backwash sequence. If the controlled process equipment is not easily accessible for manual operation and the process is critical to plant operation, local control panels may be required. These panels act as conveniently located control interfaces between process equipment (motorized valves, pumps, sensors, etc.) and automatic controllers, which are often mounted inside the panels.

Local-remote switches transfer control from the controller to the local control switches. A common example is a local backwash control panel on a filter deck, where the operator has full view of the filtering process.

The alternative approach is not to have local panels and to require operators to climb ladders in the filter pipe gallery to operate valve handwheels until the automatic controller is repaired. The logic of this approach is that the vast majority of the time, the automatic control system is operational, and local control panels add considerable cost to the system. This approach is viable only if operators are continually retrained in manual operation of the plant. Automatic controls may operate perfectly for years, and operators may forget which valves must be operated. Most operators prefer the fallback manual control switch approach.

Information Flow to Advanced Computer Applications. The real-time database in a treatment plant's process monitoring computer contains vital information for other departments within the utility, including maintenance, the laboratory, engineering, and management. For instance, water quality data can be shared with a laboratory information management system and equipment run times accessed from a maintenance management system. Today's control systems are commonly integrated with applications in other departments. Selection of a control system should include the ability to integrate with common information interfaces, such as SQL. Specifications should include detailed requirements for each interface needed.

Control System Software

This section describes general control system software. Additional details are included in the "Control System Programming and Configuration" section.

DCS and SCADA Systems. Distributed control systems (DCSs) and supervisory control and data acquisition (SCADA) systems originally solved two different problems.

The DCS was developed for in-plant applications where high-speed networking was available. DCSs are a tightly integrated software and hardware solution for plant control. Control strategies can typically be developed on a DCS through the use of graphical configuration tools. DCSs typically have powerful built-in control strategies and track data quality. IEC 61131 programming is often supported. New control logic can sometimes be tested and debugged using live plant data without disrupting the process, and then loaded into the DCS unit at the process. DCSs are historically proprietary in both hardware and software, although most vendors now use commercially available computers and networking equipment and provide industry standard interfaces to other equipment. Most systems were designed to provide redundancy and fault tolerance in all components, since the operation of the plant relied on the system.

SCADA systems were originally developed for connection to remote RTUs over low-speed data lines. Thus, they excel at remote communications. The system databases resided in two places: in the RTU and in the SCADA server, as opposed to DCSs which only have the database in the process control units.

Today, SCADA system and DCS features can be very similar, each having incorporated many of the strengths of the other. Both have been successfully implemented in water treatment plant environments.

Custom SCADA/DCS Software. Control systems in the water industry are commonly procured through a low-bid process. A specification is developed, which each bidder must meet. Many times, vendors will bid on a job that will require them to modify their control system software to meet the specifications. This does not refer to programming the controllers or configuring the screens or database, but refers to changes to the base SCADA or DCS software. It is important to understand the potential ramifications of this. If custom software is required, it should be understood that extended troubleshooting may occur after installation, even if complete testing is done at the factory. Systems cannot be fully tested until they are operational in the field. Since this software has never been tested in a field environment, it may have unexpected results. Cutovers of live systems are of particular concern in this regard. Custom software changes to manufacturer's standard SCADA/DCS code should only be permitted where no other solution is available.

Control System Hardware

This section describes control system hardware options for implementing the water treatment plant applications described above.

Programmable Logic Controllers. Programmable logic controllers are industrial-grade, special-purpose microcomputers with input/output (I/O) subsystems for monitoring and controlling processes and equipment. PLCs were originally developed in the late 1960s for factory applications to replace relay logic, where changes to the logic were costly. PLCs have evolved to include communications capabilities and process control abilities. Most PLCs are programmed in ladder logic or support IEC 61131 programming standards. PLCs are commonly used in water treatment plant applications, sometimes with multiple PLCs networked on high-speed data links to the SCADA/DCS workstations.

Remote Terminal Units. RTUs were originally designed to be installed at remote sites and linked to a central station host by low-speed telephone or radio. They were originally utilized to communicate status between the SCADA host and field equipment, with limited or no control capability. The differences between PLCs and RTUs were significant years ago, but diminished as the products evolved. Many RTUs include control abilities, as PLCs now include communications capabilities. However, some general distinctions can still be made. Although a serial communications adapter can be added to a PLC for modem communications, RTUs tend to have more sophisticated communication capabilities, such as multiple communications port data access from multiple other sites, serial data store-and-forward, and report-by-exception. Many are well suited for low-power solar applications.

RTUs can sometimes accommodate I/O from several different vendors and can use the same industry standard communication protocols as PLCs. Some RTUs have built-in operator interface displays and can accommodate external displays.

In-Plant Networks. Network design has become a significant part of modern control system design. Networks at a plant are generally one of three types: business networks, process control networks, and device networks. The process control network and device networks are the realm of the controls engineer, while the business networks are generally designed by others.

Process Control Networks. The process control network is used to connect several SCADA and DCS workstations together as well as connecting the workstations to the plant controllers (shown as PLCs in Figure 20.10).

FIGURE 20.10 Treatment plant networks.

The network design must consider several topics, including:

- *Network technology.* Several different types of networks can be used, and the selection may be driven by the SCADA/DCS vendor. Most vendors today utilize off-the-shelf networking platforms such as Ethernet or FDDI; however, some utilize proprietary networks or only support one networking platform. Older data communications approaches include multidrop systems such as RS-485/Modbus.

- *Deterministic network design.* Control systems are near-real-time systems. This must be considered when plant data networks are designed. *Deterministic* refers to the ability to determine or predict the delays associated with sending data across the network. Predictability is important for two reasons: SCADA systems can interpret a long delay as a communications failure, and process control loops can become unstable or will require detuning if faced with long or unknown delays. Non–switched Ethernet networks are generally nondeterministic. The delays cannot be predicted because they depend upon network loading which is variable. Switched Ethernet is effectively deterministic because there is no traffic contention, and therefore the response time is predictable. Care must be taken to avoid nondeterministic network designs. This is particularly important when workstations and controllers are mixed on the same network. Potential solutions include one or more of the following: prioritization schemes such as quality of service; separation of workstation and PLC networks as appropriate; and utilization of switched networks.

- *Redundancy and fault tolerance.* Today's networks allow for much more robust and flexible designs. Redundant routing paths, load sharing, and ring architectures are common and can provide various levels of fault tolerance. Availability calculations can be performed to validate when redundancy is appropriate.

- *Performance.* Since the control system specifications include system performance requirements (such as process control screen call-up times) and each vendor's products will have different network needs to meet the performance requirements, the specifications should require the vendor to provide a network that will support the system performance. However, it is preferable to include minimum requirements such as bandwidth and fault recovery times. Required performance is a function of both process needs and ergonomics. For many water treatment process elements, a response time of seconds or even minutes is adequate to maintain control. However, a delay of more than a second or two through the HMI can result in operator frustration and dissatisfaction with the system.

- *Network diameter.* Different technologies have different distances at which they can operate without the addition of bridges, switches, or routers. This needs to be considered in determining the network architecture, technologies, and plant cabling layout.

Device Networks. Field bus technology began penetrating the process control market during the 1990s. Field buses are expected to play a key role in future control systems as the technology continues to evolve. New products are being designed with field bus interfaces that provide self-diagnostics, condition monitoring, and device data that can be strategic to improving both management of plant assets and process plant availability.

The development of digital communications to field devices has been driven by a desire to reduce cabling costs and to make use of a wealth of additional information now available from "smart" field devices. Querying a smart field sensor can, for example, provide the primary measurement information on sensor failure, instrument manufacturer, make, model, software version, calibration range, date of last calibration, etc. Manufacturers are also adding *condition monitoring* to their smart devices to provide early warn-

TABLE 20.2 Common Field Bus Standards*

Field bus	Remarks	Data speed	Control strategy programmable at device?	Maximum distance	Targeted field devices
HART	Hybrid protocol combining digital data with a conventional 4 to 20 mA	1.2 kb/s	No	100 m	Analog devices
Profibus (PA)	Originally developed by Siemens	31.25 kb/s	No	1,900 m per segment	Analog devices
Foundation Fieldbus (H1)	From ISA, SP50 committee. H1 bus devices can be loop-powered. Includes a graphical programming language for control strategies	31.25 kb/s	Yes	Up to 1,900 m	Analog devices
AS-i bus	Connects simple on/off devices at low cost—bit data		No	300 m	Discrete (on/off) devices
Devicenet	Developed by Allen Bradley—byte data. Targeted for multiple discrete devices, e.g., motor starters and VFDs	125–500 kb/s	No	Up to 500 m	Discrete (on/off) devices

*All buses shown are multidrop.

ing of failure. For example, a smart valve positioner may alarm problems such as poor valve seating or excessive wear. This allows plant staff to avoid emergency shutdowns and shift the maintenance to a period of scheduled plant downtime.

A plant designer is faced with numerous field bus choices, the most popular being Hart, Devicenet, Profibus, and Foundation Fieldbus. Table 20.2 details the attributes of five common buses. Although field buses have the potential to deliver significant benefits and cost savings, this should be balanced against the added field communications complexity and the need for increased maintenance staff skills.

Both Profibus and Foundation Fieldbus define separate device level and host level networks. Foundation Fieldbus is unique in that it allows control strategies to be programmed in the smart field device as well as the linking device. For example, a smart flow control valve positioner may have the flowmeter signal as an input and may be programmed with a PID flow control loop. This improves reliability since failure of the controller (shown as a PLC in Figure 20.10) would leave the flow loop still operational. For digital communications with discrete (on/off) field devices, Devicenet or ASi-bus is commonly chosen. A treatment plant may employ more than one bus standard.

OPC. The primary objective for OPC (*o*bject linking and embedding for *p*rocess *c*ontrol) was to develop an industry standard method to exchange data between Windows-based process control systems (or devices). This can reduce the high cost of developing custom interfaces or gateways each time two systems need to exchange information. Although OPC-enabled applications are becoming common for process control, currently there are very few business applications (e.g., planned maintenance, document management) that recognize the standard. OPC is supported by most major control system manufacturers and is therefore a preferred way for an automation system to share data between various vendors' systems, especially where different communications protocols are used.

Control Center Equipment. In addition to the computer equipment in the control center, there is commonly a video projection or display system as well as console furniture. These items must be coordinated with the architect to ensure adequate space is provided. Video wall units need careful consideration for cooling and space behind the wall. Console furniture is specifically designed for control rooms. This furniture comes in numerous different configurations to maximize the utility and ergonomics for the operator. Room lighting should be coordinated with the console layout to minimize glare on the screens in the console.

Miscellaneous Control System Design Considerations

Premature Obsolescence. How can I get state-of-the-art equipment when my design is done 5 years before my plant is brought on-line? For computer technology, specifying items such as workstations, networking equipment, and video display walls can be difficult due to the long construction times for treatment plants. For example, after 5 years' time, computer models can become obsolete and unsupported. Replacement computers may not be compatible with older operating systems. Approaches to reduce this problem include postponing submittals for this equipment as late as possible; writing change orders to upgrade to the latest equipment (this is a less attractive approach, considering industry trends at the time of writing, because the contractor bids to the equipment when it is expensive but submits on it years later when costs have come down); including an allowance for a set dollar amount in the contract as well as methods for determining the costs at the time of submittal; or building the plant with the understanding that the workstations, software, and networks may need upgrading shortly afterward.

Power Reliability and Backup Power. Commercial power failures during critical process operations such as filter backwashing may result in undesirable process conditions, such as draining filter media dry. Plant controllers, computers, control system networks, and critical instruments should be powered from a plant emergency-backed-up power source, as well as individual uninterruptible power supplies (UPSs). It is recommended that the UPSs back up long enough to ride through any plant generator start-up or other plant emergency power transition. Controller software must be designed to ensure that the process recovers in proper states when power returns.

UPSs should include a bypass switch to allow commercial power to feed directly to the load if the UPS electronics fail.

Control System Security. Control system security has received a tremendous amount of attention lately and is a dynamic field. Standards groups, such as the ISA SP99 group, are currently developing control system security standards; however, these will take time. Appropriate standards, when available, should be specified and required of any new system. Security design goals of user transparency, defense in depth, cost of ownership, and ease of maintenance should be considered with business requirements and budgets.

There are numerous factors that the designer should consider in regard to security including, but not limited to, the following:

- *Physical security of SCADA/control components.* Any location that has a SCADA connection is a potential target, particularly unguarded remote sites. Servers, workstations, network hubs, switches, patch panels, PLCs, and communications equipment should be kept in locked, access-controlled areas. Access to these areas should be monitored and logged. Network wiring should be encased in conduits and should not be accessible to unauthorized personnel. Identify and assess remote telephone/computer network/ fiber optic cables that could be tapped; radio and microwave links that are exploitable; computer terminals that could be accessed; and wireless local-area network access points. The security of the site must be adequate to detect or prevent unauthorized access.
- *Remote access.* Remote access for maintenance personnel and vendors should be tightly controlled. Modems and remote access devices should incorporate appropriate security provisions. Outside computers and networks that have remote access to the system should be considered as part of the system security. Compromises on those systems can lead to a compromise of the control system.
- *Interconnection to business networks and the Internet.* When connecting to external networks, network access controls (such as firewalls and router packet filters) should be installed to limit and tightly control access to the control system. Multiple layers of control should be incorporated (defense in depth).
- *SCADA network and operating system security.* The network architecture should be reviewed by a certified security professional prior to design completion and during the submittal process. The design should include appropriate security devices such as a firewall, intrusion detection scheme, and virtual private network (VPN) devices, where appropriate.
- *SCADA software security.* Consideration should be given to configuration of the security features of the SCADA or DCS software. Limit access to the operating system to only those requiring it. Two-factor authentication methods should be utilized (passwords and either biometrics or secure ID cards).
- *Disaster recovery.* System backups and restoration should be specified that incorporate industry-accepted approaches.
- *Reliability.* Minimize the consequences of tampering by identifying and eliminating single points of failure and consider the ramifications of failure of outside services such as power and communications.

- *Acceptance testing.*   There is no way to physically inspect a workstation and know that it is secure from hackers. System vendors should be required to test their systems for network and workstation vulnerabilities and remediate appropriate security holes. All network and workstation-based testing should be performed by certified security professionals.
- *System maintenance.*   Security updates and patches for all system components should be applied when they are released. Always make a backup prior to and after applying a patch or update.

Power Line Transient Voltage Surge Suppression.   All electronic instrumentation is sensitive to power line disturbances. Power line transient voltages are spikes in the voltage and current waveforms whose amplitude may exceed twice normal values. The energy contained in these transients is proportional to the amplitude (volts) and duration (microseconds) and must be filtered by components sized for the anticipated worst-case transient energy.

A *transient voltage surge suppression* (TVSS) unit is recommended for the power feeder serving a UPS system. The TVSS helps protect the UPS electronics and loads when the UPS is bypassed. Any instrumentation and control equipment not fed by the UPS should be powered from a TVSS-protected power feeder.

Uninterruptible power systems can also protect equipment against harmonics and power line disturbances produced by adjustable-frequency drives, and undervoltage and transient conditions resulting from starting large motors or other equipment.

Communications Equipment.   Space should be set aside adjacent to the main control room for communications equipment. This can include fiber-optic, networking equipment, radio equipment, etc. This should include space for all in-plant needs as well as potential future remote telemetry needs. Non-fiber-optic data highway cables should be fitted with properly grounded communication line TVSS modules.

Signal Errors and Mitigation Measure.   Signal errors introduced by process and electrical noise can render the best plant monitoring and control system ineffective. Proper installation of process transmitters and signal loops is essential for providing clean, stable signals.

Signal Errors from Process Noise.   Modern process transmitters incorporate filtering and averaging to smooth signal output. However, care must be exercised to install primary devices and signal transmitters to mitigate the effects of process noise. For example, turbidimeters and ultrasonic transit-time flowmeters are sensitive to entrained air bubbles. Sample color also affects readings of some turbidimeters. Because pressure transmitters measure minute movements of diaphragms, they should not be mounted directly on vibrating equipment. Vibration also affects many vortex flowmeters. Propeller flowmeters mounted too close to control valves or bends read erratically, and little can be done to smooth the electronic signal output without reducing accuracy and destabilizing flow control loops.

Signal Errors from Ground Loops.   A ground loop is an unintentional connection between two points in a signal loop by means of a ground path. Ideally, signal outputs from process variable transmitters and signal inputs at the other end of the loop are electrically isolated from ground. For example, the common wire from a 24-V direct-current loop power supply is often grounded. The power supply may feed a pressure transmitter and a digital panel meter with a grounded 120-V ac power supply at the other end of the loop. A difference in earth potential between the two ends of the loop of only 10 mV will cause a ground current of 1 mA to flow in a loop with 10-Ω (ft) resistance. That represents a 6% error in the 16-mA dc transmitter span.

Ground loops can be avoided by using signal isolators—electronic amplifiers with no direct connections between signal input and output lines. Isolators can also be used as signal repeaters for driving multiple loads from one signal source.

Signal Errors from Electrical Noise. It is false economy to specify an accurate signal transmitter and allow electrical noise to couple into signal wiring. Any energized conductor of electricity is surrounded by an electric field. If the conductor is carrying current, a magnetic field is also present. Signal wiring is influenced by the electromagnetic fields produced by adjacent power wiring. Institute of Electrical and Electronics Engineers (IEEE) Standard 518 provides general guidelines for separation of power and signal wiring conductors, assuming signal circuits are in galvanized rigid steel conduits.

Signal loops should be wired in shielded, twisted-pair cables. Twisting conductors has the effect of canceling voltage generated by the transformer action of nearby magnetic fields, and the foil shielding acts as a barrier to electric fields.

Control System Programming and Configuration

Process monitoring and control system products are general-purpose tools, not designed specifically for water treatment. This section describes the process of configuring control system software to acquire data and perform control actions for a water treatment plant.

Database Development. The real-time database includes records for all analog (continuously varying) inputs, analog outputs, discrete (on/off) inputs and outputs, and digital data register points from field serial instruments and devices. Each record has numerous fields to fill in such as alphanumeric description (e.g., filter 1 flow rate), range limits (e.g., 0 to 2,500), engineering units (e.g., gpm), alarm limits, etc. In addition to the real-time database, additional databases need configuration including all calculation and control database points, archiving databases, and the security configuration. Configuring the databases is tedious and time-consuming, but much of it is repetitive.

Control Strategy Development. Control strategies are detailed sets of control actions intended to operate a process unit or area. Control strategies are programmed in one of a few ways. On a DCS, the strategies are traditionally implemented using control blocks. Example blocks include PID, pump control, analog input, etc. The blocks are connected together to create the strategies.

The principal PLC programming language is *relay ladder logic* (RLL). PLCs were originally developed for discrete logic. Over the years analog function blocks, such as analog input averaging and PID control, have been added. Although the actual program takes up a relatively small amount of PLC memory space, RLL documentation for a typical filter backwash PLC program can be an inch thick, filled with cross-references between rungs at opposite ends of the program. PLC manufacturers provide higher-level programming tools that generate RLL code executed by the PLC but are easier for the programmer to use.

In recent years, other PLC languages have taken hold. Many PLCs support the IEC 61131 standard. IEC 61131 is an international PLC programming standard that provides a PLC-independent programming language that can be run on many different PLCs. IEC 61131 programming languages include Ladder Diagram (relay ladder logic), Sequential Function Charts, Function Block Diagram, Structured Text, and Instruction List.

RTUs are seldom programmed in ladder logic; they are generally programmed in C, Basic, or proprietary programming languages. RTU manufacturers provide higher-level programming tools, which generate C or other code executed by the RTU.

FIGURE 20.11 Example of process graphic.

Process Graphic Display Development. Process graphic displays are typically the primary means for the operators to control the system. The displays are typically created to provide an intuitive depiction of the process being controlled and its current state. Most systems use a point-and-click approach to navigating the system. Double-clicking on a pump may (depending on how it is configured) bring a pop-up that allows the operator to control the pump or access additional information about it including historical data, O&M information, or a link into the records from a maintenance management system.

The example screen shown in Figure 20.11 depicts a pressure-reducing station. In this example, alarms will display in the upper left, station status shows on the upper right, and hot buttons at the bottom link to other screens. Operators control the station by changing setpoints shown on the screen. A pump can be controlled in computer manual mode by double-clicking on the pump to be manually controlled.

Developing user-friendly graphic screens is an art. There needs to be balance between the number of points on the screen and the number of screens in the system. Symbols, layout, identifiers, navigation, and actions need to be consistent.

In addition to the SCADA/DCS screens, remote operator interfaces for stand-alone process units can have graphics screens that require development.

Alarm Configuration. The alarm subsystem is used to notify operators of abnormal situations. Typical alarm systems today include tone/horn alarms, audio files, and dialer/pager systems. Some dialer alarm systems allow operators to perform limited control of the facilities through a menu-driven voice interface.

Minimizing nuisance alarms is a fundamental issue surrounding the alarm system. Normal practice is to require the alarm condition to be sustained for a time and to establish dead bands to prevent unneeded repeated alarms. In addition, situations that cause multiple alarms, such as power failures (which can cause hundreds), should disable secondary alarms so that the operator gets only the actual alarm. Some alarm systems include the ability to give recovery instructions to the operator, which can require significant time to set up. Example alarms include

- Pump/motor fails to start, fails to stop
- Valve fails to open, fails to close
- Valve position away from setpoint
- Controller at high/low limit
- Motor overload
- Motor winding/bearing high temperature
- Torque overload
- High vibration
- Lubrication failure (low oil flow)
- Water quality limit alarms
- High/low levels
- Instrumentation failure
- Power failure
- Intrusion/security related

Report Development. Management and operational report development is a key component to control system configuration. Control systems typically have a report package included or can use third-party report packages. Common reports include daily and monthly production reports as well as water quality reports. These reports can be set up to meet regulatory reporting requirements, such as the Interim Enhanced Surface Water Treatment Rule (IESWTR) turbidity reporting requirements.

Trend Displays. Trending packages allow the operator to plot historical information over time to see variations. Standard trends of the most critical plant processes are typically configured. Operators can tailor individual trends to meet their preferred ways to operate the plant.

Application and Device Interfaces. More and more, modern control systems integrate with other applications and instruments over a network. High-level applications such as LIMS, CMMS, Optimization Programs, Legacy or Adjacent Process (Transmission and Distribution, or Source of Supply) SCADA systems, or GIS are commonly interfaced to water treatment plant SCADA systems.

Device networks, such as Foundation Fieldbus and Profibus, allow network connections to instrumentation and other controllers. As described previously, this can allow more data to be transferred without the need for additional I/O.

To support this level of integration, device drivers or other interfaces sometimes need to be programmed and configured.

Specifications and Design Documents

This section reviews some of the primary documents associated with the instrumentation and controls (I&C) system.

Process and Instrumentation Diagrams. Process and instrumentation diagrams (P&IDs) are the primary design document of the instrumentation and controls engineer. P&IDs include schematic representations of piping, valving, and other process-related equipment, with all associated instrumentation and control components. P&IDs should be developed as early as possible and should be continually refined as the design progresses. P&IDs

20.28 CHAPTER TWENTY

are used to define the types and quantities of I&C devices as well as the general control strategies being used.

Process equipment is generally drawn at the bottom of the sheet with heavy lines. Instrumentation and control components are shown as circles, or bubbles, and signal and control loop wiring is shown as dashed lines. P&ID drawing format should use Instrument Society of America (ISA) symbol and tag numbering standards.

Figure 20.12 shows part of an automated filter backwash control system P&ID. Diamonds represent control interlocks, which are generally shown in greater detail on elementary control diagrams for the control panels.

FIGURE 20.12 Excerpt from automated filter backwash control P&ID.

Elementary Control Diagrams. Elementary control diagrams are "ladder-format" electrical schematics for motor starter controls and control panels. Power supply to the control circuit is represented by vertical rails at the left and right sides of each diagram. The diagram shows power flow from left to right, through relay and switch contacts, to energize motor contactor and relay coils, indicator lights, and solenoids.

Elementary diagrams usually show how control logic is wired to operate on/off control devices in various control modes. They can also depict power supply and signal connections to analog control devices. These diagrams are generated in close coordination with the electrical design engineer.

Interconnection Diagrams. Interconnection diagrams show the conduit and wiring that connect major control system components, such as control panels, sensors, and valve actuators. Signal and control wiring connections are shown as single lines, grouped and labeled by logical functions. Multiple wiring groups may share the same conduit but are labeled individually.

Interconnection diagrams define the separation of power, control, and signal wiring into different conduits and show all inputs and outputs for each I&C component. These drawings are generally created by the construction contractor, not as part of the design documents.

Instrument Loop Diagrams. Instrument loop diagrams are detailed multiline representations of unique signal and control loops within a plant. Usually only one or two loops are shown on a drawing sheet. Drawing format should use ISA loop diagram format and ISA symbol and tag numbering standards. Each wire, cable, and termination are shown, marked with signal polarity and function. Isolators and signal converters are included as well as ac and dc power connections to instruments, and final control elements are shown. These drawings are frequently created by the construction contractor, not as part of the design documents.

Specifications. Equally important to instrumentation and control design drawings are detailed design specifications. Typical specifications conform to the Construction Specifications Institute (CSI) division, section, and part formats. Instrumentation and controls are usually specified in Division 13 or in some cases Division 17. A partial description of typical sections includes

- Instrumentation and controls—general (project description, quality, submittal, testing, training, and documentation requirements)
- Process monitoring and control system (control room hardware and software requirements)
- Field instrumentation (data sheets and installation requirements for all instruments)
- Control panels (sizing, fabrication, wiring, and control diagram requirements)
- Programmable logic controllers (PLC hardware, software, testing, training, and documentation requirements)
- Communications systems

Each of the above sections is divided into three parts: general requirements, products, and execution. At the end of the specifications are the most job-specific and time-consuming sections to produce:

- Control strategies (detailed monitoring/control descriptions for each control loop)
- Input/output schedules (listing of PLC/RTU I/O signals, plus computed points, such as flow totals)
- Instrument list (list of all field instruments and signal conversion devices)

INSTRUMENTATION

This section provides an overview of the instruments used in water treatment plants. It is intended to provide the reader with a basic understanding of the instruments and applications. Much more detailed information is available in several of the texts referenced in the Bibliography.

Flowmeters

There are a wide variety of meters available for monitoring the flow of water and chemicals in a water treatment facility. Each has advantages and disadvantages to be considered such as cost, reliability, accuracy, and required maintenance. (See Table 20.3.)

Orifice Plates. Orifice plates consist of a steel plate with a carefully sized hole that is mounted between pipe flanges to restrict flow. The differential pressure across the restriction is measured to determine the flow. Orifice plates are inexpensive but not very accurate and have higher nonrecoverable head loss. They may be considered in applications where pumped energy is not being added to the process and higher head loss is acceptable.

Venturi Flowmeters. Venturi flow elements provide a restriction in the pipe diameter, creating a pressure differential between the unrestricted and restricted portions of the pipe. The flow is proportional to the square root of the differential pressure. Venturi flow elements equipped with differential pressure (DP) transmitters are one of the most common types of flowmeter used in water treatment plants. Venturi meters are relatively tolerant of flow profile anomalies produced by upstream piping bends.

In filter control applications, repeatability is usually more important than absolute accuracy, and venturi meters are often close-coupled to the downstream filter flow control valve. Long-form venturi meters also have the lowest nonrecoverable head loss of all DP meters, making them ideal for gravity filter applications where flow would be restricted by higher piping head losses.

Pitot Tubes. Pitot tubes measure the velocity head and static pressure of clean liquid, gases, or steam. Pitot tubes require more straight upstream and downstream piping than venturis and orifice plates. Averaging pitot tubes have multiple pressure-sensing ports distributed across the pipe profile and thus require less straight pipe for metering runs than single-port types. Where accuracy is required, 10 to 20 diameters of straight pipe upstream and 5 to 10 diameters downstream are sometimes recommended by the manufacturers.

Averaging pitot tubes can be installed in water lines in excess of 2 in. in diameter, have accuracies from ±0.5% to ±5%, have low head loss, and can be hot-tapped into existing piping.

Differential Pressure Transmitters for Flow Measurement. All DP flowmeters are "square law" devices; that is, flow is proportional to the square root of the differential pressure. DP transmitters are typically microprocessor-based, have high accuracy over a wide range of flows, and can have built-in square root extraction with no loss in accuracy.

Typical accuracy of a DP transmitter is ±0.2% of full scale, including nonlinearity, hysteresis, and repeatability. A flowmeter element producing 100 in. (2,540 mm) of water column differential pressure at 100 mgd (380 ML per day) full scale has a transmitter error of 0.1 in. (2.5 mm). This corresponds to an error of ±0.1% of actual transmit-

TABLE 20.3 Flowmeter Characteristics

	Magnetic	Ultrasonic	Rotary element (impeller)	Open channel	Variable area (rotameter)	Orifice plate	Venturi	Coriolis meter	Thermal mass	Vortex
Typical applications	Water, chemical slurries, or solutions	Liquid, slurry, gas	Liquid, gas, vapor	Water	Water, chlorine gas, carrier water, chemical solutions	Water with high available head, gas	Water, chemical solutions, or gas	Chemical solutions or gases requiring very high accuracy	Liquid, gas	Liquid, gas, steam
Typical flows	Main process or small flows	Main process or small flows	Main process flows	Main process flows	Small flows	Main process or small flows	Main process flows	Small flows	Small flows	Main process flows
Measurement range	1 L/h–300,000 m³/h	>1 m³/h	Any desired	Any desired	1 L/h–920 m³/h (liquids) or 2,210 m³/h (gases)	Any desired		1 kg/h–750 t/h	Small to high	0.5–40 m³/h
Accuracy	±0.5%–±1% depending on velocity	±1%	±0.25% (turbine) to ±1% (impeller)	±2%–±5%	±1%–±3%	±0.5%–±2%	±1%–±2% with a transmitter‡	±0.05%–±0.15%	±1%–±2%	±0.5%–±1%
Influence of density on mass calculation	Linear	Linear	Linear	Linear	Square root	Square root		No influence	No influence	Linear
Relative loss of pressure	Low*	Low*	Middle	Low (area velocity) to high (weir)	Middle	High	Middle	Middle to high depending on viscosity	Low	High

Note: See p. 20.32 for footnotes.

TABLE 20.3 Flowmeter Characteristics (*Continued*)

	Magnetic	Ultrasonic	Rotary element (impeller)	Open channel	Variable area (rotameter)	Orifice plate	Venturi	Coriolis meter	Thermal mass	Vortex
Interference of particles	Low	High	Low	Low	High	Middle	Low	Low	Low	High
Required inflow/outflow pipe length	Middle	Long†	Long	Middle	Not necessary	Middle	Short	Not necessary	Middle	Long
Turndown	20:1	25:1	10:1	75:1 to 500:1	10:1	10:1	10:1	20:1	Up to 100:1	15:1
Cost‡	Middle	High	Low (impeller) or High (turbine)	Depends on primary element/instrument. Low for large flows	Low	Low	Middle	High	Low	Middle

*If the full pipe diameter is used. However, for either flow range or economical reasons, the pipe is frequently necked down for this type of flowmeter and a small pressure drop results.

†The length of pipe can be greatly reduced if multiple transducer pairs are used in order to better determine the velocity profile across the meter under less uniform flow conditions.

‡Information was obtained from Figure 9.2 and the text of *Online Monitoring for Drinking Water Utilities.*

Source: American Water Works Association Research Foundation and CRS Proaqua, *Online Monitoring for Drinking Water Utilities*, Denver, Colo., 2002.

ted flow rate. The same meter measuring 10 mgd (38 ML per day) produces only 1 in. (25 mm) of DP. The transmitter produces the same DP error, which corresponds to an error of 10% of actual transmitted flow rate. A 10% error is significant if the flow signal is used for billing purposes or for pacing chemical addition, but it becomes less significant if the meter is used only for flow balancing.

Good engineering practice limits use of DP-type meters based upon this rangeability. Higher rangeability is possible using parallel DP transmitters with split ranges, although calibration time and parts count are doubled.

Electromagnetic Flowmeters. Electromagnetic flowmeters measure flow rate based on the velocity of the fluid through a known cross-sectional area of a pipe. Magnetic coils induce electric potential in the flowing fluid, which is proportional to fluid velocity. Potential is detected by electrodes in contact with the fluid. This requires a process fluid with sufficient conductivity.

Spool piece magnetic flowmeters are nonintrusive, have good rangeability, are highly accurate, have minimal head loss, and require metering runs of only 3 to 5 diameters upstream and 2 to 3 diameters downstream of the electrode plane for most situations. Large magnetic flowmeters can be heavy, and the spool pieces are not field-repairable.

Magnetic flowmeters are also available with "hot-tap" elements for installation on existing pipes. Probe meters that mount through the pipe wall are useful for smaller pipes but require longer straight metering runs than spool piece magmeters. Single-element magnetic probes must extend far enough into an existing pipe to be away from the viscous boundary layer of the flow profile. The magnetic sphere of influence is only a few inches around the tip of the probe, limiting accuracy in large pipes. Where accuracy is required, at least 10 to 15 diameters of straight pipe upstream and 5 diameters downstream are recommended for multielement probe-style magmeters.

Multielement insertion magmeters install across a pipe diameter and provide better flow averaging, slightly higher than that of an averaging pitot tube. Multielement probes are available in lengths up to 60 in. (1,500 mm). The probe is supported only from the insertion end, so line velocity must not exceed 10 ft/s (300 cm/s).

Ultrasonic Flowmeters. Ultrasonic transit-time flowmeters are nonintrusive and are available with clamp-on transducers that transmit acoustic energy through the walls of existing pipes. A pair of staggered transducers on opposite sides of the pipe measure differences in signal transit time upstream and downstream, and fluid velocity is inferred from this difference.

Accuracy is limited by the velocity profile over the cross section of the pipe, because the computed velocity is an average taken over only one narrow path. Multipath transit-time ultrasonic meters are used on larger pipes to improve accuracy. Dual-path, dual-beam meters are available that use two transducer pairs mounted orthogonally. The signal from each pair is transmitted through the pipe wall, reflected off the opposite interior pipe wall, and detected on the originating side. Because the transducers are offset along the pipe length, this effectively yields four paths.

Where accuracy is required, at least 10 to 20 diameters of straight pipe upstream and 5 diameters downstream are recommended for single-path ultrasonic meters. For multipath ultrasonic flowmeters, these requirements are significantly reduced. Where flow measurement is downstream of pumps, add 30 to 50 diameters to allow air bubbles to dissipate. Additional pipe diameters are also required downstream of control valves.

Clamp-on Doppler meters are also available that operate on the principle of measuring sonic reflections from suspended particles. These are suitable for slurries and liquids when air bubbles are present. Accuracy is not as high as that provided by transit-time meters.

Vortex-Shedding Flowmeters. Flags wave in a breeze because of the vortices alternately shedding from the sides of the flagpole. In the same manner, vortex meters infer fluid velocity from the rate at which vortices are shed behind a "bluff body." Vortex meters measure minute deflections in the rear portion of the bluff body, utilizing sensors such as ultrasonic, capacitive, thermal, and piezoelectric, as well as strain gages.

Spool piece meters are available up to 12 in. (300 mm) in diameter. Insertion vortex flowmeters can be installed in larger line sizes. Turndown ratios can be as high as 30:1.

Vortex meters have wide rangeability, moderate head losses, and relatively high cost. As with most other flowmeters, installed accuracy depends on the piping configuration. Where accuracy is required, at least 15 to 50 diameters of straight pipe upstream and 5 diameters downstream are recommended for spool piece vortex flowmeters.

Weirs and Flumes. Weirs and flumes are open-channel flowmeters that infer flow rate from the fluid surface level near a constriction. Weirs are dams with rectangular, V-notch, trapezoidal, or other geometrically shaped openings. Flumes are specially fabricated channels that shape the flow through a throat. Water level upstream of weirs and flumes is measured using floats, ultrasound, bubbler tubes, submerged hydrostatic sensors, or capacitance probes. The mathematical conversion from level to flow rate is nonlinear and is generally handled by signal transmitter electronics.

Weirs and flumes are inherently averaging flow elements, tolerant of upstream and downstream velocity profile anomalies. Accuracies are generally lower than those with most closed-conduit flowmeters.

Propeller and Turbine Meters. Propeller meters are electromechanical devices converting the rotary motion of a spiral-bladed propeller to pulses or electric current proportional to fluid velocity. Both spool piece and insertion styles are available. Propeller meters are inherently sensitive to swirl, so extreme caution must be exercised to avoid piping configurations that induce swirl. Where accuracy is required, 10 to 100 diameters of straight pipe upstream and 5 diameters downstream are recommended. Straightening vanes can be installed in the piping if swirl is present.

Typical accuracy of propeller and turbine meters is $\pm 0.25\%$ to $\pm 1\%$. Because these meters are susceptible to dirt and solids, they should be used only on clean fluids. They are not suggested for raw surface water metering. Compound turbine meters have two differently sized flow elements in separate chambers, yielding wide rangeability.

Flowmeter Piping Requirements. As a rule, flowmeters exhibit the accuracies described on the manufacturers' data sheets only when installed in ideal piping configurations. Careful consideration of upstream and downstream configurations is critical. Upstream and downstream straight-pipe requirements vary significantly with the type of flowmeter, the type of devices mounted upstream (valves or pumps), bends that traverse multiple planes, etc. The best piping configuration to have upstream is a long run of straight pipe, preceded by a 90° elbow. Additional pipe diameters are required downstream of pumps and control valves and in metering runs with bends in more than one plane.

Unfortunately, long, straight flowmeter runs are not always available at a water treatment plant. Flow disturbance mitigation measures such as straightening vanes and tube bundles can be installed upstream of the meter. Otherwise, manufacturers can provide estimated accuracies based upon the available straight pipe runs. The designer must then decide if the level of accuracy is acceptable or if an alternative meter or piping arrangement is necessary.

Level Switches and Transmitters

The liquid level control equipment used in water treatment systems ranges from simple float switches to electronic sensors.

Float-Type Level Switches. Float-type level switches are used to provide discrete level sensing. A rising liquid level lifts a buoyant float, tripping an internal switch. Floats come in different types, the most recognized being a set of floats connected to a cable that is lowered into a tank. Where turbulence is not present, cables can be tethered from the top of a pump wetwell, basin, or reservoir. In some models, a weight is molded into the cable near the float to hold the cable taut.

Where turbulence is present, as in a pumping wetwell, the cable should be clamped to a vertical mounting rail on the sidewall of the wetwell. In water storage reservoirs, cable-suspended level transmitters should be installed away from inlet and outlet piping. Some float switches contain mercury which poses a potential drinking water concern; however, non-mercury-containing float switches are available.

Conductance Probe Level Switches. Typical conductance probe level switches are cable-suspended, plastic-shrouded electrodes or straight probes connected to a sensitive electronic switching circuit. This circuit limits the electric energy to the probe electrodes for safety purposes. A reference electrode is required for nonmetallic tanks, such as underground concrete reservoirs. This electrode must be submerged below all measuring electrodes.

The electronics module can be located up to several hundred feet from the electrodes. A typical application is for reservoir high- and low-level alarm monitoring. Electrode materials must be resistant to chlorinated water.

Submersible Hydrostatic Reservoir Level Transmitters. In situations where top access to a tank or reservoir is available, the level can be sensed by cable-suspended, submersible pressure-sensing elements that measure the hydrostatic level of the fluid above the sensor. Some manufacturers place the signal processing and transmitting electronics at the top end of the cable; others place the electronics in a sealed enclosure at the sensor.

Variations in atmospheric (barometric) pressure affect the hydrostatic reading, so a DP sensor is used with the reference side connected to a capillary vent tube, which is an integral part of the cable. If this vent tube is open to atmosphere at the top of the cable, condensation eventually fills part of the tube, causing measurement errors. Some manufacturers connect a desiccant cartridge to the vent tube to dry the air. This cartridge has an expected life of a year or two and is easily forgotten and not replaced.

The preferred design uses an elastic bladder that flexes minutely with variations in atmospheric pressure at the top of the vent tube, while preserving a factory-dried volume of air in the tube. Level ranges are available from 0 to 15 psig (0 to 103 kPa) to 0 to 300 psig (0 to 2,068 kPa), depending on the manufacturer's design. Typical accuracy is $\pm 0.3\%$ of span. One manufacturer provides cable lengths up to 1,000 ft (305 m) between the submersible transmitter and the electronics/breather bag enclosure.

Cable-suspended level transmitters used in large water reservoirs should be installed away from inlet and outlet piping. Care must be taken to avoid folding the cable/vent tubing because of the minimum cable bending radius.

Continuous Purge Bubbler Level Transmitters. Continuous purge bubbler level transmitters measure hydrostatic level based on the pneumatic backpressure on a tube inserted into the fluid. A sensing tube is clamped to the side of a basin or reservoir, or a weighted hose is dropped into the reservoir through a roof hatch. Two small-diameter pneumatic impulse lines are then run from the top of the sensing tube (or hose). Air is fed from a compressed air supply at a small constant flow through one of the impulse lines to bubble air out of the bottom of the sensing tube. A gauge pressure transmitter measures the static backpressure on the second impulse line. A single impulse line can be used, but the pressure drop in the impulse line due to airflow can result in level measurement errors. The sensing line should be protected from sunlight and other sources of heat that may affect the sensing line pressure.

The accuracy of the bubbler system is comparable with that of the cable-suspended submersible level transmitters described above. Bubbler sensing tubes are virtually maintenance-free in source or treated water. However, dual compressors are typically utilized to ensure a reliable air source. Bubbler compressors, purge regulators, gauges, and component enclosures make bubbler systems more costly, require more room, and need more maintenance than submersible level transmitters.

Periodic Purge Bubbler Level Transmitters. This bubbler system employs an inverted chamber open at the bottom to trap air at the end of the bubbler tube. Without purging, air in the chamber dissolves gradually into the surrounding water, causing the water level to rise with no sensed change in backpressure. The unit is designed to have the tube and chamber charged with compressed air every hour or so to prevent inaccurate readings. The system is simpler mechanically than a continuous purge type, because the compressor is smaller and does not require an accumulator tank or a purge regulator. The controls are more complex, however, because the transmitted level signal must be held during the purge cycle. Integrated compressor/transmitter/controllers are available to simplify periodic purge system installation.

Ultrasonic Level Transmitters. Ultrasonic level sensing is noncontacting and uses transit time of ultrasonic pulses bounced off a liquid or solid surface to infer the distance to the surface. Available range spans are between 2 in. (5 cm) and 200 ft (60 m). Ultrasonic level transmitters are usually microprocessor-based, and some have built-in signal characterization for computing volumes, or flows, through weirs or flumes. Accuracy ranges from $\pm 0.25\%$ to 2% of span depending on conditions.

Ultrasonic transducers should be mounted to eliminate unwanted reflections based upon the manufacturer's installation requirements. Allowance must be made for a sensing dead zone between the maximum surface elevation and the transducer. This zone can vary from roughly 12 in. (0.3 m) for a 15-ft (4.6-m) span to 72 in. (1.8 m) for a 200-ft (60-m) span. A transducer mounting well may be required to provide the proper distance between the transducer and the maximum surface elevation.

Ultrasonic level transmitters are more expensive than contacting-type transmitters and are typically used for applications such as chemical storage tanks with top access where contact with the fluid is not desirable.

Radio-Frequency Capacitance Level Transmitters. Liquid level can be measured by changes in capacitance measured at a partially submerged coated probe or steel cable. This method uses radio-frequency excitation between the probe and a grounded vessel, such as the water in a metal storage tank, to infer the height of the fluid surface. Accuracies are typically $\pm 1\%$ of span. This corresponds to fairly significant errors of ± 3.6 in. (9.1 cm) in a 30-ft-deep (9-m) reservoir and ± 24 in. (61 cm) in a 200-ft-deep (61-m) water well. Capacitance-type level transmitters are usually microprocessor-based and have built-in signal characterization for computing volumes, or flows, through weirs or flumes.

Differential Pressure Transmitters for Level Measurement in Closed Vessels. Differential pressure transmitters can be used to measure the hydrostatic level in closed vessels, such as hydropneumatic pressure tanks and surge tanks. Two pressure taps are required: a "wet leg" tap at the bottom of the vessel and a "dry leg" tap at the top. Static pressure within the vessel is measured equally on both legs of the DP measurement cell, so its effects on the measurement are canceled out. As described, DP transmitters are microprocessor based, and typical accuracy is ±0.1% of full scale, including nonlinearity, hysteresis, and repeatability.

Pressure Switches and Transmitters

Pressure switches used in water systems range from simple off/on applications to electronic equipment that monitors and reacts to a wide range of pressure differentials.

Bourdon Tube Switches and Transmitters. Bourdon tubes are hollow brass or stainless steel curved chambers that flex as internal pressure is applied. This flexure is linked to an indicating dial or to an electronic linear variable differential transformer sensor for signal transmission. Accuracy is typically ±0.5% of full scale, plus stability effects of ±0.25% of full scale for 6 months. Allowable overpressure is typically 130% of full scale. Bourdon tubes are not recommended on pumped treated water pipeline applications because of potential overpressure damage from pipeline water hammer.

A Bourdon tube can also be used to actuate either a mercury or a snap-action switch. Accuracy is typically ±0.5% of full scale. Allowable overpressure is typically 130% of full scale.

Capacitive Pressure Transmitters. Capacitive pressure transmitters sense pressure variations as minute deflections of as little as 0.001 in. Deflections change the capacitance of a solid-state circuit connected for signal transmission. Accuracy is typically ±0.2% of full scale, and allowable overpressure is typically 2,000 psi (13,800 kPa).

Capacitive pressure transmitters are commonly microprocessor-based and are available with digital communications capability. This data communication can be used to remotely set the range, zero, span, damping, and other parameters, using a handheld programmer or computer connected anywhere on the current loop. Several pressure transmitter vendors have their own proprietary data communications protocol.

These smart transmitters offer two advantages. One is that a water system can purchase one spare transmitter that can then be reranged as needed to replace transmitters with different pressure ranges. The other is that calibration of a pressure transmitter can be accomplished using a handheld programming unit, rather than a screwdriver. The pressure at the transmitter can be easily bled to zero for zero calibration, but a full-scale pressure source is not always available. If the pressure being measured is known, the signal output of the transmitter can be forced to the correct value with the programmer. Given the high accuracy and linearity of these transmitters, this two-point calibration is often sufficient.

Pressure Transmitter Accessories. Pressure and differential pressure transmitters should have shutoff valves and sediment traps installed as close as possible to the process piping. A sediment trap can be made from a vertical pipe nipple extending down from a tee or cross, with a cap or blowdown valve at the bottom. For highly corrosive applications, an isolation diaphragm must be installed between the process liquid and the transmitter.

Transmitter calibration is usually accomplished by connecting a nitrogen bottle and deadweight tester, or a hand-operated compressor, to the transmitter impulse (sensing) tubing. Valving should be provided to isolate the process pressure, relieve impulse line pressure for zero calibration, and connect to the calibration pressure source. Zeroing manifolds are one-piece multivalve accessories designed to accomplish these tasks for DP transmitters.

Temperature Switches and Transmitters

Temperature switches are typically filled-system bulb type. The switch is isolated from the process temperature by a filled capillary tube. For equipment monitoring applications, a brass or stainless steel bulb is mounted inside a thermowell. Typical accuracy is ±2.5% of full scale.

Temperature transmitters for water applications usually employ either resistance temperature detector (RTD) or thermocouple (TC) sensors. RTDs have a slightly higher accuracy but are slightly slower-responding than TC sensors. Both types of sensors are installed in probes mounted directly to the transmitters. A thermowell (tapered metal sleeve screwed into a fitting welded to the process piping) allows the probe to be removed without shutting down the process. Thermowells add to the response time of the sensor because of their inherent thermal inertia. Water temperatures in tanks and pipelines do not vary rapidly, so thermowells are used for ease of maintenance

Temperature transmitters are microprocessor-based and are available with digital communications capability. Accuracy is typically ±0.2% of full scale. Smart transmitters allow a relatively simple two-point calibration based on an ice bath and the known temperature of the water.

Analytical Instrumentation

Highly accurate bench-type laboratory instrumentation is used at water treatment plants to verify compliance with safe drinking water regulations. Critical water quality parameters can also be monitored on-line 24 h per day for automated reporting and process control.

Chlorine Analyzers and Detectors. Chlorine gas dissolves in water to form hypochlorous acid and hydrochloric acid. The presence of hypochlorous acid and hypochlorite ion in solution is termed *free available chlorine*. If ammonia is present in the liquid being chlorinated, a reaction can take place that results in the formation of chloramines, termed *combined available chlorine*. Because there is usually limited ammonia in clean water, free chlorine analyzers are used for most applications.

Chlorine residual measurement can be affected by water temperature, pH, and total dissolved solids. Sample treatment before chlorine residual analysis generally includes a wye-type strainer for removing solids, and some analyzers require the addition of a reagent upstream of the analyzer. Several different measurement methods are used for chlorine residual, including colorimetric, amperometric, and polarographic. Amperometric and polarographic analyzers are similar in operation and use two dissimilar metals held in a solution or electrolyte. A voltage is applied to the two metals. The amount of current produced is proportional to the amount of chlorine in the sample. Colorimetric analyzers involve the reaction of the sample with a reagent that results in a color change proportional to chlorine concentration.

Amperometric Wet Chemistry Chlorine Residual Analyzers. The wet chemistry chlorine residual analyzer is an amperometric device that includes pretreatment or conditioning of the sample. A pH buffering reagent and potassium iodide reagent are typically added to the sample before analysis. The conditioned sample is then analyzed using an electrode assembly, with the current developed in proportion to the chlorine concentration. One drawback to this type of analyzer is that the reagent is stored in bottles that require periodic replenishing. This analyzer can provide good repeatability and accuracy, because the samples are conditioned before the analysis. The analyzer is sensitive to changes in total dissolved solids (TDS), but this is not typically a problem in clean water, with normally low or relatively stable TDS.

Membrane-Type Chlorine Residual Analyzers. The membrane-type chlorine residual analyzer is an amperometric device designed not to require sample preconditioning for most applications. The sensor consists of electrodes mounted in an electrolyte solution, with the electrolyte separated from the sample by a permeable membrane. Chlorine molecules permeate the membrane and enter the electrolyte. The resulting change in chlorine concentration at the electrolyte causes a change in the electrical potential between the electrodes.

The potential depends on temperature, so most analyzers are provided with temperature compensation. Although these sensors can be mounted directly in the process flow stream, varying process head conditions can affect the transfer rate across the membrane and could cause variations in the chlorine residual readings. Flow-through assemblies in sample lines are recommended because they can maintain constant head conditions on the membrane.

Colorimetric Chlorine Residual Analyzers. Colorimetric analyzers use a reagent that reacts with either free or combined chlorine to produce a color change. The intensity of the color is proportional to the chlorine concentration, and an optical sensor assembly determines the light absorbance of the sample as compared with an untreated sample. The wavelength of the light source is selected to match the characteristics of the reagent.

Drawbacks of this type of analyzer include requirement of reagent solution, interference effects of high turbidity or sample color, and slow response time. Benefits include generally low cost and simple calibration. These analyzers have typical accuracies of $\pm 5\%$ of range and sensitivity down to 0.5 mg/L of chlorine. They are mostly used for filtered water applications where samples are not highly colored.

Chlorine Leak Detectors. Chlorine gas leak detectors are required in facilities where gaseous chlorine is handled or is likely to become confined in a closed space. They are typically either reduction oxidation or galvanic cell-based devices. Leak detectors are provided with low-level warning and high-level alarm contacts. Some detectors have a refillable electrolyte electrochemical sensor or an amperometric membrane sensor in a sealed assembly. Reagents are not required for the membrane-type sensor, but maintenance consists of replacement of the sensor assembly approximately every 2 years.

Turbidity Detectors. The measurement of the concentration of turbidity is a common tool for monitoring plant operation. Turbidity is often measured as the source water enters the plant, at various points in the treatment process, at the effluent of each filter, in the filter backwash water, and at the final plant effluent.

Surface Scatter Turbidimeters. Surface scatter turbidimeters consist of a light source and receiving photocell detector. Light is reflected off the surface of the sample and measured by detector cells positioned at incident angles to the surface and light source. Detectors are typically oriented directly above the sample to measure the amount of light reflected at a 90° angle to the sample. Any particles or turbidity-causing components affect the amount of reflected light detected.

The surface scatter analyzer has a higher sensitivity than the transmittance type and has the advantage that the sensor and light source are not in contact with the sample. Where entrained air is present in the sample, bubbles scatter light, as do particles. For this reason, a bubble trap should be mounted in the sample line ahead of the analyzer. Surface scatter units are typically accurate to within 5% to 10% of reading in the 0.1 to 5,000 nephelometric turbidity units (ntu) range. Surface scatter turbidimeters are recommended for source water and filter influent.

Transmittance Turbidimeters. Transmittance-type detectors are used in clean water applications. A light source directed through the sample and a photocell detector measures the amount of light not absorbed or scattered by the suspended solids in the sample. Transmittance-type detectors are designed for use from 0.1 to 100 ntu. Bubble traps are recommended and are built into some models.

Accuracy is typically within 2% to 5% of reading in the 0.1- to 100-ntu range. Drawbacks include maintenance required to clean sensors. Transmittance turbidimeters are recommended for filter effluent and plant effluent applications.

Special-purpose backwash turbidimeters are available for monitoring filter washwater clarity. These provide readings in 0% to 100% transmittance, not directly equated to ntu readings. The submersible sensor is mounted on or near the washwater trough. The amount of washwater required to clean a filter can be minimized by automatically controlling the wash cycle based on the water's transmittance value.

Laser Turbidimeters. Laser turbidimeters provide much lower range than other turbidimeters. These instruments provide accuracies of $\pm 5\%$ of the reading with a range of 0 to 1,000 mntu.

Particle Counters. Particle counters measure the number and size of particles in a volume of water by means of a laser light source directed through the sample and a photoelectric detector that measures the amount of transmitted light. The amount of light blockage is equated to particle size. Most analyzers are equipped with multiple channels calibrated to different particle sizes.

Typical particle counters have sensitivity to suspended particles in the range of 1 to 800 μm. The sensors are typically calibrated over a smaller measurement range, with the ranges selected to match the specific size particle of interest. For clean water applications, particle counters may be calibrated to detect the following particle distribution: 0 to 1, 1 to 2, 2 to 5, 5 to 10, 10 to 15, 15 to 20, 20 to 40 μm. This way the analyzers provide information on a range of particle quantities of several size distributions. *Cryptosporidium* and *Giardia* are normally found in the 3- to 12-μm range. A common range used in application is 2 to 15 μm.

As particle counters provide the number of certain-sized particles in a volume of water, the flow rate through the instrument must be controlled. Most instruments use some form of piped weir arrangement to achieve this. Particle counts are normally expressed as a number per milliliter of water.

Because particle counters determine the size and quantity of several particle size ranges, they produce a large quantity of data. As a result, transmitters can produce several 4- to 20-mA signals. Particle counters are available with communications ports that allow the analyzer output data to be transferred digitally to controllers. A custom software driver may be required to accomplish the data transfer.

Particle counters can detect particle concentrations that are not detectable with turbidimeters. For this reason, particle counters can be used in filtration applications where turbidimeters are reaching their detection limits. Particle counters have also been used to monitor source and finished water quality, as well as coagulation and sedimentation processes.

Other Analyzers and Detectors. Many automatic analyzers are available that can continuously monitor some parameters, relieving the operator of the tedious task of sampling and running analyses, as well as providing better control of plant operations.

Fluoride Residual Measurement. Continuous on-line fluoride analyzers are based on amperometric techniques. On-line analyzers typically consist of a sample pump, sample conditioning, a constant-head sample flow assembly, and a selective ion electrode. The electrical potential developed is in proportion to the fluoride concentration.

Sample conditioning normally includes adding a buffering reagent. The reagent is stored in bottles that require replenishing on a periodic basis. Typical reagent storage volume is adequate for 20 days of on-line operation. Sample temperature is also controlled for improved accuracy.

Measurement of pH. A glass membrane electrode and a reference electrode are used to measure pH. When they are submerged in the process fluid, an electrical potential develops between the sensor and the reference electrode. This potential is dependent on the free hydrogen ion concentration.

Sensors are available in several mounting configurations, submersible style for open-channel applications and insertion type for mounting in pipelines. A common configuration consists of the insertion-type pH probe mounted in a sample sidestream along with other instruments such as chlorine residual and temperature.

Streaming Current Detectors. Streaming current detectors are primarily used for on-line determination of dosage rates for coagulants. Traditional tests for determining coagulant dosing rates are usually analytical procedures such as jar tests.

Streaming current detectors measure the net effective charge of suspended particles in the sample. Samples are usually collected downstream of the coagulant dosing point, so the analysis determines the measurement on the unreacted particles only. The resulting signal can be used for feedback control of the coagulant dose.

Conductivity Sensors. Conductivity detectors are used to determine specific conductance of a solution. The analyzer is an electrical device that determines the ability of the solution to carry an electric current. The two most common conductivity analyzers are the electrode type and the electrodeless type. The electrode type has two electrodes suspended in the sample to pass current through the sample. The electrodeless type uses induction coils linked by the sample solution. The amount of induction through the solution is affected by solution conductivity. Conductivity differs from pH in that all ions in solution affect the conductivity, but pH is affected only by hydrogen ions.

The concentration of ions in the sample solution is the main factor that affects the conductivity measurement, but temperature and ion type can also affect the conductivity determination. Solutions with few suspended solids or with relatively low conductivity usually require the electrode-type analyzer. Solutions with high suspended solids, high conductivity, or high corrosivity usually require the electrodeless type.

Sample Requirements for Analytical Measurement. Process time lag is one of the most difficult control problems to solve with analytical instrumentation. Control loops that rely on analytical signals, such as chlorine residual, operate best when the sample is analyzed as quickly as possible. If the analyzer is located some distance away from the process, bypassed sample pumping loops should be used. This arrangement pumps a relatively large volume through the sample piping to keep line velocities high. Most of the sample is bypassed back to the process by means of a drain, and a small portion is diverted to the analyzer through a wye strainer.

Analyzer process time lag is also affected by changing plant flow rate. A sample measured 30 s downstream of the disinfection point at maximum plant flow may be measured 5 min later at low flows. If the plant is subject to extremely wide variations in flow rate, it may be necessary to add a special process sample point for low-flow conditions.

BIBLIOGRAPHY

American Water Works Association. *Instrumentation and Computer Integration of Water Utility Operations.* Denver, Colo.: American Water Works Association, 1993.

American Water Works Association. *Instrumentation and Control,* Manual M2. Denver, Colo.: American Water Works Association, 2001.

American Water Works Association Research Foundation and CRS Proaqua. *Online Monitoring for Drinking Water Utilities,* Denver, Colo., 2002.

Falco, J., K. Stouffer, A. Wavering, and F. Proctor. *NIST, IT Security for Industrial Control Systems.* http://www.isd.mel.nist.gov/documents/falco/ITSecurityProcess.pdf.

Feit, Dr. S. *Local Area High Speed Networks.* Indianapolis, Ind.: Macmillan Technical Publishing, 2000.

Franklin, G. F. *Feedback Control of Dynamic Systems.* Reading, Mass.: Addison-Wesley Publishing, 1986.

Gaushell, D. J. "Expert System Development for Water Treatment Plant Backwash." *Proceedings 1992 AWWA Computer Conference.* Denver, Colo.: American Water Works Association, 1992.

Kubel, D. B., "My SCADA System Is Secure and Other Security Myths." *Proceedings WEFTEC® 2003, Annual Conference and Exhibition.* Alexandria, Va.: Water Environment Federation, October 11–15, 2003.

Liptak, B. J. *Instrument Engineer's Handbook.* Radnor, Pa.: Chilton, 1985.

Skrentner, R. G. *Instrumentation Handbook for Water and Wastewater Treatment Plants.* Ann Arbor, Mich.: Lewis Publishers, 1988.

CHAPTER 21
ELECTRICAL SYSTEMS

John H. Wiedeman
Wiedeman & Singleton, Inc.
Atlanta, Georgia

INTRODUCTION

The primary objective for a water treatment plant electrical system is safety and reliability. Cost-effectiveness is important when safety and reliability have been established. The electrical system in a typical water treatment plant comprises approximately 5% to 10% of the construction cost. A robust design of the electrical system with spare conduit where needed is desirable. Water treatment plants are usually permanent facilities. Many have been in place more than 100 years. For that reason, high-quality wiring and equipment are cost-effective in most cases. Electrical system design is an iterative process that involves the overall water system, treatment considerations, architecture, structures, security, and HVAC.

The design, installation, testing, and specifications for electrical equipment are thoroughly covered by codes and standards. These codes and standards are no substitute for professional judgment. Electrical engineers use many of these standards in specifying myriad items of conduit, wiring, and equipment.

Electrical engineers are responsible for detailed design of the system. Nonelectrical engineers and managers can aid the design process by being familiar with the NEMA handbook and other standards. Civil and environmental engineers usually specify and select treatment and pumping equipment, most of which operates with electric motors and controllers. The design process involves input from all professionals and managers involved with the project.

CODES AND STANDARDS

The National Fire Protection Association (NFPA) has published and updated the *National Electrical Code* (NEC) since 1911. The NEC 2002 handbook is the 9th edition. The code handbook is illustrated with 1191 pages and contains tables and examples. This code has been adopted by most jurisdictions and covers wiring, protection, equipment for general use, special occupancies, special equipment, special conditions, and communications systems.

Seismic design of electrical fixtures and equipment is subject to provisions of the *International Building Code* (IBC) if adopted by the political jurisdiction. The 2002 IBC has a method of seismic design with applicable provisions to electrical fixtures and equipment. As a minimum, fixtures, wiring, and equipment should be designed to withstand the design seismic event without collapse. In significant seismic areas, certain facilities including water treatment and distribution are known as *lifeline* and require special attention to remain in service after a seismic event. The structural engineer is the logical source of seismic design structural requirements for electrical fixtures and equipment. *Unified Facilities Guide Specifications,* published as *Military Specifications Section 16070A,* entitled *Seismic Protection for Electrical Facilities* provides references and detailed seismic design information.

Electrical equipment is generally designed in accordance with standards of the *National Electrical Manufacturers Association* (NEMA). Electric motors have a variety of uses and are coded in accordance with their starting and torque characteristics. Wiring design depends on the number of amperes used by the motor. There is no standard of full-load amperage that applies to every electric motor. Low-rpm induction motors sometimes have a lower power factor and higher full-load amperage than typical electric motors. This has a direct effect on the wiring, size of starter, and protective equipment required. The NEC handbook, Appendix D, has sample calculations for conductor size and overload, short-circuit, and ground fault protection when the motor full-load amperage and service factor are known.

ASTM, ANSI/IEEE, and the telecommunications industry (TIA and EIA) publish other related standards. There are 13 ANSI/IEEE national standards published beginning in 1972. The AWWA publishes *Instrumentation and Control* (M2) and policy statements in *Grounding of Electric Circuits on Water Pipe* and *Electric Power Reliability for Public Water Supply and Wastewater Utilities.*

The NFPA has seven publications concerning electrical equipment in addition to the National Electric Code:

Recommended Practice for Electrical Equipment Maintenance

Standard for the Protection of Electronic Computer/Data Processing Equipment

Recommended Practice on Static Electricity

Electrical Standard for Industrial Machinery

Standard for Emergency and Standby Power Systems

Standard on Stored Electrical Energy

Emergency and Standby Power Systems

Standard for the Installation of Lightning Protection Systems

Fire protection publications primarily involve other issues, but include references to electrical equipment.

There are 23 pages of UL/ANSI-approved standards most of which apply to electrical conduit and equipment. *Standard for Flexible Metal Conduit* and *Standard for Enclosures for Electrical Equipment* are typical items that apply to water treatment systems. *Standards for Power-Operated Pumps for Anhydrous Ammonia and LP Gas* and *Standard for Electric Gardening Appliances* are typical of standards of limited use to water treatment design and installation.

The Illuminating Society of North America publishes *Lighting Handbook.* The Canadian Standards Association (CSA) establishes and maintains laboratories for the examination and testing of electrical devices. The International Electrotechnical Commission (IEC) is a European organization similar in scope to the Underwriters Laboratories and CSA in North America.

ELECTRICAL DISTRIBUTION SYSTEMS

Public and private electric utilities normally supply electricity for all water treatment plant operations with the possible exception of heating. Electric heating is an alternative to natural gas or oil. Standby pumping can be provided directly by engine-driven pumps or separate generators. On-site power generation can be specifically for water treatment plant operations or with the option to sell power to an electric grid. Most electric utilities are required to publish a rate schedule for purchased power. A few water treatment plants have hydraulic turbine power used for generating electricity or directly connected to pumping equipment. Wind or solar energy is also a possibility. There are a few water treatment plants that use steam turbines for pumping and other purposes.

Electric utilities and their reliability depend on the source of supply, alternative supplies, and the type of distribution system. With deregulation of power in many areas, the source may vary, and probably will change with time. Power distribution and supply in many cases are from separate utilities. In some cases the supply, energy transmission, and energy distribution are from three or more separate utilities.

Electric utilities have standards for voltage, types of transformers, and customer utilization. In most cases the electric utility supplies the transformers for secondary metered service. In some cases primary metered service with the water treatment plant furnishing the transformers may be desirable. In either case the electric utility will usually provide data for the primary service and impedance for secondary transformers. The electrical engineer can use these data to determine short-circuit protection for the system and starting requirements to limit voltage dip.

Service conditions vary with the utility. In some cases these voltage short-circuit and starting standards are arbitrary, and there is no appeal. In most cases the utility will make an engineering analysis of water treatment plant needs. Service reliability, service voltage, transformer reliability, short-circuit or interrupting capacity, starting requirements, and regulations involving redistribution to auxiliary water treatment plant facilities are the major items that must be agreed to with the utility. All have a major effect on water treatment plant electrical system design.

All electric utilities are federally regulated with the exception of intrastate service in Texas. Eastern and western power is out of phase with each other and is not compatible. Equipment operates satisfactorily with either source. Regulation of utilities varies from state to state. In some states power is strictly regulated. Other states have virtually no regulation. State and federal regulation often varies between public power and private utilities. The water treatment plant manager should determine the extent of utility regulation by the state, and whether alternate sources are available. It is important to select a specific utility, if choices are available, at an early stage of design.

Almost all utilities supply water treatment plants with three-phase power. The water treatment plant uses three-phase power for most motors. Single-phase power is used for lighting and is optional for small motors. Utility service voltage is usually at the utilization voltage for pumping equipment. Currently this is usually 208 V for small plants, 480 V for medium plants, and 4 kV for large plants. Power is usually transformed to 480/277 V and/or 120/208 V for various uses such as lighting, small motors, water treatment equipment, and instrumentation.

Some utilities will provide 4-kV and 480-V service with two banks of transformers with primary service at additional cost. This may be desirable for continuity of service. With separate service the surges, harmonics, and voltage dip on plant facilities caused by operation of pumping equipment are reduced.

Existing water plants may utilize 2.4 kV for pumping and 120/240 V for lights and other requirements. A few pumping systems, particularly in textile manufacturing areas, operate at 550 V. Most systems in the United States operate at 60 Hz (60 cycles per second). Worldwide systems often operate at 50 Hz and a few at other frequencies. Electric motors can adjust to frequency changes, but not substantial changes in voltage. Some motors have dual voltage capability with different terminals for different voltages.

Failure and replacement of utility transformers are a major reliability factor. In some cases the utility maintains spare transformers for installation in the event of a failure, and in other cases dual transformers are provided, usually at additional cost. In some existing facilities three single-phase transformers are used with 2.4-kV systems. Service can be maintained in "open delta" if one transformer fails.

Voltage regulation depends on the utility system. As loads on the utility increase during the day, there is greater voltage drop in the system. The voltage drop often is compensated for by regulating transformers that boost voltage as loads increase. If there is substantial daily voltage variation at the water treatment plant, this can usually be adjusted by the utility. Transformer reliability and voltage regulation are items that should be discussed with the electric utility for new and existing facilities.

Discussions with the electric utility or utilities to determine reliability and methods to improve reliability are the best method to assess the amount and type of standby power needed.

Standby power requirements are based on the length of a power outage that can be tolerated and the reliability of the power utility. The length of a tolerable power outage is a feature of the water supply and distribution system. The number of water treatment plants on the system, the amount of ground and elevated storage, and the ability of the distribution system to use these facilities affect the electrical standby design of a particular water treatment plant. Some systems receive water from other sources normally or in an emergency. The percentage of industrial and domestic load and the use patterns are also important. These features determine the length of time an outage can be tolerated.

The cost and availability of energy are important in a determination of water treatment plant generated power. It may be economical to generate power on peak at the water treatment plant, and if so, this has a major effect on design. Peak power generation usually involves a reduction in power use in the water treatment plant instead of actually selling power to the electric utility.

Many water treatment plants have planned extensions as demand increases, usually flocculation, sedimentation, and filtration. It is desirable to provide for these additions in the original design of the electrical system on a cost-effective basis.

Legally required standby systems are typically installed to serve loads such as communication systems, exit lighting, ventilation, and smoke removal. In some cases electric power for fire fighting and rescue operations is required. Some water treatment plants include assembly areas for the public which require special standby systems.

TRANSFORMERS

Utility transformers provide the required change in voltage for proper utilization by the water treatment plant. Transformers have separate high-voltage and low-voltage coils that effectively prevent high voltage from entering the low-voltage system. It is important to have the correct voltage and to have a system with the load balanced for each of the three phases. Utility transformers are usually provided with taps to optimize voltage used by the water treatment plant. Several voltages are often required within the water treatment plant; usually they are provided by separate transformers installed in the water treatment plant, either the dry type or oil-filled.

Dry-type transformers are frequently used since conventional oil-filled transformers require special vaults or outdoor mounting Transformers with less combustible or noncombustible oil are also used. The NEC requires dry-type transformers of over 112.5 kVA to be in a transformer room, with some exceptions. Conventional oil-filled transformers can accept substantial overload for a short period; dry-type transformers cannot. It is desirable for the designer to be conservative in estimating loads when using dry-type transformers. Transformer taps for voltage adjustment are available for transformers provided by the water treatment plant, but are often omitted.

Autotransformers have a single coil and are economical for small changes in voltage but do not provide protection for higher voltage entering the low-voltage side. A common single-phase application is to boost 208-V power to 240 V. A common three-phase application is to boost 480-V service to 550 V where existing motors and controllers operate on 550 V. The NEC contains information on special protection for autotransformers.

MOTORS AND CONTROLLERS

Pump motor and controller design is a major electrical feature of a water treatment plant. The motor controller is the safety switch, starter, overload protection, and short-circuit protection. The motor controller often is located in a cabinet with safety switch, pushbutton or three-position switch, contactors for starting and operating, overload protection, control power transformer, and relays. A circuit breaker for low-voltage equipment often includes the safety switch, overload, and short-circuit protection. Medium-voltage equipment is often provided with current-limiting fuses instead of circuit breakers. Several separate enclosures were widely used for motor control devices at one time. Grouped separate devices are allowed by the NEC.

Motor and controller design is heavily influenced by short-circuit or interrupting capacity. This determines the capacity of protective devices to interrupt dangerous power surges caused by wiring failure, and at the same time it determines the starting characteristics of motors. Traditionally motor starting was determined by rule. Now voltage drop on the utility system is used in most cases since electric utilities have computer programs for analysis. A robust utility system has less voltage drop and requires stronger reinforcement for short-circuit and ground fault protection, but allows higher starting currents.

Electric motors require additional power to overcome inertia of the motor and pumps on starting. The amount of power required depends on the pumping equipment and the characteristics of the electric motor. The additional power requirement for starting, known as *inrush*, causes voltage to temporarily drop locally and on the utility system. The motor must supply enough torque to start the pump. Torque varies with the individual pump, generally related to specific speed. Motor inrush varies with motor design. Pump motors are usually low-torque devices. Starting equipment and motor design can be adjusted to fit the requirements of the utility system.

Depending on the utility system, motors of 2,500 hp or larger can be started "across the line" without methods of reducing inrush. On other systems 100-hp motors may require special starting conditions. The number of successive starts anticipated is often used to determine allowable voltage drop on the utility system and should be estimated prior to contacting the utility.

Motors used for pumping are usually three-phase and are induction, synchronous, or wound rotor motors. The induction motor is the most common. Excitation (magnetizing) for the induction motor is provided by the motor design. Synchronous motors are often cost-effective when motor horsepower exceeds rpm, and they require separate excitation from a dc source. Synchronous motors start as induction motors, and when nearly full speed is reached, dc power is supplied to provide synchronous speed. The wound rotor motor application is usually for low-inrush starting characteristics, and at one time it was widely used for variable-speed applications. Variable-frequency drives are usually cost-effective for variable-speed application in new installations of less than 600 V. Speed control can also be provided by magnetic couplings between the motor and equipment. New installations rarely use wound rotor motors or magnetic couplings for speed control.

NEMA motor designs are B through E. Design B is usually used for pump and general-purpose motors. Designs C and D have higher starting torque and are used for equipment such as air compressors. Design E is a high-efficiency design with locked rotor amperage higher than that of design B. Locked rotor code letters A through V indicate the ratio of locked rotor current to running current. Code A has the lowest ratio of locked rotor current to running current. The maximum locked rotor current for a design B motor is code G, or 600% of full-load current. Code E or F motors are sometimes a cost-effective method to reduce voltage dip. The starting torque available is reduced for low-inrush motors.

With large motors, the motor torque and inrush are calculated based on the starting characteristics of the pump. If the inrush exceeds requirements of the utility, then various types of starters are used to reduce inrush. Starters reduce inrush with variable frequency, autotransformers, reactors, or special motor winding. Variable-frequency or wound rotor starting produces the minimum inrush. Variable-frequency is often cost-effective for motors of less than 600 V, and the wound rotor is often cost-effective for voltages of more than 600 V.

For large motors it may be necessary to use wound rotor step starting depending on the utility system to reduce voltage drop to a tolerable level. The wound rotor motor is more expensive and less efficient. The designer may find that smaller units that can use induction motors are cost-effective with normal methods of limiting inrush on starting.

Starters and motor control equipment require considerable space which must be provided in water treatment plant design. The space required depends on the type of starting, type of protection devices, reliability factor, auxiliary power, type of motor, and motor control. Minimum working space to maintain electrical equipment is included in NEC Article 110.26. Space requirements vary with the manufacturer. Additional space as a safety factor is often desirable unless the equipment has been purchased or the source is known.

Synchronous motors require separate excitation that is provided by conversion of alternating current to direct current. This can be done by a dc generator directly connected to the motor shaft, brushless excitation, or a separate ac/dc converter (rectifier). The most common method of supplying dc power for a synchronous motor is by a remote rectifier.

Variable-frequency drives and variable-frequency starting are in widespread use. These devices are particularly suitable for less than 600 V, but may not be cost-effective at higher voltages. Motor speed is directly proportional to frequency (hertz), and for a pump, horsepower is proportional to the cube of the change in speed. Frequency of the power source can be adjusted by a solid-state system that converts first ac to dc power and then back to a different ac waveform at a frequency adjustable by the user. To reduce inrush for a pumping unit, a speed is selected that will produce sufficient torque for the pump to start, probably at shutoff head. Under these conditions the motor inrush is very low. The system that changes frequency is then by-passed, and the pump operates at normal speed.

Often variable-frequency devices are used for starting only and can be smaller than those for constant operation. With continuous operation at variable speed, the equipment is larger and more expensive. Variable-speed systems for constant operation often have provisions to switch the variable-frequency system out of the circuit when operating at or near full speed.

Motors and starting requirements for equipment in water treatment plants vary. An air compressor and pump, as extreme examples, have different starting and operating characteristics.

Single-phase motors are available up to 15 hp but in water treatment plant practice are usually limited to 2 hp. The constant-speed single-phase motor has two windings. One winding is used for starting. There are three basic constant-speed starting systems: resistance, shaded pole, and capacitor. The single-phase motor requires less wiring and a simpler starting system than a three-phase motor in small sizes. Ventilation fans and similar equipment are typical applications for constant-speed single-phase motors. Series-wound single-phase motors operate at variable speed, either based on load or controlled by inserting resistance into the circuit with a constant load. Laboratory equipment, sampling, mixing, and chemical feed are typical water treatment plant applications for series-wound single-phase motors.

Direct-current motors have an application for variable-speed operation, but in modern practice there are other more cost-effective methods, such as variable-frequency control with induction motors. Direct-current motors can operate directly on battery power.

Protection devices are outlined in Article 230 of the NEC for services and in Article 430 for motors, motor circuits, and controllers. Grounding is important and complex and is included in the NEC.

Pump motor starters are often in a motor control center which has a single circuit breaker or fuse connected to a common bus bar. Each motor has a separate disconnect and protection. A more reliable arrangement is two main breakers or fuses feeding two separate bus bars that can be connected in the event of failure of one of the breakers. This is known as a *double-ended* control center.

Circuit breakers and fuses come in various sizes and capacities. The designer determines the capacity needed so that fire will not occur as a result of a fault (short-circuit). The capacity depends on the utility system and to a lesser extent wiring capacity. With a high-capacity system, a combination of fuses and circuit breakers is required. For a lower-capacity system, circuit breakers are sufficient. The system is coordinated so that the circuit breaker protecting that particular section of wiring will open on a fault and no other parts of the system will be affected. Each level of protection requires a higher fault to open. In actual practice, a serious fault often causes all breakers in series to open even with coordination. If this occurs, all breakers can be reclosed except for the branch breaker involved with the fault.

Relays are used for control and protection of motors. Contactors usually are high-capacity electromagnetic actuated devices used to actually start the motor. The relay does not carry heavy motor loads, and it is used to control the motor and associated devices. The relay is a electromagnet or silicon controlled rectifier (SCR) which closes a contact. A *programmable logic controller* (PLC) is a solid-state device often used to control motors and auxiliary devices through low-voltage relays.

Phase reversal is very rare in modern utility systems, and it is generally caused by faulty reconnection of broken distribution wiring. When a phase reversal occurs, motors run in the opposite direction. Phase reversal protection relays can be used for protection and should be considered in all installations based on risk.

Relays are often used to limit the number of successive starts of a motor. The manufacturer's recommendations concerning repeated starts should be solicited. The starting cabinet may include various relays and contactors for valves used to reduce hydraulic surges. Power to devices in the starter cabinet should not be active when the cabinet is opened, and switches are usually specified for that purpose.

Pump motor operation usually requires 120-V single-phase power that is provided either by a separate 120-V circuit for all pumps and auxiliaries or by an individual small power transformer for each pump and controls. The individual small power transformer is the most reliable option, but is more expensive.

All electric motors must be cooled. Cooling of pump motors is a major issue in most water treatment plants and water intakes. Air cooling is the most common method, but there are many water-cooled motors that use the water pumped for cooling. Air is circulated in a water jacket in the usual water-cooled motor. Water is normally controlled by a thermostat and provided from a potable supply; or an external air-to-water heat exchanger can be used in suitable climates.

With an air-cooled motor, the ambient temperature during operation is important for the motor and controller. Room ambient temperature is a function of outside temperature and ventilation. The motor controller and associated relays are the most sensitive to ambient temperatures and are often located in a separately cooled area. Design of the motor cooling system is iterative with the architect, civil and environmental engineer, and HVAC engineer. It is very important that the motor and controller ambient temperature during operation be accurately specified and controlled. The ventilation system can be critical to adequate standby power operation. If a standby engine is located in a pump room, heat from that source should be considered. An external heat exchanger for a standby engine is desirable for ventilation purposes.

Standby power for pumping can be provided by a prime mover connected to a pump, by a separate switched circuit from a generator, or by a generator that synchronizes with the incoming power system. Single or multiple units can be used. Each has different costs and utilities. Battery power or stored pneumatic power can be used to start auxiliary engines. Auxiliary engines can be diesel, natural gas, liquid petroleum gas, or gasoline. Each has reliability and safety issues. Some systems plan for only short outages and use standby pumping only. Other systems use standby power for treatment operations and pumping. This is a reliability issue discussed in previous paragraphs. Management should determine the standby power method as a policy issue based on engineering studies.

LIGHTING

Indoor, outdoor, and emergency lighting is a major feature of water treatment plant design. Common voltages are 120 or 277 V. In exceptional cases where circuit runs are very long, 480 V may be used. Each type of lighting has different efficiency, different life, and different characteristics. Within each type of lighting, efficiency varies. The selection of lighting type, voltage, and efficiency is an engineering, maintenance, and management decision. Natural lighting and reflective surfaces reduce power consumption during daylight hours and usually involve architectural design. The civil and environmental engineer, electrical engineer, and architect must all be involved to provide the best lighting system.

Other than in administrative spaces, which may warrant some aesthetic consideration, plant lighting should be functional and utilitarian. Application and light source categorize industrial lighting fixtures. Applications include indoor, outdoor, wet location, corrosion-resistant. Common indoor light sources include tungsten-halogen, incandescent, fluorescent, high-pressure sodium, metal halide, and mercury vapor. The most efficient one and one with the longest life is high-pressure sodium, and the least efficient is incandescent. Fluorescent lightbulbs with internal starters fit into most incandescent fixtures and provide substantial reduction in power consumption and increased life at higher initial cost.

Fluorescent or incandescent fixtures do well in spaces that are intermittently occupied because they provide instant illumination when they are switched on. Although high-pressure sodium, metal halide, and mercury vapor (HID) sources are highly efficient, they require time to warm up when first turned on. Therefore they are the best choice for areas where the lights can be left on for long periods or continuously. As an option, these fixtures can be purchased with a delay-relay quartz lamp that will supply safety lighting until the HID lamp reaches approximately 50% brightness. Incandescent fixtures are used for spotlights, gallery areas where there are piping conflicts, or areas where dimmer controls are desirable. Fluorescent bulbs cannot be substituted where dimmer controls are used unless the fixture is purchased with electronic dimming ballast.

Minimum lighting requirements are published and vary widely depending on the use, from laboratory to storage areas and galleries. Outlets for temporary lighting should be considered in galleries and other areas that may require equipment maintenance to supplement normal low-level lighting.

In considering lighting, the designer must not forget the plant exterior. Task lighting should be provided for equipment that requires nighttime operation or maintenance. General site illumination of at least 0.5 to 1 footcandle (fc) should be provided to allow workers to safely see their way around the site and for security. A photocell or time clock that operates automatically usually controls site lighting.

An efficient way to light a site is by high-mast lighting. This consists of a group of high-wattage fixtures mounted on tall poles with a lowering mechanism so that lamp replacement can be done on the ground. It takes only a few poles to provide lighting over a large plant site. One caution, however, is that they should not be used close to a residential area because they will be bothersome to neighbors.

The use of extension lights around a water treatment plant presents a definite hazard, so the design engineer should make a special effort to provide lighting that will minimize the need for extension lighting.

Life safety manuals are published by the National Fire Protection Association and include emergency lighting, exit lighting, and fire and smoke alarms. Security issues are also involved in lighting. Often outside lighting is controlled by systems that identify intruders.

Security issues affect electrical design. Intruder alarms, automatic doors and gates, and communication systems are involved. Communication systems are included in the NEC.

Heating a water treatment plant can use direct electric heat or, with natural gas or fuel oil, control via thermostats and fans for unit heaters. Solar energy is a consideration in some areas, primarily for heating, and usually it requires circulation of liquid. In some areas, electric roof heating coils are used to prevent snow accumulation. Also, in areas where freezing temperatures are expected, electric heat trace cables are used to prevent freezing of flowing liquids in pipes.

Wiring design is based on sizing in the NEC including voltage drop. It is important to operate electrical equipment at the proper voltage and to keep voltage drop in wiring to acceptable levels. Paragraph 215-2 of the NEC discusses the results of low voltage. The paragraph states in part:

> An applied voltage of 10 percent below rating can result in a decrease in efficiency of substantially more than 10 percent. For example, fluorescent light output would be reduced by 15 percent while incandescent light output would be reduced by 50 percent.

LIGHTNING PROTECTION

There are two areas of protection from lightning: the electrical system and the structure. Lightning rods are used in some areas and not in others, depending on climate and geology. Protection from electrical surges caused by lightning is universal. Surge arresters and protective capacitors are used. Voltage surges with peaks of several thousand volts even on 120-V systems are not uncommon. These surges occur because of induced voltages in power and transmission lines resulting from lightning spikes in the vicinity of the line. Surges also occur at the water treatment plant as a result of switching inductive circuits. Surge arresters are part of the electric entrance service. The basic standards used to investigate surge arresters are ANSI/IEEE 62.1 and 62.11. Maximum protection is achieved where the surge protection device is located as close as practical to the equipment to be protected. When a surge passes through an arrester, a wave is reflected in both directions on the conductors connected to the surge arrester. The magnitude of the reflected wave increases as the distance from the arrester increases. Pump motors should be protected by surge arresters and protective capacitors located near the pump. A range of techniques are available to condition power line disturbance surges for computer systems.

Information on protection of the structure from lightning, if required, is included in NFPA 780-1997, entitled *Standard for the Installation of Lightning Protection Systems.*

Instrumentation and communications systems that have electrical connections must be protected from surges as provided in ANSI/IEEE 62.1 and 62.11.

Fiber optics are often the best choice to protect instrumentation and communications systems. Fiber-optic cables are nonconductive, conductive, or composite. Conductive cables include metal for strength. Composite cables include conductors and fiber optics. Each has special marking and different NEC requirements.

POWER FACTOR CORRECTION

Power factor is defined as the cosine of the angle between the voltage and the current in an electrical system. If the electrical load is resistive, the voltage and current are in phase, the angle is zero, and the power factor is at its maximum value of 1.0. Most types of electrical equipment (motors, ballasted lights, etc.) are a combination of resistive and reactive loads, which results in a power factor of less than 1.0. The consequence of low power factor is that the system must carry higher currents than would otherwise be required to supply the same load at a higher power factor. Often power requirements are expressed in kilovoltamperes (kVA) which includes the power factor. The notation kW (kilowatts) does not include the power factor. Many utilities have additional charges for power factor. Capacitors installed at the water treatment plant can improve power factor and may be cost-effective.

ACCEPTANCE TESTING

Testing of electrical equipment includes acceptance testing and testing for maintenance purposes. Acceptance testing occurs prior to placing new equipment into service. Acceptance testing of major cable insulation, ground fault equipment, and emergency equipment is required. Acceptance testing of pumps and motors is highly desirable. Maintenance testing involves major cable insulation and standby equipment.

Cable insulation should be tested at higher than normal voltage and at normal voltage. Testing at normal voltage establishes a benchmark for maintenance testing. Insulation deteriorates with time, particularly for service at more than 600 V; a record of insulation testing a normal voltage will allow prediction of insulation failure based on trends established. It is desirable to have sufficient conductors to remove one set from service or to have a spare conduit so that entrance conductors can be replaced. It is likely that the entrance service life will be less than the life of the water treatment plant.

Pump motor rotation should be tested electrically prior to operation so that the pumping equipment will not initially rotate in the opposite direction. It is desirable to provide a hydraulic bypass so that all pumping equipment can be tested at peak flow conditions. Electric utilities have demand charges, but normally allow testing of equipment without instituting demand charges for specific times of the day. Normally wire to water efficiency is tested. Depending on conditions the starting current requirement can be tested so that reduced voltage, wound rotor resistance banks, or soft-start equipment can be adjusted for peak performance.

MAINTENANCE

Cables

The vast majority of wire and cable failures occur at the terminations because of heating caused by a loose connection. An excellent way to spot this problem is by infrared (IR) testing, which uses a camera with film sensitive to light in the infrared region. The picture colors, or shades, give an accurate indication of temperature. IR testing requires some specialized knowledge, so it is usually best to contract with a testing firm to perform the work. In addition to poor cable connections, the test can detect abnormal temperatures in switch gear busing and other equipment. It is recommended that a water treatment plant have IR testing once per year, or at least every 3 years.

Emergency Equipment

A method of testing emergency equipment under load is required. The system must be tested periodically and a written record of testing and maintenance provided.

Standby Equipment

In addition to engine maintenance, generators should be operated at least once per month. Periodic operation, or exercise, is the only way to be confident that the generator will be available when it is needed. Time clocks can be incorporated into the generator transfer equipment to automatically exercise the unit.

Particular attention is required to batteries used for starting or control of the engine. Chargers are available that not only charge the battery, but also perform diagnostics and initiate an alarm if the battery is nearly out of life.

Switch Gear and Fault Protective Device

As noted, molded-case circuit breakers and fuses do not require maintenance. Power circuit breakers do require maintenance, and maintenance should be performed once a year. As molded-case breakers, fuses, and power circuit breakers become older, it becomes increasingly important to maintain an adequate inventory of spare units and parts, both because they are more liable to fail and because it becomes harder to obtain replacements.

Medium-voltage vacuum switch gear is well accepted in the market, but there is no ready means to check for loss of vacuum. A vacuum circuit breaker with no vacuum cannot clear a fault. However, the vacuum bottle will usually contain the arc long enough for a backup breaker to react. The damage resulting from a breaker failing to clear is usually less with vacuum equipment than with older air magnetic equipment.

Relay Testing

While one is performing annual switch gear maintenance, it is important to check the protective relays and trip devices. Simply pressing a test button to see if a device works normally performs only a partial test. A more positive method is called *injection testing*. With this, a high current is injected, usually through the current transformers. This is as close as possible to a "live" test of the relay and breaker operation.

Transformer

Dry transformers do not require maintenance except that they should be periodically cleaned or vacuumed to remove any dust and spider webs from inside the units. Oil-filled transformers should routinely be checked to make sure the oil level is being maintained at the proper point. The transformer oil should also be periodically checked for dielectric strength. The recommended interval for dielectric testing is typically 1 year, although few plants test this often, if at all. Depending on how critical the transformer is, testing at 3- to 5-year intervals would seem to be a reasonable compromise.

CHAPTER 22
DESIGN RELIABILITY FEATURES

John E. Spitko, Jr., P.E.
Boucher & James, Inc.
Doylestown, Pennsylvania

INTRODUCTION

A reliable public water supply is essential to the health and welfare of the customers served. Reliability provisions in water treatment plant designs can ensure the production of an adequate quantity of high-quality water at all times. The selection of reliability provisions during design is dictated by regulatory requirements and by the consequences of loss of all or part of the plant for a time, or by the consequences of treated water that does not meet drinking water standards (1, 2).

The design of water treatment facilities should be based on the premise that failure of any single plant component must not prevent the plant from operating at the design flow or from meeting drinking water standards. Sufficient operational flexibility should be included in the design to handle a variety of water quality problems where raw water quality is variable. Reserve or redundant capacity should be provided in each unit process, so that process efficiency can be maintained even when a single treatment unit within the unit process must be removed from service altogether.

RELIABILITY AND REDUNDANCY CONCEPTS

Reliability and redundancy, although related to each other in plant design, have different meanings.

- *Reliability* refers to the inherent dependability of a piece of equipment, a unit process, or the overall treatment process train in achieving the design objective.
- *Redundancy* refers to the provision of
  1. Standby equipment or unit process
  2. More than one process in the process train that performs the same or similar function as another process in the same process train

3. Additional (or "reserve") process capacity
4. Flexibility, enabling the operator to perform the same function while using a different arrangement of existing plant units

As a simplified example of the relationship between redundancy and reliability (1, 17), assume that the reliability of a single rapid-mix basin (expressed as a fraction) is 0.9. Using the equation

$$R_s = 1 - (1 - r)^m$$

where R_s = system reliability
r = unit reliability (i.e., probability of being on-line)
m = number of components in parallel

we find that the provision of two full-size rapid-mix basins in parallel increases the reliability of the unit process to 0.99, or 99 percent. Provision of three full-size rapid-mix basins exponentially increases the process reliability to 0.999, or 99.9 percent.

Standby Units

Providing two half-capacity basins might be considered more reliable than a single full-size basin. One full-size basin would have a reliability of 0.9, as indicated above, with a 0.1 probability of failure (1 − 0.9). Two half-size basins would only have a reliability of 0.81 to provide *100 percent* of the design detention time, but *would* have a reliability of 0.99 to provide at least *50 percent* of the design detention time.

But two half-sized basins would not be as reliable as providing a second full-size rapid-mix basin. A second full-size basin would provide sufficient redundancy to sustain process efficiency should an entire basin be lost for any reason. Increased redundancy increases reliability.

As another example, in this case with respect to raw water supply reliability, multiple wells would have significantly greater reliability than only one or two wells.

Reserve Process Capacity

As another example, three settling basins might each be designated to handle 2 mgd (7.6 ML/day) flow, but each is sized for a maximum surface overflow rate equivalent to 3 mgd (11 ML/day). The facilities would still provide a capacity of 6 mgd (23 ML/day) even if one basin were lost. In effect, this is the same as providing a standby process unit having the capacity of one-half of design flow.

One or More Processes Perform Same Function

Overall plant reliability may also be improved when more than one process in the system is able to perform a given function. Both sedimentation and filtration remove particulates. Reliability in the function of removing particulates is enhanced by providing sedimentation ahead of filtration. Complex plants may have two or more processes capable of several treatment functions, resulting in a very high overall reliability of the treatment system. The concept of maintaining multiple barriers to the passage of microbes into a distribution system is very important (34). Large-membrane plants are designed with this in mind, in one case by locating chlorine disinfection downstream of nanofiltration (27). Functional treatment reliability becomes even more important as water reclamation progresses (37).

Flexibility

Overall water treatment plant reliability is also influenced by the degree of flexibility of operation inherent in a given design. Flexibility can have a dramatic effect on a plant's capability to produce high-quality finished water. For example, having the flexibility to add potassium permanganate, ozone, or chlorine dioxide, as well as powdered activated carbon, to the treatment process at the onset of a taste and odor problem will increase the capability to eliminate the problem in the finished water. Similarly, having the flexibility to bypass a unit or a malfunctioning mechanical component enhances the reliability of the system.

Whereas the reliability obtained through the use of redundant units, conservative sizing, or functional unit arrangement presumes the use of the same treatment processes, reliability obtained through flexibility does not. A flexible system offers an operator a set of choices from which he or she may select those that are best suited to the needs of any particular time. Beyond this, flexibility can also offer an important element of redundancy. Redundancy does not always have to be "in kind," that is, extra "carbon copy" units. Often redundancy can be economically obtained with no sacrifice of reliability through multipurpose equipment. As an example, a backup washwater supply for an elevated storage tank system may often be economically obtained through appropriate valving and orifice plates in the piping interconnected with high-service pumps.

OVERALL WATER SUPPLY SYSTEM RELIABILITY

Overall public water supply system reliability for a public water supply system with multiple sources of supply and/or treatment will impact design decisions about each individual component of the system. For example,

- *Interconnections/in-system finished water storage.* When a system has significant interconnection to another public water supply system, or has several days of in-system treated water storage (especially if elevated, able to flow to customers via gravity), reliability of an individual treatment facility may not be as important as when an entire system is served by one water treatment plant from one source of supply. For example, in a system with significant interconnections to outside sources or with large in-system storage, backup power for most water treatment functions may not be necessary.
- *Multiple sources of supplies.* In a system with several different sources of raw water supply, for example, numerous wells each with treatment prior to the connection with the distribution system, the reliability of each well water treatment plant will be less critical than in a system that relies upon a few high-production wells all feeding raw water to a single groundwater treatment plant.

Categories of Component Failure

There are many categories for failure of components and systems. Examples (7) include:

- Process
- Electrical

- Mechanical
- Hydraulic
- Instrumentation/computer hardware
- Software error (i.e., human programming error)
- Human error (operational judgment/action error)
- Natural disasters
- Human-caused disasters

These are discussed further in this chapter.

Failure Analysis

The science of failure analysis has progressed significantly in such fields as microelectronics and geotechnical engineering, and in such structural engineering specialties as offshore drilling rig failure analysis and evaluation. It is not within the scope of this chapter to go into detail in this subject. Suffice it to say that goals of the use of probability and failure analysis are to avoid overdesign and in turn save cost in the completed facility, as well as improve reliability. The reader is referred to references for further reading (12–14, 17, 28–32). One important concept worthy of mention here, however, is that of instantaneous failure rate (FR) or hazard rate function; it provides an instantaneous rate of failure (31). Failure rates are sometimes characterized as increasing (IFR), decreasing (DFR), or constant (CFF) (31). In attempting to develop as reliable a design as possible, it is helpful to consider information contained in Figure 22.1 and Table 22.1 (31). Figure 22.1 illustrates what is commonly referred to as the "bathrub curve," beginning with a DFR, followed by a period of CFR, followed late in the design life by IFR, as defined

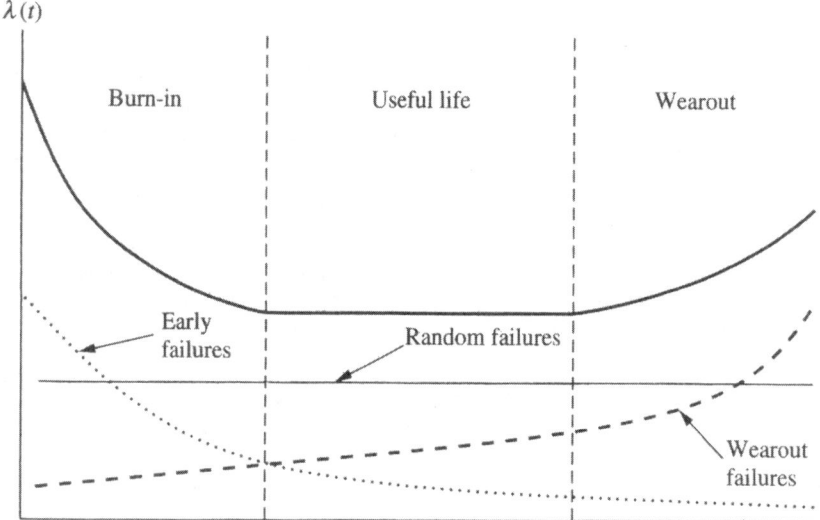

FIGURE 22.1 The bathtub curve. (*Source: Charles E. Ebeling,* An Introduction to Reliability and Maintainability Engineering, *McGraw-Hill, 1997; reprinted with permission of The McGraw-Hill Companies.*)

TABLE 22.1 The Bathtub Curve

		Characterized by	Caused by	Reduced by
Burn-in		DFR	Manufacturing defects: welding flaws, cracks, defective parts, poor quality control, contamination, poor workmanship	Burn-in testing Screening Quality control Acceptance testing
Useful life		CFR	Environment Random loads Human error "Acts of God" Chance events	Redundancy Excess strength
Wearout		IFR	Fatigue Corrosion Aginig Friction Cyclical loading	Derating Preventive maintenance Parts replacement Technology

Source: Charles E. Ebeling, *An Introduction to Reliability and Maintainability Engineering,* McGraw-Hill, New York, 1997; reprinted with permission of The McGraw-Hill Companies.

above. The third column of Table 22.1 provides some time-tested design-related means by which to reduce the potential for failures, hence increasing the reliability of a facility.

DESIGN CONCEPTS

Unit Process Configuration

Reliability of Individual Unit Processes. Unit processes that combine two or more process functions are inherently less reliable than single-purpose unit processes. For example, flocculation-clarifiers suffer from a deficiency known as the A-B syndrome, as illustrated in Figure 22.2. If unit A must be removed from service because of malfunction of the flocculation equipment, and during this time unit B has a malfunction in the clarifier (e.g., the sludge col-

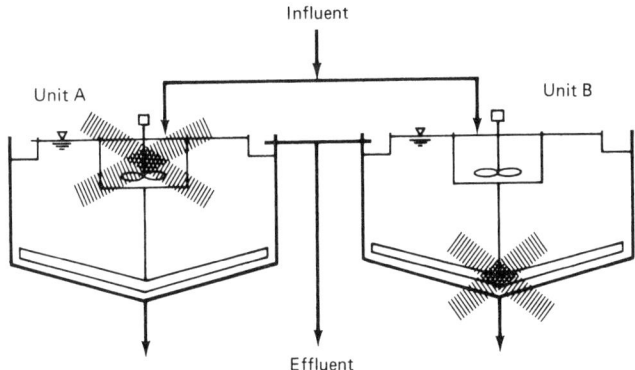

FIGURE 22.2 A-B syndrome—malfunction of two interrelated unit processes.

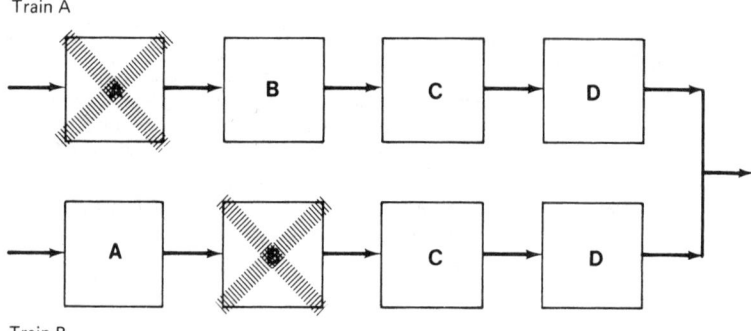

FIGURE 22.3 A-B syndrome—malfunction of two independent production trains.

lector jams), then both units could conceivably be out of service at the same time. If the flocculation and clarification steps are built as separate units, then at most only one flocculator and one clarifier would be out of service under these same conditions.

Reliability of Process Train. Another example of the A-B syndrome exists in the plant with two production trains, A and B, each with several of the unit processes shown in Figure 22.3 (note that there are no interties between the two trains). The trains operate in parallel and are totally independent of each other. If process A malfunctions in train A and process B malfunctions in train B, the entire plant may be out of service, or at least its efficiency would be impaired. Figure 22.4 illustrates an alternative design that is much more reliable, because parallel trains are interconnected between each unit process and the malfunctioning process can be bypassed. This minimizes the impact of any single unit failure on plant performance, and it nearly eliminates the possibility of both trains being out of service simultaneously.

Gravity Flow versus Pumping

Perhaps nothing in the world is as reliable as gravity. At the outset of design, careful consideration should be given to the use of gravity flow to minimize the amount of in-plant

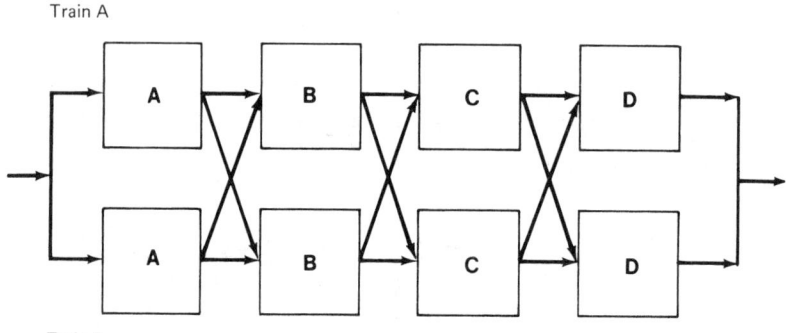

FIGURE 22.4 Interconnected parallel production trains minimize impact of single unit failure.

pumping needed. When one is evaluating alternative locations for a new plant or alternative process layouts, priority should be given to plant locations or plant layouts in which gravity flow can occur through the unit processes that are most critical to producing finished water.

Gravity Flow of Raw Water through Plant. For a new plant, if a new impoundment is to be built, or if an existing impoundment is to be utilized, try to locate the plant so that the hydraulic grade line from the impoundment will allow gravity flow to and through the plant. If influent pumping is not used and instead gravity is used, a major source of potential unreliability will be avoided (pumps, valves, piping, instrumentation, electric power outages, backup power, etc.).

Avoid Influent Pumping If Possible. If influent pumping cannot be avoided, one alternative approach is to utilize a raw water storage basin upstream of the water treatment plant, holding a significant amount of raw water pumped from the raw water source. Gravity flow from this point on will increase reliability by making the plant less susceptible to shutdown during a power failure.

Utilize elevated backwash water storage so that backwashing can continue to occur for a reasonable duration even if backwash lift pumps malfunction or a power failure occurs.

Locate Chemical Feed Equipment So That Gravity Feed to the Process Is Possible. Chemicals such as alum, lime, soda ash, and activated carbon are often better controlled and fed through gravity systems, even though this requires elevated chemical storage facilities.

Not all chemicals can be gravity-fed, however. Chlorine gas, for instance, must usually be dissolved in water prior to being fed to the process, and the solution process requires pumping of the injector water.

Individual Component System Reliability

Mechanical Reliability and Electrical Component Reliability. Reliability for mechanical and electrical equipment is influenced by several factors:

- Design that is specified with proper fault alarms and protectives to alert operators to problems (e.g., motor winding high temperature, water in oil seal, high level of vibration, low-voltage and phase loss, etc.)
- Manufacturers' experience with design of this equipment (i.e., don't be the first to specify "serial no. 1" of a new piece of equipment without very strong justification)
- Ambient operating environment (e.g., wet, dry, corrosive, etc.)
- Availability of spare parts and responsive service

Instrumentation Reliability. Design of control systems continues to make the transition away from analog controls and hardwired relays. As this continues, solid-state variable-speed drives (VSDs), industrial-grade computers known as programmable logic controllers (PLCs), human-machine interfaces (HMIs), and other digital controls have become reliable (7, 9, 33). In the past, some applications of PLCs were plagued by insufficient attention to power supply requirements (lower voltage, power spikes, etc.), which caused reliability problems with the solid-state electronics. Now, with high-quality power readily provided from uninterruptible power supplies, PLCs and other solid-state controls are reliable.

Reliability for instrumentation and SCADA is influenced by several factors (7):

- Installation techniques
- Power and grounding
- Ambient environment
- Programming techniques (software)
- Training
- Availability and repairability
- Degree of cyber protection (firewalls, password access, encryption, security policies, and protocols) (24, 25, 36)
- Change in management as a new instrumentation/SCADA system is implemented, so that new systems meet with acceptance by personnel, and are better maintained and understood, resulting in more reliable systems (33)

One utility operation responsible for a number of facilities separated by considerable distance (26) developed a centralized approach, using a scalable wide-area SCADA system that can monitor and remotely control multiple plants via the Internet. To improve reliability of this mode of centralized control, several strategies were used, such as independent verification steps for critical functions, such as maintaining chlorine residual, multitier alarm systems, and several others.

As mentioned above, proper training for operations and maintenance staff is needed since much of the solid state instrumentation appears to the uninitiated as a black box. Sometimes, all that is needed to temporarily bring a PLC-based system back on-line after a problem is to replace a board from an inventory kept at the plant. Spare boards and other spare parts should be specified by the designer so that they are on hand for the operations staff to utilize right from the outset of facility operation. However, even for the simple performance of this board replacement, specific training of O & M personnel is needed since the proper course of action may not be immediately self-evident to a typical operator.

For increased reliability of PLC-based control of large systems, control of processes such as filtration are divided among several PLCs, linked by a data highway. As an example, 40 filters at a plant in Wylie, Texas, were retrofitted with eight PLCs, each controlling five filters, with a ninth PLC provided in the plant control room to providing monitoring and alarming functions (9).

DESIGN PRINCIPLES AND PRACTICES

Security and Disaster Considerations

Natural Causes. In planning for the location, design, and operation of reliable plant facilities, the potential for natural disasters—tornadoes, hurricanes, earthquakes, fires, and floods—should be considered. Such considerations may affect the selection of location for and the design of the treatment plant.

The effects of flooding can be minimized by choosing an appropriate design elevation for the plant facilities, a location out of the 100-year flood boundary, or by designing protective walls, gasketed watertight doors, and other features to keep water out of critical locations within the plant.

The disastrous effects of earthquakes can be counteracted to some extent by choosing structural design criteria that are appropriate for the plant location, as discussed in Chapters 18 and 23. Some types of soils susceptible to liquefaction within high seismic hazard zones may have to be avoided altogether for construction of water supply facilities (5). Alternatively, a plant site may be evaluated using *failure modes and effects analysis* (FMEA) and other techniques, such as decision analysis, to reach an economical foundation design solution, such as partial ground improvement (28).

Human-Caused. Consideration of human-caused disasters must also affect plant design. Such disasters can originate either from outside the plant or from within. Examples of such disasters are chlorine leaks, strikes, civil disorders, sabotage, vandalism, explosions and fires, and airplane crashes. The occurrence of many of these events cannot be reckoned with in plant design, except from the standpoint of vulnerability. Sabotage, vandalism, aircraft accidents, and fires or explosions originating outside the plant can be assessed in this manner. Control of access to the plant and provision of plant security can alleviate, to some extent, the potential for acts of sabotage or vandalism. Careful coordination with local fire and police, even during design, will be necessary so that they may be equipped to gain quick access to facilities, without being faced with locked gates and other impediments, to assist plant personnel in a security threat situation. Well-placed video security systems are now a *must* in water facility designs.

Most water utilities are mandated to perform vulnerability assessments. Such evaluations can be performed during design, to address (24)

- Assets that are critical
- Service to critical customers (hospitals, military facilities)
- Dependence upon other infrastructures (electricity, transportation, other utilities)
- Contractual obligations
- Single points of failure (e.g., critical aqueducts, water storage facilities, transmission systems, aquifers, and the like)
- Chemical hazards

Locating the plant carefully or providing an alternative supply of drinking water can reduce the risk of being without a supply because of an airplane crash within the plant area. The impact of strikes and civil disorders, such as truck blockades, can be reduced by adequate storage of chemicals and other materials on the plant site. On-site generation of some critical chemicals, such as chlorine, also reduces a plant's dependence on truck or rail shipments. Although a supply of sodium chloride is still necessary, maintaining a large inventory of sodium chloride is simpler and safer than storing chlorine in liquid or gaseous form.

Vulnerability to explosions and fires within the plant can be reduced by establishing proper design safeguards with respect to the handling of hazardous materials. A thorough safety program established by the utility's management for the protection of operation and maintenance personnel is also a critical element in preventing in-plant accidents (4).

At the beginning of the design process, a vulnerability analysis of the plant and the site should be performed. Its results can provide critical direction for the designers and the utility, and can certainly reduce the exposure of a plant to natural and human-caused disasters. AWWA's manual *Emergency Planning for Water Utility Management* (3) is an excellent guide to assessing the need for many of these contingencies. Other AWWA and USEPA resources provide additional guidance for eliminating or minimizing vulnerability of water supply facilities (23, 24).

Supply Reliability

One of the primary concerns in system design is the adequacy of the raw water source to provide the quantity of necessary raw water to the plant. The available flow or yield of a river, impoundment, or well field must be sufficient to deliver the design flow of the plant plus estimated losses through the treatment process. Some utilities have two or more sources of supply, for example, a well field in addition to river intakes. This not only improves reliability, but also offers the possibility of improved performance in colder climates, where surface water becomes more difficult to treat as the temperature drops. Blending warmer groundwater with cold surface water generally will result in water that can be treated more reliably.

Another feature that can add to the reliability of the supply is multiple ports on intake structures or multiple screens on a submerged well-screen type of intake. With a river supply, multiple ports can eliminate the problem of an intake structure becoming clogged with ice or debris. Multiple ports on an intake in an impounding reservoir make available waters of different quality, and thereby improve the capability for producing water of the best possible quality. In cold climate locations, multiple ports, or other special design features of surface water intakes, should be considered to maintain inlet velocity below 0.5 ft/s, to minimize frazil ice formation.

Reliability of the supply water quality can also be enhanced by off-stream raw water storage reservoirs. With a polluted river supply, off-stream storage with several hours capacity will allow a plant to reduce its exposure to chemical spills or toxic discharges upstream from the intake. A surface water supply of rapidly varying water quality may be more reliably treated if a raw water storage reservoir is utilized upstream of treatment (6, 10). This feature would also allow the surface water intake to be taken out of service for maintenance.

Process Reliability

A complete discussion of individual process and equipment reliability characteristics would require a complete study and comparison. Such an effort is not undertaken here; however, a general overview of the subject can provide valuable insight.

The inherent reliability characteristics of a particular unit process are only one consideration in the selection process. Certainly process efficiency, among other factors, is extremely important. However, as the number of separate component units provided to accomplish a specific function decreases, reliability becomes increasingly important. When only two units of a particular process are provided, reliability is the prime consideration. Many states require a minimum of two component units for each of the unit processes that are primarily responsible for meeting drinking water standards. For a surface supply, such processes might include rapid mix, flocculation, clarification, and filtration as a minimum. Processes that would not normally be required to have at least two components for reliability purposes could include aeration for carbon dioxide removal, waste washwater handling facilities, and sludge processing, unless the size of the facility dictated otherwise. Other examples include certain chemical feeds, such as fluoride for fluoridation and hexametaphosphate for corrosion and scale control.

Table 22.2 lists the minimum requirements for selected treatment processes, as set by the Commonwealth of Pennsylvania and the 10-States/1-Province Great Lakes/Upper Mississippi River Board for surface water supplies.

The exclusion from Table 22.2 of requirements for appurtenant parts or accessories to the processes is not meant to minimize the need for adequate reliability. Among these appurtenant parts are flowmeters, valves, chemical feed equipment, pumps, instrumentation

TABLE 22.2 Minimum Requirements for Surface Water

	Recommended standards for water works (15)	Pennsylvania (19)
Rapid mix	2 units	2 units
Flocculation	2 units	2 units
Clarification	2 units	2 units
Filtration	2 units [Where only two units are provided, each shall be capable of meeting the plant design capacity (normally the projected maximum daily demand) at the approved filtration rate. Where more than two filter units are provided, the filters shall be capable of meeting the plant design capacity at the approved filtration rate with one filter removed from service.]	2 units [Where only two units are provided, each shall be capable of meeting the plant design capacity (normally the projected maximum daily demand). Where more than two filter units are provided, the filters shall be capable of meeting the plant design capacity at the approved filtration rate with one filter removed from service.]
Chlorinators	2 units	2 units
Coagulant feed	2 units	2 units
Ozone generator	For primary disinfection—sized so system "does not operate at peak capacity for extended periods"	For primary disinfection—sized so system "does not operate at peak capacity for extended periods"
Ozone contactor	For primary disinfection—2 units	For primary disinfection—2 units

and controls, liquid-level controls, piping, turbidity and other process monitoring equipment, and the plant service water system. Fundamental to the design of water treatment facilities should be the philosophy that failure of any single structure, piece of mechanical or electrical equipment, element of pipe, or valve should not put the entire facility out of service.

An example of the importance of a single plant component to reliability is illustrated by the August 31, 1975, incident in Trenton, New Jersey. The inability to close a single cone valve led to a series of events that flooded the high-lift pump station. As a result, 211,000 people were without drinking water and many businesses were closed for a 10-day period. The monetary loss was estimated at $5 million to $10 million. Thus the importance to system reliability of each component of a plant facility cannot be overstated.

Reliability of Plant Utilities

Electric Power. The reliability of a water treatment plant is critically affected by the reliability of its power source. Without power, all mechanical and electrical equipment ceases to function. The Institute of Electrical and Electronics Engineers (IEEE) Standard 446 contains recommendations and standards for the selection of emergency and standby power systems (16, 20). The results of one electric power service reliability study indicate that two independent power sources provide a high level of reliability (16). This same study compares the reliabilities of power sources and types of in-plant distribution systems. The reliability of sources is listed in descending order (16):

1. Two independent 13.8-kV sources
2. Single 115-kV source with two feeders to 13.8 kV
3. Single 13.8-kV source

The reliability of types of in-plant distribution systems is listed in descending order (16):

1. Secondary selective system
2. Primary selective system
3. Simple radial system

Most states require that each water treatment facility be served by two separate, independent sources of power (15, 19). If two independent power feeds from a power utility are not readily or economically available, standby power, such as gasoline, natural gas, or diesel-engine-powered generators, may be necessary to ensure the plant's ability to operate during a power outage.

Standby power supplies must be sized for the starting voltage dip and current surge of starting large motors. Switch gear/breaker low-voltage trip settings may be set at perhaps 85% of full voltage, and unexpected motor trips during an already tense situation during a power outage may contribute to difficulties already being experienced at the time.

For a natural gas–fueled engine generator or pump drive, the worst-case lowest gas pressure in the public utility's main and the actual size and capacity of the gas main must be considered in specifying the fuel feed system for the engine driving the standby device.

If standby power generation is being considered for only critical portions of a plant, the following areas should be considered (21):

- Permit critical processes such as disinfection to continue.
- Permit heating of areas where there is potential for freezing (and continued operation of vital heat tracing).
- Permit ventilation of hazardous areas.
- Provide egress illumination for safe exit.

As an alternative to standby power generation, engine-driven or dual-drive (electric motor/engine-driven) high service pumping may be most energy-efficient and lowest in initial capital cost. This alternative may be especially applicable if a plant is designed so that water flows through and is filtered via gravity flow, with no pumping required, so that need for on-site generated power would be at a minimum except for high service pumping. If standby high service pumping is handled by engine-driven or dual-drive pumps during a power outage, then on-site generation for other power demands may not be needed.

In addition to the power supply itself, dual transformer substations, duplicate primary feeders, and other duplicate components may be necessary to maintain the integrity of the power supply system. A standby source of power is of little value if failure of a plant's only primary substation transformer can prevent the power from being delivered to the equipment.

If a facility is unstaffed, its components and systems need to be more reliable than those for a staffed facility. One question which comes up during design is whether to design equipment and controls to automatically restart after power failure. Similarly, another question concerns the means of switching to an alternate source or to the emergency power source (16). Automatic, rapid transfers are more costly and more complex. At a continuously staffed facility, if a power outage of 15 to 30 min is acceptable, only a manual transfer switch may be necessary. If an outage of several minutes is acceptable, motorized primary system switching may be acceptable. Rapid switching (15 s) would require an automatic transfer switch.

Some loads such as computers and PLC control systems may require power at all times, utilizing uninterruptible power supply (UPS).

Power distribution equipment should not be located in areas below ground level which might be subject to flooding.

Remote monitoring and control systems, SCADA, can make it easier to operate a system that is not completely accessible at one location. This is an example of designing to make a system easier to operate and therefore more reliable (22).

Chemical Storage. An adequate inventory of chemicals maintained on the plant site is critical to reliability. The inventory should be sufficient to ensure the production of water at the design flow for a reasonable time. In some instances, access to the plant may dictate inventory needs (e.g., the plant's only access road is located in a flood-prone area). In others, the inventory may be dictated by the need to survive a truckers' strike. Location of the nearest source of the chemical may also affect the storage requirements.

Location and protection of chemical storage from security threats is to be considered during design of facilities.

Spare parts and chemical storage are both tools to avoid dependence on parts and chemical suppliers. The proximity, length of experience in business, and extent and reputation of service of an organization should factor into the selection of equipment and chemicals, in order to contribute to the reliability of a design.

On-site generation of a chemical may offer advantages in terms of reliability. The use of ozone for disinfection or taste and odor control does not depend on delivery of any chemicals to the plant. Except for power and air, the generation process is self-contained. As mentioned earlier, on-site generation of hypochlorite may also offer some advantages in terms of reliability, in that transport and storage of a hazardous chemical are eliminated. Again, as with power supply, system integrity is critical. An adequate supply of chemical on the plant site is of little benefit if a single conveyor, transfer pump, feeder, or other component can prevent a critical chemical from being applied.

Hydraulic Reliability

Provisions for control of hydraulic transients and design of piping to withstand calculated potential transients are critical. Provisions to allow venting of trapped air in plant piping high points are also critical.

Process Performance

Reliability of process designed for a water treatment plant can be better ensured by thorough, well-planned pilot work and by thorough, up-front raw water quality analysis. This is especially true when raw water quality is problematic and variable, and finished water goals are especially demanding (6).

OPERATION AND MAINTENANCE

Design for Ease of Maintenance and Operation

There are some very basic rules to follow to design for ease of maintenance (21).

Practically all equipment has to be moved or removed at some time for maintenance or repair. Proper lifting and openings or hatches for access should be provided for any heavy or awkward pieces of equipment.

Working room around equipment and protection of that working area outdoors from the elements will make it easier for O & M staff to maintain equipment, making the equipment systems inherently more reliable because they are probably going to be more carefully maintained. Heat tracing and insulation of water-filled piping located out doors will contribute to reliability. In locations that experience extremes of cold weather, covers over walkways and basins may be considered necessary for reliable operation and maintenance (10).

Up-to-Date, As-Built Drawings Contribute to Reliable Operation

It goes without saying that up-to-date, as-built drawings will contribute to the reliability of a water treatment plant. One trend in the future seems to be on-line equipment data sheets, parts lists, O&M manuals, all accessible through operator interface terminals throughout a water treatment plant. As budgetary pressures force limits to staff hiring levels, making information needed for repair easily available to those who need it will contribute to reliability. Some operating utilities are now adopting practices that have been common in the chemical and process industries—GIS-type or three-dimensional CADD-based O&M manuals and other information—again, accessible at locations throughout the plant, as needed. If such measures are desired by an operations staff, they must be taken into consideration during the design stage.

Specifying Computerized Maintenance Systems

It is an accepted fact that computerized maintenance systems, if rigorously followed, will improve treatment plant reliability (8). The design engineer can assist the O&M staff in establishing a computerized maintenance management system by requiring the contractor to submit sufficient types of information, such as manufacturers' recommendations for maintenance frequencies. An organized computerized maintenance management system moves plant maintenance out of the mode of responding to breakdowns and into the mode of performing most maintenance before breakdowns occur (8).

Proper Documentation for Training Purposes

Proper operation and maintenance is crucial to plant reliability, and proper training of operators will prevent many process failures. Giving adequate attention to maintenance will aid in the prevention of equipment malfunction; inadequate maintenance will guarantee malfunction (7).

Safe Design Contributes to Reliability

A safety program is a necessity at each water treatment plant facility. Safety practices of plant personnel are critical not only to their own health, but also to the operation of the plant. Often, accidents that disable plant personnel also cause damage to machinery, equipment, or materials that may put part of the plant out of service. AWWA's manual *Safety Practices for Water Utilities* (4) offers information on the development of a safety program that is expected to improve reliability.

REFERENCES

1. Hamann, Carl L., and Gene Suhr. "Reliability and Redundancy: Dual Protection for Water Treatment Plants." *J. AWWA* 72(4):182–186, 1980.
2. American Water Works Association and American Society of Civil Engineers. Chap. 22, "Design Reliability Features." In *Water Treatment Plant Design,* 3d ed. New York: McGraw-Hill, 1998.
3. American Water Works Association. *Emergency Planning for Water Utility Management, Manual of Water Supply Practices No. M19.* Denver, Colo.: AWWA, 2001.
4. American Water Works Association. *Safety Practices for Water Utilities, Manual No. M3.* Denver, Colo.: AWWA, 2002.
5. Anton, Walter F., Ronald M. Polivka, and Laurel Harrington. "Seismic Vulnerability Assessment of Seattle Water Department's Water System Facilities." In *Proc. of the 3rd U.S. Conference on Lifeline Earthquake Engineering*, August 22–23, 1991. Los Angeles, Calif.: American Society of Civil Engineers, 1991, pp. 375–384.
6. Burazer, Jane, and Steven N. Foellmi. "Tucson Water Plant Meets Community Expectations and Tough Quality Limits." *Water/Engineering and Management* 140(8):20–22, 1993.
7. Cervi, Brian. "The General Electric Genius I/O and Increasing Reliability in Water and Sewage Control." In *39th Annual Convention of the Western Canada Water and Wastewater Association*, October 21–23, 1987, Saskatoon, Saskatchewan, pp. 49–61.
8. Egleston, Philip C., Jr., Mark P. Steinert, and Kenneth P. Hammer. "Plant Improvements and Reliability Are Aided by Computerized Maintenance." In *Proc. of the 1992 Industrial Power Conference*, March 29–31, 1992. New Orleans, La.: American Society of Mechanical Engineers, 1992, pp. 25–38.
9. Moorman, Patrick. "Water Treatment Plant Implements Automatic Filter Backwash." In *Proceedings AWWA Seminar on Computers in the Water Industry*, April 14–17, 1991. Houston, Tex.: American Water Works Association, 1991, pp. 863–865.
10. Shorney, F. L., W. P. Grobmyer, and G. Taylor. "Zero Discharge Water Treatment Design Has High Reliability." *Power Engineering* 89(10):50–53, October 1985.
11. U.S. Environmental Protection Agency. *Design Criteria for Mechanical, Electrical and Fluid System and Component Reliability of Wastewater Treatment Facilities.* EPA 430-99-74-001. Washington, D.C.: Government Printing Office, 1974.
12. Cullinane, M. John, Jr., James Goodrich, and Ian Goulter. "Water Distribution System Evaluation." In Larry Mays, ed., *Reliability Analysis of Water Distribution Systems*, American Society of Civil Engineers, 1989.
13. Anonymous. 1978. "Risk Analysis Offers New Engineering Tool." *Automotive Engineering* 86(7):40–44, July 1978.
14. Jenny, Jon A. 1969. "Effect of Partial Failure Modes on Reliability Analysis." *IEEE Trans. on Reliability* R-18(4):175–180, November 1969.
15. Great Lakes–Upper Mississippi River Board of State and Provincial Public Health and Environmental Managers. *Recommended Standards for Water Works.* Albany, N.Y.: Health Education Services, 2003.
16. Krause, Terry L., and Kenneth L. Zacharias. Chap. 7, "Support Systems." In *Design of Municipal Wastewater Treatment Plants.* Water Environment Federation (MOP 8) and American Society of Civil Engineers (Manual No. 76), 1992, pp. 278–292.

17. Kennedy, John B., and Adam M. Neville. Chap. 5, "Probability, Probability Distributions, and Expectation." In *Basic Statistical Methods for Engineers and Scientists*, 3d ed. New York: Harper & Row, 1986.
18. American Water Works Association. *Minimizing Earthquake Damage*. Denver, Colo.: AWWA, 1994.
19. Pennsylvania Department of Environmental Resources, Division of Water Supplies. *Public Water Supply Manual and Revisions*. Harrisburg, Pa., 1998.
20. Institute of Electrical and Electronics Engineers. *IEEE Recommended Practice for Emergency and Standby Power Systems for Industrial and Commercial Applications (IEEE Orange Book)*, ANSI/IEEE Standard 446. New York, 1987.
21. Krishna, Gopal T. K. "Non-Sanitary Engineering Features Can Help Ensure a Plant's Success." *Water Engineering & Management* 134(10):32–37, October 1987.
22. Hill, Bruce. "Water/Wastewater Facilities Get Better with Electronics." *Water Engineering & Management*. June 1988.
23. American Water Works Association. *Water System Security: A Field Guide*, Catalog no. 20501. Denver, Colo.: AWWA, 2002.
24. United States Environmental Protection Agency, Office of Water. *Vulnerability Fact Sheet*. EPA 816-F-02-025. November 2002. www.epa.gov/ogwdw/security/index.html.
25. Brown, Alan S. "SCADA vs. the Hackers." *Mechanical Engineering*, December 2002. www.memagazine.org/backissues/dec02/features/feat_toc.html.
26. Shariff, Riyaz, Audrey Cudrak, Woytek Stachowski, Rudy Welz, and Stephen Stanley. "Centralized Remote Operation of Multiple Water and Wastewater Plants." In *2002 Information Management & Technology Conference Proceedings*, April 14–17, 2002. Kansas City, Mo.: American Water Works Association.
27. Thompson, Donald M., Dave Prah, Paul Feldman, and Gary Dernlan. "Design Decision-Making for Large Membrane Plants: Palm Beach County, Florida Case Study." In *1999 Annual Conference Proceedings*, June 20–24, 1999. Chicago, Ill.: American Water Works Association, 1999.
28. Vick, Steven G. *Degrees of Belief—Subjective Probability and Engineering Judgement*. Reston, Va.: American Society of Civil Engineers, 2002, pp. 169–175.
29. Tang, Wilson H., and A. H-S. Ang. *Probability Concepts in Engineering Design*, vol. I: *Basic Principles*. New York: John Wiley & Sons, Inc., 1975.
30. Avontuur, G. C. *Reliability Analysis in Mechanical Engineering Design*. Delft, Netherlands: Delft University Press, 2000, chaps. 1 and 2. http://gerard-avontuur.tripod.com.
31. Ebeling, Charles E. *An Introduction to Reliability and Maintainability Engineering*. New York: McGraw-Hill, 1997.
32. O'Connor, Patrick D. T. *Practical Reliability Engineering*, 4th ed. New York: Halsted Press, 2002.
33. Scholze, R. J., and H. H. Zaghloul. *Assessment of SCADA Technology Applications to Automate U.S. Army Water and Wastewater Sanitary Systems*, ERDC/CERL-TR-01-68. Champaign, Ill.: U.S. Army Construction Engineering Research Lab, November 2001.
34. Sakaji, R. H., R. Hultquist, A. Olivieri, J. Soller, R. Trussel, and J. Crook. "Whither the Multiple Treatment Barrier?" In *1998 Water Reuse Conference Proceedings*, February 1–4, 1998. Lake Buena Vista, Fla.: American Water Works Association, 1998, pp. 95–108.
35. U.S. Army Corps of Engineers. *Introduction to Probability and Reliability Methods for Use in Geotechnical Engineering*. Technical Letter no. 1110-2-547, September 30, 1997.
36. Phillips, Bill, Richard Emanuel, Ron Booth, and Linnon Parker. "Understanding and Minimizing Security Risk for Computer Information Systems." In *2002 Information Management & Technology Conference Proceedings*, April 14–17, 2002. Kansas City, Mo.: American Water Works Association, 2002.
37. Riley, Craig L. "Too Much or Too Little: Public Health Perspectives of Water Reclamation Reliability." In *2002 Water Sources Conference Proceedings*. January 27–30, 2002. Las Vegas, Nev.: American Water Works Association, 2002.

CHAPTER 23
SITE SELECTION AND PLANT ARRANGEMENT

Darryl J. Corbin
MWH
Dallas, Texas

Siting water treatment plants is becoming more complicated as public awareness and environmental activism increase. From the public's perspective, water treatment plants are often perceived in the same light as wastewater treatment plants, solid waste transfer stations, incinerators, and compost facilities. They are often categorized as a LULU—locally unwanted land use, or locally unacceptable land use.

WATER TREATMENT PLANT SITE SELECTION

In this chapter, site selection and plant arrangement are portrayed as sequential tasks: First find the site, then determine the best plant arrangement for that site. It should be recognized, however, that the two steps are not strictly sequential and are closely interrelated. One must have a preliminary plant layout before the site selection process can begin. Without a preliminary plant layout, it would be difficult to evaluate potential sites with regard to the shape of the land parcel, site topography, aesthetic effects, and so on.

Having a preliminary plant layout is also an important requirement for any public participation efforts undertaken during the site selection process. Artist renderings and other graphics that show how the proposed plant will look are an important tool for educating the public about the project. The preliminary plant layout should be developed with due consideration to likely public opposition, because the plant layout can greatly affect public acceptance of the site selection. Feedback from the public often results in changes to the preliminary plant arrangement to address real or perceived concerns. Public concerns about the plant appearance and layout must be addressed during the site selection phase. If not, sites that might otherwise be acceptable to the public will be dismissed.

Site Selection Process

It is easy for those involved in planning a new water treatment plant to misjudge the reaction of the public. Both the owner and the design engineer tend to focus on technical issues. From their perspective, the primary function of the plant is to produce a finished water that is economical, reliable, and safe. Therefore sites that are conducive to these goals are favored.

The general public, on the other hand, takes for granted that drinking water should be economical, reliable, and safe. Aesthetic considerations are the foremost concern of the neighboring communities. By getting the public involved at an early stage, the owner gains an appreciation of the issues of most concern to the neighboring communities. The owner can then take those concerns into consideration in the site selection process, as well as undertake measures to allay the concerns of the residents.

Multiple Treatment Plant Sites. Generally speaking, a water treatment plant and all its ancillary facilities should be located on one site. If a suitable site cannot be found to accommodate the entire treatment facility, consideration can be given to siting the primary water treatment processes at one location and putting the sludge treatment facility at another location. However, this approach should be used only as a last resort, because pumping or hauling sludge over long distances is troublesome and expensive.

The cost of constructing and operating one large water treatment plant is normally less than that of building and operating two or more smaller plants. However, for cities with a very large service area and multiple water sources located far apart, it may be appropriate to provide multiple water treatment plants to save pumping costs and for increased reliability.

When one is designing a new water treatment plant for a city that already has one or more existing plants, there is little opportunity to eliminate facilities at the new plant by sharing operations with existing plants. Some ancillary facilities such as administrative areas, meeting rooms, and spare parts storage can be centrally located, but in general each plant must have its own full complement of treatment units and support facilities.

Factors That Influence the Site Selection Process. One of the first things the project manager should do when beginning the design of a new treatment plant is to make a list of factors that may influence the siting effort, noting the relative level of importance of each factor with regard to the particular site selection at hand. Table 23.1 is a list of factors that can influence the site selection process. Some of these are within the designer's control, but most are not. Understanding these factors before the site selection process begins in earnest is important because it allows the designer to anticipate potential roadblocks and plan countermeasures. For example, if an election is near, political considerations may have a big influence, and efforts to predict and deal with the political aspects would take on a high priority.

Selecting a Specific Site. Given the technical, regulatory, financial, and political factors impinging on the site selection, identifying the optimum site is a complex and difficult process. The key to a successful siting is having a sound, methodical plan for carrying out the selection process. Table 23.2 shows a generic plan for performing a site selection process. The specific details of each step should be adapted to suit the particular needs of each project.

Public Participation. An essential element of the site selection process is public participation. The extent of public participation depends on the type of project and location, but it is rare that the siting of a water treatment plant can be undertaken with no involvement by the general public. Furthermore, for municipal water treatment plants, the active involvement of public officials and general citizens should be viewed as a desirable component of proper government decision making.

In past years, the public was more trusting of the judgment of public officials and engineers with regard to the need for and design of capital improvements. Water utility staff

TABLE 23.1 Factors That Influence the Site Selection Process

Factor	Characteristics that complicate the process
Project size and complexity	Large facility Multiple components Technically complex
Level of impact (real or perceived)	Major environmental impacts Close to residential areas Human health concerns Unfavorable press
Political agenda	Election year Involvement by different political bodies More than one community involved
Client experience	No previous siting experience Unsuccessful previous experience
Project costs	High costs
Regulatory drivers	No specific regulatory siting requirements No consent decree Major permitting effort required
Role of public	High level of public interest
Geographic base	Large geographic area Multiple communities

Source: Adapted from Camp Dresser & McKee Inc., *Guidelines for Site Selection of "Locally Unwanted Land Uses."* Internal document by The Engineering Center, Cambridge, Mass., 1993.

and their consulting engineer typically assessed the need for a new water treatment plant, evaluated the most appropriate type of plant, and selected the location before going public. In recent years, the trend has been toward public distrust of government, environmental activism, and moves to limit government spending. Water utilities have not escaped this growing public skepticism. In the current political climate, there is a greater need for public participation in site selection and plant design.

Public participation involves two components: one-way public education and two-way public involvement in decision making. The goal of public education is to provide information on the water treatment plant and the planning process to a wide audience. Public education efforts may include direct mail, project websites, news releases in print media, the use of broadcast media, and public meetings.

Public participation in the planning process typically involves public-opinion surveys, public meetings and public hearings, and citizen advisory committees or task forces. Surveys and questionnaires are not suited for complex issues and generally have limited value for plant site selection. Public meetings and hearings provide an opportunity to present information to a large group of interested citizens and to obtain general input and opinions from them. Public meetings are often legally required. However, they have limited effectiveness in decision making because of the size and format of the meetings.

Advisory committees and task forces composed of well-informed citizens who represent the community are better able to address detailed decision making. Care must be taken to ensure that such committees are representative of the community. For relatively noncontroversial issues, public interest and participation are generally low, and it is likely that the voluntary participants in advisory committees have some special interest and do not repre-

TABLE 23.2 General Steps of Site Selection Process

Step	Description
1. Develop work plan	The work plan defines the project and goals, technical approach, public involvement, task outlines, schedule, budget, etc.
2. Identify project components	Identify the elements of the project that must be sited. Prepare preliminary design information on these components such as size and layout.
3. Identify players and their roles	Identify the various neighborhoods, political groups, politicians, government agencies, and special interest groups who have an interest in the project.
4. Public participation	Involve the public in the selection process to the extent appropriate for the type of project. Public participation involves both public education and public involvement.
5. Develop siting criteria	Select and define the siting criteria that impact the project.
6. Identify candidate sites	Using broad criteria, identify candidate sites that meet minimum project requirements.
7. Coarse screening	Coarse screening eliminates sites that do not meet a few critical constraints. This step short-lists the sites worthy of further study.
8. Detailed multi-attribute analysis	Apply the siting criteria to the remaining sites and evaluate the sites using qualitative or quantitative analysis methods.
9. Select preferred site	Based on the ranking of sites generated by step 8 and the input of the owner and public, select the preferred site.
10. Mitigation and compensation	Apply mitigation (measures to reduce the project impact to an acceptable level) and compensation (extraordinary measures beyond mitigation negotiated with the public).

sent the community as a whole. However, as long as the decisions of the advisory committee are made public and there is some mechanism for the general public to intervene if it is perceived that the advisory committee is not properly representing the community, little else can be done. Also, the involvement of citizens with special interests is appropriate from the viewpoint that the consent of these citizens is crucial to the success of the project, even though their opinions may not represent the entire population.

Becker (1993) conducted a survey of water utilities that have had experience with public advisory committees and offered the following advice about using such committees:

- Be open and honest with the committee.
- Clearly define goals and objectives at the beginning.
- Make it clear what decision-making power, if any, the committee has.
- Invite opponents to join the committee.
- Prepare an agenda.
- Seek professional assistance, if necessary, to coordinate the group.
- Do not be afraid of criticism.
- Be prepared to compromise.
- Establish a time frame.
- Expect the process to take longer than anticipated.

Siting Criteria. In the site selection process outlined in Table 23.2, step 5 is to develop a list of siting criteria on which the candidate sites will be evaluated. This list should be long enough to include all important factors, considering the area and communities involved, but not so long that it makes the evaluation unwieldy.

A comprehensive list of potential siting criteria is shown in Table 23.3. Chapter 30, Water Treatment Plant Security, should be reviewed for additional information. In general, a criterion will be relevant to a particular siting decision if (1) it is an important consideration to at least one of the stakeholders and (2) this characteristic varies between candidate sites. For example, in an urban area with no tourism to speak of, impact on scenic areas might not be a significant concern to public officials or to citizens and therefore would not be included as a siting criterion. On the other hand, in an area with many picturesque locales and a thriving tourist industry, impact on scenic areas is obviously a major concern. If, however, all the candidate plant sites would have an equal impact in terms of aesthetic intrusion, then impact on scenic areas is not a distinguishing criterion and would not be included as a siting criterion, although it would be relevant with regard to mitigation and compensation measures.

It is usually desirable to involve the public in choosing relevant siting criteria. If a citizen task force is formed, one of its responsibilities could be to identify the siting criteria to be used in the site selection process. In addition, city and state agency regulatory requirements may dictate that specific criteria be included in the siting analysis.

Size Requirements for a Plant Site

The size of the site required for construction of a new water treatment plant generally depends on the size of the structures, the allowance for future expansion, site topography, and the need for a buffer zone around the structures.

Determining Minimum Site Area. A key siting criterion is the minimum site area required to accommodate the necessary plant facilities. It is difficult to generalize on the site area because many factors influence the minimum area required. Table 23.4 lists the main factors that determine how much area is required and includes generalizations about plant characteristics that necessitate more land.

The site areas for 21 conventional water treatment plants varying in capacity from 8 to 160 mgd (30 to 600 ML per day) are shown in Figure 23.1. The area shown includes the land required for the basic process facilities and buildings, including rapid mix, flocculation, and sedimentation basins; filters; clearwell; high-service pump station; operations/administration building; access roads; and parking. Not included is the area required for facilities such as source water reservoirs, finished water reservoirs, lagoons, and sludge drying beds that are not always required. The areas shown do not apply to nonconventional water treatment plants such as membrane plants.

According to Kawamura (1991), the minimum required site area for the basic process facilities of a conventional water treatment plant is

$$A \geq Q^{0.6}$$

where A = area in acres and Q = ultimate plant capacity (mgd). This equation has been plotted in Figure 23.1, and it is apparent that Kawamura's formula is a good approximation for the required area for conventional water treatment plants. The minimum site area required for a new water treatment plant using conventional treatment can be approximated by taking the area from the figure for the basic process facilities. Additional area may have to be allowed for routing of pipelines, large pumping facilities, site grading and

TABLE 23.3 Checklist of Siting Criteria

Size requirements
  Area of the site
  Space for future expansion
  Shape of the land parcel

Environmental issues
  Land use
    Compatibility with surrounding land uses
    Compatibility with surrounding zoning
  Aesthetics and visual effects
    Impact on the existing character
    Visibility and prominence from key locations
    Impact of night lighting on neighbors
    Natural screening
  Noise level objections
    Construction noise
    Operational noise
    Traffic to and from plant
  Impact on air quality
    Plant odors
    Accidental releases of toxic gas
  Hazardous chemical considerations
    Impact of chemical spills and releases
    Transportation route to plant
  Traffic and transportation issues
    Construction traffic
    Operational traffic
    Accidents and safety
  Archaeological and historical issues
    On-site archaeological and paleontological resources
    On-site historical and architectural resources
    Potential incompatibility with off-site resources
  Ecological issues
    Effect on terrestrial habitat
    Effect on aquatic habitat
    Presence of rare or endangered species
    Presence of wetlands
  Natural resource issues
    Loss of trees
    Impact on scenic areas
    Loss of agriculture, pasture, fisheries
    Impact on mineral and energy resources

Technical issues
  Elevation of the site
    Pumping considerations
  Convenience of location
    Proximity to source water
    Proximity to service area
    Proximity to existing transmission mains
  Transportation and site access considerations
    Proximity to major highways
    Highway restrictions
    Proximity to railroads
  Utility services availability
    Proximity to power source
    Availability of second power source
    Proximity to existing sewer
    Capacity of existing services
  Disposal of residuals
    Options for waste disposal
  Topography and soil conditions
    Site topography
    Slope constraints
    Grading of roads
    Erosion protection
  Subsurface conditions
    Difficulty of excavation
    Low-bearing-capacity soil
    Excessive settlement potential
    Expansive soil conditions
    Depth to water table
  Site drainage considerations
    Susceptibility to flooding
    Effects on existing drainage pattern
    Effects on amount of site runoff
    Effects on surface water quality
    Effects on groundwater quality
    Need for stormwater detention
  Seismic activity potential
    Proximity of earthquake faults

Institutional issues
  Land availability
    Existing land ownership
    Land acquisition issues
    Easement acquisition for source water pipeline
    Easement acquisition for finished water pipelines
  Political issues
    Political agenda
    Public opposition
    Regulatory issues
    Special agency or public policy goals
    Permits and approvals
  Security issues

Financial considerations
  Costs that will be involved
    Land costs
    Construction costs
    Mitigation costs
    Compensation costs
    Operating costs
    Opportunity costs

TABLE 23.4 Factors That Influence Required Site Area

Factor	Characteristics that tend to require less land	Characteristics that tend to require more land
Plant capacity	Small capacity	Large capacity
Plant facilities		
Type of plant	Membrane treatment	Conventional treatment
Type of construction	Prefabricated package plants	Custom-designed, in-ground plants
Sedimentation process	High-rate tube or plate settlers Reactor clarifiers	Horizontal flow basins
Filtration process	High-rate gravity filters Pressure filtration	Slow sand filters
Waste washwater handling	Waste to sewer Settling basin with plate settlers	Recycle equalization basin Conventional settling basin Lagoons
Sludge dewatering	Mechanical dewatering	Gravity thickening Sludge drying beds Vacuum drying beds
Other on-site facilities		Low-service pump station Source water storage Presedimentation basins Finished water storage High-service pump station
Plant layout		
Structural configuration	Common-wall construction	Separate structures
Site topography	Flat site	Steep irregular site
Buffering and screening	No buffer No screening	Buffer required Screening required

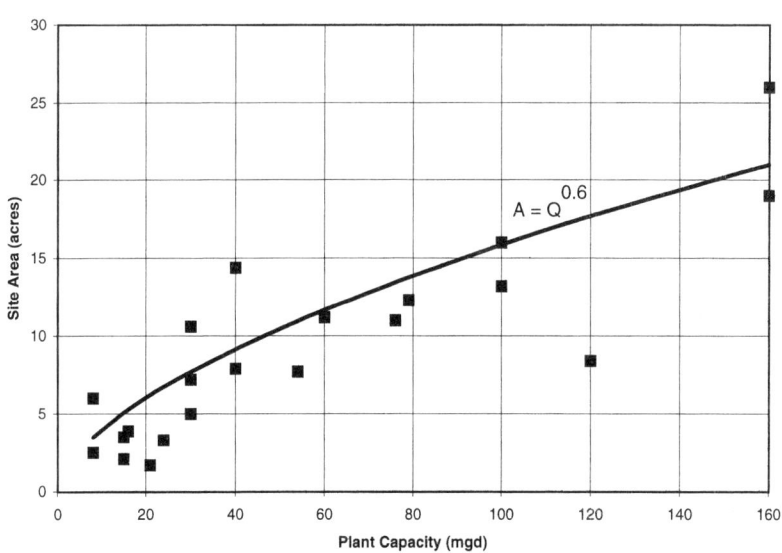

FIGURE 23.1 Minimum site area for conventional treatment plants.

stormwater handling, enhanced security concerns, etc. Additional space may also have to be added if any of the following facilities are required:

- Reservoirs or large clearwells
- Sludge disposal facilities, including backwash clarifiers, sludge thickening and/or dewatering facilities, lagoons, and associated roadways
- Provisions for future expansion
- Buffer zones
- Other special requirements

Also, it should be kept in mind that the previous site area guidelines are based on conventional treatment plants. Plants that use high-rate treatment technologies such as membranes or sand ballasted floc may fit on smaller plant areas. On the other hand, some of these high-rate processes require conventional treatment to precondition the water for the high-rate treatment, in which case the site requirements may be more than those for a conventional plant.

Providing for Future Expansion. Although new water treatment plant facilities are typically designed for a 50-year time frame, the plant site is likely to be used for water treatment purposes for centuries. Water treatment plants are rarely relocated because of the huge investments in land, treatment basins, reservoirs, source water mains, and finished water transmission mains. When water treatment plant equipment and structures reach the end of their useful life, they are typically rehabilitated, upgraded, or rebuilt at the same location. It is important, then, to provide ample room at the plant site for future expansion, not only to accommodate the projected increase in treatment capacity, but also, if possible, for expansion beyond the immediate planning period and for unanticipated plant improvements, such as additional treatment processes required to meet future regulations.

Need for a Buffer Area. In localities where aesthetic considerations are an important issue, it may be necessary to provide a buffer zone between the plant facilities and neighboring residences, recreational areas, and other sensitive surroundings. This buffer could substantially increase the required site area.

If visual intrusion is the primary concern, the buffer zone should provide a screen of undisturbed natural vegetation, new landscaping, or earth berms to hide the plant facilities. Such screening will also control noise intrusion into surrounding areas. Odors are generally not a problem at water treatment plants, but if sludge drying beds or other related processes are likely to generate objectionable odors, a buffer zone may be required to distance the source of the odors from neighboring residences.

Environmental Issues Impacting Site Selection

Environmental impact is discussed in detail in Chapter 24, but aspects that pertain to site selection are reviewed here.

Aesthetic Issues. In most instances, the issue that generates the strongest public concern over a new water treatment plant is the aesthetic impact on the neighborhood. Residents are often concerned that the facility will have the appearance of an industrial complex and that their property values will be reduced. Indeed, many existing water treatment plants do look like an industrial complex with exposed tanks, pumps, piping, maintenance vehicles, and other equipment visible to passersby. Attempts to locate a water treatment plant in a sensitive area are likely to engender strong public resistance, despite assurances about aesthetic compatibility, because of the public's general skepticism of public officials.

Aesthetic concerns are not limited to the appearance of the plant during the day. Water treatment plants must be well lighted at night, both for security reasons and to enable plant operations to continue 24 hours per day. This lighting can have a negative impact on neighbors, particularly near residential areas.

There are two ways to approach the issue of aesthetic impact. One is to treat water treatment plants as an undesirable land use and to seek neighbors who will be least impacted and least resistant to the intrusion. The other is to use mitigation measures such as innovative architectural treatments, landscaping, and screening to improve the aesthetics and make the plant more acceptable. The former approach restricts feasible plant sites to industrial and similarly zoned areas or to remote areas. The latter approach results in high costs for constructing and maintaining the architectural and landscaping treatments necessary to make the plant acceptable to a larger portion of the community.

Noise Level Concerns. It is difficult to evaluate the impact of noise in a community because of the different perceptions of how an objectionably noisy environment is defined. People respond differently to the same sound under different circumstances. Noise levels that are considered acceptable in a work environment may be deemed unacceptable in a residential setting. A noise that would not be bothersome during the day may be tolerated less at night. For a water treatment plant located on the edge of a small lake, even a low noise level may be offensive to residents on the opposite shoreline, because of the ability of sound to carry across water.

There are three chief origins of noise at a water treatment plant: construction activities, routine operation, and traffic noise. In general, residents find construction noise the least offensive of the three, because they know it is temporary. Nevertheless, construction noise may affect the surrounding population for a year or more and may be quite offensive. Because most construction noise is generated by engines such as earthmoving equipment, cranes, trucks, and compressors, the greatest potential abatement measure is to ensure that all engines have suitable exhaust mufflers, intake silencers, and engine enclosures. Appropriate construction hours should also be established to minimize noise generation during evenings and weekends.

The main sources of routine operational noise at water treatment plants are pumps, generators, compressors, blowers, and personnel paging systems. Suitable noise abatement techniques include

- Using low-noise emission equipment
- Enclosing the equipment in a sound-attenuating enclosure
- Erecting a sound wall, berm, or heavy landscaping between the source of noise and the neighbors

Traffic noise is discussed under a later section on traffic and transportation issues.

Air Quality Concerns. The Clean Air Act (CAA) sets limits on the emission of hazardous air pollutants. Volatile organic chemicals (VOCs), radionuclides, and other hazardous substances that may be emitted from water treatment plants are included under CAA requirements. The preliminary design should address the direct and indirect impact of the water treatment plant on air quality and recommend measures to mitigate any adverse effects.

Water treatment plants typically do not generate strong odors. However, citizens familiar with odors from wastewater treatment plants and solid waste facilities may have fears about potential odors. The most common source of objectionable odors at a water treatment plant is residuals handling.

Sludge drying beds and other sludge dewatering processes may generate odors. Sedimentation basins that are dewatered can also generate strong odors if the basins are not cleaned promptly and the sludge is allowed to go septic. Aeration towers can potentially cause odors in plants treating groundwater and some surface waters. Chemical storage and feed systems should not generate odors if they are operated properly.

Hazardous Chemicals Concerns. Water treatment plants use a number of hazardous chemicals. The transportation of these chemicals to the plant and the bulk storage of chemicals at the plant present a risk to the neighborhood, as well as to the plant staff. Accidental release of gaseous chemicals such as chlorine presents a significant health risk to adjacent communities. Safety concerns should influence both site selection and mitigation measures. (See Chapter 15.)

Traffic and Transportation Issues. The generation of truck traffic, both during construction and during normal operation, is an issue that can bring about a great deal of public opposition. Community concerns include public safety, disruption of neighborhood activities, and aesthetic matters. The degree to which the extra truck traffic impacts residential areas is an important siting criterion.

Mitigation measures that can be taken to reduce the impact of both construction traffic and routine delivery trucks must be carefully considered. One method is to designate truck routes that bypass residential streets and neighborhood commercial centers. Another method is to establish service schedules to avoid rush hours and the presence of schoolchildren on the streets.

Traffic noise from trucks traveling to, from, and around the plant, both during and after construction, can be minimized by the following measures:

- Identify the route to the plant with the least impact on the community.
- Locate the plant entrance where it will result in the least traffic noise.
- Avoid steep roads within the plant site.
- Erect sound walls, berms, or heavy landscaping adjacent to critical roads.

Archeological and Historical Site Considerations. Sites with structures or objects that are listed or eligible for listing on the National Register of Historical Places should be avoided, if at all possible. The National Historic Preservation Act (NHPA) requires that federal agencies consider the impact of any federal capital works on sites with historical, architectural, archeological, or cultural significance. Although the NHPA does not preclude development on historic sites if procedural requirements are met, public reaction and citizen lawsuits can delay or block projects from proceeding.

Ecological Issues. The importance of ecological issues has greatly increased in recent years. The presence of any rare or endangered species on or near a candidate site is a red flag. In fact, many would consider this to be an exclusionary criterion. That is, if a rare or endangered species is discovered at a site, that site is eliminated from further consideration.

Experts familiar with the Endangered Species Act (ESA) and other pertinent regulations should be consulted to assess candidate sites in terms of ecological issues and to assist with permitting once a final site has been selected.

Floodplain and Wetland Considerations. Floodplains and wetlands are also red flags for water treatment plant sites. Drainage and flooding are discussed later in this chapter. Wetlands are discussed in Chapter 24.

Technical Issues Impacting Site Selection

Technical issues that must be considered in site selection include the elevation of the site; the proximity to the water source, distribution system, and utility service; and transportation access.

Elevation and Pumping Considerations. The elevation of a plant site directly affects the pumping arrangement and pumping energy costs. Energy conservation is an important consideration because energy costs are a substantial portion of plant operating costs. Siting a water treatment plant at a low elevation results in a low head at the low-service pump station, but a high head at the high-service pump station. Conversely, a relatively high site results in a high head at the low-service pump station and a lower head at the high-service pumps.

For a plant located below a dam storage, it may be possible to gravity-flow into the plant and eliminate the low-service pump station entirely. Similarly, if the plant is located at the highest elevation in the region, it may be feasible to eliminate the high-service pump station and gravity-feed to the distribution system. The engineer must analyze the variables such as low-service pump size, diameter of source water pipeline, high-service pump size, diameter of finished water transmission mains, service area elevations, future service area limits, and power costs to determine the optimum configuration for the particular circumstances.

The most common situation is to locate the water treatment plant near the source water and to use low-head pumps to lift the source water to the plant and high-head pumps to deliver the finished water to the service area. One advantage of this approach is that less energy is expended in pumping water consumed by in-plant needs, such as for backwashing filters.

Proximity to the Water Source and Distribution System. The location of the water treatment plant in relation to the source water and the consumers has important implications for operating the source water pipeline, the plant itself, and the finished water transmission mains. Generally speaking, it is best to locate the water treatment plant near the source water, near the center of the distribution system, or somewhere directly between the two, to avoid duplicating the source water line and finished water lines. Whether it is better to locate the plant near the source water or near the distribution system can be determined only after a proper assessment of the advantages and disadvantages of the two approaches for the particular project circumstances.

If the plant is located close to the source water, there will be a short source water pipeline and a long finished water transmission main. Having a short source water pipeline simplifies pipeline maintenance operations. Source water pipelines typically require more maintenance than distribution system pipelines because of problems such as slime growth and deposits of silt and zebra mussel shells. Also, having a remote water source results in a remote pumping station, along with associated screening equipment, chemical systems, civil features, and structures. The site will require its own security considerations as well as power and other utilities.

Regular pigging or chlorination is often necessary to maintain design carrying capacity. Another advantage is that the long finished water transmission main can be used to provide service to customers along the pipeline route; or if there are no customers drawing from the transmission main, the residence time in the transmission main can be used for primary disinfection credit.

A disadvantage of having a long finished water transmission main is the difficulty of maintaining a disinfectant residual. Long pipeline residence times may result in loss of disinfectant residual or excessive formation of disinfectant by-products. Also, if the water source is a remote location, siting the treatment plant at the water source may present difficulties in obtaining utility services, attracting plant staff, and getting deliveries.

Transportation and Site Access Considerations. The vast majority of water treatment plants receive regular chemical deliveries by truck, so an important site selection criterion is the proximity to major highways. Sites that are accessible only through residential streets are undesirable. If streets to the site have any restrictions on the type of traffic or cargo type, that is another concern. The site must be accessible to trucks hauling hazardous chemicals, sludge, and whatever other cargo types are anticipated to and from the plant.

A plant must be accessible 365 days per year under all weather conditions. Therefore the plant access road should have no low-level river crossings or stretches subject to flooding. In areas subject to severe winter conditions, access roads to the site and inside the site should have limited grades and gentle curves to allow vehicular access during icy conditions and to facilitate snow removal.

Sites with available service from a railroad spur offer the option of receiving chemicals by rail. The lower unit cost of bulk deliveries by rail is an important consideration for plants with high chemical usage. Sites bordering lakes or rivers may also have the option of receiving chemical deliveries by barge.

Proximity to Utility Services. Water treatment plants must have a source of electric power, telephone communications, and sewerage service. It is also desirable to have a supply of natural gas. Dual electric power sources are normally required for reliability, or an on-site backup generator will be required.

Note that some or all of the cost of installing utility service lines from the nearest source to the plant site may be absorbed by the utility company. In this case, the cost of the utility connection incurred by the utility does not enter into the site selection analysis. If, however, the utility recovers the cost of the connection by way of a surcharge on the unit price of the service, ultimately the owner pays for the utility connection and the cost should be reflected in the site analysis.

In assessing the proximity of utility services, the required capacity of the service must be considered. Water treatment plants can have large power requirements, particularly plants with ozonation systems and large high-service pumps. Similarly, plants that discharge sludge to the sewer system can release large volumes of waste, and it may be difficult to locate a nearby sewer with enough available capacity to accept the additional flow.

Topography and Soil Condition Considerations

Important considerations in site selection are the topograpy of the land, soil conditions, site drainage, and potential for seismic activity.

Site Topography. Ideally, the site selected for a water treatment plant should have a uniform, moderate slope of 2% to 8%. This provides enough grade for surface drainage and to accommodate the hydraulic profile through the plant. A site steeper than 10% will probably require extensive cut and fill. Also, if process basins are aligned down the slope rather than across the slope, the vertical arrangement of the process basins will be steeper than required by the hydraulic profile. The extra drop between process units represents wasted pumping and energy consumption.

Sites with long slopes, steep grades, and highly erodible soils present special problems with erosion. The design of site drainage, landscaping, and other aspects of the civil design will have to be tailored to minimize erosion. During construction, special provisions will be necessary to prevent discharge of silt-laden soil to surrounding areas.

Sites with steep slopes also have the problem of potential landslides, particularly in view of the fact that leaking process basins or yard piping can saturate surrounding soils and increase the tendency for slope failures. Where soil conditions and topography are conducive to landslides, mitigation measures such as horizontal drainage, retaining walls, buttress fills, and tiebacks can be used. The cost of these measures should be factored into the site selection.

Subsurface Conditions. Subsurface conditions have a significant impact on the construction cost. Bedrock has an excellent bearing capacity, but if the bedrock is shallow, rock excavation will greatly inflate construction costs. On the other hand, soils that are easy to excavate often have a low bearing capacity, necessitating the use of expensive foundation types such as piles. A high-groundwater condition also increases construction costs because of the need for dewatering the soil during construction, waterproofing to keep basement areas dry, and additional design provisions to resist flotation of empty basins.

Local knowledge of subsurface conditions should be used to estimate construction cost differences resulting from differing subsurface conditions. These costs should be included in the evaluation of alternative sites. Where cost differences arising from differing subsurface conditions are a major factor in the site selection, it may be advisable to have borings drilled at short-listed sites to gain a better understanding of subsurface conditions and refine estimated costs.

Site Drainage and Flooding Considerations. Sites that receive stormwater runoff from a large upstream catchment require a more extensive site drainage system than those where only a small amount of stormwater drains onto the site. In jurisdictions where stormwater detention systems are required, evaluating differences in site drainage should include an assessment of the relative costs of providing the necessary detention basins and treatment system.

Many water treatment plants are located adjacent to a lake or river, because this provides close proximity to the source water. The disadvantage of siting a plant near a river is the risk of flooding. Protective measures such as earth levees can fail, leaving the community without a drinking water supply for an extended period. Sites well above the 100-year flood level offer increased reliability as well as avoid the cost of mitigation measures.

In addition to problems that flooding poses for site access and plant operation, any proposed construction in a floodplain triggers an array of permitting and regulatory requirements involving the U.S. Environmental Protection Agency (USEPA), the Federal Emergency Management Agency (FEMA), the U.S. Army Corps of Engineers, and other federal and state agencies. If construction in a floodplain cannot be avoided, adequate time should be included in the project schedule to deal with regulatory and permitting issues.

Potential for Seismic Activity. Damage from seismic activity results from ground shaking, ground displacement, slope failure, liquefaction of soil, lateral soil spreading, and forces due to liquid movement in the tanks (sloshing). Although all sites under consideration are likely to fall within the same seismic region, the topography and soil conditions at some sites may be more vulnerable to earthquake activity than at others. Sites at which seismically induced slope movement or liquefaction of subsurface soil materials is likely to threaten essential structures should be eliminated from consideration. Sites on or near an active fault should also be avoided, because the large movements that can occur at faults cannot be accommodated in the design of structures or piping.

Institutional Issues to Consider in Site Selection

In addition to the many technical issues that must be considered in selecting a site, there are many institutional reasons why a piece of land may or may not be available for a plant site.

Land Acquisition and Ownership. The number and type of owners of the proposed site can have a major impact on the difficulty of acquiring the land. In the ideal case, one individual owns the whole parcel and is amenable to sale of the property. Multiple ownership makes acquisition of the land more difficult and time-consuming, particularly when parts of the land are used for residential housing or similar uses.

The assessment of land acquisition issues should include not only the plant site itself, but also easements for the source water pipeline and finished water pipelines. In the ideal situation, pipelines can be installed within an existing utility easement, power company easement, railroad tract, or similar corridor. The least desirable situation is having to construct pipelines through a densely populated area.

Political Agenda. Local politics is always an important factor in the site selection process. It is important to understand the political climate and local agendas. There are two approaches to dealing with potential political problems. One is to involve the pertinent officials, groups, and agencies in the site selection process to gain their acceptance of the project goals and the selection procedures and to obtain their support. The other is to recognize the political reality of the situation and account for the political influences in the decision making. Both approaches should be used.

Permits and Approvals. If the owner of the water treatment plant is a city or county, it is generally preferable to choose a site within the owner's area of jurisdiction. This makes it easier to obtain permits and approvals (or to waive permitting requirements altogether) when dealing with one's own departments.

Security and Vulnerability. Drinking water system facilities have been identified as potential targets of criminal or terrorist groups, as these facilities are critical from both a public health and an economic standpoint. Therefore, planning for the implementation of security features is important. During the planning phase, performance of a preliminary vulnerability assessment of the new facility will help identify appropriate security components at the plant site. Required security features that may impact site selection and should be considered during this process are buffer zones around critical assets, intrusion detection systems, temporal or permanent barriers, and fences, all of which might require a relatively significant amount of area. In some cases, redundancy of certain critical assets deemed to represent single points of failure may be required, which could increase the site area needed for the facility. Additionally, the location of the site and access routes in relation to emergency response capabilities should be evaluated so that sites in areas of difficult accessibility are avoided.

Financial Considerations in Selecting a Site

At the site selection stage, the prime objective is to evaluate the relative merits of the candidate sites to identify the optimum one. Detailed cost estimates for budgeting purposes are not normally required at this stage because project details are too incomplete to estimate costs with any degree of certainty.

It is usually sufficient to use comparative costs of alternative sites, with common costs being intentionally omitted. For example, equipment costs can be neglected as long as the same process equipment will be used at all the sites. On the other hand, structural costs should be included if subsurface conditions or the structural configuration varies from site to site. The engineer should clearly denote in project documentation that the costs are for comparative purposes only and do not include all project components, so that readers do not develop an unrealistic impression of the total project cost.

Capital Costs and Operating Costs. Examples of capital costs that vary between sites include land cost, new source water pipeline to the site, new finished water pipelines, pump stations, excavation and structural foundations, civil site work, utility connections, and mitigation measures to address concerns of neighbors. Operating costs relevant to the siting decision include energy costs for pumping and residuals disposal costs.

Taxes to Other Municipalities. If an agency purchases land and constructs a treatment plant in another municipality, that agency may be subject to local property taxes, even if that agency is itself a municipality. Such property taxes can be a significant annual cost. The plant owner should investigate options for reducing or eliminating the tax burden, such as offering payment instead of taxes. Compensation could be nonmonetary, such as an agreement to provide drinking water to city buildings.

Opportunity Costs. An opportunity cost is a potential benefit that is lost or sacrificed when the course of action makes it necessary to give up a competing course of action. Opportunity costs are relevant to the siting decision and should be included, even though there is no outlay associated with the opportunity cost. Potential future benefits forgone by selecting the site should be considered where such benefits differ between sites.

An example of an opportunity cost is property tax. Construction of a water treatment plant on a site means that land is no longer available for other uses. Suppose a city is evaluating various sites within the city limits and one of the candidate sites is located in an area with high property taxes. If that site is selected, the city will lose the income it would have derived had the land been used for commercial purposes. Therefore the loss of future taxes (or more specifically, the loss of future taxes in relation to the other candidate sites) should be considered.

Sunk Costs. Sunk costs are those costs that have already been incurred, and they should have no relevance to the decision making. They cannot be avoided, regardless of which course of action is taken. For example, if candidate sites include a tract the city owns, the purchase price of that land would be irrelevant to the siting decision. Of course, the fact that the land is already owned would favor that site, because acquisition cost would be zero, and the other sites would all have an acquisition cost. However, the original purchase cost, be it small or large, should have no bearing on the decision.

It is common for decision makers to erroneously include sunk costs in the decision process. To use the example above, if the city had paid an exorbitant sum for the land, there would likely be pressure by some officials to use that particular tract for the plant site to avoid "wasting" the previous investment. However, if the owned tract has certain undesirable features, a proper analysis neglecting the sunk costs may well determine that the correct course of action is to build the plant elsewhere.

Other sunk costs that should not enter into siting decisions are the cost of existing intake structures and pump stations, pipelines, reservoirs, and all other previous capital investments in the water supply system.

Net Present Value Approach. Capital costs generally are incurred at the beginning of a project's life when the facility is initially constructed. Operating costs are ongoing, periodic costs. Both are relevant to the siting decision. The most common method of comparing two or more alternatives with differing capital and operating costs is the net present value approach.

Each of the capital and annual costs is converted to an equivalent present value. These present values are added to yield the net present value, which can then be compared with other alternatives. Net present value represents the amount of money that, if invested now for a definite period, would provide the sums necessary for constructing the project and for keeping it in operation for that period.

The *present value* of a single capital cost is

$$P = \frac{F}{(1 + i)^n}$$

where P = present value
F = future value (i.e., estimated capital cost) at year n
i = annual interest rate
n = number of years

For the initial construction cost, normally $n = 0$ and $P = F$.

The present value of a series of equal costs (i.e., an annuity) is

$$P = A \times \frac{(1 + i)^n - 1}{i(1 + i)^n}$$

where P = present value
A = periodic (annual) cost
i = annual interest rate
n = number of years

Texts on engineering economics or managerial accounting (Garrison, 1988) should be referenced for more information on net present worth calculations.

Inflation Effects on Net Present Value. A common error made in capital investment analyses is to make improper adjustments for inflation. It is common to adjust the interest rate for inflationary effects. Although current inflation is in the 3% range, over the last 20 years inflation has averaged almost 6%. Many engineers believe it is appropriate to account for inflation in long-term capital planning. However, the proper way to account for the effects of inflation is widely misunderstood, even in business texts (Garrison, 1988; Hanke, Carver, and Bugg, 1975).

Inflation can be handled in one of two ways. First, "real" dollars (also termed *uninflated* or *constant dollars*) and "real" (inflation-free) interest rates can be used in calculating net present worth. Cash flows are stated in uninflated (now) dollars. For example, if this year's labor cost was $100,000 and inflation is 5%, the projected real annual cash flow for future years would be $100,000, $100,000, $100,000, $100,000, etc.

SITE SELECTION AND PLANT ARRANGEMENT

Alternatively, *actual* (inflated) dollars and *combined* (inflated, market, or stated) interest rates can be used. Cash flows are stated in inflated (then) dollars. For the example above, the projected actual annual cash flow would be $100,000, $105,000, $110,250, $115,762, etc.

Much of the confusion occurs because the real interest rate is not the market interest rate published by financial institutions. Indeed, the real interest rate is a fictitious interest rate, frequently referred to in business texts but seldom clearly explained. Market interest rates (especially long-term treasuries such as municipal bonds) are combined interest rates because they have inflationary effects already built in. The real interest rate can be calculated from the formula

$$i = \frac{i_c - f}{1 + f} \cong i_c - f$$

where i = real (inflation-free) interest rate
 i_c = combined (market) interest rate
 f = inflation rate

Note that it makes no difference whether the effects of inflation are included in capital investment analyses, as long as there is consistent treatment of the cash flows and interest rate. Net present value and the analysis outcome will be the same whether or not inflation is accounted for (Garrison, 1988; Canada and White, 1980).

Although this may seem surprising at first, on reflection it makes sense. If cash flows and interest rate both include the effects of inflation, inflationary effects cancel out. The net present value is the same as if inflation had been ignored. If the reader needs more information, Canada and White (1980) provide a comprehensive and clear description of how to take inflation into account in capital project evaluations.

Although either of the two approaches described above provides the same result, it is generally considered preferable to use real dollars and real interest rates, because we intuitively think of future costs in terms of today's dollars. The USEPA requires that quantitative evaluations submitted in support of projects funded or controlled by the USEPA be performed without consideration of inflation. Remember, however, that to perform an inflation-free analysis, the market (combined) interest rate should not be used and the real interest rate must be computed by using the above equation.

Some texts address the situation in which the inflation movements of some component costs are projected to be different from the inflation of other component costs. For example, the escalation rate for energy costs is sometimes assumed to be different from the overall inflation rate. In this case, it is necessary to account for inflation effects. Goodman (1984) presents equations to handle differing inflation effects. However, given all the various assumptions and uncertainties inherent in long-term capital investment analyses, it is hard to justify the added complexity of such approaches.

Mitigation and Compensation.   The last step in the site selection process listed in Table 23.2 was to apply mitigation and compensation measures. These are approaches for increasing the regulatory and public acceptance of projects.

Mitigation refers to specific measures taken to reduce or eliminate undesirable impacts of the project. For example, if the proposed plant site is close to a scenic lake, the owner may agree to undertake extensive landscaping to conceal all structures and facilities from the view of residents. Mitigation measures may be required by regulatory agencies as a condition of permitting, or they may be imposed by public pressure.

Compensation is an incentive or tradeoff intended to make the project and its impacts more acceptable to the community. For example, a city may offer to build a neighborhood park. Building a park does not lessen the impact of building a new plant in the community, but it makes the construction of the plant more palatable to the community. Compensation measures are not normally imposed by regulatory agencies. They are typically used to gain political and public acceptance of a project once mitigation possibilities have been exhausted.

Compensation should be used with caution, because it may be perceived that the owner is trying to buy the decision. Compensation is typically reserved for negotiations with the host neighborhood once the site selection process is near the end. Estimated costs of any mitigation and compensation measures anticipated at each site should be factored into the site evaluation process.

Site Evaluation and Selection Methodology

After all technical, institutional, and financial issues regarding available sites have been compiled, it is then necessary to perform an evaluation and screening to determine the best site.

General Considerations. The general site selection process described in Table 23.2 includes a four-stage evaluation methodology to select the preferred site:

1. Identify the candidate sites.
2. Perform coarse screening.
3. Conduct detailed multiattribute analysis.
4. Select the preferred site.

Such a staged approach is appropriate for a siting problem involving a large area and many candidate sites. If only a small area and few potential sites are involved, an abbreviated evaluation process could be used. Identifying candidate sites is characterized as an evaluation step because some minimum site requirements must be used to assess suitability.

In identifying candidate sites, the goal is to search the entire region under consideration and locate all potential sites. Only a few broad siting criteria should be used because the emphasis at this stage is to identify all possibilities. Long-time residents with a detailed knowledge of the local area may be able to identify candidate sites, using an informal review of land use maps, aerial photographs, and other reference materials.

This approach may be adequate where the area under consideration is relatively small and where political concerns are slight. For more complex siting decisions, a formal sieving analysis as discussed below is more comprehensive and objective.

Sieving Techniques. A rigorous technique for identifying candidate sites is to perform a sieving analysis using multiple-overlay mapping (Noble, 1992), also called *land suitability analysis* (Lane, 1983). Using a few broad siting criteria, land suitability maps are prepared for the entire region under consideration. For each criterion, a map is prepared that indicates unsuitable areas. Individual maps are then overlaid to form a composite land suitability map. Areas on the composite map that have no shading meet all selected siting criteria.

In its simplest form, overlay mapping is performed using manual techniques. Transparent maps are superimposed on top of each other, and the resulting visual composite image is used. Alternatively, overlay mapping is an ideal application for computers using geographical information systems (GIS) software. It is increasingly common for cities to maintain their mappable data in GIS. Depending on the size of the GIS, data such as land use, zoning, ownership, wetlands, soil types, and sewer and water line locations may be included. If a computerized GIS has been developed, it enables the siting team to quickly and easily apply siting criteria over a large area.

The advantage of using an overlay technique, rather than selecting a few candidate sites based on local knowledge and judgment, is that all potential sites are identified. Also, the technique is objective and easy to understand, which are important considerations in obtaining support for controversial siting decisions. The overlay technique is also an effective tool for communicating and explaining the site selection analysis to public officials and the public. GIS, in particular, is an excellent presentation tool.

An example of appropriate siting criteria for identifying candidate sites for a 50 mgd (190 ML per day) plant might be as follows:

- The site must be contained within the municipal boundary.
- The site must be located no more than 5 mi (8 km) from the source water.
- The site should be located no more than 1 mi (1.6 km) from a major highway.
- The area of the site must be greater than 10 acres (4 ha).
- There must be no wetlands on the site.

These criteria are simple enough to be applied using a typical GIS.

Coarse Screening. Coarse screening reduces the number of candidate sites identified in the previous step to a manageable number. Because the amount of effort required to evaluate all candidate sites using all relevant siting criteria would be excessive in most cases, a few simple siting criteria are used for coarse screening. If the number of candidate sites identified in the previous step is relatively small, this intermediate step may not be required.

An example of coarse screening criteria for a 50 mgd (190 ML per day) plant might be these:

- The narrowest dimension must not be less than 500 ft (152 m).
- There is no evidence of rare or endangered species at the site.
- There is no evidence of historic resources at the site.
- The lowest point on site is not less than 5 ft (1.5 m) above the 100-year flood elevation.
- The site slope is not greater than 15%.

These criteria probably could not be evaluated by using GIS methods as in the previous step. Candidate sites would be reviewed individually, and those not meeting the additional criteria eliminated from further consideration. The coarse screening criteria should be ones that can be assessed relatively quickly. Screening criteria tend to be "fatal flaws"—either the site passes or it does not.

If only one or two sites pass the coarse screening, it may be necessary to "loosen up" the screening criteria so that fewer sites are culled and a greater number remain for further investigation. Ideally, coarse screening should leave three to five sites remaining for detailed evaluation. If fewer sites remain, either the screening criteria were excessively stringent or the region has few suitable sites and the choice of site will probably be obvious. If eight or more sites remain, additional screening criteria should be used.

Multiattribute Analysis Techniques. The sites that pass the coarse screening should be rated in terms of the remaining siting criteria. Ratings normally include both monetary and nonmonetary attributes. Evaluating monetary criteria requires that estimates be made of capital costs and operating costs, which is why screening is used to limit the number of sites reviewed in detail. Examples of the criteria examined at this stage include differences in pumping requirements, foundation costs, and proximity of existing services.

The most basic analysis approach is to tabulate the pros and cons of each of the sites still under consideration and to make a reasoned judgment as to the best one. This approach is commonly used because although engineers are skilled at determining the monetary and nonmonetary attributes of each site, they are often not knowledgeable in the use of multiattribute analysis techniques to interpret the resulting ratings. In instances where one site has the preponderance of favorable ratings and clearly dominates all others, a more sophisticated technique is not necessary. However, where two or more sites have a similar overall desirability, more advanced techniques are warranted.

The simplest quantitative technique is a weighted score system. Each site is scored as follows:

$$S_A = W_1 \times R_{A,1} + W_2 \times R_{A,2} + W_3 \times R_{A,3} + W_4 \times R_{A,4} + \cdots$$

where S_A = total score for site A
W_1 = weight of first siting criterion
$R_{A,1}$ = rating of first siting criterion for site A

Ratings are assigned as a number on a common scale, for example, 0 to 10, where 10 is best, 7.5 is good, 5 is fair, and so on.

This method can handle both qualitative and quantitative siting criteria. For qualitative criteria, for example, impact on scenic areas, judgment is used to arrive at a rating. For quantitative criteria, for example, foundation costs, costs are first estimated for each site, and then these costs are converted to a rating as follows:

$$R_{i,j} = \frac{P_{\max,j} - P_{i,j}}{P_{\max,j}} \times 10$$

where $R_{i,j}$ = rating for criterion j at site i (i.e., a rating of 0 to 10)
$P_{i,j}$ = estimated cost (net present value) of criteria j at site i
$P_{\max,j}$ = estimated cost (net present value) of criteria j for worst case

By using the above equation, the site with the highest cost will be assigned a rating of 0 (worst), any site with no cost for the criteria will get a rating of 10 (best), and sites with intermediate costs will be distributed linearly between 0 and 10.

Although this approach appears straightforward and easy to apply, a major difficulty arises in assigning criteria weights. The various stakeholders involved in the site selection have different perspectives and are unlikely to share the same opinion on which criteria are most important. Techniques for handling criteria weighting are discussed in the following section.

More sophisticated analysis approaches such as matrix methods, utility function techniques, and linear programming methods are available. Given the degree of uncertainty and judgment inherent in plant siting decisions, the simple weighted score system described previously is usually adequate. However, if political or other considerations dictate that a more rigorous analysis be used, advanced decision science texts such as those by Knowles (1989), Samson (1988), or Lang and Merino (1993) should be consulted.

Criteria Weighting. Criteria weights are determined on the basis of judgment, and different stakeholders often have different opinions on which criteria are most important.

The owner will likely weight cost and reliability criteria highly; environmental groups will rate environmental issues above all others; and local activist groups will focus on local impacts. Assuming a task force consisting of various representatives of stakeholder groups has been formed, one approach is for the members of the task force to discuss the various siting criteria and try to come to a consensus on weightings.

If that does not work, another approach is to average individual weightings of stakeholders. Each member is asked to score each criterion using a common scale, and then all scores are averaged to produce an aggregate weight for the criterion. Other, more advanced, techniques for developing criteria weights, such as forced choice methods, are outside the scope of this text but are described in advanced decision-making texts.

Choosing Appropriate Siting Criteria. The discussion so far has assumed that appropriate siting criteria were identified before the screening and evaluation steps. Siting criteria should be carefully selected to ensure proper evaluations.

As a first step, the siting criteria identified as impacting the siting decision should include all regulation-imposed requirements. Relevant federal, state, and local regulations must be adhered to unless it is expected that variances will be obtained. In the next step, criteria that do not discriminate among sites should not be included, even though they may be important to the facility design. For example, seismic activity has important implications for plant design, but if all sites have equal seismic activity, there is no value in including this criterion in the analysis.

The final siting criteria should be independent of one another, so that double counting does not occur. For instance, if there is a concern that operational noise may be offensive to nearby residences, the criterion could be "operational noise" or "compatibility with surrounding land uses," but not both. To assign a poor rating for both criteria would be two strikes for one problem. The listing of potential siting criteria presented in Table 23.3 includes both general and specific issues. The intent is for the designer to select the particular criteria that best suit the siting decision, but to avoid including multiple criteria describing the same problem.

Analysis of Risk and Uncertainty. An inherent characteristic of the planning process is uncertainty—uncertainty of future water demands, uncertainty of existing conditions, uncertainty of construction costs, uncertainty of future interest rates, and so on. These uncertainties have important implications for site selection. For example, if subsurface conditions are uncertain (which is normally the case), then whether a particular site is assumed to have favorable or unfavorable subsurface conditions may have a major bearing on the evaluation of that site.

At the simplest level, engineers deal with uncertainty by including contingencies for unexpected costs. Another approach is to perform sensitivity analyses to determine if the outcome of the evaluation is affected by whether optimistic or pessimistic assumptions are used. Advanced risk analysis or probabilistic analysis techniques are appropriate for complex evaluations under conditions of uncertainty, distinct from the deterministic analysis methods described earlier, which assume that uncertainty can be ignored.

With risk analysis, instead of making a single assumption about an outcome (for example, foundation costs will be $200,000), the planner describes potential outcomes in terms of probability distributions, for example, 75% probability that foundation costs will be greater than $150,000, a 50% probability that foundation costs will be greater than $200,000, and 25% probability that foundation costs will be greater than $250,000. The details of how to perform sensitivity analyses and risk analyses are beyond the scope of this text. Texts on management science and decision analysis should be consulted for descriptions of how to apply these techniques, if the nature of the project warrants such an approach.

Goodman (1984) offers the following methods of dealing with risk and uncertainty:

- Collect more detailed data to reduce uncertainty.
- Use evaluation criteria with better-known characteristics.
- Use more refined analytical techniques.
- Reduce the irreversible or irretrievable commitment of resources.
- Increase safety margins in design.
- Perform a sensitivity analysis.

Decision Software. Within the last decade, significant advancements have been achieved by software developers who now offer powerful products to support decision-making activities. These specialized software packages simplify the analysis and modeling of complex multicriteria decisions and typically provide advanced features such as sensitivity analyses. An Internet search will identify a variety of such software products. While decision software can greatly speed up the analysis of complex decisions, caution should be exercised since overreliance on software output and use by inexperienced operators can lead to flawed conclusions.

ARRANGEMENT OF WATER TREATMENT PLANT FACILITIES

Once the plant site has been selected, the next step is to determine the arrangement of the treatment process units on the site. By this stage, specific process elements and support facilities have been selected and sized. Careful consideration must be given to properly laying out the process units, buildings, and roads. A poor plant layout can have a negative impact on the plant treatment effectiveness, maintenance operations, construction and operating costs, ease of future plant expansion, and plant appearance.

The final plant layout evolves from the combined efforts of several interested groups. The key people typically involved in plant layout decisions are

- Consulting engineers, who make recommendations on mechanical, structural, civil, and electrical issues
- The architect, who advises on the layout and design of buildings
- The plant superintendent and operators, who indicate operational preferences
- The owner's staff, who provide general input
- Public citizens (through advisory committees and similar avenues), who provide input on aesthetic issues
- State regulatory agencies, which review the construction plans to ensure that the layout complies with all minimum requirements for public water supply systems
- Environmental regulators, who review plans with respect to impacts on sensitive environmental features (such as wetlands)

Factors Impacting Plant Arrangement

Each site has unique characteristics that impact the plant arrangement. In addition, preferences of the owner's staff have a strong influence on the layout. Table 23.5 summarizes

SITE SELECTION AND PLANT ARRANGEMENT

TABLE 23.5 Major Factors Affecting Plant Layout

Site topography
Hydraulic profile
Degree of redundancy desired
Degree of flexibility desired
Climatic conditions
Architectural considerations
Operations and maintenance procedures
Provisions for future expansion of the plant
Provisions for future additional treatment processes
Types of chemicals and chemical feed systems
Types of sludge handling processes

the major factors influencing plant layout and requiring careful consideration by the engineer. These factors are discussed in greater detail in the remainder of this chapter.

Basic Layout Types

Most nonpackage water treatment plants are custom-designed to suit specific requirements of the process train, site, and owner. Every layout is different, but there are three basic patterns into which all plant layouts tend to fall: the linear layout, campus layout, and compact layout (Figure 23.2). Each layout type has advantages and disadvantages, as discussed below.

Linear Layout. For a long time, the linear layout was the favored layout type. Water treatment plants were commonly arranged in essentially a linear fashion, unless site space limitations precluded it. The first edition of this text, published in 1969, advised that "with very few exceptions, a linear construction of a plant is preferable if the plant must be enlarged in the future."

One appeal of the linear layout is conceptual simplicity. The first impulse for laying out process basins on an open site is to arrange them in sequential order. The first process unit is located at the highest point on the site, and subsequent processes are arranged in order down the slope. This approach simplifies conveying water between process basins. Successive basins are connected by short pipes or channels.

Alternatively, basins can be constructed with a common end wall, in which gated ports are installed. The other advantage normally associated with the linear layout is ease of expansion. Plant capacity can be increased by simply constructing additional process trains adjacent to existing ones.

A major disadvantage of the linear layout is the unfavorable location of chemical feed points and mechanical equipment. These areas require a lot of operator attention, and the linear plant layout results in two centers of attention: one at either end of the plant. Water treatment chemicals are generally fed in the region near the rapid mix basins (e.g., predisinfectants, coagulants, flocculants, and pH control chemicals) and near the filters (e.g., filter aids, postdisinfectants, fluoride, and pH control chemicals). With a linear plant layout using conventional horizontal-flow sedimentation basins, rapid mix basins and filters

FIGURE 23.2 Basic plant layout types.

are widely separated because of the flocculation and sedimentation basins. Consequently, either the chemical storage and feed facilities must be located at one end of the plant with long runs of chemical feed piping to serve the other end, or separate chemical feed facilities have to be provided.

Major mechanical equipment associated with process units also tends to be clustered near rapid mix basins and filters. Screening equipment, rapid mixers, flocculator drives, and sedimentation basin sludge collector drives are typically all located in the vicinity of the rapid mix basins. The plant control room and any transfer pumps and high-service pumps are often located near the filters at the outlet end of the plant.

As for the segregated chemical facilities, this separation of mechanical equipment is undesirable from an operations viewpoint. It is not uncommon for large, linear water treatment plants to have two crews of operators—one based at the head of the plant to handle the chemicals and mechanical equipment in that area and the other group operating the filters and chemical feed systems at the other end of the plant. Plant operation is much more convenient if the major mechanical equipment and chemical feed points are clustered at one location.

Campus Layout. The campus layout is so named because plant facilities are spread out over the site and resemble a university campus. Buildings and basins tend to be separate from one another and are each surrounded by a clear area. The campus arrangement is appropriate where physical separation of structures is desirable, for example, in areas with uneven subsidence or in areas subject to earthquakes. The main advantages of the campus layout are

- Reduced structural complexity
- Better resistance to earthquake movement
- Better accommodation of differential settlement
- Allows vehicular access (e.g., trucks and cranes) to all sides of process basins
- Relatively simple to make modifications to connecting piping

The disadvantages are

- Access within the plant is less convenient because of longer travel distance.
- It is necessary to have buried utilities and chemical feed pipes between basins.
- Entire plant takes up more room.
- Long piping connecting the units results in higher head loss.
- Construction cost is higher because it does not allow common wall construction.

Compact Layout. The primary goal of the compact layout, also called the *cluster layout* (Kawamura, 1991) or *integrated layout* (Corbin et al., 1992), is to consolidate areas requiring frequent operator attention. The compact plant layout, shown schematically in Figure 23.2, accomplishes the clustering of chemical feed points and mechanical equipment by locating filters and clearwells near the head of the plant. As discussed previously, the large separation of chemical application points and mechanical equipment inherent in the linear plant layout occurs because of the relatively long length of the sedimentation basins.

By moving filters from the outlet end of the sedimentation basins to a location near the rapid mix basins and routing the settled water to the filters using a channel or pipe, facilities needing frequent operator attention are clustered together. Figure 23.3 shows an example of a compact plant layout. Filters are positioned adjacent to the rapid mix basins,

FIGURE 23.3 Aerial view of compact water treatment plant, Eagle Mountain WTP, Fort Worth, Texas.

locating the motor-operated equipment for the rapid mix basins, flocculation basins, sedimentation basins, and filters in close proximity to the operations building.

Another characteristic of the compact layout is that the major process units and plant facilities are normally integrated into a single, compact structure with common wall construction, rather than having separate structures as in the campus layout. In this compact arrangement, rapid mix basins, flocculation basins, sedimentation basins, filters, clearwell, chemical facilities, operations building, and administration building may all be grouped into a single complex.

Enclosed galleries and tunnels can be constructed between process units at the lower level. Such tunnels allow covered operator access between processes and provide a convenient location for routing piping and electric cables. Incorporating administration offices and operations facilities into a single building immediately adjacent to the filters means that maintenance facilities are central to the items of major equipment.

The plant shown in Figure 23.3 has the bulk chemical storage area adjacent to the operations building. Chemical feed equipment is located in the lower level of the operations building. This arrangement provides for short suction lines between the bulk tanks and the day tanks and metering pumps. Chemical discharge pipes to the various application points are also relatively short, because all chemical application points are clustered in the same general location as the mechanical equipment. Unlike the linear plant layout, there should be no need for running any chemical feed pipes the full length of the sedimentation basin.

The advantages and disadvantages of the compact layout are summarized in Table 23.6. One of the disadvantages of the compact layout is that the settled water must be transferred from the outlet end of the sedimentation basin back to the filters located near the inlet end of the sedimentation basin. One way to do this is to construct a settled wa-

TABLE 23.6 Advantages and Disadvantages of a Compact Plant Layout

Advantages	Disadvantages
Mechanical	
Major items of equipment are consolidated into one central area. Chemical feed lines are short.	Removal of equipment for off-site maintenance may be more difficult because of restricted access.
Operations	
Short walking distance to all equipment. Mechanical equipment is close to maintenance facilities. Access via covered galleries and tunnels. Required number of plant staff is reduced.	Careful equipment layout is required to ensure adequate space for maintenance activities.
Structural	
Reduced excavation volumes. Reduced concrete quantities because of common wall construction.	Larger slab penetrations because electrical conduits are concentrated in a smaller area. Extra provisions required for concrete shrinkage and expansion and for soil movement.
Electrical	
Reduced lengths of electrical and instrumentation conduits. Smaller cable size because of shorter cables. Cables can be installed in cost-efficient cable trays instead of in buried conduits. Cables routed indoors are more accessible.	Careful planning is critical because of the concentration of cables into a smaller area. Greater potential for conflicts between disciplines because other services (e.g., plumbing, HVAC) are also crowded into a smaller area.
Hydraulic	
Reduced hydraulic losses caused by less connecting piping between basins.	Long channel or pipe may be required to convey settled water to filters.
Architectural	
May be easier to develop an attractive architectural theme for a single mass than for several separate structures.	Mixed use of building (operations and administration) complicates building design.
General	
Less land area required. Fewer internal access roads and less paving. Less underground piping. Smaller area for security measures.	More difficult to comply with fire code requirements for access/egress because of the increased number of hallways, stairways, etc. Increased need for safety precautions arising from close proximity of bulk chemical storage to administration building.
Environmental	
Impacts are more concentrated. Generally more opportunities for buffering.	Impacts cover more area.

Source: Adapted from D. J. Corbin et al.: *Journal AWWA* 84(8):36, 1992.

ter return channel from the outlet end of the sedimentation basins back down to the filters. This channel can be made with an intermediate floor to provide an access tunnel running the full length of the sedimentation basin underneath the settled water channel.

An alternative to constructing a settled water return channel is to use a two-tray sedimentation basin. In this arrangement, the sedimentation basin has an intermediate suspended floor, with flow in one direction on the lower half and returning in the opposite direction in the upper half. The plan area required for the sedimentation basins can be reduced significantly.

The theoretical settling area of the two-tray basin is twice the plan area. Because the turbulence generated at the 180° turn reduces settling efficiency, a plan area somewhat larger than one-half that of a conventional sedimentation basin is required. Two-tray sedimentation basins are often economical only when available land is limited. However, engineers should not be too quick to dismiss them because the additional cost of the intermediate tray is offset by cost savings arising from the reduced volume of excavation, reduced length of walkways and handrails, reduced wall thickness and reinforcement (because walls are supported at midheight by the intermediate floor), and deletion of the settled water return channel.

Another frequent criticism of compact plant layouts is the difficulty of making plant expansions. The integrated nature of the compact layout means that modifications are more difficult and disruptive than for a linear or campus layout. However, if the initial design includes adequate provisions for future modifications and expansion, this is not a major limitation.

Using a compact layout typically results in a lower capital cost than using a campus layout because of savings in excavation and backfill, structural concrete, process piping, chemical piping, and electrical and instrumentation conduits. According to Ferguson (1991), savings for a 54 mgd (204 ML per day) plant with compact layout compared with a campus design is less than 5% of the total construction cost.

Operational Issues to Consider in Plant Arrangement

Some of the most important considerations in planning a plant arrangement are for the features that will make the plant reliable and convenient for operations and maintenance.

Need for Redundancy. The issue of plant redundancy is discussed in detail in Chapter 22. The impact of plant redundancy requirements on the plant layout is reviewed here.

Redundancy can be provided by adding standby units or by adding more treatment capacity. Typically, redundancy for equipment items such as pumps is obtained by installing standby units. Treatment basins are often oversized so that when one basin is out of service, the remaining basins can handle the flow. The size (i.e., loading rate) and number of process basins should reflect redundancy requirements. Also, the piping or channels connecting process basins should include sufficient valves and gates to allow some basins to be removed from service while the other basins are kept in operation.

Maintaining Flexibility of Operations. Operational flexibility is needed to handle a variety of operating conditions. Sufficient flexibility must be built into the plant arrangement to handle all likely variations in plant flow rate, source water quality, temperature conditions, and treatment goals. Flexibility is not the same as reliability, although flexibility can improve overall plant reliability. For example, having cross-linked interconnections between successive process basins improves both flexibility and reliability. Consider a water treatment plant with two process trains comprising rapid mix, flocculation, sedimentation, and filtration. If the trains have linked interconnections after each process, this offers the flexibility of taking one flocculation basin out of service at low flows to avoid having an excessively long period of flocculation, while not affecting rapid mixing or sedimentation. In addition, the interconnections enhance reliability, because if one flocculator fails, the inoperative basin can be bypassed, rather than taking the complete train out of service.

Providing for Convenience of Routine Operations. Plant layout can have a dramatic impact on the convenience of routine operations. A compact layout that centralizes facilities needing operator attention can significantly reduce the staffing requirements compared with a poorly arranged layout that separates operator work areas over a large area.

The daily duties of plant operators include

- Monitoring equipment (looking for signs of developing problems such as leakage, vibration, unusual noises, overheating)
- Monitoring treatment processes (observing floc formation, water levels)
- Controlling treatment processes (filter backwashing, sludge drawoff)
- Routine sampling and testing

To perform these routine tasks, the operator has to walk around the plant to the various locations of mechanical equipment, process basins, sample points, and so on. If the plant is arranged for convenient access, operators are more likely to regularly perform these routine tasks. If the equipment and plant processes are spread over a wide area, operators are discouraged from making frequent trips to perform the required monitoring and control tasks, particularly during inclement weather.

Providing Access for Equipment Maintenance. Convenient access to all equipment is necessary for proper maintenance. The layout of basins and buildings should be designed with careful consideration of all necessary maintenance activities. Refer to Chapter 25 for a detailed discussion of operations and maintenance considerations.

A special concern in compact plant layouts is how to handle media replacement in filters and GAC contactors. Where the campus layout typically has room for vacuum trucks to drive up to one side of the filters or contactors, the process units in compact layouts are often difficult to access. Appropriate provisions for removing and replacing media, such as installing a permanent suction header to each of the filters, should be provided if direct access is impossible.

Providing Access for Disabled Persons. The Americans with Disabilities Act (ADA) sets out provisions for access by disabled persons. These access requirements may apply both to the general public (visitors to the plant) and to plant employees. State and local officials knowledgeable in the ADA requirements should be consulted to determine what portions of the plant must comply with the ADA and what access provisions are required.

Meeting the ADA requirements can have a significant impact on the plant layout. To provide wheelchair access to the upper level of aboveground process basins, it may be necessary to install long ramps or provide elevators. A vertical rise of several feet requires an unobstructed total ramp length of well over 100 ft (30 m). If such ramps are not planned at an early stage of the design, it may be difficult to add them later. It is common practice to provide short containment walls around storage areas housing liquid chemical bulk tanks, with stair access over the containment wall. If wheelchair access into the chemical storage area is deemed necessary, the dimensions of the storage area may have to be significantly increased to accommodate a ramp.

Control of Noise. Aside from reducing the impact of noise on adjacent properties as discussed earlier, measures to control noise within the plant must be undertaken to avoid occupational health problems for plant employees. The Occupational Safety and Health Administration (OSHA) defines maximum noise levels for working areas. The best way to deal with noise is to isolate noisy equipment from routine work areas. Available measures to reduce the sound level for plant personnel include (U.S. Environmental Protection Agency, 1976) these:

- Locate noise-generating equipment as far away as possible from routine work areas.
- Erect sound walls, berms, and heavy landscaping in the area surrounding outside noise areas.
- Install sound-attenuating enclosures around pumps, blowers, compressors, generators, ejectors, and other types of equipment that operate at high speeds.
- Install properly designed silencers and mufflers on equipment to reduce noise.
- Install isolation fittings in piping and equipment foundations to reduce the transmission of sound through rigid piping and structures.
- Whenever possible, select equipment that meets OSHA requirements for 8 h of continuous sound exposure.
- Provide personal noise protection devices for all employees where other measures are not adequate.

Design of Internal Roads and Parking. Site layout should include permanent roadway access to all points in the plant where deliveries are made or where materials are loaded onto trucks for transport off-site, as well as adequate parking space for employees, visitors, and city vehicles.

Design criteria for internal roads generally provide for an adequate turning radius for semitrailer chemical delivery trucks. In addition, the local fire department should be consulted with regard to specific requirements for fire lanes and equipment turnarounds.

A pavement width of 25 ft (7.6 m) is appropriate for two-way traffic; a pavement width of 18 ft (5.5 m) is adequate for low-traffic roadways.

Parking should be adequate for the needs of both plant employees and anticipated visitors. Visitor parking should be near the main entrance, both for the convenience of visitors and to control visitor access. If tour groups are expected, allowance must be made for bus parking.

In areas with cold climates, roadway design should be appropriate for snow conditions. Alignment, gradient, and pavement design should be suitable for operation of snow removal equipment, and snow-dumping areas should be provided to eliminate windrows that catch drifting snow.

Providing for Chemical Delivery. The internal road layout must provide adequate space for chemical delivery and handling. Unloading bulk chemicals can take an hour or more, and the delivery trucks should not block access to other parts of the plant. Spill containment must also be provided at locations where liquid chemicals are unloaded, because the unloading operation presents a significant risk of spillage. Suitable materials handling facilities should be included at the chemical delivery area. A monorail or hoist is highly desirable for unloading and transporting drums, carboys, and other large containers. A truck dock and forklift can also be used to load and unload pallets and drums. If chlorine is delivered in ton cylinders, a monorail is essential to lift and move containers to and from the storage area.

Providing for Plant Visitors. Most water treatment plants are publicly owned and generally allow or encourage visits by schoolchildren and public groups. Conference rooms may also be used by outside groups for meetings. The layout of the internal roadways and parking areas, location of the administration building, security measures, and signage should all be designed to ensure that visitors do not inadvertently end up at hazardous locations in the plant.

The designer should keep in mind that visitors who are not familiar with plant layout can easily become confused about where to go if roads and parking areas are not clearly designated. The best approach is for the administration building to have an obvious main entrance and for it to be located close to the front gate.

Providing for Site Security. Site security is important for two reasons: (1) for the safety of trespassers and (2) to protect against vandalism and sabotage. Because children are particularly attracted to tanks and unfamiliar processes, keeping neighborhood children off the property is necessary both to safeguard against injuries to trespassers and to avoid the possibility of lawsuits brought about if a child should be injured.

Protection against vandalism, terrorism, and sabotage is required primarily to avoid loss of the drinking water supply resulting from equipment damage and secondarily to prevent deliberate contamination of the drinking water supply. Security measures that should be considered include perimeter fencing, motorized gates, security lighting, closed-circuit video cameras, motion detection systems, and telephone communications with the front gate. See Chapter 30 for additional information.

Technical Issues to Consider in Plant Arrangement

Technical issues provide some of the most important considerations in plant arrangement.

Impact of the Hydraulic Profile. One of the early steps in developing a site layout is to calculate the preliminary hydraulic profile through the proposed process basins. (See Chapter 16.) Once the hydraulic profile has been established, the top and bottom elevations of the main process basins can be determined. These basin elevations and how they relate to site topography are important considerations in locating the basins on the site.

The more treatment processes that are required, the greater the total head loss across the plant. If intermediate pumping is to be avoided, a site is needed with adequate slope for flow to occur by gravity. If necessary, upstream processes can be raised out of the ground to generate the necessary hydraulic head, but this increases construction cost, makes maintenance operations less convenient, and has a negative impact on the external aesthetics of the plant.

Optimum placement of structures relative to ground level depends on the site topography, soil conditions, and hydraulic profile; but as a general rule, it is desirable for plant treatment units to be approximately one-half underground and one-half aboveground. When process basins are located largely underground, the excavation costs, buoyancy problems, and depth of yard piping all increase. Where basins are mostly aboveground, access onto the basins becomes inconvenient, connecting piping becomes shallow, and basins become much more aesthetically obtrusive.

It should be emphasized that at plant sites with a steep grade, it is wasteful to provide large elevation drops between basins. In instances where process basins are arranged in a linear layout down a steep grade, there may be several feet or more of water elevation difference from one basin to the next.

The usual rationalization is that, because the site has plenty of grade, the large drop between basins does not matter. However, every foot that water falls through the plant represents another foot of pumping head, either at the low-service pump station or at the high-service pump station. Therefore the hydraulic profile through the plant should be kept as flat as possible, even if the site has a steep grade, to keep pumping energy costs as low as possible.

At least one plant process basin should include provision for overflow discharge to a watercourse. The overflow weir may be anywhere between the rapid mix basin and the filters, as long as the hydraulics are designed to ensure that the overflow weir will be the first place to overtop under both high-flow and low-flow conditions. The overflow weir and the grading of the ground away from the weir should be designed to accommodate the worst-case scenario of the plant at maximum source water inflow with the filters completely stopped.

In addition to providing an emergency overflow onto the ground, it is desirable to provide overflows between adjacent treatment units. For example, the dividing walls between adjacent sedimentation basins or filters can be designed with the top elevation lower than surrounding walkways. Thus if there is a valve operator malfunction or similar local failure that results in a rising water level in one unit, it can overflow into an adjacent basin rather than cause a spill onto the plant grounds.

Conveyance of Water between Process Units. A common problem in water treatment design is improper hydraulic balance between parallel treatment units. It has been reported that over 70% of all existing plants have a problem with unequal flow split between basins (Kawamura, 1991). Ideally, basin layout should provide balancing of flows between parallel treatment basins without the need for automatic control devices such as throttling valves or adjustable weirs. In addition, connecting piping or channels should include bypass facilities for operational flexibility.

Many old plants use pipes or channels to convey water from flocculation basins to sedimentation basins. When water velocity in these pipes and channels is relatively high [greater than 3 ft/s (90 cm/s)], the turbulence created at entrances, exits, and bends tends to break up the floc. When water velocity is maintained too low [less than 0.5 ft/s (15 cm/s)], large floc may have a tendency to settle out. Wherever possible, connecting pipes and channels should be eliminated, and flocculated water transferred directly through a common-wall baffle. If integrated basins are not feasible, open channels should be used to allow easy access for cleaning settled floc.

Single Application Points for Chemicals. It is highly desirable to provide a single chemical application point both before the pretreatment process units and before the filters. The application point should be a single pipe or channel through which all flow passes, as opposed to parallel conveyance systems. A single chemical stream can therefore be applied at that one location. This simplifies the chemical feed equipment considerably, because it is not necessary to use multiple metering pumps or flow-splitting devices in the chemical feed lines. Also, it eliminates problems of uneven chemical dosing due to uneven flow split of the chemical streams or of the plant flow.

One caveat is that the diffuser at the single chemical application point should be designed to disperse the chemicals uniformly into the flow. Alternatively, there should be sufficient turbulence and mixing in the piping or channel downstream of the application point to ensure a homogeneous blend before the flow divides between the downstream units. Otherwise, the downstream units receive water with varying chemical concentrations and treatment is degraded.

Separation of Filtered Water and Nonfiltered Water. State regulations for the design of water supply systems generally prohibit the separation of filtered and nonfiltered water by a single wall, because a crack or leak in the wall could result in contamination of filtered water. Nonfiltered water includes settled water channels feeding to the filters, all water in the filter box above the bottom of the media, and waste washwater channels and conduits.

Complying with this requirement sometimes becomes a challenge in compact plant layout design. It is not as severe a constraint in linear and campus layouts where common-wall construction is not used to the same degree. In instances where a compact layout configuration would result in common-wall construction between filtered and nonfiltered water, one option is to construct two walls (or two floors if vertical separation is the concern) with a narrow drainage space between them.

Finished Water Storage. Water treatment plants located close to the source water and remote from the service area should generally be designed to provide a minimum quantity of finished water storage at the plant. It is preferable to locate finished water storage (either ground storage or elevated reservoirs) near the center of the service area. A clearwell is normally all that is needed at the plant to serve as a wet well for the high-service pumps. It also often serves as the backwash water source.

Treatment plants located within the service area and with sufficiently high elevation are usually designed with finished water storage reservoirs at the plant site. In that case, additional land area is necessary to accommodate the reservoirs. There are a number of different reservoir types: ground storage or elevated, buried or exposed, square or round, concrete or steel.

Plants designed to use an elevated tank for supplying filter backwash water should have the tank located as close as possible to the filters, because large-diameter piping is typically required to accommodate the required backwash flow. The volume of the backwash storage tank should be sufficient to hold at least two full backwashes, and the height of the tank should be sufficient to produce the maximum design backwash rate at the minimum usable tank level.

Source Water Storage. Source water storage may be needed in instances where the source water delivery system is undependable, or if the daily source water delivery is constrained by the capacity of the delivery system or by contractual conditions. However, it is generally more desirable to use finished water storage to address such problems rather than to construct source water storage.

Source water storage serves only to provide a safeguard against interruptions or restrictions to source water delivery. The same storage volume provided as finished water storage serves to maintain finished water delivery not only when source water supply is restricted, but also when water demand is extreme or when water treatment capacity is constrained. Source water storage can be in open basins, so its storage is significantly cheaper than covered finished water storage. In cases where source water delivery concerns necessitate an inordinately large storage volume, the best option may be a combination of source water and finished water storage.

Providing for Lagoons. If the residuals handling system includes sludge lagoons or drying beds, they should be located where they cause the least aesthetic concern. Lagoons and drying beds are generally feasible only when the site has a large expanse of flatland. See Chapter 17 for details on sludge handling methods.

Location of Hazardous Chemicals. Careful consideration should be given to the storage location of hazardous chemicals. As a rule, chemical storage should be close to the chemical feed equipment to reduce the length of suction piping. However, hazardous chemicals should not be close to administrative offices. The resourceful use of screen walls and careful location of windows and air intakes can enable the designer to keep hazardous chemicals close to the operations area and yet isolate the danger from administrative personnel. The requirement to provide containment in the event of a massive chemical spill also influences the choice of where to locate hazardous chemicals.

If chemical storage tanks are visible from the street, serious consideration should be given to constructing a wall or berm to screen the tanks from view, particularly for the case of pressure vessels. Screens improve aesthetics and protect the tanks from drive-by vandalism.

Flooding. The importance of protecting treatment plant facilities from flooding is obvious. Aside from flooding caused by rainfall and runoff, the designer should address flooding caused by internal sources. Careful consideration must be given particularly to the possibility of flooding of the filter pipe gallery and basement level from a pipe rupture, or overtopping the filters or clearwell.

Although these events are infrequent, serious flooding of water treatment plants has occurred enough times in the past that the possibility should not be ignored. If site grading permits, the filter pipe gallery can be designed so that it will be free-draining. At flat sites where the gallery is substantially below ground level, the amount of mechanical and electrical equipment that could be flooded in the gallery should be kept to a minimum.

Climate Considerations. Cold climates require special measures that should be included in the plant design. Filters and other process basins must be covered, because cold weather can damage process equipment. It should also be kept in mind that cold weather damage of equipment can result from viscosity changes in lubricants, condensation freezing, and ice incrustation.

Snow mostly affects access around the plant site. The effects of drifting snow can be minimized by the use of trees, shrubs, and snow fences. Site design should also attempt to locate major roads parallel with the normal wind direction, and not locate roads directly upwind or downwind of large obstructions (WEF and ASCE, 1991). Structures should also be oriented, if possible, so that frequently used doors are on the sides of the building toward the upwind end.

Strong winds can induce circulating currents in open basins and interfere with settling. Long, shallow sedimentation basins are particularly prone to wind effects. The effects of strong wind can be counteracted by covering the basins with roofs, installing floating covers, constructing windbreaks beside the basins, or installing submerged baffles to impede circulating currents. Wind effects on long sedimentation basins are minimized by locating basins with their longitudinal axis aligned with the direction of the prevailing wind (AWWA/ASCE/CSSE, 1990; James M. Montgomery, Consulting Engineers, 1985).

Architectural Considerations in Plant Design

After the many operational and technical issues have determined the basic arrangement of a new plant, there should be architectural consideration of the design.

Building Orientation. Careful placement of water treatment plant buildings can greatly improve plant appearance. Buildings can be located to hide unattractive views of the facility. (See Chapter 18 for further information on architectural matters.) Site conditions and climate should also be considered in laying out the buildings on the site. Building design and orientation should be planned to minimize heat loss in winter and heat gain in summer and to use the potential for natural lighting.

Security. Building envelopes should be secure against intrusion, and sensitive rooms within buildings should receive additional security consideration. Balancing intrusion protection with convenient and efficient operation and maintenance access is often a difficult task. It may be beneficial to discuss these issues with security specialists and/or law enforcement personnel.

It is recommended that windows in control rooms and offices be arranged so that authorized personnel in the buildings can visually monitor activity in critical or vulnerable plant areas, if not in the entire treatment plant site.

Landscaping. Landscaping should be designed to be low-maintenance. An attempt should be made to preserve as much existing vegetation as possible and to use native plantings, because they are hardier than imports. Trees and shrubs should not be planted near open treatment basins to prevent leaves falling into the basins.

Areas to be mowed should have a slope of no more than 3 horizontal to 1 vertical. Steeper slopes should be planted with ground cover that does not need mowing and prevents erosion. Alternatively, steep slopes can be lined with a concrete slab, masonry pavers, or other suitable slope protection system.

Provision for Future Changes in Plant Design

Provisions should be allowed in a plant design for both expansion and process changes.

Capacity Expansion. It is important that the initial plant layout allow for future capacity expansion. Many old water treatment plants have been expanded three, four, or more times. In some instances, plants have been enlarged far beyond the expectations of the original designer, and the resulting hodgepodge of process units results in poor hydraulic conditions, scattered chemical application points, and considerable operator inconvenience.

The long-range master plan for the plant site should allow space for future expansion up to and beyond the projected long-term treatment needs. Leaving space for expansion beyond the projected needs is prudent because *long-term* demand projections typically look ahead 20 to 50 years, but most water treatment plants will be in use 100 years after original construction.

Provisions for future expansion should include not only adequate space for future basins, but also provisions to aid in the construction of future expansions. For example, blind flanges on the end of piping runs, knockout walls, reinforcing bar extensions, and similar provisions can be easily provided during original construction and will be of great assistance during future expansion. Channels to be extended should be constructed with stop plank grooves in the channel walls, so that the plant can be operated during construction of the expansion.

Future Process Additions. Aside from future process basins required to expand plant capacity, plant layout (including hydraulic design) should allow for future additional treatment processes required by new regulations. As water quality regulations become more stringent, many plants are finding it necessary to add treatment processes such as GAC contactors and disinfectant contact basins.

In addition to regulatory changes, the emergence of new or improved treatment processes sometimes causes plants to install new processes to support or replace existing ones. It is much easier and less costly to add new processes when the original designer makes allowances for such a possibility. This is especially true for compact plant layouts.

Support Facilities to Consider in Plant Design

Support facilities are those facilities that are necessary to keep the water treatment processes in operation or to accommodate the staff who operate and maintain the plant, but that are not directly involved in the treatment of water. Some of the principal facilities of this type are the maintenance workshop and spare parts storage, administrative offices and meeting rooms, lunchroom and locker rooms, and laboratory. Issues concerning the location of such support facilities within the plant site are reviewed below. See Chapter 18 for detailed information on the architectural design of plant support facilities.

Maintenance Facilities. Maintenance facilities should be located near the most equipment-intensive area. It is desirable to provide a monorail or bridge crane in the maintenance area for lifting and moving heavy items. The monorail or bridge crane should travel out to a loading dock to facilitate pickup or delivery of parts and equipment. It is also desirable for the monorail to extend into the pump rooms, pipe galleries, and other such areas containing heavy equipment to simplify movement of equipment into the maintenance area.

Laboratory Facilities. The water treatment plant laboratory is used for process control, testing for regulatory agencies, collection of historical data, and cost control. In many instances, the laboratory is also used to test distribution system samples and to perform other chemical, bacteriological, and physical testing for the utility.

The laboratory should be at a convenient, central location because operating personnel will be using the laboratory many times each day. Laboratory equipment such as analytical balances is sensitive to vibration, and the laboratory should not be located near large pumps or other heavy operating equipment. Also, heating and air conditioning systems for the laboratory should be independent of the main plant systems to prevent process area gases and fumes from being circulated into the laboratory.

The size and layout of plant laboratories vary considerably. Depending on the type of analyses to be performed on site, the laboratory could be quite small to relatively large. An alternative approach used in some large plants is to have two laboratories on the site: a small one located central to plant operations for conducting routine plant process control tests, and a full-service laboratory for complex analytical work and outside testing services located elsewhere on the site.

BIBLIOGRAPHY

AWWA/ASCE/CSSE. *Water Quality and Treatment,* 4th ed. New York: McGraw-Hill, 1990.

Becker, Joan. "Survey Says Water Utility Advisory Councils Are a Success." *Journal AWWA* 85(11):58, 1993.

Camp Dresser & McKee Inc. *Guidelines for Site Selection of "Locally Unwanted Land Uses."* Internal document by The Engineering Center, Cambridge, Mass., 1993.

Canada, J. R., and J. A. White. *Capital Investment Decision Analysis for Management and Engineering.* Englewood Cliffs, N.J.: Prentice-Hall, 1980.

Corbin, D. J., R. D. Monk, C. J. Hoffman, and S. F. Crumb, Jr. "Compact Treatment Plant Layout." *Journal AWWA* 84(8):36, 1992.

Ferguson, K. G. "Compact Water Treatment Plant Design." In *Proceedings of AWWA 1991 Annual Conference.* Denver, Colo.: American Water Works Association, 1991.

Garrison, R. H. *Managerial Accounting—Concepts for Planning, Control, Decision Making.* Plano, Tex.: Business Publications, 1988.

Goodman, A. S. *Principles of Water Resources Planning.* Englewood Cliffs, N.J.: Prentice-Hall, 1984.

Great Lakes–Upper Mississippi River Board of State Public Health and Environmental Managers. *Recommended Standards for Water Works.* Albany, N.Y.: Health Education Services, 1992.

Hanke, S. H., P. H. Carver, and P. Bugg. "Project Evaluation during Inflation." *Water Resources Research* 11(4):511, 1975.

Hudson, H. E., Jr. *Water Clarification Processes: Practical Design and Evaluation.* New York: Van Nostrand Reinhold, 1981.

James M. Montgomery, Consulting Engineers. *Water Treatment: Principles and Design.* New York: John Wiley & Sons, 1985.

Kawamura, Susumu. *Integrated Design of Water Treatment Facilities.* New York: John Wiley & Sons, 1991.

Knowles, T. W. *Management Science—Building and Using Models.* Homewood, Ill.: Irwin, 1989.

Lane, W. N. "Community and Regional Planning." In *Standard Handbook for Civil Engineers,* 3d ed. Edited by F. S. Merritt. New York: McGraw-Hill, 1983.

Lang, H. J., and D. Merino. *The Selection Process for Capital Projects.* New York: John Wiley & Sons, 1993.

Noble, George. *Siting Landfills and Other LULUs.* Lancaster, Pa.: Technomic Publishing, 1992.

Samson, D. *Managerial Decision Analysis.* Homewood, Ill.: Irwin, 1988.

U.S. Environmental Protection Agency. *Direct Environmental Factors at Municipal Wastewater Treatment Works.* EPA 430/9-76-003. Washington, D.C.: Government Printing Office, 1976.

Water Environment Federation (WEF) and American Society of Civil Engineers (ASCE). *Design of Municipal Wastewater Treatment Plants.* WEF Manual of Practice No. 8, ASCE Manual and Report on Engineering Practice No. 76. Washington, D.C.: Water Environment Federation, 1991.

CHAPTER 24
ENVIRONMENTAL IMPACT AND PROJECT PERMITTING

Philip C. Kennedy and Jane W. Wheeler
CDM
Cambridge, Massachusetts

Environmental impacts must be considered at several junctures in the planning and design of a water treatment plant project. First, as described in Chapter 23, environmental issues play an important role in site selection. Once the site is selected and design commences, plant and site design must incorporate environmental controls that will minimize the impacts to the human and natural environment during plant construction and operation.

This chapter presents an overview of the environmental issues associated with plant construction and operation, describes general permitting requirements, and discusses the integration of environmental review into permitting, design, construction, and operation. This framework will enable the design engineer to be cognizant of the basic environmental review steps and to incorporate them in the planning and design phases of the project, before construction begins. Table 24.1 is a list of possible environmental review steps in a typical project.

ENVIRONMENTAL ISSUES ASSOCIATED WITH PLANT CONSTRUCTION

The environmental issues associated with construction of water treatment plants are similar to those of any medium-size or large construction project. The site features, neighboring land uses and ecological conditions, specific facility components and design, and level of community interest are all factors in determining the nature and extent of environmental impact review. The issues can range from minor considerations of land disturbance to more significant environmental impacts caused by truck traffic or major alterations to sensitive ecological areas.

The environmental impacts of constructing water treatment plants are typically of greater potential magnitude than the environmental impacts of operations. Construction impacts are characterized by higher amounts of trucks, noise, and dust than there will be later, when the plant is operating. However, construction impacts are of shorter duration.

TABLE 24.1 Typical Environmental Review Steps

Project phase	Environmental review
Planning	
Site selection	Environmental siting criteria consideration of resource protection areas
Public hearings	Preliminary impact assessments
Initial approvals	State or municipal level environmental impact regulations
Design	
Technical design	Development of design criteria
Permits and approvals	Impact assessments and development of mitigation measures
Bidding	
Bid documents	Incorporation of permit conditions and performance standards into bid documents
Construction	
Facility construction	Environmental monitoring and reporting
Operations	
Facility operations	Environmental monitoring and reporting
	Maintaining permit compliance

Operational impacts normally involve long-term issues related to design, daily operational performance, and maintenance of the facility—attributes that generally do not have the potential to significantly impact surrounding areas if a plant is well designed, well constructed, and properly maintained and operated. A general comparison of construction versus operational environmental issues is given in Table 24.2.

Before construction, the project designer must analyze the potential environmental impacts associated with a proposed change in land use of the site and the general project compatibility with surrounding land uses. This is particularly important when construction is proposed on a previously undisturbed site or within a nonindustrial area.

TABLE 24.2 Comparison of Construction and Operational Environmental Issues

Attribute	Construction	Operations
Study area	Site and adjacent areas	Site and adjacent areas
	Truck routes	Area of groundwater influence
	Noise	Area of surface water influence
	Site compatibility	Delivery truck routes
Frequency	Daytime hours	24 h
Duration	Temporary	Long term
	6 months to 2 years	20 to 30 years
Magnitude	Ongoing disturbances for duration	Occasional disturbances
Mitigation	Timing	Performance standards
	Scheduling	Design features
	Physical barriers	
	Equipment location	
Special concerns	Traffic safety	Chemical releases
	Fuel spills	Solids disposal

Determining the Major Environmental Issues

The process of identifying and evaluating issues of concern consists of four general steps:

1. *Evaluate site conditions and features.* This includes areas of particular environmental sensitivity such as wetlands or floodplains. The evaluation should include review of state and local regulations and ordinances that protect natural resources and adjacent land uses, because they provide excellent guides for identifying the types of site features that need to be considered before construction begins. These features should then be located on a site plan.

2. *Identify major project components.* This might include some off-site activities. Typical items are access and egress, staging areas, utility connections, pumping stations, distribution mains, source water intakes, groundwater well installations, dredging, dewatering areas, and solids disposal areas.

3. *Determine the plant arrangement.* The arrangement of the major plant components on the site must be initially determined.

4. *Develop a general construction plan.* The plan should include the basic construction phases and the approximate timing of the various construction activities, a description of the sequencing of those activities to determine which are concurrent activities, and the magnitude and duration of construction activities.

From the above steps, the environmental issues associated with construction can be readily identified. These are the issues that will most likely be the subject of permit application filings, environmental impact evaluations, and public concern about the proposed facilities.

The type of facility, size and location of the facility, and the site conditions are all important variables affecting the real environmental issues associated with facility construction. Table 24.3 shows some of the potential environmental impacts by construction phase and facility component.

There are also numerous potential environmental impacts associated with the selected site. For example, if the site has wetland areas, is within a flood zone, or has particular historic or archeological significance, those issues could be extremely important.

Specific Environmental Issues during Construction

The principal environmental issues that must be considered during construction are site preparation, facility construction, and landscaping.

Site Preparation. Although site preparation is not always distinct from facility construction, it generally includes access road and staging area preparation, site clearing and grubbing, preliminary excavation, and placement of fill. Management of stormwater and dewatering discharges is important to ensure that erosion and resulting siltation of water bodies are minimized both during site preparation and during construction in general.

Environmental impacts associated with staging areas are generally temporary, although careful site selection is required to avoid sensitive areas. Access roads may be temporary or long-term, depending on whether the road is for use during construction only or for facility access during operations. In either case, sensitive siting is important. In determining locations for access roads and staging areas, consideration should be given to disturbing as few natural features as possible. Consideration must also be given to the proximity of sensitive residential activities such as homes, schools, and hospitals to

TABLE 24.3 Potential Environmental Impacts by Construction Phase or Aspect

Construction phase	Land use	Traffic	Air quality	Noise	Ecology	Aesthetics	Vibration	Public safety	Water quality	Wetlands
Site preparation										
Access roads and staging areas	X	X	X	X	X					X
Site clearing and grubbing	X	X	X	X	X	X				X
Excavation			X	X			X			X
Placement of fill		X	X	X						
Facility construction										
Materials delivery	X	X	X	X						
Dewatering systems								X	X	X
Placing concrete		X		X						
Intake or groundwater wells					X				X	X
Landscaping					X	X				

24.4

ensure that traffic conflicts and disruption from noise and dust are minimized during construction.

A certain amount of clearing and grubbing is needed on most undeveloped land. Site design should take into account any unique natural features that can be preserved, such as especially large or rare trees. There may also be vegetation or features that can be preserved to act as a natural aesthetic buffer between the plant and public view after construction. Potential impacts resulting from clearing and grubbing include construction traffic, equipment noise and emissions, dust, aesthetics, and ecological disruption.

The extent of excavation required depends on site-specific conditions. Generally, the more excavation required, the greater the anticipated impact in terms of construction traffic, noise, dust generation, and vibration, particularly if rock drilling or blasting is required. The duration of the site preparation construction is also an important consideration and is often a tradeoff with the magnitude of impact. A compressed construction schedule on a particular site, for example, may result in greater daily noise and traffic impacts than those of a longer schedule.

Facility Construction. Site preparation and facility construction both involve site disruption and on-site activities that may pose conflicts with neighboring land uses. The greatest environmental impacts associated with on-site facility construction are generally air quality (dust) and noise caused by construction traffic and aesthetics. The installation of water mains to connect with the source water and to connections within the water distribution system will have additional off-site impacts.

If a new intake is to be installed, it will undoubtedly impact a watercourse and will require additional piping. The method of intake installation and its location are critical factors in determining the extent and magnitude of environmental impacts on both aquatic species and downstream users. Environmental regulators will generally restrict construction activities in a river to times of low flow and will require that the river be restored to its original bottom contours to minimize ecological impacts.

Landscaping. Landscaping is often a mitigation measure, although it can have impacts of its own, depending on the amount required and its compatibility with the area aesthetics. Planting ornamental trees in a natural setting, for example, may have a negative aesthetic impact rather than a positive impact. However, in general, landscaping provides an opportunity to restore and even enhance the natural attributes of the site, as well as provide visual screening from neighbors.

ENVIRONMENTAL ISSUES ASSOCIATED WITH PLANT OPERATIONS

There are numerous environmental issues associated with plant operations, primarily related to the various waste streams and by-products of water treatment plant processes, but also related to the impacts on the supply source as a result of water withdrawal.

The type of treatment process that will be used will obviously affect the type and extent of the plant's operational environmental impacts. In particular, the chemicals that will be used in the treatment process and the waste stream characteristics of that process are significant in determining the environmental impact. For example, the waste streams associated with filtration processes may contain higher levels of metals than those of some other processes, which could be of concern if the resulting residuals are applied to agricultural land.

Water Withdrawals

Whether water withdrawal will be from a groundwater supply or a surface water supply may have significant environmental ramifications. The cone of influence surrounding a well may change the hydrology and, as a result, the ecology of the surrounding area. Surface water withdrawals have the potential to affect downstream ecology and uses and to entrap fish.

Surface Water Withdrawals. Some of the environmental impacts that may be caused by new or increased withdrawals from a surface water supply include

- Change in water temperature and quality characteristics as a result of river or reservoir depth changes
- Change in pollutant dilution as a result of flow changes in a river
- Ecological effects related to depth changes and reduction in flow caused by river withdrawal
- For river withdrawals in particular, entrainment of fish eggs and larvae as they are drawn into the intake, and impingement of animals and debris on the intake structure
- Potential conflicts with recreational or other human uses of the water body

Through proper design and operation, a balance can be achieved between the need to withdraw water and the goal of minimizing impacts on the water body. For example, selecting a proper type of screen and minimizing the flow velocity around an intake will help prevent entrapment of fish. Placing easy-to-read signs upstream of a river intake and buoys to mark a channel passage can help mitigate potential conflicts with boaters.

Groundwater Withdrawals. Well withdrawals can also affect surrounding ecology and human activities. Examples include

- Movement of the interface between saltwater and freshwater in the aquifer in coastal or near-coastal areas
- Reduction of groundwater available for nearby domestic and agricultural well systems
- A reduction in the aesthetics and recreational value of surface water bodies in the area
- Increased activity of sinkholes
- Changes in the composition and health of terrestrial and aquatic ecosystems

The type and extent of the drawdown impacts that will occur around a well vary widely, depending on many site-specific factors. Conducting pump tests and modeling the likely drawdown of groundwater levels at proposed withdrawal rates can help predict the magnitude of the change and determine what the optimal pumping rate will be, while minimizing impacts to the surrounding area.

Chemical Delivery, Handling, and Storage

Some of the chemicals used in water treatment are commonly found in household products; for instance, sodium hypochlorite is bleach, and aqueous ammonium hydroxide is household ammonia. These chemicals, when properly handled and stored, are safe. However, because the chemicals are often concentrated and used in bulk at water treatment plants, delivery, handling, and storage are particularly critical and are sometimes the subject of public concern because of potential environmental and safety risks. Use of chlorine gas, for example, often generates public concern because of the inhalation dangers if there should be a leak.

A number of criteria for addressing public health concerns are described in *Guiding Principles for Chemical Accident Prevention, Preparedness, and Response,* published by the Organization for Economic Cooperation and Development (OECD, 1992). These guiding principles establish "general guidance for the safe planning, construction, management, operation and review of safety performance of hazardous installations in order to prevent accidents involving hazardous substances." Although the principles define hazardous installations as fixed plants or sites, many are also applicable to transporting hazardous substances. These guiding principles are used as the basis for much of the following discussion.

Proximity of Plant to Populous Areas. The guiding principles suggest separating people from hazardous substances, which can be achieved in a number of ways. One obvious method is to site the facility away from high population densities and particularly from very sensitive persons, such as children and the elderly. However, other siting constraints may preclude this. Furthermore, this method may not adequately address transportation of chemicals and worker safety. Therefore an emergency response plan should, as required, address the areas likely to require evacuation or protection in the event of a chemical spill.

Transportation. Safe delivery of chemicals to the plant is as critical as safe chemical storage and use at the plant. Delivery modes (e.g., rail or truck) and routes are important factors to consider in transporting hazardous materials. To minimize public health risks, the ideal truck transportation route for hazardous substances is a large, little-traveled roadway in good condition or a rail line with few roadway crossings that passes through undeveloped land. If such a route is not available, a route should be selected that will be as short as possible and will avoid accident-prone areas and high-density populations.

People involved with delivery and receipt of chemicals can also assist in spill prevention by being present during off-loading of materials, checking that receiving containers can accommodate the quantity delivered, and verifying that all disconnects are made before the truck pulls away.

Engineering and Design Controls. Once chemicals have been delivered to the plant, a number of safety measures are recommended to help prevent and contain spills. Preventive measures begin with designing the facility to reduce the possibility of releases. Good engineering and design controls can range from such straightforward methods as locating all chemicals either indoors or under cover and storing reactive chemicals separately, to more advanced pollution control technologies, such as air scrubbers.

Recommended engineering controls should include installation of sprinkler systems, construction of storage spaces with fire-resistant concrete, and providing separate containment areas for each type/category of chemical. Providing enclosed areas with proper ventilation for unloading chemicals is another design control that should be incorporated in the facility.

Oil Storage

There are several examples of oil use and storage at a water treatment plant. For example, to ensure that a water treatment plant can fully function during a power outage, a backup power system (e.g., emergency/backup generators) is required. In many cases backup generators are diesel-fuel-operated; therefore, a plant would need a bulk supply of diesel fuel (e.g., aboveground storage tanks, underground storage tanks). Also, smaller quantities of oil such as lubricants and engine oil to support equipment operation require the storage of virgin oil and used oil in containers (e.g., 55-gal drums). In addition, other water treatment plant activities requiring the storage of oil include the operation of oil-filled electrical equipment (e.g., transformers) and oil-water separators. These activities entail oil handling and storage that must be managed appropriately to prevent a release of oil from reaching the environment.

Methods for preventing oil releases or spills from occurring include both administrative and engineering controls. Administrative controls might include standard operating procedures (SOPs) and policies intended to direct human behavior in such a way that oil spills can be avoided. For example, a fuel delivery SOP might require a plant operator to be present during the entire offloading of a bulk shipment of diesel fuel to provide extra assurance that fuel transfers are directly overseen by the plant. Or, a procedure that addresses visual inspections of oil storage and use areas conducted by plant personnel would enable the plant to identify potential risks proactively (for example, identifying a minor oil leak at an early stage is preferred to discovering a large spill). Engineering controls, on the other hand, such as high-level alarms, automatic leak detection systems, and secondary containment systems (e.g., dikes, berms, or containment pallets for drums) provide a last line of defense in the event that administrative controls fail.

Federal and state regulatory requirements may be applicable to a water treatment plant depending on the kind and quantities of oil stored. For example, in the United States if a plant stores more than 1,320 gal of oil, aboveground, in containers that are 55 gal or greater in capacity, then the plant is required to prepare and implement an oil spill prevention plan (under 40 CFR 112.7). Some U.S. states (e.g., Massachusetts and New York) have additional oil storage management and spill prevention requirements.

Residuals Treatment and Disposal

The primary focus of the regulatory agencies and water treatment plant owners and operators is generally on providing high-quality water to meet federal and state drinking water standards. However, attention must also be paid to the treatment and final disposal of water treatment plant waste sludges, or residuals, which were, in the recent past, discharged directly into a nearby water body or to a local wastewater treatment facility. Both of these practices, however, are receiving increased attention from the environmental and regulatory communities.

Potential localized effects of stream discharges include increased suspended solids and turbidity, creation of sludge deposits, and increased iron and aluminum concentrations. If water treatment residuals are discharged to a wastewater treatment plant, there may be an adverse impact on the plant's ability to meet discharge permit limits. As a result, alternative disposal methods are now being increasingly considered, including spreading residuals on agricultural or forestland, and landfill disposal or use as landfill cover material after thickening and dewatering.

Before land application or landfill disposal will be allowed by state or local authorities, it must be shown that groundwater will not be adversely affected. The *toxic characteristic leaching procedure* (TCLP) is commonly used to simulate the climatic leaching action expected to occur in a landfill. The TCLP identifies 8 metals and 25 organic compounds (pesticides, herbicides, etc.) with the potential to leach into groundwater.

With respect to land application, concern has also been expressed regarding the lack of organic value and the high metals content of the residuals, and the potential effect on crops and groundwater. However, lime-softening or lime-treated sludges have been shown to have positive effects on agricultural and vegetative uses of land (Robinson and Witko, 1991). The general rule in evaluating land application alternatives is to consider the ultimate land use and the residuals quality. Residuals with a high metals content, for example, will probably not be appropriate for application on agricultural land.

A 20% solids content is generally required for landfill disposal of residuals. In addition to mechanical dewatering, drying of residuals in on-site lagoons is still a common practice to achieve the appropriate solids content for landfill disposal. When suitably dry, the residuals can be scraped out of the lagoons and transported to a landfill for final disposal. But the practice of drying in lagoons also has environmental concerns, including odors (described as musty, or earthy) and land requirements for drying. In addition, the overall quality of the residuals is an important consideration in the feasibility of landfill disposal because the characteristics of the residuals will determine whether a sanitary landfill, hazardous waste disposal site, or low-level radioactive waste disposal site is most appropriate.

For use as landfill cover material, residuals could be dried to about 60% solids and applied directly, or dried to about 40% and mixed with soil. The residuals must be shown to be nonhazardous by a TCLP test and must have no free liquids. Some landfill operators and state regulators will not accept residuals for landfill cover because they are concerned that the material will be gelatinous when wet and dusty when dry.

Although there are many considerations in determining the most appropriate method of residuals treatment and disposal, environmental issues are gaining more significance as there is better understanding of the impacts on the receiving water body, treatment plant, landfill, and agricultural land. It appears likely that direct discharge to receiving streams, wastewater plants, and landfill disposal will become increasingly disfavored, and land application and other beneficial disposal alternatives will gain greater acceptance.

Wastewater Treatment and Disposal

Typical sources of liquid waste discharges from water treatment plants include filter backwash water, supernatant from sludge dewatering, emergency bypass water, water from washing equipment and flushing pipelines, and employee sewage. Disposal of employee sewage must be to an on-site septic system or to a nearby sewer. Water from equipment washing and pipeline flushing can be similarly disposed of, although in some cases it is discharged to a nearby watercourse.

Filter backwash, supernatant, and bypass water have historically been discharged to receiving waters, but in many cases these wastewaters are chlorinated, which has raised concern because of the potential toxicity effects to aquatic biota and potential health hazards. The impacts on the receiving water are primarily dependent on the concentration of chlorine residuals, which is in turn governed by the combined effects of the mixing process with the receiving waters and the time-dependent decay of the chlorine (Stanley and Smith, 1991).

There are two potential environmental hazards associated with chlorine residual discharges. The first hazard is toxicity to humans through downstream drinking water intakes. However, this concern appears to be minimal because the residual chlorine concentration is much less than the U.S. Environmental Protection Agency (USEPA) ambient water quality criteria for chlorine (Stanley and Smith, 1991). The second hazard is the potential impact of the chlorinated discharge on aquatic biota. Chlorine toxicological studies have shown that the effects on aquatic life are determined by many physical, chemical, and biological factors, including pH, alkalinity, hardness, temperature, species sensitivity to chlorine, and the speciation of the chlorine (Turner and Chu, 1983). These factors were all considered by USEPA in establishing limits for allowable total residual chlorine in receiving waters to be met by water treatment plants. As a result, any chlorinated discharges will require a permit from USEPA.

Stormwater Management

Even after construction is completed, proper management of stormwater is necessary. Appropriate controls must be incorporated into the plant design to ensure that predischarge runoff rates from the site are maintained and that direct discharges to wetlands and surface water bodies are controlled. If, for example, a new plant is constructed on a previously undeveloped site, the increased rate of runoff could be significant as a result of the increase in impervious surfaces. Typical stormwater control measures include best management practices (BMPs) such as treatment swales, vegetated buffer strips, and detention or retention basins. The purpose of these BMPs is twofold: (1) to mitigate the effects that an increase in runoff may have in a particular watershed and (2) to implement controls on potential sources of nonpoint pollution.

Air Emissions

Water treatment processes can create air emissions and odors, which must be considered during plant design. Examples of emission sources at a treatment plant include ozonation, chlorination, emergency power generation, space heating, and solids handling.

During ozonation, some ozone off-gas is released. Because ozone is a strong irritant and will harm vegetation, the off-gas must be treated by means of a catalytic or thermal ozone destruction unit. These units are effective in reducing ozone emissions to below national or state ambient air quality standards. It is also important to protect plant workers from ozone exposure. Preventive measures include maintaining the ozone contact tanks under negative pressure to prevent leaks, and placing ozone monitors in working spaces designed to shut down the entire system if a leak is detected.

Power for normal plant operations is generally provided by the local electric utility; however, many plants have their own diesel generators to supply electricity for use during commercial power failure. Diesel generator exhaust is a source of carbon monoxide, nitrogen oxides, sulfur dioxide, and particulate matter, so the release of these pollutants may pose an environmental concern if they are in large enough quantities. Similarly, the plant space heat boilers may release pollutants that may be considered objectionable.

Although not as noxious as wastewater residuals, residuals from the water treatment processes may be a source of odors, in particular, the organic matter removed from the treated water. If the residuals are discharged directly to a sewer with no open-air storage or drying, odors are usually not an issue. However, if the residuals are stored and dried on-site, then off-site odors could be a concern.

If source water is of high quality and the resulting residuals have a low organic content, odors are generally not a problem. In most cases, the odors are of an earthy or musty nature and cannot be detected beyond 100 yd (90 m) of the residuals under the worst conditions. A 1995 survey conducted for Camp Dresser & McKee showed no significant odor problems reported by any of the respondents. Some of the responses included these:

- In Englewood, Colorado, residuals are collected in large thickening lagoons and then are mechanically dewatered and stockpiled on the site for as long as 2 years. Some of the residuals may have been in the lagoon for up to 15 years. No odor complaints have been received from any of the adjacent neighbors.

ENVIRONMENTAL IMPACT AND PROJECT PERMITTING 24.11

- A facility in upper Michigan discharges alum residuals several times a year to a freeze-thaw lagoon. The residuals are excavated and disposed of each spring. In the past 20 years, no complaints have been received from residents living less than 200 yd (180 m) away. The plant manager did report, however, that on a hot windy day, the odor may be detectable 100 yd (90 m) away.
- A plant in Weymouth, Massachusetts, cleaned 6 to 8 ft of residuals out of five lagoons that had not been cleaned in 30 years. The residuals were excavated by dredging and dried on site. After drying, the residuals were hauled to a landfill in Connecticut. No odors were reported from this operation, which took place in summer.

Solid and Hazardous Waste Management

A variety of wastes can be expected from the operation of a water treatment plant, and determining how to properly manage these wastes will largely depend on the type (e.g., nonhazardous or hazardous) and amount of waste generated and where the plant is located, as applicable waste requirements often vary from one state or local jurisdiction to another. It is important that a plant manage its wastes properly because improper waste management could present a risk to the environment and to public health and safety. In addition, improper waste management could pose a financial risk for the municipality or county that owns and/or operates the water treatment plant if that plant is identified as a potentially responsible party in a contaminated site investigation.

A key step toward mitigating such risks is to first understand what wastes are generated by a given water treatment plant. A waste inventory should be conducted that is comprehensive and considers the full range of wastes potentially generated at the plant including, but not limited to, building custodial/administrative wastes (e.g., office paper, cardboard wastes, fluorescent lamps, batteries, etc.), equipment maintenance-related wastes (e.g., used oil, oily rags, oil-water separator solids, waste solvents, air drying system filters), laboratory wastes (spent chemicals), and process wastes (e.g., used scrubber filters, scrubber system wastewater). It is the responsibility of the plant to determine whether its wastes are hazardous or nonhazardous, and regulatory guidance should be consulted to make these determinations.

Hazardous wastes have been found to be potentially harmful to the environment and, therefore, must be managed in accordance with applicable container, storage area management, shipping, and record-keeping requirements. For generators of large quantities of hazardous waste, additional waste minimization, record keeping, reporting and hazardous waste training would be required. Those plants generating smaller quantities of hazardous wastes are subject to less stringent requirements for hazardous waste management than the generators of larger quantities. Federal and state regulations specify what constitutes the hazardous waste generator status of a facility (e.g., large-quantity generator, small-quantity generator, or conditionally exempt small-quantity generator), which is based on the volume (in pounds) of hazardous waste generated by a facility per month, or the volume of hazardous waste stored on site at any given time. Once a water treatment plant manager knows its hazardous waste generator status, then procedures and practices should be developed and implemented to ensure that hazardous waste is properly managed according to applicable regulatory requirements.

In addition to hazardous waste considerations described above, a water treatment plant should be aware of potentially applicable requirements for nonhazardous, special, or universal wastes. Examples of these requirements include Massachusetts requirements for the management of solid waste; Massachusetts provisions for handling asbestos waste (i.e., a special waste); and federal and state requirements for the management of used fluorescent lamps, pesticide waste, used batteries, and used mercury-containing devices and thermostats, which are universal wastes.

PROJECT PERMITTING

Numerous federal, state, and local environmental laws, regulations, and policies govern water treatment plant construction activities and operations. The focus of this discussion is on federal permits and approvals because there is a great deal of variability between the regulatory compliance and permitting processes in individual states and municipalities. It is important to note, however, that there is also some variation in implementation of federal requirements from state to state. It is therefore important that the designer carefully review the state permitting process for each particular project. Table 24.4 summarizes the most pertinent federal environmental regulations and specific triggering activities.

Federal Environmental Regulations

Federal environmental regulations can generally be categorized as affecting either plant construction or plant operations.

Construction Permitting. Construction permit requirements are primarily dictated by the site and surrounding area characteristics. Work in wetlands, for example, will generally require a permit from the U.S. Army Corps of Engineers. However, the type and extent of activity within that wetland will also determine permit jurisdiction, permit review procedures, and the relative ease or difficulty in obtaining the permit.

National Environmental Policy Act. Review under the National Environmental Policy Act (NEPA) is triggered by any federal grant or permit. If, for example, the Army Corps of Engineers determines through its review of a permit application that a proposed project will have significant environmental effects, then an environmental impact statement (EIS) under NEPA may be required. NEPA does not define an environmental "impact." The Council on Environmental Quality (CEQ) regulations define this term and require a discussion of indirect as well as direct, and beneficial as well as negative, impacts. Alternatives to the proposed action must also be presented. The alternatives analysis can be very broad and may encompass such issues as water treatment plant siting, alternative technologies, alternative supply sources when a new source is involved, and plant capacity. Mitigation of identified negative impacts must also be presented.

Generally, an EIS is required only for large, environmentally disruptive projects such as construction of a water supply lagoon that may impact a protected endangered species. By minimizing environmental impacts up front in water treatment plant site selection and preliminary design, the potential need for an EIS is reduced.

NPDES General Permit. A permit under the National Pollution Discharge Elimination System (NPDES) will be required for stormwater discharge during plant construction, if more than 1 acre is affected by construction activities and the discharge will occur to a water body or wetland. The 1-acre threshold is cumulative; that is, if total land disturbance associated with a water treatment plant project (including all project components) equals or exceeds 1 acre, a permit must be obtained. The general permit application must be submitted to USEPA at least 7 days before the start of construction, and a stormwater pollution prevention plan (SWPPP) must be prepared and kept on file. The SWPPP identifies potential sources of pollution, describes practices to reduce pollutants in stormwater, and ensures compliance with permit conditions. Monitoring and reporting

TABLE 24.4 Typical Permits, Regulations, and Triggering Activities

Federal permit and regulations	Responsible agency/authority	Triggering activity
Title III of Superfund Amendment and Reauthorization Act	USEPA	Tracking of hazardous chemicals and release reporting
Approval under CFR Title 49	U.S. Dept. of Transportation	Classification of hazardous materials; packaging and transport
Resource Conservation and Recovery Act (RCRA) of 1976 and Comprehensive Environmental Response, Compensation, and Liability Act (CERCLA) of 1980	USEPA	Review triggered by hazardous characteristics of residuals
Approval under 40 CFR 257	USEPA	Land application of residuals
Chemical accident prevention provisions under 40 CFR Part 68	USEPA	Potential accidental airborne releases of substances that pose a public health environmental threat
Title V Operating Permit Program under 40 CFR Part 70	USEPA	Air emissions exceeding certain thresholds
National Environmental Policy Act (NEPA) review to determine whether EIS is needed (42 USC 554321 et seq)	USEPA	Any federal action (including grants or permits); need for EIS depends on the magnitude of perceived impacts
National Pollutant Discharge Elimination System Permit under Section 402 of the Clean Water Act (40 CFR 122-124)	USEPA	Discharges of stormwater, treated wastewater flows, or other process flows into surface waters or wetlands; stormwater discharge from construction sites >1 acre in size
Section 404 of the Clean Water Act	U.S. Army Corps of Engineers	Discharges of dredged or fill material into U.S. waters (including wetlands); jurisdiction includes incidental discharges associated with excavation activities
Section 10 of the Rivers and Harbors Act of 1899 (33 CFR 322)	U.S. Army Corps of Engineers	Structures or work in or affecting navigable U.S. waters (navigable waters are generally defined as U.S. waters that are subject to the ebb and flow of the tide shoreward to the mean high-water mark or are used or have been used in the past to transport interstate or foreign commerce)

of stormwater discharges are also required. When construction is completed, a notice of termination must be sent to USEPA.

U.S. Army Corps of Engineers Permits. Under Section 404 of the Clean Water Act, a permit may be required from the Army Corps of Engineers for any filling of wetlands or water bodies. For structures or work affecting navigable waters, such as construction of a source water intake in a navigable river, a permit under Section 10 of the Rivers and

Harbors Act would be required. The precise permitting requirements vary between Army Corps of Engineers districts and between states; therefore it is important to understand the particular requirements within the state where the construction is to take place. Programmatic general permits or nationwide permits apply in some states, which allow certain activities involving minimal wetlands or water body disturbance to proceed without, or with limited, Army Corps of Engineers involvement.

Operations Permitting and Ongoing Compliance. Federal programs regulating chemical delivery, handling, and storage; residuals disposal; wastewater discharge and disposal; and air emissions are briefly summarized below.

Chemical Delivery, Handling, and Storage. There is potential for regulation of chemical delivery and handling by at least two federal agencies and programs. Title III of the Superfund Amendment and Reauthorization Act (SARA) of 1986 (also referred to as the Emergency Planning and Community Right-to-Know Act, or EPCRA) deals with the tracking of hazardous chemicals and release reporting. EPCRA notification allows local fire departments, local emergency planning committees, and state emergency response commissions to be informed about the existence of large quantities of hazardous materials (or smaller quantities of extremely hazardous substances) at facilities (including water treatment plants) in their community or state. The information required to be reported is used to facilitate emergency response planning at both local and state levels. The goal of the reporting system is to inform and prepare communities hosting these facilities for potential hazards and emergency situations that could occur if there was an accidental release of hazardous chemicals.

More recent federal programs include the Occupational Safety and Health Administration (OSHA) Process Safety Management (PSM) program under CFR 29, Part 1910.119, and the USEPA Risk Management Program (RMP) under CFR Title 49, Part 68. Although both programs pertain to the handling, use, or storage of listed toxic and flammable substances in quantities exceeding thresholds, the PSM program deals with worker safety, and the RMP focuses on potential impacts to the public and the environment.

Transportation of hazardous materials is regulated by the U.S. Department of Transportation (USDOT) under CFR Title 49. Title 49, Parts 171 through 180, provides for the classification of hazardous materials and for the marking, labeling, and placarding (for bulk shipments) of hazardous materials packages and containers. These regulations also contain requirements for communications, emergency response, and training of persons who handle or transport hazardous materials. Other parts of the regulations cover the specific requirements for each transportation mode (rail, air, water, and public highway).

Residuals Disposal Regulations. The regulation of residuals disposal depends on the ultimate disposal method, and the ability to comply with applicable regulations may determine which disposal method is the most feasible. If residuals are to be discharged to a watercourse, the Clean Water Act (CWA) is applicable. Under the CWA, an NPDES permit would be required from USEPA for any discharges of sludge or semisolids to navigable waters under CFR Title 40, Part 122. However, the need for compliance with water quality standards would prohibit these discharges in most cases. NPDES permit limits are determined by the appropriate water quality standards and criteria, which are based on protecting human and aquatic life.

Landfill disposal of residuals is potentially regulated under the Resource Conservation and Recovery Act (RCRA) of 1976 and the Comprehensive Environmental Response, Compensation, and Liability Act (CERCLA) of 1980, if the residuals are considered to be hazardous. Under the Toxicity Characteristics Leaching Procedure (TCLP), it is possible for water treatment plant residuals to exceed the threshold limits and therefore to be considered hazardous (Robinson and Witko, 1991). Many states require that residuals to be landfilled be a minimum of 20% solids and be able to pass the TCLP and also a filter test to determine whether there are any free liquids emanating from the residuals.

The RCRA and CERCLA guidelines for land disposal of residuals specify that disposal activities must be conducted in a manner that minimizes impact to the environment. Specifically, RCRA regulates the residuals quality and disposal activity, and CERCLA provides for the remediation of improperly operated or abandoned disposal sites. RCRA also encourages recycling of materials as an alternative to ultimate disposal.

Land application of residuals is loosely regulated by USEPA under 40 CFR 257, which deals primarily with wastewater treatment residuals as opposed to water treatment residuals.

Wastewater Disposal Regulations. The type of regulation applicable to process and sanitary wastewater disposal depends on the ultimate disposal plan. Discharge to an existing sewer would not trigger federal permitting, but would involve approval by the appropriate sewer agency. Federal approval would not be required for discharge to a septic system, but state or local approval would probably be required.

However, process wastewater discharge to a nearby watercourse may require an NPDES individual permit from USEPA under CFR Title 40, Part 122, and it may also be required that the discharge receive treatment before discharge, depending on the particular state and the discharge characteristics. In Massachusetts, New Hampshire, and Maine, for example, some water treatment plant discharges are exempt from the need to obtain an NPDES individual permit, and only a general permit may be required. Whether the discharge would be allowed is also determined by the ability of the discharge to meet the water quality criteria for the receiving watercourse.

Air Emissions Regulations. Air emissions may be regulated under several programs, depending on the amount and type of emission. A number of emerging USEPA clean-air rules may affect the design and operation of a water treatment plant. For example, on a case-by-case basis, state and local fire codes could require installation of scrubbers at water treatment plants to control accidental releases of chlorine from gas feed systems. Also, combustion of fossil fuels to operate boilers and emergency engines can result in emissions that may have to be permitted and controlled in the future.

One pertinent air quality regulatory program is 40 CFR Part 68, Chemical Accident Prevention Provisions, which addresses accidental airborne releases of substances that pose a significant threat to the public or the environment. Another is the Title V Operating Permit Program under the Clean Air Act (40 CFR Part 70), which is triggered if certain emissions thresholds are exceeded. A Title V permit is required if emissions exceed the following thresholds in tons per year (tpy):

- 100 tpy of a criteria pollutant (SO_2, NO_x, ozone, CO, lead)
- 10 tpy of a toxic compound [volatile organic compound (VOC)]
- 25 tpy of total VOCs

In determining whether any of these thresholds is being exceeded, all emission sources at a water treatment plant must be added.

State and Local Regulations

Construction and operation of a new water treatment plant are likely to require approvals from a number of state and local regulatory authorities. In fact, because there is increased deregulation at the federal level, much of the regulatory burden now rests with the states. Most states have received federally delegated authority to issue permits and enforce various federally mandated programs. Some states require environmental review and filing of environmental impact documentation at the planning stages of a project, which often mirrors the federal NEPA process in scope and process. New York and Massachusetts are two such states that have their own environmental impact review process that must be completed before project construction can begin.

If a local authority has no direct permitting responsibility, input is still provided through the state permitting process. For example, in Massachusetts, the local board of health in a community where land application of residuals is proposed is given the first opportunity to comment on a permit application submitted to the state, before the state takes any formal action on the application.

Many states have their own wetlands and surface water regulations that will require a permit if a new plant will affect these resources during construction or operation. Because federal approvals are also often required for activities affecting these resources, lack of consistency between state and federal requirements may be an issue that must be anticipated and addressed during the permitting process.

At the local level, zoning approvals are often required, as well as approvals from the local planning board, building inspector, and so on. Many communities have specially mapped districts, in addition to zoning, to control activities in floodplains, for example.

INTEGRATION OF ENVIRONMENTAL ISSUES AND REGULATORY COMPLIANCE INTO PLANT DESIGN, CONSTRUCTION, AND OPERATION

With the increase in environmental awareness and regulatory requirements in recent years, a project that might have breezed through the permitting process 10 or 15 years ago may find itself lost in the permitting maze today. As a result, it is extremely important to establish and follow a carefully planned permitting strategy, even before plant design begins. The strategy should include consideration of environmental issues, because these are the driving forces behind the permitting process. The overall steps in development and implementation of a permitting strategy are presented in Table 24.5 and summarized below.

TABLE 24.5 Permitting Strategy

1. Identify all project components.
2. Assess the existing site conditions.
3. Identify all required permits and approvals.
4. Identify the major permitting issues.
5. Develop a project-specific permitting approach.
6. Compile all necessary project information and data (coordinated with design).
7. Prepare and submit permit applications.
8. Track the review process.
9. Negotiate the permit conditions.
10. Obtain project approval by the agencies.
11. Follow through with permit conditions.
12. Know permit expiration dates.

Developing a Permitting Plan

The proposed permitting strategy should be outlined in a written plan that identifies specific permits and permit interrelationships and defines the overall approach for obtaining permits. The plan should establish the links between specific project activities or environmental issues and permit requirements (i.e., permit triggering activities). Finally, the permitting plan should present a schedule for obtaining permits, including the approximate lead time needed to prepare applications, appropriate dates for filing applications, and dates when permits are needed. Establishing a realistic schedule is particularly important when there is a state or federal consent order involved, which specifies interim dates for project activities such as bid dates, construction start, and completion.

The permitting plan also provides a good opportunity to review preliminary design decisions and to make appropriate design modifications to eliminate or reduce permitting requirements. Identifying potential modifications before design is finalized may avoid costly later design changes and project delays caused by a lengthy permitting process.

Carrying Out the Permitting Plan

The permitting plan should be updated throughout design as necessary and should be used as a checkpoint for design decisions. Using the schedule provided in the plan, applications should be prepared and submitted to the review agencies. However, the process does not end here. Tracking the applications through the agencies and the review process is just as critical as timely preparation and submittal of those applications. Knowing who the reviewer is and establishing early contact with that individual are the first step toward a smooth review process.

Once the application is submitted, maintaining contact with the reviewer is important to ensure that the application is not buried under others and that technical questions are not left unanswered. If technical questions are answered during the application review process, they are less likely to appear as permit conditions. Another approach to ensuring reasonable permit conditions is to actively negotiate the conditions with the agency before the permit is issued. This strategy may streamline the process by producing a set of permit conditions that are reasonable to both the applicant and the agency, thereby avoiding time-consuming appeals.

Compliance during Construction and Operation

Most construction-related permits become part of the contract documents so that the prospective contractor can accurately bid the project. The selected contractor then becomes a party to permit compliance. However, relying solely on a contractor to meet the terms of a permit is not advisable because it is the applicant or project owner who is ultimately responsible for permit compliance.

As a result, the permit tracking system should not end with permit issuance, but should follow through construction and operation, as applicable. Even a construction-related permit may have conditions that extend beyond construction completion. Maintaining catch basins and other stormwater controls is one example of a condition that may be effective beyond construction.

The permit tracking system should take into account permit expiration dates and regulatory changes. If, for example, a permit program is revised, a new application may be required before the existing permit expires. When a permit expires, an extension is often easier to obtain than the original permit, as long as regulatory changes have not occurred and conditions in the original permit have been met, although a smooth process is not necessarily guaranteed.

Conclusions

The key to successful permitting is to establish a strategy early in the project and to implement it throughout design, construction, and operation. Obviously, some revisions to the strategy will be necessary as the project progresses, but a permitting plan can still provide a road map to help streamline and track the permitting process.

BIBLIOGRAPHY

Camp Dresser & McKee Inc. (CDM). *Wachusett Reservoir Water Treatment Plan. Final Environmental Impact Report.* Cambridge, Mass.: CDM, 1995.

Camp Dresser & McKee Inc. *Taunton River Water Supply EIR.* Cambridge, Mass.: CDM, 1996.

Hadden, Deborah, and William Keough. *Predicting Public Water Supply Well Operation, AWWA Annual Conference.* Denver, Colo.: American Water Works Association, 1994.

Organization for Economic Cooperation and Development. *Guiding Principles for Chemical Accident Prevention, Preparedness, and Response.* Paris, France: OECD, 1992.

Robinson, M. P., and J. B. Witco. "Overview of Issues and Current State-of-the-Art Water Treatment Plant Waste Management Programs." *AWWA Conference Proceedings,* 1991.

Stanley, S. J., and D. W. Smith. "Modeling Chlorinated Discharges from Water Treatment Plants: A Case Study." *Canada Journal of Civil Engineering* 18:985, 1991.

Turner, A., and A. Chu. "Chlorine Toxicity as a Function of Environmental Variables and Species Tolerance." In *Water Chlorination: Environmental Impact and Health Effects,* vol. 4, book 2. Ann Arbor, Mich.: Ann Arbor Science, 1983.

Vicory, A. H., and L. Weaver. "Controlling Discharges of Water Plant Wastes to the Ohio River." *Journal AWWA* 4(84):122, 1984.

CHAPTER 25
OPERATIONS AND MAINTENANCE CONSIDERATIONS DURING PLANT DESIGN

Richard A. Rohan, P.E.
MWH
Houston, Texas

Stephen F. Baker
MWH
Broomfield, Colorado

DESIGN CONSIDERATIONS FOR PLANT OPERATIONS AND MAINTENANCE

While the engineering design team is considering the unit processes—mechanical, electrical, instrumentation, architectural, and structural details of a new or upgraded plant—it must also consider how to make the plant safe, efficient, economical, and easy to operate and maintain. These are considered to be *operability-* and *maintainability-*related issues. Although the guidelines here provide a comprehensive compilation of operability and maintainability issues, it is important that the design team consult with the plant operations and maintenance (O&M) staff during the design process to ensure their thoughts are considered.

Plant O&M staff involvement should increase as the plant start-up date nears. Figure 25.1 shows the recommended level of O&M staff involvement as a project moves through design and construction. O&M involvement during start-up is discussed in greater detail in Chapter 27.

Plant O&M staff participation could cause "scope creep" (work outside of the agreed upon scope of work) of the design if the design process is not managed properly. Clear and concise guidelines should be developed and agreed upon by all parties so that the O&M staff involvement is clearly defined. The guidelines should control the design process.

Designers should have a clear understanding of issues that O&M staff will most likely bring to the design process. If a designer knows and understands basic design issues, many O&M staff issues can be addressed in the design before the O&M review takes place; or the designer can have an explanation ready why certain O&M issues may not have been

FIGURE 25.1 O&M involvement.

addressed. The next two sections contain lists of design aspects that should be considered for O&M.

Design Considerations for Plant Operations

Considerations for operations encompass a wide range of issues that, if properly addressed, will enhance a facility's operability. Some of the more common issues include these:

- Safety
- Plant security
- Electrical power supply including alternative sources
- Control systems and level of automation
- Ergonomic design of pathways and access and egress areas
- Operator sampling station locations
- On-site laboratory facilities
- Operator control ergonomics
- Unit process–specific issues
- Cathodic protection systems
- Operations computerized software support

Safety Considerations. When designing a water treatment plant, designers must comply with local building codes and regulations. Many local government agencies have adopted the Uniform Building Code for construction of buildings. Other regulations and codes that must be complied with are the Uniform Fire Code, National Fire Protection

O&M CONSIDERATIONS DURING PLANT DESIGN   25.3

Association standards, and state and federal OSHA standards. By designing the plant according to these codes and regulations, the plant should be a safe place to work; but the designer should also consider the following items that may not be addressed in the codes and regulations:

- At least two self-contained breathing apparatuses (SCBAs) should be provided outside each area that can contain toxic gases or be depleted of oxygen (e.g., chlorine feed room, chlorine storage room, and indoor recarbonation basin). Take care to locate these SCBAs in a place that will be free of toxic gases in the event of a toxic gas release.
- Noise arresters or earplugs should be provided at the entrance to areas that have high noise levels (e.g., pump rooms, diesel generator room, and near air compressors).
- Plant telephones or intercoms should be installed in areas accessible to controls and equipment, but away from noise. If it is not possible to find a quiet area, install materials to control noise around communications equipment.
- All tunnels or galleries should have an exit at least every 150 ft (46 m).
- Hose bibs should be provided at intervals of no more than 50 ft (15 m) in tunnels, galleries, surface walkways, and areas where frequent washdown is required.
- Electrical receptacles should be installed at no more than 100-ft (30-m) intervals in tunnels, galleries, surface walkways, and areas where frequent use of portable electrical tools or equipment is required.
- The contract documents should specify that all drains be tested and traps charged with water before plant acceptance.
- The contract documents should specify that all chemical and process piping be color-coded and also labeled (for color-blind staff).
- Confined spaces should be avoided if possible; however if they are unavoidable, confined spaces must be given special consideration to meet all applicable regulations. It is advisable to use confined-space experts to review designs and to ensure that all requirements are met.
- Provide all necessary safety equipment for safe entry into a confined space.
- Provide warning signs for confined spaces, noting that entry is *not* permitted without proper authorization and stating the conditions for entry (e.g., SCBA is worn, safety harness is worn, second person present).
- Provide safety grates below hatch covers over submersible pumps.

Plant Security Considerations. Security at water treatment plants is an important issue. Controlling access to the plant as well as off-site storage, pumping, and distribution systems is an ever-increasing concern from the standpoint of protecting public health, plant personnel, and plant facilities and equipment. Recommendations to aid in designing a secure facility are listed below. Also review Chapter 30 for additional information.

- Incorporate security measures based on *detection, delay,* and *response.* Install intrusion detection equipment at all plant and remote sites where access to the water supply or supply equipment is possible. Design buildings so that rapid interior access is prevented. Design alarm systems so appropriate security protection personnel can safely and rapidly respond to security issues.
- Provide security fence to encompass the water treatment plant site, if possible.
- Provide gates with magnetic card readers and/or the ability for the operator to open the gate from the water treatment plant control room.

- Locate critical plant equipment in locked rooms within areas that require keyed or card access. This will provide at least two barriers to possible intruders.
- Minimize the number of windows at the water treatment facilities.
- Provide a system of locks at the plant so that a master key or card can open all of them. Plant personnel can be issued keys or cards that give them different levels of access. For example, the plant manager would have a master key or card providing access to all secured areas, O&M supervisors would have access to most secured areas, and all other staff would have access only to the areas that are necessary to do their work.
- Install motion detectors inside the perimeter fences to activate a signal in the central control room to alert plant personnel of unauthorized entry into the plant area.
- Provide adequate lighting for all outside areas of the plant for security and safety.
- Supply O&M personnel with two-way radios for in-plant communications.
- Provide security cameras at entrances to the plant site and to key facilities on the plant site.
- Security measures can be expensive to design and construct. Check the client's security requirements before estimating the design and construction costs.

Power and Control Systems. Adequate electric power and good control systems are extremely important to the successful operation of a modern water treatment plant. One of the main concerns with electric power is reliability of the power source. A water treatment facility should have two separate power feeds into the plant or a standby generator that is capable of supplying power when the regular power supply is lost. Electrical equipment must be protected from damage due to excessive heat, condensation, flooding, dust, and dirt. Control issues center on the location of controls, optimum level of automation, and human–machine interfacing. The following electrical power and control issues should be considered:

- Location of motor control centers (MCCs) and control stations (CSs) should provide adequate room, lighting, and protection from the elements for O&M access.
- MCCs and CSs must not be located in areas that are subject to flooding or water spray in the event of leaks. This may mean relocating pumps and pipes or constructing barrier walls to protect the gear.
- Access and egress to rooms where MCCs and/or CSs are located should be secure, but with entrances and exits that are easily accessible to authorized personnel.
- When practical, critical motor control centers should be decentralized. For example, if there are two sedimentation basins, it is preferable to furnish a separate control center for each one. If one MCC is taken out of service, only one sedimentation basin will be affected.
- Mechanical equipment with electric motors should be elevated on pedestals or pads. Keeping equipment elevated above the floor reduces the possibility of damage and simplifies maintenance and cleanup.
- The use of manually operated valves on automatic systems should be limited to secondary isolation of equipment units and/or processes, or for maintenance isolation, such as pressure gauges and in-line sensors. Any valve that needs to modulate or open and close frequently for process control should be automated accordingly.
- Alarm float switches should be installed near the floor in all equipment areas that are subject to flooding.

- Control systems should be provided to operate in both automatic and manual modes. Facilities that have automatic and manual control strategies provide full operator flexibility and are not labor-intensive. There should be full manual control capabilities outside of the programmable logic controller (PLC) if a PLC is provided with the equipment. This will allow the equipment to be operated even if the PLC should fail.
- Local manual control panels should preferably be placed at the equipment to allow operators to visually observe equipment operation. This is important when one is dealing with filters, chemical feed systems, sludge pumping systems, and all systems with moving mechanical equipment.
- Prepare P&ID drawings as the first step in the design process, and involve plant staff early and often in the review.
- The designer should prepare control loop descriptions early in the design process in a structured format listing equipment, manual control, PLC control, safety interlocks, and human–machine interface (HMI).
- All valves should be provided with number tags. If valves are not properly identified, it becomes difficult to develop easy-to-follow standard operating procedures (SOPs). If valves are not identified by tag numbers, a lengthy written description must be provided to accurately locate the valve.
- The design of motor control centers and control stations should provide adequate room for O&M access.
- Motor control centers and control stations must have entrances and exits that are easily accessible, should an emergency arise that requires personnel to have quick access or egress.
- Sample taps and flowmeters should be provided on all sidestreams to facilitate sampling and accounting for plant uses.
- The designer should write a draft controls section of the O&M manual during design for review and comment by plant staff.

Level of Automation. Designers must consider and sometimes determine the level of automation based on the client's O&M staffing plan. Where staffing is not a limiting factor, lower levels of automation may be acceptable. Lower levels of automation may mean that some automated systems can be started locally or require frequent local observation rather than installing high-tech software for automatic operation and/or remote monitoring equipment. Conversely, low staffing levels will mean that higher levels of automation are required where operators can monitor a number of systems from remote sites, such as from the central control room or workstations located throughout the facility.

Alarm pager systems allow operators to leave monitoring stations to aid in protecting the plant while allowing "less attended" operations.

Plant automation is typically comprised of up to three modes of operation: automatic, semiautomatic, and manual. Each mode is defined as:

- *Automatic.* Once electric power is applied, the equipment system can operate entirely on its own from complete start-up to complete shutdown without operator intervention.
- *Semiautomatic.* Certain components of the equipment can function automatically, but other components of the system require operator intervention. For example, a sump pump is started and stopped by float switches, but the pump discharge valve must be manually opened and closed. If the discharge valve is not open when power is applied to the pump, the system cannot function properly.

- *Manual.* All equipment components must be started and stopped through operator intervention.

Ergonomic Design. Treatment plants need to be easy and safe for O&M staff to move around. Designers should consider how O&M staff will move around the facility and access plant systems. The following list provides ergonomic issues that should be considered:

- Walkways around the site should go from one process area to another by the most direct path possible. For example, if elevated but separate process tanks are provided, elevated walkways that allow O&M staff to travel from one tank to another should be considered.
- Consider the path an operator might take while conducting plant rounds and/or collecting samples; then optimize the route. For example, consolidate common samples, locate sampling stations along the common route, and minimize backtracking.
- Place PLCs and remote monitoring stations at strategic locations so operators can check in as they conduct their rounds.
- Provide visual access to equipment whenever possible. This may include expanding walkway width and installing handrails between flocculation basins, sedimentation basins, filters, and other process units where critical mechanical and/or instrumentation is installed.

Operator Sampling Stations. Sampling stations should be located strategically around the plant for process control and regulatory compliance sampling. Each sampling station should provide

- Protection from extreme weather conditions.
- Wash-up facilities (important for plant waste streams such as sedimentation basin residuals and filter backwash)
- Proper drains
- Adequate lighting

On-Site Laboratory. The laboratory should be designed to obtain information to maximize the full intended use for a water treatment facility. This includes areas for sample collection staging, sample preparation, sample analysis, bench-scale testing, storage of analytical chemicals, and sample disposal. Design considerations that will help provide a safe and functional laboratory are as follows:

- Involve laboratory personnel in planning and design review for the laboratory.
- If possible, provide separate rooms for sample preparation, storage, and analyses. This may not be feasible for small facilities.
- Separate rooms should be provided for atomic adsorption spectrophotometers and gas chromatograph analyzers.
- The microbiological section of the laboratory should be in a separate room.
- Set the work flow path so that samples brought in from the plant are taken directly to the laboratory. Avoid passing through areas such as administration areas, lunch rooms, reception areas, and maintenance shops.
- Provide a separate office for laboratory administrative duties.
- Provide sufficient counter space for analyses and bench testing.

- Provide storage room for analytical chemicals and laboratory supplies.
- Provide sufficient cabinet space around the laboratory for glassware and analytical apparatus.
- Provide storage cabinets for flammable and hazardous chemicals that comply with requirements of the Uniform Building Code and Uniform Fire Code.
- Provide more than one entrance to the laboratory, so there is more than one escape route for emergencies.
- Provide deep sinks with appropriate faucets at strategic points around the laboratory.
- Provide floor drains in the laboratory rooms.
- Provide fume hoods and sample refrigerators.
- Provide analytical equipment, laboratory supplies, and reagents. These items should not be specified without consulting the plant's laboratory staff. Reagents should not be ordered until a suitable period prior to start-up of the facility. This is critical for reagents with a short shelf life.

Operator Control Room Ergonomics. Control room ergonomics are an important group of features that can improve O&M efficiencies. Some specific ergonomic features are as follows:

- Sound attenuation to reduce ambient noise generated from various conversations that normally take place in control rooms
- Adequate space in the control room to create an office for administrative work such as monthly report compilation, confidential meetings and telephone conversations, and schedule preparation
- Extra counter space, telephone receptables, and power receptables so that guests (such as contract maintenance personnel, consultants, or other plant personnel) can dial in to communicate with their home computer networks
- Sufficient storage space to store relevant O&M documents for easy retrieval when needed
- Adequate security to keep the control room off limits to all but authorized personnel

Unit Process–Specific Issues. The following process areas have O&M issues that designers should consider.

Pilot Plant Studies. Whenever possible, pilot plant studies should be conducted using the new water treatment facility's raw water. Pilot studies provide crucial information for process selection, chemical selection, dosing requirements, and chemical application points. If possible, the pilot plant should remain and be incorporated into the new treatment facility for use. This will allow operators to perform treatment enhancement and cost-saving ideas on a small scale before implementing the idea on a full plant scale.

Source Water Intakes. Some issues that will help O&M of rapid mix basins are as follows:

- Intake facilities should be designed to be as accessible as possible to plant staff. If source water pumps are located more than a short distance from the main plant, they should be fully automated and controlled and monitored from a main control room.
- Provide on-line analyzers for turbidity, pH, and other selected physical and chemical parameters to give real-time information to operators so process adjustments can be made as quickly as possible.

- Security measures should be sufficient to detect, delay, and respond to any potential threat.

Rapid Mixing Basins. These are some issues that will help O&M of rapid mix basins:

- Provide variable-speed mixers, which will allow the plant operator to adjust the intensity of mixing. This will generally allow better process control when there are changes in source water quality, water temperature, and flow rates.
- The method of introducing coagulants and coagulant aids into the rapid mix basin is important to the overall performance of the basin. They should be introduced quickly and uniformly with the source water as it enters the basin. The location of the point of application and how it is applied can have a significant impact on treatment efficiency.
- Make provisions for entering the basin for maintenance when it has been taken out of service and drained. Personnel should be able to enter without removing the mixer.
- When possible, design the basin floor so that residuals can be easily moved to a sump for withdrawal, when the basin has been taken out of service and dewatered for cleaning.
- Make provisions for lowering and removing submersible pumping equipment to facilitate maintenance.

Flocculation Basins. Here are some issues that will help O&M of flocculation basins:

- Paddles or mixers in the flocculation basin should be provided with variable-speed drives so that operators have the ability to change the mixing energy to meet different water conditions and thus produce an effluent with the best settling characteristics.
- Flocculation basins should be designed with sump drains and sluice gates in baffle walls to make it easier to drain and wash the basins.
- Sump should be placed at the effluent end or lowest point in the basin floor.

Sedimentation Basins. Some issues that will help O&M of sedimentation basins are as follows:

- Installation of automated sludge removal in sedimentation basins is usually desirable for reducing operating costs and maintaining good process performance. The sludge valves should be opened and closed by a timer set by the operator to maintain the optimum sludge level in the basin.
- Providing a davit next to the sedimentation basins and installing permanent ladders in the basins help facilitate operations and maintenance activities.

Recarbonation Basins. Some issues that will help O&M of recarbonation basins are as follows:

- Recarbonation basins may be located indoors, especially in northern climates. Enclosed recarbonation basins must be provided with exhaust and fresh air supply fans that are operated by one switch located at the enclosure entrance.
- Provide two self-contained breathing devices, located next to the enclosure entrance.
- Provide carbon dioxide analyzers and alarms to continuously monitor carbon dioxide levels in the enclosure.
- Provide a window so that personnel can observe the entire room without entering the room.

Filters. These are some issues that will help O&M of filters:

- Filters are process units that should preferably have local control. Remote control may be provided to a central point, such as the operations control room, but local control at each filter is necessary to facilitate O&M functions. Filter backwashing should be observed by operators at the filter, at least part of the time, or problems that occur in the filters may not be detected for extended periods of time. Typical problems that may be observed are formation of mudballs, cracking, jetting, incomplete backwashing, broken nozzles, and broken underdrain blocks.

- If it is not always possible for an operator to leave the control room to observe a filter backwash, closed-circuit television (CCTV) monitoring may be used to monitor filter backwashing. Good lighting would be required for CCTV monitoring to be effective.

- If the filters are provided with fully automatic control, they should also have semiautomatic and manual control for flexibility of O&M.

- Provisions should be made for continuously monitoring filter head loss and effluent turbidity at each filter because they are key parameters that tell the operator how the filter is performing.

- Filter operating data should be recorded on electronic chart recorders or by logging the data on a computer or SCADA system. Adequate records of filter operation will be important to the plant operators in determining how the filters are performing. The data will be important in the future, when the data can be reviewed for making long-term process decisions and future upgrades.

- Controls must be provided for varying the backwash water flow rate and extending or shortening the backwashing cycle. Operators will need to control the backwash water flow rate and adjust it to the optimum rate as water temperatures change. This is especially critical for surface water plants located in northern climates where wide variations in water temperature can significantly change the optimum backwash rate. The operator should be able to make changes in the backwash cycle both at the central control panel, if one has been provided, and at the local filter control panel.

Chemical Feed Systems. Chemical feed systems normally comprise storage tanks, day tanks, transfer pumps, suction and discharge piping, chemical injection points, and adjustable metering pumps. The following features should be included in chemical feed systems to assist operations and maintenance functions:

- Provide a calibration cylinder for each chemical feed pump.
- Provide a pulsation dampener on the suction and discharge sides of each chemical feed pump.
- Provide adequate pipe supports for metering pump suction and discharge piping. The supports should allow for movement of piping caused by pulsation effect of the pumps.
- Provide flexible connectors on the suction and discharge side of the pumps to minimize transfer of vibration from pump to pipe.
- Whenever possible, install chemical distribution piping in troughs with removable grating for ease of access and repair.
- Provide frequent unions or flanged fittings in chemical piping for easy disassembly for repair or replacement.
- For lime slurry piping, provide frequent cleanout connections on solid piping, and install flexible hosing with quick-connect fittings wherever possible, to allow easy cleaning and lime scale removal.

- Provide adequate roadways adjacent to the plant to allow bulk chemical delivery trucks to turn completely around near the chemical unloading stations.
- Provide chemical storage tanks with level monitoring devices, and allow ample workspace around them for chemical testing.
- Provide for adequate mixing of chemicals with process water by feeding through ejectors or by using motorized or static mixers.
- Liquid chemical systems should generally be selected over dry bulk systems unless transportation costs of the liquid are prohibitive. Liquid chemicals are normally easier to store and feed, and there are no dust problems to contend with. In addition, liquid chemical feed systems usually have a lower capital cost and are simpler to operate and maintain.
- If any chemicals are to be purchased in bags or drums, the bid documents should include specifications for proper dollies, platform trucks, hydraulic lift trucks, hydraulic pallet trucks, or forklifts for handling the chemicals.
- If dry bulk chemical systems are to be used, a dust control system must be provided that is large enough to handle dust created by the vacuum unloading system or blower on the delivery truck. The blowers on bulk delivery trucks usually deliver approximately 700 ft^3/min (1,190 m^3/h) of air at up to 10 psig (69 kPa). The capacity should be confirmed by calling chemical companies to confirm the size of blower on the truck delivering the chemical. If a vacuum system is used, call the equipment manufacturer for the size of the dust collector system required. The dust control system must have more capacity than the unloader system to contain the dust from the unloading process.
- Provide chemical tank remote-fill connections that are distinctive for each specific chemical, to prevent the wrong chemical from being introduced into a storage tank.
- Avoid installing chemical solution piping overhead wherever possible. If overhead pipes leak, it creates a hazard to personnel walking underneath. Chemical pipes should be mounted on the walls or placed in a pipe trench.
- Avoid using threaded stainless steel (typically 316 or 304) pipe. Welded or flanged stainless steel pipe is much easier to install and maintain than threaded. Installing threaded stainless steel so that it is leak-free is difficult, and the threads gall easily. When galling occurs, the pipe cannot be taken apart. If it is suitable for the application, use plastic pipe instead.
- Install chemical transfer and dosing pumps on concrete stands inside chemical containment areas, but at sufficient height that the pumps do not come in contact with the leaked chemical if the entire contents of the storage tank is expelled. Provide sufficient space around the pumps to allow safe calibration tests and maintenance.
- Provide safe and easy access to the chemical storage containment areas.
- Place safety showers and eyewash stations directly outside of chemical storage containment areas.
- Ensure that electrical and control panels are mounted outside of the chemical storage containment area.
- Check with chemical and pipe manufacturers before selection of storage tank and piping materials for liquid chemicals.
- For liquid sodium hypochlorite in schedule 80 PVC piping, specify flanged diaphragm valves, and use solvent and adhesive for solvent-welded pipe that is compatible with sodium hypochlorite solution.

Cathodic Protection Systems. Cathodic protection systems are an important addition to minimize damage to metal structures, equipment, and pipes in a water treatment facility. These design issues can facilitate the O&M of cathodic protection systems:

- Cathodic protection systems should be clearly identified so that they can be monitored by operations staff. Location markers should be specified for locating cathodic protection anodes, cables, and monitoring stations for buried pipe.
- The contract for purchase of cathodic systems should specify vendor training for O&M personnel.

Computer Programs to Assist Operations and Maintenance. Numerous software packages are available to assist in the operations, maintenance, and administration of a water treatment facility. Most types of software can be purchased to run on personal computers (PCs).

Operations Software and Hardware. Operations software usually implies treatment process control software. Several off-the-shelf programs are available to optimize pumping rates, settling rates, chemical dosing, filtration rates, and backwashing. These programs will assist the operators in optimizing processes but are not the only controlling factor; the operator will still need to apply general process history, knowledge, and skill. Most off-the-shelf programs can be customized for a specific facility. Plant-specific process control software can also be designed by an outside consultant or someone on the owner's staff.

Hardware necessary to operate the process control programs should include a high-speed PC with as much random access memory (RAM) and hard disk memory as possible. The amount of RAM and hard disk memory available on new computers is growing rapidly every year. The reason these changes occur is that technological advances accommodate new, more advanced software.

If data will be accessed from other computers, the main computer for the treatment plant should have a modem with proper communication software. The computer monitor should be at least 17 in., and preferably 21 in., to minimize eye strain and allow for proper inspection of on-screen data. If a distributed control system (DCS) is used, operations software can be incorporated in the same way.

Maintenance Software and Hardware. Off-the-shelf computerized maintenance management systems (CMMS) can easily be tested and purchased for use with the facility maintenance program. To ensure that the program will be well suited for its uses, the program selected should be designed for the water treatment industry and not customized from another trade. A good CMMS will be capable of inventorying all equipment, scheduling preventive maintenance tasks, balancing workloads, controlling spare parts, inventorying and tracking consumables, tracking corrective maintenance, printing work orders, recording run times, and following labor hours. The program selected should be easy to use and customize.

The hardware necessary to operate the CMMS should include a computer similar to the one used for process control. If graphics abilities are included in the CMMS package, the maximum amount of RAM should be ordered with the PC. The hard disk storage should be the maximum amount that is currently available.

Laboratory Software and Hardware. Laboratory software and hardware are highly specialized. The software is usually for data collection and management. In some cases, the analytical equipment itself will have the software and storage capabilities to analyze, record, and maintain data. One laboratory computer should also have software loaded for word processing and spreadsheets, and should include the ability to download data from the analytical equipment to the PC. Laboratory data can be collected, evaluated, and stored by the DCS, if a DCS is used.

Administration Software and Hardware. The uses for administration software include spreadsheets for budgets. Spreadsheets can be used to set up and monitor all the facility's budgets. The use of spreadsheets is extremely beneficial for budgetary "what if" scenarios. Spreadsheets are useful for manipulating and evaluating operations data. Scheduling programs are available for looking at working and training schedules. Treatment plant word processing requirements may include form letters, mailing labels, public relations, public presentations, and manual production and editing. Several companies produce training software for educating water system personnel.

Design Reviews. During the design of a water treatment plant, the plans should be reviewed by persons who have several years of operations and maintenance experience. This type of review can be beneficial in making the plant user-friendly to O&M personnel. These people could be on the designer's staff and/or personnel from the owner's staff.

Design Considerations for Plant Maintenance

Maintenance is considered the less glamorous part of O&M. However, that does not diminish the importance of maintenance. A good maintenance program in some respects is more important than operations. Basically, if it isn't maintained, it probably won't operate. Good maintenance keeps equipment available for service and extends the life of the equipment. The electric motor shown in Figure 25.2 is more than 70 years old. It is cur-

FIGURE 25.2 Electric pump drive functional after more than 70 years.

rently still in use driving a high-service pump. There are three main factors that contribute to its longevity:

- It is in a good operating environment. The area is clean and fairly dry.
- The motor is operated regularly.
- Last, but not least, the motor has been properly maintained since its original installation. The motor has plenty of free space around it; therefore, it is easy to maintain.

If a water treatment plant is easy and efficient to maintain, it follows that it should also be economical to maintain. Equipment should be located so that it can be easily installed, maintained, and removed. Areas where equipment is installed must be well lighted, must have easy access to power, must be well ventilated, and must have moisture kept to a minimum.

Excess Equipment. Having "excess" equipment in a treatment plant is a lot like having an automobile with many extras that are not essential, but nice to have. From a maintenance perspective, the more extras there are, the more things there are to maintain.

An example of excess equipment would be a pumping facility in which no more than three pumps of the same size are required to operate to meet the water demand for the next 10 years, but the plant has been designed and constructed with six pumps in anticipation of the 20-year water demand. If one pump is counted as standby, there are now two extra pumps that are not necessary for the operation of the facility, but must be maintained. Instead of installing all the pumps now, install the fifth and sixth pumps later when the water demand requires them. This will minimize the number of pumps to be maintained in the facility.

Accessibility of Equipment. A common problem in the design of many water treatment plants is that insufficient room has been provided to work on equipment. Convenient access to all equipment is necessary for proper maintenance. Figure 25.3 shows equipment that is in poor condition. The figure also shows that access to the equipment is difficult and working space around the equipment is limited. If equipment is not accessible and working space is cramped, the equipment probably will not be maintained. These factors enhance equipment accessibility:

- The layout of basins and buildings should be designed with careful consideration of all necessary maintenance activities. This applies to elevated equipment as well as equipment installed on the floor. There should be space to place ladders or scaffolds to gain access to all elevated equipment.
- A good rule of thumb is to provide a minimum of 3 ft (1 m) of clear space around each item of equipment that requires maintenance. When laying out areas such as pump stations, designers should remember to allow for associated appurtenances such as seal water piping, control panels, electrical conduit, and ventilation ducts. It is common for designers to show mechanical, electrical, and building utilities on separate drawings. So, unless all drawings are coordinated, it is possible for a designer to provide what appears to be adequate clearance around mechanical and electrical equipment without taking into account the space taken up by ancillary items.
- Progressive cavity pumps require extra room to remove the rotor. To ensure there is adequate space, leave the length of the rotor plus about 1 ft (0.3 m) to the nearest wall, column, or other equipment in front of the barrel of the pump.
- Maintenance personnel should be able to get to all equipment without climbing over other equipment or crawling over or under piping. Equipment should be placed so that

FIGURE 25.3 Poor access to equipment results in poor maintenance.

it can be disassembled and reassembled in place or can be removed and replaced without removing other equipment or lifting it over other equipment. The pumps shown in Figure 25.4 are examples of equipment with good accessibility.

- Equipment should be located so that grease fittings, oil fittings and drain lines, oil filters, and air filters are easy to reach. Drive belts, drive chains, and guards should be easy to remove and replace. If equipment is mounted high above the floor, the area around it should be unencumbered so that a ladder can be safely placed to perform maintenance.
- Piping with good accessibility is shown in Figure 25.5. The piping is layered on the supports with enough distance between the layers to repair, remove, and replace the piping.
- Accessways into tanks and vaults need to be sized so that maintenance personnel can comfortably enter the portholes wearing an SCBA. Portholes should be located far enough away from walls, ceilings, and columns that personnel can enter and exit without difficulty.

Installation of Cranes and Hoists. Water treatment plants should be designed to include installation of cranes and hoists for maintenance of heavy equipment. A bridge crane, monorail, or hoist beam is often provided in larger plants. For equipment that can be lifted with a 2-ton (1,800-kg) chain hoist, lifting rings can be strategically installed in the building structure above each unit so that a chain hoist can be easily attached for use. The pump shown in Figure 25.6 has a monorail and chain hoist located directly above to facilitate removal and replacement of the pump and motor. The pump shown in Figure 25.6 is undergoing maintenance. This is the reason the electric motor terminations are disconnected and the belt guard is removed.

FIGURE 25.4 Pumping equipment with adequate access.

FIGURE 25.5 Process piping supports.

FIGURE 25.6 Pump with monorail and chain hoist.

Power and Lighting Considerations. Two important items in performing maintenance are having power receptacles located near where the work is performed and having adequate lighting in the work area. Power receptacles should be located so that maintenance personnel need to carry only a 50-ft (15-m) extension cord to obtain power for tools and test equipment. Adequate lighting is absolutely necessary to perform maintenance efficiently and safely.

Tunnel and Gallery Considerations. Galleries in water treatment plants are underground areas where equipment, process piping, and utility piping are located. Tunnels are underground areas or corridors where only process and utility piping is located. Some design issues for tunnels and galleries are as follows:

- Galleries should be designed so that there is a clear passageway and adequate space to install, maintain, and remove the equipment. There should be hatches and elevators so that equipment can easily be taken in and out of the galleries. All piping in galleries and tunnels should be located so that there is access for maintenance.
- Pipe chases and tunnels are areas where lighting is usually inadequate for maintenance. To conserve on energy during normal circumstances, one-half of the lighting should be put on a separate switch for normal conditions, but extra lighting will be available when maintenance is performed. Another consideration would be to use motion-activated

lighting to facilitate the safe passage of O&M personnel through the tunnels and galleries. After a period of time with no movement in the tunnel or gallery, the lights would be turned off automatically, conserving electricity.

- Galleries and tunnels are generally prone to collecting water. Wet spots on floors caused by condensation and leaks pose a significant slipping hazards, and biological growths that form due to wet and humid conditions may make floors even more slippery. It is essential to provide good drainage to minimize wet floors. Galleries and tunnels should have floor drains, gutters, pitched floors, and sumps with automatic pumps in order to keep floors as dry as possible.
- There should also be a ventilation system for galleries and tunnels that will provide enough air circulation to dry the floors if moisture does occur, as well as ventilating the area. In plants located in high-humidity climates where at times the water temperature will be considerably colder than the ambient air temperature, it can be expected that there will be considerable condensation from piping in galleries and tunnels. Insulating the pipes can sometimes mitigate the condensation problem, but makes it more difficult and expensive to make repairs. If the pipe ever needs to be worked on, the insulation must be removed and replaced after making the repair. Another solution to keep an area dry is to install a dehumidification system.

Maintenance Shop Considerations. The design of the treatment plant maintenance shop should take into account the level of maintenance that will be done by plant personnel and the number of persons who might be working in the shop. Types of work that should be considered when designing the shop are

- Cleaning equipment prior to disassembly
- Overhaul of major equipment
- Vehicle maintenance
- Machining (lathe, press, punch, shears, milling cutters, drill press, etc.)
- Sheet metal work
- Welding (arc, submerged arc, cutting torch, etc.)
- Certified equipment calibration
- Tool storage
- Spare parts storage
- Hazardous materials storage
- Lubricant storage
- Flammable materials storage

If it is possible that complete overhaul of major equipment will be performed by plant personnel, the shop should be large and well equipped and should have large hoists and cranes to handle the equipment. The doors should be large enough to accommodate the largest equipment.

If vehicle maintenance is to be performed in the maintenance shop, some of the special requirements include

- A vehicle lift and lubrication pit
- Tire changing equipment
- Ventilation equipment for removal of exhaust fumes
- Engine analyzing equipment
- Equipment for holding spent lubricants for disposal

Certified calibration of instrumentation and test equipment requires special equipment and a clean, isolated working environment. It is important to have test equipment recertified to National Institute of Standards and Technology (NIST) criteria.

In all but the smallest maintenance shops, secure rooms or cages should be provided for tool and spare parts storage. The rooms must be sized to contain all the tools and spare parts used by the plant and to provide enough extra space for expansion.

PREPARING A PLANT FOR START-UP

The designer must start making preparations for plant start-up far in advance, to make sure that all permits, funding, staffing, and equipment checkout are completed when the plant is ready to begin operation (see Chapter 27).

Obtaining Operating Permits

There are numerous federal, state, and local government regulations and permitting requirements that must be complied with before a water treatment plant can be placed in operation. This is an area of concern not only at the beginning of the design of a water treatment plant, but also once the plant is constructed and placed in operation. Permits and regulations are discussed in detail in Chapter 24.

Preparing an Operations and Maintenance Budget

Several O&M budgets should be considered during the design phase of a water treatment facility. Initial considerations taken by the designer regarding these budgets can help the owner minimize ongoing O&M expenses. These budgets include staffing, chemicals, utilities, ongoing training, equipment, and other relevant O&M budgets. The fiscal effect of each of these budgets on the owner should be carefully considered by the designer.

Staffing Budget. Initial staffing costs for a newly constructed facility include recruitment costs such as costs for advertising the availability of the positions, interviewing costs, travel expenses, initial indoctrination and training, and possibly relocation expenses. The cost for recruitment of the higher management may include the services of a professional firm to locate potential candidates.

Once the facility is initially staffed, the staffing budget must consider employee wages, fringe benefits, and the continuing costs of restaffing due to employee turnover. Fringe benefit costs generally range from 30% to 50% of salaries. Employee turnover can be influenced by such factors as the area unemployment rate and wages and benefits provided by other employers in the area. The turnover rate can be estimated by using historical administrative records from the owner or from other water treatment plants in the area.

If the facility's employees will belong to one or more unions, labor rates should be established by contract before the new facility goes on-line so that the rates can be accurately applied to the O&M budget.

Chemical Budget. The chemical budget for a water treatment facility can be a significant portion of the total budget. To project the chemical budget, it is necessary to list all chemicals to be used in the treatment processes and the projected quantities to be used annually. If possible, three price quotes should be obtained for each chemical, including shipping costs, along with the availability from each supplier.

Utilities Budget. The budget for the utilities used by a water treatment facility is usually significant, with electrical consumption being the single largest operating cost. If the electric utility offers special off-peak rates, power use should be closely monitored. This means, if possible, backwashing or using other electrically intensive equipment during off-peak periods rather than peak periods. Other means of electrical savings can include the use of motors with higher power factors and efficiency ratings. The electric utility should be contacted to discuss other means of electrical savings. Other utilities to be considered in the budget include natural gas, sewer, and telephones.

Training Budget. Training not only is necessary for all employees who are initially hired to start up the plant, but also should continue during the lifetime of the facility. Most states have mandated training requirements for operator certification and safety. Training courses may or may not be provided locally. The training costs that should be included in the budget include salary for employees while they are attending training, as well as travel, registration, and training materials. It may also include the costs of hiring a professional trainer to provide on-the-job staff training.

Training includes new employee orientation and current employee proficiency training. Because the water supply industry is highly technical and ever changing, every opportunity should be taken to send facility employees to attend seminars, workshops, and conferences to upgrade their skills and to keep up professional contacts. On-line training opportunities are now also an option.

Equipment Budget. Equipment costs should be budgeted in several separate categories. Smaller new equipment will be purchased during the operation of the facility to replace broken equipment, add new types of equipment, and upgrade the facility. These costs tend to increase with the age of the plant.

There should be a spare parts category in the budget to cover costs of maintaining an inventory of critical and hard-to-locate parts. Another budget category is consumables necessary to keep equipment in operation, such as oil, grease, nuts, and bolts. Finally, although some tools will be initially purchased, a budget allowance should be included for constantly replacing and upgrading the tools required for plant maintenance.

Large equipment is usually a budgeted capital cost. This could include such items as vehicles, construction equipment, new pumps and generators, and high-capacity process equipment. Replacement of large equipment is usually a one-time expense, so it is budgeted by setting aside a certain amount annually in an equipment fund to purchase new equipment or replace old equipment that has failed or is a planned replacement because it is either out of date or no longer economically practical to maintain.

Other Budget Items. Other items that might have to be considered in the overall O&M budgets include payments to outside contractors for jobs such as sludge hauling, grounds keeping, painting, and maintenance of HVAC systems, if these jobs are not performed by plant staff. The costs of public relations projects such as flyers, brochures, posters, and newsletters should be planned into the future. Other items that should be provided for in the budget include personnel awards, professional journals, professional memberships, uniforms, office supplies, and office equipment.

Preparing a Staffing Plan

The designer must consider plant staffing early in the plant design, both in terms of creating a plant that can be efficiently operated by a minimum staff and to decide what degree of automation should be included in the design.

Staffing Criteria. Following are some general statements regarding how the design of a facility corresponds to the staffing requirements.

- The more automated a water treatment plant is, the fewer personnel are required to operate the plant. Although a large fully automated plant will require significantly fewer personnel for operation than the same size plant that is not automated, the automated plant will require additional instrumentation technicians to handle the increased repair and preventive maintenance of the instrumentation and controls.
- The more processes there are in a water treatment plant, the more personnel will be required to operate and maintain it. For instance, a direct filtration and chlorine disinfection plant would require fewer personnel to operate and maintain than the same size conventional filtration plant with lime-soda softening, recarbonation, ozonation, and solids dewatering.
- The more spread out a water treatment plant and distribution system are, the more personnel are required to operate and maintain them. For instance, water systems that have their source of supply, treatment plant, booster stations, and reservoirs all distributed many miles apart will have significantly greater personnel costs than a water system that is relatively compact.

The estimated number of personnel it would take to operate and maintain conventional water treatment plants with 5 and 50 mgd (19 and 190 ML per day) capacities is shown in Table 25.1. The estimated number of personnel is listed for semiautomated and fully automated facilities.

TABLE 25.1 Estimated Staffing Requirements for 50 and 5 mgd Water Treatment Plants

Position	50 mgd (190 m^3/day) water treatment plant		5 mgd (19 m^3/day) water treatment plant	
	Semiautomatic	Fully automatic	Semiautomatic	Fully automatic
Plant manager	1	1	1	1
Operations supervisor	1	1	0	0
Maintenance supervisor	1	1	0	0
Operator	15	5	5	1
Mechanical technician	3	3	1	1
Electronics technician	2	2	1*	1*
Instrument technician	2	3	0	1
Laboratory technician	2	2	1	1
Buildings maintenance	1	1	1†	1†
Grounds maintenance	1	1	0	0

*This position is split between electrical and instrumentation duties.
†This position is split between janitorial and grounds-keeping duties.

During the preliminary design stage, the designer should begin looking at the level of staffing required to operate and maintain the water treatment plant. This is valuable in helping the owner decide which of several alternative designs is most suitable. One of the issues that a staffing estimate resolves is whether it is economically feasible to go to a fully automated plant, or whether it is better to use less automation and more operating personnel. A fully automated plant normally represents a higher capital cost and lower annual O&M cost; a less automated plant generally has a lower capital cost and higher O&M cost.

The owner is usually responsible for staffing the new facility, but the designer will help estimate the staffing needs. In some instances, the owner may even have the designer locate, interview, and hire the staff. The designer needs to be aware of the job descriptions, job needs, and special consideration for each key position, to ensure that the right people are selected for the right job and the right number of people are on hand to operate and maintain the plant, once it is ready to begin operation.

Supervisory Positions. In the following discussions, supervisory positions are based on a medium to large water treatment plant.

Plant Manager. The plant manager's primary goals are to ensure that the quality of the final product continuously meets all state and federal requirements and that this is accomplished cost-effectively. It is the plant manager's responsibility to direct the management, operations, maintenance, and laboratory personnel toward this overall goal. The plant manager's office should be centrally located within the facility to allow easy access to plant supervisory personnel.

The plant manager should preferably be hired early in the design phase and included in meetings and discussions between the owner and designer regarding process design, specifications, construction, operations, and maintenance. During construction of the plant, the plant manager should be on-site to become familiar with the plant as it is being constructed and to help with on-site inspection during construction.

Operations Supervisor. The operations supervisor is the person directly responsible for controlling plant treatment processes. Duties include scheduling operator shifts, supervising operators, controlling the treatment processes, ensuring that sampling and monitoring are performed, providing operator training, and reporting on plant operations to the plant manager. The operations supervisor's office should be located centrally within the facility and should be easily accessible to operations personnel and close to plant processes.

Maintenance Supervisor. The maintenance supervisor oversees all mechanical, electrical, instrumentation, building, and grounds maintenance activities for the plant. Duties include scheduling and supervising preventive and corrective maintenance tasks and reporting on plant maintenance to the plant manager. Typically, the maintenance supervisor has an office located within the maintenance facility for better communication with maintenance personnel.

Laboratory Supervisor. The laboratory supervisor is responsible for supervising all process controls and analyses required by regulations. The supervisor's office should preferably be located next to or in the laboratory. The plant laboratory is an integral part of operations; therefore, the laboratory itself should be located near the central operations control area within the plant so that samples can easily be transported to it and the operations supervisor has easy access to the process-related analytical data.

Operations Personnel. Operations personnel are responsible for the day-to-day operation of the plant. They take samples, make process modifications, operate equipment, take and record operating data, make rounds to visually inspect processes and equipment, make entries into daily logs, keep chemical inventories, and perform various other duties related to operating the plant to meet finished water quality goals. There are usually one or two control room operators whose duties include monitoring indications of the various instruments in the control room and making changes in the processes as required. The plant may have other operators assigned to areas around the plant whose duties are related to those areas (e.g., chemical feed building, filters, and sludge dewatering).

Maintenance Personnel. Maintenance personnel are responsible for taking care of equipment and facilities at a water treatment plant. Their goal is to ensure that the plant equipment and facilities are kept in good condition, thereby extending the useful life of the equipment and facilities. To accomplish that goal, maintenance personnel must be well trained and motivated. In addition, they should have sufficient tools, spare parts, and a well-planned and implemented maintenance management system. Duties include maintenance of mechanical, electrical, and instrumentation systems as well as maintenance of buildings and grounds. Examples of each type of maintenance are given below:

- *Mechanical maintenance.* Plant maintenance technicians are responsible for preventive and corrective maintenance on mechanical equipment. This equipment includes but is not limited to air compressors, elevators, bridges, cranes, pumps, valves, and piping. Plant maintenance technicians are also responsible for all routine equipment alignment, but usually do not perform work involving electric power and instrumentation and control (I&C).

- *Electrical maintenance.* Electricians are responsible for preventive and corrective maintenance on electrical equipment. This includes electric motors, motor control centers, emergency generators, uninterruptible power systems, unit substations, protective relays, lighting, distribution panels, and lighting panels. Electricians usually do not perform work on mechanical and I&C equipment. Electricians may be required to do equipment alignments if they have had to remove and replace or uncouple an electric motor in the course of their work.

- *Instrumentation maintenance.* Instrumentation technicians are responsible for preventive and corrective maintenance on I&C equipment. This equipment includes electrical, pneumatic, hydraulic, and thermal control systems. Examples of types of equipment instrumentation technicians will work on are equipment control panels, low-voltage control systems, compressed air control systems, and programmable logic controllers (PLCs). Instrumentation technicians usually do not perform work on electric power and mechanical equipment.

- *Building maintenance.* Plant personnel responsible for building maintenance perform work to ensure that the buildings are kept clean and in good repair and that the environment in the buildings is comfortable. Building maintenance work includes but is not limited to painting, masonry, concrete, glazing, carpentry, plumbing, HVAC systems, security systems, and cleaning.

- *Grounds maintenance.* Plant grounds personnel are responsible for maintaining all the outside area of a treatment plant. Grounds maintenance work includes but is not limited to seeding and cutting lawns, planting and pruning shrubbery and trees, installing and repairing fences, installing and maintaining irrigation systems, fertilizing, roadway repair, and snow removal.

Laboratory Personnel. Laboratory personnel perform both chemical and biological testing of water. They perform process control and regulatory testing on samples collected by operations or by laboratory personnel. Results of data generated by laboratory staff are used by operations to control equipment and processes to meet plant water quality goals.

Plant Start-up Preparations

Plans for starting up a new water treatment plant should be initiated once design is well underway. This is accomplished through meetings with those who will be directly involved with start-up. Once construction is underway, the O&M manuals must be produced to aid operations and maintenance personnel in running a new facility in an efficient and cost-effective manner. These manuals should cover everything from the designer's intent for the plant processes to a suggested list of spare parts that should be ordered and kept on hand.

Training of all employees is an essential part of the successful start-up of a new treatment plant. Even if new employees have previous experience in the water supply field, they will still need training in the operation of the specific equipment and processes in a new plant.

As a water treatment plant nears the end of construction, it is time to start equipment checkout. Equipment checkout should be thoughtfully and carefully planned. For a detailed discussion of training and equipment checkout and start-up, see Chapter 27.

CHAPTER 26
WATER TREATMENT PLANT CONSTRUCTION COST ESTIMATING

Ronald F. Cilensek
CH2M HILL
Alexandria, Virginia

INTRODUCTION

It has been said that construction cost estimating is more an art than a science. This is true only to the extent that application of science also requires the use of sound judgment and experience. While a thorough understanding of the process mechanical elements of a water treatment plant would appear to be the most important aspect in estimating its cost, it is only one discipline of a multidisciplined project.

All cost estimates will contain a certain amount of experienced judgment or educated guesswork concerning the various cost elements, which comprise the estimate. An order-of-magnitude or preliminary cost estimate may be based on all judgment, while a definitive level estimate may only require judgment for those cost elements with historical cost fluctuations.

As an example, an order-of-magnitude estimate may require sound judgment in the identification of a previous project with similar scope of work characteristics that will be factored to estimate the cost of the project in question. A definitive level estimate may only require judgment concerning specialized concrete forming labor with historically wide labor cost fluctuations, with all other elements being based on verified vendor quotes or historical cost data with minimal deviations.

The various types of cost estimates used are based on the level of project design, which defines the known scope of work. The scope of work can range from only a daily treatment flow rate and major process facilities, to a complete set of plans and specifications for construction. The cost estimate that will be produced will also reflect a certain contingency and accuracy range based on the scope of work reflected by the level of design.

LEVEL OF ESTIMATES

AACE International (formerly American Association of Cost Engineers) identifies three basic levels of cost estimates: order-of-magnitude, preliminary, and definitive. A fourth type of estimate called a conceptual level has also been identified, and it is most commonly developed between the order-of-magnitude and preliminary level estimates. As their names imply, these estimates correspond to the various phases of design through which a project progresses (see Table 26.3).

Order-of-Magnitude

An order-of-magnitude estimate is usually based on previous project data that are factored up or down to reflect the size and treatment capacity of the project in question. Due to the numerous types and variations of water treatment plant (WTP) processes, it is recommended that a similar plant process type be used for factoring. Where the previous project cost data may not reflect the same process facilities for the entire plant in question, one can itemize those facilities that are applicable and use them in factoring the cost in a cut-and-paste application. For those facilities for which no historical cost data are available, a separate cost estimate will need to be developed.

In addition, cost indexes for the location of the plant in question in comparison to the cost data used should be identified and the factored cost adjusted accordingly. Also of note is that factoring of previous data is usually accurate for new construction only with favorable geotechnical conditions. Where the project in question involves modifications or expansion of existing facilities, or where geotechnical issues need to be addressed, factoring of previous data alone will not accurately address the costs involved.

Conceptual

A conceptual level cost estimate is based on a preliminary scope of work that may reflect some equipment and structure sizing. This type of estimate is recommended as a supplement to an order-of-magnitude estimate for projects that involve modifications to existing facilities. It is similar to an order-of-magnitude estimate in its accuracy but requires additional scope identification to reflect the cost of the modified facilities. It is also used where a more accurate cost estimate is required prior to the predesign stage, or where unusual site or geotechnical issues need to be addressed, i.e., extensive excavation support or groundwater control.

The scope of work required to modify existing facilities is one of the biggest unknowns in preparing a preliminary cost estimate for a treatment plant upgrade. However, if one can identify the basic process mechanical elements required, there are factors that can be applied to these elements that will generate costs for the related disciplines of work. The most common application of this method is to apply factors to the new process equipment to generate costs for the major work elements such as mechanical piping, electrical, and instrumentation and control (see Table 26.2).

If structural or geotechnical issues will need to be addressed, consultation with personnel experienced in these disciplines will be required. Where a thorough analysis of these discipline requirements is not practical, past experience with similar installations and an understanding of the work scope and existing conditions will allow an experienced process engineer to develop costs for these items.

Preliminary Design

The preliminary design cost estimate is considered by many as the most important estimate prepared for a project. This estimate is used to verify the basic concept of the project, and it usually determines the financial viability of the project. It also forms the baseline cost of the project from which all future cost impacts of the evolving design will be compared.

A preliminary design cost estimate is based on the predesign concept of the project. The major facilities of the project are identified, and the footprints for these structures are sized. Major equipment elements are identified and sized, usually based on specifications for a specific manufacturer. Preliminary geotechnical information is available to accurately evaluate structural considerations for the project. Major process mechanical piping and electrical equipment are identified and sized.

The basis of the preliminary design estimate is to obtain as much information as possible regarding the scope of the project. Most of this information is identified in a preliminary design report or other similar document. However, other pertinent information that could influence the cost estimate may not be identified in these reports. Site or existing facility access restrictions, sequencing of new and existing process elements for existing plant operations, and identification of temporary power sources for early-phase facility start-ups are types of information that are usually not identified in a preliminary design report. Information of this type is usually known by the various discipline design engineers who contributed to the preparation of the preliminary design report.

Communication with the various design discipline engineers is essential in preparing an accurate preliminary design cost estimate. At this stage in the project, they are the most knowledgeable with respect to their design elements and often are aware of information that could influence the cost estimate. However, it usually requires an experienced cost estimator with past treatment plant experience to know what additional information is required to accurately reflect the cost of the project.

As previously stated, the preliminary design cost estimate is usually identified as the baseline cost estimate for comparison of all future design considerations. This concept is also known as the *design to cost* principle. As the various facilities of the project proceed through final design, revised cost estimates based on the latest design information are prepared for comparison to the baseline cost estimate. Variations in these costs that exceed a predetermined amount (usually 15%) will require an evaluation to determine the cause of the variance.

Definitive Level

A definitive level cost estimate is based on the final design documents for the project. Also known as the *engineer's estimate,* this estimate reflects the project scope as represented by the contract documents from which construction contractors will base their contract price to build the project.

At this cost estimate level, all major equipment costs should be verified by vendor quotations that include all start-up and training requirements. Major process piping material, including fittings and valves, should be verified with vendor quotations. Major electrical components and process instrumentation and control systems (PICSs) should have costs assigned and confirmed by vendors. Local labor factors and market conditions should also be analyzed for application to the project in question. These are just a few of the numerous items that should be considered in the preparation of a definitive level cost estimate. It is recommended that a senior level cost estimator with prior definitive level cost estimating experience review the cost estimate for inclusion of these items, in addition to overall estimate quality control.

After receipt of bids and award of the construction contract, the definitive level estimate can be used as a basis of comparison in reviewing the contractor's schedule of valves for partial payment applications. Therefore, it is essential the engineer's estimate be prepared in the same format requested of the contractor in the bid documents.

ESTIMATING METHODOLOGIES

Most cost estimates, whether produced by a specialized cost estimating consultant or a water treatment plant design firm, are based to a certain degree on historical cost information. Where historical cost information is not available, other estimating factors must be used.

Historical Cost Data

Most firms that produce cost estimates have access to past water treatment plant bid data and other cost information generated during the construction of various projects. This information can be useful for estimating the costs of future projects if the data are compiled in a manner applicable to other projects.

One of the more significant and accurate sources of treatment plant cost data is the schedule of values required to be submitted by the construction contractor for partial payment applications. The contractor is required to submit a detailed cost breakdown of the bid proposal to enable assessment of the value of the work performed as the job progresses. This cost information is usually broken down by the Construction Specification Institute (CSI) format for the major facilities that make up the project.

Based on this past information, a database is usually set up to compile historical treatment plant costs. This database can be divided into process facilities and subdivided by CSI specification section. These sections can be compared with total process facility costs and total treatment plant costs, resulting in patterns of percentage allowances for each CSI specification section. This information can be useful for developing conceptual or preliminary cost estimates where only structural and major equipment elements are identified.

An example of this estimating application is a chlorine contact basin and pump station where only the civil, structural, and major process equipment elements of the facility are identified. These elements constitute the known costs and make up a certain percentage of the total facility cost based on historical cost data. Total facility cost can then be projected from the known cost percentage. Other unknown elements of the facility can be computed based on historical cost percentages applied to the projected facility total (Table 26.1). This application can also be used to verify detailed cost data that may be suspected of not being totally inclusive of the scope of work.

Cost Estimating Factors

Where historical cost data are unavailable or irrelevant to a new project, other cost estimating techniques can be employed. Because these other cost techniques are not as reliable as historical cost information, their use is usually confined to conceptual or preliminary design level cost estimates.

TABLE 26.1 Historical Cost Records

Item of work	Quantity	Unit	Unit cost	Total cost
Chlorine contact basin:				
Excavation and haul	24,000.0	CY	$ 10.00	$240,000
Steel sheet piling	23,000.0	SF	10.00	230,000
Dewatering—three wells with pumps	15.0	Mo	7,500.00	112,500
16" sq. precast conc. piles	34,000.0	VF	20.00	680,000
Granular backfill	2,100.0	CY	25.00	52,500
30" base slab	3,140.0	CY	200.00	628,000
20" walls	1,325.0	CY	300.00	397,500
12" walls	790.0	CY	500.00	395,000
8" walls	42.0	CY	600.00	25,200
12" elev. slabs	160.0	CY	400.00	64,000
72" × 72" sluice gates	4.0	Ea	23,000.00	92,000
Subtotal	75.0%			$2,916,700
General requirements	5.0%		$3,888,933	$194,000
Metals	5.0%		3,888,933	194,000
Finishes	2.5%		3,888,933	97,000
Mechanical/misc. equip.	10.0%		3,888,933	389,000
Electrical/I&C	2.5%		3,888,933	97,000
Total	100.0%			$3,887,700
Contingency	15.0%		$3,887,700	583,000
Total cost, chlorine contact basin				$4,470,700
Pump station:				
Excavation and haul	2,600.0	CY	$ 10.00	$ 26,000
Steel sheet piling	8,800.0	SF	10.00	88,000
Dewatering—one well with pump	9.0	Mo	2,500.00	22,500
16" sq. precast conc. piles	4,000.0	VF	25.00	100,000
Granular backfill	700.0	CY	25.00	17,500
30" wet well slab	270.0	CY	200.00	54,000
20" wet well walls	300.0	CY	300.00	90,000
12" weir walls	24.0	CY	500.00	12,000
24" sq. columns	12.0	CY	800.00	9,600
12" elev. slabs	86.0	CY	400.00	34,400
Electrical rm. slab on grade	60.0	CY	200.00	12,000
Building	3,500.0	SF	100.00	350,000
Overhead crane	1.0	LS	25,000.00	25,000
72" × 72" sluice gates	2.0	Ea	23,000.00	46,000
12,400 gpm pumps—CS	2.0	Ea	22,500.00	45,000
24,600 gpm pump—CS	1.0	Ea	56,500.00	55,500
24,600 gpm pump—AS w/AFD	1.0	Ea	70,500.00	70,500
24,600 gpm pumps—AS w/AFD	2.0	Ea	86,000.00	172,000
Subtotal	52.5%			$1,230,000
General requirements	5.0%		$2,342,857	$117,000
Metals	5.0%		2,342,857	117,000
Finishes	2.5%		2,342,857	59,000
Mechanical/misc. equip.	20.0%		2,342,857	469,000
Electrical/I&C	15.0%		2,342,857	351,000
Total	100.0%			$2,343,000
Contingency	15.0%		2,343,000	351,000
Total cost, pump station				$2,694,000

CY = cubic yards; SF = square feet; Mo = mouth; Ea = each; VF = vertical feet; LS = lump sum; CS = constant speed; AS = adjustable speed; AFD = adjustable frequency drive; I&C = instrumentation and control.

TABLE 26.2 Percentage of Total Facility Cost Based on Historical Cost Data

Equipment cost	100.0%
Equipment installation	50.0
Process mechanical piping	65.0
Instrumentation and control	20.0
Electrical	10.0
Buildings	20.0
Yard improvements	10.0
Service facilities	70.0
Engineering and supervision	35.0
Project management and overhead	40.0
Total percentage of equipment cost	420.0%
Subtotal percent of project cost for above	80.0
Additional project cost elements	
Misc. and unidentified equipment	10.0
Misc. and unidentified process mechanical	5.0
Misc. and unidentified electrical/I&C	5.0
Percent of total project cost	100.0%

Source: M. S. Peters and K. D. Timmerhaus, *Plant Design and Economics for Chemical Engineers,* 3d ed., McGraw-Hill, 1982.

One such method is applying a designated percentage factor to the known cost of the major process equipment for a project. This percentage application method can be used to estimate all other cost factors of a project, as shown in Table 26.2.

Another cost estimating factor is to apply the six-tenths rule. This method is primarily used to estimate costs of facilities based on a known cost and capacity of a similar facility. The capacity ratio of the known facility is multiplied by its cost and factored by 0.6 to account for the fixed costs associated with the construction of any facility. For example, a filter building facility with a capacity of 12 mgd has a construction cost of $18 million. Estimate the cost of a similar 15 mgd facility.

$$\text{Cost} = 18 \times \frac{(15)^{0.6}}{12} = \$20.58 \text{ million}$$

Although it is suggested that these cost factors be confined to use with order-of-magnitude or conceptual level cost estimates, they can also be used in conjunction with preliminary or definitive level estimates for minor items that do not significantly affect total facility cost. Using percentage factors for items such as site restoration, painting and finishes, and small piping less than 6 in. (150 mm) in diameter is appropriate because these costs historically do not exceed 10% of the total facility cost.

SPECIAL COST CONSIDERATIONS

Many construction projects have special circumstances that must be considered in preparing cost estimates, such as special soil or site conditions.

Geotechnical and Site Constraints

Locations for new water treatment plants, and to a lesser extent for existing plants, are usually confined to the secondary real estate areas. The economic reality is that most municipalities cannot afford to purchase prime real estate that reflects good soil conditions and location. Therefore the construction costs for these facilities will usually reflect additional costs to build water treatment facilities in less than ideal soil conditions or at locations far removed from the water distribution systems.

One of the cost factors to be considered in the siting of any new water treatment plant is the proximity of the plant to its primary water source. Most land adjacent to any water body with soil conditions advantageous for construction has already been claimed for development, particularly in urban areas. Therefore the remaining land will probably require extensive geotechnical considerations for facility construction.

Costs for geotechnical site investigation, structural design of special foundation support, and permitting of wetlands mitigation are just a few of the possible additional engineering costs required for plant siting. The additional construction costs include groundwater dewatering and control, installation of piling systems for structure support if soil conditions are not satisfactory, and construction of new wetland areas if the new facility site requires filling in of existing wetlands. These additional costs are unique to specific sites and are not usually reflected in most historical cost data and therefore must be added to any factored cost data used in order-of-magnitude or preliminary level estimates.

Remote Site Locations

There are occasions when the geotechnical and site constraint criteria for new plant facilities are so extensive at the selected site, or the cost of the land so prohibitively expensive, that construction of new plant facilities at a location far removed from the distribution system may be more economical. This situation may result in additional construction costs also not typically reflected in historical cost data.

Transportation to the site of all construction material will be affected, with material from local suppliers being the most pronounced. Costs for imported granular fill material from local quarries, ready-mix concrete, and other materials that are small in quantity but require full truckloads for delivery are examples of construction materials that will increase in cost because of additional transportation costs.

Labor costs will also increase to a certain extent, based on the distance of the project site from normal work locations. If the location is far from normal work areas, travel time may have to be paid to the workers as an incentive to travel a farther distance to work. If excess excavated material is required to be hauled off-site, additional labor and trucking costs will be incurred.

These are just a few of the possible additional costs that would be incurred in the construction of a new plant at a remote site. A thorough review of the scope of work by an experienced cost estimator or construction manager should be performed to identify the extent of these special remote-site cost impacts.

Other Considerations

Consideration must also be given to special costs that are caused by various federal regulations and requirements. Some examples are indicated below.

Americans with Disabilities Act (ADA). Recent regulations are affecting the costs for new and renovated water treatment plants with respect to requirements for accessibility to public facilities. Some common additional costs can be categorized in the following areas:

- Additional parking areas for handicapped persons
- Reconfiguration of door openings to be fully automated
- Rehabilitation of existing lavatory layouts
- Addition of exterior ramps for handicapped persons
- For multistory facilities, possible addition of an elevator

State Sales Tax. Most public-financed water treatment plant projects are exempt from state and local sales taxes. Where this is the case, a tax exemption certificate will be provided by the public agency. However, not all states follow this policy. In particular, any project for the federal government (such as the Army Corps of Engineers, the Park Service, and the Fish and Wildlife Service) will not be exempt from sales tax. The sales tax policy of the state and local municipal government where the project is to be constructed should be reviewed if there is a question of sales tax inclusion into the project cost. Where sales tax is applicable, it is usually applied only to the material cost portion of the project.

Hazardous Waste Audit and Removal. Recent trending in the market indicates that with future funding being levelized or reduced, the option to modify or rehabilitate existing facilities is another special cost consideration for the estimator. Because of the limited estimating manuals and available cost information, a cost estimator should request that an audit be conducted on the existing facility in the following areas:

- Lead paint—primarily on windows and walls
- Insulation material—particular attention to the presence of asbestos on piping systems, roofing systems, boiler packages, and ductwork systems
- Chimney stacks—condition, particularly the lack of liners
- Oil tanks and chemical tanks—particularly if there are any residuals or free products
- Flooring systems—types of materials and adhesives
- Underground storage tanks—the possibility of contaminated soil
- Transformers—any oil-filled transformers with polychlorinated biphenyls (PCBs)
- Ballasts in light fixtures—general condition
- Fireproofing of steel framing systems—presence of asbestos

A local hazardous waste remediation contractor should be consulted for a proper estimate and subsequent cost evaluation by the owner.

FINALIZING THE COST ESTIMATE

As the design of a facility progresses, the contingency allowance should be adjusted, and consideration must also be given to whether the scope of the job has changed since it was initiated.

Contingencies

Contingency is the application of a factor to the cost estimate that is intended to account for those items not specifically identified in the scope of work but which historically are found to be required. After completion of the estimate, a thorough review of the estimate must be made to access the completeness of the scope of work represented by the design information.

For order-of-magnitude or preliminary design level estimates, the scope of work presented by the design reports and communication with design personnel is usually insufficient to reflect the full scope of work for the project. For definitive level design estimates, the scope of work is highly detailed and sufficient to allow a contractor to actually build the project. In other words, the contingency applied to an estimate reflects the completeness of the scope of work as presented by the design documents for the particular level of design. A high contingency is generally applied to an estimate with a low degree of confidence in the scope of work, and a low contingency would be applicable to a detailed design document.

Recommended contingency factors that are applied to the various types of estimates are listed in Table 26.3. The theory is that as the level of project detail increases, the estimator's confidence in the scope of work reflected by the design information also increases. As more of the unknown elements of the project become identified, the contingency for these unknown elements can be reduced. The direct or known cost of the project will increase, but the contingency applied to the cost for unknown elements will decrease. The goal is for these increases and decreases to offset each other, thus keeping the total project cost within the parameters of the original cost estimate. However, when cost estimates between different levels of design are not within the same cost parameters, one of the more common explanations is *scope creep*.

Trending or Scope Creep Identification

Design changes frequently occur between the preliminary design and final design phases. It is during this phase that predesign concepts are finalized and presented to the client for final approval, or conceptually identified items are more clearly defined. Clients who did not fully understand the concepts or details previously discussed are now presented with a detailed system or concept that may not be exactly what they had in mind. In addition, designers may also want to "tweak" their design based on some new information or product data for their systems. The result is usually a change from the preliminary design concept to incorporate new ideas into the project.

TABLE 26.3 Level of Cost Estimates

Type of cost estimate	Level of accuracy	Recommended contingency
Order-of-magnitude	+50% to −30%	20% to 30%
Conceptual	+40% to −20%	20% to 15%
Preliminary design	+30% to −15%	15% to 10%
Definitive	+15% to −5%	10% to 5%

Note: The level of accuracy percentages listed above are interpreted as comparisons of the low bid with the established cost estimate. As an example, a low bid of 15% above to 5% below can be expected in comparison with a definitive level cost estimate.

The scope of work for the project now changes from what was reflected by the preliminary level or baseline cost estimate. Although many designers will identify these changes to the client, and many clients will acknowledge these as changes, a formal process should be put in place to document these scope changes when and if they cause an increase in the cost estimate. A procedure similar to the change order process found in most construction contracts is now being used by many firms to document these design scope changes for future reference. At a minimum it should include the following:

- Identification of the design change and scope
- Reason for the design change
- Effect on project baseline cost and schedule

In addition, a formal document prepared by the designer reflecting the above information should be presented to the client for review and approval.

Cost Indexes

Several cost indexes are used in the municipal water market. The one that appears to be the most accepted and regionalized is the Construction Cost Index published by *Engineering News-Record* (ENR-CCI), a McGraw-Hill weekly publication of the engineering and construction industry. The ENR-CCI reflects the cost of the basic commodities that affect the construction cost of any heavy construction or civil works project. The cost index is regionalized for 20 major U.S. cities. For those locations not in close proximity to any of the 20 listed cities, interpolation can be performed between nearby cities.

Cost Escalation

Although it is not as significant to construction costs as it was in the late 1970s and early 1980s, escalation of costs because of inflation over the life of the project is still a cost that should be taken into consideration. The most common method used is to base the cost estimate on current cost data and then escalate this cost to the midpoint of construction. This is usually computed by calculating the compounded percentage rate of material and labor cost increases to the construction midpoint, and applying this percentage to the current construction cost.

BIBLIOGRAPHY

Jelen, F. C., and J. N. Black. *Cost and Optimization Engineering,* 2d ed. New York: McGraw-Hill, 1983.

Peters, M. S., and K. D. Timmerhaus. *Plant Design and Economics for Chemical Engineers,* 3d ed. New York: McGraw-Hill, 1982.

CHAPTER 27
OPERATOR TRAINING AND PLANT START-UP

Thomas E. Arn
Malcolm Pirnie, Inc.
Phoenix, Arizona

John Gasser
Oakland, California

The design engineer's provisions for new, upgraded, or expanded water treatment facilities should go beyond the design and specification of equipment, steel, and concrete to include less tangible aspects of the total job: consideration of adequate staffing for the new facilities, personnel preparation, reference documentation, and start-up of new facilities. These aspects should be part of the total package of services provided by the engineer to comprehensively meet the client's needs. The engineer can then turn over facilities, knowing that everything possible has been done to provide the client a complete, functional unit.

TRAINING AND START-UP CONSIDERATIONS

Categories covered in this chapter that both engineer and owner must take into consideration during design and construction phases are design-phase training considerations and construction-phase training, start-up, and post–start-up considerations.

Training Considerations

Operator training is a joint responsibility shared by the engineer, contractor, vendors, and owner. Equipment suppliers (vendors), under the general responsibility of the contractor, should train plant staff in the operations and maintenance (O&M) of equipment and systems they have provided. The content of that training should be defined by the engineer in the design specifications. During construction the engineer should oversee vendor training of operators as part of start-up planning, including providing systems training on the functioning of the unit processes and treatment system as a whole. In short, the contractor or vendors should provide the training on the parts, and the engineer on the sum of the parts. The owner's responsibility is to adequately fund training and start-up programs, allocate sufficient personnel to receive training, and provide suitable training facilities.

Reference and training materials complement personnel training. These may include AWWA publications, vendors' O&M manuals, engineer's O&M manual, design specifications, record drawings, and records of operator training sessions. These latter records may be a particularly important source of firsthand, field-verified knowledge that can provide future operators with information that would otherwise have to be learned through experience.

Start-up Considerations

Start-up is the process of placing new facilities into service, ideally after construction is completed. More realistically, however, start-up is usually a series of steps that are neither as smooth nor as coordinated as intended. Planning for start-up should be conducted as part of the design phase to identify construction-related aspects that might become troublesome issues during both construction and start-up. Early in the design phase, consideration must be given to the effect of construction activities on existing facilities, such as the need for tie-ins and the effects of shutdowns. These considerations should be written into the specifications and drawings to the extent possible. Similarly, start-up planning during construction should include development of start-up plans and close coordination between contractor, engineer, and owner.

Not all design and construction projects involve all activities described in this chapter. Upgraded or expanded facilities require careful planning to maintain existing operations during construction. Completely new plants, on the other hand, present different challenges if the client's staff is unprepared or must be hired with no previous experience of stepping in and running new facilities. In all cases, planning for effective training and start-up establishes the preconditions for successful treatment operations.

DESIGN-PHASE TRAINING

The design phase of a project is the time to incorporate vendor training and O&M manual requirements into contract documents.

Vendor Training Requirements

Vendor training means that the equipment manufacturers or suppliers provide instruction to plant operators on operations and maintenance of equipment. Vendor training is usually specific to a piece of equipment or a subsystem provided by the vendor and generally does not cover an entire treatment process. Effective vendor training should follow a prescribed lesson plan and be conducted by a qualified instructor under controlled conditions.

To ensure that vendor training is as effective as possible, the engineer must define the requirements for vendor training and include these in the specifications. Table 27.1 lists provisions to be included in specifications.

General Training. Specifications should include general training requirements for distribution by the contractor to the vendors. These requirements typically include a lesson

TABLE 27.1 Vendor Training Requirements

1. *General conditions*
   General conditions include preconditions to training (e.g., equipment has been installed and tested), need for audiovisual equipment, and training scheduling (e.g., no sooner than 30 days before training).
2. *Instructor's qualifications*
   The instructor's qualifications and experience should be acceptable to the engineer. Qualifications can be evaluated by requiring the contractor to submit information on the instructor for approval in advance of scheduling the training.
3. *Submittal requirements*
   Training lesson plans, handouts, and other documents should be submitted for approval before the training.
4. *Lesson plan*
   A lesson plan that outlines the training content should be submitted for approval. The lesson plan should identify the topics to be covered, trainee trade, and duration of training. Comprehensive lesson plans may include such things as
   - Pretests and posttests
   - Target trade(s)
   - Purpose and specific learning objectives
   - Assumed trainee knowledge requirements (prerequisites)
   - Lesson plan contents, including duration of each phase
   - Identification of training aids, handouts, overheads, and slides and their use in the lesson plan
   - Location of training (classroom and hands-on training)
   - Audiovisual equipment requirements
5. *Minimum required training days*
   Each piece of equipment and associated specification section should be identified and the minimum number of training days specified. The owner should be consulted to develop this requirement. Also, the videotaping requirements, if any, and the duration of videotaping sessions should be defined.

plan and submittal requirements. Inclusion of other general training requirements may also be appropriate. Job descriptions for all trade groups who are to receive training and general videotaping requirements may be included.

Specific Training. Equipment sections of specifications typically include specific training requirements, such as

- Vendor trainer qualifications
- Number of days allotted for training
- Specific content of training, if beyond the scope of the general requirements

Vendor Operations and Maintenance Manual Requirements

A vendor may supply a single piece of equipment, a subsystem, or an entire unit treatment process. In each case, the vendor should be required by the specifications to supply written O&M characteristics covering equipment supplied. Vendor-supplied information should be supplemented by the contractor with additional information and submitted to the engineer for approval. This section defines the type of information required from the vendor and the contractor so that useful and site-specific vendor O&M manuals are provided to the owner's staff.

The vendors' responsibility is limited to the equipment and systems each has provided. Thus, if a pump vendor has supplied only the pump and none of the controls, the vendor's responsibility is to define the operations and maintenance characteristics of the pump only. Vendors normally provide recommendations for installation and mounting equipment. The contractor ultimately controls this aspect of the job unless the design specifications require the vendor to check, inspect, test, or approve equipment after it is installed.

Specifications should require the contractor to provide O&M manuals for each system or piece of equipment having its own specification section. The contractor will supplement information provided by the vendor with any additional information required by the specifications.

Documentation should include equipment design expectations and controls, for example, pump curves, showing the design point. Ladder diagrams clearly identify control logic for equipment. Design loadings; quality characteristics; operating parameters; and detailed start-up, shutdown, and monitoring procedures must also be included.

Table 27.2 lists typical vendor O&M manual requirements. These requirements fall on the contractor to satisfy because of limitations in the vendors' responsibilities, as discussed above. These requirements should be considered a minimum, and additional information may be needed in specific situations.

TABLE 27.2 Vendor Operations and Maintenance Manual Contents

1. Project name, contractor, subcontractor (if any), client, consulting engineer, equipment name, and installation location (normally included on title page)

2. Name, address, email address, and phone number of the manufacturer and the manufacturer's local representative, spare parts ordering information

3. Complete equipment identifying information, including model number, serial number, design parameters; functional information (such as pump curves) and operating limitations

4. Detailed operating instructions, including purpose and operational description; start-up, shutdown, and operational troubleshooting instructions; normal operating conditions or characteristics; abnormal or emergency operating conditions; safety shutdown and other shutdown conditions

5. Preventive maintenance instructions, including all information required to keep the equipment properly lubricated, adjusted, serviced, and maintained for efficient operation throughout its design life; recommended preventive maintenance schedule; lubricant type and lubrication requirements; comparative table of alternative lubricants from different manufacturers; list of required maintenance tools and specialized service equipment; maintenance troubleshooting instructions; illustrations as needed for clarity; corrective maintenance procedures and illustrations

6. Manufacturer's recommended spare parts list

7. Equipment warranty period

8. Approved shop drawings showing wiring, control logic, and installation details

9. Safe operating conditions; safe maintenance conditions; other safety information as may be required (e.g., material safety data sheets)

10. Information required by current regulations of appropriate government agencies such as material safety data sheets.

TABLE 27.3 Vendor Operations and Maintenance Manual Format Requirements

1. Title page
2. Table of contents
3. Enclosed in a three-ring loose-leaf or triple-post binder with stiff covers
4. Page size, $8\frac{1}{2}$ by 11 in.; high rag content; larger drawings to be folded and bound so that they can be unfolded without removal from the manual, or drawings contained in bound pouch; binding margin of $1\frac{1}{2}$ in.; reinforced binding edge
5. Text to be word-processed (or computer-printed) originals, or permanent copies
6. Dividers and indexed tabs to identify major divisions of the manual, such as operating instructions and maintenance information
7. Number of manual copies to be provided

Vendor O&M Manual Format. Widespread use of computers allows the engineer to consider alternative formats besides the standard three-ring binder O&M manual. Although this type of format may still be most useful to maintenance personnel in the field, engineers should also consider requiring the contractor to submit the vendor O&M information via computerized files. Computerized formats are more compact, they do not wear out as printed manuals do, they can be easily improved and modified, and copies of file pages can easily be printed.

If a hard-copy format is desired, the engineer needs to decide whether to incorporate all O&M manuals into binder volumes or to require the contractor to submit a separate binder for each vendor's equipment. Separate binders are more portable for field use, but they require more storage space. In all cases, multiple copies of the vendor O&M manuals should be required. Table 27.3 lists format items that the engineer should consider when specifying O&M manual requirements.

Manual Submittal Schedule. Vendor O&M manuals must be submitted as shop drawings for approval. The engineer should ensure by provisions in the specifications that the contractor will provide O&M submittals in a timely manner for approval, well in advance of start-up and vendor training. Because the system O&M manual prepared by the engineer will use some information from vendors' manuals, the latter should be submitted within a limited time after approval of the equipment shop drawing; 60 or 90 days is common.

CONSTRUCTION-PHASE TRAINING, START-UP, AND POST–START-UP

Start-up of a facility is the culmination of all preceding design and construction activities. Ideally, start-up should proceed only after all facilities have been tested and accepted, with staff properly trained, but this is often not the case. When facility expansion or upgrading is involved, start-up usually takes place in stages, with some areas of the site completed and in service while other areas are still under construction. This results in a difficult environment for plant staff and contractor alike, because both must coexist under less than ideal conditions. When start-up must be fragmented, start-up planning is an especially effective tool for coordinating activities of all parties. Planning start-up focuses attention on the responsibilities and activities of all parties.

A number of elements are important when one is planning for start-up, including

- Coordination of construction with existing operations (see Chapter 24)
- Vendor and systems training
- Facility operations and maintenance manual
- Roles and responsibilities
- Start-up planning and initiation
- Record keeping
- Post–start-up activities

Vendor and Systems Training

Training required by the expansion or new construction of a water treatment facility can be extensive. A facility may be increased in size and complexity, with concurrent changes in operations and maintenance requirements. New systems and equipment may be state of the art, requiring higher skill levels on the part of facility staff for proper operation. To meet these challenges, staff members must acquire technical, organizational, and managerial skills to operate and maintain facilities in a cost-effective, efficient manner.

Before start-up, facility staff should be given both vendor training and systems training. Vendor training typically focuses on individual equipment, and systems training encompasses a system's operation, process monitoring, and control. Systems training is provided by the engineer, preferably by operations specialists, who not only are familiar with the design, intent, and operational requirements of the facility, but also bring real-life experiences to complement their instruction.

Vendor Training. Vendor training is typically delivered by vendors' field representatives during equipment testing, usually just before start-up. The use of knowledgeable proctors to coordinate and oversee vendor classes and verify that vendors convey appropriate information is an effective means to enhance vendor training class quality.

Vendor training tends to focus on equipment maintenance because vendors' representatives are normally field maintenance personnel with extensive mechanical experience on particular equipment. As a result, their training may emphasize mechanical maintenance issues and leave gaps in operations training.

Unless an entire system is supplied by a single vendor, the instructor generally only knows details of a particular piece of equipment and may not fully understand its role as part of a total process. Systems training by the engineer provides the "big picture" for facility staff, whereas vendor training normally provides only a piece of the picture.

Classroom quality control is extremely important because such details as class location, seating arrangements, lighting, background noise, use of training aids, instructor control over the class, instructor voice projection, and ability to follow the lesson plan can affect the quality of instruction. A proctor can assist the instructor with classroom arrangements.

Training Setup. Vendor training should be scheduled with plant staff far enough in advance to allow time to plan around staff needs and to ensure that the appropriate personnel are available on the proposed date. Specifications should require sufficient advance notice by the contractor, normally 15 to 30 days before the proposed training date. If the contractor's proposed training date is not acceptable because plant staff are not available, the contractor should be advised of alternate acceptable training dates.

Tentative training dates can be handled orally with plant staff, but they should also be informed by letter of the final training date. A copy of this notice should be transmitted to the contractor. This ensures that the formal written notification has been provided to all parties. Plant staff should be reminded several days before and again one or two days before the scheduled training, to ensure maximum participation.

The training location should meet minimum requirements of trainer and trainees. Issues such as adequate seating and desk space, lighting, ability to use overhead and slide projectors, availability of paper and pencils for note taking, and audio considerations should all be taken into account by the training coordinator. The trainer should be notified if projection equipment is available on site.

Lesson Plan and Trainer's Qualifications. The lesson plan and written trainer's qualifications provide the documentation to evaluate before training delivery whether training will be effective. The trainer's qualifications may appear impressive on paper, citing years of experience in field applications; however, the quality of training also depends on the training abilities of the trainer. This is often not apparent until training begins, and the engineer can only hope that documented ability of the trainer will be matched by the trainer in person.

Proctoring Training. One of the best methods of ensuring that training will be effective is to provide a specialist familiar with training techniques and practices to witness training. Witnessing training, called *proctoring,* involves taking attendance, guiding the trainer if needed, making notes of key information, and preparing a brief letter report.

The proctor should use the lesson plan as a checklist and guide to verify that proposed training topics are covered. It is essential that all the topics be covered, and the order of coverage should generally follow the lesson plan. If the trainer deviates from the lesson, the proctor should ensure that training quality is not compromised.

Trainers who do not properly cover subjects or are not effectively presenting the subjects may need some guidance or coaching by the proctor. Guidance can be phrased in the form of questions by the proctor, to provide greater information or clarity without embarrassing the trainer. If the proctor's questions and suggestions are ineffective in bringing out information adequately, the proctor, as a representative of the engineer, has the authority to terminate the training.

The proctor should take attendance, noting whether trainees were present for the entire training period. It may be important to track the duration of training in hours, because the training hours may be used as continuing education credits in states where periodic retraining is required to maintain an operator's certification. A training-received verification form should be developed, documenting the instruction with sign-off spaces to be completed by the instructor, engineer, owner, and contractor at completion of training.

The proctor should also make abbreviated notes of the training. In some instances the trainer will cover information about equipment operations and maintenance that is not included in the vendor's O&M literature. This type of information is especially valuable to plant staff and should be carefully recorded. It should later be transmitted to plant staff in printed form so that all staff members may take advantage of this experience.

Videotaped Vendor Training Sessions. Videotapes specially prepared for training purposes can be useful tools for continuing training of plant staff, and construction specifications may require that tapes on certain topics be delivered by the contractor. The common availability of video recording equipment, however, can mislead people into believing that virtually any training session can be taped and used as an effective training tool. This is normally not the case unless the video session is carefully planned and prepared and the tapes are edited to produce a clear, interesting video.

A video recording of a training session delivered in a classroom or field setting, for example, picks up all the extraneous background distractions that go on in a training session and all the actions of the trainer, including misspoken words, hesitations, interruptions, and pauses. Unless the tape or DVD is heavily edited and processed, noise intrusion may distract the audience from the subject matter. Furthermore, live training is more interesting than videotapes, and the audience is so used to the fast pace of broadcast television that even a well-planned, well-edited tape will be of little use unless the tape moves fast enough to retain the trainees' interest. Requirements for good videotaping, therefore, include the following:

1. Meet with the trainer in advance to plan the videotaping session and define the purpose of the training, topics to be covered, depth of coverage, training location limit, and material to be covered so that the final tape lasts preferably no more than 20 minutes.
2. Proctor the training session for the plant staff to determine appropriateness of coverage for the videotaping session. Develop a feel for the trainer's effectiveness when appearing on tape.
3. Tape a training session delivered for taping only without an audience present. The trainer should repeat when needed to obtain clearly and confidently delivered information.
4. Edit and assemble the tape; add a title, credits, and date; send to the trainer's company, if required, for approval.
5. Copy the approved tape as needed and deliver to the client; retain the original footage and edited master.

This sequence should also be followed in producing videotapes of engineer systems training.

Systems Training. Systems (process) training by the engineer provides facility staff with knowledge they need to operate equipment and understand new processes. Systems training helps staff relate the "small picture," such as individual equipment or peripheral systems, to the "big picture," such as an entire treatment process. Systems training is most effective when created and delivered using the following concepts:

- Prepare technically accurate information that places process information and procedures into context.
- Work closely with treatment plant staff to identify specific operational concerns and incorporate concerns into training.
- Vary delivery techniques and materials to enliven the presentation.
- Use graphics extensively.
- Communicate effectively and interact with students during training to maintain interest and improve student knowledge retention.
- Solicit and respect trainees' views and use their feedback constructively to improve the instructor's delivery.
- Relate trainees' responsibilities to overall plant function and regulatory compliance.
- Limit individual training sessions to less than 3 hours of contact time, with one or two breaks.
- Use hands-on demonstrations or walk-throughs to accompany or follow classroom sessions.

Training Materials Development. Materials used in systems training must meet intended audience needs and clearly describe specific equipment, instrumentation, processes, and controls. Training materials can be most effectively developed subsequent to or concurrently with the preparation of the facility operations and maintenance manual. The instructor has contract documents, shop drawings, manufacturers' O&M manuals, existing standard operating procedures (if any), and reference materials to draw on for information. The instructor must distill and simplify material and present information in an interesting and practical way. Training materials must be technically sound, user-oriented, and effective. Formal lesson plans, handouts, and other instructional resources should be prepared for most systems training.

All training materials should be performance-based, keyed clearly to learning and performance objectives. Evaluation methods should focus on direct and indirect indicators of performance by trainees, along with more conventional trainer evaluation.

Systems Training Delivery. Systems overview training for all plant staff should be conducted several months before start-up to provide an overview of new treatment systems and the project as a whole. This training presents an opportunity for staff to voice concerns about future operations that inevitably develop as start-up gets closer. Some topics covered by this overview training include

- New process systems, locations, function, and purpose
- The projected schedule of vendor training and identification of equipment for which training will be provided
- A description of the start-up schedule
- Answers to any questions or concerns that staff members have regarding the project

Specific systems training by the engineer, conducted just before start-up and after vendor training, completes the formal start-up training program. This instruction emphasizes hands-on aspects and procedural instruction to teach how and why the system operates. This training is considered performance-based, and it should teach each staff member the skills necessary to achieve the best possible equipment and systems performance.

In addition, the initial month after start-up is an excellent opportunity for further instruction and confirmation of skills. Post–start-up training may be conducted either formally or informally by the engineer.

Facility Operations and Maintenance Manual

The facility O&M manual prepared by the engineer provides complete guidance to the client's plant and engineering staff on facility operation, with contents prepared to meet the needs of technical staff and plant operators. This dual audience, with its diversity of technical background and needs, imposes a challenge for the manual writer to prepare material both technically complete and comprehensible. Because plant operators rely on practical information as it applies to plant operations and maintenance, procedural information can be directed more to this group. Theory and design intent may be written more with technical personnel in mind.

Manual Preparation Schedule. The O&M manual should be prepared before construction is completed in order to support preparation of operator training materials. Client staff should be involved at an early stage in developing the O&M manual to assist with format and content. Later, during start-up and initial facility operation, information will be gathered and procedures developed that will modify information in the O&M manual. This material should be carefully recorded so that the manual can be updated based on actual operating conditions. A final manual can then be issued that accurately reflects operating characteristics of the facility.

Computer-Based Manuals. The traditional hard-copy O&M manual is gradually being replaced by computer-based, electronic manuals in various formats. Computer manuals have a variety of characteristics that make them superior in many respects to paper manuals. Some of these advantages are ease and speed of access to information, ability to readily update the manual, and the capability of storing graphic information. Internet-based applications are linking electronic O&M documentation to a variety of information sources, including Internet addresses, document imaging, other reference databases, and maintenance management and laboratory information management systems.

Another consideration is the likelihood that manufacturers' literature will become available in digital formats such as CD-ROM and DVD, and through the Internet. Computerized O&M manuals can be expected to become more common, and the engineer should give special consideration to developing the capability of the O&M manual to link to these other references.

The engineer must carefully consider factors that go into making a computer manual. It must generally be an entirely new document, not simply a reformatting of electronic files used to develop paper manuals. Considerations include the software used to develop the manual to make sure it is user-friendly. Input from the client's staff at an early stage of manual development is especially critical when one is preparing a computer manual.

O&M Manual Contents. Whether the O&M manual is computer-based or not, the engineer should determine the regulatory requirements over the contents of the manual. In addition to any such regulatory requirements, the manual should contain the following information:

- Design intent for each process and the overall plant
- Design parameters
- Water quality criteria and regulatory requirements
- Process operations and control
- Operational procedures
- Laboratory procedures
- Maintenance information, in addition to manufacturers' literature (preferably a complete maintenance management system)

A detailed listing of suggested manual contents is included in Table 27.4.

Start-up Planning and Initiation

When an existing plant is upgraded, two types of start-ups commonly occur—interim and final. Both can substantially affect O&M practices and procedures, and each requires extensive planning and coordination with existing operations. Interim start-up occurs during critical construction phases as existing systems are being upgraded and new systems are brought into service on an interim basis. Final start-up occurs when the system is substantially complete.

Interim start-up planning is a part of construction sequencing. For example, to keep a plant operational, a minimum number of filters must be in service at any given time to meet water demands, and a plan to maintain plant operations is developed to do this.

All parties must collectively plan for and implement process changes to the facility during each start-up. Several process and operations issues must be considered:

TABLE 27.4 Recommended Facility Operations and Maintenance Manual Requirements

1. *Introduction and process overview*
   Plant capacity, source water supply, and distribution system overview
   Treatment process overview
   Plant hydraulics
   Source water quality

2. *Water quality criteria and regulations*
   Water quality criteria
   Microbiological, organic, and inorganic contaminants
   Aesthetic qualities
   Regulations (such as federal Safe Drinking Water Act requirements)
   Nonregulatory quality issues
   Sampling and testing

3. *Process chapters (each process)*
   Design intent and criteria, process purpose
   Process and equipment description, operation, and control (standard operating procedures)
   Start-up, shutdown, monitoring, and troubleshooting
   Special safety considerations

4. *Maintenance*
   Organization of equipment manufacturers' literature
   Equipment identification system
   Maintenance and lubrication frequency tables
   Maintenance management system (work orders, maintenance history records, schedules, inventories, purchase orders)
   Standard maintenance procedures
   Warranties

5. *Laboratory*
   Monitoring program
   Quality assurance/quality control and chain-of-custody procedures
   List of analytical procedures and references
   Safety and materials safety data sheets (MSDSs) for laboratory chemicals
   Inventories
   Ordering information and procedures

6. *Safety*
   Supervisory and staff responsibilities
   Safety equipment
   Plantwide safety procedures
   Confined-space entry procedures
   MSDS for process chemicals

7. *Utilities*
   Electrical, gas, and water service
   Heating, ventilating, and air conditioning systems

8. *Records and reports*
   Record-keeping needs
   Reporting to regulatory bodies

- New equipment and control systems will be placed in service, such as tanks, chemical feed equipment, control valves, instrumentation, and pumping equipment. This equipment requires frequent monitoring and adjustment, which must be accounted for in the start-up plan.

- New flow patterns, channels, and tank configurations may be used, and new systems may also be placed into service.
- Contingencies should respond to potential emergencies, such as major equipment failure or instrumentation calibration and control problems.
- Additional operations personnel may be required to monitor and operate equipment during the critical initial phases of a start-up. It is commonly necessary to schedule special personnel for 24-h monitoring during start-up phases of new systems and equipment. Similarly, the contractor may need to schedule additional personnel and ensure that the vendor is on-site and available during start-up.

Start-up Plan. Start-up plans should include elements listed in Table 27.5. A formal start-up plan should be prepared, generally in the following sequence:

- Identify facilities to be placed into service and their functional objectives.
- Review, outline, and develop procedures with facility operations and maintenance personnel to maintain treatment process integrity.
- Identify specific process and equipment conversion sequences.
- Identify contractor preconditions to start up, including satisfactory functional test documentation; vendor training completion; accepted vendor O&M documentation, turned over to plant staff; and accessible and fully operational equipment.
- List additional preconditions, such as systems training requirements; necessary documentation availability; and related systems operability, including safety and auxiliary equipment.
- Establish process monitoring and control setpoints.
- Define owner, contractor, and engineer staff responsibilities, including emergencies and shift coverage.
- Identify other resources required for start-up, such as emergency requirements, special record-keeping forms, and specialized support or monitoring equipment.
- Establish the start-up schedule.
- Develop specific system start-up sequences for each unit process, including detailed procedures and operating responsibilities.

TABLE 27.5 Start-up Plan Elements

1. Overall objective of the plan
2. Facilities to be started
3. Sequence of events
4. Responsibilities of each party
5. Initial operating conditions and parameters
6. Intended final operating conditions and parameters
7. Laboratory requirements or arrangement for outside laboratory services
8. Operating procedures
9. Sampling and monitoring requirements

Start-up Workshop. A highly useful tool for working out details of a start-up plan is the use of workshops to bring responsible parties together. The resident engineer, contractor, client staff supervisors and engineer, and start-up manager normally should attend these meetings. The start-up manager should prepare draft documents for consideration, and agreed-on revisions should be incorporated into revised drafts. Ultimately, a final plan will be developed.

Start-up Sequence. From an operational point of view, start-up should ideally proceed in the following sequence:

1. Hire needed plant staff.
2. Complete vendor and system training.
3. Prepare and revise a start-up plan.
4. Complete construction.
5. Test, check out, and accept new facilities.
6. Stock needed chemicals and prepare for their use.
7. Notify regulatory agencies, as appropriate.
8. Conduct a walk-through of the facilities that are to be started before start-up.
9. Proceed with start-up.

This may be the sequence for a small project that is completed and then placed into service, but for a large project that proceeds in phases, this sequence takes place for each phase in a less than ideal manner. New plant staff, for example, are often not hired until start-up begins, because of the owner's budget limitations, and they have to be trained on-site while some of the facilities are placed into service. It is also sometimes necessary to start facilities that are not complete, and the contractor is responsible for finish work to be completed at a later time. Such disruptions may cause conflict between the contractor and plant staff.

Start-up/Operations Services Manager. An effective functional arrangement for coordinating start-up is the employment of a start-up/operations services manager. This person's responsibilities include not only start-up, but all operations services elements of a project. This function is particularly important when start-up is piecemeal and the resident engineer does not have time to coordinate training and start-up in addition to managing construction activities.

An effective operations services manager should establish good rapport with contractor and plant staff. Because conflicts or differences of opinion among these parties are almost inevitable, the manager should be able to work with the parties to resolve problems. Close communication and cooperation with the resident engineer are essential.

Record Keeping

Record keeping during start-up is often a chore because it requires careful recording of process conditions that may change frequently as the contractor works toward start-up. Nevertheless, good record keeping during start-up is important because it

- Defines what was done at a particular time and how the process responded to given conditions, which may be a useful guide for future process operation

- Establishes responsibilities for actions that may later be relevant to damage claims
- Defines the initiation of equipment warranties
- Identifies process operating problems and solutions
- Identifies equipment problems and unfinished work that need to be resolved by the contractor

Post–Start-up Considerations

Post–start-up activities depend on scheduling start-up and the engineer's responsibility for follow-up work. If part of the new facilities have been started up, the engineer will remain on site and available to assist plant staff in process troubleshooting, even though the engineer's contract does not specifically define this as a contract function. If a start-up/operations services manager is employed, this person normally will assist the owner.

When a new facility has been completed and start-up is over, post–start-up services provided by the engineer depend on the engineer's contract with the owner. Budget conditions may limit services the engineer can provide. However, post–start-up operational and troubleshooting assistance is so important that the engineer should seek to write a provision for this service into the contract.

CHAPTER 28
PILOT PLANT DESIGN AND CONSTRUCTION

Robert A. Stoops
Springfield Water and Sewer Commission
Springfield, Massachusetts

A pilot plant consists of equipment and materials used to simulate a full-scale process or processes. It is built to collect process design and operations data, but it is constructed on a smaller scale for ease of operation, installation, and manipulation. A pilot plant's size can vary from bench-scale equipment to systems that can treat millions of gallons per day.

Each unit treatment process has a limited scale-down factor. If this limit is exceeded, the pilot unit process will no longer simulate the full-scale process. Most scale-down limits have been established empirically from previous pilot studies. For example, gravity filter operations can be simulated using filter columns as small as 4 in. (10 cm) in diameter. However, backwashing studies using the same size filter do not provide quality data. This is so primarily because the excessive ratio of sidewall area to surface area creates friction between the media and sidewall, short-circuiting washwater.

Pilot plant design and operations should be kept as simple as possible. Pilot plants should be operated manually to maintain control of the processes, to determine idiosyncrasies associated with unit processes and individual equipment pieces, and to rectify small problems as they occur. Problems not manually corrected could jeopardize the relevance of significant amounts of data. For example, flow can be controlled using rate controllers and variable-speed drives on pumps. With small flows, in-line flowmeter accuracy can be off by as much as 25% or more. Without a manual check of flow rate, all data collected would have a significant factor of error that, if left unchecked, could result in a critical final design flaw. Automatic data collection is effective for establishing trends and monitoring operational and water quality data during the absence of operators.

Pilot plant design and operations must also be flexible because these plants are used to investigate the unknown and to stress the limits of a treatment process. During a study it is common to change process variables and even to change processes themselves. Flexibility in design allows the operator to quickly connect and disconnect unit processes or by-pass processes. Modular design allows for complete processes to be quickly removed or inserted into the treatment train, reducing downtime. For example, if the pilot plant is not operational for significant periods of time, specific water quality events important to the study might be missed.

Although the pilot plant comprises the tools used in performing a study, it is the design of the study that dictates many of the design characteristics of the pilot plant.

PILOT PLANT STUDIES

Before we describe the construction of pilot plants, some of the common variables encountered in a pilot plant study should be discussed.

Purposes of a Pilot Plant Study

Pilot plant studies are common, and their use has become almost necessary to meet regulatory requirements at minimum construction costs. Pilot plants are generally used to

- Compare alternative treatment processes
- Solve treatment process problems by investigating alternative process modifications
- Investigate new treatment processes
- Demonstrate confidence in recommended treatment processes
- Meet regulatory requirements
- Establish design criteria

Engineers generally turn to pilot plant studies to compare alternative treatment processes before design, construction, or plant modification. One significant advantage of using pilot plant studies in new facility design is that pilot plants allow the engineers to use site-specific design criteria rather than generic or rule-of-thumb conservative design criteria. This advantage can result in significant cost savings without sacrificing treatment quality.

New treatment processes are usually investigated at the pilot scale to establish a success rate before their recommendation. For example, the concern for controlling the formation of disinfection by-products has led to using ozone as a preoxidant and disinfectant in place of chlorine and chlorine dioxide. Although ozone has been used for many years in Europe, it has only recently been commonly used in North America. Pilot plants and pilot studies have allowed ozone to be compared as an alternative oxidant and disinfectant. Ozone concentrations, contact times, and dissolution techniques are investigated at the pilot scale and proved practical before full-scale plant design is started. In many instances, pilot plant studies produce results that provide for regulatory agencies the confidence to approve the use of new technologies and for municipalities confidence to accept them.

Pilot studies are often used to demonstrate treatment processes over extended periods and under varying water quality conditions. Seasonal fluctuations due to lake turnover and algae blooms in reservoirs, daily water quality changes in rivers due to biological respiration, runoff events, and industrial discharges are some of the occurrences that can disrupt treatment processes. If water sources are likely to experience any of these events, it may be necessary to study their effects on various treatment processes for extended periods. Regulatory agencies may require operating a pilot plant for a year or more before granting permission to use new treatment technologies.

It is important to have a complete understanding of source water quality before establishing a pilot plant study. Historical data can be used to some extent to determine the stability of supply and the need for extended operations. Exceptionally stable supplies may require pilot testing for a short time during each season. All water quality episodes that happen in the supply should be determined and investigated as part of the study. For example, if a study is conducted for only two to three weeks per season on a large reservoir, there is a chance that an algae bloom occurrence may be missed. Also, depending on the depth of the intake, an isotherm may pass through the water at the intake level for short periods, causing wide swings in pH, which may severely affect the coagulation process.

Another common use of pilot plant studies is to determine how best to correct full-scale unit process problems without shutting down the full-scale process. Once the comparability between the pilot plant process and the full-scale process has been established, individual process parameters can be isolated and tested for comparison with the full-scale process. For example, if floc carryover through a clarification tank is a problem, a pilot plant can be used to test how the problem can best be alleviated by testing various combinations of coagulant type and dosage, mixing times and energies, and flocculation time. Testing these variables on a full plant basis can be dangerous in that some changes could result in complete failure of the process or production of water of unsatisfactory quality.

Pilot Plant Study Checklist

In developing a pilot plant study, the following checklist of items must be considered before the design, construction, or acquisition of a pilot plant. Each item in the checklist will help focus on the type and size of pilot plant required.

- Define the purpose of the study.
- Identify the end product of the study.
- Collect all available background and historical data concerning water source and water quality.
- Compile a list of water treatment processes that should be considered.
- Establish the minimum and maximum design constraints of pilot plant processes.
- Define the length of operation of the study.
- Determine who will operate the pilot plant.
- Decide where the pilot plant will be located.
- Determine the flexibility required for process changes during the study.
- Decide the locations and frequency of sampling and analyses.

PILOT PLANT DESIGN

Pilot plant design generally consists of two steps: running bench-scale tests to obtain preliminary data and then final design based on the unit processes to be studied.

Bench-Scale Testing

Bench-scale tests are often conducted before the designing of a pilot plant to establish the feasibility of various treatment processes.

Purpose of Bench-Scale Tests. In the absence of existing treatability data or treatment plant operational data, bench-scale tests can be run to help select the pilot-scale treatment processes. Bench-scale tests are typically batch tests; the results are not capable of showing dynamics found in the continuous flow process stream. Bench-scale tests can, however, provide immediate and short-term results of significant value. Results from bench-scale tests can also be incorporated into design criteria for the processes.

Bench-scale equipment can be arranged to simulate an existing treatment plant operation, proposed modifications, or entirely new processes. Once the correlation between the bench-scale test and the full-scale process has been established, bench-scale apparatus can be used in assessing various treatment options. For example, rapid mixing, flocculation, and clarification can be accomplished in one vessel by simulating various mixing times and energy inputs. Clarified or flocculated water can then be passed through filter paper, membranes, or a carbon column to simulate further treatment.

Bench-scale tests are generally conducted concurrently with the pilot plant study to help determine proper pilot plant protocols. Preliminary selection of treatment chemicals, mixing times and energies, settling rates, flotation rates, and media selection can be estimated using bench-scale tests.

Bench-Scale Equipment. The most common types of equipment used for bench-scale tests for potable water include

- Six-paddle stirrer with variable-speed mixers (Figure 28.1)
- Set of 2-L plastic jars with sample taps and pinch valves
- Supply of 15-cm Whatman filter paper with approximately 0.5- to 10-μm porosity
- A 4-in. (10-cm) filter funnel and a 500-mL graduated flask

FIGURE 28.1 Typical jar test apparatus for bench-scale testing. (*Courtesy of CDM.*)

- Vacuum filter apparatus with 0.5-μm glass fiber filters
- Analytical equipment, including a turbidimeter, pH meter, spectrophotometer, and colorimeter

Other types of bench-scale equipment are available to test for settling, carbon adsorption, dissolved air flotation, and solids dewatering characteristics.

Bench-scale tests are generally conducted by adding various treatment chemicals directly to the jars while mixing. Treatment chemicals can be prepared in advance and added by syringe or pipette to the jar at the appropriate time. Rapid mixing is usually simulated for 0.5 to 1.0 min at full rpm. Flocculation can be simulated by varying the rpm from 80 to 20 over an extended period typically from 5 to 20 min. The floc is then allowed to settle for 30 to 60 min without mixing.

Samples are periodically collected from the sample tap on the jar to determine the settleability of the floc. Floc suitable for conventional settling will readily settle in less than 5 min, or sometimes even during the slow flocculation step. Floc suitable for flotation will remain suspended for over 30 min. Pin-size floc suitable for direct filtration will be barely visible and will remain suspended.

Samples collected from the jar after settling can be filtered through filter paper and analyzed for particulate removal and color reduction. True color can be used as a surrogate measurement for natural organic matter reduction. If samples are to be analyzed for total or dissolved organic carbon reduction, glass fiber filters should be used so as not to contaminate the samples with organic paper fibers.

General Design Considerations

Principal issues that must be addressed early in pilot plant design are the various hydraulic issues and provision of adequate electric power to operate the equipment that will simulate processes provided in the full-scale plant.

Hydraulic Requirements. Several hydraulic concerns should be addressed before pilot plant construction or installation. The first is where the source water should come from and how it is to be supplied to the pilot plant. It is important that the source water supplied to the pilot plant be representative of the water that will be used in the treatment plant being designed. In many cases, providing a continuous supply of source water requires modifications to existing structures or running a long length of pipe from the source to the pilot plant.

The method of disposal of both treated and untreated water from the pilot plant must also be determined. It is often difficult to locate gravity drain lines from the pilot plant, and it may be necessary to arrange for a sump or collection tank with pumps to remove the wastewater to a sewer or drain.

If possible, the pilot plant should be designed so that pumping will not be required between the unit processes. In most cases, water being processed contains solids, and running the water through a pump could change the characteristics of the solids. This could have a significant adverse impact on pilot plant results. In addition, every pump introduces another potential source of failure. Staged platforms are usually used to support tanks and other hydraulic structures so that flow is by gravity and the number of pumps is kept to a minimum.

Space Requirements. The pilot plant study often is required to be done adjacent to the water supply source. Operating full-scale water treatment plants are often the location for pilot plant studies. These facilities may or may not have the space to accommodate a pilot plant. If the pilot plant size excludes its location from within the existing buildings, temporary housing may be required. Tractor-trailer-mounted units may become attractive in the absence of existing housing.

Water tanks associated with larger flows and storage requirements may present a structural issue associated with floor loading capacities. Clearance associated with taller components such as filters, contactors, and sedimentation processes may restrict their location or use.

Underestimating the space requirements for a pilot study can cause delay and extra costs. Careful planning and coordination with the end user are required for a successful study.

Electric Power Requirements. The electrical loads required by the pilot plant must be determined in advance to avoid on-site problems and delays. The total electrical draw (amperes) and voltage requirements must be compared with available on-site electrical facilities. If special electrical work is required to supply power to the pilot plant, it may be necessary to obtain a local permit and hire a licensed electrician. One way of minimizing outside labor charges for electrical work is to provide a power panel with the pilot plant prewired to all electrical equipment. In this way, it is necessary to provide only a single source of power to the panel at the site.

Pilot plants designed for overseas installation will probably require special consideration of the electrical design. The voltage and hertz of power available differ from country to country, and it may be necessary to provide a transformer as part of the pilot plant equipment package.

Design for Study of Unit Processes

Unit processes most often studied by using a pilot plant to obtain improved design information are

- Preoxidation
- Coagulation
- Flocculation
- Clarification
- Filtration and granular activated carbon (GAC) contactors
- Ozonation contactors
- UV disinfection
- Solids collection and treatment
- Membranes

See Table 28.1 for recommended flow rate and equipment sizes when designing pilot plants.

Preoxidation. Preoxidation is commonly used for biological growth control through treatment plant source water piping and structures, as a first-stage treatment for color removal, for taste and odor reduction, for reduction of natural or manufactured organic matter, for metal precipitation, for disinfection credit, and for coagulation enhancement. The most common chemicals used are

- Chlorine gas or sodium hypochlorite
- Chlorine dioxide

TABLE 28.1 Recommended Flow Rates and/or Equipment Sizes for Evaluating Water Treatment Unit Processes at Pilot Plants for Design Criteria

Unit process	Minimum recommended flow or size for establishing design criteria	Typical flow range (gpm)	Comments
Coagulation • Static mixing • Mechanical mixing • Hydraulic mixing	2 gpm	2–50	• Velocity gradient, detention time, and other mixing parameters correlate well with full scale. • Dilution of coagulant chemicals may be required. • Chemical metering pump accuracy an issue.
Flocculation • Mechanical mixing • Hydraulic mixing	2 gpm	2–50	• Velocity gradient, detention time, and other mixing parameters correlate well with full scale. • Variable-speed mixing should be considered for changing raw water characteristics.
Ballasted flocculation	200 gpm	200–800	• Proprietary demonstration scale units available from manufacturers. • Effective sand separation units to treat small volumes are not available.
Gravity sedimentation • Horizontal flow • Circular clarifier	20 gpm	20–50	• Equivalent surface loading rates typically perform better at full scale than at pilot scale. • Difficult to simulate detention time and surface loading rates at the same time. • Size and weight of pilot units can be difficult to handle.
High-rate sedimentation • Tube settlers • Inclined plate settlers	20 gpm	20–50	• Equivalent surface loading rates typically perform better at full scale than at pilot scale. • Typically produce settled water quality comparable to full scale, but at shorter detention times.
Upflow solids contact clarification/lime softening	10 gpm	10–50	• Difficult to maintain a sludge blanket with smaller units. Heat can disrupt hydraulics. • Solids recycle recommended for lime softening, but can be difficult at low flows.

TABLE 28.1 Recommended Flow Rates and/or Equipment Sizes for Evaluating Water Treatment Unit Processes at Pilot Plants for Design Criteria (*Continued*)

Unit process	Minimum recommended flow or size for establishing design criteria	Typical flow range (gpm)	Comments
Dissolved air floatation	20 gpm	20–200	• Proprietary units available from manufacturers. • Manufacturers typically provide flocculation, DAF, and filter systems. • Tractor-trailer units require space. • Minimum two-stage flocculation attached direct to DAF tank. • High-pressure (80 psig) sidestream flow (10% of process).
Contact flocculation/clarification • Buoyant media • Roughing filters	4-in. dia. columns or larger	0.5–10	• The 4-in.-diameter columns or larger accurately predict water quality. • Production typically better at full scale.
Granular media filtration • Pressure filters • Gravity filters • GAC filter/adsorbers	4-in.-dia. columns or larger	0.5–10	• The 4- or 6-in.-dia. columns accurately predict water quality. • Production typically is better at full scale. • Sidewall effects can alter media settling, impact ripening and headloss, and uplift media during backwash. • Gravity units recommended for investigations to address air binding.
Backwashing granular filters • Air scour • Surface wash • Water	4-in.-dia. columns or larger	5–30 (1 cfm/sf air)	• Minimum 1 ft^2 of filter area to accurately simulate backwashing performance (for air and water rates and durations). • Others recommend a filter diameter of $\geq$1,000 times the media effective size.
Carbon adsorption contactors	2-in.-dia. columns or larger	0.5–10	• Empty bed contact time correlates well to full scale. • Rate of head loss accumulation correlates to full scale.
Ozone • Single- or multiple-column • Sidestream injection	4-in.-dia. columns or larger for diffusers 1 gpm	1–20	• The 4-in. columns or larger with a height of 8 ft or taller provide reasonable transfer efficiency. • Difficult to accurately dose ozone at flows lower than 1 gpm. Gas-to-liquid ratios become difficult to control.

Process	Unit size	Flow (gpm)	Comments
Ultraviolet light disinfection	1-in.-dia. cells or larger	2–700	• Ozone monitor gas flow requirements may be greater than process application requirements at low flows. This can impact control. • High voltage. • Commercially produced units available from several manufacturers. Smaller units acceptable for inactivating bacteria, viruses, and other pathogens, but can be difficult to accurately control UV dose. Larger units (demonstration scale) with more accurate monitoring and flow control are recommended for inactivation studies where actual log reduction values are required. • High voltage.
Chlorine disinfection • Free chlorine • Monochloramine	1 gpm	1–20	• Basins, piping, and other contactor configurations can accurately predict disinfection and DBP formation. • Tracer studies on pilot units are critical to full-scale correlation. • Increased surface area and/or exposure to sunlight can impact results.
Chlorine dioxide disinfection	2 gpm	2–10	• Small-scale generators are not commercially available. • Use of stock solutions may not accurately predict oxidation or by-product levels at full scale. • Plant-scale trials often provide more useful results. • Chemical storage and material compatibility issues.
Membrane filtration • Pressurized microfiltration or ultrafiltration • Immersed microfiltration or ultrafiltration	10 gpm	10–60	• Long-term studies needed for design criteria and require use of full-size membrane elements or slightly modified versions. • Flows are highly subject to vendor requirements. • Backwash intervals, flux rates, recycle flows and feedwater recycle rates, and membrane replace or recharge are all vendor-specific.
Softening and deionization membranes • Nanofiltration • Low-pressure reverse osmosis • Reverse osmosis	10 gpm	10–60	• Long-term studies commonly use 4-in.-dia. membrane elements arranged in arrays. • Single-element testers (1–5 gpm) characterize water quality. However, they do not accurately demonstrate long-term performance of the membranes. • Backwash intervals, flux rates, recycle flows, and feedwater rates are all vendor-specific.

Thanks to CDM for input to this table.

- Potassium permanganate
- Ozone

Preoxidation to be used in the full-scale plant must be simulated in the pilot plant, but chlorine gas, chlorine dioxide, and ozone will generally be expensive, difficult to feed and control, and potentially dangerous to use. Sodium hypochlorite is recommended for most pilot studies because it is safe to use and easy to feed and control. Sodium hypochlorite is commonly available in 5.25% and 12.5% solutions and can be easily pumped into the process stream. Fumes associated with hypochlorite solutions are not generally a major concern.

Potassium permanganate may also be used as a preoxidant for a pilot plant. It is purchased as a dry powder, and a batch of solution can be prepared with a known concentration. This solution can then be injected into the process stream with a metering pump. Metering the powder directly to the process stream is not practical because of the low flows used in most pilot plants.

Depending on the application, some extended reaction time must be provided in the process train when using chlorine or permanganate. The optimum chlorine concentration and reaction time can be simulated in jar tests before the pilot study. The total retention time may be longer than 30 min for some reactions. Some agitation is required during the retention time to guarantee that the oxidant is dispersed throughout the process stream. This mixing can be accomplished by directly injecting chlorine into the pipeline ahead of several elbows or bends, or by using a static mixer, venturi injector, or mixing in a tank.

Polyethylene tanks or drums can be used to provide contact time in a pilot plant. To eliminate the short-circuiting of flow that would occur in one large tank, proper contact time can be ensured by installing several smaller tanks piped in series. By connecting piping to the threaded bulkhead fittings in the tanks and using pipe unions, tanks can be easily added and eliminated from the series. Changing the flow through a process can also control the contact time; however, other processes downstream may then be adversely impacted or restricted.

Ozone is a strong oxidant and an unstable gas that has to be generated on-site. Its reaction time is typically much faster than that of other oxidants. If ozone is to be fed in the pilot plant, it can be directly injected into the process stream through a venturi injector or applied through a diffuser into a contact tank. Ozone injection can be accomplished by drawing off approximately 10% of the total process stream and pumping the flow through a venturi injector, creating a vacuum. The venturi is connected so that the vacuum draws the ozone gas and injects it into the partial process stream, and then the solution that has been formed is returned to the full process stream. The dosage is controlled by varying the flow of gas and the concentration of the gas.

Coagulation. Coagulation is the process in which an electrolyte is introduced to the process stream to reduce the net electric repulsive forces of suspended particle surfaces. Coagulation aids in removing those substances associated with turbidity, including colloidal solids, clay particles, suspended and some dissolved organic matter, bacteria, algae, color, and taste- and odor-causing compounds. Coagulation depends on the selection of a coagulant, applied coagulant dosage, pH, temperature, mixing energy, and time.

Coagulant application requires facilities and equipment to store and deliver the chemical to the process stream. Many suppliers of coagulant chemicals provide quantities of their products for a pilot plant study at no charge other than those associated with shipping. Containers larger than 5 gal are heavy and may be difficult to receive and move around the pilot plant location.

Day tanks generally function as a control mechanism to measure the volume of chemical used. The day tank can be as small as a 500-mL beaker and as large as a 50-gal drum.

The use of day tanks requires the pilot plant operator to frequently mix fresh chemicals for plant use.

If possible, coagulant chemicals should be fed at full strength. Once a coagulant has been diluted, hydrolysis immediately starts to take place. The metals begin to precipitate out of solution and render the coagulant less effective, which can have a significant impact on the performance of the coagulant. The design size of the process stream in the pilot plant and the selection of the coagulant feed pumps are critical for the coagulation process. The smaller the process stream, the more difficult it is to meter the coagulant accurately.

The coagulation process is almost instantaneous if a complete mixing system with sufficient energy is supplied. In-line static mixers can be used instead of mixing tanks and paddles. An in-line static mixer imparts a tremendous amount of energy in the short period of only about 1 or 2 s. If static mixers are used, it is recommended that chemicals be fed with a constant-rate (peristaltic) chemical feed pump. If pulse-type feed pumps are used, chemicals are injected only intermittently into the system. Peristaltic pumps are excellent for this type of application. There are peristaltic pumps that can reliably feed as little as 0.1 mL/min.

If a mixing tank is provided, a minimum hydraulic detention time of 15 s should be provided. Rapid mix times longer than 1 to 2 min may lead to floc shear. A mixing system should be designed to impart a velocity gradient G value of approximately 300 to 500 s^{-1}.

The pH of the water being treated is critical in the control of the coagulation process. If source water pH varies, continuous pH monitoring may be required. If source water pH is stable, periodic grab samples usually suffice. A streaming current or zeta-potential meter may be considered to monitor the coagulation process if source water quality changes frequently. A sample tap should be located immediately before and after the coagulation process.

Flocculation. Flocculation imparts energy to the process stream to force agitation, which causes very small suspended particles to collide and agglomerate into larger, heavier particles, or flocs. The optimum size of the floc particles varies depending on the downstream processes. For direct filtration, a fine, dense floc is required; for dissolved air flotation, a small floc is required. For best settling, a large, heavy floc is generally optimum. Mixing time and energy control the behavior of the flocculation process. The flocculation process is usually designed to provide mixing with a decreasing level of energy over time.

Flocculation time of at least 10 min is usually required for most treatment systems, and some conventional settling plants find that optimum flocculation times as long as 30 min may be required. The determination of the optimum flocculation time is often one of the primary goals of a pilot plant study; therefore the flocculation tanks should be designed to accommodate wide variations in flow and retention time. To avoid short-circuiting through the flocculation process, the tank or tanks should be constructed to provide for at least three separate hydraulic zones. This can be accomplished by providing baffle walls between zones or by providing two or more tanks in series. The time through the process can be varied somewhat by controlling the flow through the tanks, as long as other processes are not affected by the flow change. Provisions for bypassing hydraulic zones or installing or removing tanks in series may be a better method for controlling the time of flocculation.

Mixing in the flocculation tanks can be accommodated by installing vertical shaft paddle mixers designed to impart various energies to the flocculation zones. Variable-speed drives should be provided for the simulation of flocculation. As an alternative, the speed of the mixer can be determined by visual inspection or with the use of a tachometer or strobe light. Flocculation velocity gradients between 10 and 80 s^{-1} are most common.

Flocculation tanks should preferably be rectangular to promote complete mixing. If circular tanks are used, some type of stators should be incorporated to guarantee mixing.

Provisions for sampling should be provided throughout the flocculation process, specifically at points between mixing zones. Grab samples are probably best to evaluate the formation of floc. Velocities through piping should be kept low so as not to shear floc and present unrepresentative samples.

Clarification. Clarification removes much of the floc before filtration. Most clarification processes require sedimentation of the flocculated particles. Recently, dissolved air flotation has gained some popularity for its ability to remove significant amounts of floc before filtration, lower surface area requirements, and have fewer upsets such as those associated with temperature gradients or hydraulic surges. Other clarification processes include the use of sludge blankets and roughing filters. These processes can be effective but are usually pilot-tested by equipment suppliers with their own standard equipment. Conventional sedimentation remains the most common form of clarification process.

Sedimentation. Full-scale sedimentation basins have typically been divided into four specific zones:

- The inlet zone provides a smooth transition from the influent flow to the uniform steady flow desired in the sedimentation zone.
- The sedimentation zone provides volume and surface area for sedimentation to take place.
- The sludge zone receives and stores the settled floc particles.
- The outlet zone provides a smooth transition from the sedimentation zone to the effluent flow.

These four zones have equal importance at the pilot scale as well. The inlet zone must be designed to provide for a smooth transition from the flocculation process to the clarification process. High velocities and energy losses associated with small piping and fittings, valves, and orifices can shear and break flocculated particles, thus reducing the sedimentation tank efficiency. An open-channel zone providing laminar flow should be provided ahead of the chamber where upflow begins.

The sedimentation zone must be designed to provide the necessary volume and surface area to minimize floc carryover and maintain steady-state operations for extended periods. Simply scaling down the size of the tank to match the process flow requirements would result in significant short-circuiting through the tank with little sedimentation occurring. When pilot-testing sedimentation, it is recommended that standard inclined tubes or plates be used. Even when using tubes or plates, it is recommended that a safety factor of 1.5 to 2.0 be applied in designing the surface area requirements. A design overflow rate of 0.5 gpm/ft^2 (0.1 m/h), based on the surface area of the sedimentation tank, has been used successfully.

To eliminate short-circuiting through the sedimentation tank, it is critical to provide for at least 4 to 6 in. (5 to 15 cm) of standing water over the top of the tubes or plates. In addition, collector pipes or perforated pipes should be laid across the top of the clarifier to evenly distribute the flow across the entire basin.

The sludge zone of the sedimentation tank must be large enough to collect solids over extended periods. Continuous sludge collection constantly stirs up and resuspends solids, which greatly reduces the effectiveness of the process. Three to four vertical feet of storage volume should be provided to allow for the storage of the solids. The bottom of the tank can be angled in such a manner as to direct the solids to a central place for collection, with a drain valve provided for the removal of the solids. If the drain line is to be run continuously, it should be provided with a flowmeter.

The outlet of the sedimentation basin should be designed similar to the inlet so that excessive velocities and energy drops are not encountered. Flocculated particles that pass through the sedimentation basin can be sheared to a size where they may later pass through a filter. Aggressive pumping of settled water can also shear floc, so low-speed positive displacement pumps should be used if pumping is necessary.

Process flows of 5 to 15 gpm (3 to 9 L/s) can incorporate conventional sedimentation relatively easily. Flows less than 2 to 3 gpm (1 to 2 L/s) make short-circuiting an intolerable problem. On the other hand, flows greater than 20 gpm (13 L/s) result in requirements for large tanks. Temporary pilot plant sites typically have constraints that restrict the size of tanks that can be used. These constraints are usually associated with open space, access, and structural limitations.

Dissolved Air Flotation. The dissolved air flotation (DAF) process requires the use of a clarification tank in which the flocculated particles are drawn to the surface by their attachment to tiny air bubbles; they are then removed from the surface and piped to waste or provided with further treatment (Figure 28.2). Overflow rates can range from 3 to 10 gpm/ft^2 (8 to 24 m/h) of surface area. DAF also requires a sidestream process whereby approximately 10% of the total process flow is injected with air, pumped through a pressure vessel at approximately 60 to 100 psig (414 to 689 kPa) to saturate it with air, and delivered to the process stream at the inlet to the DAF tank. At the tank inlet, the pressure in the sidestream is dropped to near-atmospheric conditions, and the air in solution is released in the form of many tiny bubbles.

The DAF tank requires an inlet zone for smooth transition from the flocculation process, but the flows do not have to be as slow as those for sedimentation. The sidestream

FIGURE 28.2 Parallel DAF tanks with mechanical scrapers (tanks 50 gpm with overflow rate of 6 gpm/ft^2). (*Courtesy of CDM.*)

is injected into the process stream in the pipes directly ahead of the tank. A baffle wall inclined at approximately 60° from horizontal, pointed in the direction of flow, is required immediately downstream from the inlet of the tank to direct the flow upward. The flow is then directed to the bottom of the tank for discharge.

Perforated collector pipes located a short distance off the tank bottom can be used for collecting the DAF effluent and directing it to an adjustable weir that allows for maintaining a constant head in the tank at various flow rates. An inclined plate at the outlet end of the tank is provided as a "beach" for the ease of collecting the floating solids ("float"). Removal of the solids can be accomplished by backing up the flow and flooding the solids over the beach or by providing mechanical scraping of the solids toward the beach. Volume requirements for solids collection can be estimated by assuming the float material generated to be approximately 0.5% to 1.0% and 2% to 5% solids if collected hydraulically or mechanically, respectively.

The sidestream system requires piping and valves, centrifugal pump, rotameter, air compressor and receiver tank, solenoid valve, venturi injector, saturation tank with level switch and pressure relief, and discharge valves (Figure 28.3).

Filtration and GAC Contactors. Filtration is accomplished by passing water through a porous medium for the removal of suspended particles. GAC contactors are designed for adsorbing soluble organic matter onto the media. Most filters and contactors are designed to operate in a downflow manner, but upflow filters are occasionally used ahead of another filter in series. Filters and GAC contactors can be pilot-tested using similar material and equipment.

The process of filtration has been successfully simulated using filter columns as small as 4 in. (10 cm) in diameter. The use of smaller columns is not recommended because there is an excessive sidewall-to-surface ratio, which results in significant short-circuiting during the filter operation and excessive bed expansion or compaction caused by media-sidewall friction. In addition, the small flows required by small columns are difficult to control and maintain. The flow requirements for additional processes and for monitoring can help determine the size selection of the filter column, but in general, a 6-in. (15-cm) column is recommended.

The most popular material for construction of filter columns is clear polyvinyl chloride (PVC) because it allows the operator to visually inspect the bed while it is in operation. Visual inspection can help identify the level of media, formation of mud balls, excessive floc accumulating on the surface of the media, uneven distribution of media after excessive backwashing, and accumulation of media fines near the surface. Columns can be constructed in one section or in short sections with flanges that are bolted together. Inlets, outlets, overflows, underdrains, and manometer taps are installed relatively easily through the sidewall of the column (Figure 28.4).

The underdrain system can be attached directly to the bottom flange of the column. If a nozzle is used, the flange can be drilled and tapped to accommodate the thread size of the nozzle. If a porous plate is used, the flange can be cut to accommodate the plate, or a plate can be sandwiched between two flanges with the use of gasket material so that it will be watertight. Filter columns can either be supported off the ground on a stand or hung from a structural wall using steel angle iron and plates. Mobile pilot plants are constructed with individual filters built into cabinets with casters that allow for movement from one site to another (Figure 28.5).

Flow control through a filter column can be controlled by several methods. One common method is by use of a constant-head tank or altitude valve. Electronic flowmeters can also be used in conjunction with a modulating valve or variable-speed pump to control the flow through the filter. When flow control is provided by a pump located ahead of a filter, there will be some pressure through the filter, which will impair filter perfor-

FIGURE 28.3 Author shown adjusting flow to a DAF recycle stream pump adjusted manually based on data from flowmeters and pressure gauges. (*Courtesy of CDM.*)

mance if the pressure is excessive. This usually requires the installation of pressure sensors to monitor the pressure drop through the filter. In addition, the energy imparted to the water by the pump may change the character of any floc that has passed through the clarification process.

If possible, gravity feed to and through the filter is recommended for minimum adverse effects. The flow can be controlled at the filter outlet through a valve or pump.

FIGURE 28.4 Top section of filter column preassembled with quick-connect unions, flanged sections bolted together. (*Courtesy of CDM.*)

When a pump is used, it should be a positive displacement pump that is not affected by the suction head. Because the filter will experience head losses approaching 8 to 10 ft (2.4 to 3 m), the available suction head will vary by the same amount.

Because of the low flow rate required when filtering through a small column, available flowmeters may not be accurate, so a rotameter is often installed on the influent line to provide visual inspection of the flow rate. In addition, a sample point should be provided at the filter effluent so that the flow rate can be checked by diverting the flow into a graduated cylinder while the time is measured with a stopwatch.

Filter backwashing facilities should be provided to clean the filters at the end of a filter run. These facilities should include a storage reservoir for clean water and a pump. The pump should be sized to provide water to the filter at a rate of up to 25 gpm/ft^2 (61 m/h) after head losses through the pipe, media, and column have been considered.

FIGURE 28.5 Mobile field plant modules with filter assembly units mounted on casters, complete with filter, flow controls, turbidimeter, and head loss monitoring. (*Courtesy of CDM.*)

If air scour is also to be used, oil-free air should be provided if there are to be multiple backwashings. Air scour can be controlled through a needle valve–rotameter combination. The compressed air is usually delivered at a rate of 2 to 5 ft³/min (0.9 to 2.4 L/s) for a period of 2 to 5 min before backwashing. If the pilot plant is to be set up at an existing filtration plant, there may be sources of backwash water and air available on-site, but care should be taken to determine whether the available backwash water is chlorinated. The use of chlorinated water through GAC media can inactivate biological growth and reduce the adsorptive capacity of the carbon.

If several types or grades of media and media depths are to be investigated in the same filter column over time, several backwash outlets should be provided for different bed expansion depths. Sample taps should be provided according to the media configuration to be investigated. The taps should be installed to measure head loss at the media surface, at interfaces between media types, and in the bottom layer. Manometer tubes or pressure sensors can be used to measure the head loss through the filter column.

The backwash process cannot be simulated accurately enough for collecting design data when using filters of 4- to 6-in. (10- to 15-cm) diameter. Filters with a 2-ft (60-cm) minimum diameter have been shown to be effective in simulating the backwashing process. However, at this size, large volumes of water are required to operate the filter, and the collection of backwash solids requires several days of storage volume (Figure 28.6).

Ozonation Contactors. Ozone can be used for preoxidation as well as for disinfection after clarification or after filtration. Ozone generation requires special equipment for providing dry compressed air as well as the ozone generator itself (Figure 28.7).

Many sizes and types of ozone generators are available, and they are generally classified by the pounds of ozone produced. Most pilot plants can use generators that produce less than 2 lb/day (0.9 kg/day). The size of the generator can be determined by approximating the required dosage in advance and by knowing the process stream rate. These units are typically run with conventional 120-V, 60-Hz power requirements. Larger generators may require 220- to 480-V power.

Some generators come with desiccant air dryers built within their chassis. Others require independent air preparation equipment. Separate air compressors may be required ahead of the desiccant dryer if not provided as part of the unit. The air compressor should be able to provide oil-free air to the generator. Particulate filters and hydrocarbon filters should be installed between the compressor and ozone generator.

Small oxygen generators are also available for providing a clean oxygen stream, which allows an ozone generator to create more ozone than when using compressed air. They can produce oxygen streams that are up to 95% pure. The size of the airstream required can be approximated by knowing the percent weight ozone gas that can be produced by the generator and by computing the quantity of ozone required per day.

The required quantity of ozone feedgas and off-gas must be determined to establish the transfer efficiency of the ozone contactor. Several types of ozone monitors are available for measuring the concentration of ozone in air, and they all work primarily by measuring the adsorption of ultraviolet light by the ozonated airstream. Some monitors require a zero reference gas flow as well as the ozonated gas flow. Rotameters are generally used to provide an accurate measurement of the volume of ozone gas directed to the monitor and to the feed point in order to establish the dosage and transfer efficiency. The rotameters must be made primarily of glass and stainless steel, and gasket material should be made of Teflon or Hypalon to resist attack by the ozone gas.

Measuring the quantity of off-gas concentration is relatively difficult for small pilot plant systems because most monitors require greater airflows than are typically provided to small ozone contactors for accurate measurement. Also, the off-gas contains high concentrations of water vapor, which can destroy or inhibit the optical materials inside a monitor. To minimize the adverse impacts to the monitoring system, the off-gas can be measured periodically instead of continuously.

Ozone contactors for pilot plants with process flows less than 10 gpm (0.6 L/s) are typically made of clear PVC. Although PVC is attacked to some extent by ozone, most pilot studies are short-term and intermittent, so the material will hold up for the duration

FIGURE 28.6 Manifolded backwash collection, settling and storage tanks in foreground. Large-scale filter (2-ft diameter) in background generates backwash solids. (*Courtesy of CDM.*)

of the test. The benefit of using clear PVC is the visibility it provides to the operator. The relative volume of gas and the size of the bubbles can be monitored, and a diffuser blockage can be noticed through visual inspection. Most contactors are designed to provide a countercurrent flow whereby the water is lifted to the top of the ozone contactor and allowed to fall, while the ozone gas is supplied to the base of the column and allowed to

FIGURE 28.7 Ozone disinfection module includes oxygen generator, ozone generator, contact columns, pumps, flowmeters, residual and feedgas monitors, and power supply panel. (*Courtesy of CDM.*)

rise. This technique provides the necessary mixing action to induce the transfer of ozone into the liquid stream. Small porous-stone diffusers are generally used to disperse the ozone, and they can be attached to the bottom flange of the contactor.

Additional ozone contact time can be simulated by placing additional columns in series or by allowing the ozonated water to pass through one or more storage tanks. Efforts should be made to allow for smooth flow transitions between contact tanks because turbulence, hydraulic jumps, and free falls will drive ozone out of solution and reduce the contactor efficiency. In some cases a reduction of ozone residual may be required as part of the process. Excessive ozone residual, for example, can interfere with filtration and can consume chlorine. In such cases, the added turbulence can be used to reduce the ozone residual, and free falls of about 2 ft (0.6 m) have been shown to be effective in reducing ozone residual.

The flowthrough in an ozone contact tank can also be altered to change the contact time, but the variation in flow that can be achieved may be limited by minimum gas-to-liquid relationships and other process flow requirements. All ozone contactors should be covered and provided with an off-gas blower or fan to direct the vapors away from operators.

Sample ports should be provided throughout the contact tank system for the measurement of ozone residual. At a minimum, sample taps should be provided between each contact column or tank. In-line continuous ozone residual analyzers are available. Some work by continuously passing ozonated water across an ozone-specific membrane, and others work by agitating the ozonated water stream and stripping the ozone out of solution. The gas concentration is then measured by UV adsorption.

The depth of the water in the contact chamber will have a significant impact on the transfer of ozone into the water stream. Most full-scale plants use contactors that are 18 to 25 ft (5.5 to 7.6 m) deep. These depths may not be possible at the pilot plant scale, but the columns should be made as deep as possible.

In addition to the conventional bubble diffuser system, direct injection or sidestream ozone injection is available. These techniques may be more attractive where high ozone demands exist because the transfer is almost instantaneous.

Ozone can also be injected into the process stream through a venturi injector before a contact tank. A sidestream of approximately 10% of total flow is drawn off and pumped to a pressure of 20 to 35 psig (138 to 241 kPa) and then passed through an efficient, low-energy venturi injector, where a vacuum is formed. Ozone gas is drawn in by the vacuum of the injector, and the turbulence created by the flow creates a foam of ozonated gas and water. If the injector is mounted adjacent to the process flow, the foam can be returned to the process stream, where the mass transfer takes place.

Solids Collection and Treatment. Solids generated by the treatment of water have become a significant issue for most municipalities and water utilities. Pilot studies for the sole purpose of optimizing the solids treatment train are gaining popularity.

Proper pilot testing of mechanical thickening and dewatering of settled solids and backwash water solids requires a significant amount of sludge, in the range of several hundred gallons, for a continuous run. In general, it is best to first see if there are vendors who will analyze solids, perform jar tests, and perform studies using mechanical dewatering equipment in their laboratories. They can also help in selecting the types of polymers and optimum dosages for thickening and dewatering. Some vendors also have small pressure vessels and membranes that can simulate dewatering methods.

It is important to not change the solids characteristic of sludge before it is tested. If solids generated from clarification, flotation, and backwashing must be pumped, positive displacement pumps should be used and operated at low speeds. The number of times the materials are transferred or moved should be minimized, and the material should be stored in a manner that will deter it from going anaerobic or becoming contaminated in any way.

On-site pilot testing of mechanical dewatering equipment is commonly used to establish design criteria. Solids from a full-scale plant are provided daily to operate a pilot-scale dewatering operation. Before beginning an on-site study, samples of the solids to be treated should be sent to the vendor whose equipment will be used for preliminary evaluation. This will minimize start-up time and operational errors on site. The transfer of the solids to the pilot plant and disposal of the solids and supernatant are usually the responsibility of the facility owner.

Lagoons, land drying, and freeze-thaw drying can be simulated at a fairly small scale in small, easily constructed drying beds. The beds should have drains installed to capture the filtrate and the sidewalls designed so that the volume reduction of the solids due to dewatering can be easily measured. Covers can be made of plywood for protection from rain and snow, if necessary (Figure 28.8).

PILOT PLANT CONSTRUCTION

These are some of the options available for obtaining the facilities to run pilot studies:

- The owner can design and build the pilot plant.
- The owner can design the plant and hire a contractor build it.
- A contractor can be engaged to both design and construct the plant (turnkey operation).
- An independent engineer can be hired to design the plant, with either in-house or contract construction.

FIGURE 28.8 Water treatment residuals freeze-thaw beds with sand/gravel underdrain and leachate collection system included. Covers made of plywood are not shown. (*Courtesy of CDM.*)

- The owner can rent or lease portions of a plant, including specific unit processes that are difficult to build or are patented.
- A complete package pilot plant can be leased for the period of the pilot studies.

Much of the decision making is inevitably determined by the available money and time as well as the end goals for the project.

Alternative Types of Pilot Plants

Pilot plants generally fall under two distinct categories of permanent and temporary units. Permanent pilot plants require the owner to purchase the pilot plant equipment and are more common with the construction of new water treatment plants. A permanent pilot plant offers the treatment plant operator the opportunity to fine-tune treatment techniques without adversely impacting the full-scale plant. Successful modifications identified through pilot tests provide the confidence to make the modification to the full-scale process. Without the pilot plant, operators commonly hesitate to recommend changes for fear of worsening the treatment.

The approach to permanent pilot plant construction differs significantly from temporary plant construction. Permanent pilot plants have to meet local building, plumbing, electrical, and fire codes. In addition, bidding laws may require extensive engineering services to provide design drawings and specifications before bids can be taken and the plant constructed.

In locating a permanent pilot plant, it is extremely important to identify the area and height requirements, source water and drainage availability, and electrical requirements. For instance, laying pipes along the floor surface is not usually acceptable for permanent applications. Any changes to the main treatment plant facility to accommodate the pilot plant should be determined to be part of the pilot plant cost.

Temporary pilot plants offer the owner the same opportunities to test treatment options, but usually at a fraction of the cost of building a permanent installation. Temporary pilot plants can be constructed in many ways. Modular pilot plants are gaining in popularity and make it easy for vendors to move their equipment from one site to another. Start-up time is reduced because the intricate connections move with the plant. The only on-site work is limited to source water piping, electrical supply modifications, drainage piping, and process-to-process piping.

The customer user can lease or rent the equipment for the time necessary to perform the evaluations, and when the tests are completed, the equipment can be removed and the area where it was located can be returned to its previous condition. Also, the use of modular plants more readily allows for process changes during the study compared with permanent installations.

Process Drawings and Specifications

Process drawings are required for every pilot plant installation. The number and type of drawings and detail required for the drawings vary with the type of plant selected and change according to how the plant is obtained.

At a minimum, a floor plan or site layout drawing should be created. This drawing identifies exactly where the process equipment is to be located, keeping in mind space and access requirements, the source water source, electrical supply, and available drainage (Figure 28.9). A process schematic or process and instrumentation diagram (PID) should also be created to provide the level of detail required for a complete operational pilot plant (Figure 28.10). The level of detail provided by a process schematic cannot typically be provided on the floor plan. For simple plants, the schematic can include details of the mechanical process, electrical, instrumentation, and plumbing design criteria. More complicated plants may require individual function drawings.

An electrical single-line diagram is usually required to identify the numbers, types, and sizes of motor loads and other electrical demands, including those for instrumentation, controls, lighting, and lab equipment.

FIGURE 28.9 Pilot plant floor plan.

Specifications should be produced for all pilot plants. These usually consist of a simple yet comprehensive design criteria table that lists the equipment and materials associated with each unit process.

A permanent pilot plant may require a full set of design drawings, including architectural, structural, mechanical, electrical, plumbing, and HVAC, along with a full set of specifications. If a permanent pilot plant is to be built within an existing facility, the existing conditions must be determined in detail so that the installation of new water lines, electrical conduit, and drain lines will not conflict with existing pipe and conduit. The

FIGURE 28.10 Pilot plant schematic of train A.

structural limitations of floors may have to be considered when deciding on the location of large storage tanks or other heavy equipment.

Installation and Start-up

When planning the installation of a pilot plant, the time required for installation and start-up is often either forgotten or underestimated. Careful and proper installation and start-up of the pilot plant will inevitably produce benefits during the pilot plant study. On the other hand, improper installation or start-up can cause many problems, including the collection of useless data.

The use of modular equipment reduces the installation and start-up effort but does not eliminate it entirely. Site-specific constraints and minor process modifications can have major impacts. The use of equipment from different vendors, equipment built in different countries, and different construction materials, flow requirements, system pressures, and electrical demands are some of the problems that can cause installation nightmares.

Start-up time should include time for troubleshooting and repairs. Any weekend plumber knows how smart water can be when it comes to finding leaks. The shipping of equipment often produces some form of damage as well. Long periods of nonuse can have an adverse impact on equipment.

Whenever possible, the people who are designated to operate the pilot plant should help with the installation and start-up. The contractors who install the plant often know the strengths and weaknesses of the equipment and materials that are provided, and this information can be valuable to the operator.

Operations and maintenance manuals for each piece of equipment and unit process should be available to the user for proper start-up, and these should be studied by the operators during the installation and start-up. Once the study has begun, time is usually limited and operators may not take the time to learn how to properly maintain the equipment.

Each unit process should be independently tested for proper hydraulic, electrical, and mechanical function before testing the entire process train. Once all connections have been made, the entire system can be hydraulically tested by passing source water through it. Each chemical feed system should be calibrated under the flow and pressure conditions likely to be experienced before start-up, and preliminary jar tests can be conducted before start-up to identify likely chemical selections and dosages.

Consideration must also be given to the experience level of the installers and operator because this will affect the time of installation and start-up. When one is hiring engineers or contractors for the design, construction, and installation of a pilot plant, their experience with similar projects should be investigated before awarding of the contract. Selection of the lowest bid with least experience can spell disaster in the long run.

BIBLIOGRAPHY

American Society of Civil Engineers and American Water Works Association. *Water Treatment Plant Design,* 3d ed. New York: McGraw-Hill, 1998.

American Water Works Association. *Water Quality and Treatment,* 5th ed. New York: McGraw-Hill, 1999.

Daniel, Philippe, Michael Zafer, and Paul Meyerhofer. *Effectively Planning Pilot Studies.* Walnut Creek, Calif.: Camp Dresser and McKee Inc., 1992.

Thompson, J. C. "Overview of Pilot Plant Studies." Presented at the 1982 Annual Conference of the American Water Works Association, Denver, Colo.

CHAPTER 29
ULTRAVIOLET DISINFECTION

Paul D. Swaim, P.E.
CH2M HILL
Englewood, Colorado

Christine A. Cotton, P.E.
Malcolm Pirnie, Inc.
Tucson, Arizona

Samuel S. Jeyanayagam, Ph.D., P.E., D.E.E.
Malcolm Pirnie, Inc.
Columbus, Ohio

Ultraviolet (UV) light has been used for many decades for disinfection of drinking water supplies worldwide. In North America, the use of UV light was more widespread for wastewater disinfection in the last quarter-century, and only more recently has UV light been considered for drinking water disinfection for a wide range of potable water systems. For disinfection of drinking water supplies in North America, UV disinfection was long assumed to provide effective inactivation of bacteria and viruses, but UV light was believed to be ineffective for inactivating protozoa at UV doses that would be cost-effective for implementation.

In 1998 and 1999, new research was published demonstrating that UV disinfection was capable of cost-effectively inactivating *Cryptosporidium* oocysts (Clancy et al., 1998; Dyksen et al., 1998; Bukhari et al., 1999). Where previous studies had focused on in vitro assays to indicate oocyst viability, these new studies utilized animal infectivity assays to directly determine the ability of the pathogen to replicate and cause disease. In the following years, several more studies confirmed these findings and demonstrated the effectiveness of UV disinfection for the inactivation of *Giardia* cysts as well (Craik et al., 2000; Danielson et al., 2001; Hayes et al., 2001; Oppenheimer et al., 2002; Campbell and Wallis, 2002; Linden et al., 2002a; Mofidi et al., 2002). Concurrently, new research showed that higher UV doses were needed for inactivation of adenovirus compared to previous virus inactivation data (Thompson, 2003).

At the same time as the effectiveness of UV disinfection was being established for *Giardia* and *Cryptosporidium*, the U.S. Environmental Protection Agency (USEPA) was developing drinking water regulations to specifically target *Cryptosporidium* inactivation for surface water systems that demonstrate the presence of *Cryptosporidium* in their untreated water supplies. The effectiveness of UV disinfection for the inactivation of *Giardia* and

Cryptosporidium represented a cornerstone for cost-effective treatment to meet the new regulatory requirements. With this regulatory framework, the application of UV disinfection may be utilized as an additional disinfection barrier at water treatment plants (WTPs), with the continued use of chlorine or chloramines expected for virus inactivation and for residual disinfection throughout the distribution system.

Chapter 29 is new to this fourth edition of *Water Treatment Plant Design*. The science of UV disinfection for drinking water treatment has been rapidly developed in the last several years since the third edition was published. The objective of this new chapter is to provide a reference for transferring experience with this emergent knowledge base to assist in the successful integration of UV disinfection at WTPs.

In this chapter, the following major discussion topics are addressed in regard to the evaluation and design of UV disinfection:

- Principles of UV disinfection
- Applications for UV disinfection
- UV disinfection equipment
- Water quality and treatment effects on UV disinfection
- UV disinfection facility design and project implementation
- Construction, testing, start-up, and O&M
- Potential regulatory requirements for UV disinfection in North America
- Validation testing

PRINCIPLES OF UV DISINFECTION

In this section, the historical development of UV disinfection, the fundamental aspects of UV light, the disinfection mechanism, microbial response, and the factors impacting UV disinfection are addressed.

History of UV Disinfection

Over 100 years ago, in 1877, Downes and Blunt first observed the germicidal impact of sunlight (Downes and Blunt, 1877). The production of the mercury vapor lamp in 1901 and the use of quartz as a UV transmitting material in 1906 led to the first known full-scale use of UV disinfection of drinking water in 1910 at the Marseilles water treatment facility in France. In 1929, Gates identified a link between UV disinfection and the absorption of UV light by nucleic acid. In the 1930s, the fluorescent lamp was developed, which led to the production of germicidal tubular lamps (USEPA, 2003a). Despite these advances, cost and reliability issues hindered the further development of the UV equipment.

A resurgence of interest in UV disinfection occurred in the 1950s with research focusing on the mechanisms of UV light inactivation (Dulbecco, 1950; Kelner, 1950; Powell, 1959; Brandt and Giese, 1956). It was in Switzerland and Austria that the first reliable applications of UV light for drinking water disinfection were demonstrated in 1955 (Kruithof and van der Leer, 1990). As of 1996, there were more than 2,000 UV disinfection systems treating drinking water in Europe (USEPA, 2000), primarily treating flows less than 1 million gallons per day (mgd) (USEPA, 2003a).

In the United States and Canada, UV disinfection has been successfully used in wastewater treatment applications since the 1970s. With respect to drinking water, recent findings that *Cryptosporidium* (and *Giardia*) can be inactivated by relatively low UV doses (Bukhari et al., 1999; Clancy et al., 2000) have led to a rapid increase in interest in the use of this technology. The fact that no known measurable levels of disinfection by-products (DBPs) have been associated with the use of UV light at the doses used for inactivation of cysts and oocysts has further supported the viability of the technology (Awad et al., 1993, Malley et al., 1995). As of 2004, several installations are in operation across the United States and Canada, and many more are currently under design. In the coming years, the USEPA's acceptance of UV light as an available, feasible, and cost-effective disinfection technology for achieving compliance with new regulations is expected to trigger a significant increase in the number of UV installations in North America.

Properties of UV Light

As UV light propagates from its source, it interacts with the materials it encounters through absorption, reflection, refraction, and scattering. The most significant of these interactions are absorption and scattering, both of which are described in this section.

Definition of UV Light. On the electromagnetic spectrum, UV light lies between X-ray and visible light. The typical wavelengths of UV light are from 100 to 400 nanometers (nm). As illustrated in Figure 29.1, the UV spectrum is composed of UV-A, UV-B, UV-C, and vacuum UV. Although the entire UV wavelength range is capable of disinfection, the most effective wavelengths for UV disinfection are from 200 to 300 nm, a range that includes UV-B and UV-C light.

When atoms of an element are activated by an electric discharge (electrons) to a higher orbital state, light is emitted when they return to a lower energy state. If the difference between the energy levels is appropriate, the light emitted is in the UV range. Although many elements can be used for this purpose, mercury is the most commonly used element in UV lamps employed for water and wastewater disinfection. This is so because mercury emits light in the germicidal wavelength range. In addition, it is most volatile, with temperatures that can practically be attained by current lamp technology. In addition to mercury, UV lamps contain a filler gas, which promotes the ionization activation of mercury; the most common filler gas is argon.

FIGURE 29.1 UV light in the electromagnetic spectrum. (*Source: USEPA, 2003a.*)

UV Light Absorbance. As UV light propagates through the water, the light interacts with the water. The most significant interactions are UV light absorbance and scattering. UV absorbance (A_{254}) is a measure of the amount of light at 254 nm that is absorbed by a water sample over a specified path length. In water applications, a 1-cm pathlength is typically used. Dissolved organics, turbidity, and suspended particles can increase absorbance. The UV absorbance of water varies with wavelength, typically decreasing with increasing wavelength. Particles in water can absorb and scatter light, increasing the UV absorbance of the water.

Another method of quantifying the absorbance of UV light is UV transmittance (UVT). UVT is the percentage of light (typically at 254 nm) passing through a water sample over a specified distance (e.g., 1 cm). The relationship between UVT and A_{254} may be expressed as

$$\%\text{UVT} = 100 \times 10^{-A_{254}} \tag{29.1}$$

where %UVT = UV transmittance at specified wavelength (e.g., 254 nm) and a path length of 1 cm
A_{254} = UV absorbance at specified wavelength, based on 1-cm path length

UVT (or A_{254}) is a key design parameter for UV disinfection systems, and further discussion of this parameter is presented later in this chapter.

UV Light Scattering. Scattering of light is the change in direction of light propagation caused by interaction with a particle. Particles can cause scattering in all directions, including toward the incident light source (backscattering) (USEPA, 2003a). Turbidity is a measurement of light scatter caused by particles in water.

Microbial Response to UV Light

Specific wavelengths of UV light will inactivate microorganisms if light at these wavelengths is absorbed by the microorganism's deoxyribonucleic acid (DNA) or ribonucleic acid (RNA). However, following disinfection, some microorganisms have mechanisms that repair UV light–induced damage.

Mechanism of UV Disinfection. UV light inactivates microorganisms by damaging the microorganism's DNA or RNA. This interferes with the microorganism's ability to replicate. A microorganism that cannot reproduce is incapable of infecting (USEPA, 2003a).

According to the first law of photochemistry, only light that is absorbed can induce a chemical reaction. Work by Jagger (1967) shows that nucleic acid bases in DNA are capable of absorbing UV light in the 200- to 300-nm range, as shown in Figure 29.2. While the absorbed light can potentially inflict several types of damage, the formation of dimers (covalent bonds) between two complementing nucleotides on the same DNA strand is thought to be the most common damage resulting from exposure to UV light.

UV Dose Response of Microorganisms. Over the last decade, several investigators have reported the UV dose response of microorganisms of concern in the treatment of potable water supplies. UV dose-response relationships can be expressed as either the proportion of microorganisms *inactivated* (log inactivation, resulting in a dose-response curve with a positive slope) or the proportion of microorganisms *remaining* (log survival, resulting in a dose-response curve with a negative slope) as a function of UV dose. Of the pathogens of interest in drinking water, generally viruses are most resistant to UV disinfection

FIGURE 29.2 UV absorbance of nucleotides (left) and nucleic acid (right) at pH 7. (*Adapted from Jagger, 1967.*)

followed by bacteria, and then *Giardia* cysts and *Cryptosporidium* oocysts. The specific, proposed regulatory UV dose requirements for bench-scale inactivation of *Cryptosporidium*, *Giardia*, and viruses are described at the end of this chapter.

Microbial Repair. Microorganisms that have been exposed to UV light still retain metabolic functions, and some are able to repair the damage done by UV light to a limited degree. In some cases, the microorganism can become infective again. Repair of UV light–induced DNA damage includes photoreactivation and dark repair (Knudson, 1985). In photoreactivation (or photorepair), enzymes energized by exposure to light between 310 and 490 nm (near and in the visible range) repair damaged sections of DNA. For dark repair to occur, photoreactivating light is not necessary; however, dark repair can occur in the presence of light (USEPA, 2003a).

Microbial repair of UV-induced damage has been researched, and the fundamental conclusions of this research are summarized here (USEPA, 2003a):

- Although *Giardia* reactivation has been observed at very low UV doses (that is, 0.5 to 1.0 mJ/cm^2), photoreactivation or dark repair of *Giardia* at UV doses typical for UV disinfection applications (16 and 40 mJ/cm^2) does not occur (Linden et al., 2002a). The same study showed *Giardia* reactivation in light and dark conditions at very low UV doses (0.5 mJ/cm^2).

- *Cryptosporidium* does not regain infectivity after inactivation by UV light (Shin et al., 2001). One study showed that *Cryptosporidium* contains the capability to undergo some DNA repair (Oguma et al., 2001). However, infectivity was not restored.

- Viral DNA lacks the necessary enzymes for repair and depends on a host for reactivation (Rauth, 1965).

- Bacteria can be reactivated following UV damage. The presence of a residual disinfectant concentration (either chlorine or chloramines) in the distribution system is expected to prevent repair of UV damage in bacteria.

By-products of UV Disinfection

The formation of DBPs by UV disinfection can occur either directly through photochemical reactions or indirectly through reactions with products of photochemical reactions. In drinking water, research has focused on the impact of UV light on the formation of halogenated DBPs after subsequent chlorination, and on the transformation of organic material to more degradable components. For groundwater and filtered drinking water, UV disinfection at typical doses has been shown not to impact the formation of trihalomethanes or haloacetic acids, two categories of DBPs currently regulated by the USEPA (Malley et al., 1995; Kashinkunti et al., 2003).

Several studies have shown the formation of low levels of nonregulated DBPs (e.g., aldehydes) as a result of applying UV light to wastewater and raw drinking water sources. However, a study performed with filtered drinking water indicates no significant change in aldehydes, carboxylic acids, or total organic halides (TOX) (Kashinkunti et al., 2003). The different results can be attributed to the difference in water quality, most notably the higher concentration of organic material in raw waters and wastewaters.

Finally, the conversion of nitrate to nitrite is possible with MP lamps that emit at low wavelengths (von Sonntag and Schuchmann, 1992). However, due to the low conversion rate (about 1%; Sharpless and Linden, 2001), this should be a minimal concern in drinking water applications.

APPLICATIONS FOR UV DISINFECTION

Water utilities are faced with many treatment challenges in today's regulatory climate and the shifts in focus to new contaminants of concern. In addition, often several tools can be used to achieve the utility's objectives. UV disinfection can be used to address these treatment challenges: (1) disinfection of pathogenic microorganisms, (2) reduction of DBP formation, and (3) photolysis and advanced oxidation of contaminants.

Comparison of UV Light and Chemical Disinfection

UV disinfection is fundamentally different from chemical disinfection with respect to the inactivation mechanism, the response of microorganisms, and the factors that impact disinfection. The differences between UV light and chemical disinfection are summarized in Table 29.1.

Disinfection of Target Microorganisms

As discussed later under potential regulatory requirements, UV light can be used to inactivate different types of pathogenic microorganisms. From a regulatory perspective, UV disinfection can be used to inactivate *Giardia lamblia, Cryptosporidium parvum*, bacteria, and viruses. In addition, some utilities may have specific disinfection objectives beyond these regulatory requirements.

A comprehensive disinfection strategy provides multiple barriers to reduce microbial risk while minimizing DBP formation. UV disinfection is a tool that can contribute to a comprehensive disinfection strategy by providing a cost-effective method of inactivating target pathogens that are more resistant to more traditional disinfection methods. For UV projects, the target pathogen(s) should be clearly defined during the planning stages (USEPA, 2003a). This can ensure that the design meets the utility's and the primacy agency's expectations based on the regulatory requirements, target microorganism(s), and overall disinfection strategy. An overview of general regulatory requirements for drinking water systems is provided in Chapter 2, and an overview of the potential regulatory requirements for UV disinfection of *Giardia lamblia, Cryptosporidium parvum*, bacteria, and viruses is presented at the end of this chapter.

TABLE 29.1 Comparison of Chemical and UV Disinfection

Factor	Chemical disinfection	UV disinfection
Disinfection mechanism	Pathogen is killed by exposure to chemical (e.g., destruction of cell wall). Associated chemical reactions well understood	DNA is damaged by UV light, and pathogen replication is prevented. Cell structure is left intact. Associated optical interactions relatively more complex
DBP formation	High potential for DBP formation	No measurable DBPs formed at typical doses for disinfection
Flow-through reactor	Water, disinfectant, and pathogens flow together	Water and pathogens flow past a fixed UV light field
Detention time	Relatively long and measurable. A key variable in determination of regulatory compliance	Very short. Path taken by pathogen more important
Residual	Used, along with detention time, for determining regulatory compliance	No residual. UV intensity sensors provide a surrogate measurement. Need another measure of disinfection effectiveness
Primary factors impacting disinfection effectiveness	• Water quality • Residual concentration • Contact time • Temperature • pH • Type of chemical	• Water quality (e.g., UVT) • Characteristics of UV equipment • UV intensity and dose distribution • Power quality • Sensor performance

The UV doses necessary for *Cryptosporidium* and *Giardia* inactivation are lower than those required to inactivate viruses. Accordingly, the capital costs for inactivating *Cryptosporidium* and *Giardia* should be lower than the capital costs for virus inactivation by UV disinfection alone. One study estimated capital costs for virus inactivation by UV disinfection are approximately 50% higher than the costs associated with inactivation of *Cryptosporidium* and *Giardia* inactivation by UV disinfection (Cotton et al., 2002). Therefore, the target microorganism(s) and required inactivation level should be determined early in the planning process (USEPA, 2003a).

Reduction of Disinfection By-product Formation

To a degree, UV disinfection can be used to replace chemicals for disinfection of chlorine-resistant pathogens (i.e., *Cryptosporidium* and *Giardia*), thereby potentially reducing DBP formation. However, UV disinfection is not as efficient at inactivating viruses as more traditional, chlorine-based disinfection processes. Because of its effectiveness at inactivating viruses and the need to maintain a disinfectant residual in the distribution system, some chlorine-based disinfection (chlorine or chloramines) will be needed even if UV disinfection is implemented. To achieve a free chlorine residual, the chlorine demand in the water must be satisfied. Thus, chlorine-based disinfection by-products will still be formed if the necessary precursors are present. However, providing chemical disinfection for viruses will also contribute to the disinfection of bacteria and will control microbial repair.

Photolysis and Advanced Oxidation of Contaminants

UV light alone (i.e., photolysis) or in conjunction with specific oxidants (e.g., hydrogen peroxide) can oxidize several constituents of concern, such as tetrachloroethene (TCE), bromate, methyl tertiary-butyl ether (MTBE), taste- and odor-causing compounds, 1-4 dioxane, N-nitrosodimethylamine (NDMA), and hormonally active substances.

UV photolysis requires sufficient UV energy for chemical decomposition. Advanced oxidation processes (AOPs) are typically defined as those processes employing the indiscriminant oxidative properties of the hydroxyl radical (OH·) for destruction of target compounds. Because the hydroxyl radical has a greater oxidation potential (2.8 V) than ozone (2.07 V) or hydrogen peroxide (1.8 V) alone, the hydroxyl radical is able to oxidize target compounds at a much faster rate than other oxidants.

Advanced oxidation processes are differentiated by the rate and the method by which they produce the hydroxyl radical. AOPs include ozone and hydrogen peroxide (O_3/H_2O_2); UV light and ozone (UV/O_3); UV light and hydrogen peroxide (UV/H_2O_2); and UV light, ozone, and hydrogen peroxide ($UV/O_3/H_2O_2$). An example of how UV light is utilized to form hydroxyl radicals is shown using peroxide. Production of hydroxyl radicals with hydrogen peroxide irradiated by UV light proceeds according to the following reaction:

$$H_2O_2 + hv \rightarrow 2OH·$$

The UV energy needed to achieve this oxidation is greater than that for UV disinfection, as shown in Figure 29.3. Therefore, the cost for advanced oxidation with UV light is significantly higher than that for UV disinfection. Site-specific water quality parameters will have an effect on AOP performance and cost and should be evaluated on a site-specific basis. The presence of natural organic matter, alkalinity (Adams et al., 1994), nitrates, reduced metals, or turbidity may interfere with the efficacy of the selected AOP (Kommineni et al., 2000).

The effectiveness of UV photolysis and AOPs for emerging constituents of concern and hormonally active substances represents areas of new research, with more research anticipated in the next few years. Note that UV disinfection is the focus of this chapter, and the design elements of photolysis and advanced oxidation processes using UV light are not described further in this chapter.

UV DISINFECTION EQUIPMENT

This section outlines the key components that comprise UV disinfection equipment. As shown in Figure 29.4, the UV equipment consists of the control panel, UV reactor, and UVT monitor. The UV reactor houses the UV lamps, lamp sleeves, UV intensity sensors, cleaning mechanism (lamp sleeve wipers), and temperature sensors. Of these components, the cleaning mechanism and UVT analyzer are optional components.

UV Reactor

The UV reactor (or chamber) is where the water is exposed to UV light and where disinfection occurs. UV reactors are designed to optimize UV dose delivery, through the UV

FIGURE 29.3 Summary of UV dose ranges to achieve between 90% and 99% removal. (*Courtesy of Alex Mofidi, Metropolitan Water District of Southern California.*)

FIGURE 29.4 Schematic of UV disinfection system. (*Courtesy of Severn Trent Services.*)

intensity field and the UV reactor hydraulics. The need to optimize the UV dose delivery is balanced by the need to minimize the head loss through the reactor. The UV intensity field depends on the lamp placement and configuration, lamp power, and UVT of the water. The reactor hydraulics are affected by the flow rate, inlet and outlet piping, lamp configuration, and internal baffles. The UV reactor design is typically proprietary to the UV equipment manufacturer.

UV reactors are either open-channel or closed reactors. Closed reactors are typically used for drinking water disinfection where the water flows under pressure (i.e., no free surface) and the reactor flows full. Open-channel reactors are typically used in wastewater treatment applications.

UV Lamps

Several UV lamp technologies are available:

- Low-pressure (LP) mercury vapor lamps
- Low-pressure, high-output (LPHO) mercury vapor lamps
- Medium-pressure (MP) mercury vapor lamps

TABLE 29.2 Characteristics of Lamp Types for Typical Drinking Water UV Systems

Parameter	Low-pressure lamps	Low-pressure, high-output lamps	Medium-pressure lamps
Germicidal UV light	Nearly monochromatic at 254 nm	Nearly monochromatic at 254 nm	Polychromatic, including germicidal range (200 to 300 nm)
Mercury vapor pressure (torr)	Optimal at 0.007	0.76	300–30,000
Operating temperature (°C)	Optimal at 40	130–200	600–900
Electrical input (W/cm)	0.5	1.5–10	50–250
Germicidal UV output (W/cm)	0.2	0.5–3.5	5–30
Arclength (cm)	10–150	10–150	5–120
Electrical-to-germicidal UV conversion efficiency (%)	35–38	30–40	10–20
Relative number of lamps needed for a given dose	High	Intermediate	Low
Lifetime (h)	8,000–12,000	8,000–12,000	3,000–8,000

Source: USEPA (2003a).

- Electrodeless mercury vapor lamps
- Metal halide lamps
- Xenon lamps (pulsed UV)
- Eximer lamps
- UV lasers

Full-scale drinking water applications generally use LP, LPHO, or MP lamps, and more recently, LPHO and MP lamps have been used almost exclusively. The *pressure* in LP, LPHO, and MP lamps is associated with the vapor pressure of mercury within the lamps themselves. The remainder of the UV lamp technologies are not as fully developed and/or not validated for use in drinking water disinfection. The characteristics associated with the LP, LPHO, and MP lamps are listed in Table 29.2, and the comparative operational advantages and disadvantages of LPHO and MP lamps are summarized in Table 29.3.

The wavelength distribution of the emitted light depends on the lamp's concentration of mercury atoms, which is directly related to the mercury vapor pressure and amount of mercury (USEPA, 2003a). In low-pressure lamps, mercury is maintained at low vapor pressure (near vacuum, 2×10^{-5} to 2×10^{-3} psi) and moderate temperatures, producing essentially monochromatic UV light at 253.7 nm. At these operating conditions, liquid mercury is always present in excess. LPHO lamps also produce essentially monochromatic UV light at 253.7 nm, but with higher vapor pressures and operating temperatures to allow more germicidal UV output.

MP lamps have higher vapor pressures (near atmospheric pressure) and higher operating temperatures, allowing the emission of UV light over a broad (polychromatic) spectrum with an overall higher intensity. In these lamps, mercury is not normally present in excess during operation. With either type of lamp, proper disposal and/or recycling of the used lamps is essential, given the mercury content.

As shown in Figure 29.5, the light emitted by LP and LPHO lamps is essentially monochromatic at 253.7 nm, which is near the optimum wavelength of the microbial action spectrum, shown in Figure 29.2 as 260 nm. MP lamps emit at a wide range of wave-

TABLE 29.3 Comparison of Mercury Vapor Lamp Types for Drinking Water Applications

	LPHO lamps	MP lamps
Comparative advantages	• Higher germicidal efficiency; nearly all output at 254 nm • Smaller power draw per lamp (less reduction in dose if lamp fails) • Longer lamp life	• Higher power output • Fewer lamps for a given application • Potentially smaller reactors, smaller footprint • Automatic physical or physical/chemical sleeve cleaning typically provided
Comparative disadvantages	• More lamps needed for a given application • For very cold water temperatures, there are potential water temperature impacts on the magnitude of the lamp output • Additional chemical cleaning of the lamps may be required	• Higher operating temperature can accelerate fouling • Shorter lamp life • Lower electrical-to-germicidal UV conversion efficiency • Cooling water system may be necessary during start-up

Source: Adapted from USEPA (2003a).

FIGURE 29.5 UV output of (*a*) low-pressure and (*b*) medium-pressure lamp types. (*Source: Sharpless and Linden, 2001.*)

lengths across the germicidal action spectra (i.e., 200 to 300 nm) and at other wavelengths, and thus electrical efficiency is reduced. With either lamp type, power that is not converted to light is primarily lost as heat.

Lamp Aging. The output (intensity) from UV lamps decreases over time, resulting in a reduction in UV dose delivery as lamp aging occurs. Lamp output will decrease over time as a function of lamp hours in operation, number of on/off cycles, and power applied per unit (lamp) length. The rate of decrease in lamp output slows as the lamp ages (USEPA, 2003a). In design, the engineer typically determines a lamp fouling/aging factor, which accounts for both lamp aging and sleeve fouling (described later in this chapter).

TABLE 29.4 Comparison of Advantages of Magnetic and Electronic Ballasts

Magnetic ballast	Electronic ballast
• Less potential for power interference due to stored energy • More resistant to power surges • More resistant to high temperatures • Less prone to interference with electronic devices • Less prone to sputtering (inductive less than capacitive) • Proven technology (in use for nearly 70 years) • Less costly	• More efficient • Lighter weight • Smaller size • Less potential for heat generation • Less potential for noise • Continuous power adjustment • Longer lamp operating life

Source: USEPA (2003a).

Lamp Power Supply and Ballasts

UV lamps are typically powered by an alternating-current (AC) supply. A mercury vapor lamp has a nonlinear voltage-to-ampere characteristic (Persson and Kuusisto, 1998). Consequently, it cannot be connected directly to the AC electric service, and a ballast must be inserted into the circuit to limit the current flow through the lamp and provide reliable service.

Ballasts are either magnetic (also known as electromagnetic) or electronic. Magnetic ballasts can be inductive or capacitive and operate at the line frequency. Electronic ballasts operate at frequencies higher than that of the line voltage and involve solid-state devices or a mixture of solid-state devices, inductors, and capacitors. A comparison of the advantages of each ballast technology is presented in Table 29.4.

Lamp Sleeves

UV lamps are housed in a sleeve, which is in direct contact with the water; thus, the lamp itself is not in contact with the water. The sleeve also helps maintain the optimal lamp operating temperature and protects the lamp from breaking. For UV disinfection, lamp sleeves are most commonly constructed of quartz. The distance between the exterior of the lamp and interior of the lamp sleeve is approximately 1 cm (USEPA, 2003a).

Lamp sleeve fouling on external surfaces is caused by the reaction of compounds in the water with the lamp sleeve surface. External fouling should be removed by cleaning. In addition, extensive exposure of the lamp sleeve to UV light can cause solarization, especially if the quartz is contaminated with metal cations. Both fouling and solarization can decrease the UV transmittance of the sleeve.

Cleaning Systems

UV manufacturers have different techniques to clean the lamp sleeves. These approaches may be broadly categorized as off-line chemical cleaning (OCC) and on-line mechanical cleaning (OMC) (USEPA, 2003a).

OCC systems require shutting down the reactor and flushing the interior with a cleaning solution. Solutions used for cleaning include citric acid, phosphoric acid, or other food-grade proprietary solutions provided by the UV manufacturer. Following chemical contact, the reactor is rinsed and returned to operation. OCC is most often used in LPHO systems.

An OMC system provides in situ cleaning capability. The cleaning system contains mechanical wipers that are driven over the sleeves by electric motors or pneumatic pistons. Cleaning is performed by stainless steel brush collars or elastomeric wiper rings. In some cases, the wiper collar is assisted with cleaning chemical that is applied to the sleeve during the cleaning operation. Cleaning may be manually initiated or automatically controlled. MP systems typically use OMC systems because the higher lamp temperatures can accelerate fouling.

With either type of cleaning system, it may also be necessary to perform occasional supplemental manual cleanings of the lamp sleeves. The frequency of this task will depend on the utility's water quality and operational strategy.

UV Intensity Sensors

UV intensity sensors measure UV intensity at a point within the UV reactor. The sensor responds to changes in lamp output due to lamp power setting, lamp aging, and lamp sleeve fouling. Depending on its position, a UV intensity sensor may also respond to changes in UVT of the water being treated. UV intensity sensors are composed of optical components, a photodetector, an amplifier, a housing, and an electrical connector. Some sensors include optical filters for modifying the spectral response such that the sensor only responds to germicidal wavelengths (i.e., 200 to 300 nm). At the time of publication, sensors are specific to each manufacturer, and sensor performance is verified during validation testing (USEPA, 2003a).

UV intensity sensors can monitor the UV lamp intensity through a monitoring window (i.e., the sensor itself is dry), or they can be in direct contact with the water (i.e., wet sensor). Monitoring windows and the wetted ends of wet sensors can foul over time and may need cleaning, similar to lamp sleeves. For UV intensity sensors, frequent (e.g., monthly) calibration checks are likely to be required (USEPA, 2003a), and factory calibration may be necessary as often as annually. The dry UV sensors allow a quick calibration check with a reference sensor without taking the UV reactor out of service.

UV Transmittance Monitors

The efficiency of UV disinfection is closely linked to the water's UVT. Some manufacturers use on-line UVT monitors as an integral part of their control and monitoring strategy. In general, commercial on-line UVT monitors calculate UVT by measuring the UV intensity at various distances from a lamp inside the monitor. The difference in sensor readings at these different locations is used to calculate UVT; this calculated UVT is transmitted from the on-line monitor to the control system (USEPA, 2003a).

Water Temperature Sensors

UV lamps have high operating temperatures and are cooled by the water flowing through a reactor. The water efficiently absorbs the wasted heat and maintains operating lamp temperatures within a desirable range. (With the short contact times at disinfection doses, the impact on water temperature is negligible.) It is important to maintain the correct lamp temperature because excessively high lamp temperatures can harm the lamp. In addition, during low-flow periods or shutdowns, the water temperature must be maintained below the UV manufacturer's specified value (approximately 50° C for MP lamps) to ensure that the UV lamp is properly cooled and maintains its temperature at an acceptable level. The following conditions could potentially result in elevated lamp temperatures:

- Water level in the reactor drops, and lamp sleeves are exposed to air.
- Water stops flowing in the reactor, or flow rates drop to very low values.

Water temperature sensors monitor the operating temperature within a UV reactor and shut down the reactor if the water temperature exceeds the setpoint. Some reactors also include level sensors as another safeguard to ensure the reactor is always flooded.

Control Strategies and Dose Delivery Monitoring

The performance of operating UV equipment must be monitored to demonstrate that adequate disinfection is being achieved (USEPA, 2003a). Because the concentration of pathogenic organisms cannot be measured continuously in the UV-treated water and the dose distribution cannot be measured directly in real time, various strategies have been developed to monitor dose delivery. Any dose monitoring method must be evaluated during reactor validation, and the outputs measured during validation will be part of the monitoring requirements.

The strategy for dose monitoring depends on the manufacturer and may be proprietary. Currently, there are three fundamental approaches to monitor UV disinfection performance in a UV reactor:

1. *UV intensity setpoint approach.* With this approach, measurements made by the UV intensity sensor are used to control the UV reactor. The UV intensity sensor is located in a position that allows it to properly respond to changes in both the UV intensity output of the lamps and the UVT of the water. The UV intensity sensor output and the flow rate are used to monitor dose delivery. The setpoint value(s) for UV intensity over a range of flow rates is (are) determined during validation.

2. *UV intensity and UVT setpoint approach.* This approach is similar to the UV intensity sensor setpoint approach, except that the UV sensor is placed close to the lamp such that it only responds to changes in UV lamp output. UVT is monitored separately. For a specific flow rate, the UV intensity and UVT measurements are used to monitor dose delivery. The setpoints for UV intensity and UVT over a range of flow rates are determined during validation.

3. *Calculated UV dose approach.* With this approach, the UV intensity sensor is placed close to the lamp, which is similar to the UV intensity and UVT setpoint approach. Flow rate, UVT, and UV intensity are all monitored, and the outputs are used to calculate UV dose via a validated computational algorithm developed by the UV reactor manufacturer. Dose is then controlled relative to a setpoint.

WATER QUALITY AND TREATMENT EFFECTS ON UV DISINFECTION

The primary factors impacting UV disinfection include water quality, sleeve fouling, lamp aging, power quality, flow rate, and the UV equipment. Of these, UV equipment options were discussed in the preceding section, and power quality is addressed in the design section. Of the water quality factors, UV absorbance, particle content, constituents that foul lamp sleeves and other wetted components, and the chemicals added upstream are the water quality factors that affect UV disinfection effectiveness. In spite of these effects, the impact of water quality on dose delivery can be adequately addressed in virtually all drinking water applications if carefully considered during the design of the UV disinfection system.

UV Absorbance and Transmittance

The most important water quality parameter affecting reactor performance is UV absorbance at 254 nm, or UVT, where these two parameters are related as described in Eq. (29.1). For a given reactor, as UVT decreases, the UV intensity throughout the reactor decreases and the UV dose delivered to the microorganisms by the reactor is reduced as a result. Later in this chapter, in the section on UV disinfection implementation and design, we discuss how to incorporate the impact of UVT into UV facility design.

When one is assessing the UVT of the water, it is recommended that measurements be made without filtering the sample in the laboratory. Therefore, the measurement accounts for scattering and some absorption from particles in the water sample that may interfere with UV disinfection. If the measurement is made according to a modified version of Standard Method 5910B (APHA/AWWA/WEF, 1998), the water sample is not pH-adjusted or filtered.

Particle Content

Particle content can also impact UV disinfection performance, although the potential impacts of particles on UV disinfection performance are not as significant for drinking water applications compared to wastewater UV disinfection. Particles may scatter light and reduce the UV intensity delivered to the microorganisms. Particles may also shield microorganisms from UV light, effectively reducing disinfection performance (USEPA, 2003a).

Microorganism UV dose response is affected by particle association and clumping of microorganisms. Solids present in wastewater samples can cause a tailing or flattening of the dose-response curve at higher inactivation levels because clumping or particle association shields a fraction of the microorganisms from UV light (USEPA, 1986). Compared to wastewater, there are significantly fewer particles in filtered drinking water and unfiltered water (assuming compliance with the filtration avoidance criteria). Research by Linden et al. (2002b) indicated that the UV dose response of microorganisms added to filtered drinking waters is not altered by variation in turbidity within regulatory requirements for filtered water turbidity.

For unfiltered waters, Passantino and Malley (2001) found that source water turbidity up to 10 ntu does not impact the UV dose response of separately added (seeded) microorganisms. However, these drinking water studies can only suggest the impact of turbidity on dose response as it relates to the impact of UV light scattering by particles, rather than particle association or clumping of microorganisms, because the microorganisms were seeded and not naturally associated with the particles. Further research into the significance of particle association for unfiltered drinking water applications is in progress. It is important that instruments that are used with higher-turbidity waters measure A_{254}/UVT appropriately.

The USEPA has only proposed requirements and guidance for postfilter applications and for waters that meet the filtration avoidance criteria (i.e., unfiltered water systems).

If UV disinfection is installed upstream of filtration or in other application points, the primacy agency will need to develop specific criteria for that application (USEPA, 2003a).

Lamp Sleeve Fouling Potential

Compounds in the water can cause fouling in a UV reactor on the external surfaces of the lamp sleeves and other wetted components such as the monitoring windows of UV intensity sensors. Fouling on the lamp sleeves reduces the transmittance of UV light through the sleeve into the water, thereby reducing power efficiency and potentially system capacity. Fouling on the monitoring windows impacts UV intensity and dose monitoring.

The rate of fouling and the corresponding frequency of sleeve cleaning necessary to maintain disinfection effectiveness depend on hardness, alkalinity, lamp temperature, pH, and certain inorganic constituents such as iron and calcium. Fouling is typically caused by precipitation of compounds with low solubility or compounds where the solubility decreases as temperature increases (e.g., $CaCO_3$).

While the specific rate of fouling and the optimal cleaning protocol for any given application cannot currently be predicted quantitatively, a proper cleaning protocol can be adequately estimated for most water sources in the start-up testing following construction (without the need for pilot- or demonstration-scale testing) and then adjusted during normal operation. Extensive data have been generated from pilot-scale testing on waters of low to moderate hardness and iron content (Mackey et al., 2001; Mackey et al., 2003). At total and calcium hardness levels below 140 mg/L as $CaCO_3$ and low iron concentration (less than 0.1 mg/L), standard cleaning protocols and wiper frequencies (one sweep every 15 min to 1 h) were adequate to address the impact of sleeve fouling at the sites tested. At sites with hardness or iron concentrations that exceed these levels, it may be advantageous to evaluate fouling on a site-specific or worst-case basis to identify how best to address fouling.

Upstream Chemical Impacts

Unit processes upstream of UV reactors can have a significant impact on the UV reactor performance. The three potential ways in which upstream processes may affect UV performance are (1) to increase UVT by increasing organics removal or oxidizing organics, (2) to decrease UVT because certain treatment chemicals will absorb UV light, and (3) to affect the lamp sleeve fouling rate.

It is possible to increase filtered water UVT by increasing the coagulant dose; however, the results will be site-specific. In one study, the UVT was increased from 80% to 89% per centimeter by increasing the alum dose from 15 to 45 mg/L (Cushing et al., 2001). However, the UVT increase from an increased alum dose should be considered against the increased alum chemical costs and sludge production. UVT increases would also probably be observed if other iron coagulant and poly-aluminum chloride coagulant doses were increased.

Properly implemented, ozone disinfection prior to UV disinfection has the potential to increase the UVT due to oxidation of organic matter. If the ozone residual is adequately quenched, a decrease in UV absorbance (increase in UVT) will be observed (Malley, 2002); an example of this increase for an unfiltered water is shown in Figure 29.6. If a UVT increase is desired, then a combination of increased coagulant dose and ozone disinfection will likely give the greatest UVT increase (Cushing et al., 2001).

Most common water treatment chemicals themselves will not significantly impact UVT. The following common water treatment chemicals do not significantly affect UVT

FIGURE 29.6 Example effect of preozonation on UV absorbance if ozone is quenched prior to UV disinfection. (*Source: Malley, 2002.*)

at typical concentrations present in filtered water: alum, aluminum, ammonia, ammonium, zinc, phosphate, calcium, hydroxide, and ferric iron (Fe^{3+}) (Cushing et al., 2001). However, hypochlorite (OCl^-), ferrous iron (Fe^{2+}), permanganate, and ozone are commonly used chemicals examined that might reduce UVT (Cushing et al., 2001), as described below (USEPA, 2003a).

- Residual OCl^- has only a slight effect on UVT. For example, an OCl^- residual of 3.5 mg/L was shown to cause UVT to decrease from 91% to 90% per centimeter (Cushing et al., 2001). However, in most cases, a hypochlorite residual that high will not be flowing through the UV reactor.
- Ferrous iron does decrease UVT; however, it is unlikely that ferrous iron will be present in filtered waters because ferrous iron is only present when there is low dissolved oxygen.
- Permanganate is a strong absorber of UV light; however, it is typically added in the raw water to oxidize taste and odor or iron and manganese. Therefore, when it is applied to raw water, there should not be a significant permanganate concentration in the filtered water and in the UV reactor.
- Ozone disinfection can decrease UVT if a residual ozone concentration is present in the UV reactors. Ozone residual concentrations can be quenched, and then the UVT will not be decreased. However, two common quenching chemicals, thiosulfate and peroxide, are both strong absorbers of UV light and will decrease the UVT. Sodium bisulfite, an alternative to thiosulfate, will not significantly impact UVT.

The possible UVT variation from upstream processes should be assessed by collecting UVT data during various operating conditions (e.g., a range of alum doses) that are typically observed. Potential treatment process upsets should also be considered in the water quality analysis to determine the extent to which they impact the design UVT and cleaning regime.

Some unit processes that use metal-based coagulants may affect the rate of fouling. Mackey et al. (2001) found that iron levels less than 0.1 mg/L could be adequately cleaned by standard protocols as described previously. In addition, lime softening has been shown to reduce fouling potential (Mackey et al., 2001). Overall, the effect of upstream coagulant addition and residual metals should be considered in the fouling data monitoring as described previously.

UV DISINFECTION FACILITY PLANNING, DESIGN, AND PROJECT IMPLEMENTATION

As of 2004, it is estimated that less than 100 UV disinfection projects had been constructed at water treatment plants in North America. Although there is not a "standard" method for UV disinfection project implementation, one approach is shown in Figure 29.7.

The flowchart includes the steps necessary for UV facility design; however, the order of these steps will vary depending on a utility's specific goals and constraints (e.g., procurement laws). In the flowchart, validation testing is shown with dashed lines to indicate potential options for the timing of this key step. Of the steps shown, defining goals

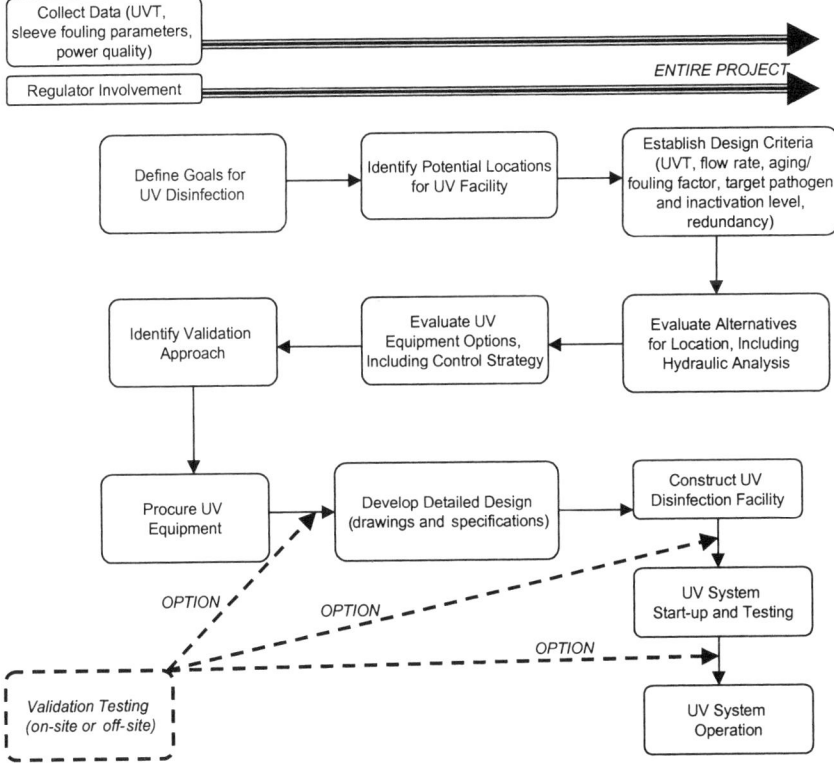

FIGURE 29.7 Flowchart for an approach for UV disinfection project implementation.

for UV disinfection was addressed previously, while additional discussion of validation testing is included in the next section. In this section, the other steps in the project implementation approach are addressed.

Data Collection

In Figure 29.7, collecting data is shown as an ongoing task for a UV disinfection project, to indicate that it is important to begin collecting data early in the project and to continue data collection throughout the duration of the project. Key components of the data collection task include UVT data, data on water quality constituents that contribute to sleeve fouling, power quality data, and conducting optional bench-scale, pilot-scale, or demonstration-scale testing if desired.

UVT Data. The collection and evaluation of UVT data are an essential step in UV project implementation. Because UVT is a direct measure of the ability of the water to transmit UV light, the required size of a UV reactor is strongly influenced by the design value selected for this parameter. Small differences in UVT can have a large influence on UV project construction costs. Thus, UVT is a critical design criterion for UV facility design.

For a sampling program, UVT samples should be collected at the point of implementation. As data are available over a longer time, the data set becomes more representative of any temporal and seasonal variations and allows the design engineer to avoid selecting an inappropriately optimistic value or incorporating unnecessary conservatism. The collection of UVT data should start as soon as possible in the implementation process and then continue through the project. Specific data needs will depend on the watershed characteristics, data variability, and site-specific events that result in organic peaks (e.g., changes in source waters, spring runoff, reservoir turnover). It may be beneficial for a utility to consider UVT variations together with seasonal water demands (USEPA, 2003a).

Sleeve Fouling Data. Early in the project, it is also important to collect water quality data for parameters that can contribute to lamp sleeve fouling. An evaluation of the data may assist in determining the appropriate value for the lamp fouling/aging factor to use as a design criterion. Discussion of the water quality parameters of concern was provided previously in this chapter.

Power Quality Data. UV lamps are extremely sensitive to power fluctuations and interruptions. For example, voltage fluctuations that exceed 10% to 15% of the normal operating conditions for as little as 2 to 5 cycles (0.03 to 0.08 s) may cause UV lamps to lose their arc. The time required for UV lamps to return to full operating status when power is restored varies with the type of lamp and can be from 2 to 10 min. Because of the UV lamp's sensitivity to power quality, a reliable power supply is critical to UV facility design.

At the start of the project, it is important to review power quality data to evaluate the potential implications for off-spec operation and the potential need for mitigating equipment such as an uninterruptible power supply (UPS) or other device. Additional discussion of this issue is provided later in this section.

Bench, Pilot, and Demonstration Testing. Bench-scale inactivation tests, bench-scale jar tests, and pilot-scale testing of UV disinfection equipment can provide some useful information as part of the implementation of UV projects, but none of these testing steps is necessary for successful implementation. Consequently, these steps are not included in the implementation flowchart (Figure 29.7). More important than any testing step is the collection of UVT data for use in facility design, as described previously.

Bench-scale inactivation tests can be used to evaluate the dose response of microorganisms, potentially to confirm the consistency of results in comparison to research findings. Jar tests may be useful for a new water supply in estimating the resulting UVT following treatment, or in evaluating the impacts of coagulation conditions on UVT for an existing supply.

Pilot testing can provide information about the operation, maintenance, and control of specific equipment. This information can include gaining an understanding of sleeve fouling for the particular water, although scale-up issues and the fact that pilot testing occurs for a limited time can limit the usefulness of the data collected. Pilot testing may also be used to generate information on the steps and time commitment associated with typical maintenance activities associated with UV equipment such as sensor calibration checks and parts replacement.

Demonstration testing, like pilot testing, can provide useful information but is not necessary as part of project implementation. Depending on the reactor size selected for demonstration, scale-up issues can be minimized or eliminated. With the correct piping and with the reactor size equivalent to the future full-scale reactors, demonstration testing can be used for validation of UV reactor performance and establishing safety factors under the USEPA requirements. Demonstration testing also provides an opportunity for plant operations and maintenance staff to become familiar with the UV system prior to commissioning of the full facility.

Regulator Involvement

UV disinfection is a relatively new technology for WTPs, and it is important that regulators from the primacy agency be involved early in decision making and throughout the implementation process. The USEPA's proposed *UV Disinfection Guidance Manual* (USEPA, 2003a) recommends regulator involvement throughout the planning and design phases of implementation. Key milestones for regulator involvement include reviews of the conceptual planning or preliminary design report, equipment procurement documents; approval of drawings and specifications; review of planning for on-site validation testing (if applicable); review of the validation testing report; and final approval upon UV facility start-up.

Define Goals for UV Disinfection

A comprehensive disinfection strategy provides multiple barriers to reduce microbial risk while minimizing DBP formation. UV disinfection can contribute to a comprehensive disinfection strategy by providing a cost-effective method of inactivating target pathogens that are more resistant to more traditional chemical disinfection methods. The specific objectives of a given UV installation should be clearly defined during the planning stages. This can ensure that the design meets the utility's and the regulator's expectations based on the regulatory requirements, target microorganism(s), and the overall disinfection strategy. The potential applications for UV disinfection were described previously. The most important goal to define is the target pathogen because viruses will require higher doses compared to *Cryptosporidium* and *Giardia* inactivation and will significantly affect the UV facility design.

Identify Potential Locations for the UV Facility

While different WTPs may have unique potential locations for the UV facility, some generalizations are possible regarding the potential locations for consideration. The USEPA's UV dose requirements, although published in draft form at present (USEPA, 2003a), apply only to postfiltration applications and for unfiltered supplies that meet the filtration avoidance criteria. Thus, although alternative upstream locations are allowed, implementation of UV upstream of filtration would represent a special case for evaluation by the primacy agency.

For installations downstream of filtration, typical potential locations to consider are

1. Individual filter effluent piping within the filter piping gallery
2. Combined filter effluent flow, by gravity, upstream of the finished water clearwell
3. Pumped (combined) filter effluent flow, upstream of the finished water clearwell
4. Pumped or gravity flow downstream of the finished water clearwell

Each of the potential locations listed above has unique associated advantages and disadvantages. If there are treatment processes downstream of filtration, such as GAC contactors, there may be additional potential locations to consider.

In developing the alternatives for the identified potential locations, it is important to coordinate with the UV manufacturers to understand reactor dimensions and sizes of ancillary equipment components such as electrical and control panels. The engineer should consider at least a conceptual level design for the UV equipment options, become familiar with the hydraulic requirements, and determine the space requirements for upstream and downstream piping. The conceptual level layout and sizing of the UV system should consider provisions for reactor isolation, flow splitting and/or flow control, flow measurement, assurance of reactor/lamp submergence, sample and drain lines, a cooling water system (if necessary), air relief, space and access for reactor and ancillary equipment maintenance, and potentially validation facilities and facilities for the management of off-specification water (as defined later). These considerations are discussed further later in this chapter.

Individual Filter Effluent Piping. A UV reactor installed on individual filter effluent piping is shown in Figure 29.8. This approach can result in eliminating the need for a new building to house the UV equipment. As such, this approach results in construction cost savings; however, O&M costs may increase because more UV reactors, and thus more individual system components (e.g., lamps, sensors, sleeves), may be necessary. Other advantages include that filter-to-waste flows could be disinfected with UV light and that head loss associated with the installation may be minimized.

Another consideration for UV installation on individual filter effluent piping is that the number of UV reactors would match the number of filters, and operation of the filter and UV reactor would be coupled. Thus, the reliable operation of each component would depend on the other. Other potential disadvantages include the following:

- Available space in the filter pipe gallery may limit design options, equipment options, access to UV equipment, optimum hydraulic conditions, and locations for ancillary components such as control panels.
- Existing piping may result in unique inlet and outlet configurations and require site-specific validation testing. In addition, it may be difficult to accommodate equipment and locate effective additive feed points for on-site validation.
- Operation of the UV equipment can be complicated during filter backwashing and filter-to-waste operation, depending on the pipe arrangement.

FIGURE 29.8 UV installation in existing filter gallery in Poughkeepsie, New York. (*Photo courtesy of Jay Maggi and the Poughkeepsie Water Treatment Facility.*)

- Head loss through the UV reactor and piping will reduce filter driving head, thereby altering operation of the filters, potentially shortening filter run times, and possibly altering operation of the finished water clearwell.
- Feasible UV equipment options may be limited.
- The complexity of construction sequencing and the duration of construction may be impacted to accommodate installation at each individual filter.

Gravity Combined Filter Effluent Flow. A UV facility installed to treat combined filter effluent is shown in Figure 29.9. This system treats gravity flow, splitting the combined flow to separate UV trains, at a location between the filters and the finished water clearwell in the process flow stream. At this location, the UV equipment would typically be installed in a separate UV building, and the number of UV reactors could be minimized. For example, three UV reactors may be needed to disinfect the combined filter effluent from six or eight filters. Additional advantages of this location include the following:

- A dedicated building would typically be provided for the UV equipment, and the building can be designed and constructed specifically to accommodate the UV equipment.
- The use of gravity flow would minimize costs associated with incorporating additional in-plant pumping.
- More options would typically be available for the types of UV equipment that can be accommodated compared to the individual filter effluent piping retrofit.
- The operation of the UV facility is more independent of the operation of other unit processes such as filters.

FIGURE 29.9 UV facility to disinfect combined filter effluent in Clayton County, Georgia. (*Photo courtesy of Clayton County, Ga., Water Authority, Freeman Road Water Production Plant.*)

- The upstream and downstream piping configurations can be designed and constructed to optimize disinfection performance.
- The upstream and downstream configuration will be more likely to be equivalent to or better than the off-site validation configurations, such that the off-site validation results may be valid. Alternatively, on-site validation equipment can be accommodated.
- O&M costs may be minimized compared to a filter gallery location because the number of reactors will typically be reduced.

On the other hand, a combined filter effluent location also has the following disadvantages:

- Construction costs will be higher for this option compared to a filter gallery retrofit because a new building will most likely be needed.
- Head loss through the UV reactor and piping will likely be highest for this application, and the head loss will reduce filter driving head, thereby altering operation of the filters, reducing filter run time, and also possibly resulting in a decrease in the available operational storage or disinfection contact time at the finished water clearwell.

Pumped Combined Filter Effluent Flow. The head loss associated with a gravity flow, combined filter effluent location will vary largely with the pipe diameter and configuration, but head loss will typically be approximately 1 to 6 ft. Reducing the filter driving head by this amount can substantially shorten filter run times, reduce unit filter run volumes, and increase filter backwash frequency. In addition, it can reduce the available operating storage or disinfection contact time in the finished water clearwell, which may not be desirable. Because of the head loss associated with the UV facility, it will not be possible for many utilities to implement UV disinfection on the filter effluent without providing intermediate pumping. Thus, the advantage of an application following intermediate pumping is that the installation of the UV facility does not negatively impact filter or clearwell operations.

The most significant disadvantages associated with this alternative are that surge and water hammer analysis may be required to protect equipment (unless the hydraulic grade line allows breaking head) and that the project, operating, and maintenance costs are increased by the cost of the new pump station. The controls associated with the UV equipment and the pumps may need to be integrated based on the requirements of both components.

Pumped or Gravity Flow Downstream of Clearwell. The fourth alternative location for UV facility is downstream of the finished water clearwell. This alternative offers the advantage of preserving filter driving head while eliminating the additional project costs associated with a new pump station. In addition, there may be more space available for installation of a new building. However, there are also significant disadvantages associated with this alternative, particularly if the location is on the discharge from high-service pumps. The potential disadvantages include the following:

- Flow rates may change frequently to accommodate downstream demand, particularly in a gravity flow system. Therefore, the design should consider measures to ensure that an adequate dose is delivered under flow conditions that are frequently changing.
- Special design attention will be required for water at distribution system pressure. Special reactor construction may be necessary.
- Surge and water hammer analysis is critical, and additional design components may be necessary.
- Clearwell volume may be affected, and available operational storage or distribution system pressures may be reduced.
- In the event of a structural failure of a lamp sleeve and lamp, glass, quartz, and mercury may be released into the water being discharged from the plant. Thus, there may be limited opportunity to capture the contaminants before delivery of the water to the end user.

Establish Design Criteria

The key design criteria for UV facilities include UVT, flow rate, lamp sleeve fouling/lamp aging factors, the target pathogen and inactivation level, and the level of redundancy. Additional criteria to consider in establishing design criteria include the potential for off-specification operation, potential future expansion, potential impacts of changes to source water quality, and influence of upstream treatment processes on water quality.

Flow Rate. The UV facility should be designed to provide adequate disinfection at the maximum instantaneous flow rate. Plant minimum and average flow rates should also be determined, with all three values determined with potential future flow increases taken into consideration.

The UV facility can be designed with additional space to accommodate the future installation of more reactors. As an alternative means of increasing UV facility capacity, it may be possible to expand the capacity of individual UV reactors through the addition of more lamps or more powerful lamps. To install different lamps or more lamps, however, revalidation of UV reactor performance may be necessary.

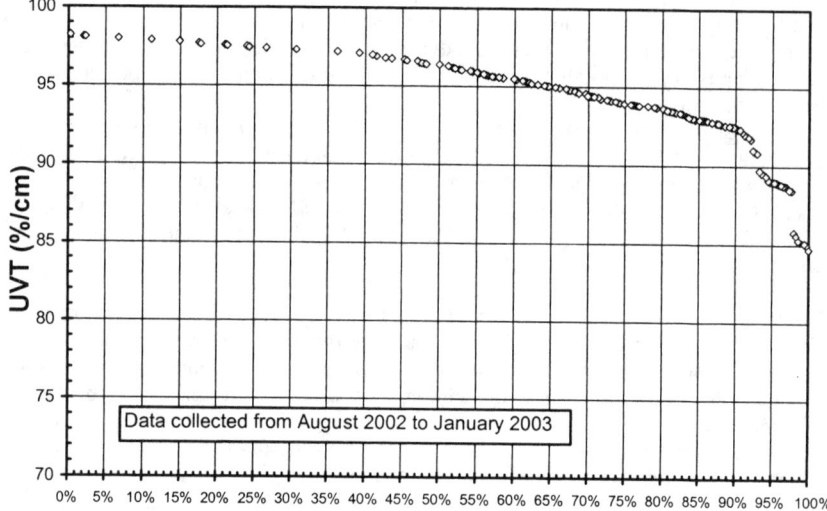

FIGURE 29.10 Example UV transmittance data from sampling program for filtered water from City of Newark, Ohio, WTP. (*Photo courtesy of City of Newark, Ohio.*)

UV Transmittance (UVT). The selection of a design criterion value for UVT should be based on as comprehensive a data set as possible, an understanding of the relationship between UVT and plant flow rates, an understanding of the impacts of upstream treatment processes, and the utility's preferred level of conservatism. The value selected should be tied to validation of the reactor to minimize off-spec operation.

A data set from sampling combined filter effluent at the Newark, Ohio, WTP is shown in Figure 29.10. From the data shown, in 95% of the samples analyzed, UVT exceeded 89%/cm, and UVT exceeded 92.5%/cm in 90% of the samples analyzed. For design of this facility, a design UVT value of 88%/cm was selected during the data collection process. This value was selected to balance conservatism and the potential for off-specification operation. From Figure 29.10, it is expected that UVT will be below the design value approximately 3% of the time. These low UVT events will most likely occur during low demand periods, so the flow rate in the UV reactor could potentially be reduced to compensate for these low UVTs.

Lamp Fouling/Aging Factor. The UV design dose should be based on the output from the UV lamps at the end of lamp life because the output decreases over time as the lamp ages. As part of the equipment procurement process, lamp aging data should be evaluated, and guarantees for the end of lamp life output and the associated lamp replacement frequency should be required for each UV manufacturer.

In addition to lamp aging, the output from the lamps that reaches the water flow will diminish over time as the result of any fouling, coating, scarring, or scratching that occurs on the quartz sleeves. Although an automatic cleaning system will help to keep the lamp sleeves relatively clean, both water quality and use characteristics can contribute to permanent sleeve fouling and reduced dose delivery.

It is recommended that lamp aging and lamp fouling factors be included as design criteria for the UV equipment. In practice, these factors are often combined into a lamp fouling/aging factor. Typical design values are 40% to 80%, meaning that the dose delivery capability with new lamps should be specified to be the design dose divided by the lamp fouling/aging factor. Therefore, a system designed with a UV dose of 40 mJ/cm^2 and a lamp fouling/aging factor of 0.7 would be capable of providing a dose of 40 divided by 0.7, or 57 mJ/cm^2, with new lamps and clean sleeves. In practice, there is a tradeoff between equipment costs and maintenance costs, and the fouling/aging factor should be selected to balance equipment costs with component replacement costs.

Target Pathogen and Inactivation Level. The selection of the target pathogen and inactivation level should be based on the defined goals for the UV process. In the United States, UV dose requirements are based on the target pathogen (i.e., *Cryptosporidium, Giardia*, or viruses), the target level of inactivation (see Table 29.7), and validation safety factors. Since UV dose cannot be directly measured at full scale, the UV dose achieved by a specific reactor can only be assessed based on validation testing data.

Redundancy. At a minimum, UV facility design should include one redundant reactor to serve as a standby when the system is operated at the design conditions. The standby reactor should be configured for duty in the event of component failure in a duty reactor. Some utilities may elect to provide additional redundancy, and additional redundancy may also be warranted for larger-capacity facilities. To minimize reactor downtime following failure of components, utilities should consider the availability of spare parts and the need for maintaining an on-site inventory to expedite the return to service of any reactor following component failure.

Evaluate Alternative Locations

Once the potential locations for the UV facility have been identified, they should be evaluated and compared so that the optimum alternative can be identified. This evaluation will include determining the site-specific advantages and disadvantages for each alternative. Elements of this evaluation include the following:

- Hydraulic analysis, development of hydraulic profiles, and identification of impacts of each alternative (reactor and associated piping and appurtenances) on the operation of the filters, finished water clearwell, and other processes
- Schematic layouts of UV equipment for each alternative including reactors, piping, valves, flowmeters, flow-splitting facilities, UVT analyzer, and electrical and control panels
- Identification of electrical requirements and necessary electrical modifications to accommodate the UV equipment for each alternative including UPS (or other) power quality correcting devices
- Schematic design of yard piping and site electrical modifications
- Identification of any locations that are not feasible with specific UV equipment because of site-specific constraints (e.g., head loss, reactor dimensions)
- Consideration of where debris/mercury from a reactor sleeve/lamp failure will go and what opportunities the operator will have to contain the debris/mercury
- Consideration of how off-specification water will be managed in the event of a power or component failure
- Consideration of validation components and water disposal if on-site validation is proposed.
- Estimates of construction and O&M costs for each alternative

If the facility is to be retrofit into existing building space as part of one of the alternatives, structural and HVAC evaluations may be necessary. As part of the evaluation of alternatives, it is essential that the different UV equipment options, the potential controls philosophy, the potential operational strategy, operations and maintenance issues, and the preferred validation approach be defined.

Evaluation of Hydraulic Considerations. The evaluation of UV equipment in conjunction with the overall plant hydraulics is critical during project planning to determine the optimum UV facility location. The head loss associated with a UV reactor at the peak flow rate can vary from as little as a few inches to as much as several feet. Head loss through the reactor depends on the lamp array, internal baffling, inlet and outlet conditions, and the reactor size. While head loss through the reactor is important, upstream and downstream piping, flow control valves, and appurtenances are also important contributors to the total head loss through the UV facility for typical installations.

For operating UV facilities, the total head loss through UV facilities varies widely. For planning purposes, in establishing a hydraulic profile through a new WTP, a minimum of 1 to 6 ft of head should be allocated for the UV facility. As the project progresses, a more detailed hydraulic analysis is essential. As with any process, the design can minimize head loss through component sizing and piping configuration, if necessary.

The upstream and downstream piping configuration is critical in the design of effective UV disinfection facilities. One objective of any design should be to avoid locating any mechanical design components where they may hinder the optimum flow characteristics through the UV reactor. In many of the first UV designs for drinking water facilities, the UV process was treated similar to a flowmeter, with the piping designed to allow a minimum of five upstream pipe diameters and three downstream pipe diameters of straight piping. The USEPA's potential regulatory requirements are still being developed, but the agency may recommend more upstream and downstream pipe diameters. As the use of computational fluid dynamics becomes more widespread as a UV design tool, these values may be able to be reduced based on site-specific evaluations. In designing UV facilities, some components that need to be carefully considered include piping elbows, tees, reducers, and other appurtenances, as well as modulating valves for flow control, if used.

Computational Fluid Dynamics Modeling. Computational fluid dynamics (CFD) modeling can be an important tool for use in the implementation of UV facility design. CFD modeling can assist in understanding the impacts of upstream and downstream piping and components, scale-up issues, characterization of potential head loss related to the location of the UV facility, and modeling of reactor entry and exit conditions. CFD modeling can be used to evaluate the following items:

- Candidate locations for the UV facility
- Candidate UV reactor designs
- Head loss estimations through the UV reactor and development of optimum piping configurations
- Effects of upstream and downstream piping configurations

An example of the output from a CFD model is shown in Figure 29.11.

An enhancement to CFD modeling involves the use of sophisticated UV dose models or light intensity distribution models that calculate the UV intensity distribution in the UV reactor, considering the reflection, refraction, and absorption of UV light within the reactor. These models allow computation of the UV dose distribution using techniques such as Lagrangian particle tracking.

FIGURE 29.11 Example velocity contour output from computational fluid dynamics model. (*Courtesy of Alex Mofidi, Metropolitan Water District of Southern California.*)

CFD modeling can provide site-specific information to help assess the performance of a UV reactor upon installation. However, CFD represents a tool to assist in implementation rather than a method for validating performance that is currently accepted. Full-scale reactor biodosimetry is the only regulator-accepted method for validation of UV reactor performance. Where CFD models are used to compute UV dose distribution, the model results will require calibration via microbial challenge testing.

Evaluate UV Equipment

For a comprehensive evaluation of alternatives, it will be necessary to contact UV manufacturers to determine applicable reactor types, reactor dimensions, maintenance clearances, piping configurations that have been validated, head loss through the reactor, electrical loads, maximum distances between control panels and reactors, and other equipment installation requirements. With this information, it may become apparent that some manufacturers' UV equipment may not be feasible for installation at one or more of the identified locations. At this stage in the project, it may also be possible to develop a short list of UV equipment options that meet the objectives for the selected location for the UV facility.

Control Strategy. The available UV equipment differs in how the UV reactor is controlled to maintain effective disinfection. It is anticipated that the USEPA's *UV Disinfection Guidance Manual* will recognize three control strategies for validation and operation. For any project, the control strategy for implementation will depend on the equipment selected and how the equipment is validated. The three control strategies consist of: (1) UV intensity setpoint, (2) UVT and UV intensity setpoint, and (3) calculated dose. A more detailed description of these control strategies was provided in the description of UV equipment.

These control strategies should be compared by the owner and engineer to determine if a particular control strategy is preferred. For each option, the potential for optimization and energy savings should be identified.

Identify the Validation Approach

Validation testing is a critical step in UV project implementation, and issues associated with validation testing are addressed later in this chapter. At this stage in project implementation, it is desirable to identify the preferred validation approach before procuring the UV equipment. The available options include off-site validation prior to equipment procurement, off-site validation after equipment procurement, and on-site validation after the construction of the UV facility. Each option has advantages and disadvantages. The logistics and cost issues require careful examination and understanding prior to selecting the optimum validation approach for each utility. If validation is to be completed after detailed design, the incorporation of additional electrical capacity and building space may be advantageous in the event that the validation testing necessitates a change in the UV reactor or facility design.

Off-Site Validation Test Centers. Manufacturers will likely validate UV reactors over a wide range of flow rates and water quality (e.g., UVT) conditions at off-site validation test centers. Until recently, the only test center dedicated to UV validation was in Germany. In 2003, two dedicated test center facilities in the United States began validation testing, and at least one manufacturer has performed off-site validation under controlled conditions to seek regulatory approval.

Off-site validation has several advantages, including simplicity and the ability to design around a UV reactor with known performance characteristics for given inlet and outlet hydraulics. However, the Long Term 2 Enhanced Surface Water Treatment Rule (LT2ESWTR) requires that the site-specific installation and operating conditions fall within the range of conditions used when the installed UV reactor is validated off-site (USEPA, 2003a). The inlet and outlet hydraulic conditions during validation will probably be selected so the UV reactors can be installed in most WTPs. If the validation conditions do not encompass the utility's design criteria or inlet and outlet piping configurations, the utility may request that the UV manufacturer revalidate the unit off-site under specific testing conditions that more closely match those of the proposed UV facility. Alternatively, on-site validation can be performed.

On-Site Validation. If on-site validation is desired, then the UV facility design should be developed to enable testing. On-site validation testing is complex and requires sufficient facilities, planning, and expertise.

The advantages of on-site validation are that the UV reactors can be validated for the exact piping hydraulic conditions, and the UVT will more accurately represent the UV facility even if a UV-absorbing chemical is added. In addition, the equipment necessary for on-site validation will provide the flexibility for future testing to optimize the UV reactor performance under specific hydraulic and water quality conditions even if they are not tested as part of the initial validation. On-site validation will also allow for future testing of different or improved technologies or equipment configurations.

On the other hand, a disadvantage of on-site validation is that the UV facility may be designed and constructed without prior validation of the performance of the UV reactors (unless on-site validation is conducted in addition to off-site validation). This may lead to the UV reactor failing to meet performance requirements, and it may be difficult to increase UV disinfection efficiency after the UV reactors are already installed. In addition, on-site validation is limited to the highest UVT available at the time of testing. Consequently, UV reactor performance characteristics cannot be determined at higher UVT, and the UV reactors may need to be operated at conditions other than optimal, possibly resulting in higher power use. Other significant disadvantages include the logistics and cost of the testing. For example, one unit must be isolated from the system to allow validation testing to occur, and a permit may be needed to discharge the nonpathogenic challenge microorganism (USEPA, 2003a).

If on-site testing is selected as the desired approach, the facility layout should accommodate piping and ancillary equipment necessary for validation. As described by Swaim et al. (2003), the requirements for on-site validation testing include the following:

- Willingness of utility and regulators to allow on-site testing
- Ability to shut down all or part of the WTP to accommodate testing
- Sufficient quantity of challenge organism
- Mechanical equipment for the challenge organism and the UVT adjustment chemical including dilution water, batch tanks, tank mixers, additive feed pumps, and feed ports
- Static mixers located between feed ports and sample ports both upstream and downstream of the reactor, if possible
- Representative sample ports on influent and effluent piping
- Ability to isolate test loop both upstream and downstream of test reactor
- Location to collect test waste stream volume
- Method for disposal and possibly chemical disinfection of the test waste stream

An equipment skid for on-site validation testing that includes tanks, tank mixers, and tubing feed pumps is shown in Figure 29.12.

FIGURE 29.12 Equipment skid for on-site validation testing. (*Photo courtesy of Clayton County, Ga., Water Authority.*)

Equipment Procurement

There are several different manufacturers of UV equipment, and these manufacturers are currently offering several different UV reactor types and models for drinking water applications. The available approaches for equipment procurement include owner prepurchase, equipment selection during preliminary design with the equipment contract assigned to a general contractor (selected later), a base bid approach with design based on one manufacturer but bids open to alternatives, and selection by the contractor during the bid phase for construction. Each approach offers advantages and disadvantages, and selecting the best approach for a specific project is up to the owner.

There is enough difference among the alternative equipment systems that it can be difficult to prepare a "generic" detailed design that can be used to solicit construction bids from all the different manufacturers. At the same time, it may not be cost-effective to develop specific designs for each potential UV system.

Many of the first drinking water UV disinfection projects developed approaches for preselection of UV equipment. On many of these projects, the UV equipment was prepurchased or the price was guaranteed through some type of contractual mechanism. Including a life-cycle cost component in the procurement documents will assist in procuring the most efficient UV system for a given application. This type of approach is described by Brauer et al. (2001). These are very important considerations and should be reviewed with the owner's purchasing agency to ensure a suitable procurement strategy is established early in the project.

Detailed Design

The next step in project implementation consists of detailed design leading to the development of bid documents for construction, including specifications and drawings. The detailed design of a UV facility may incorporate several engineering disciplines, including structural, architectural, mechanical, civil, electrical, and instrumentation. Some of the elements of a UV facility that should be considered as part of the detailed design effort include

- Flowmeter installation at a location that allows accurate measurement of flow rate to each reactor
- Effective flow splitting by either a passive flow split or the use of modulating flow control valves
- Adequate access for UV reactor components that are subject to typical O&M activities
- Structural support of UV reactor, piping, associated components, and forces
- Sample taps for the collection of representative upstream and downstream samples
- Drain valves and drain piping of adequate size, particularly if regular O&M tasks require reactor draining
- Recirculation loops or flow-to-waste capability for UV reactors that require cooling water during lamp warm-up
- Atmospheric conditions (i.e., HVAC) for UV equipment, electrical component, and control panel longevity
- Vents and/or air release, air/vacuum, or combination air valves to discharge any introduced air and to facilitate reactor draining
- Hydraulic design so that reactors stay flooded under all operating conditions
- Actuated isolation valves for automatic shutdown procedures

- Provisions to minimize and manage off-specification water
- Provisions to minimize and manage the impacts of any sleeve and lamp breakage events
- Use of appropriate materials of construction for piping materials exposed to UV light and for all components within the building environment
- Identification of construction sequencing needs

Instrumentation and Control. Detailed instrumentation and control design should include development of a P&ID drawing, control philosophy, identification of parameters to monitor locally and at SCADA stations, designation of alarms and associated actions, and identification of conditions that will result in reactor start-up and shutdown.

It may not be possible to identify all specific elements of the instrumentation and control system until the UV manufacturer is selected. The owner's specific preferences should be identified and included as part of equipment specification and procurement. The UV manufacturer will typically have a PLC-based control system developed for the reactor. This may be a standard package using a particular brand of PLC, and the designer may be limited in options for interfacing the plant SCADA system with the UV control panel if the plant SCADA system uses a different hardware base. The UV reactor control system may require that certain external signals be provided by the owner, such as the reactor flow rate.

It is important to develop the instrumentation and controls design together with the control strategy, manufacturer input, and consideration of the potential causes of off-spec operation. With this approach, steps to minimize off-spec operation and maximize energy efficiency can be incorporated into facility design.

Power Supply and Quality. In addition to typical electrical engineering design tasks, electrical design for UV facilities must consider the quality of power supply and the potential power supply and backup power options.

UV lamps are extremely sensitive to power quality fluctuations. UV lamps may lose their arc if voltage fluctuations exceed 10% to 15% of the normal operating conditions for as little as 2 to 5 cycles (0.03 to 0.08 s). The start-up and restart behavior for LPHO and MP lamps is summarized in Table 29.5.

TABLE 29.5 Start and Restart Times for Low-Pressure, High-Output, and Medium-Pressure Lamps*

Lamp type	Cold start[†]	Warm start[‡]
LPHO	2-min warm-up + 4 to 5 min to full power	2-min warm-up + 2 to 5 min to full power
	Total time: 6 to 7 min	Total time: 4 to 7 minutes
MP	No warm-up or cooldown + 5 min to full power[§]	5-min cooldown + 5 min to full power[§]
	Total time: 5 min	Total time: 10 min

*Information shown in table is compiled from UV manufacturers (Calgon Carbon, Severn Trent, Trojan, and Wedeco).
[†]A cold start occurs when UV lamps are started when they have not been operating for a significant period of time.
[‡]A warm start occurs when UV lamps are started after they have just lost their arc (e.g., due to voltage sag).
[§]Intensity of 60% is obtained after 3 min.
Source: Cotton et al. (2002).

The effects of temperature can increase or decrease the times listed in Table 29.5 and should be discussed with the UV manufacturer.

In locations that are subject to frequent power quality fluctuations or outages, the following options should be considered alone or in combination to minimize off-specification operation and ensure regulatory compliance:

1. Installation of a backup generator to allow UV equipment operation during long-term outages
2. Connection to a second, independent power source
3. Installation of power conditioning equipment or a battery-supported, uninterruptible power supply (UPS)
4. Automatic rapid flow shutoff and/or diversion to waste

The assessment of the above options should consider the different response and backup periods associated with them. For example, a backup generator cannot ensure continuous power supply to avoid UV reactor shutdown due to voltage sags, while a battery-supported UPS can. The hydraulic and process constraints must also be considered. For example, a rapid shutoff may not be practical due to upstream processes or hydraulic transients. Conversely, without proper programming, a UPS may result in lamp overheating if a power failure also results in loss of flow in a pumped system (USEPA, 2003a).

Off-Spec Operation and Provisions to Address Equipment Failure. At the time of publication, the USEPA has recommended in the proposed *UV Disinfection Guidance Manual* that UV facilities disinfecting filtered water minimize off-spec operation, that is, operation outside of the validated limits. To comply with this recommendation for system operation, it is critical that detailed design include consideration and mitigation of the potential causes of off-spec operation.

There are many potential causes of off-spec operation, such as poor power quality, unexpectedly poor water quality, discrepancies in flow splits, reactor cooldown and warm-up time following lamp failures, low UV intensity, intensity sensor failure, and flow surges that result in flow rates outside of validated limits.

Provisions to address equipment failure should be considered to mitigate excessive off-spec operation. These provisions include automatic valve closures as part of a reactor or system shutdown, increasing reactor power levels in anticipation of flow surges (such as the reintroduction of flow from a filter following completion of filter-to-waste), provision for alternative disinfection backup, and providing reactor-to-waste piping. As discussed above, selection of these options must reflect the overall plant constraints.

In evaluating the ability of the facility as designed to deal with the potential for equipment failure, one additional consideration is the potential for mercury release due to a sleeve and lamp breakage. There are a small number of documented lamp and sleeve breakage events resulting in mercury and quartz debris release, but the majority of these events were the result of construction debris (USEPA, 2003a). Any regulatory requirements for addressing this possibility will be the responsibility of the primacy agency.

There are several components of facility design that can be configured to reduce the possibility of mercury and quartz debris release. One option to reduce the potential for mercury and quartz debris to move downstream is to orient the UV reactors vertically, as shown in Figure 29.13. It is recommended that utilities using UV disinfection develop emergency operating procedures in the event of a sleeve and lamp breakage event (USEPA, 2003a).

FIGURE 29.13 UV reactor oriented vertically. (*Photo courtesy of City of Albany, New York.*)

Other Considerations for Bid Document Preparation. In addition to the items addressed above, the following topics should be addressed as part of the development of UV facility bid documents:

- Manufacturer warranties and bonds for system performance, O&M costs, and component lifetimes
- Corrective actions and design flexibility, particularly if validation will be completed following equipment procurement
- Identification of a spare parts package that will expedite the return of reactors to service while also taking into account the available space on-site for spare parts storage
- Definition of testing requirements including factory, functional, performance, and validation testing
- Requirements for shop drawing and O&M manual submittals, and requirements for manufacturer services on-site

CONSTRUCTION, TESTING, START-UP, AND O&M

Once detailed design of the UV facility has been completed, the remaining steps in project implementation consist of construction, testing, start-up, and facility operation, maintenance, monitoring, and reporting.

Construction of UV Facilities

The construction of UV disinfection facilities is similar to that of other treatment plant processes. If the UV facility will be installed without bypass piping, it will be important to follow construction sequencing requirements that limit piping modifications that affect WTP operation to the desired season. As a developing technology, equipment components are evolving, and manufacturers may propose the use of new generations of components during the construction phase. Thus, a thorough understanding of the impacts of any change is important, and field coordination may be necessary. UV equipment includes breakable components such as lamps and sleeves that require special care during handling and installation. In addition, special care is recommended to ensure that no loose items or debris are left in the process piping upstream of the UV facility.

Testing of UV Facilities

UV equipment testing should include factory testing prior to installation and functional testing and performance testing upon UV equipment installation. Validation testing, a critical step for all UV projects, is discussed subsequently in this chapter.

Factory testing consists of inspection of the equipment upon fabrication but prior to delivery to the project site. Factory testing can be certified by the manufacturer and possibly witnessed by the owner and/or engineer. Typical activities include verification of equipment and dimensions, demonstration of control functions, and simulated operation.

Functional testing is the first testing step after construction, and a typical objective is to demonstrate the effective operation of electrical, mechanical, and instrumentation/control components to confirm that all components function as intended. For functional testing, the contractor, engineer, owner, and UV manufacturer are typically involved to meet the objectives of demonstration of system functionality and certification of the installation.

Components that should be tested during functional testing include flowmeters, valves, sensors, cleaning systems, the UVT analyzer (if included), the standby generator and/or UPS (if included), and other system components related to operation of the UV facility. As part of functional testing, the operation of all UV facility components, all control system functions, and all communication links should be verified. Specific tasks should include verifying hydraulic integrity, demonstrating lamp starts and stops, demonstrating changes in power levels, simulating loss of input signals, demonstrating alarms, verifying accurate displays, and demonstrating reactor start-up and shutdown sequences.

Upon the successful completion of functional testing, an extended-period performance test is recommended. The duration of performance testing could be as little as a few days to as much as several months. The scope and responsibilities during performance testing should be developed based on the objectives of the owner and the specific project. Typical elements of performance testing include the following:

- Verification of flow distribution and head loss
- Demonstration of operation within validated setpoints in automatic control mode under a range of flow rates and operating conditions
- Verification of the accuracy of intensity sensors based on calibration checks
- Verification of the accuracy of the UVT analyzer (if applicable)
- Assessment of cleaning system effectiveness over an extended time period
- Confirmation of UPS and backup power transfer (if applicable)
- Assessment of O&M needs of the UV facility and compliance with manufacturer guarantees

UV Facility Start-up

Following construction and testing, the UV facility will be ready for system start-up. For the start-up effort, it is necessary to complete final inspections of the facility and components, develop O&M manuals, develop emergency operating procedures, and complete operator training. Upon system start-up, the UV facility will require operation and maintenance and regular monitoring.

UV Facility Operation and Maintenance. Typical O&M activities for a UV disinfection facility include the following tasks:

- General system operation, upkeep, monitoring, and record-keeping
- Regular checks of duty sensor calibration versus reference sensors (sensor checks), and sensor recalibration as necessary
- Lamp replacement and recycling of spent lamps
- Sleeve removal, inspection, and cleaning (if necessary)
- UVT analyzer recalibration and maintenance (for systems with UVT analyzer)
- Replacement of sensors, sleeves, ballasts, cleaning system components, and other replaceable parts
- Regular maintenance of ancillary equipment (e.g., UPS, flowmeters, generator)

Typical component lifetimes for UV equipment components are shown in Table 29.6. The UV equipment specifications should include manufacturer guarantees for many of or all the components shown.

The operation and maintenance costs for UV disinfection consist of costs for electricity, labor, and replacement parts. Power consumption will depend on the operating characteristics of the facility such as UVT, flow rate, control strategy, and the validated operating limits. Labor costs will also be site-specific, although the time necessary for typical O&M tasks can be estimated based on the specific characteristics of the selected equipment. When one is budgeting for annual operating costs, provisions should also be made for patent fees (based on the volume of water treated) that may be applicable.

The O&M costs for an operating 15-mgd-capacity facility were estimated in 2002 (Swaim et al.), including estimated time requirements and component replacement costs for a specific manufacturer's system. The USEPA developed cost curves for capital and O&M for facilities from less than 0.1 mgd to more than 300 mgd (USEPA, 2003b). From

TABLE 29.6 Typical Lifetimes for UV Equipment Components

Component	Typical guarantees	Potential lifetime
Low-pressure lamps	8,000 to 12,000 h	8,000 to 12,000 h
Medium-pressure lamps	3,000 to 8,000 h	3,000 to 8,000 h
Sleeves	1 to 3 years	3 to 10 years
Intensity sensors	1 year	1 to 10 years
UVT monitor	1 year	3 to 10 years
Cleaning system components	1 to 3 years	3 to 5 years
Ballasts	1 to 10 years	10 to 20 years

Source: USEPA (2003a).

the USEPA's cost curves, which were based on an assumed combined filter effluent installation in a new building, design dose of 40 mJ/cm^2, and design UVT of 88% per centimeter, the estimated annual O&M costs for a UV disinfection system were approximately $9,000 for a design flow rate of 1.0 mgd, approximately $19,000 for a design flow rate of 10 mgd, and approximately $86,000 for a design flow rate of 100 mgd (USEPA, 2003b).

UV Facility Monitoring and Reporting. During operation, UV equipment will be monitored to demonstrate operation within the validated conditions. At a minimum, utilities will monitor flow rates, lamp outages, UV intensity, UVT (if part of the control strategy), calculated dose (if part of the control strategy), and any additional parameters as required by the primacy agency. Additional parameters recommended for regular monitoring include power draw, water temperature, lamp on/off cycles, and the age of lamps and other system components.

UV facilities will be required to monitor the calibration of UV intensity sensors on a regular basis, and the results of these calibration checks will need to be recorded and reported. In the proposed *UV Disinfection Guidance Manual*, the sensor calibration checks are required on a monthly basis. From the proposed *UV Disinfection Guidance Manual*, the UV facility is required to report UV intensity, flow rate, UVT and calculated dose (if part of the control strategy), lamp outage, and other parameters as required by the primacy agency. Reporting requirements will also include the duration and volume treated during periods of off-spec operation. The specific reporting requirements will be determined by the primacy agency.

POTENTIAL REGULATORY REQUIREMENTS FOR UV DISINFECTION IN NORTH AMERICA

This section describes the proposed requirements for UV disinfection in the United States that are currently under development by the USEPA. In addition, a description is provided of the current status of regulations pertaining to the use of UV disinfection for drinking water applications in Canada.

Many utilities are not waiting for USEPA requirements and guidance to be finalized to begin planning and designing UV disinfection facilities. The information in this chapter provides a summary of proposed Long Term 2 Enhanced Surface Water Treatment Rule (LT2ESWTR) requirements, which at the time of publication are subject to change. Therefore, the reader should evaluate the most recent regulatory requirements before developing UV facility design, validation, and operation strategies.

Current and Future Regulations in the United States

The USEPA's Surface Water Treatment Rule (SWTR) provides minimum disinfection requirements for systems using surface water or groundwater under the direct influence (GWUDI) of surface water. The focus of the SWTR is on the removal/inactivation of *Giardia* cysts and viruses. The Interim Enhanced Surface Water Treatment Rule (IESWTR) and Long Term 1 Enhanced Surface Water Treatment Rule (LT1ESWTR) build upon SWTR requirements by establishing baseline treatment requirements for *Cryptosporidium* by filtration systems.

The LT2ESWTR will supplement these regulations by establishing additional treatment technique requirements for *Cryptosporidium* for systems with greater vulnerability to this pathogen. The LT2ESWTR will be implemented simultaneously with the upcoming Stage 2 Disinfectant Disinfection By-Product Rule (DDBPR).

The basis of the LT2ESWTR has been developed through the negotiations of a Federal Advisory Committee, which reached a consensus Agreement-in-Principle in September 2000. In essence, the LT2ESWTR will require additional treatment for *Cryptosporidium* by all unfiltered systems and by a subset of filtered systems with elevated source water *Cryptosporidium* levels, as demonstrated through monitoring. The LT2ESWTR was proposed in summer 2003, and the LT2ESWTR is expected to be finalized in summer 2005. UV disinfection is one of the technologies recognized for inactivation of *Cryptosporidium* and one of the few technologies accepted for high levels (2 to 3 logs) of *Cryptosporidium* inactivation. Accordingly, there are specific requirements related to UV disinfection in the proposed LT2ESWTR preamble. The proposed requirements for UV disinfection are summarized in this section, although these requirements may change in the final LT2ESWTR. Therefore, the final LT2ESWTR should be consulted in lieu of this section once the LT2ESWTR is promulgated (finalized). The LT2ESWTR has requirements for UV disinfection that pertain to the UV doses required, UV reactor validation, operation, monitoring, and reporting.

Because UV disinfection is a relatively new technology in the U.S. drinking water industry, the USEPA agreed to publish UV disinfection guidelines as the *UV Disinfection Guidance Manual*. Its purpose is to facilitate planning, design, and operation of UV installations by familiarizing primacy agencies and utilities with important design, operation, and UV reactor validation issues. The LT2ESWTR proposal draft of the *UV Disinfection Guidance Manual* was published in conjunction with the proposed LT2ESWTR in June 2003.

UV Dose Requirements. UV dose requirements will be established for up to 3-log inactivation of *Cryptosporidium* and *Giardia* (or more) and up to 4-log inactivation of viruses (USEPA, 2003a). These UV dose requirements will be for postfilter applications of UV disinfection and for unfiltered systems that meet filtration avoidance criteria. The proposed UV dose requirements are presented in Table 29.7 (USEPA, 2003a). However, note that there are safety factors that will be applied to these dose requirements to determine what UV dose must be demonstrated in the validation testing (as discussed in the next section).

The UV dose requirements represent the UV doses that would be needed for the various levels of inactivation of the specified microorganisms in *controlled, bench-scale experiments* (i.e., *collimated beam tests*). Therefore, these UV dose requirements cannot be applied to a full-scale UV reactor because of the UV dose distributions that occur in these

TABLE 29.7 Proposed UV Dose Requirements (mJ/cm^2) without Safety Factors

Microorganism	Log inactivation							
	0.5	1	1.5	2.0	2.5	3.0	3.5	4.0
Giardia	1.6	2.5	3.9	5.8	8.5	12	—	—
Cryptosporidium	1.5	2.1	3.0	5.2	7.7	11	—	—
Viruses	39	58	79	100	121	143	163	186

Source: USEPA (2003a).

systems and their associated uncertainties. Validation testing will verify reactor performance, and validation safety factors will need to be applied to account for these uncertainties in *full-scale UV reactors*. Thus, UV systems will be designed for UV doses that exceed the doses shown in Table 29.7 by a factor of safety. The actual safety factor will depend on the methodology defined by the USEPA for determining safety factors and site-specific and validation conditions.

UV Reactor Validation Testing. For a utility to receive inactivation credit for UV disinfection, the UV reactor utilized must be tested to ensure that it delivers the required UV dose. According to the USEPA (2003a),

> The demonstration [validation testing] would be required to involve the following:
> - Full scale testing of a reactor which conforms uniformly to the UV reactors used by the system
> - Inactivation of a test microorganism whose dose response characteristics have been quantified with a low pressure mercury vapor lamp

Additional explanations of the methods, terms, and approaches for validation testing are presented in a later section of this chapter. The minimum conditions that are required to be validated are flow rate, UV intensity (as measured by a UV intensity sensor), and lamp status (e.g., power level). In addition, the validated operating conditions must account for the following factors:

- Lamp aging
- Lamp sleeve fouling
- UVT of the water
- Inlet and outlet piping or channel configurations of the UV reactor
- Dose distributions arising from the velocity profiles through the reactor
- Failure of UV lamps or other critical system components
- Measurement uncertainty of on-line sensors

Validation testing must determine a set of operating conditions that can be monitored by a utility to ensure that the UV dose required for a given pathogen inactivation credit is delivered, and the utility must then monitor to demonstrate it is operating within the range of conditions under which the reactor was validated.

The USEPA's recommended validation testing procedures are described in the proposed *UV Disinfection Guidance Manual*. The USEPA validation protocol establishes a safety factor that should be applied to the UV dose requirements. These safety factors account for differences in the microbial response of the challenge and target microorganism, hydraulics, water quality issues, equipment uncertainty, and potential polychromatic biases. At the time of publication, the methodology for calculating these safety factors was under development. Refer to the final *UV Disinfection Guidance Manual* to determine the appropriate safety factor. Currently, the safety factors can vary between 1.4 and 5.1. From the proposed *UV Disinfection Guidance Manual*, the reduction equivalent doses (REDs) to be determined during validation testing were up to 12, 24, 32, and 42 mJ/cm^2 for 1.0-, 2.0-, 2.5-, and 3.0-log *Cryptosporidium* inactivation, respectively.

To determine the RED that needs to be shown in validation, the validation safety factor is multiplied by the UV dose requirements (Table 29.7).

Operating out of Validated Conditions (Off-Spec). Validation testing establishes the conditions under which the UV reactors must be operated to ensure the required dose delivery (USEPA, 2003a). In the proposed *UV Disinfection Guidance Manual*, off-specification operation is defined as occurring when a UV reactor is operating outside of its validated limits. For example, off-spec operation will occur if a UV reactor is operating with a flow rate that exceeds the maximum validated flow rate. To receive inactivation credit, the UV disinfection system is required to operate under validated conditions.

In general, the proposed *UV Disinfection Guidance Manual* defines two ways a UV system may operate outside of the validated conditions, as follows:

1. The flow, UVT, calculated dose, and/or UV intensity sensor readings are not in accordance with the validation conditions *and* water is flowing to the distribution system. This could represent inadequate disinfection.

2. All UV lamps in all reactors lose their arc (i.e., shut off) because of a power quality problem *and* water is flowing to the distribution system. This would represent no disinfection.

It is unrealistic to require UV systems to operate within the validated conditions 100% of the time. According to the proposed *UV Disinfection Guidance Manual*, unfiltered systems that use UV disinfection to meet the *Cryptosporidium* treatment requirement of the LT2ESWTR must demonstrate that at least 95% of the water delivered to the public during each month is treated by UV reactors operating within validated limits (USEPA, 2003a). Or in other words, the UV reactor cannot be operated off-specification while passing more than 5% of the water delivered to the public (USEPA, 2003a). The draft LT2ESWTR did not state an off-specification requirement for filtered systems; however, the primacy agency may establish requirements for its filtered systems, including a limit for off-specification operation.

Monitoring and Reporting Requirements. Monitoring of UV facilities must demonstrate operation within the validated conditions. Monitoring must include UV intensity, flow rate, lamp outage, and other parameters as required by the primacy agency. In addition, the UV intensity sensor calibration must be checked regularly to verify that it is within the calibrated limits established in the validation testing. Monthly reporting by utilities with UV facilities must also include the percentage of off-spec water (USEPA, 2003a).

Regulations in Canada

Water quality standards in Canada are a provincial responsibility, and the regulations vary across the country. In Table 29.8, a summary is provided of the provincial regulations that affect the decision of where UV disinfection could be considered.

Most provinces refer to the *Guidelines for Canadian Drinking Water Quality* (GCDWQ) when considering drinking water quality requirements. These guidelines are developed by a Federal-Provincial Subcommittee through Health Canada. The April 2003 edition (current at the time of writing) of the *Guidelines for Canadian Drinking Water Quality* does not specifically set requirements for *Giardia* or *Cryptosporidium* reduction. It is recommended that if past experience demonstrates that a particular raw water supply source could harbor pathogens for which *E. coli* are not good indicators (i.e., *Giardia* cysts and *Cryptosporidium* oocysts), then the source should receive treatment known to remove or inactivate these pathogens. The *Guidelines for Canadian Drinking Water Quality* are generally outcome-based so specific requirements for UV disinfection may not be forthcoming.

As shown in Table 29.8, as of 2003, many provincial regulations do not specifically address UV disinfection. However, many municipalities across Canada have been able to obtain permission from their respective regulators to proceed with UV disinfection as an

TABLE 29.8 Summary of 2003 Canadian Drinking Water Standards for Application of UV Disinfection

Province/territory	Current UV standard	Comment
British Columbia	None	Outcome-based negotiation with Drinking Water Officer. Only prescribed standard is disinfection of bacteria.
Alberta	1-log protozoa removal/inactivation Expected to be revisited once USEPA requirements are finalized	*Giardia* inactivation/removal requirement based on quantity of cysts in the source water. Minimum of 3-log inactivation/removal required for surface and GWUDI sources. *Cryptosporidium* regulations anticipated.
Saskatchewan	None	3-log *Cryptosporidium*, 3-log *Giardia,* and 4-log virus inactivation/removal required for surface and GWUDI sources unless approved filtration/disinfection provided
Manitoba	None	2-log *Cryptosporidium*, 3-log *Giardia*, and 4-log virus inactivation/removal required for surface and GWUDI sources
Ontario	UV disinfection is recognized as an acceptable alternate disinfection technology	2-log *Cryptosporidium*, 3-log *Giardia*, and 4-log virus inactivation/removal required for surface and GWUDI sources. Particle removal can be waived if proved not necessary
Quebec	UV disinfection is recognized as an acceptable alternate disinfection technology	2-log *Cryptosporidium*, 3-log *Giardia*, and 4-log virus inactivation/removal required for surface and GWUDI sources. Particle removal can be waived if proved not necessary
New Brunswick	None	3-log *Giardia* and 4-log virus inactivation/removal required for surface and GWUDI sources
Nova Scotia	None	3-log *Giardia* and 4-log virus inactivation/removal required for surface and GWUDI sources
Prince Edward Island	None	All public waterworks supply sources in the province are classified as groundwater sources. Only chlorine disinfection is required

TABLE 29.8 Summary of 2003 Canadian Drinking Water Standards for Application of UV Disinfection (*Continued*)

Province/territory	Current UV standard	Comment
Newfoundland and Labrador	None	GCDWQ pathogen requirements are followed
Nunavut	None	GCDWQ pathogen requirements are followed Surface water and GWUDI sources require chlorination as a minimum
Northwest Territories	None	GCDWQ pathogen requirements are followed. Surface water and GWUDI sources require chlorination as a minimum
Yukon Territory	None	GCDWQ pathogen requirements are followed. Surface water and GWUDI sources require chlorination as a minimum

alternative treatment technology in order to provide protection to the public from *Giardia* cysts and *Cryptosporidium* oocysts. In the absence of any Federal Canadian Guidelines, the industry has frequently drawn upon the U.S. or European standards to establish design and operational criteria.

VALIDATION TESTING

As discussed previously, full-scale validation testing of UV reactors will be required by the LT2ESWTR. The purpose of validation testing is to verify the disinfection performance of the UV reactor under specific operating (e.g., water quality, flow rate, and piping configuration) conditions. Validation testing, or biodosimetry, is the current water industry standard to assess UV reactor performance.

Biodosimetry

Biodosimetry is a biological method of determining the RED provided by a UV reactor. Biodosimetry uses full-scale testing and bench-scale analysis to verify UV reactor performance with a challenge microorganism. In the full-scale tests, the UV reactor is injected with a challenge microorganism (e.g., MS-2 bacteriophage), and the log inactivation is measured at different water quality, lamp power, and flow conditions. The purpose of the bench-scale tests is to understand the UV dose response of the challenge microorganism, which is typically accomplished by using a collimated beam test with a low-pressure lamp. The log reduction (inactivation) observed in the full-scale tests is then correlated to the RED that is measured in the collimated beam test for the same level of inactivation. Biodosimetry tests would be completed for a matrix of water quality and

FIGURE 29.14 Biodosimetry process. (*Adapted from USEPA, 2003a.*)

flows for the particular UV reactor and installation conditions, and the RED would be calculated at each condition. An overview of the biodosimetry process is shown in Figure 29.14.

Additives Used during Biodosimetry

The UV reactor performance must be validated for the range of UVT values under which the UV reactor will be operating. Because it is possible to decrease the test water UVT (as opposed to increasing it), UVT is typically decreased using a UV-absorbing compound. The common UV absorbers used in validation to date are lignin sulfonate and instant coffee.

In addition to the UV absorbing compound, a challenge microorganism has to be added to the test water to determine the log inactivation of the UV reactor, as described previously. The challenge microorganism is typically a nonpathogenic microorganism that has similar UV dose response compared to the target microorganism. Currently, MS-2 bacteriophage (MS-2) and *Bacillus subtilis* are commonly used challenge microorganisms when *Cryptosporidium* is the target microorganism. The UV-absorbing compound and the challenge microorganism chosen for validation testing affect the validation safety factor calculations described previously.

BIBLIOGRAPHY

Adams, C., P. Scanlan, and N. Secrist. "Oxidation and Biodegradability Enhancement of 1,4-Dioxane Using Hydrogen Peroxide and Ozone." *Environmental Science and Technology* 28:1812–1818, 1994.

APHA/AWWA/WEF. *Standard Methods for the Examination of Water and Wastewater*, 20th ed. Washington, D.C., 1998.

Awad, J., C. Gerba, and G. Magnuson. 1993. "Ultraviolet Disinfection for Water Reuse." *WEF Specialty Conference Proceedings, Planning, Design, and Operation of Effluent Disinfection Systems*, Whippany, N.J.

Brandt, C. L., and A. C. Giese. "Photoreversal of Nuclear and Cytoplasmic Effects of Short Ultraviolet Radiation on Paramecium caudatum." *J. Gen. Physiol.* 39(5):735–751, 1956.

Brauer, R. G., M. A. Morine, P. D. Swaim, A. Squatrito, M. Neher, and W. D. Bellamy. "Achieving Regulatory Approval of UV Disinfection: The City of Henderson/Nevada State Health Division Approach." *Proceedings IUVA Congress*, Washington, D.C., 2001.

Bukhari, Z., T. M. Hargy, J. R. Bolton, B. Dussert, and J. L. Clancy. "Medium-Pressure UV for Oocyst Inactivation." *J. AWWA* 91(3):86–94, 1999.

Campbell, A. T., and P. Wallis. "The Effect of UV Irradiation on Human Derived *Giardia lamblia* Cysts." *Water Research* 36(4):963–969, 2002.

Clancy, J. L., T. M. Hargy, M. Marshall, and J. Dyksen. "UV Light Inactivation of *Cryptosporidium* Oocysts." *J. AWWA* 90(9):92–102, 1998.

Clancy, J. L., Z. Bukhari, T. M. Hargy, J. R. Bolton, B. W. Dussert, and M. M. Marshall. "Using UV to Inactivate *Cryptosporidium*." *J. AWWA* 92(9):97–104, 2000.

Cotton, C. A., R. S. Cushing, and D. M. Owen. "The Impact of the Draft UV Disinfection Requirements on UV Facility Design and Operation." *Proceedings AWWA Annual Conference and Exposition*, New Orleans, La., 2002.

Craik, S. A., G. R. Finch, J. R. Bolton, and M. Belosevic. "Inactivation of *Giardia muris* Cysts Using Medium-Pressure Ultraviolet Radiation in Filtered Drinking Water." *Water Research* 34(18):4325–4332, 2000.

Cushing, R. S., E. D. Mackey, J. R. Bolton, and M. I. Stefan. "Impact of Common Water Treatment Chemicals on UV Disinfection." *Proceedings AWWA Annual Conference and Exposition*, Washington, D.C., 2001.

Danielson, R. E., J. Phillips, and F. Soroushian. "Inactivation of *Giardia muris*, *Bacillus subtilis* Spores, and MS2 Coliphage by Ultraviolet Light." *Proceedings IUVA Congress*, Washington, D.C., 2001.

Downes, A., and T. P. Blunt. "Researches on the Effect of Light upon Bacteria and Other Organisms." *Proceedings Royal Society London*, vol. 26, 1877.

Dulbecco, R. "Experiments on Photoreactivation of Bacteriophages Inactivated with Ultraviolet Radiation." *Journal of Bacteriology* 59:329–347, 1950.

Dyksen, J., M. Marshall, A. Gera, and J. Clancy. "Cost of Advanced UV for Inactivating Crypto." *J. AWWA* 90(9):103–111, 1998.

Hayes, S. L., C. D. Barnes, E. W. Rice, M. W. Ware, and F. W. Schaefer, III. Low Pressure Ultraviolet Studies for Inactivation of *Giardia muris* Cysts." *Proceedings IUVA Congress*, Washington, D.C., 2001.

Jagger, J. *Introduction to Research in Ultraviolet Photobiology*. Englewood Cliffs, N.J.: Prentice-Hall, 1967.

Kashinkunti, R., K. G. Linden, G. Shin, D. H. Metz, M. D. Sobsey, M. Moran, and A. Samuelson. "Achieving Multi-barrier Inactivation in Cincinnati: UV, By-products, and Biostability." *J. AWWA* 96(6):114–127, 2004.

Kelner, A. "Light-Induced Recovery of Microorganisms from Ultraviolet Radiation Injury, with Special Reference to *Escherichia coli*." *Bulletin of the New York Academy of Medicine* 26:189–199, 1950.

Knudson, G. B. "Photoreactivation of UV-Irradiated *Legionella pneumophila* and Other *Legionella* Species." *Applied and Environmental Microbiology* 49(4):975–980, 1985.

Kommineni, S., J. Zoeckler, A. Stocking, S. Liang, A. Flores, and M. Kavanaugh. "Advanced Oxidation Processes," Chap. 3 of *Treatment Technologies for Removal of MTBE from Drinking Water*. California MTBE Research Partnership, National Water Research Institute, Fountain Valley, Calif., 2000.

Kruithof, J. C., and R. C. van der Leer. "Practical Experiences with UV-Disinfection in the Netherlands." *Proceedings of the AWWA Seminar on Emerging Technologies in Practice. Proceedings AWWA Annual Conference and Exposition,* Cincinnati, Ohio, 1990.

Linden, K. G., G. A. Shin, G. Faubert, W. Cairns, and M. D. Sobsey. "UV Disinfection of *Giardia lamblia* Cysts in Water." *Environmental Science and Technology* 36(11):2519–2522, 2002a.

Linden, K. G., L. Batch, and C. Schulz. "UV Disinfection of Filtered Water Supplies: Water Quality Impacts on MS2 Dose-Response Curves." *Proceedings AWWA Annual Conference and Exposition,* New Orleans, La., 2002b.

Mackey, E. D., R. S. Cushing, and H. B. Wright. "Effect of Water Quality on UV Disinfection of Drinking Water." *Proceedings IUVA Congress,* Washington, D.C., 2001.

Mackey, E. D., R. S. Cushing, M. L. Janex, N. Picard, J. M. Laine, and J. P. Malley. "Bridging Pilot-Scale Testing to Full-Scale Design of UV Disinfection Systems," draft report. Final report to be published in 2004. Denver, Colo.: AWWA Research Foundation, 2003.

Malley, J. P. "Historical Perspective of UV Use." *Proceedings AWWA Water Quality Technology Conference,* Seattle, Wash., 2002.

Malley, J. P., J. P. Shaw, and J. R. Ropp. *Evaluation of By-products Produced by Treatment of Groundwaters with Ultraviolet Irradiation.* Denver, Colo.: AWWA Research Foundation, 1995.

Mofidi, A. A., E. A. Meyer, P. M. Wallis, C. I. Chou, B. P. Meyer, S. Ramalingam, and B. M. Coffey. "The Effect of UV Light on the Inactivation of *Giardia lamblia* and *Giardia muris* Cysts as Determined by Animal Infectivity Assay (P-2951-01)." *Water Research* 36:2098–2108, 2002.

Oguma, K., H. Katayama, H. Mitani, S. Morita, T. Hirata, and S. Ohgaki. "Determination of Pyrimidine Dimers in *Escherichia coli* and *Cryptosporidium parvum* during UV Light Inactivation, Photoreactivation and Dark Repair." *Applied and Environmental Microbiology* 67(10):4630–4637, 2001.

Oppenheimer, J., T. Gillogly, G. Stolarik, and R. Ward. "Comparing the Efficiency of Low and Medium Pressure UV Light for Inactivation of *Giardia muris* and *Cryptosporidium parvum*." *Proceedings AWWA Annual Conference and Exposition,* New Orleans, La., 2002.

Passantino, L. B., and J. P. Malley. "Impacts of Turbidity and Algal Content of Unfiltered Drinking Water Supplies on the Ultraviolet Disinfection Process." *Proceedings AWWA Annual Conference and Exposition,* Washington, D.C., 2001.

Persson, E., and D. Kuusisto. "A Performance Comparison of Electronic vs. Magnetic Ballast for Powering Gas-Discharge UV Lamps." *Proceedings of the RadTech Conference,* RadTech International, Chicago, 1998.

Powell, W. F. "Radiosensitivity as an Index of *Herpes simplex* Virus Development." *Virology* 9:1–19, 1959.

Rauth, A. M. "The Physical State of Viral Nucleic Acid and the Sensitivity of Viruses to Ultraviolet Light." *Biophysical Journal* 5:257–273, 1965.

Sharpless, C. M., and K. G. Linden. "UV Photolysis of Nitrate: Effects of Natural Organic Matter and Dissolved Inorganic Carbon, and Implications for UV Water Disinfection." *Environmental Science & Technology* 35(14):2949–2955, 2001.

Shin, G. A., K. G. Linden, M. J. Arrowood, G. Faubert, and M. D. Sobsey. "DNA Repair of UV-Irradiated *Cryptosporidium parvum* Oocysts and *Giardia lamblia* Cysts." *Proceedings IUVA Congress,* Washington, D.C., 2001.

Swaim, P. D., M. A. Morine, R. G. Brauer, M. A. Neher, J. Gebhart, and W. D. Bellamy. "Operating Data from Henderson's Full-Scale UV Disinfection Facility." *Proceedings AWWA Water Quality Technology Conference,* Seattle, Wash., 2002.

Swaim, P. D., R. G. Brauer, M. A. Morine, M. A. Neher, W. D. Bellamy, and J. P. Malley. "On-Site Validation Testing at a Full-Scale UV Disinfection Facility." *Proceedings AWWA Annual Conference and Exposition,* Anaheim, Calif., 2003.

Thompson, S. S., et al. "Detection of Infectious Human Adenovirus in Tertiary-Treated and Ultraviolet-Disinfected Wastewater." *Water Environment Research* 75:163–170, 2003.

U.S. Environmental Protection Agency. *Design Manual, Municipal Wastewater Disinfection.* EPA/625/1-86/021. Office of Research and Development and Water Engineering Research Laboratory, Cincinnati, Ohio, 1986.

U.S. Environmental Protection Agency. *Technical Memorandum: Worldwide Ultraviolet Disinfection Installations for Drinking Water.* Office of Groundwater and Drinking Water, Washington, D.C., 2000.

U.S. Environmental Protection Agency. *Ultraviolet Disinfection Guidance Manual.* Draft. EPA 815-D-03-007. Office of Groundwater and Drinking Water, Washington, D.C., 2003a.

U.S. Environmental Protection Agency. *Technologies and Costs for Control of Microbial Contaminants and Disinfection By-products,* 2003b.

von Sonntag, C., and H. P. Schuchmann. "UV Disinfection of Drinking Water and By-product Formation—Some Basic Considerations." *Journal of Water Supply Research and Technology* 41(2):67–74, 1992.

CHAPTER 30
WATER TREATMENT PLANT SECURITY

John M. Winslow
AMRIC Associates, Ltd.
East Syracuse, New York

Water treatment plant security is now a requirement for operation of a water facility. Water utilities are viewed as part of the critical infrastructure of any society. As such, they have to be protected to ensure a supply of safe drinking water, as well as sufficient water for fire protection and industry.

Any contamination of the water supply or interruption of the distribution of the water to customers will have a grave effect on the population and on business operations.

Following the terrorist attacks of September 11, 2001, the U.S. Environmental Protection Agency mandated that all water utilities serving over 3,300 people develop a risk assessment of their operations. Those assessments determined what functions and assets of the utilities were most critical and what the system vulnerabilities were. Countermeasure options were developed to reduce the risk.

Real concerns about water system security were raised when intelligence was developed that members of the al Qaeda terrorist network had discussed plans to attack the U.S. drinking water supply. While no specific information indicating that an attack was imminent or that any specific target, systems, or location had been identified, two potential attack scenarios were discussed by the terrorist.

One scenario was the disruption of water delivery through a physical attack on the water supply infrastructure, while the other involved the introduction of chemical or biological agents into water distribution systems and posttreatment facilities. The U.S. Department of Homeland Security determined that while both types of attack would be difficult, they were also feasible.

Consequently, it is now necessary for water utility systems to develop realistic physical protection systems (PPSs) to meet this threat. The PPS should be well integrated, in depth, and focused on specific utility assets and designed to protect against attack from well-defined threats. Threats come from two simple categories: insider threat and outsider threat. An insider can be the most difficult to defend against. The insider can be a disgruntled current or former employee, a contractor, or a vendor. Just by showing up for work, the insider circumvents many of the PPS features designed to protect the facility. The outsider can be a teenage vandal, a community activist, a mentally deranged customer, or a dedicated terrorist.

Modes of potential attack should be considered. Attackers can try to contaminate raw or finished water with chemical, biological, or radiological contaminants. They can physically assault the water treatment plant, pump stations, dams, and water tanks to destroy the machinery and process needed to treat and transmit the water

When one is assessing a threat, a number of questions need to be answered. What are the motives and goals of a potential attacker? Is there a history of similar attacks and a presence of potential attackers in the area? What are the most likely tactics of the attacker? When questions like these are answered, a better threat assessment can be made and the total risk to the water treatment plant determined.

Items to consider when one is designing a PPS for a water treatment plant will be the cost of the improvements; the assets needed to be protected; the vulnerabilities of the utility; the consequence of damage, destruction, or compromise of the utilities assets that are to be protected; and the risk level and threat.

A good PPS will deter people from thinking about attacking, detect an attacker as he or she attempts the attack, delay the attacker in his or her mission, and respond to the attack. Physical protective systems are composed of barriers such as chain-link fences and gates; good perimeter and plant lighting; electronic security system to detect and assess intrusions; and security procedures that would make an attack unattractive to an adversary. A solid PPS will help protect the water quality and the physical facilities themselves.

WATER QUALITY

The source of raw water can be an attractive target for an adversary. Whether it is a lake, river, or well field, many sources are remote and can offer an attacker numerous opportunities to attempt a contamination or physical attack. It is believed that large quantities of contaminant are required to successfully poison a water source such as a lake or river. However, great damage can be done to the public's perception of the utility itself, if even unsubstantiated claims of contamination are made. Steps can be taken to ensure that any claim of contamination, real or false, can be proved or disproved.

Raw water intakes in rivers and lakes are often located within an area where routine boaters can navigate. They are often marked by buoys. The elimination of the buoy markers over the intakes will make it harder for someone to easily locate the intake.

Closed-circuit television (CCTV) coverage on the area over the intake can help prove or disprove whether anyone attempted to place contaminants at that location. Tapes from the CCTV can be reviewed to disprove an allegation that the water system has been compromised, thus helping to maintain the public's confidence in the utility.

In the event the CCTV detects that a vessel is lingering near the intake, the authorities can be notified to investigate the situation. Depending on the resolution of the CCTV camera, the boat may even be identified by the registration number.

Open-air water storage tanks can be attacked by aircraft, as well as by a casual passerby with a strong throwing arm. If possible, these tanks should be covered or replaced. Steps can be taken to harden these open tanks, such as the use of secure fencing, intrusion detection sensors, CCTV, and security guards. However, those steps can be counteracted very easily by a low-flying crop duster.

BARRIERS

The first line of defense for a facility is at the perimeter. Barriers define the utility's borders and create a buffer zone with neighboring property. The barriers should control vehicle and pedestrian entry to the facility. They can act as a delay to forced entry by an intruder and can protect individual assets within the outer perimeter.

The most common perimeter barrier is a chain-link fence. The fence delineates the boundary of the facility and whether it is a water treatment plant, finished water storage tank, or a booster pumping station. However, fences will keep out only the casual passerby. A determined adversary can climb a fence in 4 s or cut through the fencing in about 10 s.

However, a sturdy, well-maintained 7- to 8-ft galvanized chain-link fence will serve two purposes. First, a well-maintained, secure fence, topped by three strands of barbwire with at least a 20-ft clear zone on the outside of the fence line, may give an indication of the type of security an adversary may face. The attacker will look for a system with overgrown vegetation growing right next to the fence line. He will look for rusted, sagging fencing and ground eroded from under the fence, enabling him to crawl under without being noticed.

More importantly, the fence can be a platform for an electronic security system's sensor. There are several exterior intrusion detection systems that are attached to the fence. When someone comes close to the fence, cuts the fence wire, or climbs over the fence, these various systems will sound an alarm.

A taut wire fence acts as both a barrier and an intrusion detection system. In a total taut wire system, the entire fence consists of taut wire. When there is a change in the tension of the wire fencing, an alarm will be triggered. Although this type of system is very reliable, it is also costly.

Gates to a facility should be kept to the minimum needed for the safe and efficient running of the facility. The most secure vehicle gates are one-piece, sliding gates that are remotely controlled. Those gates should also have a top guard of barbwire. The gates need to be inspected to ensure that a person cannot crawl underneath or through them. Utilities with double or single swinging gates should ensure that the gates are locked with high-security, tamper-resistant padlocks with a shrouded shank. Any chains used to secure the gates should be high-security chains.

Other barriers used to control vehicles are earthen berms, Jersey barriers, planters, fixed and retractable bollards, and drum-type barriers. Entrance roads should be designed so that the approaching vehicle must travel through a serpentine path as it approaches the facility. This is done to decrease the speed of the vehicle.

Doors and windows are the most common interior perimeter barriers. Doors should be secure, heavy-duty metal without windows. If not used frequently, the doors should have dead-bolt locks. Exterior doors or ones leading to critical operations areas should have secure, interior hinges.

Window barriers consist of bars, grilles, and chain-link screen to prevent entry. The security devices should be securely fastened on the inside. If windows are replaced, consideration should be given to installing glass block or wire-reinforced windows in their place.

Access hatches to clearwells as well as manholes located at critical access points should be locked.

Electrical substations at the water treatment plant should be fenced and locked. Electric control panels should be locked. Consideration should be given to obtaining backup generators to power the plant in case of attack or natural disaster.

Chemical connection or injection points should be located inside a secure area.

Lighting at the water treatment plant should be designed with security in mind. Perimeter lighting should be adequate to allow identification of intruders attacking the fence line. Illumination of the water treatment plant grounds should eliminate blind spots.

ELECTRONIC SECURITY SYSTEMS

Electronic security systems (ESSs) provide the detection element of a PPS. Properly functioning and used, the ESS will provide an early warning that an intruder is entering the water treatment plant property or compromising one of your buildings.

Intrusion detection sensors, both exterior and interior, are the most common ESS. There are numerous intrusion detection sensors on the market today. The performance of an intrusion detection sensor (IDS) is rated by the probability that the sensor will detect an intruder, the nuisance alarm rate generated by the sensor; and how vulnerable the sensor is to defeat. All three factors need to be taken into consideration in selecting an IDS.

Intrusion detection sensors are classified by several types: passive or active, covert or visible, line-of-sight or terrain-following, volumetric or line detection, and by application. Active sensors, such as active infrared and ported coaxial cable, generate energy and detect a change in the energy field when something or someone enters the field. Passive sensors detect energy coming from whatever enters the field. For example, passive infrared detects the body heat of a person entering the targeted zone.

Covert sensors are hidden from view while visible sensors can be seen on fences or on another structure such as a building or tower. Covert sensors, often buried underground as is ported coaxial cable, are harder for an intruder to recognize. Clearly seen visible sensors, such as passive infrared, can give an intruder ideas on how to circumvent the areas covered by that technology. However, the visible sensors may indicate to a potential intruder that the defenses of the facility are too hard for an attempted attack, thus acting as a deterrent.

Terrain-following sensors are most often associated with the fence line. Sensors associated with the fence will follow the natural lay of the land. For line-of-sight sensors, the sensor must be in close proximity to the intruder with nothing between the sensor and the intruder.

Volumetric sensors, such as active ultrasonic sensors, emit energy into a monitored area and sound an alarm when that area is penetrated by an object such as a person, vehicle, or animal. Line detection sensors, such as passive and active infrared, work along a line; and when that line is penetrated, an alarm is activated.

The last classification is based on how the sensors are applied. Buried line sensors are placed in line underground. Fence-associated sensors use the fence as platforms. Freestanding sensors are neither buried nor placed on the fence line, but are usually mounted on a separate support.

Exterior intrusion detection systems should give a warning that someone is attempting to penetrate the facility. As an intruder approaches the water treatment plant property from an adjacent lot, he or she should see a clear space of at least 30 ft in front of the 7-ft-chainlink fence. That fence should have three strands of barbwire on outriggers, facing out toward the intruder. The fence should be taut and in good repair. The intruder will have to cross the 30-ft clear space before climbing over, crawling under, or cutting the fence. The fence can be equipped with several options in intrusion detection sensors.

The fence can have fence vibration sensors mounted directly on the fence fabric. If a person climbs, cuts, or lifts the fence, mechanical vibrations or stress in the fence fabric, different from the usual natural disturbances such as wind, will cause an alarm.

Another fence-associated sensor is fiber-optic cable. This system uses light instead of electricity for transmission and detection. The fiber-optic cable is attached to the fence. A steady stream of light travels along the cable. When someone grabs, steps on, pulls, or cuts the cable, the stream of light is interrupted, causing an alarm. Inside the fence line, buried fiber-optic cable will trip an alarm when someone walks, runs, or crawls over the cable.

Ported coaxial buried line sensors are coaxial cables with closely spaced holes in the outer shield of the cable. Electromagnetic energy escapes from these small holes and radiates a short distance, creating an electric field. When an intruder walks, runs, or crawls into the electric field, an alarm will sound.

Another example of a freestanding sensor is the microwave sensor. These are motion detection sensors that flood a designated surveillance zone with an electric field. Human or animal movement in the field will set off an alarm.

As the intruder reaches the intended target building, she or he can enter through a door, window, or skylight or can cut a hole in the roof or wall. Doors should be secure, metal, and locked. The doors should have balanced magnetic switches that will sound an alarm when the door is opened.

Windows and skylights can also be equipped with balanced magnetic switches and/or glass-break sensors. These sensors are mounted on a stable interior wall or ceiling. When glass is broken, the sensors listen for frequencies made by the sound of the glass breaking or detect the shock wave created when glass is broken. The sensor processor filters out the normal sounds and can detect the appropriate frequency. An alarm is then sounded.

If a window has only balanced magnetic switches, the glass can be broken and a nimble intruder can crawl through without tripping an alarm. If only glass-break sensors are used, the window can be jimmied and the alarm will not sound. Windows can also be protected by steel bars or wire mesh or replaced by glass blocks. Any bars or wire mesh or screens should be attached on the interior.

As an alternative to sensors targeting the windows, interior active or passive infrared (IR) sensors can be deployed in rooms and hallways. The active infrared sensors shoot beams of infrared rays from a sending unit to a receiving unit. When something breaks the beam of IR light, an alarm will sound.

Passive infrared detects the body heat of humans as they cross the path of the beam.

If the water treatment plant has any critical components that must be reached by climbing a ladder, fiber-optic cable can be affixed to the ladder rungs. When an unauthorized person attempts to climb the ladder, he or she will grab, pull, or step on the cable. When the light within the cable is interrupted, an alarm will sound. Fiber-optic cable can also be placed around a tank hatch. When someone steps on the cable mesh, an alarm will sound.

Alarms triggered by intruders should be audible, which may deter any further action by a potential attacker.

Once an intruder gains entry to the water treatment plant, locked interior doors leading to critical operations will delay the malevolent act. Another delay feature is to erect floor-to-ceiling wire mesh security cage partitions around pumps or other critical processes.

A water utility employee, contractor, or vendor can circumvent many of the security devices just by showing up for work. The insider threat is the most difficult to defend against. Good access control is the most effective everyday security defense against an insider threat. Installing electronic access devices such as card key or proximity card readers will give management a record of who used a card and when it was used to open a specific door. An insider who knew he or she would be identified as being the person who gained access to a location where a malevolent act occurred may be deterred.

Good access control using electronic access devices can also limit the number of employees who can enter critical areas.

CCTV

An important element in security systems is the use of closed-circuit television as a tool to assess the alarms triggered by exterior or interior intrusion sensors. CCTV systems can also provide recorded evidence for use in criminal courts or administrative procedures against employees. Visible, overt CCTV cameras can also serve as a deterrent to intruders or insiders who are worried that they will be detected committing an act of sabotage.

CCTV should be used as part of an integrated security system. For instance, when an intrusion sensor detects activity at a fence, the water treatment plant operator should be able to direct a CCTV camera to the detection zone. Some integrated systems automatically focus the cameras on the location of the suspected intrusion.

The water treatment plant operator should have the ability to determine if the cause of the alarm is a human or animal. That assessment will allow for a quicker response to malevolent intrusions.

Cameras can be employed at the perimeter, at access control points, at critical operation locations, and on the interior of buildings. A CCTV system will consist of one or more cameras, lenses for those cameras, camera housing, mounting bracket, cable, multiplexer, video recorder, and monitor. The cameras can be black and white or color. The resolution of the camera will dictate the quality of the image captured. The higher the resolution, the clearer the picture and the greater the ability to recognize or identify an intruder as a specific person.

Cameras can be fixed or pan tilt zoom (PTZ) models. Fixed cameras focus on one area, while PTZ cameras have the ability to move focus from one spot to another and to bring an image up closer to view. Some cameras can operate only in good lighting, while others have the ability to capture images in a night environment. Infrared cameras will give the operator the ability to identify an image as a human walking in a totally darkened area by capturing the person through her or his body heat.

Weather conditions sometimes dictate what type of exterior camera housing is needed. In cold climates a small heater may be needed to ensure the smooth operation of the camera and lens. In extreme heat, a fan may be needed.

If more than one camera is used in a system, a video multiplexer is needed to collect the signals and send them to a single video cassette recorder (VCR).

The video recorder will capture the image on tape or digitally. VCRs need a good deal of maintenance. They are the weakest link in a CCTV system because of their mechanical nature. Because the VCRs are constantly recording, they receive a lot of wear and tear. Regular VCR heads should be cleaned after every 100 h of use. That would be about every 4 days of continuous recording. The entire VCR unit needs to be serviced every 3 months, if constantly used.

Another option is to employ video motion sensors, which change a closed-circuit television system into an alarm system as well. In this system, the sensors detect changes caused by movement of an object within the video's field of view. Parameters can be established with video motion sensors to view only a portion of the total field of view. When movement is detected in that portion, an alarm will sound.

Policy should be developed regarding the retention of videotapes. They should be kept for 30 days before reusing them. This will allow for review if some questioned incident comes to the attention of the utility.

Digital CCTV technology has advanced by leaps and bounds. As the costs for digital video cameras and recorders have come down, more and more units are being used as part of integrated physical security systems. Research has developed digital "smart" cameras. These systems not only record what is viewed, but also analyze the image. The system can be programed to look for specific objects or suspicious activity. Digital video recorders produce better images and have excellent search capability and easier integration with the overall security system.

PROCEDURES

Simple changes in policy and procedures can also help protect the water treatment plant. When chemical deliveries are made to water treatment plants, steps should be taken to ensure that the chemical order is correct and the delivery driver is actually the legitimate person sent by the chemical company.

Prior notification should be made for all visits from chemical companies. A protocol should be established in which the chemical company and the plant operators or management communicate prior to any delivery. The time of the delivery should be known beforehand as well as the identity of the driver. Photographs of the chemical company drivers could be kept by the water treatment plant for comparison to the actual driver, or photographs can be faxed to the water treatment plant prior to delivery. Deliveries should be made during daytime operations.

Each delivery manifest should be checked against the purchase order. To protect against the delivery of the wrong chemicals, consideration should be given to constructing an extra tank in which a sampling of the chemical can be tested prior to using the chemical during operations.

Thorough background checks should be conducted on prospective employees, to include criminal arrest checks. The use of specific uniforms and photo ID badges for all water treatment plant employees should be considered. Employees should be encouraged to challenge people not in uniform or without their badges. Employees should be encouraged to note and report any suspicious activity in and around the water treatment plant, such as people taking photographs or lingering in the area. Periodic training or bulletins should be given to employees to maintain situational awareness of world events and potential threats to water systems.

Protocols should be developed to handle bomb or contamination threats. Caller ID should be installed on the phones likely to be called with such threats.

Warning signs should be posted on the perimeter fence and throughout the facility, warning that the site is under video surveillance.

BIBLIOGRAPHY

Defense Advanced Research Project, Joint Program Steering Group. *Perimeter Security Sensor Technologies Handbook.* Arlington, Va., 1997.

Department of the Army. *Physical Security,* FM 3-19.30. 2001.

Department of the Navy. *Anti-Terrorism/Force Protection Considerations in Facility Design Seminar Workbook.* 2002.

Garcia, M. *The Design and Evaluation of Physical Protection Systems.* Boston, Mass.: Butterworth-Heinemann, 2001.

Gips, Michael. "Behind the Scenes: CCTV Trends." *Security Management On Line,* May 2003.

Lockhart, Pete. "Emerging Solutions in the Government Sector." *ST&D Magazine,* August 2003.

Lowere, Mark, Tony Raker, and Jim Rogers. "Do Fence Me In." *Security Management On Line,* December 2001.

Maurer, Richard. "Asset Protection Technology." *ST&D Magazine,* August 2003.

Roper, C. *Physical Security and the Inspection Process.* Boston, Mass.: Butterworth-Heinemann, 1997.

APPENDIX A
PROPERTIES AND CHARACTERISTICS OF WATER TREATMENT CHEMICALS

Jerry L. Anderson, P.E.
CH2M HILL
Louisville, Kentucky

ACTIVATED ALUMINA

See Aluminum oxide.

ACTIVATED CARBON

Activated carbon is a form of charcoal that has acquired the property of adsorbing various substances from water through treatment by a carefully controlled combustion process. It is available in two forms, powdered and granulated, and it has high affinity for adsorbing chlorine and taste- and odor-causing substances.

Activated carbon particles have a large surface area. The grade of activated carbon used for water treatment has a specific surface area ranging from 500 to 600 m^2/g. The particles appear to be solid, but are actually honeycombed with minute "tunnels" or pores on the molecular order of size. Pore dimensions are expressed in angstroms (Å). The materials responsible for taste and odor in water, such as the products of industrial waste, sewage, and plant and animal organisms, are thought to be adsorbed in pores not greater than 20 Å in diameter. In the highly activated carbon, about 75% of the surface exists as pores less than 20 Å in diameter.

Activated carbon is prepared from coal, wood, coconut char, petroleum coke, lignite, charred peach pits, and other carbonaceous materials. The carbons used in the water industry are prepared principally from paper char, hardwood charcoal, or lignite.

Granular Activated Carbon

Granular activated carbon (GAC) is used by passing the water through a relatively thick bed of carbon. Common applications are by replacing some of or all the granular media in a filter or using a separate tank filled with the carbon. It can be ground and screened to any desired size. In the size used in some water treatment plants, 100% will pass an 8-mesh screen, and at least 90% will be retained on a 12-mesh screen. Other sizes commonly available have effective sizes of 0.8 to 0.9 mm and 0.55 to 0.65 mm. GAC is covered in AWWA Standard B604.

Powdered Activated Carbon

Powdered activated carbon (PAC) is a very finely ground material; more than 90% passes a 300-mesh screen. PAC is used by adding it to the water to be treated at a point where it will be subsequently removed by sedimentation and filtration. PAC is covered in AWWA Standard B600.

ALUMINUM OXIDE

Aluminum oxide (Al_2O_3) is also known as activated alumina. It is a highly porous, granular material with a preferential adsorptive capacity for moisture from gases, vapors, and some liquids. One of the principal uses in the waterworks field is for removal of excess levels of fluorides from drinking water by percolating the water through beds of alumina. The beds can be arranged as either pressure or gravity filters, and piped for backwashing and regeneration.

When the alumina becomes saturated with fluorides, it must be regenerated. This is accomplished by first backwashing with water to remove the accumulated solids and then backwashing with a weak caustic solution to remove fluorides. The residual caustic is then neutralized with a weak acid, followed by water rinses.

Activated alumina is available in granules ranging in size from a powder to approximately 1.5 in. (38 mm) in diameter. The 14 to 8 and the 8 to $1/4$ U.S. Standard Sieve mesh are the sizes most generally used.

ALUMINUM SULFATE

The most common coagulating agent used for water treatment is aluminum sulfate (alum). Most commercial grades of alum have the formula $Al_2(SO_4)_3 \cdot 14H_2O$. It is available in lump, ground, or liquid form. Ground aluminum sulfate for use in dry feed machines should be of such size that not less than 90% will pass a National Bureau of Standards no. 10 sieve and 100% will pass a no. 4 sieve. It is a grayish-white crystalline solid completely soluble in water with a tendency to absorb moisture from the air. Under such conditions, it tends to deliquesce to a fine white powder. This does not affect its efficiency but requires a slight change in the chemical feed rate.

Liquid alum is a clear, amber-colored liquid sometimes called 50% alum. This is because 1 gal of liquid alum weighs 11.2 lb (0.5 g) and contains 5.4 lb (0.2 g) of dry aluminum sulfate. Actually, it usually contains 8.5% or more available water-soluble alumina (Al_2O_3), as compared with the 17% Al_2O_3 available in dry alum. Alum is covered in AWWA Standard B403.

AMMONIA, ANHYDROUS

Ammonia is used in water treatment to add to chlorinated water so that chloramines will be formed. In the gaseous state, ammonia is colorless and about 0.6 times as heavy as air.

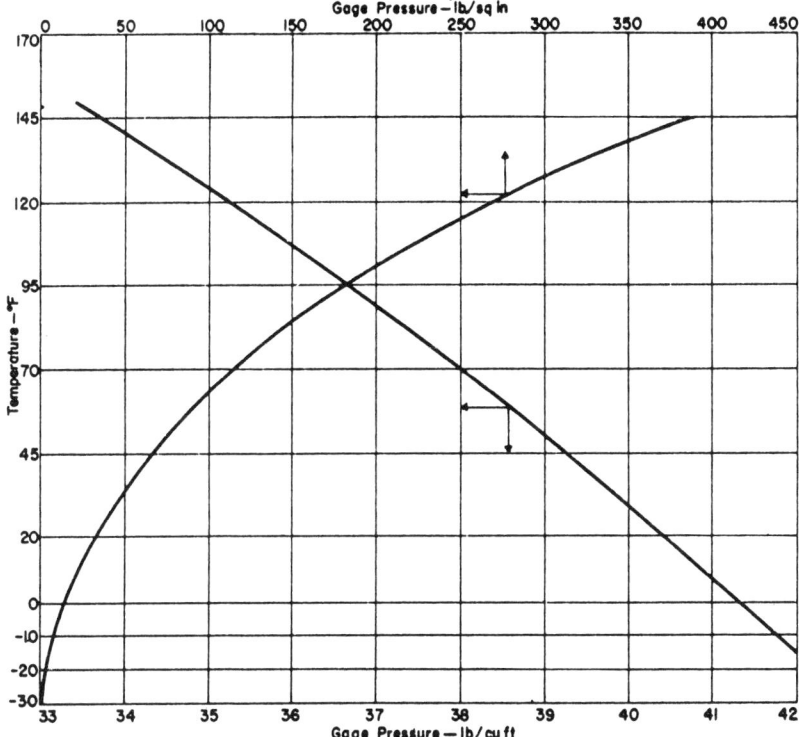

FIGURE A.1 Relationship of pressure–temperature to density–temperature in anhydrous ammonia.

The liquid, also colorless, is about 0.68 times as heavy as water. Unconfined liquid ammonia rapidly vaporizes to gas. The temperature, pressure, and density characteristics of anhydrous ammonia are shown in Figure A.1. An AWWA standard on anhydrous ammonia is currently under development.

AMMONIA, AQUA

See Ammonium hydroxide.

AMMONIUM ALUMINUM SULFATE

Also referred to as ammonium alum and crystal alum, ammonium aluminum sulfate has the form of dry, colorless, and odorless crystals or white powder. Although it has many industrial applications, it is not often used in water treatment because it has coagulation properties similar to alum but is more expensive. A characteristic unique to alum, though, is the residual ammonium ion after the coagulation process, which can be used to form chloramines.

AMMONIUM HYDROXIDE

Aqua ammonia is the ammonium hydroxide of commerce. At 60° F, it consists of a stable solution of 29.4% NH_3, with a specific gravity of 0.8974 and a density of 7.48 lb/gal (0.9 g/mL). Its boiling and freezing points are 84° F and −107° F, respectively. Its vapor pressure varies from about 4.5 lb/in.^{2} absolute (psia) (31 kPa) at 32° F (0° C) to 50 psia (344.5 kPa) at 104° F (40° C).

AMMONIUM SULFATE

Like ammonia, ammonium sulfate is sometimes used as a source of ammonia for the formation of chloramines for disinfection. Because it can vary widely in purity, purchase specifications regarding quality should be carefully written and should conform to the AWWA standard. Being a manufacturing by-product, it should be carefully examined for the presence of heavy metals. It is a white, crystalline solid that is readily soluble in water. Ammonium sulfate is covered in AWWA Standard B302.

ASCORBIC ACID

Ascorbic acid is available in white granular or powder form and has a shelf life of about 1 year; however, once a solution is prepared, the chemical decays within a few days. The stored chemical should be kept in a dark, dry area sealed from the atmosphere. The chemical is a weak acid and will decrease the pH of the dechlorinated water somewhat. Ascorbic acid can actually be beneficial to aquatic life if concentrations are not excessive. Ascorbic acid is vitamin C, a diet requirement, and is recognized as safe by the Food and Drug Administration (FDA), Department of Health and Human Services. Skin contact is not dangerous, but could produce slight irritation.

BENTONITE

Bentonite is a form of clay that swells on absorbing moisture. It is sometimes used as a coagulant aid in waters of very low turbidity to create a nucleus around which floc will form and to add weight to the floc. Many variations of bentonite are available, depending on their source, but the primary constituents are silicone and aluminum oxides.

CALCIUM CARBONATE

Calcium carbonate is also known as unburned lime, limestone, calcite, whiting, chalk, and precipitated chalk. It is widely used as an agricultural lime and is also used in neutralization, stabilization, or corrosion control in water treatment. The chemical is commercially available in a number of forms, both prepared and naturally occurring. It is only slightly soluble in water. Calcium carbonate is also the main precipitant generated as a by-product in lime softening.

CALCIUM CHLORIDE

Calcium chloride ($CaCl_2$) is manufactured as white, odorless, deliquescent flakes or pellets with a minimum strength of 77%. It is used in water treatment as a coagulant and sludge conditioner. Calcium chloride is covered in AWWA Standard B550.

CALCIUM HYDROXIDE

See Lime.

CALCIUM HYPOCHLORITE

Calcium hypochlorite is an off-white chemical that is furnished in granular, free-flowing, or compressed-tablet form. Commercial high-test calcium hypochlorite products, such as HTH, contain at least 70% available chlorine and have from less than 3% up to about 5% lime residual. All commercial calcium hypochlorites also contain minor amounts of other impurities, such as calcium carbonate and other insoluble substances. In hard water, an additional amount of calcium carbonate is formed.

Although a highly active oxidizer, calcium hypochlorite is relatively stable. Under normal storage conditions, commercial preparations lose about 3% to 5% of their available chlorine content in a year. But contact with water or the atmosphere induces a pronounced increase in the decomposition rate and greatly increases the reaction rate with organic materials.

Decomposition is exothermic and proceeds rapidly if any part of the material is heated to 350° F (176° C), yielding oxygen and chlorine as well as a powdery dust that has an irritating action. Heat is also evolved, which further supports and increases the decomposition rate. The granular form is essentially nonhydroscopic and resists moist caking tendencies, when properly stored. Although calcium hypochlorite is a stable, nonflammable material that cannot be ignited, contact with heat, acids, combustible, organic, or oxidizable materials may cause fire.

Granular calcium hypochlorite is readily soluble in water, varying from about 21.5 g/100 mL at 32° F (0° C) to 23.4 g/100 mL at 104° F (40° C). Tablet forms dissolve more slowly than the granular materials and can provide a fairly steady source of available chlorine over an 18- to 24-h period. Small tablets are often used for disinfection of new water mains. Large tablets are commonly used in swimming pool and sewage disinfection. Hypochlorites are covered in AWWAStandard B300.

CALCIUM OXIDE

See Lime.

CALCIUM THIOSULFATE

Calcium thiosulfate is a clear solution with nearly neutral pH. It reacts with both free and combined chlorine. A 30% solution of calcium thiosulfate has a pH ranging from 6.5 to

7.5. Because dechlorination produces acids, the pH may be lowered somewhat. The reaction does not scavenge oxygen or produce sulfur dioxide. The application rate should be 1.3 parts by weight of the agent per 1 part of chlorine residual to be neutralized. Actual feed rates will depend on the concentration of the batch solution, discharge rate, and chlorine residual concentration. For example, if a 10% stock solution is used, a feed rate of 0.0013 gpm is required for a 10 gpm discharge rate of 1 mg/L chlorine residual. The reaction at stoichiometric dosages requires about 5 min to take place. Overdosing will decrease the reaction time, but may produce milky-colored suspended solids that would cause turbidity violations.

Although the chemical may cause eye and skin irritation, the National Fire Protection Association (NFPA) hazard ratings are 0, 0, 0 for health, fire, and reactivity, respectively, out of a scale of 0 to 4 in each category. Hazardous reporting is not required.

CARBON

See Activated carbon.

CARBON DIOXIDE

Carbon dioxide (CO_2) gas is commonly used in water treatment to lower pH and stabilize lime-softened water. Once stabilized, further precipitation of soluble calcium carbonate is less likely to occur. In the past, CO_2 was commonly produced on-site by the burning of carbonaceous material such as natural gas. It is now typically manufactured at an industrial plant and delivered to the treatment plant, where it is stored in liquid form in a refrigerated tank under pressure.

Carbon dioxide is a clear, colorless, and odorless gas with a high specific gravity. When confined, carbon dioxide tends to seek the lowest level possible. In water, its solution produces carbonic acid. Carbon dioxide functions as an aid to coagulation, primarily by adjusting the pH or solubilities so that proper coagulation is obtained. Carbon dioxide is covered in AWWA Standard B510.

CHLORINATED LIME

Chlorinated lime is sometimes referred to as bleaching powder. It has been used as a disinfectant in treating water supplies and more particularly for swimming pools. It is a white powder prepared by chlorinating slaked lime and decomposes in water, releasing 39% available chlorine for disinfecting action.

CHLORINE

In the gaseous state, chlorine is greenish yellow and about 2.48 times as heavy as air. The liquid is amber-colored and about 1.44 times as heavy as water. Unconfined liquid chlorine rapidly vaporizes to gas with 1 volume of liquid yielding about 450 volumes of gas.

Chlorine is only slightly soluble in water, its maximum solubility being approximately 1% at 50° F (10° C). At lower temperatures, chlorine combines with water, forming chlorine "ice," a crystalline hydrate ($Cl_2 \cdot 8H_2O$).

The temperature–pressure characteristics of chlorine are shown in Figure A.2. Chlorine confined in a container may exist as a gas, a liquid, or a mixture of both. Thus any consideration of liquid chlorine includes consideration of gaseous chlorine. The vapor pressure of chlorine in a container is a function of temperature and is independent of the contained volume of chlorine; therefore gauge pressure does not necessarily indicate its state.

The volume–temperature characteristics of chlorine in a container loaded to the limit authorized by Department of Transportation (DOT) regulations are shown in Figure A.3. If a container is filled to this limit, it becomes completely full of liquid at approximately 154° F (67° C), and at temperatures beyond that point, excessive pressure that could result in rupture of the container. Safety devices are provided on cylinders and containers to relieve the excessive pressures that accompany excessive temperature elevations.

Under specific conditions, chlorine reacts with most elements, sometimes with extreme rapidity. The reaction with hydrocarbons, alcohols, and ethers can be explosive. Although chlorine itself is neither explosive nor flammable, it is capable of supporting combustion of certain substances. It should, accordingly, be handled and stored away from other compressed gases (such as anhydrous ammonia) and kept apart from turpentine, ether, finely divided metals, and hydrocarbons or other flammable materials. Chlorine is covered in AWWA Standard B301.

CHLORINE DIOXIDE

Under atmospheric conditions, chlorine dioxide is a yellow to red, unpleasant-smelling, irritating, unstable gas. It is produced by mixing a chlorine solution (chlorine gas dissolved in water or sodium hypochlorite and hydrochloric acid) with a solution of sodium chlorite (typically 25%). Because chlorine dioxide decomposes very quickly, it is usually generated on-site before its application.

To generate chlorine dioxide, the chlorine and sodium chlorite solutions are injected and mixed together and passed through a media-packed reaction column to promote mixing and a near-complete reaction (greater than 95%) to generate chlorine dioxide. Various types and designs of chlorine dioxide generators are manufactured. This material is a powerful oxidizing agent and disinfectant, used primarily in reducing or eliminating tastes and odors.

COPPER SULFATE

Copper sulfate is available in blue crystalline granules or powder. It is used most often in water treatment to control algae in impounding reservoirs. It is sometimes applied as a powder, distributed by dusting on the surface of the water. Another technique is to drag burlap bags of copper sulfate crystals through the water, effecting solution and distribution at the same time. Another method of application is to meter the copper sulfate solution into a narrow control channel at the entrance to the impounding reservoir. This chemical and its solution are poisonous and corrosive. Copper sulfate is covered in AWWA Standard B602.

FIGURE A.2 Relationship of temperature to pressure in chlorine.

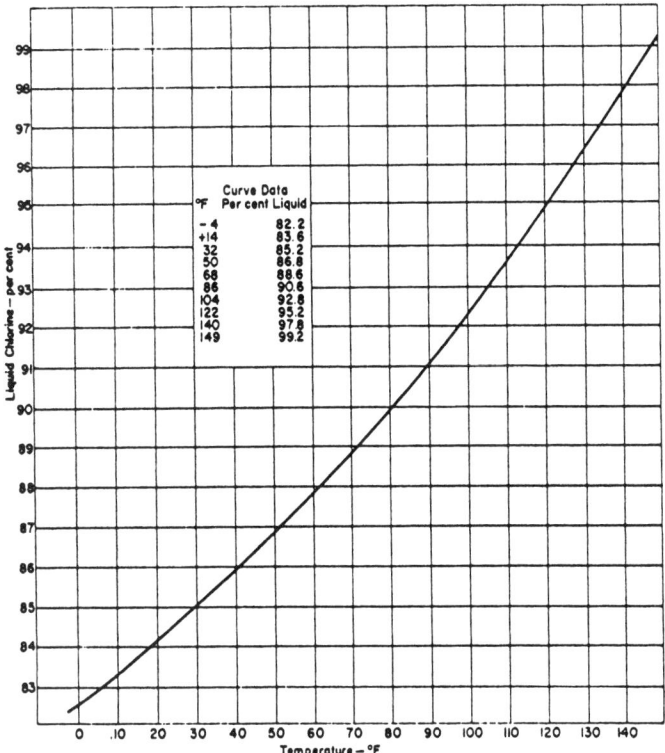

FIGURE A.3 Relationship of volume to temperature in chlorine.

DISODIUM PHOSPHATE

See Phosphates.

EPI-DMA POLYAMINES

Epichlorohydrin dimethylamine (EPI-DMA) polyamines are polyquaternary polymers produced from the reaction of dimethylamine and epichlorohydrin. Polymers may contain residual amounts of these chemicals and their by-products, so the chemical should be tested for impurities. The product is shipped in drums or bulk as an aqueous solution typically ranging from 10% to 65% by weight active polymer. EPI-DMA polyamines are covered in AWWA Standard B452.

FERRIC CHLORIDE, LIQUID

Ferric chloride is commonly used as a coagulant in potable water, sewage, and industrial waste treatment. It is usually prepared as a liquid of 37% to 40% solution, and it is shipped in 55-gal (208-L) drums or 4,000- to 8,000-gal (15,000- to 30,000-L), rubber-lined tank trucks. Ferric chloride is fed about the same as alum, but because it is more corrosive than alum, special handling materials must be used. Ferric chloride is covered in AWWA Standard B407.

FERRIC SULFATE

Ferric sulfate is produced to some degree when ferrous sulfate is chlorinated. It is available on the chemical market as a reddish-brown deliquescent solid. Ferric sulfate is soluble in water, producing a somewhat acidic solution. Ferric sulfate is covered in AWWA Standard B406.

FERROUS SULFATE

The combination of ferrous sulfate and lime forms an effective coagulant for the clarification of turbid water. Ferrous sulfate itself is a greenish-white crystalline solid that is obtained as a by-product of other chemical processes.

The AWWA standard for ferrous sulfate gives size standards for both its granular and lump forms. The standard-size granular ferrous sulfate has a tendency to cake and arch in dry feeder hoppers or storage bins. Material of a finer crystal size that contains only five molecules of water crystallization ($FeSO_4 \cdot 5H_2O$) may be obtained. This material has less tendency to cake or arch. Ferrous sulfate is also available in a liquid form. Ferrous sulfate is covered in AWWA Standard B402.

FLUOROSILICIC ACID

Fluorosilicic acid (formerly called hydrofluosilicic acid) is commonly fed at a very low dose rate to drinking water supplies to increase the fluoride level in the water. Pure acid is a colorless, transparent, fuming, corrosive liquid, with a pungent odor and an irritating action on the skin. The commercial product used in water treatment is a 20% to 30% water solution of H_2SiF_6 that has a straw or slightly reddish color owing to dissolved iron. In the amounts used in water fluoridation, the iron presents no problem. Fluorosilicic acid is covered in AWWA Standard B703.

HYDRATED LIME

See Lime.

HYDROCHLORIC ACID

Hydrochloric acid is a clear or slightly yellow, fuming, pungent liquid. It may contain iron or arsenic and is poisonous. If it is to be used in a potable water supply, care should be taken to obtain the purified form (USP). To reduce the fuming characteristics, it should be diluted (by adding the acid to water) to approximately 20% HCl. It is not often used in the treating of potable water supplies because it increases the chlorine content of the water.

HYDROFLUOSILICIC ACID

See Fluorosilicic acid.

HYDROGEN PEROXIDE

Hydrogen peroxide is an aqueous solution prepared in solution strengths ranging from 30% to 70%. In water treatment it is used as a catalyst with ozone for oxidation and for dechlorination. Hydrogen peroxide will react and neutralize free chlorine in solutions with a pH greater than 7. It reacts slowly and does not effectively neutralize combined chlorine. Oxygen will be produced in the free chlorine reaction and in closed environments while reacting with hyperchlorinated water. The solution may effervesce, requiring special provisions to handle the gas.

Hydrogen peroxide is very reactive and rated hazardous when its strength is greater than 52%. The Superfund Amendment and Reauthorization Act (SARA) Title III, Sections 311 and 312, classifies hydrogen peroxide as an immediate health hazard and a fire hazard. It is shipped in 6- to 55-gal (23- to 208-L) polyethylene drums, tank trucks, and railroad tank cars.

HYPOCHLORITES

See Calcium hypochlorite and Sodium hypochlorite.

LIME

The most common forms of lime in water treatment are quicklime and hydrated lime. Other forms of lime may also be found in specific areas. Quicklime and hydrate lime are covered in AWWA Standard B202.

Quicklime

Quicklime (CaO) results from the calcination of limestone or an equivalent, such as dried water-softening sludge of suitable analysis, and consists essentially of calcium oxide in natural association with a lesser amount of magnesium oxide. The revised AWWA standard for quicklime and hydrated lime gives detailed recommendations on impurities and particle sizes. It recommends a minimum of 90% available calcium oxide content. Quicklime is available in any particle size required by a given installation, and size is usually specified in each contract. Quicklime has some tendency to air-slake, so exposure to outside air should be kept to a minimum.

Hydrated Lime

Hydrated lime, $Ca(OH)_2$, is a very finely divided powder resulting from the hydration of quicklime with enough water to satisfy its chemical affinity. Commercial hydrated lime consists essentially of calcium hydroxide or a mixture of calcium hydroxide and magnesium hydroxide, depending on the type of quicklime slaked. Hydrated lime is a powder that should be uniform in size and should have a minimum available calcium oxide content of 68.0%. Hydrated lime tends to absorb carbon dioxide from the atmosphere, even in the slurry form; therefore its exposure to outside air should be carefully controlled.

Dolomitic Hydrated Lime

This lime is produced by hydrating dolomitic quicklime. When normal hydrating conditions are present, only the calcium oxide is hydrated and the magnesium oxide remains unchanged. This is referred to as *monohydration*.

Dolomitic Lime

Dolomitic lime contains from 35% to 40% magnesium oxide. When the magnesium content exceeds the calcium content, the chemical is termed *dolomitic magnesite*. If the magnesium oxide is greater than 5% but less than 35%, it is termed *magnesia quicklime*.

MONOSODIUM PHOSPHATE

See Phosphates.

OXYGEN, LIQUID

Liquid oxygen is pale blue, extremely cold, and nonflammable. It is a strong oxidizer and, as such, supports combustion. Oxygen will react with nearly all organic materials and metals. Materials that burn easily in air usually burn vigorously in oxygen. In water treatment, pure oxygen is used to generate a higher concentration of ozone that can be generated with prepared air.

OZONE

Ozone is a faintly blue gas with a pungent odor. It is an unstable form of oxygen composed of three-atom molecules that break down readily to normal oxygen and nascent oxygen. The latter is a powerful oxidizing agent and has germicidal action. Ozone is usually produced with on-site generators by passing high-voltage electricity through dry atmospheric air or pure oxygen between stationary electrodes. This process converts a small percentage of the oxygen in the air to ozone. It is usually injected into the water to be treated in a highly baffled mixing chamber. The ozone residual is determined by use of the indigo blue test.

PHOSPHATES

Many different forms of phosphates are marketed for corrosion control, preventing scale formation, and sequestering unwanted precipitants. Generic forms of these compounds and compound groups include sodium hexametaphosphate, tricalcium phosphate, monosodium phosphate, disodium phosphate, trisodium phosphate, sodium tripolyphosphate, polyphosphates, zinc orthophosphates, bimetallic zinc orthophosphates, and silicate phosphates. These products are available in both liquid and powder forms. They often have a very low pH, requiring special care in handling and storage. Phosphates are covered in AWWA Standards B502, B503, B504, and B505.

PHOSPHORIC ACID

Phosphoric acid (H_3PO_4) is a colorless and odorless water-white liquid used for corrosion and pH control. It is shipped in polyethylene drums, railroad tank cars, and tank trucks at concentrations ranging from 70% to 85%.

POLYALUMINUM CHLORIDE

Polyaluminum chloride (PAC) is a clear, pale amber liquid used in place of an inorganic coagulant. It is shipped in bulk and 55-gal drums. Polyaluminum chloride is covered in AWWA Standard B408.

POLYDADMAC

Polydiallyldimethylammonium chloride (polyDADMAC) is a synthetic cationic organic polyelectrolyte used as a coagulant. PolyDADMAC is usually sold as an aqueous solution between 10% and 50% by weight of active polymer in drums or bulk. PolyDADMAC is covered in AWWA Standard B451.

POLYMERS

Polymers are synthetic organic chemicals typically possessing high molecular weight. They are available in positively charged (cationic), negatively charged (anionic), or neutral (near zero net charge) forms and are manufactured as both powders and liquids. They are used as primary coagulants, coagulant aids, flocculants, filter aids, filter backwash water conditioning, and solids thickening and dewatering. Compared with other chemicals routinely used in water treatment, polymer dosages often are much lower.

When a polymer in powder form is to be used, a stock solution must first be prepared. Some liquid polymers must also be prepared in a stock solution before use, although other liquid products can be fed directly to the process with dilution water. In preparing the stock solutions, warm water up to 100° F (38° C) should be used to hasten the dissolving process, and a relatively long mixing time is usually recommended.

To protect the effectiveness of the polymer, there are typically upper and lower limits to the strength of the solution. Once the stock solution is ready to apply, it is further diluted to help ensure complete mixing with the process water. Because polymers are highly variable in their performance on various types of water, bench- or pilot-scale studies are necessary to make an appropriate selection.

POTASSIUM HYDROXIDE

Potassium hydroxide, also known as potash and caustic potash, is used in lime softening and for pH adjustment. It is available in both solid and liquid forms. As an unground solid it is shipped in 700- to 725-lb (318- to 329-kg) steel drums. Flaked, ground, crystalline, or beaded materials are shipped in drums, kegs, barrels, polyethylene-lined bags, or fiberboard drums. The liquid form typically contains 45% to 52% by weight and is shipped in carboys, steel drums, tank trucks, and railroad tank cars.

Considerable heat is generated when this product is mixed with water. As a result, when one is making solutions, protective clothing is required, and the material should only be added to water with constant stirring. Potassium hydroxide is covered in AWWA Standard B511.

POTASSIUM HYPOCHLORITE

Potassium hypochlorite is also referred to as potassium hypochlorite bleach. It has similar properties and is used as a substitute for sodium hypochlorite when the use of sodium must be avoided or minimized. As a stock solution, it is typically prepared at 10% strength. Heat, sunlight, and certain metallic contamination will affect the hypochlorite stability, so the solution should be stored away from heat and sunlight in containers of appropriate material.

POTASSIUM PERMANGANATE

This material is in the form of black or purple crystals or pellets with a blue metallic sheen. It is highly soluble in water, which allows easy application. It is usually prepared in dilute solution (1% to 4%) as needed for application. Potassium permanganate is covered in AWWA Standard B603.

QUICKLIME

See Lime.

SODA ASH

See Sodium carbonate.

SODIUM ALUMINATE

Sodium aluminate is sometimes used as an auxiliary coagulant for the removal of fine turbidity and color in soft, low-pH water. The solid form is a white or brown powder containing 70% to 80% $Na_2Al_2O_4$. The solution form is a concentrated solution containing approximately 32% $Na_2Al_2O_4$. This material is readily soluble in water, producing a noncorrosive solution. Sodium aluminate is covered in AWWA Standard B405.

SODIUM ASCORBATE

Sodium ascorbate is available in crystalline form and has a shelf life of at least 1 year, if kept in a dark, cool place. More sodium ascorbate than ascorbic acid is required to react with the same amount of chlorine, although it does not significantly depress pH.

SODIUM BISULFITE

Sodium bisulfite, also called sodium metabisulfite, is used as a dechlorination agent. The principal constituent of this product is sodium pyrosulfite, or sodium metabisulfite ($Na_2S_2O_5$). Sodium bisulfite is available as a white powder, granule, or clear liquid solution. These white, crystalline powders are readily soluble in water. Sodium metabisulfite is covered in AWWA Standard B601. A 1% solution of this agent has a pH of 4.3, so dechlorination will depress pH somewhat. Sodium bisulfite is also a strong oxygen scavenger, and slug dosages of this chemical could cause injury to aquatic organisms if surplus chemical did not react with chlorinated water. Although sodium bisulfite can cause skin, eye, and respiratory tract irritation, it is used in food and drugs as a preservative and is safe, if properly handled and used. The chemical will react with air to decompose and form sulfur dioxide gas.

SODIUM CARBONATE

Sodium carbonate, commonly called soda ash (Na_2CO_3), is commonly used in the water softening process. It is a white powder or granular material containing well above 99% sodium carbonate, or 58% as a sodium oxide equivalent. Although the product is available in two different grades, light and dense, they differ only in bulk density, size, and appearance. Quality and other properties are the same. Packaging is in 50- and 100-lb (23- and 46-kg) bags and as bulk in tank trucks and railcars. Soda ash is covered in AWWA Standard B201.

SODIUM CHLORIDE

Sodium chloride is used in water treatment to recharge cation resins of ion exchange treatment units. The AWWA standard for sodium chloride requires that the material be homogeneous and in a dry granular form, white, grayish white, pink, brown, or brownish white. The size requirements state that rock salt must be of such fineness that it passes a no. 3 sieve and 95% of it is retained on a no. 7 sieve. Evaporated salt must be of such fineness that at least 85% is retained on a no. 7 sieve. The material must also dissolve rapidly without packing. The solution formed by dissolving the salt in distilled water is to have a phenolphthalein alkalinity of zero and a hydrogen-ion concentration (pH) no higher than 8.0. Sodium chloride is covered in AWWA Standard B200.

SODIUM CHLORITE

Sodium chlorite is a dry, flaked salt with a powerful oxidizing nature. It is stable when sealed or in solution, but it is highly combustible in the presence of organic material. Technical-grade sodium chlorite (approximately 80% dry weight $NaClO_2$) is an orange-colored flaked salt that is a powerful oxidizing agent.

Sodium chlorite in contact with acid will react rapidly to evolve chlorine dioxide gas. When heated above 347° F (175° C), sodium chlorite will decompose rapidly, liberating oxygen and evolving sufficient heat to make the decomposition self-sustaining. If this decomposition is confined, as in a closed container, the effect is explosive. Therefore it should be protected at all times from exposure to heat.

Sodium chlorite dissolves easily in water at ordinary temperatures to form an orange-brown solution that is chemically stable under ordinary conditions of temperature and pressure. Sodium chlorite is available from suppliers in dry form or as a solution. Because of the hazardous nature of dry sodium chlorite, liquid solutions are the preferred chemical form. For most applications, sodium chlorite solutions of 25% by weight should be used. This concentration has been selected by the industry because it has the lowest freezing point of sodium chlorite solutions, 5° F ($-15°$ C). If dry sodium chlorite is to be used, special precautions are required for storage, handling, dust control, and worker safety.

For most chlorine dioxide generation systems, a low-alkalinity (<10 g/L as $CaCO_3$) sodium chlorite should be used. Sodium chlorite is covered in AWWA Standard B303.

SODIUM FLUORIDE

Sodium fluoride is produced by neutralizing hydrofluoric acid with either sodium carbonate or sodium hydroxide. It is available as crystals or white powder. It is highly soluble in water, which enhances its use in the fluoridation of water supplies. Sodium fluoride is covered in AWWA Standard B701.

SODIUM FLUOROSILICATE

Sodium fluorosilicate was previously called sodium silicofluoride in the United States, but the name has been changed to conform with international usage. This material is a white, free-flowing, odorless, crystalline powder. It is produced by neutralizing hydrofluosilicic acid with soda ash (Na_2CO_3) or sodium hydroxide (NaOH) and evaporating the resulting solution. Sodium silicofluoride is the most extensively used compound in the fluoridation of water supplies because of its availability and low cost. Its disadvantages lie in its very low solubility, 60 gal of water being required to dissolve 1 lb of sodium fluorosilicate. Sodium fluorosilicate is covered in AWWA Standard B702.

SODIUM HEXAMETAPHOSPHATE

This is one of several forms of what are known as glassy phosphates. It is used in water softening and boiler water treatment and is available from manufacturers in both powdered and granular forms. It is also used as a sequestering agent in municipal water supplies and is highly soluble in water. See Phosphates and Sodium polyphosphate.

SODIUM HYDROXIDE

Sodium hydroxide (NaOH) is also known as caustic soda. It is available as a solid in flake form, but more commonly in liquid solution at a concentration of 50% as NaOH. The specific gravity of 50% liquid caustic is 1.525, or a density of 12.7 lb/gal (1,522 g/L). Because it is a powerfully alkaline corrosive chemical, extreme care must be used when handling caustic soda. Body contact will result in immediate burning of tissue. Sodium hydroxide is covered in AWWA Standard B501.

SODIUM HYPOCHLORITE

Commercial sodium hypochlorite, or liquid bleach, usually contains 12.5% to 17% available chlorine at the time of manufacture and is available only in liquid form. It is typically used for disinfection at smaller water treatment plants, where use of chlorine gas is a safety concern, and for swimming pool water disinfection.

All NaOCl solutions are unstable to some degree and deteriorate more rapidly than calcium hypochlorite. The effect can be minimized by care in the manufacturing processes, controlling the alkalinity of the solution, and storing away from heat and sunlight. Greatest stability is attained with a pH close to 11.0 and in the absence of heavy metal cations. Hypochlorites are covered in AWWA Standard B300.

SODIUM METABISULFITE

Sodium metabisulfite is a dechlorination agent available as crystalline granules or liquid solution. A 1% solution of this agent has a pH of 4.3, so dechlorination will depress pH. Sodium metabisulfite is also a strong oxygen scavenger, and slug dosages of this chemical could cause injury to aquatic organisms comparable to sodium bisulfite. Dilution with water produces sulfur dioxide. Solutions degrade when exposed to oxygen, and stability increases with solution concentration.

Sodium metabisulfite is an eye, skin, mucous membrane, and respiratory tract irritant. Overexposure can produce highly toxic effects. The OSHA Permissible Exposure Limit (PEL) for sodium metabisulfite based on time-weighted average (TWA) criteria is 5 mg/m^3. The NFPA hazard rating is 3, 0, 1 for health, fire, and reactivity, respectively, out of a scale of 0 to 5 for each category.

SODIUM POLYPHOSPHATE

See Sodium hexametaphosphate.

SODIUM SILICATE

Sodium silicate (liquid) in the form known as activated silica is used as a coagulant aid. The chief advantage of activated silica is its ability to toughen the floc. Activated silica may be prepared in a number of ways. Sodium silicate solution is available as a 40° Baumé liquid containing approximately 30% SiO_2. Sodium silicate is covered in AWWA Standard B404.

SODIUM SULFITE

Sodium sulfite is a widely used dechlorination agent, available in powder and crystalline forms, and is the only agent available in tablet form. Sodium sulfite solutions are slightly alkaline. Although dechlorination produces hydrochloric acid, the reaction does not significantly affect pH. About 8 parts of sodium sulfite removes 1 part of dissolved oxygen.

Sodium sulfite is an eye, skin, mucous membrane, and respiratory tract irritant and has an NFPA hazard rating of 2, 0, 0 for health, fire, and reactivity, respectively, out of a scale of 0 to 4 in each category. Although the crystalline form is considerd to be stable, a solution of the chemical will react with air to decompose and form sulfur dioxide gas.

Sodium sulfite is also available in tablet form and can be used for dechlorinating open-channel flows. Tablets can be inserted in dechlorination mats or in dispensers to dissolve the tablets as discharged water flows through the tablets. The dispensers are filled with as many tablets as will fit. For surface water flows over mats, tablets should be placed no more than 4 in. apart. Tablets must be regularly checked and replaced as needed.

SODIUM THIOSULFATE

Sodium thiosulfate is a colorless, transparent crystal widely used for dechlorination. Dechlorination of combined chlorine may produce ammonia or ammonium ion, depending on pH. Although some hydrochloric acid is produced during dechlorination, solution pH is not affected significantly. Thiosulfate is an oxygen scavenger and reducing agent, but scavenges less oxygen than sulfite, bisulfite, or metabisulfite. The reaction ratio for dechlorination is heavily dependent on pH. Although this chemical scavenges less oxygen than do other agents, its reaction time for dechlorination is slower.

Sodium thiosulfate is a skin, eye, nose, and throat irritant, and it is moderately toxic. Its NFPA rating is 1, 0, and 0, for health, fire, and reactivity, respectively, out of a scale of 0 to 4 in each category.

SODIUM TRIPOLYPHOSPHATE

See Phosphates.

SULFUR DIOXIDE

Sulfur dioxide is used in water treatment for dechlorination. It is produced in North America by the combustion of sulfur in special burners, by burning pyrites, or as a by-product of smelting operations. In the gaseous state, sulfur dioxide is colorless and about 2.26 times as heavy as air. The liquid, also colorless, is about 1.44 times as heavy as water. Unconfined liquid sulfur dioxide rapidly vaporizes to gas.

Sulfur dioxide is about 20 times as soluble as chlorine in water. At 32° F (0° C), up to 20% by weight will dissolve to form a weak solution of sulfurous acid, or ammonium chloride in the presence of combined chlorine residual. Because its vapor pressure increases with increasing temperature, its solubility decreases with increasing temperature; at 80° F (27° C) it is soluble to less than 10%. Sulfur dioxide is covered in AWWA Standard B512. This agent can deplete dissolved oxygen (DO) in the discharge water and receiving stream.

Sulfur dioxide is a toxic chemical subject to reporting requirements of the SARA Title III, Section 313. It has an NFPA rating of 2, 0, and 0 for health, fire, and reactivity, respectively, out of a scale of 0 to 4 in each category.

SULFURIC ACID

Sulfuric acid (H_2SO_4) is a strong, corrosive, dense, oily liquid, lightly colored or dark brown, depending on purity. For use in water treatment it should be of the USP grade and free of heavy metals. It is available in a number of grades, containing from 60% to 90% H_2SO_4.

QUALITY OF PURCHASED CHEMICALS

Chemicals can be purchased in several grades. Contaminants from manufacturing or by-products can be present in poorer-quality grades. The presence of some contaminants, such as heavy metal ions, could cause water quality problems in finished water or in the disposal of treatment plant residuals. The presence of iron, nickel, copper, cobalt, and others will cause sodium hypochlorite to decompose at a faster rate. For that reason, appropriate specifications for chemical quality should be used for all purchases. AWWA has prepared standards for purchasing certain water treatment chemicals to help ensure that quality is acceptable. Table A.1 on pp. A.20 to A.29 provides an alphabetical listing of commonly used water treatment chemicals. Table A.2 on pp. A.30 and A.31 lists the process use of common water treatment chemicals.

TABLE A.1 Alphabetical Listing of Commonly Used Water Treatment Chemicals

Chemical name and formula	Common or trade name	Shipping containers	Suitable storage materials*	Available forms/descriptions	Density	Solubility (lb/gal)	Commercial strength (%)	Additional characteristics and properties
Activated carbon, powdered carbon	Aqua Nuchor, Hydrodarco, Herite	Bags, bulk	Dry: iron, steel; wet: rubber and silicon linings, type 316 stainless steel	Black granules, powder	15 to 30 lb/ft^3	Insoluble (suspension used)		1 lb/gal suspension used for storage and handling
Aluminum oxide, Al_2O_3	Activated alumina	Bags, drums	Iron, steel	Powder granules (up to 1-in. diameter)		Insoluble	100	
Aluminum sulfate, $Al_2(SO_4)_3 \cdot 14H_2O$ (dry)	Alum, filter alum, sulfate of alumina	100- to 200-lb bags, 300- to 400-lb barrels, bulk (carloads), tank truck, 228-36 tank car	Dry: iron, steel; wet: stainless steel, rubber, plastic	Ivory-colored; powder, granule, lump	38 to 45 lb/ft^3 60 to 63 lb/ft^3 62 to 67 lb/ft^3	6.2 (60° F)	17 as Al_2O_3	pH of 1% solution: 3.4
Aluminum sulfate (liquid)	50% alum	Tank cars and tank trucks	FRP,† PE,‡ type 316 stainless steel, rubber linings	Liquid	11.2 lb/gal		8.5 as Al_2O_3	Freezing point −13° C (9° F)
Ammonium aluminum sulfate, $Al_2(SO_4)_3 \cdot (NH_4)_2-SO_4 \cdot 24H_2O$	Ammonia alum, crystal alum	100-lb bags, barrels, bulk	FRP, PE, type 316 stainless steel, rubber linings	Colorless crystals or white powder	65 to 75 lb/ft^3	0.3 (32° F)	99	pH of 1% solution: 3.5

Chemical	Other names	Shipping containers	Suitable materials	Appearance	Weight	Solubility	Purity (%)	Remarks
Ammonium hydroxide, NH_4OH	Ammonia water, ammonium hydrate, aqua ammonia	Carboys, 750-lb drums, bulk	Glass lining, steel, iron, FRP, PE	Colorless liquid	7.48 lb/gal	Complete	29.4 (NH_3) max 26° Baumé	pH 14; freezing point = −107° F
Ammonium silicofluoride, $(NH_4)_2SiF_6$	Ammonium fluorosilicate	100- and 400-lb drums	Steel, iron, FRP, PE	White crystals	65 to 70 lb/ft^3	1.7 (63° F)	100	White, free-flowing solid
Ammonium sulfate, $(NH_4)_2SO_4$	Sulfate of ammonia	50- and 100-lb bags, 725-lb drums	FRP, PE, ceramic and rubber linings; iron (dry)	White or brown crystal	70 lb/ft^3	6.3 (68° F)	>99	Cakes in dry feed; add $CaSO_4$ for free flow
Anhydrous ammonia, NH_3	Ammonia	50-, 100-, 150-lb cylinders, bulk tank cars, and trucks	Shipping containers	Colorless gas	38.6 lb/ft^3	3.9 (32° F) 3.1 (60° F)	99.9+ (NH_3)	
Ascorbic acid	Vitamin C	Bags; drums	Plastic, glass	White granules and powder	1.65 specific gravity	2.75	100	Will depress pH in large doses
Bentonite	Colloidal clay, volclay, wilkinite	100-lb bags, bulk	Iron, steel, FRP, PE	Powder, pellet, mixed sizes	60 lb/ft^3	Insoluble (colloidal solution used)		Free-flowing, nonabrasive
Calcium fluoride, CaF_2	Fluorspar	Bags, drums, barrels, hopper cars, trucks	Steel, iron, FRP, PE	Powder		Very slight	85 (CaF_2), less than 5 (SiO_2)	
Calcium hydroxide, $Ca(OH)_2$	Hydrated lime, slaked lime	50-lb bags, bulk	FRP, PE, iron, steel, rubber lining	White powder, light, dense	28 to 36 lb/ft^3	0.14 (68°F) 0.12 (90°F)	85 to 99 [$Ca(OH)_2$], 63 to 73 (CaO)	Hopper agitation required for dry feed of light form

*Always contact chemical suppliers to select best materials for handling. †FRP = fiber-reinforced plastic. ‡PE = polyethylene.

TABLE A.1 Alphabetical Listing of Commonly Used Water Treatment Chemicals (*Continued*)

Chemical name and formula	Common or trade name	Shipping containers	Suitable storage materials*	Available forms/descriptions	Density	Solubility (lb/gal)	Commercial strength (%)	Additional characteristics and properties
Calcium hypochlorite, $Ca(OCl)_2 \cdot 4H_2O$	HTH, perchloron, pittchlor	5-lb cans; 100-, 300-, 800-lb drums	Glass, plastic, and rubber linings, FRP, PE	White granule, powder, tablet	52.5 lb/ft^3	1.5 at 25° C	65 (available Cl_2)	1 to 3 (available Cl_2 solution used)
Calcium oxide, CaO	Burned lime, chemical lime, quicklime, unslaked lime	80- and 100-lb bags, bulk	FRP, PE, iron, steel, rubber	Lump, pebble, granule	35 to 71 lb/ft^3	Slaked to form hydrated lime	75 to 99 (CaO)	pH of saturated solution, on detention time temp. amount of water critical for efficient slaking
Calcium thiosulfate	BSP Captor	5-gal drums; 55-gal drums	Plastic, steel	Clear solution	1.24 specific gravity	10.3	30	Weak oxygen scavenger. Reaction is pH-dependent, and reaction time is slower than that of other agents
Carbon dioxide, liquid CO_2	Carbonic anhydride	Bulk	Carbon steel (dry); type 316 stainless steel (solution)	Liquid		0.012 at 25° C	99.5	Solution is acid

Chemical	Common name	Shipping containers	Suitable handling materials	Appearance	Weight	Solubility (lb/gal water)	Commercial strength (%)	Remarks
Chlorinated lime, $2CaOCl_2 \cdot 3H_2O$	Bleaching powder, chloride of lime	100-, 300-, 800-lb drums	Glass and rubber lining, FRP, PE	White powder	48 lb/ft³		25 to 37 (available Cl_2)	Deteriorates
Chlorine, Cl_2	Chlorine gas, liquid chlorine	100-, 150-lb cylinders; 1-ton tanks; 16-, 30-, 55-ton tank cars	Shipping containers	Greenish-yellow, liquefied gas under pressure	91.7 lb/ft³	0.07 (60° F) 0.04 (100° F)	99.8 (Cl_2)	Forms HCl and HOCl when mixed with water
Chlorine dioxide, ClO_2	Chlorine dioxide	Generated as used	Glass, PVC, and rubber linings; FRP; PE	Greenish-yellow gas		0.02 (30 mu)	26.3 (available Cl_2)	Explosive under certain conditions
Copper sulfate, $CuSO_4 \cdot 5H_2O$	Blue vitriol, blue stone	100-lb bags, 450-lb barrels, drums	FRP, PE, silicon lining, iron, stainless steel	Crystal, lump, powder	75 to 90 lb/ft³ 73 to 80 lb/ft³ 60 to 64 lb/ft³	1.6 (32° F) 2.2 (68° F) 2.6 (86° F)	99 ($CuSO_4$)	25% solution pH approx. 3.0
Disodium phosphate, anhydrous $Na_2HPO_4 \cdot 12H_2O$	Basic sodium phosphate, DSP, secondary sodium phosphate	100- and 300-lb drums, 50- and 100-lb bags	Cast iron, steel, FRP, PE	White crystal, granular or powder	60 to 64 lb/ft³	0.4 (32° F) 6.4 (86° F)	64.3 (PO_4) 48 (P_2O_5)	Precipitates Ca, Mg; pH of 1% solution = 9.1; solubility is 11 g/100 g at 25° C (77° F)
Ferric chloride, $FeCl_3$ (33% to 45% solution)	Ferrichlor, iron chloride	55-gal drums, bulk	Glass, PVC, and rubber linings; FRP, PE	Dark brown, syrupy liquid	11.9 lb/gal (40%)	Complete	37 to 45 ($FeCl_3$) 20 to 21 (Fe)	

*Always contact chemical suppliers to select best materials for handling. †FRP = fiber-reinforced plastic. ‡PE = polyethylene.

TABLE A.1 Alphabetical Listing of Commonly Used Water Treatment Chemicals (*Continued*)

Chemical name and formula	Common or trade name	Shipping containers	Suitable storage materials*	Available forms/descriptions	Density	Solubility (lb/gal)	Commercial strength (%)	Additional characteristics and properties
Ferric chloride, $FeCl_3 \cdot 6H_2O$	Crystal ferric chloride	300-lb barrels	Keep in original containers	Yellow-brown lump			59 to 61 ($FeCl_3$), 20 to 21 (Fe)	Hygroscopic (store lumps and powder in tight container), no dry feed; optimum pH, 4.0 to 11.0
Ferric chloride, $FeCl_3$	Anhydrous ferric chloride	500-lb casks; 100-, 300-, 400-lb kegs; 65-, 135-, 250-lb drums	Keep in original containers	Greenish-black powder or crystals	175 lb/ft^3		98 ($FeCl_3$) 34 (Fe)	
Ferric sulfate, $Fe_2(SO_4)_3 \cdot 9H_2O$	Ferrifloc, ferrisul	100- to 175-lb bags, 400- to 425-lb drums	Glass, plastic, and rubber linings; FRP, PE, type 316 stainless steel	Red-brown powder 70 or granule 72	60 to 70 lb/ft^3	Soluble in 2 to 4 parts cold water	90 to 94 [as $Fe_2(SO_4)_3$] 25 to 26 (Fe)	Mildly hygroscopic coagulant at pH 3.5 to 11.0
Ferrous sulfate, $FeSO_4 \cdot 7H_2O$	Copperos, green vitriol	Bags, barrels, bulk	Glass, plastic, and rubber linings; FRP, PE, type 316 stainless steel	Green crystal, granule, lump	63 to 66 lb/ft^3		55 ($FeSO_4$) 20 (Fe)	Hygroscopic; cakes in storage; optimum pH 8.5 to 11.0

Chemical	Trade names	Shipping containers	Materials of construction	Appearance	Properties	Concentration (%)	Remarks
Fluorosilicic acid, H_2SiF_6	Fluorosilicic acid	Rubber-lined drums, trucks, or railroad tank cars	Rubber-lined steel, PE	Liquid	Specific gravity = 1.46 for 61% solution strength at 25° C, 1.27 for 30% sol str at 17.5° C	Approx. 1.2 (68° F)	40 (approx.)	Corrosive, etches glass
Hydrogen fluoride, HF	Hydrofluoric acid	Steel drums, tank cars	Steel, FRP, PE	Liquid			70 (HF)	Below 60% steel cannot be used
Hydrogen peroxide	Durox; Hybrite; OxyPure	Drums	Glass, stainless steel, aluminum, plastic	Clear colorless liquid	1.11 specific gravity	Complete	20–40	Not recommended for combined chlorine due to slowness of reaction
Oxygen, liquid	LOX	Dewars, cylinders, truck and rail tankers	Steel	Pale blue liquid	9.52 lb/gal at 68° F and 1 atm	3.16% by volume at 25° C	99.5	Prevent LOX from contacting grease, oil, asphalt, or other combustibles
Ozone, O_3	Ozone	Generated at site of application		Colorless gas				
Phosphoric acid, H_3PO_4		Polyethylene drums, bulk	FRP, epoxy, rubber lining, polypropylene, type 316 stainless steel	Watery white liquid	13.1 lb/gal at 75% solution		75/80/85	Freezing point: 0.5° F at 75% 40.2° F at 80% 70.0° F at 85%

*Always contact chemical suppliers to select best materials for handling. †FRP = fiber-reinforced plastic. ‡PE = polyethylene.

TABLE A.1 Alphabetical Listing of Commonly Used Water Treatment Chemicals (*Continued*)

Chemical name and formula	Common or trade name	Shipping containers	Suitable storage materials*	Available forms/descriptions	Density	Solubility (lb/gal)	Commercial strength (%)	Additional characteristics and properties
Polyaluminum chloride, $Al_{13}(OH)_{20}(SO_4)_2Cl_{15}$	SternPac	55-gal drums and bulk	FRP, PE, type 316 stainless steel, rubber linings	Pale amber liquid	10.0 lb/gal		10.3 (Al_2O_3)	Freezing point = $-12°$ C
Potassium aluminum sulfate, $K_2SO_4 \cdot Al_2(SO_4)_3 \cdot 24H_2O$	Potash alum, potassium alum	Bags, lead-lined bulk (carloads)	FRP, PE, ceramic and rubber linings	Lump, granule, powder	60 to 67 lb/ft^3	0.5 (32° F) 1.0 (68° F) 1.4 (86° F)	10 to 11 (Al_2O_3)	Low, even solubility; pH of 1% solution: 3.5
Potassium permanganate, $KMnO_4$	Purple salt	Bulk, barrels, drums	Iron, steel, FRP, PE	Purple crystals	90 to 105 lb/ft^3	Infinite	100	Danger of explosion in contact organic matters
Pyrosodium sulfite	Sodium metabisulfite, $Na_2S_2O_5$	Bags, drums, barrels	Iron, steel, FRP, PE	White crystalline powder		Complete in water	Dry 67 (SO_2), sol 33.3 (SO_2)	Sulfurous odor
Sodium aluminate, $Na_2OAl_2O_3$	Soda alum	100- to 150-lb bags, 250- to 440-lb drums, solution	Iron, FRP, PE, rubber, steel	Brown powder liquid (27° Baumé)	50 to 60 lb/ft^3	3.0 (68° F) 3.3 (86° F)	70 to 80 (Na_2) Al_2O_4 min. 32 $Na_2Al_2O_4$	Hopper agitation required for dry feed; very hygroscopic
Sodium ascorbate		Bags; drums	Plastic, FRP, stainless steel	White crystalline powder	1.5	13.3	99	Minimal effect on pH
Sodium bisulfite*	Sodium acid sulfite	Drums; bulk	Plastic, FRP, stainless steel	Milky-white liquid in solution form	1.2 to 1.4 specific gravity	Complete	28–43	Strong oxygen scavenger

Chemical	Common name	Shipping containers	Suitable handling materials	Appearance	Weight	Solubility (lb/gal water)	Commercial strength (%)	Remarks
Sodium carbonate, Na_2CO_3	Soda ash	Bags, barrels, bulk (carloads), trucks	Iron, rubber lining, steel, FRP, PE	White powder, extra light, light, dense	31.2 to 56.2 (light); 56.2 to 68.7 (dense) lb/ft³	1.5 (68° F) 2.3 (86° F)	99.4 (Na_2CO_3) 57.9 (Na_2O)	Hopper agitation required for dry feed of light and extra-light forms; pH of 1% solution: 11.3
Sodium chloride, NaCl	Common salt, salt	Bags, barrels, bulk (carloads)	Bronze, FRP, rubber	Rock, fine	50 to 60 58 to 78 lb/ft³	2.9 (32° F) 3.0 (68° F)	98 (NaCl)	Absorbs moisture
Sodium chlorite, $NaClO_2$	ADOX dry	100-lb drums	Metals (avoid cellulose materials)	Light orange powder, flake, or crystals	53 to 56 lb/ft³	3.5 (68° F)	80 ($NaClO_2$) 30 (available Cl_2)	Generates ClO_2 at pH 3.0; explosive
Sodium fluoride, NaF	Fluoride	Bags, barrels, fiber drums, kegs	Iron, steel, FRP, PE	Nile blue or white powder, light, dense	50 lb/ft³ 75 lb/ft³	0.35 (most temps.)	90 to 95 (NaF)	pH of 4% solution: 6.6
Sodium fluorosilicate, Na_2SiF_6	Sodium silicofluoride	Bags, barrels, fiber drums	Cast iron, rubber lining, steel, FRP, PE	Nile blue or yellowish-white powder	72 lb/ft³	0.03 (32° F) 0.06 (68° F) 0.12 (140° F)	99 (Na_2)	pH of 1% solution: 5.3
Sodium hexametaphosphate, $Na(PO_3)_6$	Calgon, glassy phosphate, vitreous phosphate	100-lb bags	Rubber linings, plastics, type 316 stainless steel	Crystal, flake, powder	47 lb/ft³	1 to 4.2	66 (P_2O_5 unadjusted)	pH of 0.25% solution: 6.0 to 8.3
Sodium hydroxide, NaOH	Caustic soda, soda lye	100- to 700-lb drums; bulk (trucks, tank cars)	Carbon steel, polypropylene, FRP, rubber lining	Flake, lump, liquid	95.5 lb/ft³; 12.8 lb/gal for 50% solution	2.4 (32° F) 4.4 (68° F) 4.8 (104° F)	98.9 (NaOH) 74 to 76 ($NaOH_2$)	Solid, hygroscopic; pH of 1% solution = 12.9; freezing point of 50% solution = 53° F

*Always contact chemical suppliers to select best materials for handling. †FRP = fiber-reinforced plastic. ‡PE = polyethylene.

TABLE A.1 Alphabetical Listing of Commonly Used Water Treatment Chemicals (*Continued*)

Chemical name and formula	Common or trade name	Shipping containers	Suitable storage materials*	Available forms/descriptions	Density	Solubility (lb/gal)	Commercial strength (%)	Additional characteristics and properties
Sodium hypochlorite, NaOCl	Sodium hypochlorite	5-, 13-, 50-gal carboys; 1,300- to 2,000-gal tank trucks	Ceramic, glass, plastic, and rubber linings; FRP, PE	Light yellow liquid			12 to 15 (available Cl_2)	Unstable
Sodium metabisulfite	Disodium disulfite	Drums; bulk	Plastic, FRP, stainless steel	White crystalline powder	1.48 specific gravity	3.25	98	Strong oxygen scavenger
Sodium silicate, Na_2OSiO_2	Water glass	Drums, bulk (tank trucks, tank cars)	Cast iron, rubber lining, steel, FRP, PE	Opaque, viscous liquid		Complete	38 to 42° Baumé	Variable ratio of Na_2O to SiO_2; pH of 1% solution: 12.3
Sodium sulfite, Na_2SO_3	Sulfite	Bags, drums, barrels	Cast iron, rubber lining, steel, FRP, PE	White crystalline powder	80 to 90 lb/ft^3	Complete in water	23 (SO_2)	Sulfurous taste and odor
Sodium thiosulfate	Thiosulfuric acid	Drums; bulk	Plastic, FRP, stainless steel	Monoclinic, colorless crystals	1.75 specific gravity	6.6	100	Weak oxygen scavenger. Reaction is pH-dependnet and will depress pH. Reaction time is slower than that of other agents

Chemical	Also known as	Shipping container		Physical state	Density	Solubility in water (lb/gal)	Commercial strength (%)	Remarks
Sulfur dioxide, SO_2	Sulfurous acid anhydride	100- to 150-lb steel cylinders, ton containers, tank cars, tank trucks		Colorless gas		20% at 32° F, complete in water	99 (SO_2)	Irritating gas
Sulfuric acid, H_2SO_4	Oil of vitriol, vitriol	Bottles, carboys, drums, trucks, tank cars	FRP, PE porcelain, glass, and rubber linings	Solution	81.4 lb/ft³ (59.3° Baumé)	Complete	77 (59.3 Baumé)	Approx. pH of 0.5% solution: 1.2
Tetrasodium pyrophosphate, $Na_4P_2O_7 \cdot 10H_2O$	Alkaline sodium, pyrophosphate, TSPP	125-lb kegs, 200-lb bags, 300-lb barrels	Cast iron, steel, plastics	White powder	68 lb/ft³	0.6 (80° F) 3.3 (212° F)	53 (P_2O_5)	pH of 1% solution: 10.8
Tricalcium phosphate	Fluorex	Bags, drums, bulk, barrels	Cast iron, steel, plastics	Granular	Variable	Insoluble		Also available as white powder
Trisodium phosphate, $Na_3PO_4 \cdot 12H_2O$	Normal sodium phosphate, tertiary sodium phosphate, TSP	125-lb kegs, 200-lb bags, 325-lb barrels	Cast iron, steel, plastics	Crystal coarse, medium, standard	56 lb/ft³ 58 lb/ft³ 61 lb/ft³	0.1 (32° F) 13.0 (158° F)	19 (P_2O_5)	pH of 1% solution: 11.9

*Always contact chemical suppliers to select best materials for handling. †FRP = fiber-reinforced plastic. ‡PE = polyethylene.

TABLE A.2 Process Use of Common Water Treatment Chemicals

Process	Chemicals	AWWA Standard
Coagulants and coagulant aids	Aluminum sulfate	B403-93
	Bentonite	
	Calcium carbonate	
	Calcium hydroxide	
	Calcium oxide	
	EPI-DMA polyamines	B452-90
	Ferric chloride	B407-93
	Ferric sulfate	B406-92
	Ferrous sulfate	B402-90
	Polyaluminum chloride	B408-93
	PolyDADMAC	B451-92
	Polymers	
	Sodium aluminate	B405-94
	Sodium silicate	B404-92
Dechlorination	Activated carbon, granular	B604-90
	Ion exchange resins	
	Sodium bisulfite	
	Sodium metabisulfite	B601-93
	Sulfur dioxide	B512-91
Disinfection and chlorination	Anhydrous ammonia	
	Ammonium hydroxide	
	Ammonium sulfate	B302-90
	Calcium hypochlorite	B300-92
	Chlorinated lime	
	Chlorine	B301-92
	Chlorine dioxide (sodium chlorite + Cl_2)	
	Ozone	
	Sodium chlorite	B303-88
	Sodium hypochlorite	B300-92
Fluoridation and fluoride adjustment	Activated alumina (aluminum oxide)	
	Calcium fluoride	
	Fluorosilicic acid	B703-94
	Hydrogen fluoride	
	Sodium fluoride	B704-94
	Sodium fluorosilicate	B702-94
Mineral oxidation	Chlorine	B301-92
	Chlorine dioxide (sodium chlorite + Cl_2)	
	Ozone	
	Potassium permanganate	B603-93
Organics adsorption oxidation	Activated carbon, granular	B604-90
	Chlorine dioxide	
	Ozone	
	Potassium permanganate	B603-93
pH adjustment, stabilization, and corrosion control	Calcium carbonate	
	Calcium chloride	B550-90
	Calcium hydroxide	B202-93
	Calcium oxide	B202-93
	Carbon dioxide	B510-89
	Disodium phosphate	B505-88
	Hydrochloric acid	
	Monosodium phosphate	B504-88

TABLE A.2 Process Use of Common Water Treatment Chemicals (*Continued*)

Process	Chemicals	AWWA Standard
pH adjustment, stabilization, and corrosion control (*cont.*)	Phosphoric acid	
	Potassium hydroxide	B511-90
	Sodium carbonate	
	Sodium hexametaphosphate	B502-88
	Sodium hydroxide	B501-93
	Sodium polyphosphate	B502-88
	Sodium silicate	
	Sodium tripolyphosphate	B503-89
	Sulfuric acid	
Softening	Calcium hydroxide	B202-93
	Calcium oxide	B202-93
	Sodium carbonate	B201-92
	Sodium chloride	B200-93
	Sodium hydroxide	
Taste and odor control	Activated carbon, powdered	B600-90
	Activated carbon, granular	B604-90
	Chlorine	B301-92
	Chlorine dioxide (sodium chlorite + Cl_2)	
	Copper sulfate	B602-91
	Ozone	
	Potassium permanganate	B603-93

APPENDIX B
ABBREVIATIONS COMMONLY USED IN THE WATER INDUSTRY

A

acre-foot: **acre·ft**
acutely hazardous material: **AHM**
Advanced Notice of Proposed Rule Making: **ANPRM**
aggressiveness index: **AI**
alkylbenzene sulfonate: **ABS**
alternating current: **ac**
American Association of Cost Engineers: **AACE**
American Association of State Highway and Transportation Officials: **AASHTO**
American Chemical Society: **ACS**
American Concrete Institute: **ACI**
American Gas Association: **AGA**
American Institute of Chemical Engineers: **AIChE**
American Institute of Electrical Engineers: **AIEE**
American National Standards Institute: **ANSI**
American Public Health Association: **APHA**
American Public Works Association: **APWA**
American Society of Civil Engineers: **ASCE**
American Society of Heating, Refrigeration and Air Conditioning: **ASHRAE**
American Society of Mechanical Engineers: **ASME**
American Society for Testing and Materials: **ASTM**
American Water Works Association: **AWWA**
American Water Works Association Research Foundation: **AWWARF**
Americans with Disabilities Act: **ADA**
ampere: **A**
ampere-hour: **Ah**
answer: **ans.**
area: A
asbestos cement: **A-C**
asbestos-containing material: **ACM**
assimilable organic carbon: **AOC**

APPENDIX B

Association of Metropolitan Water Agencies: **AMWA**
Association of State Drinking Water Administrators: **ASDWA**
Association of State Sanitary Engineers: **ASSE**
atomic absorption spectrophotometry: **AAS**
atomic emission spectroscopy: **AES**
atomic weight: **at wt**
available filter head: **AFH**
avoirdupois: **avdp** or **avoir**

B

background organic matter: **BOM**
barrel: **bbl** or **brl**
base (as the "length" of a triangle): B or b
becquerel (metric equivalent of curie): **Bq**
benzene-toluene-xylene (the volatiles measured in gasoline): **BTX**
best-available technology: **BAT**
billion: means 10^9 to Americans, 10^{12} to Europeans
billion electronvolts: **BeV**
billion gallons: **bil gal**
billion gallons per day: **bgd**
biochemical oxygen demand or biological oxygen demand: **BOD**
biodegradable organic matter: **BDOM**
biologically enhanced activated carbon: **BEAC**
board feet (feet board measure): **fbm**
brake horsepower: **bhp**
British thermal unit: **Btu** or **BTU**
bushel: **bu**

C

calculation error: **CE**
carbon usage rate: **CUR**
cast iron: **CI**
cathode-ray tube: **CRT**
Centers for Disease Control and Prevention: **CDC**
centimeter: **cm**
central processing unit: **CPU**
chemical oxygen demand: **COD**
chloramines: includes **monochloramine (NH_2Cl), dichloramine ($NHCl_2$), and trichloramine (NCl_3)**
chlorinated polyvinyl chloride: **CPVC**
chloroplatinate units (color indicator): **cpu**
Clean Air Act: **CAA**
Clean Water Act: **CWA**
Code of Federal Regulations: **CFR**
colony-forming units: **cfu**
color units: **cu**

community water system: **CWS**
Comprehensive Environmental Response, Compensation, and Liability Act: **CERCLA**
computerized maintenance management system: **CMMS**
Consumer Products Safety Commission: **CPSC**
contact time (disinfectant with water): *T*
counts per minute: **cpm**
cubic feet: **ft³** or **cu ft**
cubic feet per hour: **ft³/h** or **cu ft/h**
cubic feet per minute: **ft³/min** or **cu ft/min**
cubic feet per second: **ft³/s** or **cu ft/s** or **cfs**
cubic inches: **in.³** or **cu in.**
cubic meters: **m³** or **cu m**
cubic yards: **yd³** or **cu yd**
curie: **Ci**
cycles per second: **cps** (1 cps = 1 Hz)

D

deciliter: **dL**
Department of Agriculture: **DOA** or **USDA**
Department of Commerce: **DOC**
Department of Interior: **DOI**
Department of Transportation: **DOT**
diameter: *D*
diatomaceous earth (filtration): **DE**
dibromochloropropane: **DBCP**
dichlorodiphenyltrichloroethane: **DDT**
differential pressure cell: **DP cell**
differential pressure meter: **DP meter**
direct current: **DC** or **dc**
disinfectant by-products: **DBPs**
disinfectant concentration × time: *CT* or *C* × *T*
disinfectant–disinfection by-products: **D-DBP** or **D/DBP**
dissolved air flotation: **DAF**
dissolved organic carbon: **DOC**
dissolved organic halogen: **DOX**
dissolved oxygen: **DO**
distributed control system: **DCS**
ductile iron pipe: **DIP**
Ductile Iron Pipe Research Association: **DIPRA**

E

effective size (of granular media): **E.S.**
efficiency: **eff.**
electrodialysis: **ED**
electrodialysis reversal: **EDR**
electrolytic conductivity detector: **ElCD**
electromotive force: **emf**

electronic data processing: **EDP**
electronvolts: **eV**
Emergency Planning and Community Right-to-Know Act: **EPCRA**
empty bed contact time: **EBCT**
Endangered Species Act: **ESA**
Engineers Joint Council: **EJC**
Enhanced Surface Water Treatment Rule: **ESWTR**
environmental impact statement: **EIS**
Environmental Protection Agency (U.S.): **USEPA**
epichlorohydrin dimethylamine: **EPI-DMA**
ethylenediaminetetraacetic acid: **EDTA**
ethylene dibromide: **EDB**

F

failure rate: **FR**
Federal Emergency Management Agency: **FEMA**
Federal Insecticide, Fungicide, and Rodenticide Act: **FIFRA**
feet: **ft**
feet per hour: **ft/h**
feet per minute: **ft/min**
feet per second: **ft/s**
fiberglass-reinforced plastic: **FRP**
fiscal year: **FY**
flow rate: Q
foot-pound: **ft·lb**

G

gallon: **gal**
gallons per capita per day: **gpcd**
gallons per day: **gpd**
gallons per hour: **gph**
gallons per minute: **gpm**
gallons per minute per square foot: **gpm/ft^2**
gallons per second: **gps**
gallons per square foot: **gal/ft^2**
gas chromatography: **GC**
gas chromatography–mass spectrometry: **GC-MS**
General Accounting Office: **GAO**
geographical information system: **GIS**
gigaliter: **GL**
Government Printing Office: **GPO**
grains per gallon: **gpg**
granular activated carbon: **GAC**
greater than: **>**
greater than or equal to: **≥**
groundwater under the direct influence of surface water: **GWUDI**

H

haloacetic acid: **HAA**
haloacetic nitriles: **HANs**
heating, ventilating, and air conditioning: **HVAC**
hectare: **ha**
height: *h* or *H*
hertz: **Hz**
heterotrophic plate count (replaces *standard plate count*): **HPC**
high-density, cross-linked polyethylene: **HDXLPE**
high-density polyethylene: **HDPE**
high-intensity discharge (lighting): **HID**
high-pressure liquid chromatography: **HPLC**
highway-addressable remote transducer: **HART**
horsepower: **hp**
hour: **h** or **hr**
hundred (centum): **C**
hundredweight: **cwt**
hydraulic grade line: **HGL**
hydraulic grade line elevation: **HGLE**
hydraulic retention time: **HRT**

I

imperial: **imp**
inch: **in.**
inches per minute: **in./min**
inches per second: **in./s**
Information Collection Rule: **ICR**
inorganic chemical: **IOC**
inside diameter: **ID**
Institute of Electrical and Electronics Engineers: **IEEE**
Instrument Society of America: **ISA**
instrumentation and control: **I&C**
integrally molded flanged outlets: **IMFO**
International Research Center: **IRC**
Interstate Commerce Commision: **ICC**
iron pipe size: **IP**

J

Jackson turbidity units: **Jtu**
joule: **J**

K

kilogram: **kg**
kilohertz (kilocycles per second): **kHz**
kilometer: **km**
kilovoltampere: **kVA**
kilowatt: **kW**
kilowatthours: **kWh**

L

Langelier saturation index: **LSI**
length: l or L
less than: **<**
less than or equal to: **≤**
linear feet: **lin ft**
liquid oxygen: **LOX**
liquefied petroleum gas: **LPG**
liter: **L**
local emergency planning committee: **LEPC**
locally unacceptable land use: **LULU**
local operator interface (control system): **LOI**
lowest observed adverse effect level: **LOAEL**
lumen: **lm**

M

mass spectrometry: **MS**
mass transfer zone: **MTZ**
material safety data sheet: **MSDS**
maximum contaminant level: **MCL**
maximum contaminant level goal: **MCLG**
maximum permissible concentration: **MPC**
maximum tolerated dose: **MTD**
mean sea level: **MSL**
megahertz (megacycles per second): **MHz**
megaliter: **ML**
membrane filter: **MF**
meter: **m**
2-methylisoborneol: **MIB**
microfiltration: **MF**
mile: **mi**
miles per hour: **mph**
mille (thousand): **M**
milliampere direct current: **mADC**
milliampers: **mA**
milligram: **mg**

milliliter: **mL**
millimeter: **mm**
million electronvolts: **MeV**
million gallons: **mil gal** or **MG**
million gallons per day: **mgd**
minute: **min**
mixing intensity: **G**
mole: **mol**
molecular weight: **mol wt**
molecular weight cutoff: **MWCO**
moles per liter: **mol/L**
month: **mo**
most probable number: **MPN**
motor starting current: **IM**

N

nanofiltration: **NF**
nanometer: **nm**
National Academy of Science: **NAS**
National Electrical Code: **NEC**
National Electrical Manufacturers Association: **NEMA**
National Environmental Policy Act: **NEPA**
National Fire Protection Association: **NFPA**
National Institute of Occupational Safety and Health: **NIOSH**
National Pollutant Discharge Elimination System: **NPDES**
National Sanitation Foundation: **NSF International**
natural organic matter: **NOM**
nephelometric turbidity unit: **ntu**
net positive suction head: **NPSH**
New England Water Works Association: **NEWWA**
nitrogen (gaseous): **GAN**
nitrogen (liquid): **LIN**
N,N-diethyl-p-phenylenediamine (used to determine presence of chlorine): **DPD**
nominal pipe size: **NPS**
not equal to: $\neq$

O

Occupational Safety and Health Administration: **OSHA**
ohm: **Ω**
one thousand: **M**
operations and maintenance: **O&M**
Organization for Economic Cooperation and Development: **OECD**
ounce: **oz**
outside diameter: **OD**
oxidation reduction potential: **ORP**

oxygen (gas): **GOX**
oxygen (liquid): **LOX**
ozone: **O₃**

P

package: **pkg**
parts per billion: **ppb** or **μg/L** (preferred)
parts per million: **ppm** or **mg/L** (preferred)
parts per trillion: **ppt** or **ng/L** (preferred)
perchloroethylene: **PCE**
perimeter: *P*
personal computer: **PC**
picocurie: **pCi**
picocuries per liter: **pCi/L**
point of entry (water treatment device): **POE**
point of use (water treatment device): **POU**
polyaluminum chloride: **PACl**
polyaromatic hydrocarbon: **PAH**
polychlorinated biphenyl: **PCB**
polydiallyldimethylammonium chloride: **PolyDADMAC**
polytetrafluoroethylene: **PTFE** (generic name for Teflon)
polyvinyl chloride: **PVC**
Portland Cement Association: **PCA**
pound: **lb**
pounds per day: **lb/d**
pounds per square foot: **lb/ft²**
pounds per square inch: **psi** or **lb/in.²**
pounds per square inch absolute: **psia**
pounds per square inch gauge: **psig**
powdered activated carbon: **PAC**
practical quantitation level: **PQL**
presence-absence (coliform test): **P-A**
pressure-reducing valve: **PRV**
pressure-swing adsorption (manufacture of O_2): **PSA**
process and instrumentation drawing: **P&ID**
process hazard analysis: **PHA**
process safety management: **PSM**
programmable logic controller: **PLC**
proportional, plus integral, plus derivative (control system): **PID**
publicly owned treatment works: **POTW**

Q

quantity (flow rate): *Q*

R

radius (of a circle): *r*
random access memory: **RAM**
redox potential: **pE**
regulatory negotiations: **reg neg**
reinforced thermoset plastic: **RTP**
relay ladder logic (computer): **RLL**
reliable detection limit: **RDL**
remote terminal unit: **RTU**
resistance temperature detector: **RTD**
Resource Conservation and Recovery Act: **RCRA**
reverse osmosis: **RO**
revolutions per minute: **rpm**
revolutions per second: **rps**
risk management program: **RMP**
roentgen equivalent, mammal: **rem**

S

Safe Drinking Water Act: **SDWA**
second: **s**
second-feet (cubic feet per second): **ft³/s** or **cfs**
self-contained breathing apparatus: **SCBA**
specific gravity: **sp gr**
specific heat: **sp ht**
square foot: **ft²** or **sq ft**
square inch: **in.²** or **sq in.**
square meters: **m²** or **sq m**
square miles: **mi²** or **sq mi**
square yards: **yd²** or **sq yd**
standard plate count: see *heterotrophic plate count*
standard query language: **SQL**
state emergency response commission: **SERC**
Stormwater Pollution Prevention Plan: **SWPPP**
streaming current detector: **SCD**
Superfund Amendments and Reauthorization Act: **SARA**
supervisory control and data acquisition: **SCADA**
Surface Water Treatment Rule: **SWTR**
synthetic organic chemical: **SOC**

T

temperature: *T* or **temp.**
tetrachloroethylene (perchloroethylene): **PCE**
tetrafluoroethylene: **TFE**

thermal-swing adsorption (manufacture of O_2): **TSA**
thermocouple: **TC**
thousand: **thous**
thousand (mille): **M**
threshold odor number: **TON**
ton: **T** or **ton**
total dissolved solids: **TDS**
total organic carbon: **TOC**
total organic halide: **TOX**
total suspended solids: **TSS**
total trihalomethanes: **TTHMs**
toxic characteristic leaching procedure: **TCLP**
Toxic Substances Control Act: **TSCA**
transient voltage surge suppression: **TVSS**
trichloroethylene (or trichloroethene): **TCE**
trihalomethane: **THM**
trihalomethane formation potential: **THMFP**
true color units: **b**
turbidity units: see *nephelometric turbidity units*

U

ultrafiltration: **UF**
ultrahigh-molecular-weight (plastic): **UHMW**
ultraviolet: **UV**
underground injection control: **UIC**
uniformity coefficient: **uc**
uniformity coefficient (of granular media): **U.C.**
uninterruptible power system: **UPS**
unreasonable risk to health: **URTH**
U.S. Environmental Protection Agency: **USEPA**
U.S. Public Health Service: **USPHS**

V

vacuum-swing adsorption (manufacture of O_2): **VSA**
variable-frequency drive: **VFD**
variable-speed drive: **VSD**
volatile organic compound: **VOC**
volt: **V**
volt drop: **VD**
voltamperes: **VA**
volts alternating current: **VAC**
volts direct current: **VDC**
volume: *V* or **vol.**

W

Water Environment Federation (formerly Water Pollution Control Federation): **WEF**
water treatment plant: **WTP**
watt: **W**
week: **wk.**
weight: **wt**
width: *w* or *W*
World Health Organization: **WHO**

X

xenon: **Xe**

Y

yard: **yd**
year: **yr**

Z

zeta potential: **zp**

INDEX

A-B syndrome, **22.5–22.6**
Abbreviations, **11.1–11.11**
Abrasion resistance, **14.3**
Accelerator solids contact unit, **7.31**
Acceptance testing, **21.10–21.11**
Access ramp and filter, **9.8**
Accessibility of equipment, **25.13–25.14**
Accidental release scrubber systems, **15.50**
Acetaldehyde, **12.53**
Acid base exchanges, **12.42**
Acidification, **13.42**
Acronyms, **11.1–11.11**
Actiflo ballasted flocculation system, **7.24–7.25**
Activated alumina:
   arsenic removal, **12.17**
   fluoride removal, **12.22**
   protective equipment, **15.42**
Activated carbon, **14.1**, **A.1**
   abrasion resistance, **14.3**
   ash content, **14.3**
   BAC, **9.14**, **14.29**
   carbon weight, **14.2–14.3**
   durability, **14.3**
   GAC (*see* Granular activated carbon)
   GAC particle size, **14.3–14.4**
   iodine number, **14.2**
   ion exchange, **12.30**
   moisture content, **14.3**
   molasses number, **14.2**
   PAC (*see* Powdered activated carbon)
   piping, **15.35**
   storage, **15.22**
Activated carbon feeding system, **14.7**
Activated carbon regeneration system
   schematic, **14.34**
Active frazil ice, **4.23**
Active IDS, **30.4**
Active infrared (IR) sensors, **30.5**
ADA, **18.5**, **23.29–23.30**, **26.8**
Adaptive gain control, **20.7**
Adiabatic equation, **10.23**
Adjacency diagram, **18.4**
Administrative facilities, **18.5–18.6**, **18.7**

Adsorption, **2.10**
Adsorption clarifier, **7.40**, **7.41**
Adsorptive separation, **10.53–10.54**
Advanced oxidation process (AOP), **29.8**
Aeration and stripping:
   design, **5.10–5.22**
   diffuse-type aeration, **5.4–5.5**, **5.11–5.12**
   equilibrium conditions, **5.8**
   factors to consider, **5.8**
   films, **5.9**
   lime softening, **11.5–11.6**
   mechanical aeration, **5.7**
   multiple-tray aerators, **5.2**, **5.13**
   pilot testing, **5.22–5.24**
   pressure aeration, **5.7**
   PTA, **5.3**, **5.4**, **5.14–5.22**
   (*See also* Packed tower aeration)
   rate of achievement, **5.8–5.9**
   rate of transfer, **5.9–5.10**
   saturation value, **5.8**
   spray aerators, **5.2**, **5.12–5.13**
   uses, **5.1**
   waterfall aeration, **5.2–5.4**
Afterburner, **14.37**
Air binding, **8.40**
Air block exchangers, **12.39**
Air emissions, **5.20**, **24.10–24.11**, **24.15**
Air entrainment, **16.13**
Air mixing, **6.11–6.13**
Air-operated diaphragm pumps, **15.24**
Air pollution, **17.66**
Air preparation ozonation system, **10.46–10.48**
Air scour, **8.14**, **8.18**, **28.17**
Air stripping, **2.10**, **5.1**
   (*See also* Aeration and stripping)
Air-to-water ratio, **5.3**, **5.14**
Alarm systems, **20.26–20.27**
Algae, **9.3–9.4**
Algal species, **9.4**
Allen substation aerator (Memphis), **5.13**
Alum [*see* Aluminum (alum)]
Alum recovery, **17.17**
Alum sludge, **17.5–17.6**

Aluminosilicates, **12.**4
Aluminum (alum):
  characteristics, **15.**10
  MCL, **2.**7
  piping, **15.**35
  recovery, **17.**17
  sludge, **17.**5–**17.**6
  storage, **15.**12–**15.**13
  structural design, **19.**3
Aluminum oxide, **A.**2
  (*See also* Activated alumina)
Aluminum sulfate, **6.**3, **15.**42, **A.**2, **A.**20
Ambient-type vaporizer, **10.**49, **10.**50
American National Standard Practices for Respiratory Protection, **15.**47
Americans with Disabilities Act (ADA), **18.**5, **23.**29–**23.**30, **26.**8
Ammonia, **5.**10, **10.**24
  anhydrous (*see* Anhydrous ammonia)
  aqua (*see* Aqua ammonia)
  dumping, **12.**10
  removal, **12.**10
  slow sand filters, **9.**5
Ammonium alum, **A.**3
Ammonium aluminum sulfate, **A.**3, **A.**20
Ammonium hydroxide, **15.**42, **A.**4, **A.**21
Ammonium silicofluoride, **A.**21
Ammonium sulfate, **10.**25, **15.**42, **A.**4, **A.**21
Amperometric chlorine residual analyzer, **10.**33
Amperometric wet chemistry chlorine residual analyzers, **20.**38–**20.**39
Analytical instrumentation, **20.**38–**20.**41
Analyzer process time lag, **20.**41
Analyzers and detectors, **20.**38–**20.**41
Anchor bolt embedments, **19.**10
Anchor ice, **4.**23
Ancillary feed equipment, **15.**32–**15.**37
Anderson, Jerry L., **15.**1, **A.**1
Anhydrous ammonia, **A.**2–**A.**3, **A.**21
  accidental release regulatory thresholds, **15.**44
  characteristics, **15.**10
  oxidation and disinfection, **10.**24
  protective equipment, **15.**42
  shipment/storage, **15.**6
  temperature, **A.**3
  threshold planning quantities, **15.**44
Anhydrous ammonia systems, **10.**24
Anion exchange process, **12.**12–**12.**30
Anion resins, **12.**3, **12.**13, **12.**31, **12.**51–**12.**52
Anion softeners, **12.**13
Anionic polymers, **8.**3
Anthracite, **8.**5
Anthracite sand filter media, **8.**6
AOC, **14.**29–**14.**30
AOP, **29.**8
Application and device interfaces, **20.**27
Applied pressure, **13.**8

Aqua ammonia, **A.**4
  accidental release regulatory thresholds, **15.**44
  piping, **15.**35
  oxidation and disinfection, **10.**24–**10.**25
  threshold planning quantities, **15.**44
Aqueous chlorine-sodium chlorite system, **10.**36, **10.**37
Architectural design, **3.**8, **18.**1–**18.**19
  administrative facilities, **18.**5–**18.**6, **18.**7
  aesthetics/contexturalism, **18.**1–**18.**2
  architectural programming, **18.**4
  building panel systems, **18.**15
  codes and standards, **18.**2–**18.**3
  control system, **20.**15–**20.**18
  environmental considerations, **18.**3, **18.**15–**18.**16
  facilities design, **18.**5–**18.**14
  interior finishes, **18.**16–**18.**17
  laboratories, **18.**8–**18.**11
  maintenance facilities, **18.**12–**18.**14
  person with disabilities, **18.**5
  personnel facilities, **18.**6–**18.**7
  plant design, **23.**24–**23.**25
  preengineered structures, **18.**14–**18.**15
  professional organizations, **18.**17–**18.**19
  sealing of documents, **18.**3
  structures over process units, **18.**14
Army Corps of Engineers permits, **24.**13, **24.**14
Arn, Thomas E., **27.**1
Arrangement of plant facilities, **23.**22–**23.**26
  architectural considerations, **23.**24–**23.**25
  campus layout, **23.**24, **23.**25
  compact layout, **23.**24, **23.**25–**23.**28
  factors to consider, **23.**22–**23.**23
  linear layout, **23.**23–**23.**25
  operational issues, **23.**28–**23.**31
  provision for future changes, **23.**35–**23.**36
  security, **23.**31, **23.**35
  support facilities, **23.**36
  technical issues, **23.**31–**23.**34
Arrays, **13.**31–**13.**32, **13.**37
Arsenate, **12.**15–**12.**17
Arsenate dumping, **12.**17
Arsenic, **12.**15–**12.**17, **17.**14–**17.**15
"Arsenic Treatability Options and Evaluation of Residuals Management Issues," **17.**15
Arsenite, **12.**15–**12.**17
AS-i bus, **20.**21
Ascorbic acid, **A.**4, **A.**21
Ash content, **14.**3
Assimilable organic carbon (AOC), **14.**29–**14.**30
Attleboro (Mass.) water treatment plant, **8.**19
Auto mode, **20.**4
Automated filter backwash control P&ID, **20.**28
Automatic flushing systems, **13.**34

Automation, **25.**5–**25.**6
 (*See also* Process instrumentation and controls)
Autotransformers, **21.**5
Auxiliary surface wash, **8.**14–**8.**16, **8.**18
Available filter head (AFH), **16.**2, **16.**11
Average flow HGL, **16.**2
A/W ratio, **5.**3, **5.**14
AWWA chemical standards, **15.**45
AWWA manuals, **8.**37
AWWA Sourcebook, **7.**8
AWWA standards, **A.**30–**A.**31
AWWARF research reports, **8.**37

BAC, **9.**14, **14.**29
BAC adsorption, **2.**10
Background checks, **30.**7
Backing screen, **9.**22
Backmix reactor, **6.**8
Backwash turbidimeters, **20.**40
Backwash waste, **17.**9
Backwashing:
 CEB, **13.**38
 dirty water, **16.**14
 filter backwash controls, **20.**11–**20.**13
 filter backwash system, **16.**13–**16.**14
 filter backwash water sludge, **17.**4–**17.**5, **17.**20–**17.**28
 GAC, **14.**13, **14.**29, **28.**16–**28.**18
 manifold backwash collection, **28.**19
 pumps, **16.**15
 UF/MF units, **13.**38
Backwashing granular filters, **8.**13
Bacteria, viruses, cyst removal, **2.**10
Baffles, **6.**20–**6.**22
Baffling classification, **10.**8
Baffling conditions in basins, **10.**7
Baker, Stephen F., **25.**1
Ballasted flocculation, **7.**24–**7.**25
Ballasts, **29.**13
Bar racks, **4.**31, **4.**32
Barium, **12.**10–**12.**12
Barthuly, Peter J., **16.**1
Baseline filtration options (rapid sand filters), **2.**13
Basins, vaults, large conduits, channels, **19.**10–**19.**12
Basket screen, **4.**31
Batch mixing, **15.**20, **15.**22
Bathtub curve, **22.**4–**22.**5
Bauxite, **15.**42
Belt feeders, **15.**20
Belt presses, **17.**43–**17.**45, **17.**46
Belt-type gravimetric feeders, **15.**19
Bench-scale testing, **28.**3
 equipment, **28.**4–**28.**5
 membrane processes, **13.**47
 purpose, **28.**4
 residuals handling, **17.**67
 UV disinfection, **29.**20–**29.**21
Beneficial reuse, **17.**11–**17.**12
Bentonite, **15.**42, **A.**4, **A.**21
Benzene, **5.**10
Bergman, Robert A., **13.**1
Bicarbonate, **12.**14
Bid administration, **3.**10
Bid documents, **3.**10
Binding frame closures, **9.**23
Biodosimetry, **29.**43–**29.**44
Biologically activated carbon (BAC), **9.**14, **14.**29
Biosolids, **17.**11
Bleaching powder, **A.**6
Blending, **13.**33
Block underdrain, **8.**23–**8.**25, **14.**20
Bone char, **12.**22
Bourdon tube switches and transmitters, **20.**37
Bowl centrifuge, **17.**45, **17.**47
Brady, Richard D., **14.**1
Breakpoint, **10.**15
Breakthrough, **14.**13
Brick making, **17.**11–**17.**12
Brine disposal options, **17.**12
Brine reclamation, **12.**45
Brine wastes, **17.**7–**17.**9
Broder, Michael V., **9.**1
Bromamine, **10.**17
Bromate, **2.**6, **10.**9
Bromide, **10.**18
Bromochloramine, **10.**17, **10.**18
Bromodichloromethane, **5.**10, **10.**9
Bromoform, **5.**10, **10.**9
Bubble point theory, **13.**39
Bubble-type aerators, **5.**4
Bubbler system, **20.**35–**20.**36
Budget, **25.**18–**25.**19
Buffer area, **23.**8
Building cladding system, **19.**13
Building code, **15.**48
Building enclosure, **19.**13
Building maintenance, **25.**22
Building panel systems, **18.**15
Buildings/superstructures, **19.**12–**19.**13
Buried line sensors, **30.**4
Burn-in, **22.**4, **22.**5
Butterfly valve, **16.**13
By-product removal, **2.**11
Byron, John C., **9.**1

Cable insulation, **21.**10
Cable-suspended level transmitters, **20.**35
Cake resistance, **13.**35
Calcite, **A.**4
Calcium carbonate, **15.**42, **A.**4
Calcium carbonate control, **13.**16–**13.**18
Calcium chloride, **A.**5

INDEX

Calcium fluoride, **A.21**
Calcium hydroxide, **15.**35, **A.21**
Calcium hypochlorite, **A.5, A.22**
  piping, **15.**35
  protective equipment, **15.**42
  storage, **15.**23
Calcium oxide, **A.**12, **A.22**
Calcium thiosulfate, **A.**5–**A.6, A.22**
Calculated UV dose approach, **29.**15
Calendering, **9.**23
Campus plant layout, **23.**24, **23.**25
Canyon Lake, **4.**11, **4.**14
Capacitative pressure transmitters, **20.**37
Carbon, **4.**31, **6.**4
  activated (*see* Activated carbon)
  BAC, **9.**14, **14.**29
  dilution in water, **15.**27
  GAC (*see* Granular activated carbon)
  PAC (*see* Powdered activated carbon)
  slow and filtration, **9.**5
Carbon bed depth, **14.**28
Carbon dioxide, **A.6, A.22**
  Henry's constant atm, **5.**9–**5.**10
  membrane processes, **13.**42
  piping, **15.**35
  protective equipment, **15.**42
  shipment/storage, **15.**6
Carbon dust, **14.**9
Carbon loading, **14.**35
Carbon losses, **14.**30–**14.**31
Carbon reactivation system, **14.**33–**14.**37
Carbon slurries, **15.**27–**15.**29
Carbon slurry feed system, **15.**27, **15.**28
Carbon tetrachloride, **5.**10
Carbon usage rate (CUR), **14.**15, **14.**35
Carbonate hardness, **11.**2
Carbonates, **12.**13
Carboxylic acids, **12.**30
Carrying water, **15.**38
Cascade aerators, **5.**2–**5.**3
Cascade level flow control, **20.**11
Castro, Kevin, **8.**1
Catalytic reduction, **10.**58
Cathodic protection systems, **25.**11
Cation exchange process, **12.**9–**12.**12
Cation resin, **12.**3, **12.**51
Cation resin fouling, **12.**6
Cationic polymers, **6.**3
Caustic potash, **A.**14
  (*See also* Potassium permanganate)
Caustic soda, **A.**17
  characteristics, **15.**10
  dilution, **15.**38
  feeding system, **15.**16
  freezing points, **15.**15
  lime softening, **11.**4–**11.**5
  storage, **15.**13
  (*See also* Sodium hydroxide)
CCR exchangers, **12.**39–**12.**40

CCTV, **30.**2, **30.**5–**30.**6
CEB, **13.**38
Cell pairs, **13.**3
Cement manufacturing, **17.**11
Centrate, **17.**47
Centrifugal pumps, **12.**45, **15.**25, **15.**27, **15.**31
Centrifuge, **17.**45–**17.**50
Centrifuge loading, **17.**47
Certification seals, **12.**2
Cesium, **12.**27
CFD modeling, **29.**28–**29.**29
CFE turbidity limit, **8.**2
CFR, **12.**12, **22.**4
Chabasite, **12.**4
Chain-and-flight collectors, **7.**5, **7.**9, **7.**10
Chain-and-flight skimmer, **7.**39
Chain-and-flight-type cross collector, **7.**11–**7.**12
Challenges:
  integrated treatment systems, **1.**2–**1.**3
  multidiscipline teams, **1.**3–**1.**4
  new technologies, **1.**3
  project delivery, **1.**4
  regulatory uncertainties, **1.**3
Change order, **26.**10
Channels, **16.**9
Channels, conduits, etc., **19.**10–**19.**12
Chemical accident prevention provisions, **24.**15
Chemical addition, **17.**52–**17.**54
Chemical budget, **25.**19
Chemical conditioning zone, **17.**43
Chemical diffusers, **15.**39–**15.**41
Chemical feed rates, **15.**1–**15.**2
Chemical feeding systems, **15.**17–**15.**41
  activated carbon, **15.**22–**15.**23
  ancillary feed equipment, **15.**32–**15.**37
  batch mixing, **15.**20, **15.**22
  calcium hypochlorite, **15.**23
  carbon slurries, **15.**27–**15.**29
  chlorinators, **15.**18–**15.**19
  day tanks, **15.**33
  decanting feeders, **15.**37
  diffusion of chemicals, **15.**39–**15.**41
  dilution of chemicals, **15.**38
  dissolving tanks, **15.**37
  dry chemical feed equipment, **15.**19–**15.**24
  gas feed equipment, **15.**18–**15.**19
  gravimetric feeders, **15.**19
  gravity orifice feeders, **15.**37
  lime slakers, **15.**20, **15.**22
  lime slurries, **15.**29–**15.**31
  liquid chemical feed equipment, **15.**24–**15.**37
  O&M, **25.**9–**25.**10
  piping, **15.**33–**15.**37
  polymers, **15.**32
  pot feeders, **15.**37
  rotameters, **15.**37
  slurries, **15.**27–**15.**32
  soda ash, **15.**23
  sodium chloride, **15.**23–**15.**24

sodium chlorite, **15**.23
transfer pumps, **15**.32–**15**.33
utility water, **15**.37–**15**.39
volumetric feeders, **15**.19–**15**.20
Chemical fouling, **12**.52
Chemical mixing:
  air mixing, **6**.11–**6**.13
  hydraulic mixing, **6**.13
  in-line blenders, **6**.9–**6**.11
  induction mixing, **6**.13–**6**.14
  mechanical mixing, **6**.7–**6**.9
  mixing intensity, **6**.6
  rapid mixing systems, **6**.6–**6**.14
  temperature effects, **6**.6
Chemical oxidation treatment, **10**.10–**10**.13
Chemical pumps, **12**.45
Chemical reduction, **10**.58
Chemical scrubbers, **15**.49, **17**.57
Chemical stabilization, **17**.56, **A**.30–**A**.31
Chemically enhanced backwash (CEB), **13**.38
Chemicals/chemical handling, **15**.1–**15**.3
  ammonia, **15**.6
  AWWA standards, **A**.30–**A**.31
  carbon dioxide, **15**.6
  chlorine, **15**.3–**15**.6
  chlorine gas leak containment, **15**.51
  dechlorinating agents, **15**.51–**15**.52
  description of individual chemicals,
    A.1–**A**.19
  diffusion of chemicals, **15**.39–**15**.41
  dry chemicals, **15**.8–**15**.10
  environmental impact, **24**.6–**24**.7, **24**.14
  evaporators, **15**.6–**15**.7
  feeding systems, **15**.17–**15**.41
    (*See also* Chemical feeding systems)
  inventory of chemicals, **22**.13
  liquefied gases, **15**.3–**15**.8
  liquid chemicals, **15**.10–**15**.17
  location of hazardous chemicals,
    23.33–**23**.34
  new chemicals, **15**.51
  oxygen, **15**.6
  process use, **A**.30–**A**.31
  quality, **A**.19
  receiving shipments, **15**.3
  recent trends, **15**.50–**15**.52
  reliability, **22**.13
  safety, **15**.41–**15**.50
    (*See also* Safety—chemicals)
  sizing storage and feed systems, **15**.1–**15**.3
  sodium hypochlorite systems, **15**.51
  trade names, **A**.20–**A**.29
Chick, Harriette, **10**.1, **10**.13
Chick-Watson theory, **10**.13
Chloramination, **10**.14–**10**.18
Chloramines, **10**.9, **10**.14
Chloride, **2**.7
Chlorinated copperas, **15**.35
Chlorinated lime, **A**.6, **A**.23

Chlorination, **10**.13–**10**.34
  ammonium sulfate systems, **10**.25
  anhydrous ammonia systems, **10**.24
  aqueous ammonia systems, **10**.24–**10**.25
  chemicals, **A**.30
  chloramination, **10**.14–**10**.18
  chlorine chemistry, **10**.14
  chlorine gas feeders, **10**.20–**10**.22
  chlorine gas system, **10**.18
  chlorine injector, **10**.22
  chlorine piping and valving, **10**.22–**10**.24
  chlorine residual analyzers, **10**.32–**10**.33
  control system, **10**.30–**10**.32
  controls, **20**.13–**20**.14
  CT credit, **10**.34
  gas induction units, **10**.22
  membrane processes, **13**.22–**13**.23
  misting, **10**.18
  on-site generation of chorine, **10**.28–**10**.30
  sodium hypochlorite systems, **10**.25–**10**.28
  storage, evaporators, **10**.18, **10**.20
  SWTR compliance, **10**.34
  vacuum regulators, **10**.20
  (*See also* Chlorine)
Chlorination control, **10**.30–**10**.32, **20**.13–**20**.14
Chlorinators, **10**.20–**10**.22, **15**.18–**15**.19
Chlorine, **4**.31, **10**.9, **A**.6–**A**.7, **A**.23
  accidental release regulatory thresholds,
    15.44
  gaseous, **10**.13, **15**.3
  iron and manganese removal, **10**.11–**10**.12
  physical constraints, **10**.2
  piping, **10**.22–**10**.24, **15**.35
  protective equipment, **15**.42
  safety measures, **15**.48, **15**.51
  shipment/storage, **15**.3–**15**.6
  temperature, **A**.8, **A**.9
  threshold planning quantities, **15**.44
  (*See also* Chlorination)
Chlorine analyzers and detectors, **20**.38
Chlorine chemistry, **10**.14
Chlorine cylinders, **10**.18, **15**.4–**15**.6
Chlorine demand, **10**.14
Chlorine dioxide, **4**.31, **10**.9, **A**.7, **A**.23
  accidental release regulatory thresholds,
    15.44
  oxidation, **10**.13
  physical constraints, **10**.2
  piping, **15**.35
  protective equipment, **15**.42
  threshold planning quantities, **15**.44
Chlorine dioxide disinfection, **10**.35
Chlorine dioxide generation, **10**.35–**10**.37
Chlorine dioxide systems, **10**.34–**10**.39
  chlorine dioxide disinfection, **10**.35
  chlorine dioxide generation, **10**.35–**10**.37
  sodium chlorite, **10**.38–**10**.39
Chlorine gas feeders, **10**.20–**10**.22
Chlorine gas leak containment, **15**.51

Chlorine gas piping, **15**.19
Chlorine gas system, **10**.18
Chlorine injector, **10**.22
Chlorine-intolerant composite membranes, **13**.34
Chlorine leak detectors, **20**.39
Chlorine piping and valving, **10**.22–**10**.24
Chlorine residual analyzers, **10**.32–**10**.33, **20**.38–**20**.39
Chlorine scrubbing system, **15**.49–**15**.50
Chlorine tank barges, **15**.4
Chlorite, **2**.6, **10**.9
Chloroform, **5**.10, **10**.9
Chloromethane, **5**.10
Cilensek, Ronald F., **26**.1
CIP systems, **13**.38
CIP waste, **17**.9
Circular adsorbers, **14**.26
Circular sedimentation basins, **7**.19–**7**.22
Circular sludge collector units, **7**.9–**7**.10
Circulation diagram, **18**.4
Civil process design, **3**.6–**3**.7
Clackamas River, **4**.27
Clarification:
  applications, **7**.1
  ballasted flocculation, **7**.24–**7**.25
  basin depth/velocities, **7**.4
  categories, **7**.1
  circular basins, **7**.19–**7**.22
  contact, **7**.40–**7**.43
  controls/monitoring, **20**.9, **20**.10
  DAF, **7**.36–**7**.40
  design, **7**.1–**7**.22
  detention time, **7**.4
  high-rate, **7**.22–**7**.28
  overflow rates, **7**.3
  plate settlers, **7**.24–**7**.25
  pulsed blanket clarifier, **7**.32–**7**.35
  rectangular basins, **7**.4–**7**.19
  sedimentation theory, **7**.2–**7**.3
  sludge blanket, **7**.32–**7**.36
  solids contact/slurry recirculation units, **7**.28–**7**.31
  square basins, **7**.22
  tube settlers, **7**.23–**7**.24
Clean Air Act Amendments of 1990, **15**.44
Clean water HGLE, **16**.2
Cleaning-in-place, **13**.38
Climate considerations, **23**.34
Clinoptilolite, **12**.4
ClorTec on-site hypochlorite-generating process, **10**.29
Closed-circuit television (CCTV), **30**.2, **30**.5–**30**.6
Closed-loop control, **20**.5
Closed-loop system, **15**.32
Cluster layout, **23**.25
CMMS, **25**.11

Coagulant alum/ferric chloride system, **15**.14
Coagulant chemicals, **6**.2, **6**.3
Coagulant recovery, **17**.12
Coagulant sludge, **17**.5–**17**.6
Coagulants and coagulant aids, **15**.35, **A**.30
Coagulation:
  chemical selection, **6**.3–**6**.4
  coagulant chemicals, **6**.3
  color, **2**.10
  defined, **6**.1
  flocculation aids, **6**.3
  membrane technology, **6**.5
  pH, **6**.4–**6**.5
  pilot plant, **28**.7, **28**.10–**28**.11
Coal-based carbons, **14**.3
Coarse screening, **4**.31, **4**.32, **23**.19
Coarse-to-fine media, **8**.5–**8**.6
Cobalt, **12**.27
Cocurrent centrifuge, **17**.47
Cocurrent exchangers, **12**.36–**12**.37
Cocurrent regeneration, **12**.33–**12**.35
Coflow exchanger, **12**.36–**12**.37
Coflow regeneration, **12**.33–**12**.35
Cold lime softening (*see* Lime softening)
Collector wells, **4**.7–**4**.9
Color:
  architectural design, **18**.16
  DE filtration, **9**.18
  maximum containment level, **2**.7
  piping, **18**.16
  process components, **2**.10
  slow sand filtration, **9**.4
Colorimetric chlorine residual analyzers, **20**.39
Combined available chlorine, **20**.38
Combined filter effluent (CFE), **8**.2
Combined residual chlorine, **10**.14
Commercial Application and Marketing of Water Treatment Plant Residuals, **17**.12
Common multipurpose treatment measures, **2**.12
Common treatment process, **2**.10–**2**.11
Communications equipment, **20**.24
Compact plant layout, **23**.24, **23**.25–**23**.28
Compensation, **23**.17–**23**.18
"Compliance Guidance and Model Risk Management Program for Water Treatment Plants," **15**.45
Composting, **17**.11
Compound 90° bends, **16**.13
Compound chlorination control loop, **20**.14
Compound control, **20**.10
Compound-loop control, **10**.32
Compound pH control loop, **20**.10
Compound turbine meters, **20**.34
Compression zone, **17**.44–**17**.45
Computational fluid dynamics (CFD) modeling, **29**.28–**29**.29
Computer control, **20**.4
Computer manual, **20**.4

Computer software/hardware, **25.**11–**25.**12
Computerized maintenance management system (CMMS), **25.**11
Concentrate pressure, **13.**8
Concentrate staging, **13.**31–**13.**32
Concentration and sidestreams, **17.**57
Concentration polarization, **13.**8
Concentration ratio, **12.**46
Conceptual level cost estimate, **26.**2
Concrete, **19.**3
Concrete joints, **19.**5
Conductance probe level switches, **20.**35
Conductivity detectors, **20.**41
Conduit, **8.**32
Cone aerators, **5.**3
Confined spaces, **15.**46, **15.**49
Constant failure rate (CFR), **22.**4
Constant-rate filter with influent splitting and constant water level, **8.**9–**8.**10
Constant-rate filter with influent splitting and varying water level, **8.**10, **8.**11
Constant-rate filter with rate-of-flow controller, **8.**9
Construction administration, **3.**10–**3.**11
Construction approvals, **3.**11
Construction assistance:
 bid administration, **3.**10
 construction administration, **3.**10–**3.**11
 construction approvals, **3.**11
 construction finalization, **3.**12
 coordination of design disciplines, **3.**11
 equipment start-up, **3.**12
 facilities start-up, **3.**11
 operator training, **3.**12
Construction Contract Administration, **3.**6
Construction cost estimating, **26.**1–**26.**10
 ADA, **26.**8
 contingencies, **26.**9
 cost escalation, **26.**10
 cost estimating factors, **26.**4, **26.**6
 cost indexes, **26.**10
 estimating methodologies, **26.**4–**26.**6
 finalizing the estimate, **3.**9–**3.**10, **26.**8–**26.**10
 geotechnical/site constraints, **26.**7
 hazardous waste audit, **26.**8
 historical cost data, **26.**4, **26.**5
 levels of estimates, **26.**2–**26.**4
 remote site locations, **26.**7
 sales tax, **26.**8
 scope creep/change orders, **26.**9–**26.**10
Construction cost index, **26.**10
Construction finalization, **3.**12
Construction joints, **19.**5
Construction permit, **24.**12–**24.**13
Contact clarification, **7.**40–**7.**43
Contact lenses, **15.**47
Contact time, **10.**5–**10.**8
Contingencies, **26.**9

Contingency planning (process residuals), **17.**63, **17.**64–**17.**65
Continuous backwashing filter beds, **8.**16
Continuous control, **20.**4
Continuous purge bubbler level transmitters, **20.**35–**20.**36
Contract flocculation, **6.**23
Contracting out, **12.**54
Contraction joint, **19.**5
Control (*see* Process instrumentation and controls)
Control center equipment, **20.**22
Control drawdown, **4.**25
Control hierarchy, **20.**2–**20.**4
Control joints, **19.**4
Control room ergonomics, **25.**7
Control system design, **20.**15–**20.**29
 alarm systems, **20.**26–**20.**27
 application and device interfaces, **20.**27
 architectural issues, **20.**15–**20.**18
 hardware, **20.**19–**20.**22
 information flow to computer application, **20.**18
 local control panels, **20.**17–**20.**18
 operator interfaces/control centers, **20.**17
 power reliability/backup power, **20.**23
 premature obsolescence, **20.**22
 redundancy/availability, **20.**17
 security, **20.**23–**20.**24
 signal errors, **20.**24–**20.**25
 software, **20.**18–**20.**19, **20.**25–**20.**28
 specifications/design documents, **20.**27–**20.**29
 surge protection, **20.**24
 system sizing/performance issues, **20.**17
 trend displays, **20.**27
 unattended plant operation, **20.**17
Control system hardware, **20.**19–**20.**22
Control system software, **20.**18–**20.**19, **20.**25–**20.**28
Controlled-volume pumps, **15.**31
Conventional softening basins, **11.**13
Conveying water, **15.**38
Conveyor system, **17.**53
Coordination of design disciplines, **3.**6–**3.**9, **3.**11
Copper, **2.**7
Copper sulfate, **A.**7, **A.**23
 piping, **15.**35
 protective equipment, **15.**42
Corbin, Darryl J., **23.**1
Corner floc, **7.**22
Corner sweeps, **7.**10, **7.**22
Corrosion control:
 chemicals, **A.**30–**A.**31
 membrane processes, **13.**42
 process components, **2.**11
 structural design, **19.**6

Cost escalation, **26**.10
Cost estimating (*see* Construction cost estimating)
Cost indexes, **26**.10
Cotton, Christine A., **29**.1
Countercurrent centrifuge, **17**.47
Countercurrent regeneration, **12**.34–**12**.35
Countercurrent regeneration (CCR) exchanges, **12**.39–**12**.40
Covert IDS, **30**.4
Cracks, **8**.39
Cranes/hoists, **25**.14
Crane's Technical Paper No. 410, 1988, **10**.23
Crib-type intake facilities, **4**.6, **4**.8
Criteria weighting, **23**.20–**23**.21
Cross collectors/cross hoppers, **7**.11–**7**.12
Cross-flow baffle, **6**.20
Cross-flow baffle flocculate, **6**.21–**6**.22
Cross-flow velocity, **13**.37
Cross-wash, **8**.33
Crossflow tower, **5**.2
Cryogenic air separation, **10**.49–**10**.52
Cryptosporidium cyst, **8**.29, **9**.1, **9**.16
Crystal alum, **A**.3
CT credit, **10**.34
CT value, **10**.3–**10**.4
CUR, **14**.15, **14**.35
Custom-designed process units, **3**.3
Custom SCADA/DCS software, **20**.19

D/DBP Rule, **2**.5–**2**.7, **10**.9–**10**.10
DAF, **7**.36–**7**.40, **28**.8, **28**.13–**28**.14
Dampproofing, **19**.6
Darcy equation, **10**.23
Darcy-Weisbach equation, **16**.6
Darcy's law, **13**.35
Day tanks, **15**.33
DCS, **20**.18
DE, **9**.27
DE filtration (*see* Diatomaceous earth filtration)
DE recovery system, **9**.28
DE slurry, **9**.27–**9**.28
Dead band, **20**.4
Deadweight, **4**.28
Dealkalization, **12**.13–**12**.15
Decanting, **17**.36
Decanting feeder, **15**.37
Dechlorinating agents, **15**.51–**15**.52
Dechlorination, **A**.30
Decision software, **23**.22
Declining-rate filters, **8**.10, **8**.11
Decreasing failure rate (DFR), **22**.4
Deep foundations, **19**.8
Deep U-tube contactor, **10**.55
Deep wall injection, **17**.14
Definitive level cost estimate, **26**.3–**26**.4
Defluoridation, **12**.22

Degasification, **12**.36, **13**.42
Degree of hardness, **11**.2
Deionization, **12**.30–**12**.32
Delivery bags, **9**.27
Delphos, Paul J., **6**.1
Delta-P, **13**.8
Demineralization, **12**.30–**12**.32
Demineralized stream pressure, **13**.8
Demineralizers, **12**.33–**12**.42
 (*See also* Regenerating resins)
Design and construction:
 construction assistance, **3**.10–**3**.12
 final design, **3**.5–**3**.10
 preliminary design, **3**.1–**3**.5
 (*See also individual subject headings*)
Design-build, **1**.4, **3**.10
Design drawings and specifications, **3**.5–**3**.6
Design reference books, **1**.6
Design reliability features, **22**.1–**22**.16
 A-B syndrome, **22**.5–**22**.6
 bathtub curve, **22**.4–**22**.5
 chemical storage, **22**.13
 definition, **22**.1–**22**.2
 disasters, **22**.8–**22**.9
 failure analysis, **22**.4–**22**.5
 flexibility, **22**.3
 gravity flow *vs.* pumping, **22**.6–**22**.7
 hydraulic reliability, **22**.13
 mechanical/electrical reliability, **22**.7
 operation and maintenance, **22**.14–**22**.15
 plant utilities, **22**.11–**22**.13
 process performance, **22**.13
 process reliability, **22**.10–**22**.11
 reserve process capacity, **22**.2
 standby units, **22**.2
 supply reliability, **22**.10
 unit process configuration, **22**.5–**22**.6
Design reviews, **3**.9–**3**.10, **25**.12
Design standards, **3**.6
Desorption, **5**.1
 (*See also* Aeration and stripping)
Detention time, **7**.4
Deterministic network design, **20**.20
Device networks, **20**.21–**20**.22
Devicenet, **20**.21
Dewatering processes, **17**.31–**17**.48
 belt presses, **17**.43–**17**.45, **17**.46
 centrifuge, **17**.45–**17**.50
 drying beds, **17**.32–**17**.35
 filter presses, **17**.39–**17**.44
 holding phase, **17**.36
 lagoons, **17**.35–**17**.36
 loading phase, **17**.36
 mechanical dewatering, **17**.39–**17**.48
 natural dewatering, **17**.31–**17**.39
 sizing natural dewatering processes, **17**.36–**17**.39
DFR, **22**.4

Diagrams/documentation, I&C system, **20**.27–**20**.29
Diamond Valley Lake inlet/outlet tower, **4**.10, **4**.11
Diaphragm filters, **17**.41
Diaphragm pumps, **15**.24
Diatomaceous earth (DE), **9**.27
Diatomaceous earth filters, **2**.14
Diatomaceous earth (DE) filtration, **9**.16–**9**.29
 ancillary facilities, **9**.27–**9**.29
 backing screen, **9**.22
 binding frame closures, **9**.23
 central drainage chamber, **9**.22
 color, **9**.18
 DE delivery, **9**.27
 design, **9**.19–**9**.24
 filter elements, **9**.20–**9**.21
 filter system, **9**.22–**9**.23
 iron and manganese, **9**.19
 outlet connections, **9**.24
 porosity control, **9**.17–**9**.18
 pressure filters, **9**.20, **9**.25–**9**.26, **9**.28–**9**.29
 recovery, **9**.28
 slurry conveyance, **9**.27–**9**.28
 slurry storage, **9**.27
 system controls, **9**.28–**9**.29
 three-step process, **9**.17
 vacuum filter leaves, **9**.23
 vacuum filters, **9**.20, **9**.25–**9**.26, **9**.28–**9**.29
 waste disposal, **9**.28
Diatomaceous earth recovery system, **9**.28
Dibromochloromethane, **5**.10, **10**.9
Dichloramine, **10**.15
Dichloroacetic acid, **10**.9
1,2-Dichloroethane, **5**.10
Dieldrin, **5**.10
Differential pressure, **13**.8
Differential pressure transmitter, **20**.30–**20**.31, **20**.37
Diffused Air Pilot Studies, **5**.22
Diffuser aerators, **5**.5, **5**.11–**5**.12
Diffuser contactors, **10**.55–**10**.57
Diffusion of chemicals, **15**.39–**15**.41
Diffusion-type aeration, **5**.4–**5**.5, **5**.11–**5**.12
Diffusive airflow test, **13**.39
Digital CCTV technology, **30**.6
Dilution of chemicals, **15**.38
Dimethylaminoethanol, **12**.25
Dimethylethanolamine, **12**.3
Dipper-wheel feeders, **15**.31–**15**.32
Direct-current motors, **21**.7
Direct filtration, **6**.2, **6**.14, **8**.4
Direct injection contactors, **10**.57–**10**.58
Direct thermal drying, **17**.48
Dirty backwash water, **16**.14
Dirty bombs, **12**.1, **12**.27
Disabilities, persons with, **18**.5, **23**.29–**23**.30, **26**.8

Disasters, **22**.8–**22**.9
Disinfectant contact requirements, **10**.3–**10**.8
Disinfectant monitoring requirements, **10**.8–**10**.9
Disinfectants/Disinfection By-products Rule, **2**.5–**2**.7, **10**.9–**10**.10
Disinfection, **20**.9, **20**.13–**20**.15, **A**.30
 (*See also* Oxidation and disinfection)
Disinfection by-product control, **2**.10
Disk (drum) screen, **4**.31
Disodium phosphate, **A**.23
Displacement (decanting) feeder, **15**.37
Disposal:
 ion exchange system, **12**.45–**12**.46. **12**.50
 membrane processes, **13**.43–**13**.44
 off-gas, **10**.58–**10**.59
 residuals (*see* Process residuals)
Dissolved air flotation (DAF), **7**.36–**7**.40, **28**.8, **28**.13–**28**.14
Dissolved oxygen, **9**.4, **9**.5
Dissolving tanks, **15**.37
Distributed control systems (DCSs), **20**.18
Distribution of flow, **16**.11–**16**.12
Distributors, **5**.18, **5**.19
Dolomitic hydrated lime, **A**.12
Dolomitic lime, **15**.35, **A**.12
Dolomitic magnesite, **A**.12
Double-ended, **21**.7
Downflow fixed bed in parallel operation, **14**.24
Downflow fixed bed in series operation, **14**.22–**14**.24
Downflow gravel bed flocculator, **6**.24
DP transmitter, **20**.30–**20**.31, **20**.37
DPD, **10**.18
Draft-tube aerator, **5**.7
Drawing off underflow, **7**.17–**7**.18
Drawings and specifications, **3**.5–**3**.6
Drinking water standards (*see* Water quality goals)
Drinking water treatment processes, **2**.10–**2**.11
Dry chemicals, **15**.8–**15**.10
Dry chemical dust, **15**.9
Dry chemical feed equipment, **15**.19–**15**.24
Dry-type transformers, **21**.5
Drying, **17**.48, **17**.51
Drying beds, **17**.32–**17**.35
Drying bed loading sequence, **17**.41
Drying bed loading table, **17**.40
Drying bed operating layers, **17**.37
Dual-media bed, **8**.6
Dual-membrane system, **13**.12
Dumping, **12**.10
 ammonia, **12**.10
 arsenate, **12**.17
 nitrate, **12**.19–**12**.20
 selenium, **12**.26
Dust collector, **14**.37, **15**.9

Dyksen, John E., **5.**1
Dynamic ice, **4.**22

Earthquakes (seismic design), **19.**8–**19.**10
EBCT, **14.**13
Eberts, Ralph T., **19.**1
ED/EDR system, **13.**3–**13.**4, **13.**39–**13.**41
  applications, **13.**6
  differential pressure, **13.**41
  electric current, **13.**39
  electrode compartments, **13.**41
  feedwater analysis, **13.**14
  I&C equipment, **13.**44
  operating mode, **13.**41
  pH, **13.**41
  recovery, **13.**39–**13.**40
  solute removal, **13.**40
  staging, **13.**41
  temperature, **13.**40, **13.**41
  (*See also* Membrane processes)
Eductors, **15.**25
Effective size (ES), **8.**5
Effluent piping, **16.**8–**16.**9, **16.**12–**6.**13
EIS, **24.**12
Ejectors, **15.**25
Elastomers, **10.**27
Elder, Doug B., **11.**1
Electric infrared oven furnace, **14.**36
Electric motors, **21.**5–**21.**9
Electric power monitoring, **20.**15
Electric vaporizer, **10.**49, **10.**50
Electrical component reliability, **22.**7, **22.**11–**22.**13
Electrical design, **3.**8–**3.**9
Electrical distribution systems, **21.**3–**21.**4
Electrical maintenance, **25.**22
Electrical noise, **20.**24–**20.**25
Electrical systems, **21.**1–**21.**12
  acceptance testing, **21.**10–**21.**11
  codes/standards, **21.**1–**21.**2
  electrical distribution systems, **21.**3–**21.**4
  lightning protection, **21.**9–**21.**10
  maintenance, **21.**11–**21.**12
  motors and controllers, **21.**5–**21.**9
  power factor correction, **21.**10
  reliability, **22.**7, **22.**11–**22.**13
  security, **21.**9
  transformers, **21.**4–**21.**5
Electrical voltage-driven membrane processes, **13.**3–**13.**4
Electrochemical generators, **10.**36, **10.**38
Electrodialysis (ED)/electrodialysis reversal (EDR) (*see* ED/EDR system)
Electromagnetic flowmeters, **20.**31–**20.**32, **20.**33
Electronic O&M documents, **3.**12
Electronic security system (ESS), **30.**3–**30.**5
Elementary control diagrams, **20.**29

Embodied energy, **18.**15
Emergency Planning and Community Right-to-Know Act (EPCRA), **15.**44, **24.**14
Emergency Planning for Water Utility Management, **22.**9
Encrustation, **11.**16
End baffles, **6.**22
Energy monitoring, **20.**15
Energy use, **17.**66
Engineering News-Record, **26.**10
Engineer's estimate, **26.**3–**26.**4
Enhanced coagulation, **6.**2
Entamoeba histolytica cyst, **9.**16
Environmental impact, **24.**1–**24.**18
  air emissions, **24.**10–**24.**11, **24.**15
  chemical handling/storage, **24.**6–**24.**7, **24.**14
  oil storage, **24.**7–**24.**8
  permitting strategy/plan, **24.**16–**24.**18
  plant construction, **24.**1–**24.**5
  plant operations, **24.**5–**24.**12
  project permitting, **24.**12–**24.**18
  residuals treatment/disposal, **24.**8–**24.**9, **24.**14–**24.**15
  solid/hazardous waste management, **24.**11–**24.**12
  stormwater management, **24.**10
  wastewater treatment/disposal, **24.**9–**24.**10, **24.**15
  water withdrawals, **24.**6
Environmental impact statement (EIS), **24.**12
EPCRA, **15.**44, **24.**14
EPCRA threshold planning quantities, **15.**44
Epichlorohydrin dimethylamine (EPI-DMA), A.9
Equalization, **17.**21–**17.**23
Equalization basin layout, **17.**23
Equalization storage curve, **17.**22
Equilibrium, **5.**8
Equipment budget, **25.**19
Equipment start-up, **3.**12
Ergonomics, **25.**6–**25.**7
ES, **8.**5
Escalation of costs, **26.**10
ESS, **30.**3–**30.**5
ESSD wash trough with Type II MULTIWASH separator baffles, **8.**34
Estimating cost (*see* Construction cost estimating)
European-type deep-bed filters, **8.**22
Evaporators, **15.**6–**15.**7
Excess equipment, **25.**13
Excess inventory, **15.**2
Excess lime softening process, **11.**9–**11.**12
Exchange process theory, **12.**2
Exchange tank services, **12.**48, **12.**54
Exchangers, ion, **12.**33–**12.**42
  (*See also* Regenerating resins)
Exhaust emission, **5.**20–**5.**21

Exhaust system, **15**.49
Expansion joint, **19**.4
Explosion prevention, **15**.50
Exposed intake, **4**.3, **4**.4
Exposed intake design criteria, **4**.17
Exterior intrusion detection system, **30**.4
External regeneration, **12**.41
External soil loads, **19**.10
Eye protection, **15**.47
Eye wash, **15**.47–**15**.48

Face shield, **15**.47
Facilities and Operating Procedures for Chlorine Storage, **15**.5
Facilities design, **18**.5–**18**.14
Facilities start-up, **3**.11
Facility O&M manual, **27**.9–**27**.10, **27**.11
Failure analysis, **22**.4–**22**.5
Failure modes and effects analysis (FMEA), **22**.9
False-bottom underdrains, **8**.25–**8**.26
Faraday's law, **13**.39
FBRR, **17**.15, **17**.24
Feed gas preparation, **10**.44–**10**.46
Feed pressure, **13**.9
Feed rate, **15**.1–**15**.2
Feedback control, **10**.30–**10**.32
Feedback flow controller, **20**.4
Feedforward control, **10**.30–**10**.32, **20**.5–**20**.7
Feedwater acidification, **13**.17
Feedwater analysis, **13**.14
Feedwater LSI, **13**.17, **13**.18
Feedwater recovery, **13**.35–**13**.36
Feedwater suspended solids content, **13**.16
Feedwater turbidity, **13**.16
Fence vibration sensors, **30**.4
Ferric chloride, **6**.3, **A**.9–**A**.10, **A**.23, **A**.24
  characteristics, **15**.10
  piping, **15**.35
  protective equipment, **15**.42
Ferric sulfate, **15**.42, **A**.10, **A**.24
Ferrous iron, **29**.16
Ferrous iron in groundwater, **5**.22
Ferrous sulfate, **15**.42, **A**.10
Fiber-optic cable, **30**.4, **30**.5
Fiberglass, **19**.3
Fiberglass-reinforced plastic trough, **8**.34
Field bus technology, **20**.21–**20**.22
50% alum, **A**.2
50-lb (22.7-kg) plastic-lined paper bags, **9**.27
Films, **5**.9
Filter, **25**.9
Filter-adsorbers, **14**.17–**14**.21
Filter aid polymer, **8**.3–**8**.4
Filter backwash controls, **20**.11–**20**.13
Filter backwash layout, **20**.12
Filter Backwash Recycling Rule (FBRR), **17**.15, **17**.24

Filter backwash system, **8**.12, **16**.13–**16**.14
Filter backwash water sludge, **17**.4–**17**.5, **17**.20–**17**.28
Filter belts, **17**.45
Filter box elevation, **16**.2–**16**.3
Filter efficiency, **8**.38
Filter effluent piping, **16**.8–**16**.9, **16**.12–**6**.13
Filter evaluation, **8**.41
Filter flow controls, **20**.11
Filter harrowing, **9**.15–**9**.16
Filter leaf drainage chambers, **9**.22
Filter media replacement, **8**.40–**8**.41
Filter media support gravel upset, **8**.39–**8**.40
Filter media washing, **8**.11–**8**.18, **8**.39
Filter presses, **17**.39–**17**.44
Filter rate control valve, **8**.32
Filter ripening, **8**.37–**8**.38
Filter roofing system, **9**.10
Filter septum, **9**.22–**9**.23
Filter-to-waste period, **8**.40
Filter-to-waste system, **9**.10
Filter-to-waste valve, **8**.32
Filter waste discharge, **16**.14
Filtered/non-filtered water, **23**.33
Filtering Material (AWWA Standard B100), **8**.4
Filtrate, **17**.30, **17**.41
Filtration, **2**.10, **2**.14, **20**.9, **20**.11–**20**.13
Filtration techniques (*see* Diatomaceous earth filtration; Rapid sand filtration; Slow sand filtration)
Final design:
  bid documents, **3**.10
  coordination of design disciplines, **3**.6–**3**.9
  design reviews, **3**.9–**3**.10
  design standards, **3**.6
  drawings and specifications, **3**.5–**3**.6
  final cost estimates, **3**.9–**3**.10
  project coordination, **3**.6
Financial feasibility evaluation, **3**.4
Fine-bubble diffuser contactor, **10**.55–**10**.57
Fine-bubble gas diffusers, **15**.40
Fine screens, **4**.31, **4**.32
Fine-to-coarse media, **8**.5
Finger weir, **7**.7
Finished water storage, **23**.23
Fire code, **15**.48
Fire Protection for Laboratories Using Chemicals, **18**.8
Fire protection publications, **21**.2
Fire resistance rating, **19**.13
First cell log flat inactivation credits, **10**.57
Fish protection, **4**.27–**4**.28, **4**.29
Fixed beds in parallel, **14**.24
Fixed beds in series, **14**.22–**14**.24
Fixed-leaf filter leaves, **9**.24
Fixed-leaf pressure filters, **9**.20, **9**.24
Fixed-volume presses, **17**.39, **17**.41

Flammable and Combustible Liquids Code, **18**.9
Flash mixing, **6**.2
Flash mixing facility, **6**.10
Flat-faced diaphragm pumps, **15**.24
Flat gasket, **9**.24
Flat-leaf DE filter, **9**.19–**9**.24
Flexibility, **22**.3, **23**.28–**23**.29
Flies, **17**.66
Float, **17**.29
Float-type level switches, **20**.35
Floated solids removal, **7**.39–**7**.40
Floating bridge-type collector, **7**.17, **7**.18
Floating intake, **4**.11, **4**.14
Floc, **6**.2
Flocculation, **6**.2, **28**.7, **28**.11–**28**.12
Flocculation aids, **6**.2, **6**.3
Flocculation basins, **25**.8
Flocculation process design, **6**.2, **6**.14–**6**.23
   contract flocculation, **6**.23
   energy requirements, **6**.16–**6**.17
   flocculation time, **6**.15–**6**.16
   hydraulic flocculation, **6**.20–**6**.23
   incidental flocculation, **6**.14–**6**.15
   mechanical flocculators, **6**.18–**6**.20
   sludge recirculation/solids contact, **6**.23
Flocculator clarifiers, **7**.20
Flooding:
   site selection, **23**.34
   reducing effects of, **22**.8
   river intakes, **4**.10
Flotation thickening, **17**.29–**17**.30
Flow control valves, **15**.32
Flow distribution, **16**.11–**16**.12
Flowmeters, **20**.30–**20**.34
Fluidized bed furnace, **14**.36
Flumes, **20**.34
Fluoridation/fluoride adjustment, **A**.30
Fluoride, **2**.7, **12**.22
Fluoride residual measurement, **20**.41
Fluorosilicic acid, **15**.10, **15**.43, **A**.10, **A**.25
Fluorspar, **15**.43
Fluosilicic acid, **15**.35
Flushing system, **13**.34
Flux:
   defined, **13**.9
   RO/NF units, **13**.28–**13**.29
   UF/MF units, **13**.36
Flux decline, **13**.30
Flux decline curve, **13**.30
FMEA, **22**.9
Foaming agents, **2**.7
Foellmi, Steven N., **4**.1
Food locker, **18**.6
Fouling, **5**.21–**5**.22, **12**.6, **12**.52
Foundation Fieldbus, **20**.21, **20**.22
Foundation ties, **4**.28
FR, **22**.4

Frazil ice, **4**.22–**4**.25
Free available chorine, **20**.38
Free residual chlorine, **10**.13
Free residual chlorine curves, **10**.16
Freeboard, **16**.3
Freeze-assisted sand bed, **17**.34
Freezing point:
   caustic soda, **15**.15
   hypochlorite solutions, **15**.17
FRP tanks, **10**.27
Fuller, R. Gary, **3**.1
Fulton, George P., **2**.1
Future drinking water regulations, **2**.5, **2**.7
Fuzzy logic control, **20**.8

GAC (*see* Granular activated carbon)
GAC adsorption, **5**.21
GAC density, **14**.29
GAC filter-adsorbers, **14**.17–**14**.21
GAC postfilter adsorbers, **14**.22–**14**.28
   basic characteristics, **14**.25
   circular adsorbers, **14**.26
   configurations, **14**.22–**14**.25
   design, **14**.25
   fixed beds in parallel, **14**.24
   fixed beds in series, **14**.22–**14**.24
   number of adsorbers, **14**.26–**14**.28
   pulsed beds, **14**.24–**14**.25
   rectangular adsorbers, **14**.26
   underdrain system, **14**.28, **14**.31
   upflow expanded beds, **14**.24
GAC sandwich filter, **9**.14–**9**.15, **14**.21, **14**.22
GAC scrubbers, **17**.57
Galleries, **25**.16–**25**.17
GAN, **10**.50
Garnet sand, **8**.6
Gas absorption, **5**.1
   (*See also* Aeration and stripping)
Gas chlorine-sodium chlorite system, **10**.36, **10**.38
Gas feed equipment, **15**.18–**15**.19
Gas induction units, **10**.22
Gas masks, **15**.47
Gaseous chlorine, **10**.13, **15**.3
Gaseous nitrogen (GAN), **10**.50
Gaseous oxygen (GOX), **10**.49, **10**.50
Gasser, John, **27**.1
Gates, **16**.9–**16**.10
Gear pumps, **15**.24
Gelular resins, **12**.3–**12**.4, **12**.30
Generation:
   chlorine dioxide, **10**.35–**10**.37
   on-site (*see* On-site generation/regeneration)
   ozone, **10**.42–**10**.44
   (*See also* Regeneration)
Geosmin, **10**.13
Geotechnical investigation, **3**.7–**3**.8
Geotechnical report, **19**.6–**19**.7

Geotechnical/site constraints, **26.**7
Geren Island Treatment Facility, **4.**28, **4.**29
GFH, **12.**16
Giardia cyst, **8.**29, **9.**1, **9.**16
GIS, **23.**19
Glassy phosphates, **A.**17
Gloves, **15.**47
Gnats, **17.**66
Gottlieb, Michael C., **12.**1
GOX, **10.**49, **10.**50
gpm/ft^2, **8.**16
Granular Activated Carbon (AWWA Standard B604), **8.**4, **14.**2
Granular activated carbon (GAC), **14.**9–**14.**37, **A.**1–**A.**2
   adsorber volume/bed depth, **14.**13
   backwashing, **14.**13, **14.**29
   basic options, **14.**9–**14.**10
   biological activity, **14.**29–**14.**30
   breakthrough, **14.**13
   carbon bed depth, **14.**28
   carbon losses, **14.**30–**14.**31
   carbon storage/transport facilities, **14.**32–**14.**33
   CUR, **14.**15, **14.**35
   design flow rate/system hydraulics, **14.**11
   design of adsorber vessels, **14.**31–**14.**32
   EBCT, **14.**13
   effective size, **14.**3–**14.**4
   filter-adsorbers, **14.**17–**14.**21
   filtration/adsorption, **14.**10, **14.**17
   GAC sandwich filter, **9.**14–**9.**15, **14.**21, **14.**22
   gravity adsorbers, **14.**16–**14.**17
   maintenance requirements, **14.**30
   operating problems, **14.**28–**14.**30
   ozonation, **14.**31
   pilot testing, **28.**8, **28.**14–**28.**18
   postfilter adsorbers, **14.**22–**14.**28
   (*See also* GAC postfilter adsorbers)
   pressure adsorbers, **14.**16–**14.**17
   protective equipment, **15.**42
   regeneration, **14.**33–**14.**37
   removing carbon from adsorbers, **14.**29
   RSSCT, **14.**15
   slow sand filtration, **9.**14–**9.**15, **14.**21
   surface loading rate, **14.**13
Granular activated carbon (GAC) adsorption, **5.**21
Granular activated carbon (GAC) scrubbers, **17.**57
Granular ferric hydroxide (GFH), **12.**16
Granular filtration techniques, **8.**1
   (*See also* Diatomaceous earth filtration; Rapid sand filtration; Slow sand filtration)
Granular media filtration, **28.**8, **28.**14–**28.**18
Graphical information system (GIS), **23.**19

Gravel bed flocculator, **6.**23, **6.**24
Gravel-packed wells, **4.**7, **4.**9
Gravel upset, **8.**39–**8.**40
Gravimetric feeders, **15.**19
Gravity adsorbers, **14.**16–**14.**17
Gravity belt thickener, **17.**30–**17.**31
Gravity drainage zone, **17.**44
Gravity filter, **8.**20, **8.**23
Gravity filter control, **8.**9
Gravity filter piping, **8.**33
Gravity flow *vs.* pumping, **22.**6–**22.**7
Gravity orifice feeders, **15.**37
Gravity thickening, **17.**28–**17.**29
Green architecture, **18.**3
Green building rating system, **18.**3
Green sand, **12.**4
Ground loop, **20.**24
Ground Water Rule (GWR-2005), **2.**5, **2.**7, **10.**10
Grounds maintenance, **25.**22
Groundwater disinfection rule, **10.**10
Groundwater elevation, **19.**8
Groundwater under direct influence of surface water (GWUI), **10.**3
Groundwater withdrawals, **24.**6
Guidance Manual for Compliance with the Interim Enhanced Surface Water Treatment Rule Turbidity Provisions, **8.**29
Guiding Principles for Chemical Accident Prevention, Preparedness, and Response, **24.**7
GWR-2005, **2.**5, **2.**7, **10.**10
GWUI, **10.**3

Haloacetic acids (HAA5), **2.**7, **2.**8, **2.**11, **10.**9
Hard hats, **15.**47
Hardness, **11.**1
Hardness reduction, **2.**11
Hardwired interlocks, **20.**2
Harms, Leland L., **11.**1
Harrow tractor, **9.**16
Harrowing filters, **9.**15–**9.**16
HART, **20.**21
Hazard rate function, **22.**4
Hazardous chemicals (*see* Chemicals/chemical handling)
Hazardous waste, **24.**11
Hazardous waste audit, **26.**8
Hazen-Williams equation, **16.**6, **17.**18
Head loss, **8.**30
Heat drying process, **17.**50
Heating, ventilation, and air conditioning (HVAC), **15.**49
Height of transfer unit (HTU), **5.**14
Helical flow solids contact units, **11.**16
Helical-screw-type cross collector, **7.**11, **7.**12
Hemlock water filtration (Rochester), **8.**19
Henry's law, **5.**8–**5.**10

Henry's law constants, **5.9–5.10**
Hesby, James C., **10.**1
HGLE, **16.**2
HID lamp, **21.**9
Hierarchy of control, **20.**2–**20.**4
High-concentration wastes, **17.**51–**17.**52
High-energy flocculators, **6.**20
High-intensity ultrasound, **17.**26–**17.**27
High-mast lighting, **21.**9
High-pressure air preparation ozonation system, **10.**46, **10.**48
High-pressure membranes, **17.**24
High-pressure (utility) water, **15.**37–**15.**39
High-pressure zone, **17.**45
High-purity oxygen (LOX), **10.**46, **10.**49
High-rate clarification, **7.**22–**7.**28
High-rate granular media filtration (*see* Rapid sand filtration)
High-rate water wash, **8.**13, **8.**14
Historical cost data, **26.**4, **26.**5
Hoists, **25.**14
Holding phase, **17.**36
Hollow-fiber UF module, **13.**8
Hollow fine-fiber RO membrane module, **13.**6, **13.**7, **13.**32
Home water softeners, **12.**46, **12.**47
Hopper feeder, **15.**20
Horizontal groundwater collection system, **4.**7, **4.**9
Horizontal shaft flocculator, **6.**15, **6.**18
Horsley, Michael B., **11.**1
Hose mask, **15.**47
Hot-process softening, **11.**1
HTU, **5.**14
Human-caused disasters, **22.**9
HVAC system, **15.**49
Hydrated lime, **15.**43, **A.**12
Hydraulic flocculation, **6.**20–**6.**23
Hydraulic grade line (HGL), **16.**2
Hydraulic grade line elevation (HGLE), **16.**2
Hydraulic grade line maximum day flow (50 MGD) conventional rapid sand filter water treatment plant, **16.**5
Hydraulic mixing, **6.**13
Hydraulic overflow rate, **7.**3
Hydraulically balanced inlet cones, **4.**20
Hydraulics, **16.**1–**16.**15
  AFH, **16.**2, **16.**11
  computer programs, **16.**11
  design, **16.**1–**16.**5
  filter backwash system, **16.**13–**16.**14
  filter box elevation, **16.**2–**16.**3
  filter effluent piping, **16.**8–**16.**9, **16.**12–**16.**13
  filter waste discharge, **16.**14
  flow distribution, **16.**11–**16.**12
  freeboard, **16.**3
  head loss, **16.**15–**16.**11
    (*See also* Head loss)
  HGL, **16.**2
  HGLE, **16.**2
  intermediate/final design, **16.**4–**16.**5
  preliminary design, **16.**1–**16.**4
  preliminary HGL, **16.**3, **16.**4
  pumps, **16.**15
  reliability, **22.**13
  residuals handling/transportation, **17.**18
Hydrocarbons, **10.**45
Hydrochloric acid, **A.**10–**A.**11
  ion exchange, **12.**42
  piping, **15.**35
  protective equipment, **15.**43
Hydrofluorosilicic acid, **A.**10
Hydrogen fluoride, **A.**25
Hydrogen peroxide, **A.**11, **A.**25
  accidental release regulatory thresholds, **15.**44
  oxidation, **15.**51
  threshold planning quantities, **15.**44
Hydrogen sulfide, **5.**10, **13.**42
Hydrogen sulfide control, **13.**23
Hydroxides, **12.**13
Hydroxyl radical, **10.**40–**10.**41
Hypo pump, **10.**28
Hypochlorite (*see* Calcium hypochlorite; Potassium hypochlorite; Sodium hypochlorite)
Hypochlorous acid, **10.**25, **15.**13

IBC, **18.**2, **21.**2
Ice, **4.**21–**4.**25
Ice floes, **4.**22
Ice formation conditions, **4.**22
Ice jams, **4.**22
IDS, **30.**4
IEC 61131 programming, **20.**18, **20.**19, **20.**25
IESWTR, **2.**6, **10.**8–**10.**9
IFR, **22.**4
IMS cap, **8.**26
In-line blenders, **6.**9–**6.**11
In-line diffuser, **15.**40
In-line filtration, **6.**2
In-line jet mixers, **6.**9, **6.**11
In-plant networks, **20.**19–**20.**22
In-well aeration, **5.**7
Inactivation ratio, **10.**5
Inactive frazil ice, **4.**23
Incidental flocculation, **6.**14–**6.**15
Increasing failure rate (IFR), **22.**4
Indexing grid sludge removal system, **7.**12–**7.**13, **7.**14
Indirect thermal drying, **17.**48
Individual component system reliability, **22.**7–**22.**8
Induction mixing, **6.**13–**6.**14
Induction motor, **21.**5
Inflation, **23.**16–**23.**17
Influent conduit, **8.**32
Infrared furnace, **14.**36

Initial mixing, **6.**2
Injection testing, **21.**12
Inlet works and ports, **4.**16–**4.**19
Inlet zone (sedimentation basin), **28.**12
Inorganic, **2.**16
Inorganic ion exchangers, **12.**4
Inorganic organic chemical reduction, **2.**11
Inrush, **21.**5
Insects, **4.**30, **17.**66
Inside-out membranes, **17.**24
Instantaneous failure rate (FR), **22.**4
Instrument loop diagram, **20.**29
Instrumentation, **20.**30–**20.**41
Instrumentation design, **3.**9
Instrumentation maintenance, **25.**22
Intake conduit, **4.**19–**4.**21
Intake facilities:
 capacity, **4.**2, **4.**15
 chemical treatment, **4.**30–**4.**31
 component block diagram, **4.**2
 design objectives, **4.**2
 exposed systems, **4.**3, **4.**4
 fish protection, **4.**27–**4.**28, **4.**29
 flooding, **4.**30
 flotation, **4.**28
 geotechnical considerations, **4.**28–**4.**30
 ice, **4.**21–**4.**25
 inlet works and ports, **4.**16–**4.**19
 insects, **4.**30
 intake conduit, **4.**19–**4.**21
 lake and reservoir intakes, **4.**10–**4.**14
 layout, **4.**16
 location, **4.**3
 mechanical considerations, **4.**30
 O&M, **25.**7–**25.**8
 pumping facilities, **4.**21
 purpose, **4.**1
 quagga mussel, **4.**26
 racks and screens, **4.**31–**4.**33
 redundancy, **4.**2–**4.**3
 reliability, **4.**2–**4.**3
 river intakes, **4.**4–**4.**10
  (*See also* River intakes)
 submerged systems, **4.**3, **4.**4
 submergence design, **4.**15–**4.**16
 ventilation systems, **4.**30
 zebra mussel, **4.**25–**4.**26
Intake port selection, **4.**16
Integral media support (IMS) cap, **8.**26
Integrated treatment systems, **1.**2–**1.**3
Interconnected parallel production trains, **22.**6
Interconnection diagram, **20.**29
Interim Enhanced Surface Water Treatment Rule (IESWTR), **2.**6, **10.**8–**10.**9
Interim start-up planning, **27.**10
Interior finishes, **18.**16–**18.**17
Internal service loads, **19.**10
International Building Code (IBC), **18.**2, **21.**2
Interstage pressure, **13.**9

Intrusion detection sensor (IDS), **30.**4
Inventory size, **15.**2
Iodine, **12.**28
Iodine number, **14.**2
Ion exchange applications, **2.**11, **12.**1–**12.**61
 ammonia, **12.**10
 anion exchange process, **12.**12–**12.**30
 arsenic, **12.**15–**12.**17
 automation, **12.**47
 barium, **12.**10–**12.**12
 basic theory, **12.**2, **12.**4–**12.**5, **12.**6–**12.**8
 cation exchange process, **12.**9–**12.**12
 centralized treatment systems, **12.**47–**12.**48
 certification seals, **12.**2
 CFR, **12.**12
 chemical storage/dilution, **12.**44–**12.**45
 contracting out, **12.**54
 dealkalization, **12.**13–**12.**15
 demineralization, **12.**30–**12.**32
 disposal, **12.**45–**12.**46, **12.**50
 distributor design, **12.**43–**12.**44
 dumping (*see* Dumping)
 exchange tank services, **12.**48, **12.**54
 exchangers, **12.**33–**12.**42
 flow rates, **12.**32
 fluoride, **12.**22
 future trends, **12.**54
 inorganic ion exchangers, **12.**4
 lead, **12.**21–**12.**22
 media depth, **12.**33
 monitoring/validation, **12.**46
 nitrate, **12.**18–**12.**20
 operation/maintenance of equipment, **12.**48–**12.**51
 perchlorate, **12.**22–**12.**25
 piping, **12.**44
 point-of-use systems, **12.**47
 radioactive substances, **12.**27–**12.**28
 radium, **12.**12
 regeneration, **12.**33–**12.**42
 regeneration frequency, **12.**32–**12.**33
 reliability/availability, **12.**50–**12.**51
 resin, **12.**3–**12.**4, **12.**51–**12.**52
 resin cleaning, **12.**49
 resin degradation/deterioration, **12.**5–**12.**6, **12.**51–**12.**52
 resin leachables, **12.**52–**12.**53
 resin removal and replacement, **12.**46–**12.**47, **12.**49–**12.**50, **12.**53
 selenium, **12.**26–**12.**27
 sulfates, **12.**18
 suppliers, consultants, users, **12.**53–**12.**54
 terrorism, **12.**1, **12.**27
 TOC, **12.**29–**12.**30
 uranium, **12.**28–**12.**29
 vessel design, **12.**42–**12.**43
 waste, **17.**9
 waste collection and disposal, **12.**45–**12.**46
Ion exchange polymers, **12.**2–**12.**3

Ion exchange waste brine, **17.**9
Ion exchangers, **12.**33–**12.**42
Iron, **2.**7, **9.**4
Iron and manganese:
 chlorine, **10.**11–**10.**13
 DE filtration, **9.**19
 membrane processes, **13.**23–**13.**24
 oxidation, **10.**11
 reduction/sequestering (process components), **2.**11
 slow sand filtration, **9.**4
 sludge, **17.**6–**17.**7
Iron bacteria, **5.**22
Iron coagulant recovery, **17.**17
Iron-exchange resins, **15.**43
Iron oxidation, **5.**22
Iron salts coagulation sludge, **17.**6
Isothermal equation, **10.**23

Jet pumps, **15.**25
Jeyanayagam, Samuel S., **29.**1
Jointing of structures, **19.**4–**19.**5

K value, **16.**7, **16.**11
Kennedy, Philip C., **24.**1
Koch, Robert, **10.**1
Kubel, Dave, **20.**1

Laboratory, **18.**8–**18.**11, **23.**36, **25.**6–**25.**7
Laboratory personnel, **25.**22
Laboratory software/hardware, **25.**11
Laboratory supervisor, **25.**21
Lag time, **10.**32
Lagoon, **9.**16, **17.**11, **17.**13, **17.**35–**17.**36
Lagoon decant structure, **17.**37
Lagoon lining profile, **17.**36
Lake and reservoir intakes, **4.**10–**4.**14
Lake intake crib, **4.**19
Lake Norman in-lake tower, **4.**10, **4.**12
Lamp sleeves, **29.**13, **29.**17
Land application, **17.**11, **17.**13–**17.**14
Land Application of Water Treatment Plant Sludge, **17.**12
Land suitability analysis, **23.**18
Landfill, **17.**11, **17.**13, **24.**14
Landfill cover, **17.**12
Landscaping, **18.**4, **23.**35, **24.**5
Lane, Thomas J., **1.**1
Langelier saturation index (LSI), **13.**17
Larger-diameter accelerators, **7.**31
Laser turbidimeters, **20.**40
Lateral loads, **19.**10
Lateral soil pressure, **19.**8
L/d ratio, **8.**8
Leachables resin, **12.**52–**12.**53
Lead, **12.**21–**12.**22
Leaf outlet connection, **9.**24
LEED, **18.**3

LEPC, **15.**44
Level of automation, **25.**5–**25.**6
Level switches/transmitters, **20.**35–**20.**37
Lifeline, **21.**2
Lighting, **21.**8–**21.**9, **25.**16
Lightning protection, **21.**9–**21.**10
Lime, **11.**3, **A.**11–**A.**12
Lime recovery, **17.**17
Lime slakers, **15.**20, **15.**22
Lime sludge thickening, **17.**29
Lime slurries, **15.**29–**15.**31
Lime softening, **11.**1–**11.**20
 aeration, **11.**5–**11.**6
 basins, **11.**13–**11.**16
 chemical feed/layout considerations, **11.**19
 chemistry, **11.**3–**11.**5
 encrustation, **11.**16
 excess lime-soda ash process, **11.**9–**11.**12
 filtration, **11.**18
 future trends, **11.**19
 hardness reduction, **2.**11
 iron and manganese removal, **10.**13
 pellet reactors, **11.**18–**11.**19
 presedimentation, **11.**6
 pretreatment, **11.**5
 reference text, **11.**3, **11.**6, **11.**12
 residuals removal/handling, **11.**17
 residuals production, **11.**17–**11.**18
 sludge, **17.**7
 softening plants, **11.**3
 split-treatment excess lime process, **11.**12
 straight lime-soda ash process, **11.**6–**11.**9
Lime softening sludge, **17.**7
Limestone, **A.**4
LIMS, **18.**9
LIN, **10.**50
Linear plant layout, **23.**23–**23.**25
Lined lagoon, **17.**35, **17.**36
Liquefaction of soil, **19.**8–**19.**9
Liquefied gases, **15.**3–**15.**8
Liquid alum, **A.**2
Liquid bleach, **A.**17
Liquid chemical feed equipment, **15.**24–**15.**37
Liquid chemicals, **15.**10–**15.**17
Liquid chlorine, **15.**3
Liquid level control equipment, **20.**35–**20.**37
Liquid nitrogen (LIN), **10.**50
Liquid oxygen, **15.**51, **A.**12, **A.**25
 (*See also* Oxygen)
Liquid-oxygen storage tanks, **10.**46, **10.**49
Liquid petroleum gas (LPG) natural gas–type vaporizers, **10.**49, **10.**50
Loading criteria, **19.**10
Loading on downstream process, **17.**21
Loading phase, **17.**36
Lobby and reception area, **18.**5
Local control, **20.**2–**20.**4
Local control panels, **20.**17–**20.**18

Local Emergency Planning Committee (LEPC), **15.**44
Local equipment controls, **20.**2
Local manual controls, **20.**3
Local operator interface (LOI), **20.**17
Locker rooms, **18.**6
Log removal, **8.**36
Logsdon, Gary, **8.**1
LOI, **20.**17
Long-finger weir, **7.**7
Long Term 1 Enhanced Surface Water Treatment Rule (LT1ESWTR-2005), **2.**5, **2.**7, **10.**8–**10.**9
Long-Term 2 Enhanced Surface Water Treatment Rule (LT2ESWTR-2006), **2.**5, **2.**7, **10.**9, **29.**39
Los Angeles aqueduct water filtration plant, **8.**19
Loss-in-weight gravimetric feeders, **15.**19
Louisville Water Company Ohio River intake, **4.**6, **4.**8
Low-concentration wastes, **17.**51
Low-energy flocculators, **6.**20
Low-fouling RO membranes, **13.**28
Low-pressure, high-output (LPHO) lamps, **29.**10–**29.**11, **29.**33
Low-pressure air preparation ozonation system, **10.**46, **10.**47
Low-pressure (LP) lamps, **29.**10–**29.**12, **29.**37
Low-pressure membranes, **6.**2, **17.**24
Low-pressure RO/NF systems, **13.**28
Low-pressure zone, **17.**45
Low-profile, diffused air system, **5.**5, **5.**6
Low-rate water wash, **8.**13
Low-tip-speed flocculators, **6.**20
Lowther plate, **10.**43, **10.**44
LOX, **10.**46, **10.**49
LOX storage tanks, **10.**46, **10.**49
LP lamps, **29.**10–**29.**12, **29.**37
LPG/natural gas vaporizer, **10.**49, **10.**50
LPHO lamps, **29.**10–**29.**11, **29.**33
LSI, **13.**17
LT2ESWTR, **2.**5, **2.**7, **10.**9, **29.**39
Lunchrooms, **18.**6

Macroporous resins, **12.**4, **12.**30
Magnesia quicklime, **A.**12
Magnesite, **9.**18
Magnesium hardness, **11.**2
Magnesium hydroxide solids, **17.**7
Magnesium oxide, **9.**18
Magnetic flowmeters, **20.**31–**20.**32, **20.**33
Magnetic ion exchange (MIEX) resin, **6.**4
Maintenance (*see* Operations and maintenance)
Maintenance facilities, **18.**12–**18.**14, **23.**36
Maintenance personnel, **25.**22
Maintenance shop, **25.**17–**25.**18
Maintenance shop tools, **18.**12, **18.**14

Maintenance software/hardware, **25.**11
Maintenance supervisor, **25.**21
Management and operational report development, **20.**27
Manganese, **2.**7, **9.**4
  (*See also* Iron and manganese)
Manifolded backwash collection, **28.**19
Mannings equation, **16.**6, **16.**12
Manual solids removal, **7.**8
Martin, Stephen R., **8.**1
Masks, **15.**47
Masonry, **19.**3
Master plan, **2.**1–**2.**3
  contents, **2.**2
  factors to consider, **2.**2
  planning periods, **2.**3
  principal influences, **2.**2–**2.**3
  scope, **2.**2
Mat foundations, **19.**7
Material safety data sheet (MSDS), **15.**41, **15.**45–**15.**46
Maximum contaminant level (MCL), **2.**7
Maximum contaminant level goal (MCLG), **2.**8
Maze-type baffles, **6.**20
MCL, **2.**7
MCLG, **2.**8
Mechanical aerators, **5.**7
Mechanical design, **3.**8
Mechanical dewatering, **17.**39–**17.**48
Mechanical (thermal) drying, **17.**48
Mechanical flocculators, **6.**18–**6.**20
Mechanical in-line blenders, **6.**11, **6.**12
Mechanical maintenance, **25.**22
Mechanical mixing, **6.**7–**6.**9
Mechanical reliability, **22.**7
Mechanical skimming, **7.**39–**7.**40
Mechanical solids removal, **7.**8–**7.**19
Mechanical surface aerators, **5.**7
Mechanical thickening, **17.**30–**17.**31
Medium-pressure air preparation ozonation system, **10.**46, **10.**47
Medium-pressure (MP) lamps, **29.**10–**29.**12, **29.**33, **29.**37
Membrane compaction, **13.**30
Membrane filtration, **2.**14
Membrane flux decline curves, **13.**30
Membrane integrity testing, **13.**38–**13.**39
Membrane module array, **13.**31–**13.**32, **13.**37
Membrane pilot testing, **13.**47
Membrane process waste, **17.**7–**17.**9
Membrane processes, **13.**1–**13.**49
  bench testing, **13.**47
  building design, **13.**45–**13.**46
  calcium carbonate control, **13.**16–**13.**18
  cleaning system, **13.**43
  coagulation, **6.**5
  concentration of solubility salts and silica, **13.**20–**13.**22

Membrane processes (*Cont.*):
  definitions, **13.**8–**13.**11
  designer-vendor interface issues, **13.**47
  ED/EDR system (*see* ED/EDR system)
  electrical voltage-driven membranes, **13.**3–**13.**4
  feedwater analysis, **13.**14
  feedwater characteristics, **13.**12–**13.**13
  feedwater pretreatment, **13.**13–**13.**25
  floor plans, **13.**45
  high-/low-pressure membranes, **17.**25
  hydrogen sulfide control, **13.**23
  inside-out/outside-in membranes, **17.**25
  iron and manganese control, **13.**23–**13.**24
  materials of construction, **13.**46
  membrane configuration, **13.**5–**13.**8
  microbial control, **13.**22–**13.**23
  O&M design issues, **13.**45
  organics control, **13.**24
  pH, **13.**24
  pilot testing, **13.**47
  posttreatment, **13.**41–**13.**42
  pressure-driven membranes, **13.**2–**13.**3
  reliability/redundancy, **13.**46–**13.**47
  residuals disposal, **13.**43–**13.**44
  RO and NF units (*see* RO and NF units)
  scaling control, **13.**16
  selecting a process, **13.**4–**13.**5
  silica control, **13.**18–**13.**20
  sulfate scale control, **13.**18
  suspended solids control, **13.**13–**13.**16
  system flow schematic, **13.**12
  temperature control, **13.**25
  types, **13.**1
  UF and MF units (*see* UF and MF units)
  waste, **17.**7–**17.**9
Membrane system flow schematic, **13.**12
Membrane-type chlorine residual analyzers, **20.**38–**20.**39
Mesh size, **14.**3
Metering pumps, **12.**45
Methyl tertiary butyl ether, **5.**10
Methylisoborneol (MIB), **10.**13
MF (*see* UF and MF units)
MIB, **10.**13
Microbial contamination, **10.**1
  (*See also* Oxidation and disinfection)
Microbial fouling, **5.**21–**5.**22
Microfiltration (MF) (*see* UF and MF units)
Microsand, **7.**25
Microwave sensor, **30.**4
MIEX resin, **6.**4
Milk-of-lime, **15.**29
Mineral oxidation, **A.**30
  (*See also* Oxidation and disinfection)
Mini-clean, **13.**38
Missouri River intake, **4.**4, **4.**5
Misting, **10.**18

Mitigation, **23.**17
Mixed bed demineralization, **12.**32
Mixed-media filter bed, **8.**6
Mixing, **6.**2 (*see* Chemical mixing)
Mixing basins, **16.**13
Model-based control, **20.**7–**20.**8
Module, **13.**2
Modular pilot plants, **28.**23
Moisture content of gases, **10.**44
Molasses number, **14.**2
Molecular weight fingerprinting, **13.**13
Monochloramine, **10.**15, **10.**17
Monohydration, **A.**12
Monolithic wheeler bottom, **8.**27
Monosodium phosphate, **A.**13
Monroe Reservoir, **4.**17
Montgomery, James M., **8.**7, **8.**37, **16.**12, **23.**34
Motionless mixers, **6.**11
Motors and controllers, **21.**5–**21.**9
MP lamps, **29.**10–**29.**12, **29.**33, **29.**37
MSDS, **15.**41, **15.**45–**15.**46
Mudballs, **8.**39
Multiattribute analysis techniques, **23.**20
Multiple-hearth furnace, **14.**36
Multiple inlet gates, **4.**18
Multiple-overlay mapping, **23.**18–**23.**19
Multiple-pass design, **13.**32
Multiple-step regeneration, **12.**41
Multiple-tray aerators, **5.**2, **5.**13
Mutltidiscipline teams, **1.**3–**1.**4

Nanofiltration (NF) (*see* RO and NF units)
National Electrical Code (NEC), **21.**1, **21.**2
National Environmental Policy Act (NEPA), **24.**12
National Pollution Discharge Elimination System (NPDES), **17.**10, **17.**13
National Pollutant Discharge Elimination System (NPDES) permit, **8.**36, **24.**12
Natural dewatering, **17.**31–**17.**39
Natural dewatering sizing parameters, **17.**31
Natural disasters, **22.**8–**22.**9
NDP, **13.**9–**13.**10
NEC, **21.**1, **21.**2
NEPA, **24.**12
Nephelometric turbidity unit (ntu), **8.**28
Net driving pressure (NDP), **13.**9–**13.**10
Net filter production, **8.**39
Net positive suction head (NPSH), **15.**34
Net present value approach, **23.**16–**23.**17
Network diameter, **20.**21
NF (*see* RO and NF units)
90° V-notch weir, **16.**10
900-lb (408-kg) woven plastic bags, **9.**27
NIOSH Pocket Guide to Chemical Hazards, **15.**41
Nitrate, **12.**18–**12.**20
Nitrate dumping, **12.**19–**12.**20

Nitrate selective, **12.**20, **12.**23
Nitrogen, **9.**5, **10.**15
Noise, **17.**66, **20.**24–**20.**25, **23.**9, **23.**30
Noncarbonate hardness, **11.**2
Nonfiltered/filtered water, **23.**33
Nonionic (anionic) polymers, **8.**3
Nonlinear control, **20.**7
Nonmechanical Dewatering of Water Plant Residuals, **17.**39
Nonrecoverable water, **17.**57
North Santiam River, **4.**28
Nozzle-type underdrains, **8.**25
NPDES, **17.**10, **17.**13
NPDES permit, **8.**36, **24.**12
NPSH, **15.**34
NSF International, **15.**45
NSF International standards, **8.**37
NTU, **5.**14
ntu, **8.**28
Number of transfer units (NTU), **5.**14
Nutrient control, **17.**12

O ring gasket, **9.**24
Obsolescence, **20.**22
OCC system, **29.**13–**29.**14
Occupational Safety and Health Administration (OSHA), **15.**41
Odor control, **2.**10, **10.**13, **17.**54–**17.**57, **17.**66, **A.**31
Off-gas disposal, **10.**58–**10.**59
Off-line chemical cleaning (OCC), **29.**13–**29.**14
Off-line integrity testing, **13.**39
Off-site validation test center, **29.**30
Office area, **18.**5
Ohio River intake, **4.**4–**4.**6
Oil storage, **24.**7–**24.**8
O&M (*see* Operations and maintenance)
O&M manuals, **27.**3–**27.**5, **27.**9–**27.**10, **27.**11
On-line mechanical cleaning (OMC), **29.**14
On-line turbidimeters, **8.**28
On-off control, **20.**4–**20.**5
On-site generation/regeneration:
 chlorine, **10.**28–**10.**30
 GAC, **14.**33–**14.**37
 oxygen, **10.**49–**10.**54
 reliability, **22.**13
On-site hypochlorite-generating process, **10.**29
On-site validation testing, **29.**30–**29.**31
1,1,1-Trichloroethane, **5.**10
1,2-Dichloroethane, **5.**10
OPC, **20.**22
Open-air drying, **17.**48
Open-top concrete, rectangular, gravity flow contactor, **14.**26, **14.**27
Operational flexibility, **23.**28–**23.**29
Operations and maintenance (O&M):
 accessibility of equipment, **25.**13–**25.**14
 budget, **25.**18–**25.**19

cathodic protection systems, **25.**11
chemical feed systems, **25.**9–**25.**10
computer software/hardware, **25.**11–**25.**12
cranes/hoists, **25.**14
design reviews, **25.**12
ergonomics, **25.**6–**25.**7
excess equipment, **25.**13
filters, **25.**9
flocculation basins, **25.**8
laboratory, **25.**6–**25.**7
level of automation, **25.**5–**25.**6
maintenance shop, **25.**17–**25.**18
manuals, **27.**3–**27.**5, **27.**9–**27.**10, **27.**11
plant start-up, **25.**18–**25.**23
 (*See also* Plant start-up)
power and control systems, **25.**4–**25.**5
power/lighting, **25.**16
rapid mix basins, **25.**8
recarbonation basins, **25.**8
safety, **25.**2–**25.**3
sampling stations, **25.**6
security, **25.**3–**25.**4
sedimentation basins, **25.**8
source water intakes, **25.**7–**25.**8
staffing plan, **25.**20–**25.**22
tunnels/galleries, **25.**16–**25.**17
unit process—specific uses, **25.**7–**25.**11
Operations and maintenance (O&M) manuals, **3.**12
Operations personnel, **25.**21–**25.**22
Operations services manager, **27.**13
Operations software/hardware, **25.**11
Operations supervisor, **25.**21
Operator control room ergonomics, **25.**7
Operator sampling station, **25.**6
Operator training, **3.**12 (*see* Plant start-up)
Opportunity cost, **23.**15
Optimal expansion rate, **8.**14
Optimization control, **20.**4
Order-of-magnitude estimate, **26.**2
Organic fouling, **12.**6
Organics adsorption oxidation, **A.**30
Organics control, **13.**24
Orifice plates, **20.**30, **20.**31–**20.**32
Orifice trough, **7.**21
Orifice-type distributor, **5.**19
Oscillating-bridge collectors, **7.**11
Oscillating flocculator, **6.**19
Oscillating hopper feeder, **15.**20
Oscillating skimmer, **7.**39–**7.**40
OSHA, **15.**41
Osmotic changes, **12.**51
Osmotic drawback, **13.**35
Osmotic pressure, **13.**10
Otto plate, **10.**43–**10.**44
Outlet zone (sedimentation basin), **28.**12
Outside-in membranes, **17.**24
Overflow rates, **7.**3

Overlay mapping, **23.**18–**23.**19
Owner approval of change orders, **3.**11
Owner reviews, **3.**9
Oxidation and disinfection, **2.**10, **10.**1–**10.**59
  chemical oxidation treatment, **10.**10–**10.**13
  chemicals, **A.**30
  Chick-Watson theory, **10.**13
  chlorination, **10.**13–**10.**34
    (*See also* Chlorination)
  chlorine dioxide systems, **10.**34–**10.**39
    (*See also* Chlorine dioxide systems)
  contact time, **10.**5–**10.**8
  CT value, **10.**3–**10.**4
  disinfectant contact requirements, **10.**3–**10.**8
  disinfectant monitoring requirements, **10.**8–**10.**9
  disinfection by-product regulation, **10.**9–**10.**10
  groundwater disinfection rule, **10.**10
  historical overview, **10.**1–**10.**2
  iron and manganese removal, **10.**11–**10.**13
  ozone disinfection systems, **10.**39–**10.**59
    (*See also* Ozone disinfection systems)
  performance ratio, **10.**4
  reference text, **10.**2
  regulatory framework, **10.**3–**10.**10
  resin degradation, **12.**5–**12.**6, **12.**51
  SFA, **10.**4
  SWTR, **10.**3–**10.**9
  taste and odor control, **10.**13
Oxidation-reduction (redox) reaction, **10.**10
Oxygen:
  dissolved, **9.**4, **9.**5
  GOX, **10.**49, **10.**50
  Henry's constant atm, **5.**9–**5.**10
  liquid (*see* Liquid oxygen)
  LOX, **10.**46, **10.**49
  on-site generation, **10.**49–**10.**54
  shipment/storage, **15.**6
Ozonation contactors, **28.**8, **28.**18–**28.**21
Ozonator, **10.**42–**10.**44
Ozone, **A.**12, **A.**25
  accidental release regulatory thresholds, **15.**44
  GAC, **14.**31
  physical constants, **10.**2
  preoxidation studies, **28.**10
  protective equipment, **15.**43
  sidestream residual flows, **17.**26
  slow sand filters, **9.**14
  threshold planning quantities, **15.**44
  UV disinfection, **29.**17–**29.**18
Ozone chemistry, **10.**39–**10.**41
Ozone contactors, **10.**54–**10.**58
Ozone disinfection module, **28.**20
Ozone disinfection systems, **10.**39–**10.**59
  air feed systems, **10.**46–**10.**48
  construction materials, **10.**59

  feed gas preparation, **10.**44–**10.**46
  high-purity oxygen (LOX), **10.**46, **10.**49
  off-gas disposal, **10.**58–**10.**59
  on-site generation of oxygen, **10.**49–**10.**54
  ozone chemistry, **10.**39–**10.**41
  ozone contractors, **10.**54–**10.**58
  ozone generation, **10.**42–**10.**44
Ozone generation, **10.**42–**10.**44

PAC:
  polyaluminum chloride, **6.**3, **15.**51, **A.**13, **A.**26
  powdered activated carbon (*see* Powdered activated carbon)
PAC application point, **14.**5, **14.**6
PAC dosage, **14.**5
Package units, **3.**3
Packed bed exchanger, **12.**40–**12.**41
Packed column contactors, **10.**57
Packed column installation (*see* Packed tower aeration)
Packed Column Pilot Studies, **5.**22–**5.**24
Packed tower aeration (PTA), **5.**3, **5.**4, **5.**14–**5.**22
  air blower, **5.**18
  A/W ratio, **5.**14
  column design, **5.**14–**5.**15
  distributors, **5.**18, **5.**19
  exhaust emission, **5.**20–**5.**21
  fouling of packing, **5.**21–**5.**22
  liquid distribution, **5.**18
  packing materials, **5.**18, **5.**20
  pilot testing, **5.**22–**5.**24
  site considerations, **5.**18
Paddle-type flocculators, **6.**15, **6.**18, **6.**20
Painting, **18.**17
Pan tilt zoom (PTZ) cameras, **30.**6
Parallel DAF tanks, **28.**13
Particle counters, **20.**40
Particle counting, **8.**29
Particulates, **10.**45
Passive frazil ice, **4.**23
Passive IDS, **30.**4
Passive infrared (IR) sensors, **30.**5
Paste, **15.**29
Pasteur, Louis, **10.**1
Peck, Brian E., **17.**1
Pellet reactors, **11.**18–**11.**19
Pentachlorophenal, **5.**10
Percentage application method, **26.**6
Perchlorate, **12.**22–**12.**25
Performance ratio, **10.**5
Periodic purge bubbler transmitters, **20.**36
Peristaltic pumps, **15.**24
Permanent pilot plants, **28.**23
Permanganate, **29.**16
Permeate, **13.**34
Permeate drawback tank, **13.**34–**13.**35

Permeate staging, **13**.32
Permeators, **13**.6
Permit tracking system, **24**.17
Permitting strategy/plan, **24**.16–**24**.18
Person with disabilities, **18**.5
Personnel facilities, **18**.6–**18**.7
Persons with disabilities, **23**.29–**23**.30, **26**.8
Perstraction, **10**.38
Pests, **17**.66
pH:
    adjustment, **A**.30–**A**.31
    calcium carbonate, **13**.16
    coagulation process, **6**.4–**6**.5
    control of, **20**.10
    dealkalization, **12**.14, **12**.15
    ED/EDR system, **13**.41
    MCL, **2**.7
    measurement, **20**.41
    membrane processes, **13**.24
    RO/NF units, **13**.28
PHA, **15**.45
Phantom loads, **19**.12
Phase reversal protection relays, **21**.7
Phosphates, **15**.51, **A**.13
Phosphoric acid, **15**.10, **15**.51, **A**.13, **A**.25
Phosphorus, **9**.5
Physical fouling, **12**.52
Physical protection system (PPS), **30**.1–**30**.2
PI control, **20**.6–**20**.7
PID control, **20**.7
P&IDs, **20**.27–**20**.28
Pilot air stripping column, **5**.23
Pilot filters, **8**.31
Pilot plant design and construction:
    backwashing, granular filters, **28**.8, **28**.16–**28**.18
    bench-scale testing, **28**.3–**28**.5
    coagulation, **28**.7, **28**.10–**28**.11
    construction, **28**.21
    design, **28**.3–**28**.21
    dissolved air flotation (DAF), **28**.8, **28**.13–**28**.14
    electric power requirements, **28**.6
    evaluating unit processes, **28**.6–**28**.21
    flocculation, **28**.7, **28**.11–**28**.12
    granular media filtration, **28**.8, **28**.14–**28**.18
    hydraulic requirements, **28**.5
    installation and start-up, **28**.25
    ozonation contactors, **28**.8, **28**.18–**28**.21
    permanent *vs.* temporary plants, **28**.23
    preoxidation, **28**.6, **28**.10
    process drawings/specifications, **28**.23–**28**.25
    sedimentation, **28**.7, **28**.12–**28**.13
    solids collection/treatment, **28**.21
    space requirements, **28**.5–**28**.6
    (*See also* Pilot testing)
Pilot plant floor plan, **28**.24
Pilot plant studies, **25**.7, **28**.2–**28**.3

Pilot testing, **28**.6–**28**.21
    aeration and air stripping, **5**.22–**5**.24
    membrane processing, **13**.47
    residuals handling, **17**.67
    slow sand filtration, **9**.12–**9**.13
    UV disinfection, **29**.21
    (*See also* Pilot plant design and construction)
Pipe lateral underdrains, **8**.22–**8**.23
Piping:
    chlorine, **10**.22–**10**.24, **15**.19
    effluent, **16**.8–**16**.9, **16**.12–**16**.13
    feeding system, **15**.33–**15**.37
    flowmeter, **20**.34
    head loss, **16**.5–**16**.6, **16**.8–**16**.9
    ion exchange system, **12**.44
    membrane processes, **13**.46
    residuals handling /transportation, **17**.18–**17**.19
    sodium hypochloride, **10**.27
    types/chemicals, **15**.4–**15**.5
Piping gallery, **9**.11
Piston and diaphragm pumps, **15**.24
Pitot tubes, **20**.30
Plain sedimentation, **7**.1
Plant arrangement, **23**.22–**23**.36
    (*See also* Arrangement of plant facilities)
Plant hydraulic design, **3**.7
Plant maintenance (*see* Operations and maintenance)
Plant maintenance shop, **25**.17–**25**.18
Plant manager, **25**.21
Plant operations (*see* Operations and maintenance)
Plant reliability (*see* Design reliability features)
Plant siting analysis, **3**.2
Plant start-up, **25**.18–**25**.23, **27**.1–**27**.10
    budget, **25**.18–**25**.19
    construction-phase training, **27**.5–**27**.14
    design-phase training, **27**.2–**27**.5
    facility O&M manual, **27**.9–**27**.10, **27**.11
    post-start-up considerations, **27**.14
    record keeping, **27**.13–**27**.14
    staffing plan, **25**.20–**25**.22
    start-up/operations services manager, **27**.13
    start-up planning and initiation, **27**.10–**27**.13
    start-up sequence, **27**.13
    systems training, **27**.8–**27**.9
    vendor O&M manual, **27**.3–**27**.5
    vendor training, **27**.2–**27**.3, **27**.6–**27**.8
Plastic block underdrain, **8**.25
Plastic packing, **5**.20
Plate and frame filter press, **17**.41–**17**.43
Plate settlers, **7**.24–**7**.25
PLC, **20**.19, **21**.7, **22**.7–**22**.8
Plutonium, **12**.28
Polyacrylic acid antiscalants, **13**.17, **13**.18, **13**.23
Polyaluminum chloride, **6**.3, **15**.51, **A**.13, **A**.26

polyDADMAC, **A.**13
Polydiallyldimethylammonium chloride
   (polyDADMAC), **A.**13
Polymer, **A.**13–**A.**14
   anionic, **8.**3
   cationic, **6.**3
   dilution, **15.**38
   feed and metering system, **15.**32
   filter aid, **8.**3–**8.**4
   ion exchange, **12.**2–**12.**3
   sludge treatment, **17.**52–**17.**54
   uses, **15.**51
Polymer feed equipment, **17.**54
Polymer feed system schematics, **17.**55
Polymer usage, **17.**53
Porous-bottom underdrains, **8.**26, **8.**28
Ported coaxial buried line sensors, **30.**4
Ports, **16.**9–**16.**10
Positive displacement pumps, **15.**24, **15.**25,
   17.18, **17.**52
Postfilter adsorbers, **14.**22–**14.**28
   (*See also* GAC postfilter adsorbers)
Pot feeders, **15.**37
Potash, **A.**14
   (*See also* Potassium permanganate)
Potassium aluminum sulfate, **A.**26
Potassium hydroxide, **A.**14
Potassium hypochlorite, **A.**14
Potassium hypochlorite bleach, **A.**14
Potassium permanganate, **A.**14, **A.**26
   DE filtration, **9.**19
   intake facilities, **4.**31
   oxidation and disinfection, **10.**12–**10.**13
   pilot plant, **28.**10
   piping, **15.**35
   protective equipment, **15.**43
Potassium permanganate system, **15.**21
Powdered Activated Carbon (AWWA Standard
   B600), **14.**2
Powdered activated carbon (PAC), **14.**4–**14.**9,
   A.2, **A.**20
   application equipment, **14.**6–**14.**8
   application point, **14.**5, **14.**6
   black dust, **15.**9
   carbon addition, **6.**4
   characteristics, **14.**4–**14.**5
   coagulation, **6.**4
   dosage, **14.**5
   manufacturers' specifications, **14.**4
   odor, **17.**55–**17.**56
   operating problems, **14.**8–**14.**9
   protective equipment, **15.**42
   questions to ask, **14.**6
   storage/shipment, **15.**9, **15.**22
Power and control systems, **25.**4–**25.**5
Power factor correction, **21.**10
Power failures, **20.**23
Power line transient voltage surge suppression,
   20.24

Power receptacles, **25.**16
PPS, **30.**1–**30.**2
Precast wheeler bottom, **8.**27
Precipitated chalk, **A.**4
Precipitation, **12.**52
Precoat filters, **9.**16
   (*See also* Diatomaceous earth filtration)
Precoat filtration cycles, **9.**17
Precursor reduction, **2.**10
Preengineered structures, **18.**14–**18.**15
Preengineered units, **3.**3
Preliminary design:
   custom design vs. packaged units, **3.**3
   design development, **3.**3–**3.**4
   financial feasibility evaluation, **3.**4
   ongoing coordination/cost evaluations,
      3.4–**3.**5
   plant siting analysis, **3.**2
   regulatory agency coordination, **3.**1–**3.**2
   upgrading/expansion of existing facilities,
      3.2–**3.**3
Preliminary design cost estimate, **26.**3
Preliminary HGL, **16.**3, **16.**4
Premature obsolescence, **20.**22
Preoxidation, **28.**6, **28.**10
Preozonation, **9.**14
Presedimentation, **11.**6
Present value, **23.**16
Pressure adsorbers, **14.**16–**14.**17
Pressure aeration, **5.**7
Pressure decay test, **13.**39
Pressure-driven membrane process application
   guide, **13.**3
Pressure-driven membrane processes,
   13.2–**13.**3, **13.**44
Pressure drop, **12.**51
Pressure filter dewatering, **17.**39–**17.**43
Pressure filter system, **9.**20, **9.**25–**9.**26,
   9.28–**9.**29
Pressure-reducing valve (PRV), **19.**7
Pressure swing adsorption (PSA), **10.**53–**10.**54
Pressure switches/transmitters, **20.**37–**20.**38
Pressure transmitter accessories, **20.**37–**20.**38
Pressurized chlorine piping, **10.**23
Primary drinking water standards, **2.**5, **2.**6
Privatization approaches, **1.**4
Process alternatives, **2.**9–**2.**12
Process and instrumentation diagrams (P&IDs),
   20.27–**20.**28
Process control network, **20.**19–**20.**21
Process drawings/specifications, **28.**23–**28.**25
Process graphic displays, **20.**26
Process hazard analysis (PHA), **15.**45
Process instrumentation and controls,
   20.1–**20.**42
   analytical instrumentation, **20.**38–**20.**41
   chlorination controls, **20.**13–**20.**14
   clarification and sedimentation, **20.**9, **20.**10
   control, types of, **20.**4–**20.**8

control system design, **20.15–20.29**
  (*See also* Control system design)
  disinfection, **20.9**, **20.13–20.15**
  filtration, **20.9**, **20.11–20.13**
  flowmeters, **20.30–20.34**
  hierarchy of control, **20.2–20.4**
  importance, **20.1–20.2**
  instrumentation, **20.30–20.41**
  level switches/transmitters, **20.35–20.37**
  mixing, coagulation, flocculation, **20.9**, **20.10**
  pressure switches/transmitters, **20.37–20.38**
  raw water pump controls, **20.8**, **20.9**
  sludge handling and return water, **20.14–20.15**
  temperature switches/transmitters, **20.38**
Process laboratory, **18.11**
  (*See also* Laboratory)
Process lag, **20.7**
Process lag time, **20.13**
Process noise, **20.24**
Process piping supports, **25.15**
Process reliability, **22.10–22.11**
Process residuals:
  arsenic, **17.14–17.15**
  bench tests, **17.67**
  contingency planning, **17.63**, **17.64–17.65**
  disposal limitations, **17.59–17.60**
  disposal methods, **17.9–17.12**
    (*See also* Disposal)
  environmental impact, **17.65–17.66**, **24.8–24.9**, **24.11–24.12**, **24.14–24.15**
  equipment sizing, **17.62**
  estimating solids production, **17.61–17.62**
  facility drainage, **17.65**
  hydraulic profile, **17.63–17.64**
  hydraulics, **17.18**
  minimizing sludge quantities, **17.16**
  overview, **17.1–17.4**
  pilot tests, **17.67**, **28.21**
  piping, **17.18–17.19**
  pumping, **17.18**
  quantity/quality of residuals, **17.60**
  reference sources, **17.67–17.68**
  regulatory requirements, **17.12–17.15**
  residuals mass balance, **17.61**
  resource recovery potential, **17.60–17.61**
  selecting handling process, **17.65–17.67**
  sidestreams, **17.24–17.28**
  sludge treatment techniques, **17.19–17.58**
    (*See also* Sludge treatment techniques)
  sludge types, **17.4–17.9**
    (*See also* Sludge types)
  steps in process, **17.1**, **17.2**
  trucking, **17.19**
  UIC rules, **17.14**
Process safety management (PSM) program, **15.45**, **24.14**
Process time lag, **20.41**
Process train selection, **1.4**, **1.5**

Proctoring, **27.7**
Product pressure, **13.10**
Production trains, **22.6**
Profibus, **20.21**, **20.22**
Programmable logic controller (PLC), **20.19**, **21.7**, **22.7–22.8**
Progressive cavity pumps, **15.24**
Project coordination, **3.6**
Project delivery, **1.4**
Project permitting, **24.12–24.18**
Propeller and turbine meters, **20.34**
Propeller-type mechanical flash mixer, **6.7**
Propeller-type mixers, **6.8–6.9**
Proportional band, **20.6**
Proportional control, **20.6**
Proportional plus integral (PI) control, **20.6–20.7**
Proportional plus integral plus derivative (PID) control, **20.7**
Proportional-speed floating control, **20.7**
Proposed rules, **2.4**, **2.5**, **2.7**
Protective clothing, **15.47**
PRVs, **19.7**
PSA oxygen generation, **10.53–10.54**
PSM program, **24.14**
PSM regulation, **15.45**
PTA (*see* Packed tower aeration)
PTZ cameras, **30.6**
Public advisory committees, **23.3–23.4**
Public Health Security and Bioterrorism Preparedness and Response Act, **2.3**
Public Works Manual, **7.8**
Pulsed bed absorbers, **14.24–14.25**
Pulsed blanket clarifier, **7.32–7.35**
Pump motor and controller design, **21.5–21.9**
Pumping Station Design, **17.18**
Pumps, **16.15**
  backwash systems, **16.15**
  centrifugal, **15.25**, **15.27**
  jet, **15.25**
  liquid chemical feed, **15.24**
  motors/controllers, **21.5–21.9**
  piston and diaphragm, **15.24**
  progressive cavity, **15.24**
  residuals handling/transportation, **17.18**
  slurry, 15.31.27
  standby power, **21.8**
  transfer, **15.31–15.32**
Pyramid-type sludge blanket clarifier, **7.33**
Pyrosodium sulfite, **A.26**

Quagga mussel, **4.26**
Quality control reviews, **3.9**
Quench tank, **14.37**
Quicklime, **11.3**, **15.9**, **15.43**, **A.11**

Radio-frequency capacitance level transmitters, **20.36**
Radioactive substances, **12.27–12.28**
Radium, **12.12**

Radon rule, **2.5**, **2.7**
Rapid mixing, **6.2**
Rapid mixing basins, **16.**13, **25.8**
Rapid mixing systems, **6.6–6.**14
Rapid sand filtration, **8.**1–**8.**43
    air binding, **8.**40
    auxiliary equipment, **8.**32–**8.**36
    color, **2.**10
    conduits, **8.**32
    cracks/mudballs, **8.**39
    design, **8.**2–**8.**37
    filter arrangements, **8.**18–**8.**21
    filter efficiency, **8.**38
    filter evaluation, **8.**41
    filter media, **8.**4–**8.**8
    filter media replacement, **8.**40–**8.**41
    filter media washing, **8.**11–**8.**18, **8.**39
    filter operation and maintenance, **8.**37–**8.**41
    filter operational control, **8.**8, **8.**31
    filter performance monitoring, **8.**28–**8.**31
    filter ripening, **8.**37–**8.**38
    filter wash rate, **8.**31
    filtration rates, **8.**8, **8.**30
    head loss, **8.**29
    length of filter run, **8.**31
    mechanism of filtration, **8.**2
    net filter production, **8.**39
    particle counting, **8.**29
    pilot filters, **8.**31
    pretreatment, **8.**3–**8.**4, **8.**39
    problems, **8.**39–**8.**41
    regulatory requirements, **8.**2–**8.**3
    restart after shutdown, **8.**40
    standards, **8.**36–**8.**37
    support gravel upset, **8.**39–**8.**40
    turbidity, **8.**28–**8.**29
    underdrain systems, **8.**21–**8.**28
    valves/actuators, **8.**32
    washwater troughs, **8.**32–**8.**36
Rapid small-scale column test (RSSCT), **14.**15
Rate controller, **16.**8
Raw water pump controls, **20.**8, **20.**9
Reactivation rate, **14.**35
Recalcination of spent lime, **17.**16–**17.**17
Recarbonation, **11.**4, **11.**7, **11.**10
Recarbonation basin, **11.**16, **25.**8
Recessed chamber filter presses, **17.**41
Recommended Standards for Water Works, **8.**36, **8.**40, **11.**16, **15.**49
Record keeping, **27.**13–**27.**14
Recoverable water, **17.**57, **17.**58
Recovery (*Y*), **13.**10
Rectangular adsorbers, **14.**26, **14.**27
Rectangular sedimentation basins, **7.**4–**7.**19
    basin dimensions, **7.**5
    carriage-type collectors, **7.**11
    circular collector equipment, **7.**9–**7.**10
    cross collectors/cross hoppers, **7.**11–**7.**12

    floating bridge hydraulic systems, **7.**17, **7.**18
    indexing grid system, **7.**12–**7.**13, **7.**14
    inlet zone, **7.**5
    manual solids removal, **7.**8
    mechanical solids removal, **7.**8–**7.**19
    outlet design, **7.**5–**7.**8
    track-mounted hydraulic systems, **7.**14–**7.**17
    underflow control, **7.**17
Recycle/disposal of liquid wastes, **17.**57–**17.**58
Recycle flotation, **17.**29
Recycle limitation, **17.**21
Recycle of waste washwater, **8.**36
Redox reaction, **10.**10
Redundancy, **22.**1–**22.**2, **23.**28
    (*See also* Design reliability features)
Redundant diffusers, **15.**41
Regenerating resins, **12.**33–**12.**42
    CCR exchangers, **12.**39–**12.**40
    cocurrent exchangers, **12.**36–**12.**37
    cocurrent regeneration, **12.**33–**12.**35
    coflow exchanger, **12.**36–**12.**37
    coflow regeneration, **12.**33–**12.**35
    countercurrent regeneration, **12.**34–**12.**35
    degasification, **12.**36
    external regeneration, **12.**41
    frequency, **12.**32–**12.**33
    multiple-step regeneration, **12.**41
    packed bed exchanger, **12.**40–**12.**41
    salt regeneration, **12.**41
    split-flow exchangers, **12.**37–**12.**38
    throughfare regeneration, **12.**33
Regeneration:
    GAC, **14.**33–**14.**37
    ion exchange system, **12.**33–**12.**42
    (*See also* Regenerating resins)
    thermal, **14.**35–**14.**37
    (*See also* Generation)
Regeneration furnaces, **14.**35–**14.**36
Regulations (*see* Water quality goals)
Regulatory agency coordination, **3.**1–**3.**2
Regulatory agency reviews, **3.**9
Regulatory approval of change orders, **3.**11
Regulatory conflict, **2.**9
Regulatory uncertainties, **1.**3
Reinforced concrete trough, **8.**34
Reinforced Thermoset Plastic Corrosion Resistant Equipment, **15.**11
Reinjection, **10.**58
Relay ladder logic (RLL), **20.**25
Relay testing, **21.**12
Reliability (*see* Design reliability features)
Remote site locations, **26.**7
Remote terminal units (RTUs), **20.**19
Repair shops, **18.**12
Report development, **20.**27
Reserve process capacity, **22.**2
Reservoir intakes, **4.**10–**4.**14
Reset windup, **20.**7

INDEX

Residuals disposal limitations, **17**.59–**17**.60
Residuals handling/disposal (*see* Process residuals)
Residuals mass balance, **17**.61
Resin, **12**.3–**12**.4, **12**.51–**12**.52
Resin cleaning, **12**.49
Resin degradation/deterioration, **12**.5–**12**.6, **12**.51–**12**.52
Resin leachables, **12**.52–**12**.53
Resin removal and replacement, **12**.46–**12**.47, **12**.49–**12**.50, **12**.53
Respirators, **15**.47
Restart after shutdown (filters), **8**.40
Return water pumping, **20**.15
Reverse deionization, **12**.31
Reverse osmosis (RO) (*see* RO and NF units)
Rising corner floc, **7**.22
Risk analysis, **23**.21–**23**.22
Risk management program (RMP), **15**.45, **24**.14
River channel degradation, **4**.4
River intakes, **4**.4–**4**.10
  flooding, **4**.10
  location, **4**.9–**4**.10
  Missouri River intake, **4**.4, **4**.5
  Ohio River intake, **4**.4, **4**.6
  silt/bed load, **4**.10
  siphon well intakes, **4**.6, **4**.7
RLL, **20**.25
RMP program, **15**.45, **24**.14
RO and NF units, **13**.2–**13**.3, **13**.25–**13**.35
  applications, **13**.6
  arrays, **13**.31–**13**.32
  automatic flushing systems, **13**.34
  blending, **13**.33
  cleaning system, **13**.43
  computer performance projections, **13**.33–**13**.34
  concentrate staging, **13**.31–**13**.32
  design equations, **13**.26
  energy recovery, **13**.34
  feed pressure, **13**.29–**13**.31
  feedwater analysis, **13**.14
  flux, **13**.28–**13**.29
  hollow-fiber RO unit, **13**.6, **13**.7
  I&C equipment, **13**.44–**13**.45
  permeate drawback tank, **13**.34–**13**.35
  permeate staging, **13**.32
  pretreatment, **13**.15
  recovery considerations, **13**.27
  seawater desalination system, **13**.13
  solute rejection/passage, **13**.27–**13**.28
  spiral-wound RO unit, **13**.5–**13**.6, **13**.7
  temperature, **13**.29
RO hollow-fiber membrane modules, **13**.6, **13**.7, **13**.32
RO seawater desalination system, **13**.13
Road subgrade, **17**.12

Rock anchors, **4**.28
Rohan, Richard A., **25**.1
Roll feeder, **15**.20
Rotameters, **15**.37
Rotary kiln furnace, **14**.36
Rotary paddle feeders, **15**.20
Rotary surface wash systems, **8**.15
Rotating skimmer, **7**.40
Roughing filters, **9**.13–**9**.14
RSSCT, **14**.15
RTUs, **20**.19
Rubber-seated butterfly valve, **8**.32
Rules under development, **2**.4, **2**.5, **2**.7
Russell, Jerry S., **17**.1

Safe Drinking Water Act (SDWA), **2**.4
Safety, **25**.2–**25**.3
Safety—chemicals, **15**.41–**15**.50
  building code, **15**.48
  chlorine, **15**.48, **15**.51
  confined spaces, **15**.46
  containment and treatment systems, **15**.49–**15**.50
  explosion prevention, **15**.50
  eye wash, **15**.47–**15**.48
  fire code, **15**.48
  masks, **15**.47
  MSDS, **15**.41, **15**.45–**15**.46
  OSHA, **15**.41
  overview, **15**.42–**15**.43
  protective clothing, **15**.47
  reference material, **15**.41
  regulations/codes, **15**.41–**15**.45
  respirators, **15**.47
  safety screens, **15**.48
  safety showers, **15**.47–**15**.48
  scrubbers, **15**.49–**15**.50
  ventilation, **15**.48–**15**.49
Safety Color Code, **18**.16
Safety Practices for Water Utilities, **22**.15
Safety screens, **15**.48
Safety showers, **15**.47–**15**.48
Sales tax, **26**.8
Salt regeneration, **12**.30, **12**.41
Sampling stations, **25**.6
Sand drying beds, **17**.32–**17**.34
Saturation value, **5**.8
SCADA systems, **20**.18
Scaling control, **13**.16
Schedule 80 seamless steel piping, **10**.23
Schellhase, Todd F., **19**.1
Scheme for the Identification of Piping System, **18**.16
Schmutzdecke, **9**.2
Scope creep, **25**.1, **26**.9–**26**.10
ScourGuard baffled wash trough, **8**.34
Screens, **4**.31–**4**.33
Screw feeder, **15**.20

Scrubbers/scrubbing:
  accidental release scrubber systems, **15.**50
  carbon removal, **15.**10
  chemical scrubbers, **15.**49, **17.**57
  chlorine scrubbing system, **15.**49–**15.**50
  GAC scrubbers, **17.**57
  membrane processes, **13.**42
  odor control, **17.**56–**17.**57
  regeneration of GAC, **14.**37
  vortex scrubbers, **15.**10
  water scrubbers, **17.**56
SDI, **13.**16
SDSI, **13.**17, **13.**18
SDWA, **2.**4
Sealing of documents, **18.**3
Seawater desalination system, **13.**13
Secondary drinking water standards, **2.**5, **2.**7, **10.**2–**10.**3
Section baffle flocculator, **6.**20–**6.**22
Section maze flocculator, **6.**20, **6.**21
Security, **30.**1–**30.**7
  barriers, **30.**2–**30.**3
  CCTV, **30.**2, **30.**5–**30.**6
  control system, **20.**23–**20.**24
  electrical system, **21.**9
  electronic security system, **30.**3–**30.**5
  O&M, **25.**3–**25.**4
  plant arrangement, **23.**31, **23.**35
  PPS, **30.**1–**30.**2
  procedures, **30.**6–**30.**7
  water quality, **30.**2
  (*See also* Terrorism)
Sedimentation, **7.**2–**7.**3, **20.**10–**20.**11
  filter backwash recycle, **17.**23, **17.**24
  gravity, **28.**7
  monitored and controlled items, **20.**9
  pilot plant, **28.**12–**28.**13
  reference text, **7.**2
Sedimentation basins/tanks, **7.**2, **25.**8, **28.**12–**28.**13
Sedimentation zone (sedimentation basin), **28.**12
Segregated flow analysis (SFA), **10.**4
Seismic design, **19.**8–**19.**10
Seismic Protection for Electrical Facilities, **21.**2
Selecting a site (*see* Site selection)
Selenium, **12.**26–**12.**27
Selenium dumping, **12.**26
Self-backwashing filters, **8.**12, **8.**13, **8.**20
Semiquantitative fit test, **15.**47
Sequential control, **20.**4
SERC, **15.**44
7M vessel, **13.**32
SFA, **10.**4
Sharp-crested weir, **16.**10
Shasta Dam, **4.**17
Shocking the filter, **16.**14
Shop and storage building, **18.**13

Short-length weir, **16.**10
Shrinkage cracking, **19.**6
Sidestream residual flow, **17.**24–**17.**28
Sieving techniques, **23.**18–**23.**19
Signal errors, **20.**24–**20.**25
Silica control, **13.**18–**13.**20
Silica fouling of anion resins, **12.**52
Silica sand, **8.**6
Silt density index (SDI), **13.**16
Silver, **2.**7
Simplified Procedures for Water Examination, **18.**9
Single chemical application point, **23.**32
Single end weir, **7.**7
Single gravity filters, **8.**20
Single-medium anthracite beds, **8.**5
Single-phase motors, **21.**7
Single-stage excess lime softening, **11.**9
Single-stage lime softening process, **11.**7
Single-unit solids contact type of flocculator-clarifier, **6.**23
Siphon well intakes, **4.**6, **4.**7
Site selection, **23.**1–**23.**22
  environmental issues, **23.**8–**23.**11
  financial considerations, **23.**15–**23.**18
  institutional issues, **23.**14
  public participation, **23.**2–**23.**4
  site evaluation/selection methodology, **23.**18–**23.**22
  siting criteria, **23.**5, **23.**6
  size requirements, **23.**5–**23.**8
  steps in process, **23.**4
  technical issues, **23.**11–**23.**12
  topography/soil conditions, **23.**12–**23.**14
Site surveying, **3.**7
Site topography, **23.**12–**23.**13
Six-tenths rule, **26.**6
6M vessel, **13.**31–**13.**32
Skimming, **7.**39–**7.**40
Slaking, **15.**20
Slow sand filtration, **2.**14, **9.**2–**9.**16
  access, **9.**8–**9.**9
  algae, **9.**3–**9.**4
  automation, **9.**12
  color, **9.**4
  design, **9.**5–**9.**13
  dissolved oxygen, **9.**4–**9.**5
  drainage, **9.**10–**9.**11
  effluent water quality, **9.**5
  filter box design, **9.**5–**9.**6
  filter fabric, **9.**15
  filter harrowing, **9.**15–**9.**16
  filter rates, **9.**11–**9.**12
  filter-to-waste system, **9.**10
  flow control, **9.**11
  GAC sandwich filter, **9.**14–**9.**15, **14.**21, **14.**22
  historical overview, **9.**2
  iron and manganese, **9.**4

lighting, **9.9–9.**10
nutrients, **9.**5
pilot testing, **9.**12–**9.**13
preozonation, **9.**14
roof system, **9.**10
roughing filters, **9.**13–**9.**14
schmutzdecke, **9.**2
source water quality considerations, **9.**3
source water storage, **9.**8
treatment performance, **9.**5
turbidity, **9.**3
underdrain and support system, **9.**6–**9.**7
ventilation, **9.**10
SLR, **17.**38
Sludge blanket clarification, **7.**32–**7.**36
Sludge disposal alternatives:
  beneficial reuse, **17.**11–**17.**12
  burial in landfill, **17.**11, **17.**13
  discharge to lagoon, **17.**11, **17.**13
  discharge to natural waterway, **17.**10, **17.**13
  discharge to sanitary sewer, **17.**10, **17.**13
  disposal of brine, **17.**12
Sludge handling/disposal (*see* Process residuals)
Sludge hopper arrangements (rectangular basins), **7.**13
Sludge loading rate (SLR), **17.**38
Sludge recirculation, **6.**23
Sludge removal:
  circular sedimentation basins, **7.**21–**7.**22
  contact clarification, **7.**42–**7.**43
  plate settlers, **7.**28
  rectangular sedimentation basins, **7.**8–**7.**19
  sludge blanket clarification, **7.**36
  slurry recirculation units, **7.**30
  tube settlers, **7.**24
Sludge treatment techniques, **17.**19–**17.**58
  backwash wastewater, **17.**20–**17.**28
  chemical addition, **17.**52–**17.**54
  controls, **20.**14–**20.**15
  dewatering, **17.**31–**17.**48
    (*See also* Dewatering processes)
  drying, **17.**48, **17.**51
  equalization, **17.**21–**17.**23
  handling and transport, **17.**51–**17.**52
  high-concentration wastes, **17.**51–**17.**52
  low-concentration wastes, **17.**51
  MF/UF membranes, **17.**25–**17.**26
  odor control, **17.**54–**17.**57
  ozone, **17.**26
  recycle/disposal of liquid wastes, **17.**57–**17.**58
  sedimentation, **17.**23–**17.**24
  sidestreams, **17.**24–**17.**28
  thickening, **17.**23–**17.**24, **17.**28–**17.**31, **17.**58
  ultrasound, **17.**26–**17.**27
  UV light, **17.**24–**17.**25
Sludge types, **17.**4–**17.**9
  alum sludge, **17.**5–**17.**6

brine wastes, **17.**7–**17.**9
coagulant sludge, **17.**5–**17.**6
filter backwash water sludge, **17.**4–**17.**5, **17.**20–**17.**28
ion exchange waste brine, **17.**9
iron and manganese precipitates, **17.**6–**17.**7
iron salts coagulation sludge, **17.**6
membrane process waste, **17.**7–**17.**9
softening plant sludge, **17.**7
Sludge zone (sedimentation basin), **28.**12
Slurries:
  carbon, **15.**27–**15.**29
  defined, **15.**27
  feed systems, **15.**27–**15.**32
  lime, **15.**29–**15.**31
Slurry conveyance, **9.**27–**9.**28
Slurry pumps, **15.**31
Slurry recirculation units, **7.**28–**7.**31
Slurry storage, **9.**27
Soda ash, **11.**4, **15.**23, **A.**15
Sodium aluminate, **15.**43, **A.**15, **A.**26
Sodium ascorbate, **A.**15, **A.**26
Sodium bisulfite, **15.**43, **A.**15, **A.**26
Sodium carbonate, **A.**15, **A.**27
  piping, **15.**36
  protective equipment, **15.**43
  storage, **15.**23
Sodium chloride, **A.**15–**A.**16, **A.**27
  nitrate removal, **12.**20
  piping, **15.**36
  protective equipment, **15.**43
  storage, **15.**23
Sodium chlorite, **A.**16, **A.**27
  oxidation and disinfection, **10.**38–**10.**39
  piping, **15.**36
  storage, **15.**23
Sodium dioxide, **15.**43
Sodium fluoride, **A.**16, **A.**27
  piping, **15.**36
  protective equipment, **15.**43
Sodium fluorosilicate, **15.**43, **A.**16, **A.**27
Sodium hexametaphosphate, **13.**23, **15.**36, **A.**13, **A.**17, **A.**27
Sodium hydroxide, **A.**27
  piping, **15.**36
  protective equipment, **15.**43
  (*See also* Caustic soda)
Sodium hypochlorite, **A.**17, **A.**28
  characteristics, **15.**10
  chlorination, **10.**25
  degradation, **10.**26
  elastomers, **10.**27
  feeding systems, **15.**51
  hypochlorite feed, **10.**27–**10.**28
  piping, **10.**27, **15.**36
  preoxidation studies, **28.**10
  protective equipment, **15.**43
  purchasing, **10.**25–**10.**26
  storage, **10.**26–**10.**27, **15.**13, **15.**15

Sodium hypochlorite systems, **15.**51
Sodium metabisulfite, **A.**15, **A.**17, **A.**28
Sodium polyphosphate, **15.**43, **A.**17
Sodium silicate, **A.**18, **A.**28
  piping, **15.**36
  protective equipment, **15.**43
Sodium silicofluoride, **15.**36, **A.**16
Sodium sulfite, **15.**43, **A.**18, **A.**28
Sodium thiosulfate, **A.**18, **A.**28
Sodium tripolyphosphate, **A.**13
Softening, **11.**1–**11.**3
  (*See also* Ion exchange applications; Lime softening; Membrane processes)
Softening plant sludge, **17.**7
Softening process basins, **11.**13–**11.**16
Soil liquefaction, **19.**8–**19.**9
Soils report, **19.**6
Solar drying bed, **17.**34
Solid bowl centrifuge, **17.**45, **17.**47
Solids contact, **6.**23
Solids contact clarifiers, **6.**2
Solids contact reactor unit, **6.**22
Solids contact/slurry recirculation units, **7.**28–**7.**31
Solids contact softening basins, **11.**13–**11.**16
Solids handling/disposal (*see* Process residuals)
Solids removal (*see* Sludge removal)
Solute passage, **13.**10
Solute rejection (removal), **13.**11
Source water storage, **23.**23
Spare parts, **18.**14
Spatially varied flow, **16.**9
Spill containment, **15.**11–**15.**12
Spiral-wound RO membrane module, **13.**5–**13.**6, **13.**7
Spitko, John E., Jr., **22.**1
Split-flow exchangers, **12.**37–**12.**38
Split-treatment excess lime process, **11.**12
Splitter boxes, **15.**32
Spray aerators, **5.**2
Spread footings, **19.**7–**19.**8
Square sedimentation basins, **7.**22
Stacked DAF, **7.**38
Staffing budget, **25.**18
Staffing plan, **25.**20–**25.**22
Stage 1 DBPR, **2.**6, **10.**9
Stage 2 Disinfectants/Disinfection By-products Rule (Stage 2 D/DBPR-2006), **2.**5, **2.**7, **10.**10
Stainless steel, **19.**3
Standard 60 (NSF International), **8.**37
Standard 61 (NSF International), **8.**37
Standard for the Installation of Lightning Protection Systems, **21.**10
Standard Methods for the Examination of Water and Wastewater, **18.**9
Standard Specification for Highway Bridges, **19.**12
Standards (*see* Water quality goals)

Standby power for pumping, **21.**8
Standby units, **22.**2
Start-up/operations services manager, **27.**13
Start-up workshop, **27.**13
State Emergency Response Commission (SERC), **15.**44
State sales tax, **26.**8
Static ice, **4.**22
Static in-line mixers, **6.**11, **6.**12
Steam-type vaporizer, **10.**49, **10.**50
Steel, **19.**3
Stiff and Davis saturation index (SDSI), **13.**17, **13.**18
Stoops, Robert A., **28.**1
Storage of chemicals (*see* Chemicals/chemical handling)
Stormwater management, **24.**10
Stormwater pollution prevention plan (SWPPP), **24.**12
Straight lime softening, **11.**6–**11.**9
Streaming current detectors, **20.**41
Strong base acrylic resins, **12.**52
Strong base anion resins, **12.**52
Strong base dealkalizer, **12.**14
Structural design, **3.**8, **19.**1–**19.**13
  basins, vaults, large conduits, channels, **19.**10–**19.**12
  buildings/superstructures, **19.**12–**19.**13
  codes and standards, **19.**2
  configuration of structures, **19.**3–**19.**5
  corrosion protection, **19.**6
  environmental considerations, **19.**5–**19.**6
  foundation considerations, **19.**6–**19.**8
  freezing considerations, **19.**5
  geotechnical report, **19.**6–**19.**7
  jointing of structures, **19.**4–**19.**5
  loading criteria, **19.**10
  mat foundations, **19.**7
  material selection, **19.**2–**19.**3
  public access considerations, **19.**5
  seismic design, **19.**8–**19.**10
  watertightness, **19.**6
Structural loads, **19.**10
Structural watertightness, **19.**6
Subbituminous coals, **14.**3
Submerged aerators, **5.**7
Submerged intake, **4.**3, **4.**4, **4.**11
Submerged intake screens, **4.**13
Submerged orifice trough, **7.**21
Submersible hydrostatic reservoir level transmitters, **20.**35
Subsurface conditions, **23.**13
Sulfate, **2.**7, **12.**18
Sulfate rule, **2.**5, **2.**7
Sulfate scale control, **13.**18
Sulfur dioxide, **15.**36, **A.**19, **A.**29
Sulfuric acid, **A.**19, **A.**29
  accidental release regulatory thresholds, **15.**44
  piping, **15.**36

protective equipment, **15.**43
threshold planning quantities, **15.**44
Sunk costs, **23.**15–**23.**16
Superpulsator sludge blanket clarifier, **7.**34
Supervisory control and data acquisition (SCADA) systems, **20.**18
Supervisory positions, **25.**21–**25.**22
Supply reliability, **22.**10
Support gravel upset, **8.**39–**8.**40
Surface aerators, **5.**7
Surface agitator and arrangements, **8.**15
Surface ice, **4.**22
Surface loading rate, **7.**29, **14.**13
Surface scatter turbidimeters, **20.**39–**20.**40
Surface wash systems, **8.**14–**8.**16, **8.**18
Surface Water Treatment Rule (SWTR), **2.**6, **8.**36, **10.**3–**10.**9, **29.**38
Surface water withdrawals, **24.**6
Surge protection, **20.**24
Suspended solids control, **13.**13–**13.**16
Sustainable architecture, **18.**3
Swaim, Paul D., **29.**1
Switched Ethernet, **20.**20
Switches and transmitters:
    DP transmitter, **20.**30–**20.**31, **20.**37
    level, **20.**35–**20.**37
    pressure, **20.**37–**20.**38
    temperature, **20.**38
SWPPP, **24.**12
SWTR, **2.**6, **8.**36, **10.**3–**10.**9, **29.**38
SWTR changes, **10.**8–**10.**9
Synchronous motor, **21.**5–**21.**6
Synthetic organics, **2.**16
Systems training, **27.**8–**27.**9

T_{10}, **10.**57
T screen (fish protection), **4.**28
Tanks, vaults, etc., **19.**10–**19.**12
Taste and odor control, **2.**10, **10.**13, **A.**31
TCLP, **24.**8, **24.**18
Technical library, **18.**6
Technology Transfer Handbook: Management of Water Treatment Plant Residuals, **17.**68
Temperature:
    chlorine, **A.**8, **A.**9
    ED/EDR system, **13.**40, **13.**41
    hypochlorite solutions, **15.**17
    membrane processes, **13.**25
    RO/NF units, **13.**29
    UF/MF units, **13.**36
Temperature switches/transmitters, **20.**38
Temporary pilot plants, **28.**23
10% finer size, **8.**5
Ten States Standards, **8.**36, **9.**8, **11.**16
Terminal head loss, **8.**30
Terrain-following sensors, **30.**4
Terrorism:
    dirty bombs, **12.**1, **12.**27
    removal of radioactive substances, **12.**27

(*See also* Security)
Tetrachloroethylene, **5.**10
Tetrachloroferrate, **12.**24
Tetrasodium pyrophosphate, **A.**29
Textbook:
    overview, **1.**5–**1.**6
    purpose, **1.**5
Thames Water, **9.**14
Thermal/catalytic destruct systems, **10.**58–**10.**59
Thermal drying, **17.**48
Thermal regeneration, **14.**35–**14.**37
Thermal swing adsorption (TSA), **10.**53
Thickening, **17.**23–**17.**24, **17.**28–**17.**31, **17.**58
316L stainless steel piping, **13.**46
Throughfare regeneration, **12.**33
Time demand regeneration, **12.**46
Title III of SARA, **15.**44
Title V operating permit program, **24.**15
TMP (transmembrane pressure), **13.**11, **13.**35
TOC, **12.**29–**12.**30
Toluene, **5.**10
Ton chlorine containers, **15.**6
Topographic and property surveys, **3.**7
Total hardness, **11.**2
Total trihalomethanes (TTHM), **2.**7, **2.**8, **2.**11, **10.**9
Tower intake, **4.**10–**4.**13
Toxaphene, **5.**10
Toxic characteristic leaching procedure (TCLP), **24.**8, **24.**18
Toxic cloud (Bhopal, India), **15.**44
Track-mounted hydraulic sludge removal system, **7.**14–**7.**17
Trade percent, **10.**26, **15.**13
Training (*see* Plant start-up)
Training budget, **25.**19
Transfer pumps, **15.**32–**15.**33
Transformers, **21.**4–**21.**5
Transient voltage surge suppression (TVSS), **20.**24
Transmembrane osmotic pressure, **13.**11
Transmembrane pressure (TMP), **13.**11, **13.**35
Transmittance turbidimeters, **20.**40
Transmitters (*see* Switches and transmitters)
Trash racks, **4.**31, **4.**32
Traveling water screens, **4.**31–**4.**33
Tray aerators, **5.**2, **5.**13
Treated water pump controls, **20.**15
Treatment comparisons/evaluations, **2.**15
Treatment obsolescence, **2.**8
Treatment options:
    common multipurpose treatment measures, **2.**12
    common treatment process, **2.**10–**2.**11
    less common treatment, **2.**14–**2.**15
    process alternatives, **2.**9–**2.**12
    treatment comparisons/evaluations, **2.**15
    treatment train alternatives, **2.**12–**2.**14

Treatment train alternatives, **2.12–2.14**
Trend displays, **20.**27
Trenton, New Jersey, reliability incident, **22.**11
Tributylamine, **12.**25
Tricalcium phosphate, **A.**29
Trichloramine, **10.**15
Trichloroacetic acid, **10.**9
1,1,1-Trichloroethane, **5.**10
Tricholoroethylene, **5.**10
Triethylamine, **12.**25
Trimethylamine, **12.**3, **12.**25, **12.**53
Triple-media bed, **8.**6
Tripropylamine, **12.**25
Trislot, **9.**22
Trisodium phosphate, **A.**29
Trough-type distributor, **5.**19
Troughs, **16.**9
Truck unloading system, **17.**52
TSA, **10.**53
TTHM, **2.**7, **2.**8, **2.**11, **10.**9
Tube settlers, **7.**23–**7.**24
Tube-type ozone generator, **10.**43, **10.**44
Tubular diaphragm pumps, **15.**24
Tunnels/galleries, **25.**16–**25.**17
Turbidimeters, **20.**39–**20.**40
Turbidity:
 feedwater, **13.**16
 rapid sand filtration, **8.**28–**8.**29
 slow sand filtration, **9.**3
 sludge, **17.**5
 units of measure, **8.**28
 UV light, **29.**4
Turbidity detectors, **20.**39
Turbidity-particulate reduction, **2.**10
Turbine contactors, **10.**57
Turbine-type mechanical flash mixer, **6.**8
Turntable drives, **7.**21
TVSS, **20.**24
Two-bed demineralization, **12.31–12.32**
Two-film theory of gas transfer, **10.**54
2:1 array, **13.**31
Two-pass RO seawater desalination system, **13.**13
Two-stage excess lime softening process, **11.9–11.10**
Two-stage filtration, **2.**14, **8.**4
Two-tray sedimentation basin, **7.**6
Type I styrenic resins, **12.**52
Type II anion resins, **12.**3, **12.**13, **12.**31
Type II styrenic resins, **12.**52

UC, **8.**5
UF and MF units, **13.2–13.3**, **13.35–13.39**
 applications, **13.**6
 arrays, **13.**37
 backwashing, **13.**38
 CEB, **13.**38
 CIP, **13.**38
 configuration, **13.**35
 cross-flow velocity, **13.**37
 design equations, **13.**35
 feedwater analysis, **13.**14
 flux, **13.**36
 hollow-fiber units, **13.**8
 I&C equipment, **13.44–13.45**
 membrane integrity tests, **13.38–13.39**
 pumping requirement, **13.**37
 recovery, **13.35–13.36**
 sidestream residual flows, **17.24–17.25**
 solute rejection, **13.**36
 temperature, **13.**36
 (*See also* Membrane processes)
UIC rules, **17.**14
Ultrafiltration (UF) (*see* UF and MF units)
Ultrasonic flowmeters, **20.31–20.32**, **20.**33
Ultrasonic level transmitters, **20.**36
Ultrasound, **17.26–17.27**
Ultraviolet disinfection, **29.1–29.47**
 absorbance, **29.**4, **29.**16
 advanced oxidation, **29.**8
 ballasts, **29.**13
 bid documents, **29.**35
 biodosimetry, **29.43–29.44**
 by-products, **29.5–29.6**
 Canada, **29.41–29.43**
 CFD modeling, **29.28–29.29**
 chemical disinfection, compared, **29.**7
 cleaning systems, **29.13–29.14**
 construction, **29.**36
 data collection, **29.20–29.21**
 definitions, **29.**3
 design criteria, **29.25–29.27**
 detailed design, **29.32–29.35**
 disinfection of target microorganisms, **29.6–29.7**
 dosage ranges/requirements, **29.**9, **29.39–29.40**
 dose monitoring, **29.**15
 facility monitoring/reporting, **29.**38, **29.**41
 flow rate, **29.**25
 goals, **29.**21
 historical review, **29.2–29.3**
 hydraulics, **29.**28
 instrumentation and control, **29.**33
 intensity sensors, **29.**14
 lamp fouling/aging factor, **29.**12, **29.26–29.27**
 lamp sleeves, **29.**13, **29.**17
 lamps, **29.10–29.13**, **29.**33
 location of UV facility, **29.22–29.25**, **29.27–29.29**
 microbial response to UV light, **29.4–29.5**
 off-spec operation, **29.**34, **29.40–29.41**
 O&M activities, **29.37–29.38**
 ozone disinfection, **29.**17, **29.**18
 particle content, **29.16–29.17**

photolysis, **29**.8
power supply and quality, **29**.33–**29**.34
project implementation, **29**.19–**29**.35
reduction of disinfection by-product formation, **29**.7–**29**.8
redundancy, **29**.27
regulations, **29**.38–**29**.43
regulator involvement, **29**.21
scattering, **29**.4
schematic of disinfection system, **29**.9
sidestream residual flows, **17**.24–**17**.25
target pathogen/inactivation level, **29**.27
testing, **29**.36
upstream chemical impacts, **29**.17–**29**.19
UV reactor, **29**.8, **29**.10, **29**.40–**29**.41
UVT, **29**.4, **29**.16, **29**.26
UVT monitors, **29**.14
validation testing, **29**.30–**29**.31, **29**.40–**29**.41, **29**.43–**29**.44
water temperature sensors, **29**.14–**29**.15
Unattended plant operation, **20**.17
Unburned lime, **A**.4
Uncertainty, **23**.21–**23**.22
Underdrain system:
  GAC postfilter adsorbers, **14**.28, **14**.31
  rapid sand filters, **8**.21–**8**.28
  slow sand filtration, **9**.6–**9**.7
Underflow control, **7**.17
Underground injection control (UIC) rules, **17**.14
Ungraded media, **8**.5
Uniformity coefficient (UC), **8**.5, **14**.4
Uniformity graded deep-bed filters, **8**.7–**8**.8
Uninterruptible power supply (UPS), **20**.23, **20**.24
Unit process configuration, **22**.5–**22**.6
Up-to-date, as-built drawings, **22**.14
Upflow expanded beds, **14**.24
Upflow gravel bed flocculator, **6**.24
Upflow water wash with air scour, **8**.14, **8**.18
Upflow water wash with surface wash, **8**.14–**8**.16, **8**.18
Upflow water wash without auxiliary scour, **8**.13–**8**.14, **8**.16–**8**.17
Uplift forces, **4**.28
UPS, **20**.23, **20**.24
Uranium, **12**.28–**12**.29
U.S. Army Corps of Engineers permits, **24**.13, **24**.14
USEPA drinking water regulations in development, **2**.5, **2**.7
USEPA Membrane Filtration Guidance Manual, **13**.39
USEPA primary drinking water standards, **2**.5, **2**.6
USEPA secondary drinking water standards, **2**.5, **2**.7, **10**.2–**10**.3
USFilter Kruger "Actiflo" ballasted flocculation system, **7**.24–**7**.25

Utilities budget, **25**.19
Utility transformers, **21**.4–**21**.5
Utility water, **15**.37–**15**.39
UV (*see* Ultraviolet disinfection)
UV absorbance, **29**.4, **29**.16
UV intensity and UVT setpoint approach, **29**.15
UV intensity sensors, **29**.14
UV intensity setpoint approach, **29**.15
UV lamps, **29**.10–**29**.13, **29**.33
UV photolysis, **29**.8
UV reactor, **29**.8, **29**.10, **29**.40–**29**.41
UV reactor oriented vertically, **29**.35
UV reactor validation testing, **29**.40
UV transmittance (UVT), **29**.4, **29**.16, **29**.26
UVT monitors, **29**.14

V-notch weir, **16**.10
Vacuum-assisted drying beds, **17**.34–**17**.35
Vacuum filter leaves, **9**.23
Vacuum filter systems, **9**.20, **9**.26–**9**.27, **9**.28–**9**.29
Vacuum leaf mountings, **9**.24
Vacuum regulators, **10**.20
Vacuum swing adsorption (VSA), **10**.53–**10**.54
Vacuum-type hollow-fiber MF, **13**.8
Valentine chloramine stability coefficient, **10**.17
Validation testing, **29**.30–**29**.31, **29**.40–**29**.41, **29**.43–**29**.44
Value engineering, **3**.5, **3**.9
Valve actuating systems, **8**.32
Valves:
  chlorine, **10**.22–**10**.24
  flow control, **15**.32
  head loss, **16**.6–**16**.7
  liquid chemical feed, **15**.24–**15**.25, **15**.26
  PRVs, **19**.7
  rapid sand filters, **8**.32
Vapor-phase GAC system, **5**.21
Vaporizers, **10**.49, **10**.50
Variable-volume presses, **17**.41
Vaults, tanks, etc., **19**.10–**19**.12
Velocity cap, **4**.27
Vendor O&M manual, **27**.3–**27**.5
Vendor training, **27**.2–**27**.3, **27**.6–**27**.8
Ventilation, **15**.48–**15**.49
Venturi flowmeters, **20**.30
Venturi tube, **8**.30
Vertical end suction pumps, **19**.9
Vertical flocculators, **6**.19–**6**.20
Vertical leaf filters, **9**.21
Vertical loads, **19**.10
Vertical paddle flocculator compartment, **6**.18
Vertical velocity problem, **7**.8
Vibrating feeders, **15**.20
Video motion sensors, **30**.5–**30**.6
Videotaped vendor training sessions, **27**.7–**27**.8

Vinyl chloride, **5.**10
Visual rotameters, **15.**37
Vitamin C, **A.**4, **A.**21
Vitrified clay tile underdrain, **8.**24
Volatile organic compound (VOC), **13.**42
Volatile organic reduction, **2.**10
Volatile organics, **2.**16
Voltage-driven electrodialysis process, **13.**4
Volumetric feeders, **15.**19–**15.**20
Volumetric sensors, **30.**4
Vortex scrubbers, **15.**10
Vortex-shedding flowmeters, **20.**31–**20.**32, **20.**34
VSA oxygen generation, **10.**53–**10.**54
Vulnerability assessment, **22.**9

Wash rates, **8.**16–**8.**18
Washwater trough-sizing diagram, **8.**35
Washwater troughs, **8.**32–**8.**36
Waste DE slurry, **9.**28
Waste disposal (*see* Process residuals)
Waste handling system hydraulic profile, **17.**63–**17.**64
Waste inventory, **24.**11
Wastewater Engineering: Treatment, Disposal, and Reuse, **17.**18
Wastewater lagoons, **9.**16
Wastewater treatment/disposal, **24.**9–**24.**10, **24.**15
Water block CCR regeneration, **12.**39
Water flux (*see* Flux)
Water hardness, **11.**1–**11.**2
Water Quality and Treatment:
 chemical treatment, **15.**1
 coagulation, **6.**1
 exchange process theory, **12.**2
 flocculation, **6.**1
 lime softening, **11.**3, **11.**6, **11.**12
 oxidation and disinfection, **10.**2
 sedimentation theory, **7.**2
Water quality goals:
 finalized regulations, **2.**4–**2.**5, **2.**6
 primary standards, **2.**5, **2.**6
 regulations under development, **2.**5, **2.**7
 secondary standards, **2.**5, **2.**7
 treatment obsolescence, **2.**8
Water scrubber, **17.**56
Water softening, **11.**1–**11.**3
 (*See also* Ion exchange applications; Lime softening; Membrane processes)

Water system control architecture, **20.**16
Water temperature sensors, **29.**14–**29.**15
Water treatment chemicals (*see* Chemicals/chemical handling)
Water Treatment Plant Design, **1.**6
Water treatment plant residuals (*see* Process residuals)
Water Treatment Plant Waste Management, **17.**58, **17.**68
Water treatment waste disposal injection wells, **17.**14
Water-type vaporizer, **10.**49, **10.**50
Water Utility Capital Financing, **3.**4
Water viscosity and water temperature, **6.**6
Water wash, **8.**13–**8.**18
Water withdrawals, **24.**6
Waterborne disease, **10.**1
 (*See also* Oxidation and disinfection)
Waterfall aeration, **5.**2–**5.**4
Waterstops, **19.**5
Watertightness, **19.**6
Watson, H. E., **10.**13
Wausau, Wisconsin, **5.**21
Weakly basic anion resins, **12.**31
Wearout, **22.**4, **22.**5
Wedge wire underdrain (GAC filters), **14.**19
Wedgewater process, **17.**34
Wedgewire beds, **17.**34
Weir, **16.**10–**16.**11, **20.**34
Weir mixers, **6.**13
Weir trough, **7.**21
Well injection, **17.**14
Well intakes, **4.**7
Wesner, George M., **6.**1
West Hartford, **9.**15
Wheeler, Jane W., **24.**1
Wheeler bottom, **8.**27
Wiedeman, John H., **21.**1
Willis, John F., **7.**1
Winslow, John M., **30.**1
Wire cloths (water filtration), **9.**23
Woodward Reservoir, **4.**11
Wound rotor motor, **21.**6

Y, **13.**10

Zebra mussel, **4.**25–**4.**26
Zeolites, **12.**4
Zimmerman, Phillip C., **18.**1
Zinc, **2.**7